Periodic Table of the Elements

Main groups

Handwritten notes: *electronegativity too increase* · E_i *increase* · *radius → decrease* · (left margin) *decrease* · E_i · *radius* · *increase*

1A 1	2A 2		3B 3	4B 4	5B 5	6B 6	7B 7	8B 8	8B 9	8B 10	1B 11	2B 12	3A 13	4A 14	5A 15	6A 16	7A 17	8A 18
1 H 1.00794																		2 He 4.00260
3 Li 6.941	4 Be 9.01218												5 B 10.81	6 C 12.011	7 N 14.0067	8 O 15.9994	9 F 18.998403	10 Ne 20.1797
11 Na 22.98977	12 Mg 24.305												13 Al 26.98154	14 Si 28.0855	15 P 30.97376	16 S 32.066	17 Cl 35.453	18 Ar 39.948
19 K 39.0983	20 Ca 40.078		21 Sc 44.9559	22 Ti 47.88	23 V 50.9415	24 Cr 51.996	25 Mn 54.9380	26 Fe 55.847	27 Co 58.9332	28 Ni 58.69	29 Cu 63.546	30 Zn 65.39	31 Ga 69.72	32 Ge 72.61	33 As 74.9216	34 Se 78.96	35 Br 79.904	36 Kr 83.80
37 Rb 85.4678	38 Sr 87.62		39 Y 88.9059	40 Zr 91.224	41 Nb 92.9064	42 Mo 95.94	43 Tc (98)	44 Ru 101.07	45 Rh 102.9055	46 Pd 106.42	47 Ag 107.8682	48 Cd 112.41	49 In 114.82	50 Sn 118.710	51 Sb 121.757	52 Te 127.60	53 I 126.9045	54 Xe 131.29
55 Cs 132.9054	56 Ba 137.33		57 *La 138.9055	72 Hf 178.49	73 Ta 180.9479	74 W 183.85	75 Re 186.207	76 Os 190.2	77 Ir 192.22	78 Pt 195.08	79 Au 196.9665	80 Hg 200.59	81 Tl 204.383	82 Pb 207.2	83 Bi 208.9804	84 Po (209)	85 At (210)	86 Rn (222)
87 Fr (223)	88 Ra 226.0254		89 †Ac 227.0278	104 Db (261)	105 Jl (262)	106 Rf (263)	107 Bh (262)	108 Hn (265)	109 Mt (268)									

Transition metals · Main groups

*Lanthanide series

58 Ce 140.12	59 Pr 140.9077	60 Nd 144.24	61 Pm (145)	62 Sm 150.36	63 Eu 151.96	64 Gd 157.25	65 Tb 158.9254	66 Dy 162.50	67 Ho 164.9304	68 Er 167.26	69 Tm 168.9342	70 Yb 173.04	71 Lu 174.967

†Actinide series

90 Th 232.0381	91 Pa 231.0359	92 U 238.0289	93 Np 237.048	94 Pu (244)	95 Am (243)	96 Cm (247)	97 Bk (247)	98 Cf (251)	99 Es (252)	100 Fm (257)	101 Md (258)	102 No (259)	103 Lr (260)

[a]The larger labels are common American usage. The smaller labels are those recommended by the International Union of Pure and Applied Chemistry.

CHEMISTRY

CHEMISTRY

John McMurry
Cornell University

Robert C. Fay
Cornell University

 Prentice Hall, *Englewood Cliffs*, New Jersey 07632

Library of Congress Cataloging-in-Publication Data

McMurry, John.
  Chemistry / John McMurry, Robert C. Fay.
    p.   cm.
  Includes index.
  ISBN 0-13-350281-3
    1. Chemistry.   I. Fay, Robert C., (date).   II. Title.
  QD33.M137   1995   94-37065
  540--dc20   CIP

Editor in Chief: *Paul F. Corey*
Editorial Director: *Tim Bozik*
Development Editor: *Stephen Deitmer*
Director of Production and Manufacturing: *David W. Riccardi*
Managing Production Editor: *Kathleen Schiaparelli*
Production Project Coordinator: *Barbara DeVries*
Marketing Manager: *Kelly McDonald*
Associate Editor: *Mary Hornby*
Manufacturing Buyer: *Trudy Pisciotti*
Creative Director: *Paula Maylahn*
Text Designer: *Lisa A. Jones*
Page Layout: *Natasha Sylvester*
Cover Designer: *Jerry Votta*
Cover Art: © *Kenneth Edward/BioGrafx*
Photo Editor: *Lorinda Morris-Nantz*
Photo Researchers: *Diane Austin/Tobi Zausner*
Editorial Assistants: *Veronica A. Wade and Nancy Bauer*
Copy Editor: *Barbara Liguori*
Art Studio: *Academy ArtWorks, Inc.*
Text Composition: *Better Graphics, Inc.*

Photo credits appear on the last page of this book
and are an extension of the copyright page.

 © 1995 by Prentice-Hall, Inc.
A Simon & Schuster Company
Englewood Cliffs, New Jersey 07632

Printed in the United States of America
10 9 8 7 6 5 4 3 2 1

ISBN 0-13-350281-3

Prentice-Hall International (UK) Limited, *London*
Prentice-Hall of Australia Pty. Limited, *Sydney*
Prentice-Hall Canada Inc., *Toronto*
Prentice-Hall Hispanoamericana, S.A., *Mexico*
Prentice-Hall of India Private Limited, *New Delhi*
Prentice-Hall of Japan, Inc., *Tokyo*
Simon & Schuster Asia Pte. Ltd., *Singapore*
Editors Prentice-Hall do Brasil, Ltds., *Rio de Janeiro*

ℬRIEF CONTENTS

CONTENTS

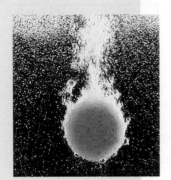

9 ➤ GASES: THEIR PROPERTIES AND BEHAVIOR   *324*

10 ➤ LIQUIDS, SOLIDS, AND CHANGES OF STATE   *362*

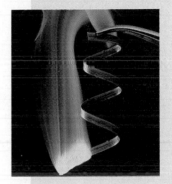

14 ▶ HYDROGEN, OXYGEN, AND WATER   536

15 ▶ AQUEOUS EQUILIBRIA: ACIDS AND BASES   570

16 ➤ APPLICATIONS OF AQUEOUS EQUILIBRIA *616*

17 ➤ THERMODYNAMICS: ENTROPY, FREE ENERGY, AND EQUILIBRIUM *664*

18 ➤ ELECTROCHEMISTRY *704*

19 ➤ THE MAIN-GROUP ELEMENTS *750*

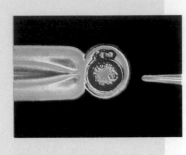

PREFACE

Novelist Kurt Vonnegut (a former chemistry major at Cornell) has a scene in his classic book *Cat's Cradle* in which Francine Pefko, secretary to the famous surface chemist Dr. Nilsak Horvath, is bemoaning her job:

> "I take dictation from Dr. Horvath and it's just like a foreign language. I don't think I'd understand it—even if I was to go to college."
>
> "If there's something you don't understand, ask Dr. Horvath to explain it. He's very good at explaining. Dr. Hoenikker used to say that any scientist who couldn't explain to an eight-year-old what he was doing was a charlatan."

The case may be overstated, but the underlying sentiment is valid: Scientists *should* be able to explain their subject in a clear, direct, and understandable way to anyone who wants to learn. And that has been our goal in writing this book—to produce the most readable and effective teaching text possible.

We have paid an extraordinary amount of attention to the writing in this book, trying to make it as easy as possible for a reader to understand and learn chemistry. Beginning with atomic structure, proceeding next to bonding and molecules, then to bulk physical properties of substances, and ending with a study of chemical properties, we have told a cohesive story about chemistry. Transitions between topics are smooth, explanations are lucid, and tie-ins to earlier material are frequent. Every attempt has been made to explain chemistry in a visual, intuitive way so that it can be understood by all who give it an honest effort.

Insofar as possible, distractions within the text are minimized. Each chapter is broken into numerous sections to provide frequent breathers, and each section has a consistent format. Sections generally begin with an explanation of their subject, move to an Example problem that shows how to work with the material, and end with practice Problems for the reader to work through. Each chapter ends with a brief Interlude that describes an interesting application or extension of the chapter subject, a Summary, a list of Key Words, and a large number of Additional Problems.

The Problems in the text have received almost as much attention as the writing. In addition to the usual dimensional-analysis method for solving numerical problems, the idea of first getting an approximate "ballpark" answer of the right magnitude is introduced. This method serves not only as a test of understanding specific concepts but also as a useful device for developing general intellectual skills. Additional end-of-chapter problems begin with a series of visual, non-numerical "Understanding Key Concepts" problems that test a true understanding of the material. (Some of them are real challenges; just because a problem doesn't require math doesn't mean it's easy.)

We hope that this book, the final result of eight year's work, has met the goals we set for it and that students find it to be friendly, accessible, and above all effective in teaching chemistry.

Many of our colleagues contributed their expertise to the book by reading, and criticizing the manuscript to improve our work. We thank our Cornell colleagues and the following reviewers:

Anneke S. Allen	*Wichita State University*
Herbert Beall	*Worcester Polytechnic Institute*
Rathindra N. Bose	*Kent State University*
Albert W. Burgstahler	*University of Kansas*
James Coke	*The University of North Carolina–Chapel Hill*
Michael Denniston	*DeKalb College-Central Campus*
Norman Duffy	*Kent State University*
Dale D. Ensor	*Tennessee Technological University*
Joanne M. Follweiler	*Lafayette College*
Charles Fritchie, Jr.	*Tulane University*
Roy Garvey	*North Dakota State University*
Stephen J. Hawkes	*Oregon State University*
Harry G. Hecht	*South Dakota State University*
Colin D. Hubbard	*University of New Hampshire*
Kenneth J. Klabunde	*Kansas State University*
Peter Lykos	*Illinois Institute of Technology*
Gary W. Morrow	*University of Dayton*
Heidi Reese	*New York University*
Clyde Riley	*University of Alabama–Huntsville*
B. Ken Robertson	*University of Missouri–Rolla*
Victor Rodwell	*Purdue University*
Steven Ruis	*American River College*
Gene D. Schaumberg	*Sonoma State University*
Richard L. Schowen	*The University of Kansas*
Henry Shanfield	*University of Houston*
Klaus H. Theopold	*University of Delaware*
Bruno M. Vittimberga	*University of Rhode Island*
Charles A. Wilkie	*Marquette University*
Shelby D. Worley	*Auburn University*
Charles M. Wynn	*Eastern Connecticut State University*
William H. Zoller	*University of Washington*

Additionally, our thanks go to Robert Pribush of Butler University, who prepared the annotations for the Annotated Instructor's Edition, Thomas Herrinton of the University of San Diego, who wrote the Test Item File, and Theodore Sakano of Rockland Community College, who wrote the Instructor's Resource Manual.

Prentice Hall has been most supportive throughout the process of writing and producing CHEMISTRY. We especially thank Paul Corey, Editor-in-Chief, and Tim Bozik, Editorial Director, for their guidance. Thanks also go to Paul Banks for his editorial work early in the production process, Raymond Mullaney, Editor in Chief of Book Development, Steve Deitmer, Development Editor, and Barbara DeVries,

Production Project Coordinator. Kelly McDonald, Marketing Manager, Gary June, Director of Marketing, and Meghan Dacey, Advertising Copywriter, have each played an important role. We are grateful to Lisa Jones, who designed the text, and Paula Maylahn, Creative Director, for their valuable part in the finished product, and to the many other people at Prentice Hall who helped make the book a reality.

And finally, our sincere thanks to our families.

John McMurry

Robert C. Fay

$\mathcal{N}$OTE FOR STUDENTS

We have similar goals. Your goal is to learn chemistry; our goal is to do everything possible to help you learn. It took some work on our part, and it's going to take some work on your part, but the following suggestions should prove helpful.

- **Don't read the text immediately.** As you begin each new chapter, look it over first. Read the introductory paragraphs, find out what topics will be covered, and then turn to the end of the chapter and read the Summary. You'll be in a much better position to learn new material if you first have a general idea of where you're going.
- **Work the problems.** There are no shortcuts here; working problems is the only way to learn chemistry. The in-text examples show you how to approach the material, the problems at the end of each section provide immediate reinforcement, and the end-of-chapter problems give additional drill. Answers to most odd-numbered problems are given in an appendix at the back of the book, and full answers for all problems are provided in an accompanying *Solutions Manual*.
- **Ask questions.** Faculty members and teaching assistants are there to help you. Most of them will turn out to be genuinely nice people with a sincere interest in helping you learn.

Prentice Hall provides a number of student-oriented supplements that will aid in your study of general chemistry and help prepare you for subsequent courses. The key items are listed below. Check with your bookstore for more details.

- **Chemistry Explorer Software** This is an interactive simulation program based on worked problems and examples from the text. It allows students to manipulate variables and physical parameters in performing experiments to observe how these manipulations affect the results. It also provides data analysis tools, such as spreadsheets and graphs.
- **Solutions Manual** by Joseph Topich of Virginia Commonwealth University, contains worked-out solutions to all in-chapter, end-of-chapter problems and critical-thinking problems. Available to students and instructors.
- **Selected Solutions Manual** also by Joseph Topich contains worked-out solutions to over half of the text's problems. The answers to this set of problems appear in the Answers to Selected Problems.
- **Study Guide** by Donna Jean Fredeen of Southern Connecticut State University contains learning goals, chapter outlines, and self tests with solutions.
- **Math Review Toolkit** by Gary Long, Virginia Polytechnic Institute, is available at no additional charge. Part One of the Toolkit consists of a chapter-by-chapter guide to the mathematics used throughout the book. Part Two is a guide to career planning and chemistry. It highlights the value of chemistry training in

business and other careers not specifically related to chemistry. Part Three looks at the special requirements of writing in chemistry, focusing particularly on the lab notebook.

➤PRENTICE HALL/THE NEW YORK TIMES
THEMES OF THE TIMES

The New York Times and Prentice Hall are sponsoring *Themes of the Times*, a program designed to enhance student access to current information of relevance in the classroom.

Through this program, the core subject matter provided in the text is supplemented by a collection of time-sensitive articles from one of the world's most distinguished newspapers, *The New York Times*. These articles demonstrate the vital, ongoing connection between what is learned in the classroom and what is happening in the world around us.

To enjoy the wealth of information of *The New York Times* daily, a reduced subscription rate is available. For information, call toll-free: 1 800 631 1222.

Prentice Hall and *The New York Times* are proud to cosponsor *Themes of the Times*. We hope it will make the reading of both textbooks and newspapers a more dynamic, involving process.

ABOUT THE AUTHORS

John McMurry *(left)*
Robert Fay *(right)*

JOHN E. McMURRY, educated at Harvard and Columbia, has taught approximately 11,000 students in general and organic chemistry over a 27-year period. A Professor of Chemistry at Cornell University since 1980, Dr. McMurry previously spent 13 years on the faculty at the University of California at Santa Cruz. He has received numerous awards, including the Alfred P. Sloan Fellowship (1969–71), the National Institute of Health Career Development Award (1975–80), the Alexander von Humboldt Senior Scientist Award (1986–87), and the Max Planck Research Award (1991).

ROBERT C. FAY, Professor of Chemistry at Cornell University, has been teaching general and inorganic chemistry at Cornell since 1962. Known for his clear, well organized lectures, Dr. Fay was the 1980 recipient of the Clark Distinguished Teaching Award. He has also taught as a visiting professor at Harvard University and at the University of Bologna (Italy). A Phi Beta Kappa graduate of Oberlin College, Fay received his Ph.D. from the University of Illinois. He has been an NSF Science Faculty Fellow at the University of East Anglia and the University of Sussex (England) and a NATO/Heineman Senior Fellow at Oxford University.

CHEMISTRY

chapter 1

CHEMISTRY: MATTER AND MEASUREMENT

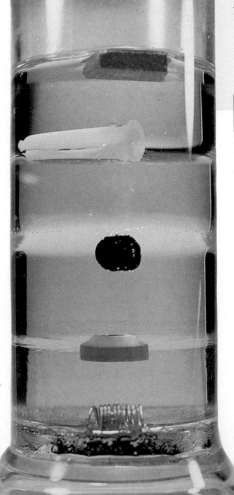

L ife has changed more in the past 200 years than in all the previously recorded span of human history. The earth's population has increased more than fivefold since 1800, and life expectancy has nearly doubled because of our ability to synthesize medicines, control diseases, and increase crop yields. Methods of transportation have changed from horseback to automobiles and airplanes because of our ability to harness the energy in petroleum. Many goods are now made of polymers and ceramics instead of wood and metal because of our ability to manufacture materials with properties unlike any found in nature.

In one way or another, all these changes involve **chemistry**, the study of the composition, properties, and transformations of matter. Chemistry is responsible for the changes that take place in nature, and chemistry is in many ways responsible for the profound social changes of the past two centuries. In addition, chemistry is central to the current revolution in molecular biology that is exploring the details of how life is genetically controlled. No educated person today can understand the world around them without a basic knowledge of chemistry.

Different materials have different densities, ranging from wood at the top of this column, to mercury at the bottom.

1.1 ►APPROACHING CHEMISTRY: EXPERIMENTATION

By opening this book, you have already decided that chemistry is worth looking into. Perhaps you want to know more about how medicines are made, how fertilizers and pesticides work, how living organisms function, how new high-temperature ceramics are used in space vehicles, or how microelectronic circuits are etched onto silicon chips. How do you approach chemistry?

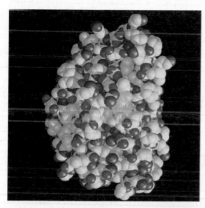

Computer modeling of potential drugs is a powerful new technique in pharmaceutical research.

The high-temperature ceramic tiles on the space shuttle protect it from damage during reentry into the atmosphere.

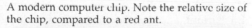

A modern computer chip. Note the relative size of the chip, compared to a red ant.

One way you might begin is by looking and thinking. Look carefully at what you see around you, and then try to arrive at logical explanations for your observations. You would certainly observe, for instance, that different substances have different forms and appearances: Some substances are gases, some are liquids, and some are solids; some are hard and shiny, but others are soft and dull. You'd also observe that different substances behave differently: Iron rusts but gold does not; copper conducts electricity but sulfur doesn't. How can these and an infinite number of other observations be explained?

Gold, one of the most valuable of elements, is prized for its beauty and resistance to corrosion.

Iron, though widely used as a structural material, corrodes easily.

Though useful, the looking-and-thinking approach is not sufficient. The natural world is too complex to be understood by thinking alone; a more active approach is needed. Specific questions must be posed, and experiments must be carried out to find their answers. Only when the results of many experiments are known can we then apply logic to devise an interpretation, or *hypothesis*, that explains the results. The hypothesis, in turn, can be used to make more predictions and to suggest more experiments until a consistent explanation, or **theory**, of known observations is finally arrived at.

It's important to keep in mind as you study chemistry or any other science that scientific theories are just logical interpretations of experimental results. They aren't handed down from above; they're arrived at by trial and error and by the hard work of thousands of people over many years. Some theories may eventually be modified or replaced by better ones if new experiments uncover results that accepted theories can't explain. Scientific theories just represent the best explanations we can come up with at the present time.

1.2 ▶ CHEMISTRY AND THE ELEMENTS

Everything you see around you is formed from one or more of 109 presently known elements. An **element** is a fundamental substance that can't be chemically changed or broken down into anything simpler. Silver, mercury, and sulfur are common examples, as listed in Table 1.1.

TABLE 1.1	Names and Symbols of Some Common Elements[a]						
Al	aluminum	**Cl**	chlorine	**Mn**	manganese	**Cu**	copper (cuprum)
Ar	argon	**F**	fluorine	**N**	nitrogen	**Fe**	iron (ferrum)
Ba	barium	**He**	helium	**O**	oxygen	**Pb**	lead (plumbum)
B	boron	**H**	hydrogen	**P**	phosphorus	**Hg**	mercury (hydrargyrum)
Br	bromine	**I**	iodine	**Si**	silicon	**K**	potassium (kalium)
Ca	calcium	**Li**	lithium	**S**	sulfur	**Ag**	silver (argentum)
C	carbon	**Mg**	magnesium	**Zn**	zinc	**Na**	sodium (natrium)

[a] The names in parentheses are the Latin names from which the symbols for the elements are derived.

Samples of mercury, silver, and sulfur

Only about 90 of the 109 presently known elements occur naturally. The remaining ones have been produced artificially by nuclear chemists using high-energy particle accelerators. Furthermore, not all of the 90 naturally occurring elements are equally abundant. Hydrogen is thought to account for approximately 75% of the mass in the universe; oxygen and silicon together account for 75% of the earth's crust; and oxygen, carbon, and hydrogen make up more than 90% of the human body (Figure 1.1). By contrast, it has been estimated that there are probably less than 20 grams of the unstable, radioactive element francium (Fr) dispersed over the entire earth at any one time.[1]

---

[1] Francium is a short-lived radioactive element, atoms of which are continually being formed and destroyed in natural radiochemical processes. We'll discuss radioactivity in Chapter 22.

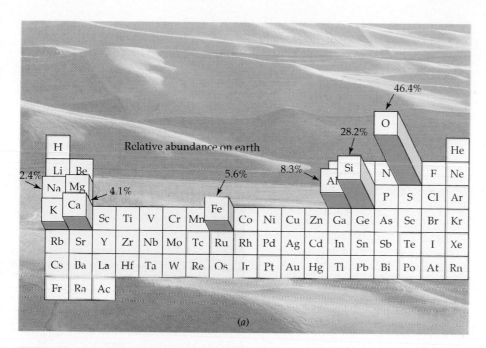

(a)

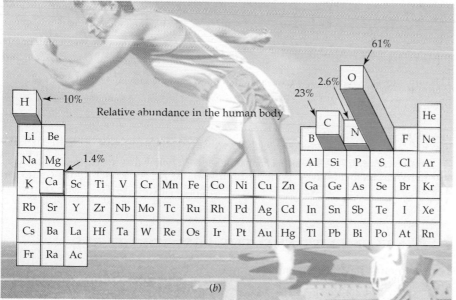

(b)

FIGURE 1.1 Estimated elemental composition by mass percent of **(a)** the earth's crust and **(b)** the human body. Oxygen is the most abundant element in both. (Only the major constituents are shown in each case. Small amounts of many other elements are also present.)

Chemists refer to specific elements using a shorthand code of one-, two-, or three-letter symbols. As shown in Table 1.1, the first letter of an element's symbol is always capitalized, and the second or third letters, if any, are lower case. Many of the symbols are just the first one or two letters of the element's English name: H = hydrogen, C = carbon, Al = aluminum, and so forth. Other symbols derive from Latin or other languages: Na = sodium (Latin, *natrium*), Pb = lead (Latin, *plumbum*), W = tungsten (German, *wolfram*). The names, symbols, and other information about all 109 elements are given inside the front cover, organized in a tabular form called the *periodic table*.

┌ **PROBLEM 1.1** Look at the periodic table of elements inside the front cover, and find the symbols for these elements: **(a)** copper (used in electrical wires) **(b)** platinum (used in automobile emission control devices) **(c)** plutonium (used in nuclear weapons)

Copper wire.

This automobile catalytic converter uses platinum to catalyze the reaction of unburned fuel with air.

┌ **PROBLEM 1.2** Look at the alphabetical list of elements inside the front cover, and tell what elements these symbols represent: **(a)** Ag **(b)** Rh **(c)** Re **(d)** Cs **(e)** Ar **(f)** As

1.3 ►ELEMENTS AND THE PERIODIC TABLE

Ten elements have been known since the beginning of recorded history: antimony (Sb), carbon (C), copper, (Cu), gold (Au), iron (Fe), lead (Pb), mercury (Hg), silver (Ag), sulfur (S), and tin (Sn). The first "new" element to be found in several thousand years was arsenic (As), discovered in about 1250. In fact, only 24 elements were known up to the time of the American Revolution in 1776. A graph showing the decades when different elements were discovered is given in Figure 1.2.

The first tabulation of the "chemically simple" substances that we today call elements appeared in a treatise published in 1789 by the French scientist Antoine Lavoisier. As the pace of discovery quickened in the late 1700s and early 1800s, chemists began to look for similarities among elements that might allow them to draw general conclusions. Particularly important among the early successes was Johann Döbereiner's observation in 1829 that there were several *triads*, or groups of three elements, that

FIGURE 1.2 A chart showing decades in which new elements were discovered. Ten elements have been known since antiquity. The photo shows Marie Curie, discoverer of the element radium.

appeared to have similar chemical and physical properties. Calcium (Ca), strontium (Sr), and barium (Ba) form one such triad; chlorine (Cl), bromine (Br), and iodine (I) form another; and lithium (Li), sodium (Na), and potassium (K) form a third. By 1843, 16 such triads were known and chemists had begun to search for an explanation.

Numerous attempts were made in the mid-1800s to account for the similarities among groups of elements, but the great breakthrough came in 1869 when the Russian chemist Dmitri Mendeleev published the forerunner of the modern **periodic table**, shown in Figure 1.3. In this modern version, elements are placed on a grid with 7 horizontal rows called **periods** and 18 vertical columns called **groups**. When organized in this way, *the elements in a given group have similar chemical properties*. Lithium, sodium, potassium, and the other elements in group 1A behave similarly. Beryllium, magnesium, calcium, and the other elements in group 2A behave similarly. Fluorine, chlorine, bromine, and the other elements in group 7A behave similarly, and so on throughout the table.

Few people now question the overall form of the modern periodic table, but chemists in different countries have historically used different conventions for labeling the columns. To resolve these difficulties, a newly adopted standard calls for numbering the columns from 1 to 18 going from left to right. This new standard has not yet found wide acceptance, however, and we'll continue to use the U.S. system of numbers and capital letters—group 7A instead of group 17, for example. Labels for the new system are shown in parentheses in Figure 1.3.

One further note: There are actually *32* columns in the table rather than 18, but to make the table fit manageably on a page, the 14 elements following lanthanum (the *lanthanides*) and the 14 following actinium (the *actinides*) are pulled out and shown below the others.

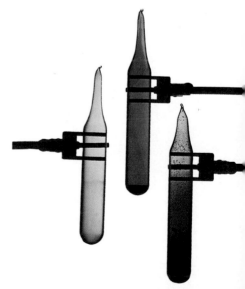

Samples of chlorine, bromine, and iodine, one of Döbereiner's triads of elements with similar chemical properties.

1 H																	2 He
3 Li	4 Be											5 B	6 C	7 N	8 O	9 F	10 Ne
11 Na	12 Mg											13 Al	14 Si	15 P	16 S	17 C	18 Ar
19 K	20 Ca	21 Sc	22 Ti	23 V	24 Cr	25 Mn	26 Fe	27 Co	28 Ni	29 Cu	30 Zn	31 Ga	32 Ge	33 As	34 Se	35 Br	36 Kr
37 Rb	38 Sr	39 Y	40 Zr	41 Nb	42 Mo	43 Tc	44 Ru	45 Rh	46 Pd	47 Ag	48 Cd	49 In	50 Sn	51 Sb	52 Te	53 I	54 Xe
55 Cs	56 Ba	57 La	72 Hf	73 Ta	74 W	75 Re	76 Os	77 Ir	78 Pt	79 Au	80 Hg	81 Tl	82 Pb	83 Bi	84 Po	85 At	86 Rn
87 Fr	88 Ra	89 Ac	104 Db	105 Jl	106 Rf	107 Bh	108 Hn	109 Mt									

58 Ce	59 Pr	60 Nd	61 Pm	62 Sm	63 Eu	64 Gd	65 Tb	66 Dy	67 Ho	68 Er	69 Tm	70 Yb	71 Lu
90 Th	91 Pa	92 U	93 Np	94 Pu	95 Am	96 Cm	97 Bk	98 Cf	99 Es	100 Fm	101 Md	102 No	103 Lr

FIGURE 1.3 The modern form of the periodic table. Each element is identified by a one- or two-letter symbol and is characterized by an *atomic number*. The table begins with hydrogen (H), atomic number 1, in the upper left-hand corner, and continues to meitnerium (Mt), atomic number 109. The 14 elements following lanthanum (La, atomic number 57) and the 14 elements following actinium (Ac, atomic number 89) are pulled out and shown below the others.

1.4 ►SOME CHARACTERISTICS OF THE ELEMENTS

We'll see repeatedly that the periodic table of the elements is the most important organizing principle of chemistry. The time you take now to familiarize yourself with the overall layout and organization of the periodic table will pay you back later on. Notice in Figure 1.4, for example, that there is a regular progression in the size of the 7 periods (rows). The first period has only 2 elements, hydrogen (H) and helium (He); the second and third periods have 8 elements each; the fourth and fifth periods have 18 elements each; and the sixth and seventh (incomplete) periods have 32 elements each. We'll see in Chapter 5 that this regular progression in the periodic table reflects a similar regularity in atomic structure. Notice also that not all groups (columns) in the periodic table have the same number of elements. The 2 larger groups on the left and the 6 larger groups on the right of the table are called the **main groups**, the 10 smaller ones in the middle of the table are called the **transition-metal groups**, and the 14 shown separately at the bottom of the table are called the **inner transition-metal groups**.

Groups of elements in the periodic table often show remarkable similarities in their properties. Look at the following 4 groups, for example:

- *Group 1A—Alkali metals*: Lithium (Li), sodium (Na), potassium (K), rubidium (Rb), and cesium (Cs) are shiny, soft, low-melting metals. All react rapidly (often violently) with water to form products that are highly alkaline, or basic—hence the name *alkali metals*. Because of their high reactivity, the alkali metals are never found in nature in the pure state but only in combination with other elements.

- *Group 2A—Alkaline earth metals*: Beryllium (Be), magnesium (Mg), calcium (Ca), strontium (Sr), barium (Ba), and radium (Ra) are also lustrous, silvery metals, but all are less reactive than their neighbors in group 1A. Like the alkali metals, the alkaline earths are never found in nature in the pure state.

Sodium, one of the alkali metals, reacts violently with water to yield hydrogen gas and an alkaline (basic) solution.

Magnesium, one of the alkaline earth metals, burns in air.

FIGURE 1.4 Variation in the size of different periods. The first period has 2 elements, the second and third periods have 8 elements each, the fourth and fifth periods have 18 elements each, and the sixth and seventh periods have 32 elements each.

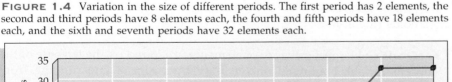

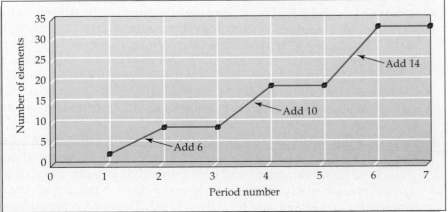

Bromine, a halogen, is a corrosive dark red liquid at room temperature.

- *Group 7A—Halogens*: Fluorine (F), chlorine (Cl), bromine (Br), and iodine (I) are corrosive and nonmetallic. All are found in nature only in combination with other elements, such as with sodium in table salt (sodium chloride, NaCl). In fact, the group name *halogen* is taken from the Greek word *hals*, meaning salt.

- *Group 8A—Noble gases*: Helium (He), neon (Ne), argon (Ar), krypton (Kr), xenon (Xe), and radon (Rn) are gases of very low reactivity. Helium, neon, and argon don't combine with any other elements; krypton and xenon combine with very few.

Neon, one of the noble gases, is used in neon lights.

Although the resemblances aren't as pronounced as they are within a single group, *neighboring* groups of elements also behave similarly in many ways. Thus, as indicated in Figure 1.3, the periodic table is often divided into three major classes of elements—metals, nonmetals, and semimetals, or metalloids (metal-like):

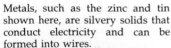

Metals, such as the zinc and tin shown here, are silvery solids that conduct electricity and can be formed into wires.

- *Metals*, the largest category of elements, are found on the left side of the periodic table, bounded on the right by a zigzag line running from boron (B) at the top to astatine (At) at the bottom. The metals are easy to characterize by their appearance: All except mercury are solid at room temperature, and most have the silvery shine we normally associate with metals. In addition, metals are generally malleable rather than brittle, can be twisted and drawn into wires without breaking, and are good conductors of heat and electricity.

- *Nonmetals* are found on the right side of the periodic table and are also easy to characterize by their appearance. Eleven of the 17 nonmetals are gases, 1 is a liquid (bromine), and only 5 are solids at room temperature (carbon, phosphorus, sulfur, selenium, and iodine). None are silvery in appearance, and several are brightly colored. The solid nonmetals are brittle rather than malleable, and they tend to be poor conductors of heat and electricity.

- *Semimetals* are 7 of the elements adjacent to the zigzag boundary between metals and nonmetals (boron, silicon, germanium, arsenic, antimony, tellurium, and astatine). As you might expect, their properties fall between those of their neighbors on either side. Though most are silvery in appearance, they are brittle rather than malleable and tend to be poor conductors of heat and electricity. Silicon, for example, is a widely used *semiconductor*, a substance whose electrical conductivity is intermediate between that of a metal and an insulator.

Sulfur, red phosphorus, and iodine (clockwise from bottom) are typical nonmetals. All are brittle, and none conducts electricity.

⌐ **PROBLEM 1.3** Identify these elements as metals, nonmetals, or semimetals:
(a) Ti **(b)** Te **(c)** Se **(d)** Sc **(e)** At **(f)** Ar ⌐

1.5 ►EXPERIMENTATION AND MEASUREMENT

As noted previously, chemistry is an experimental science. If our experiments are to be reproducible, we must be able to describe fully the substances we're working with—their amounts, sizes, temperatures, and so forth. Thus, one of the most important requirements in chemistry is that we have a way to *measure* things.

Under an international agreement concluded in 1960, scientists throughout the world now use the **International System of Units** for measurement, abbreviated **SI** for the French *Système Internationale d'Unités*. Based on the well-known metric system, which is used in all industrialized countries of the world except the United States, there are only 7 fundamental SI units (Table 1.2). These seven base units, along with others derived from them, suffice for all scientific measurements. We'll look at 3 of the most common units in this chapter—those for mass, length, and temperature—and discuss others as the need arises in later chapters.

Boron, a semimetal, is used in making this tennis racket.

TABLE 1.2 The Seven Fundamental SI Units of Measure		
Physical Quantity	**Name of Unit**	**Abbreviation**
Mass	Kilogram	kg
Length	Meter	m
Temperature	Kelvin	K
Amount of substance	Mole	mol
Time	Second	s
Electric current	Ampere	A
Luminous intensity	Candela	cd

One problem with any system of measurement is that the sizes of the base units sometimes turn out to be inconveniently large or small. For example, a chemist describing the diameter of a sodium atom (0.000 000 000 372 m) would find the meter (m) to be an inconveniently large unit, but an astronomer measuring the average distance from the earth to the sun (150,000,000,000 m) would find the meter to be inconveniently small. For this reason, SI units are modified through the use of prefixes when they refer to either smaller or larger quantities. Thus, the prefix *milli-* means one-thousandth, and a *milli*meter (mm) is 1/1000 of 1 meter; similarly, the prefix *kilo-* means one thousand, and a *kilo*meter (km) is 1000 meters. (Note that the SI unit for mass already contains the *kilo-* prefix.) A list of prefixes is shown in Table 1.3, with the most commonly used ones given in red.

· TABLE 1.3	Some Prefixes for Multiples of SI Units		
Factor	**Prefix**	**Symbol**	**Example**
$1\ 000\ 000\ 000 = 10^9$	giga	G	1 gigameter (Gm) $= 10^9$ m
$1\ 000\ 000 = 10^6$	mega	M	1 megameter (Mm) $= 10^6$ m
$1\ 000 = 10^3$	kilo	k	1 kilogram (kg) $= 10^3$ g
$100 = 10^2$	hecto	h	1 hectogram (hg) $= 100$ g
$10 = 10^1$	deka	da	1 dekagram (dag) $= 10$ g
$0.1 = 10^{-1}$	deci	d	1 decimeter (dm) $= 0.1$ m
$0.01 = 10^{-2}$	centi	c	1 centimeter (cm) $= 0.01$ m
$0.001 = 10^{-3}$	milli	m	1 milligram (mg) $= 0.001$ g
$0.000\ 001 = 10^{-6}$	micro	μ	1 micrometer (μm) $= 10^{-6}$ m
$0.000\ 000\ 001 = 10^{-9}$	nano	n	1 nanosecond (ns) $= 10^{-9}$ s
$0.000\ 000\ 000\ 001 = 10^{-12}$	pico	p	1 picosecond (ps) $= 10^{-12}$ s

Notice how numbers that are either very large or very small are referred to in Table 1.3 using an exponential format called **scientific notation.** For example, the number 55,000 is written in scientific notation as 5.5×10^4, and the number 0.003 20 as 3.20×10^{-3}. You might want to review Appendix A if you are uncomfortable with scientific notation or if you need to brush up on how to do mathematical manipulations on numbers with exponents.

Notice also in Table 1.3 that all measurements of physical quantities contain *both* a number and a unit label. A number alone is not much good without a unit to define it. If you asked a friend how far it was to the nearest tennis court, the answer "3" alone wouldn't tell you much—3 blocks? 3 kilometers? 3 miles?

┌ **PROBLEM 1.4** Express the following quantities in scientific notation.
(a) The diameter of a sodium atom, 0.000 000 000 372 m
(b) The distance from the earth to the sun, 150,000,000,000 m

┌ **PROBLEM 1.5** What units do these abbreviations stand for?
(a) μg   **(b)** dm   **(c)** ps   **(d)** kA   **(e)** mmol

1.6 ➤MEASURING MASS

Mass is defined as the amount of *matter* in an object. **Matter,** in turn, is a catchall term used to describe anything physically real—anything you can touch, taste, or smell. (Stated more scientifically, matter is anything that has mass.) Mass is measured in SI units by the **kilogram (kg;** 1 kg = 2.205 U.S. lb). Because the kilogram is too large for many purposes in chemistry, the familiar metric **gram (g;** 1 g = 0.001 kg), the **milligram (mg,** 1 mg = 0.001 g

This pile of 400 pennies weighs about 1 kg.

$= 10^{-6}$ kg), and the **microgram** (μg; 1 μg = 0.001 mg = 10^{-6} g = 10^{-9} kg) are more commonly used. One gram is a bit less than half the mass of a U.S. penny.

$$1 \text{ kg} = 1000 \text{ g} = 1{,}000{,}000 \text{ mg} = 1{,}000{,}000{,}000 \ \mu g \qquad (2.205 \text{ lb})$$

$$1 \text{ g} = 1000 \text{ mg} = 1{,}000{,}000 \ \mu g \qquad (0.03527 \text{ oz})$$

$$1 \text{ mg} = 1000 \ \mu g$$

The standard kilogram, against which all other masses are compared, is defined as the mass of a cylindrical bar of platinum–iridium alloy stored in a vault in a suburb of Paris, France. There are 40 copies of this bar distributed throughout the world, with two (Numbers 4 and 20) now stored at the U.S. National Institute of Standards and Technology near Washington, D.C.

The terms "mass" and "weight," though often used interchangeably, have quite different meanings. *Mass* measures the amount of matter in an object, whereas *weight* measures the pull of gravity on an object by the earth or other celestial body. Clearly, the amount of matter in an object is independent of its location. Your body has the same mass whether you're on earth or on the moon. Just as clearly, though, the weight of an object *does* depend on its location. You might weigh 140 lb on earth, but you would weigh only about 23 lb on the moon, which has a lower gravity than the earth.

At the same location on earth, two objects with identical masses have identical weights; that is, the objects experience an identical pull of the earth's gravity. Thus, the *mass* of an object can be measured on a balance by comparing the *weight* of the object to the weight of a reference standard of known mass. Much of the confusion between mass and weight is simply due to a language problem: We speak of "weighing" when we really mean that we are measuring mass by comparing two weights. Figure 1.5 shows two of the balances normally used for measuring mass in the laboratory.

FIGURE 1.5 (a) The single-pan balance has a sliding counterweight that is adjusted until the weight of the object in the pan is just balanced. **(b)** Modern electronic balances whose workings are concealed by the case.

(a)

(b)

FIGURE 1.6 A comparison of different measures of length. One centimeter is about 0.4 in.

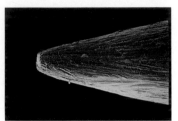

The length of these bacteria on the tip of a pin is about 5×10^{-7} m.

1.7 ►MEASURING LENGTH

The **meter (m)** is the standard unit of length in the SI system. Although originally defined in 1790 as being one ten-millionth of the distance from the equator to the North Pole, the meter was redefined in 1889 as the distance between two thin lines on a bar of platinum–iridium alloy stored outside Paris, France. To accommodate an increasing need for precision, the meter was redefined again in 1983 as equal to the distance traveled by light in a vacuum in 1/299,792,458th of a second. Although this new definition isn't as easy to understand as the distance between two scratches on a bar, it has the great advantage that it can't be lost, damaged, or destroyed.

One meter is 39.37 inches, about 10% longer than an English yard and much too large for most measurements in chemistry. Other, more commonly used measures of length are the **centimeter (cm;** 1 cm = 0.01 m, a bit less than half an inch), the **millimeter (mm;** 1 mm = 0.001 m, about the thickness of a U.S. dime), the **micrometer (μm;** 1 μm = 10^{-6} m), the **nanometer (nm;** 1 nm = 10^{-9} m), and the **picometer (pm;** 1 pm = 10^{-12} m). Thus, a chemist might refer to the diameter of a sodium atom as 372 pm (3.72×10^{-10} m). Figure 1.6 shows a comparison between inches and centimeters.

$$1 \text{ m} = 100 \text{ cm} = 1000 \text{ mm} = 1,000,000 \text{ μm} = 1,000,000,000 \text{ nm} \ (1.0936 \text{ yd})$$

$$1 \text{ cm} = 10 \text{ mm} = 10,000 \text{ μm} = 10,000,000 \text{ nm} \ (0.3937 \text{ in.})$$

$$1 \text{ mm} = 1000 \text{ μm} = 1,000,000 \text{ nm}$$

1.8 ►DERIVED UNITS: MEASURING VOLUME

Look back at the seven fundamental SI units given in Table 1.2 and you'll find that measures for such familiar quantities as area, volume, speed, and pressure are missing. All are examples of *derived* quantities rather than fundamental quantities because they can be expressed in terms of one or more of the seven base units (Table 1.4).

Volume, the amount of space occupied by an object, is measured in SI units by the **cubic meter (m^3),** defined as the amount of space occupied by a cube one meter on each edge (Figure 1.7).

The cubic meter, equivalent to 264.2 U.S. gallons, is much too large for normal use in chemistry, and smaller, more convenient measures are com-

TABLE 1.4	Some Derived Quantities	
Quantity	**Definition**	**Derived Unit (Name)**
Area	Length times length	m^2
Volume	Area times length	m^3
Density	Mass per unit volume	kg/m^3
Speed	Distance per unit time	m/s
Acceleration	Change in speed per unit time	m/s^2
Force	Mass times acceleration	$(kg \cdot m)/s^2$ (newton, N)
Pressure	Force per unit area	$kg/(m \cdot s^2)$ (pascal, Pa)
Energy	Force times distance	$(kg \cdot m^2)/s^2$ (joule, J)

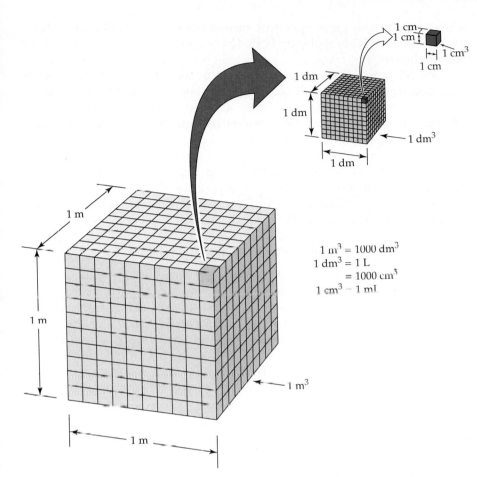

FIGURE 1.7 A cubic meter is the volume of a cube 1 m on each edge. Each cubic meter contains 1000 cubic decimeters (liters), and each cubic decimeter contains 1000 cubic centimeters (milliliters).

$$1 \text{ m}^3 = 1000 \text{ dm}^3$$
$$1 \text{ dm}^3 = 1 \text{ L}$$
$$= 1000 \text{ cm}^3$$
$$1 \text{ cm}^3 = 1 \text{ mL}$$

FIGURE 1.8 Some laboratory equipment used for measuring liquid volume.

(a) 100 mL graduated cylinder
(b) 50 mL volumetric flask
(c) 5 mL syringe
(d) 50 mL buret

monly employed. Both the **cubic decimeter** (1 **dm³** = 0.001 m³), equal in size to the more familiar metric **liter** (**L**), and the **cubic centimeter** (1 **cm³** = 0.001 dm³ = 10⁻⁶ m³), equal in size to the metric **milliliter** (**mL**), are particularly convenient. Slightly larger than 1 U.S. quart, a cubic decimeter (liter) has the volume of a cube 1 dm on each edge. Similarly, a cubic centimeter (milliliter) has the volume of a cube 1 cm on each edge (Figure 1.7).

$$1 \text{ m}^3 = 1000 \text{ dm}^3 = 1{,}000{,}000 \text{ cm}^3 \text{ (264.2 gal)}$$

$$1 \text{ dm}^3 = 1 \text{ L} = 1000 \text{ mL (1.057 qt)}$$

Volume measurement, particularly for liquids, is often necessary in laboratory work. Figure 1.8 shows some of the more frequently used pieces of equipment.

1.9 ➤ MEASURING TEMPERATURE

Just as the kilogram and the meter are slowly replacing the pound and the yard as common units for mass and length measurement in the United States, the **Celsius degree** (°C) is slowly replacing the Fahrenheit degree (°F)

The coldest temperature ever recorded on the earth's surface was −128.6°F at Vostok Station, Antarctica.

as the common unit for temperature measurement. In scientific work, however, the **kelvin (K)** is replacing both. (Note that we say only "kelvin," not "kelvin degree.")

For all practical purposes, the kelvin and the Celsius degree are the same size—both are 1/100th of the interval between the freezing point of water and the boiling point of water at atmospheric pressure. The only difference between the two units is that the numbers assigned to various points on the scales differ. Whereas the Celsius scale assigns a value of 0°C to the freezing point of water and 100°C to the boiling point of water, the Kelvin scale assigns a value of 0 K to the coldest possible temperature, −273.15°C, sometimes called *absolute zero*. Thus, 0 K = −273.15°C, and 273.15 K = 0°C. For example, a warm spring day with a temperature of 25°C has a kelvin temperature of 25 + 273.15 = 298 K.

$$\text{Temperature in K} = \text{temperature in °C} + 273.15$$

$$\text{Temperature in °C} = \text{temperature in K} - 273.15$$

In contrast to the Kelvin and Celsius scales, the Fahrenheit scale specifies an interval of 180° between the freezing point (32°F) and the boiling point (212°F) of water. Thus, it takes 180 Fahrenheit degrees to cover the same range encompassed by only 100 Celsius degrees (or kelvins), and a Fahrenheit degree is therefore only 100/180; = 5/9 as large as a Celsius degree. Figure 1.9 gives a comparison of Kelvin, Celsius, and Fahrenheit scales.

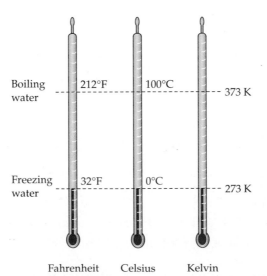

FIGURE 1.9 A comparison of Kelvin, Celsius, and Fahrenheit temperature scales. One Fahrenheit degree is 5/9 the size of a kelvin or Celsius degree.

Two corrections need to be made to convert between Fahrenheit and Celsius scales—one to adjust for the difference in degree size and one to adjust for the difference in zero point. The size correction is made by remembering that a Celsius degree is 9/5ths the size of a Fahrenheit degree (or, conversely, that a Fahrenheit degree is 5/9ths the size of a Celsius degree). The zero-point correction is made by remembering that the freezing point is higher by 32 on the Fahrenheit scale than it is on the Celsius scale. Thus, if you want to convert from Celsius to Fahrenheit, you do a size correction (multiply °C by 9/5) and then a zero-point correction (add 32); if you want to convert from Fahrenheit to Celsius, you find out how many Fahrenheit degrees there are above freezing (by subtracting 32) and then do a size correction (multiply by 5/9). The following formulas show the conversion methods:

$$°F = \left(\frac{9}{5} \times °C\right) + 32 \qquad °C = \frac{5}{9} \times (°F - 32)$$

EXAMPLE 1.1

The normal body temperature of a healthy adult is 98.6°F. Express this value on both Celsius and Kelvin scales.

SOLUTION   There are two ways to do this and every other problem in chemistry. One way is to think things through to be sure you understand what's going on, and the other way is to blindly apply the right formula. Let's try both ways:

The thinking approach: A temperature of 98.6°F corresponds to (98.6 − 32) = 66.6 Fahrenheit degrees above the freezing point of water. Since each Fahrenheit degree is only 5/9ths as large as a Celsius degree, 66.6 Fahrenheit degrees above freezing are equal to (66.6 × 5/9) = 37.0 Celsius degrees above freezing (0°C), or 37.0°C. The same number of degrees above freezing on the Kelvin scale (273.15 K) corresponds to a temperature of 310.15 K.

The formula approach: Set up an equation using the temperature-conversion formula:

$$°C = \frac{5}{9} \times (°F - 32) = \frac{5}{9} \times (98.6 - 32) = 37.0° C$$

Since the answers obtained by the two approaches agree, you can feel fairly confident that your thinking processes are following the right lines and that you understand the material. (If the answers *did not* agree, you'd be alerted to a fundamental misunderstanding somewhere.)

> **PROBLEM 1.6** The melting point of table salt is 1474°F. What temperature is this on the Celsius and Kelvin scales?

> **PROBLEM 1.7** Carry out the indicated temperature conversions.
> **(a)** −78°C = ? K　　　　　**(b)** 158°C = ?°F　　　　　**(c)** 375 K = ?°F

> **PROBLEM 1.8** At a certain point, the Celsius and Fahrenheit scales "cross," giving the same numerical value on both. At what temperature does this crossover occur?

1.10 ►ACCURACY, PRECISION, AND SIGNIFICANT FIGURES IN MEASUREMENT

Any measurement is only as good as the skill of the person doing the work and the reliability of the equipment used. Although most of us tend to use the words accuracy and precision interchangeably, there is an important distinction between the two terms. **Accuracy** refers to how close to the true value a given measurement is, whereas **precision** refers to how well a

The melting point of sodium chloride is 1474°F.

This tennis ball has a mass of about 54 g.

number of independent measurements agree with one another. To see the difference, imagine that you weigh a tennis ball whose mass is 54.44178 g. Assume that you take three independent measurements on each of three different types of balance to give the data shown in the following table.

Measurement #	Bathroom Scale	Lab Balance	Analytical Balance
1	0 kg	54.4 g	54.4419 g
2	0 kg	54.7 g	54.4417 g
3	1 kg	54.1 g	54.4417 g
(Avg.)	(0.3 kg)	(54.4 g)	(54.4418 g)

If you use a bathroom scale, your measurement (avg. = 0.3 kg) is neither accurate nor precise because the scale is not very sensitive. If you take the ball into your chemistry lab and weigh it on a single-pan balance, your measurement (avg. = 54.4 g) is fairly accurate, but the data are not very precise because they vary widely (from 54.1 g to 54.7 g). Going one step further by using an expensive analytical balance such as those found in research laboratories, your measurement (avg. = 54.4418 g) is both precise and accurate.

To indicate the precision of a measurement, the value recorded should use all the digits known with certainty, plus one additional estimated digit. The total number of digits in the measurement is called the number of **significant figures**. For example, the mass of the tennis ball as determined on the single-pan balance (54.4 g) has three significant figures, whereas the mass determined on the analytical balance (54.4418 g) has six significant figures. All digits but the last are known with certainty; the final digit is only a best guess, and we generally assume that it has an error of plus or minus one (±1).

Finding the number of significant figures in a measurement is usually easy but can be troublesome if zeros are present. Look at the following four quantities:

1. 4.803 cm   [four significant figures: 4, 8, 0, 3]
2. 0.00661 g   [three significant figures: 6, 6, 1]
3. 55.220 K   [five significant figures: 5, 5, 2, 2, 0]
4. 34,200 m   [anywhere from three (3, 4, 2) to five (3, 4, 2, 0, 0) significant figures]

The following rules cover the different situations that can arise:

1. *Zeros in the middle of a number are like any other digit; they are always significant.* Thus, 4.803 cm has four significant figures.
2. *Zeros at the beginning of a number are not significant*; they act only to locate the decimal point. Thus, 0.00661 g has three significant figures. (Note that 0.00661 g can be rewritten as 6.61 mg or as 6.61×10^{-3} g.)
3. *Zeros at the end of a number and* after *the decimal point are always significant.* The assumption is that these zeros wouldn't be shown unless they were significant. Thus, 55.220 K has five significant figures.

4. *Zeros at the end of a number and before the decimal point may or may not be significant.* We can't tell whether they are part of the measurement or whether they act only to locate the decimal point. Thus, 34,200 m may have three, four, or five significant figures.

The fourth rule shows why it is helpful to write numbers in scientific notation rather than ordinary notation. Doing so makes it possible to indicate the number of significant figures. Thus, writing the number 34,200 as 3.42×10^4 indicates three significant figures, but writing it as 3.4200×10^4 indicates five significant figures.

One further point about significant figures: Certain numbers, such as those obtained when counting objects, are *exact* and have an effectively infinite number of significant figures. For example, a week has *exactly* 7 days, not 6.9 or 7.0 or 7.1. Similarly, 1 ft has *exactly* 12 in., not 11.9 or 12.0, or 12.1.

EXAMPLE 1.2

How many significant figures does each of the following measurements have?
(a) 0.036653 m   (b) 7.2100×10^3 g   (c) 72,100 km   (d) $25.03

SOLUTION   (a) 5 (by rule 2); (b) 5 (by rule 3); (c) 3, 4, or 5 (by rule 4); (d) $25.03 is an exact number

┌ **PROBLEM 1.9** A 1.000 mL sample of acetone, a common solvent used as a paint remover, was placed in a small bottle whose mass was known to be 38.0015 g. The following values were obtained when the acetone-filled bottle was weighed: 38.7798 g, 38.7795 g, and 38.7801 g. How would you characterize the precision and accuracy of these measurements if the actual mass of the acetone was 0.7791 g?

┌ **PROBLEM 1.10** How many significant figures does each of the following quantities have? Explain your answers.

(a) 76.600 kg   (b) 4.50200×10^3 g   (c) 3000 nm   (d) 0.00300 mL   (e) 18 students

1.11 ►ROUNDING NUMBERS

It often happens, particularly when doing arithmetic on a pocket calculator, that a quantity appears to have more significant figures than are really justified. For example, you might calculate the gas mileage of your car by finding that it takes 11.70 gallons of gasoline to drive 278 miles:

$$\text{Mileage} = \frac{\text{miles}}{\text{gallons}} = \frac{278 \text{ mi}}{11.70 \text{ gal}} = 23.760\,684 \text{ mi/gal (mpg)}$$

Although the answer on the pocket calculator has eight digits, your measurement is really not so precise as it appears. In fact, your answer is precise to only three significant figures and should be **rounded off** to 23.8 mi/gal.

How do you decide how many figures to keep and how many to ignore? The full answer to this question is a bit complex and involves a mathematical treatment of the data known as *error analysis*. For many pur-

Calculators often give more significant figures than are justified.

poses, though, a simplified procedure using two easy-to-remember rules is sufficient. These two rules give only an approximate value of the actual error, but that approximation is often good enough:

1. *In carrying out a multiplication or division, the answer can't have more significant figures than either of the original numbers.* This is just a common-sense rule. After all, if you don't know the number of miles you drove to better than three significant figures (278 could mean 277, 278, or 279), you certainly can't calculate your mileage to more than the same number of significant figures.

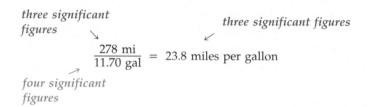

2. *In carrying out an addition or subtraction, the answer can't have more digits to the right of the decimal point than either of the original numbers.* For example, if you have 3.18 L of water and you add 0.01315 L more, you now have 3.19 L. Again, this rule is just common sense. If you don't know the volume you started with past the second decimal place (it could be 3.17, 3.18, or 3.19), you can't know the total of the combined volumes past the same decimal place.

$$
\begin{array}{ll}
3.18??? & \leftarrow \textit{ends two places past decimal point} \\
+\ 0.01315 & \leftarrow \textit{ends five places past decimal point} \\
\hline
3.19??? & \leftarrow \textit{ends two places past decimal point}
\end{array}
$$

Once you decide how many digits to drop from your answer, the rules for rounding off numbers are straightforward:

1. *If the first digit you remove is less than 5, round down by dropping it and all following digits.* Thus, 5.663 **507** becomes 5.66 when rounded off to three significant figures because the first of the dropped digits (3) is less than 5.
2. *If the first digit you remove is 6 or greater, round up by adding 1 to the digit on the left.* Thus, 5.663 **507** becomes 5.7 when rounded off to two significant figures because the first of the dropped digits (6) is greater than 5.
3. *If the first digit you remove is 5 and there are more nonzero digits following, round up.* Thus, 5.663 **507** becomes 5.664 when rounded to four significant figures because there is a nonzero digit (7) after the 5.
4. *If the digit you remove is a 5 with nothing following, you arbitrarily round up if the preceding digit is odd and arbitrarily round down if the preceding digit is even.* Thus, 4.747 **5** becomes 4.748 when rounded to four significant figures because there is nothing after the 5 and the digit to its left is odd (7). Conversely, 4.746 **5** becomes 4.746 when rounded to four significant figures because there is nothing after the 5 and the digit to its left is even (6).

EXAMPLE 1.3

If it takes 9.25 hours to fly from London, England to Chicago, a distance of 3952 miles, what is the average speed of the airplane in miles per hour?

SOLUTION   First, set up an equation dividing the number of miles flown by the number of hours:

$$\text{Average speed} = \frac{3952 \text{ mi}}{9.25 \text{ h}} = 427.243\,24 \text{ mi/h}$$

Next, decide how many significant figures should be in your answer. Since the problem involves a division, and since one of the quantities you started with (9.25 hours) has only three significant figures, the answer must also have three significant figures.

Finally, round off your answer. Since the first digit to be dropped (2) is less than 5, 427.243 24 must be rounded off to 427 mi/h. Notice that in doing the calculation, you use all figures, significant or not, and then round off the final answer.

PROBLEM 1.11 Round off each of the following quantities to the number of significant figures indicated in parentheses.

(a) 3.774 499 L (4)   (b) 255.0974 K (3)   (c) 55.265 kg (4)

PROBLEM 1.12 Carry out these calculations, expressing each result in the correct number of significant figures.

(a) 24.567 g + 0.044 78 g EQ ? g   (b) 4.6742 g ÷ 0.003 71 L = ? g/L
(c) 0.378 mL + 42.3 mL − 1.5833 mL = ? mL

1.12 ►CALCULATIONS: CONVERTING FROM ONE UNIT TO ANOTHER

Many scientific activities involve numerical calculations—measuring, weighing, preparing solutions—and it's often necessary to convert a quantity from one unit to another. Converting between units isn't difficult; we all do it every day. For example, if you swim 15 lengths of a 50-m pool, you have to convert between the distance unit *length* and the distance unit *meter* to find that you have swum 750 m (15 lengths times 50 meters per length).

$$15 \text{ lengths} \times \frac{50 \text{ meters}}{1 \text{ length}} = 750 \text{ meters}$$

This swimmer has to convert from lengths to meters to find out how far she has swum.

Converting from one scientific unit to another is just as easy.

The simplest way to carry out calculations involving different units is to use the **dimensional-analysis method**. In this method, a quantity described in one unit is converted into an equivalent quantity described in a different unit by using a **conversion factor** to express the specific relationship between units:

$$\text{Starting quantity} \times \text{conversion factor} = \text{equivalent quantity}$$

As an example, we know from Section 1.6 that 1 kg is equal to 2.205 lb. Writing this relationship as a fraction restates it in the form of a conversion factor, either kilograms per pound or pounds per kilogram:

Conversion factors between pounds and kilograms

$$\text{Since } 1 \text{ kg} = 2.205 \text{ lb, then}$$

$$\frac{1 \text{ kg}}{2.205 \text{ lb}} = \frac{2.205 \text{ lb}}{1 \text{ kg}} = 1$$

Note that this and all other conversion factors are numerically equal to 1 because the value of the quantity above the division line (the numerator) is equal to the value of the quantity below the division line (the denominator). Thus, multiplying by a conversion factor is equivalent to multiplying by 1.

These two quantities are the same   *These two quantities are the same*

$$\frac{1 \text{ kg}}{2.205 \text{ lb}} \quad \text{or} \quad \frac{2.205 \text{ lb}}{1 \text{ kg}}$$

The key to the dimensional-analysis method of problem solving is that units are treated like numbers and can thus be multiplied and divided just as numbers can. The idea when solving a problem is to set up an equation so that all unwanted units cancel, leaving only the desired units. For example, if you knew your weight in pounds, say 138 lb, and wanted to find it in kilograms, you could set up an equation multiplying the weight in pounds by the conversion factor in kilograms per pound.

$$138 \text{ lb} \times \frac{1 \text{ kg}}{2.205 \text{ lb}} = 62.6 \text{ kg}$$

Starting quantity   *Conversion factor*   *Equivalent quantity*

The unit ''lb''cancels from the left side of the equation since it appears both above and below the division line, and the only unit that remains is ''kg.''

The dimensional-analysis method gives the right answer only if the equation is set up so that the unwanted units cancel. If the equation is set up in any other way, the units won't cancel properly, and you won't get the right answer. Thus, if you were to multiply your weight in pounds by the incorrect conversion factor pounds per kilogram, you would end up with an incorrect answer expressed in meaningless units:

$$??? \quad 138 \text{ lb} \times \frac{2.205 \text{ lb}}{1 \text{ kg}} = 304 \text{ lb}^2/\text{kg} \quad ???$$

The main drawback to using the dimensional-analysis method is that it's easy to get the "right" answer without really understanding what you're doing. It's therefore best to estimate a ballpark answer before doing the detailed calculation. If your rough guess isn't close to the detailed answer, there is a misunderstanding somewhere, and you should think the problem through again. Even if you don't ballpark the full solution, it's important to be sure that your detailed answer makes sense. If, for example, you were trying to calculate the volume of a human cell and you came up with the answer 5.3 cm^3, you should realize that such an answer couldn't possibly be right; cells are too tiny to be distinguished with the naked eye, but a volume of 5.3 cm^3 is about the size of a walnut.

Examples 1.4–1.6 show how to estimate ballpark answers. To conserve space, we'll continue this approach in only the next few chapters, but you should make it a routine part of your problem solving.

EXAMPLE 1.4

Many of the German autobahns have no speed limit, and it's not unusual to see a Porsche or Mercedes flash by at 190 km/h. What is this speed in mi/h?

BALLPARK SOLUTION   Common sense says that the answer is probably large, perhaps over 100 mi/h. A better estimate is to realize that, since 1 km − 0.6214 mi, a mile is larger than a kilometer and it takes only about 2/3 as many miles as kilometers to measure the same distance. Thus, 190 km/h is about 120 mi/h.

DETAILED SOLUTION   Use the dimensional-analysis method to set up an equation so that the km units cancel:

$$\frac{190 \text{ km}}{1 \text{ h}} \times \frac{0.6214 \text{ mi}}{1 \text{ km}} - 118 \frac{\text{mi}}{\text{h}}$$

The ballpark solution and the detailed solution agree.

EXAMPLE 1.5

The Mercedes that just went by at 190 km/h is using gas at a rate of 16 L per 100 km. What is this rate in miles per gallon?

BALLPARK SOLUTION   Common sense says that the mileage is probably low, perhaps in the range of 10 to 15 mi/gal. This is a harder problem than the previous one to estimate, though, because it requires several different conversions. It's therefore best to think the problem through one step at a time. A rate of 16 L per 100 km is the same as 100 km per 16 L, or approximately 7 km/L. Since 1 km is about 0.6 mi, 7 km/L is about 4 mi/L. Furthermore, 1 L is approximately 1 qt, or 1/4 gal. Thus, the Mercedes is getting about 4 mi per 1/4 gal or about 16 mi/gal.

DETAILED SOLUTION   It's best to do multiple conversions one step at a time until you get used to dealing with them. You might start by converting the distance from kilometers to miles and the fuel consumption from liters to gallons:

$$100 \text{ km} \times \frac{0.6214 \text{ mi}}{1 \text{ km}} = 62.14 \text{ mi} \qquad 16 \text{ L} \times \frac{1 \text{ gal}}{3.78 \text{ L}} = 4.2328 \text{ gal}$$

Dividing the distance by the amount of fuel then gives the mileage:

$$\frac{62.14 \text{ mi}}{4.2328 \text{ gal}} = 14.68 \frac{\text{mi}}{\text{gal}}$$

which must be rounded off to 15 mi/gal. Note that extra digits are carried through the intermediate calculations, and only the final answer is rounded off.

Alternatively, when you become more confident in working conversion problems, you might set up one large equation so that all unwanted units cancel:

$$\frac{100 \text{ km}}{16 \text{ L}} \times \frac{3.78 \text{ L}}{1 \text{ gal}} \times \frac{0.6214 \text{ mi}}{1 \text{ km}} = 14.68 \frac{\text{mi}}{\text{gal}}$$

$$= 15 \frac{\text{mi}}{\text{gal}} \text{ (rounded off)}$$

However you do the problem, the ballpark solution and detailed solution agree.

EXAMPLE 1.6

The volcanic explosion that destroyed the Indonesian island of Krakatau on August 27, 1883, released an estimated 4.3 mi^3 of debris into the atmosphere. In SI units, how many cubic meters were released?

BALLPARK SOLUTION   One meter is much less than one mile, so it takes a large number of cubic meters to fit into 1 mi^3, and the answer is going to be very large. In fact, 1 mi^3 is the volume of a cube 1 mi on each edge, and 1 km^3 is the volume of a cube 1 km on each edge. Since a kilometer is only about 0.6 mile, 1 km^3 is about $(0.6)^3 = 0.2$ times as large as 1 mi^3. Thus, each cubic mile contains about 5 km^3, and 4.3 mi^3 contains about 20 km^3. Each cubic kilometer, in turn, contains $(1000 \text{ m})^3$ $= 10^9 \text{ m}^3$. Thus, the volume of debris from the Krakatau explosion was about $20 \times 10^9 \text{ m}^3$, or $2 \times 10^{10} \text{ m}^3$.

DETAILED SOLUTION   First convert cubic miles into cubic kilometers:

$$4.3 \text{ mi}^3 \times \left[\frac{1 \text{ km}}{0.6214 \text{ mi}} \right]^3 = 17.92 \text{ km}^3$$

Then convert cubic kilometers into cubic meters:

$$17.92 \text{ km}^3 \times \left[\frac{1000 \text{ m}}{1 \text{ km}} \right]^3 = 1.792 \times 10^{10} \text{ m}^3$$

$$= 1.8 \times 10^{10} \text{ m}^3 \text{ (rounded off)}$$

The ballpark solution and the detailed solution agree.

☐ **PROBLEM 1.13** Estimate the answers to the following problems:
(a) The melting point of gold is 1064°C. What is this temperature in °F?
(b) About how large in cubic centimeters is the volume of a red blood cell if the cell has the shape of a disk with a diameter of 6×10^{-6} m and a height of 2×10^{-6} m?

☐ **PROBLEM 1.14** How many yards are there in a marathon (26 mi, 385 yd)? How many meters?

☐ **PROBLEM 1.15** Gemstones are weighed in *carats*, with 1 carat = 200 mg. What is the mass in grams of the Hope Diamond, the world's largest blue diamond at 44.4 carats? What is this mass in ounces?

☐ **PROBLEM 1.16** Batrachotoxin, the active component of the South American arrow poison obtained from the golden frog, *Phyllobates terribilis*, is so poisonous that a single frog contains enough toxin (1100 μg) to kill 2300 people. Expressed in scientific notation, how many grams does it take to kill one person?

The golden frog, *Phyllobates terribilis*, contains one of the most potent toxins known.

1.13 ➤PROPERTIES OF MATTER: DENSITY

Any characteristic that can be used to describe or identify matter is called a **property**. Size, mass, odor, color, and temperature are all well-known examples. Still other properties of matter include such characteristics as melting point, solubility, and chemical behavior. For example, we might list some properties of sodium chloride (table salt) by saying that it melts at 801°C, that it dissolves in water, and that it undergoes a chemical reaction when it comes into contact with a silver nitrate solution.

Properties are often classified as either *physical* or *chemical* (Table 1.5). **Physical properties** are those characteristics like color, mass, melting point, and temperature that can be determined without changing the chemical makeup of the sample. **Chemical properties** are those that *do* change the chemical makeup of the sample. For example, the rusting that occurs when a bicycle is left out in the rain is due to the chemical combination of oxygen with iron to give the new substance iron oxide. Rusting is therefore a chemical property of iron.

Properties can also be classified as either *intensive* or *extensive*, depending on whether their value changes with the size of the sample. **Intensive properties**, like temperature and melting point, have values that do not depend on the amount of sample. Thus, a small ice cube at 0°C might have the same temperature as a massive iceberg. **Extensive properties**, like mass, length, and volume, have values that *do* depend on the sample size. An ice cube is much smaller than an iceberg and has less mass.

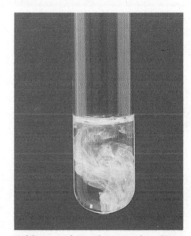

Addition of a solution of sodium chloride to a solution of silver nitrate yields a white precipitate of solid silver chloride.

TABLE 1.5	Some Examples of Physical and Chemical Properties		

Physical Properties		Chemical Properties
Temperature	Mass	Rusting (of iron)
Color	Odor	Combustion (of coal)
Boiling point	Solubility	Tarnishing (of silver)
Heat capacity	Hardness	Hardening (of cement)
Electrical conductance	Density	Explosion (of dynamite)

Both the weight and the pillow have the same mass, but the weight has a higher density because its volume is smaller.

TABLE 1.6	Densities of Some Common Materials		
Substance	**Density (g/mL)**	**Substance**	**Density (g/mL)**
Ice (0°C)	0.917	Human fat	0.94
Water (4.0°C)	1.0000	Cork	0.22–0.26
Gold	19.31	Table sugar	1.59
Helium (25°C)	0.000 164	Balsa wood	0.12
Air (25°C)	0.001 185	Earth	5.54

The intensive physical property that relates the mass of an object to its volume is called **density**. Density, which is simply the mass of an object divided by its volume, is often expressed in derived units of g/cm^3 or g/mL. Thus, if we know the density of a substance, we know how much a given volume will weigh. The densities of some common materials are given in Table 1.6.

$$\text{Density} = \frac{\text{mass (g)}}{\text{volume (mL or cm}^3)}$$

Most substances change in volume when heated or cooled, and densities are therefore temperature dependent. For example, at 3.98°C, a 1.0000 mL container holds exactly 1.0000 g of water (density = 1.0000 g/mL). As the temperature is raised, however, the volume occupied by the water expands so that only 0.9584 g fits in the 1.0000 mL container at 100°C (density = 0.9584 g/mL). When reporting a density, the temperature must also be specified.

Icebergs float because solid ice is less dense than liquid water.

Although most substances expand when heated and contract when cooled, water behaves differently. Water contracts when cooled from 100°C to 3.98°C, but below this temperature it begins to expand again. Thus, the density of liquid water is at its maximum of 1.0000 g/mL at 3.98°C, but decreases to 0.999 87 g/mL at 0°C (Figure 1.10). When freezing occurs, the density drops still further to a value of 0.917 g/cm^3 for ice at 0°C. Ice and any other substance with a density less than 1 g/cm^3 will float on water, but any substance with a density greater than 1 g/cm^3 will sink.

Knowing the density of a substance can be very useful, because it is often easier to measure the volume of a liquid than its mass. Suppose, for example, that you needed 1.5 g of ethyl alcohol. Rather than try to weigh out exactly the right amount, it would be much easier to look up the density of ethyl alcohol (0.7893 g/mL at 20°C) and measure out the correct volume with a syringe or graduated cylinder.

$$1.5 \text{ g ethyl alcohol} \times \frac{1 \text{ mL}}{0.7893 \text{ g}} = 1.9 \text{ mL ethyl alcohol}$$

A precise amount of a liquid is easily measured by syringe if the density of the liquid is known.

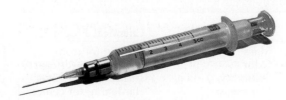

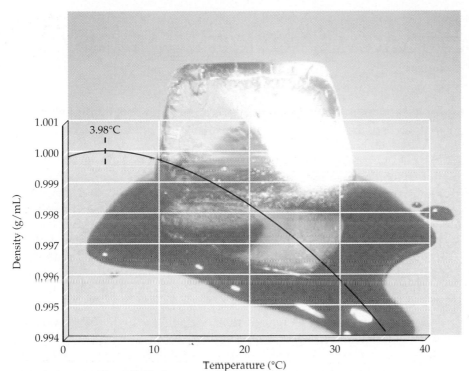

FIGURE 1.10 The density of water at different temperatures. Density reaches a maximum value of 1.000 g/mL at 3.98°C.

EXAMPLE 1.7

What is the volume (in cm^3) of 1.000 lb of gold?

BALLPARK SOLUTION   One pound equals about 450 g. Since 20 g of gold has a volume of about 1 cm^3 (Table 1.6), 100 g of gold has a volume of 5 cm^3, and 450 g has a volume of about 22 cm^3.

DETAILED SOLUTION   First convert pounds into grams by setting up an equation:

$$1.000 \text{ lb gold} \times \frac{454 \text{ g}}{1 \text{ lb}} = 454 \text{ g gold}$$

Then, use the density of gold as a conversion factor to calculate volume:

$$454 \text{ g gold} \times \frac{1 \text{ cm}^3}{19.31 \text{ g}} = 23.5 \text{ cm}^3 \text{ gold}$$

PROBLEM 1.17 What is the density of glass in g/cm^3 if a sample weighing 27.43 g has a volume of 12.40 cm^3?

PROBLEM 1.18 Chloroform, a substance once used as an anesthetic agent, has a density of 1.483 g/mL at 20°C. How many mL would you use if you needed 9.37 g?

interlude—CHEMICALS, TOXICITY, AND RISK

Life is not risk-free—we all take many risks each day, beginning when we get out of bed in the morning and turn on the lights. (200 people are electrocuted in home accidents each year in the United States.) We may decide to ride a bike rather than drive, even though there is a 10 times greater likelihood per mile of dying on a bicycle than in a car. We may decide to smoke cigarettes, even though smoking increases our chance of getting cancer by 50 percent. Making judgments that affect our health is something we do every day without thinking about it.

What about risks from "chemicals?" News reports sometimes make it seem that our food is covered with pesticide residues and filled with dangerous additives, that our land is polluted by toxic waste dumps, and that our medicines are unsafe. How bad are the risks from chemicals, and how are the risks evaluated?

First, it's important to realize that *everything* is made of chemicals. There is no such thing as a "chemical-free" food, cleanser, cosmetic, or anything else. Second, it's important to realize that there is no meaningful distinction between "natural" substances and "synthetic" ones; a chemical is a chemical is a chemical. Many naturally occurring substances—botulism toxin, for example—are extraordinarily toxic, and many synthetic substances—polyethylene, for example—are virtually harmless.

Risk evaluation of chemicals is generally carried out by exposing test animals, usually mice or rats, to a chemical and then monitoring for signs of harm. To limit the expense and time needed for testing, the amounts administered are hundreds or thousands of times larger than those a human might normally encounter. Test animals are exposed to different amounts of the chemical to generate an S-shaped *dose-response curve* such as that in Figure 1.11, from which toxicity is determined. The *acute chemical toxicity* (as opposed to long-term toxicity) observed in animal tests is then reported as an LD_{50} *value*, the amount of a substance per kilogram of body weight that is a lethal dose for 50 percent of the test animals. LD_{50} values of various substances are shown in Table 1.7.

Even with an LD_{50} value established in test animals, the risk of human exposure to a given substance is still hard to assess. If a substance is harmful to rats, is it necessarily harmful to humans? How can a large dose for a small animal be translated into a small dose for a large human? As pointed out by the sixteenth-century Swiss physician Paracelsus, "The dose makes the

The annual risks of the average automobile driver and a professional mountaineer are about the same.

TABLE 1.7	Some LD_{50} Values		
Substance	LD_{50} (g/kg)	Substance	LD_{50} (g/kg)
Aflatoxin B_1	4×10^{-4}	Sodium chloride	4
Aspirin	1.1	Sodium cyanide	1.5×10^{-2}
Chloroform	3.2	Sodium cyclamate	17
Ethyl alcohol	10.6	Water	500

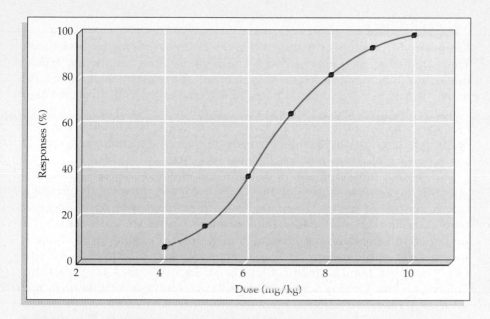

FIGURE 1.11 Typical S-shaped dose-response curve for a group of test animals exposed to a chemical. The response levels off at both low and high doses.

poison." All substances are toxic to some organisms to some extent, and the difference between help and harm is often a matter of degree. Common table salt (sodium chloride), for example, is essential to health yet has an LD_{50} of 4 g/kg in rats. Vitamin A is necessary for vision yet causes cancer at high doses. Even water is toxic at a high enough dose, with an LD_{50} of 500 g/kg. Thus, an important aspect of risk management involves establishing *threshold* values, an exposure level below which the risk from a substance is assumed to be negligible.

A threshold exposure value is obtained from a dose-response curve by extrapolating the curve to lower doses until an arbitrarily low response point is reached, often 0.0001 percent, or 1 person in 1 million. Unfortunately, the mathematical uncertainties involved in the extrapolation are enormous, and the final result may be off by a factor of 10,000 or more. Even if the extrapolation is accurate, the benefit produced by the chemical may far outweigh the additional danger to 1 in 1 million persons.

How we evaluate risk is strongly influenced by familiarity. The presence of chloroform in municipal water supplies at a barely detectable level of 0.000 000 01 percent has caused concern in many cities, yet chloroform has a lower acute toxicity than aspirin, a much more familiar substance. Many foods contain natural ingredients far more toxic than synthetic food additives or pesticide residues, but the ingredients are ignored because the foods are familiar. Peanut butter, for example, contains tiny amounts of aflatoxin, an enormously more potent cancer threat than sodium cyclamate, an artificial sweetener that has been banned because of its "risk."

All decisions involve trade-offs. Does the benefit of a pesticide that will increase the availability of food outweigh the health risk to 1 person in 1 million who are exposed to it? Do the beneficial effects of a new drug outweigh a potentially dangerous side effect in 0.001 percent of users? The answers are not always obvious, but it is the responsibility of legislators and well-informed citizens to keep their responses on a factual level rather than an emotional one.

Which is more toxic?

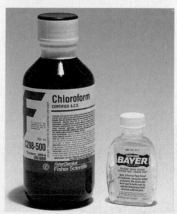

SUMMARY

Chemistry is the study of **matter** and of the changes that matter undergoes. These studies are best approached by posing questions, conducting experiments, and then devising **theories** to interpret the experimental results. All matter is formed from one or more of 109 presently known **elements**—fundamental substances that can't be chemically broken down further. Elements are symbolized by one-, two-, or three-letter abbreviations and are organized into a **periodic table**. Elements in the same column, or **group**, of the periodic table show similar chemical properties.

Accurate **measurement** of physical quantities is crucial to accurate scientific experimentation, and the units used are those of the Système Internationale (**SI units**). There are seven fundamental SI units, together with other derived units: **Mass**, the amount of matter an object contains, is measured in **kilograms (kg)**; **length** is measured in **meters (m)**; **volume** is measured in **cubic meters (m³)**; and **temperature** is measured in **kelvins (K)**. The more familiar metric **liter (L)** is also still used for measuring volume, and the **Celsius degree** (°C) is still used for measuring temperature.

Since many experiments involve numerical calculations, it's often necessary to manipulate and convert different units of measure. The simplest way to carry out such conversions is to use the **dimensional-analysis** method, in which an equation is set up so that unwanted units cancel and only the desired units remain. It's also important when measuring physical quantities or carrying out calculations to indicate the precision of the measurement by **rounding-off** the result to the correct number of **significant figures**.

The characteristics, or **properties**, that are used to describe matter can be classified in several ways. **Physical properties** are those that can be determined without changing the chemical composition of the sample, whereas **chemical properties** are those that *do* involve a chemical change in the sample. **Intensive properties** are those whose values do not depend on the size of the sample, whereas **extensive properties** are those whose values *do* depend on sample size. **Density**, the intensive property that relates mass to volume, is one of the most important and useful physical properties.

1. Where on the following outline of a periodic table are the indicated elements or groups of elements?
   (a) the alkali metals   (b) the halogens
   (c) the alkaline-earth metals (d) the transition metals
   (e) hydrogen   (f) helium

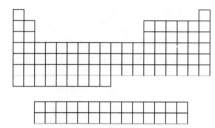

2. Show approximately where on the following outline of a periodic table the dividing line between metals and nonmetals falls.

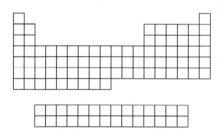

3. Is the element marked in red on the following periodic table likely to be a gas, a liquid, or a solid? What is the atomic number of the element in blue? Name at least one other element that is likely to be chemically similar to the element in green.

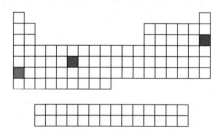

4. Characterize each of the following dartboards according to the accuracy and precision of the results.

(a)   (b)

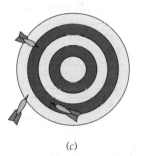

(c)

5. Assume that at some time in the future an antigravity machine is invented. If a person weighing 135 lb (61.2 kg) steps inside the machine and sets it for 85 lb of antigravity force, what will the person weigh? What will the person's mass be? What would happen to the person if the machine was set for 165 lb of antigravity force?

ADDITIONAL PROBLEMS

Problems 1.1–1.18 appear within the chapter.

ELEMENTS AND THE PERIODIC TABLE

1.19 How many elements are presently known? About how many occur naturally?

1.20 What are the rows called and what are the columns called in the periodic table?

1.21 How many groups are there in the periodic table? How are they labeled?

1.22 What common characteristics do elements within a group of the periodic table have?

1.23 Where in the periodic table are the main-group elements found? Where are the transition-metal groups found?

1.24 Where in the periodic table are the metallic elements found? Where are the nonmetallic elements found?

1.25 What is a semimetal, and where in the periodic table are semimetals found?

1.26 List several general properties of **(a)** alkali metals **(b)** noble gases **(c)** halogens

1.27 Without looking at a periodic table, list as many alkali metals as you can (there are five common ones.)

1.28 Without looking at a periodic table, list as many halogens as you can (there are four common ones).

1.29 Without looking at a periodic table, list as many noble gases as you can (there are six).

1.30 What are the symbols for the following elements?
(a) americium (used in smoke detectors)
(b) germanium (used in semiconductors)
(c) technetium (used in biomedical imaging)
(d) arsenic (used in pesticides)
(e) cadmium (used in rechargeable batteries)
(f) iridium (used for hardening alloys)

1.31 Give the names corresponding to the following symbols.
(a) Te **(b)** Re **(c)** Be **(d)** Ar **(e)** Pu

1.32 What is wrong with each of the following statements?
(a) The symbol for tin is Ti.
(b) The symbol for manganese is Mg.
(c) The symbol for potassium is Po.
(d) The symbol for helium is HE.

1.33 What is wrong with each of the following statements?
(a) The symbol for carbon is ca.
(b) The symbol for sodium is So.
(c) The symbol for nitrogen is Ni.
(d) The symbol for chlorine is Cr.

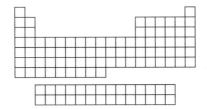

UNITS AND SIGNIFICANT FIGURES

1.34 Why is accurate measurement crucial in science?

1.35 What is the difference between a physical quantity like length and a number?

1.36 What SI units are used for measuring these quantities?
(a) mass
(b) length
(c) temperature
(d) volume

1.37 What is the difference between a derived SI unit and a base SI unit? Give an example of each.

1.38 What SI prefixes correspond to the following multipliers?
(a) 10^3 **(b)** 10^{-6} **(c)** 10^9 **(d)** 10^{-12} **(e)** 10^{-2}

1.39 What is the difference between mass and weight?

1.40 Which is larger, a Fahrenheit degree or a Celsius degree? By how much?

1.41 What is the difference between a kelvin and a Celsius degree?

1.42 What is the difference between a cubic decimeter (SI) and a liter (metric)?

1.43 Why is it often necessary to round off the results of numerical calculations?

1.44 Which of the following statements use exact numbers?
(a) 1 ft = 12 in.
(b) The population of Mexico City is 8,988,230.
(c) The world record for the 1 mile run, set by Britain's Steve Cram in 1985, is 3 minutes 46.31 seconds.

1.45 What is the difference in mass between a nickel that weighs 4.8 g and a nickel that weighs 4.8673 g?

1.46 Bottles of wine sometimes carry the notation "Volume = 75 cL." What does the unit cL mean?

1.47 What do the following abbreviations stand for?
(a) dL
(b) dm
(c) μm
(d) nL

1.48 How many picograms are there in 1 mg? in 35 ng?

1.49 How many microliters are there in 1 L? in 20 mL?

1.50 Carry out the following conversions:
(a) 5 pm = _____ cm = _____ nm
(b) 8.5 cm^3 = _____ m^3 = _____ mm^3
(c) 65.2 mg = _____ g = _____ pg

1.51 Which is larger and by approximately how much?
(a) a liter or a quart
(b) a mile or a kilometer
(c) a gram or an ounce
(d) a centimeter or an inch

1.52 How many significant figures are there in each of the following measurements?
(a) 35.0445 g **(b)** 59.0001 cm **(c)** 0.030 03 kg
(d) 0.004 50 m **(e)** 67,000 m^2 **(f)** 3.8200×10^3 L

1.53 The Vehicle Assembly Building at the John F. Kennedy Space Center in Cape Canaveral, Florida, is the largest building in the world, with a volume of 3,666,500 m³. Express this volume in scientific notation.

1.54 Express the following measurements in scientific notation.
(a) 453.32 mg
(b) 0.000 0421 mL
(c) 667,000 g

1.55 Convert the following measurements from scientific notation to ordinary notation:
(a) 3.221×10^{-3} mm (b) 8.940×10^5 m
(c) $1.350\ 82 \times 10^{-12}$ m³

1.56 The diameter of the earth at the equator is 7,926.381 mi. Round off this quantity to four significant figures; to two significant figures. Express the answers in scientific notation.

1.57 Round off the following quantities to the number of significant figures indicated in parentheses:

(a) 35,670.06 m (4, 6)
(b) 68.507 g (2, 3)
(c) 4.995×10^3 cm (3)
(d) $2.309\ 85 \times 10^{-4}$ kg (5)

1.58 Express the results of the following calculations with the correct number of significant figures:
(a) 4.884×2.05
(b) $94.61 \div 3.7$
(c) $3.7 \div 94.61$
(d) $5,502.3 + 24 + 0.01$
(e) $86.3 + 1.42 - 0.09$
(f) 5.7×2.31

1.59 Express the results of the following calculations with the correct number of significant figures.
(a) $\dfrac{3.41 - 0.23}{5.2333} \times 0.205$
(b) $\dfrac{5.556 \times 2.3}{4.223 - 0.08}$

UNIT CONVERSIONS

1.60 Carry out the indicated conversions.
(a) How many grams of meat are in a quarter-pound hamburger (0.25 lb)?
(b) How tall in meters is the Sears Tower in Chicago (1454 ft)?
(c) How large in m² is the land area of Australia (2,941,526 mi²)?

1.61 Convert the following quantities into SI units with the correct number of significant figures:
(a) 5.4 in.
(b) 66.31 lb
(c) 0.5521 gal
(d) 65 mi/h
(e) 978.3 yd³

1.62 The volume of water used for crop irrigation is measured in units of acre-feet, where 1 acre-foot is the amount of water needed to cover 1 acre of land to a depth of 1 ft.
(a) If there are 640 acres per square mile, how many ft³ of water are in 1 acre-foot?
(b) How many acre-feet are in Lake Erie (total volume = 116 mi³)?

1.63 The height of a horse is usually measured in hands instead of in feet, where one hand equals 1/3 ft (exactly).
(a) How tall in cm is a horse of 18.6 hands?
(b) What is the volume in m³ of a box measuring $6 \times 2.5 \times 15$ hands?

1.64 Amounts of substances dissolved in solution are often expressed as mass per unit volume. For example, normal human blood has a cholesterol concentration of about 200 mg/100 mL. Express this concentration in the following units.
(a) mg/L
(b) µg/mL
(c) g/L
(d) ng/µL

1.65 How much total blood cholesterol in grams (Problem 1.64) does a person have if the normal blood volume in the body is 5 L?

1.66 Weights in England are commonly measured in *stones*, where 1 stone = 14 lb. What is the weight in pounds of a person who weighs 8.65 stones?

1.67 Among many alternative units that might be considered as a measure of time is the *shake* rather than the second. Based on the expression "faster than a shake of a lamb's tail," we'll define 1 shake as equal to 2.5 $\times 10^{-4}$ s. If a car is traveling at 55 mi/h, what is its speed in cm/shake?

1.68 Administration of digitalis, a drug used to control atrial fibrillation in heart patients, must be carefully controlled because even a modest overdosage can be fatal. To take differences between patients into account, drug dosages are prescribed in terms of mg/kg body weight. Thus, a child and an adult differ greatly in weight, but both receive the same dosage per kilogram of body weight. At a dosage of 20 mg/kg body weight, how many mg of digitalis should a 160 lb patient receive?

TEMPERATURE

1.69 The normal body temperature of a goat is 39.9°C, and that of an Australian spiny anteater is 22.2°C. Express these temperatures in °F.

1.70 Of the 90 elements that occur naturally on earth, only three are liquid at normal temperatures: mercury (melting point −38.87°C), gallium (melting point 29.78°C), and bromine (melting point −7.2°C). Convert these melting points from °C to °F.

1.71 Tungsten, the element used to make filaments in light bulbs, has a melting point of 6170°F. Convert this temperature to °C and to K.

1.72 Suppose you were dissatisfied with both Celsius and Fahrenheit units and wanted to design your own temperature scale based on ethyl alcohol (ethanol). On the Celsius scale, ethanol has a melting point of −117.3°C and a boiling point of 78.5°C, but on your new scale calibrated in units of °E, you define ethanol to melt at 0°E and boil at 200°E. Answer the following questions.
   (a) How does your new Ethanol degree compare in size with a Celsius degree?

(b) How does an Ethanol degree compare in size with a Fahrenheit degree?
(c) What are the melting and boiling points of water on the Ethanol scale?
(d) What is normal human body temperature (98.6°F) on the Ethanol scale?
(e) If the outside thermometer read 130°E, how would you dress to go out?

1.73 Answer parts **(a)** through **(d)** of Problem 1.72 assuming that your new temperature scale is based on ammonia, NH_3. On the Celsius scale, ammonia has a melting point of −77.7°C and a boiling point of −33.4°C, but on your new scale calibrated in units of °A, you define ammonia to melt at 0°A and boil at 100°A.

DENSITY

1.74 What is density, and why do more dense substances like gold appear heavier than less dense substances like styrofoam?

1.75 Aspirin has a density of 1.40 g/cm^3. What is the volume in cm^3 of an aspirin tablet weighing 250 mg? Of a tablet (?) weighing 500 lb?

1.76 Gaseous hydrogen has a density of 0.0899 g/L at 0°C, and gaseous chlorine has a density of 3.214 g/L at the same temperature. How many liters of each would you need if you wanted 1.0078 g hydrogen and 35.45 g chlorine?

1.77 What is the density of lead in g/cm^3 if a rectangular bar measuring 0.50 cm in height, 1.55 cm in width, and 25.00 cm in length has a mass of 220.9 g?

1.78 What is the density of lithium metal in g/cm^3 if a cylindrical wire with a diameter of 2.40 mm and a length of 15.0 cm has a mass of 0.3624 g?

1.79 When an irregularly shaped chunk of silicon weighing 8.763 g was placed in a graduated cylinder containing 25.00 mL of water, the water level in the cylinder rose to 28.76 mL. What is the density of silicon in g/cm^3?

1.80 When the experiment outlined in Problem 1.79 was repeated using a chunk of sodium metal, rather than silicon, an explosion occurred. Was this due to a chemical property or a physical property of sodium?

GENERAL PROBLEMS

1.81 What is the difference between a hypothesis and a theory?

1.82 What is the difference between a physical property and a chemical property?

1.83 Give four examples of physical properties.

1.84 Give the symbol for each of the following elements.
   (a) selenium (used in photocopiers)
   (b) rhenium (used for hardening alloys)
   (c) cobalt (used in magnets)
   (d) rhodium (used in catalytic converters)

1.85 Give the names corresponding to the following symbols.
   (a) Se
   (b) Sb
   (c) Ba
   (d) Cs
   (e) Np

1.86 What SI prefixes correspond to the following multipliers?
   (a) 10^6
   (b) 10^{-9}
   (c) 10^1
   (d) 10^2

1.87 What do the following abbreviations stand for?
   (a) cL
   (b) m^2
   (c) K
   (d) nm

1.88 How many significant figures are in each of the following measurements?
   (a) 105.331 mg
   (b) 4050 km
   (c) 405.0 km
   (d) 0.022 107 kg

1.89 Sodium chloride has a melting point of 1074 K and a boiling point of 1686 K. Convert these temperatures to °C and to °F.

1.90 The mercury in thermometers freezes at −38.9°C. What is this temperature in °F?

1.91 The density of chloroform, a widely used organic solvent, is 1.4832 g/mL at 20°C. How many mL would you use if you wanted 112.5 g of chloroform?

1.92 The density of sulfuric acid, H_2SO_4, is 15.28 lb/gal. What is the density of sulfuric acid in g/mL?

1.93 Sulfuric acid (Problem 1.92) is produced in larger amount than any other chemical—80.31 × 10⁹ lb in 1993. What is the volume of this amount in liters?

1.94 The caliber of a gun is expressed by measuring the diameter of the gun barrel in hundredths of an inch. A "22" rifle, for example, has a barrel diameter of 0.22 in. What is the barrel diameter of a .22 rifle in millimeters?

chapter 2 ATOMS, MOLECULES, AND IONS

People have always been fascinated by changes, particularly by those that are useful or beneficial. In the ancient world, the change that occurred when a stick of wood burned, gave off heat, and turned to a small pile of ash was especially important. Similarly, the change that occurred when a reddish lump of rock (iron ore) was heated with charcoal and produced a gray metal (iron) useful for making weapons, tools, and other implements was of enormous value. Observing such changes eventually caused Greek philosophers to think about what different materials might be composed of and led to the idea of fundamental substances that we today call elements.

At the same time that the ancient Greeks were pondering the question of elements, they were also musing about related matters: What is an element made of? Is matter continuously divisible into ever smaller and smaller pieces, or is there an ultimate limit? Can you cut a piece of gold in two, take one of the pieces and cut *it* in two, and so on indefinitely, or is there a point at which you must stop? Although most thinkers, including Plato and Aristotle, thought that matter is continuous, the Greek philosopher Democritus disagreed. Democritus proposed that elements are composed of tiny particles that we now call **atoms**, from the Greek word *atomos*, meaning indivisible. Little else was learned about elements and atoms until the birth of modern experimental science nearly 2000 years later.

Most substances, including the granite in Yosemite National Park's Half-dome, are mixtures rather than pure compounds.

2.1 ►CONSERVATION OF MASS AND THE LAW OF DEFINITE PROPORTIONS

The Englishman Robert Boyle (1627–1691) is generally credited with being the first person to study chemistry as a separate intellectual discipline and with being the first to carry out rigorous chemical experiments. Through a careful series of researches into the nature and behavior of gases, Boyle provided clear evidence for the atomic character of matter. In addition, Boyle was the first to clearly define an element as a substance that can't be chemically broken down further and to suggest that a substantial number of different elements might exist.

Progress in the field of chemistry was slow in the 100 years following Boyle, and it was not until the work of Joseph Priestley (1733–1804) and Antoine Lavoisier (1743–1794) that the next great leap was made. Priestley isolated the gas *oxygen*[1] in 1774, and Lavoisier soon showed that oxygen is the key substance in combustion. Furthermore, Lavoisier demonstrated by careful measurements that when combustion is carried out in a closed container, the mass of the combustion products is exactly equal to the mass of the starting reactants. For example, when hydrogen gas burns and combines with oxygen to yield water (H_2O), the mass of the water formed is equal to the mass of the hydrogen and oxygen consumed. Called the *law of mass conservation*, this principle is a cornerstone of chemical science.

LAW OF MASS CONSERVATION   Mass is neither created nor destroyed in chemical reactions.

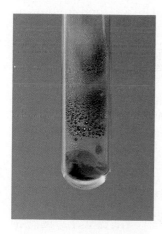

Pure oxygen was first made by heating HgO.

It's easy to demonstrate the truth of the law of mass conservation by carrying out a simple experiment like that shown in Figure 2.1. If weighed amounts of mercuric nitrate [$Hg(NO_3)_2$] and potassium iodide (KI) are dissolved in water and the solutions are mixed, an immediate chemical reaction occurs, leading to formation of an insoluble orange solid, mercuric iodide (HgI_2). Filtering the reaction mixture removes the mercuric iodide, and evaporation of the water leaves a deposit of potassium nitrate (KNO_3).

---

[1] Priestley discovered that pure oxygen can be prepared by heating mercuric oxide, HgO, according to the equation $2\,HgO \longrightarrow O_2 + 2\,Hg$.

FIGURE 2.1 An illustration of the law of conservation of mass. In any chemical reaction, the combined masses of the products always equals the combined masses of the reactants. In this sequence of photos, **(a)** weighed amounts of solid KI and solid $Hg(NO_3)_2$ are taken. **(b)** After dissolving the reactants in water, the solutions are mixed to give a precipitate of solid HgI_2, which is removed by filtration. The aqueous solution that remains is then evaporated to yield solid KNO_2. **(c)** When the two products are weighed, the mass of the reactants KI and $Hg(NO_3)_2$ is equal to the mass of the products HgI_2 and KNO_2.

When the two products are weighed, it's found that their combined masses exactly equal the combined masses of the two starting substances

The combined masses of these two starting materials . . .   *. . . equal the combined masses of these two products.*

$$Hg(NO_3)_2 + 2\,KI \longrightarrow HgI_2 + 2\,KNO_3$$

Mercuric nitrate   Potassium iodide   Mercuric iodide   Potassium nitrate

Further investigations in the decades following Lavoisier led the French chemist Joseph Proust (1754–1826) to formulate a second fundamental chemical principle that we now call the *law of definite proportions*.

LAW OF DEFINITE PROPORTIONS   Different samples of a pure chemical substance always contain the same proportion of elements by mass.

Every sample of water (H_2O) contains 1 part hydrogen and 8 parts oxygen by mass; every sample of carbon dioxide (CO_2) contains 1 part carbon and 2.7 parts oxygen by mass, and so on. Elements do not combine chemically in random proportions.

2.2 ►DALTON'S ATOMIC THEORY AND THE LAW OF MULTIPLE PROPORTIONS

How can the law of mass conservation and the law of definite proportions be explained? Why do elements behave as they do? The answers were provided by the English schoolteacher John Dalton (1766–1844), who in 1808 published a new theory of matter. Dalton reasoned as follows:

1. *Elements are made of tiny particles called atoms.* Although Dalton didn't know what atoms might look like, he nevertheless felt that they were necessary to explain why there were so many different elements.

2. *Each element is characterized by the mass of its atoms. All atoms of the same element have the same mass, but atoms of different elements have different*

masses. Dalton realized that there must be some feature that distinguishes the atoms of one element from those of another. Because Proust's law of definite proportions showed that elements always combine in specific mass ratios, Dalton reasoned that the distinguishing feature between atoms of different elements must be *mass.*

3. *Chemical combination of elements to make different substances occurs when atoms join together in small, whole-number ratios.* This statement follows from the law of definite proportions because random combinations of elements are not observed.

4. *Chemical reactions only rearrange the way that atoms are combined; the atoms themselves are not changed.* Dalton realized that the law of mass conservation is valid only if atoms are chemically indestructible. If the same numbers and kinds of atoms are present in both reactants and products, then the masses of reactants and products must also be the same.

These samples of sulfur and carbon both contain the same number of atoms.

For any theory to be successful, it must not only explain known observations, it must also lead to the prediction of events or results yet unknown. Dalton realized that his atomic theory does exactly this: It predicts what has come to be called the *law of multiple proportions.*

Read again the third statement in Dalton's atomic theory: "Chemical combination . . . occurs when atoms join together in small, whole-number ratios." According to this statement, it might be possible for the same elements to combine in *different* ratios to give *different* substances. For example, nitrogen and oxygen can combine either in a 7:16 mass ratio to make a substance we know today as NO_2, or in a 7:8 mass ratio to make a substance we know as NO. Clearly, the first substance contains exactly twice as much oxygen as the second one.

LAW OF MULTIPLE PROPORTIONS	If two elements combine in different ways to form different substances, the mass ratios are small, whole-number multiples of each other.

NO_2: 16 g oxygen per 7 g nitrogen   O:N mass ratio = 16:7

NO: 8 g oxygen per 7 g nitrogen   O:N mass ratio = 8:7

Comparison of O:N ratios in NO_2 and NO:

$$\frac{\text{O:N mass ratio in } NO_2}{\text{O:N mass ratio in NO}} = \frac{(16 \text{ g O})/(7 \text{ g N})}{(8 \text{ g O})/(7 \text{ g N})} = 2$$

As another example, 1.00 g of oxygen can combine with 0.0625 g of hydrogen to yield hydrogen peroxide (H_2O_2) or with 0.125 g of hydrogen to yield water (H_2O). Again, the hydrogen reacts in a multiple of 0.125/0.0625, or 2:

Comparison of H:O ratios in H_2O and H_2O_2:

$$\frac{\text{H:O mass ratio in } H_2O}{\text{H:O mass ratio in } H_2O_2} = \frac{(0.125 \text{ g H})/(1.00 \text{ g O})}{(0.0625 \text{ g H})/(1.00 \text{ g O})} = 2.00$$

These results make sense only if we assume that matter is composed of discrete atoms, which combine with each other in specific and well-defined ways (Figure 2.2).

When ignited, hydrogen gas reacts violently with oxygen to yield water.

FIGURE 2.2 An illustration of Dalton's law of multiple proportions. Atoms of nitrogen and oxygen can combine in specific proportions to make either NO or NO_2. NO_2 contains exactly twice as many atoms of oxygen per atom of nitrogen as NO does.

EXAMPLE 2.1

Methane and ethane are both constituents of natural gas. A sample of methane contains 11.40 g of carbon and 2.85 g of hydrogen, whereas a sample of ethane contains 4.47 g of carbon and 1.49 g of hydrogen. Show that the two substances obey the law of multiple proportions.

SOLUTION   First, find the C:H mass ratio in each compound.

$$\text{Methane: C:H mass ratio} = \frac{(11.40 \text{ g C})}{(2.85 \text{ g H})} = 4.00$$

$$\text{Ethane: C:H mass ratio} = \frac{(4.47 \text{ g C})}{(1.49 \text{ g H})} = 3.00$$

The two C:H ratios are clearly small, whole-number multiples of each other:

$$\frac{\text{C:H mass ratio in methane}}{\text{C:H mass ratio in ethane}} = \frac{4.00}{3.00} = \frac{4}{3}$$

⌐ **PROBLEM 2.1** Substances 1 and 2 are colorless gases obtained by combination of sulfur with oxygen. Substance 1 results from combination of 6.00 g of sulfur with 5.99 g of oxygen, and substance 2 results from combination of 8.60 g of sulfur with 12.88 g of oxygen. Show that the mass ratios in the two substances are simple multiples of each other.

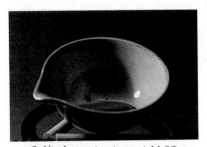

Sulfur burns in air to yield SO_2.

This photograph of the carbon atoms in graphite was taken by a scanning tunneling microscope, a remarkable device that lets scientists actually "see" individual atoms. (For more detail, see the *Interlude* at the end of this chapter.)

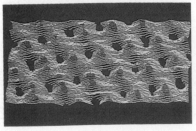

2.3 ▶THE STRUCTURE OF ATOMS: ELECTRONS

Dalton's atomic theory is fine as far as it goes, but it leaves unanswered the question, What is an atom made of? Dalton himself had no way of answering this question, and it was not until nearly a century later that experiments by the English physicist J. J. Thomson (1856–1940) provided some clues.

Thomson's experiments involved the use of *cathode-ray tubes*, the early predecessors of today's television and computer-display tubes. As illustrated in Figure 2.3, a cathode-ray tube is a glass tube from which the air has been removed and in which two thin pieces of metal called *electrodes* have been sealed. When a sufficiently high voltage is applied across the electrodes, an electric current flows through the tube from the negatively charged electrode (the *cathode*) to the positively charged electrode (the *anode*). If the tube is not fully evacuated but still contains a small amount of

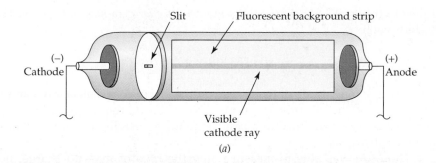

Slit

Fluorescent background strip

(−)
Cathode

(+)
Anode

Visible
cathode ray

(a)

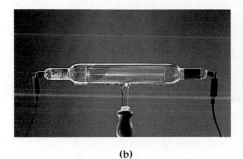

(b)

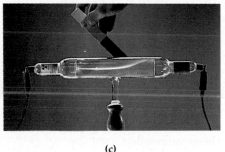

(c)

FIGURE 2.3 (a) Schematic of a cathode-ray tube; **(b)** an actual tube. A stream of rays (electrons) emitted from the negatively charged cathode passes through a slit and is detected by a fluorescent strip. The electron beam ordinarily travels in a straight line **(b)**, but it is deflected if either a magnetic field **(c)** or an electric field is present.

air or other gas, the flowing current is visible as a glow called a **cathode ray**. Furthermore, if the anode has a hole in it and the end of the tube is coated with a phosphorescent substance like zinc sulfide, the rays pass through the hole and strike the tube end, where they are visible as a bright spot of light. (In fact, this is exactly what happens in a television set.)

Experiments by a number of physicists in the 1890s had shown that cathode rays can be deflected by bringing either a magnet or an electrically charged plate near the tube (Figure 2.3). Because the beam is produced at a *negative* electrode and is deflected toward a *positive* plate, Thomson proposed that cathode rays must consist of tiny negatively charged particles, which we now call **electrons**. Furthermore, because electrons are emitted from electrodes made of many different metals, *all* these different substances must contain electrons.

Thomson reasoned that the amount of deflection of an electron beam by a nearby magnetic or electric field depends on three factors:

1. *The strength of the deflecting magnetic or electric field.* The stronger the magnet or the higher the voltage on the charged plate, the greater the deflection.
2. *The size of the negative charge on the electron.* The larger the charge on the particle, the greater the interaction with the magnetic or electric field, and the greater the deflection.
3. *The mass of the electron.* The lighter the particle, the greater the deflection. (Just as it's easier to deflect a Ping-Pong ball than a bowling ball.)

By carefully measuring the amount of deflection caused by electric and magnetic fields of known strength, Thomson was able to calculate the ratio of the electron's electric charge to its mass—its *charge-to-mass ratio, e/m*. The modern value is

$$\frac{e}{m} = 1.758\ 819 \times 10^8 \ \text{C/g}$$

where e is the magnitude of the charge on the electron in *coulombs* (C) and m is the mass of the electron in grams.[2] Note that e, the magnitude of the electron's charge, is defined as a *positive* quantity, making the actual charge on the electron $-e$.

Thomson was able to measure only the *ratio* of mass to charge, not mass itself, and it was left to the American R. A. Millikan (1868–1953) to devise a method for measuring the mass of an electron (Figure 2.4). In Millikan's experiment, a fine mist of oil was sprayed into a box, and the tiny droplets were allowed to fall between two horizontal plates. Observation of the spherical drops through a telescopic eyepiece made it possible to determine how rapidly they fell through the air, from which their mass could be calculated. The drops were then given a negative charge by irradiation with X rays. By applying a voltage to the plates, with the upper plate positive, it was possible to counteract the downward fall of the charged drops and keep them suspended.

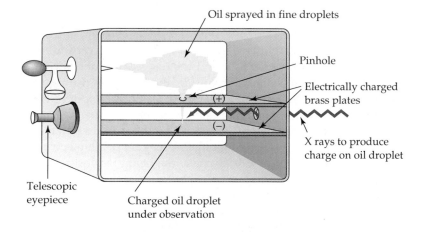

FIGURE 2.4 Millikan's oil-drop experiment. The falling oil droplets are given a negative charge, which makes it possible for them to be suspended between two electrically charged plates. Knowing the mass of the drop and the voltage on the plates makes it possible to calculate the charge on the drop, as explained in the text.

With the voltage on the plates and the mass of the drops thus known, Millikan was able to show that the charge on a drop was always a small, whole-number multiple of e, whose modern value is $1.602\ 177 \times 10^{-19}$ C. Substituting this e value into Thomson's charge-to-mass ratio gives the mass m of the electron as $9.109\ 390 \times 10^{-28}$ g:

Because
$$\frac{e}{m} = 1.758\ 819 \times 10^8 \ \text{C/g}$$

then
$$m = \frac{e}{1.758\ 819 \times 10^8 \ \text{C/g}} = \frac{1.602\ 177 \times 10^{-19}\ \cancel{\text{C}}}{1.758\ 819 \times 10^8 \ \frac{\cancel{\text{C}}}{\text{g}}}$$

$$= 9.109\ 390 \times 10^{-28} \ \text{g}$$

[2] We'll say more about coulombs and electrical charge in a later chapter.

2.4 ➤THE STRUCTURE OF ATOMS: RUTHERFORD'S NUCLEAR MODEL

Think about the consequences of Thomson's cathode-ray experiments. Because matter is electrically neutral overall, the fact that the atoms in an electrode can give off *negatively* charged particles (electrons) must mean that those same atoms also contain *positively* charged particles. The search for those positively charged particles and for an overall picture of atomic structure led to a landmark experiment published in 1911 by the British physicist Ernest Rutherford (1871–1937).

Rutherford's work involved the use of **alpha (α) particles**, a type of emission that had previously been observed to be given off by a number of naturally occurring radioactive elements, including radium, polonium, and radon. Rutherford knew that alpha particles are about 7000 times more massive than electrons and that they have a *positive* charge that is twice the magnitude of but opposite in sign to the charge on an electron.

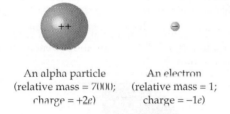

An alpha particle
(relative mass = 7000;
charge = +2e)

An electron
(relative mass = 1;
charge = −1e)

When Rutherford directed a beam of alpha particles at a thin metal foil, he found that almost all the particles passed through the foil undeflected but that a very small number (about 1 of every 20,000) were deflected at large angles (Figure 2.5).

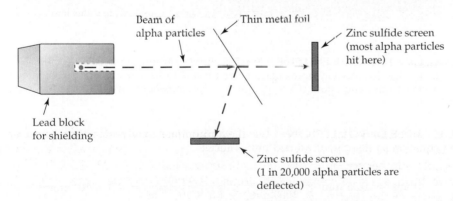

Beam of alpha particles

Thin metal foil

Zinc sulfide screen
(most alpha particles hit here)

Lead block for shielding

Zinc sulfide screen
(1 in 20,000 alpha particles are deflected)

FIGURE 2.5 An illustration of the Rutherford scattering experiment. When a beam of alpha particles is directed at a thin metal foil, most particles pass through the foil undeflected, but a small number are deflected at large angles.

Rutherford explained his results by proposing that a metal atom must be almost entirely empty space and have its mass concentrated in a tiny central core that he called the **nucleus** (Figure 2.6). If the nucleus contains the atom's positive charges and most of its mass, and if the electrons move in space a relatively large distance away, then it is clear why the observed scattering results are obtained: Most alpha particles encounter empty space as they fly through the foil. Only when a positive alpha particle chances to come near a massive positive nucleus is it repelled strongly enough to make it bounce backward.

FIGURE 2.6 The Rutherford model of an atom. Positive charge is concentrated in a small, dense core called the *nucleus*, and electrons move around the nucleus at a relatively large distance.

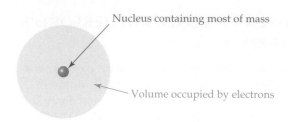

Nucleus containing most of mass

Volume occupied by electrons

The relative size of the nucleus in an atom is the same as that of a pea in the middle of this stadium.

Modern measurements show that an atom has a diameter of roughly 10^{-10} m and that a nucleus has a diameter of about 10^{-15} m. It's difficult to imagine from these numbers alone, though, just how small a nucleus really is. For comparison purposes, if an atom were the size of a large domed stadium such as the Houston Astrodome, the nucleus would be approximately the size of a small pea in the center of the playing field.

EXAMPLE 2.2

Ordinary "lead" pencils actually are made of a form of carbon called graphite. If the diameter of a carbon atom is 1.5×10^{-10} m and a pencil line is 0.35 mm wide, how many atoms wide is the line?

How many atoms wide is this line?

BALLPARK SOLUTION   Since a single carbon atom is about 10^{-10} m wide, it takes 10^{10} carbon atoms placed side by side to stretch 1 m, 10^7 carbon atoms to stretch 1 mm, and about 0.3×10^7 (or 3×10^6; 3 *million*) carbon atoms to stretch 0.35 mm.

DETAILED SOLUTION   Use the appropriate conversion factors to set up an equation so that the unwanted units cancel:

$$\text{Atoms} = 0.35 \text{ mm} \times \frac{1 \text{ m}}{1000 \text{ mm}} \times \frac{1 \text{ atom}}{1.5 \times 10^{-10} \text{ m}} = 2.3 \times 10^6 \text{ atoms}$$

The ballpark answer and the exact answer agree.

┌ **PROBLEM 2.2** The gold foil Rutherford used had a thickness of approximately 0.000 2 in. If a single gold atom has a diameter of 2.9×10^{-8} cm, how many atoms thick was Rutherford's foil?

┌ **PROBLEM 2.3** Here's a gee-whiz calculation: A small speck of carbon the size of a pinhead contains about 10^{19} atoms, and the circumference of the earth at the equator is 40,075 km. How many times around the earth would the carbon atoms extend if they were laid side by side?

2.5 ►THE STRUCTURE OF ATOMS: PROTONS AND NEUTRONS

Further experiments by Rutherford and others in the period from 1910 to 1930 ultimately showed that a nucleus is composed of two kinds of particles, called *protons* and *neutrons*. **Protons** have a mass of $1.672\,623 \times 10^{-24}$ g (about 1836 times greater than that of an electron) and are positively charged. Because the charge on a proton is opposite in sign but equal in size to that on an electron, the numbers of protons and electrons in a neutral atom are equal. **Neutrons** ($1.674\,929 \times 10^{-24}$ g) are almost identical in mass to protons but carry no charge. Thus, the number of neutrons in a nucleus is not directly related to the numbers of protons and electrons. Table 2.1 compares the three fundamental subatomic particles.

	Mass		**Charge**	
Particle	**(grams)**	**(amu[a])**	**(coulombs)**	**(e)**
Electron	$9.109\,390 \times 10^{-28}$	$5.485\,799 \times 10^{-4}$	$-1.602\,177 \times 10^{-19}$	-1
Proton	$1.672\,623 \times 10^{-24}$	$1.007\,276$	$+1.602\,177 \times 10^{-19}$	$+1$
Neutron	$1.674\,929 \times 10^{-24}$	$1.008\,664$	0	0

TABLE 2.1   A Comparison of Subatomic Particles

[a] The atomic mass unit (amu) is defined in the next section.

Thus far, we have described atoms only in general terms and have not yet answered the most important question, What is it that makes one atom different from another? How, for example, does an atom of gold differ from an atom of carbon? The answer to this question is really quite simple: *Elements differ from one another according to the number of protons their atoms contain*, a value called the *atomic number (Z)* of the element. Hydrogen, atomic number 1, has one proton and one electron; helium, atomic number 2, has two protons and two electrons; carbon, atomic number 6, has six protons and six electrons; and so on up to element number 109.

Atomic number (Z) = Number of protons in atom's nucleus
                      = Number of electrons surrounding atom's nucleus

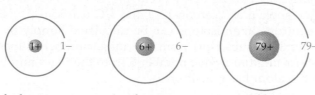

A hydrogen atom   A carbon atom   A gold atom
(1 proton; 1 electron) (6 protons; 6 electrons) (79 protons; 79 electrons)

In addition to protons, the nuclei of all but the simplest atom (hydrogen) also contain neutrons. The sum of the number of protons (Z) and the number of neutrons (N) in an atom is called the atom's *mass number (A)*.

Mass number of atom = Number of protons + number of neutrons in atom
$$A = Z + N$$

For example, most hydrogen atoms have one proton and no neutrons, and therefore have mass number 1. Most helium atoms have two protons and two neutrons, and therefore have mass number 4. Most carbon atoms have six protons and six neutrons, and therefore have mass number 12, and so on. Except for hydrogen, atoms always contain at least as many neutrons as protons, although there is no simple way to predict how many neutrons a given atom will have.

Notice in the previous paragraph that we said *most* hydrogen atoms have mass number 1, *most* helium atoms have mass number 4, and *most* carbon atoms have mass number 12. In fact, different atoms of the same element can have different mass numbers depending on how many neutrons they have. Atoms with identical atomic numbers but different mass numbers are called **isotopes**. Hydrogen, for example, has three isotopes (Figure 2.7). All hydrogen atoms have one proton in their nucleus (otherwise they wouldn't be hydrogen), but most have no neutrons. These hydrogen atoms, called *protium*, have mass number 1. In addition, a small fraction of hydrogen atoms, called *deuterium*, have one neutron and mass number 2. Still other hydrogen atoms, called *tritium*, have two neutrons and mass number 3. (Tritium is an unstable, radioactive isotope that does not occur naturally but is made artificially in nuclear reactors.) As other examples, there are 13 known isotopes of carbon, of which only two occur naturally, and there are 19 known isotopes of uranium.

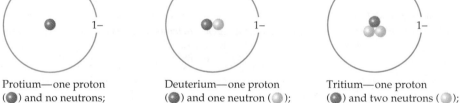

FIGURE 2.7 The three isotopes of hydrogen. All have atomic number 1, but they differ in their mass numbers—that is, in the numbers of neutrons in their nuclei.

Protium—one proton (●) and no neutrons; mass number = 1

Deuterium—one proton (●) and one neutron (○); mass number = 2

Tritium—one proton (●) and two neutrons (○); mass number = 3

Specific isotopes are represented by symbols in which the mass number is written as a left superscript and the atomic number is written as a left subscript. Thus, protium is represented as ^1_1H, deuterium as ^2_1H, and tritium as ^3_1H. Similarly, the two naturally occurring isotopes of carbon are represented as $^{12}_6C$ (spoken as "carbon-12") and $^{13}_6C$ (carbon-13). Note that the number of neutrons in an isotope can be calculated simply by subtracting the atomic number (subscript) from the mass number (superscript). For example, subtracting the atomic number 6 from the mass number 12 of $^{12}_6C$ tells us that the atom has 6 neutrons.

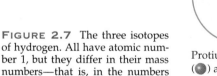

Mass number (number of protons plus neutrons)

$^{12}_6C$ ← Symbol of element

Atomic number (number of protons or electrons)

It's important to realize that the number of neutrons in an atom has little effect on the chemical properties of the atom. The chemistry of an element is largely determined by the number of protons in the nucleus (or, stated more properly, by the number of electrons surrounding the nucleus). All three isotopes of hydrogen behave almost identically in their chemical reactions.

EXAMPLE 2.3

The isotope of uranium used in nuclear power plants and atomic weapons is $^{235}_{92}U$ (uranium-235). How many protons, neutrons, and electrons does an atom of $^{235}_{92}U$ have?

SOLUTION   The subscript 92 in the symbol $^{235}_{92}U$ indicates that the element uranium has 92 protons and 92 electrons. The number of neutrons is equal to the difference between mass number (235) and atomic number (92): $235 - 92 = 143$ neutrons.

Uranium-235 is used as fuel in this nuclear power plant.

EXAMPLE 2.4

Element X is toxic to humans in high concentration but is essential to life at low concentrations. If an atom of element X contains 24 protons and 28 neutrons, identify the element and write the symbol for the isotope in the standard format.

SOLUTION   The atomic number is the number of protons in the atom's nucleus. A look at the periodic table shows that the element with atomic number 24 is chromium, Cr. The particular isotope of chromium in the present instance has a mass number of $24 + 28 = 52$ and is written $^{52}_{24}Cr$.

$\ulcorner$ **PROBLEM 2.4**   The isotope $^{75}_{34}Se$ is used medically for diagnosis of pancreatic disorders. How many protons, neutrons, and electrons does an atom of $^{75}_{34}Se$ have?

$\ulcorner$ **PROBLEM 2.5**   Chlorine, one of the elements in common table salt (sodium chloride), has two main isotopes with mass numbers 35 and 37. Look up the atomic number of chlorine, tell how many neutrons each isotope contains, and give the standard symbol for each.

$\ulcorner$ **PROBLEM 2.6**   Tell how many of each subatomic particle the following isotopes contain, and give the standard symbol for each.

(a) nitrogen-15   **(b)** iodine-131   **(c)** plutonium-242

$\ulcorner$ **PROBLEM 2.7**   An atom of element X contains 47 protons and 62 neutrons. Identify the element, and write the symbol for the isotope in the standard format.

2.6 ➤ATOMIC WEIGHT

Atoms are so tiny that even the smallest dust speck visible to the naked eye contains about 10^{16} atoms. Thus, the mass of a single atom in grams is much too small a number for convenience. We therefore use a unit called an **atomic mass unit (amu)**, also known as a *dalton*. One amu is defined as exactly 1/12th the mass of an atom of $^{12}_{6}C$ and is equal to 1.6605×10^{-24} g:

Mass of one $^{12}_{6}C$ atom = 12 amu (exactly)

$$1 \text{ amu} = \frac{\text{mass of one } ^{12}_{6}C \text{ atom}}{12} = 1.6605 \times 10^{-24} \text{ g}$$

The mass of a $^{12}_{6}C$ atom, defined to be exactly 12 amu, is the standard for expressing the values of all other atomic masses. Thus, a ^1_1H atom, which has a mass 1.007 825/12 times that of a $^{12}_{6}C$ atom, has a mass of 1.007 825 amu. Similarly, a $^{235}_{92}U$ atom, which has a mass 235.043 924/12 times that of a $^{12}_{6}C$ atom, has a mass of 235.043 924 amu, and so forth.

Since the mass of an atom's electrons is negligible compared with the mass of its protons and neutrons, defining 1 amu as $\frac{1}{12}$ the mass of a $^{12}_{6}C$ atom means that both protons and neutrons have a mass of almost exactly 1 amu (Table 2.1). Thus, the mass of an atom in amu is numerically close to the atom's mass number and is called its **atomic weight**. (The term *atomic mass* is really more accurate, but chemists have spoken of "atomic weight" for so long that its usage is unlikely to change.)

Look at the periodic table inside the front cover and you'll see the atomic weights of the elements given to several decimal places. (The unit amu is understood but not usually specified.) Surprisingly, though, no elements have atomic weights with whole-number values. Even carbon has its atomic weight listed as 12.011 amu, rather than exactly 12 amu. One reason for this apparent discrepancy is that most elements occur naturally as *mixtures* of different isotopes, and the atomic-weight values listed in tables are weighted averages for these isotope mixtures.

Carbon, for example, occurs on earth as a mixture of two isotopes, $^{12}_{6}C$ (98.89%) and $^{13}_{6}C$ (1.11%).[3] Although the atomic weight of any *individual* carbon atom is either 12 amu (a carbon-12 atom) or 13.0034 amu (a carbon-13 atom), the *average* atomic weight of a large collection of carbon atoms is 12.011 amu:

Avg. atomic weight = (fraction of $^{12}_{6}C$)(mass of $^{12}_{6}C$) + (fraction of $^{13}_{6}C$)(mass of $^{13}_{6}C$)

= (0.9889 × 12 amu) + (0.0111 × 13.0034 amu)

= 11.8668 amu + 0.1443 amu = 12.011 amu

The advantage of using average atomic weights is that it allows us to *count* a large number of atoms by *weighing* a sample of the element. For example, we can calculate that a small speck of carbon weighing 1.00 mg (1.00×10^{-3} g) contains 5.01×10^{19} carbon atoms:

$$1.00 \times 10^{-3} \, \cancel{g} \times \frac{1 \, \cancel{amu}}{1.6605 \times 10^{-24} \, \cancel{g}} \times \frac{1 \, C \, atom}{12.011 \, \cancel{amu}} = 5.01 \times 10^{19} \, C \, atoms$$

We'll have many occasions in future chapters to take advantage of this relationship between mass of sample, atomic weight, and number of atoms.

EXAMPLE 2.5

Chlorine has two naturally occurring isotopes: $^{35}_{17}Cl$ with an abundance of 75.77% and an atomic weight of 34.969 amu, and $^{37}_{17}Cl$ with an abundance of 24.23% and an atomic weight of 36.966 amu. What is the average atomic weight of chlorine?

---

[3] A third carbon isotope, $^{14}_{6}C$, is produced in small amounts in the upper atmosphere when cosmic rays strike $^{14}_{7}N$. The natural abundance of $^{14}_{6}C$ is so low, however, that it can be ignored when calculating an average atomic weight.

SOLUTION   The atomic weight of an element is equal to the sum of the masses of each isotope times the fraction of that isotope:

$$\text{Atomic weight} = (\text{fraction of } {}^{35}_{17}\text{Cl})(\text{mass of } {}^{35}_{17}\text{Cl}) + (\text{fraction of } {}^{37}_{17}\text{Cl})(\text{mass of } {}^{37}_{17}\text{Cl})$$

$$= (0.7577 \times 34.969 \text{ amu}) + (0.2423 \times 36.966 \text{ amu}) = 35.45 \text{ amu}$$

An alternative way of solving the problem is to imagine that you have a large number of chlorine atoms, say 10,000. Of the 10,000 atoms, 7577 of them (75.77%) would have an atomic weight of 34.969 amu, and 2423 of them (24.23%) would have an atomic weight of 36.966 amu. Thus, the total weight of the 10,000 atoms would be

$$\left(7577 \text{ atoms} \times 34.969 \frac{\text{amu}}{\text{atom}}\right) + \left(2423 \text{ atoms} \times 36.966 \frac{\text{amu}}{\text{atom}}\right) = 354,500 \text{ amu}$$

and the average atomic weight would be

$$\frac{354,500 \text{ amu}}{10,000 \text{ atoms}} = 35.45 \frac{\text{amu}}{\text{atom}}$$

PROBLEM 2.8   Copper metal has two naturally occurring isotopes: copper-63 (69.17%; atomic weight 62.94 amu) and copper-65 (30.83%; atomic weight 64.93 amu). Calculate the atomic weight of copper, and check your answer in a periodic table.

PROBLEM 2.9   Based on your answer to Problem 2.8, how many atoms of copper are there in a penny that weighs 2.15 g? (1 amu = 1.6605×10^{-24} g)

2.7 ▸ COMPOUNDS AND MIXTURES

Although there are only 90 naturally occurring elements, there are obviously far more than 90 different substances on earth. Just look around, and you'll find several hundred. These other substances form when atoms from two or more different elements combine, creating new materials with properties completely unlike those of their constituent elements. For example, when atoms of sodium (a soft, lustrous metal) combine with atoms of chlorine (a toxic green gas), the harmless white solid called sodium chloride is formed. Similarly, when two atoms of hydrogen combine with one atom of oxygen, water is formed. The transformations are said to be **chemical reactions**, and the substances formed are called **compounds**.

A compound is written by giving its **chemical formula**, which lists the symbols of the individual constituent elements and indicates the number of atoms of each element with a subscript. If no subscript is given, the number 1 is understood. Thus, water is written as H_2O, sodium chloride as NaCl, and sucrose (table sugar) as $C_{12}H_{22}O_{11}$. A chemical reaction is written in a standard format with the starting substances (called *reagents* or *reactants*) on the left, the products on the right, and an arrow between them to indicate a transformation. Although this manner of writing is usually called a **chemical equation**, the word "equation" is not strictly correct because the substances on either side of the arrow are not identical. Only the numbers and kinds of *atoms* are the same on both sides of the reaction arrow; the compounds themselves are different.

Two oxygen atoms   Four hydrogen atoms   Two water molecules
                                                 (H_2O - a chemical compound)

$$O_2 + 2\,H_2 \longrightarrow 2\,H_2O$$

Notice how compounds differ from mixtures. *Mixtures* are formed simply by blending two or more substances together in some random proportion without chemically changing the individual substances in the mixture. Thus, hydrogen gas and oxygen gas can be mixed together in any ratio without changing them (as long as there is no flame nearby to initiate reaction), just as a spoonful of sugar and a spoonful of salt can be mixed. *Compounds*, however, always have a constant composition because they are formed when atoms combine chemically in specific ways to yield new substances.

Crystalline quartz is a pure compound (SiO_2), but granite is a solid mixture of many compounds.

Mixtures can be further classified as either homogeneous or heterogeneous (Figure 2.8). **Homogeneous mixtures**, usually called **solutions**, have a constant composition throughout. Air is a gaseous solution of oxygen

FIGURE 2.8 A scheme for the classification of matter.

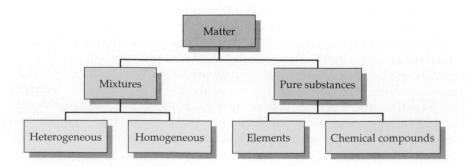

and nitrogen, seawater is a liquid solution of (primarily) sodium chloride dissolved in water, and brass is a solid solution of copper and zinc. **Heterogeneous mixtures** have regions with differing compositions. Salt with sugar, water with gasoline, and dust with air, are all heterogeneous mixtures.

With liquids it's often possible to distinguish between a homogeneous mixture and a heterogeneous one simply by looking at them. Homogeneous solutions are transparent, and heterogeneous mixtures are often cloudy. We'll look further at the nature and properties of liquid solutions in Chapter 11.

The ocean is a homogeneous mixture of (primarily) NaCl and water; milk is a heterogeneous mixture.

2.8 ►MOLECULES, IONS, AND CHEMICAL BONDS

Imagine what must happen when two atoms approach each other at the beginning of a chemical reaction. Because the electrons of an atom occupy a much greater volume than the nucleus, it's the *electrons* that actually make the contact when atoms collide. Thus, it's the electrons that form the **chemical bonds** that join atoms together in compounds. There are two fundamental kinds of chemical bonds between atoms—*covalent bonds* and *ionic bonds*. As a general rule, covalent bonds occur between two nonmetallic elements, and ionic bonds occur between a metal and a nonmetal.

Covalent Bonds: Molecules

A **covalent bond**, the most common kind of chemical bond, results when two atoms *share* several (usually two) of their electrons. A simple way to think about covalent bonds is to imagine what happens when two people want the same object. If both people hold on to the object tightly, they are effectively bound together. Neither person can walk away from the other as long as they both hold on. Similarly with atoms: When two atoms both hold on to some shared electrons, the atoms are bonded together.

The unit of matter that results when two or more atoms are joined by covalent bonds is called a **molecule**. A hydrogen chloride molecule (HCl) results when a hydrogen atom shares two electrons with a chlorine atom. A water molecule (H_2O) results when two hydrogen atoms share two electrons each with a single oxygen atom. An ammonia molecule (NH_3) results when three hydrogen atoms share two electrons each with a nitrogen atom, and so on. To visualize these and other molecules, it helps to imagine the individual atoms as spheres that stick together to give a discrete molecule with a specific three-dimensional shape (Figure 2.9).

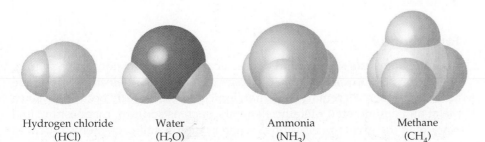

Hydrogen chloride (HCl) Water (H_2O) Ammonia (NH_3) Methane (CH_4)

FIGURE 2.9 Visualizing some simple molecules. The individual atoms (spheres) are joined together with specific geometries by sharing electrons in a covalent bond.

We sometimes represent molecules by giving their **structural formula**, which shows the specific connections between atoms and therefore gives more information than the chemical formula alone. Ethyl alcohol, for example, has the chemical formula C_2H_6O and the following structural formula:

$$C_2H_6O$$
(chemical formula)

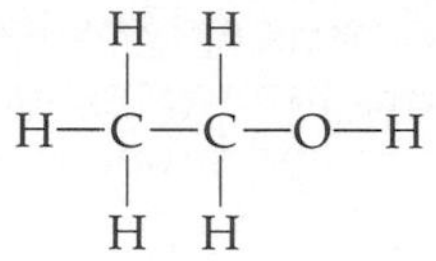

$$\begin{array}{ccc} & H & H \\ & | & | \\ H - & C - & C - O - H \\ & | & | \\ & H & H \end{array}$$
(structural formula)

Ethyl alcohol

The structural formula of ethyl alcohol uses lines between atoms to show that the two carbon atoms are covalently bonded to each other, that the oxygen atom is bonded to one of the carbon atoms, and that the six hydrogen atoms are distributed three to one carbon, two to the other carbon, and one to the oxygen.

Even some *elements* exist as molecules rather than as atoms. Hydrogen, oxygen, nitrogen, fluorine, chlorine, and bromine all exist as diatomic (two-atom) molecules whose two atoms are held together by covalent bonds. We therefore have to write them as such when using any of these elements in a chemical reaction.

H_2	O_2	N_2	F_2	Cl_2	Br_2
Hydrogen	Oxygen	Nitrogen	Fluorine	Chlorine	Bromine

EXAMPLE 2.6

Ethane, C_2H_6, has a structural formula in which each carbon atom is bonded to three hydrogens and to the other carbon. Draw this structural formula using lines between atoms to represent specific covalent bonds.

SOLUTION

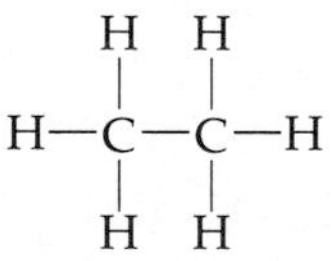

$$\begin{array}{ccc} & H & H \\ & | & | \\ H - & C - & C - H \\ & | & | \\ & H & H \end{array}$$
Ethane

⌜ **PROBLEM 2.10** Draw the structural formula of methylamine, CH_5N, a substance responsible for the odor of rotting fish. The carbon atom is bonded to the nitrogen atom and to three hydrogens. The nitrogen atom is bonded to the carbon and to two hydrogens. ⌟

Ionic Bonds

An **ionic bond**, the second kind of chemical bond, results not from a sharing of electrons but from a complete *transfer* of one or more electrons from one atom to another. As remarked earlier, ionic bonds generally occur between a metal and a nonmetal. Metallic elements, such as sodium, magnesium, and zinc, tend to give up electrons, whereas nonmetallic elements, such as oxygen, nitrogen, and chlorine, tend to accept electrons.

Imagine what happens when sodium metal comes in contact with chlorine gas. The sodium atom gives an electron to chlorine, resulting in the formation of two charged particles, called **ions**. Because sodium *loses* one electron, it loses one negative charge and becomes an Na^+ ion. Such positive ions are called **cations** (**cat**-ions). Conversely, because chlorine *gains* an electron, it gains a negative charge and becomes a Cl^- ion. Such negative ions are called **anions** (**an**-ions).

Sodium is a reactive metal, chlorine is a toxic green gas, and sodium chloride is a harmless white solid.

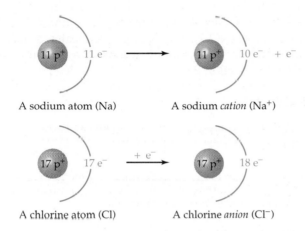

A similar reaction takes place when magnesium and chlorine (Cl_2) come in contact to form $MgCl_2$. Magnesium transfers *two* electrons to two chlorine atoms, yielding the doubly charged Mg^{2+} cation and two Cl^- anions.

$$Mg + Cl_2 \longrightarrow Mg^{2+} + Cl^- + Cl^- \ (MgCl_2)$$

As you might expect, positively charged cations like Na^+ and Mg^{2+} are strongly attracted by electrical forces to negatively charged anions like Cl^-, an attraction that we call an ionic bond. Unlike what happens when covalent bonds are formed though, we can't really talk about discrete Na^+Cl^- *molecules* under normal conditions. We can speak only of an **ionic solid**, in which equal numbers of Na^+ and Cl^- ions are packed together in a regular way (Figure 2.10). In a crystal of salt, for instance, each Na^+ ion is

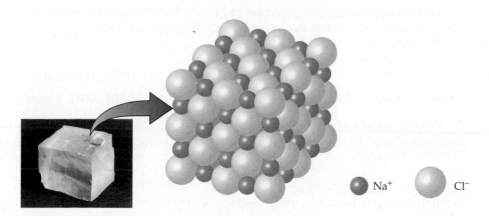

$\bullet$ Na^+ $\bigcirc$ Cl^-

FIGURE 2.10 The arrangement of Na^+ ions and Cl^- ions in a crystal of sodium chloride. Each Na^+ ion is surrounded by six neighboring Cl^- ions, and each Cl^- ion is surrounded by six neighboring Na^+ ions. Thus, there is no discrete "molecule" of NaCl. Instead, the entire crystal is an ionic solid.

surrounded by six nearby Cl^- ions, and each Cl^- ion is surrounded by six nearby Na^+ ions, but we can't specify what pairs of ions "belong" to each other as we can with atoms in covalent molecules.

Charged, covalently bonded groups of atoms, called **polyatomic ions**, also exist—for example, ammonium ion (NH_4^+), hydroxide ion (OH^-), nitrate ion (NO_3^-), and the doubly charged sulfate ion (SO_4^{2-}). You can think of these polyatomic ions as charged molecules because they consist of specific numbers and kinds of atoms joined together by covalent bonds in a definite way. When writing the formulas of substances that contain more than one of these ions, parentheses are placed around the entire polyatomic unit. The formula $Ba(NO_3)_2$, for example, indicates a substance made of Ba^{2+} cations and NO_3^- polyatomic anions in a 1:2 ratio. We'll say more about these ions in Section 2.10.

EXAMPLE 2.7

Which of the following compounds might be ionic and which molecular?
(a) BaF_2 **(b)** SF_4 **(c)** PH_3 **(d)** CH_3OH

SOLUTION Compound (a) is composed of a metal (barium) and a nonmetal (fluorine) and might therefore be ionic. Compounds (b)–(d) are composed entirely of nonmetals and therefore are probably molecular.

PROBLEM 2.11 Which of the following compounds might be ionic and which molecular?
(a) LiBr **(b)** $SiCl_4$ **(c)** BF_3 **(d)** CaO

2.9 ▶ACIDS AND BASES

Among the many ions we'll be discussing in the remainder of this book, the two most important are the hydrogen cation (H^+) and the hydroxide anion (OH^-). Since a hydrogen *atom* contains one proton and one electron, a hydrogen *cation* is simply a proton. A hydroxide ion, by contrast, is a polyatomic anion in which an oxygen atom is covalently bonded to a hydrogen atom. Although much of Chapter 15 is devoted to the chemistry of H^+ and OH^- ions, it's worthwhile now taking a preliminary look at these two species.[4]

The importance of the H^+ cation and OH^- anion is that they are fundamental to the concept of acids and bases. In fact, one useful definition of an **acid** is a substance that provides H^+ ions when dissolved in water, and one definition of a **base** is a substance that provides OH^- ions when dissolved in water.

An **acid:** A substance that provides H^+ ions in water   HCl, HNO_3, H_2SO_4, H_3PO_4

A **base:** A substance that provides OH^- ions in water   **NaOH, KOH, Ba(OH)$_2$**

Hydrochloric acid (HCl), nitric acid (HNO_3), sulfuric acid (H_2SO_4), and phosphoric acid (H_3PO_4) are among the most common acids. When any of

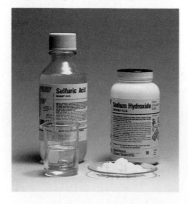

Sulfuric acid, H_2SO_4, is a commonly used acid; sodium hydroxide, NaOH, is a common base.

---

[4] The word *species* is frequently used in chemistry as a catchall term to refer to any particle, whether atom, ion, or molecule.

these substances is dissolved in water, H^+ ions are formed along with the corresponding anion. For example, HCl yields H^+ ions and Cl^- ions when it dissolves. We sometimes attach the designation (*aq*) to show that the ions are present in aqueous solution. In the same way, we often attach the designation (*g*) for gas, (*l*) for liquid, or (*s*) for solid to indicate the state of other reactants or products. Pure HCl, for example, is a gas, HCl(*g*), and pure HNO_3 is a liquid, $HNO_3(l)$.

Different acids can provide different numbers of H^+ ions depending on their structure. Hydrochloric acid and nitric acid provide one H^+ ion each per molecule, sulfuric acid can provide two H^+ ions per molecule, and phosphoric acid can provide three H^+ ions per molecule.

Hydrochloric acid: $\quad HCl(g) \xrightarrow{\text{dissolve in } H_2O} H^+(aq) + Cl^-(aq)$

Nitric acid: $\quad HNO_3(l) \xrightarrow{\text{dissolve in } H_2O} H^+(aq) + NO_3^-(aq)$

Sulfuric acid: $\quad H_2SO_4(l) \xrightarrow{\text{dissolve in } H_2O} H^+(aq) + HSO_4^-(aq)$

$\quad HSO_4^-(aq) \xrightarrow{} H^+(aq) + SO_4^{2-}(aq)$

Phosphoric acid: $\quad H_3PO_4(l) \xrightarrow{\text{dissolve in } H_2O} H^+(aq) + H_2PO_4^-(aq)$

$\quad H_2PO_4^-(aq) \xrightarrow{} H^+(aq) + HPO_4^{2-}(aq)$

$\quad HPO_4^{2-}(aq) \xrightarrow{} H^+(aq) + PO_4^{3-}(aq)$

Sodium hydroxide (NaOH; also known as *lye* or *caustic soda*), potassium hydroxide (KOH; also known as *caustic potash*), and barium hydroxide [$Ba(OH)_2$] are examples of bases. When any of these compounds dissolves in water, OH^- anions go into solution along with the corresponding metal cation. Sodium hydroxide and potassium hydroxide provide one OH^- ion each, and barium hydroxide provides two OH^- ions, as indicated by its formula, $Ba(OH)_2$.

Sodium hydroxide: $\quad NaOH(s) \xrightarrow{\text{dissolve in } H_2O} Na^+(aq) + OH^-(aq)$

Potassium hydroxide: $\quad KOH(s) \xrightarrow{\text{dissolve in } H_2O} K^+(aq) + OH^-(aq)$

Barium hydroxide: $\quad Ba(OH)_2(s) \xrightarrow{\text{dissolve in } H_2O} Ba^{2+}(aq) + 2\,OH^-(aq)$

⌐ **PROBLEM 2.12** Which of the following compounds are acids and which are bases? Explain your answers. **(a)** HF   **(b)** $Ca(OH)_2$ **(c)** LiOH   **(d)** HCN   ⌐

2.10 ➤NAMING CHEMICAL COMPOUNDS

In the early days of chemistry, when few pure compounds were known, newly discovered ones were often given fanciful names—morphine, quick lime, muriatic acid, and barbituric acid (named by its discoverer in honor of his friend Barbara) to cite a few. Today, with more than 11 million pure compounds known, there would be chaos unless a systematic method for naming compounds were used. Ideally, every chemical compound can be given a name that not only defines it uniquely but also allows chemists (and computers) to tell the compound's chemical makeup.

Different kinds of compounds are named by different rules. Ordinary salt, for example, is named *sodium chloride* because of its formula NaCl, but common table sugar ($C_{12}H_{22}O_{11}$) is named β-*D-fructofuranosyl-α-D-glucopyranoside* because of special rules for carbohydrates. We'll begin in this section by seeing how to name simple binary compounds—those made of only two elements—and then introduce additional rules in later chapters as the need arises.

Naming Binary Ionic Compounds

Binary ionic compounds are named by identifying the positive ion first and then the negative ion. The positive ion takes the same name as the element, but the negative ion takes only the first part of its name from the element and then adds the ending *-ide*. For example, KBr is named potassium bromide—*potassium* for the K^+ ion, and *bromide* for the negative Br^- ion derived from the element *brom*ine. Table 2.2 shows some common main-group ions, and Table 2.3 shows some common transition-metal ions.

LiF	CaBr$_2$	AlCl$_3$
Lithium fluoride	Calcium bromide	Aluminum chloride

There are several interesting features about Table 2.2. Note, for instance, that metals tend to form cations and nonmetals tend to form anions, as mentioned previously in Section 2.8. Note also that elements

---

TABLE 2.2   Main-Group Cations and Anions[a]

(1) 1A	(2) 2A		(13) 3A	(14) 4A	(15) 5A	(16) 6A	(17) 7A	(18) 8A
H^+ H^- Hydride								
Li^+	Be^{2+}				N^{3-} Nitride	O^{2-} Oxide	F^- Fluoride	
Na^+	Mg^{2+}		Al^{3+}			S^{2-} Sulfide	Cl^- Chloride	
K^+	Ca^{2+}		Ga^{3+}			Se^{2-} Selenide	Br^- Bromide	
Rb^+	Sr^{2+}		In^{3+}	Sn^{2+} Sn^{4+}		Te^{2-} Telluride	I^- Iodide	
Cs^+	Ba^{2+}		Tl^+ Tl^{3+}	Pb^{2+} Pb^{4+}				

[a] Cation names are the same as the element they're derived from; anion names have an *-ide* ending.

TABLE 2.3 **Common Transition-Metal Ions[a]**

(3) 3B	(4) 4B	(5) 5B	(6) 6B	(7) 7B	(8)	(9) 8B	(10)	(11) 1B	(12) 2B
Sc^{3+}	Ti^{3+}	V^{3+}	Cr^{2+} Cr^{3+}	Mn^{2+}	Fe^{2+} Fe^{3+}	Co^{2+}	Ni^{2+}	Cu^{2+}	Zn^{2+}
Y^{3+}					Ru^{3+}	Rh^{3+}	Pd^{2+}	Ag^{+}	Cd^{2+}
La^{3+}									Hg^{2+}

[a] Only ions that exist in aqueous solution are shown.

within a group often form similar kinds of ions and that the charge on the ion depends on the group number. Main-group metals usually form cations whose charge is equal to the group number. Group 1A elements form monopositive ions (M^+, where M is a metal), group 2A elements form doubly positive ions (M^{2+}), and group 3A elements form triply positive ions (M^{3+}). Main-group nonmetals usually form anions whose charge is equal to the group number minus eight. Thus, group 6A elements form doubly negative ions ($6 - 8 = -2$), group 7A elements form mononegative ions ($7 - 8 = -1$), and group 8A elements form no ions at all ($8 - 8 = 0$).

Note in both Tables 2.2 and 2.3 that some metals can form more than one kind of cation. Iron, for example, can form two different cations, the doubly charged Fe^{2+} ion and the triply charged Fe^{3+} ion. In naming these ions, it's necessary to distinguish between them by using a Roman numeral in parentheses to indicate the number of charges. Thus, $FeCl_2$ is iron(II) chloride, and $FeCl_3$ is iron(III) chloride. Alternatively, an older method distinguishes between the ions by using the Latin name of the element (*ferrum*) together with the ending *-ous* for the ion with lower charge and *-ic* for the ion with higher charge. Thus, $FeCl_2$ is sometimes called ferrous chloride, and $FeCl_3$ is called ferric chloride. Though still in use, this older naming system is slowly being phased out and will not be used in this book.

Fe^{2+}	Fe^{3+}	Sn^{2+}	Sn^{4+}
Ferrous ion	Ferric ion	Stannous ion	Stannic ion
(From the Latin *ferrum* = iron)		(From the Latin *stannum* = tin)	

In any compound, you can always figure out the number of positive charges on a cation by counting the number of negative charges on the associated anion(s). In $FeCl_2$, for example, the iron ion must be Fe(II) because there are two Cl^- ions associated with it. Similarly, in AlF_3 the aluminum ion is Al(III) because there are three F^- anions associated with it. As a general rule, a Roman numeral is needed for transition-metal compounds to avoid ambiguity. If in doubt about whether or not a Roman numeral is needed in a name, it is better to include it than to leave it out.

Crystals of ferrous chloride tetrahydrate ($FeCl_2 \cdot 4H_2O$) are green; crystals of ferric chloride hexahydrate ($FeCl_3 \cdot 6H_2O$) are mustard yellow.

EXAMPLE 2.8

Give systematic names for the following compounds:

(a) $BaCl_2$ (b) $CrCl_3$ (c) PbS (d) Fe_2O_3

SOLUTION   Refer to Tables 2.2 and 2.3 if you are unsure about charges on the ions.

(a) barium chloride   No Roman numeral is necessary because barium, a group 2A element, forms only Ba^{2+}.

(b) chromium(III) chloride   The Roman numeral III is necessary to specify the +3 charge on chromium (a transition metal).

(c) lead(II) sulfide   The sulfide anion (S^{2-}) has a double negative charge, so the lead cation must be doubly positive.

(d) iron(III) oxide   The three oxide anions (O^{2-}) have a total negative charge of −6, so the two iron cations must have a total charge of +6. Thus, each is Fe(III).

EXAMPLE 2.9

Write formulas for the following compounds:

(a) magnesium hydride   (b) tin(IV) oxide   (c) iron(III) sulfide

SOLUTION   Refer to Tables 2.2 and 2.3 to find the charges on ions.

(a) MgH_2 Magnesium (group 2A) forms only a +2 cation, so there must be two hydride ions (H^-) to balance the charge.

(b) SnO_2 Tin(IV) has a +4 charge, so there must be two oxide ions (O^{2-}) to balance the charge.

(c) Fe_2S_3 Iron(III) has a +3 charge and sulfide ion a −2 charge (S^{2-}), so there must be two irons and three sulfurs.

┌ **PROBLEM 2.13** Give systematic names for the following compounds:

(a) CsF (b) K_2O (c) CuO (d) BaS (e) $BeBr_2$

┌ **PROBLEM 2.14** Write formulas for the following compounds:

(a) vanadium(III) chloride   (b) manganese(IV) oxide
(c) copper(II) sulfide   (d) aluminum oxide

Naming Binary Molecular Compounds

Binary molecular compounds are named by assuming that one of the two elements in the molecule is more cationlike and the other element is more anionlike. As with ionic compounds, the cationlike element takes the name of the element itself, and the anionlike element takes an *-ide* ending. The compound CO, for example, is called *carbon monoxide*.

CO: Carbon is more cationlike because it is farther left in the periodic table, and oxygen is more anionlike because it is farther right. The compound is therefore named *carbon monoxide*.

We'll see a quantitative way to decide which element is more cationlike and which is more anionlike in Section 7.7, but it is usually possible to decide by looking at the relative positions of the elements in the periodic table. The farther left in the periodic table the element occurs, the more likely it is to have cation character; the farther right the element occurs, the more likely it is to have anion character. Look at the following examples to see how this generalization applies:

CO_2 carbon dioxide (C is in group 4A; O is in group 6A)
SF_4 sulfur tetrafluoride (S is in group 6A; F is in group 7A)
N_2O_4 dinitrogen tetroxide (N is in group 5A; O is in group 6A)

As the examples show, numerical prefixes must often be included in the names of binary molecular compounds to specify the numbers of atoms of each element present. The compound CO, for example, is called carbon *monoxide*, and CO_2 is called carbon *dioxide*. Table 2.4 lists the most common prefixes. Note that when the prefix ends in *a* or *o* and the element begins with a vowel (*oxide*, for instance), the *a* or *o* on the prefix is dropped to avoid having two vowels together in the name. Thus, we write carbon *monoxide* rather than carbon *monooxide* and dinitrogen *tetroxide* rather than dinitrogen *tetraoxide*. Note also that the *mono* prefix is not used for the atom named first. NO_2, for instance, is called nitrogen dioxide rather than mononitrogen dioxide.

TABLE 2.4   Numerical Prefixes for Naming Compounds

Prefix	Meaning
mono-	1
di-	2
tri-	3
tetra-	4
penta-	5
hexa-	6
hepta-	7
octa-	8

EXAMPLE 2.10

Give systematic names for the following compounds.

(a) PCl_3 (b) N_2O_3 (c) P_2O_5

SOLUTION   Look at a periodic table to see which element in each compound is more cationlike (farther to the left) and which is more anionlike (farther to the right). Then name the compound using the appropriate numerical prefix from Table 2.4.

(a) phosphorus trichloride   (b) dinitrogen trioxide   (c) diphosphorus pentoxide

⌐ **PROBLEM 2.15** Give systematic names for the following compounds.

(a) NCl_3 (b) P_4O_6 (c) S_2F_2

⌐ **PROBLEM 2.16** Write formulas for the following names.
(a) disulfur dichloride   **(b)** iodine monochloride   **(c)** nitrogen triiodide   ◺

Naming Compounds with Polyatomic Ions

Ionic compounds with polyatomic ions (Section 2.8) are named in the same way as binary ionic compounds: First the cation is identified and then the anion. For example, $Ba(NO_3)_2$ is called *barium nitrate* because Ba^{2+} is the cation and the NO_3^- polyatomic anion has the name *nitrate*. Unfortunately, there is no systematic way of naming polyatomic ions, so it's necessary to memorize the names, formulas, and charge numbers of the most common ones (Table 2.5). The ammonium ion NH_4^+ is the only cation on the list; all the others are anions.

TABLE 2.5	Some Common Polyatomic Ions		
Formula	**Name**	**Formula**	**Name**
Cation		**Singly charged anions (continued)**	
NH_4^+	Ammonium	MnO_4^-	Permanganate
Singly charged anions		NO_2^-	Nitrite
$C_2H_3O_2^-$	Acetate	NO_3^-	Nitrate
CN^-	Cyanide	**Doubly charged anions**	
ClO^-	Hypochlorite	CO_3^{2-}	Carbonate
ClO_2^-	Chlorite	CrO_4^{2-}	Chromate
ClO_3^-	Chlorate	$Cr_2O_7^{2-}$	Dichromate
ClO_4^-	Perchlorate	O_2^{2-}	Peroxide
$H_2PO_4^-$	Dihydrogen phosphate	HPO_4^{2-}	Hydrogen phosphate
HCO_3^-	Hydrogen carbonate (or bicarbonate)	SO_3^{2-}	Sulfite
		SO_4^{2-}	Sulfate
HSO_4^-	Hydrogen sulfate (or bisulfate)	$S_2O_3^{2-}$	Thiosulfate
		Triply charged anion	
OH^-	Hydroxide	PO_4^{3-}	Phosphate

Several points about the ions in Table 2.5 need special mention. First, note that the names of most polyatomic anions end in *-ite* or *-ate*; only hydroxide (OH^-), cyanide (CN^-), and peroxide (O_2^{2-}) take the *-ide* ending.

These blue crystals of copper sulfate pentahydrate contain the Cu^{2+} ion and the polyatomic SO_4^{2-} ion.

Second, note that several of the ions form a series of **oxoanions**, in which an atom of the same element is combined with different numbers of oxygen atoms—hypochlorite (ClO^-), chlorite (ClO_2^-) chlorate (ClO_3^-), and perchlorate (ClO_4^-), for example. When there are only two oxoanions in the series, as with sulfite (SO_3^{2-}) and sulfate (SO_4^{2-}), the ion with fewer oxygens takes the *-ite* ending and the ion with more oxygens takes the *-ate* ending.

SO_3^{2-} sul*ite* ion (fewer oxygens) SO_4^{2-} sul*fate* ion (more oxygens)
NO_2^- nit*rite* ion (fewer oxygens) NO_3^- nit*rate* ion (more oxygens)

When there are more than two oxoanions in the series, the prefix *hypo-* (meaning "less than") is used for the ion with the fewest oxygens, and the prefix *per-* (meaning "more than") is used for the ion with the most oxygens.

ClO^- *hypo*chlorite ion (less oxygen than chlorite)
ClO_2^- chlorite ion
ClO_3^- chlorate ion
ClO_4^- *per*chlorate ion (more oxygen than chlorate)

Third, note that several pairs of ions are related by the presence or absence of a hydrogen atom. The carbonate anion (CO_3^{2-}) differs from the hydrogen carbonate anion (HCO_3^-) by the absence of H^+, and the sulfate anion (SO_4^{2-}) differs from the hydrogen sulfate anion (HSO_4^-) by the absence of H^+. The ion that has the additional hydrogen is often referred to using the prefix *bi-*, although this usage is now discouraged; for example, sodium bicarbonate for $NaHCO_3$.

HCO_3^- hydrogen carbonate (*bi*carbonate) ion CO_3^{2-} carbonate ion
HSO_4^- hydrogen sulfate (*bi*sulfate) ion SO_4^{2-} sulfate ion

EXAMPLE 2.11

Give systematic names for the following compounds.
(a) $LiNO_3$ **(b)** $KHSO_4$ **(c)** $CuCO_3$ **(d)** $Fe(ClO_4)_3$

SOLUTION

(a) lithium nitrate   Lithium (group 1A) forms only the Li^+ ion and does not need a Roman numeral.

(b) potassium hydrogen sulfate   Potassium (group 1A) forms only the K^+ ion.

(c) copper(II) carbonate   The carbonate ion has a -2 charge, so copper must be $+2$. A Roman numeral is needed because copper, a transition metal, can form more than one kind of ion.

(d) iron(III) perchlorate   There are three perchlorate ions, each with a -1 charge, so the iron must have a $+3$ charge.

EXAMPLE 2.12

Write formulas for the following compounds.
(a) potassium hypochlorite   **(b)** silver(I) chromate   **(c)** iron(III) carbonate

SOLUTION

(a) KClO   Potassium forms only the K^+ ion, so only one ClO^- is needed.

(b) Ag_2CrO_4 The polyatomic chromate ion has a -2 charge, so two Ag^+ ions are needed.

(c) $Fe_2(CO_3)_3$ Iron(III) has a $+3$ charge and the polyatomic carbonate ion has a -2 charge, so there must be two iron ions and three carbonate ions. The entire polyatomic carbonate ion is set off in parentheses, because there is more than one of them.

┌ **PROBLEM 2.17**   Give systematic names for the following compounds.

(a) $Ca(ClO)_2$ (b) $Ag_2S_2O_3$ (c) NaH_2PO_4 (d) $Sn(NO_3)_2$ (e) $Pb(C_2H_3O_2)_4$

┌ **PROBLEM 2.18**   Write formulas for the following compounds.

(a) lithium phosphate   (b) magnesium hydrogen sulfate
(c) manganese(II) nitrate   (d) chromium(III) sulfate

Naming Acids

Most acids are **oxoacids**. That is, they contain oxygen in addition to hydrogen and another element. When dissolved in water, an oxoacid yields one or more H^+ ions and a polyatomic oxoanion, such as one of those listed in Table 2.6.

The names of oxoacids are related to the names of the corresponding oxoanions as described previously, with the *-ite* or *-ate* ending of the anion name replaced by *-ous acid* or *-ic acid*, respectively. In other words, the acid with fewer oxygens has an *-ous* ending, and the related acid with more oxygens has an *-ic* ending. The compound HNO_2, for example, is called *nitrous acid* because it has fewer oxygens and yields the nitr*ite* ion (NO_2^-) when dissolved in water. HNO_3 is called *nitric acid* because it has more oxygens and yields the nitr*ate* ion (NO_3^-) when dissolved in water.

Nitr*ous acid* gives nitr*ite ion*
$$HNO_2(l) \xrightarrow{\text{dissolve in water}} H^+(aq) + NO_2^-(aq)$$

Nitr*ic acid* gives nitr*ate ion*
$$HNO_3(l) \xrightarrow{\text{dissolve in water}} H^+(aq) + NO_3^-(aq)$$

In a similar way, hypochlor*ous* acid yields the hypochlor*ite* ion, chlor*ous* acid yields the chlor*ite* ion, chlor*ic* acid yields the chlor*ate* ion, and perchlor*ic* acid yields the perchlor*ate* ion. These and other examples are shown in Table 2.6.

In addition to the oxoacids, there are a small number of other acids, such as HCl, that do not contain oxygen. Although the pure, gaseous compound HCl is named hydrogen chloride, the aqueous solution is named *hydrochloric acid*, HCl(*aq*). This example is typical: The prefix *hydro-* and the suffix *-ic acid* are used in such cases.

Hydrogen chloride $HCl(g) \xrightarrow{\text{dissolve in water}} HCl(aq)$ *Hydrochloric acid*

Hydrogen bromide $HBr(g) \xrightarrow{\text{dissolve in water}} HBr(aq)$ *Hydrobromic acid*

TABLE 2.6 Some Common Oxoacids and Their Anions			
Oxoacid		**Oxoanion**	
HNO_2	Nitr*ous* acid	NO_2^-	Nitrite ion
HNO_3	Nitr*ic* acid	NO_3^-	Nitrate ion
H_3PO_4	Phosphor*ic* acid	PO_4^{3-}	Phosphate ion
H_2SO_3	Sulfur*ous* acid	SO_3^{2-}	Sulfite ion
H_2SO_4	Sulfur*ic* acid	SO_4^{2-}	Sulfate ion
$HClO$	*Hypo*chlor*ous* acid	ClO^-	Hypochlorite ion
$HClO_2$	Chlor*ous* acid	ClO_2^-	Chlorite ion
$HClO_3$	Chlor*ic* acid	ClO_3^-	Chlorate ion
$HClO_4$	*Per*chlor*ic* acid	ClO_4^-	Perchlorate ion

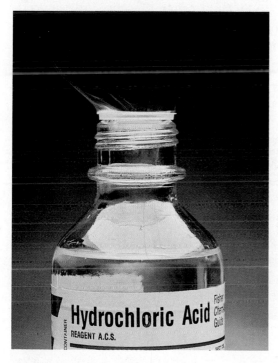

Pure HCl is a gas; hydrochloric acid
is an aqueous solution of HCl.

EXAMPLE 2.13

Name the following acids.
(a) HBrO   (b) HCN(*aq*)

SOLUTION   (a) This compound is an oxoacid that yields hypobromite ion
(BrO^-) when dissolved in water. Its name is hypobromous acid.
(b) This compound is not an oxoacid but yields cyanide ion when dissolved in water.
Its name is hydrocyanic acid. (As a pure gas, HCN is named hydrogen cyanide.)

⌐ **PROBLEM 2.19** Name the following acids.
(a) HIO_4 (b) $HBrO_2$ (c) H_2CrO_4

interlude—ARE ATOMS REAL?

The atomic theory of matter lies at the heart of chemistry. Every chemical reaction and every physical law that describes the behavior of matter is explained by chemists in terms of atoms. How do we know, though, that atoms are real and that our explanations have a factual basis? The best answer to that question is that we can now actually "see" individual atoms with a remarkable device called a *scanning tunneling microscope*, or *STM*. Invented in 1981 by a research team at the IBM Corporation, this special microscope has achieved magnifications of up to 10 million (!), allowing chemists for the first time to look directly at atoms. Figure 2.11 shows a computer-enhanced representation of individual atoms lined up on the face of a crystal of silicon.

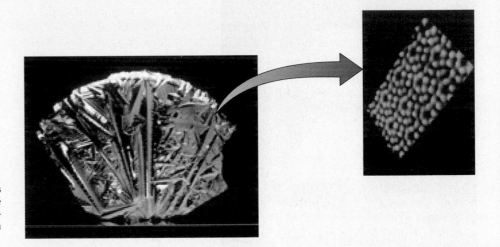

FIGURE 2.11 Individual atoms on the face of a crystal of silicon are seen at a magnification of 10 million in this view produced by a scanning tunneling microscope.

The principle behind the operation of an STM is shown in Figure 2.12. A sharp tungsten probe, whose tip is only one or two atoms across, is brought near the surface of the sample, and a small voltage is applied. When the tip comes within a few atomic diameters of the sample, a small electric current flows from the sample to the probe in a process called *electron tunneling*. The strength of the current flow is extremely sensitive to the distance between sample and probe, varying by as much as 1000-fold over a distance of just 100 pm (about one atomic diameter.) Passing the probe across the sample while moving it up and down over individual atoms to keep current flow constant, gives a two-dimensional map of the probe's path. By then moving the probe back and forth in a series of closely spaced parallel tracks and storing the data in a computer, a three-dimensional image of the surface can be constructed.

The kind of image "seen" by scanning tunneling microscopy is quite different from the kind of image we normally see with our eyes. A normal visual image results when light from the sun or other source reflects off an

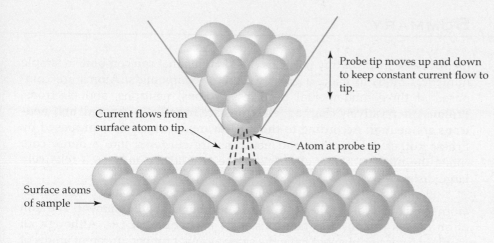

Current flows from
surface atom to tip.

Atom at probe tip

Probe tip moves up and down
to keep constant current flow to
tip.

Surface atoms
of sample

FIGURE 2.12 A scanning tunneling microscope works by moving an extremely fine probe along the surface of a sample, applying a small voltage, and measuring a current flow between atoms in the sample and the atom at the tip of the probe. By raising and lowering the moving probe to keep current flow constant as the tip passes over atoms, a map of the surface can be obtained.

object, strikes the retina in our eye, and is converted into electrical signals that are processed by the brain. The image obtained with a scanning tunneling microscope, by contrast, is a three-dimensional, computer-generated data plot that uses tunneling current to mimic depth perception. The nature of the computer-generated image depends on the identity of the molecules or atoms on the surface, on the precision with which the probe tip is made, on how the data are manipulated, and on other experimental variables. Thus, the image is not a true "picture" of the surface.

Most of the present uses of the scanning tunneling microscope involve studies of surface chemistry. Processes such as the deposition of monomolecular layers on smooth surfaces can be studied, the nature of industrial catalysts can be probed, and events accompanying the corrosion of metals can be examined. Even more exciting are some potential applications of scanning tunneling microscopy. Although the technology is yet too new to be fully developed, the possibility exists that complex molecular structures can be determined with the STM. At that point, it may even become possible to determine the structure of DNA strands in human genes, thereby helping to unravel the mysteries of heredity.

Summary

Elements are made of tiny particles called **atoms** that can combine in simple numerical ratios to make a great variety of compounds. Atoms are composed of three fundamental particles: protons, neutrons, and electrons. **Protons** are positively charged, **electrons** are negatively charged, and **neutrons** are neutral. According to the nuclear model of an atom proposed by Ernest Rutherford, protons and neutrons are clustered into a dense core called the **nucleus**, while electrons move around the nucleus a relatively large distance away.

Elements differ from one another according to how many protons their atoms contain, a value called the **atomic number (Z)** of the element. The sum of an atom's protons and neutrons is its **mass number** (*A*). Although all atoms of a specific element have the same atomic number, different atoms of an element can have different mass numbers depending on how many neutrons they have. Atoms with identical atomic numbers but different mass numbers are called **isotopes**. Atomic masses are measured using the **atomic mass unit** (**amu**), defined as 1/12th the mass of a ^{12}C atom. Because both protons and neutrons have a mass of approximately 1 amu, the total mass of an atom in amu (the **atomic weight**) is numerically close to the atom's mass number. The atomic weight values listed in tables are weighted averages for isotope mixtures.

Most substances on earth are **compounds**, formed when atoms of two or more elements combine in a **chemical reaction**. The atoms in compounds are held together by one of two fundamental kinds of **chemical bonds**. **Covalent bonds** form when two atoms share several electrons to give a new unit of matter called a **molecule**. **Ionic bonds** form when one atom completely transfers one or more electrons to another atom, resulting in the formation of **ions**. Positively charged ions (**cations**) are strongly attracted to negatively charged ions (**anions**) by electrical forces.

The hydrogen ion (H^+) and the **polyatomic** hydroxide ion (OH^-) are among the most important ions in chemistry because they are fundamental to the idea of acids and bases. According to one common definition, **acids** are substances that yield H^+ ions when dissolved in water, and **bases** are substances that yield OH^- ions when dissolved in water.

All chemical compounds can be named systematically by following a series of rules. Binary ionic compounds are named by identifying first the positive ion and then the negative ion. Binary molecular compounds are similarly named by identifying the cationlike and anionlike elements. Naming compounds with polyatomic ions involves memorizing the names and formulas of common ones.

1. If yellow spheres represent sulfur atoms, and blue spheres represent oxygen atoms, which of the following drawings depicts a collection of SO_2 molecules?

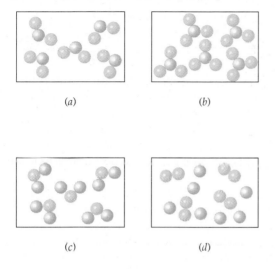

(a)   (b)

(c)   (d)

2. Using solid dots to represent hydrogen atoms and open circles to represent oxygen atoms, make drawings similar to those in Problem 1 to represent a collection of H_2O molecules and a collection of H_2O_2 molecules.

3. Which of the following drawings represents an Na atom? a Ca^{2+} ion? an F^- ion?

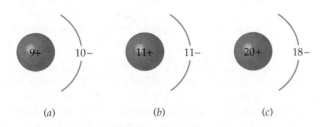

(a)   (b)   (c)

4. Make drawings similar to those in Problem 3 to represent a Ne atom, a Br^- ion, and an Au^{3+} ion.

5. Assume that the mixture of substances in drawing (a) undergoes a reaction. Which of drawings (b) through (d) represents a product mixture consistent with the law of mass conservation?

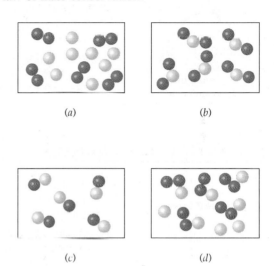

(a)   (b)

(c)   (d)

6. Give names and symbols for the following atoms.
   (a) An atom with atomic number 31
   (b) An atom with 24 protons in its nucleus
   (c) An atom with 13 electrons

7. The subscript giving the atomic number of an atom is often left off an isotope symbol. For example, $^{13}_{6}C$ is often written simply as ^{13}C. Why is this allowable?

8. Iodine has a *lower* atomic weight that tellurium (126.90 for iodine versus 127.60 for tellurium) even though it has a *higher* atomic number (53 for iodine versus 52 for tellurium). Explain how this is possible.

9. In dimethyl ether, C_2H_6O, three hydrogen atoms are bonded to each carbon, and the carbons are bonded to the oxygen. Draw a structural formula for dimethyl ether.

10. Give molecular formulas corresponding to each of the following ball-and-stick representations. Red repre- sents oxygen, gray represents carbon, blue represents nitrogen, and light green represents hydrogen.

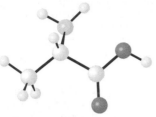

Alanine
(an amino acid)

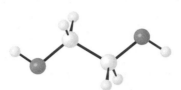

Ethylene glycol
(automobile antifreeze)

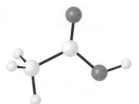

Acetic acid
(vinegar)

ADDITIONAL PROBLEMS

Problems 2.1–2.19 appear within the chapter.

ATOMIC THEORY

2.20 State the law of definite proportions in your own words.

2.21 State the law of mass conservation in your own words.

2.22 Explain how Dalton's atomic theory accounts for the law of mass conservation and the law of definite proportions.

2.23 What is the law of multiple proportions, and how is it predicted by Dalton's atomic theory?

2.24 Benzene, ethane, and ethylene are just three of a large number of *hydrocarbons*—compounds that contain only carbon and hydrogen. Show how the following data are consistent with the law of multiple proportions.

Compound	Mass of carbon in 5.00 g sample	Mass of hydrogen in 5.00 g sample
Benzene	4.61 g	0.39 g
Ethane	4.00 g	1.00 g
Ethylene	4.29 g	0.71 g

2.25 In addition to carbon monoxide (CO) and carbon dioxide (CO_2), there is a third compound of carbon and oxygen called carbon suboxide. If a 2.500 g sample of carbon suboxide contains 1.32 g of C and 1.18 g of O, show that the law of multiple proportions is followed.

2.26 The atomic weight of carbon (at wt = 12.011) is approximately 12 times that of hydrogen (at wt = 1.008).

(a) Show how you can use this knowledge to calculate possible formulas for benzene, ethane, and ethylene (Problem 2.24).

(b) Show how your answer to part (a) is consistent with the modern formulas for benzene (C_6H_6), ethane (C_2H_6), and ethylene (C_2H_4).

2.27 What is a possible formula for carbon suboxide (Problem 2.25)?

2.28 (a) If the average mass of a hydrogen atom is 1.67×10^{-24} g, what is the mass in grams of 6.02×10^{23} hydrogen atoms? How does your answer compare numerically with the atomic weight of hydrogen?

(b) If the average mass of an oxygen atom is 26.558×10^{-24} g, what is the mass in grams of 6.02×10^{23} oxygen atoms? How does your answer compare numerically with the atomic weight of oxygen?

(c) If the atomic weight of an element is X, what is the mass in grams of 6.02×10^{23} atoms of the element?

2.29 A binary compound of zinc and sulfur contains 67.1% zinc by mass. What is the ratio of zinc and sulfur atoms in the compound?

2.30 There are two binary compounds of titanium and chlorine. One compound contains 31.04% titanium by mass, and the other contains 74.76% chlorine by mass. What are the ratios of titanium and chlorine atoms in the two compounds?

2.31 What is a cathode-ray tube, and how does it work?

ELEMENTS AND ATOMS

2.32 What are the names of the three subatomic particles, what are their masses in amu, and what is the unit charge on each?

2.33 What is the difference between an atom's atomic number and its mass number?

2.34 What is the difference between an element's atomic number and its atomic weight?

2.35 What is an isotope?

2.36 Give names and symbols for the following elements.
(a) An element with atomic number 6
(b) An element with 18 protons in its nucleus
(c) An element with 23 electrons

2.37 Carbon-14 and nitrogen-14 both have the same mass number yet they are different elements. Explain.

2.38 The radioactive isotope cesium-137 was produced in large amounts in fallout from the 1985 nuclear power plant disaster at Chernobyl, Ukraine. Write the symbol for this isotope in standard format.

2.39 Write standard symbols for the following isotopes.
(a) radon-220 (b) polonium-210
(c) gold-197

2.40 Write symbols for the following isotopes.
(a) $Z = 58$ and $A = 140$
(b) $Z = 27$ and $A = 60$

2.41 How many protons, neutrons, and electrons are in each of the following atoms?
(a) $^{15}_{7}N$ (b) $^{60}_{27}Co$
(c) $^{131}_{53}I$

2.42 How many protons and neutrons are in the nucleus of the following atoms?
(a) ^{27}Al (b) ^{32}S (c) ^{64}Zn (d) ^{207}Pb

2.43 Identify the following elements:
(a) $^{24}_{12}X$ (b) $^{58}_{28}X$

2.44 Identify the following elements:
(a) $^{202}_{80}X$ (b) $^{195}_{78}X$

2.45 Naturally occurring boron consists of two isotopes, ^{10}B (19.9%) with an atomic mass of 10.0129, and ^{11}B (80.1%) with an atomic mass of 11.00931. What is the atomic weight of boron? Check your answer by looking at a periodic table.

2.46 Naturally occurring silver consists of two isotopes, ^{107}Ag (51.84%) with an atomic mass of 106.9051, and ^{109}Ag (48.16%) with an atomic mass of 108.9048. What is the atomic weight of silver? Check your answer by looking at a periodic table.

2.47 Magnesium has three naturally occurring isotopes, ^{24}Mg (23.985 amu) with 78.99% abundance, ^{25}Mg (24.986 amu) with 10.00% abundance, and a third with 11.01% abundance. Look up the average atomic weight of magnesium and then calculate the atomic mass of the third isotope.

2.48 The best balances commonly found in laboratories can weigh amounts as small as 10^{-5} g. If you were to count out carbon atoms at the rate of two each second, how long would it take you to count a pile of atoms large enough to weigh?

COMPOUNDS AND MIXTURES, MOLECULES AND IONS

2.49 What is the difference between a compound and a mixture? Give an example of each.

2.50 What is the difference between a homogeneous mixture and a heterogeneous one? Give an example of each.

2.51 Which of the following mixtures are homogeneous and which are heterogeneous?
(a) muddy water (b) concrete
(c) housepaint (d) a soft drink

2.52 What is the difference between an atom and a molecule? Give an example of each.

2.53 What is the difference between a molecule and an ion? Give an example of each.

2.54 What is the difference between a covalent bond and an ionic bond? Give an example of each.

2.55 The symbol CO stands for carbon monoxide, but the symbol Co stands for the element cobalt. Explain.

2.56 Correct the error in each of the following statements.
(a) The formula of ammonia is NH3.
(b) Molecules of potassium chloride have the formula KCl.
(c) Cl^- is a cation.
(d) CH_4 is a polyatomic ion.

2.57 How many protons and electrons are in the following *ions*?
(a) Be^{2+} (b) Rb^+
(c) Se^{2-} (d) Au^{3+}

2.58 What is the identity of the element X in the following ions?
(a) X^{2+}, a cation that has 36 electrons
(b) X^-, an anion that has 36 electrons

2.59 The structural formula of isopropyl alcohol, better known as "rubbing alcohol," is shown. What is the chemical formula of isopropyl alcohol?

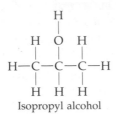

Isopropyl alcohol

2.60 Butane, the fuel used in disposable lighters, has the formula C_4H_{10}. The carbon atoms are connected in the sequence C—C—C—C, and each carbon has a total of four covalent bonds. Draw a structural formula for butane.

2.61 Isooctane, the substance in gasoline from which the term *octane rating* derives, has the formula C_8H_{18}. Each carbon has a total of four covalent bonds, and the atoms are connected in the following sequence. Draw a structural formula for isooctane.

$$\begin{array}{c} \quad\ \ C \qquad\quad C \\ \quad\ \ | \qquad\quad | \\ C-C-C-C-C \\ \quad\ \ | \\ \quad\ \ C \end{array}$$

ACIDS AND BASES

2.62 Which of the following compounds are acids and which are bases?
(a) HI   (b) CsOH   (c) H_3PO_4
(d) $Ba(OH)_2$ (e) H_2CO_3

2.63 Identify the anion that results when each of the acids in Problem 2.62 dissolves in water.

2.64 Identify the cation that results when each of the bases in Problem 2.62 dissolves in water.

NAMING COMPOUNDS

2.65 Write formulas for the following binary compounds.
(a) potassium chloride   (b) tin(II) bromide
(c) calcium oxide   (d) barium chloride
(e) aluminum hydride

2.66 Write formulas for each of the following compounds.
(a) calcium acetate   (b) iron(II) cyanide
(c) sodium dichromate   (d) chromium(III) sulfate
(e) mercury(II) perchlorate

2.67 Name the following ions.
(a) Ba^{2+} (b) Cs^+ (c) V^{3+} (d) HCO_3^-
(e) NH_4^+ (f) Ni^{2+} (g) NO_2^- (h) ClO_2^-
(i) Mn^{2+} (j) ClO_4^-

2.68 Name the following binary molecular compounds.
(a) CCl_4 (b) SiO_2 (c) N_2O (d) N_2O_3

2.69 Give the formulas and charges of the following ions.
(a) sulfite ion   (b) phosphate ion
(c) zirconium(IV) ion   (d) chromate ion
(e) acetate ion   (f) thiosulfate ion

2.70 What are the charges on the positive ions in the following compounds?
(a) $Zn(CN)_2$ (b) $Fe(NO_2)_3$ (c) $Ti(SO_4)_2$
(d) $Sn_3(PO_4)_2$ (e) Hg_2S (f) MnO_2
(g) KIO_4 (h) $Cu(C_2H_3O_2)_2$

2.71 Name each of the compounds in Problem 2.70.

2.72 Name each of the following compounds.
(a) $MgSO_3$ (b) $Co(NO_2)_2$ (c) $Mn(HCO_3)_2$
(d) $ZnCrO_4$ (e) $BaSO_4$ (f) $KMnO_4$
(g) $Al_2(SO_4)_3$ (h) $LiClO_3$

2.73 Fill in the missing information to give formulas for the following compounds.
(a) $Na_?SO_4$ (b) $Ba_?(PO_4)_?$ (c) $Ga_?(SO_4)_?$

2.74 Write formulas for each of the following compounds.
(a) sodium peroxide   (b) aluminum bromide
(c) chromium(III) sulfate

2.75 Name each of the compounds in Problem 2.73.

GENERAL PROBLEMS

2.76 Germanium has five naturally occurring isotopes: ^{70}Ge, 20.5 %, 69.924 amu; ^{72}Ge, 27.4 %, 71.922 amu; ^{73}Ge, 7.8 %, 72.923 amu; ^{74}Ge, 36.5 %, 73.921 amu; and ^{78}Ge, 7.8 %, 75.921 amu. What is the atomic weight of germanium?

2.77 Name the following compounds.
(a) $NaBrO_3$ (b) H_3PO_4 (c) H_3PO_3 (d) V_2O_5

2.78 Write formulas for the following compounds.
(a) calcium hydrogen sulfate
(b) tin(II) oxide
(c) ruthenium(III) nitrate
(d) ammonium carbonate
(e) hydriodic acid
(f) beryllium phosphate

2.79 Ammonia, NH_3, and hydrazine, N_2H_4, are both binary compounds of nitrogen and hydrogen. Based on the law of multiple proportions, how many grams of hydrogen would you expect 2.34 g of nitrogen to combine with to yield ammonia? to yield hydrazine?

2.80 If 3.670 g of nitrogen combines with 0.5275 g of hydrogen to yield compound X, how many grams of nitrogen would combine with 1.575 g of hydrogen? Is X ammonia or hydrazine (Problem 2.79)?

2.81 Tellurium, a group 6A element, forms the oxoanions TeO_4^{2-} and TeO_3^{2-}. What are the likely names of these ions? To what other group 6A oxoanions are they analogous?

2.82 Give the formulas and the likely names of the acids derived from the tellurium-containing oxoanions in Problem 2.81.

2.83 Identify the following atoms or ions.
   (a) a halogen anion with 54 electrons
   (b) a metal cation with 79 protons and 76 electrons
   (c) a noble gas with $A = 84$

2.84 Prior to 1961, the atomic mass unit was defined as 1/16th the mass of the atomic weight of oxygen. That is, the atomic weight of oxygen was defined as exactly 16. What was the mass of a ^{12}C atom prior to 1961 if the atomic weight of oxygen on today's scale is 15.9994 amu?

2.85 What was the mass in amu of a ^{40}Ca atom prior to 1961 if its mass on today's scale is 39.9626 amu? (See Problem 2.84.)

chapter 3 FORMULAS, EQUATIONS, AND MOLES

I t's sometimes possible when beginning the study of chemistry to lose sight of the fact that *reactions* are at the heart of the science. New words, ideas, and principles are sometimes introduced so quickly that the central concern of chemistry—the change of one substance into another—gets lost in the rush.

In this chapter, we'll begin learning how to describe chemical reactions. We'll start the process by learning about the conventions for writing reactions and about the necessary mass relationships between reactants. Since most chemical reactions are carried out using solutions rather than pure materials, we'll also discuss units for describing the concentration of a solution and will see how to determine the value of the concentration. Finally, we'll look at how chemical formulas are determined and how molecular masses are measured.

Chemists need to put the right amounts of substances into their reactions, just as chefs need to put the right amount of ingredients into their cooking.

3.1 ►BALANCING CHEMICAL EQUATIONS

The previous two chapters have provided several examples of reactions: hydrogen reacting with oxygen to yield water, sodium reacting with chlorine to yield sodium chloride, mercuric nitrate reacting with potassium iodide to yield mercuric iodide, and so forth. We can write these reactions in the following format:

4 H's and 2 O's on this side ⟶ 4 H's and 2 O's on this side
$$2 H_2 + O_2 \rightarrow 2 H_2O$$

2 Na's and 2 Cl's on this side ⟶ 2 Na's and 2 Cl's on this side
$$2 Na + Cl_2 \rightarrow 2 NaCl$$

1 Hg, 2 N's, 6 O's, 2 K's, and 2 I's on this side ⟶ 1 Hg, 2 N's, 6 O's, 2 K's, and 2 I's on this side
$$Hg(NO_3)_2 + 2 KI \rightarrow HgI_2 + 2 KNO_3$$

Look carefully at how these reactions are written. Since we know from Section 2.8 that hydrogen, oxygen, and chlorine exist as covalent H_2, O_2, and Cl_2 *molecules* rather than as isolated atoms, we must write them as such in the chemical equations. Now look at the atoms on each side of the reaction arrow. In each reaction, the numbers and kinds of atoms on both sides of the arrow are the same, and the equations are therefore said to be **balanced**.

The requirement that an equation be balanced is a direct consequence of the mass conservation law: All chemical equations must be balanced because mass is neither created nor destroyed in chemical reactions. The numbers and kinds of atoms in the products must be the same as in the reactants.

Balancing a chemical equation involves finding out how many *formula units* of each different substance take part in the reaction. (A **formula unit**, as the name implies, is one unit—an atom, ion, or molecule—corresponding to a given formula. One formula unit of NaCl, for example, is one Na^+ ion and one Cl^- ion. One formula unit of H_2O is one H_2O molecule.) The balancing process is carried out in four steps using a mixture of common sense and trial and error.

1. Write the unbalanced equation using the correct chemical formulas for all reactants and products. In the reaction of hydrogen with oxygen to yield water, for example, we begin by writing:

$$H_2 + O_2 \rightarrow H_2O$$

2. Find suitable **coefficients**—numbers placed before each formula to indicate how many formula units of each substance are required to balance the equation. *Only these coefficients can be changed when balancing an equation; the formulas themselves can't be changed.* Again taking the reaction of hydrogen with oxygen as an example, we can balance the equation by adding coefficients of 2 to both H_2 and H_2O. By so doing, we now have four hydrogen atoms and two oxygen atoms on each side of the equation:

add these *coefficients* to balance the equation
$$2 H_2 + O_2 \rightarrow 2 H_2O$$

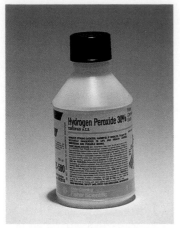

Hydrogen peroxide is used as a bleach and a sterilant.

It might seem easier at first glance to balance the equation simply by adding a subscript 2 to the oxygen atom in water, thereby changing H_2O into H_2O_2. That is not allowed, however, because the resulting equation would no longer describe the same reaction. The substances H_2O (water) and H_2O_2 (hydrogen peroxide) are two entirely different compounds. Water is a substance used for drinking and swimming; hydrogen peroxide is a substance used for bleaching hair.

NOT ALLOWED!
When this subscript is added, we get a completely different reaction.

$$H_2 + O_2 \rightarrow H_2O_2$$

3. Reduce the coefficients to their smallest whole-number values, if necessary, by dividing them by a common divisor.
4. Check your answer by making sure that the numbers and kinds of atoms are the same on both sides of the equation.

Let's work through some examples to see how equations are balanced.

EXAMPLE 3.1

Propane, C_3H_8, is a colorless, odorless gas often used as a cooking fuel in campers and rural homes. Write a balanced equation for the combustion reaction of propane with oxygen to yield carbon dioxide and water.

SOLUTION　*Step 1* Write the unbalanced equation using correct chemical formulas for all substances:

$$C_3H_8 + O_2 \rightarrow CO_2 + H_2O \quad \text{(unbalanced)}$$

Step 2 Find coefficients to balance the equation. It's usually best to start with the most complex substance—in this case C_3H_8—and to think about one element at a time. Look first at the unbalanced equation, and note that there are 3 carbon atoms on the left side of the equation but only 1 on the right side. If we add a coefficient of 3 to CO_2 on the right, the carbons balance:

$$C_3H_8 + O_2 \rightarrow 3\,CO_2 + H_2O \quad \text{(balanced for C)}$$

Next, look at the number of hydrogen atoms. There are 8 hydrogens on the left but only 2 (in H_2O) on the right. By adding a coefficient of 4 to the H_2O on the right, the hydrogens balance:

$$C_3H_8 + O_2 \rightarrow 3\,CO_2 + 4\,H_2O \quad \text{(balanced for C and H)}$$

Finally, look at the number of oxygen atoms. There are 2 on the left but 10 on the right. By adding a coefficient of 5 to the O_2 on the left, the oxygens balance:

$$C_3H_8 + 5\,O_2 \rightarrow 3\,CO_2 + 4\,H_2O \quad \text{(balanced for C, H, and O)}$$

Step 3 Make sure the coefficients are reduced to their smallest whole-number values. In fact, our answer is already correct, but we might have arrived at a different answer through trial and error:

Liquefied propane (LP) gas is used as a fuel in campers and rural homes.

$$2\ C_3H_8 + 10\ O_2 \rightarrow 6\ CO_2 + 8\ H_2O$$

Although the preceding equation is balanced, the coefficients are not the smallest whole numbers. It's necessary to divide all coefficients by 2 to reach the final equation.

Step 4 Check your answer. Count the numbers and kinds of atoms on both sides of the equation to make sure they're the same:

3 C's, 8 H's, and 10 O's   3 C's, 8 H's, and 10 O's
on this side   on this side

$$C_3H_8 + 5\ O_2 \rightarrow 3\ CO_2 + 4\ H_2O$$

EXAMPLE 3.2

The major ingredient in ordinary safety matches is potassium chlorate ($KClO_3$), a substance that can act as a source of oxygen in chemical reactions. Its reaction with ordinary table sugar (sucrose, $C_{12}H_{22}O_{11}$), for example, occurs violently to yield potassium chloride, carbon dioxide, and water. Write a balanced equation for the reaction.

SOLUTION *Step 1* Write the unbalanced equation, making sure the formulas for all substances are correct:

$$KClO_3 + C_{12}H_{22}O_{11} \rightarrow KCl + CO_2 + H_2O \quad \text{(unbalanced)}$$

Step 2 Find coefficients to balance the equation by starting with the most complex substance (sucrose) and considering one element at a time. Since there are 12 C's on the left and only 1 on the right, we can balance for carbon by adding a coefficient of 12 to CO_2:

$$KClO_3 + C_{12}H_{22}O_{11} \rightarrow KCl + 12\ CO_2 + H_2O \quad \text{(balanced for C)}$$

Potassium chlorate reacts violently with sucrose to yield KCl, CO_2, and H_2O.

Since there are 22 H's on the left and only 2 on the right, we can balance for hydrogen by adding a coefficient of 11 to H_2O:

$$KClO_3 + C_{12}H_{22}O_{11} \rightarrow KCl + 12\ CO_2 + 11\ H_2O \quad \text{(balanced for C, H)}$$

Since there are 35 O's on the right but only 14 on the left (11 in sucrose and 3 in $KClO_3$), 21 oxygens must be added on the left. We can do this without disturbing the C and H balance by adding 7 more $KClO_3$'s, giving a coefficient of 8 for $KClO_3$:

$$8\ KClO_3 + C_{12}H_{22}O_{11} \rightarrow KCl + 12\ CO_2 + 11\ H_2O \quad \text{(balanced for C, H, O)}$$

Potassium and chlorine can both be balanced by adding a coefficient of 8 to KCl:

8 K, 8 Cl, 12 C,   8 K, 8 Cl, 12 C,
22 H, and 35 O   22 H, and 35 O

$$8\ KClO_3 + C_{12}H_{22}O_{11} \rightarrow 8\ KCl + 12\ CO_2 + 11\ H_2O$$

Steps 3 and 4 The coefficients in the balanced equation are already reduced to their smallest whole-number values, and a check shows that the numbers and kinds of atoms are the same on both sides of the equation.

⌐ **PROBLEM 3.1**   Potassium chlorate, $KClO_3$, decomposes when heated to yield potassium chloride and oxygen, a reaction used to provide oxygen for the emergency breathing masks in airliners. Balance the equation.

⌐ **PROBLEM 3.2**   Balance the following equations:
(a) $C_6H_{12}O_6 \longrightarrow C_2H_6O + CO_2$ (fermentation of sugar to yield ethyl alcohol)
(b) $Fe + O_2 \longrightarrow Fe_2O_3$ (rusting of iron)
(c) $NH_3 + Cl_2 \longrightarrow N_2H_4 + NH_4Cl$ (synthesis of hydrazine for rocket fuel)

3.2 ►CHEMICAL SYMBOLS ON DIFFERENT LEVELS

What does it mean when we write a chemical formula or equation? Answering this question isn't as easy as it sounds because the same chemical symbol has different meanings under different circumstances. Chemists use symbols to represent both a small-scale, *microscopic* level and a large-scale, *macroscopic* level, and they often tend to slip back and forth between the two levels without realizing the confusion this can cause for newcomers to the field.

On the microscopic level, chemical symbols represent the behavior of individual atoms and molecules. Atoms and molecules are much too small to be seen, but we can nevertheless portray their microscopic behavior if we read the equation $2\,H_2 + O_2 \rightarrow 2\,H_2O$ to mean "Two molecules of hydrogen react with one molecule of oxygen to yield two molecules of water." It's this microscopic world that we deal with when trying to understand reactions, and it's often helpful to visualize molecules as a collection of spheres stuck together. In trying to understand how H_2 reacts with O_2, for example, you might try picturing H_2 and O_2 molecules as made of two spheres pressed together and a water molecule as made of three spheres (Figure 3.1).

FIGURE 3.1 A useful way of visualizing the reaction of hydrogen with oxygen to yield H_2O is to think of the individual molecules as spheres (atoms) pressed together.

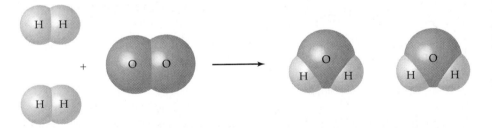

On the macroscopic level, formulas and equations represent the large-scale behavior of atoms and molecules that gives rise to observable properties. In other words, the symbols H_2, O_2, and H_2O can represent not just single molecules but vast numbers of molecules that collectively have a set of measurable physical properties. A huge collection of H_2O molecules appears to us as a colorless liquid that freezes at 0°C and boils at 100°C, but a *single* H_2O molecule in isolation has no such properties. Clearly, it's this macroscopic behavior we deal with in the laboratory when we weigh out specific amounts of reactants, place them in a flask, and observe visible changes.

What does a chemical formula or equation mean at any particular time? It means whatever you want it to mean depending on the context. The symbol H_2O can mean either one tiny, invisible molecule or a vast collection of molecules you can swim in.

3.3 ►AVOGADRO'S NUMBER AND THE MOLE

Imagine a laboratory experiment—perhaps the reaction of ethylene (C_2H_4) with hydrogen chloride (HCl) to prepare ethyl chloride (C_2H_5Cl), a colorless, low-boiling liquid that doctors and athletic trainers use as a spray-on anesthetic agent.

$$C_2H_4(g) + HCl(g) \rightarrow C_2H_5Cl(l)$$
Ethylene   Ethyl chloride
                                (an anesthetic)

Ethyl chloride is often used as a spray-on anesthetic for athletic injuries.

Because molecules are so small and the number of molecules needed to make a visible sample is so large, the experiment must involve a huge number of ethylene molecules reacting with a huge number of hydrogen chloride molecules.

When referring to the large numbers of molecules or ions that take part in a visible chemical reaction, it's convenient to use a special unit. Just as the unit "dozen" refers to the number 12 and the unit "gross" refers to the number 144, the unit **mole**,[1] abbreviated mol, refers to the number 6.022 × 10^{23}. This number is called **Avogadro's number** and is abbreviated N_A.[2] Thus, 12 molecules of HCl is a dozen, 144 molecules of HCl is a gross, and 6.022 × 10^{23} molecules of HCl (or the same number of any other object, molecule, or ion) is a mole.

1 mol HCl = 6.022 × 10^{23} HCl molecules

1 mol Na^+ ions = 6.022 × 10^{23} Na^+ ions

1 mol Ping-Pong balls = 6.022 × 10^{23} Ping-Pong balls

Why are the mole and Avogadro's number so important in chemistry? The answer comes from the relationship between *numbers* of molecules and *masses* of molecules. Take the reaction of ethylene with hydrogen chloride, for example. The coefficients in the balanced equation say that a 1:1 number ratio of the two reactants is needed:

1 C_2H_4 molecule + 1 HCl molecule $\rightarrow$ 1 C_2H_5Cl molecule

1 dozen C_2H_4 molecules + 1 dozen HCl molecules $\rightarrow$ 1 dozen C_2H_5Cl molecules

1 mol C_2H_4 molecules + 1 mol HCl molecules $\rightarrow$ 1 mol C_2H_5Cl molecules

You can't count the reactant molecules, though, so you have to weigh them to make sure you have the correct numbers of each. In other words, you must convert a *number* ratio into a *mass* ratio.

Mass ratios are determined by calculating the *molecular weights* of the substances involved in a reaction. The **molecular weight**, abbreviated **mol wt** or **MW**, of a molecule is the sum of the atomic weights (at. wt) of all atoms in a molecule. [The more general term **formula weight** (form. wt) is

---

[1] In precise terms, 1 mole is defined as the quantity of a substance that contains as many molecules or formula units as there are atoms in exactly 12 g of carbon-12.

[2] Because Avogadro's number tells the number of objects *per mole*, it has the units mol^{-1}.

the sum of atomic weights of all atoms in one formula unit of *any* substance, whether ionic or molecular.] For example, the molecular weight of ethylene is 28.0 amu, the molecular weight of hydrogen chloride is 36.5 amu, the molecular weight of ethyl chloride is 64.5 amu, and the formula weight of NaCl is 58.5 amu. (These numbers are rounded off to one decimal place for convenience; the actual values are more precise.)

Molecular weight = the sum of atomic weights of the atoms in a molecule

For ethylene, C_2H_4:
at. wt of 2 C = 2 × 12.0 amu = 24.0 amu
at. wt of 4 H = 4 × 1.0 amu   = 4.0 amu
mol wt of C_2H_4 = 28.0 amu

For hydrogen chloride, HCl:
at. wt of H   = 1.0 amu
at. wt of Cl = 35.5 amu
mol wt of HCl = 36.5 amu

For ethyl chloride, C_2H_5Cl:
at. wt of 2 C = 2 × 12.0 amu = 24.0 amu
at. wt of 5 H = 5 × 1.0 amu   = 5.0 amu
at. wt of Cl   = 35.5 amu
mol wt of C_2H_5Cl = 64.5 amu

For NaCl:
at. wt of Na   = 23.0 amu
at. wt of Cl   = 35.5 amu
form. wt of NaCl = 58.5 amu

Because the mass ratio of a *single* HCl molecule to a *single* ethylene molecule is 36.5:28.0, the mass ratio of *any* number of HCl molecules to the same number of ethylene molecules is also 36.5:28.0. In other words, a 36.5:28.0 *mass* ratio of HCl and ethylene always guarantees a 1:1 *number* ratio. *Samples of different substances contain the same number of molecules (or formula units) whenever their mass ratio is the same as their molecular-weight ratio* (Figure 3.2).

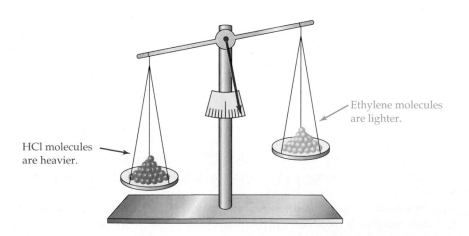

HCl molecules are heavier.

Ethylene molecules are lighter.

FIGURE 3.2 Equal numbers of HCl and ethylene molecules always have a mass ratio equal to the ratio of their molecular weights, 36.5:28.0.

Once the molecular weights of the reactant molecules have been determined, Avogadro's number and the mole come into play. *One mole of any substance (Avogadro's number of formula units) has a mass equal to its molecular or formula weight in grams.* For example, 1 mol (6.022×10^{23} molecules) of HCl weighs 36.5 g, 1 mol of ethylene weighs 28.0 g, and 1 mol of ethyl chloride weighs 64.5 g. Each amount is called the **molar mass** of the substance. To give you an idea of the amounts involved, molar amounts of some common substances are shown in Figure 3.3.

FIGURE 3.3 These samples of table sugar, lead shot, potassium dichromate, mercury, water, copper, sodium chloride, and sulfur contain 1 mol each.

Molar mass = mass of one mole of substance
              = mass of 6.02×10^{23} molecules (formula units) of substance
              = molecular (formula) weight of substance in grams

Mol wt of HCl = 36.5 amu	Molar mass of HCl = 36.5 g	$(6.02 \times 10^{23}$ HCl molecules$)$
Mol wt of C_2H_4 = 28.0 amu	Molar mass of C_2H_4 = 28.0 g	$(6.02 \times 10^{23}$ C_2H_4 molecules$)$
Mol wt of C_2H_5Cl = 64.5 amu	Molar mass of C_2H_5Cl = 64.5 g	$(6.02 \times 10^{23}$ C_2H_5Cl molecules$)$
Form. wt of NaCl = 58.5 amu	Molar mass of NaCl = 58.5 g	$(6.02 \times 10^{23}$ NaCl formula units$)$

In effect, molar mass serves as a conversion factor between numbers of molecules and mass. If you know how many molecules you have, you can calculate their mass; if you know the mass of a sample, you can calculate how many molecules you have. Note, though, that it's always necessary when calculating or using a molar mass to specify the formula of the particles you're talking about. For example, 1 mol of hydrogen *atoms*, H, has a molar mass of 1.0 g/mol, but 1 mol of hydrogen *molecules*, H_2, has a molar mass of 2.0 g/mol.

Whenever you see a balanced chemical equation, the coefficients tell how many moles of each substance are needed for the reaction. You can then use molar mass to calculate reactant masses. If you saw the following balanced equation for the industrial synthesis of ammonia, for example, you would know that 3 mol of H_2 (3 mol $\times$ 2.0 g/mol = 6.0 g) are required for reaction with 1 mol of N_2 (28.0 g) to yield 2 mol of NH_3 (2 mol $\times$ 17.0 g/mol = 34.0 g).

This number of moles of hydrogen . . .   . . . reacts with this number of moles of nitrogen . . .   to yield this number of moles of ammonia.

$$3 \; H_2 + 1 \; N_2 \rightarrow 2 \; NH_3$$

EXAMPLE 3.3

What is the molecular weight of sucrose, $C_{12}H_{22}O_{11}$? What is the molar mass of sucrose in g/mol?

SOLUTION   The molecular weight of a substance is the sum of the atomic weights of the constituent atoms. First, list the elements present in the molecule, and then look up the atomic weight of each (we'll round off to one decimal place for convenience):

$$C \text{ (12.0 amu); } H \text{ (1.0 amu); } O \text{ (16.0 amu)}$$

Next, multiply the atomic weight of each element by the number of times that element appears in the chemical formula, and then total the results.

$$
\begin{aligned}
C_{12} \ (12 \times 12.0 \text{ amu}) &= 144.0 \text{ amu} \\
H_{22} \ (22 \times 1.0 \text{ amu}) &= 22.0 \text{ amu} \\
O_{11} \ (11 \times 16.0 \text{ amu}) &= 176.0 \text{ amu} \\
\hline
C_{12}H_{22}O_{11} \text{ Mol wt} &= 342.0 \text{ amu}
\end{aligned}
$$

The molar mass of sucrose is its molecular weight in grams: 342.0 g/mol.

PROBLEM 3.3 Calculate the formula weight or molecular weight of the following substances.

(a) Fe_2O_3 (rust)   **(b)** H_2SO_4 (sulfuric acid)
(c) $C_6H_8O_7$ (citric acid)   **(d)** $C_{16}H_{18}N_2O_4S$ (penicillin G)

PROBLEM 3.4 The commercial production of iron from iron ore involves the reaction of Fe_2O_3 with CO to yield metallic iron plus carbon dioxide:

$$Fe_2O_3(s) + CO(g) \rightarrow Fe(s) + CO_2(g)$$

Balance the equation, and predict how many moles of CO will react with 0.5 mol of Fe_2O_3.

Iron is produced by reduction of iron ore with carbon monoxide.

PROBLEM 3.5 The molar mass of HCl is 36.5 g/mol, and the average mass per HCl molecule is 36.5 amu. Use the fact that 1 amu = 1.6605×10^{-24} g to show how Avogadro's number can be calculated.

3.4 ➤STOICHIOMETRY: CHEMICAL ARITHMETIC

We saw in the previous section that the coefficients in a balanced equation describe how many moles of each substance are needed for a reaction. To do actual laboratory work, though, it's necessary to convert between moles and grams to be sure that the correct amounts of reactants are used. In referring to these mole/mass relationships between reactants and products, we use the word **stoichiometry** (stoy-key-**ahm**-uh-tree; from the Greek *stoicheion*, "element," and *metron*, "measure"). Let's look again at the reaction of ethylene with HCl to see how stoichiometric relationships can be used.

$$C_2H_4(g) + HCl(g) \rightarrow C_2H_5Cl(l)$$

Let's assume that we have 15.0 g of ethylene and we need to know how many grams of HCl to use in the reaction. According to the balanced equation, 1 mol of HCl is required for each mole of ethylene. To find out how many grams of HCl are required to react with 15.0 g of ethylene, we first have to find out how many moles of ethylene are in 15.0 g. We do this gram-to-mole conversion by calculating the molar mass of ethylene and using that value as a conversion factor:

Molecular weight of C_2H_4 = (2 × 12.0 amu) + (4 × 1.0 amu) = 28.0 amu

Molar mass of C_2H_4 = 28.0 g/mol

Moles of C_2H_4 = 15.0 g ethylene × $\dfrac{1 \text{ mol ethylene}}{28.0 \text{ g ethylene}}$ = 0.536 mol ethylene

Now that we know how many moles of ethylene we have (0.536 mol), we also know from the balanced equation how many moles of HCl we need (0.536 mol), and we have to do a mole-to-gram conversion to find the mass of HCl required. Once again, we do the conversion by calculating the molecular weight of HCl and using molar mass as a conversion factor:

Molecular weight of HCl = 1.0 amu + 35.5 amu = 36.5 amu

Molar mass of HCl = 36.5 g/mol

Grams of HCl = 0.536 mol C_2H_4 × $\dfrac{1 \text{ mol HCl}}{1 \text{ mol } C_2H_4}$ × $\dfrac{36.5 \text{ g HCl}}{1 \text{ mol HCl}}$ = 19.6 g HCl

Thus, 19.6 g of HCl is needed to react with 15.0 g of ethylene.

Look carefully at the sequence of steps used in the calculation just completed. The important point is that *moles* (numbers of molecules) are used to balance equations but that *grams* are used to weigh reactants in the laboratory. Moles tell us *how many molecules* of each reactant we need, but grams tell us *how much mass* of each reactant we need.

Moles: numbers of molecules or formula units

Grams: mass of reactants

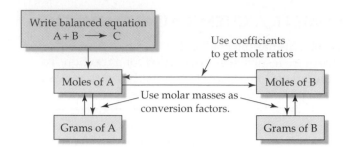

FIGURE 3.4 A flow diagram summarizing conversions between moles and grams for chemical reactions. Moles are used to balance equations and to tell how many molecules of each reactant are needed; grams tell how much mass of each reactant is needed.

The flow diagram in Figure 3.4 illustrates the necessary conversions. Note that you can't go directly from the number of grams of one reactant to the number of grams of another reactant. You *must* first convert to moles.

EXAMPLE 3.4

How many moles of sucrose are in an average tablespoon of sugar weighing 2.85 g?

BALLPARK SOLUTION   Since the molecular weight of sucrose (calculated in Example 3.3) is 342.0 amu, 1 mol of sucrose weighs 342.0 g. Thus, 2.85 g of sugar is a bit less than 0.01 mol.

DETAILED SOLUTION   The problem gives the mass of sucrose and asks for a mass-to-mole conversion. Use the molar mass of sucrose as a conversion factor, and set up an equation so that the unwanted unit cancels:

$$2.85 \text{ g sucrose} \times \frac{1 \text{ mol sucrose}}{342.0 \text{ g sucrose}} = 0.008\ 33 \text{ mol sucrose}$$

The ballpark answer and the detailed answer agree.

Sodium bicarbonate is the main ingredient in this Alka-Seltzer tablet.

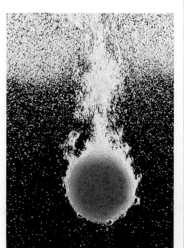

EXAMPLE 3.5

How many grams are in 0.0626 mol of $NaHCO_3$, the main ingredient in Alka-Seltzer tablets?

SOLUTION   The problem gives the number of moles of $NaHCO_3$ and asks for a mole-to-mass conversion. First, calculate the formula weight and molar mass of $NaHCO_3$:

$$\text{Form. wt of } NaHCO_3 = 23.0 \text{ amu} + 1.0 \text{ amu} + 12.0 \text{ amu} + (3 \times 16.0 \text{ amu}) = 84.0 \text{ amu}$$

$$\text{Molar mass of } NaHCO_3 = 84.0 \text{ g/mol}$$

Next, use the molar mass as a conversion factor, and set up an equation so that the unwanted unit cancels:

$$0.0626 \text{ mol } NaHCO_3 \times \frac{84.0 \text{ g } NaHCO_3}{1 \text{ mol } NaHCO_3} = 5.26 \text{ g } NaHCO_3$$

EXAMPLE 3.6

Aqueous solutions of sodium hypochlorite (NaOCl), best known as household bleach, are prepared by reaction of sodium hydroxide with chlorine:

$$2 \text{ NaOH}(aq) + \text{Cl}_2(g) \rightarrow \text{NaOCl}(aq) + \text{NaCl}(aq) + \text{H}_2\text{O}(l)$$

How many grams of NaOH are needed to react with 25.0 g Cl_2?

Household bleach is made by reacting chlorine gas with aqueous sodium hydroxide.

SOLUTION   Finding the relationship between numbers of reactant molecules always requires working in moles. Thus, the first job is to find out many moles of Cl_2 are in 25.0 g of Cl_2. This gram-to-mole conversion is done in the normal way, using the molar mass of Cl_2 (70.9 g/mol) as the conversion factor:

$$25.0 \text{ g Cl}_2 \times \frac{1 \text{ mol Cl}_2}{70.9 \text{ g Cl}_2} = 0.353 \text{ mol Cl}_2$$

Next, look at the balanced equation. Each mole of Cl_2 reacts with 2 mol of NaOH, so 0.353 mol of Cl_2 reacts with $2 \times 0.353 = 0.706$ mol of NaOH. With the number of moles of NaOH known, carry out a mole-to-gram conversion using the molar mass of NaOH (40.0 g/mol) as a conversion factor.

$$\text{Grams NaOH} = 0.353 \text{ mol Cl}_2 \times \frac{2 \text{ mol NaOH}}{1 \text{ mol Cl}_2} \times \frac{40.0 \text{ g NaOH}}{1 \text{ mol NaOH}} = 28.2 \text{ g NaOH}$$

We find that 25.0 g of Cl_2 reacts with 28.2 g of NaOH.

The problem can also be worked by combining the steps and setting up one large equation:

$$\text{Grams NaOH} = 25.0 \text{ g Cl}_2 \times \frac{1 \text{ mol Cl}_2}{70.9 \text{ g Cl}_2} \times \frac{2 \text{ mol NaOH}}{1 \text{ mol Cl}_2} \times \frac{40.0 \text{ g NaOH}}{1 \text{ mol NaOH}}$$

$$= 28.2 \text{ g NaOH}$$

Aspirin is manufactured by reaction of acetic anhydride with salicylic acid.

⌐ PROBLEM 3.6   Aspirin has the formula $\text{C}_9\text{H}_8\text{O}_4$. How many moles of aspirin are in a tablet weighing 500 mg? How many molecules?

⌐ PROBLEM 3.7   Aspirin is prepared by reaction of salicylic acid ($\text{C}_7\text{H}_6\text{O}_3$) with acetic anhydride ($\text{C}_4\text{H}_6\text{O}_3$) according to the following equation:

$$\text{C}_7\text{H}_6\text{O}_3 + \text{C}_4\text{H}_6\text{O}_3 \rightarrow \text{C}_9\text{H}_8\text{O}_4 + \text{C}_2\text{H}_4\text{O}_2$$

Salicylic   Acetic   Aspirin   Acetic acid
acid   anhydride

How many grams of acetic anhydride are necessary to react with 4.5 g of salicylic acid? How many grams of aspirin will result? How many grams of acetic acid are formed as a byproduct?

3.5 ▶YIELDS OF CHEMICAL REACTIONS

In the examples of chemical stoichiometry worked out in the preceding section, we made the unstated assumption that all reactions "go to completion." That is, we assumed that *all* molecules present in a reactant mixture are converted to products. In fact, few reactions behave so nicely. Most of the time, a large majority of molecules react in the specified way, but other processes, called *side reactions*, also occur. Thus, the amount of product actually formed is usually somewhat less than the amount that theory predicts.

The amount of product actually formed in a reaction divided by the amount theoretically possible and multiplied by 100% is called the reaction's **percent yield**. For example, if a given reaction *could* provide 6.9 g of a product according to its stoichiometry but actually provides only 4.7 g, then its percent yield is $4.7/6.9 \times 100\% = 68\%$.

$$\textbf{Percent yield} = \frac{\text{Actual yield of product}}{\text{Theoretical yield of product}} \times 100\%$$

The following example shows how to calculate a reaction's percent yield.

EXAMPLE 3.7

Ethyl acetate ($C_4H_8O_2$), a colorless liquid used as a solvent in such familiar products as nail-polish remover, is prepared commercially by reaction of acetic acid ($C_2H_4O_2$) with ethyl alcohol (C_2H_6O). How many grams of ethyl acetate would you obtain from 45.0 g of acetic acid if the percent yield of the reaction is 87%?

$$C_2H_4O_2(l) + C_2H_6O(l) \rightarrow C_4H_8O_2(l) + H_2O(l)$$
Acetic acid   Ethyl alcohol   Ethyl acetate

SOLUTION   Always begin stoichiometry problems by calculating the molar masses of the reactants and products:

Acetic acid:
  mol wt $C_2H_4O_2$ = (2 × 12.0 amu) + (4 × 1.0 amu) + (2 × 16.0 amu) = 60.0 amu

Molar mass of acetic acid = 60.0 g/mol

Ethyl acetate:
  mol wt $C_4H_8O_2$ = (4 × 12.0 amu) + (8 × 1.0 amu) + (2 × 16.0 amu) = 88.0 amu

Molar mass of ethyl acetate = 88.0 g/mol

Next, find how many moles of acetic acid are in 45.0 g by using molar mass as a conversion factor:

$$45.0 \text{ g acetic acid} \times \frac{1 \text{ mol acetic acid}}{60.0 \text{ g acetic acid}} = 0.750 \text{ mol acetic acid}$$

Because we started with 0.750 mol acetic acid and because the balanced equation indicates that each mole of acetic acid yields 1 mol of ethyl acetate, we can theoretically form 0.750 mol of product. We must therefore find how many grams of ethyl acetate are in 0.750 mol:

$$0.750 \; \text{mol ethyl acetate} \times \frac{88.0 \text{ g ethyl acetate}}{1 \text{ mol ethyl acetate}} = 66.0 \text{ g ethyl acetate}$$

Finally, we have to multiply the theoretical amount of product by the observed yield (87% = 0.87) to find how much ethyl acetate is actually formed:

$$66.0 \text{ g ethyl acetate} \times 0.87 = 57 \text{ g ethyl acetate}$$

☐ PROBLEM 3.8 Ethyl alcohol is prepared industrially by the reaction of ethylene, C_2H_4, with water. What is the percent yield of the reaction if 4.6 g of ethylene gives 4.7 g of ethyl alcohol?

$$\underset{\text{Ethylene}}{C_2H_4(g)} + H_2O(l) \longrightarrow \underset{\text{Ethyl alcohol}}{C_2H_6O(l)}$$

☐ PROBLEM 3.9 Dichloromethane (CH_2Cl_2), a solvent used in the decaffeination of coffee beans, is prepared by reaction of methane (CH_4) with chlorine. How many grams of dichloromethane result from reaction of 1.85 kg of methane if the yield is 43.1%?

$$\underset{\text{Methane}}{CH_4(g)} + \underset{\text{Chlorine}}{2 \; Cl_2(g)} \longrightarrow \underset{\text{Dichloromethane}}{CH_2Cl_2(l)} + 2 \; HCl(g)$$

Dichloromethane is used as a solvent to remove caffeine from coffee beans.

3.6 ►REACTIONS WITH LIMITING AMOUNTS OF REACTANTS

Because chemists usually write balanced equations, it's easy to get the impression that reactions are always carried out using exactly the right proportions of reactants. In fact, this is often not the case. Many reactions are carried out using an excess amount of one reactant—more than is actually needed according to stoichiometry. Look, for example, at the industrial synthesis of ethylene glycol, $C_2H_6O_2$, a substance used both as an automobile antifreeze and as a starting material for the preparation of polyester polymers. More than 2 million tons of ethylene glycol are prepared each year in the United States by reaction of ethylene oxide, C_2H_4O, with water at high temperature:

The antifreeze being added to this car is prepared by reaction of water with ethylene oxide.

$$\underset{\text{Ethylene oxide}}{C_2H_4O} + H_2O \xrightarrow{\text{heat}} \underset{\text{Ethylene glycol}}{C_2H_6O_2}$$

Because water is so cheap and so abundant, it doesn't make sense to worry about using exactly 1 mol of water for each mole of ethylene oxide. It's much easier to use an excess of water to be certain that enough is present to consume entirely the more valuable ethylene oxide reactant. Of course, when an excess of water is present, only the amount required by stoichiometry reacts with ethylene oxide. The excess water acts only as a spectator and takes no role in the reaction.

Whenever the numbers of reactant molecules actually used are different from those given in the balanced equation, an excess of one reactant is left over after the reaction is finished. Thus, the extent to which a chemical reaction takes place depends on the reactant present in limiting amount— the **limiting reactant**. The situation is analogous to what happens if there are five people in a room but only three chairs. Only three people can sit while the other two stand, because the number of people sitting is limited by the number of available chairs. In the same way, if five water molecules (or 5 mol) come in contact with 3 ethylene oxide molecules (or 3 mol), only three water molecules undergo a reaction. The other two are merely spectators because the number of water molecules that react is limited by the number of available ethylene oxide molecules.

$$5 \text{ H}_2\text{O} + 3 \text{ ethylene oxide} \rightarrow 3 \text{ ethylene glycol} + 2 \text{ H}_2\text{O (unreacted)}$$

       ↖ in excess   ↖ limiting reactant

The following example shows how to tell if a limiting amount of one reactant is present and how to calculate the amount of nonlimiting reactant consumed.

EXAMPLE 3.8

Cisplatin, an anticancer agent used for the treatment of solid tumors, is prepared by the reaction of ammonia with potassium tetrachloroplatinate:

$$\underset{\substack{\text{Potassium} \\ \text{tetrachloroplatinate}}}{\text{K}_2\text{PtCl}_4} + 2 \text{ NH}_3 \rightarrow \underset{\text{Cisplatin}}{\text{Pt(NH}_3)_2\text{Cl}_2} + 2 \text{ KCl}$$

Assume that 10.0 g K_2PtCl_4 reacts with 10.0 g NH_3. **(a)** Which reactant is limiting and which is in excess? **(b)** How many grams of the excess reactant are consumed? **(c)** How many grams of cisplatin are formed?

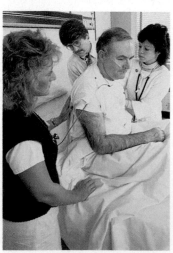

Some kinds of cancers can be treated with cisplatin, $\text{Pt(NH}_3)_2\text{Cl}_2$.

SOLUTION   **(a)** Complex stoichiometry problems should be taken slowly and carefully, one step at a time. The first step is to find how many moles of each reactant are present:

Form. wt of K_2PtCl_4 = (2 × 39.1 amu) + 195.1 amu + (4 × 35.5 amu) = 415.3 amu

Molar mass of K_2PtCl_4 = 415.3 g/mol

Moles K_2PtCl_4 = 10.0 g K_2PtCl_4 × $\dfrac{1 \text{ mol K}_2\text{PtCl}_4}{415.3 \text{ g K}_2\text{PtCl}_4}$ = 0.0241 mol K_2PtCl_4

Mol wt of NH_3 = 14.0 amu + (3 × 1.0 amu) = 17.0 amu

Molar mass of NH_3 = 17.0 g/mol

Moles NH_3 = 10.0 g NH_3 × $\dfrac{1 \text{ mol NH}_3}{17.0 \text{ g NH}_3}$ = 0.588 mol NH_3

Next, look at the coefficients in the balanced equation to see the required mole ratio of the two reactants. Since 2 mol of NH_3 are required for reaction with 1 mol of K_2PtCl_4, 0.0482 mol of NH_3 is required for reaction with 0.0241 mol of K_2PtCl_4. We have 0.588 mol of NH_3 available, though, so a large excess of NH_3 is left over, and K_2PtCl_4 is the limiting reactant:

Moles of K_2PtCl_4 consumed: 0.0241 mol K_2PtCl_4

Moles of NH_3 consumed: 2×0.0241 mol = 0.0482 mol NH_3

Moles of NH_3 not consumed: $(0.588 - 0.0482)$ mol NH_3 = 0.540 mol NH_3

(b) To find out the mass of NH_3 consumed, a mole-to-gram conversion is necessary:

$$\text{Grams of } NH_3 \text{ consumed} = 0.0482 \text{ mol } NH_3 \times \frac{17.0 \text{ g } NH_3}{1 \text{ mol } NH_3} = 0.819 \text{ g } NH_3$$

(c) To find out how much cisplatin is formed, we again have to check the balanced equation to find that 1 mol of cisplatin is formed for each mole of K_2PtCl_4 consumed. Thus, 0.0241 mol of cisplatin is formed from 0.0241 mol of K_2PtCl_4. To determine the mass of cisplatin produced, we must calculate its molar mass and then carry out a mole-to-gram conversion:

Mol wt of $Pt(NH_3)_2Cl_2$ = 195.1 amu + $(2 \times 17.0$ amu$)$ + $(2 \times 35.5$ amu$)$ = 300.1 amu

Molar mass of $Pt(NH_3)_2Cl_2$ = 300.1 g/mol

$$\text{Grams } Pt(NH_3)_2Cl_2 = 0.0241 \text{ mol } Pt(NH_3)_2Cl_2 \times \frac{300.1 \text{ g } Pt(NH_3)_2Cl_2}{1 \text{ mol } Pt(NH_3)_2Cl_2}$$
$$= 7.23 \text{ g } Pt(NH_3)_2Cl_2$$

☐ **PROBLEM 3.10** We know from everyday experience that flames are extinguished when the oxygen supply is used up. Imagine that you have a tank containing 700 g of propane (C_3H_8). How many grams of propane would remain if only 65 g of O_2 was available for combustion? (See Example 3.1 for the balanced equation.)

Lithium oxide is used aboard the space shuttle to remove water from the air.

☐ **PROBLEM 3.11** Lithium oxide is used aboard the space shuttle to remove water from the air supply according to the equation

$$Li_2O(s) + H_2O(g) \rightarrow 2 \text{ LiOH}(s)$$

Which reactant is limiting if there are 80 kg of water to be removed and 65 kg of Li_2O available? How many kg of the excess reactant remain?

☐ **PROBLEM 3.12** After lithium hydroxide is produced aboard the space shuttle by reaction of Li_2O with H_2O (Problem 3.11), it is used to remove exhaled carbon dioxide from the air supply according to the equation:

$$LiOH(s) + CO_2(g) \rightarrow LiHCO_3(s)$$

How many grams of CO_2 can 500 g of LiOH absorb?

3.7 ▶CONCENTRATIONS OF REACTANTS IN SOLUTION: MOLARITY

For a chemical reaction to occur, the reacting molecules or ions must come into close contact with each other. This means that the reactants must have considerable mobility, which in turn means that most chemical reactions are carried out in the liquid state or in solution rather than in the solid state. It's therefore necessary to have a standard means for describing exact quantities of reactants in solution.

Because we need to know numbers of moles to do stoichiometry calculations for chemical reactions, the most generally useful means of expressing a solution's concentration is to use **molarity** (**M**), the number of moles of a substance (the *solute*) dissolved in each liter of solution. For example, a solution made by dissolving 1.00 mol (58.5 g) of NaCl in enough water to give 1.00 L of solution has a concentration of 1.00 mol/L, or 1.00 **M**. The molarity of any solution is found by dividing the number of moles of solute by the number of liters of solution.

$$\text{Molarity (M)} = \frac{\text{Moles of solute}}{\text{Liters of solution}}$$

Note that a solution of known molarity is prepared by dissolving the solute in enough solvent to give a *final* solution volume of 1.00 L, not by dissolving it in a *starting* volume of 1.00 L. If a starting volume of 1.00 L were used, the final solution volume would be a bit larger than 1.00 L because of the additional volume of the solute. In practice, the appropriate amount of solute is weighed out and placed in a *volumetric flask*, as shown in Figure 3.5. Enough solvent is then added to dissolve the solute, and further solvent is added until an accurately calibrated final volume is reached. The solution is then shaken until it is uniformly mixed.

Molarity can be used as a conversion factor to connect a solution's volume with the number of moles of solute. If we know the molarity and volume of a solution, we can calculate the number of moles of solute. If we know the number of moles of solute and the molarity of the solution, we can find the solution's volume. Examples 3.9 and 3.10 show how the calculations are done.

FIGURE 3.5 Preparing a solution of known molarity. (a) A measured number of moles of solute is placed in a volumetric flask. (b) Enough solvent is added to dissolve the solute. (c) Further solvent is carefully added until the calibration mark on the neck of the flask is reached. The solution is then shaken until uniform.

(a)

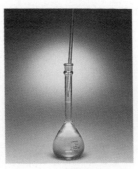
(b)

(c)

$$\text{Molarity} = \frac{\text{Moles of solute}}{\text{Volume of solution (L)}}$$

$$\begin{array}{c}\text{Moles of}\\\text{solute}\end{array} = \text{Molarity} \times \begin{array}{c}\text{Volume of}\\\text{solution}\end{array} \qquad \begin{array}{c}\text{Volume}\\\text{of solution}\end{array} = \frac{\text{Moles of solute}}{\text{Molarity}}$$

A sulfuric acid solution is prepared by diluting a known amount of H_2SO_4 to a precisely calibrated volume.

EXAMPLE 3.9

What is the molarity of a solution made by dissolving 2.355 g of sulfuric acid (H_2SO_4) in water and diluting to a final volume of 50.0 mL?

SOLUTION   Since molarity is defined as the number of moles of solute per liter of solution, it's first necessary to find the number of moles of sulfuric acid in 2.355 g.

Mol wt of H_2SO_4 = (2 × 1.0 amu) + 32.1 amu + (4 × 16.0 amu) = 98.1 amu

Molar mass of H_2SO_4 = 98.1 g/mol

$$2.355 \text{ g } H_2SO_4 \times \frac{1 \text{ mol } H_2SO_4}{98.1 \text{ g } H_2SO_4} = 0.0240 \text{ mol } H_2SO_4$$

Next, divide the number of moles of solute by the volume of the solution in liters:

$$\frac{0.0240 \text{ mol } H_2SO_4}{0.0500 \text{ L}} = 0.480 \text{ M}$$

The solution has a sulfuric acid concentration of 0.480 M.

EXAMPLE 3.10

Hydrochloric acid is sold commercially as a 12.0 M solution. How many moles of HCl are in 300 mL of 12.0 M solution?

SOLUTION   You can calculate the number of moles of solute by multiplying the molarity of the solution by its volume:

$$\text{Moles of HCl} = (\text{molarity of solution}) \times (\text{volume of solution})$$

$$= \frac{12.0 \text{ mol HCl}}{1 \text{ L}} \times 0.300 \text{ L} = 3.60 \text{ mol HCl}$$

There are 3.60 mol of HCl in 300 mL of 12.0 M solution.

PROBLEM 3.13 How many moles of solute are present in the following solutions?

(a) 125 mL of 0.20 M $NaHCO_3$ (b) 650 mL of 2.50 M H_2SO_4?

PROBLEM 3.14 How many grams of solute would you use to prepare the following solutions?

(a) 500 mL of 1.25 M NaOH   (b) 1.50 L of 0.250 M glucose ($C_6H_{12}O_6$)

PROBLEM 3.15 How many mL of a 0.20 M glucose ($C_6H_{12}O_6$) solution are needed to provide a total of 25.0 g of glucose?

PROBLEM 3.16 The concentration of cholesterol ($C_{27}H_{46}O$) in normal blood is approximately 0.005 M. How many grams of cholesterol are in 750 mL of blood?

This artery has become blocked by cholesterol.

3.8 ►DILUTING CONCENTRATED SOLUTIONS

For convenience, chemicals are sometimes bought and stored as concentrated solutions that must be diluted before use. Aqueous hydrochloric acid, for example, is sold commercially as a 12.0 M solution yet is most commonly used in the laboratory after dilution with water to a final concentration of 6.0 M or 1.0 M.

Concentrated solution + solvent → dilute solution

The key fact to remember when diluting a concentrated solution is that the number of moles of *solute* remains the same; only the *volume* is changed by adding more solvent. Because the number of moles of solute can be calculated by multiplying molarity times volume, we can set up the following equation:

$$\text{Moles of solute} = M_i\left(\frac{mol}{L}\right) \times V_i(L) = M_f\left(\frac{mol}{L}\right) \times V_f(L)$$

where M_i is the initial molarity, V_i is the initial volume, M_f is the final molarity, and V_f is the final volume after dilution. Rewriting this equation in a more useful form shows that the molar concentration after dilution (M_f) can be found by multiplying the initial concentration (M_i) by the ratio of initial and final volumes (V_i/V_f):

$$M_f = M_i \times \frac{V_i}{V_f}.$$

Suppose, for example, that we dilute 50.0 mL of a solution of 2.00 M H_2SO_4 to 200.0 mL. The solution volume *increases* by a factor of four (from 50 mL to 200 mL), so the concentration of the solution must *decrease* by a factor of four (from 2.00 M to 0.500 M):

$$M_f = 2.00 \text{ M} \times \frac{50.0 \text{ mL}}{200.0 \text{ mL}} = 0.500 \text{ M}.$$

In practice, dilutions are carried out as shown in Figure 3.6. The volume to be diluted is withdrawn with a pipet, placed in an empty volumetric flask of the chosen volume, and diluted to the calibration mark on the flask.

FIGURE 3.6 The procedure for diluting a concentrated solution. (a) The volume to be diluted is withdrawn using a pipet and placed in an empty volumetric flask. (b) Solvent is then added to reach the calibration mark on the flask, and the solution is thoroughly mixed (c) by shaking it.

(a)   (b)   (c)

EXAMPLE 3.11

How would you prepare 500.0 mL of 0.2500 M sodium hydroxide solution starting from a concentration of 1.000 M?

BALLPARK SOLUTION   Because the concentration of the final solution is one-fourth that of the initial solution, the volume of the initial solution must increase by a factor of four. Thus, to prepare 500.0 mL of solution, we should start with 500.0/4 = 125.0 mL.

DETAILED SOLUTION   The problem gives initial and final concentrations (M_i and M_f) and final volume (V_f) and asks for the initial volume (V_i) that we need to dilute. Rewriting the equation $M_i \times V_i = M_f \times V_f$ as $V_i = (M_f/M_i) \times V_f$ gives the answer:

$$V_i = \frac{M_f}{M_i} \times V_f = \frac{0.2500 \text{ M}}{1.000 \text{ M}} \times 500.0 \text{ mL} = 125.0 \text{ mL}$$

We therefore need to pipet 125.0 mL of 1.000 M NaOH solution into a 500.0 mL volumetric flask and fill to the mark with water.

PROBLEM 3.17 What is the final concentration if 75.0 mL of a 3.50 M glucose solution is diluted to a volume of 400.0 mL?

PROBLEM 3.18 Sulfuric acid is normally purchased at a concentration of 18.0 M. How would you prepare 250 mL of 0.500 M aqueous H_2SO_4?

3.9 ►SOLUTION STOICHIOMETRY

We remarked in Section 3.7 that molarity can serve as a conversion factor between numbers of moles of solute and the volume of a solution. If we know the volume and molarity of a solution, we can calculate the number of moles of solute. If we know the number of moles of solute and molarity, we can find the volume.

As indicated by the flow diagram in Figure 3.7, a knowledge of molarity is critical for carrying out stoichiometry calculations on substances in solution. Molarity makes it possible to calculate the volume of one solution needed to react with a given volume of another solution. This sort of calculation is particularly important in acid–base chemistry, as shown in Example 3.12.

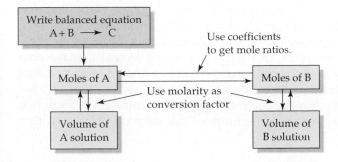

FIGURE 3.7 A flow diagram summarizing the use of molarity as a conversion factor between moles and volume in stoichiometry calculations.

Neutralization of sodium bicarbonate with acid leads to release of CO_2 gas, visible in this fizzing solution.

EXAMPLE 3.12

Stomach acid, a dilute solution of HCl in water, can be neutralized by reaction with sodium bicarbonate, $NaHCO_3$, according to the equation:

$$HCl(aq) + NaHCO_3(aq) \rightarrow NaCl(aq) + H_2O(l) + CO_2(g)$$

How many mL of 0.125 M $NaHCO_3$ solution are needed to neutralize 18.0 mL of 0.100 M HCl?

BALLPARK SOLUTION   The balanced equation shows that HCl and $NaHCO_3$ react in a 1:1 ratio, and we are told that the concentrations of the two solutions are about the same. Thus, the volume of the $NaHCO_3$ solution must be about the same as that of the HCl solution.

DETAILED SOLUTION   Since we need to know the numbers of moles to solve stoichiometry problems, first find how many moles of HCl are in 18.0 mL of a 0.100 M solution by multiplying volume times molarity.

$$\text{Moles HCl} = 18.0 \text{ mL} \times \frac{1 \text{ L}}{1000 \text{ mL}} \times \frac{0.100 \text{ mol}}{1 \text{ L}} = 1.80 \times 10^{-3} \text{ mol HCl}$$

Next, check the coefficients of the balanced equation to find that each mole of HCl reacts with 1 mol of $NaHCO_3$, and then calculate how many mL of 0.125 M $NaHCO_3$ solution contains 1.80×10^{-3} mol.

$$1.80 \times 10^{-3} \text{ mol HCl} \times \frac{1 \text{ mol NaHCO}_3}{1 \text{ mol HCl}} \times \frac{1 \text{ L solution}}{0.125 \text{ mol NaHCO}_3} = 0.0144 \text{ L solution}$$

Thus, 14.4 mL of the 0.125 M $NaHCO_3$ solution is needed to neutralize 18.0 mL of the 0.100 M HCl solution.

⌐ **PROBLEM 3.19** What volume of 0.250 M H_2SO_4 is needed to react with 50.0 mL of 0.100 M NaOH? The equation is

$$H_2SO_4(aq) + 2 \text{ NaOH}(aq) \rightarrow Na_2SO_4(aq) + 2 \text{ H}_2O(l)$$

⌐ **PROBLEM 3.20** What is the molarity of an HNO_3 solution if 68.5 mL is needed to react with 25.0 mL of 0.150 M KOH solution?

3.10 ➤TITRATION

There are two ways to make a solution of known molarity. The first and most obvious way is to make the solution carefully in the first place, using an accurately weighed amount of solute diluted with solvent to an accurately calibrated volume. Often, though, it's more convenient to make up a solution quickly, using an estimated amount of solute and an estimated final volume, and then determine the solution's exact molarity by *titration*.

Titration is a procedure for determining the concentration of a solution by allowing a carefully measured volume of the solution to react with a second solution of known concentration. By finding the volume of the second solution that reacts with the known volume of the first solution, the concentration of the first solution can be calculated. (Of course, it's necessary that the reaction go to completion and reach a yield of 100%.)

To see how titration works, let's imagine that we have an HCl solution of unknown concentration. We measure out a known volume—say, 20.0 mL—and add a small amount of an *indicator* such as phenolphthalein, a compound that is colorless in acidic solution but that turns a pink color when in basic solution. Next, we fill a buret with an NaOH solution of known concentration—say 0.100 M—and we slowly add the NaOH to the HCl until the phenolphthalein just begins to turn pink, indicating that all the HCl has reacted. By then reading from the buret to find the volume of the 0.100 M NaOH solution that has reacted with the 20.0 mL of HCl, we can calculate the concentration of the HCl. If, for example, 48.6 mL of 0.100 M NaOH was needed, then the HCl concentration is 0.243 M:

$$NaOH(aq) + HCl(aq) \rightarrow H_2O(l) + NaCl(aq)$$

Moles of HCl = moles of NaOH

$$\text{Moles of NaOH} = 0.0486 \ \cancel{\text{L NaOH}} \times \frac{0.100 \ \text{mol NaOH}}{1 \ \cancel{\text{L NaOH}}} = 0.00486 \ \text{mol NaOH}$$

$$\text{Moles of HCl} = 0.00486 \ \cancel{\text{mol NaOH}} \times \frac{1 \ \text{mol HCl}}{1 \ \cancel{\text{mol NaOH}}} = 0.00486 \ \text{mol HCl}$$

$$\text{HCl molarity} = \frac{0.00486 \ \text{mol HCl}}{0.0200 \ \text{L HCl}} = 0.243 \ \text{M HCl}$$

The entire sequence is shown in Figure 3.8.

(a)

(b)

(c)

FIGURE 3.8 Titration of an acid solution of unknown concentration with a base solution of known concentration. (a) A measured volume of acid solution is placed in a flask, and phenolphthalein indicator is added. (b) A base solution of known concentration is added from a buret until (c) the indicator changes color to signal that all the acid has reacted. Reading the volume of base solution added from the buret makes it possible to calculate the concentration of the acid solution.

⌐ **PROBLEM 3.21** A 25.0 mL sample of vinegar (dilute acetic acid, $HC_2H_3O_2$) is titrated and found to react with 94.7 mL of 0.200 M NaOH. What is the molarity of the acetic acid solution? The reaction is

$$NaOH(aq) + HC_2H_3O_2(aq) \rightarrow NaC_2H_3O_2(aq) + H_2O(l)$$

3.11 ►PERCENT COMPOSITION
AND EMPIRICAL FORMULAS

Whenever a new compound is made in the laboratory or is found in nature, it must be analyzed to find what elements it contains and how much of each element is present—that is, to find its *composition*. The **percent composition** of a compound is expressed by identifying the elements present and giving the mass percent of each. For example, a white solid might contain 43.6% phosphorus and 56.4% oxygen by mass. In other words, a 100 g sample of the compound contains 43.6 g of phosphorus atoms and 56.4 g of oxygen atoms.

Knowing a compound's percent composition makes it possible to calculate its chemical formula. First, we find the relative numbers of moles of each element in the compound by using the molar masses of the elements as conversion factors and carrying out gram-to-mole conversions. For example, arbitrarily taking 100 g of the white solid whose composition is 43.6% P and 56.4% O, we find that the 100 g contains:

$$43.6 \text{ g P} \times \frac{1 \text{ mol P}}{31.0 \text{ g P}} = 1.41 \text{ mol P}$$

$$56.4 \text{ g O} \times \frac{1 \text{ mol O}}{16.0 \text{ g O}} = 3.52 \text{ mol O}$$

With the relative numbers of moles of P and O known, we next find their ratio by dividing the larger number of moles (3.52 mol) by the smaller number (1.41 mol):

$$\frac{\text{Moles of O}}{\text{Moles of P}} = \frac{3.52 \text{ mol O}}{1.41 \text{ mol P}} = \frac{2.50 \text{ mol O}}{1 \text{ mol P}}$$

The O:P mole ratio of 2.50:1 means that we can write $P_1O_{2.50}$ as a temporary formula for our white solid. Multiplying the subscripts by small integers in a trial-and-error procedure until whole numbers are found then gives the final formula. In the present instance, multiplication of the subscripts by 2 does the job:

$$P_{(1 \times 2)}O_{(2.50 \times 2)} = P_2O_5$$

A formula such as P_2O_5 that is determined from data about percent composition is called an **empirical formula** because it tells only the *ratios* of atoms in a compound. The **molecular formula**, which tells the actual numbers of atoms in a molecule, can be either the same as the empirical formula or a multiple of it such as P_4O_{10} or P_6O_{15}. To determine the molecular formula, it is necessary to know the molecular weight of the substance. In the present instance, the molecular weight of the phosphorus/oxygen compound is 283.9 amu, which is a simple multiple of the empirical formula weight for P_2O_5 (141.9 amu). To find the multiple, calculate the ratio of the molecular weight to the empirical formula weight:

$$\text{Multiple} = \frac{\text{Molecular weight}}{\text{Empirical formula weight}} = \frac{283.9}{141.9} = 2.001, \text{ or } 2$$

Then multiply the subscripts in the empirical formula by this multiple to obtain the molecular formula. In our example, the compound is $P_{(2\times2)}O_{(2\times5)}$, or P_4O_{10}.

Just as we can derive an empirical formula from the percent composition, we can also derive the percent composition from an empirical (or molecular) formula. Aspirin, for example, has the molecular formula $C_9H_8O_4$ and thus has a C:H:O mole ratio of 9:8:4. We can convert this mole ratio into a mass ratio, and thus into percent composition, by carrying out mole-to-gram conversions on a 1 mol sample of compound:

$$1 \text{ mol aspirin} \times \frac{9 \text{ mol C}}{1 \text{ mol aspirin}} \times \frac{12.0 \text{ g C}}{1 \text{ mol C}} = 108 \text{ g C}$$

$$1 \text{ mol aspirin} \times \frac{8 \text{ mol H}}{1 \text{ mol aspirin}} \times \frac{1.01 \text{ g H}}{1 \text{ mol H}} = 8.08 \text{ g H}$$

$$1 \text{ mol aspirin} \times \frac{4 \text{ mol O}}{1 \text{ mol aspirin}} \times \frac{16.0 \text{ g O}}{1 \text{ mol O}} = 64.0 \text{ g O}$$

Dividing the mass of each element by the total mass and multiplying by 100% then gives the percent composition.

$$\text{Total mass of 1 mol aspirin} = 108 \text{ g} + 8.08 \text{ g} + 64.0 \text{ g} = 180 \text{ g}$$

$$\% \text{ C} = \frac{108 \text{ g C}}{180 \text{ g}} \times 100\% = 60.0\%$$

$$\% \text{ H} = \frac{8.08 \text{ g H}}{180 \text{ g}} \times 100\% = 4.49\%$$

$$\% \text{ O} = \frac{64.0 \text{ g O}}{180 \text{ g}} \times 100\% = 35.6\%$$

The answer can be checked by confirming that the sum of the mass percentages is 100%: 60.0% + 4.49% + 35.6% = 100.1%

Examples 3.13 and 3.14 show more conversions between percent composition and empirical formulas.

EXAMPLE 3.13

Vitamin C (ascorbic acid) contains 40.92% C, 4.58% H, and 54.50% O. What is the empirical formula of ascorbic acid?

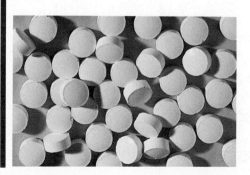

Vitamin C is an organic acid.

SOLUTION   Assume that you have 100.0 g of ascorbic acid, and then carry out gram-to-mole conversions to find the number of moles of each element:

$$40.92 \text{ g C} \times \frac{1 \text{ mol C}}{12.0 \text{ g C}} = 3.41 \text{ mol C}$$

$$4.58 \text{ g H} \times \frac{1 \text{ mol H}}{1.01 \text{ g H}} = 4.53 \text{ mol H}$$

$$54.50 \text{ g O} \times \frac{1 \text{ mol O}}{16.0 \text{ g O}} = 3.41 \text{ mol O}$$

Dividing each of the three numbers by the smallest one (3.41 mol) gives a C:H:O mole ratio of 1:1.33:1 and a temporary formula of $C_1H_{1.33}O_1$. Multiplying the subscripts by small integers in a trial-and-error procedure until whole numbers are found then gives the empirical formula $C_3H_4O_3$.

Multiply subscripts by 2: $C_{(2 \times 1)}H_{(2 \times 1.33)}O_{(2 \times 1)} = C_2H_{2.66}O_2$

Multiply subscripts by 3: $C_{(3 \times 1)}H_{(3 \times 1.33)}O_{(3 \times 1)} = C_3H_4O_3$

EXAMPLE 3.14

Glucose, or blood sugar, has the molecular formula $C_6H_{12}O_6$. What is the empirical formula, and what is the percent composition of glucose?

SOLUTION   Dividing the subscripts in the molecular formula $C_6H_{12}O_6$ by 6 gives an empirical formula of CH_2O for glucose.

The percent composition of glucose can be calculated either from the molecular formula or from the empirical formula. Using the molecular formula, for example, the C:H:O mole ratio of 6:12:6 can be converted into a mass ratio by assuming that we have 1 mol of compound and carrying out mole-to-gram conversions:

$$1 \text{ mol glucose} \times \frac{6 \text{ mol C}}{1 \text{ mol glucose}} \times \frac{12.0 \text{ g C}}{1 \text{ mol C}} = 72.0 \text{ g C}$$

$$1 \text{ mol glucose} \times \frac{12 \text{ mol H}}{1 \text{ mol glucose}} \times \frac{1.01 \text{ g H}}{1 \text{ mol H}} = 12.1 \text{ g H}$$

$$1 \text{ mol glucose} \times \frac{6 \text{ mol O}}{1 \text{ mol glucose}} \times \frac{16.0 \text{ g O}}{1 \text{ mol O}} = 96.0 \text{ g O}$$

Dividing the mass of each element by the total mass and multiplying by 100% gives the percent composition.

Total mass of 1 mol of glucose = 72.0 g + 12.1 g + 96.0 g = 180.1 g

$$\% \text{ C} = \frac{72.0 \text{ g C}}{180.1 \text{ g}} \times 100\% = 40.0\%$$

$$\% \text{ H} = \frac{12.1 \text{ g H}}{180.1 \text{ g}} \times 100\% = 6.72\%$$

$$\% \text{ O} = \frac{96.0 \text{ g O}}{180.1 \text{ g}} \times 100\% = 53.3\%$$

The sum of the mass percentages is 100%.

Dimethylhydrazine is used as a liquid rocket propellant.

☐ PROBLEM 3.22 What is the empirical formula, and what is the percent composition of dimethylhydrazine, $C_2H_8N_2$, a colorless liquid used as a rocket fuel?

☐ PROBLEM 3.23 What is the empirical formula of an ingredient in Bufferin tablets that has the percent composition C 14.25%, O 56.93%, Mg 28.83%?

3.12 ►DETERMINING EMPIRICAL FORMULAS: ELEMENTAL ANALYSIS

One of the most common analytical methods used to determine empirical formulas, particularly for compounds containing carbon and hydrogen, is *combustion analysis*. In this method, a compound of unknown composition is burned with oxygen to produce the volatile combustion products CO_2 and H_2O, which are separated and have their masses determined by an automated instrument called a gas chromatograph. Methane (CH_4), for instance, burns according to the balanced equation

$$CH_4(g) + 2\ O_2(g) \rightarrow CO_2(g) + 2\ H_2O(g)$$

To understand how combustion analysis works, imagine that we have a sample of a pure substance—say, naphthalene (ordinary household moth balls). We weigh out a known amount of the sample, burn it in pure oxygen, and then analyze the products. Let's say that 0.330 g of naphthalene reacts with O_2 and that 1.133 g of CO_2 and 0.185 g of H_2O are formed. The first thing to find out is the number of moles of carbon and hydrogen in the CO_2 and H_2O products so that we can calculate the number of moles of each element in the naphthalene sample. We do this by carrying out gram-to-mole conversions.

Moles of C in 1.133 g CO_2

$$= 1.133 \; \cancel{\text{g } CO_2} \times \frac{1 \; \cancel{\text{mol } CO_2}}{44.01 \; \cancel{\text{g } CO_2}} \times \frac{1 \; \text{mol C}}{1 \; \cancel{\text{mol } CO_2}} = 0.025\,74 \; \text{mol C}$$

Moles of H in 0.185 g H_2O

$$= 0.185 \; \cancel{\text{g } H_2O} \times \frac{1 \; \cancel{\text{mol } H_2O}}{18.02 \; \cancel{\text{g } H_2O}} \times \frac{2 \; \text{mol H}}{1 \; \cancel{\text{mol } H_2O}} = 0.0205 \; \text{mol H}$$

Although not necessary in this instance since naphthalene contains only carbon and hydrogen, it's possible to make sure that all the mass is accounted for and that no other elements are present. To do so, we carry out mole-to-gram conversions to find the number of grams of C and H in the starting sample:

$$\text{Mass of C} = 0.025\,74 \; \cancel{\text{mol C}} \times \frac{12.01 \; \text{g C}}{1 \; \cancel{\text{mol C}}} = 0.3091 \; \text{g C}$$

$$\text{Mass of H} = 0.0205 \; \cancel{\text{mol H}} \times \frac{1.01 \; \text{g H}}{1 \; \cancel{\text{mol H}}} = 0.0207 \; \text{g H}$$

$$\text{Total mass of C and H} = 0.3091 \; \text{g} + 0.0207 \; \text{g} = 0.3298 \; \text{g}$$

Because the total mass of the C and H in the products (0.3298 g) is the same as the mass of the starting sample (0.330 g), we know that no other elements are present in naphthalene.

With the relative number of moles of C and H known, we divide the larger number of moles by the smaller number to get a C:H mole ratio of 1.26:1 and the temporary formula $C_{1.26}H_1$.

$$\frac{\text{Moles of C}}{\text{Moles of H}} = \frac{0.025\,74 \; \text{mol C}}{0.0205 \; \text{mol H}} = \frac{1.26 \; \text{mol C}}{1 \; \text{mol H}}$$

Multiplying the subscripts by small integers in a trial-and-error procedure until whole numbers are found gives the final formula C_5H_4. (Of course, the subscripts may not always be *exact* integers because of small experimental errors in the data, but the discrepancies should be small.)

Multiply subscripts by 2: $C_{(1.26 \times 2)}H_{(1 \times 2)} = C_{2.52}H_2$

Multiply subscripts by 3: $C_{(1.26 \times 3)}H_{(1 \times 3)} = C_{3.78}H_3$

Multiply subscripts by 4: $C_{(1.26 \times 4)}H_{(1 \times 4)} = C_{5.04}H_4$ (both subscripts are integers)

A flow diagram for the overall process of determining an empirical formula from analytical data is shown in Figure 3.9.

As mentioned in the previous section, analysis can provide only an empirical formula. To determine the molecular formula, it's also necessary to know the substance's molecular weight. In the present problem, the

FIGURE 3.9 A flow diagram for determining an empirical formula from combustion analysis of a compound containing C and H.

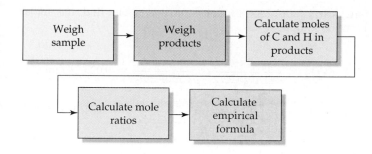

molecular weight of naphthalene is 128.2 amu, or twice the empirical formula weight of C_5H_4 (64.1 amu).

$$\text{Multiple} = \frac{\text{Molecular weight}}{\text{Empirical formula weight}} = \frac{128.2}{64.1} = 2$$

Thus, the molecular formula of naphthalene is $C_{(2\times5)}H_{(2\times4)}$, or $C_{10}H_8$.

Example 3.16 shows a combustion analysis when the sample contains oxygen in addition to carbon and hydrogen. Because oxygen yields no combustion products, its presence in a molecule can't be directly detected by this method. Rather, the presence of oxygen must be inferred by subtracting the calculated masses of C and H from the total mass of the sample.

EXAMPLE 3.15

We calculated in Example 3.13 that ascorbic acid (vitamin C) has the empirical formula $C_3H_4O_3$. If the molecular weight of ascorbic acid is 176 amu, what is the molecular formula?

SOLUTION Calculate a formula weight for the empirical formula, and compare it with the molecular weight of ascorbic acid:

$$\text{Form. wt } C_3H_4O_3 = (3 \times 12.0 \text{ amu}) + (4 \times 1.0 \text{ amu}) + (3 \times 16.0 \text{ amu}) = 88.0 \text{ amu}$$

The empirical formula weight (88.0 amu) is half the molecular weight of ascorbic acid (176 amu), so the subscripts in the empirical formula must be multiplied by 2:

$$\text{Mol wt for ascorbic acid} = 2 \times 88.0 \text{ amu} = 176.0 \text{ amu}$$

$$\text{Molecular formula for ascorbic acid} = C_{(2\times3)}H_{(2\times4)}O_{(2\times3)} = C_6H_8O_6$$

The unmistakable aroma of these running shoes is due to caproic acid.

EXAMPLE 3.16

Caproic acid, the substance responsible for the aroma of dirty gym socks and running shoes, contains carbon, hydrogen, and oxygen. On combustion analysis, a 0.450 g sample of caproic acid gives 0.418 g of H_2O and 1.023 g of CO_2. What is the empirical formula of caproic acid? If the molecular weight of caproic acid is 116.2 amu, what is the molecular formula?

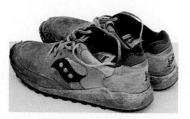

SOLUTION   First, carry out gram-to-mole conversions to find the molar amounts of C and H in the sample.

$$\text{Moles of C} = 1.023 \text{ g CO}_2 \times \frac{1 \text{ mol CO}_2}{44.01 \text{ g CO}_2} \times \frac{1 \text{ mol C}}{1 \text{ mol CO}_2} = 0.023\,24 \text{ mol C}$$

$$\text{Moles of H} = 0.418 \text{ g H}_2\text{O} \times \frac{1 \text{ mol H}_2\text{O}}{18.02 \text{ g H}_2\text{O}} \times \frac{2 \text{ mol H}}{1 \text{ mol H}_2\text{O}} = 0.0464 \text{ mol H}$$

Next, carry out mole-to-gram conversions to find the number of grams of C and H in the starting sample:

$$\text{Mass of C} = 0.023\,24 \text{ mol C} \times \frac{12.01 \text{ g C}}{1 \text{ mol C}} = 0.2791 \text{ g C}$$

$$\text{Mass of H} = 0.0464 \text{ mol H} \times \frac{1.01 \text{ g H}}{1 \text{ mol H}} = 0.0469 \text{ g H}$$

Subtracting the masses of C and H from the mass of the starting sample indicates that 0.124 g is unaccounted for.

$$0.450 \text{ g} - (0.2791 \text{ g} + 0.0469 \text{ g}) = 0.124 \text{ g}$$

Since we are told that oxygen is also present in the sample, the "missing" mass must be due to oxygen, which can't be detected by combustion. We therefore need to carry out a gram-to-mole conversion to find the number of moles of oxygen in the sample:

$$\text{Moles of O} = 0.124 \text{ g O} \times \frac{1 \text{ mol O}}{16.00 \text{ g O}} = 0.007\,75 \text{ mol O}$$

Knowing the relative numbers of moles of all three elements, C, H, and O, we next divide the three numbers of moles by the smallest number (0.007 75 mol of oxygen) to arrive at a C:H:O ratio of 3:6:1.

$$\frac{\text{Moles of C}}{\text{Moles of O}} = \frac{0.023\,24 \text{ mol C}}{0.007\,75 \text{ mol O}} = \frac{3.00 \text{ mol C}}{1 \text{ mol O}}$$

$$\frac{\text{Moles of H}}{\text{Moles of O}} = \frac{0.0464 \text{ mol H}}{0.007\,75 \text{ mol O}} = \frac{5.99 \text{ mol H}}{1 \text{ mol O}}$$

The empirical formula of caproic acid is therefore C_3H_6O, and the empirical formula weight is 58.1 amu. Since the molecular weight of caproic acid is 116.2, or twice the empirical formula weight, the molecular formula of caproic acid must be $C_{(2\times3)}H_{(2\times6)}O_{(2\times1)}$, or $C_6H_{12}O_2$.

⌐ **PROBLEM 3.24**   Ethyl alcohol, the "alcohol" in beverages, contains carbon, hydrogen, and oxygen. On combustion analysis, 1.00 g of ethyl alcohol yields 1.174 g of H_2O and 1.910 g of CO_2. What is the empirical formula of ethyl alcohol?   ⌐

3.13 ▶DETERMINING MOLECULAR WEIGHTS: MASS SPECTROMETRY

We said in the previous section that elemental analysis gives only an empirical formula. To determine a compound's molecular formula, it's also necessary to know the compound's molecular weight. How is molecular weight determined?

The most common method of determining both atomic and molecular weights is with an instrument called a *mass spectrometer*, shown schematically in Figure 3.10. The sample is first vaporized and then injected as a dilute gas into an evacuated chamber, where it is bombarded with a beam of high-energy electrons. The electron beam knocks other electrons from the sample molecules, which become positively charged ions. These ions are then accelerated by an electric field and passed between the poles of a strong magnet that deflects them through a curved, evacuated pipe.

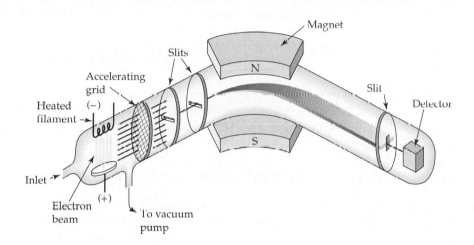

FIGURE 3.10 Schematic illustration of a mass spectrometer. Sample molecules are ionized by collision with a high-energy electron beam and are then passed between the poles of a magnet, where they are deflected according to their mass. The deflected ions pass through a slit into a detector assembly.

The radius of deflection of a charged ion M^+ as it passes between the magnet poles depends on its mass, with lighter ions deflected more strongly than heavier ones. By varying the strength of the magnetic field, it's possible to focus ions of different masses through a slit and onto a detector assembly, where they are identified. So precise are modern mass spectrometers that molecular weights can often be measured to seven significant figures. A $^{12}C_{10}{}^{1}H_8$ molecule of the naphthalene sample we analyzed in Section 3.12, for example, has a molecular weight of 128.0626 amu as measured by mass spectrometry.

┌ **PROBLEM 3.25** Convert the following empirical formulas into molecular formulas.

(a) benzene: empirical formula CH; mol wt = 78 amu
(b) acetylene: empirical formula CH; mol wt = 26 amu
(c) acetic acid (vinegar): empirical formula CH_2O; mol wt = 60 amu

┌ **PROBLEM 3.26** Convert the following percent compositions into molecular formulas.

(a) diborane: H 21.86%, B 78.14%; mol wt = 27.7 amu
(b) trioxane: H 6.71%, C 40.00%, O 53.28%; mol wt = 90.08 amu

interlude—GOUT AND KIDNEY STONES: PROBLEMS IN SOLUBILITY

All the foods we eat must be chemically metabolized in the body, the energy those foods contain must be extracted, and the waste products must be removed. The general classes of substances found in food are familiar: proteins, carbohydrates, and fats. Less familiar are the *nucleic acids*, large molecules that are present in all living cells and that contain an organism's genetic information. How do our bodies handle nucleic acids?

One of the major pathways for the chemical breakdown of nucleic acids in the body is their conversion to a substance called *uric acid*, $C_5H_4N_4O_3$, so named because it was first isolated in 1776 from urine. Normal people excrete about 0.5 g of uric acid every day. Unfortunately, the amount of uric acid that dissolves in water (or urine) is fairly low—only about 0.07 mg/mL, or 4×10^{-4} mol/L, at the normal body temperature of 37°C. When too much uric acid is produced by the body, its concentration in blood and urine rises, and the excess amount is deposited in the joints and kidneys.

Gout is a disorder of nucleic acid metabolism that primarily affects middle-aged men (only 5% of gout patients are women). It is characterized by an increased uric acid concentration in blood, leading to the deposit of uric acid crystals in soft tissue around the joints, particularly in the hands and at the base of the big toe. Deposits of the sharp, needlelike crystals cause an extremely painful inflammation that can lead ultimately to arthritis and even to bone destruction.

Just as increased uric acid concentration in blood can lead to gout, increased concentration in urine can result in the formation of *kidney stones*, small crystals of uric acid deposited in the kidney. Although often quite small, kidney stones cause excruciating pain when they pass through the ureter, the duct that carries urine from the kidney to the bladder. In some cases, complete blockage of the ureter occurs. Medical treatment for excessive uric acid production involves both dietary modification and drug therapy. Liver, sardines, asparagus, and other foods rich in nucleic acid precursors should be avoided, and drugs such as allopurinol can be taken to lower uric acid production.

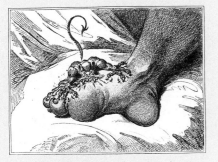

An old cartoon of a gouty foot.

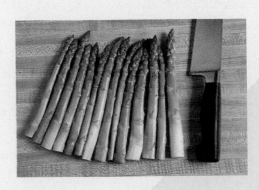

Sufferers from gout are advised to stay away from this.

SUMMARY

All chemical equations must be **balanced**—that is, the numbers and kinds of atoms on both sides of the reaction arrow must be the same because mass is neither created nor destroyed in reactions. A balanced equation tells the number ratio of reactants in a reaction and the number ratio of products. A great many molecules are involved in a reaction on a visible scale, but the number ratio of reactants is always that described by the balanced equation.

Just as atomic weight specifies an atom's relative mass, **molecular weight** specifies a molecule's relative mass. Molecular weight is the sum of the atomic weights of all atoms in the molecule. (The analogous term **formula weight** is used for ionic and other nonmolecular substances.) When referring to the large numbers of molecules or ions that take part in a visible chemical reaction, it is convenient to use a special unit called the **mole**, abbreviated mol. One mole of any object, atom, molecule, or ion contains **Avogadro's number** of formula units, $6.022 \times 10^{23} \text{ mol}^{-1}$. For work in the laboratory, it's necessary to weigh reactants rather than just know mole ratios. Thus, it's necessary to convert between numbers of moles and numbers of grams by using **molar mass** as the conversion factor. The molar mass of any substance is the amount in grams numerically equal to the substance's molecular or formula weight. Carrying out chemical calculations using these relationships is called **stoichiometry**.

The amount of product actually formed in a reaction is often less than the amount theoretically possible. By dividing the actual amount by the theoretical amount and multiplying by 100%, we can calculate a reaction's **percent yield**. Often, reactions are carried out with an excess of one reactant beyond that called for by the balanced equation. In such cases, the extent to which the reaction takes place depends on the reactant present in limiting amount, the **limiting reactant**.

The concentration of a substance in solution is usually expressed in **molarity (M)**, which is defined as the number of moles of a substance (the *solute*) dissolved per liter of solution. A solution's molarity acts as a conversion factor between solution volume and number of moles of solute, making it possible to carry out stoichiometry calculations on solutions. Often, chemicals are stored as concentrated aqueous solutions that are diluted before use. The key fact to remember when carrying out a dilution is that only volume is changed by adding solvent; the amount of solute is unchanged.

The chemical makeup of a substance is described by its **percent composition**—the percentage of the substance's mass due to each of its constituent elements. Elemental analysis is used to calculate a substance's **empirical formula**, which gives the smallest whole-number ratio of atoms of the elements in the compound. To determine the **molecular formula**, it is also necessary to know the substance's molecular weight since the molecular formula may be a simple multiple of the empirical formula. Molecular formulas are usually determined by mass spectrometry.

KEY WORDS

1. Box **(a)** represents a solution of particles at a given concentration. Which of the remaining boxes represents the solution after it has been diluted by doubling the volume of solvent?

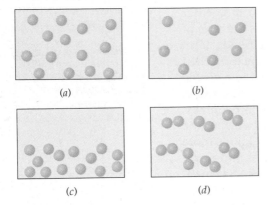

(a) (b)

(c) (d)

2. Reaction of A (green spheres) with B (blue spheres) is shown in the following schematic equation. Which equation best describes the stoichiometry of the reaction?

(a) $A_2 + 2B \longrightarrow A_2B_2$ **(b)** $10A + 5B_2 \longrightarrow 5A_2B_2$
(c) $2A + B_2 \longrightarrow A_2B_2$ **(d)** $5A + 5B_2 \longrightarrow 5A_2B_2$

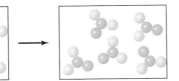

3. Balance the equation for the combustion reaction of ethylene with oxygen.

$$C_2H_4 + O_2 \rightarrow CO_2 + H_2O \text{ (unbalanced)}$$

Write the reaction in a format similar to that in Problem 2, using red dots to represent oxygen, black dots to represent carbon, and light green dots to represent hydrogen.

4. If blue dots represent nitrogen atoms and red dots represent oxygen atoms, which box represents reactants and which represents products for the reaction $2NO(g) + O_2(g) \longrightarrow 2NO_2(g)$?

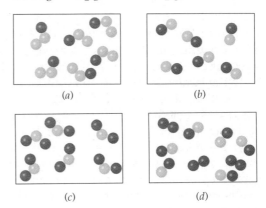

(a) (b)

(c) (d)

5. Methionine, an amino acid used by organisms to make proteins, can be represented by the following ball-and-stick molecular model. Write the formula for methionine, and give its molecular weight. (Red represents oxygen, gray represents carbon, blue represents nitrogen, yellow represents sulfur, and light green represents hydrogen.)

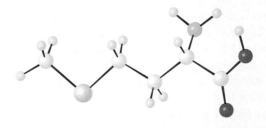

ADDITIONAL PROBLEMS

Problems 3.1–3.26 appear within the chapter.

BALANCING EQUATIONS

3.27 Which of the following equations are balanced?
　(a) The thermite reaction, used in welding:
　　$Al + Fe_2O_3 \longrightarrow Al_2O_3 + Fe$
　(b) The photosynthesis of glucose from CO_2:
　　$6CO_2 + 6H_2O \longrightarrow C_6H_{12}O_6 + 6O_2$

　(c) The separation of gold from its ore: $Au + 2NaCN + O_2 + H_2O \longrightarrow NaAu(CN)_2 + 3NaOH$

3.28 Balance any of the equations in Problem 3.27 that need it.

3.29 The unbalanced equation for the combustion of

hydrogen to yield water is $H_2 + O_2 \longrightarrow H_2O$. Why can't this equation be balanced by adding a subscript 2 to the H_2O oxygen: $H_2 + O_2 \longrightarrow H_2O_2$?

3.30 Balance the following equations.
  (a) $Mg + HNO_3 \longrightarrow H_2 + Mg(NO_3)_2$
  (b) $CaC_2 + H_2O \longrightarrow Ca(OH)_2 + C_2H_2$
  (c) $O_2 + S \longrightarrow SO_3$
  (d) $UO_2 + HF \longrightarrow UF_4 + H_2O$

3.31 Balance the following equations.
  (a) The explosion of ammonium nitrate:
    $NH_4NO_3 \longrightarrow N_2 + O_2 + H_2O$
  (b) The spoilage of wine into vinegar:
    $C_2H_6O + O_2 \longrightarrow C_2H_4O_2 + H_2O$
  (c) The burning of rocket fuel:
    $C_2H_8N_2 + N_2O_4 \longrightarrow N_2 + CO_2 + H_2O$

MOLECULAR WEIGHTS AND MOLES

3.32 What are the molecular (formula) weights of the following substances?
  (a) Hg_2Cl_2 (calomel, used at one time as a bowel purgative)
  (b) $C_4H_8O_2$ (butyric acid, the odor of rancid butter)
  (c) CF_2Cl_2 (a chlorofluorocarbon that destroys the atmospheric ozone layer)

3.33 How many grams are there in a mole of each of the following substances?
  (a) Ti   **(b)** Br_2 **(c)** Hg   **(d)** H_2O

3.34 How many moles are there in a gram of each of the following substances?
  (a) Cr   **(b)** Cl_2 **(c)** Au   **(d)** NH_3

3.35 How many moles of ions are there in 2.5 mol of NaCl?

3.36 How many moles of cations are in 1.45 mol of K_2SO_4?

3.37 How many moles of anions are in 35.6 g of AlF_3?

3.38 What is the molecular weight of chloroform if 0.0275 mol weighs 3.28 g?

3.39 What is the molecular weight of cholesterol if 0.5731 mol weighs 221.6 g?

3.40 Iron(II) sulfate, $FeSO_4$, is prescribed for the treatment of anemia. How many moles of $FeSO_4$ are present in a standard 300 mg tablet? How many atoms of iron?

3.41 The "lead" in lead pencils is actually almost pure carbon, and the mass of a period mark made by a lead pencil is about 0.0001 g. How many carbon atoms are in the period?

3.42 Calculate the molecular (formula) weights of the following substances.
  (a) $Be_3Al_2Si_6O_{18}$ (emerald)
  (b) $C_{17}H_{19}NO_3$ (morphine)
  (c) $Na_2Cr_2O_7$ (sodium dichromate)
  (d) $NaC_5H_8NO_4$ (monosodium glutamate)

3.43 An average cup of coffee contains about 125 mg of caffeine, $C_8H_{10}N_4O_2$. How many moles of caffeine are in a cup? How many molecules?

3.44 Let's say that an average egg has a mass of 45 g. What mass would a mole of eggs have?

3.45 How many moles does each of the following samples contain?
  (a) 1.0 g of lithium   **(b)** 1.0 g of gold
  (c) 1.0 g of penicillin G, $C_{16}H_{17}N_2O_4SK$

3.46 What is the mass in grams of each of the following samples?
  (a) 0.0015 mol of sodium   **(b)** 0.0015 mol of lead
  (c) 0.0015 mol of diazepam (Valium), $C_{16}H_{13}ClN_2O$

STOICHIOMETRY CALCULATIONS

3.47 Titanium metal is obtained from the mineral rutile, TiO_2. How many kg of rutile are needed to produce 100 kg of Ti?

3.48 Iron metal can be produced from the mineral hematite, Fe_2O_3, by reaction with carbon. How many kg of iron are present in 105 kg of hematite?

3.49 In the preparation of iron from hematite (Problem 3.48), Fe_2O_3 reacts with carbon: $Fe_2O_3 + C \longrightarrow Fe + CO_2$.
  (a) Balance the equation.
  (b) How many moles of carbon are needed to react with 525 g of hematite?
  (c) How many grams of carbon are needed to react with 525 g of hematite?

3.50 An alternate method for preparing pure iron from Fe_2O_3 (Problem 3.48) is by reaction with carbon monoxide: $Fe_2O_3 + CO \longrightarrow Fe + CO_2$.

  (a) Balance the equation.
  (b) How many grams of CO are needed to react with 3.02 g of Fe_2O_3?
  (c) How many grams of CO are needed to react with 1.68 mol of Fe_2O_3?

3.51 Magnesium metal burns in oxygen to form magnesium oxide, MgO.
  (a) Write a balanced equation for the reaction.
  (b) How many grams of oxygen are needed to react with 25.0 g of Mg? How many grams of MgO will result?
  (c) How many grams of Mg are needed to react with 25.0 g of O_2? How many grams of MgO will result?

3.52 Ethylene gas, C_2H_4, reacts with water at high temperature to yield ethyl alcohol, C_2H_6O.
  (a) How many grams of ethylene are needed to

react with 0.133 mol of H_2O? How many grams of ethyl alcohol will result?

(b) How many grams of water are needed to react with 0.371 mol of ethylene? How many grams of ethyl alcohol will result?

3.53 Pure oxygen was first made by heating mercury(II) oxide:

$$HgO \xrightarrow{\text{heat}} Hg + O_2$$

(a) Balance the equation.

(b) How many grams of Hg and how many grams of O_2 are formed from 45.5 g of HgO?

(c) How many grams of HgO would you need to obtain 33.3 g of O_2?

3.54 Silver metal reacts with chlorine (Cl_2) to yield silver chloride. If 2.00 g of Ag reacts with 0.657 g of Cl_2, what is the empirical formula of silver chloride?

3.55 Aluminum reacts with oxygen to yield aluminum oxide. If 5.0 g of Al reacts with 4.45 g of O_2, what is the empirical formula of aluminum oxide?

3.56 Titanium dioxide (TiO_2), the substance used as the pigment in white paint, is prepared industrially by reaction of $TiCl_4$ with O_2 at high temperature.

$$TiCl_4 + O_2 \xrightarrow{\text{heat}} TiO_2 + 2 Cl_2$$

How many kg of TiO_2 can be prepared from 5.60 kg of $TiCl_4$?

LIMITING REACTANTS AND REACTION YIELD

3.57 You have 1.39 mol of H_2 and 3.44 mol of N_2. How many grams of ammonia (NH_3) can you make, and how many grams of which reactant will be left over?

$$3 H_2 + N_2 \rightarrow 2 NH_3$$

3.58 Hydrogen and chlorine react to yield hydrogen chloride: $H_2 + Cl_2 \longrightarrow 2 HCl$. How many grams of HCl are formed from reaction of 3.56 g of H_2 with 8.94 g of Cl_2? Which reactant is limiting?

3.59 How many grams of the dry-cleaning solvent ethylene chloride, $C_2H_4Cl_2$, can be prepared by reaction of 15.4 g of ethylene, C_2H_4, with 3.74 g of Cl_2?

$$C_2H_4 + Cl_2 \rightarrow C_2H_4Cl_2$$

3.60 How many grams of each product result from the following reactions?

(a) $(1.3 \text{ g NaCl}) + (3.5 \text{ g AgNO}_3) \rightarrow (x \text{ g AgCl}) + (y \text{ g NaNO}_3)$

(b) $(2.65 \text{ g BaCl}_2) + (6.78 \text{ g H}_2SO_4) \rightarrow (x \text{ g BaSO}_4) + (y \text{ g HCl})$

3.61 How many grams of which reactant is left over in each of the reactions shown in Problem 3.60?

3.62 Limestone ($CaCO_3$) reacts with hydrochloric acid according to the equation $CaCO_3 + 2 HCl \longrightarrow$

$CaCl_2 + H_2O + CO_2$. If 1.00 mol of CO_2 has a volume of 22.4 L, how many liters of gas are formed by reaction of 2.35 g of $CaCO_3$ with 2.35 g of HCl? Which reactant is limiting?

3.63 Acetic acid (vinegar, $C_2H_4O_2$) reacts with isopentyl alcohol ($C_5H_{12}O$) to yield isopentyl acetate ($C_7H_{14}O_2$), a fragrant substance with the odor of bananas. If 1.87 g of acetic acid reacts with 2.31 g of isopentyl alcohol to give 2.96 g of isopentyl acetate, what is the percent yield of the reaction?

$$C_2H_4O_2 + C_5H_{12}O \rightarrow C_7H_{14}O_2 + H_2O$$

3.64 If the yield from the reaction of acetic acid with isopentyl alcohol (Problem 3.63) is 45%, how many grams of isopentyl acetate are formed from 3.58 g of acetic acid and 4.75 g of isopentyl alcohol?

3.65 Cisplatin [$Pt(NH_3)_2Cl_2$], a compound used in cancer treatment, is prepared by reaction of ammonia with potassium tetrachloroplatinate:

$$K_2PtCl_4 + 2 NH_3 \rightarrow 2 KCl + Pt(NH_3)_2Cl_2$$

How many grams of cisplatin are formed from 55.8 g of K_2PtCl_4 and 35.6 g of NH_3 if the reaction takes place in 95% yield based on the limiting reactant?

MOLARITY, SOLUTION STOICHIOMETRY, DILUTION, AND TITRATION

3.66 How many moles of solute are present in each of the following solutions?

(a) 35.0 mL of 1.200 M HNO_3

(b) 175 mL of 0.67 M glucose ($C_6H_{12}O_6$)

3.67 How many grams of solute would you use to prepare each of the following solutions?

(a) 250 mL of 0.600 M ethyl alcohol (C_2H_6O)

(b) 167 mL of 0.200 M boric acid [$B(OH)_3$]

3.68 How many mL of a 0.45 M $BaCl_2$ solution contain 15.0 g of $BaCl_2$?

3.69 The sterile saline solution used to rinse contact lenses is made by dissolving 400 mg of NaCl in 100 mL of water. What is the molarity of the solution?

3.70 The concentration of glucose ($C_6H_{12}O_6$) in normal blood is approximately 90 mg per 100 mL. What is the molarity of the glucose?

3.71 *Ringer's solution*, used in the treatment of burns and wounds, is prepared by dissolving 4.30 g of NaCl, 0.150 g of KCl, and 0.165 g of $CaCl_2$ in water and diluting to a volume of 500 mL. What is the molarity of each of the component ions in the solution?

3.72 A bottle of 12.0 M hydrochloric acid has only 35.7 mL left in it. What will the HCl concentration be if the solution is diluted to 250 mL?

3.73 A flask containing 450 mL of 0.500 M HBr was accidentally knocked to the floor. How many grams of K_2CO_3 would you need to put on the spill to neutralize the acid according to the equation $2 \, HBr$ $(aq) + K_2CO_3(aq) \longrightarrow 2 \, KBr(aq) + CO_2(g) + H_2O(l)$?

3.74 Oxalic acid, $C_2H_2O_4$, is a toxic substance found in spinach. What is the molarity of a solution made by dissolving 12.0 g of oxalic acid in 400 mL of water?

3.75 How many mL of 0.100 M KOH would you need to titrate 25.0 mL of the oxalic acid solution made in Problem 3.74 according to the equation

$$C_2H_2O_4(aq) + 2 \, KOH(aq) \rightarrow K_2C_2O_4(aq) + 2 \, H_2O(l)$$

3.76 Potassium permanganate ($KMnO_4$) reacts with oxalic acid ($C_2H_2O_4$) in aqueous sulfuric acid according to the equation

FORMULAS AND ELEMENTAL ANALYSIS

3.78 Urea, a substance used as a fertilizer, has the formula CH_4N_2O. What is its percent composition?

3.79 Calculate percent compositions for each of the following substances.
  (a) malachite, a copper-containing mineral: $Cu_2(OH)_2CO_3$
  (b) acetaminophen, a headache remedy: $C_8H_9NO_2$
  (c) prussian blue, an ink pigment: $Fe_4[Fe(CN)_6]_3$

3.80 What is the empirical formula of stannous fluoride, a compound added to toothpaste to protect teeth against decay, if its percent composition is 24.25% F, 75.75% Sn?

3.81 What are the empirical formulas of each of the following substances?
  (a) ibuprofen, a headache remedy: 75.69% C, 15.51% O, 8.80% H
  (b) tetraethyllead, the "lead" in gasoline: 29.71% C, 64.06% Pb, 6.23% H
  (c) zircon, a diamondlike mineral: 34.91% O, 15.32% Si, 49.76% Zr

3.82 Combustion analysis of 45.62 mg of toluene, a common solvent, gives 35.67 mg of H_2O and 152.5 mg of CO_2. What is the empirical formula of toluene?

3.83 Coniine, a toxic substance isolated from poison hemlock, contains only carbon, hydrogen, and nitrogen. Combustion analysis of a 5.024 mg sample yields 13.90 mg of CO_2 and 6.048 mg of H_2O. What is the empirical formula of coniine?

GENERAL PROBLEMS

3.90 Give the percent compositions of each of the following substances.
  (a) glucose, $C_6H_{12}O_6$
  (b) sulfuric acid, H_2SO_4
  (c) potassium permanganate, $KMnO_4$
  (d) saccharin, $C_7H_5NO_3S$

3.91 What are the empirical formulas of substances with the following percent compositions?
  (a) aspirin: 4.48% H, 60.00% C, 35.52% O
  (b) ilmenite (a titanium-containing ore): 31.63% O, 31.56% Ti, 36.81% Fe
  (c) sodium thiosulfate (photographic "fixer"): 30.36% O, 29.08% Na, 40.56% S

$$2 \, KMnO_4 + 5 \, C_2H_2O_4 + 3 \, H_2SO_4 \rightarrow$$
$$2 \, MnSO_4 + 10 \, CO_2 + 8 \, H_2O + K_2SO_4$$

How many mL of a 0.250 M $KMnO_4$ solution are needed to react with 3.225 g of oxalic acid?

3.77 Copper reacts with dilute nitric acid according to the equation

$$3 \, Cu(s) + 8 \, HNO_3(aq) \rightarrow$$
$$3 \, Cu(NO_3)_2(aq) + 2 \, NO(g) + 4 \, H_2O(l)$$

If a copper penny weighing 3.045 g is dissolved in a small amount of nitric acid and the resultant solution is diluted to 50.0 mL with water, what is the molarity of the $Cu(NO_3)_2$?

3.84 Ethylene glycol, commonly used as automobile antifreeze, contains only carbon, hydrogen, and oxygen. Combustion analysis of a 23.46 mg sample yields 20.42 mg of H_2O and 33.27 mg of CO_2. What is the empirical formula of ethylene glycol? What is its molecular formula if it has a molecular weight of 62.0 amu?

3.85 The molecular weight of ethylene glycol (Problem 3.84) is 62.0689 amu when calculated using the atomic weights found in a standard periodic table, yet the molecular weight determined experimentally by high-resolution mass spectrometry is 62.0368 amu. Explain the discrepancy.

3.86 Cytochrome c is an iron-containing enzyme found in the cells of all aerobic organisms. If cytochrome c is 0.43% Fe by weight, what is its minimum molecular weight?

3.87 Nitrogen fixation in the root nodules of peas and other leguminous plants is carried out by the molybdenum-containing enzyme *nitrogenase*. What is the molecular weight of nitrogenase if the enzyme contains two molybdenum atoms and is 0.0872% Mo by weight?

3.88 Disilane, Si_2H_x, is analyzed and found to contain 90.28% by weight silicon. What is the value of x?

3.89 A certain metal sulfide, MS_2, is used extensively as a high-temperature lubricant. If MS_2 is 40.06% by mass sulfur, what is the identity of the metal M?

3.92 Balance the following equations:
  (a) $SiCl_4 + H_2O \rightarrow SiO_2 + HCl$
  (b) $P_2O_5 + H_2O \rightarrow H_3PO_4$
  (c) $CaCN_2 + H_2O \rightarrow CaCO_3 + NH_3$
  (d) $NO_2 + H_2O \rightarrow HNO_3 + NO$

3.93 What is the molecular weight of cholesterol if 0.5731 mol weighs 221.6 g?

3.94 Calculate molecular (formula) weights of the following substances.
  (a) geraniol, $C_{10}H_{18}O$ (found in rose oil)
  (b) sodium borate, $Na_2B_4O_7$ (borax)

chapter 4 REACTIONS IN AQUEOUS SOLUTIONS

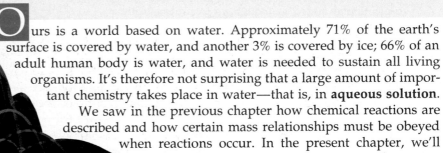

Ours is a world based on water. Approximately 71% of the earth's surface is covered by water, and another 3% is covered by ice; 66% of an adult human body is water, and water is needed to sustain all living organisms. It's therefore not surprising that a large amount of important chemistry takes place in water—that is, in **aqueous solution**. We saw in the previous chapter how chemical reactions are described and how certain mass relationships must be obeyed when reactions occur. In the present chapter, we'll continue our look at chemical reactions by seeing how different reactions can be classified and by learning some of the general ways in which reactions take place.

The reaction of potassium metal with water is so violent that the hydrogen gas produced burst into purple flame.

4.1 ➤SOME WAYS THAT CHEMICAL REACTIONS OCCUR

Every chemical reaction occurs for a reason—a reason that we'll explore in more detail in Chapter 8 and that has to do with the energies of reactants and products. It's often convenient, though, to think of reactions as being "driven" from starting material to product by some "driving force" that pushes them along. In the reaction of mercury(II) nitrate with sodium iodide, for example, the driving force is the precipitation of solid, insoluble mercury(II) iodide, whose formation drags the product out of solution.

$$Hg(NO_3)_2(aq) + 2\ NaI(aq) \rightarrow 2\ NaNO_3(aq) + HgI_2(s)$$

Reaction of aqueous $Hg(NO_3)_2$ with aqueous NaI leads to a precipitate of orange HgI_2.

When classified according to how they occur, reactions can be grouped into three general categories, each with its own kind of driving force: precipitation reactions, acid–base neutralization reactions, and oxidation–reduction reactions. Let's look briefly at an example of each before studying the three kinds of reactions in more detail in subsequent sections.

1. **Precipitation reactions** are processes in which an insoluble solid precipitate forms and drops out of solution, thereby removing material from the aqueous solution and driving the reaction along. Most precipitations take place when the anions and cations of two ionic compounds change partners. For example, an aqueous solution of lead(II) nitrate reacts with an aqueous solution of potassium iodide to yield an aqueous solution of potassium nitrate plus an insoluble yellow precipitate of lead iodide:

$$Pb(NO_3)_2(aq) + 2\ KI(aq) \rightarrow 2\ KNO_3(aq) + PbI_2(s)$$

Reaction of aqueous $Pb(NO_3)_2$ with aqueous KI gives a yellow precipitate of PbI_2.

2. **Acid–base neutralization reactions** are processes in which an acid reacts with a base to yield water plus an ionic compound called a **salt**. We'll look at both acids and bases in more detail in Section 4.5, but you might recall for the moment that we previously defined acids as compounds that produce H^+ ions when dissolved in water and bases as compounds that produce OH^- ions when dissolved in water (Section 2.9). Thus, the driving force behind a neutralization reaction is the production of the stable covalent water molecule, whose formation removes H^+ and OH^- ions from solution. The reaction between hydrochloric acid and sodium hydroxide to yield sodium chloride plus water is a typical example:

$$HCl(aq) + NaOH(aq) \rightarrow H_2O(l) + NaCl(aq)$$

3. **Oxidation–reduction reactions**, or **redox reactions**, are processes in which one or more electrons are transferred between reaction partners—atoms, molecules, or ions. The driving force is a decrease in electrical potential analogous to what happens when a live electrical wire is grounded and electrons flow from the wire to ground. As a result of this transfer of electrons, charges on atoms in the various reactants change. When metallic magnesium reacts with iodine vapor, for instance, a magnesium atom gives an electron to each of two iodine atoms, forming an Mg^{2+} ion and two I^- ions. The charge on the magnesium changes from 0 to +2, and the charge on each iodine changes from 0 to −1:

$$Mg(s) + I_2(g) \rightarrow MgI_2(s)$$

Reaction of magnesium metal with iodine vapor yields magnesium iodide, MgI_2.

┌ **PROBLEM 4.1** Classify each of the following processes as a precipitation, acid–base neutralization, or redox reaction.

(a) $AgNO_3(aq) + KCl(aq) \longrightarrow AgCl(s) + KNO_3(aq)$
(b) $2\ P(s) + 3\ Br_2(l) \longrightarrow 2\ PBr_3(l)$
(c) $Ca(OH)_2(aq) + 2\ HNO_3(aq) \longrightarrow 2\ H_2O(l) + Ca(NO_3)_2(aq)$

4.2 ➤ELECTROLYTES IN AQUEOUS SOLUTION

We all know through experience that both table salt (NaCl) and sugar (sucrose) dissolve in water. The solutions that result, though, are quite different. When sucrose, a molecular substance, dissolves in water, the solution that results contains neutral sucrose *molecules* surrounded by water. When NaCl, an ionic substance, dissolves in water, the solution contains Na^+ and Cl^- *ions*. Because of the presence of the ions, the NaCl solution conducts electricity but the sucrose solution does not.

$$C_{12}H_{22}O_{11}(s) \xrightarrow{H_2O} C_{12}H_{22}O_{11}(aq)$$
$$\text{Sucrose}$$

$$NaCl(s) \xrightarrow{H_2O} Na^+(aq) + Cl^-(aq)$$

The electrical conductivity of an aqueous NaCl solution is easy to demonstrate using a battery, a flashlight bulb, and several pieces of wire

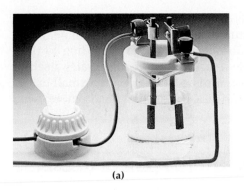

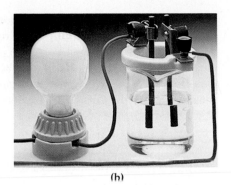

(a) (b)

FIGURE 4.1 A simple device to demonstrate the electrical conductivity of an ionic solution. **(a)** A solution of NaCl conducts electricity, thereby completing the circuit and allowing the bulb to light. **(b)** A solution of sucrose does not conduct electricity and complete the circuit. The bulb therefore remains dark.

connected as shown in Figure 4.1. When the wires are dipped into an aqueous NaCl solution, the positively charged Na^+ ions move through the solution toward the wire connected to the negatively charged terminal of the battery. At the same time, the negatively charged Cl^- ions move toward the wire connected to the positively charged terminal of the battery. The resulting movement of ions allows a current to flow, and the bulb lights. When the wires are dipped into an aqueous sucrose solution, however, the bulb remains dark.

Substances such as NaCl or KBr, which dissolve in water to produce ionic solutions, are called **electrolytes**. Substances such as sucrose or ethyl alcohol, which do not produce ions in aqueous solution, are called **nonelectrolytes**. Not surprisingly, most electrolytes are ionic compounds, but some are molecular. Hydrogen chloride, for example, is a molecular compound when pure but **dissociates** (splits apart) to give H^+ and Cl^- ions when it dissolves in water.

$$HCl(g) \xrightarrow{H_2O} H^+(aq) + Cl^-(aq)$$

Ionic compounds, such as NaCl, and molecular compounds that dissociate completely into ions when dissolved in water are said to be **strong electrolytes**. Molecular compounds that dissociate incompletely are **weak electrolytes**. Acetic acid, for example, is a weak electrolyte because only about 1% of the acetic acid molecules in a 1.0 M solution dissociate into H^+ and $CH_3CO_2^-$ ions. The remaining 99% of acetic acid molecules are undissociated. As a result, a 1.0 M solution of acetic acid is only weakly conducting.

$$\underset{\text{Acetic acid}}{CH_3COOH(l)} \overset{H_2O}{\rightleftharpoons} H^+(aq) + \underset{\text{Acetate ion}}{CH_3CO_2^-(aq)}$$

Note that we use a forward-and-backward double arrow $\rightleftharpoons$ when we write the acetic acid dissociation to indicate that the process takes place simultaneously in both directions. That is, the dissociation process is a dynamic one in which an *equilibrium* is established between the forward and backward reactions. At any given moment, some acetic acid molecules are dissociating while some acetate ions and hydrogen ions are recombining to form acetic acid. The balance between the two opposing reactions defines the exact concentrations of the various species in solution. We'll learn much more about chemical equilibria in future chapters.

A brief list of some common substances classified according to their electrolyte strength is given in Table 4.1. Note that pure water itself is classified as a weak electrolyte because it dissociates only to a small extent into H^+ and OH^- ions. We'll explore the dissociation of water in more detail in Chapter 15.

TABLE 4.1	Electrolyte Classification of Some Common Substances		
Strong Electrolytes	**Weak Electrolytes**	**Nonelectrolytes**	
HCl	CH_3COOH	CH_3OH (methyl alcohol)	
$HClO_4$	HF	C_2H_5OH (ethyl alcohol)	
HNO_3	H_2O	$C_{12}H_{22}O_{11}$ (sucrose)	
H_2SO_4	NH_3	Most compounds of carbon	
KBr		(organic compounds)	
NaCl			
NaOH			
Other ionic compounds			

EXAMPLE 4.1

What is the total molar concentration of ions in a 0.350 M solution of the strong electrolyte Na_2SO_4?

SOLUTION   Writing an equation for dissolving Na_2SO_4 in water shows that three moles of ions—two moles of Na^+ and one mole of SO_4^{2-}—are formed from one mole of Na_2SO_4.

$$Na_2SO_4(s) \xrightarrow{H_2O} 2\ Na^+(aq) + SO_4^{2-}(aq)$$

The molar concentration of ions is therefore 1.05 M.

$$\frac{3\ \text{mol ions}}{1\ \text{mol Na}_2\text{SO}_4} \times \frac{0.350\ \text{mol Na}_2\text{SO}_4}{1\ \text{L soln}} = 1.05\ \text{M}$$

PROBLEM 4.2 What is the molar concentration of Br^- ions in a 0.225 M aqueous solution of $FeBr_3$?

4.3 ►AQUEOUS REACTIONS AND NET IONIC EQUATIONS

The equations we've been writing up to this point have all been **molecular equations**. That is, all the substances involved in reactions have been written using their full formulas as if they were *molecules*. In the precipitation reaction of lead(II) nitrate with potassium iodide mentioned in Section 4.1, for example, only the parenthetical (*aq*) indicated that the reaction actually takes place in aqueous solution, and nowhere was it indicated that ions are involved:

A molecular equation:

$$Pb(NO_3)_2(aq) + 2\ KI(aq) \rightarrow 2\ KNO_3(aq) + PbI_2(s)$$

In fact, lead nitrate, potassium iodide, and potassium nitrate are strong electrolytes that dissolve in water to yield solutions of *ions*. Thus, it is more accurate to write the reaction as an **ionic equation**, in which all the ions are explicitly shown:

An ionic equation:

$$Pb^{2+}(aq) + 2\ NO_3^-(aq) + 2\ K^+(aq) + 2\ I^-(aq) \rightarrow 2\ K^+(aq) + 2\ NO_3^-(aq) + PbI_2(s)$$

A look at this ionic equation shows that the NO_3^- and K^+ ions undergo no change during the reaction. They appear on both sides of the reaction arrow and act merely as **spectator ions**, which are present but play no role. The actual reaction, when stripped to its essentials, can be described more simply by writing a **net ionic equation**, in which the spectator ions are removed. A net ionic equation focuses only on the ions undergoing change—the Pb^{2+} and I^- ions in this instance—and ignores everything extraneous:

An ionic equation:

$$Pb^{2+}(aq) + 2\ \cancel{NO_3^-}(aq) + 2\ \cancel{K^+}(aq) + 2\ I^-(aq) \rightarrow 2\ \cancel{K^+}(aq) + 2\ \cancel{NO_3^-}(aq) + PbI_2(s)$$

A net ionic equation:

$$Pb^{2+}(aq) + 2\ I^-(aq) \rightarrow PbI_2(s)$$

Leaving spectator ions out of a net ionic equation does not imply that their presence is irrelevant. Clearly, if a reaction occurs by mixing a solution of Pb^{2+} ions with a solution of I^- ions, then those solutions must also contain additional ions to balance the charge in each: The Pb^{2+} solution must contain an anion, and the I^- solution must contain a cation. Leaving these other ions out of the net ionic equation merely implies that the specific *identity* of the spectator ions is not important; *any* ions could fill the same role.

Zinc metal reacts with aqueous hydrochloric acid to give hydrogen gas and aqueous Zn^{2+} ion.

EXAMPLE 4.2

Aqueous hydrochloric acid reacts with zinc metal to yield hydrogen gas and aqueous zinc chloride. Write a net ionic equation for the process.

$$2\ HCl(aq) + Zn(s) \rightarrow H_2(g) + ZnCl_2(aq)$$

SOLUTION   First, write the ionic equation, listing all the species present in solution. Both HCl (a molecular compound; Table 4.1) and $ZnCl_2$ (an ionic compound) are strong electrolytes, which exist as ions in solution.

$$2\,H^+(aq) + 2\,Cl^-(aq) + Zn(s) \rightarrow H_2(g) + Zn^{2+}(aq) + 2\,Cl^-(aq)$$

Next, find the ions that are present on both sides of the reaction arrow (the spectator ions) and cancel them, leaving the net ionic equation. In the present instance, the chloride ions cancel from both sides.

$$2\,H^+(aq)\ 2\,\cancel{Cl^-}(aq) + Zn(s) \rightarrow H_2(g) + Zn^{2+}(aq) + 2\,\cancel{Cl^-}(aq)$$

Net ionic equation: $2\,H^+(aq) + Zn(s) \rightarrow H_2(g) + Zn^{2+}(aq)$

PROBLEM 4.3   Write net ionic equations for the following reactions:
(a) $2\,AgNO_3(aq) + Na_2CrO_4(aq) \longrightarrow Ag_2CrO_4(s) + 2\,NaNO_3(aq)$
(b) $H_2SO_4(aq) + MgCO_3(s) \longrightarrow H_2O(l) + CO_2(g) + MgSO_4(aq)$

4.4 ▸PRECIPITATION REACTIONS AND SOLUBILITY RULES

To predict whether a precipitation reaction will occur on mixing two substances together, you must know the **solubilities** of the potential products—how much of each compound will dissolve in a given amount of solvent at a given temperature. If a substance has a low solubility in water, then it is likely to precipitate from an aqueous solution. If a substance has a high solubility in water, then no precipitate will form.

Solubility is a complex matter, and it's not always possible to make correct predictions. As a rule of thumb, though, the following solubility guidelines apply:

1. *These compounds are usually soluble:*

   • Compounds that contain an alkali metal (group 1A) cation—Li^+, Na^+, K^+, Rb^+, and Cs^+
   • Compounds that contain an ammonium cation, NH_4^+
   • Compounds that contain a chloride (Cl^-), bromide (Br^-), or iodide (I^-) anion, except for Ag^+, Hg_2^{2+}, and Pb^{2+} compounds
   • Compounds that contain a nitrate (NO_3^-), chlorate (ClO_3^-), perchlorate (ClO_4^-), acetate ($CH_3CO_2^-$), or sulfate (SO_4^{2-}) anion, except for Ba^{2+}, Hg_2^{2+}, and Pb^{2+} sulfates

2. *These compounds are usually insoluble:*

   • Compounds that contain a hydroxide (OH^-) or oxide (O^{2-}) anion, except for the alkali metal and Ba^{2+} compounds
   • Compounds that contain a carbonate (CO_3^{2-}), phosphate (PO_4^{3-}), chromate (CrO_4^{2-}), or sulfide (S^{2-}) anion, except for those of the alkali metals and the NH_4^+ ion

Using these guidelines makes it possible not only to predict whether a precipitate will form when solutions of two ionic compounds are mixed but also to prepare a specific compound by purposefully carrying out a precipitation. If, for example, you wanted to prepare a sample of silver carbonate, Ag_2CO_3, you could mix a solution of a soluble silver compound, such as $AgNO_3$, with a solution of a soluble carbonate, such as Na_2CO_3. Both starting compounds are soluble in water, as is $NaNO_3$. Silver carbonate is the only insoluble combination of ions and will therefore precipitate from solution.

$$2\ AgNO_3(aq) + Na_2CO_3(aq) \rightarrow Ag_2CO_3(s) + 2\ NaNO_3(aq)$$

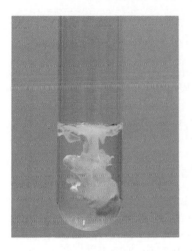

Reaction of aqueous $AgNO_3$ with aqueous Na_2CO_3 gives a precipitate of Ag_2CO_3.

EXAMPLE 4.3

Predict whether a precipitation reaction will occur when aqueous solutions of $CdCl_2$ and $(NH_4)_2S$ are mixed. Write the net ionic equation.

SOLUTION The idea in this problem is to identify the two potential products and to predict the solubility of each. In the present instance, $CdCl_2$ and $(NH_4)_2S$ might give CdS and $2\ NH_4Cl$. Of the two possible products, the solubility guidelines predict that CdS, a sulfide, is insoluble and that NH_4Cl, an ammonium salt, is soluble. Thus, a precipitation reaction will occur:

$$Cd^{2+}(aq) + S^{2-}(aq) \rightarrow CdS(s)$$

EXAMPLE 4.4

How might you use a precipitation reaction to prepare a sample of $CuCO_3$? Write the net ionic equation.

SOLUTION To prepare a precipitate of $CuCO_3$, a soluble Cu^{2+} salt must react with a soluble CO_3^{2-} salt. A look at the solubility guidelines suggests that $Cu(NO_3)_2$ and Na_2CO_3 (among many other possibilities) might work:

$$Cu^{2+}(aq) + CO_3^{2-}(aq) \rightarrow CuCO_3(s)$$

Reaction of aqueous Cd^{2+} with aqueous S^{2-} gives a precipitate of CdS.

Reaction of aqueous $Cu(NO_3)_2$ with aqueous Na_2CO_3 gives a precipitate of $CuCO_3$.

⌐ **PROBLEM 4.4** Use the solubility guidelines to predict the solubility of each of the following compounds:

(a) $CdCO_3$ **(b)** MgO **(c)** Na_2S **(d)** $PbSO_4$ **(e)** $(NH_4)_3PO_4$ **(f)** $HgCl_2$

⌐ **PROBLEM 4.5** Predict whether a precipitation reaction will occur in each of the following situations. Write a net ionic equation for each reaction that does occur.

(a) $NiCl_2(aq) + (NH_4)_2S(aq) \longrightarrow$ **(b)** $Na_2CrO_4(aq) + Pb(NO_3)_2(aq) \longrightarrow$
(c) $AgClO_4(aq) + CaBr_2(aq) \longrightarrow$

⌐ **PROBLEM 4.6** How might you use a precipitation reaction to prepare a sample of $Ca_3(PO_4)_2$? Write the net ionic equation. ⌐

4.5 ►ACIDS, BASES, AND NEUTRALIZATION REACTIONS

We've mentioned the subject of acids and bases on several previous occasions, but now let's look more carefully at just what it is that makes an acid an acid, and a base a base. In 1777 the French chemist Antoine Lavoisier proposed that all acids contain a common element, oxygen; in fact, the word "oxygen" is derived from a Greek phrase meaning "acid former." That idea had to be modified, however, when the English chemist Sir Humphrey Davy (1778–1829) showed in 1810 that muriatic acid (hydrochloric acid) contains no oxygen, only hydrogen and chlorine. Davy's studies thus suggested that the common element in acids is *hydrogen*, not oxygen.

The relation between the presence of hydrogen in a compound and acidic behavior was clarified by the Swedish chemist Svante Arrhenius (1859–1927) in 1887. Arrhenius proposed that acids are substances that dissociate in water to produce hydrogen ions (H^+) and that bases are substances that dissociate in water to yield hydroxide ions (OH^-):

$$HA(aq) \rightarrow H^+(aq) + A^-(aq)$$
An acid

$$MOH(aq) \rightarrow M^+(aq) + OH^-(aq)$$
A base

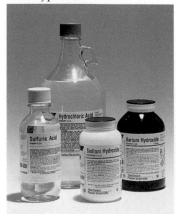

Some typical acids and bases.

In these equations, HA is a general formula for an acid—for example, HCl or HNO_3—and MOH is a general formula for a metal hydroxide—for example, NaOH or KOH.

Although convenient to write and to use in equations, the symbol $H^+(aq)$ does not really represent the structure of the ion present in aqueous solution. As just a bare proton, H^+ is much too reactive to exist by itself. Rather, the H^+ attaches to a water molecule, giving the more stable **hydronium ion**, H_3O^+. We'll sometimes write $H^+(aq)$ for convenience, particularly when balancing equations, but will more often write H_3O^+ or $H_3O^+(aq)$ to represent a generalized aqueous acid solution.

Different acids dissociate to different extents in aqueous solutions. Those acids that dissociate completely and are strong electrolytes are **strong acids**; those acids that dissociate incompletely and are weak electrolytes are **weak acids**. We've already seen in Table 4.1, for example, that HCl, $HClO_4$, HNO_3, and H_2SO_4 are strong electrolytes and therefore strong acids, whereas CH_3COOH and HF are weak electrolytes and therefore weak acids.

It should also be reiterated, as pointed out in Section 2.9, that different acids have different numbers of acidic hydrogens and yield different numbers of H_3O^+ ions in solution. Sulfuric acid, for instance, can dissociate twice, and phosphoric acid can dissociate three times:

Sulfuric acid: $H_2SO_4(l) + H_2O(l) \rightarrow HSO_4^-(aq) + H_3O^+(aq)$

$HSO_4^-(aq) + H_2O(l) \rightarrow SO_4^{2-}(aq) + H_3O^+(aq)$

Phosphoric acid: $H_3PO_4(l) + H_2O(l) \rightarrow H_2PO_4^-(aq) + H_3O^+(aq)$

$H_2PO_4^-(aq) + H_2O(l) \rightarrow HPO_4^{2-}(aq) + H_3O^+(aq)$

$HPO_4^{2-}(aq) + H_2O(l) \rightarrow PO_4^{3-}(aq) + H_3O^+(aq)$

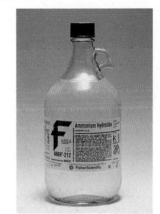

In the case of sulfuric acid, the first dissociation of an H^+ is complete—all the H_2SO_4 molecules lose one H^+—but the second is incomplete. In the case of phosphoric acid, none of the three dissociations is complete.

Bases, like acids, can also be either strong or weak, depending on the extent to which they dissociate and produce OH^- ions in aqueous solutions. Most metal hydroxides, such as NaOH and $Ba(OH)_2$, are strong electrolytes and strong bases, but ammonia (NH_3) is a weak electrolyte and a weak base. The weakly basic behavior of ammonia is due to the fact that it reacts to a small extent with water to yield NH_4^+ and OH^- ions. In fact, aqueous solutions of ammonia are often called "ammonium hydroxide," although this is really a misnomer since the concentrations of NH_4^+ and OH^- ions are low.

This bottle should be labeled "Aqueous Ammonia" rather than "Ammonium Hydroxide".

$$NH_3(g) + H_2O(l) \rightleftharpoons NH_4^+(aq) + OH^-(aq)$$

As with the dissociation of acetic acid (Section 4.2), the reaction of ammonia with water takes place only to a small extent (about 1%). Most of the ammonia remains unreacted, and we therefore write the reaction with a double arrow to show that a dynamic equilibrium between forward and backward reactions exists.

Table 4.2 summarizes the names, formulas, and classification of some common acids and bases.

TABLE 4.2	Some Common Acids and Bases					
Strong acid	$HClO_4$	Perchloric acid	NaOH	Sodium hydroxide	Strong base	
	H_2SO_4	Sulfuric acid	KOH	Potassium hydroxide		
	HBr	Hydrobromic acid	$Ba(OH)_2$	Barium hydroxide		
	HCl	Hydrochloric acid	$Ca(OH)_2$	Calcium hydroxide		
	HNO_3	Nitric acid				
Weak acid	H_3PO_4	Phosphoric acid	NH_3	Ammonia	Weak base	
	HF	Hydrofluoric acid				
	CH_3COOH	Acetic acid				

When acids and bases are mixed in the right proportion, both acidic and basic properties disappear because of a **neutralization reaction**, which produces water and a salt. The anion of the salt (A$^-$) comes from the acid, and the cation of the salt (M$^+$) comes from the base.

A neutralization reaction $\underset{\text{Acid}}{HA(aq)} + \underset{\text{Base}}{MOH(aq)} \rightarrow H_2O(l) + \underset{\text{Salt}}{MA(aq)}$

Because salts are generally strong electrolytes in aqueous solution, we can write the neutralization process of a strong acid with a strong base as an ionic equation:

$$H^+(aq) + A^-(aq) + M^+(aq) + OH^-(aq) \rightarrow H_2O(l) + M^+(aq) + A^-(aq)$$

Canceling the ions that appear on both sides of the ionic equation gives the net ionic equation that holds for the reaction of any strong acid with any strong base in water:

$$H^+(aq) + \cancel{A^-}(aq) + \cancel{M^+}(aq) + OH^-(aq) \rightarrow H_2O(l) + \cancel{M^+}(aq) + \cancel{A^-}(aq)$$

$$H^+(aq) + OH^-(aq) \rightarrow H_2O(l)$$

or $H_3O^+(aq) + OH^-(aq) \rightarrow 2\,H_2O(l)$

For the reaction of a weak acid with a strong base, a similar neutralization occurs, but we must write the molecular formula of the acid rather than simply H$^+$(aq), because the acid dissociation in water is incomplete. In the reaction of HF with KOH, for example, we write the net ionic equation as

$$HF(aq) + OH^-(aq) \rightarrow H_2O(l) + F^-(aq)$$

EXAMPLE 4.5

Write both an ionic equation and a net ionic equation for the neutralization reaction of aqueous HBr and aqueous Ba(OH)$_2$.

SOLUTION   Aqueous hydrobromic acid contains a solution of H$^+$ ions and Br$^-$ ions. Aqueous barium hydroxide contains a solution of Ba^{2+} and OH$^-$ ions. Thus, we have a mixture of four different ions on the reactant side:

$$H^+(aq) + Br^-(aq) + Ba^{2+}(aq) + 2\,OH^-(aq) \rightarrow$$

The reactant side of the equation contains both H$^+$ ions and OH$^-$ ions, which will react to yield H$_2$O, so we can write the balanced ionic equation as

$$2\,H^+(aq) + 2\,Br^-(aq) + Ba^{2+}(aq) + 2\,OH^-(aq) \rightarrow 2\,H_2O(l) + 2\,Br^-(aq) + Ba^{2+}(aq)$$

Canceling the spectator ions Br$^-$(aq) and Ba^{2+}(aq) from both sides of the ionic equation gives the net ionic equation

$$2\,H^+(aq) + 2\,\cancel{Br^-}(aq) + \cancel{Ba^{2+}}(aq) + 2\,OH^-(aq) \rightarrow 2\,H_2O(l) + 2\,\cancel{Br^-}(aq) + \cancel{Ba^{2+}}(aq)$$

$$2\,H^+(aq) + 2\,OH^-(aq) \rightarrow 2\,H_2O(l)   \text{or}   H^+(aq) + OH^-(aq) \rightarrow H_2O(l)$$

The reaction of HBr with Ba(OH)$_2$ thus involves the combination of a proton (H$^+$) from the acid with OH$^-$ from the base to yield water and a salt (BaBr$_2$).

⌐ PROBLEM 4.7 Balance and write both the ionic equation and net ionic equation for each of the following acid-base reactions.

(a) $2 \, CsOH(aq) + H_2SO_4(aq) \longrightarrow$
(b) $Ca(OH)_2(aq) + 2 \, CH_3COOH(aq) \longrightarrow$

4.6 ➤OXIDATION–REDUCTION (REDOX) REACTIONS

Magnesium metal burns in air with an intense white light to form solid magnesium oxide. Red phosphorus reacts with liquid bromine to form liquid phosphorus tribromide. Purple aqueous permanganate ion (MnO_4^-) reacts with aqueous Fe^{2+} ion to yield Fe^{3+} and pale pink Mn^{2+}.

$$2 \, Mg(s) + O_2(g) \rightarrow 2 \, MgO(s)$$

$$2 \, P(s) + 3 \, Br_2(l) \rightarrow 2 \, PBr_3(l)$$

$$MnO_4^-(aq) + 5 \, Fe^{2+}(aq) + 8 \, H^+(aq) \rightarrow Mn^{2+}(aq) + 5 \, Fe^{3+}(aq) + 4 \, H_2O(l)$$

Magnesium metal burns in air.

Elemental phosphorus reacts spectacularly with liquid bromine.

Aqueous potassium permanganate is frequently used as an oxidizing agent.

Although these and many thousands of other reactions appear unrelated, all are examples of *oxidation–reduction* reactions.

Historically, the word *oxidation* referred to the combination of an element with oxygen to yield an oxide, and the word *reduction* referred to the removal of oxygen from an oxide to yield the element. Such oxidation–reduction processes have been crucial to the development of human civilization and still have enormous commercial value. The oxidation (rusting) of iron metal by reaction with moist air has been known for millennia and is still a serious problem that causes enormous damage to buildings, bridges, and automobiles. The reduction of iron ore (Fe_2O_3) with charcoal (C) to make iron metal has been carried out since prehistoric times and is still used today in the initial stages of steelmaking.

$$4\ Fe(s)\ +\ 3\ O_2(g) \rightarrow 2\ Fe_2O_3(s) \quad \text{rusting of iron: an } \textit{oxidation} \text{ of Fe}$$

$$2\ Fe_2O_3(s)\ +\ 3\ C(s) \rightarrow 4\ Fe(s)\ +\ 3\ CO_2(g) \quad \text{manufacture of iron: a } \textit{reduction} \text{ of Fe}$$

Today, the words oxidation and reduction have taken on a much broader meaning. An **oxidation** is now defined as the loss of one or more electrons by a substance—element, compound, or ion. Conversely, a **reduction** is the gain of one or more electrons by a substance. Thus, an **oxidation–reduction reaction**, or **redox reaction**, is a process in which *electrons are transferred from one substance to another.*

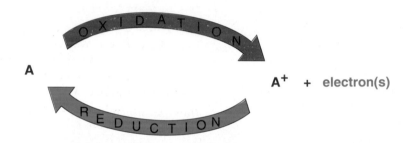

How can you tell when a redox reaction is taking place? The answer is that we assign to each atom in a substance a value called an **oxidation number** (or *oxidation state*), which provides a measure of whether the atom is neutral, electron-rich, or electron-poor. By comparing the oxidation number of an atom before and after reaction, we can tell whether the atom has gained or lost one or more electrons. Note that *oxidation numbers don't necessarily imply ionic charges.* They are simply a convenient device to help us keep track of electrons in redox reactions.

The rules for assigning oxidation numbers are straightforward:

1. **An atom in an uncombined element has an oxidation number of 0.** The oxidation number of an H atom in gaseous H_2 is 0, of an Na atom in metallic sodium is 0, and so on.

2. **An atom in a monatomic ion has an oxidation number identical to its charge.** The oxidation number of Na^+ is +1, of Ca^{2+} is +2, of Cl^- is −1, of O^{2-} is −2, and so on. You might want to review Section 2.10 to see the charges on some of the more common ions.

3. **An atom in a polyatomic ion or in a molecular compound has an oxidation number that represents a hypothetical charge *as if the atom were a monatomic ion.*** In the hydroxide ion (OH^-), for example, the oxygen atom has an oxidation number of −2, as if it were a monatomic O^{2-} ion, and the hydrogen atom has an oxidation number of +1, as if it were H^+. In general, the farther left and lower down an element is in the periodic table, the more likely it is that the atom will be "cation-like". Metals, therefore, usually have positive oxidation numbers. The farther right and higher up an element is in the periodic table, the more likely it

is that the atom will be "anion-like". Nonmetals, such as O, S, N, and the halogens, usually have negative oxidation numbers.

(a) **Hydrogen can be either +1 or −1.** When bonded to a metal, such as Na or Ca, hydrogen has oxidation number −1. When bonded to a nonmetal, such as C, N, O, or Cl, hydrogen has oxidation number +1.

(b) **Oxygen usually has an oxidation number of −2.** The major exception is in compounds called *peroxides*, which contain the O_2^{2-} ion. Each oxygen atom in the O_2^{2-} ion has an oxidation number of −1.

(c) **Halogens usually have an oxidation number of −1.** The major exception is in compounds of chlorine, bromine, or iodine in which the halogen atom is bonded to oxygen. In such cases, the oxygen has an oxidation number of −2, and the halogen has a positive oxidation number. In Cl_2O, for example, each Cl atom has an oxidation number of +1 and the O atom has an oxidation number of −2.

4. **The sum of the oxidation numbers must be 0 for a neutral compound and must equal the net charge for a polyatomic ion.** This rule is particularly useful in helping to find the oxidation number of an atom in difficult cases. The general idea is to assign oxidation numbers to the "easy" atoms first and then find the oxidation number of the "difficult" atom by subtraction. For example, suppose we need to find the oxidation number of the sulfur atom in sulfuric acid (H_2SO_4). We know that each H atom is +1 and each O atom is −2, so the S atom must have an oxidation number of +6 for the compound to have no net charge:

$H_2SO_4 = (2\ H^+)\ (S?)\ (4\ O^{2-})$ $2(+1) + (?) + 4(-2) = 0$ net charge
                                            $? = 0 - 2(+1) - 4(-2) = +6$

$H_2SO_4 = (2\ H^+)\ (S^{6+})\ (4\ O^{2-})$ $2(+1) + (+6) + 4(-2) = 0$ net charge

To find the oxidation number of the chlorine in the perchlorate anion (ClO_4^-), we know that each oxygen is −2, so the Cl atom must have an oxidation number of +7 for there to be a net charge of −1:

$ClO_4^- = (Cl?)\ (4\ O^{2-})$ $? + 4(-2) = -1$ net charge
                                   $? = -1 - 4(-2) = +7$

$ClO_4^- = (Cl^{7+})\ (4\ O^{2-})$ $(+7) + 4(-2) = -1$ net charge

To find the oxidation number of the nitrogen atom in the ammonium cation (NH_4^+), we know that each H atom is +1, so the N atom must have an oxidation number of −3 for there to be a net charge of +1:

$NH_4^+ = (N?)\ (4\ H^+)$ $? + 4(+1) = +1$ net charge
                             $? = +1 - 4(+1) = -3$

$NH_4^+ = (N^{3-})\ (4\ H^+)$ $(-3) + 4(+1) = +1$ net charge

Once oxidation numbers are assigned, it's clear why the various reactions mentioned at the beginning of this section are redox processes. Take the rusting of iron, for example. Two of the starting substances, Fe and O_2, are elements, and both therefore have an oxidation number of 0. In the product, however, the oxygen atoms have an oxidation number of -2, and the iron atom has an oxidation number of $+3$. Thus, Fe has undergone a change from 0 to $+3$ (an oxidation), and O has undergone a change from 0 to -2 (a reduction). Note that the total number of electrons given up by the substance being oxidized (4 Fe $\times$ 3 electrons/Fe = 12 electrons) is the same as the number gained by the substance being reduced (6 O $\times$ 2 electrons/O = 12 electrons).

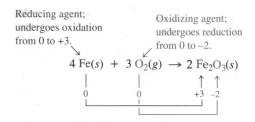

A similar analysis can be carried out for the production of iron metal from its ore. The iron atom is reduced because it goes from an oxidation number of $+3$ in the starting material (Fe_2O_3) to 0 in the product (Fe). At the same time, the carbon atom is oxidized because it goes from an oxidation number of 0 in the starting material (C) to $+4$ in the product (CO_2). The oxygen atoms undergo no change because they have an oxidation number of -2 in both starting material and product. The total number of electrons given up by the substance being oxidized (3 C $\times$ 4 electrons/C = 12 electrons) is the same as the number gained by the substance being reduced (4 Fe $\times$ 3 electrons/Fe = 12 electrons).

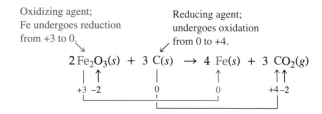

As these examples show, oxidations and reductions always occur together. Whenever one substance loses one or more electrons (is oxidized), another substance must gain those electrons (be reduced). The substance that *causes* a reduction by giving up an electron—the iron atom in the reaction of Fe with O_2 and the carbon atom in the reaction of C with Fe_2O_3—is called a **reducing agent**. The substance that causes an oxidation by accepting an electron—the oxygen atom in the reaction of Fe with O_2 and the iron atom in the reaction of C with Fe_2O_3—is called an **oxidizing agent**. The reducing agent is itself oxidized when it gives up electrons, and the oxidizing agent is itself reduced when it accepts electrons.

Reducing agent   loses one or more electrons;
undergoes oxidation;
oxidation number on atom increases

> **Oxidizing agent**   gains one or more electrons;
> undergoes reduction;
> oxidation number on atom decreases

We'll see in later chapters that redox reactions are common for almost every element in the periodic table except for the noble-gas elements, in group 8A. In general, metals act as reducing agents, and reactive nonmetals such as O_2 and the halogens act as oxidizing agents.

Different metals can give up different numbers of electrons in oxidation–reduction reactions. Lithium, sodium, and the other group 1A elements give up only one electron and become monopositive ions with oxidation numbers of +1. Beryllium, magnesium, and the other group 2A elements, however, give up two electrons and become dipositive ions. The transition metals in the middle of the periodic table can give up a variable number of electrons to yield more than one kind of ion depending on the circumstances. Titanium, for example, can react with chlorine to yield either $TiCl_3$ or $TiCl_4$. Because a chloride ion always has a −1 oxidation number, the titanium atom in $TiCl_3$ must have a +3 oxidation number and the titanium atom in $TiCl_4$ must be +4.

EXAMPLE 4.6

Assign oxidation numbers to each atom in the following substances.

(a) CdS   (b) AlI_3 (c) $S_2O_3{}^{2-}$ (d) $Na_2Cr_2O_7$

SOLUTION

(a) CdS   (b) AlI_3 (c) $S_2O_3{}^{2-}$ (d) $Na_2Cr_2O_7$
 ↑ ↑   ↑ ↑   ↑ ↑   ↑   ↑   ↑
+2 −2   +3 −1   +2 −2   +1 +6 −2

(a) The sulfur atom in S^{2-} has an oxidation number of −2, so Cd must be +2.
(b) H bonded to a metal has the oxidation number 1, so Al must be +3.
(c) O usually has the oxidation number −2, so S must be +2 for the anion to have a net charge of −2: $(2 \, S^{2+})(3 \, O^{2-}) = 2(+2) + 3(-2) = -2$ net charge
(d) Na is always +1, and oxygen is −2, so Cr must be +6 for the compound to be neutral: $(2 \, Na^+)(2 \, Cr^{6+})(7 \, O^{2-}) = 2(+1) + 2(+6) + 7(-2) = 0$ net charge

EXAMPLE 4.7

Assign oxidation numbers to all atoms, tell in each case which substance is undergoing an oxidation and which a reduction, and identify the oxidizing and reducing agents.

(a) $Ca(s) + 2 \, H^+(aq) \longrightarrow Ca^{2+}(aq) + H_2(g)$
(b) $2 \, Fe^{2+}(aq) + Cl_2(aq) \longrightarrow 2 \, Fe^{3+}(aq) + 2 \, Cl^-(aq)$

SOLUTION   (a) The neutral elements Ca and H_2 have oxidation numbers of 0; Ca^{2+} is +2 and H^+ is +1.

$$Ca(s) + 2 \, H^+(aq) \longrightarrow Ca^{2+}(aq) + H_2(g)$$
$$\quad 0 \qquad\quad +1 \qquad\qquad +2 \qquad\quad 0$$

Ca is oxidized (its oxidation number increases from 0 to +2), and H^+ is reduced (its oxidation number decreases from +1 to 0). The reducing agent is the substance that gives away an electron, thereby going to a higher oxidation number, and the oxidizing agent is the substance that accepts an electron, thereby going to a lower oxidation number. In the present case, calcium is the reducing agent, and H^+ is the oxidizing agent.

(b) The neutral element Cl_2 has an oxidation number of 0; the monatomic ions have oxidation numbers equal to their charge.

$$2 \ Fe^{2+}(aq) \ + \ Cl_2(aq) \rightarrow 2 \ Fe^{3+}(aq) \ + \ 2 \ Cl^-(aq)$$
$$\uparrow \qquad \uparrow \qquad \uparrow \qquad \uparrow$$
$$+2 \qquad 0 \qquad +3 \qquad -1$$

Fe^{2+} is oxidized (its oxidation number increases from +2 to +3); Cl_2 is reduced (its oxidation number decreases from 0 to −1). Fe^{2+} is the reducing agent and Cl_2 is the oxidizing agent.

⌐ PROBLEM 4.8 Assign an oxidation number to each atom in the following compounds.

(a) $SnCl_4$ **(b)** CrO_3 **(c)** $VOCl_3$ **(d)** V_2O_3 **(e)** HNO_3 **(f)** $FeSO_4$

⌐ PROBLEM 4.9 Aqueous copper(II) ion reacts with aqueous iodide ion to yield solid copper(I) iodide and aqueous iodine. Write the balanced net ionic equation, assign oxidation numbers to all species present, and identify the oxidizing and reducing agents.

⌐ PROBLEM 4.10 Tell in each case which substance is undergoing an oxidation and which a reduction, and identify the oxidizing and reducing agents.

(a) $SnO_2(s) \ + \ 2 \ C(s) \longrightarrow Sn(s) \ + \ 2 \ CO(g)$
(b) $Sn^{2+}(aq) \ + \ 2 \ Fe^{3+}(aq) \longrightarrow Sn^{4+}(aq) \ + \ 2 \ Fe^{2+}(aq)$

4.7 ➤ THE ACTIVITY SERIES OF THE ELEMENTS

Among the simplest of all redox processes is the reaction of an aqueous cation, usually a metal ion, with a free element. The products are a different ion and a different element. Iron metal reacts with aqueous copper(II) ion, for example, to give iron(II) ion and copper metal:

$$Fe(s) \ + \ Cu^{2+}(aq) \rightarrow Fe^{2+}(aq) \ + \ Cu(s)$$

The iron nail reduces the Cu^{2+} ion and becomes coated with metallic Cu.

Similarly, magnesium metal reacts with aqueous acid to yield magnesium ion and hydrogen gas:

$$Mg(s) + 2\,H^+(aq) \rightarrow Mg^{2+}(aq) + H_2(g)$$

Whether a reaction occurs between a given ion and a given element depends on the relative ease with which the various species gain or lose electrons—that is, on the relative ease with which the species are reduced or oxidized. By noting the results from a succession of different reactions, it's possible to organize an **activity series**, which ranks the elements in order of their reducing ability in aqueous solution (Table 4.3).

Magnesium metal reacts with aqueous acid to give hydrogen gas and Mg^{2+} ion.

TABLE 4.3	The Activity Series of the Elements	

	Oxidation Reaction	
Strongly reducing	Li → Li$^+$ + e$^-$ K → K$^+$ + e$^-$ Ba → Ba^{2+} + 2 e$^-$ Ca → Ca^{2+} + 2 e$^-$ Na → Na$^+$ + e$^-$	These elements react rapidly with aqueous acid or with liquid H_2O to release H_2 gas.
	Mg → Mg^{2+} + 2 e$^-$ Al → Al^{3+} + 3 e$^-$ Mn → Mn^{2+} + 2 e$^-$ Zn → Zn^{2+} + 2 e$^-$ Cr → Cr^{3+} + 3 e$^-$ Fe → Fe^{2+} + 2 e$^-$	These elements react rapidly with aqueous acid or with steam to release H_2 gas.
	Co → Co^{2+} + 2 e$^-$ Ni → Ni^{2+} + 2 e$^-$ Sn → Sn^{2+} + 2 e$^-$	These elements react rapidly with aqueous acid to release H_2 gas.
	H$_2$ → 2 H$^+$ + 2 e$^-$	
Weakly reducing	Cu → Cu^{2+} + 2 e$^-$ Ag → Ag$^+$ + e$^-$ Hg → Hg^{2+} + 2 e$^-$ Pt → Pt^{2+} + 2 e$^-$ Au → Au^{3+} + 3 e$^-$	These elements do not react with aqueous acid to release H_2.

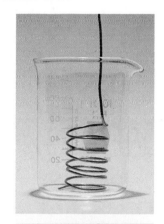

The copper wire reduces aqueous Ag^+ ion and becomes coated with metallic silver.

Those elements at the top of Table 4.3 give up electrons readily and are stronger reducing agents, whereas those elements at the bottom of Table 4.3 give up electrons less readily and are weaker reducing agents. As a result, *any element higher in the activity series will react with the ion of any element lower in the activity series.* Because copper is above silver, for example, we expect copper metal to give electrons to Ag^+ ions:

$$Cu(s) + 2\,Ag^+(aq) \rightarrow Cu^{2+}(aq) + 2\,Ag(s)$$

This gold ring does not react with aqueous Sn^{2+}.

Conversely, because gold is below tin in the activity series, we do *not* expect gold metal to give electrons to Sn^{2+} ion:

$$2 \, Au(s) + 3 \, Sn^{2+}(aq) \xrightarrow{\,\,\,//\,\,\,} 2 \, Au^{3+}(aq) + 3 \, Sn(s) \quad \textbf{\textit{Does not occur}}$$

The position of hydrogen in the activity series is particularly important because it indicates which metals react with aqueous acid (H^+) to release H_2 gas. Those metals at the top of the series—the alkali metals in group 1A and alkaline earth metals in group 2A—are such powerful reducing agents that they even react with pure water, in which the concentration of H^+ is very low:

$$2 \, Na(s) + 2 \, H_2O(l) \rightarrow 2 \, Na^+(aq) + 2 \, OH^-(aq) + H_2(g)$$

Metallic sodium reacts violently with water, generating H_2 gas.

Metallic calcium reacts only slowly with water at room temperature.

By contrast, those metals in the middle of the series react only with aqueous acids but not with water, and those metals at the bottom of the series react with neither aqueous acid nor water:

$$Fe(s) + 2 \, H^+(aq) \rightarrow Fe^{2+}(aq) + H_2(g)$$

$$Cu(s) + H^+(aq) \rightarrow No \ reaction$$

Notice that the most reactive metals (the top of the activity series) are on the left of the periodic table, whereas the least reactive metals (the bottom of the activity series) are in the transition-metal groups closer to the right side of the table. We'll see the reasons for this behavior in Chapter 6.

EXAMPLE 4.8

Predict whether the following reactions will occur.
(a) $Hg^{2+}(aq) + Zn(s) \longrightarrow Hg(l) + Zn^{2+}(aq)$
(b) $2 H^+(aq) + 2 Ag(s) \longrightarrow H_2(g) + 2 Ag^+(aq)$

SOLUTION   Check the activity series in Table 4.3 to find the relative reactivities of the metals. **(a)** Zinc is above mercury in the activity series, so this reaction will occur. **(b)** Silver is below hydrogen in the activity series, so this reaction will not occur.

⌐ **PROBLEM 4.11**   Predict whether the following reactions will occur.
(a) $2 H^+(aq) + Pt(s) \longrightarrow H_2(g) + Pt^{2+}(aq)$
(b) $Ca^{2+}(aq) + Mg(s) \longrightarrow Ca(s) + Mg^{2+}(aq)$

4.8 ►BALANCING REDOX REACTIONS BY THE OXIDATION-NUMBER METHOD

Many redox reactions can be balanced by the common-sense method described in Section 3.1, but other reactions are so complex that a more systematic approach is needed. There are two such systematic approaches often used for balancing oxidation–reduction reactions: the *oxidation-number method* and the *half-reaction method*. Different chemists prefer different methods, so we'll discuss both, beginning in this section with the oxidation-number method.

The key to the **oxidation-number method** of balancing redox equations is to realize that the net change in the total of all oxidation numbers must be zero. That is, any *increase* in oxidation number for the oxidized atoms must be exactly matched by a corresponding *decrease* in oxidation number for the reduced atoms. Take the reaction of potassium permanganate, $KMnO_4$, with sodium bromide in aqueous acid, for example. An aqueous acidic solution of the purple permanganate anion, MnO_4^-, is reduced by Br^- to yield the pale pink Mn^{2+} ion along with Br_2. The unbalanced ionic equation for the process is

$$MnO_4^-(aq) + Br^-(aq) \rightarrow Mn^{2+}(aq) + Br_2(aq)$$

The first step is to balance the equation for all atoms other than O and H. In the present instance, a coefficient of 2 is needed for Br^- on the left:

Add this coefficient
to balance for Br.
$$MnO_4^-(aq) + 2 Br^-(aq) \rightarrow Mn^{2+}(aq) + Br_2(aq)$$

Next, assign oxidation numbers to all atoms, including O and H, in both starting materials and products:

$$MnO_4^-(aq) + 2 Br^-(aq) \rightarrow Mn^{2+}(aq) + Br_2(aq)$$
$$+7\ -2 \quad -1 \quad +2 \quad 0$$

Now, decide which of the atoms have changed their oxidation number and have thus been either oxidized or reduced. In the present instance, manganese has been reduced from +7 to +2 (gaining five electrons), bromine has been oxidized from −1 to 0 (losing one electron), and oxygen has not changed its oxidation number.

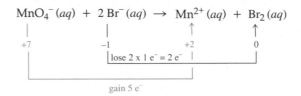

The next step is to find the net increase in oxidation number for the oxidized atoms and the net decrease in oxidation number for the reduced atoms. Then, multiply the net increase and the net decrease by suitable factors so that the two become equal. In the present instance, the net increase in oxidation number is 2 (as two Br^- ions go from −1 to 0), and the net decrease in oxidation number is 5 (as Mn goes from +7 to +2). Multiplying the net increase by 5 and the net decrease by 2 will make them equal at 10. Thus, we must multiply the coefficients of the manganese species by 2 and the coefficients of the Br species by 5.

Increase in oxidation number: $2 \times [Br^{-1} \rightarrow Br^0]$ Net increase = +2

Decrease in oxidation number: $Mn^{+7} \rightarrow Mn^{+2}$ Net decrease = −5

Multiply by these coefficients to make the net increase and net decrease in oxidation number equal.

$$(2 \times 1)\, MnO_4^-(aq) + (5 \times 2)\, Br^-(aq) \rightarrow 2\, Mn^{2+}(aq) + 5\, Br_2(aq)$$

Finally, because we know that the reaction is occurring in acidic solution, we balance the equation for oxygen by adding H_2O to the side with less O and then balance for hydrogen by adding H^+ to the side with less H. In this example, $8\ H_2O$ must be added to the right side to balance for O, and $16\ H^+$ must then be added to the left side to balance for H. If everything has been done correctly, the final net ionic equation will result. The answer can be checked by noting that the equation is balanced both for atoms and for charge, with 4+ on both sides.

Add these H^+ ions to balance H.

Add these water molecules to balance O.

$$2\, MnO_4^-(aq) + 10\, Br^-(aq) + 16\, H^+(aq) \rightarrow 2\, Mn^{2+}(aq) + 5\, Br_2(aq) + 8\, H_2O(l)$$

Charge: $(2 \times 1-) + (10-) + (16+) = 4+$ Charge: $(2 \times 2+) = 4+$

To summarize, there are six steps needed to balance a redox reaction by the oxidation-number method:

1. Write the unbalanced ionic equation.
2. Balance the equation for all atoms other than H and O.
3. Assign oxidation numbers to all atoms.
4. Decide which atoms have changed oxidation number and by how much.
5. Find the total net increase in oxidation number for the oxidized atoms and the total net decrease in oxidation number for the reduced atoms. Multiply the net increase and the net decrease by suitable numbers so that the two values become equal.
6. Balance the equation for oxygen by adding water to the side with less O, and then balance for hydrogen by adding H^+ to the side with less H. Check your answer by making sure that the equation is balanced both for atoms and for charge.

Example 4.9 shows how the oxidation-number method can be used for a reaction carried out in basic solution. The procedure is exactly the same as when balancing a reaction in acidic solution, but at the end OH^- ions must be added to neutralize any H^+ that appears in the equation.

EXAMPLE 4.9

A purple solution of aqueous potassium permanganate ($KMnO_4$) reacts with aqueous sodium sulfite (Na_2SO_3) in basic solution to yield the green manganate ion (MnO_4^{2-}) and sulfate ion (SO_4^{2-}). The unbalanced ionic equation is

$$MnO_4^-\ (aq) + SO_3^{2-}\ (aq) \rightarrow MnO_4^{2-}(aq) + SO_4^{2-}(aq)$$

Balance the equation by the oxidation-number method.

Reduction of the purple MnO_4^- ion with SO_3^{2-} in basic solution yields the deep green MnO_4^{2-} ion.

SOLUTION   *Steps 1 and 2.* The unbalanced ionic equation is already balanced for atoms other than O and H.

Step 3. Assign oxidation numbers to all atoms:

$$MnO_4^-(aq) + SO_3^{2-}(aq) \rightarrow MnO_4^{2-}(aq) + SO_4^{2-}(aq)$$

$$\begin{array}{cccc} \uparrow\ \uparrow & \uparrow\ \uparrow & \uparrow\ \uparrow & \uparrow\ \uparrow \\ +7\ -2 & +4\ -2 & +6\ -2 & +6\ -2 \end{array}$$

Step 4. Decide which atoms have changed oxidation number, and by how much. Manganese has been reduced from +7 to +6 (gaining one electron), and sulfur has been oxidized from +4 to +6 (losing two electrons):

$$MnO_4^-(aq) + SO_3^{2-}(aq) \rightarrow MnO_4^{2-}(aq) + SO_4^{2-}(aq)$$

+7   +4   +6   +6

lose 2 e⁻

gain 1 e⁻

Step 5. Find the total net increase in oxidation number for the oxidized atoms and the total net decrease in oxidation number for the reduced atoms:

Increase in oxidation number: $S^{4+} \rightarrow S^{6+}$ Net increase = +2

Decrease in oxidation number: $Mn^{7+} \rightarrow Mn^{6+}$ Net decrease = −1

Now, choose coefficients that make the net increase equal to the net decrease. In the present instance, the net decrease must be multiplied by 2, meaning that a coefficient of 2 is needed for both Mn species:

Add this coefficient to make the net increase
and net decrease in oxidation numbers equal.

$$2\ MnO_4^-(aq) + SO_3^{2-}(aq) \rightarrow 2\ MnO_4^{2-}(aq) + SO_4^{2-}(aq)$$

Step 6. Balance the equation for O by adding 1 H_2O on the left, and then balance for H by adding 2 H^+ on the right.

Add this water molecule   Add these H⁺ ions to
to balance for O.   balance for H.

$$2\ MnO_4^-(aq) + SO_3^{2-}(aq) + H_2O\ (l) \rightarrow 2\ MnO_4^{2-}(aq) + SO_4^{2-}(aq) + 2\ H^+(aq)$$

At this point, the equation is fully balanced. Even so, it's not correct because it assumes an *acidic* solution (the 2 H^+ on the right), while we were told that the reaction occurs in *basic* solution. We can correct the situation by adding 2 OH^- ions to each side of the equation. The 2 OH^- on the right will "neutralize" the 2 H^+ ions, giving 2 H_2O:

2 OH⁻ ions have been
added to both sides.   These water molecules result from
neutralizing 2 H⁺ with 2 OH⁻.

$$2\ MnO_4^-(aq) + SO_3^{2-}(aq) + H_2O(l) + 2\ OH^-(aq) \rightarrow$$
$$2\ MnO_4^{2-}(aq) + SO_4^{2-}(aq) + 2\ H_2O(l)$$

Finally, we can cancel one H_2O molecule that occurs on both sides of the equation, giving the final balanced net ionic equation:

$$2\ MnO_4^-(aq) + SO_3^{2-}(aq) + 2\ OH^-(aq) \rightarrow 2\ MnO_4^{2-}(aq) + SO_4^{2-}(aq) + H_2O\ (l)$$

Charge: $(2 \times 1-) + (2-) + (2 \times 1-) = 6-$ Charge: $(2 \times 2-) + (2-) = 6-$

The final equation is balanced both for atoms and for charge, with 6− on both sides.

┌ **PROBLEM 4.12** Balance the following ionic equation by the oxidation-number method. The reaction takes place in acidic solution.

$$Cr_2O_7{}^{2-}(aq) + I^-(aq) \rightarrow Cr^{3+}(aq) + IO_3{}^-(aq)$$

┌ **PROBLEM 4.13** Balance the following ionic equation by the oxidation-number method. The reaction takes place in basic solution.

$$MnO_4{}^-(aq) + Br^-(aq) \rightarrow MnO_2(s) + BrO_3{}^-(aq)$$ ⌐

4.9 ▶BALANCING REDOX REACTIONS BY THE HALF-REACTION METHOD

An alternative to the oxidation-number method for balancing redox reactions is the **half-reaction method**. The key to this method is to realize that the overall reaction can be broken into two parts, or *half-reactions*. One half-reaction describes the oxidation part of the process, and the other half-reaction describes the reduction part. Each half is balanced separately, and the two halves are then added to obtain the final equation. Let's look at the reaction of aqueous potassium dichromate ($K_2Cr_2O_7$) with aqueous sodium chloride to see how the method works. The reaction occurs in acidic solution according to the unbalanced ionic equation:

$$Cr_2O_7{}^{2-}(aq) + Cl^-(aq) \rightarrow Cr^{3+}(aq) + Cl_2(aq)$$

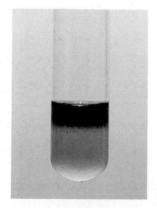

The orange dichromate ion is reduced by Cl^- to the green Cr^{3+} ion.

The first step is to decide which atoms have been oxidized and which have been reduced. Though helpful for understanding what's going on, it's not necessary to assign exact oxidation numbers to carry out this step. For the reaction of aqueous dichromate ion with aqueous chloride ion, the chloride atom is oxidized (from −1 to 0), and the chromium atom is reduced (from +6 to +3). Thus, we can write two unbalanced **half-reactions** that show the separate steps:

Oxidation half-reaction: $Cl^-(aq) \rightarrow Cl_2(aq)$

Reduction half-reaction: $Cr_2O_7{}^{2-}(aq) \rightarrow Cr^{3+}(aq)$

With the two half-reactions identified, each is then balanced separately. Begin by balancing for all atoms other than hydrogen and oxygen. The oxidation half-reaction needs a coefficient of 2 before the Cl^-, and the reduction half-reaction needs a coefficient of 2 before the Cr^{3+}.

Add this coefficient to balance for Cl.

Oxidation: $2\ Cl^-(aq) \rightarrow Cl_2(aq)$

Add this coefficient to balance for Cr.

Reduction: $Cr_2O_7^{2-}(aq) \rightarrow 2\ Cr^{3+}(aq)$

Next, balance both half-reactions for oxygen by adding H_2O to the side with less O, and then balance for hydrogen by adding H^+ to the side with less H. The oxidation half-reaction has no H or O, but the reduction reaction needs 7 H_2O on the product side and 14 H^+ on the reactant side:

Oxidation: $2\ Cl^-(aq) \rightarrow Cl_2(aq)$

Add this coefficient to balance for H.   Add this coefficient to balance for O.

Reduction: $Cr_2O_7^{2-}(aq) + 14\ H^+(aq) \rightarrow 2\ Cr^{3+}(aq) + 7\ H_2O(l)$

Now, balance both half-reactions for charge by adding electrons (e^-) to the side with the greater positive charge. The oxidation half-reaction must have 2 e^- added to the *product* side, and the reduction half-reaction must have 6 e^- added to the *reactant* side:

Add these electrons to balance for charge.

Oxidation: $2\ Cl^-(aq) \rightarrow Cl_2(aq) + 2\ e^-$

Add these electrons to balance for charge.

Reduction: $Cr_2O_7^{2-}(aq) + 14\ H^+(aq) + 6\ e^- \rightarrow 2\ Cr^{3+}(aq) + 7\ H_2O(l)$

With both half-reactions now balanced, we need only to multiply them by suitable coefficients so that the electron count is the same in both. Since the reduction half-reaction has 6 e^- but the oxidation half-reaction has only 2 e^-, the entire oxidation half-reaction must be multiplied by 3:

Multiply by this coefficient to equalize the numbers of electrons in the two half-reactions

Oxidation: $3 \times [2\ Cl^-(aq) \rightarrow Cl_2(aq) + 2\ e^-]$

or $6\ Cl^-(aq) \rightarrow 3\ Cl_2(aq) + 6\ e^-$

Reduction: $Cr_2O_7^{2-}(aq) + 14\ H^+(aq) + 6\ e^- \rightarrow 2\ Cr^{3+}(aq) + 7\ H_2O(l)$

Adding the two half-reactions together and canceling the species that occur on both sides (only the electrons in this example) then gives the final balanced equation:

$$6\ Cl^-(aq) \rightarrow 3\ Cl_2(aq) + 6\ e^-$$

$$\underline{Cr_2O_7{}^{2-}(aq) + 14\ H^+(aq) + 6\ e^- \rightarrow 2\ Cr^{3+}(aq) + 7\ H_2O(l)}$$

$$Cr_2O_7{}^{2-}(aq) + 14\ H^+(aq) + 6\ Cl^-(aq) \rightarrow 3\ Cl_2(aq) + 2\ Cr^{3+}(aq) + 7\ H_2O(l)$$

Charge: $(2-) + (14+) + (6 \times 1-) = 6+$ Charge: $(2 \times 3+) = 6+$

The answer can be checked by noting that it is balanced both for atoms and for charge, with $6+$ on both sides. To summarize, there are six steps in balancing a redox reaction by the half-reaction method.

1. Write the unbalanced ionic equation.
2. Decide which atoms are oxidized and which are reduced, and write two unbalanced half-reactions for the separate processes.
3. Balance both half-reactions for all atoms other than H and O.
4. Balance both half-reactions for O by adding H_2O to the side with less O, and then balance for H by adding H^+ to the side with less H.
5. Balance both half-reactions for charge by adding electrons (e^-) to the side with the greater positive charge. Then multiply the half-reactions by suitable numbers to make the electron count the same in both.
6. Add the two balanced half-reactions, and cancel the electrons and any other species that appear on both sides of the equation. Check to see that both atoms and charges balance, and that the coefficients are reduced to the simplest whole numbers.

   Example 4.11 shows how to use the method for balancing a reaction that takes place in basic solution. As in the oxidation-number method, we first balance the reaction for an *acidic* solution and then add OH^- ions in the last step to neutralize H^+.

EXAMPLE 4.10

Write unbalanced half-reactions for the following ionic equations.
(a) $Mn^{2+}(aq) + ClO_3{}^-(aq) \longrightarrow MnO_2(s) + ClO_2(aq)$
(b) $Cr_2O_7{}^{2-}(aq) + Fe^{2+}(aq) \longrightarrow Cr^{3+}(aq) + Fe^{3+}(aq)$

SOLUTION   Look at each equation to see which atoms are being oxidized (increasing in oxidation number) and which are being reduced (decreasing in oxidation number).

(a) Oxidation: $Mn^{2+}(aq) \longrightarrow MnO_2(s)$ (Mn goes from $+2$ to $+4$)
    Reduction: $ClO_3{}^-(aq) \longrightarrow ClO_2(aq)$ (Cl goes from $+5$ to $+4$)
(b) Oxidation: $Fe^{2+}(aq) \longrightarrow Fe^{3+}(aq)$ (Fe goes from $+2$ to $+3$)
    Reduction: $Cr_2O_7{}^{2-}(aq) \longrightarrow Cr^{3+}(aq)$ (Cr goes from $+6$ to $+3$)

EXAMPLE 4.11

Aqueous sodium hypochlorite (NaOCl; household bleach) is a strong oxidizing agent that reacts with chromite ion [$Cr(OH)_4{}^-$] in basic solution to yield chromate ion ($CrO_4{}^{2-}$) and chloride ion. The ionic equation is

$$OCl^-(aq) + Cr(OH)_4{}^-(aq) \rightarrow CrO_4{}^{2-}(aq) + Cl^-(aq)$$

Balance the equation using the half-reaction method.

SOLUTION

Step 1. We have already been given the unbalanced ionic equation.

Step 2. Chromium is oxidized (from +3 to +6), and chlorine is reduced (from +1 to −1). Thus, we can write the following two half-reactions:

$$\text{Oxidation half-reaction:}\quad Cr(OH)_4^-(aq) \rightarrow CrO_4^{2-}(aq)$$

$$\text{Reduction half-reaction:}\quad ClO^-(aq) \rightarrow Cl^-(aq)$$

Step 3. The half-reactions are already balanced for atoms other than O and H.

Step 4. Balance both half-reactions for O by adding H_2O to the side of each that has less O, and then balance both for H by adding H^+ to the side of each that has less H.

$$\text{Oxidation:}\quad Cr(OH)_4^-(aq) \rightarrow CrO_4^{2-}(aq) + 4\ H^+(aq)$$

$$\text{Reduction:}\quad ClO^-(aq) + 2\ H^+(aq) \rightarrow Cl^-(aq) + H_2O(l)$$

Step 5. Balance both half-reactions for charge by adding electrons to the side with the greater positive charge:

$$\text{Oxidation:}\quad Cr(OH)_4^-(aq) \rightarrow CrO_4^{2-}(aq) + 4\ H^+(aq) + 3\ e^-$$

$$\text{Reduction:}\quad ClO^-(aq) + 2\ H^+(aq) + 2\ e^- \rightarrow Cl^-(aq) + H_2O(l)$$

Now, multiply the half-reactions by numbers that will make the electron count in each the same. The oxidation half-reaction must be multiplied by 2, and the reduction half-reaction must be multiplied by 3:

$$\text{Oxidation:}\quad 2 \times [Cr(OH)_4^-(aq) \rightarrow CrO_4^{2-}(aq) + 4\ H^+(aq) + 3\ e^-]$$
$$\text{or } 2\ Cr(OH)_4^-(aq) \rightarrow 2\ CrO_4^{2-}(aq) + 8\ H^+(aq) + 6\ e^-$$

$$\text{Reduction:}\quad 3 \times [ClO^-(aq) + 2\ H^+(aq) + 2\ e^- \rightarrow Cl^-(aq) + H_2O(l)]$$
$$\text{or } 3\ ClO^-(aq) + 6\ H^+(aq) + 6\ e^- \rightarrow 3\ Cl^-(aq) + 3\ H_2O(l)$$

Step 6. Add the balanced half-reactions:

$$2\ Cr(OH)_4^-(aq) \rightarrow 2\ CrO_4^{2-}(aq) + 8\ H^+(aq) + 6\ e^-$$
$$\underline{3\ ClO^-(aq) + 6\ H^+(aq) + 6\ e^- \rightarrow 3\ Cl^-(aq) + 3\ H_2O(l)}$$
$$2\ Cr(OH)_4^-(aq) + 3\ ClO^-(aq) + 6\ H^+(aq) + 6\ e^- \rightarrow$$
$$2\ CrO_4^{2-}(aq) + 3\ Cl^-(aq) + 3\ H_2O(l) + 8\ H^+(aq) + 6\ e^-$$

Now, cancel the species that appear on both sides of the equation:

$$2\ Cr(OH)_4^-(aq) + 3\ ClO^-(aq) \rightarrow 2\ CrO_4^{2-}(aq) + 3\ Cl^-(aq) + 3\ H_2O(l) + 2\ H^+(aq)$$

Finally, since we know that the reaction takes place in basic solution, we must add $2\ OH^-$ ions to both sides of the equation to neutralize the $2\ H^+$ ions on the right. The final net ionic equation is

$$2\ Cr(OH)_4^-(aq) + 3\ ClO^-(aq) + 2\ OH^-(aq) \rightarrow 2\ CrO_4^{2-}(aq) + 3\ Cl^-(aq) + 5\ H_2O(l)$$

Charge: $(2 \times 1-) + (3 \times 1-) + (2 \times 1-) = 7-$ Charge: $(2 \times 2-) + (3 \times 1-) = 7-$

The atoms and the charges both balance, with 7− on both sides.

⌐ **PROBLEM 4.14** Write unbalanced half-reactions for the following ionic equations.

(a) $MnO_4^-(aq) + IO_3^-(aq) \longrightarrow MnO_2(s) + IO_4^-(aq)$
(b) $NO_3^-(aq) + SO_2(aq) \longrightarrow SO_4^{2-}(aq) + 2\ NO_2(g)$

⌐ **PROBLEM 4.15** Balance the following ionic equation by the half-reaction method. The reaction takes place in acidic solution.

$$NO_3^-(aq) + Cu(s) \rightarrow NO(g) + Cu^{2+}(aq)$$

⌐ **PROBLEM 4.16** Balance the following ionic equation by the half-reaction method. The reaction takes place in basic solution.

$$Fe(OH)_2(aq) + O_2(g) \rightarrow Fe(OH)_3(aq)$$

4.10 ➤REDOX TITRATIONS

We saw in Section 3.10 that the concentration of an acid or base solution can be determined by *titration*. A measured volume of acid (or base) solution of unknown concentration is placed in a flask, and a base (or acid) solution of known concentration is slowly added from a buret. By measuring the volume of the added solution needed for a complete reaction (as signaled by an indicator), the unknown concentration can be calculated.

A similar procedure can be carried out to determine the concentration of many oxidizing or reducing agents using a **redox titration**. All that's necessary is that the substance whose concentration you want to determine undergo an oxidation or reduction reaction in 100% yield and that there be a color change to signal when the reaction is complete. The color change might be due to the one of the substances undergoing reaction or to some added redox indicator. Let's imagine, for instance, that we have a potassium permanganate solution whose concentration we want to find. Aqueous $KMnO_4$ reacts with oxalic acid, $H_2C_2O_4$, in acidic solution according to the net ionic equation.

$$5\ H_2C_2O_4(aq) + 2\ MnO_4^-(aq) + 6\ H^+(aq) \rightarrow 10\ CO_2(g) + 2\ Mn^{2+}(aq) + 8\ H_2O(l)$$

The reaction takes place in 100% yield and is accompanied by a sharp color change as the intense purple color of the MnO_4^- ion is replaced by the pale pink color of Mn^{2+}.

To see how the redox titration process works, let's carefully weigh a precise amount of $H_2C_2O_4$—say, 0.2585 g—and dissolve it in approximately 100 mL of 0.5 M H_2SO_4. The exact volume isn't important because we aren't concerned with the concentration of the oxalic acid solution, only with the total amount of dissolved $H_2C_2O_4$. Next, we place an aqueous $KMnO_4$ solution of unknown concentration in a buret and slowly add it to the $H_2C_2O_4$ solution. The purple permanganate color initially disappears as reaction occurs, but we continue the addition until a purple color persists, indicating that all the $H_2C_2O_4$ has reacted and that MnO_4^- ion is no longer being reduced. At this point, we might find that 22.35 mL of the $KMnO_4$ solution has been added (Figure 4.2).

FIGURE 4.2 The redox titration of oxalic acid, $H_2C_2O_4$, with $KMnO_4$. **(a)** A precise amount of oxalic acid is weighed and **(b)** dissolved in aqueous sulfuric acid. **(c)** Aqueous $KMnO_4$ of known concentration is then added from a buret until **(d)** the purple color persists, indicating that all the $H_2C_2O_4$ has been oxidized.

To calculate the molarity of the $KMnO_4$ solution, we need to find the number of moles of $KMnO_4$ present in the 22.35 mL of solution used for titration. We do this by first calculating the number of moles of oxalic acid that react with the permanganate ion. A gram-to-mole conversion is needed, using the molar mass of $H_2C_2O_4$ (90.04 g/mol) as the conversion factor:

$$\text{Moles } H_2C_2O_4 = 0.2585 \text{ g } H_2C_2O_4 \times \frac{1 \text{ mol } H_2C_2O_4}{90.04 \text{ g } H_2C_2O_4} = 2.871 \times 10^{-3} \text{ mol } H_2C_2O_4$$

According to the balanced equation, 5 mol of oxalic acid react with 2 mol of permanganate ion. Thus, we can calculate the number of moles of $KMnO_4$ that react with 2.871×10^{-3} mol of $H_2C_2O_4$:

$$\text{Moles } KMnO_4 = 2.871 \times 10^{-3} \text{ mol } H_2C_2O_4 \times \frac{2 \text{ mol } KMnO_4}{5 \text{ mol } H_2C_2O_4}$$
$$= 1.148 \times 10^{-3} \text{ mol } KMnO_4$$

Knowing both the number of moles of $KMnO_4$ that react (1.148×10^{-3} mol) and the volume of the solution (22.35 mL) then makes it possible to calculate molarity:

$$\text{Molarity} = \frac{1.148 \times 10^{-3} \text{ mol } KMnO_4}{22.35 \text{ mL}} \times \frac{1000 \text{ mL}}{1 \text{ L}} = 0.051\ 36 \text{ M}$$

The molarity of the $KMnO_4$ solution as determined by redox titration is 0.051 36 M.

EXAMPLE 4.12

The concentration of an aqueous I_3^- solution can be determined by titration with aqueous sodium thiosulfate, $Na_2S_2O_3$, in the presence of a starch indicator that turns from deep blue to colorless when all the I_3^- has reacted. The net ionic equation is

$$2\ S_2O_3^{2-}(aq) + I_3^-(aq) \rightarrow S_4O_6^{2-}(aq) + 3\ I^-(aq)$$

What is the molar concentration of I_3^- if 24.55 mL of 0.102 M $Na_2S_2O_3$ is needed for complete reaction with 10.00 mL of the I_3^- solution?

SOLUTION   We first need to find the number of moles of thiosulfate ion needed for the titration:

$$24.55 \text{ mL} \times \frac{1 \text{ L}}{1000 \text{ mL}} \times \frac{0.102 \text{ mol } S_2O_3^{2-}}{1 \text{ L}} = 2.50 \times 10^{-3} \text{ mol } S_2O_3^{2-}$$

According to the balanced equation, 2 mol of $S_2O_3^{2-}$ ion react with 1 mol of I_3^- ion. Thus, we can find the number of moles of I_3^- ion.

$$2.50 \times 10^{-3} \text{ mol } S_2O_3^{2-} \times \frac{1 \text{ mol } I_3^-}{2 \text{ mol } S_2O_3^{2-}} = 1.25 \times 10^{-3} \text{ mol } I_3^-$$

The I_3^- ion turns a deep blue color in the presence of starch.

Knowing both the number of moles (1.25×10^{-3} mol) and the volume (10.00 mL) then lets us calculate molarity:

$$\frac{1.25 \times 10^{-3} \text{ mol } I_3^-}{10.00 \text{ mL}} \times \frac{10^3 \text{ mL}}{1 \text{ L}} = 0.125 \text{ M}$$

The molarity of the I_3^- solution is 0.125 M.

◻ **PROBLEM 4.17** The concentration of Fe^{2+} ion in aqueous solution can be determined by redox titration with bromate ion, BrO_3^-, according to the net ionic equation

$$6\ Fe^{2+}(aq) + BrO_3^-(aq) + 6\ H^+(aq) \rightarrow 6\ Fe^{3+}(aq) + Br^-(aq) + 3\ H_2O(l)$$

What is the molar concentration of Fe^{2+} if 31.50 mL of 0.105 M $KBrO_3$ is required for reaction with 10.00 mL of the Fe^{2+} solution?

interlude—PHOTOGRAPHY: A SERIES OF REDOX REACTIONS

Photography is so common that most people never give a moment's thought to how remarkable the process really is. Ordinary photographic film consists of a celluloid strip that has been coated with a gelatin emulsion containing very tiny crystals, or "grains" of a silver halide, usually AgBr. There is a considerable amount of art as well as science to making the film, and the recipes major manufacturers use are well-guarded secrets. When exposed to light, the surfaces of the AgBr grains turn dark because of a light-induced redox reaction in which Br^- transfers an electron to Ag^+, producing atoms of elemental silver (Figure 4.3). Those areas of the film exposed to the brightest light have the largest number of silver atoms, and those areas exposed to the least light have the smallest number.

$$2\ AgBr(s) \xrightarrow{\text{light}} 2\ Ag(s)\ +\ Br_2(g)$$

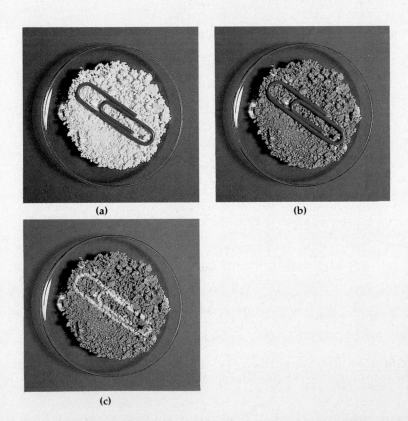

(a)

(b)

(c)

FIGURE 4.3 (a) When the paper clip sitting on powdered AgCl crystals is exposed to a bright light, the crystals around the clip darken (b) to provide an image (c).

Perhaps surprisingly in view of what a finished photograph looks like, only a few hundred out of many trillions of Ag^+ ions in each grain are reduced to Ag atoms, and the latent image produced on the film is still invisible at this point. The key to silver halide photography is the *developing* process, in which the latent image is amplified. By mechanisms still not understood in detail, the presence of a relatively tiny number of Ag atoms on the surface of an AgBr grain sensitizes the remaining Ag^+ ions in the grain toward further reduction when the film is exposed to the organic reducing agent hydroquinone. Those grains that have been exposed to the strongest light and thus have more Ag atoms reduce and darken quickly, while those grains with fewer Ag atoms reduce and darken more slowly. By carefully monitoring the amount of time allowed for reduction of the AgBr grains with hydroquinone, it is possible to amplify the latent image on the exposed film and make it visible. At the desired point, further developing is stopped by transferring the film from the basic hydroquinone solution to a weakly acidic one.

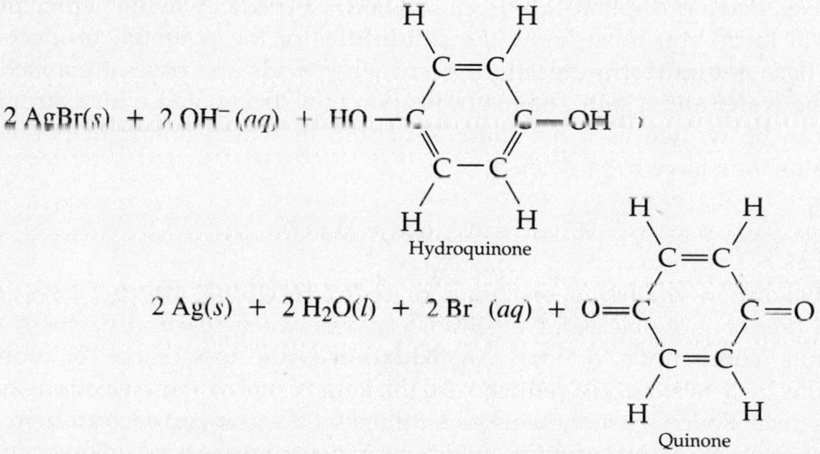

$$2\,AgBr(s)\ +\ 2\,OH^-(aq)\ +\ \text{Hydroquinone} \longrightarrow$$

Hydroquinone

$$2\,Ag(s)\ +\ 2\,H_2O(l)\ +\ 2\,Br^-(aq)\ +\ \text{Quinone}$$

Quinone

Once the image is fully formed, the film is *fixed* by washing away the remaining unreduced AgBr so that the film is no longer sensitive to light. Although AgBr is insoluble in water, it is made soluble by reaction with a solution of sodium thiosulfate, $Na_2S_2O_3$, called *hypo* by photographers.

$$AgBr(s)\ +\ 2\,S_2O_3^{2-}(aq) \rightarrow Ag(S_2O_3)_2^{3-}(aq)\ +\ Br^-(aq)$$

At this point, all that remains on the film is a negative image formed by a layer of black, finely divided silver metal, a layer that is denser and darker in those areas exposed to the most light but lighter in those areas exposed to the least light. To convert this negative image into the final printed photograph, the entire photographic procedure has to be repeated a second time. Light is passed through the negative image onto special photographic paper that is coated with the same kind of gelatin–AgBr emulsion used on the original film. Developing the photographic paper with hydroquinone and fixing the image with sodium thiosulfate reverses the image again, and the final positive image is produced. The whole process from film to print is a complex one, but it is carried out billions of times and consumes over 3 million pounds of silver each year.

KEY WORDS

SUMMARY

Many reactions, particularly those that involve ionic compounds, take place in **aqueous solution**. Substances whose aqueous solutions contain ions and therefore conduct electricity are called **electrolytes**. Ionic compounds such as NaCl and molecular compounds that **dissociate** completely into ions when dissolved in water are **strong electrolytes**. Molecular substances such as acetic acid that dissociate incompletely are **weak electrolytes**, and substances such as sucrose that do not produce ions in aqueous solution are **nonelectrolytes**. **Acids** dissociate in aqueous solutions to yield an anion and a **hydronium ion**, H_3O^+. Those acids that dissociate completely are **strong acids**; those acids that dissociate incompletely are **weak acids**.

There are three important types of aqueous reactions: precipitation reactions, acid–base neutralization reactions, and oxidation–reduction reactions. **Precipitation reactions** occur when solutions of two ionic substances are mixed and a precipitate falls from solution. To predict whether a precipitate will form, you must know the **solubilities** of the potential products. **Acid–base neutralization reactions** occur when acids and bases are mixed, yielding water and a **salt**. The neutralization of a strong acid with a strong base can be written as a **net ionic equation**, in which nonparticipating, **spectator ions** have been deleted:

$$H^+(aq) + OH^-(aq) \rightarrow H_2O(l)$$

Oxidation–reduction reactions, or **redox reactions**, are processes in which one or more electrons are transferred between reaction partners—elements, compounds, or ions. An **oxidation** is the loss of one or more electrons by a substance; a **reduction** is the gain of one or more electrons by a substance. Redox reactions can be identified by assigning to each atom in a substance an **oxidation number**, which provides a measure of whether the atom is neutral, electron-rich, or electron-poor. Comparison of the oxidation number of an atom before and after reaction shows whether the atom has gained or lost electrons.

Oxidations and reductions must occur together. Whenever one substance loses one or more electrons (is oxidized), another substance must gain those electrons (be reduced). The substance that causes a reduction by giving up an electron is called a **reducing agent**. The substance that causes an oxidation by accepting an electron is called an **oxidizing agent**. The reducing agent is itself oxidized when it gives up electrons, and the oxidizing agent is itself reduced when it accepts electrons.

Among the simplest of redox processes is the reaction of an aqueous cation, usually a metal ion, with a free element to give a different ion and a different element. Noting the results from a succession of different reactions makes it possible to organize an **activity series**, which ranks the elements in order of their reducing ability in aqueous solution.

Redox reactions can be balanced using either the **oxidation-number method** or the **half-reaction method**. The concentration of an oxidizing agent or a reducing agent in solution can be determined by a **redox titration**.

1. Convert each of the following descriptions into a balanced net ionic equation.
   (a) An aqueous solution of sodium sulfate reacts with an aqueous solution of barium chloride to yield a precipitate of barium sulfate.
   (b) An aqueous solution of perchloric acid reacts with an aqueous solution of potassium carbonate to yield aqueous potassium perchlorate, carbon dioxide gas, and water.

2. Assume that you are given an unidentified solid substance. Describe how you can tell whether the substance is a strong electrolyte, a weak electrolyte, or a nonelectrolyte.

3. Use the following reactions to arrange the elements A through D in order of their decreasing ability as reducing agents.

   $C + B^+ \longrightarrow C^+ + B$ $A^+ + D \longrightarrow$ no reaction

   $C^+ + A \longrightarrow$ no reaction $D + B^+ \longrightarrow D^+ + B$

4. Which of the following reactions would you expect to occur according to the activity series you established in Problem 3?
   (a) $A^+ + C \longrightarrow A + C^+$ (b) $A^+ + B \longrightarrow A + B^+$

5. Assume that an aqueous solution of a cation, represented by a red dot, is allowed to mix with a solution of an anion, represented by a yellow dot. Three possible outcomes are represented by boxes (1)–(3).

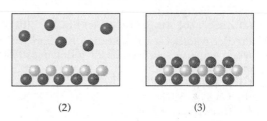

(1)

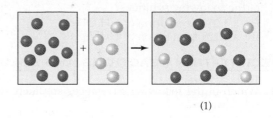

(2)   (3)

Which outcome corresponds to each of the following reactions?
(a) $2\,Na^+(aq) + CO_3^{2-}(aq) \longrightarrow$
(b) $Ba^{2+}(aq) + CrO_4^{2-}(aq) \longrightarrow$
(c) $2\,Ag^+(aq) + SO_4^{2-}(aq) \longrightarrow$

ADDITIONAL PROBLEMS

Problems 4.1–4.17 appear within the chapter.

AQUEOUS REACTIONS AND NET IONIC EQUATIONS

4.18 Classify each of the following processes as a precipitation, acid–base neutralization, or redox reaction.
   (a) $Hg(NO_3)_2(aq) + 2\,NaI(aq) \longrightarrow 2\,NaNO_3(aq) + HgI_2(s)$
   (b) $2\,HgO(s) \xrightarrow{\text{heat}} 2\,Hg(l) + O_2(g)$
   (c) $H_3PO_4(aq) + 3\,KOH(aq) \longrightarrow K_3PO_4(aq) + 3\,H_2O(l)$

4.19 Classify each of the following processes as a precipitation, acid–base neutralization, or redox reaction.
   (a) $S_8(s) + 8\,O_2(g) \longrightarrow 8\,SO_2(g)$
   (b) $NiCl_2(aq) + Na_2S(aq) \longrightarrow NiS(s) + 2\,NaCl(aq)$
   (c) $2\,CH_3COOH(aq) + Ba(OH)_2(aq) \longrightarrow (CH_3COO)_2Ba(aq) + 2\,H_2O(l)$

4.20 Write net ionic equations for the reactions listed in Problem 4.18.

4.21 Write net ionic equations for the reactions listed in Problem 4.19.

4.22 Convert each of the following descriptions into a balanced net ionic equation.
   (a) A piece of solid sodium reacts with water to give an aqueous solution of sodium hydroxide and hydrogen gas.
   (b) An aqueous solution of silver nitrate reacts with an aqueous solution of hydrochloric acid to yield a precipitate of silver chloride.

4.23 What is the difference between a strong electrolyte and a weak electrolyte?

4.24 Classify each of the following substances as either a strong or a weak electrolyte.
   (a) HBr   (b) HF
   (c) $NaClO_4$ (d) $(NH_4)_2CO_3$
   (e) NH_3

4.25 What is the total molar concentration of ions in each of the following solutions?
   (a) a 0.750 M solution of K_2CO_3
   (b) a 0.355 M solution of $AlCl_3$

4.26 What is the total molar concentration of ions in each of the following solutions?

    (a) a 1.250 M solution of CH_3OH

    (b) a 0.225 M solution of $HClO_4$

4.27 Explain why a solution of HCl in water conducts electricity but a solution of HCl in chloroform, $CHCl_3$, does not.

PRECIPITATION REACTIONS AND SOLUBILITY RULES

4.28 Which of the following substances are likely to be soluble in water?

    (a) Ag_2O **(b)** $Ba(NO_3)_2$ **(c)** $SnCO_3$ **(d)** Fe_2O_3

4.29 Which of the following substances are likely to be soluble in water?

    (a) ZnS **(b)** $Au_2(CO_3)_3$ **(c)** $PbCl_2$ **(d)** MnO_2

4.30 Predict whether a precipitation reaction will occur when aqueous solutions of the following substances are mixed.

    (a) $NaOH + HClO_4$

    (b) $FeCl_2 + KOH$

    (c) $(NH_4)_2SO_4 + NiCl_2$

4.31 Predict whether a precipitation reaction will occur when aqueous solutions of the following substances are mixed.

    (a) $MnCl_2 + Na_2S$

    (b) $HNO_3 + CuSO_4$

    (c) $Hg(NO_3)_2 + Na_3PO_4$

4.32 How would you prepare the following substances by a precipitation reaction?

    (a) $PbSO_4$ **(b)** $Mg_3(PO_4)_2$ **(c)** $ZnCrO_4$

4.33 How would you prepare the following substances by a precipitation reaction?

    (a) $Al(OH)_3$ **(b)** FeS **(c)** $CoCO_3$

4.34 Assume that you have an aqueous mixture of $NaNO_3$ and $AgNO_3$. How could you use a precipitation reaction to separate the two metal ions?

4.35 Assume that you have an aqueous mixture of $BaCl_2$ and $CuCl_2$. How could you use a precipitation reaction to separate the two metal ions?

4.36 Assume that you have an aqueous solution of an unknown salt. Treatment of the solution with dilute $NaOH$, Na_2SO_4, and KCl produces no precipitate. Which of the following cations might the solution contain?

    (a) Ag^+ **(b)** Cs^+ **(c)** Ba^{2+} **(d)** NH_4^+

4.37 Assume that you have an aqueous solution of an unknown salt. Treatment of the solution with dilute $BaCl_2$, $AgNO_3$, and $Cu(NO_3)_2$ produces no precipitate. Which of the following anions might the solution contain?

    (a) Cl^- **(b)** NO_3^- **(c)** OH^- **(d)** SO_4^{2-}

ACIDS, BASES, AND NEUTRALIZATION REACTIONS

4.38 What is the difference between a strong acid and a weak acid? Between a strong base and a weak base?

4.39 Assume that you are given a solution of an unknown acid or base. How can you tell whether the unknown substance is acidic or basic?

4.40 Why do we use a double, forward-and-backward arrow to show the dissociation of a weak acid or weak base in aqueous solution?

4.41 Write balanced ionic equations for the following reactions.

    (a) Aqueous perchloric acid is neutralized by aqueous calcium hydroxide.

    (b) Aqueous sodium hydroxide is neutralized by aqueous acetic acid.

4.42 Write balanced ionic equations for the following reactions.

    (a) Aqueous hydrofluoric acid is neutralized by aqueous calcium hydroxide.

    (b) Aqueous magnesium hydroxide is neutralized by aqueous nitric acid.

4.43 Balance and write net ionic equations for the following reactions.

    (a) $LiOH(aq) + HI(aq) \longrightarrow$

    (b) $HBr(aq) + Ca(OH)_2(aq) \longrightarrow$

4.44 Balance and write net ionic equations for the following reactions.

    (a) $Fe(OH)_3(s) + H_2SO_4(aq) \longrightarrow$

    (b) $HClO_3(aq) + NaOH(aq) \longrightarrow$

REDOX REACTIONS AND OXIDATION NUMBERS

4.45 Where in the periodic table are the best reducing agents found? The best oxidizing agents?

4.46 Where in the periodic table are the most easily reduced elements found? The most easily oxidized?

4.47 Tell in each of the following instances whether the substance gains electrons or loses electrons in a redox reaction.

    (a) an oxidizing agent *gains*

    (b) a reducing agent *loses*

    (c) a substance undergoing oxidation *loses*

    (d) a substance undergoing reduction *gains*

4.48 Tell for each of the following substances whether the oxidation number increases or decreases in a redox reaction.

    (a) an oxidizing agent

    (b) a reducing agent

(c) a substance undergoing oxidation

(d) a substance undergoing reduction

4.49 Assign oxidation numbers to each element in the following compounds.

(a) NO_2 (b) SO_3 (c) $COCl_2$
(d) CH_2Cl_2 (e) $KClO_3$ (f) HNO_3

4.50 Assign oxidation numbers to each element in the following compounds.

(a) $VOCl_3$ (b) $CuSO_4$ (c) CH_2O
(d) Mn_2O_7 (e) OsO_4 (f) H_2PtCl_6

4.51 Assign oxidation numbers to each element in the following ions.

(a) ClO_3^- (b) SO_3^{2-} (c) $C_2O_4^{2-}$ (d) NO_2^- (e) BrO^-

4.52 Assign oxidation numbers to each element in the following ions.

(a) $Cr(OH)_4^-$ (b) $S_2O_3^{2-}$ (c) NO_3^-
(d) MnO_4^{2-} (e) HPO_4^{2-}

4.53 Which element is oxidized and which is reduced in each of the following reactions?

(a) $Ca(s) + Sn^{2+}(aq) \longrightarrow Ca^{2+}(aq) + Sn(s)$
(b) $ICl(s) + H_2O(l) \longrightarrow HCl(aq) + HOI(aq)$

4.54 Which element is oxidized and which is reduced in each of the following reactions?

(a) $Si(s) + 2 Cl_2(g) \longrightarrow SiCl_4(l)$
(b) $Cl_2(g) + 2 NaBr(aq) \longrightarrow Br_2(aq) + 2 NaCl(aq)$

4.55 Use the activity series of metals (Table 4.3) to predict the outcome of each of the following reactions. If no reaction occurs, write N.R.

(a) $Na^+(aq) + Zn(s) \longrightarrow$
(b) $HCl(aq) + Pt(s) \longrightarrow$
(c) $Ag^+(aq) + Au(s) \longrightarrow$
(d) $Au^{3+}(aq) + Ag(s) \longrightarrow$

4.56 Use the following reactions to arrange the elements A through D in order of their redox activity.

$A + B^+ \longrightarrow A^+ + B$; $C^+ + D \longrightarrow$ no reaction;
$B + D^+ \longrightarrow B^+ + D$; $B + C^+ \longrightarrow B^+ + C$

4.57 Tell which of the following reactions you would expect to occur according to the activity series you established in Problem 4.56.

(a) $A^+ + C \longrightarrow A + C^+$ (b) $A^+ + D \longrightarrow A + D^+$

BALANCING REDOX REACTIONS

4.58 Classify each of the following unbalanced half-reactions as either an oxidation or a reduction.

(a) $NO_3^-(aq) \longrightarrow NO(g)$
(b) $Zn(s) \longrightarrow Zn^{2+}(aq)$
(c) $Ti^{3+}(aq) \longrightarrow TiO_2(s)$
(d) $Sn^{4+}(aq) \longrightarrow Sn^{2+}(aq)$

4.59 Classify each of the following unbalanced half-reactions as either an oxidation or a reduction.

(a) $O_2(g) \longrightarrow OH^-(aq)$
(b) $H_2O_2(aq) \longrightarrow O_2(g)$
(c) $MnO_4^-(aq) \longrightarrow MnO_4^{2-}(aq)$
(d) $CH_3OH(aq) \longrightarrow CH_2O(aq)$

4.60 Balance the half-reactions in Problem 4.58, assuming that they occur in acidic solution.

4.61 Balance the half-reactions in Problem 4.59, assuming that they occur in basic solution.

4.62 Write unbalanced oxidation and reduction half-reactions for the following processes.

(a) $Te(s) + NO_3^-(aq) \longrightarrow TeO_2(s) + NO(g)$
(b) $H_2O_2(aq) + Fe^{2+}(aq) \longrightarrow Fe^{3+}(aq) + H_2O(l)$

4.63 Write unbalanced oxidation and reduction half-reactions for the following processes.

(a) $Mn(s) + NO_3^-(aq) \longrightarrow Mn^{2+}(aq) + NO_2(g)$
(b) $Mn^{3+}(aq) \longrightarrow MnO_2(s) + Mn^{2+}(aq)$

4.64 Balance the following half-reactions.

(a) (acidic) $Cr_2O_7^{2-}(aq) \longrightarrow Cr^{3+}(aq)$
(b) (basic) $CrO_4^{2-}(aq) \longrightarrow Cr(OH)_4^-(aq)$
(c) (basic) $Bi^{3+}(aq) \longrightarrow BiO_3^-(aq)$
(d) (basic) $ClO^-(aq) \longrightarrow Cl^-(aq)$

4.65 Balance the following half-reactions.

(a) (acidic) $VO^{2+}(aq) \longrightarrow V^{3+}(aq)$

(b) (basic) $Ni(OH)_2(s) \longrightarrow Ni_2O_3(s)$
(c) (acidic) $NO_3^-(aq) \longrightarrow NO_2(g)$
(d) (basic) $Br_2(aq) \longrightarrow BrO_3^-(aq)$

4.66 Write balanced net ionic equations for the following reactions in basic solution.

(a) $MnO_4^-(aq) + IO_3^-(aq) \longrightarrow MnO_2(s) + IO_4^-(aq)$
(b) $Cu(OH)_2(s) + N_2H_4(aq) \longrightarrow Cu(s) + N_2(g)$
(c) $Fe(OH)_2(s) + CrO_4^{2-}(aq) \longrightarrow Fe(OH)_3(s) + Cr(OH)_4^-(aq)$
(d) $H_2O_2(aq) + ClO_4^-(aq) \longrightarrow ClO_2^-(aq) + O_2(g)$

4.67 Write balanced net ionic equations for the following reactions in basic solution.

(a) $S_2O_3^{2-}(aq) + I_2(aq) \longrightarrow S_4O_6^{2-}(aq) + I^-(aq)$
(b) $Mn^{2+}(aq) + H_2O_2(aq) \longrightarrow MnO_2(s)$
(c) $Zn(s) + NO_3^-(aq) \longrightarrow NH_3(aq) + Zn(OH)_4^{2-}(aq)$
(d) $Bi(OH)_3(s) + Sn(OH)_3^-(aq) \longrightarrow Bi(s) + Sn(OH)_6^{2-}(aq)$

4.68 Write balanced net ionic equations for the following reactions in acidic solution.

(a) $Zn(s) + VO^{2+}(aq) \longrightarrow Zn^{2+}(aq) + V^{3+}(aq)$
(b) $Ag(s) + NO_3^-(aq) \longrightarrow Ag^+(aq) + NO_2(g)$
(c) $Mg(s) + VO_4^{3-}(aq) \longrightarrow Mg^{2+}(aq) + V^{2+}(aq)$
(d) $I^-(aq) + IO_3^-(aq) \longrightarrow I_3^-(aq)$

4.69 Write balanced net ionic equations for the following reactions in acidic solution.

(a) $MnO_4^-(aq) + C_2H_5OH(aq) \longrightarrow Mn^{2+}(aq) + CH_3COOH(aq)$
(b) $H_2O_2(aq) + Cr_2O_7^{2-}(aq) \longrightarrow O_2(g) + Cr^{3+}(aq)$
(c) $Sn^{2+}(aq) + IO_4^-(aq) \longrightarrow Sn^{4+}(aq) + I^-(aq)$
(d) $PbO_2(s) + Cl^-(aq) \longrightarrow PbCl_2(s) + O_2(g)$

REDOX TITRATIONS

4.70 Iodine, I_2, reacts with aqueous thiosulfate ion in neutral solution according to the balanced equation

$$I_2(aq) + 2\,S_2O_3{}^{2-}(aq) \rightarrow S_4O_6{}^{2-}(aq) + 2\,I^-(aq)$$

How many grams of I_2 are present in a solution if 35.20 mL of 0.150 M $Na_2S_2O_3$ solution is needed to titrate the I_2 solution?

4.71 How many mL of 0.250 M $Na_2S_2O_3$ solution are needed to titrate 2.486 g of I_2 according to the equation in Problem 4.70?

4.72 Titration with potassium bromate, $KBrO_3$, solutions can be used to determine the concentration of As(III) in aqueous medium. What is the molar concentration of As(III) in a solution if 22.35 mL of 0.100 M $KBrO_3$ is needed to titrate 50.00 mL of the As(III) solution? The balanced equation is

$$3\,H_3AsO_3(aq) + BrO_3{}^-(aq) \rightarrow$$
$$Br^-(aq) + 3\,H_3AsO_4(aq)$$

4.73 Standardized solutions of $KBrO_3$ are frequently used in redox titrations. The necessary solution can be made by dissolving $KBrO_3$ in water and then titrating it with an As(III) solution. What is the molar concentration of a $KBrO_3$ solution if 28.55 mL of the solution is needed to titrate 1.550 g of As_2O_3? See Problem 4.72 for the balanced equation. (As_2O_3 dissolves in aqueous acid solution to yield H_3AsO_3: $As_2O_3 + 3\,H_2O \longrightarrow 2\,H_3AsO_3$).

4.74 The metal content of iron in ores can be determined by a redox procedure in which the sample is first oxidized with Br_2 to convert all the iron to Fe^{3+} and then titrated with Sn^{2+} to reduce the Fe^{3+} to Fe^{2+}. The balanced equation is

$$2\,Fe^{3+}(aq) + Sn^{2+}(aq) \rightarrow 2\,Fe^{2+}(aq) + Sn^{4+}(aq)$$

What is the mass percent Fe in a 0.1875 g sample if 13.28 mL of a 0.1015 M Sn^{2+} solution is needed to titrate the Fe^{3+}?

4.75 The Sn^{2+} solution used in Problem 4.74 can be standardized by titrating it with a known amount of Fe^{3+}. What is the molar concentration of an Sn^{2+} solution if 23.84 mL is required to titrate 1.4855 g of Fe_2O_3?

4.76 Alcohol levels in blood can be determined by a redox titration with potassium dichromate according to the balanced equation

$$C_2H_5OH(aq) + 2\,Cr_2O_7{}^{2-}(aq) + 16\,H^+(aq) \rightarrow$$
$$2\,CO_2(g) + 4\,Cr^{3+}(aq) + 11\,H_2O(l)$$

What is the blood alcohol level in mass percent if 8.76 mL of 0.049 88 M K_2CrO_7 is required for titration of a 10.002 g sample of blood?

4.77 Calcium levels in blood can be determined by adding oxalate ion to precipitate calcium oxalate, CaC_2O_4, followed by dissolving the precipitate in aqueous acid and titrating the oxalic acid ($H_2C_2O_4$) with $KMnO_4$:

$$5\,H_2C_2O_4(aq) + 2\,MnO_4{}^-(aq) + 6\,H^+(aq) \rightarrow$$
$$10\,CO_2(g) + 2\,Mn^{2+}(aq) + 8\,H_2O(l)$$

How many milligrams of Ca^{2+} are present in 10.0 mL of blood if 21.08 mL of 0.000 988 M $KMnO_4$ solution is needed for the titration?

GENERAL PROBLEMS

4.78 Balance the equations for the following reactions in basic solution.
  (a) $[Fe(CN)_6]^{3-}(aq) + N_2H_4(aq) \longrightarrow$
    $[Fe(CN)_6]^{4-}(aq) + N_2(g)$
  (b) $SeO_3{}^{2-}(aq) + Cl_2(g) \longrightarrow SeO_4{}^{2-}(aq) + Cl^-(aq)$
  (c) $CoCl_2(s) + HO_2{}^-(aq) \longrightarrow Co(OH)_3(aq) + Cl^-(aq)$

4.79 Citric acid, $H_3C_6H_5O_7$, a substance widely found in citrus fruit, has three weakly acidic hydrogens. Write a net ionic equation for the neutralization reaction of citric acid with calcium hydroxide.

4.80 Which of the following compounds would you expect to be water soluble?
  (a) TiO_2 **(b)** $SnCl_4$
  (c) Ag_2S **(d)** $Pd(NO_3)_2$

4.81 An alternative procedure to that given in Problem 4.74 for determining the amount of iron in a sample is to convert the iron to Fe^{2+} and then titrate with a solution of $Ce(NH_4)_2(NO_3)_6$.

$$Fe^{2+}(aq) + Ce^{4+}(aq) \rightarrow Fe^{3+}(aq) + Ce^{3+}(aq)$$

What is the mass percentage of iron in a sample if titration of 1.2284 g of the sample requires 57.91 mL of 0.1018 M $Ce(NH_4)_2(NO_3)_6$?

4.82 Assign oxidation numbers to each atom in the following substances.
  (a) ethane, C_2H_6, a constituent of natural gas
  (b) borax, $Na_2B_4O_7$, a mineral used in laundry detergents
  (c) $Mg_2Si_2O_6$, a mineral

4.83 Balance the equations for the following reactions in acidic solution.

(a) $PbO_2(s) + Mn^{2+}(aq) \longrightarrow Pb^{2+}(aq) + MnO_4^-(aq)$

(b) $As_2O_3(s) + NO_3^-(aq) \longrightarrow H_3AsO_4(aq) + HNO_2(aq)$

(c) $Br_2(aq) + SO_2(g) \longrightarrow Br^-(aq) + HSO_4^-(aq)$

(d) $NO_2^-(aq) + I^-(aq) \longrightarrow I_2(s) + NO(g)$

4.84 Which of the following ions can be reduced to elemental forms by reaction with iron?

(a) Ni^{2+} (b) Au^{3+}

(c) Zn^{2+} (d) Ba^{2+}

chapter 5 PERIODICITY AND ATOMIC STRUCTURE

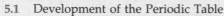

The periodic table is the most important organizing principle in chemistry. If you know the chemical and physical properties of any one element in a group, or vertical column, of the periodic table, you can make a good guess at the chemical and physical properties of every other element in the same group and even of elements in neighboring groups.

To see why it's called the *periodic* table, look at the graph of atomic radius versus atomic number shown in Figure 5.1. The graph shows a clearly periodic, rise-and-fall pattern. Beginning on the left with atomic number 1 (hydrogen), the size of the atoms increases to a maximum at atomic number 3 (lithium), then decreases to a minimum, then increases again to a maximum at atomic number 11 (sodium), then decreases, and so on. It turns out that all the maxima occur for atoms of group 1A elements—Li (atomic number $Z = 3$), Na ($Z = 11$), K ($Z = 19$), Rb ($Z = 37$), Cs ($Z = 55$), and Fr ($Z = 87$)—and that the minima occur for atoms of either the group 7A or group 8A elements.

There's nothing unique about the periodicity of atomic radii shown in Figure 5.1. Any of several dozen other physical or chemical properties can be plotted in a similar way with similar results. We'll look at several examples of such periodicity in this and the next chapter.

The light produced by this antique bulb is an example of blackbody radiation, the visible glow emitted by a heated object.

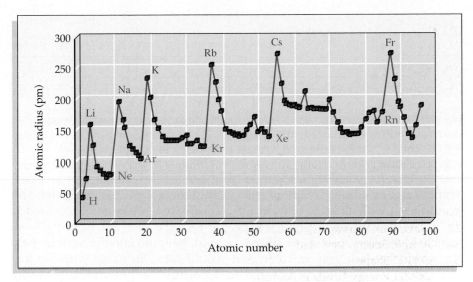

FIGURE 5.1 A graph of atomic radius in picometers (pm) versus atomic number shows a clear rise-and-fall pattern of periodicity. The maxima occur for atoms of group 1A elements (Li, Na, K, Cs, Fr); the minima occur for atoms of either the group 7A or 8A elements.

5.1 ►DEVELOPMENT OF THE PERIODIC TABLE

In many ways, the creation of the periodic table by Dmitri Mendeleev in 1869 is an ideal example of how a scientific theory comes into being. At first there is only random information—a large number of elements and many observations about their properties and behavior. As more and more facts become known, people try to organize the data in ways that make sense, until ultimately a consistent hypothesis emerges. Any good hypothesis must do two things: It must explain known facts, and it must make predictions about phenomena yet unknown. If the predictions are tested and found true, then the hypothesis is a good one and will stand until additional facts require it to be modified or discarded.

Mendeleev's hypothesis about how known chemical information could be organized passed all tests. Not only did the periodic table arrange the data in a useful and consistent way to explain known facts about chemical reactivity, it also led to several remarkable predictions that were later found to be accurate. Taking the chemistry of the elements as his primary organizing principle, Mendeleev listed the known elements by atomic weight and grouped them together according to their chemical reactivity. On so doing, he realized that there were several "holes" in the table, some of which are shown in Figure 5.2. The chemical behavior of aluminum (atomic weight ≈ 27.3) is similar to that of boron (atomic weight ≈ 11), but there was no element known at the time that fit into the slot below aluminum. In the same way, silicon (atomic weight ≈ 28) is similar in many respects to carbon (atomic weight ≈ 12), but there was no element known that fit below silicon.

Looking at the holes in the table, Mendeleev predicted that two then-unknown elements existed and would be found at some future time. Furthermore, he predicted with remarkable accuracy what the properties of these unknown elements would be. The element immediately below alumi-

H = 1							
Li = 7	Be = 9.4			B = 11	C = 12	N = 14	O = 16	F = 19
Na = 23	Mg = 24			Al = 27.3	Si = 28	P = 31	S = 32	Cl = 35.5
K = 39	Ca = 40	?, Ti, V, Cr, Mn, Fe, Co, Ni, Zn		? = 68	? = 72	As = 75	Se = 78	Br = 80

FIGURE 5.2 A portion of Mendeleev's periodic table giving the atomic weights known at the time and showing some of the "holes" representing unknown elements. There is an unknown element (which turned out to be gallium, Ga) underneath aluminum (Al) and another unknown element (which turned out to be germanium, Ge) underneath silicon (Si).

Gallium (top) is a shiny, low-melting metal. Germanium (bottom) is a hard, gray semimetal.

num, which he called *eka*-aluminum from a Sanskrit word meaning "first", would have an atomic weight near 68, would have a low melting point, and would react with chlorine to form a trichloride XCl_3. Gallium, discovered in 1875, has exactly these predicted properties. The element below silicon, which Mendeleev called *eka*-silicon, would have an atomic weight near 72, would be dark gray in color, and would form an oxide with the formula XO_2. Germanium, discovered in 1886, fits the description perfectly (Table 5.1).

The success of these and other predictions convinced other chemists of the usefulness of Mendeleev's periodic table and led to its wide acceptance. Even Mendeleev made some mistakes, though. He was completely unaware of the existence of the group 8A elements—He, Ne, Ar, Kr, Xe, and Rn—because none were known at the time. All are colorless, odorless gases with little or no chemical reactivity, and none were discovered until 1894 when argon was first isolated.

TABLE 5.1   A Comparison of Predicted and Observed Properties for Gallium (*eka*-Aluminum) and Germanium (*eka*-Silicon)

	Mendeleev's Prediction		Observed	
Gallium (*eka*-aluminum)	Atomic weight	~68	Atomic weight	69.72
	Density	~5.9	Density	5.91
	Melting point	Low	Melting point	29.8°
	Formula of oxide	X_2O_3	Formula of oxide	Ga_2O_3
	Formula of chloride	XCl_3	Formula of chloride	$GaCl_3$
Germanium (*eka*-silicon)	Atomic weight	~72	Atomic weight	72.59
	Density	~5.5	Density	5.35
	Color	Dark gray	Color	Light gray
	Formula of oxide	XO_2	Formula of oxide	GeO_2
	Formula of chloride	XCl_4	Formula of chloride	$GeCl_4$

5.2 ▶ LIGHT AND THE ELECTROMAGNETIC SPECTRUM

What fundamental property of atoms is responsible for the periodic variations we observe in atomic radii and in so many other characteristics of the elements? This question occupied the thoughts of chemists for more than 50 years after Mendeleev, and it was not until well into the 1920s that the answer was established. To understand how the answer slowly emerged, we must look first at the nature of visible light and other forms of radiant energy. Historically, studies of the interaction of radiant energy with matter have provided invaluable insights into atomic and molecular structure.

Although they appear quite different to our senses, visible light, infrared radiation, microwaves, radio waves, X rays, and other forms of radiant energy are all different kinds of **electromagnetic radiation**. Collectively, they make up the **electromagnetic spectrum**, shown in Figure 5.3.

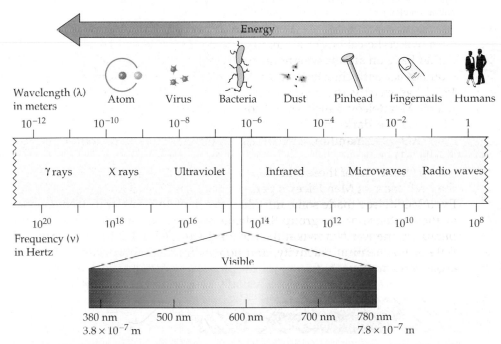

FIGURE 5.3 The electromagnetic spectrum consists of a continuous range of wavelengths and frequencies, from radio waves at the low-frequency end to gamma rays at the high-frequency end. The familiar visible region accounts for only a small portion near the middle of the spectrum. Note that waves in the X-ray region have a length that is approximately the diameter of atoms (10^{-10} m).

Electromagnetic radiation traveling through a vacuum has properties similar to those of an ocean wave traveling through water, and we therefore speak of radiant energy as having wavelike properties. Like an ocean wave, electromagnetic radiation is characterized by a *frequency*, a *wavelength*, and an *amplitude*. If you could look at a sideways, cutaway view of an ocean wave moving through the water, you would see a regular rise-and-fall pattern (Figure 5.4). The **frequency** (v; Greek nu) of a wave is simply the number of peaks (wave maxima) that pass by a fixed point per unit time, usually expressed in units of reciprocal seconds ($1/s$ or s^{-1}), or **hertz**, (**Hz**; $1 \text{ Hz} = 1 \text{ s}^{-1}$). The **wavelength** ($\lambda$; Greek lambda) of the wave is the length from one wave maximum to the next. The **amplitude** of the wave is its height measured from the middle point between peak and trough. (Physically, what we perceive as the *intensity* of radiant energy is proportional to the square of the wave amplitude. A very feeble beam and a blinding glare of light may have the same wavelength and frequency, but they will differ greatly in amplitude.)

Like electromagnetic waves, ocean waves are characterized by a wavelength, a frequency, and an amplitude.

FIGURE 5.4 Electromagnetic waves are characterized by a wavelength, a frequency, and an amplitude. **(a)** Wavelength (λ) is the distance between two successive wave maxima, and frequency (ν) is the number of whole waves that pass a fixed point per unit time. Amplitude is the height of the maximum measured from the center. **(b)** What we perceive as different kinds of electromagnetic radiation are simply waves with different wavelengths and frequencies.

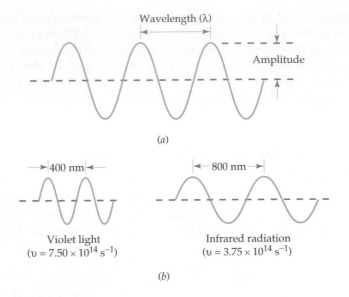

Wavelength (λ)

Amplitude

(a)

400 nm

800 nm

Violet light
($\nu = 7.50 \times 10^{14}\ \text{s}^{-1}$)

Infrared radiation
($\nu = 3.75 \times 10^{14}\ \text{s}^{-1}$)

(b)

Multiplying the wavelength of a wave in meters (m) by its frequency in reciprocal seconds (s^{-1}) gives the speed of the wave in meters per second (m/s). The rate of travel of all electromagnetic radiation in a vacuum is a constant value, commonly called the *speed of light* and abbreviated *c*. It is one of the most accurately known of all physical constants, with a numerical value of $2.997\ 924\ 58 \times 10^{8}$ m/s, usually rounded off to 3.00×10^{8} m/s:

$$\text{wavelength} \times \text{frequency} = \text{speed}$$

$$\lambda\ (\text{m}) \times \nu\ (\text{s}^{-1}) = c\ (\text{m/s})$$

which can be rewritten as: $\lambda = \dfrac{c}{\nu}$ or $\nu = \dfrac{c}{\lambda}$

This equation says that frequency and wavelength are inversely related: Electromagnetic radiation with a long wavelength has a low frequency, and radiation with a short wavelength has a high frequency.

EXAMPLE 5.1

The light blue glow given off by mercury street lamps has a wavelength of 436 nm. What is its frequency?

The bluish light from this mercury street lamp has a wavelength of 436 nm.

SOLUTION   We are given a wavelength and need to find the corresponding frequency. Wavelength and frequency are inversely related by the equation $\lambda\nu = c$, which can be solved for ν. Don't forget to convert from nanometers to meters.

$$\nu = \frac{c}{\lambda} = \frac{3.00 \times 10^{8}\ \frac{\text{m}}{\text{s}}}{436\ \text{nm}} \times \frac{10^{9}\ \text{nm}}{\text{m}}$$

$$= 6.88 \times 10^{14}\ \text{s}^{-1} = 6.88 \times 10^{14}\ \text{Hz}$$

The frequency of the light is $6.88 \times 10^{14}\ \text{s}^{-1}$, or 6.88×10^{14} Hz.

┌ **PROBLEM 5.1** What is the frequency of a gamma ray with $\lambda = 3.56 \times 10^{-11}$ m? Of a radar wave with $\lambda = 10.3$ cm?

┌ **PROBLEM 5.2** What is the wavelength in meters of an FM radio wave with a frequency $\nu = 102.5$ MHz? Of a medical X ray with $\nu = 9.55 \times 10^{17}$ Hz?

┌ **PROBLEM 5.3** Look at Figure 5.3 and tell which color of visible light has the highest frequency? Which color has the longest wavelength?

5.3 ►ELECTROMAGNETIC RADIATION AND ATOMIC SPECTRA

The light that we see from the sun or from a light bulb is "white" light, meaning that it consists of an essentially continuous distribution of wavelengths spanning the entire visible region of the electromagnetic spectrum. When a narrow beam of white light is passed through a glass prism, the different wavelengths travel through the glass at different rates. As a result, the white light is separated into its component colors, ranging from red at the long-wavelength end of the spectrum (700 nm) to violet at the short-wavelength end (400 nm) [Figure 5.5(a)]. A similar separation into colors occurs when light travels through water droplets in the air, forming rainbows.

(a)   (b)

FIGURE 5.5 **(a)** When a narrow beam of ordinary white light is passed through a glass prism, different wavelengths travel through the glass at different rates and appear as a continuous broad band of different colors. **(b)** A similar effect occurs when light passes through water droplets in the air, forming rainbows.

What do visible light and other kinds of electromagnetic radiation have to do with atomic structure? The answer involves the fact that individual atoms give off visible light when heated or otherwise excited energetically, thereby providing a clue to their atomic makeup.[1] Unlike the white light from the sun, though, the light given off by an energetically excited atom is not a continuous distribution of wavelengths. When passed first through a narrow slit and then through a prism, the visible light given off by an excited atom is found to consist of only a few wavelengths rather than a

[1] Actually, energetically excited atoms give off electromagnetic radiation at many different wavelengths, not just in the visible range. We'll develop this point shortly.

The brilliant colors of these fireworks are due to the emission of light by excited metal atoms.

full rainbow of colors, giving a series of discrete lines separated by blank areas—a **line spectrum**. Excited sodium atoms produced by heating NaCl or some other sodium salt in the flame of a Bunsen burner give off yellow light [Figure 5.6(a)], potassium atoms give off purple light, barium atoms give off green light, and so on. In fact, the brilliant colors of fireworks are produced by adding mixtures of different metal salts to explosive powder.

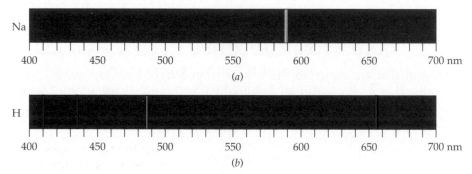

FIGURE 5.6 (a) The visible line spectrum of energetically excited sodium atoms consists of a single yellow line. (b) The visible line spectrum of excited hydrogen atoms consists of four lines.

Excited neon atoms emit orange light, and excited hydrogen atoms give off a bluish light.

Soon after the initial discovery that energetic atoms emit light of specific wavelengths, chemists began cataloging the line spectra of various elements. They rapidly found that each element has its own unique spectral "signature," and they began using the results as a method for identifying elements present in minerals and other substances. Not until the work of the Swiss schoolteacher Johann Balmer in 1885, though, was a pattern discovered in atomic line spectra. Working with a glass tube containing hydrogen gas at low pressure, Balmer noticed that the gas emitted a bluish glow when an electric discharge was passed through it, giving a spectrum with four lines against a black background, as shown in Figure 5.6(b). The wavelengths of the four lines are 656.3 nm (red), 486.1 nm (green), 434.0 nm (blue), and 410.1 nm (indigo).

After thinking about his data and trying by trial and error to organize it in various ways, Balmer discovered that the wavelengths of all four lines in the hydrogen spectrum could be expressed by the equation

$$\frac{1}{\lambda} = R\left[\frac{1}{4} - \frac{1}{n^2}\right] \quad \text{or} \quad \nu = R \cdot c\left[\frac{1}{4} - \frac{1}{n^2}\right]$$

where R is a constant (now called the *Rydberg constant*) equal to 1.097×10^{-2} nm^{-1} and n is an integer greater than 2. The red spectral line at 656.3 nm, for example, results from Balmer's equation when $n = 3$:

$$\frac{1}{\lambda} = [1.097 \times 10^{-2} \text{ nm}^{-1}]\left[\frac{1}{4} - \frac{1}{3^2}\right] = 1.524 \times 10^{-3} \text{ nm}^{-1}$$

$$\lambda = \frac{1}{1.524 \times 10^{-3} \text{ nm}^{-1}} = 656.3 \text{ nm}$$

Similarly, a value of $n = 4$ gives the green line at 486.1 nm, a value of $n = 5$ gives the blue line at 434.0 nm, and so on. Solve Balmer's equation yourself to make sure.

Subsequent to the discovery of the Balmer series of lines in the *visible* region of the electromagnetic spectrum, it was found that a host of other spectral lines are also present in *nonvisible* regions of the electromagnetic spectrum. Hydrogen, for example, shows a series of spectral lines called the *Lyman series* in the far ultraviolet region and still other series (the *Paschen*, *Brackett*, and *Pfund series*) in the infrared region.

By adapting Balmer's equation, the Swedish physicist Johannes Rydberg was able to show that every line in the entire spectrum of hydrogen can be fit by the generalized Balmer-Rydberg equation

Balmer-Rydberg equation
$$\frac{1}{\lambda} = R\left[\frac{1}{m^2} - \frac{1}{n^2}\right] \quad \text{or} \quad \nu = R \cdot c\left[\frac{1}{m^2} - \frac{1}{n^2}\right]$$

where m and n are integers with $n > m$. If $m = 1$, then the Lyman series of lines results. If $m = 2$ so that $m^2 = 4$, then Balmer's series of visible lines results. If $m = 3$, then the Paschen series is described, and so forth for still larger values of m. Some of these other spectral lines are calculated in Example 5.2.

EXAMPLE 5.2

What are the two longest wavelength lines (in nm) in the Lyman series of the hydrogen spectrum?

SOLUTION The Lyman series is given by the Balmer-Rydberg equation with $m = 1$ and $n > 1$.

$$\frac{1}{\lambda} = R\left[\frac{1}{m^2} - \frac{1}{n^2}\right] \quad \text{where } m = 1$$

The longest wavelength lines are those where $n = 2$ and $n = 3$.
    Solving the equation first for $n = 2$ gives

$$\frac{1}{\lambda} = R\left[\frac{1}{1^2} - \frac{1}{2^2}\right] = (1.097 \times 10^{-2} \text{ nm}^{-1}) \times \left(1 - \frac{1}{4}\right) = 8.228 \times 10^{-3} \text{ nm}^{-1}$$

or
$$\lambda = \frac{1}{8.228 \times 10^{-3} \text{ nm}^{-1}} = 121.5 \text{ nm}$$

Solving the equation next for $n = 3$ gives

$$\frac{1}{\lambda} = R\left[\frac{1}{1^2} - \frac{1}{3^2}\right] = (1.097 \times 10^{-2} \text{ nm}^{-1}) \times \left(1 - \frac{1}{9}\right) = 9.751 \times 10^{-3} \text{ nm}^{-1}$$

or
$$\lambda = \frac{1}{9.751 \times 10^{-3} \text{ nm}^{-1}} = 102.6 \text{ nm}$$

The two longest wavelength lines in the Lyman series are at 121.5 nm and 102.6 nm.

EXAMPLE 5.3

What is the shortest wavelength line (in nm) in the Lyman series of the hydrogen spectrum?

SOLUTION   The Lyman series is given by the Balmer-Rydberg equation with $m = 1$ and $n > 1$. The shortest wavelength line occurs when n is infinitely large so that $1/n^2$ is zero. That is, if $n = \infty$, then $1/n^2 = 0$. Thus, the equation becomes

$$\frac{1}{\lambda} = R\left[\frac{1}{1^2} - \frac{1}{\infty^2}\right] = (1.097 \times 10^{-2} \text{ nm}^{-1}) \times (1 - 0) = 1.097 \times 10^{-2} \text{ nm}^{-1}$$

or
$$\lambda = \frac{1}{1.097 \times 10^{-2} \text{ nm}^{-1}} = 91.2 \text{ nm}$$

⌐ **PROBLEM 5.4**   Although unknown to Balmer at the time, his equation can be extended beyond the visible portion of the electromagnetic spectrum to include lines in the ultraviolet. What is the wavelength in nm of ultraviolet light in the Balmer series corresponding to a value of $n = 7$?

⌐ **PROBLEM 5.5**   What is the longest wavelength line (in nm) in the Paschen series for hydrogen?

⌐ **PROBLEM 5.6**   What is the shortest wavelength line (in nm) in the Paschen series for hydrogen?

5.4 ➤PARTICLELIKE PROPERTIES OF ELECTROMAGNETIC RADIATION: THE PLANCK EQUATION

The red glow given off by this molten iron is an example of blackbody radiation.

The existence of atomic line spectra and the fit of the hydrogen spectrum to the Balmer-Rydberg equation implied the existence of a general underlying principle about atomic structure, yet it was many years before that principle was found. One of the key discoveries came in 1900, when the German physicist Max Planck proposed a theory to explain a seemingly unrelated phenomenon called *blackbody radiation*—the visible glow that solid objects give off when heated. The reddish glow from the heating element in an electric stove and the white light emitted by the hot filament in a light bulb are two examples of blackbody radiation.

Experimentally, it's found that the intensity of blackbody radiation varies with the wavelength of the emitted light. When an object such as an iron bar is heated, it first begins to glow a dull red but then changes to a brighter orange and ultimately to a blinding white as its temperature increases. Thus, longer wavelengths around 700 nm (red) have a lower intensity, and shorter wavelengths have a higher intensity. If this trend continues, one might expect the intensity to keep rising indefinitely as the wavelength becomes ever shorter and enters the ultraviolet region. In fact, though, the intensity of blackbody radiation does not continue rising indefinitely at shorter wavelengths. Instead, the intensity reaches a maximum and then falls rapidly at wavelengths shorter than about 500 nm (Figure 5.7).

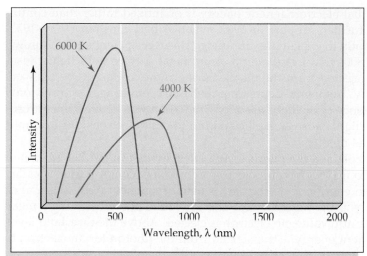

Intensity →

6000 K

4000 K

Wavelength, λ (nm)

0   500   1000   1500   2000

FIGURE 5.7 The dependence of the intensity of blackbody radiation on wavelength at two different temperatures. Intensity increases from right to left on the curve as wavelength decreases, reaches a maximum, and then drops off to zero as the wavelength continues to decrease.

To explain the observation that the intensity of blackbody radiation does not continue to rise indefinitely as the wavelength decreases, Planck concluded that the energy radiated by a heated object can't be continuously variable but instead must be *quantized*. That is, radiant energy is not emitted in amounts of any allowable size but only in discrete units, or **quanta**. Furthermore, the amount of energy, *E*, associated with each quantum depends on the frequency of the emitted radiation, *ν*, according to the equation

$$E = h\nu$$

or, since $\nu = c/\lambda$,

$$E = \frac{hc}{\lambda}$$

The symbol *h* is a fundamental physical constant that we now call **Planck's constant** and that has the value $h = 6.626 \times 10^{-34}$ J·s. For example, one quantum of red light with a frequency $\nu = 4.62 \times 10^{14}$ s^{-1} (wavelength λ = 649 nm) has an energy in joules[2] of 3.06×10^{-19} J:

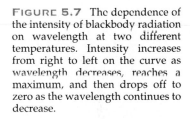

Medical X rays have short wavelengths on the order of 1 nm; FM radio waves have wavelengths near 10 m.

$$E = h\nu = (6.626 \times 10^{-34} \text{ J·s}) \times (4.62 \times 10^{14} \text{ s}^{-1}) = 3.06 \times 10^{-19} \text{ J}$$

Note that higher frequencies and shorter wavelengths correspond to higher energy radiation, while lower frequencies and longer wavelengths correspond to lower energy. Blue light (λ ≈ 450 nm), for instance, has a shorter wavelength and is more energetic than red light (λ ≈ 650 nm). Similarly, an X ray (λ ≈ 1 nm) has a shorter wavelength and is more energetic than an FM radio wave (λ ≈ 10^{10} nm, or 10 m).

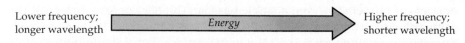

Lower frequency; longer wavelength

Energy

Higher frequency; shorter wavelength

[2] As noted in Section 1.8, the joule, abbreviated J, is the SI unit for energy (1 J = 1 kg·m^2/s^2). The joule is a fairly small amount of energy—it takes 100 J to light a 100 watt light bulb for 1 second.

The idea that electromagnetic energy is quantized rather than continuous received further support in 1905, when Albert Einstein (1879–1955) used it successfully to explain the **photoelectric effect**. Scientists had known since the late 1800s that irradiating a clean metal surface with light causes electrons to be ejected from the metal. Furthermore, different metals were known to show frequency-dependent behavior, emitting electrons only when the frequency of the light used for irradiation was above some threshold value. Blue light causes metallic sodium to emit electrons, for example, but red light has no effect.

Einstein explained the photoelectric effect by assuming that a beam of light behaves as if it were composed of a stream of small particles called **photons**, whose energy, E, is related to their frequency, v, by the Planck equation, $E = hv$. If the frequency (or energy) of the photon striking a metal is below a minimum value, no electron is ejected. Above the threshold level, however, sufficient energy is transferred from the photon for an electron to overcome the attractive forces holding it to the metal (Figure 5.8).

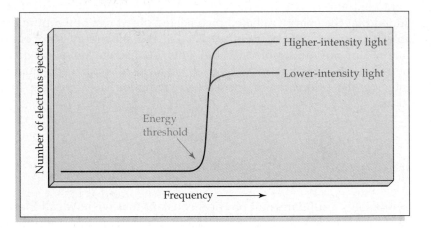

FIGURE 5.8 The photoelectric effect. A plot of the number of electrons ejected from a metal surface versus light frequency shows a threshold value. Increasing the intensity of the light while keeping the frequency the same increases the number of electrons ejected but does not change the threshold value.

Note that the energy of an individual photon depends only on its frequency, not on the intensity (amplitude) of the light beam. The intensity of a light beam is a measure of the *number* of photons in the beam, not of the *energies* of those photons. A low-intensity beam of high-energy photons might easily knock a few electrons loose from a metal, but a high-intensity beam of low-energy photons might not be able to knock loose a single electron.

As a rough analogy, think of throwing balls of different masses at a glass window. A thousand Ping-Pong balls (lower energy) would only bounce off the window, but a single baseball (higher energy) thrown at the same velocity as the Ping-Pong balls would break the glass. In the same way, low-energy photons bounce off the metal surface, but a single photon at or above a certain threshold energy can ''break'' the metal and dislodge an electron.

The main conclusion from both Planck's and Einstein's work was that the behavior of light and other forms of electromagnetic radiation is more complex than had been formerly believed. *In addition to behaving as waves, light energy can also behave as small particles.* The idea isn't really so strange if you think of light as analogous to matter. The amount of energy corresponding to one quantum of light is almost inconceivably small, just as the amount of matter in one atom is inconceivably small, but the principle is the same: *Both matter and energy occur only in discrete units.* Just as there can be either 1 or 2 hydrogen atoms but not 1.5 or 1.8 atoms, there can be 1 or 2 photons of light but not 1.5 or 1.8.

Once the quantized nature of electromagnetic radiation is accepted, part of the puzzle of atomic line spectra is explained. Energetically excited atoms evidently are not able to emit light of continuously varying wavelengths and therefore don't give a continuous spectrum. The atoms are somehow constrained to emit light quanta (photons) of only a few specific energies, and they therefore give a line spectrum. Why this should be so is the next question to answer.

Stairs are quantized, changing height only in discrete amounts. A ramp, by contrast, changes height continuously.

EXAMPLE 5.4

What is the energy in kJ of 1 mol of photons of FM radio waves with $\nu = 102.5$ MHz?

SOLUTION   The energy of a photon with frequency ν can be calculated with the Planck equation $E = h\nu$:

$$E = h\nu = (6.626 \times 10^{-34} \text{ J} \cdot \text{s})(102.5 \times 10^6 \text{ s}^{-1}) = 6.792 \times 10^{-26} \text{ J}$$

To find the energy of one mole of photons, the energy of one photon must be multiplied by Avogadro's number:

$$\left(6.792 \times 10^{-26} \frac{\text{J}}{\text{photon}}\right)\left(6.022 \times 10^{23} \frac{\text{photon}}{\text{mol}}\right) = 4.090 \times 10^{-2} \text{ J/mol}$$

$$= 4.090 \times 10^{-5} \text{ kJ/mol}$$

PROBLEM 5.7   In Example 5.3, we calculated the shortest wavelength line in the Lyman series for hydrogen. What is the energy in kJ of 1 mol of photons with this wavelength?

PROBLEM 5.8   The biological effects of a given dose of electromagnetic radiation generally become more serious as the energy of the radiation increases: Infrared radiation has a pleasant warming effect; ultraviolet radiation causes tanning and burning; and X rays can cause considerable tissue damage. What energies in kJ/mol are associated with the following wavelengths: infrared radiation with $\lambda = 1.55 \times 10^{-6}$ m? ultraviolet light with $\lambda = 250$ nm? X rays with $\lambda = 5.49$ nm?

Ultraviolet radiation from the sun can lead to the tissue damage we associate with sunburn.

5.5 ▶WAVELIKE PROPERTIES OF MATTER: THE DE BROGLIE EQUATION

The analogy between matter and radiant energy developed by Planck and Einstein in the early 1900s was further extended in 1924 by the French physicist Louis de Broglie. de Broglie suggested that if *light* can behave in some respects like *matter*, then perhaps *matter* can behave in some respects like *light*. That is, perhaps *both* light and matter are wavelike as well as particlelike.

To help construct his argument about the wavelike behavior of matter, de Broglie used the now-famous equation $E = mc^2$, which had been proposed by Einstein in 1905 to predict the mass (m) of photons of wavelength λ emitted during the radioactive decay of radium and other elements.

Since $$E = mc^2$$

then $$m = \frac{E}{c^2}$$

Einstein had seen that, since $E = hc/\lambda$ according to the Planck equation, it is possible to substitute for E to derive a relationship between mass and wavelength:

$$m = \frac{E}{c^2} = \frac{hc/\lambda}{c^2} = \frac{h}{\lambda c}$$

de Broglie suggested that a similar equation might be applied to an *electron* instead of a photon by replacing the speed of the photon (c) by the speed of the electron (v). The resultant **de Broglie equation** thus allows calculation of the "wavelength" of an electron or of any other particle or object of mass m moving at velocity v:

de Broglie equation $$m = \frac{h}{\lambda v} \quad \text{or} \quad \lambda = \frac{h}{mv}$$

For example, the mass of an electron is 9.11×10^{-31} kg, and the velocity v of an electron in a hydrogen atom is 2.18×10^6 m/s (about 1% of the speed of light). Thus, the de Broglie wavelength of an electron in a hydrogen atom is 3.34×10^{-10} m, or 334 pm:

$$\lambda = \frac{h}{mv} = \frac{6.626 \times 10^{-34} \, \frac{\text{kg} \cdot \text{m}^2}{\text{s}}}{(9.11 \times 10^{-31} \, \text{kg})\left(2.18 \times 10^6 \, \frac{\text{m}}{\text{s}}\right)} = 3.34 \times 10^{-10} \text{ m}$$

Note that Planck's constant, which is usually expressed in units of joule seconds (J · s), is expressed for the present purposes in units of (kg · m^2)/s [1 J = 1 (kg · m^2)/s^2]. Note also that the calculated de Broglie wavelength of the electron in a hydrogen atom, 334 pm, is larger than the diameter of an isolated H atom itself (about 240 pm). We'll return to this point in the next section.

What does it mean to say that light and matter act both as waves and as particles? The answer is "not much," at least not on the everyday human scale. The problem in trying to understand the dual wave/particle description of light and matter is that our common sense isn't up to the task. Our intuition has been developed from personal experiences, using our eyes and other senses to tell us how light and matter are "supposed" to behave. We have no personal experience on the atomic scale, though, and thus have no common-sense way of dealing with the behavior of light and matter at that level. On the atomic scale, where distances and masses are so tiny, light and matter behave in a manner different from what we are used to.

The dual wave/particle description of light and matter is really just a mathematical *model*. Since we can't see atoms and observe their behavior directly, the best we can do is to construct a set of mathematical equations that correctly account for atomic properties and behavior. The wave/particle description does this extremely well, even though it is not easily understood using day-to-day experience.

EXAMPLE 5.5

What is the de Broglie wavelength in meters of a pitched baseball with a mass of 120 g and a speed of 100 mph (44.7 m/s)?

SOLUTION   The de Broglie relationship says that wavelength λ of an object with mass m moving at a velocity v can be calculated by the equation $\lambda = h/mv$.

$$\lambda = \frac{h}{mv} = \frac{6.626 \times 10^{-34} \frac{\text{kg} \cdot \text{m}^2}{s}}{(0.120 \text{ kg})\left(44.7 \frac{m}{s}\right)} = 1.24 \times 10^{-34} \text{ m}$$

The batter has no excuse. He didn't miss because of the ball's de Broglie wavelength.

The de Broglie wavelength of the baseball is 1.24×10^{-34} m.

PROBLEM 5.9   What is the de Broglie wavelength in meters of a small car with a mass of 1150 kg traveling at a speed of 55.0 mi/h (24.6 m/s)?

5.6 ►THE QUANTUM MECHANICAL DESCRIPTION OF THE HYDROGEN ATOM

With the quantized nature of energy and the wavelike nature of matter now established, let's return to the problem of atomic structure. Several models of atomic structure were proposed in the late nineteenth and early twentieth centuries. A model proposed in 1914 by the Danish physicist Neils Bohr (1885–1962), for example, described the hydrogen atom as a nucleus with an electron circling it in a specific path. Unfortunately, the Bohr model failed for atoms with more than one electron.

The breakthrough in our understanding of atomic structure came in 1926, when the Austrian physicist Erwin Schrödinger (1887–1961) proposed what has come to be called the *wave mechanical model*, or **quantum mechanical model**, of the atom. The fundamental idea behind Schrödinger's quantum mechanical model is that it is best to abandon the notion of an electron as a small particle moving around the nucleus in a defined path and to concentrate instead on the electron's wavelike properties. (Remember that the de Broglie wavelength of the electron in a hydrogen atom is greater than the diameter of the atom.) In fact, it was shown by Werner Heisenberg (1901–1976) in 1927 that it is *impossible* to know precisely where the electron is and what path it follows.

The **Heisenberg uncertainty principle** can be understood in simple terms by imagining what would happen if we tried to determine the position of an electron at a given moment. For us to "see" the electron, light photons of an appropriate frequency would have to interact with and bounce off the electron before our "eye" could detect it. But the interaction

Werner Heisenberg. Where is this man? His whereabouts aren't known with certainty.

process would result in a transfer of energy from the photon to the electron, thereby increasing the energy of the electron and making it move faster. Thus, the very act of determining the electron's position would make that position change.

In mathematical terms, Heisenberg's principle states that the uncertainty in the electron's position, Δx, times the uncertainty in its momentum, Δmv, is equal to or greater than the quantity $h/4\pi$:

$$\textit{Heisenberg uncertainty principle} \qquad (\Delta x)(\Delta mv) \geq \frac{h}{4\pi}$$

The equation says that we can never know both the position and the velocity of an electron (or of any other object) beyond a certain level of precision. If we know the electron's *velocity* with a high degree of certainty (Δmv is small), then the *position* of the electron must be uncertain (Δx must be large). Conversely, if we know the *position* of the electron exactly (Δx is small), then we can't know its *velocity* (Δmv must be large). As a result, an electron will always appear as something of a blur whenever we attempt to make any physical measurements of its position and velocity.

A brief calculation can help put the conclusions of the uncertainty principle more clearly. As mentioned in the previous section, the mass m of an electron is 9.11×10^{-31} kg and the velocity v of an electron in a hydrogen atom is 2.18×10^6 m/s. If we assume that the velocity is known precisely to within 10%, or 0.2×10^6 m/s, then we can calculate that the uncertainty in the electron's position in a hydrogen atom is greater than 3×10^{-10} m, or 300 pm:

$$\text{If } (\Delta x)(\Delta mv) \geq \frac{h}{4\pi}, \quad \text{then } (\Delta x) \geq \frac{h}{(4\pi)(\Delta mv)}$$

$$(\Delta x) \geq \frac{6.626 \times 10^{-34}\,\frac{\cancel{\text{kg}} \cdot \cancel{\text{m}^2}}{\cancel{\text{s}}}}{(4)(3.1416)(9.11 \times 10^{-31}\,\cancel{\text{kg}})\left(0.2 \times 10^6\,\frac{\cancel{\text{m}}}{\cancel{\text{s}}}\right)}$$

$$(\Delta x) \geq 3 \times 10^{-10}\text{ m}$$

But since the diameter of a hydrogen atom is only 240 pm, *the uncertainty in the electron's position is approximately the size of the atom itself!*

When the mass m of an object is relatively large, as is true in daily life, then Δx and Δv in the Heisenberg relationship can *both* be very small. We therefore have no apparent problem in measuring both position and velocity for visible objects. The problem arises only on the atomic scale. Example 5.6 gives a sample calculation.

Schrödinger's quantum mechanical model of atomic structure is framed mathematically in the form of a *wave equation* similar to that used to describe ordinary wave motion. The solutions (there are many) to the wave equation are called **wave functions**, or **orbitals**, and are represented by the symbol ψ (Greek *psi*). Each wave function has a specific energy associated with it, and each contains information about an electron's position in three-dimensional space. The best way to think about a wave function is to regard

Even the motion of very fast objects can be captured in daily life. On the atomic scale, however, velocity and position can't both be known precisely.

it as a quantity whose square, ψ^2, gives the *probability* of finding an electron within a given region in space. (Remember: Heisenberg showed that we can never be completely certain about an electron's position; the electron will always look like a blur.)

$$\underset{\text{equation}}{\text{Wave}} \xrightarrow{\text{solve}} \underset{\substack{\text{or orbital } (\psi)}}{\text{Wave function}} \longrightarrow \underset{\substack{\text{electron in a region}\\ \text{of space } (\psi^2)}}{\textbf{Probability of finding}}$$

Let's take the simplest wave function for the hydrogen atom. The lowest energy level available to an electron in the hydrogen atom is associated with the wave function corresponding to what is called a 1*s* orbital. As displayed in Figure 5.9 in three formats, an electron in a 1*s* orbital will be found within a spherical region of space surrounding the nucleus. Parts (a) and (b) in Figure 5.9 are two-dimensional plots that show how far from the nucleus the electron is most likely to be found. Part (c) gives a three-dimensional view of the spherical region occupied by the electron. Note that the value of ψ^2 is greatest near the nucleus, indicating that this is where the electron is most likely to be found. The probability of finding the electron drops off rapidly as distance from the nucleus increases, although it never goes all the way to zero, even at large distances. As a result, there is no definite boundary to the atom and no definite "size." For purposes like that of Figure 5.9(c), however, we usually imagine a boundary surface enclosing the volume where an electron spends *most* (say, 95%) of its time.

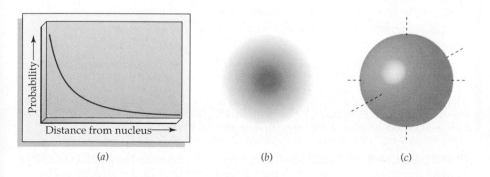

(a)   (b)   (c)

FIGURE 5.9 Three representations of a 1*s* orbital. The plot in **(a)** represents the probability of finding the electron as a function of its distance from the nucleus. The variable density of the dots in **(b)** represents the probability of the electron's being at those locations, and the spherical surface in **(c)** arbitrarily encloses the volume in which the electron spends 95% of its time.

The wave mechanical model of atomic structure just described always raises many questions in the minds of people exposed to it for the first time: How does an electron get from one place to another in an atom? What path does it follow? The problem with these and similar questions is that they're unanswerable because they assume *particlelike* behavior for the electron. The wave mechanical model, however, specifically rejects a particlelike description of the electron and assumes instead a description of an electron as being *wavelike*. The best we can do is to define a volume of space around the nucleus where there is a high probability of finding an electron. We can say nothing about the electron's path or movement.

EXAMPLE 5.6

Assume that you are traveling at a speed of 90 km/h in a small car with a mass of 1250 kg. If the uncertainty in the velocity of the car is 1% (Δv = 0.9 km/h), what is the uncertainty in meters in the position of the car? How does this compare with the uncertainty in the position of an electron in a hydrogen atom?

How certain is the driver of the car's position?

SOLUTION   The Heisenberg relationship says that the uncertainty in an object's position, Δx, times the uncertainty in its momentum, Δmv, is equal to or greater than the quantity $h/4\pi$. In the present instance, we need to find Δx when Δv is known:

$$(\Delta x) \geq \frac{h}{(4\pi)\,(\Delta mv)}$$

$$(\Delta x) \geq \frac{6.626 \times 10^{-34}\ \frac{\text{kg}\cdot\text{m}^2}{\text{s}}}{(4)\,(3.1416)\,(1250\ \text{kg})\left(0.9\ \frac{\text{km}}{\text{h}}\right)\left(\frac{1\,\text{h}}{3600\ \text{s}}\right)\left(\frac{1000\ \text{m}}{1\ \text{km}}\right)}$$

$$(\Delta x) \geq 2 \times 10^{-37}\ \text{m}$$

The uncertainty in the position of the car is 2×10^{-37} m, far smaller than the uncertainty in the position of an electron in a hydrogen atom (3×10^{-10} m), and far too small a value to have any measurable consequences.

⌐ **PROBLEM 5.10** Calculate the uncertainty in meters in the position of a 120 g baseball thrown at a velocity of 45 m/s if the uncertainty in the velocity is 2%. ⌐

5.7 ▶WAVE FUNCTIONS AND QUANTUM NUMBERS

The behavior of each electron in an atom is described mathematically by a wave function, or *orbital*, and each wave function contains a set of three variables, called **quantum numbers** n, l, and m_l. These quantum numbers describe the energy level of an orbital and define the shape and orientation of the region in space where the electron is most likely to be found.

1. The **principal quantum number (n)** is a positive integer ($n = 1, 2, 3, 4, \ldots$) on which the size and energy level of the orbital primarily depend. For hydrogen and other one-electron atoms such as He^+, the energy of an orbital depends *only* on n. For atoms with more than one electron, the energy level of an orbital depends both on n and on the value of the l quantum number.

   As the value of n increases, the number of allowed orbitals increases and the size of the orbitals becomes larger, thus allowing an electron to be farther from the nucleus. Because it takes energy to separate a negative charge from a positive charge, this increased distance between the electron and the nucleus means that the energy of the electron in the orbital increases as the quantum number n increases.

   We often speak of orbitals as being grouped according to the principal quantum number n into successive layers or **shells** around the atom. Those orbitals with $n = 3$, for example, are said to be in the third shell.

Electrons occupy layers in atoms in analogy with the layers in this onion.

2. The **angular-momentum quantum number (l)** defines the three-dimensional shape of the orbital. For an orbital whose principal quantum number is n, the angular-momentum quantum number l can have any integral value from 0 to $n - 1$:

$$\text{If } n = 1, \quad \text{then } l = 0$$

$$\text{If } n = 2, \quad \text{then } l = 0, \text{ or } 1$$

$$\text{If } n = 3, \quad \text{then } l = 0, 1, \text{ or } 2$$
$$\vdots$$

Thus, within each shell, there are n different shapes for orbitals.

Just as it's convenient to think of orbitals as being grouped into shells according to the principal quantum number n, we often speak of orbitals as being grouped into **subshells** according to the angular-momentum quantum number l. Different subshells are usually referred to by letter rather than by number, following the order s, p, d, f, g.[3]

quantum number l:	0	1	2	3	4	...
subshell notation:	s	p	d	f	g	...

As an example, we might speak of an orbital with $n = 3$ and $l = 2$ as being a $3d$ orbital: 3 to represent the third shell and d to represent the $l = 2$ subshell.

---

[3] The letters s, p, d, and f arose historically from the use of the words *sharp, principal, diffuse,* and *fundamental* to describe various lines in atomic spectra. After f, successive subshells are designated alphabetically: $\ldots f$, g, h, and so on.

3. The **magnetic quantum number (m_l)** defines the spatial orientation of the orbital along a standard set of coordinate axes. For an orbital whose angular-momentum quantum number is l, the magnetic quantum number m_l can have any integral value from $-l$ to $+l$:

$$\text{If } l = 0, \text{ then } m_l = 0$$

$$\text{If } l = 1, \text{ then } m_l = -1, 0, \text{ or } +1$$

$$\text{If } l = 2, \text{ then } m_l = -2, -1, 0, +1, \text{ or } +2$$
$$\vdots$$

Thus, within each group of orbitals having the same principal quantum number n and the same shape (same value of l), there are $2l + 1$ different spatial orientations for those orbitals. We'll explore this point further in the next section.

A summary of the allowed combinations of quantum numbers for the first four shells is given in Table 5.2.

TABLE 5.2	Allowed Combinations of Principal, Angular-Momentum, and Magnetic Quantum Numbers for the First Four Shells				
n	l	m_l	Orbital Notation	Number of Orbitals in Subshell	Number of Orbitals in Shell
1	0	0	1s	1	1
2	0	0	2s	1	4
	1	−1, 0 +1	2p	3	
3	0	0	3s	1	9
	1	−1, 0 +1	3p	3	
	2	−2, −1, 0 +1, +2	3d	5	
4	0	0	4s	1	16
	1	−1, 0 +1	4p	3	
	2	−2, −1, 0 +1, +2	4d	5	
	3	−3, −2, −1, 0 +1, +2, +3	4f	7	

The energy levels of various orbitals are shown in Figure 5.10. As noted earlier in this section, the energy levels of different orbitals in a hydrogen atom depend only on the principal quantum number n, but the energy levels of orbitals in multielectron atoms depend on both n and l. In other words, orbitals in all subshells within a shell have the same energy for hydrogen but have slightly different energies for other atoms. In fact, there is even some crossover of energies from one shell to another. A 3d orbital in a multielectron atom has a higher energy than a 4s orbital, for instance.

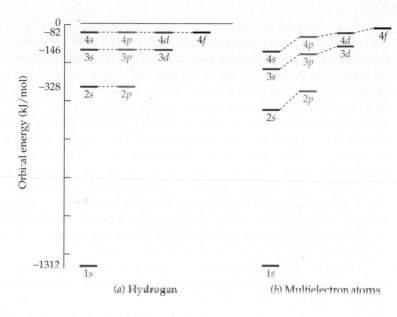

FIGURE 5.10 Orbital energy levels for **(a)** hydrogen and **(b)** multielectron atoms. The differences between energy levels of various subshells in **(b)** are exaggerated for clarity. Note, though, that there is some "crossover" of energies from one shell to another. A 3*d* orbital has a higher energy than a 4*s* orbital, for instance.

EXAMPLE 5.7

Identify the shell and subshell of an electron with the quantum numbers $n = 3$, $l = 1$, $m_l = 1$.

SOLUTION The principal quantum number $n = 3$ indicates that the orbital is in the third shell, and the angular-momentum quantum number $l = 1$ indicates that the orbital is of the *p* type. Thus, the orbital has the designation 3*p*. The magnetic quantum number m_l is related only to the spatial orientation of the orbital.

PROBLEM 5.11 Extend Table 5.2 to show allowed combinations of quantum numbers when $n = 5$. How many orbitals are there in the fifth shell?

PROBLEM 5.12 Why can't an electron have the following quantum numbers?

(a) $n = 2$, $l = 2$, $m_l = 1$ **(b)** $n = 3$, $l = 0$, $m_l = 3$ **(c)** $n = 5$, $l = -2$, $m_l = 1$

PROBLEM 5.13 Give orbital notations for electrons with the following quantum numbers.

(a) $n = 2$, $l = 1$, $m_l = 1$ **(b)** $n = 4$, $l = 3$, $m_l = -2$ **(c)** $n = 3$, $l = 2$, $m_l = -1$

PROBLEM 5.14 Give the possible combinations of quantum numbers for the following orbitals.

(a) a 3*s* orbital   **(b)** a 2*p* orbital   **(c)** a 4*d* orbital

5.8 ►ATOMIC SPECTRA REVISITED

Now that we've seen how atoms are constructed according to the wave-mechanical model, let's return briefly to the subject of atomic line spectra (Section 5.3). How does the wave-mechanical model account for the discrete lines in a line spectrum?

Each electron in an atom occupies an orbital, and each orbital has a specific energy. Thus, *the energies available to electrons are quantized* and can have only the specific values associated with the orbital they occupy. When an atom is heated in a flame or electric discharge, the added energy causes an electron to jump from a lower-energy orbital to a higher-energy orbital. In a hydrogen atom, for example, the electron might jump from the 1s orbital to a second-shell orbital, to a third-shell orbital, or to *any* higher-shell orbital depending on the amount of energy added. The energetically excited atom is relatively unstable, though, and the electron rapidly returns to a lower-energy level accompanied by *emission* of an amount of energy equal to the energy difference between the higher and lower orbitals. *Since the energies of the orbitals are quantized, the amount of energy emitted is also quantized.* Thus, we observe the emission of only specific frequencies of radiation (Figure 5.11). By measuring the frequencies emitted by excited hydrogen atoms, we can calculate the energy differences between orbitals.

It turns out that the variables m and n in the Balmer-Rydberg equation for hydrogen (Section 5.3) correspond to the principal quantum numbers of the two orbitals involved in the electronic transition. The variable n corresponds to the principal quantum number of the outer-shell orbital that the transition is *from*, and the variable m corresponds to the principal quantum number of the inner-shell orbital that the transition is *to*. When $m = 1$ (the Lyman series), for example, the frequencies of emitted light correspond to energy differences between outer-shell orbitals and the first-shell orbital. When $m = 2$ (the Balmer series), the frequencies correspond to energy differences between outer-shell orbitals and the second-shell orbitals, as shown in Figure 5.11(b).

FIGURE 5.11 The origin of atomic line spectra. **(a)** As an energetically excited electron falls from an outer-shell orbital to an inner-shell orbital, it emits electromagnetic radiation whose frequency corresponds to the energy difference between the two orbitals. **(b)** The different spectral series correspond to electronic transitions from outer-shell orbitals to different inner-shell orbitals.

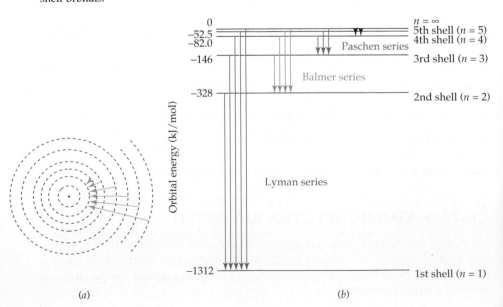

$$\frac{1}{\lambda} = R\left[\frac{1}{m^2} - \frac{1}{n^2}\right]$$

shell the transition is to ╱　　╲ shell the transition is from

Notice in Figure 5.11(b) that as n becomes larger and approaches infinity, the energy difference between the n shell and the first shell converges to a value of 1312 kJ/mol. That is, 1312 kJ/mol of energy is released when an electron comes from a great distance (the "infinite" shell) and reacts with H^+ to give a hydrogen atom with an electron in the first shell.

$$H^+ + e^- \rightarrow H + \text{energy (1312 kJ/mol)}$$

Since the energy *released* on adding an electron to H^+ is equal in magnitude but opposite in sign to the energy *absorbed* on removing an electron from a hydrogen atom, we can also say that *1312 kJ/mol of energy is required to remove an electron from a hydrogen atom.* We'll see in the next chapter that the amount of energy necessary to remove an electron from an element provides an important clue about the element's chemical reactivity.

What is true for hydrogen is also true for all other atoms: All atoms show atomic line spectra when energetically excited electrons fall from higher-energy orbitals in outer shells to lower-energy orbitals in inner shells. As you might expect, though, these spectra become very complex for multi-electron atoms in which different orbitals within a shell no longer have identical energies and in which a large number of electronic transitions are possible.

EXAMPLE 5.8

What is the energy difference in kJ/mol between the first and second shells of the hydrogen atom if the first emission in the Lyman series occurs at $\lambda = 121.57$ nm?

SOLUTION　　The first line in the Lyman series corresponds to the emission of light as an electron falls from the second shell to the first shell, and the energy of that light is equal to the energy difference between shells. Knowing the wavelength of the light, you can calculate the energy of one photon using the Planck equation, $E = hc/\lambda$:

$$E = \frac{hc}{\lambda} = \frac{(6.626 \times 10^{-34}\text{ J}\cdot s)\left(3.00 \times 10^8 \frac{m}{s}\right)\left(10^9 \frac{nm}{m}\right)}{121.57\ nm} = 1.64 \times 10^{-18}\text{ J}$$

Multiplying by Avogadro's number gives the answer in J/mol.

$$(1.64 \times 10^{-18}\text{ J}) \times (6.022 \times 10^{23}\text{ mol}^{-1}) = 9.88 \times 10^5\text{ J/mol} = 988\text{ kJ/mol}$$

The energy difference between the first and second shells of the hydrogen atom is 988 kJ/mol, a large difference indeed.

⌐ **PROBLEM 5.15** Show how to calculate in kJ/mol the energy necessary to remove an electron from the first shell of a hydrogen atom. ($R = 1.097 \times 10^{-2}\text{ nm}^{-1}$)

5.9 ▶ORBITAL SHAPES

We said in Section 5.7 that the shapes of orbitals are defined by the designations $s, p, d, f, \ldots$ corresponding to the angular-momentum quantum number l. Of the various possible shapes, the s, p, d, and f orbitals are the most important because these are the only ones actually used by known elements. Let's look at each of the four individually.

s Orbitals

All s orbitals are spherical, meaning that the probability of finding an electron in one of these orbitals depends only on the distance of the electron from the nucleus, not on direction. Furthermore, because there is only one possible orientation of a sphere in space, every s orbital has $m_l = 0$ and there is only one s orbital per shell. There are, however, significant differences among the s orbitals in the various shells. For one thing, the size of the s orbitals increases in successively higher shells, implying that an electron in an outer-shell s orbital is farther from the nucleus on average than an electron in an inner-shell s orbital. For another thing, the electron probability distribution within the sphere is not continuous in shells beyond the first one. As shown in Figure 5.12, a 2s orbital is essentially a sphere within a sphere. There are *two* regions of maximum probability, separated by a surface of zero probability called a **node**. Similarly, a 3s orbital has three regions of maximum probability separated by two nodes.

The concept of a node—a surface of zero electron probability separating regions of nonzero electron probability—is difficult to grasp because it raises the question, How does an electron get from one region of nonzero

FIGURE 5.12 (Top) Electron probability distribution plots and (bottom) cutaway dot-density representations of 1s, 2s, and 3s orbitals showing regions of maximum electron probability. All three orbitals are spherically symmetric, but the 2s orbital has buried within it a spherical surface of zero probability, and the 3s orbital has within it two spherical surfaces of zero probability.

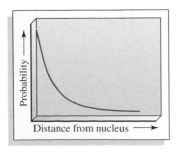

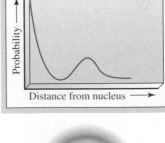

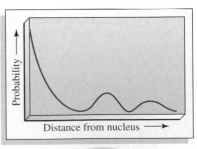

1s orbital

2s orbital

3s orbital

probability to another if it's not allowed to be at the node? The question is misleading, though, because it assumes particlelike behavior for the electron rather than wavelike behavior. In fact, *nodes are an intrinsic property of waves*, from moving water waves in the ocean to the standing (stationary) wave generated by vibrating a rope, as shown in Figure 5.13. A node simply corresponds to the region where the wave has zero amplitude.

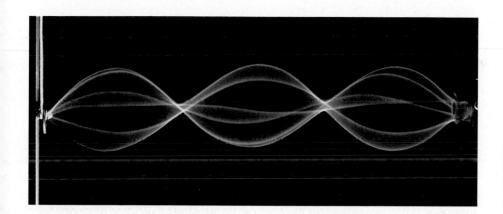

FIGURE 5.13 When a rope is fixed at one end and vibrated rapidly at the other, a standing (stationary) wave is generated. The zero-amplitude regions are called *nodes*.

p Orbitals

p Orbitals are dumbbell shaped rather than spherical, with the electron distribution concentrated in identical lobes on either side of the nucleus and separated by a nodal plane cutting through the nucleus (Figure 5.14). Since

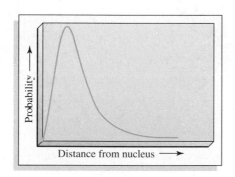

(a)

FIGURE 5.14 **(a)** An electron probability plot of a 2*p* orbital, and **(b)** computer-generated representations of all three 2*p* orbitals, each of which is dumbbell shaped and oriented in space along one of the three coordinate axes *x*, *y*, or *z*. Each *p* orbital has two lobes of high electron probability separated by a nodal plane passing through the nucleus.

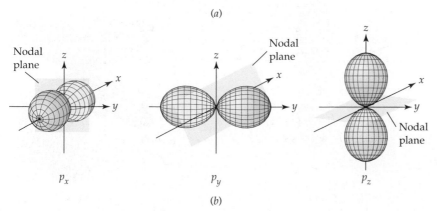

(b)

there are three allowable values of m_l when $l = 1$, each shell beginning with the second has three *p* orbitals, which are oriented in space at 90° angles to one another along the three coordinate axes *x*, *y*, and *z*. The three *p* orbitals in the second shell, for example, are designated $2p_x$, $2p_y$, and $2p_z$. As you might expect, *p* orbitals in the third and higher shells are larger than those in the second shell and extend farther from the nucleus. Their shape is roughly the same, however.

d and *f* Orbitals

The five *d* orbitals in the third and higher shells differ from their *s* and *p* counterparts because they have two different shapes. Four of the five *d* orbitals are cloverleaf shaped and have four lobes of maximum electron probability separated by two nodal planes through the nucleus (Figure 5.15). The fifth *d* orbital is similar in shape to a p_z orbital but has an additional donut-shaped region of electron probability centered in the *xy* plane. In spite of their different shapes, all five *d* orbitals in a given shell have the same energy.

You may have noticed by now that the number of nodal planes through the nucleus and the overall geometric complexity of the orbitals increases with the *l* quantum number of the subshell: *s* orbitals have one lobe and no nodal planes; *p* orbitals have two lobes and one nodal plane; *d*

FIGURE 5.15 **(a–e)** Computer-generated representations of the five 3*d* orbitals. Four of the orbitals are shaped like a cloverleaf, and the fifth is shaped like an elongated dumbbell inside a donut. **(f)** Also shown is one of the seven 4*f* orbitals.

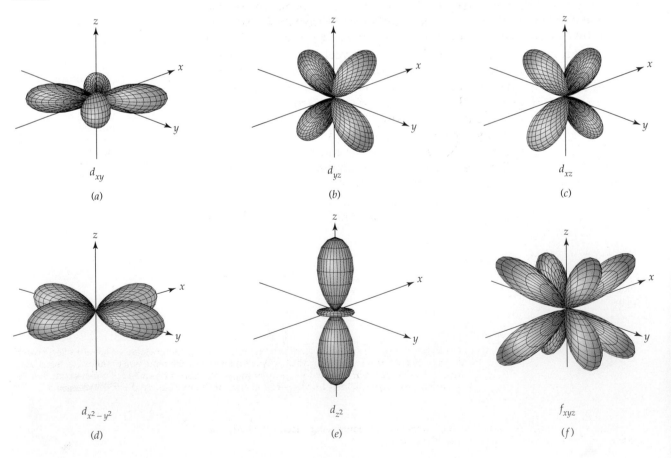

d_{xy}

(a)

d_{yz}

(b)

d_{xz}

(c)

$d_{x^2-y^2}$

(d)

d_{z^2}

(e)

f_{xyz}

(f)

orbitals have four lobes and two nodal planes. The seven *f* orbitals are more complex still, having eight lobes of maximum electron probability separated by three nodal planes through the nucleus. (Figure 5.15 shows one of the seven 4*f* orbitals.) Fortunately, most of the elements we'll deal with in the following chapters don't use *f* orbitals, so we won't worry much about them.

PROBLEM 5.16 How many nodal planes through the nucleus do you think a *g* orbital has?

5.10 ▶ ORBITAL ENERGY LEVELS IN MULTIELECTRON ATOMS

We said in Section 5.7 that the energy level of an orbital in a hydrogen atom is determined by its principal quantum number *n*. Within a shell, all hydrogen orbitals have the same energy independent of their other quantum numbers. The situation is different in multielectron atoms, however, because the energy level of an orbital in these atoms depends not only on the shell but also on the subshell. The *s*, *p*, *d*, and *f* orbitals within a given shell in a multielectron atom have slightly different energies, as shown previously in Figure 5.10. In fact, there's even some crossover of energies between shells: The 4*s* orbital is lower in energy than the 3*d* orbitals, for instance.

The difference in energy between subshells in multielectron atoms results from electron–electron repulsions. In hydrogen, the only electrostatic interaction is the attraction of the positive nucleus for the negative electron, but in multielectron atoms there are many different electrostatic interactions to consider.[4] Not only are there the *attractions* of the nucleus for each separate electron, there are also the *repulsions* between every electron and each of its neighboring electrons. These repulsions act to push the electrons apart, with the result that an individual electron is held less tightly to the nucleus than it would be in the absence of other electrons. Part of the electron–nucleus attraction is thus canceled by the electron–electron repulsion, an effect we describe by saying that the electrons are *shielded* from the nucleus by the other electrons (Figure 5.16). The net nuclear charge actually felt by an electron is called the **effective nuclear charge**, Z_{eff}, and is often substantially lower than the actual nuclear charge Z.

$$Z_{eff} = Z_{actual} - \text{electron shielding}$$

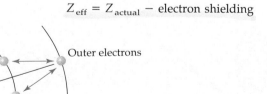

Outer electrons

FIGURE 5.16 The origin of electron shielding and Z_{eff}. Outer electrons are attracted toward the nucleus by the nuclear charge but are pushed away by the repulsion of inner electrons. As a result, the nuclear charge actually felt by an outer electron is diminished, and we say that the outer electrons are shielded from the full charge of the nucleus by the inner electrons.

[4] The word *electrostatic* refers to the interaction of nonmoving electrical charges as opposed to the moving charges in the flow of electricity.

How does electron shielding lead to energy differences among orbitals within a given shell? The answer results from the differences among orbital shapes. Let's compare a 2s and a 2p orbital, for instance. The s orbitals are spherical and have a large probability density near the nucleus, while p orbitals are dumbbell shaped and have a node at the nucleus. An electron in a 2s orbital can therefore penetrate closer to the nucleus than an electron in a 2p orbital can and feels less of a shielding effect from other electrons. A 2s electron thus feels a higher Z_{eff}, is more tightly held by the nucleus, and has a lower energy than a 2p electron does. In the same way, a 3p electron penetrates closer to the nucleus, feels a higher Z_{eff}, and has a lower energy than a 3d electron does. More generally, for any given value of the principal quantum number n, a lower value of l corresponds to a higher value of Z_{eff} and to a lower energy for the orbital.

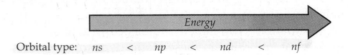

Orbital type: ns $<$ np $<$ nd $<$ nf

The concept of different effective nuclear charges for electrons in different orbitals is a very useful one that we'll return to on several occasions to explain various chemical phenomena. The concept is inexact, though, and numerous methods have been devised for calculating Z_{eff} values. The values obtained from an equation proposed by J. C. Slater in 1932 are shown in Figure 5.17 for the highest energy electron in each of the first 18 elements. There is an obvious periodicity to the data, with minimum Z_{eff} values for the group 1A elements (H, Li, and Na), and maximum Z_{eff} values for the group 8A elements (Ne and Ar). In addition, there is a regular increase in Z_{eff} across a row of the periodic table from left to right. We'll account for this periodicity in Section 5.15.

FIGURE 5.17 Calculated Slater values of effective nuclear charge (Z_{eff}) for the highest-energy electron in each of the first 18 elements. There is an obvious periodicity to the data, with group 1A elements at minima and group 8A elements at maxima.

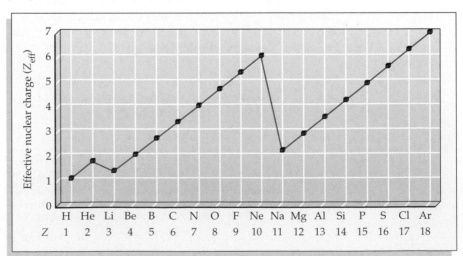

5.11 ►ELECTRON SPIN AND THE PAULI EXCLUSION PRINCIPLE

The three quantum numbers n, l, and m_l define the energy, shape, and spatial orientation of orbitals, but they don't quite tell the whole story. When the line spectra of multielectron atoms are studied in detail, it turns out that lines actually occur as very closely spaced *pairs*. (You can see this pairing if you look closely at the visible spectrum of sodium in Figure 5.6.) Thus, there are twice as many electron energy levels as simple quantum mechanics predicts, and a fourth quantum number is required. Denoted m_s, this fourth quantum number is related to a property called *electron spin*.

In certain respects, electrons behave as if they were spinning around an axis, much as the earth spins daily. Unlike the earth, though, electrons are free to spin in either a clockwise or a counterclockwise direction. As a result, the **spin quantum number** m_s can have either of two values, $+1/2$ or $-1/2$ (Figure 5.18). A spin of $+1/2$ is usually represented by an up arrow (↑), and a spin of $-1/2$ is represented by a down arrow (↓). Note that the value of m_s is independent of the other three quantum numbers, unlike the values of n, l, and m_l, which are interrelated.

The importance of the spin quantum number m_s comes when we assign electrons to specific orbitals in multielectron atoms. According to the **Pauli exclusion principle**, proposed by the Austrian physicist Wolfgang Pauli (1900–1958) in 1925, no two electrons in an atom can have the same four quantum numbers. In other words, the set of four quantum numbers associated with an electron acts as a unique "address" for each electron in an atom, and no two electrons can have the same address.

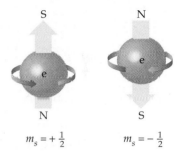

FIGURE 5.18 Electrons behave in certain respects as if they were tiny charged spheres spinning around an axis. This spin gives rise to a tiny magnetic field and to a fourth quantum number, m_s, which can have a value of either $+1/2$ or $-1/2$.

> *Pauli exclusion principle:* No two electrons in an atom have the same four quantum numbers.

Think about the consequences of the Pauli exclusion principle. Because any electrons in a given orbital must have the same values for three quantum numbers, n, l, and m_l, they must have different values for the fourth quantum number m_s. But because there are only two possible values of m_s, $+1/2$ and $-1/2$, each orbital can hold only two electrons, which must have opposite spins. An atom with x number of electrons therefore has at least $x/2$ occupied orbitals, though it might have more if some of its orbitals are only half-filled.

5.12 ►ELECTRON CONFIGURATIONS OF MULTIELECTRON ATOMS

All the parts are now in place to provide a complete electronic description for every element. Knowing the relative energies of the various orbitals, we can predict for each element which orbitals are occupied by electrons—the element's **electron configuration**.

A set of three simple rules called the **aufbau principle** (from the German word for "building up") guides the filling order of orbitals. In general, each successive electron added to an atom occupies the lowest-energy orbital available. The resultant lowest-energy electron configuration is called the **ground-state configuration** of the atom. (It may happen, of course, that several orbitals will have the same energy level—for example, the three *p* orbitals or the five *d* orbitals in a given shell. Such orbitals are said to be **degenerate**.)

Rules of the aufbau principle:

1. Fill the lowest-energy orbitals first before filling successively higher energy orbitals. (The ordering of energy levels for orbitals is shown in Figure 5.10.)
2. Put only two electrons of opposite spin into any one orbital. This is just a restatement of the Pauli exclusion principle (Section 5.11), emphasizing that no two electrons in an atom can have the same four quantum numbers.
3. If two or more degenerate orbitals are available, put one electron in each until all are half full, a statement called **Hund's rule**. Only then should you add a second electron to one of the orbitals. Furthermore, the electrons in each of the singly occupied orbitals must have the same value of the spin quantum number.

Hund's rule:	If two or more orbitals with the same energy are available, put one electron (with the same spin quantum number) in each until all are half full.

Hund's rule is just a matter of common sense. Because electrons repel each other, it makes sense that they should remain as far apart as possible. Clearly, they can remain farther apart and be lower in energy if they are in different orbitals describing different spatial regions than if they're in the same orbital occupying the same region.

Let's look at some examples to see how the rules of the aufbau principle are applied.

*   **Hydrogen:** Hydrogen has only one electron, which must go into the lowest-energy, 1*s* orbital. Thus, we say that the ground-state electronic configuration of hydrogen is $1s^1$, where the superscript indicates the number of electrons in the specified orbital.

$$\text{H: } 1s^1$$

*   **Helium:** Helium has two electrons, both of which fit into the lowest-energy, 1*s* orbital. The two electrons have opposite spins.

$$\text{He: } 1s^2$$

*   **Lithium and beryllium:** With the 1*s* orbital full, both the third and fourth electrons go into the next available, 2*s* orbital.

$$\text{Li: } 1s^2\,2s^1 \qquad \text{Be: } 1s^2\,2s^2$$

- **Boron through neon:** The six elements from boron through neon have their three $2p$ orbitals filled successively. Since these three $2p$ orbitals have the same energy, they are degenerate and are filled according to Hund's rule. In carbon, for example, the two $2p$ electrons are in different orbitals, which can be arbitrarily specified as $2p_x$, $2p_y$, or $2p_z$ when writing the electron configuration. Similarly for nitrogen, whose three $2p$ electrons must be in three different orbitals. Although not usually noted in the written electronic configuration, the electrons in each of the singly occupied carbon and nitrogen $2p$ orbitals must have the same value of the spin quantum number—both either $+1/2$ or $-1/2$.

  For clarity, we sometimes specify electron configurations using orbital-filling diagrams in which electrons are represented by arrows. The two values of the spin quantum numbers are indicated by having the arrow point either up or down. An up-down pair indicates that an orbital is filled, while a single up (or down) arrow indicates that an orbital is half-filled. Note in the diagrams for carbon and nitrogen that the degenerate $2p$ orbitals are half-filled rather than filled, according to Hund's rule, and that the electron spin is the same in each.

B: $1s^2\,2s^2\,2p^1$ or $\underset{1s}{\Updownarrow}\quad\underset{2s}{\Updownarrow}\qquad\underset{\;\;2p}{\uparrow\quad\underline{}\quad\underline{}}$

C: $1s^2\,2s^2\,2p_x^{\,1}\,2p_y^{\,1}$ or $\underset{1s}{\Updownarrow}\quad\underset{2s}{\Updownarrow}\qquad\underset{\;\;2p}{\uparrow\quad\uparrow\quad\underline{}}$

N: $1s^2\,2s^2\,2p_x^{\,1}\,2p_y^{\,1}\,2p_z^{\,1}$ or $\underset{1s}{\Updownarrow}\quad\underset{2s}{\Updownarrow}\qquad\underset{\;\;2p}{\uparrow\quad\uparrow\quad\uparrow}$

From oxygen through neon, the three $2p$ orbitals are successively filled. For fluorine and neon, it's no longer necessary to distinguish among the different $2p$ orbitals, so we can simply write $2p^5$ and $2p^6$.

O: $1s^2\,2s^2\,2p_x^{\,2}\,2p_y^{\,1}\,2p_z^{\,1}$ or $\underset{1s}{\Updownarrow}\quad\underset{2s}{\Updownarrow}\qquad\underset{\;\;2p}{\Updownarrow\quad\uparrow\quad\uparrow}$

F: $1s^2\,2s^2\,2p^5$ or $\underset{1s}{\Updownarrow}\quad\underset{2s}{\Updownarrow}\qquad\underset{\;\;2p}{\Updownarrow\quad\Updownarrow\quad\uparrow}$

Ne: $1s^2\,2s^2\,2p^6$ or $\underset{1s}{\Updownarrow}\quad\underset{2s}{\Updownarrow}\qquad\underset{\;\;2p}{\Updownarrow\quad\Updownarrow\quad\Updownarrow}$

- **Sodium and magnesium:** The $3s$ orbital is filled next, giving sodium and magnesium the ground-state electron configurations shown. Note that we often write the configurations in a shorthand version by giving the symbol of the noble gas in the previous row to indicate electrons in filled shells and then specifying only those electrons in unfilled shells.

neon configuration

Na: $1s^2\,2s^2\,2p^6\,3s^1$ or $[\text{Ne}]\,3s^1$

Mg: $1s^2\,2s^2\,2p^6\,3s^2$ or $[\text{Ne}]\,3s^2$

- **Aluminum through argon:** The $3p$ orbitals are now filled according to the same rules used previously for filling the $2p$ orbitals of boron through neon. Rather than explicitly identify which of the degenerate $3p$ orbitals are occupied in Si, P, and S, we'll simplify the writing by giving just the total number of electrons in the subshell. For example, we'll write $3p^2$ for silicon rather than $3p_x^1\, 3p_y^1$.

Al: [Ne] $3s^2\, 3p^1$	**Si:** [Ne] $3s^2\, 3p^2$	**P:** [Ne] $3s^2\, 3p^3$
S: [Ne] $3s^2\, 3p^4$	**Cl:** [Ne] $3s^2\, 3p^5$	**Ar:** [Ne] $3s^2\, 3p^6$

FIGURE 5.19   Outer shell ground state electron configurations of the elements.

1 H $1s^1$								
3 Li $2s^1$	4 Be $2s^2$							
11 Na $3s^1$	12 Mg $3s^2$							
19 K $4s^1$	20 Ca $4s^2$	21 Sc $4s^23d^1$	22 Ti $4s^23d^2$	23 V $4s^23d^3$	24 Cr $4s^13d^5$	25 Mn $4s^23d^5$	26 Fe $4s^23d^6$	27 Co $4s^23d^7$
37 Rb $5s^1$	38 Sr $5s^2$	39 Y $5s^24d^1$	40 Zr $5s^24d^2$	41 Nb $5s^14d^4$	42 Mo $5s^14d^5$	43 Tc $5s^24d^5$	44 Ru $5s^14d^7$	45 Rh $5s^14d^8$
55 Cs $6s^1$	56 Ba $6s^2$	57 La $6s^25d^1$	72 Hf $6s^24f^{14}5d^2$	73 Ta $6s^24f^{14}5d^3$	74 W $6s^24f^{14}5d^4$	75 Re $6s^24f^{14}5d^5$	76 Os $6s^24f^{14}5d^6$	77 Ir $6s^24f^{14}5d^7$
87 Fr $7s^1$	88 Ra $7s^2$	89 Ac $7s^26d^1$	104 Rf $7s^25f^{14}6d^2$	105 Ha $7s^25f^{14}6d^3$	106 Sg $7s^25f^{14}6d^4$	107 Uns $7s^25f^{14}6d^5$	108 Uno $7s^25f^{14}6d^6$	109 Une $7s^25f^{14}6d^7$

58 Ce $6s^24f^2$	59 Pr $6s^24f^3$	60 Nd $6s^24f^4$	61 Pm $6s^24f^5$	62 Sm $6s^24f^6$	63 Eu $6s^24f^7$	64 Gd $6s^24f^75d^1$	65 Tb $6s^24f^9$	66 Dy $6s^24f^{10}$	67 Ho $6s^24f^{11}$	68 Er $6s^24f^{12}$	69 Tm $6s^24f^{13}$	70 Yb $6s^24f^{14}$	71 Lu $6s^24f^{14}5d^1$
90 Th $7s^26d^2$	91 Pa $7s^25f^26d^1$	92 U $7s^25f^36d^1$	93 Np $7s^25f^46d^1$	94 Pu $7s^25f^6$	95 Am $7s^25f^7$	96 Cm $7s^25f^76d^1$	97 Bk $7s^25f^9$	98 Cf $7s^25f^{10}$	99 Es $7s^25f^{11}$	100 Fm $7s^25f^{12}$	101 Md $7s^25f^{13}$	102 No $7s^25f^{14}$	103 Lr $7s^25f^{14}6d^1$

- **Elements past argon:** Following the filling of the $3p$ subshell in argon, the first crossover in the orbital-filling order is encountered. Rather than continue filling the third shell by populating the $3d$ orbitals, the next two electrons for potassium and calcium go into the $4s$ subshell. Only then does filling of the $3d$ subshell occur to give the first transition-metal series from scandium through zinc.

$$\textbf{K: } [Ar]\ 4s^1 \qquad \textbf{Ca: } [Ar]\ 4s^2 \qquad \textbf{Sc: } [Ar]\ 4s^2\ 3d^1 \rightarrow \textbf{Zn: } [Ar]\ 4s^2\ 3d^{10}$$

The experimentally determined ground-state electron configurations of all 109 known elements are shown in Figure 5.19.

						2 He $1s^2$
		5 B $2s^22p^1$	6 C $2s^22p^2$	7 N $2s^22p^3$	8 O $2s^22p^4$	9 F $2s^22p^5$	10 Ne $2s^22p^6$

			13 Al $3s^23p^1$	14 Si $3s^23p^2$	15 P $3s^23p^3$	16 S $3s^23p^4$	17 Cl $3s^23p^5$	18 Ar $3s^23p^6$
28 Ni $4s^23d^8$	29 Cu $4s^13d^{10}$	30 Zn $4s^23d^{10}$	31 Ga $4s^23d^{10}4p^1$	32 Ge $4s^23d^{10}4p^2$	33 As $4s^23d^{10}4p^3$	34 Se $4s^23d^{10}4p^4$	35 Br $4s^23d^{10}4p^5$	36 Kr $4s^23d^{10}4p^6$
46 Pd $4d^{10}$	47 Ag $5s^14d^{10}$	48 Cd $5s^24d^{10}$	49 In $5s^24d^{10}5p^1$	50 Sn $5s^24d^{10}5p^2$	51 Sb $5s^24d^{10}5p^3$	52 Te $5s^24d^{10}5p^4$	53 I $5s^24d^{10}5p^5$	54 Xe $5s^24d^{10}5p^6$
78 Pt $6s^14f^{14}5d^9$	79 Au $6s^14f^{14}5d^{10}$	80 Hg $6s^24f^{14}5d^{10}$	81 Tl $6s^24f^{14}5d^{10}6p^1$	82 Pb $6s^24f^{14}5d^{10}6p^2$	83 Bi $6s^24f^{14}5d^{10}6p^3$	84 Po $6s^24f^{14}5d^{10}6p^4$	85 At $6s^24f^{14}5d^{10}6p^5$	86 Rn $6s^24f^{14}5d^{10}6p^6$

Why are electron configurations so important, and what do they have to do with the periodic table? The answers become clear when you look closely at Figure 5.19. Focussing only on the electrons in the outermost shell, or **valence shell**, *the elements in each group of the periodic table have similar valence-shell electron configurations* (Table 5.3). The group 1A elements, for example, all have an s^1 valence-shell configuration; the group 2A elements have an s^2 valence-shell configuration; the group 3A elements have an s^2p^1 valence-shell configuration; and so on across every column of the periodic table (except for a small number of anomalies). Furthermore, because the valence-shell electrons are the most loosely held and therefore most important in determining an element's properties, similar electron configurations explain why the elements in a given group of the periodic table have similar chemical behavior.

TABLE 5.3	Valence-Shell Electron Configurations of Main-Group Elements
Group	**Valence-Shell Electron Configuration**
1A	ns^1
2A	ns^2
3A	$ns^2\,np^1$
4A	$ns^2\,np^2$
5A	$ns^2\,np^3$
6A	$ns^2\,np^4$
7A	$ns^2\,np^5$
8A	$ns^2\,np^6$

The periodic table can be divided into four regions, or *blocks*, of elements according to the orbitals being filled (Figure 5.20). The group 1A and 2A elements on the left side of the table are called the **s-block elements** because they result from the filling of an s orbital; the group 3A–8A elements on the right side of the table are the **p-block elements** because they result from the filling of p orbitals; the transition-metal **d-block elements** in the middle of the table result from the filling of d orbitals; and the lanthanide/actinide **f-block elements**, detached at the bottom of the table, result from the filling of f orbitals.

Thinking of the periodic table as outlined in Figure 5.20 provides a simple way to remember the order of orbital filling. Beginning at the top left corner of the periodic table and going across successive rows gives the correct orbital-filling order automatically. The first row of the periodic table, for instance, contains only the two s-block elements H and He, so we fill the first available s orbital ($1s$). The second row begins with two s-block elements (Li and Be) and continues with six p-block elements (B through Ne), so we fill the next available s orbital ($2s$) and then the first available p orbitals ($2p$). The third row is similar to the second row, so we fill the $3s$ and $3p$ orbitals. The fourth row again starts with two s-block elements (K and Ca) but is then followed by ten d-block elements (Sc through Zn) and six p-block elements (Ga through Kr). Thus, the order of orbital filling is $4s$ followed by

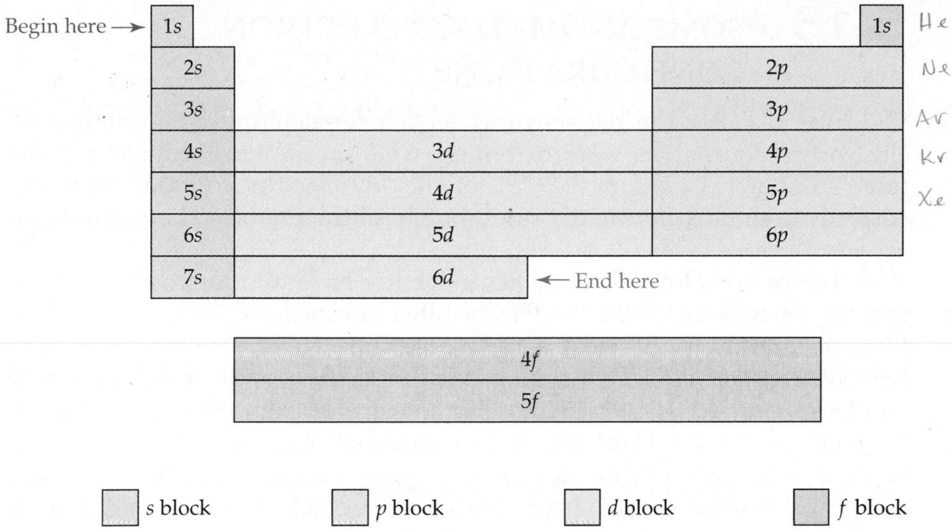

Begin here → | 1s | | | | | | 1s | He

2s				2p	Ne
3s				3p	Ar
4s	3d		4p	Kr
5s	4d		5p	Xe
6s	5d		6p	
7s	6d	← End here		

| 4f |
| 5f |

☐ s block   ☐ p block   ☐ d block   ☐ f block

FIGURE 5.20 The regions of the periodic table that correspond to filling the different blocks of orbitals. Beginning at the top left and going across successive rows of the periodic table provides a method for remembering the order of orbital filling: $1s \rightarrow 2s \rightarrow 2p \rightarrow 3s \rightarrow 3p \rightarrow 4s \rightarrow 3d \rightarrow 4p$, and so on.

the first available *d* orbitals (3*d*) followed by 4*p*. Continuing through successive rows of the periodic table gives the entire filling order:

$$1s \rightarrow 2s \rightarrow 2p \rightarrow 3s \rightarrow 3p \rightarrow 4s \rightarrow 3d \rightarrow 4p \rightarrow 5s \rightarrow 4d \rightarrow$$
$$5p \rightarrow 6s \rightarrow 4f \rightarrow 5d \rightarrow 6p \rightarrow 7s \rightarrow 5f \rightarrow 6d$$

EXAMPLE 5.9

Give the ground-state electron configuration of arsenic, $Z = 33$. Draw an orbital-filling diagram to indicate the electrons as up or down arrows.

SOLUTION   Think of the periodic table as having *s*, *p*, *d*, and *f* blocks of elements, as shown in Figure 5.20. Start with hydrogen at the upper left and fill orbitals until 33 electrons have been added. Remember that only two electrons can go into an orbital and that each one of a set of degenerate orbitals must be half-filled before any one can be completely filled.

As: $1s^2 \, 2s^2 \, 2p^6 \, 3s^2 \, 3p^6 \, 4s^2 \, 3d^{10} \, 4p^3$ or   [Ar] $4s^2 \, 3d^{10} \, 4p^3$

An orbital-filling diagram indicates the electrons in each orbital as arrows. Note that the three 4*p* electrons all have the same spin:

As: [Ar] $\underset{4s}{\underline{\uparrow\downarrow}}$ $\underline{\uparrow\downarrow}$ $\underline{\uparrow\downarrow}$ $\underset{3d}{\underline{\uparrow\downarrow}}$ $\underline{\uparrow\downarrow}$ $\underline{\uparrow\downarrow}$ $\underline{\uparrow}$ $\underset{4p}{\underline{\uparrow}}$ $\underline{\uparrow}$

⌐ **PROBLEM 5.17** Give expected ground-state electron configurations for the following atoms, and draw orbital-filling diagrams for parts (a)–(c).
(a) Ti ($Z = 22$)   **(b)** Zn ($Z = 30$)   **(c)** Sn ($Z = 50$)   **(d)** Pb ($Z = 82$)

⌐ **PROBLEM 5.18** What do you think is a likely ground-state electron configuration for the sodium *ion*, Na^+, formed by loss of an electron from a neutral sodium atom?

5.13 ►SOME ANOMALOUS ELECTRON CONFIGURATIONS

The rules discussed in the previous section for determining ground-state electron configurations work extremely well but are not completely accurate. As shown in Figure 5.19, 92 of the 109 electron configurations are correctly accounted for by the rules, but 16 of the predicted configurations are incorrect.

The reasons for the anomalies generally have to do with the unusual stability of both half-filled and fully filled subshells. To take an example from the first transition-metal series, chromium, which we would expect to have the configuration [Ar] $4s^2 3d^4$, actually has the experimentally observed configuration [Ar] $4s^1 3d^5$. By moving an electron from the $4s$ orbital to an energetically similar $3d$ orbital, chromium trades one filled subshell ($4s^2$) for two half-filled subshells ($4s^1 3d^5$). In the same way, copper, which we would expect to have the configuration [Ar] $4s^2 3d^9$, actually has the configuration [Ar] $4s^1 3d^{10}$. By transferring an electron from the $4s$ orbital to a $3d$ orbital, copper trades one filled subshell ($4s^2$) for a different filled subshell ($3d^{10}$) and gains a half-filled subshell ($4s^1$).

Not surprisingly, most of the anomalous electron configurations shown in Figure 5.19 occur in elements heavier than $Z = 40$, where the energy differences between subshells are small. In all cases, the transfer of an electron from one subshell to another lowers the total energy of the atom because of a decrease in electron–electron repulsions.

⌐ **PROBLEM 5.19** Look at the electron configurations in Figure 5.19 and identify the 16 elements that you think are anomalous.   ⌐

5.14 ►ELECTRON CONFIGURATIONS OF IONS

We've seen on several occasions that the metallic elements on the left side of the periodic table have a tendency to give up electrons to form cations, while the halogens and a few other nonmetallic elements on the right side of the table have a tendency to accept electrons to form anions. What are the ground-state electron configurations of the resultant ions?

For main-group elements, the electrons that are given up by a metal in forming a cation come from the highest-energy occupied orbital, while the electrons that are accepted by a nonmetal in forming an anion go into the lowest-energy unoccupied orbital according to the aufbau principle. When a sodium atom ($1s^2 2s^2 2p^6 3s^1$) gives up an electron, for example, the valence-shell $3s$ electron is lost, giving an Na^+ ion with the stable, noble-gas electronic configuration of neon ($1s^2 2s^2 2p^6$). Similarly, when a chlorine atom ($1s^2 2s^2 2p^6 3s^2 3p^5$) accepts an electron, the electron fills the remaining vacancy in the $3p$ subshell to give a Cl^- ion with the noble-gas electronic configuration of argon ($1s^2 2s^2 2p^6 3s^2 3p^6$).

$$\textbf{Na: } 1s^2\, 2s^2\, 2p^6\, 3s^1 \xrightarrow{-e^-} \textbf{Na}^+\text{: } 1s^2\, 2s^2\, 2p^6$$

$$\textbf{Cl: } 1s^2\, 2s^2\, 2p^6\, 3s^2\, 3p^5 \xrightarrow{+e^-} \textbf{Cl}^-\text{: } 1s^2\, 2s^2\, 2p^6\, 3s^2\, 3p^6$$

What is true for sodium is also true for the other elements in group 1A: All form positive ions by losing their valence-shell s electron, and all the resultant ions have noble-gas electron configurations. Similarly for the elements in group 2A: All form a doubly positive ion by losing both their valence-shell s electrons. An Mg atom ($1s^2\, 2s^2\, 2p^6\, 3s^2$), for example, goes to an Mg^{2+} ion with the neon configuration $1s^2\, 2s^2\, 2p^6$ by loss of its two $3s$ electrons.

Group 1A Atom: [noble gas] ns^1 $\xrightarrow{-e^-}$ **Group 1A Ion$^+$:** [noble gas]

Group 2A Atom: [noble gas] ns^2 $\xrightarrow{-2\,e^-}$ **Group 2A Ion^{2+}:** [noble gas]

Just as the group 1A and 2A metals *lose* the appropriate number of electrons to yield ions with noble-gas configurations, the group 6A and group 7A nonmetals *gain* the appropriate number of electrons. The halogens in group 7A gain one electron to form singly charged anions with noble-gas configurations, and the elements in group 6A gain two electrons to form doubly charged anions with noble-gas configurations. Oxygen ($1s^2\, 2s^2\, 2p^4$), for example, gives the O^{2-} ion with the neon configuration ($1s^2\, 2s^2\, 2p^6$).

Group 6A Atom: [noble gas] $ns^2\, np^4$ $\xrightarrow{+2\,e^-}$ **Group 6A Ion^{2-}:** [noble gas] $ns^2\, np^6$

Group 7A Atom: [noble gas] $ns^2\, np^5$ $\xrightarrow{+e^-}$ **Group 7A Ion$^-$:** [noble gas] $ns^2\, np^6$

The formulas and electron configurations of the most common main-group ions are shown in Table 5.4.

TABLE 5.4	Some Common Main-Group Ions and Their Noble-Gas Electron Configurations				
Group 1A	**Group 2A**	**Group 3A**	**Group 6A**	**Group 7A**	**Electron Configuration**
H^-					[He]
Li^+	Be^{2+}				[He]
Na^+	Mg^{2+}	Al^{3+}	O^{2-}	F^-	[Ne]
K^+	Ca^{2+}	Ga^{3+}	S^{2-}	Cl^-	[Ar]
Rb^+	Sr^{2+}	In^{3+}	Se^{2-}	Br^-	[Kr]
Cs^+	Ba^{2+}	Tl^{3+}	Te^{2-}	I^-	[Xe]

The situation is a bit different for the formation of ions from the transition-metal elements than it is for the main-group elements. Transition metals form cations by first losing their valence-shell s electrons and then losing d electrons. As a result, all the valence electrons in transition-metal cations occupy the d orbitals. Iron, for example, forms the Fe^{2+} ion by losing its two $4s$ electrons, and forms the Fe^{3+} ion by losing two $4s$ electrons and one $3d$ electron:

$$\textbf{Fe: } [Ar]\, 4s^2\, 3d^6 \xrightarrow{-2\,e^-} \textbf{Fe}^{2+}\textbf{: } [Ar]\, 3d^6$$

$$\textbf{Fe: } [Ar]\, 4s^2\, 3d^6 \xrightarrow{-3\,e^-} \textbf{Fe}^{3+}\textbf{: } [Ar]\, 3d^5$$

It may seem strange that ion formation removes the 4s electrons *before* the 3d electrons, whereas building up the periodic table adds the 3d electrons *after* the 4s electrons. Note, though, that the two processes are not the reverse of each other: Progressing from one element to the next in building up the periodic table adds one electron to the valence shell and also adds one positive charge to the nucleus, whereas ionization removes an electron from the valence shell but does not alter the nucleus.

⌐ **PROBLEM 5.20** Predict the ground-state electron configuration for each of the following ions, and explain your answers.
(a) Ra^{2+} **(b)** La^{3+} **(c)** Ti^{4+} **(d)** N^{3-} ⌐

5.15 ►ELECTRON CONFIGURATIONS AND PERIODIC PROPERTIES: ATOMIC RADII

One of the many periodic properties of the elements that can be explained by electron configurations is size, or atomic radius. You might wonder, though, how we can talk about a definite "size" for an atom, having said in Section 5.6 that the electron clouds around atoms have no specific boundaries. What is usually done is to assume that the radius of an atom is half the distance between the nuclei of two identical atoms when they are covalently bonded together. In the Cl_2 molecule, for example, the distance between the two chlorine nuclei is 198 pm; in diamond (elemental carbon), the distance between two carbon nuclei is 154 pm. Thus, we say that the atomic radius of chlorine is half the Cl–Cl distance, or 99 pm, and the atomic radius of carbon is half the C–C distance, or 77 pm.

It's possible to check the accuracy of atomic radii by making sure that the assigned values are additive. For instance, since the atomic radius of Cl is 99 pm and the atomic radius of C is 77 pm, the distance between Cl and C nuclei when those two atoms are bonded together ought to be roughly 99 pm + 77 pm, or 176 pm. In fact, the measured distance between carbon and chlorine in the chloroform molecule ($CHCl_3$) is 178 pm, remarkably close to the expected value.

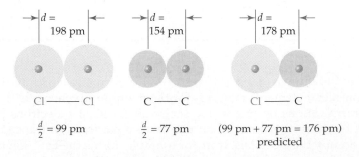

As shown pictorially in Figure 5.21 and graphically in Figure 5.1, a comparison of atomic radius versus atomic number shows a periodic rise-and-fall pattern. Atomic radius *increases* down a group of the periodic table (Li → Na → K → Rb → Cs, for example) but *decreases* across a row of the table (Na → Mg → Al → Si → P → S → Cl, for example).

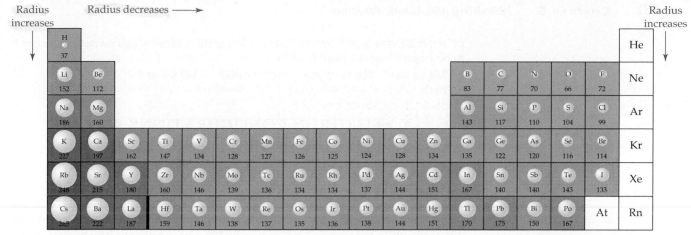

H																	He
37																	
Li	Be											B	C	N	O	F	Ne
152	112											83	77	70	66	72	
Na	Mg											Al	Si	P	S	Cl	Ar
186	160											143	117	110	104	99	
K	Ca	Sc	Ti	V	Cr	Mn	Fe	Co	Ni	Cu	Zn	Ga	Ge	As	Se	Br	Kr
227	197	162	147	134	128	127	126	125	124	128	134	135	122	120	116	114	
Rb	Sr	Y	Zr	Nb	Mo	Tc	Ru	Rh	Pd	Ag	Cd	In	Sn	Sb	Te	I	Xe
248	215	180	160	146	139	136	134	134	137	144	151	167	140	140	143	133	
Cs	Ba	La	Hf	Ta	W	Re	Os	Ir	Pt	Au	Hg	Tl	Pb	Bi	Po	At	Rn
265	222	187	159	146	138	137	135	136	138	144	151	170	175	150	167		

FIGURE 5.21 Atomic radii of the elements in picometers.

The increase in atomic radius that occurs down a group of the periodic table occurs because successively larger valence-shell orbitals are occupied. In Li, for example, the outermost occupied shell is the second one ($2s^1$); in Na it's the third one ($3s^1$); in K it's the fourth one ($4s^1$); and so on through Rb ($5s^1$), Cs ($6s^1$), and Fr ($7s^1$). Because larger shells are occupied, the atomic radii also become larger.

The decrease in atomic radius from left to right across the periodic table is accounted for by a change in effective nuclear charge, Z_{eff}, for the valence-shell electrons (Section 5.10). On going across the third period from Na to Ar, for example, each additional electron adds to the same shell (from $3s^1$ for Na to $3s^2 3p^6$ for Ar). Because electrons in the same shell are at approximately the same distance from the nucleus, they are relatively ineffective at shielding each other. At the same time, though, the nuclear charge increases from +11 for Na to +18 for Ar. Thus, the *effective* nuclear charge for the valence-shell electrons increases across the period from 2.20 for Na to 6.75 for Ar, drawing all the valence-shell electrons closer to the nucleus and shrinking the atomic radii (Figure 5.22).

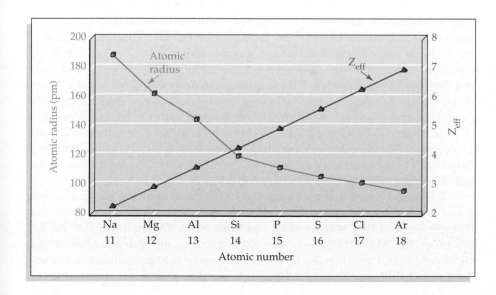

FIGURE 5.22 A plot of both atomic radius and Z_{eff} for the valence-shell electrons versus atomic number. As Z_{eff} increases, the valence-shell electrons are attracted more strongly to the nucleus, and the atomic radius therefore decreases.

181

⌐ **PROBLEM 5.21** Which atom in each of the following pairs would you expect
to be larger? Explain your answers.

(a) Mg or Ba   **(b)** W or Au   **(c)** Si or Sn   **(d)** Ce or Lu

5.16 ►ELECTRON CONFIGURATIONS AND
PERIODIC PROPERTIES: IONIC RADII

Just as there are systematic differences in the sizes of atoms, there are also
systematic differences in the sizes of ions. As shown in Figure 5.23 for the
elements of groups 1A and 2A, atoms shrink dramatically when an electron
is removed to form a cation. The radius of an Na atom, for example, is
186 pm, while that of an Na^+ cation is 102 pm. Similarly, the radius of an
Mg atom is 160 pm, and that of an Mg^{2+} cation is 72 pm.

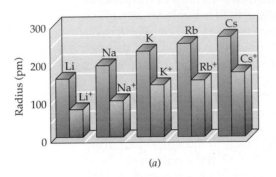   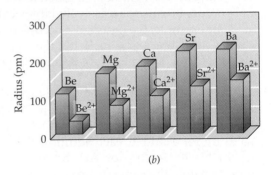

(a)   (b)

FIGURE 5.23 The relative radii of **(a)** the group 1A atoms and their cations; **(b)** the group
2A atoms and their cations. The cations are much smaller than the neutral atoms both because
the principal quantum number of the valence-shell electrons is smaller in the cations and
because Z_{eff} is higher.

The shrinkage that occurs when an electron is removed from an atom
is explained both by the fact that the electrons are removed from larger,
valence-shell orbitals and by an increase in effective nuclear charge. In going
from a neutral Na atom to a charged Na^+ cation, for example, the electron
configuration changes from $1s^2\ 2s^2\ 2p^6\ 3s^1$ to $1s^2\ 2s^2\ 2p^6$. The valence shell of
the Na *atom* is the *third* one, but the valence shell of the Na^+ *cation* is the
second one. Thus, the Na^+ ion has a smaller valence shell than the Na atom
and therefore a smaller size. Also important in accounting for the shrinkage
in size is that the effective nuclear charge felt by the valence-shell electrons is
greater in the Na^+ cation than in the neutral atom. The Na atom has
11 protons and 11 electrons, but the Na^+ cation has 11 protons and only
10 electrons. The smaller number of electrons in the cation means that they
shield each other to a lesser extent than the electrons in the neutral atom do,
and thus they feel a stronger pull toward the nucleus.

The same effects felt by the group 1A elements when a single electron
is lost are felt by the group 2A elements when two electrons are lost. For
example, loss of two valence-shell electrons from an Mg atom ($1s^2\ 2s^2\ 2p^6$
$3s^2$) gives the Mg^{2+} cation ($1s^2\ 2s^2\ 2p^6$). The smaller valence shell of the
Mg^{2+} cation and the increase in effective nuclear charge combine to cause a
dramatic shrinkage. Not surprisingly, a similar shrinkage is encountered

whenever any of the metal atoms on the left-hand two-thirds of the periodic table is converted into a cation.

Just as atoms shrink when converted to cations by loss of an electron, they expand when converted to anions by gain of an electron. As shown in Figure 5.24 for the group 7A elements (halogens), the expansion is a dramatic one. Chlorine, for example, expands from 99 pm for the neutral atom to 184 pm for the chloride anion.

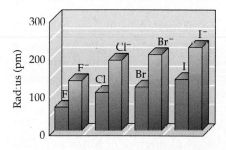

FIGURE 5.24 The relative radii of the group 7A atoms (halogens) and their anions. The anions are much larger than the neutral atoms because of additional electron–electron repulsions and a decrease in Z_{eff}.

The expansion that occurs when a group 7A atom gains an electron to yield an anion can't be accounted for by a change in the quantum number of the valence shell, since the added electron simply completes the p subshell of an already occupied shell: [Ne] $3s^2\,3p^5$ for a Cl atom to [Ne] $3s^2\,3p^6$ for a Cl$^-$ anion, for example. Thus, the expansion is due entirely to a decrease in effective nuclear charge and to an increase in electron-electron repulsions that occurs when an extra electron is added.

▢ **PROBLEM 5.22** Which atom or ion in each of the following pairs would you expect to be larger? Explain your answers.

(a) O or O^{2-} **(b)** O or S   **(c)** Fe or Fe^{3+} **(d)** H or H$^-$

interlude—THE AURORA BOREALIS: ATOMIC SPECTRA ON A GRAND SCALE

The aurora borealis, or northern lights.

Every so often on a clear evening, residents in Alaska, Canada, and other far-northern parts of the world are treated to a breathtaking display of celestial fireworks—the so-called northern lights, or *aurora borealis*. The lights appear in many different forms—as curtains, arcs, rays, and gauzy patches—and have many different colors. All, however, result from the emission of light by energetically excited atoms in the upper atmosphere, the same kind of phenomenon that gives rise to atomic line spectra.

The aurora borealis is caused by a chain of events that begins on the surface of the sun with a massive solar flare. These flares eject a solar "gas" of energetic protons and electrons that travel at about 900 km/s and that after 2 days reach earth, where they are attracted toward the north and south magnetic poles. (The Southern Hemisphere has its own display of lights called the *aurora australis*.) The energetic electrons are deflected by the earth's magnetic field into a series of sheet-like beams, much as iron filings scattered around a magnet are deflected into a series of lines by the magnet's field of force. The electrons then collide with O_2 and N_2 molecules in the upper atmosphere, exciting them, ionizing them, and breaking them apart into O and N atoms.

The energetically excited atoms, ions, and molecules generated by collisions with electrons emit energy of characteristic wavelengths when they decay to their ground states. O_2^+ ions emit a crimson-red light around 630 nm; N_2^+ ions emit violet and blue light at 391.4 nm and 470.0 nm; and O atoms emit a greenish-yellow light at 557.7 nm and a deep red light at 630.0 nm.

Protons in the solar gas are also responsible for part of the auroral display as they too collide with oxygen atoms when they descend into the upper atmosphere. The protons pull electrons from the oxygen atoms, yielding excited O^+ ions and H atoms that give off still more colors when they return to their ground states. The hydrogen atoms, in particular, emit all the wavelengths of visible light in the Balmer series.

Northern lights are seen almost every night by observers within about 2000 km of the north magnetic pole and are visible in the northern parts of the United States several times a year. They have even been seen as far south as Mexico during times of massive solar disturbances.

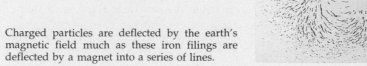

Charged particles are deflected by the earth's magnetic field much as these iron filings are deflected by a magnet into a series of lines.

Understanding the nature of atoms and molecules begins with an understanding of light and other kinds of **electromagnetic radiation** that make up the **electromagnetic spectrum**. An electromagnetic wave travels through a vacuum at the speed of light (c) and is characterized by its **frequency** (ν), **wavelength** (λ), and **amplitude**. Unlike the white light of the sun, which consists of a nearly continuous distribution of wavelengths, the light emitted by an excited atom consists of only a few discrete wavelengths, a so-called **line spectrum**. The observed wavelengths correspond to the specific energy differences between energies of different orbitals.

Atomic line spectra arise because the energy of electromagnetic radiation, like mass, is **quantized** and occurs only in discrete units, or **quanta.** Just as light behaves in some respects like a stream of small particles (**photons**), so electrons and other tiny units of matter behave in some respects as if they were waves. The wavelength of a particle of mass m traveling at a velocity v is given by the **de Broglie equation**, $\lambda = h/mv$, where h is a physical constant called **Planck's constant**.

The **quantum-mechanical** model describes an atom by a mathematical equation similar to that used to describe wave motion. The behavior of each electron in an atom is characterized by a **wave function**, or **orbital**, the square of which defines the probability of finding the electron in various regions of space. Each wave function has a set of three variables, called **quantum numbers**, that describe the energy, shape, and spatial orientation of the corresponding orbital. The principal quantum number n defines the size of the orbital; the angular-momentum quantum number l defines the shape of the orbital; and the magnetic quantum number m_l defines the spatial orientation of the orbital. In a one-electron hydrogen atom the energy of an orbital depends only on n. In a multielectron atom, the energy of an orbital depends on both n and l. In addition, a fourth quantum number, m_s, specifies the electron spin as either up or down.

Orbitals can be grouped into successive **shells** according to their principal quantum number n. Within a shell, orbitals are grouped into s, p, d, and f **subshells** according to their angular-momentum quantum numbers. An s orbital is spherical, a p orbital is dumbbell shaped, and four of the five d orbitals are cloverleaf shaped.

The **ground-state electron configuration** of a multielectron atom is arrived at by following the rules of **aufbau principle**:

1. Fill the lowest energy orbitals first.
2. Put only two electrons of opposite spin into any one orbital (the **Pauli exclusion principle**).
3. If two or more orbitals are equal in energy (**degenerate**), fill each one half full before fully filling any one of them (**Hund's rule**).

The **periodic table** created by Dmitri Mendeleev in 1869 is the most important organizing principle of chemistry. Elements in each group of the periodic table have similar **valence-shell electron configurations** and therefore have similar properties. For example, atomic radii show a periodic rise-and-fall pattern according to their position in the table. Atomic radii decrease from left to right across a period and increase down a group.

UNDERSTANDING KEY CONCEPTS

1. Test your knowledge of the periodic table by indicating where on the following outline elements that meet the following descriptions appear.
   (a) elements with the valence-shell ground-state electron configuration $ns^2 np^5$
   (b) an element whose fourth shell contains two p electrons
   (c) an element with the ground-state electron configuration [Ar] $4s^2 3d^{10} 4p^5$

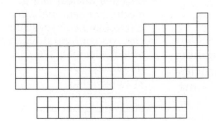

2. What atom has the following orbital-filling diagram?

   [Ar] $\underset{4s}{\uparrow\downarrow}$ $\quad$ $\underset{3d}{\uparrow\downarrow \; \uparrow\downarrow \; \uparrow\downarrow \; \uparrow\downarrow \; \uparrow\downarrow}$ $\quad$ $\underset{4p}{\uparrow \quad \_\_ \quad \_\_}$

3. Draw orbital-filling diagrams of the sort shown in the previous problem for the following atoms. Show each electron as an up or down arrow, and use the abbreviation of the preceding noble gas to represent core electrons.
   (a) K   (b) Mo   (c) Sn   (d) Ir

4. Which of the following three spheres is most likely to represent a K^+ ion, which a K atom, and which a Cl^- ion?

   $r = 227$ pm $\qquad$ $r = 184$ pm $\qquad$ $r = 133$ pm

5. What is the atomic number and expected ground-state electron configuration of the yet undiscovered element directly below Rn in the periodic table?

6. At what atomic number is the filling of a g orbital likely to begin?

7. Identify each of the following orbitals, and give n and l quantum numbers for each.

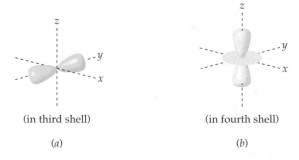

   (in third shell) $\qquad\qquad$ (in fourth shell)

   (a) $\qquad\qquad\qquad\qquad$ (b)

ADDITIONAL PROBLEMS

Problems 5.1 to 5.22 apear within the chapter.

ELECTROMAGNETIC RADIATION

5.23 Which color of light has the higher frequency, red light or violet light? Which has the higher wavelength? Which has the higher energy?

5.24 Which has the higher frequency, infrared light or ultraviolet light? Which has the higher wavelength? Which has the higher energy?

5.25 What is the wavelength in meters of ultraviolet light with $\nu = 5.5 \times 10^{15}$ s^{-1}?

5.26 What is the frequency of a microwave with $\lambda = 4.33 \times 10^{-3}$ m?

5.27 What is the energy in joules of one photon of microwave radiation with $\nu = 2.4 \times 10^{11}$ s^{-1}? What is the energy in kJ/mol of Avogadro's number (1 mol) of these photons?

5.28 Calculate the energies of the following waves in kJ/mol, and tell which member of each pair has the higher value.
   (a) an FM radio wave at 99.5 MHz and an AM radio wave at 115.0 KHz?

(b) an X ray with $\lambda = 3.44 \times 10^{-9}$ m and a microwave with $\lambda = 6.71 \times 10^{-2}$ m?

5.29 The MRI (magnetic resonance imaging) body scanners used in hospitals operate with 400 MHz radiofrequency energy. How much energy does this correspond to in kJ/mol?

5.30 Launched in August 1977, the unmanned space probe *Voyager 2* reached the planet Neptune in August 1989. How many seconds did it take a transmission from *Voyager 2* to reach Earth from Neptune, a distance of 4.5×10^9 km?

5.31 What is the wavelength in meters of photons with the following energies? In what region of the electromagnetic spectrum does each appear?
   (a) 90.5 kJ/mol   (b) 8.05×10^{-4} kJ/mol
   (c) 1.83×10^3 kJ/mol

5.32 What is the energy of each of the following photons in kJ/mol?
   (a) $\nu = 5.97 \times 10^{19}$ s^{-1}

(b) $\nu = 1.26 \times 10^6 \text{ s}^{-1}$
(c) $\lambda = 2.57 \times 10^2 \text{ m}$

5.33 What is the wavelength in meters of a proton (mass $= 1.673 \times 10^{-24}$ g) that has been accelerated to 25% of the speed of light?

5.34 What is the de Broglie wavelength in meters of a baseball weighing 145 g and traveling at 156 km/h?

5.35 What is the de Broglie wavelength in meters of a mosquito weighing 1.55 mg and flying at 1.38 m/s?

5.36 At what speed in m/s would a 145 g baseball have to be traveling to have a de Broglie wavelength of 0.500 nm?

ATOMIC SPECTRA

5.37 Explain how atomic line spectra can be used to identify the elements present in a material of unknown composition.

5.38 According to the equation for the Balmer line spectrum of hydrogen, a value of $n = 3$ gives a red spectral line at 656.3 nm, a value of $n = 4$ gives a green line at 486.1 nm, and a value of $n = 5$ gives a blue line at 434.0 nm. Calculate the energy in kJ/mol of the radiation corresponding to each of these spectral lines.

5.39 According to the values cited in Problem 5.38, the wavelength differences between lines in the Balmer series become smaller as n becomes larger. In other words, the wavelengths converge toward a minimum value as n becomes very large. At what wavelength in nm do the lines converge?

5.40 The wavelength of light at which the Balmer series converges (Problem 5.39) corresponds to the amount of energy required to completely remove an electron from the second shell of a hydrogen atom. Calculate this energy in kJ/mol.

5.41 Lines in the Brackett series of the hydrogen spectrum are caused by emission of energy accompanying the fall of an electron from outer shells to the fourth shell. The lines can be calculated using the Balmer-Rydberg equation:

$$\frac{1}{\lambda} = R\left[\frac{1}{m^2} - \frac{1}{n^2}\right]$$

where $m = 4$, $R = 1.097 \times 10^{-2}$ nm^{-1}, and n is an integer greater than 4. Calculate the wavelengths in nm and energies in kJ/mol of the first two lines in the Brackett series. In what region of the electromagnetic spectrum do they fall?

5.42 Sodium atoms emit light with a wavelength of 330 nm when an electron moves from a $4p$ orbital to a $3s$ orbital. What is the energy difference between the orbitals in kJ/mol?

ORBITALS AND QUANTUM NUMBERS

5.43 What are the four quantum numbers, and what does each specify?

5.44 What is the Heisenberg uncertainty principle, and how does it affect our description of atomic structure?

5.45 Why do we have to use an arbitrary value such as 95% to determine the spatial limitations of an orbital?

5.46 Draw representations of the three $2p$ orbitals. How do they differ?

5.47 What is meant by the term effective nuclear charge, Z_{eff}?

5.48 How does electron shielding in multielectron atoms give rise to energy differences among $3s$, $3p$, and $3d$ orbitals?

5.49 What l quantum numbers are associated with d subshells? With f subshells?

5.50 How many nodal surfaces does a $4s$ orbital have? Draw a cutaway representation of a $4s$ orbital showing the nodes and the regions of maximum electron probability.

5.51 Give the allowable combinations of quantum numbers for each of the following electrons.
(a) a $4s$ electron   **(b)** a $3p$ electron
(c) a $5f$ electron   **(d)** a $5d$ electron

5.52 Give the orbital designations of electrons with the following quantum numbers.
(a) $n = 3$, $l = 0$, $m_l = 0$
(b) $n = 2$, $l = 1$, $m_l = -1$
(c) $n = 4$, $l = 3$, $m_l = -2$
(d) $n = 4$, $l = 2$, $m_l = 0$

5.53 Tell which of the following combinations of quantum numbers are not allowed. Explain your answers.
(a) $n = 3$, $l = 0$, $m_l = -1$
(b) $n = 3$, $l = 1$, $m_l = 1$
(c) $n = 4$, $l = 4$, $m_l = 0$

5.54 What is the maximum number of electrons in an atom whose highest-energy electrons have the principal quantum number $n = 5$? $n = 6$?

5.55 What is the maximum number of electrons in an atom whose highest-energy electrons have the principal quantum number $n = 4$ and the angular-momentum quantum number $l = 0$?

5.56 Use the Heisenberg uncertainty principle to calculate the uncertainty in meters in the position of a honeybee weighing 0.68 g and traveling at a velocity of 0.85 m/s. Assume that the uncertainty in the velocity is 0.1 m/s.

ELECTRON CONFIGURATIONS

5.57 Why does the number of elements in successive periods of the periodic table increase by the progression 2, 8, 18, 32?

5.58 Explain the following terms.
(a) the aufbau principle
(b) the Pauli exclusion principle
(c) Hund's rule

5.59 Test your knowledge of the periodic table by indicating where on the blank outline elements that meet the following descriptions appear.
(a) elements with electrons whose largest principal quantum number is $n = 4$
(b) elements with the valence-shell ground-state electron configuration $ns^2 np^3$
(c) elements that have only one unpaired p electron
(d) the d-block elements
(e) the p-block elements

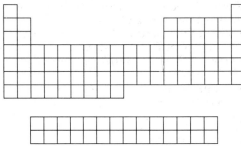

5.60 Which two of the four quantum numbers determine the energy level of an orbital in a multielectron atom?

5.61 Order the orbitals for a multielectron atom in each of the following lists according to increasing energy.
(a) $4d, 3p, 2p, 5s$ (b) $2s, 4s, 3d, 4p$
(c) $6s, 5p, 3d, 4p$

5.62 According to the aufbau principle, which orbital is filled immediately *after* each of the following in a multielectron atom?
(a) $4s$ (b) $3d$ (c) $5f$ (d) $5p$

5.63 According to the aufbau principle, which orbital is filled immediately *before* each of the following?
(a) $3p$ (b) $4p$ (c) $4f$ (d) $5d$

5.64 Give the expected ground-state electron configurations for the following elements.
(a) Ti   (b) Ru   (c) Sn   (d) Sr   (e) Se

5.65 Give the expected ground-state electron configurations for atoms with the following atomic numbers.
(a) $Z = 55$ (b) $Z = 41$ (c) $Z = 80$ (d) $Z = 62$

5.66 Draw orbital-filling diagrams for the following atoms. Show each electron as an up or down arrow, and use the abbreviation of the preceding noble gas to represent core electrons.
(a) Rb   (b) W   (c) Ge   (d) Zr

5.67 Draw orbital-filling diagrams for atoms with the following atomic numbers. Show each electron as an up or down arrow, and use the abbreviation of the preceding noble gas to represent core electrons.
(a) $Z = 25$ (b) $Z = 56$ (c) $Z = 28$ (d) $Z = 47$

5.68 Order the electrons in the following orbitals according to their shielding ability: $4s, 4d, 4f$

5.69 If an element with $Z = 118$ is discovered, what ground-state electron configuration do you think it will have?

5.70 What is the atomic number and expected ground-state electron configuration of the yet undiscovered element directly below Fr in the periodic table?

5.71 What is the atomic number and expected ground-state electron configuration of the yet undiscovered element directly below Bi in the periodic table?

5.72 How many unpaired electrons are present in each of the following ground-state atoms?
(a) O   (b) Si   (c) K   (d) As

5.73 Identify the following atoms:
(a) It has the ground-state electron configuration [Ar] $3d^{10} 4s^2 4p^1$.
(b) It has the ground-state electron configuration [Kr] $4d^{10}$.

5.74 What are the probable ground-state electron configurations of the following ions.
(a) La^{3+} (b) Ag^+ (c) Sn^{2+}

5.75 Identify the element whose 2+ ion has the ground-state electron configuration [Ar] $3d^{10}$.

5.76 There are two elements in the transition-metal series Sc through Zn that have four unpaired electrons in their 2+ ions. Identify them.

ATOMIC RADII AND PERIODIC PROPERTIES

5.77 Why do atomic radii increase down a group of the periodic table?

5.78 Why do atomic radii decrease from left to right across a period of the periodic table?

5.79 Order the following atoms according to increasing atomic radius. S, F, O.

5.80 Which atom in each of the following pairs has a

larger radius?
(a) Na or K   (b) V or Ta   (c) V or Zn   (d) Li or Ba

5.81 The amount of energy required to remove an electron from a neutral atom to give a cation is called the atom's *ionization energy*. Which would you expect to have the larger ionization energy, Na or Mg? Explain.

5.82 The amount of energy released when an electron is added to a neutral atom to give an anion is called the atom's *electron affinity*. Which would you expect to have the larger electron affinity, C or F? Explain.

5.83 In which group of the periodic table will element 120 fall if and when it is discovered?

5.84 Cu^+ has an ionic radius of 77 pm, but Cu^{2+} has an ionic radius of 73 pm. Explain.

5.85 The following ions all have the same number of electrons: Ti^{4+}, Sc^{3+}, Ca^{2+}, S^{2-}. Order them according to their expected size, and explain your answer.

GENERAL PROBLEMS

5.86 Use the Balmer equation to calculate the wavelength in nm of the spectral line for hydrogen when $n = 6$. What is the energy in kJ/mol of the radiation corresponding to this line?

5.87 Use the generalized Balmer-Rydberg equation for hydrogen line spectra to calculate the wavelengths in nm corresponding to the third through sixth lines in the Lyman series.

5.88 Lines in the Pfund series of the hydrogen spectrum are caused by emission of energy accompanying the fall of an electron from outer shells to the fifth shell. Use the Balmer-Rydberg equation to calculate the wavelengths in nm and energies in kJ/mol of the two longest wavelength lines in the Pfund series. In what region of the electromagnetic spectrum do they fall?

5.89 What is the shortest wavelength (in nm) in the Pfund series (Problem 5.88) for hydrogen?

5.90 What is the wavelength (in meters) of photons with the following energies? In what region of the electromagnetic spectrum does each appear?
(a) 142 kJ/mol   (b) 4.55×10^{-2} kJ/mol
(c) 4.81×10^4 kJ/mol

5.91 What is the energy of each of the following photons in kJ/mol?
(a) $\nu = 3.79 \times 10^{11}$ s^{-1}
(b) $\nu = 5.45 \times 10^4$ s^{-1}
(c) $\lambda = 4.11 \times 10^{-5}$ m

5.92 What is the wavelength in meters of a proton (mass $= 1.673 \times 10^{-24}$ g) that has been accelerated to 5.3% of the speed of light?

5.93 What is the de Broglie wavelength in meters of a hummingbird weighing 12.6 g and flying at 12.6 m/s?

5.94 The laser light used in compact disc players has $\lambda = 780$ nm. In what region of the electromagnetic spectrum does this light appear? What is the energy of this light in kJ/mol?

5.95 Excited rubidium atoms emit red light with $\lambda = 795$ nm. What is the energy difference in kJ/mol between orbitals that give rise to this emission?

5.96 Draw orbital-filling diagrams for the following atoms. Show each electron as an up or down arrow, and use the abbreviation of the preceding noble gas to represent core electrons.
(a) Sr   (b) Cd   (c) has $Z = 22$ (d) has $Z = 34$

5.97 One method for calculating Z_{eff} is to use the equation

$$Z_{eff} = \sqrt{\frac{(E)(n^2)}{1312 \text{ kJ/mol}}}$$

where E is the energy necessary to remove an electron from an atom and n is the principal quantum number of the electron. Use this equation to calculate Z_{eff} values for the highest-energy electrons in potassium ($E = 418.8$ kJ/mol) and krypton ($E = 1350.7$ kJ/mol).

5.98 One watt (W) is equal to 1 J/s. Assuming that 5.0% of the energy output of a 75 W light bulb is visible light and that the average wavelength of the light is 550 nm, how many photons are emitted by the light bulb each second?

5.99 Microwave ovens work by irradiating food with microwave radiation, which is absorbed and converted into heat. Assuming that radiation with $\lambda = 15.0$ cm is used and that all the energy is converted to heat, how many photons are necessary to raise the temperature of a 350 mL cup of water from 20°C to 95°C?

5.100 Photochromic sunglasses, which darken when exposed to light, contain a small amount of colorless $AgCl(s)$ embedded in the glass. When irradiated with light, metallic silver atoms are produced and the glass darkens: $AgCl(s) \longrightarrow Ag(s) + Cl$. Escape of the chlorine atoms is prevented by the rigid structure of the glass, and the reaction therefore reverses as soon as the light is removed. If 310 kJ/mol of energy is required to make the reaction proceed, what wavelength of light is necessary?

chapter **6** IONIC BONDS AND SOME
MAIN-GROUP CHEMISTRY

Clearly, there must be some force that holds atoms together in chemical compounds. Otherwise, the atoms would simply fly apart, and no compounds could exist. As we saw briefly in Section 2.8, the forces that hold atoms together are called *chemical bonds* and are of two types—ionic bonds and covalent bonds. In this and the next two chapters, we'll look at the nature of chemical bonds and at the energy changes that accompany their forming and breaking. We'll begin in the present chapter with a look at the ionic bonds formed between halogens and main-group metals.

This beautiful example of ram's-horn selenite gypsum is made of Ca^{2+} ions and $SeO_4{}^{2-}$ ions.

6.1 ►IONIZATION ENERGY

We saw in the previous chapter that the absorption of light energy by an atom leads to a change in electron configuration. A valence-shell electron is promoted from a lower-energy orbital to a higher-energy orbital with a larger principal quantum number n. If enough energy is absorbed, the electron can be removed completely from the atom, leaving behind an ion. The amount of energy necessary to remove the outermost electron from an isolated neutral atom in the gaseous state is called the atom's **ionization energy**, abbreviated E_i. For 1 mol of hydrogen atoms, 1312.0 kJ of energy is needed.

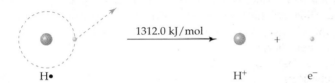

As shown by the plot in Figure 6.1, ionization energies of the various elements differ widely, from a low of 375.7 kJ/mol for cesium to a high of 2372.3 kJ/mol for helium. Furthermore, there is a clear periodicity to the data. The minimum E_i values correspond to the group 1A alkali metals, the maximum E_i values correspond to the group 8A noble gas elements, and a gradual increase in E_i occurs from left to right across any one period of the periodic table—from Na to Ar, for example.

The periodicity evident in Figure 6.1 is easily explained by looking at electron configurations. All the noble-gas elements in group 8A have filled valence subshells, either s (for helium) or both s and p for the other noble gases. As we saw in Section 5.15, electrons in a filled valence subshell feel a relatively high effective nuclear charge, Z_{eff}, because electrons within the same shell don't shield each other very strongly. As a result, the electrons

FIGURE 6.1 Ionization energies of the first 92 elements. There is an obvious periodicity to the data, with maximum values for the noble-gas elements and minimum values for the alkali metals. Note that all the values are positive, meaning that energy is always required to remove an electron from an atom.

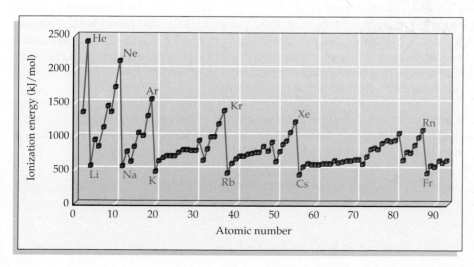

are held tightly to the nucleus, the radius of the atom shrinks, and the energy necessary to remove an electron is relatively high. The alkali-metal elements in group 1A, by contrast, have only a single *s* electron in their valence shell. This single valence electron is well shielded from the nucleus by the inner-shell electrons, called the **core electrons,** resulting in a low Z_{eff}. The electron is thus held loosely, and the energy necessary to remove it is relatively low.

The three-dimensional display of ionization energies in Figure 6.2 shows that there are other trends in the data beyond the obvious periodicity. One such trend is that ionization energies gradually decrease down a column in the periodic table, from Li to Cs and from He to Rn, for example. As atomic number increases down a column, both the principal quantum number of the valence-shell electrons and their average distance from the nucleus also increase. As a result, the valence-shell electrons are less tightly held, and E_i is lower.

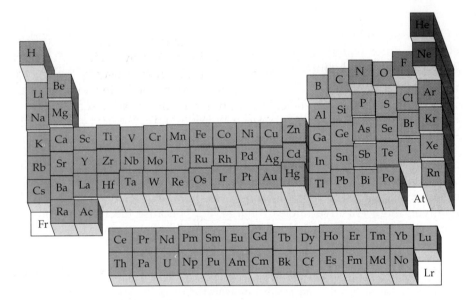

FIGURE 6.2 A three-dimensional display showing how the first ionization energies increase from left to right across a row and decrease from top to bottom down a group of the periodic table. The elements at the lower left therefore have the smallest E_i values, and the elements at the upper right have the highest.

Still another feature of the E_i data is that minor irregularities occur from left to right across a row of the periodic table. As shown in the close-up look at E_i values of the first 20 elements in Figure 6.3, the E_i of beryllium is higher than that of its neighbor boron, and the E_i of nitrogen is higher than that of its neighbor oxygen. Similarly, magnesium has a higher E_i than aluminum, and phosphorus has a slightly higher E_i than sulfur.

The unusually high E_i values for the group 2A elements Be, Mg, and so forth can be explained by looking at their electron configurations. Compare beryllium with boron, for example. An *s* electron is removed on ionization of beryllium, but a *p* electron is removed on ionization of boron.

$$\text{Be } (1s^2 2s^2) \rightarrow \text{Be}^+ (1s^2 2s) + \text{e}^- \quad (E_i = 899.4 \text{ kJ/mol})$$
s electron removed

$$\text{B } (1s^2\, 2s^2\, 2p^1) \rightarrow \text{B}^+ (1s^2\, 2s^2) + \text{e}^- \quad (E_i = 800.6 \text{ kJ/mol})$$
p electron removed

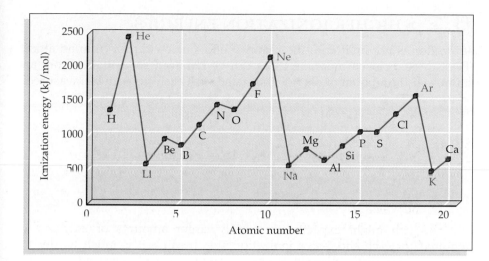

Because a $2s$ electron penetrates closer to the nucleus than a $2p$ electron does, it is held more tightly and is harder to remove. Thus, the E_i of beryllium is higher than that of boron. An alternative way of saying the same thing is to note that the $2p$ electron of boron is shielded somewhat by the $2s$ electrons, feels a smaller Z_{eff}, and is thus more easily removed.

The unusually low E_i values for group 6A elements can also be explained by their electron configurations. In comparing nitrogen with oxygen, for example, the nitrogen electron is removed from a half-filled orbital, whereas the oxygen electron is removed from a filled orbital.

half-filled orbital

$$N\ (1s^2\ 2s^2\ 2p_x^{\ 1}\ 2p_y^{\ 1}\ 2p_z^{\ 1}) \rightarrow N^+\ (1s^2\ 2s^2\ 2p_x^{\ 1}\ 2p_y^{\ 1}) + e^- \quad (E_i = 1402.3\ \text{kJ/mol})$$

filled orbital

$$O\ (1s^2\ 2s^2\ 2p_x^{\ 2}\ 2p_y^{\ 1}\ 2p_z^{\ 1}) \rightarrow O^+\ (1s^2\ 2s^2\ 2p_x^{\ 1}\ 2p_y^{\ 1}\ 2p_z^{\ 1}) + e^- \quad (E_i = 1313.9\ \text{kJ/mol})$$

Because electrons repel each other and try to stay as far apart as possible, electrons that are forced together in a filled orbital are slightly higher in energy than those in a half-filled orbital, so it is slightly easier to remove one. Thus, oxygen has a lower E_i than nitrogen.

EXAMPLE 6.1

Arrange the three elements Se, Cl, and S in order of increasing ionization energy.

SOLUTION   Ionization energy generally increases from left to right across a row of the periodic table and decreases from top to bottom down a column. Chlorine should therefore have a higher E_i than its neighbor sulfur, and selenium should have a lower E_i than sulfur. Thus, the order is Cl > S > Se.

PROBLEM 6.1 Using only the periodic table and not the data in Figure 6.1, predict which element in each of the following pairs has the higher ionization energy.
(a) K or Br   **(b)** S or Te   **(c)** Ga or Se   **(d)** Ne or Sr

6.2 ►HIGHER IONIZATION ENERGIES

Ionization is not limited to the removal of a single electron from an atom. Two, three, or even more electrons can be removed sequentially from an atom, and the amount of energy associated with each step can be measured.

$$M + energy \rightarrow M^+ + e^- \qquad \text{First ionization energy } (E_{i1})$$

$$M^+ + energy \rightarrow M^{2+} + e^- \qquad \text{Second ionization energy } (E_{i2})$$

$$M^{2+} + energy \rightarrow M^{3+} + e^- \qquad \text{Third ionization energy } (E_{i3})$$

.
.
.

As you might expect, successively larger amounts of energy are required for each successive ionization step because it is much harder to remove a negatively charged electron from a positively charged ion than from a neutral atom. Interestingly, though, the energy differences between successive steps vary dramatically from one element to another. Removing the second electron from sodium takes nearly 10 times as much energy as removing the first one (4562 versus 496 kJ/mol), but removing the second electron from magnesium takes only twice as much energy as removing the first one (1451 versus 738 kJ/mol). Similarly large jumps are found for the other elements, as can be seen by following the zigzag line in Table 6.1. Magnesium has a large jump between its second and third ionization energies, aluminum has a large jump between its third and fourth ionization energies, silicon has a large jump between its fourth and fifth ionization energies, and so on.

The large jumps in ionization energies that coincide with the zigzag line in Table 6.1 are yet another consequence of electron configuration. It is relatively easy to remove an electron from a *partially* filled valence shell, but

TABLE 6.1	Successive Ionization Energies (kJ/mol) for Third-Row Elements							
E_i Number	Na	Mg	Al	Si	P	S	Cl	Ar
E_{i1}	496	738	578	787	1012	1000	1251	1520
E_{i2}	4562	1451	1817	1577	1903	2251	2297	2665
E_{i3}	6912	7733	2745	3231	2912	3361	3822	3931
E_{i4}	9543	10,540	11,575	4356	4956	4564	5158	5770
E_{i5}	13,353	13,630	14,830	16,091	6273	7013	6540	7238
E_{i6}	16,610	17,995	18,376	19,784	22,233	8495	9458	8781
E_{i7}	20,114	21,703	23,293	23,783	25,397	27,106	11,020	11,995

it's relatively difficult to remove an electron from an atom or ion that has a filled valence shell and a noble-gas electron configuration. In other words, there is high degree of stability associated with filled s and p sublevels, which corresponds with having eight electrons (an *octet*) in the valence shell of an atom or ion. Sodium ([Ne] $3s^1$) loses only one electron easily, magnesium ([Ne] $3s^2$) loses only two electrons easily, aluminum ([Ne] $3s^2 3p^1$) loses only three electrons easily, and so on across the row. We'll explore further the stability of valence-shell electron octets in Section 6.10.

8 electrons in
outer (2nd) shell

$$\text{Na } (1s^2\ 2s^2\ 2p^6\ 3s^1) \quad \rightarrow \text{Na}^+ (1s^2\ 2s^2\ 2p^6) + e^-$$

$$\text{Mg } (1s^2\ 2s^2\ 2p^6\ 3s^2) \quad \rightarrow \text{Mg}^{2+} (1s^2\ 2s^2\ 2p^6) + 2\ e^-$$

$$\text{Al } (1s^2\ 2s^2\ 2p^6\ 3s^2\ 3p^1) \rightarrow \text{Al}^{3+} (1s^2\ 2s^2\ 2p^6) + 3\ e^-$$

$$\text{Cl } (1s^2\ 2s^2\ 2p^6\ 3s^2\ 3p^5) \rightarrow \text{Cl}^{7+} (1s^2\ 2s^2\ 2p^6) + 7\ e^-$$

Because valence-shell electrons rather than core electrons are the most easily lost during ionization, valence-shell electrons are also the most likely to be lost or shared during chemical reactions. We'll see repeatedly in later chapters that *the valence-shell electron configuration of an atom controls the atom's chemistry*.

EXAMPLE 6.2

Which would you expect to have the larger fifth ionization energy, Ge or As?

SOLUTION   As their positions in the periodic table indicate, the group 4A element germanium has four valence-shell electrons and thus four relatively low ionization energies, whereas the group 5A element arsenic has five valence-shell electrons and five low ionization energies. Germanium therefore has a higher E_{i5} than arsenic.

PROBLEM 6.2 (a) Which would you expect to have the larger third ionization energy, Be or N? **(b)** Which would you expect to have the larger fourth ionization energy, Ga or Ge?

PROBLEM 6.3 Three atoms have the following electron configurations:
(a) $1s^2\ 2s^2\ 2p^6\ 3s^2\ 3p^1$ **(b)** $1s^2\ 2s^2\ 2p^6\ 3s^2\ 3p^5$ **(c)** $1s^2\ 2s^2\ 2p^6\ 3s^2\ 3p^6\ 4s^1$
Which of the three has the highest E_{i1}? Which has the lowest E_{i4}?

6.3 ►ELECTRON AFFINITY

Just as it's possible to measure the energy necessary to *remove* an electron from an atom to form a cation, it's also possible to measure the energy necessary to *add* an electron to an atom to form an anion. An element's **electron affinity,** abbreviated E_{ea}, is defined as the energy change that occurs when an electron is added to an isolated atom in the gaseous state. By convention, electron affinity has a negative value when energy is released. The more negative the E_{ea}, the greater the tendency of the atom to accept an electron, and the more stable the anion that results. By contrast, an atom that forms an unstable anion should, in principle, have a positive value of E_{ea}, but no experimental measurement can be made in such circumstances. All we can say is that the E_{ea} for such an atom is greater than zero. The E_{ea} of hydrogen, for instance, is -72.8 kJ/mol, meaning that energy is released and the H^- anion is stable. The E_{ea} of calcium, however, is greater than 0 kJ/mol, meaning that energy is absorbed and the Ca^- anion is unstable.

$$H\ (1s^1) + e^- \rightarrow H^-\ (1s^2) + 72.8\ kJ/mol \quad (E_{ea} = -72.8\ kJ/mol)$$

$$Ca\ (.\ .\ .\ 4s^2) + e^- + energy \rightarrow Ca^-\ (.\ .\ .\ 4s^2\ 4p^1) \quad (E_{ea} > 0\ kJ/mol)$$

As was true for ionization energies, electron affinities show a periodicity that is related to the electron configurations of the various elements (Figure 6.4). The data in Figure 6.4 show that addition of an electron to an atom is energetically favorable, except for the alkaline earths and noble gases. Group 7A elements have the most negative electron affinities, corresponding to the largest release of energy, whereas group 2A and group 8A elements have positive electron affinities, corresponding to an absorption of energy.

FIGURE 6.4 Electron affinities for elements 1–57 and 72–86. A negative value means that energy is released when an electron adds to an atom, while a value of zero means that energy is absorbed but that the exact amount can't be measured. Note that the group 2A elements (alkaline earths) and the group 8A elements (noble gases) have E_{ea} values near zero, while the group 7A elements (halogens) have large negative E_{ea}'s. Electron affinities have not been measured for elements 58–71.

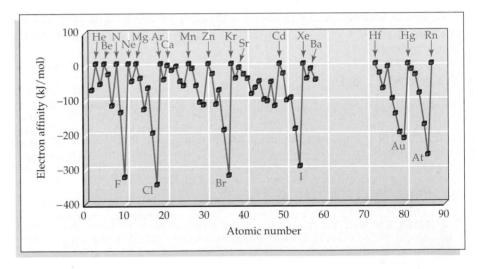

The sign and magnitude of an element's electron affinity are caused by an interplay of several offsetting factors. Attraction between the additional electron and the atomic nucleus favors a negative E_{ea}, but the increase in electron-electron repulsions that results from addition of the extra electron favors a positive E_{ea}. The large negative E_{ea}'s of the halogens (F, Cl, Br, I) occur because these elements have a high effective nuclear charge, yet also have room in their valence shell for the additional electron. The resulting anions have a noble-gas electron configuration with filled s and p sublevels, and the attraction between the additional electron and the atomic nucleus is high. The positive E_{ea}'s of the noble-gas elements (He, Ne, Ar, Kr, Xe) occur because the s and p sublevels are full, and the additional electron must therefore go into the next higher shell, where it is shielded from the nucleus and feels a relatively low Z_{eff}. The attraction of the nucleus for the added electron is therefore small and is outweighed by the additional electron–electron repulsions.

A halogen: $Cl\,(\ldots 3s^2\,3p^5) + e^- \rightarrow Cl^-\,(\ldots 3s^2 3p^6) + energy$ $(E_{ea} = -348.6\;kJ/mol)$

A noble gas: $Ar\,(\ldots 3s^2\,3p^6) + e^- + energy \rightarrow Ar^-\,(\ldots 3s^2\,3p^6 4s^1)\;(E_{ea} > 0\;kJ/mol)$

In looking for other trends in the data shown in Figure 6.4, the near-zero E_{ea}'s of the alkaline-earth metals (Be, Mg, Ca, Sr, Ba) are particularly striking. Atoms of these elements have filled s subshells, which means that the additional electron must go into a p subshell. The higher energy of the p subshell, together with a relatively low Z_{eff} for elements on the left side of the periodic table, means that the alkaline earths accept an electron reluctantly and have E_{ea} values near zero.

An alkaline earth: $Mg\,(\ldots 3s^2) + e^- + energy \rightarrow Mg^-\,(\ldots 3s^2 3p^1)\;(E_{ea} > 0\;kJ/mol)$

EXAMPLE 6.3

Why does nitrogen (atomic number 7) have a less favorable (more positive) E_{ea} than its neighbors on either side, C and O?

SOLUTION   The magnitude of an element's E_{ea} depends heavily on the element's valence-shell electron configuration. The electron configurations of C, N, and O are as follows:

    Carbon: $1s^2\,2s^2\,2p_x^{\,1}\,2p_y^{\,1}$ Nitrogen: $1s^2\,2s^2\,2p_x^{\,1}\,2p_y^{\,1}\,2p_z^{\,1}$
    Oxygen: $1s^2\,2s^2\,2p_x^{\,2}\,2p_y^{\,1}\,2p_z^{\,1}$

Carbon has only two electrons in its $2p$ subshell and can readily accept another in its vacant $2p_z$ orbital. Nitrogen, however, has a half-filled $2p$ subshell, and the additional electron must pair up in a $2p$ orbital, where it feels a repulsion from the electron already present. Thus, the E_{ea} of nitrogen is more positive than that of carbon. Oxygen also must add an electron to an orbital that already has one electron, but the additional stabilizing effect of increased Z_{eff} across the periodic table evidently counteracts the effect of electron repulsion, resulting in a lower E_{ea} for O than for N.

PROBLEM   Use electron configurations to explain why manganese (atomic number 25) has a less negative E_{ea} than its neighbors on either side.

Sodium metal burns in a chlorine atmosphere to yield solid sodium chloride.

6.4 ➤IONIC BONDS AND THE FORMATION OF IONIC SOLIDS

What would happen if an element that gives up an electron easily (that is, has a low ionization energy) were to come in contact with an element that accepts an electron easily (that is, has a highly negative electron affinity)? The element with a low E_i can transfer an electron to the element with a negative E_{ea}, yielding a cation and an anion. Sodium, for example, reacts with chlorine to give Na^+ ions and Cl^- ions.

$$Na + Cl \rightarrow Na^+\;Cl^-$$

$1s^2\,2s^2\,2p^6\,3s^1$ $1s^2\,2s^2\,2p^6\,3s^2\,3p^5$ $1s^2\,2s^2\,2p^6$ $1s^2\,2s^2\,2p^6\,3s^2\,3p^6$

The oppositely charged Na^+ and Cl^- ions that result when sodium transfers an electron to chlorine are attracted together by electrostatic forces, and we say that they are joined by an **ionic bond.** It's important to realize, though, that a visible sample of sodium chloride does not consist of individual pairs of Na^+ and Cl^- ions. Solid NaCl consists of a three-dimensional network of ions in which each Na^+ is surrounded by and attracted to *many* Cl^- ions, and each Cl^- is surrounded by and attracted to many Na^+ ions (Figure 6.5).

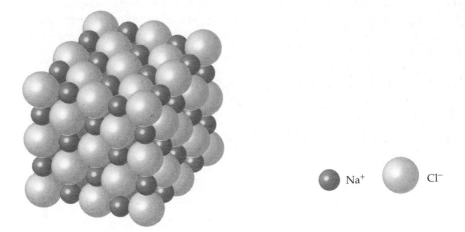

FIGURE 6.5 The arrangement of ions in a sodium chloride crystal. Each Na^+ ion is surrounded by six nearest-neighbor Cl^- ions, and each Cl^- ion is surrounded by six nearest-neighbor Na^+ ions.

What about the energy change, ΔE, that occurs when sodium and chlorine react to yield Na^+ and Cl^- ions? It's apparent from E_i and E_{ea} values that the amount of energy released when a chlorine atom accepts an electron ($E_{ea} = -348.6$ kJ/mol) is not large enough to offset the amount absorbed when a sodium atom loses an electron ($E_i = +495.8$ kJ/mol):

E_i for Na	$= +495.8$ kJ/mol
E_{ea} for Cl	$= -348.6$ kJ/mol
ΔE	$= +147.2$ kJ/mol (unfavorable)

Thus, the net energy change for the reaction of sodium and chlorine atoms would be unfavorable by $+147.2$ kJ/mol, and no reaction would occur, unless some other factor were involved. This additional factor, which overcomes the unfavorable energy change of electron transfer, is the large gain in stability due to the formation of ionic bonds.

Although the actual reaction of sodium with chlorine occurs all at once rather than in a stepwise manner, it's possible to make energy calculations more easily if we imagine a series of hypothetical steps for which experimentally measured energy values can be obtained. There are five contributions that must be taken into account to calculate the overall energy release during the formation of NaCl from solid sodium metal and gaseous chlorine molecules:

$$Na(s) + 1/2\ Cl_2(g) \rightarrow NaCl(s)$$

1. Conversion of solid Na metal into isolated, gaseous Na atoms without changing the temperature, a process called *sublimation*. Since energy must be added to disrupt the forces holding atoms together in a solid, the heat required for sublimation has a positive value: 107.3 kJ/mol for Na.

$$Na(s) \rightarrow Na(g)$$
$$+107.3 \text{ kJ/mol}$$

2. Dissociation of gaseous Cl_2 molecules into individual Cl atoms. Energy must be added to break molecules apart before reaction can occur, and the energy required for bond breaking therefore has a positive value: 243 kJ/mol for Cl_2 (or 122 kJ/mol for 1/2 Cl_2). We'll look further into the energetics of bond dissociation in Section 8.10.

$$1/2\ Cl_2(g) \rightarrow Cl(g)$$
$$+122 \text{ kJ/mol}$$

3. Ionization of isolated Na atoms into Na^+ ions plus electrons. This is simply the first ionization energy of sodium and has a positive value: 495.8 kJ/mol.

$$Na(g) \rightarrow Na^+(g) + e^-$$
$$+495.8 \text{ kJ/mol}$$

4. Formation of Cl^- ions from Cl atoms by addition of an electron. This is simply the electron affinity of chlorine and has a negative value: −348.6 kJ/mol.

$$Cl(g) + e^- \rightarrow Cl^-(g)$$
$$-348.6 \text{ kJ/mol}$$

5. Formation of solid NaCl from isolated Na^+ and Cl^- ions. The energy change for this step is a measure of the overall electrostatic attractions between ions in the solid. It is the amount of energy released when isolated ions condense to form a solid, and it has a negative value: −787 kJ/mol for NaCl.

$$Na^+(g) + Cl^-(g) \rightarrow NaCl(s)$$
$$-787 \text{ kJ/mol}$$

Net reaction: $\quad Na(s) + 1/2\ Cl_2(g) \rightarrow NaCl(s)$
Net energy change: $\quad$ **−411 kJ/mol**

A pictorial way of viewing the five hypothetical steps in the reaction between sodium and chlorine is shown Figure 6.6. Called a **Born-Haber cycle**, the figure shows how each step contributes to the overall energy change and how the net process is the sum of the individual steps. It is clear from the diagram that steps 1, 2, and 3 absorb energy (have positive values), while steps 4 and 5 release energy (have negative values). The largest

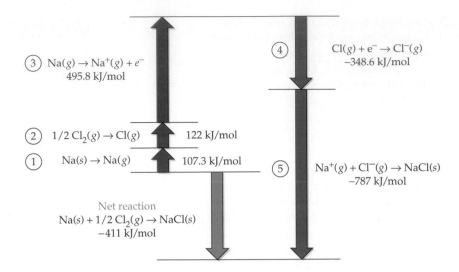

③ $Na(g) \rightarrow Na^+(g) + e^-$
495.8 kJ/mol

④ $Cl(g) + e^- \rightarrow Cl^-(g)$
−348.6 kJ/mol

② $1/2\ Cl_2(g) \rightarrow Cl(g)$　122 kJ/mol

① $Na(s) \rightarrow Na(g)$　107.3 kJ/mol

⑤ $Na^+(g) + Cl^-(g) \rightarrow NaCl(s)$
−787 kJ/mol

Net reaction
$Na(s) + 1/2\ Cl_2(g) \rightarrow NaCl(s)$
−411 kJ/mol

FIGURE 6.6 A Born-Haber cycle for the formation of NaCl from Na(s) and $Cl_2(g)$. The sum of the energy changes for the five hypothetical steps is equal to the net energy change for the overall reaction. Note that the most favorable step is the formation of solid NaCl from gaseous Na^+ and Cl^- ions (step 5).

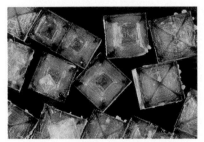

These large NaCl crystals consist of Na^+ and Cl^- ions held together by electrostatic attraction.

contribution by far is step 5, which takes into account the electrostatic interactions between ions. Were it not for this large amount of stabilization of the solid due to ionic bonding, no reaction would take place.

The sum of the electrostatic interaction energies between ions in a solid is called the **lattice energy (U)** of the solid. By convention, the lattice energy refers to the *breakup* of a crystal into individual ions. It therefore has a positive value because energy is required to separate the electrical charges. The formation of a crystal from ions, however, is the reverse of the breakup, and step 5 in the Born-Haber cycle of Figure 6.6 therefore has the value $-U$.

$NaCl(s) \rightarrow Na^+(g) + Cl^-(g)$　　$U = +787$ kJ/mol (energy absorbed)

$Na^+(g) + Cl^-(g) \rightarrow NaCl(s)$　　$-U = -787$ kJ/mol (energy released)

The magnitude of a lattice energy is described by Coulomb's law and is equal to a constant k times the charges on the ions, z_1 and z_2, divided by the distance between their centers, d:

$$-U = k \times \frac{z_1 z_2}{d}$$

The value of the constant k depends on the geometric arrangement of the ions in the specific compound and is different for different substances. Lattice energies are largest when the distance d between ions is small and when one or both of the charges z_1 and z_2 are large.

A small distance d means that the ions are close together, which means in turn that they have small ionic radii. Thus, if z_1 and z_2 are held constant, the largest lattice energies belong to compounds formed from the smallest ions, as shown in Table 6.2. Within a series of compounds that have the same anion but different cations, the order of lattice energies is the same as the order of decreasing cation size. Among LiF, NaF, and KF, for example, the

TABLE 6.2	Lattice Energies of Some Ionic Solids (kJ/mol)				
	Anion				
Cation	F^-	Cl^-	Br^-	I^-	O^{2-}
Li^+	1036	853	807	757	2925
Na^+	923	787	747	704	2695
K^+	821	715	682	649	2360
Be^{2+}	3505	3020	2914	2800	4443
Mg^{2+}	2957	2524	2440	2327	3791
Ca^{2+}	2630	2258	2176	2074	3401
Al^{3+}	5215	5492	5361	5218	15,916

order of decreasing ionic size is $K^+ > Na^+ > Li^+$, and the order of lattice energies is LiF > NaF > KF. Similarly, within a series of compounds that have the same cation but different anions, the order of lattice energies is the same as the order of decreasing anion size. Among LiF, LiCl, LiBr, and LiI, for example, the order of decreasing anion size is $I^- > Br^- > Cl^- > F^-$, and the order of lattice energies is LiF > LiCl > LiBr > LiI.

Table 6.2 also shows that compounds of ions with higher charges have greater lattice energies than compounds of ions with lower charges. In comparing NaI, MgI_2, and AlI_3, for example, the order of charges on the cations is $Al^{3+} > Mg^{2+} > Na^+$, and the order of lattice energies is $AlI_3 > MgI_2 > NaI$.

EXAMPLE 6.4

Which has the higher lattice energy, NaCl or CsI?

SOLUTION The magnitude of a substance's lattice energy is affected both by the charges on its constituent ions and by the sizes of those ions. The higher the charges on the ions and the smaller the sizes of the ions, the greater the lattice energy. All four ions, Na^+, Cs^+, Cl^-, and I^-, are singly charged, but they differ in size. Since Na^+ is smaller than Cs^+, and Cl^- is smaller than I^-, the distance between ions is smaller in NaCl than in CsI. Thus, NaCl has the higher lattice energy.

PROBLEM 6.5 Calculate the energy change in kJ/mol if potassium atoms react with fluorine atoms to yield isolated K^+ and F^- ions. Is the reaction favorable (negative overall energy change) or unfavorable (positive overall energy change)? (E_{ea} for F = -328 kJ/mol; E_i for K = $+418.8$ kJ/mol)

PROBLEM 6.6 Calculate the net energy change in kJ/mol that takes place on formation of KF(s) from the elements: $K(s) + 1/2 \ F_2(g) \rightarrow KF(s)$. The following information is needed:
Heat of sublimation for K = 89.2 kJ/mol E_{ea} for F = -328 kJ/mol
Bond dissociation energy for F_2 = 158 kJ/mol E_i for K = 418.8 kJ/mol
Lattice energy for KF = 821 kJ/mol

PROBLEM 6.7 Predict which substance in each of the following pairs has the higher lattice energy: **(a)** KCl or RbCl   **(b)** CaF_2 or BaF_2 **(c)** CaO or KI

6.5 ▸THE ALKALI METALS (GROUP 1A)

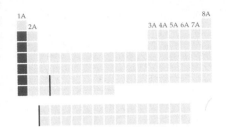

Now that we have discussed ionization energies, electron affinities, and ionic bonding, let's look at the chemistry of some of the elements that form ionic bonds. The alkali-metals in group 1A—Li, Na, K, Rb, Cs, and Fr—have the lowest ionization energies of all the elements because of their valence-shell ns^1 electron configurations (Figure 6.1). They therefore lose this ns^1 electron easily in chemical reactions to yield +1 ions and are thus among the most powerful reducing agents in the periodic table (Sections 4.6–4.7). In fact, the entire chemistry of the alkali metals is dominated by their ability to lose an electron to other elements or compounds.

As their group name implies, the alkali metals are *metallic*. They have a bright, silvery appearance, are malleable, and are good conductors of electricity. Unlike the more common metals such as iron, though, the alkali metals are all soft enough to cut with a dull knife, have low melting points and densities, and are so reactive that they must be stored under oil to prevent their instantaneous reaction with oxygen and water. None is found in the elemental state in nature; they occur only in salts. Their properties are summarized in Table 6.3.

TABLE 6.3 Properties of the Alkali Metals

Name	Melting Point (°C)	Boiling Point (°C)	Density (g/cm³)	1st Ionization Energy (kJ/mol)	Abundance on Earth (%)	Atomic Radius (pm)	Ionic (M⁺) Radius (pm)
Lithium	181	1342	0.53	520.2	0.002 0	152	68
Sodium	98	883	0.97	495.8	2.36	186	102
Potassium	63	760	0.86	418.8	2.09	227	138
Rubidium	39	686	1.53	403.0	0.009 0	248	147
Cesium	28	669	1.87	375.7	0.000 10	265	167
Francium	—	—	—	~400	Trace	—	—

A sample of lithium metal.

Occurrence and Uses of the Alkali Metals

First isolated in 1817 from the mineral petalite, $LiAlSi_4O_{10}$, lithium was named from the Greek word *lithos* (stone) because of its common occurrence in rocks. Most lithium today is obtained from the mineral spodumene, $LiAlSi_2O_6$, large deposits of which occur scattered throughout the world in the United States, Canada, Brazil, and the former USSR. The major industrial use of lithium is in all-purpose automotive greases, but lithium salts such as Li_2CO_3 also have a variety of specialized applications, including use as a pharmaceutical agent for the treatment of manic-depressive behavior.

All-purpose automobile greases are one of the main commercial uses of lithium.

Sodium, the sixth most abundant element in the earth's crust, was first prepared in 1807 from caustic soda, NaOH, after which it was named. Sodium occurs throughout the world in vast deposits of NaCl (halite, or rock salt), $Na_2CO_3 \cdot NaHCO_3 \cdot 2H_2O$ (trona)[1], $NaNO_3$ (saltpeter), and $Na_2SO_4 \cdot 10H_2O$ (mirabilite)[1], all laid down by evaporation of ancient seas. In addition, the world's oceans are approximately 3% by weight NaCl. Uses of sodium and its salts span nearly the entire range of processes in the modern chemical industry. Glass, rubber, pharmaceutical agents, and many other substances use sodium or its salts in their production.

A sample of sodium metal.

Ordinary soft glass is made by melting silica (SiO_2) with sodium carbonate.

Potassium, the eighth most abundant element in the earth's crust, was first prepared in 1807, at the same time as sodium. The name of the element is derived from *potash*, a made-up word for K_2CO_3, which had been isolated from the "pot ashes" left over from wood fires. (Today, the term *potash* is often used as a general name for all water-soluble, potassium-containing minerals; the name *caustic potash* is used for KOH.) Potassium is found primarily in deposits of KCl (sylvite) and $KCl \cdot MgCl_2 \cdot 6H_2O$ (carnalite), most of which are in Canada and the former USSR. The major use of potassium salts is as a plant fertilizer.

A sample of potassium.

Potassium salts are much used as fertilizers.

[1] Dots are sometimes used when writing complex formulas such as those for trona [$Na_2CO_3 \cdot NaHCO_3 \cdot 2H_2O$] and mirabilite [$Na_2SO_4 \cdot 10H_2O$] to specify the overall composition of a substance without indicating exactly how the various parts separated by the dots are bound together.

Samples of rubidium and cesium.

Rubidium and cesium, the two most chemically reactive of the alkali metals, were both detected as impurities in other substances by chance observation of their characteristic colored spectral lines (Section 5.3). Rubidium, named from the Latin *rubidius* (deepest red), occurs with lithium in the mineral lepidolite, $K_2Li_3Al_5Si_6O_{20}(OH,F)_4$, and is obtained as a byproduct of lithium manufacture. Cesium, named from the Latin *caesius* (sky blue), also occurs with lithium in many minerals and is found in pollucite ($Cs_4Al_4Si_9O_{26} \cdot H_2O$). Neither rubidium nor cesium has any major commercial importance.

Francium, the heaviest of the group 1A elements, is highly radioactive, and no visible amount of the element has ever been prepared. Little is known about its properties from direct observation, but its behavior would presumably be similar to that of the other alkali metals.

Production of Alkali Metals

All the alkali metals are produced commercially by reduction of their chloride salts, although the exact procedure differs for each element. Both lithium metal and sodium metal are produced by *electrolysis,* a process in which an electric current is passed through the molten salt. Although the details of the process won't be discussed until Sections 18.11 and 18.12, the fundamental idea of electrolysis is simply to convert electrical energy into the chemical energy needed to break down an ionic compound into its elements. A high reaction temperature is necessary to keep the salt liquid.

$$2 \text{ LiCl}(l) \xrightarrow[\substack{450° \text{ C}}]{\substack{\text{electrolysis} \\ \text{in KCl}}} 2 \text{ Li}(l) + Cl_2(g)$$

$$2 \text{ NaCl}(l) \xrightarrow[\substack{580° \text{ C}}]{\substack{\text{electrolysis} \\ \text{in CaCl}_2}} 2 \text{ Na}(l) + Cl_2(g)$$

Potassium, rubidium, and cesium metals are produced by chemical reduction rather than by electrolysis. Sodium is the reducing agent used in potassium production, and calcium is the reducing agent used for preparing rubidium and cesium.

$$\text{KCl}(l) + \text{Na}(l) \underset{}{\overset{850°C}{\rightleftharpoons}} \text{K}(g) + \text{NaCl}(l)$$

$$2 \text{ RbCl}(l) + \text{Ca}(l) \underset{}{\overset{750°C}{\rightleftharpoons}} 2 \text{ Rb}(g) + \text{CaCl}_2(l)$$

$$2 \text{ CsCl}(l) + \text{Ca}(l) \underset{}{\overset{750°C}{\rightleftharpoons}} 2 \text{ Cs}(g) + \text{CaCl}_2(l)$$

All three of the above reductions appear contrary to the activity series described in Section 4.7, according to which sodium is not a strong enough reducing agent to react with K^+, and calcium is not a strong enough reducing agent to react with either Rb^+ or Cs^+. At the high reaction temperatures, however, *equilibria* are established in which small amounts of the products are formed. These products are then removed from the reaction mixture by distillation, thereby driving the reactions toward more product formation. We'll explore the general nature of such chemical equilibria in Chapter 13.

Reactions of Alkali Metals

REACTION WITH HALOGENS   The alkali metals react rapidly with the group 7A elements (halogens) to yield colorless, crystalline ionic salts called *halides*.

$$2 \, M + X_2 \rightarrow 2 \, MX$$

where M = an alkali metal (Li, Na, K, Rb, Cs)
X = a halogen (F, Cl, Br, I)

The reactivity of the alkali metals increases as their ionization energy decreases, giving a reactivity order Cs > Rb > K > Na > Li. Cesium is the most reactive, combining almost explosively with the halogens.

REACTION WITH HYDROGEN AND NITROGEN   Alkali metals react with hydrogen gas to form a series of white crystalline compounds called *hydrides*, MH, in which the hydrogen has an oxidation number of -1. The reaction is sluggish at room temperature and requires heating to melt the alkali metal before reaction takes place.

$$2 \, M(s) + H_2(g) \rightarrow 2 \, MH(s)$$

where M = an alkali metal (Li, Na, K, Rb, Cs)

A similar reaction takes place between lithium and nitrogen gas to form lithium nitride, Li_3N, but the other alkali metals do not react with nitrogen.

$$6 \, Li(s) + N_2(g) \rightarrow 2 \, Li_3N(s)$$

REACTION WITH OXYGEN   All the alkali metals react rapidly with oxygen, but they give different kinds of products. Lithium reacts with O_2 to yield the *oxide*, Li_2O; sodium reacts to yield the *peroxide*, Na_2O_2; and the remaining alkali metals, K, Rb, and Cs, form either peroxides or *superoxides*, MO_2, depending on the reaction conditions and on how much oxygen is present. The reasons for the differences have to do largely with the differences in stability of the various products and on the way in which the ions pack together in crystals. Note that the alkali metal cations have a $+1$ oxidation number in all cases, but that the oxidation numbers of the oxygen atoms in the O^{2-}, O_2^{2-}, and O_2^{-} anions vary.

$4 \, Li(s) + O_2(g) \rightarrow 2 \, Li_2O(s)$ an *oxide*; oxidation number of O = -2

$2 \, Na(s) + O_2(g) \rightarrow Na_2O_2(s)$ a *peroxide*; oxidation number of O = -1

$K(s) + O_2(g) \rightarrow KO_2(s)$ a *superoxide*; oxidation number of O = $-1/2$

Potassium superoxide, KO_2, is a particularly valuable compound because of its use in self-contained breathing devices to remove moisture and CO_2 from exhaled air, generating oxygen in the process.

$$2 \, KO_2(s) + H_2O(g) \rightarrow KOH(s) + KOOH(s) + O_2(g)$$

$$4 \, KO_2(s) + 2 \, CO_2(g) \rightarrow 2 \, K_2CO_3(s) + 3 \, O_2(g)$$

REACTION WITH WATER   The most well-known and dramatic reaction of the alkali metals is with water to yield hydrogen gas and an alkali metal hydroxide, MOH. In fact, it's this reaction that gives the elements their group name: The solution of metal hydroxide that results from adding an alkali metal to water is *alkaline*, or basic.

$$2\,M(s) + 2\,H_2O(l) \rightarrow 2\,M^+(aq) + 2\,OH^-(aq) + H_2(g)$$

where M = Li, Na, K, Rb, or Cs

Lithium undergoes the reaction with vigorous bubbling as hydrogen is released, sodium reacts rapidly with evolution of heat, and potassium reacts so violently that the hydrogen produced bursts instantly into flame. Rubidium and cesium react almost explosively.

Like all reactions of the alkali metals, the reaction with water is a redox process in which the metal M loses an electron and is oxidized to M^+. At the same time, a hydrogen from water gains an electron and is reduced to H_2 gas, as can be seen by assigning oxidation numbers to the various substances in the usual way.

$$2\,Na(s) + 2\,H_2O(l) \rightarrow 2\,Na^+(aq) + 2\,OH^-(aq) + H_2(g)$$

$$\begin{array}{ccccccc} 0 & +1 & -2 & +1 & -2 & +1 & 0 \end{array}$$

The alkali metals all react with water to generate H_2 gas. Lithium reacts vigorously, sodium reacts violently, and potassium reacts almost explosively.

REACTION WITH AMMONIA   The alkali metals react very slowly with ammonia, NH_3, to yield H_2 gas plus a metal *amide*, MNH_2. The reaction is exactly analogous to that between an alkali metal and water.

$$2\,M(s) + 2\,NH_3(l) \rightarrow 2\,M^+(soln) + 2\,NH_2^-(soln) + H_2(g)$$

where M = Li, Na, K, Rb, or Cs

The reaction is so slow at low temperature that it's possible for the alkali metals to dissolve in liquid ammonia at −33°C, forming deep blue solutions of metal cations and dissolved electrons. As you might expect, these solutions have extremely powerful reducing properties.

$$M(s) \xrightarrow[\text{solvent}]{\text{liquid NH}_3} M^+(soln) + e^-(soln)$$

where M = Li, Na, K, Rb, or Cs

Sodium metal dissolves in liquid ammonia to yield a blue solution of Na^+ ions and solvent-surrounded electrons.

┌ **PROBLEM 6.8** Assign oxidation numbers to the oxygen atoms in the following compounds.

(a) Li_2O **(b)** K_2O_2 **(c)** CsO_2

┌ **PROBLEM 6.9** Write products for, and balance the following equations. If no reaction takes place, write N.R.

(a) $Cs(s) + H_2O(l) \longrightarrow ?$ **(b)** $Na(s) + N_2(g) \longrightarrow ?$ **(c)** $Rb(s) + O_2(g) \longrightarrow ?$
(d) $K(s) + NH_3(g) \longrightarrow ?$ **(e)** $Rb(s) + H_2(g) \longrightarrow ?$

6.6 ➤THE ALKALINE-EARTH METALS (GROUP 2A)

The alkaline-earth elements in group 2A—Be, Mg, Ca, Sr, Ba, and Ra—are similar to the alkali metals in many respects but differ in that they have ns^2 valence-shell electron configurations and can therefore lose *two* electrons in redox reactions. Alkaline-earth metals are thus powerful reducing agents and form ions with a +2 charge.

The Born-Haber cycle for the reaction of magnesium with chlorine in Figure 6.7 provides a graphic view of the energy changes involved in the redox reaction of an alkaline-earth element. As in the reaction of sodium and chlorine to form NaCl, shown previously in Figure 6.6, there are five contributions to the overall energy change in the reaction of magnesium with chlorine to form $MgCl_2$. First, solid magnesium metal must be converted into isolated gaseous magnesium atoms (sublimation). Second, the bond in chlorine molecules must be broken to yield two chlorine atoms. Third, magnesium atoms must lose *two* electrons to form the dipositive Mg^{2+} ion. Fourth, the two chlorine atoms formed in step 2 must accept electrons to form two Cl^- ions. Fifth, the gaseous ions must combine to form the ionic solid, $MgCl_2$. As the Born-Haber cycle indicates, it is the large contribution from ionic attractions (the negative of the lattice energy) that releases enough energy to drive the entire process.

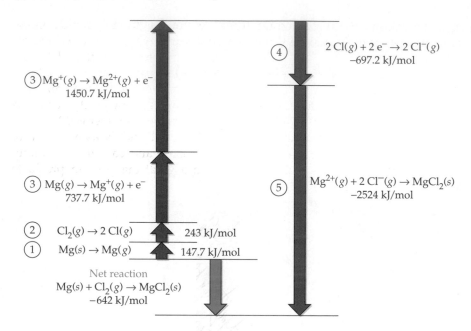

FIGURE 6.7 A Born-Haber cycle for the formation of $MgCl_2$ from the elements. The large contribution from ionic bonding in the solid (step 5) provides more than enough energy to remove two electrons from magnesium (step 3).

② $Cl_2(g) \rightarrow 2\ Cl(g)$ 243 kJ/mol
① $Mg(s) \rightarrow Mg(g)$ 147.7 kJ/mol
Net reaction
$Mg(s) + Cl_2(g) \rightarrow MgCl_2(s)$
−642 kJ/mol

③ $Mg(g) \rightarrow Mg^+(g) + e^-$
737.7 kJ/mol

③ $Mg^+(g) \rightarrow Mg^{2+}(g) + e^-$
1450.7 kJ/mol

④ $2\ Cl(g) + 2\ e^- \rightarrow 2\ Cl^-(g)$
−697.2 kJ/mol

⑤ $Mg^{2+}(g) + 2\ Cl^-(g) \rightarrow MgCl_2(s)$
−2524 kJ/mol

Though harder than their neighbors in group 1A, the alkaline-earth elements are still relatively soft, silvery metals. They tend, however, to have higher melting points and densities than alkali metals, as shown in Table 6.4. Alkaline-earth elements are less reactive toward oxygen and water than alkali metals but are nevertheless found in nature only in salts, not in the elemental state.

Occurrence and Uses of the Alkaline-Earth Metals

Beryllium was first detected in 1798 in the gemstones beryl and emerald ($Be_3Al_2Si_6O_{18}$) and was subsequently prepared in pure form in 1828 by the reduction of $BeCl_2$ with potassium. It is obtained today from large commercial deposits of beryl in Brazil and southern Africa. Though beryllium compounds are extremely toxic, particularly when inhaled as dust, the metal is nevertheless useful in forming alloys. Addition of a few percent beryllium to copper or nickel results in hard, corrosion-resistant alloys that are used in airplane engines and precision instruments.

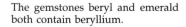

The gemstones beryl and emerald both contain beryllium.

TABLE 6.4	Properties of the Alkaline-Earth Metals						
Name	Melting Point (°C)	Boiling Point (°C)	Density (g/cm³)	1st Ionization Energy (kJ/mol)	Abundance on Earth (%)	Atomic Radius (pm)	Ionic (M^{2+}) Radius (pm)
Beryllium	1278	2970	1.848	899.4	0.000 28	112	44
Magnesium	649	1107	1.738	737.7	2.33	160	66
Calcium	839	1484	1.550	589.8	4.15	197	99
Strontium	769	1384	2.540	549.5	0.038	215	112
Barium	725	1640	3.510	502.9	0.042	222	134
Radium	700	1140	~5.0	509.3	Trace	223	143

Compounds of magnesium, the seventh most abundant element in the earth's crust, have been known since ancient times, although the pure metal was not prepared until 1808. The element is named after the Magnesia district in Thessally, Greece, where large deposits of talc [$Mg_3Si_4O_{10}(OH)_2$] are found. There are a large number of magnesium-containing minerals, including dolomite ($CaCO_3 \cdot MgCO_3$) and magnesite ($MgCO_3$), and the world's oceans provide a nearly infinite supply, since sea water is 0.13% Mg. When alloyed with aluminum, magnesium is used widely as a structural material because of its high strength, low density, and ease in machining. Airplane fuselages, automobile engines, and a great many other products are made with magnesium.

A sample of magnesium

Magnesium metal is used as a structural material in applications where keeping weight low is important.

Calcium is the fifth most abundant element in the earth's crust, owing largely to the presence of huge $CaSO_4 \cdot 2H_2O$ (gypsum) and $CaCO_3$ deposits in ancient sea beds. Limestone, marble, chalk, and coral are all slightly different forms of $CaCO_3$. Though the metal was not obtained in pure form until 1808, compounds of calcium have been known for millennia. Lime (CaO), for example, was prepared by the Romans by heating $CaCO_3$ and was used as a mortar in their constructions. In fact, the name *calcium* is derived from the Latin word *calx*, meaning "lime". The primary industrial use of calcium metal is as an alloying agent to harden aluminum. Calcium compounds such as lime and gypsum are used for many purposes throughout the chemical and construction industries. Portland cement, for example, contains approximately 70% CaO.

A sample of calcium.

Dolomite mountains in Italy and coral reefs in the Caribbean are just two of the many natural occurrences of calcium carbonate.

A barium "cocktail" is given to patients prior to a stomach X ray to help in visualizing the gastrointestinal tract.

Strontium was discovered near and named after the small town of Strontian, Scotland, in 1787. There are no commercial uses for the pure metal, but the carbonate salt, $SrCO_3$, is used in the manufacture of glass for color TV picture tubes. Barium is found principally in the minerals witherite ($BaCO_3$) and barite ($BaSO_4$), after which it is named. Though water-soluble salts of barium are extremely toxic, barium sulfate is so insoluble that it is used in medicine as a contrast medium for stomach and intestinal X rays. Like strontium, barium metal has no commercial uses, but various compounds are used in glass manufacture and in oil-well drilling.

Radium, the heaviest of the group 2A elements, occurs with uranium and was isolated as its chloride salt from the mineral pitchblende in 1898 by Marie and Pierre Curie. Radium is highly radioactive, and no more than a few kilograms of the pure metal have ever been produced. Though used for many years as a radiation source for cancer radiotherapy, better sources are now available, and there are no longer any commercial uses for radium.

Production of Alkaline-Earth Metals

Like the alkali metals, the pure alkaline-earth elements are produced commercially by reduction of their salts, either chemically or through electrolysis. Beryllium is prepared by reduction of BeF_2 with magnesium, and magnesium is prepared by electrolysis of its molten chloride salt.

$$BeF_2(l) + Mg(l) \xrightarrow{1300°C} Be(l) + MgF_2(l)$$

$$MgCl_2(l) \xrightarrow[750°C]{\text{electrolysis}} Mg(l) + Cl_2(g)$$

Calcium, strontium, and barium are all made by high-temperature reduction of their oxides with aluminum metal.

$$3\ MO(l) + 2\ Al(l) \xrightarrow{\text{high temp.}} 3\ M(l) + Al_2O_3(s)$$

$$\text{where M = Ca, Sr, Ba}$$

Magnesium metal burns in air with a brilliant white glare.

Reactions of Alkaline-Earth Metals

The alkaline-earth metals undergo the same kinds of redox reactions that the alkali metals do, but they lose two electrons rather than one to yield dipositive ions, M^{2+}. Because their first ionization energy is larger than that of their alkali-metal neighbors (Figure 6.1), the group 2A metals tend to be somewhat less reactive than alkali metals. The general reactivity trend is Ba > Sr > Ca > Mg > Be.

Alkaline earths react with halogens to yield ionic halide salts, MX_2, and with oxygen to form oxides, MO.

$$M + X_2 \rightarrow MX_2 \qquad \text{where M = Be, Mg, Ca, Sr, or Ba}$$
$$\text{X = F, Cl, Br, or I}$$

$$2\ M + O_2 \rightarrow 2\ MO$$

Beryllium and magnesium are relatively unreactive toward oxygen at room temperature, but both burn with a brilliant white glare when ignited by a flame. Calcium, strontium, and barium are reactive enough that they are best stored under oil to keep them from contact with air. As is true for the heavier alkali metals, strontium and barium also form peroxides, MO_2.

With the exception of beryllium, the alkaline-earth elements react with water to yield metal hydroxides, $M(OH)_2$, but the reaction is sluggish. Magnesium undergoes reaction only at temperatures above 100°C, and calcium and strontium react slowly with liquid water at room temperature. Only barium reacts vigorously.

$$M(s) + 2\ H_2O(l) \rightarrow M^{2+}(aq) + 2\ OH^-(aq) + H_2(g)$$

where M = Mg, Ca, Sr, or Ba

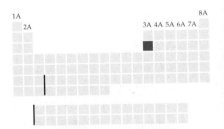

Calcium metal reacts very slowly with water at room temperature.

EXAMPLE 6.5

Calcium metal reacts with hydrogen gas at high temperature to give calcium hydride. Predict the formula of the product, and write the balanced equation.

SOLUTION   Metal hydrides contain hydrogen with a −1 oxidation number. Since calcium always has a +2 oxidation number, there must be two H^- ions per Ca^{2+} ion, and the formula of calcium hydride must be CaH_2.

$$Ca(s) + H_2(g) \rightarrow CaH_2(s)$$

PROBLEM 6.10   Predict the products of the following reactions, and balance the equations.
(a) $Be(s) + Br_2(l) + ?$ **(b)** $Sr(s) + H_2O(l) \longrightarrow ?$ **(c)** $Mg(s) + O_2(g) \longrightarrow ?$

PROBLEM 6.11   Write a balanced equation for the preparation of beryllium metal by the reduction of beryllium chloride with potassium.

PROBLEM 6.12   What product do you think is formed by reaction of magnesium with sulfur, a group 6A element? What is the oxidation number of sulfur in the product?

6.7 ➤ALUMINUM (GROUP 3A)

The elements in group 3A—B, Al, Ga, In, and Tl—are the first of the *p*-block elements (Section 5.12) and have the valence-shell electron configuration $ns^2\ np^1$. With the exception of boron, which behaves as a semimetal rather than a metal, the group 3A elements are silvery in appearance, good conductors of electricity, and relatively soft. Gallium, in fact, has a melting point of only 30°C. Although properties of the entire group are listed in Table 6.5, we'll concentrate for now on the most common element, aluminum, and look at the others in Chapter 19.

Aluminum, the most abundant metal in the earth's crust at 8.3%, takes its name from alum, $KAl(SO_4)_2 \cdot 12H_2O$, a salt that has been used medicinally since Roman times. In spite of its abundance, the metal nevertheless proved difficult to isolate in pure form and was so valuable in the mid-nineteenth century that aluminum cutlery was used by Napoleon for state dinners. Not until 1886 did an economical manufacturing process become available.

	Melting Point (°C)	Boiling Point (°C)	Density (g/cm³)	1st Ionization Energy (kJmol)	Abundance on Earth (%)	Atomic Radius (pm)	Ionic (M³⁺) Radius (pm)
Name							
Boron	2300	3650	2.35	800.6	0.001	83	—
Aluminum	660	2467	2.699	577.6	8.32	143	51
Gallium	30	2403	5.904	578.8	0.0015	135	62
Indium	157	2080	7.31	558.3	0.000 01	167	81
Thallium	304	1457	11.85	589.3	0.000 04	170	95

TABLE 6.5   Properties of the Group 3A Elements

When dedicated in 1885, the Washington Monument was capped by a pyramid of pure aluminum, a precious substance at the time.

Aluminum occurs in many common minerals and clays, as well as in gemstones. Ruby and sapphire are both impure forms of Al_2O_3 that receive their color from the presence of small amounts of other elements (Cr in ruby; Fe and Ti in sapphire). Most aluminum is currently obtained from bauxite, $Al_2O_3 \cdot xH_2O$, which occurs in large deposits in Australia, the United States, Jamaica, and elsewhere. The preparation of Al from ores is extremely energy-intensive, requiring high temperatures and large amounts of electrical current to carry out the electrolysis of Al_2O_3. We'll examine the process in more detail in Section 18.12.

$$2\ Al_2O_3(soln) \xrightarrow[\substack{\text{electrolysis in} \\ Na_3AlF_6 \\ 980°C}]{} 4\ Al(l) + 3\ O_2(g)$$

The gemstones ruby and sapphire are both impure forms of Al_2O_3.

Following the trend exhibited by the group 1A and group 2A metals, aluminum behaves as a reducing agent. It undergoes redox reactions by losing all three valence-shell electrons to yield Al^{3+} ions. For example, it reacts with the halogens to yield colorless halides AlX_3, with oxygen to yield an oxide Al_2O_3, and with nitrogen to yield a nitride AlN.

$$2\ Al + 3\ X_2 \rightarrow 2\ AlX_3 \quad \text{where X = F, Cl, Br, or I}$$

$$4\ Al + 3\ O_2 \rightarrow 2\ Al_2O_3$$

$$2\ Al + \ N_2 \rightarrow 2\ AlN$$

Reactions with the halogens occur vigorously at room temperature and release large amounts of heat. Reaction with oxygen is also vigorous at room temperature, yet we know that aluminum can be used in a huge array of consumer products without evident corrosion from the air. The explanation for this apparent inconsistency is that aluminum metal reacts rapidly with oxygen only on its *surface*. In so doing, it forms a thin, hard, oxide coating that does not flake off and that protects the underlying metal from contact with air.

Aluminum is less reactive than the group 1A and 2A metals and does not react with water. It does, however, react with both acidic and basic solutions to give aluminum ions and release H_2 gas.

Acid solution: $2\ Al(s) + 6\ H^+(aq) \rightarrow 2\ Al^{3+}(aq) + 3\ H_2(g)$

Basic solution: $2\ Al(s) + 2\ OH^-(aq) + 6\ H_2O(l) \rightarrow 2\ Al(OH)_4^-(aq) + 3\ H_2(g)$

Aluminum metal reacts with liquid bromine in a spectacular display of sparks.

▷ **PROBLEM 6.13** Identify the oxidizing agent and the reducing agent in the reaction of aluminum metal with $H^+(aq)$.

▷ **PROBLEM 6.14** Aluminum reacts with sulfur to give a sulfide in the same way that it reacts with oxygen to give an oxide. Identify the product, and write a balanced equation for the reaction.

6.8 ▶THE HALOGENS (GROUP 7A)

The halogens in group 7A—F, Cl, Br, I, and At—are completely different from the elements we've been discussing up to this point. The halogens are nonmetals rather than metals, and they have a tendency to gain rather than lose electrons when they enter into redox reactions because of their $ns^2\ np^5$ electron configurations. In other words, the halogens are powerful oxidizing agents, characterized by large negative electron affinities and large positive ionization energies. Some of their properties are shown in Table 6.6.

The halogens are too reactive to occur in nature as the free elements. Instead, they are found only as their anions in various salts and minerals. Even the name "halogen" implies reactivity, since it comes from the Greek words *hals* (salt) and *gennan* (to form). Thus, a halogen is literally a "salt-former."

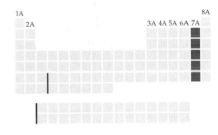

Name	Melting Point (°C)	Boiling Point (°C)	Density (g/cm³)	Electron Affinity (kJ/mol)	Abundance on Earth (%)	Atomic Radius (pm)	Ionic (X⁻) Radius (pm)
Fluorine	−220	−188	1.516(*l*)	−328	0.062	72	133
Chlorine	−101	−35	2.030(*l*)	−349	0.013	99	181
Bromine	−7	59	3.12(*l*)	−325	0.000 3	114	196
Iodine	114	184	4.930	−295	0.000 05	133	220
Astatine	—	—	—	−270	Trace	—	—

TABLE 6.6   Properties of the Halogens

Teflon, a fluorine-containing polymer, is used because of its nonstick properties for coating this frying pan.

Occurrence and Uses of The Halogens

Fluorine, a corrosive, pale yellow gas, is the 13th most abundant element in the earth's crust, more common than such well-known elements as sulfur (16th), carbon (17th), and copper (26th). It is far too reactive to occur in the free state but is found in several common minerals including fluorspar, or fluorite (CaF_2), and fluorapatite [$Ca_5(PO_4)_3F$]. Note that fluorine is spelled with the *u* first as in *flu* rather than with the *o* first as in *flour*.

In spite of its toxicity and difficulty in handling, fluorine is widely used in the manufacture of polymers such as Teflon, $(C_2F_4)_n$. Fluorine is also important in the production of UF_6, used in the separation of uranium isotopes for nuclear power plants.

Chlorine, like fluorine, is a toxic, reactive, greenish-yellow gas that must be handled with great care. Though chlorine is only the twentieth most abundant element in crustal rocks, there are vast additional amounts of chloride ion in the world's oceans (1.9% by mass). The free element was first prepared in 1774 by oxidation of NaCl with MnO_2, though not until 1810 was it recognized that the product of the reaction was indeed a new element rather than a compound.

$$MnO_2(s) + 2\ Cl^-(aq) + 4\ H^+(aq) \rightarrow Mn^{2+}(aq) + 2\ H_2O(l) + Cl_2(g)$$

Industrial applications of chlorine center around its uses as a reagent for the preparation of numerous chlorinated organic chemicals such as PVC [poly(vinyl chloride)] plastic and the solvents chloroform ($CHCl_3$), methylene chloride (CH_2Cl_2), and ethylene dichloride ($C_2H_4Cl_2$). Large amounts of chlorine are also used as a bleach during paper manufacture and as a disinfectant for swimming pools and municipal water supplies.

Chlorine is a toxic greenish-yellow gas.

Bromine is a volatile, reddish liquid.

Chlorine is frequently added as a disinfectant to the water in swimming pools.

As a fourth-row element, bromine is a volatile, reddish liquid rather than a gas like fluorine and chlorine. Its fumes are quite toxic, however, and it causes particularly painful burns when spilled on bare skin. Bromine was first isolated in 1826 by the 23-year-old French chemist A.-J. Balard by oxidation of KBr with MnO_2 in a manner similar to that used for the synthesis of chlorine.

$$MnO_2(s) + 2\ Br^-(aq) + 4\ H^+(aq) \rightarrow Mn^{2+}(aq) + 2\ H_2O(l) + Br_2(aq)$$

The primary uses of bromine are as its silver salt, AgBr, in photographic emulsions and as a reagent for preparing brominated organic compounds. Fuel additives, pesticides, fungicides, and flame retardants are but a few of the many kinds of compounds manufactured from bromine.

Iodine is a volatile purple solid with a beautiful metallic sheen. As the least reactive halogen, iodine is safe to handle and is widely used as a skin disinfectant. It was first prepared in 1811 from seaweed ash, but commercially useful deposits of the iodine-containing minerals lautarite ($CaIO_3$) and dietzeite [$7Ca(IO_3)_2 \cdot 8CaCrO_4$] were subsequently found in Chile. Iodine is used in the preparation of numerous organic compounds, including dyes and pharmaceutical agents, but there is no one single use of major importance.

Iodine is a volatile purple solid.

Astatine in group 7A, like francium in group 1A, is a radioactive element that occurs only in minute amounts in nature. No more than about 5×10^{-8} g has ever been prepared at one time, and little is known about its chemistry.

Production of the Halogens

All the free halogens are produced commercially by oxidation of their anions. Fluorine and chlorine are both produced by electrolysis, fluorine from a molten 1:2 mixture of KF and HF, and chlorine from molten NaCl.

$$2\ HF(l) \xrightarrow[\text{100°C}]{\text{electrolysis}} H_2(g) + F_2(g)$$

$$2\ NaCl(l) \xrightarrow[\text{580°C}]{\text{electrolysis}} 2\ Na(l) + Cl_2(g)$$

Bromine and iodine are both prepared by oxidation of their aqueous halide solutions with chlorine, followed by distillation. Naturally occurring aqueous solutions of bromide ion with concentrations of up to 5000 ppm are found in Arkansas and in the Dead Sea in Israel. Iodide ion solutions of up to 100 ppm concentration are found in Oklahoma and Michigan.

$$2\ Br^-(aq) + Cl_2(g) \rightarrow Br_2(l) + 2\ Cl^-(aq)$$

$$2\ I^-(aq) + Cl_2(g) \rightarrow I_2(s) + 2\ Cl^-(aq)$$

The Dead Sea in Israel has a particularly high concentration of bromide ion.

Reactions of the Halogens

The halogens are among the most reactive elements in the periodic table. Fluorine, in fact, forms compounds with every element except the noble gases He, Ne, and Ar. As noted several times already in this chapter, the dominant reaction of the halogens is as strong oxidizing agents in redox reactions. That is, their high electron affinities allow halogens to accept electrons from other atoms to yield halide anions, X^-.

REACTION WITH METALS   Halogens react with practically every metal in the periodic table to yield metal halides. With the alkali and alkaline-earth metals, the formula of the halide product is easily predictable. With transition metals, though, more than one product can sometimes form depending on the reaction conditions and on the amounts of reactants

present. Iron, for example, can react with Cl_2 to form either $FeCl_2$ or $FeCl_3$. Without knowing a good deal more about transition-metal chemistry, it's not possible to make predictions at this point. The reaction can be generalized as

$$2\,M + n\,X_2 \rightarrow 2\,MX_n \quad \text{where M = a metal}$$
$$X = \text{F, Cl, Br, or I}$$

Unlike the metallic elements, halogens become *less* reactive going down the periodic table. Thus, their reactivity order is $F_2 > Cl_2 > Br_2 > I_2$. Fluorine often reacts extremely vigorously, chlorine and bromine somewhat less so, and iodine often sluggishly.

REACTION WITH HYDROGEN   The halogens react with hydrogen gas to yield hydrogen halides, HX.

$$H_2(g) + X_2 \rightarrow 2\,HX(g) \quad \text{where X = F, Cl, Br, or I}$$

Fluorine reacts explosively as soon as the two gases come in contact. Chlorine also reacts explosively once the reaction is initiated by a spark or by ultraviolet light, but the mixture of gases is stable in the dark. Bromine and iodine react more slowly.

The hydrogen halides are extremely valuable chemical reagents because they behave as acids when dissolved in water. Hydrogen fluoride is a weak acid, dissociating only to a small extent in aqueous solution, but the other hydrogen halides are strong acids. As one of the few substances that reacts with glass, HF is frequently used for the etching or fogging of glass. The aqueous solution of HCl, called hydrochloric acid or muriatic acid, is used throughout the chemical industry in a vast number of processes, from pickling steel (removing its iron oxide coating) to dissolving animal bones for producing gelatin.

$$HX(g) \xrightarrow[\text{in H}_2\text{O}]{\text{dissolve}} H^+(aq) + X^-(aq)$$

Gaseous HF is one of the few substances that reacts with glass.

REACTION WITH OTHER HALOGENS   Since all the halogens exist as diatomic molecules, X_2, it's not surprising that a variety of covalent **interhalogen compounds XY** exist, where X and Y are different halogens. Iodine reacts with chlorine, for example, to yield iodine chloride, ICl, and bromine reacts with fluorine to yield bromine fluoride, BrF. These reactions can be thought of as redox processes in which the lighter, more reactive element is the oxidizing agent and the heavier, less reactive element is the reducing agent. In the reaction of iodine with chlorine, for example, Cl_2 acts as the oxidizing agent and is itself reduced to a -1 oxidation state, while I_2 acts as the reducing agent and is itself oxidized to a $+1$ oxidation state.

$$I_2(s) + Cl_2(g) \rightarrow 2\,ICl(s)$$

$$Br_2(l) + F_2(g) \rightarrow 2\,BrF(g)$$

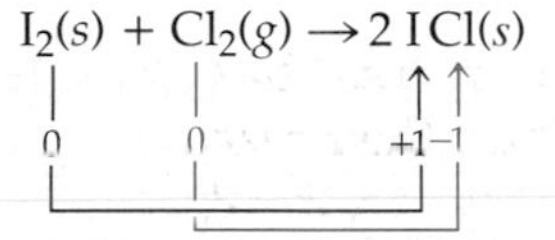

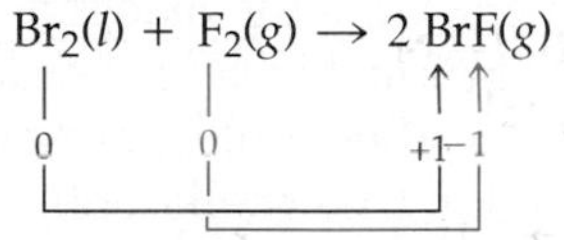

As a general rule, the properties of interhalogen compounds are intermediate between those of their parent elements. ICl, for example, is a red solid that melts near room temperature, and BrF is a brownish gas that condenses to a liquid near room temperature. All six possible diatomic interhalogen compounds are known, and all act as strong oxidizing agents in redox reactions.

In addition to the diatomic interhalogen compounds, a number of polyatomic substances with formulas XY_3, XY_5, and XY_7 are also known. ClF_3, BrF_5, and IF_7 are typical examples. Once again, the less reactive halogen has a positive oxidation state and the more reactive halogen has a negative oxidation state. In BrF_5, for example, the oxidation number of Br is $+5$ and that of F is -1.

EXAMPLE 6.6

The UF_6 used in producing nuclear fuels is prepared by reaction of uranium metal with chlorine trifluoride. Tell which atoms have been oxidized and which reduced, and balance the equation.

$$U(s) + ClF_3(l) \rightarrow UF_6(l) + ClF(g)$$

SOLUTION   First assign oxidation numbers to the various elements, and decide which atoms have undergone a change. Those that have increased in oxidation number have been oxidized, and those that have decreased in oxidation number have been reduced.

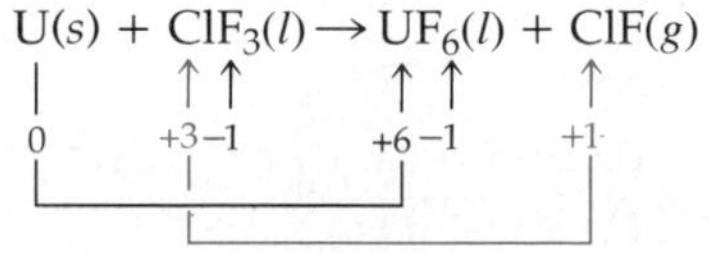

$$U(s) + ClF_3(l) \rightarrow UF_6(l) + ClF(g)$$

Uranium is oxidized from 0 to $+6$, and chlorine is reduced from $+3$ to $+1$.

Balance the equation, either by inspection or by one of the redox methods discussed in Sections 4.8 and 4.9.

$$U(s) + 3\,ClF_3(l) \rightarrow UF_6(l) + 3\,ClF(g)$$

PROBLEM 6.15 Write the products of the following reactions, and balance the equations.

(a) $Br_2(l) + Cl_2(g) \longrightarrow$?   **(b)** $Al(s) + F_2(g) \longrightarrow$?   **(c)** $H_2(g) + I_2(s) \longrightarrow$?

PROBLEM 6.16 Bromine reacts with sodium iodide to yield iodine and sodium bromide. Identify the oxidizing and reducing agents, and write the balanced equation.

6.9 ►THE NOBLE GASES (GROUP 8A)

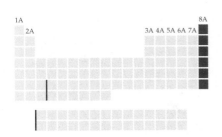

The noble gases in group 8A—He, Ne, Ar, Kr, Xe, and Rn—are different still from the elements we've been discussing. They are neither metals like most elements nor reactive nonmetals like the halogens; rather they are colorless, odorless, unreactive gases. Their stable $ns^2 np^6$ valence-shell electron configurations make it difficult for the noble gases either to gain or to lose electrons, and the elements don't normally enter into redox reactions.

Though sometimes referred to as "rare gases" or "inert gases," these older names are not really accurate because the group 8A elements are neither rare nor completely inert. Argon, for instance, makes up nearly 1% by volume of dry air, and there are several dozen known compounds of krypton and xenon, though none occur naturally. Some properties of the noble gases are shown in Table 6.7.

Occurrence and Uses of the Noble Gases

The natural abundance of the noble-gas elements depends on where you look. In the universe, helium is the second most abundant element, accounting for about 25% of its total mass (hydrogen accounts for the other 75%). On earth, though, the fact that they're gases means that the abundance of the group 8A elements in crustal rocks is very low. Helium occurs as a minor constituent of natural gas, and argon makes up about 1% of air. The remaining noble gases occur in small amounts in the air (Table 6.7), from which they are obtained by liquefaction and distillation.

Their relative lack of reactivity means that the only commercial uses of the noble gases are in applications that require an inert (unreactive) atmosphere. Argon is used for arc welding to protect the metal from oxygen;

TABLE 6.7	Properties of the Noble Gases				
Name	Melting Point (°C)	Boiling Point (°C)	1st Ionization Energy (kJ/mol)	Electron Affinity[a] (kJ/mol)	Abundance in Air (vol %)
Helium	−272.2	−268.9	2372.3	21	5.2×10^{-4}
Neon	−248.7	−246.1	2080.6	29	1.8×10^{-3}
Argon	−189.2	−185.7	1520.4	35	0.93
Krypton	−156.6	−152.3	1350.7	39	1.1×10^{-4}
Xenon	−111.9	−107.1	1170.4	41	9×10^{-6}
Radon	−71	−61.8	1037	41	Trace

[a]Electron affinities are calculated rather than experimental values.

Titanium metal must be welded under an atmosphere of argon because of its sensitivity to atmospheric oxygen.

helium is used in deep-sea diving gas as a replacement for nitrogen so that divers can avoid the "bends" that can accompany rapid depressurization when they return to the surface. Liquid helium is widely used in scientific research as a cooling agent for extremely low-temperature studies because it has the lowest boiling point (4.2 K) of any substance known.

Radon, the heaviest of the noble gases, has been much publicized in recent years because of fear that low-level exposures to it increase the risk of cancer. Like astatine and francium, its neighbors in the periodic table, radon is a radioactive element with only a minute natural abundance. It is produced by radioactive decay of the radium present in small amounts in many granitic rocks, and it can slowly seep into basements, where it remains unless vented. If then breathed into the lungs, it can cause radiation damage.

Radioactive radon gas has been found in the basement of many homes and other buildings.

Reactions of the Noble Gases

Helium, neon, and argon undergo no chemical reactions and form no known compounds; krypton and xenon react only with fluorine. Depending on the reaction conditions and on the amounts of reagents present, xenon can form three different fluorides, XeF_2, XeF_4, and XeF_6. All three xenon fluorides are powerful oxidizing agents and undergo a wide variety of redox reactions.

$$Xe(g) + F_2(g) \rightarrow XeF_2(s)$$

$$Xe(g) + 2\ F_2(g) \rightarrow XeF_4(s)$$

$$Xe(g) + 3\ F_2(g) \rightarrow XeF_6(s)$$

The lack of reactivity of the noble gases is a consequence of their unusually high ionization energies (Figure 6.1) and their unusually low electron affinities (Figure 6.3), which result from their valence-shell electron configurations.

┌ **PROBLEM 6.17** Assign oxidation numbers to the elements in the following compounds of xenon.

(a) XeF_2 (b) XeF_4 (c) $XeOF_4$

6.10 ➤ THE OCTET RULE

Let's list some general conclusions about main-group chemistry that we can draw from the data in the preceding five sections:

1. Group 1A elements tend to lose their ns^1 valence-shell electron, thereby adopting the electron configuration of the noble-gas element in the previous row of the periodic table.
2. Group 2A elements tend to lose both of their ns^2 valence-shell electrons and adopt a noble-gas electron configuration.
3. Group 3A elements tend to lose all three of their $ns^2\,np^1$ valence-shell electrons and adopt a noble-gas electron configuration.
4. Group 7A elements tend to gain one electron, changing from $ns^2\,np^5$ to $ns^2\,np^6$, thereby adopting the configuration of the neighboring noble-gas element in the same row.
5. Group 8A (noble-gas) elements are essentially inert; they rarely gain or lose electrons.

All these observations can be gathered into a single statement called the **octet rule**:

OCTET RULE: *Main-group elements tend to undergo reactions that leave them with eight valence electrons.* That is, main-group elements react so that they attain a noble-gas electron configuration with filled s and p sublevels in their valence electron shell.

There are many exceptions to the octet rule—after all, it's called the octet *rule*, not the octet *law*—but it is nevertheless useful for making predictions about unknown reactions and for providing insights about chemical bonding.

Why does the octet rule work? What factors determine whether an atom is likely to gain or to lose electrons? Clearly, electrons are most likely to be lost if they are held loosely in the first place—that is, if they feel a relatively low effective nuclear charge, Z_{eff}. Valence-shell electrons in the group 1A, 2A, and 3A metals, for example, are shielded from the nucleus by core electrons. They feel a low Z_{eff}, and they are therefore lost relatively easily. Once the next lower noble-gas configuration is reached, though, loss of an additional electron is much more difficult, since it must come from an inner shell, where it feels a high Z_{eff}.

Conversely, electrons are most likely to be gained if they can be held tightly by a high Z_{eff}. Valence-shell electrons in the group 6A and 7A elements, for example, are poorly shielded. They feel high values of Z_{eff}, and they aren't lost easily. The high Z_{eff} thus makes possible the gain of one or more additional electrons into vacant valence-shell orbitals. Once the noble-gas configuration is reached, though, there is no longer an available low-energy orbital. Any additional electron would have to be placed in a higher-energy orbital, where it would feel only a low Z_{eff}.

Eight is therefore the "magic number" for valence-shell electrons. Taking electrons away *from* a filled octet is difficult because they are tightly held by a high Z_{eff}; adding more electrons *to* a filled octet is difficult because, with *s* and *p* sublevels full, there is no low-energy orbital available.

When the octet rule fails, it generally does so for elements toward the right side of the periodic table (groups 3A–8A) that are in the third row and lower (Figure 6.8). The reason is straightforward and has to do with the electron configurations of these elements: With few exceptions, the main-group elements that occasionally break the octet rule have vacant, low-energy *d* orbitals, which allow them to accommodate more than the usual number of electrons. Phosphorus, for example, has the electron configuration [Ne] $3s^2 3p^3$ and has a vacant $3d$ subshell that is only slightly higher in energy than the $3s$ and $3p$ levels. As a result, phosphorus is occasionally able to add more than the three electrons predicted by the octet rule.

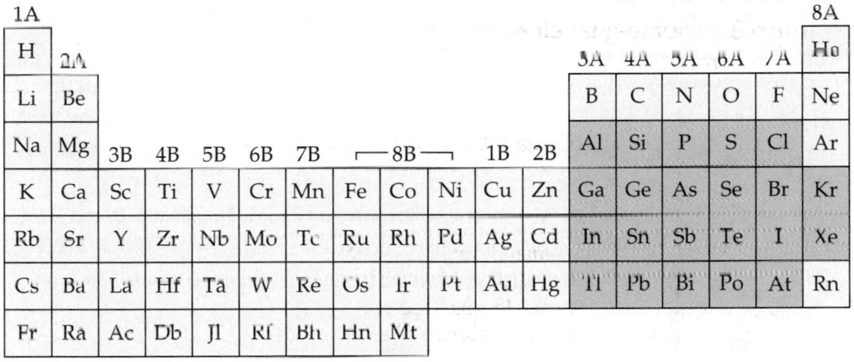

FIGURE 6.8 The octet rule occasionally fails for the shaded main-group elements because these elements, all of which are in the third row or lower, can use low-energy unfilled *d* orbitals to expand their valence shell beyond the normal octet.

EXAMPLE 6.7

We saw in Section 6.5 that lithium reacts with nitrogen to yield Li_3N. What noble-gas configuration does the nitrogen atom in Li_3N have?

SOLUTION   The nitrogen atom in Li_3N has an oxidation number of -3 and has gained three electrons over the neutral atom, giving it a valence-shell octet with the neon configuration:

N configuration: $(1s^2 2s^2 2p^3)$ N^{3-} configuration: $(1s^2 2s^2 2p^6)$

PROBLEM 6.18 What noble-gas configurations are the following elements likely to adopt in redox reactions?
(a) Rb   **(b)** Ba   **(c)** Ga   **(d)** F

PROBLEM 6.19 Although we've not talked in this chapter about group 6A elements, what are they likely to do in redox reactions—gain or lose electrons? How many?

interlude—SALT

If you're like most people, you feel a little guilty about reaching for the salt shaker at mealtime. The notion that high salt intake and high blood pressure go hand in hand is surely the most highly publicized piece of nutritional lore to appear in recent decades.

Salt has not always been held in such disrepute. Historically, salt has been prized since the earliest recorded times as a seasoning and a food preservative. "Salt" words and phrases are used in many languages to reflect the importance of salt as a life-giving and life-sustaining substance. We refer to a kind and generous person as "the salt of the earth," for instance, and we speak of being "worth one's salt." In Roman times, soldiers were paid in salt; the English word "salary" is derived from the Latin word for paying salt wages (*salarium*).

Salt is perhaps the easiest of all minerals to obtain and purify. The simplest method, used for thousands of years throughout the world in coastal climates where sunshine is abundant and rainfall is scarce, is to evaporate sea water. Though the exact amount varies depending on the source, sea water contains an average of about 3.5% by mass of dissolved substances, most of which is sodium chloride. It has been estimated that evaporation of all the world's oceans would yield approximately *4.5 million cubic miles* of NaCl.

Only about 10% of current world salt production comes from evaporation of sea water. Most salt is obtained by mining the vast deposits of *halite*, or *rock salt*, formed by evaporation of ancient inland seas. These salt beds vary in thickness up to hundreds of meters and vary in depth from a few

Ocean water is evaporated by the sun in these massive salt pans.

Underground salt mining has been carried out for more than a thousand years.

meters to thousands of meters below the earth's surface. Salt mining has gone on for at least 3400 years, and the Wieliczka mine in Galicia, Poland, has been worked continuously from A.D. 1000 to the present.

Let's get back now to the dinner table. What about the link between dietary salt intake and high blood pressure? There's no doubt that most people in industrialized nations have a relatively high salt intake, and there's no doubt that high blood pressure among industrialized populations is on the rise. What's not so clear is whether the two observations are related by cause and effect. The case against salt has been made largely by comparing widely diverse populations with different dietary salt intakes—by comparing the health of modern Americans with that of inhabitants of the Amazon rain forest, for example. Obviously, though, industrialization brings with it far more changes than simply an increase in dietary salt intake, and many of these other changes may be far more important contributors to hypertension than salt is.

The largest and most definitive study of the connection between salt and high blood pressure was published in the summer of 1991 by the Intersalt Cooperative Research Group, made up of scientists from 32 countries. This group found that there is indeed an increase in blood pressure with increasing salt intake but that the increase is minimal. If Americans were to cut their salt intake from the present average of 9 g per day to an average of 3 g per day (a near impossibility), the blood pressure of the average person would decline by 1%. Certain *individuals* would, of course, benefit far more substantially, but most people would see no reduction at all.

What should an individual do? The best answer, as in so many things, is to use moderation and common sense. Cut your salt intake if you can or if you have extreme hypertension, but don't spend a lot of time worrying about it.

223

KEY WORDS

Born-Haber cycle, *199*
core electron, *192*
electron affinity, *195*
ionic bond, *198*
ionization energy, *191*
interhalogen compound
   XY, *216*
lattice energy, *200*
octet rule, *220*

SUMMARY

The amount of energy necessary to remove a valence electron from an isolated neutral atom is called the atom's **ionization energy**, E_i. Ionization energies are lowest for metallic elements on the left side of the periodic table and highest for nonmetallic elements on the right side, implying that metals can act as electron donors (reducing agents) in chemical reactions.

The amount of energy released or absorbed when an electron adds to an isolated neutral atom is called the atom's **electron affinity**, E_{ea}. By convention, a negative E_{ea} corresponds to a release of energy, and a positive E_{ea} corresponds to an absorption of energy. Electron affinities are most negative for group 7A elements and most positive for group 2A and 8A elements. As a result, the group 7A elements can act as electron acceptors (oxidizing agents) in chemical reactions.

Main-group metallic elements in groups 1A, 2A, and 3A undergo redox reactions with halogens in group 7A, during which the metal loses an electron to the halogen. The product, a metal halide salt such as NaCl, consists of metal cations and halide anions electrostatically attracted to one another by **ionic bonds**. The sum of the interaction energies among all ions in a crystal is called the crystal's **lattice energy**.

In general, redox reactions of main-group elements can be described by the **octet rule**, which states that these elements tend to undergo reactions so as to attain a noble-gas electron configuration with filled s and p sublevels in their valence shell. Elements on the left side of the periodic table tend to give away electrons until a noble-gas configuration is reached, elements on the right side of the table tend to accept electrons until a noble-gas configuration is reached, and the noble gases themselves are essentially unreactive.

UNDERSTANDING KEY CONCEPTS

1. Little is known about the chemistry of astatine (At) from direct observation, but reasonable predictions can be made.
   (a) Is astatine likely to be a gas, a liquid, or a solid?
   (b) What color is astatine likely to have?
   (c) Is astatine likely to react with sodium? If so, what would be the formula of the product?

2. Test your knowledge of the periodic table by indicating approximately where on the blank outline the following elements appear.
   (a) main groups   **(b)** halogens
   (c) alkali metals   **(d)** noble gases
   (e) alkaline earths   **(f)** group 3A elements
   (g) lanthanides

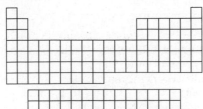

3. Which of the following drawings is more likely to represent an ionic compound and which a covalent compound?

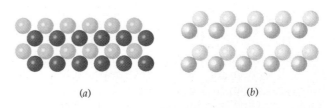

   (a)   *(b)*

4. Circle the approximate part or parts of the periodic table where the following elements appear.
   (a) elements with the smallest values of E_{i1}
   (b) elements with the largest atomic radii
   (c) elements with the most negative values of E_{ea}

224

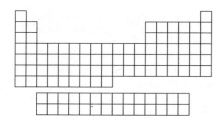

E_{ea} for Cl $= -348.6$ kJ/mol
Heat of sublimation for Li $= +159.4$ kJ/mol
E_{i1} for Li $= +520$ kJ/mol
Bond dissociation energy for $Cl_2 = +243$ kJ/mol
Lattice energy for LiCl $= +853$ kJ/mol

6. Look at the properties of the alkali metals summarized in Table 6.3, and predict reasonable values for the melting point, boiling point, density, and atomic radius of francium.

5. Given the following values for the formation of LiCl from its elements, draw a Born-Haber cycle similar to that shown in Figure 6.6.

ADDITIONAL PROBLEMS

Problems 6.1–6.19 appear within the chapter.

IONIZATION ENERGY AND ELECTRON AFFINITY

6.20 Do ionization energies have a positive sign or a negative sign? Explain.

6.21 Which group of elements in the periodic table has the largest E_{i1}, and which group has the smallest? Explain.

6.22 Without looking at Figure 6.1, which element would you expect to have the smallest ionization energy in the periodic table? Which the largest?

6.23 Which electron is most likely to be lost from elements with the following electron configurations?
(a) [Ne] $3s^2 3p^4$ **(b)** [Ar] $4s^2 3d^{10} 4p^1$ **(c)** [Ar] $4s^2 3d^1$

6.24 **(a)** Which has the smaller second ionization energy, K or Ca? **(b)** Which has the larger third ionization energy, Ga or Ca?

6.25 **(a)** Which has the smaller fourth ionization energy, Sn or Sb? **(b)** Which has the larger sixth ionization energy, Se or Br?

6.26 Three atoms have the following electron configurations:
(a) $1s^2 2s^2 2p^6 3s^2 3p^3$ **(b)** $1s^2 2s^2 2p^6 3s^2 3p^6$
(c) $1s^2 2s^2 2p^6 3s^2 3p^6 4s^2$
Which of the three has the highest E_{i2}? Which has the smallest E_{i7}?

6.27 Identify the electron most likely to be lost from the following elements.
(a) strontium **(b)** bromine **(c)** lanthanum

6.28 Identify the second electron to be lost from the elements listed in Problem 6.27.

6.29 Which element in each of the following sets has the lowest first ionization energy, and which has the highest?
(a) Li, Ba, K **(b)** B, Be, Cl **(c)** Ca, C, Cl

6.30 What elements meet the following descriptions?
(a) has highest E_{i3} **(b)** has highest E_{i7}

6.31 Use the data in Table 6.1 to calculate the total amount of energy in kJ/mol necessary to remove all valence-shell electrons from each of the following elements.
(a) Na **(b)** Mg **(c)** Al **(d)** Cl

6.32 Which element in each of the following pairs has the more negative electron affinity?
(a) F or Fe **(b)** Ne or Na **(c)** Ba or Br

6.33 What is the relationship between the electron affinity of a univalent cation such as Na^+ and the ionization energy of the neutral atom?

6.34 Which do you think has the more negative electron affinity, Na^+ or Na? Na^+ or Cl?

6.35 Which do you think has the more negative electron affinity, Br or Br^-?

6.36 Why is energy usually released when an electron is added to a neutral atom but absorbed when an electron is removed from a neutral atom?

6.37 Why does ionization energy increase steadily across the periodic table from group 1A to group 8A, whereas electron affinity increases irregularly from group 1A to group 7A and then falls dramatically for group 8A?

LATTICE ENERGY AND IONIC BONDS

6.38 Order the following compounds according to their expected lattice energies.
(a) LiCl **(b)** KCl **(c)** KBr **(d)** $MgCl_2$

6.39 Order the following compounds according to their expected lattice energies.
(a) $AlBr_3$ **(b)** $MgBr_2$ **(c)** LiBr **(d)** CaO

6.40 What five factors contribute to the overall energy change in the reaction of lithium with bromine to form LiBr?

6.41 Calculate the energy change in kJ/mol if lithium atoms lose an electron to bromine atoms to form isolated Li^+ and Br^- ions. E_i for Li is 520 kJ/mol, E_{ea} for Br is -325 kJ/mol.

6.42 Find the lattice energy of LiBr in Table 6.2, and calculate the energy change in kJ/mol for the formation of solid LiBr from the elements. (The sublimation energy for Li is +159.4 kJ/mol, the bond dissociation energy of Br_2 is +224 kJ/mol, and the energy necessary to convert $Br_2(l)$ to $Br_2(g)$ is 30.9 kJ/mol.)

6.43 Look up the lattice energies in Table 6.2, and then calculate the energy change in kJ/mol for the formation of the following substances from their elements.

   (a) LiF (The sublimation energy for Li is +159.4 kJ/mol, E_i for Li is 520 kJ/mol, E_{ea} for F is −328 kJ/mol, and the bond dissociation energy of F_2 is +158 kJ/mol.)

   (b) CaF_2 (The sublimation energy for Ca is +178.2 kJ/mol, E_{i1} = +589.5 kJ/mol, and E_{i2} = +1145 kJ/mol.)

6.44 Born-Haber cycles, such as that shown in Figure 6.6, are called *cycles* because they form closed loops. If any five of the six energy changes in the cycle are known, the value of the sixth can be calculated. Use the following five values to calculate a lattice energy in kJ/mol for sodium hydride, NaH.
E_{ea} for H = −72.8 kJ/mol
Heat of sublimation for Na = +107.3 kJ/mol
E_{i1} for Na = +495.8 kJ/mol
Bond dissociation energy for H_2 = +435.9 kJ/mol
Net energy change for the formation of NaH from its elements = −60 kJ/mol

6.45 Calculate a lattice energy for CaH_2 in kJ/mol using the following information:
E_{ea} for H = −72.8 kJ/mol
Heat of sublimation for Ca = +178.2 kJ/mol
E_{i1} for Ca = +589.8 kJ/mol
Bond dissociation energy for H_2 = +435.9 kJ/mol
E_{i2} for Ca = +1145 kJ/mol
Net energy change for the formation of CaH_2 from its elements = −186.2 kJ/mol

6.46 Calculate the overall energy change in kJ/mol for the formation of CsF from its elements using the following data:
E_{ea} for F = −328 kJ/mol
Heat of sublimation for Cs = +76.1 kJ/mol
E_{i1} for Cs = +375.7 kJ/mol
Bond dissociation energy for F_2 = +158 kJ/mol
E_{i2} for Cs = +2422 kJ/mol
Lattice energy for CsF = +740 kJ/mol

6.47 The estimated lattice energy for CsF_2 is +2347 kJ/mol. Use the data given in Problem 6.46 to calculate an overall energy change in kJ/mol for the formation of CsF_2 from its elements. Does the overall reaction absorb energy or release it?

6.48 In light of your answers to problems 6.46 and 6.47, which compound is more likely to form in the reaction of cesium with fluorine, CsF or CsF_2?

6.49 Calculate overall energy changes in kJ/mol for the formation of CaCl and $CaCl_2$ from their elements. The following data are needed.
E_{ea} for Cl = −348.6 kJ/mol
Heat of sublimation for Ca = +178.2 kJ/mol
E_{i1} for Ca = +589.8 kJ/mol
Bond dissociation energy for Cl_2 = +243 kJ/mol
E_{i2} for Ca = +1145 kJ/mol
Lattice energy for $CaCl_2$ = +2258 kJ/mol
Lattice energy for CaCl = +717 kJ/mol (estimated)

6.50 In light of your answers to Problem 6.49, which compound is more likely to form in the reaction of calcium with chlorine, CaCl or $CaCl_2$?

6.51 Use the data and the result in Problem 6.44 to draw a Born-Haber cycle for the formation of NaH from its elements.

6.52 Use the data and the result in Problem 6.43a to draw a Born-Haber cycle for the formation of LiF from its elements.

MAIN-GROUP CHEMISTRY

6.53 Indicate where in the periodic table the following elements appear.

   (a) Li   **(b)** Al   **(c)** Ba   **(d)** Br
   (e) Ne   **(f)** O   **(g)** Fr   **(h)** He

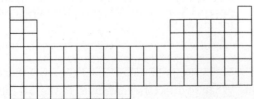

6.54 Which of the elements in groups 7A (F, Cl, Br, I) and 8A (He, Ne, Ar, Kr, Xe) are gases, which are liquids, and which are solids at room temperature?

6.55 Give at least one important use for each of the following elements.

   (a) lithium   **(b)** potassium
   (c) strontium   **(d)** helium

6.56 Tell how each of the following elements is produced commercially.

   (a) sodium   **(b)** aluminum
   (c) argon   **(d)** bromine

6.57 Why does chemical reactivity increase from top to bottom in group 1A but decrease from top to bottom in group 7A?

6.58 Give a brief statement of the octet rule and an explanation of why it works.

6.59 Which main-group elements occasionally break the octet rule?

6.60 Write balanced equations for the reaction of potassium with the following reagents. If no reaction occurs, write N.R.

(a) H_2 (b) H_2O (c) NH_3 (d) Br_2 (e) N_2 (f) O_2

6.61 Write balanced equations for the reaction of calcium with the following reagents. If no reaction occurs, write N.R.

(a) H_2 (b) H_2O (c) He (d) Br_2 (e) O_2

6.62 Write balanced equations for the reaction of chlorine with the following reagents. If no reaction occurs, write N.R.

(a) H_2 (b) Ar (c) Br_2 (d) N_2

6.63 Aluminum metal can be prepared by reaction of $AlCl_3$ with Na. Write a balanced equation for the reaction, and tell which atoms have been oxidized and which have been reduced.

6.64 The widely used antacid called milk of magnesia is an aqueous suspension of $Mg(OH)_2$. How would you prepare $Mg(OH)_2$ from magnesium metal?

6.65 What is the maximum amount in grams of pure iodine that you could obtain from 1.00 kg of the mineral lautarite ($CaIO_3$)?

6.66 Assume that you wanted to prepare a small volume of pure hydrogen by reaction of lithium metal with water. How many grams of lithium would you need to prepare 455 mL of H_2 if the density of hydrogen is 0.0893 g/L?

6.67 How many grams of strontium bromide are formed if 5.65 g of strontium undergoes complete reaction with excess bromine?

6.68 As a general rule, more reactive halogens can oxidize the anions of less reactive halogens. Predict the products of the following reactions, and identify the oxidizing and reducing agents in each. If no reaction occurs, write N.R.

(a) $2 Cl^-(aq) + F_2(g) \longrightarrow$?
(b) $2 Br^-(aq) + I_2(s) \longrightarrow$?
(c) $2 I^-(aq) + Br_2(aq) \longrightarrow$?

6.69 Identify the oxidizing agent and the reducing agent in each of the following reactions.

(a) $Mg(s) + 2 H^+(aq) \longrightarrow Mg^{2+}(aq) + H_2(g)$
(b) $Kr(g) + F_2(g) \longrightarrow KrF_2(s)$
(c) $I_2(s) + 3 Cl_2(g) \longrightarrow 2 ICl_3(l)$

6.70 Identify the oxidizing agent and the reducing agent in each of the following reactions.

(a) $2 XeF_2(s) + 2 H_2O(l) \longrightarrow 2 Xe(g) + 4 HF(aq) + O_2(g)$
(b) $NaH(s) + H_2O(l) \longrightarrow Na^+(aq) + OH^-(aq) + H_2(g)$
(c) $2 TiCl_4(l) + H_2(g) \longrightarrow 2 TiCl_3(s) + 2 HCl(g)$

GENERAL PROBLEMS

6.71 The ionization energy (E_i) of an atom can be measured by a technique known as *photoelectron spectroscopy*, in which light of wavelength λ is directed at an atom, causing an electron to be ejected. The kinetic energy of the ejected electron (E_k) is measured by determining its velocity, v ($E_k = 1/2\ mv^2$), and E_i is then calculated using the conservation of energy principle. That is, the energy of the incident light is equal to E_i plus E_k. What is the ionization energy of rubidium atoms in kJ/mol if light with $\lambda = 58.4$ nm produces electrons with a velocity of 2.450×10^6 m/s? The mass, m, of an electron is 9.109×10^{-31} kg.

6.72 What is the ionization energy of potassium in kJ/mol if light with $\lambda = 142$ nm produces electrons with a velocity of 1.240×10^6 m/s? (See Problem 6.71.)

6.73 Calculate overall energy changes in kJ/mol for the formation of MgF and MgF_2 from their elements. The following data are needed.

E_{ea} for F $= -328$ kJ/mol
Heat of sublimation for Mg $= +147.7$ kJ/mol
E_{i1} for Mg $= +737.7$ kJ/mol
Bond dissociation energy for $F_2 = +158$ kJ/mol
E_{i2} for Mg $= +1450.7$ kJ/mol
Lattice energy for $MgF_2 = +2952$ kJ/mol
Lattice energy for MgF $= 930$ kJ/mol (estimated)

6.74 In light of your answers to Problem 6.73, which compound is more likely to form in the reaction of magnesium with fluorine, MgF_2 or MgF?

6.75 Give at least one important use of each of the following elements.

(a) sodium (b) magnesium (c) fluorine

6.76 Tell how each of the following elements is produced commercially.

(a) fluorine (b) calcium (c) chlorine

6.77 Write balanced equations for the reaction of lithium with the following reagents. If no reaction occurs, write N.R.

(a) H_2 (b) H_2O (c) NH_3 (d) Br_2 (e) N_2 (f) O_2

6.78 Write balanced equations for the reaction of fluorine with the following reagents. If no reaction occurs, write N.R.

(a) H_2 (b) Na (c) Br_2 (d) NaBr

6.79 Many early chemists noted a *diagonal relationship* among elements in the periodic table, whereby a given element is sometimes more similar to the element below and to the right than it is to the element directly below. Lithium is more similar to magnesium than to sodium, for example, and boron is more similar to silicon than to aluminum. Use your knowledge about the periodic trends of such properties as atomic radii and Z_{eff} to see if you can explain the existence of diagonal relationships.

chapter 7 COVALENT BONDS AND MOLECULAR STRUCTURE

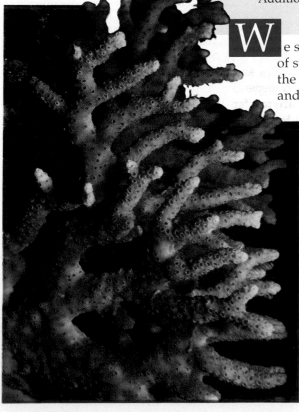

We saw in the last chapter that bonds are formed between the atoms of such dissimilar elements as the alkali metals and the halogens by the transfer of electrons. The metal atom loses one or more electrons and becomes a cation, while the halogen atom gains an electron and becomes an anion. The resultant ions are held together by electrostatic attraction, forming what chemists call ionic bonds.

How, though, do bonds form between atoms of the same or similar elements? How can we describe the bonds in such substances as H_2, Cl_2, CO_2, and the literally millions of other non-ionic compounds? Simply put, the answer is that the bonds in such compounds are formed by the *sharing* of electrons between atoms rather than by the complete transfer of electrons from one atom to another. The resultant shared-electron bond is called a **covalent bond** and is the most important kind of bond in all of chemistry. We'll explore the nature of covalent bonding in this chapter.

Most of this coral plant consists of the ionic compound $CaCO_3$. The tip of the plant, however, is a living organism, full of organic molecules.

7.1 ➤ THE COVALENT BOND

The same electrostatic forces that are responsible for the formation of ionic bonds between ions are also responsible for the formation of covalent bonds between neutral atoms. Let's look at the H–H bond in the H_2 molecule as an example. As two hydrogen atoms come close together, electrostatic interactions begin to develop between them. The two positively charged nuclei repel each other and the two negatively charged electrons repel each other, but each proton attracts both electrons and each electron attracts both protons. If the attractive forces are stronger than the repulsive forces, then the hydrogen atoms stay together and a covalent bond is formed. In essence, the electrons act as a kind of "glue" to bind the two nuclei together into an H_2 molecule. Both nuclei are simultaneously attracted to the same electrons and are therefore held together (Figure 7.1).

The two teams are effectively bonded together because both are tugging on the same rope.

As you might imagine, the magnitudes of the various attractive and repulsive forces between protons and electrons in a covalent bond depend on how close the atoms are to each other. If the hydrogen atoms are too far apart, the attractive forces are small and no bond exists. If the hydrogen atoms press too closely together, the repulsive interaction between the nuclei becomes so strong that it pushes the atoms apart. Thus, there is an optimum point where net attractive forces are maximized and where the H–H molecule is most stable. This optimum distance between nuclei is

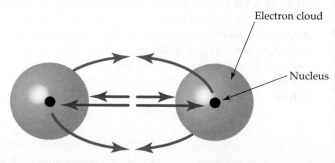

Electron cloud

Nucleus

FIGURE 7.1 A covalent H–H bond is the net result of attractive and repulsive electrostatic forces. The proton–electron attractions (blue arrows) are greater than the proton–proton and electron–electron repulsions (red arrows), resulting in a net attractive force that holds the atoms together to form an H_2 molecule.

called the **bond length** and is 74 pm in the H_2 molecule. On a graph of energy versus internuclear distance, the bond length corresponds to the minimum-energy, most stable arrangement (Figure 7.2).

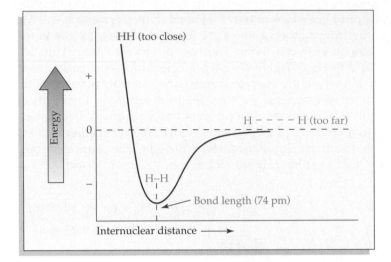

FIGURE 7.2 A graph of potential energy versus internuclear distance for the H_2 molecule. When the hydrogen atoms are too far apart, attractions are weak and no bonding occurs; when the atoms are too close, strong repulsions occur; and when the atoms are optimally separated, the energy is at a minimum.

Every covalent bond has its own characteristic length that leads to maximum stability. As you might expect, bond lengths are roughly predictable from a knowledge of atomic radii (Section 5.15). For example, since the atomic radius of hydrogen is 37 pm and the atomic radius of chlorine is 99 pm, the H–Cl bond length in a hydrogen chloride molecule is about 37 pm + 99 pm = 136 pm. (The actual value is 127 pm.)

7.2 ➤ STRENGTHS OF COVALENT BONDS

Look again in Figure 7.2 at the graph of energy versus internuclear distance for the H_2 molecule, and note how the H_2 molecule is lower in energy than two separate hydrogen atoms. This result is just one example of a general tendency throughout nature for systems to find as low an energy level as possible—after all, water always runs downhill, and a released spring always uncoils. In the same way, when pairs of hydrogen atoms bond together, they end up as lower-energy H_2 molecules and release 436 kJ/mol of energy in the process. Looked at from the other direction, 436 kJ/mol of energy must be *added* to H_2 molecules to split them apart into hydrogen atoms.

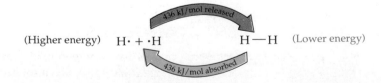

(Higher energy)   H· + ·H   H — H   (Lower energy)

The amount of energy necessary to break a chemical bond in an isolated molecule in the gaseous state—and thus the amount of energy released when the bond forms—is called the **bond dissociation energy,** *D*. Bond dissociation energies always have a positive value because energy must be supplied to break a bond. Conversely, the amount of energy released on forming a bond has a negative value.

Every bond in every molecule has its own specific bond dissociation energy. Not surprisingly, though, bonds between the same pairs of atoms usually have similar bond dissociation energies. For example, carbon–carbon bonds usually have *D* values of approximately 350 to 380 kJ/mol regardless of the exact structure of the molecule.

Ethane
D = 376 kJ/mol

Propane
D = 356 kJ/mol

Butane
D = 352 kJ/mol

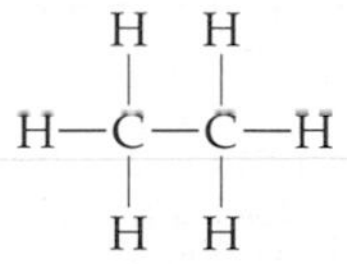

As a result, we can construct a useful table of *average* bond dissociation energies (Table 7.1) to compare different kinds of bonds. Keep in mind, though, that the actual value in a specific molecule might vary by ±10% from the average.

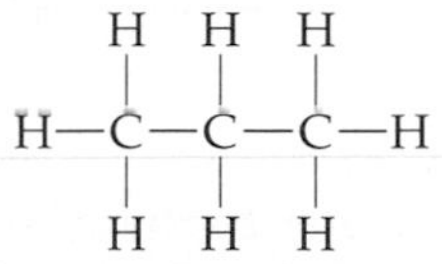

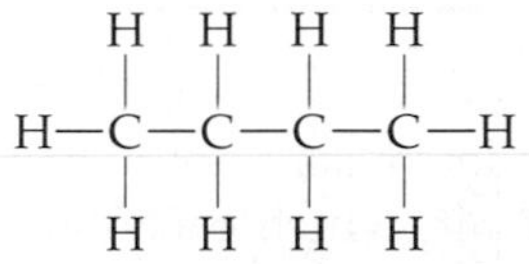

TABLE 7.1 Average Bond Dissociation Energies, *D* (kJ/mol)[a]

H—H	436[a]	C—H	410	N—H	390	O—H	460	F—F	158[a]
H—C	410	C—C	350	N—C	300	O—C	350	Cl—Cl	243[a]
H—F	570[a]	C—F	450	N—F	270	O—F	180	Br—Br	193[a]
H—Cl	432[a]	C—Cl	330	N—Cl	200	O—Cl	200	I—I	151[a]
H—Br	366[a]	C—Br	270	N—Br	240	O—Br	210	S—F	310
H—I	298[a]	C—I	240	N—I	—	O—I	220	S—Cl	250
H—N	390	C—N	300	N—N	240	O—N	200	S—Br	210
H—O	460	C—O	350	N—O	200	O—O	180	S—S	225
H—S	340	C—S	260	N—S	—	O—S	—		

[a] Bond dissociation energies for diatomic molecules are exact.

The bond dissociation energies listed in Table 7.1 cover a wide range, from a low of 151 kJ/mol for the I–I bond to a high of 570 kJ/mol for the H–F bond. As a rule of thumb, though, most of the bonds commonly encountered in naturally occurring molecules—C–H, C–C, C–O—have values in the range 350 to 400 kJ/mol.

7.3 ➤A COMPARISON OF IONIC AND COVALENT COMPOUNDS

Look at the comparison of physical properties between NaCl and HCl shown in Table 7.2. Sodium chloride, a white ionic solid, has a melting point of 801°C and a boiling point of 1413°C. Hydrogen chloride, a colorless gas, has a melting point of −115°C and a boiling point of −84.9°C. What accounts for such large differences?

Ionic compounds are high-melting solids because of the kinds of bonds they have. As discussed previously in Section 6.4, a visible sample of sodium

TABLE 7.2 Some Physical Properties of NaCl and HCl		
Property	**NaCl**	**HCl**
Formula weight	58.44	36.46
Physical appearance	White solid	Colorless gas
Type of bond	Ionic	Covalent
Melting point	801°C	−115°C
Boiling point	1413°C	−84.9°C

chloride consists not of NaCl molecules but of a three-dimensional network of ions in which each Na$^+$ is attracted to many surrounding Cl$^-$ ions and each Cl$^-$ is attracted to many surrounding Na$^+$ ions. For sodium chloride to melt or boil, every ionic attraction in the entire crystal must be overcome, a process that requires a large amount of energy.

Sodium chloride, an ionic compound, is a white, crystalline solid. Hydrogen chloride, a molecular compound, is a gas at room temperature.

Covalent compounds, by contrast, are low-melting solids, liquids, or even gases. A sample of a covalent compound such as hydrogen chloride consists of discrete HCl molecules. The covalent bond within an *individual* molecule may be very strong, but the attractive forces between *different* molecules are relatively weak. As a result, relatively little energy is required to overcome intermolecular forces and cause a covalent compound to melt or boil.

7.4 ►LEWIS STRUCTURES

One of the most convenient ways to picture the sharing of electrons between atoms is to use what are known as *electron-dot structures*, or *Lewis structures*, named after G. N. Lewis of the University of California at Berkeley. A **Lewis structure** represents an atom's valence electrons by dots and indicates by the placement of the dots how the valence electrons are distributed in a molecule. A hydrogen molecule, for example, is written showing a pair of dots between the hydrogen atoms, indicating that the hydrogens share the pair of electrons in a covalent bond:

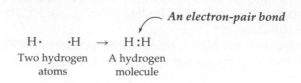

An electron-pair bond

H·   ·H   →   H:H
Two hydrogen   A hydrogen
      atoms   molecule

By sharing two electrons in a covalent bond, each hydrogen effectively has one electron pair and a stable filled-shell configuration. As with ions (Section 6.10), a filled valence shell for each atom in a molecule leads to maximum stability.

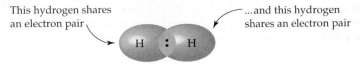

Atoms other than hydrogen also form covalent bonds by sharing electron pairs, and the Lewis structures of the resultant molecules are easily drawn by assigning the correct number of valence electrons to each atom. Group 3A atoms (such as boron) have three valence electrons, group 4A atoms (such as carbon) have four valence electrons, and so on across the periodic table. The group 7A element fluorine has seven valence electrons, and a Lewis structure for the F_2 molecule shows how a covalent bond can form:

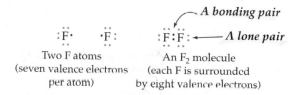

Six of the seven valence electrons in each fluorine atom are paired in three filled atomic orbitals and are not shared in bonding. The seventh fluorine electron, however, is unpaired and can be used in forming a covalent bond to another fluorine. Each atom in the resultant F_2 molecule thus achieves a filled valence-shell octet. The three pairs of nonbonding electrons on each fluorine are called **lone pairs**, and the shared electrons are called a **bonding pair**.

The tendency of main-group atoms to fill their s and p subshells when they form bonds—the *octet rule* discussed in Section 6.10—is an important guiding principle that makes it possible to predict the formulas and Lewis structures of a great many molecules. As a general rule, an atom shares as many of its valence-shell electrons as possible, either until it has no more to share or until it reaches an octet configuration. For second-row elements in particular, the following guidelines apply:

- Group 3A elements, such as boron, have three valence electrons and can therefore form three electron-pair bonds, as in borane, BH_3.

$$\cdot \ddot{B}\cdot + 3\,H\cdot \rightarrow H\!:\!\ddot{B}\!:\!H \qquad \text{Borane}$$

Only six electrons around boron

- Group 4A elements, such as carbon, have four valence electrons and form four bonds, as in methane, CH_4.

$$\cdot \dot{C}\cdot + 4\,H\cdot \rightarrow H\!:\!\ddot{C}\!:\!H \qquad \text{Methane}$$

- Group 5A elements, such as nitrogen, have five valence electrons and form three bonds, as in ammonia, NH_3.

$$\cdot \overset{\cdot\cdot}{N} \cdot \ + \ 3\,H\cdot \ \rightarrow \ H:\overset{\overset{\textstyle H}{\cdot\cdot}}{N}:H \quad \text{Ammonia}$$

- Group 6A elements, such as oxygen, have six valence electrons and form two bonds, as in water, H_2O.

$$\cdot \overset{\cdot\cdot}{O} \cdot \ + \ 2\,H\cdot \ \rightarrow \ H:\overset{\cdot\cdot}{O}:H \quad \text{Water}$$

- Group 7A elements (halogens), such as fluorine, have seven valence electrons and form one bond, as in hydrogen fluoride, HF.

$$:\overset{\cdot\cdot}{F}\cdot \ + \ H\cdot \rightarrow \ H:\overset{\cdot\cdot}{\underset{\cdot\cdot}{F}}: \quad \text{Hydrogen fluoride}$$

- Group 8A elements (noble gases), such as neon, rarely form covalent bonds because they already have valence-shell octets.

$$:\overset{\cdot\cdot}{\underset{\cdot\cdot}{Ne}}: \quad \textit{(does not form covalent bonds)}$$

These conclusions are summarized in Table 7.3.

TABLE 7.3	Covalent Bonding for Second-Row Elements	
Group	**Number of Bonds**	**Example**
3A	3	BH_3
4A	4	CH_4
5A	3	NH_3
6A	2	H_2O
7A	1	HF
8A	0	—

Not all covalent molecules are described as simply as those just discussed. In molecules such as O_2, N_2, and many others, the atoms share more than one pair of electrons, leading to the formation of *multiple* covalent bonds. The only way, for example, that both oxygen atoms in the O_2 molecule can have valence-shell octets is for them to share four electrons (two pairs), giving a **double bond**. Similarly, the nitrogen atoms in the N_2 molecule share six electrons (three pairs), giving a **triple bond**. In speaking of such compounds, we often use the term **bond order** to refer to the number of electron pairs shared between atoms. Thus, the oxygen–oxygen bond in the O_2 molecule[1] has a bond order of 2, and the nitrogen–nitrogen bond in the N_2 molecule has a bond order of 3.

$$\cdot\overset{\cdot\cdot}{O}\cdot \ + \ \cdot\overset{\cdot\cdot}{O}\cdot \rightarrow \ :\overset{\cdot\cdot}{O}::\overset{\cdot\cdot}{O}: \quad \begin{array}{l}\textit{Two electron pairs}\\ \textit{— a double bond}\end{array}$$

$$:\overset{\cdot}{N}\cdot \ + \ \cdot\overset{\cdot}{N}: \rightarrow \ :N:::N: \quad \begin{array}{l}\textit{Three electron pairs}\\ \textit{— a triple bond}\end{array}$$

[1] Although the O_2 molecule does have a double bond, the Lewis structure shown for O_2 is incorrect in other respects, as explained in Section 7.14.

As you might expect, multiple bonds are both shorter and stronger than their corresponding single-bond analogs because there are more shared electrons holding the nuclei together. Compare, for example, the O=O double bond in O_2 with the O–O single bond in H_2O_2 (hydrogen peroxide) and the N≡N triple bond in N_2 with the N–N single bond in N_2H_4 (hydrazine):

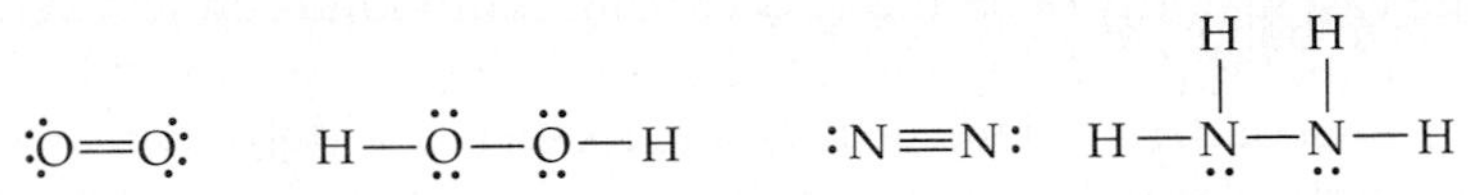

:O̤=O̤:	H—O̤—O̤—H	:N≡N:	H—N—N—H (with H above each N)
Bond length: 121 pm	148 pm	110 pm	145 pm
Bond strength: 498 kJ/mol	213 kJ/mol	945 kJ/mol	275 kJ/mol

One final point about covalent bonds involves the origin of the bonding electron pair. Although most covalent bonds result when each of two atoms contributes one electron, bonds can also form when one atom donates *both* electrons (a lone pair) to another atom that has a vacant valence orbital. The ammonium ion (NH_4^+), for example, forms when two electrons from the nitrogen atom of ammonia bond to H^+. Such bonds are called **coordinate covalent bonds**.

An ordinary covalent bond: each atom donates one electron

$$H\cdot + \cdot H \rightarrow H:H$$

A coordinate covalent bond: the nitrogen atom donates both electrons

$$H^+ + \; :\overset{H}{\underset{H}{N}}:H \rightarrow \left[H:\overset{H}{\underset{H}{N}}:H \right]^+$$

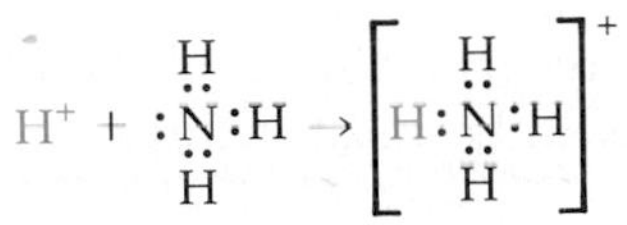

Note that the nitrogen atom in the ammonium-ion product has more than the usual number of bonds—four instead of three—but that it still has an octet of valence electrons. Nitrogen, oxygen, phosphorus, and sulfur form such coordinate covalent bonds frequently.

EXAMPLE 7.1

Draw a Lewis structure for phosphine, PH_3.

SOLUTION   Phosphorus, a group 5A element, has five valence electrons and can achieve a valence-shell octet by forming three bonds:

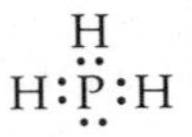

$$H:\overset{H}{\underset{\cdot\cdot}{P}}:H \quad \text{Phosphine}$$

☐ PROBLEM 7.1 Draw Lewis structures for the following molecules.
(a) H_2S, a poisonous gas produced by rotten eggs
(b) $CHCl_3$, chloroform

☐ PROBLEM 7.2 Draw a Lewis structure for the hydronium ion, H_3O^+, and show how a coordinate covalent bond is formed by the reaction of H_2O with H^+.

7.5 ►LEWIS STRUCTURES OF POLYATOMIC MOLECULES

Compounds of Second-Row Elements: C, H, N, O

The vast majority of naturally occurring compounds on which life is based—proteins, fats, carbohydrates, and many others—contain only hydrogen and a few second-row elements: carbon, oxygen, and nitrogen. Lewis structures are particularly easy to draw for such compounds because the octet rule almost always applies and because the number of bonds formed by each element is usually predictable. If, for example, you were asked to figure out a Lewis structure for ethane, C_2H_6, you could take the two carbons, each with four valence electrons, and the six hydrogens, each with one valence electron, and then assemble the pieces so that each hydrogen forms one bond and each carbon forms four bonds. There is only one possibility:

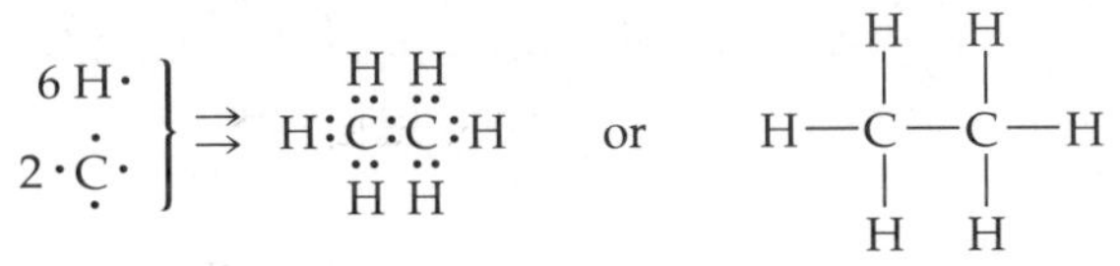

$$6\,H\cdot \atop 2\cdot\ddot{C}\cdot \Bigg\} \rightrightarrows \quad H\!:\!\ddot{C}\!:\!\ddot{C}\!:\!H \quad \text{or} \quad H\!-\!\overset{\displaystyle H}{\underset{\displaystyle H}{C}}\!-\!\overset{\displaystyle H}{\underset{\displaystyle H}{C}}\!-\!H$$

Ethane, C_2H_6

Note the alternative ways of showing the ethane structure. Because it is time consuming to draw all the dots in a Lewis structure, chemists generally indicate a covalent bond by a line. By convention, *a line drawn between two atoms represents an electron pair in a covalent bond.* Similarly, two lines between atoms represent four electrons (two pairs) in a double bond, and three lines represent six electrons (three pairs) in a triple bond. The following problems give you more practice with Lewis structures.

EXAMPLE 7.2

Draw a Lewis structure for hydrazine, N_2H_4.

SOLUTION   Draw the individual atoms with their valence electrons, and assemble them so that each hydrogen forms one bond and each nitrogen forms three bonds:

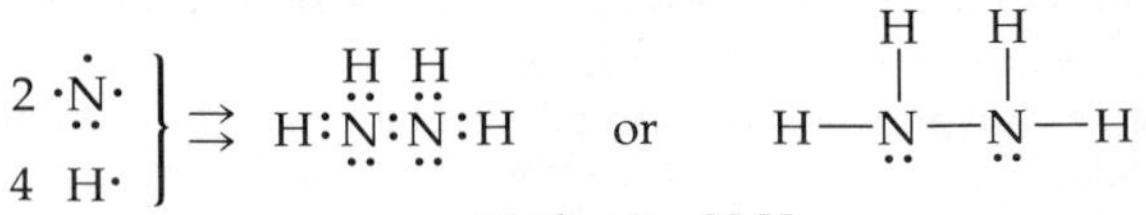

$$2\cdot\ddot{N}\cdot \atop 4\,H\cdot \Bigg\} \rightrightarrows \quad H\!:\!\ddot{N}\!:\!\ddot{N}\!:\!H \quad \text{or} \quad H\!-\!\overset{\displaystyle H}{\underset{\displaystyle\cdot\cdot}{N}}\!-\!\overset{\displaystyle H}{\underset{\displaystyle\cdot\cdot}{N}}\!-\!H$$

Hydrazine, N_2H_4

EXAMPLE 7.3

Draw a Lewis structure for carbon dioxide, CO_2.

SOLUTION   Draw the individual atoms with their valence electrons, and assemble them so that the carbon forms four bonds and each oxygen forms two bonds. This can happen only if there are two carbon–oxygen double bonds.

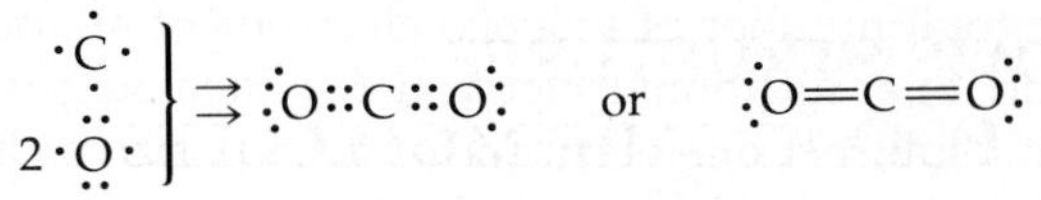

$$\left. \begin{array}{c} \cdot\ddot{C}\cdot \\ 2\cdot\ddot{O}\cdot \end{array} \right\} \rightarrow \ddot{O}::C::\ddot{O} \quad \text{or} \quad \ddot{O}=C=\ddot{O}$$

Carbon dioxide, CO_2

EXAMPLE 7.4

Draw a Lewis structure for the deadly gas hydrogen cyanide, HCN.

SOLUTION   Draw the individual atoms with their valence electrons, and assemble them so that the hydrogen forms one bond, the carbon forms four bonds, and the nitrogen forms three bonds. This can happen only if there is a carbon–nitrogen triple bond.

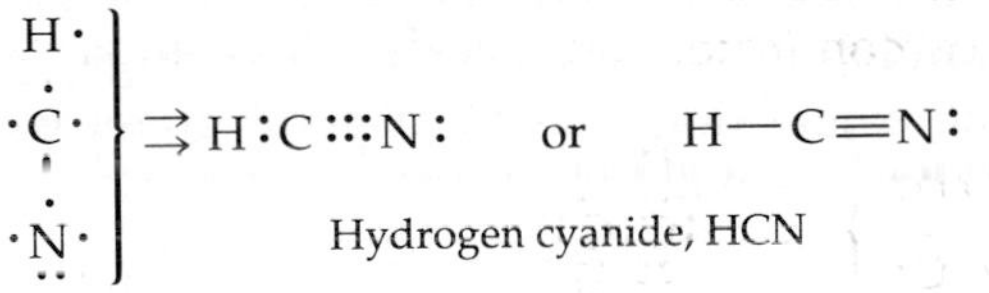

$$\left. \begin{array}{c} H\cdot \\ \cdot\dot{C}\cdot \\ \cdot\dot{N}\cdot \end{array} \right\} \rightarrow H:C:::N: \quad \text{or} \quad H-C\equiv N:$$

Hydrogen cyanide, HCN

┌ **PROBLEM 7.3** Draw Lewis structures for the following molecules.
(a) propane, C_3H_8 **(b)** hydrogen peroxide, H_2O_2 **(c)** methylamine, CH_5N

┌ **PROBLEM 7.4** Draw Lewis structures for the following molecules.
(a) ethylene, C_2H_4 **(b)** acetylene, C_2H_2 **(c)** phosgene, Cl_2CO

┌ **PROBLEM 7.5** There are two molecules with the formula C_2H_6O. Draw Lewis structures for both.

Compounds With Elements Beyond the Second Row

The simple method of drawing Lewis structures that works so well for most compounds of second-row elements sometimes breaks down for compounds that contain elements from lower than the second row in the periodic table. Because elements in the third row and lower have unfilled valence-shell d orbitals, they are able to expand their valence shell beyond the normal octet of electrons, forming more than the "normal" number of bonds predicted by their group number. In bromine trifluoride, for example, the bromine atom forms three electron-pair bonds rather than one and has ten valence electrons rather than eight:

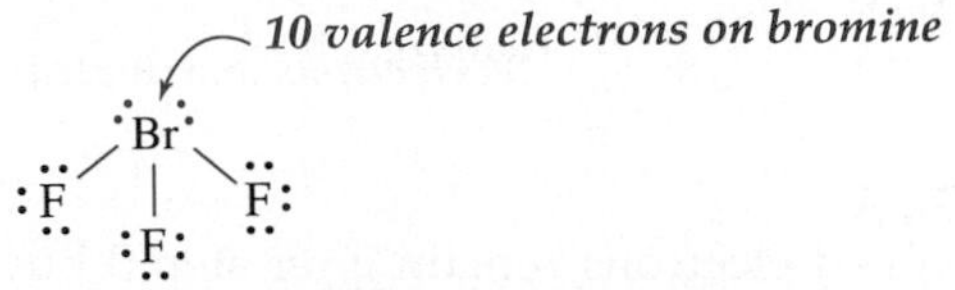

10 valence electrons on bromine

Bromine trifluoride, BrF_3

A general method of drawing Lewis structures that works for any compound is to use the following steps:

Step 1. Find the total number of valence electrons of all atoms in the molecule. Add one additional electron for each negative charge in an anion and subtract one electron for each positive charge in a cation. In SF_4, for example, the total is 34 (6 from sulfur and 7 from each of 4 fluorines). In OH^-, the total is 8 (6 from oxygen, 1 from hydrogen, and 1 for the negative charge). In NH_4^+, the total is 8 (5 from nitrogen, 1 from each of 4 hydrogens, minus 1 for the positive charge).

Step 2. Decide what the connections are between atoms, and draw a line to represent each bond. Often, you'll be told the connections; other times you'll have to guess. Remember that hydrogen and the halogens usually form only one bond, elements in the second row usually form the number of bonds given in Table 7.3, and elements in the third row and lower often expand their valence shells and occur as the central atom in a cluster. If, for example, you were asked to predict the connections in SF_4, a good guess would be that each fluorine forms one bond to sulfur, which expands its valence shell to accommodate the necessary number of bonds.

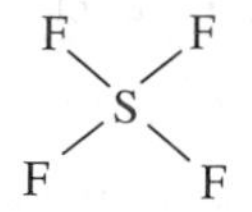

Sulfur tetrafluoride, SF_4

Step 3. Find the number of valence electrons remaining by subtracting the number used in bonding from the total number calculated in step 1. Assign as many of these remaining electrons as necessary to the terminal atoms (other than hydrogen) so that each has an octet. In SF_4, 8 of the 34 total valence electrons are used in covalent bonding, leaving $34 - 8 = 26$. Twenty-four of these 26 are assigned to the four terminal fluorine atoms to reach an octet configuration for each:

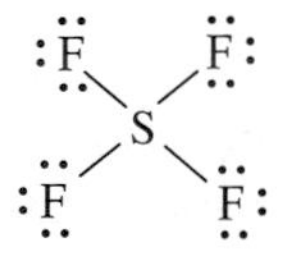

32 electrons distributed

Step 4. If unassigned electrons remain after step 3, place them on the central atom. In SF_4, 32 of the 34 electrons have been assigned, leaving the final two to be placed on the central S atom:

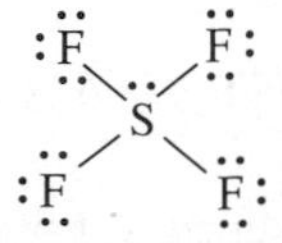

34 electrons distributed

Step 5. If no unassigned electrons remain after step 3 but the central atom does not yet have an octet, move one or more lone pairs of electrons from a neighboring atom to form a multiple bond, either double or triple. O, N, C, and S often form multiple bonds.

The following examples show how to use these rules.

EXAMPLE 7.5

Write a Lewis structure for phosphorus pentachloride, PCl_5.

SOLUTION   First, count the total number of valence electrons. Phosphorus has 5, and each chlorine has 7, for a total of 40. Next, decide on the connections between atoms and draw a line to indicate each bond. Since chlorine normally forms only one bond, it's likely in the case of PCl_5 that all five chlorines are bonded to phosphorus, which expands its valence shell:

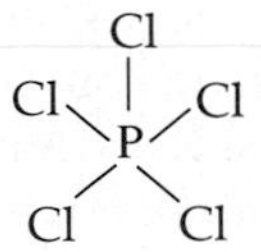

$$\begin{array}{c} \text{Cl} \\ | \\ \text{Cl} \diagdown \, \diagup \text{Cl} \\ \text{P} \\ \diagup \quad \diagdown \\ \text{Cl} \quad \text{Cl} \end{array}$$

Ten of the 40 valence electrons are necessary for the five P–Cl bonds, leaving 30 to be distributed so that each chlorine has an octet. All 30 remaining valence electrons are used up in this step, giving the structure

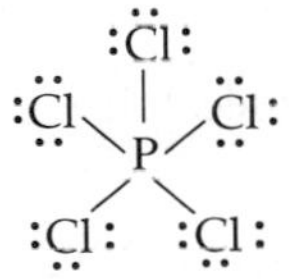

$$\begin{array}{c} :\ddot{\text{Cl}}: \\ | \\ :\ddot{\text{Cl}} \diagdown \, \diagup \ddot{\text{Cl}}: \\ \text{P} \\ \diagup \quad \diagdown \\ :\ddot{\text{Cl}}: \quad :\ddot{\text{Cl}}: \end{array}$$ Phosphorus pentachloride, PCl_5

EXAMPLE 7.6

Write a Lewis structure for formaldehyde, CH_2O.

SOLUTION   First, count the total number of valence electrons. Carbon has 4, each hydrogen has 1, and the oxygen has 6, for a total of 12. Next, decide on the probable connections between atoms, and draw a line to indicate each bond. In the case of formaldehyde, both hydrogens and the oxygen are bonded to carbon:

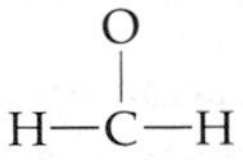

$$\begin{array}{c} \text{O} \\ | \\ \text{H} - \text{C} - \text{H} \end{array}$$

Six of the 12 valence electrons are used for bonds, leaving 6 for assignment to the terminal oxygen atom.

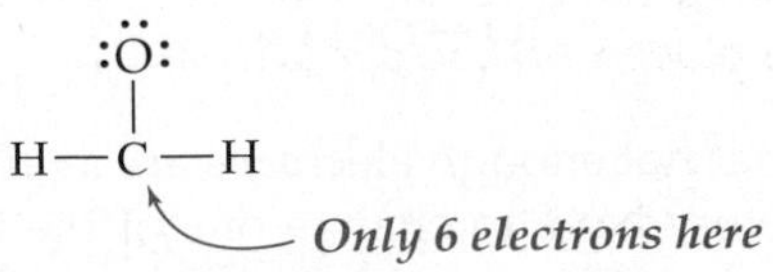

$$\begin{array}{c} :\ddot{\text{O}}: \\ | \\ \text{H} - \text{C} - \text{H} \end{array}$$
— *Only 6 electrons here*

At this point, all the valence electrons are assigned, yet the central carbon atom does not have an octet. We therefore convert two of the oxygen electrons from a lone pair into a bonding pair, generating a carbon-oxygen double bond and satisfying the octet rule for both oxygen and carbon.

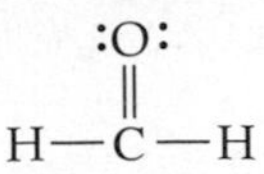

$$\begin{array}{c} :\text{O}: \\ \| \\ \text{H} - \text{C} - \text{H} \end{array}$$ Formaldehyde, CH_2O

EXAMPLE 7.7

Write a Lewis structure for xenon tetrafluoride, XeF_4.

SOLUTION First, count the total number of valence electrons. Xenon has 8, and each fluorine has 7, for a total of 36. Next, decide on the probable connections between atoms, and draw a line for each bond. In the case of XeF_4, it's likely that the four fluorines are bonded to xenon, a fifth-row atom, giving

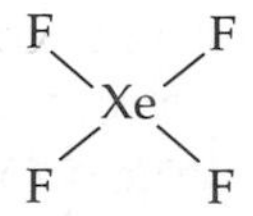

With 8 of the 36 valence electrons used in bonds, distribute as many of the remaining 28 electrons as necessary so that each of the terminal fluorine atoms has an octet. Four electrons still remain, so we assign them to xenon to give the final Lewis structure:

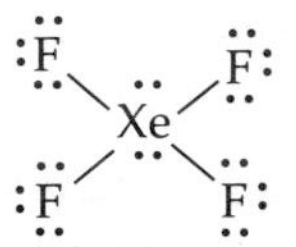

Xenon tetrafluoride, XeF_4

┌ **PROBLEM 7.6** Carbon monoxide, CO, is a deadly gas produced by incomplete combustion of fuels. Write a Lewis structure for CO.

┌ **PROBLEM 7.7** Write a Lewis structure for each of the following molecules.
(a) $AlCl_3$ **(b)** ICl_3 **(c)** $XeOF_4$ **(d)** HOBr

┌ **PROBLEM 7.8** Write a Lewis structure for each of the following ions.
(a) OH^- **(b)** H_3S^+ **(c)** HCO_3^-

7.6 ►LEWIS STRUCTURES AND RESONANCE

The steps given in the previous section for drawing Lewis structures lead to an interesting problem in some cases. Let's look at ozone, O_3, for instance. Step 1 says that there are 18 valence electrons in the molecule, and steps 2 through 4 let us draw the following structure:

$$:\ddot{O}—\ddot{O}—\ddot{O}:$$

We find at this point that not enough electrons are available to give all atoms an octet, and we therefore have to change one of the terminal oxygen lone pairs of electrons into a bonding pair, giving the central oxygen an octet. But do we take the lone pair from the "right-hand" oxygen or the "left-hand" one? Both possibilities lead to acceptable Lewis structures:

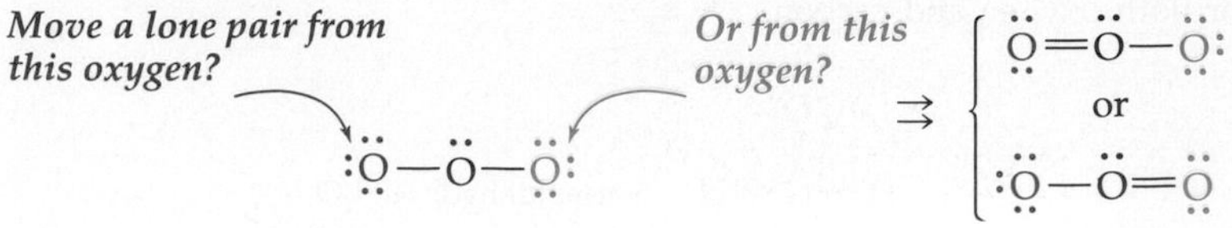

Move a lone pair from this oxygen?

Or from this oxygen?

Which of the two Lewis structures for O_3 is correct? In fact, *neither* is correct by itself. Whenever it's possible to write more than one Lewis structure for a molecule, the actual electronic structure is an *average* of the various possibilities, called a **resonance hybrid**. Ozone doesn't have one O=O double bond and one O–O single bond as the individual Lewis structures imply; rather, ozone has two *equivalent* O–O bonds that we can think of as having a bond order of 1.5, midway between pure single bonds and pure double bonds. Both bonds have an identical length of 128 pm.

We can't draw a single Lewis structure that indicates the equivalence of the two O–O bonds in O_3 because the conventions we use for indicating electron placement aren't good enough. Instead, the idea of resonance between two or more Lewis structures is indicated by drawing the individual Lewis structures and using a double-headed "resonance arrow" to show that both contribute to the resonance hybrid. A straight, double-headed arrow always indicates resonance; it is never used for any other purpose.

This double-headed arrow means that the structures on either side are contributors to a resonance hybrid.

$$:\ddot{O}—\ddot{O}=\ddot{O} \longleftrightarrow \ddot{O}=\ddot{O}—\ddot{O}:$$

The fact that single Lewis structures can't be written for all molecules indicates that Lewis structures are oversimplified and don't always give an accurate representation of the electron distribution in a molecule. There's a more accurate way of describing electron distributions called molecular orbital theory, which we'll look into shortly. This theory is more complex, however, and chemists still rely to a great extent on Lewis structures despite their deficiencies.

Ozone can be generated by passing a stream of oxygen through an electric discharge.

EXAMPLE 7.8

The nitrate ion, NO_3^-, has three equivalent oxygens, and its electronic structure is a resonance hybrid of three Lewis structures. Draw them.

SOLUTION   There are a total of 24 valence electrons in the nitrate ion: 5 from nitrogen, 6 from each of 3 oxygens, and 1 for the negative charge. The three equivalent oxygens are all bonded to nitrogen:

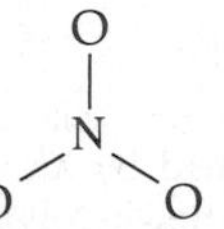

6 of 24 valence electrons assigned

Distributing the remaining 18 valence electrons among the three terminal oxygen atoms completes the octet of each oxygen but leaves nitrogen with only six electrons.

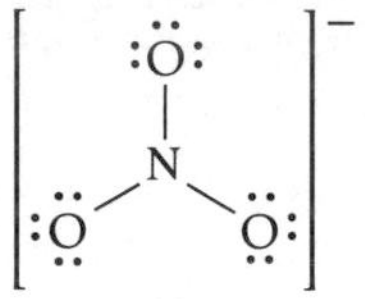

To give nitrogen an octet, one of the oxygen atoms must use a lone pair to form an N–O double bond. But which one? There are three possibilities, and thus three Lewis structures for the nitrate ion:

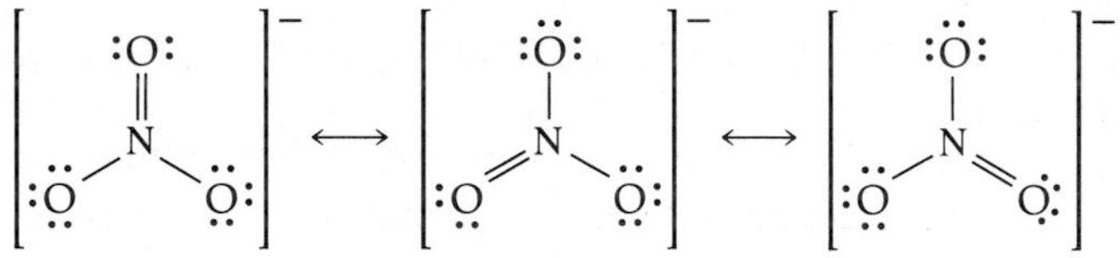

PROBLEM 7.9   Nitrous oxide, N_2O, is called "laughing gas" and is sometimes used by dentists as an anesthetic. Assuming the connections N–N–O, draw two Lewis structures for N_2O.

PROBLEM 7.10   Draw the indicated number of Lewis structures for each of the following molecules or ions:

(a) SO_2 (two structures)   **(b)** CO_3^{2-} (three structures)   **(c)** $SOCl_2$ (two structures)

7.7 ➤POLAR COVALENT BONDS: ELECTRONEGATIVITY

Without actually saying so, we've left the impression in previous sections that chemical bonding is an either/or proposition: A particular bond is either ionic or covalent. In fact, bonding is a continuous spectrum of possibilities, from an ionic bond between positive and negative ions on the one hand, to a nonpolar covalent bond with a symmetrical electron distribution between atoms on the other hand. Between these two extremes are the large majority of bonds, in which the bonding electrons are attracted somewhat more strongly by one atom than by the other but are not completely transferred. Such bonds are said to be **polar covalent bonds** (Figure 7.3). The Greek letter delta (δ) is used to symbolize the resultant partial charges on the atoms, either partial positive (δ^+) or partial negative (δ^-).

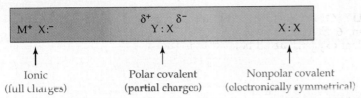

Ionic	Polar covalent	Nonpolar covalent
(full charges)	(partial charges)	(electronically symmetrical)

FIGURE 7.3 The continuum in bond types from ionic to nonpolar covalent. Polar covalent bonds lie between the two extremes. They are characterized by an unsymmetrical electron distribution in which the bonding electrons are attracted somewhat more strongly by one atom than by the other. The symbol δ (Greek delta) means *partial* charge, either partial positive (δ^+) or partial negative (δ^-).

As examples of various points along the bonding spectrum, let's compare three substances, NaCl, HCl, and Cl_2:

- **NaCl**   The bond in sodium chloride is a largely ionic one between Na^+ and Cl^-. (Even here, though, experiments show that the NaCl bond is only about 80% ionic and that complete electron transfer from Na to Cl does not occur.)

$$Na^+ \; Cl^- \qquad \text{An ionic bond}$$

- **HCl**   The bond in a hydrogen chloride molecule is polar covalent. The chlorine atom attracts the bonding electron pair more strongly than hydrogen does, resulting in an unsymmetrical distribution of electrons. Chlorine thus has a partial negative charge, and hydrogen has a partial positive charge.

$$\overset{\delta^+}{H}-\overset{\delta^-}{Cl} \qquad H:Cl \qquad \text{A polar covalent bond}$$

The bonding electrons are attracted more strongly by Cl than by H.

- **Cl_2**   The bond in a chlorine molecule is nonpolar covalent, with the bonding electrons symmetrically centered between the two identical chlorine atoms and attracted equally to both.

$$Cl-Cl \qquad \text{A nonpolar covalent bond}$$

Bond polarity is due to differences in **electronegativity (EN)**, the ability of an atom in a molecule to attract the shared electrons in a bond. As shown graphically in Figure 7.4, metallic elements on the left of the periodic table attract electrons only weakly, whereas the halogens and other elements in the upper right of the table attract electrons strongly. The alkali metals are the least electronegative elements; fluorine, oxygen, nitrogen, and chlorine are the most electronegative. Figure 7.4 also indicates that electronegativity generally decreases down the periodic table within a group.

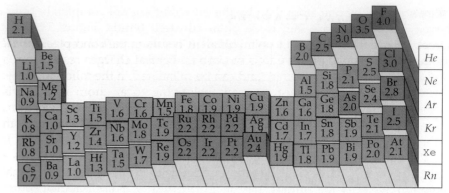

FIGURE 7.4 Electronegativity trends in the periodic table. Electronegativity increases from left to right and generally decreases from top to bottom.

Since electronegativity measures the ability of an atom in a molecule to attract shared electrons, it seems logical that it should be related to electron affinity (E_{ea}, Section 6.3) and ionization energy (E_i, Section 6.1). Electron affinity, after all, is a measure of the tendency of an isolated atom to gain an electron, and ionization energy is a measure of the tendency of an isolated atom to lose an electron. In fact, one of the ways in which electronegativities were first calculated was by taking an average of the absolute values of E_{ea} and E_i and then expressing the results on a unitless scale, with fluorine assigned a value of 4.0.

How can we use a knowledge of electronegativities to make predictions about bond polarity? A general but somewhat arbitrary guideline is that bonds between atoms with the same or similar electronegativities are nonpolar covalent, bonds between atoms whose electronegativities differ by more than 2 units are substantially ionic, and bonds between atoms whose electronegativities differ by less than 2 units are polar covalent. Thus, we can be reasonably sure that a C–Cl bond in chloroform, $CHCl_3$, is polar covalent, while an Na^+Cl^- bond in sodium chloride is largely ionic.

Chlorine: EN = 3.0
Carbon: EN = 2.5

$\triangle$EN = 0.5

Chlorine: EN = 3.0
Sodium: EN = 0.9

$\triangle$EN = 2.1

PROBLEM 7.11 Use the electronegativity values in Figure 7.4 to predict which of the following compounds have polar covalent bonds and which have ionic bonds.

(a) $SiCl_4$ **(b)** CsBr   **(c)** $FeBr_3$ **(d)** CH_4

PROBLEM 7.12 Order the following compounds according to the increasing ionic character of their bonds: CCl_4, $MgCl_2$, $TiCl_3$, Cl_2O

7.8 ►FORMAL CHARGES

Closely related to the idea of polar covalent bonds is the concept of *formal charges* on specific atoms in Lewis structures. Formal charges result from a kind of electron "bookkeeping" and can be calculated in the following way: Find the number of valence electrons belonging to an atom in a Lewis structure, and compare that value with the number of valence electrons belonging to the isolated atom. If the numbers aren't the same, then the atom in the molecule has either gained or lost electrons and has a formal charge. If the atom in a molecule has more electrons than the isolated atom, it has a negative formal charge; if it has fewer electrons, it has a positive formal charge.

$$\textbf{Formal charge} = \begin{pmatrix} \text{Number of} \\ \text{valence electrons} \\ \text{in free atom} \end{pmatrix} - \begin{pmatrix} \text{Number of} \\ \text{valence electrons} \\ \text{in bound atom} \end{pmatrix}$$

In counting the number of valence electrons in a bound atom, it's necessary to make a distinction between bonding electrons and nonbonding, lone-pair electrons. For bookkeeping purposes, an atom can be thought of as "owning" all its nonbonding electrons but only half of its bonding electrons, since the bonding electrons are shared with another atom. Thus, we can rewrite the definition of formal charge as

$$\textbf{Formal charge} = \begin{pmatrix} \text{Number of} \\ \text{valence electrons} \\ \text{in free atom} \end{pmatrix} - \frac{1}{2}\begin{pmatrix} \text{Number of} \\ \text{bonding} \\ \text{electrons} \end{pmatrix} - \begin{pmatrix} \text{Number of} \\ \text{nonbonding} \\ \text{electrons} \end{pmatrix}$$

Look at the atoms in the ammonium ion, $NH_4{}^+$, for example. Each of the four identical hydrogen atoms has two valence electrons (in its covalent bond to nitrogen), and the nitrogen atom has eight valence electrons (two from each of its four N–H bonds):

$$\begin{bmatrix} \text{H} \\ \text{H:N:H} \\ \text{H} \end{bmatrix}^+$$

Ammonium ion:
 eight valence electrons around nitrogen
 two valence electrons around each hydrogen

For bookkeeping purposes, each hydrogen "owns" half of its two bonding electrons, or one, while the nitrogen atom owns half of its eight bonding electrons, or four. Since an isolated hydrogen atom has one electron and since the hydrogens in the ammonium ion each still own one electron, they have neither gained nor lost electrons and thus have no formal charge. An isolated nitrogen atom, however, has five valence electrons. Since the nitrogen atom in $NH_4{}^+$ owns only four valence electrons, it has a formal charge of +1. The sum of the formal charges on all the atoms (+1 in this example) is, of course, equal to the overall charge on the ion.

For hydrogen:
Isolated hydrogen valence electrons	1
Bound hydrogen bonding electrons	2
Bound hydrogen nonbonding electrons	0

$$\left[\begin{array}{c} H \\ H \!:\! N \!:\! H \\ H \end{array} \right]^{+}$$

Formal charge $= 1 - \dfrac{1}{2}(2) - 0 = 0$

For nitrogen:
Isolated nitrogen valence electrons	5
Bound nitrogen bonding electrons	8
Bound nitrogen nonbonding electrons	0

Formal charge $= 5 - \dfrac{1}{2}(8) - 0 = +1$

Example 7.9 at the end of this section shows more formal charge calculations.

The real value of formal charge calculations is that they make it possible to evaluate the relative importance of different resonance structures for a molecule and thus to gain a deeper understanding of a molecule's properties. Take nitrous oxide, for instance, a colorless gas used occasionally as an anesthetic. We can write two Lewis structures for N_2O, one of which has a negative formal charge on the oxygen atom and the other of which has a negative formal charge on the terminal nitrogen atom. (Check for yourself that the formal charges are correct.)

$$:N\!\equiv\!\overset{+}{N}\!-\!\overset{..}{\underset{..}{O}}\!:^{-} \longleftrightarrow \ ^{-}\overset{..}{\underset{..}{N}}\!=\!\overset{+}{N}\!=\!\overset{..}{\underset{..}{O}}$$

Which of the two Lewis structures is "better"? That is, which of the two is the more important contributor to the N_2O resonance hybrid and thus approximates the actual electronic structure of N_2O more closely? Since we saw in the previous section that oxygen is a more electronegative element than nitrogen, the Lewis structure that places a negative formal charge on oxygen is probably lower in energy than the Lewis structure that has a negative formal charge on nitrogen. Thus, the actual electronic structure of N_2O is probably closer to that of the more stable Lewis structure.

$$:N\!\equiv\!\overset{+}{N}\!-\!\overset{..}{\underset{..}{O}}\!:^{-} \longleftrightarrow \ ^{-}\overset{..}{\underset{..}{N}}\!=\!\overset{+}{N}\!=\!\overset{..}{\underset{..}{O}}$$

This Lewis structure . . . is better than . . . this one.

EXAMPLE 7.9

Calculate formal charges on each atom in the following Lewis structure for SO_2.

$$:\overset{..}{\underset{..}{O}}\!-\!\overset{..}{S}\!=\!\overset{..}{O}\!:$$

SOLUTION

For sulfur:
Isolated sulfur valence electrons	6
Bound sulfur bonding electrons	6
Bound sulfur nonbonding electrons	2

Formal charge $= 6 - \dfrac{1}{2}(6) - 2 = +1$

For singly bound oxygen:	Isolated oxygen valence electrons	6
	Bound oxygen bonding electrons	2
	Bound oxygen nonbonding electrons	6

$$\text{Formal charge} = 6 - \frac{1}{2}(2) - 6 = -1$$

For doubly bound oxygen:	Isolated oxygen valence electrons	6
	Bound oxygen bonding electrons	4
	Bound oxygen nonbonding electrons	4

$$\text{Formal charge} = 6 - \frac{1}{2}(4) - 4 = 0$$

These calculations say that the sulfur atom of SO_2 has a formal charge of $+1$ and the singly bonded oxygen atom has a formal charge of -1. We might therefore write the Lewis structure for SO_2 as

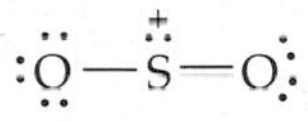

$$:\overset{..}{\underset{..}{O}} - \overset{+}{\underset{..}{S}} - \overset{..}{O}:$$

PROBLEM 7.13 Calculate formal charges in the three resonance structures for the nitrate ion in Example 7.8.

PROBLEM 7.14 Calculate formal charges for the following Lewis structures:

(a) cyanate ion: $\left[:\overset{..}{N} :: C :: \overset{..}{\underset{..}{O}} : \right]^{-}$ **(b)** ozone: $:\overset{..}{\underset{..}{O}} : \overset{..}{\underset{..}{O}} :: \overset{..}{\underset{..}{O}} :$

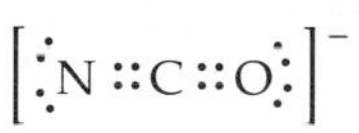

7.9 ➤MOLECULAR SHAPES: THE VSEPR MODEL

Look at the following computer-generated models of water, ammonia, and methane. Each of these molecules—and every other covalent molecule as well—has a specific three-dimensional shape. Often, particularly for biologically important molecules, three-dimensional shape plays a crucial role in determining the molecule's chemistry.

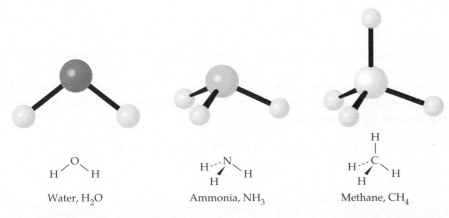

Water, H_2O Ammonia, NH_3 Methane, CH_4

Like so many other properties, molecular shapes are determined by the numbers of valence electrons around the atoms. The approximate shape of a molecule can often be predicted by using what is called the **valence-shell electron-pair repulsion (VSEPR) model**. Electrons in bonds and in lone pairs can be thought of as occupying charge clouds that repel each other and stay as far apart as possible, thus causing molecules to assume specific shapes. There are only two steps to remember in applying the VSEPR method:

Step 1. Count the number of electron "charge clouds" surrounding the atom of interest. A charge cloud is simply a group of electrons, either in a bond or in a lone pair. Thus, the number of charge clouds is the total number of bonds and lone pairs. Multiple bonds count the same as single bonds because we're interested only in the *number* of charge clouds, not in how many electrons each contains.

Step 2. Predict the geometry around each atom by assuming that the atom's charge clouds are oriented in space so that they are as far away from each other as possible. How they achieve this orientation depends on their number. Let's look at the possibilities.

TWO CHARGE CLOUDS   If there are only two charge clouds, as occurs on the beryllium atom of $BeCl_2$ (electrons in two single bonds) and on the carbon atom of CO_2 (electrons in two double bonds), the clouds are farthest apart when they point in opposite directions. Thus, both $BeCl_2$ and CO_2 are linear molecules with **bond angles** of 180°.

These molecules are linear, with bond angles of 180°.

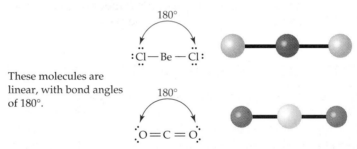

THREE CHARGE CLOUDS   When there are three charge clouds, as occurs on the boron atom of BF_3 (three single bonds) and the sulfur atom of SO_2 (one single bond, one double bond, and one lone pair), the clouds are farthest apart when they lie in the same plane and point to the corners of an equilateral triangle. Thus, a BF_3 molecule is trigonal planar, with F–B–F bond angles of 120°. Similarly, an SO_2 molecule has a trigonal planar arrangement of its three charge clouds on sulfur, but one point of the triangle is occupied by a lone pair and two points by oxygen atoms. The molecule therefore has a bent rather than a linear geometry, with an O–S–O bond angle of approximately 120°.

A BF_3 molecule is trigonal planar, with bond angles of 120°.

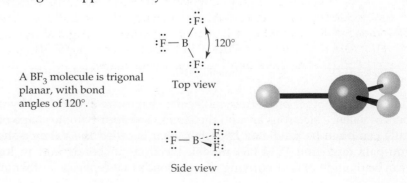

Top view

Side view

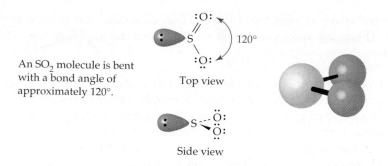

An SO_2 molecule is bent with a bond angle of approximately 120°.

FOUR CHARGE CLOUDS   When there are four charge clouds, as occurs on the central atoms in CH_4 (four single bonds), NH_3 (three single bonds and one lone pair), and H_2O (two single bonds and two lone pairs), the clouds are farthest apart if they extend toward the corners of a regular tetrahedron. As illustrated in Figure 7.5, a *regular tetrahedron* is a geometric solid whose four identical faces are equilateral triangles. The central atom lies at the center of the tetrahedron, the charge clouds point toward the four corners, and the angle between two lines drawn from the center to any two corners is 109.5°. Note how the three-dimensional shape of a tetrahedral molecule is shown: Normal lines are assumed to be in the plane of the paper; the dashed line recedes behind the plane of the paper away from the viewer; and the solid wedged line protrudes out of the paper toward the viewer.

A regular tetrahedron

A tetrahedral molecule

(a)   (b)   (c)

FIGURE 7.5 The tetrahedral geometry of an atom surrounded by four charge clouds. The atom is located in the center of the tetrahedron, and the four charge clouds point toward the four corners. In **(c)**, the solid lines are in the plane of the paper, the dashed line recedes behind the plane of the paper away from the viewer, and the wedged line protrudes out of the paper toward the viewer.

Because valence electron octets are so common, particularly for second-row elements, the atoms in a great many molecules have geometries based on the tetrahedron. In CH_4, for example, the carbon atom has tetrahedral geometry with H–C–H bond angles of 109.5°. In NH_3, the nitrogen atom has a tetrahedral arrangement of its four charge clouds, but one corner of the tetrahedron is occupied by a lone pair, resulting in a trigonal pyramidal shape for the molecule. Similarly, H_2O has two corners of the tetrahedron occupied by lone pairs and thus has a bent shape.

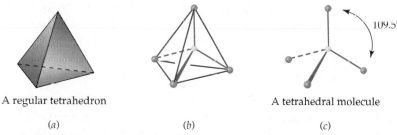

A methane molecule is tetrahedral, with bond angles of 109.5°.

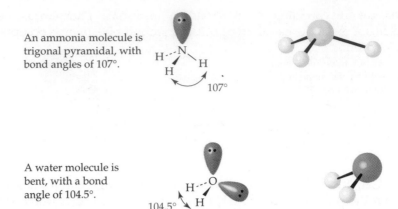

An ammonia molecule is trigonal pyramidal, with bond angles of 107°.

A water molecule is bent, with a bond angle of 104.5°.

Note that the H–N–H bond angle in ammonia (107°) and the H–O–H bond angle in water (104.5°) are close to, but not exactly equal to, the ideal 109.5° tetrahedral value. The angles are diminished somewhat from their ideal value because of the presence of lone pairs. Charge clouds holding lone-pair electrons spread out more than charge clouds holding bonding electrons because they aren't confined to the space between two atoms. As a result, the somewhat enlarged lone-pair charge clouds tend to compress the bond angles in the rest of the molecule.

FIVE CHARGE CLOUDS Five charge clouds, such as are found on the central atoms in PCl_5, SF_4, ClF_3, and I_3^-, are oriented toward the corners of a geometric figure called a *trigonal bipyramid*. Three clouds lie in a plane and point toward the corners of an equilateral triangle, the fourth cloud points directly up, and the fifth cloud points down:

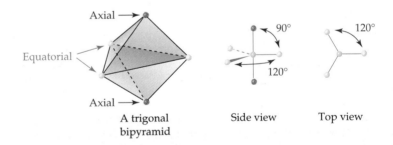

A trigonal bipyramid   Side view   Top view

Trigonal bipyramidal geometry differs from the linear, trigonal planar, and tetrahedral geometries discussed previously because it has two kinds of positions—*equatorial* positions (around the "equator" of the bipyramid) and *axial* positions (along the "axis" of the bipyramid). The three equatorial positions in the horizontal plane are at angles of 120° to each other and at an angle of 90° to the axial positions. The two axial positions are at angles of 180° to each other and at an angle of 90° to the equatorial positions.

Different substances with a trigonal bipyramidal arrangement of charge clouds adopt different shapes depending on whether the five charge clouds contain bonding or nonbonding electrons. Phosphorus penta-chloride, for example, has all five positions around phosphorus occupied by chlorine atoms:

A PCl_5 molecule is trigonal bipyramidal.

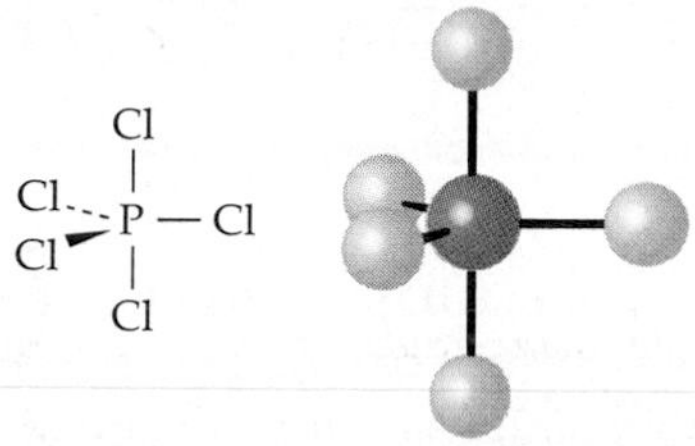

The sulfur atom in SF_4 is bonded to four other atoms and has one nonbonding electron lone pair. Because an electron lone pair spreads out and occupies more space than a bonding pair, the nonbonding electrons in SF_4 occupy an equatorial position where they are close to (90° away from) only two charge clouds. Were they instead to occupy an axial position, they would be close to three charge clouds. As a result, SF_4 has a shape often described as that of a seesaw. (The two axial bonds form the board, and the two equatorial bonds form the legs of the seesaw. You have to tilt your head 90° to see it.)

An SF_4 molecule is see-saw shaped.

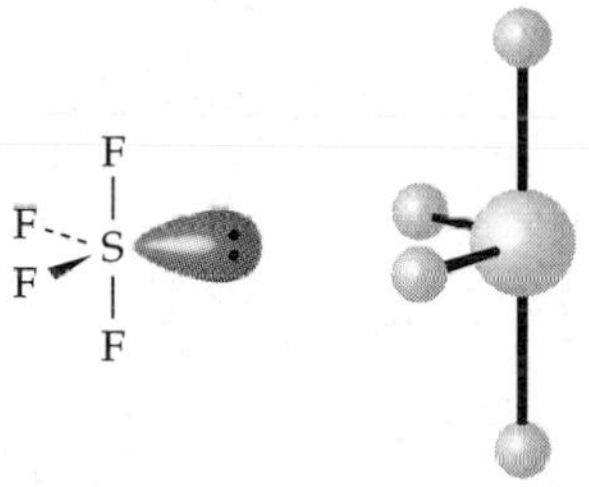

The chlorine atom in ClF_3 is bonded to three other atoms and has two nonbonding electron lone pairs. The lone pairs both occupy equatorial positions, resulting in a T shape for the ClF_3 molecule. (As with the seesaw, you have to tilt your head 90° to see the T.)

A ClF_3 molecule is T shaped.

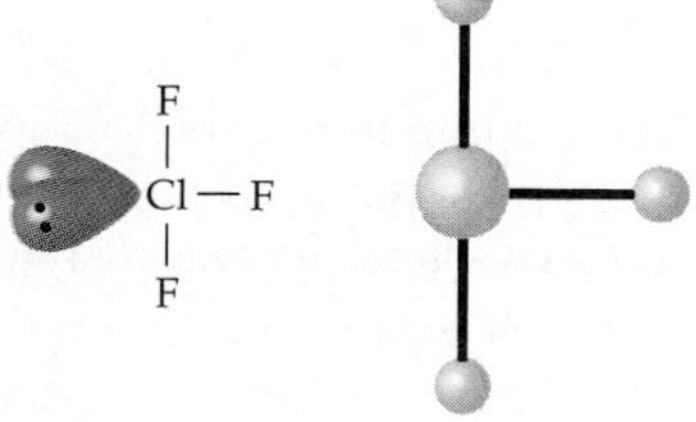

The central iodine atom in the I_3^- ion is bonded to two other atoms and has three lone pairs. All three lone pairs occupy equatorial positions, resulting in a linear structure for I_3^-.

An I$_3^-$ ion is linear.

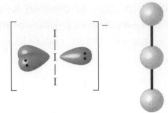

SIX CHARGE CLOUDS

SIX CHARGE CLOUDS   Six charge clouds around an atom are oriented toward the six corners of a regular *octahedron*, a geometric solid whose eight faces are equilateral triangles. All six positions are equivalent, and the angle between any two adjacent positions is 90°.

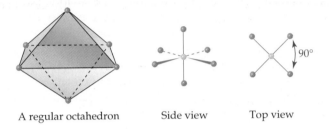

A regular octahedron   Side view   Top view

As was true in the case of five charge clouds, there are different shapes possible for molecules having atoms with six charge clouds, depending on whether the clouds contain bonding or nonbonding electrons. Sulfur hexafluoride, for example, has all six positions around sulfur occupied by fluorine atoms:

An SF$_6$ molecule is octahedral.

The antimony atom in the SbCl$_5^{2-}$ ion also has six charge clouds, but it is bonded to only five atoms and has one nonbonding electron lone pair. As a result, the ion has a *square pyramidal* shape—a pyramid with a square base:

An SbCl$_5^{2-}$ ion has a square pyramidal shape.

The xenon atom in XeF_4 is bonded to four atoms and has two lone pairs. As you might expect, the lone pairs orient as far away from each other as possible to minimize electronic repulsions, giving the molecule a *square planar* shape:

An XeF_4 molecule has a square planar shape.

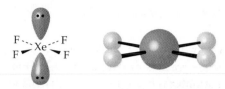

All the geometries discussed above for 2 to 6 charge clouds around an atom are summarized in Table 7.4.

GEOMETRY OF LARGER MOLECULES   The geometries around individual atoms in larger molecules can also be predicted from the rules summarized in Table 7.4. For example, each of the two carbon atoms in ethylene ($H_2C=CH_2$) has three charge clouds, giving rise to trigonal planar geometry for each carbon. The molecule as a whole is planar, with H–C–C and H–C–H bond angles of approximately 120°.

The ethylene molecule is planar, with bond angles of 120°

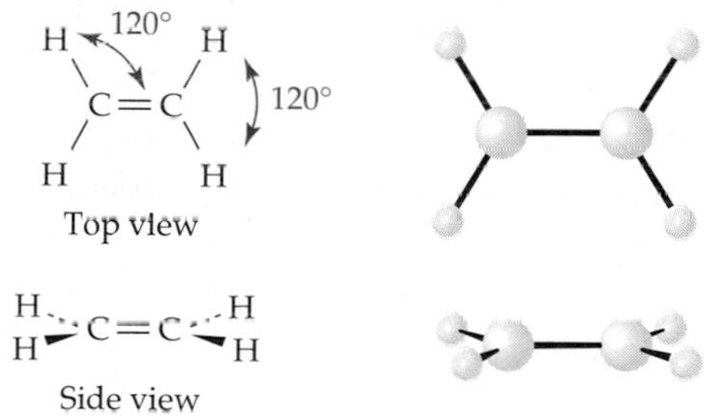

Top view

Side view

Carbon atoms bonded to four other atoms are each at the center of a tetrahedron. As shown below for ethane, $H_3C–CH_3$, the two tetrahedrons are joined together so that the center atom of one is a corner atom of the other.

The ethane molecule has tetrahedral carbon atoms, with bond angles of 109.5°.

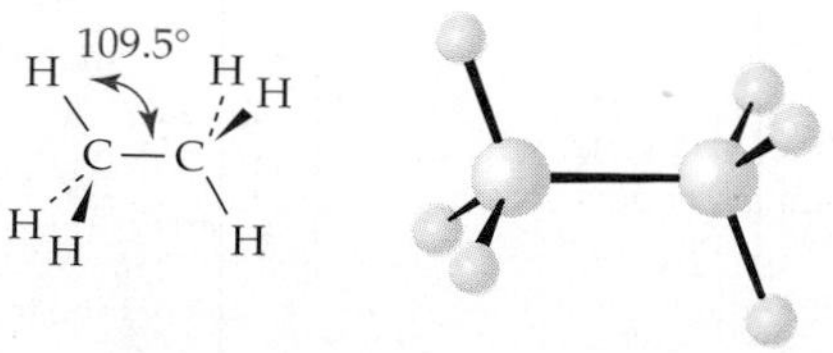

| TABLE 7.4 | Molecular Geometry Around Atoms With 2, 3, 4, 5, and 6 Charge Clouds | | | | |

Number of Bonds	Number of Lone Pairs	Number of Charge Clouds	Molecular Geometry		Example
2	0	2	 Linear	*lener* *Overall*	Cl—Be—Cl
3	0	3	Triangular planar	*trignal*	F, F >B—F
2	1		Bent		O, O >S—:
4	0	4	 Tetrahedral	*tetra hedral*	H—C(H)(H)H
3	1		Trigonal pyramidal		H—N—H, H
2	2		Bent		H, H—O—:
	0	5	 Trigonal bipyramidal	*Trigonal bipyramidal*	Cl, Cl, Cl >P—Cl, Cl
	1		See-saw		F, F >S—:, F, F
	2		T shaped		:Cl—F, F
	3		Linear		[:I—I—I:]⁻

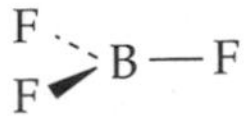

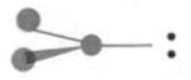

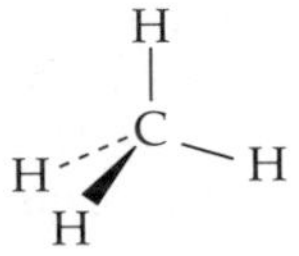

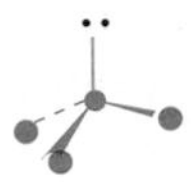

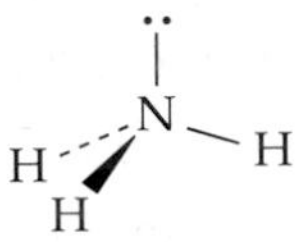

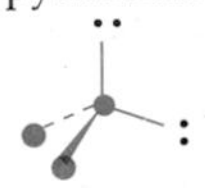

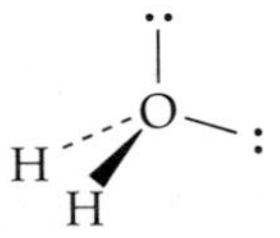

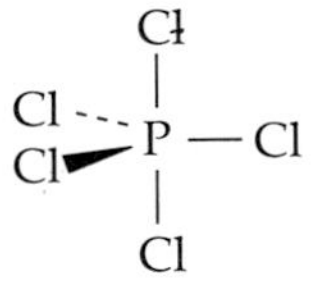

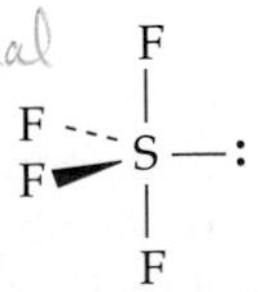

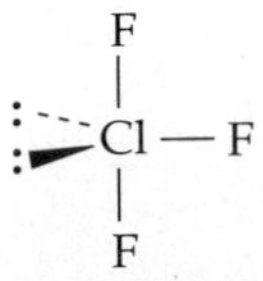

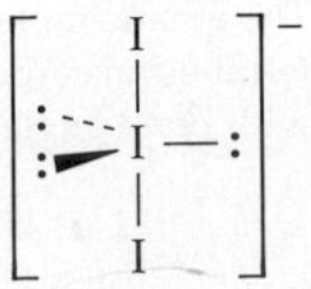

TABLE 7.4   (continued)

Number of Bonds	Number of Lone Pairs	Number of Charge Clouds	Molecular Geometry	Example
6	0		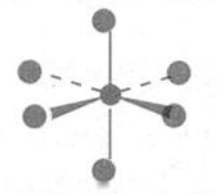 Octahedral	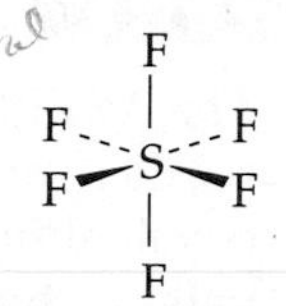 SF_6 structure
5	1	6	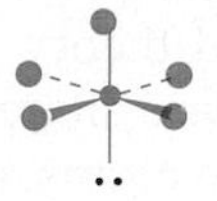 Square pyramidal	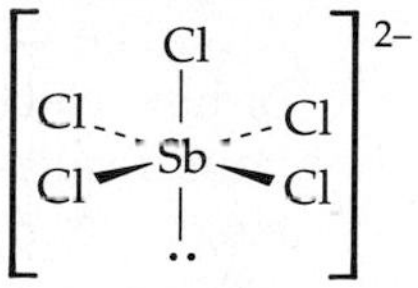 $[SbCl_5]^{2-}$ structure
4	2		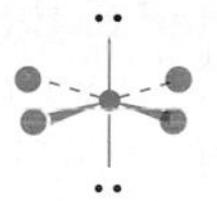 Square planar	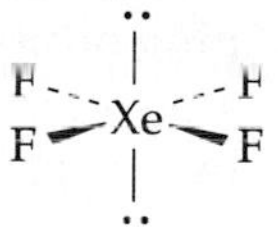 XeF_4 structure

EXAMPLE 7.10

Predict the shape of BrF_5.

SOLUTION   First, draw a Lewis structure for BrF_5, showing that the central bromine atom has six charge clouds (five bonds and one lone pair):

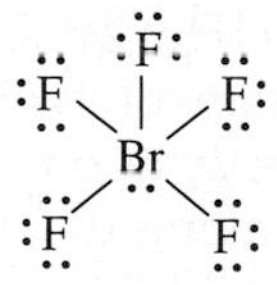

Bromine pentafluoride

Six charge clouds implies an octahedral arrangement; five attached atoms and one lone pair give BrF_5 a square pyramidal shape:

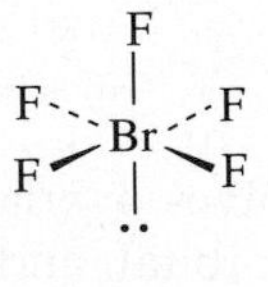

⌐ **PROBLEM 7.15** Predict the shapes of the following molecules or ions.

(a) O_3 (b) H_3O^+ (c) XeF_2 (d) PF_6^- (e) $XeOF_4$
(f) AlH_4^- (g) BF_4^- (h) $SiCl_4$ (i) ICl_4^- (j) $AlCl_3$

⌐ **PROBLEM 7.16** Wood alcohol, so called because it is obtained by heating wood in the absence of air, has the formula CH_3OH. What is its shape?

┌ **PROBLEM 7.17**   Acetic acid, CH_3COOH, is the primary nonaqueous constituent of vinegar. Draw a Lewis structure for acetic acid, and show its overall shape.
(The two carbons are connected, and both oxygens are connected to the same
carbon.)

7.10 ➤ VALENCE BOND THEORY

The VSEPR model discussed in the previous section provides a way to
predict molecular shapes but says nothing about the electronic nature of
covalent bonds. To describe bonding, a quantum mechanical method called
valence bond theory has been developed.

The valence bond theory provides an easily visualized orbital picture
of how electron pairs are shared in a covalent bond. In essence, a covalent
bond results when two atoms approach each other closely enough so that a
singly occupied valence orbital on one atom *overlaps* a singly occupied
valence orbital on the other atom. The now-paired electrons in the overlapping orbitals are attracted to the nuclei of both atoms and thus bond the two
atoms together. In the H_2 molecule, for instance, the H–H bond results from
the overlap of two singly occupied hydrogen 1s orbitals:

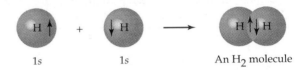

In the valence bond model, the strength of a covalent bond depends on
the amount of orbital overlap: The greater the overlap, the stronger the
bond. This, in turn, means that bonds formed by overlap of other than s
orbitals have a directionality to them. In the F_2 molecule, for instance, each
fluorine atom has the electron configuration $[He]\ 2s^2\ 2p_x^2\ 2p_y^2\ 2p_z^1$, meaning
that the F–F bond results from the overlap of two singly occupied 2p
orbitals. The two p orbitals must point directly at one another for optimum
overlap to occur, and the F–F bond forms along the orbital axis:

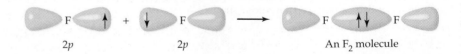

In HCl, the covalent bond involves overlap of a hydrogen 1s orbital (non-
directional) with a chlorine 3p orbital, and forms along the p-orbital axis.

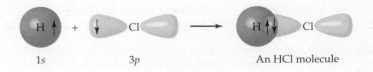

The key ideas of valence bond theory can be summarized in a few
short statements:

1. Covalent bonds are formed by overlap of atomic orbitals, each of which contains one electron of opposite spin.

2. Each of the bonded atoms maintains its own atomic orbitals, but the electron pair in the overlapping orbitals is shared by both atoms

3. The greater the amount of orbital overlap, the stronger the bond. This leads to a directional character to the bond when other than s orbitals are involved.

7.11 ►HYBRIDIZATION AND sp^3 HYBRID ORBITALS

How does valence bond theory describe the electronic structure of poly-atomic molecules, and how does it account for their geometry? Let's look, for example, at a simple tetrahedral molecule such as methane, CH_4. There are several problems we haven't yet dealt with.

Carbon has the ground-state electron configuration [He] $2s^2\ 2p_x^1\ 2p_y^1$, meaning that it has four valence electrons, two of which are paired in a $2s$ orbital and two of which are unpaired in different $2p$ orbitals.[2] The first problem is this: How can carbon form four bonds if two of its valence electrons are already paired and only two unpaired electrons are available for sharing? The answer is that an electron must be promoted from the lower-energy $2s$ orbital to the vacant, higher-energy $2p$ orbital, giving an *excited-state configuration* [He] $2s^1\ 2p_x^1\ 2p_y^1\ 2p_z^1$, which has *four* unpaired electrons.

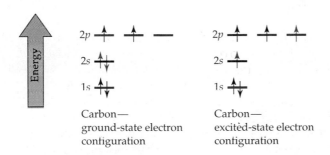

Carbon—
ground-state electron
configuration

Carbon—
excited-state electron
configuration

A second problem is more difficult to resolve: If excited-state carbon uses two kinds of orbitals for bonding, $2s$ and $2p$, how can it form four *equivalent* bonds? Furthermore, if the three $2p$ orbitals in carbon are at angles of 90° to each other, and if the $2s$ orbital has no directionality, how can we account for the observation that carbon forms bonds with tetrahedral angles of 109.5°? The answers to these questions were provided in 1931 by Linus Pauling, who introduced the idea of *hybrid orbitals*.

Pauling showed that the quantum mechanical wave functions for s and p atomic orbitals derived from the Schrödinger equation (Section 5.6) can be mathematically combined to form a new set of equivalent wave functions known as **hybrid atomic orbitals**. When one s orbital combines with three

---

[2] In describing an electron configuration, the assignment of unpaired electrons to specific p orbitals is arbitrary. For carbon, the configurations [He] $2s^2\ 2p_x^1\ 2p_y^1$, [He] $2s^2\ 2p_x^1\ 2p_z^1$, and [He] $2s^2\ 2p_y^1\ 2p_z^1$ are equally satisfactory.

p orbitals, as occurs in an excited-state carbon atom, four equivalent new orbitals, called **sp^3 hybrids**, result.[3]

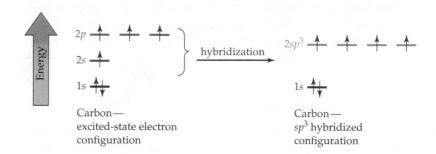

Carbon—
excited-state electron
configuration

Carbon—
sp^3 hybridized
configuration

Each of the four equivalent sp^3 hybrid orbitals has two lobes like an atomic *p* orbital, but one of the lobes is much larger than the other. The four large lobes are oriented toward the four corners of a tetrahedron (Figure 7.6).

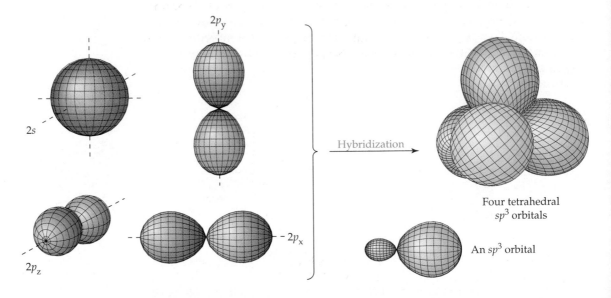

Four tetrahedral
sp^3 orbitals

An sp^3 orbital

FIGURE 7.6 The formation of four sp^3 hybrid orbitals by combination of an atomic *s* orbital with three atomic *p* orbitals. Each sp^3 hybrid orbital has two lobes, one of which is larger than the other. The four large lobes are oriented toward the corners of a tetrahedron.

The shared electrons in a covalent bond made with a strongly directed hybrid orbital spend most of their time in the region between the two bonded nuclei. As a result, covalent bonds made with sp^3 hybrid orbitals are often strong ones. In fact, the energy released on forming the four strong C–H bonds in CH_4 more than compensates for the energy required to produce the excited state of carbon. Figure 7.7 shows how the C–H bonds in methane can form by overlap of carbon sp^3 hybrid orbitals with hydrogen $1s$ orbitals.

[3] Note that the superscript 3 in the name sp^3 tells how many *p* atomic orbitals are combined to construct the hybrid orbitals, not how many electrons occupy each orbital.

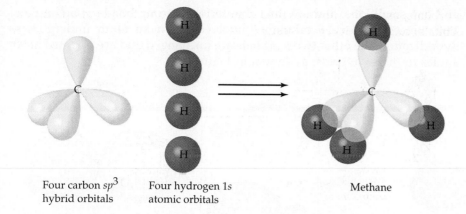

Four carbon sp^3
hybrid orbitals

Four hydrogen $1s$
atomic orbitals

Methane

FIGURE 7.7 The bonding in methane. Each of the four C–H bonds results from overlap of a singly occupied carbon sp^3 hybrid orbital with a singly occupied hydrogen $1s$ orbital.

The same kind of sp^3 hybridization that accounts for the tetrahedral shape of methane also accounts for the trigonal pyramidal shape of ammonia, the bent shape of water, and indeed the shape around all atoms that VSEPR theory predicts to have a tetrahedral arrangement of four charge clouds. A tetrahedral arrangement of charge clouds always implies sp^3 hybridization.

Methane, CH_4

Ammonia, NH_3

Water, H_2O

PROBLEM 7.18 Describe the bonding in ethane, C_2H_6. Tell what kinds of orbitals on each atom overlap to form the C–H and C–C bonds.

7.12 ►OTHER KINDS OF HYBRID ORBITALS

Each of the different geometries shown in Table 7.4—whether based on two, three, four, five, or six charge clouds—can be accounted for by a specific kind of hybridization. Let's look at each in turn.

sp HYBRIDIZATION

Atoms that have two charge clouds undergo hybridization by combination of one atomic s orbital with one p orbital, resulting in two **sp hybrid** orbitals that are linearly oriented. As with sp^3 hybrids, sp hybrids have one large lobe and one small lobe, and are thus able to form strong bonds at an angle of 180°. Since only one *p* orbital is involved when an atom undergoes *sp* hybridization, the other two *p* orbitals are unchanged and are oriented at 90° angles to the *sp* hybrids, as shown in Figure 7.8.

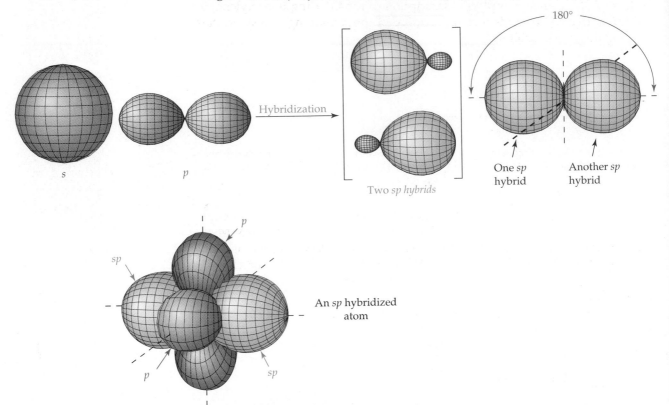

FIGURE 7.8   The formation of two *sp* hybrid orbitals by combination of one *s* and one *p* orbital. The hybrid orbitals are oriented at an angle of 180° to each other.

The presence of the two unhybridized *p* orbitals on an *sp* hybridized atom has some interesting consequences. Look, for example, at acetylene, H–C≡C–H, a colorless gas used in welding. Each carbon atom in the acetylene molecule has linear geometry and is *sp* hybridized. When two *sp* hybridized carbon atoms approach each other with their *sp* orbitals aligned head-on for bonding, the unhybridized *p* orbitals on each carbon also approach each other and form bonds, but in a parallel, sideways manner rather than head-on. Such sideways bonding, in which the shared electrons occupy a region above and below a line connecting the two nuclei, is called a **pi (π) bond**. By contrast, a bond that is formed by head-on overlap and that has its shared electrons centered about the axis between the two nuclei is called a **sigma (σ) bond**.

Sideways overlap
—a π bond

Head-on overlap
—a σ bond

In acetylene, two *p* orbitals are aligned in an up/down manner, and two are aligned in an in/out manner. Thus there are *two* mutually perpendicular π bonds that form in acetylene by *p*-orbital overlap, along with one σ bond formed by overlap of the *sp* orbitals. The net result is a triple bond (Figure 7.9). To then complete a description of the bonding in acetylene, in addition, two C–H bonds form by overlap of the remaining two *sp* orbitals with hydrogen 1*s* orbitals.

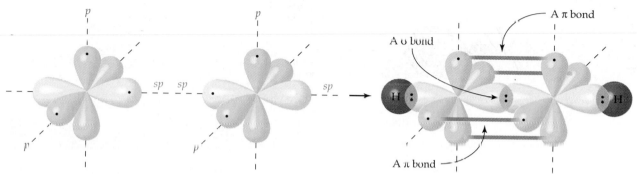

FIGURE 7.9 Formation of a triple bond by two *sp* hybridized atoms. A σ bond forms by head-on overlap of two *sp* orbitals, and two mutually perpendicular π bonds form by sideways overlap of *p* orbitals.

┌ **PROBLEM 7.19** Draw an orbital picture of the C and N atoms, and describe the bonding in the hydrogen cyanide molecule, H–C≡N. ┘

sp² HYBRIDIZATION

Atoms with three charge clouds undergo hybridization by combination of one atomic *s* orbital with two *p* orbitals, resulting in three *sp²* **hybrid** orbitals that lie in a plane and are oriented toward the corners of an equilateral triangle. One *p* orbital remains unchanged and is oriented at a 90° angle to the plane of the *sp²* hybrids, as shown in Figure 7.10.

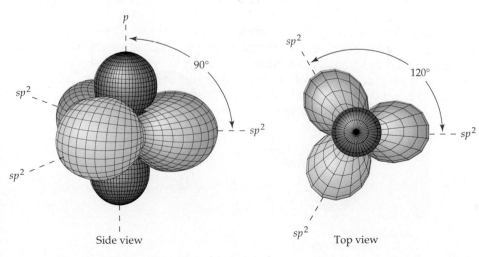

Side view

Top view

FIGURE 7.10 The formation of *sp²* hybrid orbitals by combination of one *s* orbital and two *p* orbitals. The three hybrids lie in a plane at angles of 120° to one another. One unhybridized *p* orbital remains, oriented at a 90° angle to the plane of the *sp²* orbitals.

One of the simplest examples of sp^2 hybridization and of its geometric consequences is found in ethylene, $H_2C=CH_2$, a colorless gas used as starting material for the industrial preparation of polyethylene. Each carbon atom in ethylene has three charge clouds and is sp^2 hybridized. When two sp^2 hybridized carbon atoms approach each other with sp^2 orbitals aligned head-on for σ bonding, the unhybridized p orbitals on each carbon overlap in a π bond, resulting in a net carbon–carbon double bond. The two remaining sp^2 hybrids on each carbon form bonds to hydrogen to complete the $H_2C=CH_2$ structure (Figure 7.11).

FIGURE 7.11   The structure of ethylene. The carbon–carbon double bond consists of one σ bond from the overlap of sp^2 hybrid orbitals and one π bond from the overlap of p orbitals. The overall shape of the molecule is planar (flat).

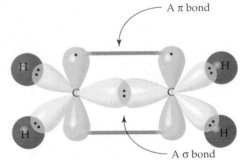

sp^3d HYBRIDIZATION

Atoms with five charge clouds, such as the phosphorus in PCl_5, undergo hybridization by combination of five atomic orbitals. Since a given shell has a total of only four s and p orbitals, the need to use five orbitals implies that valence-shell d orbitals must be involved. Thus, only atoms of the third row or lower form the necessary hybrids. Hybridization of five atomic orbitals occurs by a combination of one s orbital, three p orbitals, and one d orbital, giving five **sp^3d hybrid** orbitals in a trigonal bipyramidal arrangement (Figure 7.12). Three of the hybrid orbitals lie in a plane at angles of 120°, with the remaining two orbitals perpendicular to the plane, one above and one below.

FIGURE 7.12 The five sp^3d hybrid orbitals and their trigonal bipyramidal geometry.

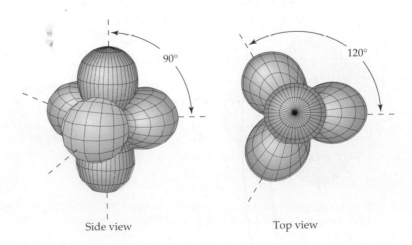

Side view   Top view

sp^3d^2 HYBRIDIZATION

Atoms with six charge clouds, such as the sulfur in SF_6, undergo hybridization by combination of six atomic orbitals. This again implies that valence-shell d orbitals are involved and that only atoms in the third row or lower form the necessary hybrids. Hybridization occurs by a combination of one s orbital, three p orbitals, and two d orbitals, resulting in six **sp^3d^2 hybrid** orbitals with an octahedral arrangement (Figure 7.13). All six orbitals are equivalent, and the angle between any two adjacent orbitals is 90°.

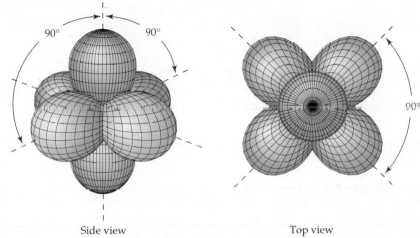

Side view   Top view

FIGURE 7.13 The six sp^3d^2 hybrid orbitals and their octahedral geometry.

A summary of the five most important kinds of hybridization and the geometric arrangement that each corresponds to is given in Table 7.5. Remember: A given geometry implies the necessary hybridization. If you know the observed geometry around an atom, you know the hybridization of the atom.

TABLE 7.5   Hybrid Orbitals and Their Geometry

Number of Charge Clouds	Geometry	Hybridization
2	Linear	sp
3	Trigonal planar	sp^2
4	Tetrahedral	sp^3
5	Trigonal bipyramidal	sp^3d
6	Octahedral	sp^3d^2

EXAMPLE 7.11

Describe the hybridization of the carbon atoms in allene, $H_2C=C=CH_2$, and make a rough sketch of the molecule showing its hybrid orbitals.

SOLUTION   Draw a Lewis structure to find the number of charge clouds on each atom. Because the central carbon atom has two charge clouds (two double bonds), it has a linear geometry and is *sp* hybridized. Since the two terminal carbon atoms have three charge clouds each (one double bond and two C–H bonds), they have trigonal planar geometry and are *sp²* hybridized. The central carbon uses its *sp* orbitals to form two σ bonds at 180° angles and uses its two unhybridized *p* orbitals to form π bonds, one to each of the terminal carbons. Each terminal carbon atom uses an *sp²* orbital for σ bonding, a *p* orbital for π bonding, and its two remaining *sp²* orbitals for C–H bonds.

$H_2C{=}C{=}CH_2$

(The carbon orbitals shown in green are *sp²* hybrids; those shown in blue are unhybridized *p* orbitals.)

⌐ PROBLEM 7.20 Describe the hybridization of the carbon atom in formaldehyde, $H_2C{=}O$, and make a rough sketch of the molecule showing the orbitals involved in bonding.

⌐ PROBLEM 7.21 Describe the hybridization of the central iodine atom in I_3^-, and make a rough sketch of the ion showing the orbitals involved in bonding.

⌐ PROBLEM 7.22 Describe the hybridization of the sulfur atom in SF_2, SF_4, and SF_6 molecules.

7.13 ➤MOLECULAR ORBITAL THEORY OF BONDING: THE HYDROGEN MOLECULE

The valence bond model of covalent bonding is easy to visualize and leads to a satisfactory picture for most molecules. It does, however, have some problems. Perhaps the most serious flaw in the valence bond model is that it sometimes leads to an incorrect electronic description. For this reason, another bonding description called the *molecular orbital* (MO) model is often used. The molecular orbital model is more complex than the valence bond model, particularly for larger molecules, but sometimes gives a more satisfactory accounting of chemical and physical properties.

To introduce some of the basic ideas of the **molecular orbital theory**, let's look again at orbitals. The idea of orbitals derives from quantum mechanics, in which the square of the wave function gives the probability of finding an electron within a given region of space. The kinds of orbitals that we've been concerned with to this point are called *atomic orbitals* because they are characteristic of individual atoms:

ATOMIC ORBITAL   A wave function whose square gives the probability of finding the electron within a given region of space *in an atom*

Atomic orbitals on the same atom can combine to form hybrids, and atomic orbitals on different atoms can overlap to form covalent bonds, but the orbitals and the electrons in them remain localized on specific atoms.

Molecular orbital theory takes a different approach to bonding by considering the molecule as a whole rather than concentrating on individual atoms. Thus, a *molecular* orbital is to a *molecule* what an *atomic* orbital is to an *atom*.

MOLECULAR ORBITAL   A wave function whose square gives the probability of finding the electron within a given region of space *in a molecule*

Like atomic orbitals, molecular orbitals have specific energy levels and specific shapes, and they can be occupied by a maximum of two electrons with opposite spins. The energy and shape of a molecular orbital depend on the size and complexity of the molecule and can thus be fairly complicated, but the fundamental analogy between atomic and molecular orbitals is a good one.

Let's look at the molecular orbital description of the simple diatomic molecule H_2 to see some general features of MO theory. Imagine what might happen when two isolated hydrogen atoms approach each other and begin to interact. The 1s orbitals begin to blend together, and the electrons spread out over both atoms. Molecular orbital theory says that there are two ways for the orbital interaction to occur—an additive way and a subtractive way. The additive interaction leads to formation of a molecular orbital that is roughly egg-shaped, whereas the subtractive interaction leads to formation of a molecular orbital that contains a node between atoms (Figure 7.14).

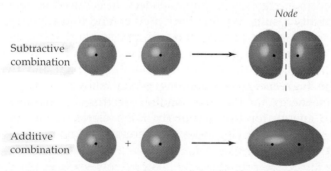

FIGURE 7.14 Formation of molecular orbitals in the H_2 molecule. The additive combination of two atomic 1s orbitals leads to formation of a lower-energy, bonding molecular orbital. The subtractive combination leads to formation of a higher-energy, antibonding molecular orbital that has a node between the nuclei.

The additive combination, denoted σ, is lower in energy than the two isolated 1s orbitals and is called a **bonding molecular orbital** because any electrons it contains spend most of their time in the region between the two nuclei, helping to bond the atoms together. The subtractive combination, denoted σ^* (spoken as "sigma star"), is higher in energy than the two isolated 1s orbitals and is called an **antibonding molecular orbital** because any electrons it contains can't occupy the central region between the nuclei and can't contribute to bonding. Diagrams of the following sort are used to show the energy relationships of the various orbitals. The two isolated H atomic orbitals are shown on either side, and the two H_2 molecular orbitals are shown in the middle. Each of the starting hydrogen atomic orbitals holds one electron, which pair up and occupy the low-energy bonding MO after covalent bond formation.

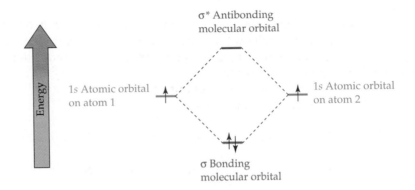

Similar MO diagrams can be drawn, and predictions about stability can be made for related diatomic molecules such as H_2^- and He_2. For example, we might imagine constructing the H_2^- ion by bringing together a neutral H· atom with one electron and an H:⁻ anion with two electrons. Since the resultant H_2^- ion has three electrons, two of them will occupy the low-energy bonding σ MO and one will occupy the high-energy antibonding σ* MO according to the energy diagram in Figure 7.15. Two electrons are lowered in energy while only one electron is raised in energy, so a net gain in stability results. We therefore predict (and find experimentally) that the H_2^- ion is a stable species, although less stable than the H_2 molecule.

What about He_2? A hypothetical He_2 molecule has four electrons, two of which occupy the low-energy bonding orbital and two of which occupy the high-energy antibonding orbital shown in Figure 7.15. The decrease in energy for the two bonding electrons is counteracted by the increase in energy for the two antibonding electrons, leading to no net change in stability. The He_2 molecule therefore has no net bonding energy and is not stable.

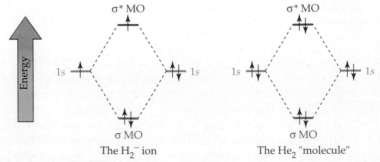

FIGURE 7.15 MO diagrams for the stable H_2^- ion and the unstable He_2 molecule.

Note that bond orders—the number of electron pairs shared between atoms (Section 7.4)—are easy to calculate from MO diagrams by subtracting the number of antibonding electrons from the number of bonding electrons and dividing by two.

$$\text{Bond order} = \frac{\text{number of bonding electrons} - \text{number of antibonding electrons}}{2}$$

The H_2 molecule, for example, has a bond order of 1 since it has two bonding electrons and no antibonding electrons. Similarly, the H_2^- ion has a bond order of $1/2$, and the hypothetical He_2 molecule has a bond order of 0, which explains why He_2 is unstable.

The most important conclusions about the molecular orbital theory of bonding can be summarized as follows:

1. Molecular orbitals are to molecules what atomic orbitals are to atoms. Molecular orbitals describe regions of space in a molecule where electrons are most likely to be found, and they have a specific size, shape, and energy level.
2. Molecular orbitals are formed by combining atomic orbitals. The number of molecular orbitals formed is the same as the number of atomic orbitals combined.
3. Molecular orbitals that are lower in energy than the starting atomic orbitals are bonding; MOs higher in energy than the starting atomic orbitals are antibonding.
4. Electrons occupy molecular orbitals beginning with the MO of lowest energy. Only two electrons occupy each orbital.
5. Bond order can be calculated by subtracting the number of electrons in antibonding MOs from the number in bonding MOs and dividing by two.

⌐ PROBLEM 7.23 Construct an MO diagram for the He_2^+ ion. Is the ion likely to be stable? What is its bond order?

7.14 ➤ MOLECULAR ORBITAL THEORY OF OTHER DIATOMIC MOLECULES

Having now looked at the fundamental ideas of molecular orbital theory and having seen how the theory accounts for bonding in the H_2 molecule, let's move up a level in complexity by looking at the bonding in several second-row diatomic molecules—N_2, O_2, and F_2. The valence bond model developed in Section 7.10 predicts that the nitrogen atoms in N_2 are triply bonded and have one lone pair each, that the oxygen atoms in O_2 are doubly bonded and have two lone pairs each, and that the fluorine atoms in F_2 are singly bonded and have three lone pairs each:

$$:N{\equiv}N:$$

(one σ bond
and two π bonds)

$$\overset{..}{:}O{=}O\overset{..}{:}$$

(one σ bond
and one π bond)

$$\overset{..}{:}\overset{..}{F}{-}\overset{..}{F}\overset{..}{:}$$

(one σ bond)

Unfortunately, this simple valence bond picture can't be right because it predicts that the electrons in all three molecules are *spin-paired*. In other words, Lewis structures indicate that all the occupied atomic orbitals in the three molecules contain two electrons each. It's easy to demonstrate experimentally, however, that the O_2 molecule has two electrons that are *not* spin-paired and that these electrons therefore must be in different, singly occupied orbitals.

Experimental evidence for the electronic structure of O_2 rests on the observation that substances with unpaired electrons are attracted by magnetic fields and are thus said to be **paramagnetic**. The more unpaired electrons the substance has, the stronger the paramagnetic attraction. Substances whose electrons are all spin-paired, by contrast, are weakly repelled by magnetic fields and are said to be **diamagnetic**. Both N_2 and F_2 are diamagnetic, just as predicted by Lewis structures, but O_2 is paramagnetic, as shown in Figure 7.16. When liquid O_2 is poured over the poles of a strong magnet, the O_2 sticks to the poles.

Why is O_2 paramagnetic? Although Lewis structures and valence bond theory fail in their descriptions, MO theory explains the experimental results nicely. In an MO description of O_2, two separate oxygen atoms come together and their atomic orbitals interact to form molecular orbitals. As shown in Figure 7.17, four kinds of interactions occur, leading to the formation of four bonding molecular orbitals and four antibonding ones. The $2s$ orbitals interact giving σ_{2s} and σ^*_{2s} MOs, the two $2p$ orbitals that lie on the internuclear axis interact head-on to give σ_{2p} and σ^*_{2p} MOs, and the two remaining pairs of $2p$ orbitals that are perpendicular to the internuclear axis interact in a sideways manner to give two equivalent π_{2p} and two equivalent π^*_{2p} MOs oriented 90° apart.

FIGURE 7.16 Liquid O_2 sticks to the poles of a magnet because it has unpaired electrons and is paramagnetic.

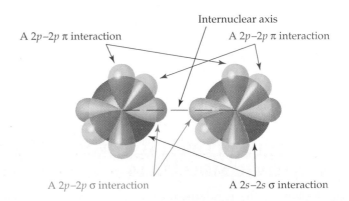

Internuclear axis

A $2p$–$2p$ π interaction   A $2p$–$2p$ π interaction

A $2p$–$2p$ σ interaction   A $2s$–$2s$ σ interaction

FIGURE 7.17 Four kinds of interactions occur among valence-shell atomic orbitals when two oxygen atoms approach each other: one σ_{2s} interaction, one σ_{2p} interaction along the internuclear axis, and two π_{2p} interactions perpendicular to the internuclear axis.

As we saw with H_2 in Section 7.13, the σ_{2s} and σ^*_{2s} MOs in N_2 and O_2 result from the interaction of s atomic orbitals. Similarly, the σ_{2p} and σ^*_{2p} MOs result from head-on interaction of p orbitals, and the π_{2p} and π^*_{2p} MOs result from sideways interaction of p atomic orbitals. In each case, additive interaction of the atomic orbitals leads to the bonding MO, and subtractive interaction leads to the antibonding MO containing a node between nuclei. The shapes of the resultant σ_{2p}, σ^*_{2p}, π_{2p}, and π^*_{2p} MOs are shown in Figure 7.18, and the relative energy levels of the eight molecular orbitals in N_2, O_2, and F_2 are shown in Figure 7.19.

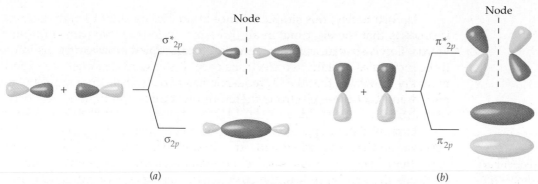

FIGURE 7.18 Formation of **(a)** σ_{2p} and σ^*_{2p} MOs by head-on interaction of two p atomic orbitals, and **(b)** π_{2p} and π^*_{2p} MOs by sideways interaction. In each case, the bonding MO concentrates electron density between atomic nuclei, whereas the antibonding MO has a node between nuclei.

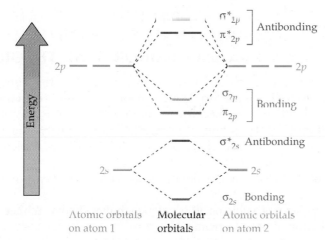

FIGURE 7.19 An energy diagram for molecular orbitals of second-row diatomic molecules. There are eight MOs, four of them bonding and four antibonding.

When appropriate numbers of electrons are added to fill the molecular orbitals, the results shown in Figure 7.20 are obtained. Both N_2 and F_2 have all their electrons spin-paired, but O_2 has two unpaired electrons in the degenerate π^*_{2p} orbitals (*degenerate* orbitals are those that have the same energy).[4] Both N_2 and F_2 are therefore diamagnetic, while O_2 is paramagnetic.

It should be pointed out that MO diagrams like that in Figure 7.19 require a reasonable amount of experience and (often) mathematical calculation to generate. MO theory is therefore less easy to visualize and to understand on an intuitive level than valence bond theory is.

PROBLEM 7.24 Use the MO diagram in Figure 7.19 to predict electronic structures for B_2 and C_2 molecules. Is either of these substances paramagnetic?

---

[4] Actually, the MO diagram for O_2 and F_2 should have the σ_{2p} orbital a bit lower in energy than the π_{2p} orbitals, but this does not affect the overall conclusions.

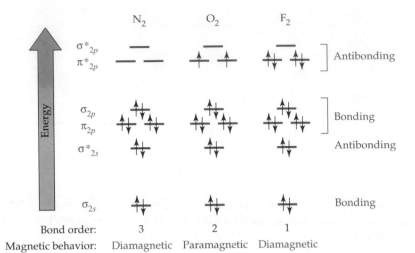

FIGURE 7.20 Molecular orbital diagrams for the second-row diatomic molecules N_2, O_2, and F_2. The O_2 molecule has two unpaired electrons in its two degenerate π^*_{2p} orbitals and is therefore paramagnetic.

7.15 ➤COMBINING VALENCE BOND THEORY AND MOLECULAR ORBITAL THEORY

Whenever two different theories are used to explain the same concept, the question naturally comes up, Which theory is better? Sometimes the question isn't easy to answer because it depends on what is meant by "better." Valence bond theory is better because of its simplicity, but MO theory is better because of its accuracy. Best of all, though, is a blend of the two theories that combines the strengths of both.

Lewis structures and valence bond theory have two main problems:

1. For molecules such as O_2, valence bond theory makes an incorrect prediction about electronic structure.
2. For molecules such as O_3, no single Lewis structure is adequate, and the concept of resonance involving two or more Lewis structures must be added (Section 7.6).

The first problem occurs rarely, but the second is much more common.

To deal with resonance, chemists often use a blend of valence bond and MO theories. Take ozone, for instance. We said in Section 7.6 that O_3 can be described as a resonance hybrid of two equivalent Lewis structures. One structure has a lone pair of electrons in the p orbital on the left-hand oxygen atom and has a π bond to the right-hand oxygen. The other structure has a lone pair of electrons in the p orbital on the right-hand oxygen and a π bond to the left-hand oxygen. The actual structure of O_3 is an average of the two resonance forms in which four π electrons occupy the entire region encompassed by the overlapping set of three p orbitals. The only difference between the resonance structures is in the placement of electrons. The atoms themselves are in the same positions in both, and the geometries of both are the same (Figure 7.21).

Valence bond theory is perfectly adequate for describing the σ bonds in O_3 but has trouble in describing the π bonding of the p atomic orbitals whose four electrons are spread out, or *delocalized*, over the molecule. Yet this is exactly what MO theory does best—describe bonds in which elec-

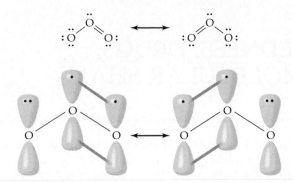

FIGURE 7.21 Ozone is a hybrid of two resonance forms that differ only in the placement of electrons.

trons are delocalized throughout a molecule. Thus, a blend of valence theory and MO theory describes O_3 nicely. The σ electrons are localized between pairs of nuclei and are best described in valence-bond terminology as resulting from σ overlap of hybrid orbitals on the central oxygen atom with a p orbital on each of the terminal oxygens. The π electrons are delocalized over the molecule and are best described by MO theory.

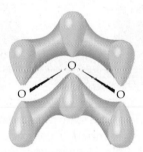

Lowest-energy π molecular orbital

The carbonate ion, CO_3^{2-}, and all other structures for which a resonance description is needed can be described in a similar fashion: The σ-bond framework can be thought of as localized and best described by valence bond theory, while the π-bond framework is delocalized and best described by π molecular orbitals that spread out over the molecule.

CO_3^{2-} ion

p orbitals

Lowest-energy π molecular orbital

⌐ **PROBLEM 7.25** Draw two resonance structures for the formate ion, HCO_2^-, and sketch its lowest-energy π molecular orbital showing how the π electrons are delocalized over both oxygen atoms. ⌐

interlude—HANDEDNESS, DRUGS, AND MOLECULAR SHAPE

Why does your right glove fit only on your right hand and not on your left hand? Why do the threads on a light bulb twist only in one direction so that you have to turn the bulb clockwise to screw it in? The reason is that the glove and the light bulb have a *handedness* to them. When the right-handed glove is held up to a mirror, the reflected image looks like a left-handed glove. (Try it.) When the light bulb with clockwise threads is reflected in a mirror, the threads in the mirror image twist in a counterclockwise direction.

A right hand fits only into a right-handed glove with a complementary shape, not into a left-handed glove.

Molecules too can have a handedness and can thus exist in mirror-image forms, one right-handed and one left-handed. Take, for example, the main classes of biomolecules found in living organisms: carbohydrates (sugars), proteins, fats, and nucleic acids. These and most other biomolecules are handed, and usually only one of the two possible mirror-image forms occurs naturally in a given organism. The other form can often be made in the laboratory but does not occur naturally.

The biological consequences of molecular handedness can be dramatic. Look at the structures of dextromethorphan and levomethorphan, for instance. The right-handed form, dextromethorphan, is a common cough

The gray spheres in these computer-generated molecular structures represent carbon atoms, the light green spheres represent hydrogens, the red spheres represent oxygens, and the blue spheres represent nitrogens.

Mirror

Levomethorphan
(a narcotic analgesic)

Dextromethorphan
(a cough suppressant)

suppressant found as an ingredient in many over-the-counter cold medicines; the left-handed form, levomethorphan, is a powerful narcotic pain-reliever (analgesic) similar in its effects to morphine. The two substances are chemically identical except for their handedness, yet their biological properties are completely different.

As another example of the effects of molecular handedness, look at the substance called *carvone*. The left-handed form of carvone occurs in mint plants and has the characteristic odor of spearmint, while the right-handed form occurs in several herbs and has the odor of caraway seeds. Again, the two structures are the same except for their handedness, yet they have entirely different odors.

These two plants both produce carvone, but the mint plant yields the left-handed form, while the caraway plant yields the right-handed form.

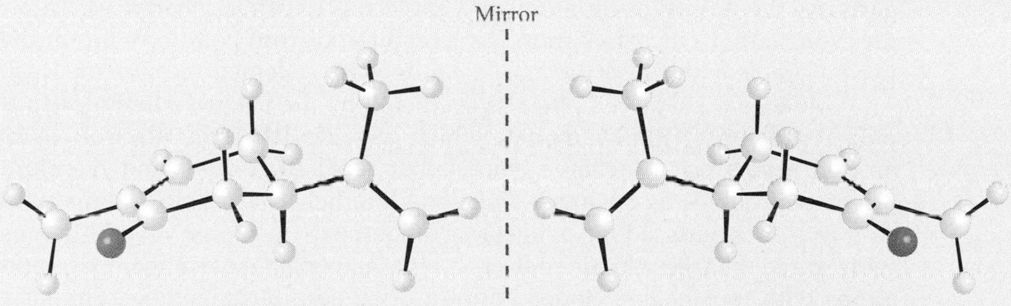

Mirror

"Left-handed" carvone
(odor of spearmint)

"Right-handed" carvone
(odor of caraway)

Why do different mirror-image forms of molecules have different biological properties? The answer goes back to the question about why a right glove fits only on the right hand: A right hand in a right glove is a perfect match because the two shapes are complementary. Putting the same right hand into a left glove produces a mismatch because the two shapes are *not* complementary. In the same way, handed molecules such as dextromethorphan and carvone have specific shapes that match only complementary-shaped receptor sites in the body. The mirror-image forms of the molecules can't fit into the receptor sites and thus don't elicit the same biological response.

Precise molecular shape is of crucial importance to every living organism. Almost every chemical interaction in living systems is governed by complementarity between handed molecules and their receptors.

273

Covalent bonds result from the sharing of electrons between atoms. Every covalent bond has a specific **bond length** that leads to optimum stability and a specific **bond dissociation energy** that describes the strength of the bond. Energy is released when bonds are formed and is absorbed when bonds are broken. As a general rule, an atom will share as many of its valence-shell electrons as possible, either until it has no more to share or until it reaches an octet. Atoms in the third and lower rows of the periodic table can expand their valence shells beyond the normal octet by using valence-shell d orbitals in covalent bond formation.

Lewis structures represent an atom's valence electrons by dots and show the two electrons in a **single bond** as a pair of dots shared between atoms or as a single line. In the same way, a **double bond** is represented as four dots or two lines between atoms, and a **triple bond** is represented as six dots or three lines between atoms. Occasionally, a molecule can be represented by more than one satisfactory Lewis structure. In such cases, no single Lewis structure is adequate by itself. The actual electronic structure of the molecule is said to be a **resonance hybrid** of the different individual Lewis structures. A double-headed arrow is used to indicate resonance:

$$\ddot{\text{O}}{-}\ddot{\text{O}}{=}\ddot{\text{O}} \quad \longleftrightarrow \quad \ddot{\text{O}}{=}\ddot{\text{O}}{-}\ddot{\text{O}}$$

In a bond between dissimilar atoms, such as that in HCl, one atom often attracts the bonding electrons more strongly than the other, giving rise to a **polar covalent bond**. Bond polarity is due to differences in **electronegativity**, the ability of an atom in a molecule to attract shared electrons. Electronegativity increases from left to right and from bottom to top in the periodic table. Fluorine is the most electronegative element; cesium the least.

Molecular shape can often be predicted by the **valence-shell electron-pair repulsion (VSEPR) model**, which assumes that the electron pairs around atoms occupy charge clouds that repel each other and therefore orient themselves as far away from one another as possible. Atoms with two charge clouds adopt a linear arrangement of charge clouds, atoms with three charge clouds adopt a trigonal planar arrangement, and atoms with four charge clouds adopt a tetrahedral arrangement. Similarly atoms with five charge clouds are trigonal bipyramidal, and atoms with six charge clouds are octahedral.

According to **valence bond theory**, covalent bond formation occurs by the overlap of singly occupied atomic orbitals, either in a head-on manner to form a **σ bond** or in a sideways manner to form a **π bond**. The observed geometry of covalent bonding can be accounted for by assuming that s, p, and d atomic orbitals combine to generate **hybrid orbitals** that are strongly oriented in specific directions: *sp* **hybrid orbitals** have linear geometry, *sp²* **hybrid orbitals** have trigonal planar geometry, *sp³* **hybrid orbitals** have tetrahedral geometry, *sp³d* **hybrid orbitals** have trigonal bipyramidal geometry, and *sp³d²* **hybrid orbitals** have octahedral geometry.

Molecular orbital theory sometimes gives a more accurate picture of electronic structure than the valence bond model. Formed from the mathematical combination of atomic orbitals, **molecular orbitals** are wave functions whose square gives the probability of finding an electron in a given region of space in a molecule. Combination of two atomic orbitals gives two molecular orbitals, a **bonding MO** that is lower in energy than the starting atomic orbitals and an **antibonding MO** that is higher in energy than the starting atomic orbitals. Molecular orbital theory is particularly useful for describing delocalized π bonding in molecules.

UNDERSTANDING KEY CONCEPTS

1. What is the geometry around the central atom in each of the following molecular models?

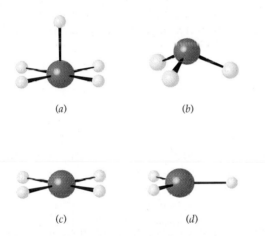

(a)   (b)

(c)   (d)

2. What is the geometry around the central atom in each of the following molecular models? The views are a bit unusual, but they are good practice for visualizing molecules.

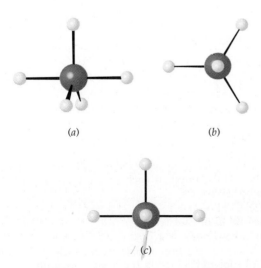

(a)   (b)

(c)

3. Sketch your own rough ball-and-stick views similar to those in Problem 1 for each of the following geometries.
   (a) T shaped   **(b)** seesaw shaped   **(c)** octahedral

4. Identify each of the following sets of hybrid orbitals.

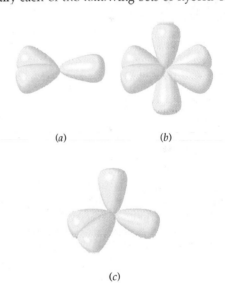

(a)   (b)

(c)

5. Make a sketch similar to that in Problem 4 to represent the following sets of hybrid orbitals.
   (a) sp **(b)** sp^3d

6. The odor of cinnamon oil is due to cinnamaldehyde, C_9H_8O. What is the hybridization of each carbon atom in cinnamaldehyde? How many σ bonds and how many π bonds does cinnamaldehyde have?

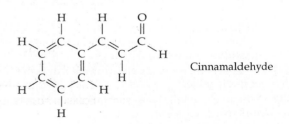

Cinnamaldehyde

275

ADDITIONAL PROBLEMS

Problems 7.1–7.25 appear within the chapter.

LEWIS STRUCTURES

7.26 Which of the following substances contains an atom that does not obey the octet rule?
  (a) $AlCl_3$ (b) PCl_3 (c) PCl_5 (d) $SiCl_4$

7.27 Draw Lewis structures for the following molecules or ions.
  (a) CBr_4 (b) NCl_3 (c) C_2H_5Cl
  (d) BF_4^- (e) O_2^{2-} (f) NO^+

7.28 Draw Lewis structures for the following molecules, which contain atoms from the third row or lower.
  (a) $TiCl_3$ (b) KrF_2 (c) ClO_2 (d) PF_5 (e) H_3PO_4

7.29 Draw as many resonance structures as you can for each of the following chlorine-containing ions.
  (a) ClO_2^- (b) ClO_3^- (c) ClO_4^-

7.30 Draw as many resonance structures as you can for the following nitrogen-containing compounds.
  (a) N_2O (b) NO   (c) NO_2 (d) N_2O_3 ($ONNO_2$)

7.31 Which of the following molecules do not need to be described as resonance hybrids?
  (a) phosgene, $COCl_2$
  (b) thionyl chloride, $SOCl_2$
  (c) sulfuryl chloride, SO_2Cl_2
  (d) diazomethane, H_2CN_2

7.32 Oxalic acid, $H_2C_2O_4$, is a poisonous substance found in uncooked spinach leaves. If oxalic acid has a C–C single bond, draw its Lewis structure.

7.33 Draw a Lewis structure for carbon dioxide. How many double bonds does CO_2 have?

7.34 Which of the following pairs of structures represent resonance forms, and which do not?

(a) $H-C\equiv N-\ddot{\underset{..}{O}}:$ and $H-C=\overset{..}{N}-\ddot{\underset{..}{O}}:$

(b)

(c)

(d)

7.35 Draw two resonance structures for methyl isocyanate, CH_3NCO, a toxic gas accidentally released into the atmosphere in December 1984 in Bhopal, India.

7.36 Identify the third-row elements that form the following ions.

7.37 Identify the fourth-row elements that form the following compounds.

7.38 Write Lewis structures for molecules with the following connections.

7.39 Write Lewis structures for molecules with the following connections.

ELECTRONEGATIVITY AND POLAR COVALENT BONDS

7.40 Without looking at Figure 7.4, order the following elements according to their increasing electronegativity: Li, Br, Pb, K, Mg, C.

7.41 What general trends in electronegativity occur in the periodic table?

7.42 Which of the following substances are largely ionic and which covalent?
  (a) HF   (b) HI   (c) $PdCl_2$ (d) BBr_3 (e) NaOH

7.43 Use the electronegativity data in Figure 7.4 to predict which bond in each of the following pairs is more polar.
  (a) C–H or C–Cl
  (b) Si–Li or Si–Cl
  (c) N–Cl or N–Mg

7.44 Show the direction of polarity for each of the bonds in Problem 7.43 using the δ^+/δ^- notation.

FORMAL CHARGES

7.45 Draw a Lewis structure for carbon monoxide, CO, and assign formal charges to both atoms.

7.46 Calculate formal charges for the atoms in the resonance forms of N_2O that you drew in Problem 7.30(a). Which resonance form do you think is the more significant contributor to the resonance hybrid? Explain.

7.47 Assign formal charges to the atoms in the following Lewis structures.

(a)

$$H-\overset{\overset{\displaystyle H}{|}}{\underset{\displaystyle ..}{N}}-\overset{..}{\underset{..}{O}}-H$$

(b)

$$\left[H-\overset{..}{\underset{..}{N}}-\overset{\overset{\displaystyle H}{|}}{\underset{\underset{\displaystyle H}{|}}{C}}-H \right]^-$$

(c)

$$:\overset{..}{\underset{..}{Cl}}-\overset{\overset{\displaystyle :\overset{..}{O}:}{|}}{\underset{\underset{\displaystyle :\overset{..}{Cl}:}{|}}{P}}-\overset{..}{\underset{..}{Cl}}:$$

7.48 Assign formal charges to the atoms in the following Lewis structures. Which of the two do you think is the more important contributor to the resonance hybrid?

(a)

$$H\overset{..}{O}-\overset{\overset{\displaystyle :O:}{\|}}{S}-\overset{..}{O}H$$

(b)

$$H\overset{..}{O}-\overset{\overset{\displaystyle :\overset{..}{O}:}{|}}{S}-\overset{..}{O}H$$

7.49 Assign formal charges to the atoms in the following Lewis structures. Which of the two do you think is the more important contributor to the resonance hybrid?

(a)

$$\overset{\displaystyle H}{\underset{\displaystyle H}{>}}C=N=\overset{..}{\underset{..}{N}}:$$

(b)

$$\overset{\displaystyle H}{\underset{\displaystyle H}{>}}C-\overset{..}{N}=\overset{..}{\underset{..}{N}}:$$

THE VSEPR MODEL

7.50 What geometric arrangement of charge clouds do you expect for atoms that have the following numbers of charge clouds?
(a) 3 (b) 5 (c) 2 (d) 6

7.51 How many charge clouds are there around the central atom in a molecule that has the following geometry?
(a) tetrahedral (b) octahedral
(c) bent (d) linear
(e) square pyramidal (f) trigonal pyramidal

7.52 Give an example of a molecule or ion that has the following shape.
(a) linear (b) tetrahedral (c) bent

7.53 How many charge clouds are there around the central atom in a molecule that has the following geometry?
(a) seesaw (b) square planar
(c) trigonal bipyramidal (d) T shaped
(e) trigonal planar (f) linear

7.54 What shape do you expect for each of the following molecules?
(a) H_2Se (b) $SiCl_4$ (c) O_3 (d) GaH_3

7.55 What shape do you expect for each of the following molecules?
(a) XeO_4 (b) SO_2Cl_2 (c) OsO_4 (d) SeO_2

7.56 What shape do you expect for each of the following molecules or ions?
(a) SbF_5 (b) IF_4^+ (c) SeO_3^{2-} (d) CrO_4^{2-}

7.57 Predict the shape around the central atom in each of the following molecules or ions.
(a) NO_3^- (b) NO_2^+ (c) NO_2^-

7.58 What shape do you expect for each of the following anions?
(a) PO_4^{3-} (b) MnO_4^- (c) SO_4^{2-} (d) SO_3^{2-} (e) ClO_4^-

7.59 What shape do you expect for each of the following cations?
(a) XeF_3^+ (b) SF_3^+ (c) ClF_2^+ (d) CH_3^+

7.60 What bond angles do you expect for each of the following?
(a) the F–S–F angle in SF_2
(b) the H–N–N angle in N_2H_2
(c) the F–Kr–F angle in KrF_4
(d) the Cl–N–O angle in NOCl

7.61 There are two possible shapes for diimide, H–N=N–H. Draw them both. Are they resonance forms? Explain.

7.62 Acrylonitrile is used as the starting material for manufacturing acrylic fibers. Predict values for all bond angles in acrylonitrile.

$$H_2C=\overset{\overset{\displaystyle H}{|}}{C}-C\equiv N \qquad \text{Acrylonitrile}$$

7.63 Predict values for all bond angles in dimethylsulfoxide, a powerful solvent used in veterinary medicine to treat inflammation.

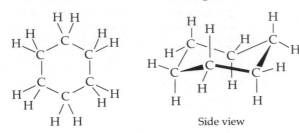

$$H_3C-\underset{..}{\overset{:\overset{:O:}{\|}}{S}}-CH_3 \qquad \text{Dimethylsulfoxide}$$

7.64 Explain why cyclohexane, a substance that contains a six-membered ring of carbon atoms, is not flat but instead has a puckered, nonplanar shape. Predict the values of the C–C–C bond angles.

Cyclohexane   Side view

7.65 Like cyclohexane (Problem 7.64), benzene also contains a six-membered ring of carbon atoms, but it is flat rather than puckered. Explain. Predict the values of the C–C–C bond angles.

Benzene

HYBRID ORBITALS AND MOLECULAR ORBITAL THEORY

7.66 What is the difference in spatial distribution between electrons in a π bond and electrons in a σ bond?

7.67 What is the difference in spatial distribution between electrons in a bonding MO and electrons in an antibonding MO?

7.68 What hybridization do you expect for atoms that have the following numbers of charge clouds?
(a) 2   (b) 5   (c) 6   (d) 4

7.69 What hybridization do you expect for the central atom in a molecule that has the following geometry?
(a) tetrahedral   (b) octahedral   (c) bent
(d) linear   (e) square pyramidal

7.70 What hybridization do you expect for the central atom in a molecule that has the following geometry?
(a) seesaw   (b) square planar
(c) trigonal bipyramidal   (d) T shaped
(e) trigonal planar

7.71 What hybridization would you expect for the underlined atom in each of the following molecules?
(a) H₂C=O   (b) B̲H₄⁻   (c) X̲eOF₄   (d) S̲O₃

7.72 What hybridization would you expect for the underlined atom in each of the following ions?
(a) B̲rO₃⁻   (b) HC̲O₂⁻   (c) C̲H₃⁺   (d) C̲H₃⁻

7.73 Urea is excreted as a waste product in animal urine. What is the hybridization of the C and N atoms in urea, and what are the values of the various bond angles?

$$H_2N-\overset{\overset{\text{O}}{\|}}{C}-NH_2 \qquad \text{Urea}$$

7.74 Aspirin has the following connections among atoms. Complete the Lewis structure for aspirin, and tell how many σ bonds and how many π bonds the molecule contains.

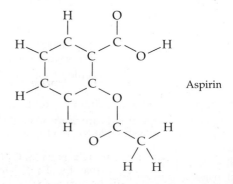

Aspirin

7.75 Use the MO diagram in Figure 7.19 to describe the bonding in O₂⁺, O₂, and O₂⁻. Which of the three should be stable? What is the bond order of each? Which contain unpaired electrons?

7.76 Use the MO diagram in Figure 7.19 to describe the bonding in N₂⁺, N₂, and N₂⁻. Which of the three should be stable? What is the bond order of each? Which contain unpaired electrons?

7.77 Calcium carbide, CaC_2, reacts with water to produce acetylene, C_2H_2, and is sometimes used as a convenient source of that substance. Use the MO diagram in Figure 7.19 to describe the bonding in the carbide anion, C_2^{2-}. What is its bond order?

7.78 Make a sketch showing the location and geometry of the *p* orbitals in the allyl cation. Describe the bonding in this cation using a localized valence bond model for σ bonding and a delocalized MO model for π bonding.

Allyl cation

GENERAL PROBLEMS

7.79 Predict the hybridization of each carbon atom in aspirin (Problem 7.74).

7.80 Boron trifluoride reacts with dimethyl ether to form a coordinate covalent bond:

Boron trifluoride   Dimethyl ether

(a) Assign formal charges to the B and O atoms in the product.

(b) Describe the geometry and hybridization of the B and O atoms in both reactants and product.

7.81 What is the hybridization of the B and N atoms in borazine, what are the values of the B–N–B and N–B–N bond angles, and what is the overall geometry of the molecule?

Borazine

7.82 Look at the molecular orbital diagram for O_2 (Figure 7.20), and then suppose that the Pauli exclusion principle could somehow be changed to allow three electrons per orbital rather than two. What would the bond order of O_2 be?

7.83 Benzyne, C_6H_4, is a stable, though highly energetic and reactive, molecule. What hybridization do you expect for the two triply bonded carbon atoms? What are the "theoretical" values for the C–C≡C bond angles? Why do you suppose benzyne is so reactive?

Benzyne

7.84 The dichromate ion, $Cr_2O_7^{2-}$, has neither Cr–Cr nor O–O bonds. Write a Lewis structure, and predict the geometry around the Cr atoms.

7.85 Propose structures for examples of molecules that meet the following descriptions.

(a) contains a C atom that has two π bonds and two σ bonds

(b) contains an N atom that has one π bond and two σ bonds

(c) contains an S atom that has a coordinate covalent bond

7.86 What is the geometry about the C, N, and S atom in each of the molecules described in Problem 7.85?

7.87 Write a Lewis structure for chloral hydrate, also known in detective novels as "knock-out drops."

Chloral hydrate

7.88 Draw as many resonance structures as you can for the diphosphate ion, $P_2O_7^{4-}$, a biologically important ion with the following connections:

Diphosphate ion

7.89 Draw a molecular orbital diagram for Li_2. What is the bond order? Is the molecule likely to be stable? Explain.

chapter 8 THERMOCHEMISTRY: CHEMICAL ENERGY

Why do chemical reactions occur? As mentioned briefly in the previous chapter, the answer involves *stability*. For a reaction to take place spontaneously, the products of the reaction must be more stable than the starting reactants.

But what is "stability," and what does it mean to say that one substance is more stable than another? The most important factor in determining the stability of a substance is the amount of energy it contains. Substances with larger amounts of energy are generally less stable and more reactive, while substances with smaller amounts of energy are generally more stable and less reactive. We'll explore some different forms of energy in this chapter and look at the subject of **thermochemistry**—the heat changes that take place during reactions.

The water held back by the Glen Canyon Dam contains potential energy that is used to drive power-generating turbines when it falls.

8.1 ►HEAT AND ENERGY

Energy. The word is familiar but is surprisingly hard to define in simple, nontechnical terms. The best working definition is that **energy** is the capacity to do work or supply heat. The water falling over a dam, for example, contains energy that can be used to turn a turbine and generate electricity. A tank of propane gas contains energy that, when released on burning, can heat a house or camper.

$$\text{Energy} = \text{work} + \text{heat}$$

The potential energy stored in the chemical bonds of propane molecules changes into kinetic energy of molecular motion (heat) when the propane undergoes a combustion reaction.

Energy can be classified as either *kinetic* or *potential*. **Kinetic energy** (E_K) is the energy of motion. A moving object with mass m and velocity v has an amount of kinetic energy given by the equation

$$E_K = \frac{1}{2}mv^2$$

The larger the mass of an object and the larger its velocity, the larger the amount of kinetic energy. Thus, water that has fallen over a dam from a great height has a greater velocity and far more kinetic energy than the same amount of water that has fallen only a short way.

Potential energy (E_P), by contrast, is stored energy—stored either in an object because of its position or in a molecule because of its chemical composition. The water sitting in a reservoir behind the dam contains potential energy because of its position high above the stream at the bottom of the dam. When the water is allowed to fall, its potential energy is converted into the kinetic energy of motion. Propane and other substances used as fuels contain potential energy in their chemical bonds. When these substances undergo reaction with oxygen during combustion, some of this potential energy is released as heat.

The SI unit for energy follows from the expression for kinetic energy, $E_K = (1/2) mv^2$. Imagine, for instance, that your body has a mass of 50.0 kg (about 110 lb) and that you are going downhill on a bicycle at a speed of 10.0 m/s (about 22 mi/h). Substituting your mass and velocity into the equation for kinetic energy shows that your kinetic energy is 2500 (kg·m²)/s².

$$E_K = \frac{1}{2}mv^2 = \frac{1}{2}(50.0 \text{ kg})\left(10.0 \frac{\text{m}}{\text{s}}\right)^2 = 2500 \frac{\text{kg} \cdot \text{m}^2}{\text{s}^2} = 2500 \text{ J}$$

One $(kg \cdot m^2)/s^2$, the SI unit for energy, is given the name *joule* (J) after the English physicist James Prescott Joule. The joule is a fairly small amount of energy—it takes roughly 100,000 J to heat a coffee cup of water from room temperature to boiling—so kilojoules (kJ) are more frequently used in chemistry.

The 75 watt light bulb in this lamp uses energy at the rate of 75 J/s. Unfortunately, only about 5% of that energy appears as light; the remainder appears as heat.

In addition to the SI unit joule, some chemists still use the unit *calorie* (cal, with a lowercase *c*), where 1 cal was originally defined as the amount of energy necessary to raise the temperature of 1 g of water by 1°C (specifically, from 14.5°C to 15.5°C). One calorie is now defined as exactly 4.184 J.

$$1 \text{ cal} = 4.184 \text{ J (exactly)}$$

Nutritionists also use the somewhat confusing unit *Calorie* (Cal, with a capital *C*), where 1 Cal is equal to 1000 calories, or 1 kilocalorie (kcal).

$$1 \text{ Cal} = 1000 \text{ cal} = 1 \text{ kcal} = 4.184 \text{ kJ}$$

The energy value, or caloric content, of food is measured in Calories. Thus, the statement that a banana contains 70 Calories means that 70 Cal (70 kcal, or 290 kJ) of energy is released when the banana is used by the body for fuel.

The potential energy in this food is released when your body uses it for fuel.

⌐ **PROBLEM 8.1** This book weighs about 2 kg. If it fell from your desk onto your foot, what would its kinetic energy in joules be at the moment of impact if its velocity at that point was 4 m/s?

⌐ **PROBLEM 8.2** What is the kinetic energy in kilojoules of a 2300 lb car moving at 55 mi/h?

8.2 ➤ENERGY CHANGES
AND ENERGY CONSERVATION

Let's pursue a bit further the relationship between potential energy and kinetic energy. According to the **law of conservation of energy**, energy can be neither created nor destroyed. It can only be converted from one form into another. To take the example of falling water again, the water in a reservoir has potential energy because of its height but has no kinetic energy because it isn't moving ($v = 0$). As the water starts to fall over the dam, though, its height (and potential energy) decrease while its velocity (and kinetic energy) increase. The total of potential energy plus kinetic energy always remains constant (Figure 8.1). When the water reaches the bottom and dashes against the rocks or drives a turbine, its kinetic energy is converted to still other forms of energy—perhaps into heat that raises the temperature of the water, or into moving the turbine and then into electricity.

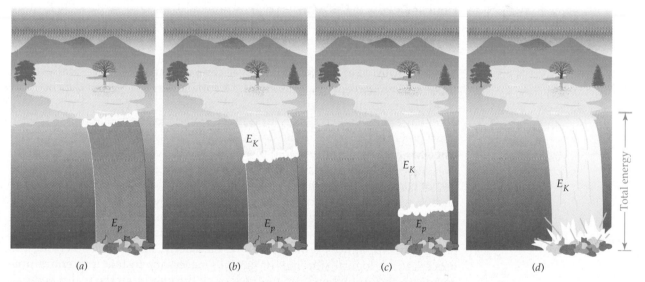

FIGURE 8.1 Conservation of energy. The total amount of energy contained by the water in a reservoir is constant. **(a)** At the top of the dam, the energy is entirely potential (E_P). **(b)**, **(c)** As the water falls over the dam, its velocity continues to increase, and its potential energy is converted into kinetic energy (E_K). **(d)** Just before the bottom of the dam, the energy is entirely kinetic. When the water dashes against the rocks at the bottom of the dam, the kinetic energy heats the water.

The conversion of kinetic energy into heat when water falls over a dam and strikes the rocks at the bottom illustrates several other important points about energy. One point is that energy has many forms. **Thermal energy**, for

The kinetic energy of falling water is used to turn huge turbines, thereby generating electricity.

example, seems different from the kinetic energy of falling water, yet it's really quite similar. **Heat** is the energy transferred from one object to another as the result of a temperature difference between them. **Temperature**, in turn, is a measure of the kinetic energy of molecular motion. An object has a low temperature and we perceive it as "cold" if its atoms or molecules are moving slowly; it has a high temperature and we perceive it as "hot" if its atoms or molecules are moving rapidly and are colliding forcefully with the thermometer or other measuring device.

Chemical energy is another kind of energy that seems somehow different from that of the water in a reservoir, yet again is really quite similar. Chemical energy is a kind of potential energy in which the chemical bonds of molecules act as the "storage" medium. Just as water releases its potential energy when it falls to a more stable position, chemicals release their potential energy in the form of heat or light when they undergo reactions and form more stable products. We'll explore this topic shortly.

A second point illustrated by the water falling over a dam involves the law of conservation of energy. To keep track of all the energy involved, it's necessary to take into account the entire chain of events that ensue from the falling water: the heating of the rocks at the bottom of the dam, the driving of the turbines and electrical generators, the transmission of the electrical power, the appliances powered by the electricity, and so on. Carrying the process to its logical extreme, in fact, it's necessary to take the entire universe into account when keeping track of all the energy in the water because the energy lost in one form always shows up elsewhere in another form. This implies that the amount of energy in the universe is constant and that all events, whether chemical or physical, natural or manmade, only change part of the overall energy in the universe from one form to another. So important is this statement that it is known as the **first law of thermodynamics**:

FIRST LAW OF THERMODYNAMICS   The energy of the universe is constant

8.3 ►INTERNAL ENERGY AND THE FIRST LAW OF THERMODYNAMICS

When keeping track of the energy changes that occur in a chemical reaction, it's helpful to think of the reaction as separate from the world around it. Those things we focus on in an experiment—the starting reactants and the final products—are collectively called the *system*, while everything in the rest of the universe—the reaction flask, the room, the building, and so on—is called the *surroundings* (Figure 8.2). If somehow the system were truly isolated from its surroundings so that no energy transfer could occur between the two, then the total **internal energy** E of the system—the sum of all the kinetic and potential energies for each particle in the system—would be constant throughout the reaction. In fact, this assertion is just a restatement of the first law of thermodynamics:

FIRST LAW OF THERMODYNAMICS RESTATED	The total internal energy of an isolated system is constant

In practice, of course, it's not possible to isolate a chemical reaction from its surroundings. After all, the chemicals are in physical contact with the walls of the flask or container, and the container itself might be in contact with the laboratory bench. What's important is not that the system be isolated but that we be able to measure accurately any energy that enters the system from the surroundings or leaves the system and flows to the surroundings. That is, we must be able to measure any *change* in the internal energy of the system, represented by ΔE (read "delta E"). The energy change ΔE represents the difference in energy between final and initial states:

$$\Delta E = E_{final} - E_{initial}$$

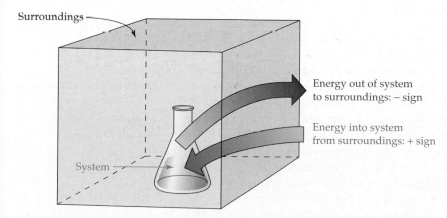

Surroundings

Energy out of system
to surroundings: – sign

Energy into system
from surroundings: + sign

System

FIGURE 8.2 When energy changes are measured in chemical reactions, the *system* is the reaction mixture being studied, and the *surroundings* are the flask, the room, and the rest of the universe. The energy change is defined as the difference between final and initial states ($\Delta E = E_{final} - E_{initial}$). Any energy that flows out from the system to the surroundings has a negative sign (since E_{final} is smaller than $E_{initial}$), and any energy that flows in from the surroundings to the system has a positive sign (since E_{final} is larger than $E_{initial}$).

By convention, *energy changes are measured from the point of view of the system*. Any energy that flows *out* of the system *to* the surroundings has a negative sign because the system has lost it. Any energy that flows *into* the system *from* the surroundings has a positive sign because the system has gained it. If, for instance, we were to burn 1 mol of methane in the presence of oxygen, 890 kJ of energy would be released and transferred as heat from the system to the surroundings ($\Delta E = -890$ kJ). This energy flow could be detected and measured by placing the reaction vessel in a water bath and noting the temperature of the bath before and after reaction.

$$CH_4(g) + 2\,O_2(g) \rightarrow CO_2(g) + 2\,H_2O(l) + 890\,kJ\ \text{energy released}$$

The experiment tells us that the products of the reaction, CO_2 and $2\,H_2O$, have 890 kJ less internal energy than the reactants, CH_4 and $2\,O_2$, even though we don't know the exact values at the beginning ($E_{initial}$) and end (E_{final}) of the reaction.

The internal energy of a system depends on many things: chemical identity, sample size, temperature, pressure, physical state (gas, solid, or liquid), and so forth. What the internal energy does *not* depend on is the system's past history. It doesn't matter what the system's temperature or physical state was yesterday or an hour ago, and it doesn't matter how the chemicals were made; all that matters is the present condition of the system. Thus, internal energy is said to be a **state function**.

STATE FUNCTION   A function or property whose value depends only on the present state (condition) of the system, not on the path used to arrive at that condition.

Let's illustrate the idea of a state function by making a trip across country, starting on the west coast in Santa Cruz, California, and ending on the east coast in Boston, Massachusetts. The straight-line distance between the initial and final points (about 2720 mi) is a state function because how we get from one place to the other is irrelevant; it doesn't matter whether we drive, fly, or crawl. Most of the other characteristics of the trip, however, are not state functions. The total distance traveled, for instance, is not a state function because it depends on the exact route taken, whether in a straight line or through Miami, Florida, for example. Similarly, the amount of fuel consumed, the time taken, and the cost of the trip are not state functions (Figure 8.3).

FIGURE 8.3 The point-to-point distance in a cross-country trip is a state function because it does not depend on how the trip is made. Other variables of the trip such as time required and total distance covered are not state functions because they depend on the route and manner of transportation.

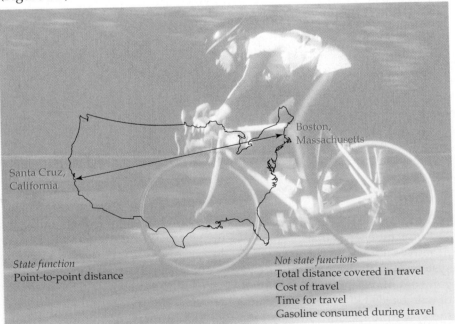

Santa Cruz, California

Boston, Massachusetts

State function
Point-to-point distance

Not state functions
Total distance covered in travel
Cost of travel
Time for travel
Gasoline consumed during travel

The cross-country trip in Figure 8.3 illustrates a particularly important point about all state functions—their *reversibility*. Imagine that after traveling from Santa Cruz to Boston, you turn around and go back to Santa Cruz. Since your final position is now identical to your initial position, the value of the point-to-point state function is 0 mi. *For any state function, the overall change is zero if the system returns to its original condition.* For non-state functions, however, the overall change is nonzero even if the path returns the system to its original condition. We don't recover the gasoline, the cost, or the time spent on a trip when we return home.

PROBLEM 8.3 Which of the following are state functions and which are not?
(a) The temperature of an ice cube
(b) The volume of an aerosol can
(c) The amount of work expended in a 10 mi bike ride

8.4 ►EXPANSION WORK

Work, like energy, comes in many forms and is hard to define in everyday terms. In physics, **work** (w) is defined as the distance (d) moved times the force (F) that opposes the motion:

$$\text{Work} = \text{distance} \times \text{force}$$

$$w = d \times F$$

When you run up stairs, for instance, your leg muscles provide a force sufficient to overcome the opposing force of gravity. When you swim, you have to provide a force sufficient to push water out of the way and to overcome the effects of frictional drag.

It may look like play, but the swimmer is doing a lot of physical work to push the water out of the way.

The most common type of work encountered in chemical systems is the *expansion work* (also called *pressure-volume*, or *PV work*) done as the result of a volume change in the system. Take the combustion of propane (C_3H_8) with oxygen, for instance. The balanced equation says that seven molecules of product come from six molecules of reactant:

$$\underbrace{C_3H_8(g) + 5\,O_2(g)}_{\text{Six molecules of gas}} \rightarrow \underbrace{3\,CO_2(g) + 4\,H_2O(g)}_{\text{Seven molecules of gas}}$$

If the reaction takes place inside a container outfitted with a movable piston, the increased pressure caused by the greater amount of gas in the product will force the piston outward against the pressure of the atmosphere, moving air molecules aside and thereby doing work (Figure 8.4).

FIGURE 8.4 A volume change during a reaction can do work on the surroundings by forcing a piston outward against air pressure P. The amount of work is equal to the pressure times the volume change $P\Delta V$, where ΔV is the area of the piston A times the distance the piston moves d.

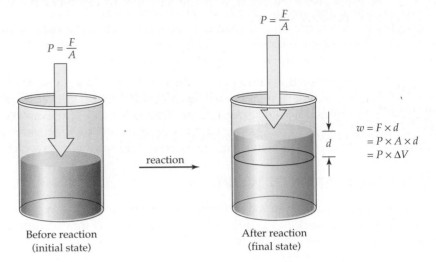

A simple calculation gives the exact amount of work done by the expanding gas. Since pressure is equal to force per unit area of the piston, the force opposing the piston's motion is the atmospheric pressure (P) times the piston's area (A).

$$\text{Force on piston} = \text{pressure on piston} \times \text{area of piston}$$

$$F = P \times A$$

If the piston is moved out a distance d by the expanding gas, then the amount of work done is distance times force, or distance times pressure times area:

$$w = d \times F = d \times (P \times A)$$

This equation can be put into a more useful form by noticing that the distance the piston moves times the area of the piston is simply the volume change in the system, $\Delta V = d \times A$. Thus, we can say that the amount of work done is equal to the pressure on the piston times the volume change.

$$w = P \times \Delta V$$

Of course, if there is no volume change, then $\Delta V = 0$ and there is no work.

If the pressure P is given in atmospheres and the volume change ΔV is given in liters, then the amount of work done has the unit liter-atmosphere (L · atm), where 1 atm = 101 × 10³ kg/(m · s²). Thus, 1 L · atm = 101 J.

$$1\ \text{L} \cdot \text{atm} = (1\text{L})\left(\frac{10^{-3}\text{m}^3}{1\text{L}}\right)\left(101 \times 10^3 \frac{\text{kg}}{\text{m} \cdot \text{s}^2}\right) = 101 \frac{\text{kg} \cdot \text{m}^2}{\text{s}^2} = 101\ \text{J}$$

What about the sign of the work done during an expansion of the system? Since the work is being done *by* the system *on* the surroundings, energy must be leaving the system and the sign of the energy change must be negative according to the convention previously established. In other words, we must add a minus sign to the equation to represent the work done *by* an expanding system:

$$w = -P\Delta V$$

Of course if a reaction takes place with a *contraction* in volume, then the ΔV term has a negative sign and work has a positive sign because the work is done *on* the system *by* the surroundings. This is the case, for example, in the industrial synthesis of ammonia by the reaction of hydrogen with nitrogen, in which four reactant gas molecules yield only two product gas molecules:

$$\underbrace{3\ H_2(g) + N_2(g)}_{\substack{\text{Four gas} \\ \text{molecules}}} \rightarrow \underbrace{2\ NH_3(g)}_{\substack{\text{Two gas} \\ \text{molecules}}}$$

EXAMPLE 8.1

Calculate the work in joules done by the system during a reaction in which the volume expands from 12.0 L to 14.5 L against an external pressure of 5.0 atm.

SOLUTION   Expansion work done by a chemical reaction is calculated by the formula $w = -P\Delta V$. In this instance, $P = 5.0$ atm and $\Delta V = (14.5 - 12.0)\ L = 2.5$ L. Thus,

$$w = -(5.0\ \text{atm})(2.5\ \text{L}) = -12.5\ \text{L} \cdot \text{atm}$$

$$(-12.5\ \text{L} \cdot \text{atm})\left(101\ \frac{J}{\text{L} \cdot \text{atm}}\right) = -1.3 \times 10^3\ J = -1.3\ kJ$$

The negative sign indicates that the expanding system does work on the surroundings.

▸ **PROBLEM 8.4**   Calculate the work in joules done by the system during a synthesis of ammonia in which the volume contracts from 8.6 to 4.3 L at a constant external pressure of 44 atm. In which direction does the energy flow? What is the sign of the energy change?

8.5 ►ENERGY AND ENTHALPY

We've seen up to this point that a system can exchange energy with its surroundings either by transferring heat or by doing work. Using the symbol q to represent heat, we can represent the total energy change of a system ΔE as

$$\Delta E = q + w = q + (-P\Delta V)$$

We can rearrange this equation to represent the amount of heat transferred as

$$q = \Delta E + P\Delta V$$

Let's look at two ways in which a chemical reaction might be carried out. On the one hand, a reaction might be carried out in a closed container with a constant volume, so that $\Delta V = 0$. In such a case, no PV work is done, and the energy change in the system is due entirely to heat transfer, which we write as q_v to indicate heat at constant volume:

$$q_v = \Delta E \quad \text{(at constant volume; } \Delta V = 0\text{)}$$

Alternatively, a reaction might be carried out at constant pressure in an open flask or other apparatus that allows the volume of the system to change freely. In such a case, $\Delta V \neq 0$ and the energy change in the system is due to both heat transfer and PV work. We indicate the heat transfer at constant pressure by the symbol q_p:

$$q_p = \Delta E + P\Delta V \quad \text{(at constant pressure)}$$

Most chemical reactions are carried out in open vessels at atmospheric pressure.

Because such processes are so important in chemistry, the heat change in a reaction carried out at constant pressure is given a special symbol, ΔH, called the *enthalpy change* of the reaction. The **enthalpy (H)** of a system is the name given to the quantity $E + PV$.

$$q_p = \Delta E + P\Delta V = \Delta H \quad \text{(where } H = E + PV\text{)}$$

Note that only the enthalpy *change* during a reaction is important. As with the internal energy, E, we don't need to know the absolute value of the system's enthalpy before and after a reaction. We only need to know the difference between final and initial states.

$$\Delta H = H_{\text{products}} - H_{\text{reactants}}$$

Enthalpy is a state function, whose value depends only on the current state of the system, not on the path taken to arrive at that state.

How big a difference is there between ΔE, the heat flow at constant volume, and ΔH, the heat flow at constant pressure? Let's look again at the combustion reaction of propane, C_3H_8, as an example. When the reaction is carried out at constant volume, no PV work is possible and all the energy is released as heat—2043 kJ per mole of propane reacted ($\Delta E = -2043$ kJ). When the same reaction is carried out at constant pressure, however, only 2041 kJ of heat is released per mole of propane ($\Delta H = -2041$ kJ). The difference, 2 kJ/mol, occurs because at constant pressure, a small amount of expansion work is done by the system against the atmosphere as 6 mol of gaseous reactants are converted into 7 mol of gaseous products.

$$C_3H_8(g) + 5\,O_2(g) \rightarrow 3\,CO_2(g) + 4\,H_2O(g) \qquad \Delta E = -2043 \text{ kJ}$$
Propane
$$\Delta H = -2041 \text{ kJ}$$
$$P\Delta V = +2 \text{ kJ}$$

What is true of the propane combustion reaction is also true of most other reactions: The difference between ΔH and ΔE is almost always so small that the two quantities are nearly equal. Chemists usually measure and think in terms of ΔH, though, because most reactions are carried out at constant atmospheric pressure in loosely covered vessels.

Note that the value given for the enthalpy change ΔH represents the amount of heat released when molar amounts of reactants undergo reaction in the direction written as represented by coefficients of the balanced equation. In the propane combustion example, 1 mol of propane gas reacting with 5 mol of oxygen gas to give 3 mol of CO_2 gas and 4 mol of water vapor releases 2041 kJ of energy. The actual amount of heat released in a *specific* reaction depends on the actual amounts of reactants. Thus, combustion of 0.5000 mol of propane with 2.500 mol of O_2 would release 0.5000×2041 kJ = 1021 kJ.

Note also that the physical states of reactants and products must be specified as solid (*s*), liquid (*l*), gaseous (*g*), or aqueous (*aq*) when enthalpy changes are reported. The enthalpy change for the combustion of propane is $\Delta H = -2041$ kJ if water is produced as a gas but $\Delta H = -2217$ kJ if water is produced as a liquid.

$$C_3H_8(g) + 5\ O_2(g) \rightarrow 3\ CO_2(g) + 4\ H_2O(g) \qquad \Delta H = -2041 \text{ kJ}$$

$$C_3H_8(g) + 5\ O_2(g) \rightarrow 3\ CO_2(g) + 4\ H_2O(l) \qquad \Delta H = -2217 \text{ kJ}$$

The difference between the values of ΔH for the two reactions—176 kJ—arises because the physical change of liquid water to gaseous water requires energy. If liquid water is produced, ΔH is larger (more negative), but if gaseous water is produced, 44.0 kJ/mol is needed for the vaporization, and ΔH is smaller (less negative).

$$H_2O(l) \rightarrow H_2O(g) \qquad \Delta H = 44.0 \text{ kJ}$$

$$\text{or } 4\ H_2O(l) \rightarrow 4\ H_2O(g) \qquad \Delta H = 176 \text{ kJ}$$

In addition to specifying the physical state of reactants and products when reporting an enthalpy change, it is also necessary to specify the pressure and temperature. To insure that all measurements are reported in the same way so that different reactions can be compared, a standard set of conditions called the **thermodynamic standard state** has been defined.

THERMODYNAMIC   298.15 K (25°C); 1 atm pressure of each gas;
STANDARD STATE   1 M concentration (for solutions)
CONDITIONS

Measurements made under these conditions are indicated by addition of a superscript zero (°) to the symbol of the quantity reported. Thus, heats of reactions measured under standard-state conditions are called **standard enthalpies of reaction** and are indicated by the symbol $\Delta H°$. The combustion reaction of propane, for example, might be written

$$C_3H_8(g) + 5\ O_2(g) \rightarrow 3\ CO_2(g) + 4\ H_2O(g) \qquad \Delta H° = -2041 \text{ kJ}$$

EXAMPLE 8.2

The reaction of nitrogen with hydrogen to make ammonia has $\Delta H° = -92.2$ kJ.

$$N_2(g) + 3\,H_2(g) \rightarrow 2\,NH_3(g) \qquad \Delta H° = -92.2 \text{ kJ}$$

What is the value of ΔE in kJ if the reaction is carried out at a constant pressure of 40.0 atm and the volume change is -1.12 L?

SOLUTION First rearrange the equation $\Delta H = \Delta E + P\Delta V$ to the form $\Delta E = \Delta H - P\Delta V$ and then substitute the appropriate values for ΔH, P, and ΔV:

$$\Delta E = \Delta H - P\Delta V$$

where
$$\Delta H = -92.2 \text{ kJ}$$

$$P\Delta V = (40.0 \text{ atm})(-1.12 \text{ L}) = -44.8 \text{ L} \cdot \text{atm}$$

$$= (-44.8 \text{ L} \cdot \text{atm})\left(101\,\frac{\text{J}}{\text{L} \cdot \text{atm}}\right) = -4520 \text{ J} = -4.52 \text{ kJ}$$

$$\Delta E = (-92.2 \text{ kJ}) - (-4.52 \text{ kJ}) = -87.7 \text{ kJ}$$

Note that ΔE is *smaller* (less negative) than ΔH for this reaction because the volume change is negative. Since the products have less volume than the reactants, a small amount of expansion work is done *by* the surroundings *on* the system.

◤ **PROBLEM 8.5** The reaction between hydrogen and oxygen to yield water has $\Delta H° = -484$ kJ.

$$2\,H_2(g) + O_2(g) \rightarrow 2\,H_2O(g) \qquad \Delta H° = -484 \text{ kJ}$$

Explosions cause the essentially instantaneous release of large amounts of gas, thereby doing *PV* work.

How much *PV* work is done, and what is the value of ΔE in kJ for the reaction of 0.50 mol of H_2 with 0.25 mol of O_2 at atmospheric pressure if the volume change is -5.6 L?

◤ **PROBLEM 8.6** The explosion of 2.00 mol of solid trinitrotoluene (TNT) with a volume of approximately 274 mL produces gases with a volume of 448 L at room temperature. How much *PV* work in kJ is done during the explosion?

$$2\,C_7H_5N_3O_6(s) \rightarrow 12\,CO(g) + 5\,H_2(g) + 3\,N_2(g) + 2\,C(s)$$

8.6 ▶ENTHALPIES OF PHYSICAL AND CHEMICAL CHANGE

Every change by a system involves either a gain or a loss of enthalpy. The change can be either physical, such as the melting of a solid to a liquid, or chemical, such as the combustion of propane. Let's briefly look at examples of both kinds.

Enthalpies of Physical Change

Imagine what would happen if you were to start with a block of ice at a low temperature—say, $-100°C$—and slowly increase its enthalpy by adding heat to it. As shown in Figure 8.5, the initial input of heat causes the temperature of the ice to rise until it reaches 0°C. Additional heat then causes the ice to melt without raising its temperature as the added energy goes into overcoming the forces that hold H_2O molecules rigidly together in the ice crystal. The amount of heat required for melting is called the enthalpy of fusion, or **heat of fusion**, ΔH_{fusion}, and has a value of 6.01 kJ/mol for H_2O.

Once the ice has melted, further input of heat raises the water's temperature until it reaches 100°C, and additional heat input then causes the water to boil. Once again, energy is necessary to overcome the forces holding molecules together in the liquid, and the temperature does not rise again until all the liquid has been converted into vapor. The amount of heat required for evaporation is called the enthalpy of vaporization, or **heat of vaporization**, ΔH_{vap}, and has a value of 40.7 kJ/mol at 100°C for H_2O.

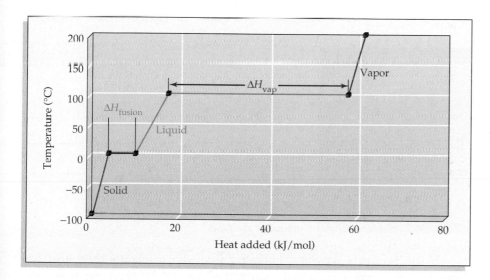

FIGURE 8.5 The effect of adding heat to a block of ice. From a starting temperature of −100°C, the initial heat input raises the temperature of the ice to 0°C, at which point additional heat causes the ice to melt without raising its temperature. Further input of heat raises the temperature of the liquid water to 100°C, where a second change of physical state takes place and the liquid is converted into vapor.

Another kind of physical change is **sublimation**—the direct conversion of a solid to a vapor without going through a liquid state. Solid CO_2 (dry ice), for example, changes directly from solid to vapor without first melting to a liquid. Since enthalpy is a state function, the enthalpy change on going from solid to vapor must be constant regardless of the path taken. Thus, a substance's enthalpy of sublimation, or **heat of sublimation**, ΔH_{subl}, at a given temperature is equal to the sum of the heat of fusion and heat of vaporization at the same temperature (Figure 8.6).

Dry ice (solid CO_2) sublimes directly from solid to gas at atmospheric pressure.

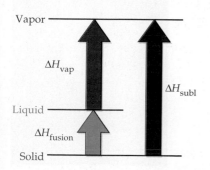

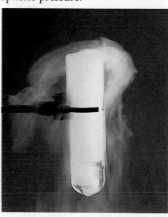

FIGURE 8.6 Since enthalpy is a state function, the enthalpy change from solid to vapor does not depend on the path taken between the two states. Therefore, $\Delta H_{subl} = \Delta H_{fusion} + \Delta H_{vap}$ when the temperature is constant.

Enthalpies of Chemical Change

We saw in Section 8.5 that when a reaction is carried out at constant pressure, the enthalpy change is equal to the heat flow into or out of the system. For this reason, enthalpies of chemical change are often called **heats of reaction**. If the products have *more* enthalpy than the reactants, then heat has flowed into the system from the surroundings and ΔH has a positive sign. Such reactions are said to be **endothermic** (*endo* means "within"; heat flows in). The reaction of 1 mol of barium hydroxide octahydrate with ammonium chloride, for example, absorbs 80.3 kJ of heat from the surroundings [$\Delta H° = +80.3$ kJ]. The surroundings become so cold that water freezes around the outside of the flask (Figure 8.7).

$$Ba(OH)_2 \cdot 8H_2O(s) + 2\ NH_4Cl(s) \rightarrow$$

$$BaCl_2(aq) + 2\ NH_3(aq) + 10\ H_2O(l)   \Delta H° = +80.3\ kJ$$

FIGURE 8.7 The reaction of barium hydroxide octahydrate with ammonium chloride is so strongly endothermic and draws in so much heat from the surroundings that it causes water to freeze around the outside of the flask.

If the products have *less* enthalpy than the reactants, then heat has flowed out from the system to the surroundings and ΔH has a negative sign. Such reactions are said to be **exothermic** (*exo* means "out"; heat flows out). The so-called thermite reaction of aluminum with iron(III) oxide, for example, releases so much heat [$\Delta H° = -851.5$ kJ] that it has been used in construction to weld iron (Figure 8.8).

$$2\ Al(s) + Fe_2O_3(s) \rightarrow 2\ Fe(s) + Al_2O_3(s)   \Delta H° = -851.5\ kJ$$

FIGURE 8.8 The thermite reaction of aluminum with iron(III) oxide is so strongly exothermic and releases so much heat to the surroundings that the products are molten.

As noted in the previous section, the $\Delta H°$ values given for an equation assume that the equation is balanced for the number of moles of reactants and products, that all substances are in their standard states, and that the physical state of each substance is as specified. The actual amount of heat released in a reaction depends on the amounts of reactants, as illustrated in Example 8.3.

It should also be emphasized that $\Delta H°$ values refer to the reaction going *in the direction written*. If a reaction is written backward, the sign of $\Delta H°$ must be changed. Because of the reversibility of state functions (Section 8.3), the enthalpy change for any reaction is equal in magnitude but opposite in sign to that for the reverse reaction. For example, the reaction of iron with aluminum oxide to yield aluminum and iron oxide (the reverse of the thermite reaction) would be endothermic and have $\Delta H° = +851.5$ kJ:

$$2\ Fe(s) + Al_2O_3(s) \rightarrow 2\ Al(s) + Fe_2O_3(s) \qquad \Delta H° = +851.5\ \text{kJ}$$

EXAMPLE 8.3

How much heat in kJ is evolved when 5.00 g of aluminum reacts with a stoichiometric amount of Fe_2O_3?

$$2\ Al(s) + Fe_2O_3(s) \rightarrow 2\ Fe(s) + Al_2O_3(s) \qquad \Delta H° = -851.5\ \text{kJ}$$

SOLUTION   The balanced equation tells how much heat is evolved from the reaction of 2 mol of Al. To find out how much heat is evolved from the reaction of 5.00 g of Al, we first have to find out how many moles of aluminum are in 5.00 g. Looking in the periodic table, we see that the molar mass of Al is 26.98 g/mol, so 5.00 g of Al is equal to 0.185 mol:

$$5.00\ \text{g Al} \times \frac{1\ \text{mol Al}}{26.98\ \text{g Al}} = 0.185\ \text{mol Al}$$

Since 2 mol of Al releases 851.5 kJ of heat, 0.185 mol of Al releases 78.8 kJ of heat:

$$0.185\ \text{mol Al} \times \frac{851.5\ \text{kJ}}{2\ \text{mol Al}} = 78.8\ \text{kJ}$$

PROBLEM 8.7   How much heat in kJ is evolved or absorbed in each of the following reactions?

(a) Combustion of 15.5 g of propane:

$$C_3H_8(g) + 5\ O_2(g) \rightarrow 3\ CO_2(g) + 4\ H_2O(l) \qquad \Delta H° = -2217\ \text{kJ}$$

(b) Reaction of 4.88 g of barium hydroxide octahydrate with ammonium chloride:

$$Ba(OH)_2 \cdot 8H_2O(s) + 2\ NH_4Cl(s) \rightarrow$$
$$BaCl_2(aq) + 2\ NH_3(aq) + 10\ H_2O(l) \qquad \Delta H° = +80.3\ \text{kJ}$$

8.7 ►CALORIMETRY AND HEAT CAPACITY

The amount of heat transferred during a reaction can be measured by a device called a *calorimeter*, shown schematically in Figure 8.9. At its simplest, a calorimeter is just an insulated vessel with a stirrer, a thermometer,

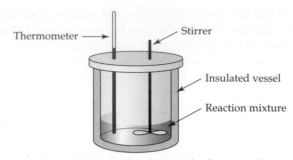

FIGURE 8.9 A calorimeter for measuring heat evolved in a reaction at constant pressure (ΔH). The reaction takes place inside an insulated vessel outfitted with a loose-fitting top, a thermometer, and a stirrer. Measuring the temperature change that accompanies the reaction makes it possible to calculate ΔH.

and a loose-fitting lid to keep the contents at atmospheric pressure. The reaction is carried out inside the vessel, and the heat evolved or absorbed is measured by the temperature change. Since the pressure inside the calorimeter is constant, the temperature measurement makes it possible to calculate the enthalpy change ΔH during the reaction.

A somewhat more complicated device called a *bomb calorimeter* is used to measure heats of combustion reactions. The sample is placed in a small cup and sealed under an oxygen atmosphere inside a steel "bomb" that is itself placed in an insulated, water-filled container (Figure 8.10). The reactants are ignited electrically, and the evolved heat is measured by reading the temperature change of the surrounding water. Since the reaction takes place at constant volume, the measurement provides a value for ΔE rather than ΔH.

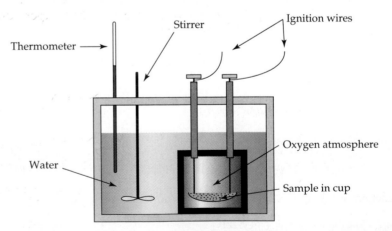

FIGURE 8.10 Diagram of a bomb calorimeter for measuring heat evolved at constant volume in a combustion reaction (ΔE). The reaction is carried out inside a steel bomb, and the heat evolved is transferred to the surrounding water, where the temperature rise is measured.

How can the temperature change inside a calorimeter be used to calculate ΔH (or ΔE) for a reaction? When a calorimeter and its contents absorb a given amount of heat, the temperature rise that results depends on the calorimeter's **heat capacity** (C). Heat capacity is the amount of heat required to raise the temperature of an object or substance a given amount, a relationship that can be expressed by the equation

$$C = \frac{q}{\Delta T}$$

where q is the quantity of heat transferred and ΔT is the temperature change ($\Delta T = T_{final} - T_{initial}$). The greater the heat capacity, the greater the amount of heat needed to produce a given temperature change. A bathtub full of water, for instance, has a greater heat capacity than a coffee cup full, and it therefore takes far more heat to warm the tubful than the cupful. The exact amount of heat is simply the heat capacity times the temperature rise.

$$q = C \times \Delta T$$

Heat capacity is an extensive property, one whose value depends on both the size of an object and its composition. To compare different substances, it is useful to define a quantity called **specific heat**, the amount of heat necessary to raise the temperature of exactly 1 g of a substance by exactly 1°C. The amount of heat necessary to raise the temperature of a given object, then, is the specific heat times the mass of the object times the rise in temperature.

$$q = (\text{specific heat}) \times (\text{mass of substance}) \times \Delta T$$

Example 8.5 shows how specific heats are used in calorimetry calculations.
    Closely related to specific heat is the **molar heat capacity (C_m)**, defined as the amount of heat necessary to raise the temperature of 1 mol of a substance by 1°C. The amount of heat necessary to raise the temperature of a given number of moles of a substance is thus

$$q = (C_m) \times (\text{moles of substance}) \times \Delta T$$

Values of specific heats and of molar heat capacities for some common substances are given in Table 8.1.

TABLE 8.1	Specific Heats and Molar Heat Capacities for Some Common Substances at 25°C	
Substance	**Specific Heat** J/(g · °C)	**Molar Heat Capacity** J/(mol · °C)
Air	1.01	29.1
Aluminum	0.902	24.35
Copper	0.385	24.44
Gold	0.129	25.42
Iron	0.450	25.10
Mercury	0.140	27.98
NaCl	0.864	50.50
Water(s)[a]	2.03	36.57
Water(l)	4.179	75.29

[a] At −11°C

As indicated in Table 8.1, the specific heat of liquid water is considerably higher than that of most other substances, and a large transfer of heat is therefore necessary either to cool or to warm a given amount of water. One consequence of this high specific heat is that large lakes or other bodies of water tend to moderate the air temperature in the surrounding areas. Another consequence is that the human body, which is about 60% water, is able to maintain a steady internal temperature under changing outside conditions.

Large masses of water moderate the temperature of the surroundings because of their high heat capacity.

EXAMPLE 8.4

What is the specific heat of silicon if it takes 192 J to raise the temperature of 45.0 g of Si by 6.0°C?

SOLUTION   A substance's specific heat is the amount of energy necessary to raise the temperature of 1 g of the substance by 1°C. Since 192 J raises the temperature of 45.0 g of Si by 6.0°C, 192/45.0 = 4.27 J will raise the temperature of 1 g of Si by 6.0°C, and 4.27/6.0 = 0.71 J will raise the temperature of 1 g of Si by 1°C.

$$\text{Specific heat of Si} = \frac{192 \text{ J}}{(45.0 \text{ g})(6.0°C)} = 0.71 \text{ J/(g} \cdot °C)$$

The specific heat of this silicon sample is 0.71 J/(g · °C).

EXAMPLE 8.5

Aqueous silver ion reacts with aqueous chloride ion to yield a white precipitate of solid silver chloride:

$$Ag^+(aq) + Cl^-(aq) \rightarrow AgCl(s)$$

When 10.0 mL of 1.00 M $AgNO_3$ solution is added to 10.0 mL of 1.00 M NaCl solution at 25.0°C in a calorimeter, a white precipitate of AgCl forms, and the temperature of the aqueous mixture increases to 32.6°C. Assuming that the specific heat of the aqueous mixture is 4.18 J/(g · °C), that the density of the mixture is 1.00 g/mL, and that the calorimeter itself absorbs a negligible amount of heat, calculate ΔH in kJ for the reaction.

The reaction of aqueous Ag^+ with aqueous Cl^- to yield solid AgCl is an exothermic process.

SOLUTION   Since the temperature rises during the reaction, heat must flow from the reactants to the product solution and ΔH must therefore be negative. The amount of heat evolved during the reaction is equal to the amount of heat absorbed by the mixture and can be calculated in the following way:

Heat evolved = (specific heat) × (mass of mixture) × (temperature change)

Specific heat = 4.18 J/(g · °C)

Mass = 20.0 mL × $1.00 \dfrac{g}{mL}$ = 20.0 g

Temperature change = 32.6°C − 25.0°C = 7.6°C

Heat evolved = $4.18 \dfrac{J}{g \cdot °C}$ × 20.0 g × 7.6°C = 6.4×10^2 J

Next, calculate how many moles of AgCl were formed. According to the balanced equation, the number of moles of AgCl produced is equal to the number of moles of Ag^+ (or Cl^-) reacted:

Moles of Ag^+ = 10.0 mL × $\dfrac{1.00 \text{ mol } Ag^+}{1000 \text{ mL}}$ = 1.00×10^{-2} mol Ag^+

Moles of AgCl = 1.00×10^{-2} mol AgCl

Finally, calculate ΔH for the reaction. Don't forget that its sign is negative, because heat is evolved.

Heat evolved per mole of AgCl = $\dfrac{6.4 \times 10^2 \text{ J}}{1.00 \times 10^{-2} \text{ mol AgCl}}$ = 64 kJ/mol AgCl

Therefore, ΔH = −64 kJ

PROBLEM 8.8   Assuming that Coca Cola has the same specific heat as water [4.18 J/(g · °C)], calculate the amount of heat in kJ transferred when one can (about 350 g) is cooled from 25°C to 3°C.

PROBLEM 8.9   What is the specific heat of lead if it takes 96 J to raise the temperature of a 75 g block by 10°C?

⌐ **PROBLEM 8.10** When 25 mL of 1.0 M H_2SO_4 is added to 50 mL of 1.0 M NaOH at 25.0°C in a calorimeter, the temperature of the aqueous solution increases to 33.9°C. Assuming that the specific heat of the solution is 4.18 J/(g · °C), that its density is 1.00 g/mL, and that the calorimeter itself absorbs a negligible amount of heat, calculate ΔH in kJ for the reaction. ⌐

8.8 ►HESS'S LAW

Having talked in general terms about the energy changes that occur during chemical reactions, let's now look in some detail at a specific example. In particular, let's look at the *Haber process*, the industrial method by which some 17 million tons of ammonia are produced each year in the United States for use as fertilizer. The reaction of hydrogen with nitrogen to make ammonia is exothermic, with $\Delta H° = -92.22$ kJ.

$$3 H_2(g) + N_2(g) \rightarrow 2 NH_3(g) \qquad \Delta H° = -92.22 \text{ kJ}$$

If we dig into the details of the reaction, we find that it's not as simple as it looks. In fact, the overall reaction occurs by a series of steps, with hydrazine (N_2H_4) produced at an intermediate stage:

$$2 H_2(g) + N_2(g) \xrightarrow{} \underset{\text{Hydrazine}}{N_2H_4(g)} \xrightarrow{H_2} \underset{\text{Ammonia}}{2 NH_3(g)}$$

The enthalpy change for the conversion of hydrazine to ammonia can be measured as $\Delta H° = -187.6$ kJ, but if we wanted to measure $\Delta H°$ for the formation of hydrazine from hydrogen and nitrogen, we would have difficulty because the reaction doesn't go cleanly. Some of the hydrazine is converted into ammonia while some of the starting nitrogen still remains.

Fortunately, there's a way around the difficulty—a way that makes it possible to measure an energy change indirectly when a direct measurement can't be made. The trick is to realize that because enthalpy change is a state function, its value is the same no matter what path is taken between two states. Thus, the sum of the enthalpy changes for the individual reactions in a sequence must equal the enthalpy change for the overall reaction, a statement known as **Hess's law**:

HESS'S LAW   The overall enthalpy change for a reaction is equal to the sum of the enthalpy changes for the individual steps in the reaction.

Reactants and products in the individual steps can be added and subtracted like algebraic quantities in determining the overall equation. In the synthesis of ammonia, for example, the sum of the two steps (1) and (2) is equal to the overall reaction (3). Thus, the sum of the enthalpy changes for the two steps (1) and (2) is equal to the enthalpy change for the overall reaction (3). With this knowledge, we can calculate the enthalpy change for step (1). Figure 8.11 shows the situation pictorially.

Reaction
(1) $2 H_2(g) + N_2(g) \rightarrow \cancel{N_2H_4(g)}$ $\Delta H°_1 = ?$
(2) $\cancel{N_2H_4(g)} + H_2(g) \rightarrow 2 NH_3(g)$ $\Delta H°_2 = -187.6$ kJ
(3) $3 H_2(g) + N_2(g) \rightarrow 2 NH_3(g)$ $\Delta H°_3 = -92.22$ kJ

Since $\qquad\qquad \Delta H_1^\circ + \Delta H_2^\circ = \Delta H_3^\circ$

then $\qquad\qquad \Delta H_1^\circ = \Delta H_3^\circ - \Delta H_2^\circ$

$$= (-92.22 \text{ kJ}) - (-187.6 \text{ kJ}) = +95.4 \text{ kJ}$$

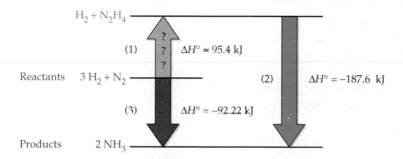

FIGURE 8.11 A representation of the enthalpy changes for the synthesis of ammonia from nitrogen and hydrogen. If ΔH° values for step (2) and for the overall reaction (3) are known, then ΔH° for step (1) can be calculated. That is, the enthalpy change for the overall reaction (3) is equal to the sum of the enthalpy changes for the individual steps (1) and (2).

EXAMPLE 8.6

Methane, the main constituent of natural gas, burns in oxygen to yield carbon dioxide and water:

$$CH_4(g) + 2 O_2(g) \rightarrow CO_2(g) + 2 H_2O(l)$$

Use the following information to calculate ΔH° in kJ for the combustion of methane:

$$CH_4(g) + O_2(g) \rightarrow CH_2O(g) + H_2O(g) \qquad \Delta H^\circ = -284 \text{ kJ}$$

$$CH_2O(g) + O_2(g) \rightarrow CO_2(g) + H_2O(g) \qquad \Delta H^\circ = -518 \text{ kJ}$$

$$H_2O(l) \rightarrow H_2O(g) \qquad\qquad\qquad\qquad \Delta H^\circ = 44.0 \text{ kJ}$$

SOLUTION It takes some trial and error, but the idea is to combine the individual reactions so that their sum is the desired reaction. Of course, all the reactants [$CH_4(g)$ and $O_2(g)$] must appear on the left, and all the products [$CO_2(g)$ and $H_2O(l)$] must appear on the right. Intermediate products such as $CH_2O(g)$ and $H_2O(g)$ must occur on *both* the left and the right so that they cancel. Note that the conversion $H_2O(g) \rightarrow H_2O(l)$ must be written backward from the direction given, and the sign of ΔH° must be changed. As noted in Section 8.6, the enthalpy change for any reaction is equal in magnitude but opposite in sign to that for the reverse reaction. Note also that ΔH° for the condensation of water vapor must be multiplied by 2 in accord with the balanced equation for the overall reaction.

$$
\begin{array}{ll}
CH_4(g) + O_2(g) \rightarrow \cancel{CH_2O(g)} + \cancel{H_2O(g)} & \Delta H^\circ = -284 \text{ kJ} \\
\cancel{CH_2O(g)} + O_2(g) \rightarrow CO_2(g) + \cancel{H_2O(g)} & \Delta H^\circ = -518 \text{ kJ} \\
2 \, [\cancel{H_2O(g)} \rightarrow H_2O(l)] & 2 \, [\Delta H^\circ = -44.0 \text{ kJ}] = -88.0 \text{ kJ} \\
\hline
CH_4(g) + 2 O_2(g) \rightarrow CO_2(g) + 2 H_2O(l) & \Delta H^\circ = -890 \text{ kJ}
\end{array}
$$

EXAMPLE 8.7

Water gas is the name for the industrially important mixture of CO and H_2 prepared by passing steam over hot charcoal at 1000°C.

$$C(s) + H_2O(g) \rightarrow CO(g) + H_2(g) \qquad \text{"water gas"}$$

The hydrogen is purified and used as a starting material for preparing ammonia, NH_3. Use the following information to calculate $\Delta H°$ in kJ for the water-gas reaction:

$$C(s) + O_2(g) \rightarrow CO_2(g) \qquad \Delta H° = -393.5 \text{ kJ}$$

$$2 H_2(g) + O_2(g) \rightarrow 2 H_2O(g) \qquad \Delta H° = -483.6 \text{ kJ}$$

$$2 CO(g) + O_2(g) \rightarrow 2 CO_2(g) \qquad \Delta H° = -566.0 \text{ kJ}$$

SOLUTION As in the previous example, the idea is to find a combination of the individual reactions whose sum is the desired reaction. In this instance, it is necessary to multiply the second and third steps by 1/2 to make the overall equation balance. In so doing, of course, the enthalpy changes for those steps must also be multiplied by 1/2. (Alternatively, we could multiply the first step by 2 and then divide the final result by 2.) Note that $CO_2(g)$ and $O_2(g)$ cancel because they appear on both the right and left sides of reactions.

$$
\begin{array}{ll}
C(s) + \cancel{O_2(g)} \rightarrow \cancel{CO_2(g)} & \Delta H° = -393.5 \text{ kJ} \\
1/2 \, [2 \, \cancel{CO_2(g)} \rightarrow 2 \, CO(g) + \cancel{O_2(g)}] & 1/2 \, [\Delta H° = 566.0 \text{ kJ}] = 283.0 \text{ kJ} \\
1/2 \, [2 \, H_2O(g) \rightarrow 2 \, H_2(g) + \cancel{O_2(g)}] & 1/2 \, [\Delta H° = 483.6 \text{ kJ}] = 241.8 \text{ kJ} \\
\hline
C(s) + H_2O(g) \rightarrow CO(g) + H_2(g) & \Delta H° = 131.3 \text{ kJ}
\end{array}
$$

The water-gas reaction is endothermic by 131.3 kJ/mol.

PROBLEM 8.11 The industrial degreasing solvent methylene chloride, CH_2Cl_2, is prepared from methane by reaction with chlorine:

$$CH_4(g) + 2 Cl_2(g) \rightarrow CH_2Cl_2(g) + 2 HCl(g)$$

Use the following data to calculate $\Delta H°$ in kJ for the reaction.

$$CH_4(g) + Cl_2(g) \rightarrow CH_3Cl(g) + HCl(g) \qquad \Delta H° = -98.3 \text{ kJ}$$

$$CH_3Cl(g) + Cl_2(g) \rightarrow CH_2Cl_2(g) + HCl(g) \qquad \Delta H° = -104 \text{ kJ}$$

PROBLEM 8.12 Recalculate $\Delta H°$ for the formation of methylene chloride (Problem 8.11) assuming it is formed as a liquid rather than a gas. For CH_2Cl_2, $\Delta H_{vap} = 29.0$ kJ/mol. (ΔH_{vap} is the amount of energy necessary to convert a liquid into a gas.)

PROBLEM 8.13 Draw a diagram similar to that in Figure 8.11 depicting the energy changes for the reaction in Problem 8.11.

8.9 ▸ STANDARD HEATS OF FORMATION

Where do the $\Delta H°$ values we've been using in the previous sections come from? There are so many chemical reactions—well over 100 million are known— that it's obviously not possible to measure $\Delta H°$ for every one of them. A better way is needed.

The most efficient way to make do with the smallest number of experimental measurements is to use what are called **standard heats of formation**, symbolized ΔH_f°.

STANDARD HEAT OF FORMATION	The enthalpy change ΔH_f° for the hypothetical formation of 1 mol of a substance in its standard state from the most stable forms of its constituent elements in their standard states.

Note several points about this definition. First, the "reaction" to form a substance from its constituent elements can be (and often is) hypothetical. We can't combine carbon and hydrogen in the laboratory to make methane, for instance, yet the heat of formation for methane is still defined as the enthalpy change for the hypothetical reaction

$$C(s) + 2\ H_2(g) \rightarrow CH_4(g) \qquad \Delta H_f^\circ = -74.8\ kJ$$

Second, note that all the elements in the reaction must be in their most stable, standard-state form at 298.15 K (25°C) and 1 atm pressure. Carbon, for example, is most stable as solid graphite (not diamond) under these conditions, and hydrogen is most stable as gaseous H_2 (not H). Table 8.2 gives some standard heats of formation for several common substances, and Appendix B gives a more detailed list.

TABLE 8.2   Standard Heats of Formation for Some Common Substances at 25°C

Substance	Formula	ΔH_f° (kJ/mol)	Substance	Formula	ΔH_f° (kJ/mol)
Acetylene	$C_2H_2(g)$	226.7	Hydrogen chloride	$HCl(g)$	−92.3
Ammonia	$NH_3(g)$	−46.1	Iron(III) oxide	$Fe_2O_3(s)$	−824.2
Carbon dioxide	$CO_2(g)$	−393.5	Magnesium carbonate	$MgCO_3(s)$	−1095.8
Carbon monoxide	$CO(g)$	−110.5	Methane	$CH_4(g)$	−74.8
Ethanol	$C_2H_5OH(l)$	−277.7	Nitric oxide	$NO(g)$	90.2
Ethylene	$C_2H_4(g)$	52.3	Water(g)	$H_2O(g)$	−241.8
Glucose	$C_6H_{12}O_6(s)$	−1260	Water(l)	$H_2O(l)$	−285.8

No elements are listed in Table 8.2 because *the most stable forms of all elements in their standard state have $\Delta H_f^\circ = 0$*. Remember that all our calculations are based on enthalpy *changes*, not on actual enthalpy values. Defining ΔH_f° as zero for all elements thus establishes a kind of thermochemical "sea level," or reference point, from which all changes can be measured.

How can standard heats of formation be used for thermochemical calculations? *The standard enthalpy change for any chemical reaction is found simply by subtracting the total heats of formation of all reactants from the total heats of formation of all products.* (Don't forget, though, to multiply each heat of formation by the coefficient of that substance in the balanced equation.)

$$\Delta H^\circ = H_{(products)}^\circ - H_{(reactants)}^\circ$$

Then to find $\Delta H°$ for the reaction

$$\underbrace{a\text{A} + b\text{B} + \ldots}_{\substack{\text{Subtract the sum of the}\\\text{heats of formation for}\\\text{these reactants}\ldots}} \rightarrow \underbrace{c\text{C} + d\text{D} + \ldots}_{\substack{\ldots\text{from the sum of the}\\\text{heats of formation for}\\\text{these products.}}}$$

$$\Delta H°_{\text{reaction}} = [c \cdot \Delta H°_f(\text{C}) + d \cdot \Delta H°_f(\text{D}) + \ldots] - [a \cdot \Delta H°_f(\text{A}) + b \cdot \Delta H°_f(\text{B}) + \ldots]$$

For example, let's calculate $\Delta H°$ for the fermentation of glucose to make ethyl alcohol (ethanol), the reaction that occurs during the production of alcoholic beverages:

$$C_6H_{12}O_6(s) \rightarrow 2\, C_2H_5OH(l) + 2\, CO_2(g)$$

Fermentation of the sugar in these grapes yields the ethyl alcohol in wine.

Using the data in Table 8.2 gives the following answer:

$$\Delta H°_{\text{reaction}} = [2 \times \Delta H°_f(\text{ethanol})] + [2 \times \Delta H°_f(\text{CO}_2)] - \Delta H°_f(\text{glucose})$$

$$= (2\ \text{mol} \times -277.7\ \text{kJ/mol}) + (2\ \text{mol} \times -393.5\ \text{kJ/mol}) - (1\ \text{mol} \times -1260\ \text{kJ/mol})$$

$$= -82\ \text{kJ}$$

The fermentation reaction is exothermic by 82 kJ.

Why does this calculation "work"? It works because fundamentally it's just an application of Hess's law. That is, we can add the individual equations corresponding to the heat of formation for each substance in the reaction to arrive at the enthalpy change for the overall reaction:

Reaction

(1)	$C_6H_{12}O_6(s) \rightarrow 6\,C(s) + 6\,H_2(g) + 3\,O_2(g)$	$-\Delta H°_f = +1260\ \text{kJ}$
(2)	$2\,[2\,C(s) + 3\,H_2(g) + 1/2\,O_2(g) \rightarrow C_2H_5OH(l)]$	$2\,[\Delta H°_f = -277.7\ \text{kJ}]$
(3)	$2\,[C(s) + O_2(g) \rightarrow CO_2(g)]$	$2\,[\Delta H°_f = -393.5\ \text{kJ}]$
Net	$C_6H_{12}O_6(s) \rightarrow 2\,C_2H_5OH(l) + 2\,CO_2(g)$	$\Delta H°_f = -82\ \text{kJ}$

Note that reaction 1 represents the formation of glucose from its elements written backward, so the sign of $\Delta H°_f$ is reversed. Note also that reactions 2 and 3, which represent the formation of ethyl alcohol and carbon dioxide, respectively, are multiplied by 2 to arrive at the correct net equation.

When we use heats of formation to calculate standard reaction enthalpies, what we're really doing is referencing the enthalpies of both products and reactants to the same point—their constituent elements. By being referenced to the same point, the product and reactant enthalpies are thereby referenced to each other, and their difference is the reaction enthalpy (Figure 8.12). Examples 8.8 and 8.9 give further instances of how to use standard heats of formation.

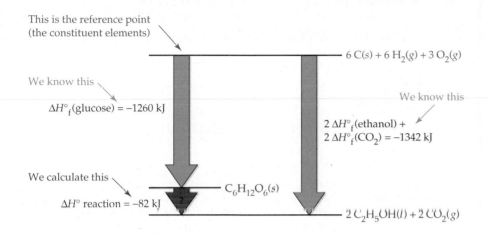

This is the reference point
(the constituent elements)

$6\ C(s) + 6\ H_2(g) + 3\ O_2(g)$

We know this

$\Delta H°_f(\text{glucose}) = -1260\ \text{kJ}$

We know this

$2\ \Delta H°_f(\text{ethanol}) + 2\ \Delta H°_f(CO_2) = -1342\ \text{kJ}$

We calculate this

$\Delta H°\ \text{reaction} = -82\ \text{kJ}$

$C_6H_{12}O_6(s)$

$2\ C_2H_5OH(l) + 2\ CO_2(g)$

FIGURE 8.12 The standard reaction enthalpy, $\Delta H°_{\text{reaction}}$, for the formation of ethyl alcohol and CO_2 from glucose is the difference between the total standard heats of formation of products and reactants. Since the different heats of formation are referenced to the same point (the constituent elements), they are referenced to each other.

EXAMPLE 8.8

Calculate $\Delta H°$ in kJ for the synthesis of lime, CaO, from limestone, $CaCO_3$, an important step in the manufacture of cement.

$$CaCO_3(s) \rightarrow CaO(s) + CO_2(g)$$

$$\Delta H°_f[CaCO_3(s)] = -1206.9\ \text{kJ/mol}$$

$$\Delta H°_f[CaO(s)] = -635.1\ \text{kJ/mol}$$

$$\Delta H°_f[CO_2(g)] = -393.5\ \text{kJ/mol}$$

Limestone is converted into lime by heating in this kiln.

SOLUTION Subtract the heat of formation of the reactant from the total heats of formation of the products:

$$\Delta H°_{\text{reaction}} = [\Delta H°_f(CaO) + \Delta H°_f(CO_2)] - \Delta H°_f(CaCO_3)$$

$$= (-635.1\ \text{kJ}) + (-393.5\ \text{kJ}) - (-1206.9\ \text{kJ}) = 178.3\ \text{kJ}$$

The reaction is endothermic by 178.3 kJ.

Acetylene burns at a high temperature, making the reaction useful for welding.

EXAMPLE 8.9

Oxyacetylene welding torches burn acetylene gas, C_2H_2. Use the information in Table 8.2 to calculate $\Delta H°$ in kJ for the combustion reaction of acetylene to yield $CO_2(g)$ and $H_2O(g)$.

SOLUTION   First write the balanced equation.

$$2\ C_2H_2(g) + 5\ O_2(g) \rightarrow 4\ CO_2(g) + 2\ H_2O(g)$$

Next, look up the appropriate heats of formation in Table 8.2.

$$\Delta H_f°[C_2H_2(g)] = 226.7\ \text{kJ/mol} \qquad \Delta H_f°[H_2O(g)] = -241.8\ \text{kJ/mol}$$

$$\Delta H_f°[CO_2(g)] = -393.5\ \text{kJ/mol}$$

Finally, carry out the calculation, making sure to multiply each $\Delta H_f°$ by the coefficient given in the balanced equation. Remember: $\Delta H_f°(O_2) = 0$ kJ/mol.

$$\begin{aligned}\Delta H_{\text{reaction}}° &= \{[4 \times \Delta H_f°(CO_2)] + [2 \times \Delta H_f°(H_2O)]\} - [2 \times \Delta H_f°(C_2H_2)]\\ &= (4 \times -393.5\ \text{kJ}) + (2 \times -241.8\ \text{kJ}) - (2 \times 226.7\ \text{kJ}) = -2511\ \text{kJ}\end{aligned}$$

The combustion of 2 mol of acetylene according to the balanced equation is exothermic by 2511 kJ(!)

⌐ **PROBLEM 8.14** Use the information in Table 8.2 to calculate $\Delta H°$ in kJ for the reaction of ammonia with O_2 to yield nitric oxide, NO, a step in the Ostwald process for the commercial production of nitric acid.

⌐ **PROBLEM 8.15** Use the information in Table 8.2 to calculate $\Delta H°$ in kJ for the photosynthesis of glucose from CO_2 and liquid water, a reaction carried out by all green plants.

8.10 ▸BOND DISSOCIATION ENTHALPIES

The procedure described in the previous section for determining heats of reactions from heats of formation is extremely useful, but it has a major flaw: To use the method, it's necessary to know $\Delta H_f°$ for every substance in a reaction. This implies, in turn, that vast numbers of measurements are needed, since there are over 10 million known chemical compounds. In practice, though, only a few thousand $\Delta H_f°$ values have been determined.

For those reactions where insufficient $\Delta H_f°$ data are available to allow an exact calculation of $\Delta H°$, it's often possible to get an approximate answer by using the average bond dissociation energies (D) discussed previously in Section 7.2. Although we didn't identify them as such at the time, bond dissociation energies such as those listed in Table 7.1, p. 231, are really just standard enthalpy changes, $\Delta H°$, for the corresponding bond-breaking reactions:

For the reaction X—Y $\rightarrow$ X + Y $\Delta H° = D = $ Bond dissociation energy

When we say, for example, that Cl_2 has a bond dissociation energy $D = 243$ kJ/mol, we mean that the standard enthalpy change for the reaction

Cl–Cl → 2 Cl is $\Delta H° = 243$ kJ/mol. Bond dissociation enthalpies are always positive because energy must always be put into a bond to break it.

Applying Hess's law, we can calculate an approximate enthalpy change for any reaction simply by subtracting the total energy of bonds formed in the products from the total energy of bonds broken in the reactants:

$$\Delta H°_{reaction} = D \text{ (bonds broken)} - D \text{ (bonds formed)}$$

In the reaction of H_2 with Cl_2 to yield HCl, for example, one Cl–Cl bond and one H–H bond are broken, while two H–Cl bonds are formed:

bonds broken   bond formed

$$H{-}H + Cl{-}Cl \rightarrow 2\ H{-}Cl$$

According to the data in Table 7.1, the bond dissociation energy of Cl_2 is 243 kJ/mol, that of H_2 is 436 kJ/mol, and that of HCl is 432 kJ/mol. We can thus calculate an approximate overall $\Delta H°$ for the reaction of −185 kJ.

$$\Delta H°_{reaction} = D \text{ (bonds broken)} - D \text{ (bonds formed)}$$
$$= (D_{Cl_2} + D_{H_2}) - (2 \times D_{HCl})$$
$$= (1\text{ mol} \times 243\text{ kJ/mol}) + (1\text{ mol} \times 436\text{ kJ/mol}) - (2\text{ mol} \times 432\text{ kJ/mol})$$
$$= -185\text{ kJ}$$

EXAMPLE 8.10

Use the information in Table 7.1 to find an approximate $\Delta H°$ in kJ for the industrial synthesis of chloroform by reaction of methane with Cl_2

$$CH_4(g) + 3\ Cl_2(g) \rightarrow CHCl_3(g) + 3\ HCl(g)$$

SOLUTION The balanced equation says that three C–H bonds and three Cl–Cl bonds are broken, while three C–Cl and three H–Cl bonds are formed. The bond dissociation energies from Table 7.1 are:

Bonds broken: C–H $D = 410$ kJ/mol   Cl–Cl $D = 243$ kJ/mol

Bonds formed: C–Cl $D = 330$ kJ/mol   H–Cl $D = 432$ kJ/mol

Next, carry out the appropriate calculation, subtracting the total energies of the bonds formed from the total energies of the bonds broken:

$$\Delta H°_{reaction} = D \text{ (bonds broken)} - D \text{ (bonds formed)}$$
$$= [(3 \times D_{Cl_2}) + (3 \times D_{CH})] - [(3 \times D_{HCl}) + (3 \times D_{CCl})]$$
$$= [(3\text{ mol} \times 243\text{ kJ/mol}) + (3\text{ mol} \times 410\text{ kJ/mol})]$$
$$\quad - [(3\text{ mol} \times 432\text{ kJ/mol}) + (3\text{ mol} \times 330\text{ kJ/mol})]$$
$$= -327\text{ kJ}$$

The reaction is exothermic by approximately 327 kJ.

PROBLEM 8.16 Use the information in Table 7.1 to calculate an approximate $\Delta H°$ value in kJ for the industrial synthesis of ethyl alcohol from ethylene. (The strength of the C=C bond is 720 kJ/mol.) $C_2H_4(g) + H_2O(g) \rightarrow C_2H_5OH(g)$

PROBLEM 8.17 Use the information in Table 7.1 to calculate an approximate $\Delta H°$ value in kJ for the synthesis of hydrazine from ammonia:
$2\ NH_3(g) + Cl_2(g) \rightarrow N_2H_4(g) + 2\ HCl(g)$

8.11 ►FUEL EFFICIENCY AND HEATS OF COMBUSTION

Surely the most familiar of all exothermic reactions is the one that takes place every time we turn up a thermostat, drive a car, or light a match—the burning, or **combustion**, of a fuel by reaction with oxygen to yield H_2O, CO_2, and heat. With the exception of hydrogen itself, all common fuels are organic compounds, whose energy derives ultimately from the sun through the photosynthesis of carbohydrates in green plants.

Though the details are complex, the net result of the photosynthesis reaction is the conversion of carbon dioxide and water into glucose, $C_6H_{12}O_6$. Glucose, once formed, is converted into cellulose and starch, which in turn act as structural materials for plant life and as food sources for animals. The conversion is highly endothermic and therefore requires a large input of solar energy.

$$6\ CO_2(g) + 6\ H_2O(l) \rightarrow C_6H_{12}O_6(s) + 6\ O_2(g) \qquad \Delta H° = 2816\ \text{kJ}$$

It has been estimated that the total amount of solar energy absorbed by the earth's vegetation is approximately 10^{19} kJ each year, an amount sufficient to synthesize 5×10^{14} kg of glucose per year.

Those fuels we use most—coal, natural gas, and petroleum—are the decayed remains of organisms from previous geologic eras. Both coal and petroleum are enormously complex mixtures of compounds. Coal is primarily of vegetable origin, and many of the compounds it contains are structurally similar to graphite (pure carbon). Petroleum is a viscous liquid mixture of *hydrocarbons*—compounds of carbon and hydrogen—that are primarily of marine origin.

Much coal lies near the surface and is obtained by strip mining.

Coal is burned directly as it comes from the mine, but petroleum must be *refined* before use. Refining involves the distillation of crude oil into fractions that are separated on the basis of their boiling points. So-called straight-run gasoline (bp 30–200°C) consists of compounds with 5 to 11 carbon atoms per molecule; kerosene (bp 175–300°C) contains compounds in the C_{11}–C_{14} range; gas oil (bp 275–400°C) contains C_{14}–C_{25} substances; and lubricating oils contain whatever remaining compounds will distill. Left over is an undistillable tarry residue of asphalt (Figure 8.13).

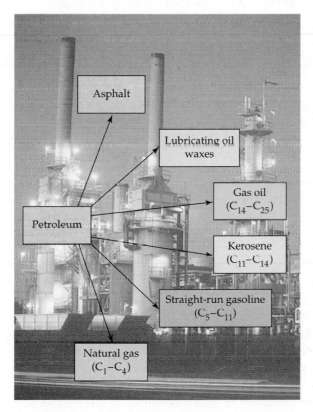

The amount of energy released on burning a substance is called its **heat of combustion**, or *combustion enthalpy*, ΔH_c°. It is simply the standard enthalpy change for the reaction of the substance with oxygen. Hydrogen, for example, has $\Delta H_c^\circ = -285.8$ kJ, and methane has $\Delta H_c^\circ = -890.3$ kJ. Note that the H_2O product is assumed to be liquid, since this is its standard-state form at 298 K.

$$H_2(g) + 1/2\ O_2(g) \rightarrow H_2O(l) \qquad \Delta H_c^\circ = -285.8 \text{ kJ}$$

$$CH_4(g) + 2\ O_2(g) \rightarrow CO_2(g) + 2\ H_2O(l) \qquad \Delta H_c^\circ = -890.3 \text{ kJ}$$

To compare the efficiency of different fuels, it's more useful to calculate combustion enthalpies per gram or per milliliter of substance rather than per mole (Table 8.3). For applications where weight is important, as in rocket engines, hydrogen is an ideal fuel because its combustion enthalpy per gram is the highest of any known fuel. For applications where volume is impor-

Hydrogen is an ideal fuel for rocket engines because of its high energy content per gram.

TABLE 8.3	Thermochemical Properties of Some Fuels		
	Combustion Enthalpy		
Fuel	**(kJ/mol)**	**(kJ/g)**	**(kJ/mL)**
Hydrogen, H_2	−285.8	−141.8	−9.9[a]
Ethanol, C_2H_5OH	−1367	−29.7	−23.4
Graphite, C	−393.5	−32.8	−73.8
Methane, CH_4	−890.3	−55.5	−30.8[a]
Methanol, CH_3OH	−726.4	−22.7	−17.9
Octane, C_8H_{18}	−5470	−47.9	−33.6
Toluene, C_7H_8	−3910	−42.3	−36.7

[a] Calculated for the compressed liquid at 0°C

tant, as in automobiles, a mixture of hydrocarbons such as those in gasoline is most efficient because hydrocarbon combustion enthalpies per milliliter are relatively high. Octane and toluene are representative examples.

As the world's petroleum deposits become ever more scarce, other sources of energy will have to be found to replace the hydrocarbon fuels we rely on. Hydrogen, though it burns cleanly and is relatively nonpolluting, has two drawbacks: low availability and low combustion enthalpy per mL. Ethanol and methanol look like the best current choices for alternative fuels because both can be produced relatively cheaply and have reasonable combustion enthalpies. Ethanol can be produced from wood by the breakdown of cellulose to glucose and subsequent fermentation. Methanol is produced directly from natural gas.

$$CH_4(g) + H_2O(g) \rightarrow CO(g) + 3\ H_2(g)$$

$$CO(g) + 2\ H_2(g) \rightarrow CH_3OH(l)$$

PROBLEM 8.18 Liquid butane (C_4H_{10}), the fuel used in many disposable lighters, has $\Delta H_f^\circ = -147.5$ kJ/mol and a density of 0.579 g/mL. Use Hess's law to calculate butane's enthalpy of combustion in kJ/mol, kJ/g, and kJ/mL.

8.12 ➤AN INTRODUCTION TO ENTROPY

We've said on several occasions that chemical and physical processes occur spontaneously only if they go "downhill" energetically so that the final state is more stable and has less energy than the initial state. In other words, energy must be *released* for a process to occur spontaneously. At the same time, though, we've said that some processes occur perfectly well, yet *absorb* heat. The endothermic reaction of barium hydroxide octahydrate with ammonium chloride shown in Figure 8.7, for example, absorbs 80.3 kJ of heat ($\Delta H^\circ = +80.3$ kJ) and leaves the surroundings so cold that water freezes around the outside of the flask.

$$Ba(OH)_2 \cdot 8H_2O(s) + 2\ NH_4Cl(s) \rightarrow$$
$$BaCl_2(aq) + 2\ NH_3(aq) + 10\ H_2O(l) \qquad \Delta H^\circ = +80.3\ kJ$$

What's going on? How can the spontaneous reaction of barium hydroxide octahydrate with ammonium chloride *release* energy yet *absorb* heat? The answer is that, in the context of a chemical reaction, the words "energy" and "heat" don't refer to exactly the same thing. There is some other factor in addition to heat that determines whether or not "energy" is released and thus determines whether or not a reaction takes place spontaneously. We'll just take a brief look at this additional factor at the moment and return for a more in-depth study in Chapter 17.

Before exploring the situation further, it's important to understand what the word *spontaneous* means in chemistry, for it's not the same as in everyday language. In chemistry, a **spontaneous** process is one that proceeds on its own without any continuous external influence. The change need not happen quickly, like a spring uncoiling or a rock rolling downhill. It can also happen slowly, like the gradual rusting away of an abandoned car. A *nonspontaneous* process, by contrast, takes place only in the presence of some continuous external influence. Energy must be continuously expended to re-coil a spring or to push a rock uphill. In general, the reverse of a spontaneous process is always nonspontaneous.

As another example of a process that takes place spontaneously yet absorbs heat, think about what happens when an ice cube melts. At a temperature slightly above 0°C, the ice spontaneously absorbs heat from the surroundings to turn from solid into liquid water without changing temperature.

What do the melting ice cube and the reaction of barium hydroxide octahydrate have in common? *The common feature of these and all other spontaneous processes that absorb heat is an increase in molecular disorder or randomness of the system.* The eight water molecules rigidly held in the $Ba(OH)_2 \cdot 8H_2O$ crystal break loose and become free to move about in the aqueous liquid product; similarly, the rigidly held H_2O molecules in the ice lose their crystalline ordering and move around with much more freedom in liquid water.

The amount of molecular disorder or randomness in a system is called the system's **entropy**, denoted S. Entropy has the units J/K (*joules* per kelvin, not kilojoules) and is a quantity that can be determined for pure substances (Section 17.3). The larger the value of S, the greater the disorder or randomness of the particles in the substance. Gases, for example have more randomness and higher entropy than liquids, and liquids have more randomness and higher entropy than solids.

Sledding downhill is a spontaneous process that continues once started. Hauling the sled back uphill is a nonspontaneous process that requires a continuous input of energy.

Molecular disorder increases when solid ice melts to liquid water.

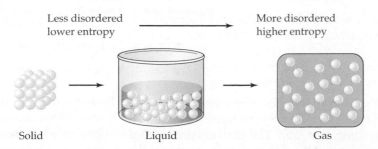

Less disordered
lower entropy   ⟶   More disordered
higher entropy

Solid   Liquid   Gas

Changes in entropy are represented by ΔS. When randomness increases, as it does when barium hydroxide octahydrate reacts or ice melts,

ΔS has a positive value because disorder has come into the system. The reaction of $Ba(OH)_2 \cdot 8H_2O(s)$ with $NH_4Cl(s)$ has $\Delta S° = +428$ J/K, and the melting of ice has $\Delta S° = +22.0$ J/(K · mol). When randomness decreases, as it does when water freezes, ΔS is negative because the system has become less disordered. The freezing of water, for example, has $\Delta S° = -22.0$ J/(K · mol). (As with $\Delta H°$, the superscript ° is used in $\Delta S°$ to refer to the standard entropy change in a reaction where all products and reactants are in their standard states.)

$$\Delta S = S_{final} - S_{initial}$$

If ΔS is positive, then the system has become more disordered

If ΔS is negative, then the system has become less disordered

The tendency of a system to increase its randomness (positive ΔS) is the second important factor in determining the spontaneity of a chemical or physical change in addition to ΔH. To decide whether a process is spontaneous, *both* enthalpy and entropy changes must be taken into account:

Spontaneous process:	favored by decrease in H (negative ΔH)
	favored by increase in S (positive ΔS)
Nonspontaneous process:	favored by increase in H (positive ΔH)
	favored by decrease in S (negative ΔS)

Clearly, the two factors don't always have to operate in the same direction. It's possible for a process to be *disfavored* by enthalpy (endothermic, positive ΔH) yet still be spontaneous because it is strongly *favored* by entropy (positive ΔS). The melting of ice [$\Delta H° = +6.01$ kJ/mol; $\Delta S° = +22.0$ J/(K · mol)] is just such a process, as is the reaction of barium hydroxide octahydrate with ammonium chloride ($\Delta H° = +80.3$ kJ; $\Delta S° = +428$ J/K). In the latter case, 3 mol of solid reactants produce 10 mol of liquid water, 2 mol of dissolved ammonia, and 3 mol of dissolved ions (1 mol of Ba^{2+} and 2 mol of Cl^-), with a consequent large increase in the amount of molecular disorder:

$$\underbrace{Ba(OH)_2 \cdot 8\ H_2O(s) + 2\ NH_4Cl(s)}_{\text{3 mol solid reactants}} \rightarrow \underset{\substack{\text{3 mol} \\ \text{dissolved ions}}}{BaCl_2(aq)} + \underset{\substack{\text{2 mol dissolved} \\ \text{molecules}}}{2\ NH_3(aq)} + \underset{\substack{\text{10 mol} \\ \text{liquid} \\ \text{water molecules}}}{10\ H_2O(l)}$$

$$\Delta H° = +80.3 \text{ kJ} \leftarrow \text{unfavorable}$$

$$\Delta S° = +428 \text{ J/K} \leftarrow \text{favorable}$$

It's also possible for a process to be favored by enthalpy (exothermic, negative ΔH) yet be nonspontaneous because it is strongly disfavored by entropy (negative ΔS). The conversion of liquid water to ice is nonspontaneous above 0°C, for example, because the process is disfavored by entropy ($\Delta S° = -22.0$ J/mol) even though it is favored by enthalpy ($\Delta H° = -6.01$ kJ/mol).

EXAMPLE 8.11

Predict whether $\Delta S°$ is likely to be positive or negative for each of the following reactions.

(a) $H_2C{=}CH_2(g) + Br_2(g) \longrightarrow CH_2BrCH_2Br(l)$
(b) $2\ C_2H_6(g) + 7\ O_2(g) \longrightarrow 4\ CO_2(g) + 6\ H_2O(g)$

SOLUTION   **(a)** The amount of disorder in the system decreases when two reactant gas molecules combine to give one liquid product molecule, so the reaction has a negative $\Delta S°$.
(b) The amount of disorder in the system increases when nine reactant gas molecules give ten product gas molecules, so the reaction has a positive $\Delta S°$.

PROBLEM 8.19   Ethane, C_2H_6, can be prepared by the reaction of acetylene, C_2H_2, with hydrogen:

$$C_2H_2(g) + 2\ H_2(g) \rightarrow C_2H_6(g)$$

Would you expect $\Delta S°$ for the reaction to be positive or negative? Explain.

8.13 ➤ AN INTRODUCTION TO FREE ENERGY

How do we weigh the different contributions of heat (enthalpy) and randomness (entropy) to the overall spontaneity of a process? To take both factors into account when deciding about the spontaneity of a chemical reaction or other process, we define a quantity called the **Gibbs free-energy change (ΔG)**, $\Delta G = \Delta H - T\Delta S$.

$$\underset{\substack{\text{free-energy}\\\text{change}}}{\Delta G} - \underset{\substack{\text{heat of}\\\text{reaction}}}{\Delta H} - \underset{\substack{\text{temperature}\\\text{(in kelvins)}}}{T}\,\underset{\substack{\text{entropy}\\\text{change}}}{\Delta S}$$

The value of the free-energy change ΔG is a general criterion for the spontaneity of a chemical reaction or physical process. If ΔG has a negative value, the process is spontaneous; if ΔG has a value of 0, the process is at equilibrium and is neither spontaneous nor nonspontaneous; and if ΔG has a positive value, the process is nonspontaneous:

$$\Delta G < 0 \qquad \text{process is spontaneous}$$

$$\Delta G = 0 \qquad \text{process is at equilibrium}$$

$$\Delta G > 0 \qquad \text{process is nonspontaneous}$$

Figure 8.14 summarizes the possible combinations of ΔH and $-T\Delta S$ to give ΔG.

The fact that the $T\Delta S$ term in the Gibbs free-energy equation is temperature dependent implies that some processes might be either spontaneous or nonspontaneous depending on the temperature. For example, the melting of ice is spontaneous above 0°C because the increase in $T\Delta S$ (posi-

FIGURE 8.14 The possible combinations of ΔH and $-T\Delta S$. A blue down arrow indicates that a process is favored, and a red up arrow indicates that a process is not favored. Processes (a–c) are spontaneous: (a) Both ΔH and $-T\Delta S$ are favorable; (b) a favorable ΔH term outweighs an unfavorable $-T\Delta S$ term; (c) a favorable $-T\Delta S$ term outweighs an unfavorable ΔH term. Processes (d–e) are at equilibrium: (d) a favorable ΔH term is balanced by an unfavorable $-T\Delta S$ term; (e) an unfavorable ΔH term is balanced by a favorable $-T\Delta S$ term. Processes (f–h) are nonspontaneous: (f) both ΔH and $-T\Delta S$ are unfavorable; (g) a favorable ΔH term is outweighed by an unfavorable $-T\Delta S$ term; (h) a favorable $-T\Delta S$ term is outweighed by an unfavorable ΔH term.

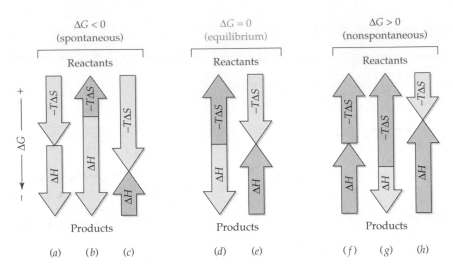

tive ΔS) of liquid over solid water outweighs the unfavorable heat absorption (endothermic, positive ΔH) during the process. Freezing is spontaneous below 0°C, however, because the release of heat (exothermic, negative ΔH) outweighs the decrease in $T\Delta S$ (negative ΔS).

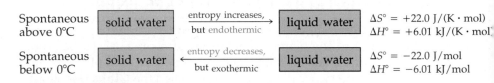

| Spontaneous above 0°C | solid water | entropy increases, but endothermic $\longrightarrow$ | liquid water | $\Delta S° = +22.0 \text{ J/(K·mol)}$ $\Delta H° = +6.01 \text{ kJ/(K·mol)}$ |
| Spontaneous below 0°C | solid water | $\longleftarrow$ entropy decreases, but exothermic | liquid water | $\Delta S° = -22.0 \text{ J/mol}$ $\Delta H° = -6.01 \text{ kJ/mol}$ |

The melting of ice is favored by entropy but disfavored by enthalpy. The freezing of water is disfavored by entropy but favored by enthalpy. The entropy and enthalpy terms for both processes balance each other at 0°C.

As a general rule, the enthalpy term (ΔH) usually dominates the entropy term ($T\Delta S$) at low temperature and controls spontaneity, but at higher temperature, the $T\Delta S$ term can dominate. Thus, an endothermic process that is nonspontaneous at low temperature can become spontaneous at higher temperature. This is exactly what happens in the ice/water transition. Below 0°C, nothing happens to the ice because the unfavorable ΔH term dominates the favorable $T\Delta S$ term, making ΔG positive. Exactly at 0°C, the two terms are balanced, making $\Delta G = 0$. Above 0°C, the ice spontaneously melts because the favorable $T\Delta S$ term now dominates the unfavorable ΔH term, making ΔG negative.

$$\Delta G = \Delta H - T\Delta S$$

At $-10°C$ (263 K): $\Delta G = 6.01 \dfrac{kJ}{mol} - (263\ K)\left(0.0220\dfrac{kJ}{K\cdot mol}\right) = +0.22\ kJ/mol$

At $0°C$ (273 K): $\Delta G = 6.01 \dfrac{kJ}{mol} - (273\ K)\left(0.0220\dfrac{kJ}{K\cdot mol}\right) = 0.00\ kJ/mol$

At $+10°C$ (283 K): $\Delta G = 6.01 \dfrac{kJ}{mol} - (283\ K)\left(0.0220\dfrac{kJ}{K\cdot mol}\right) = -0.22\ kJ/mol$

An example of a chemical reaction in which temperature controls spontaneity is that of carbon with water to yield carbon monoxide. The reaction has an unfavorable ΔH term (positive) but a favorable $T\Delta S$ term (positive) because randomness increases when a solid and a gas are converted into two gases:

$$C(s) + H_2O(g) \rightarrow CO(g) + H_2(g) \qquad \begin{aligned} \Delta H° &= +131\ kJ \\ \Delta S° &= +134\ J/K \end{aligned}$$

No reaction occurs if carbon and water are mixed at room temperature (300 K) because the unfavorable ΔH term dominates the favorable $T\Delta S$ term. If the reaction is carried out at a very high temperature, however, it becomes spontaneous because the favorable $T\Delta S$ term begins to dominate the unfavorable ΔH term at approximately 978 K (705°C). Below 978 K, ΔG has a positive value; at 978 K, $\Delta G = 0$; and above 978 K, ΔG has a negative value:

$$\Delta G = \Delta H - T\Delta S$$

At 695°C (968 K): $\Delta G = 131\ kJ - (968\ K)\left(0.134\ \dfrac{kJ}{K}\right) = +1\ kJ$

At 705°C (978 K): $\Delta G = 131\ kJ - (978\ K)\left(0.134\ \dfrac{kJ}{K}\right) = 0\ kJ$

At 715°C (988 K): $\Delta G = 131\ kJ - (988\ K)\left(0.134\ \dfrac{kJ}{K}\right) = -1\ kJ$

The reaction of carbon with water is, in fact, the first step of an industrial process used to manufacture methanol (CH_3OH). As supplies of natural gas and oil are used, this reaction may become important for the manufacture of synthetic fuels.

A process is at equilibrium when it is balanced between spontaneous and nonspontaneous—that is, when $\Delta G = 0$ and it is energetically unfavorable to go either way. Thus, at the equilibrium point, we can set up the equation

$$\Delta G = \Delta H - T\Delta S = 0 \quad \text{at equilibrium}$$

Solving this equation for T gives

$$T = \dfrac{\Delta H}{\Delta S}$$

which makes it possible to calculate the temperature at which a changeover in behavior between spontaneous and nonspontaneous occurs. Using the known values of ΔH and ΔS for the melting of ice, for instance, we find that the temperature at which liquid water and solid ice are in equilibrium is 273 K or 0°C:

$$T = \frac{\Delta H}{\Delta S} = \frac{6.01 \text{ kJ}}{0.0220 \text{ kJ/K}} = 273 \text{ K} = 0°C$$

(It's no surprise, of course, to find that the ice/water equilibrium point is 0°C, the melting point of ice.)

In the same way, the temperature at which the reaction of carbon with water changes between spontaneous and nonspontaneous is 978 K, or 705°C:

$$T = \frac{\Delta H}{\Delta S} = \frac{131 \text{ kJ}}{0.134 \text{ kJ/K}} = 978 \text{ K}$$

Below 978 K the reaction of carbon with water is nonspontaneous, at 978 K the reaction is at equilibrium, and above 978 K the reaction is spontaneous.

EXAMPLE 8.12

Quicklime, CaO, is produced by heating limestone, $CaCO_3$, to drive off CO_2 gas. Is the reaction spontaneous under standard-state conditions? Calculate the temperature at which the reaction becomes spontaneous.

$$CaCO_3(s) \rightarrow CaO(s) + CO_2(g) \qquad \Delta H° = 178.3 \text{ kJ}; \Delta S° = 160 \text{ J/K}$$

SOLUTION   The spontaneity of the reaction at a given temperature can be found by determining whether ΔG is positive or negative. At 25°C (298 K), we have:

$$\Delta G = \Delta H - T\Delta S = 178.3 \text{ kJ} - (298 \text{ K})\left(0.160 \frac{\text{kJ}}{\text{K}}\right) = +130.6 \text{ kJ}$$

Since ΔG is positive at this temperature, the reaction is nonspontaneous.

The changeover point between spontaneous and nonspontaneous can be found by setting $\Delta G = 0$ and rearranging the equation:

$$T = \frac{\Delta H}{\Delta S} = \frac{178.3 \text{ kJ}}{0.160 \text{ kJ/K}} = 1114 \text{ K}$$

The reaction becomes spontaneous at 1114 K (841°C).

⌐ **PROBLEM 8.20** Which of the following reactions are spontaneous under standard-state conditions, and which are nonspontaneous?
(a) $AgNO_3(aq) + NaCl(aq) \rightarrow AgCl(s) + NaNO_3(aq)$ $\qquad \Delta G° = -55.7 \text{ kJ}$
(b) $2 \text{ C}(s) + 2 \text{ H}_2(g) \rightarrow C_2H_4(g)$ $\qquad \Delta G° = 68.1 \text{ kJ}$

⌐ **PROBLEM 8.21** Is the Haber process for the industrial synthesis of ammonia spontaneous or nonspontaneous under standard-state conditions? At what temperature in °C does the changeover occur?

$$N_2(g) + 3 \text{ H}_2(g) \rightarrow 2 \text{ NH}_3(g) \qquad \Delta H° = -92.2 \text{ kJ}; \Delta S° = -199 \text{ J/K}$$

THE EVOLUTION OF ENDOTHERMY—*interlude*

Most people cringe at the thought of a cold shower. Imagine what it must be like for fish, which spend their lives immersed in bone-chilling water. Water is such an efficient heat conductor and cools bodies so rapidly that most fish don't even try to keep warm: Of the approximately 30,000 known species of bony fishes, all but a handful are coldblooded. Only tunas, mackerel, billfish (marlin, swordfish), and a few others are warmblooded.

The active lives of mammals and birds require a high metabolic rate, which these animals achieve by capturing the energy released during exo-thermic metabolic reactions and using that energy to maintain a high body temperature. Fish, however, function at a much lower metabolic rate than mammals and are therefore able to live with a lower, more variable body temperature. Thus, from an evolutionary point of view, the development of warmbloodedness (endothermy) in a few fish species is an extraordinary occurrence, and the reasons for that development are not fully understood. Recently obtained evidence indicates that endothermy has evolved independently in three different fish lineages, probably as a means for the fish to expand their habitat into colder water. The warmblooded bluefin tuna, for example, migrates annually between tropical and polar waters, and is at home in both.

The different ways in which endothermy has developed in different fishes may provide valuable clues about how it developed in mammals and birds. Most important, it appears that warmbloodedness is not necessarily an all-or-nothing proposition; it may well have evolved in steps. Both billfish and the butterfly mackerel, for instance, keep only their eyes and brains warm; the rest of their bodies are coldblooded. Both species accomplish their partial warmbloodedness through the use of a special heater muscle in the eye. The muscle operates at a high metabolic rate and acts as a heat exchanger to warm the blood passing though it. A special network of blood vessels then distributes this heated blood directly to the eye and brain.

Tunas, by contrast, have evolved a mechanism of endothermy completely different from that of mackerel and billfish. In most fish, the powerful aerobic muscle used for swimming is located just beneath the skin, but in tuna, the muscle is positioned centrally in the body, where it is thermally insulated from the outside water. The heat produced by this muscle during swimming is captured by circulating blood and transported to the cranium through the vascular system. Exactly how all this information about fish relates to the development of endothermy in mammals is not yet clear, but it appears that warmbloodedness is a much more complex process than once believed.

Before being caught, this marlin was one of only a handful of fish in the ocean that are partially warm-blooded.

KEY WORDS

SUMMARY

Energy is of two types: potential and kinetic. **Kinetic energy** (E_K) is the energy of motion; its value depends on both the mass m and velocity v of an object according to the equation $E_K = (1/2)mv^2$. **Potential energy** (E_P) is the energy stored in an object because of its position or in a chemical substance because of its composition. **Heat** is the energy transferred between two objects as the result of a temperature difference, whereas **temperature** is a measure of the kinetic energy of molecular motion. According to the **law of conservation of energy**, also known as the **first law of thermodynamics**, energy can be neither created nor destroyed, and the total energy of an isolated system must therefore remain constant. The total **internal energy** (E) of a system—the sum of all kinetic and potential energies for each particle in the system—is a **state function** because its value depends only on the present condition of the system, not on how that condition was reached.

Work (w) is defined as the distance moved times the force that opposes the motion. In chemistry, most work is expansion work (PV work) done as the result of a volume change during a reaction when air molecules have to be moved aside. The amount of work done by an expanding gas is given by the equation $w = -P\Delta V$, where P is the pressure on the system and ΔV is the change in volume of the system.

The total amount of internal energy change that takes place during a reaction is the sum of the heat transferred (q) and the work done ($-P\Delta V$). The equation

$$\Delta E = q + (-P\Delta V) \quad \text{or} \quad q = \Delta E + P\Delta V = \Delta H$$

where ΔH is called the **enthalpy change** of the system, is one of the fundamental equations of thermochemistry. In general, the $P\Delta V$ term is much smaller than the ΔE term so that the total internal energy change of a reacting system is approximately equal to ΔH, also called the **heat of reaction**. Reactions that have a negative ΔH are said to be **exothermic** because heat is leaving the system; reactions that have a positive ΔH are said to be **endothermic** because heat is being absorbed by the system.

Since ΔH is a state function, its value is the same no matter what path is taken between starting reactants and final products. Thus, the sum of the enthalpy changes for the individual steps in a reaction is the same as the overall enthalpy change for the entire reaction, a relationship known as **Hess's law**. Using this law, it is possible to calculate overall enthalpy changes for individual steps that can't be measured directly. Hess's law also makes it possible to calculate the enthalpy change of any reaction if the standard heats of formation (ΔH_f°) are known for the reactants and products. The **standard heat of formation** is the enthalpy change for the hypothetical formation of 1 mol of a substance in its standard state from the most stable form of the constituent elements in their **standard states** (1 atm and 298 K).

In addition to enthalpy, **entropy** (S)—a measure of the amount of molecular randomness in a substance—is also important in determining whether or not a given process will occur spontaneously. Together, changes in enthalpy and entropy define a quantity called the **Gibbs free-energy change** (ΔG) according to the equation $\Delta G = \Delta H - T\Delta S$. The value of ΔG is a general criterion for whether or not a reaction will take place spontaneously. If ΔG is negative, the reaction is spontaneous; if ΔG is positive, the reaction is nonspontaneous.

UNDERSTANDING KEY CONCEPTS

1. Imagine a reaction that results in a change in both volume and temperature:

$-P(V_2 - V_1)$

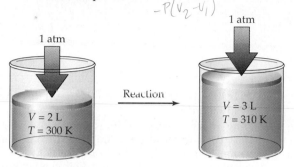

(a) Has any work been done? If so, is its sign positive or negative?
(b) Has there been an enthalpy change? If so, what is the sign of ΔH? Is the reaction exothermic or endothermic?

2. Redraw the following diagram to represent the product mixture when (a) work has been done *on* the system and (b) when work has been done *by* the system.

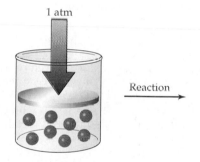

3. Acetylene, C_2H_2, reacts with H_2 in two steps to yield ethane, CH_3CH_3:

1. $HC\equiv CH + H_2 \rightarrow CH_2=CH_2 \quad \Delta H° = -174.4 \text{ kJ}$
2. $CH_2=CH_2 + H_2 \rightarrow CH_3CH_3 \quad \Delta H° = -137.0 \text{ kJ}$

Net $HC\equiv CH + 2 H_2 \rightarrow CH_3CH_3 \quad \Delta H° = -311.4 \text{ kJ}$

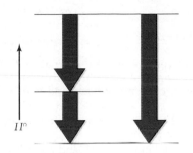

Hess's law diagram

Which arrow in the Hess's law diagram corresponds to which step, and which arrow corresponds to the net reaction? Where are the reactants located on the diagram, and where are the products located?

4. Draw a Hess's law diagram similar to that in Problem 3 for the reaction of ethyl alcohol (CH_3CH_2OH) with oxygen to yield acetic acid (CH_3COOH).

1. $CH_3CH_2OH + 1/2 O_2 \rightarrow CH_3CHO + H_2O \quad \Delta H° = -200.4 \text{ kJ}$
2. $CH_3CHO + 1/2 O_2 \rightarrow CH_3COOH \quad \Delta H° = -292.2 \text{ kJ}$

Net $CH_3CH_2OH + O_2 \rightarrow CH_3COOH + H_2O \quad \Delta H° = -492.6 \text{ kJ}$

5. Tell whether the entropy changes for the following processes are likely to be positive or negative.
(a) The fizzing of a newly opened can of soda
(b) The growth of a plant from seed

6. When a bottle of perfume is opened, odor-causing molecules mix with air and slowly diffuse throughout the entire room. Is ΔG for the diffusion process positive, negative, or zero? What about ΔH and ΔS for the diffusion?

ADDITIONAL PROBLEMS

Problems 8.1–8.21 appear within the chapter.

HEAT, WORK, AND ENERGY

8.22 What is the difference between heat and temperature? Between work and energy? Between kinetic energy and potential energy?

8.23 Which has more kinetic energy, a 1400 kg car moving at 115 km/h or a 12,000 kg truck moving at 38 km/h?

8.24 Assume that the kinetic energy of a 1400 kg car moving at 115 km/h (Problem 8.23) could be con-verted entirely into heat. What amount of water could be heated from 20°C to 50°C by the car's energy?

8.25 Calculate (in joules) the work done by a chemical reaction if the volume increases from 3.2 L to 3.4 L against a constant external pressure of 3.6 atm. What is the sign of the energy change?

8.26 The addition of H_2 to carbon–carbon double bonds is an important reaction used in the preparation of margarine from vegetable oils. If 50.0 mL of H_2 and 50.0 mL of ethylene (C_2H_4) are allowed to react at 1.5 atm, the product ethane (C_2H_6) has a volume of 50.0 mL. Calculate the amount of *PV* work done and tell the direction of the energy flow.

$$C_2H_4(g) + H_2(g) \rightarrow C_2H_6(g)$$

ENERGY AND ENTHALPY

8.27 What is the difference between internal energy change ΔE and enthalpy change ΔH? Which of the two is measured at constant pressure and which at constant volume?

8.28 What is the sign of ΔH for an exothermic reaction? For an endothermic reaction?

8.29 Under what circumstances are ΔE and ΔH nearly equal?

8.30 The enthalpy change for the reaction of 50.0 mL of ethylene with 50.0 mL of H_2 at 1.5 atm pressure (Problem 8.26) is $\Delta H = -0.31$ kJ. What is the value of ΔE?

8.31 Assume that a particular reaction evolves 244 kJ of heat and that 35 kJ of *PV* work is done on the system. What are the values of ΔE and ΔH for the system? For the surroundings?

8.32 Which of the following has the highest enthalpy content and which the lowest at a given temperature: $H_2O(s)$, $H_2O(l)$, or $H_2O(g)$? Explain.

8.33 Used in welding metals, the reaction of acetylene with oxygen has $\Delta H = -1255.5$ kJ.

$$C_2H_2(g) + 5/2\ O_2(g) \rightarrow$$
$$H_2O(g) + 2\ CO_2(g) \qquad \Delta H = -1255.5\ kJ$$

How much *P-V* work is done in kJ and what is the value of ΔE in kJ for the reaction of 6.50 g of acetylene at atmospheric pressure if the volume change is -2.80 L?

8.34 The familiar "ether" used as an anesthetic agent is diethyl ether, $C_4H_{10}O$. Its heat of vaporization is $+26.5$ kJ/mol at its boiling point. How much energy in kJ is required to convert 100 mL of diethyl ether at its boiling point from liquid to vapor if its density is 0.7138 g/mL?

8.35 How much energy in kJ is required to convert 100 mL of water at its boiling point from liquid to vapor, and how does this compare with the result calculated in Problem 8.34 for diethyl ether? [ΔH_{vap} (H_2O) = $+40.7$ kJ/mol]

8.36 Aluminum metal reacts with chlorine with a spectacular display of sparks.

$$2\ Al(s) + 3\ Cl_2(g) \rightarrow 2\ AlCl_3(s) \qquad \Delta H° = -1408.4\ kJ$$

How much heat in kJ is released on reaction of 5.00 g of Al?

8.37 How much heat in kJ is evolved or absorbed in the reaction of 1.00 g of Na with H_2O? Is the reaction exothermic or endothermic?

$$2\ Na(s) + 2\ H_2O(l) \rightarrow$$
$$2\ NaOH(aq) + H_2(g) \qquad \Delta H° = -368.4\ kJ$$

8.38 How much heat in kJ is evolved or absorbed in the following reactions? Identify each reaction as either exothermic or endothermic.

(a) Reaction of 2.50 g of Fe_2O_3 with enough carbon monoxide to produce iron metal:

$$Fe_2O_3(s) + 3\ CO(g) \rightarrow$$
$$2\ Fe(s) + 3\ CO_2(g) \qquad \Delta H° = -24.8\ kJ$$

(b) Reaction of 233.0 g of calcium oxide with enough carbon to produce calcium carbide:

$$CaO(s) + 3\ C(s) \rightarrow$$
$$CaC_2(s) + CO(g) \qquad \Delta H° = 464.8\ kJ$$

CALORIMETRY AND HEAT CAPACITY

8.39 What is the difference between heat capacity and specific heat?

8.40 Sodium metal is sometimes used as a cooling agent in heat exchange units because of its relatively high molar heat capacity of 28.2 J/(mol · °C). What is the specific heat of sodium in J/(g · °C)?

8.41 Titanium metal is used as a structural material in many high-tech applications such as in jet engines. What is the specific heat of titanium in J/(g · °C) if it takes 89.7 J to raise the temperature of a 33.0 g block by 5.20°C? What is the molar heat capacity of titanium in J/(mol · °C)?

8.42 When 1.045 g of CaO is added to 50.0 mL of water at 25.0°C in a calorimeter, the temperature of the water increases to 32.3°C. Assuming that the specific heat of the solution is 4.18 J/(g · °C), that its density is 1.00 g/mL, and that the calorimeter itself absorbs a negligible amount of heat, calculate ΔH for the reaction in kJ.

$$CaO(s) + H_2O(l) \rightarrow Ca(OH)_2(aq)$$

HESS'S LAW AND HEATS OF FORMATION

8.43 How is the standard state of an element defined?

8.44 What is a compound's standard heat of formation?

8.45 What is Hess's law, and why does it "work"?

8.46 The standard heat of formation of CuO(s) is $\Delta H_f^\circ = -157$ kJ. How much energy in kJ is needed to convert 2.00 mol of CuO into Cu + O_2 under standard-state conditions?

8.47 Sulfuric acid (H_2SO_4), the most widely produced chemical in the world, is made by a two step oxidation of sulfur to sulfur trioxide, SO_3, followed by reaction with water. Calculate ΔH_f° for SO_3 in kJ/mol, given the following data:

$$S(s) + O_2(g) \rightarrow SO_2(g) \qquad \Delta H^\circ = -296.8 \text{ kJ}$$

$$SO_2(g) + 1/2 \; O_2(g) \rightarrow SO_3(g) \qquad \Delta H^\circ = -98.9 \text{ kJ}$$

8.48 The standard enthalpy change for the reaction of $SO_3(g)$ with $H_2O(l)$ to yield $H_2SO_4(aq)$ is $\Delta H^\circ = -227.8$ kJ. Use the information in Problem 8.47 to calculate ΔH_f° for $H_2SO_4(aq)$ in kJ/mol. (For $H_2O(l)$, $\Delta H_f^\circ = -285.8$ kJ/mol.)

8.49 Draw a diagram of the sort used in Figure 8.11 to depict the energy changes in Problem 8.47.

8.50 Acetic acid (CH_3COOH), whose aqueous solutions are known as *vinegar*, is prepared by reaction of ethyl alcohol (CH_3CH_2OH) with oxygen:

$$CH_3CH_2OH(l) + O_2(g) \rightarrow CH_3COOH(l) + H_2O(l)$$

Use the following data to calculate ΔH° in kJ for the above reaction.

$$\Delta H_f^\circ(CH_3CH_2OH) = -277.7 \text{ kJ/mol}$$

$$\Delta H_f^\circ(CH_3COOH) = -484.5 \text{ kJ/mol}$$

$$\Delta H_f^\circ(H_2O) = -285.8 \text{ kJ/mol}$$

8.51 Calculate ΔH_f° for benzene, C_6H_6, in kJ/mol from the following data:

$$2 \; C_6H_6(l) + 15 \; O_2(g) \rightarrow$$
$$12 \; CO_2(g) + 6 \; H_2O(l) \qquad \Delta H^\circ = -6534 \text{ kJ}$$

$$\Delta H_f^\circ(CO_2) = -393.5 \text{ kJ/mol}$$

$$\Delta H_f^\circ(H_2O) = -285.8 \text{ kJ/mol}$$

8.52 Styrene (C_8H_8), the precursor of polystyrene polymers, has a standard heat of combustion of -4395.2 kJ/mol. Write a balanced equation for the combustion reaction, and calculate ΔH_f° for styrene in kJ/mol. [$\Delta H_f^\circ(CO_2) = -393.5$ kJ/mol; $\Delta H_f^\circ(H_2O) = -285.8$ kJ/mol]

8.53 Propene and cyclopropane have the same formula, C_3H_6. Propene has $\Delta H_f^\circ = 20.0$ kJ/mol, cyclopropane has $\Delta H_f^\circ = 53.3$ kJ/mol, and their densities are similar. Which of the two is the more efficient fuel? Explain your answer.

8.54 Methyl *tert*-butyl ether (MTBE), $C_5H_{12}O$, a gasoline additive used to boost octane ratings, has $\Delta H_f^\circ = -313.6$ kJ/mol. Write a balanced equation for its combustion reaction and calculate its standard heat of combustion in kJ.

8.55 Methyl *tert*-butyl ether (Problem 8.54) is prepared by reaction of methanol ($\Delta H_f^\circ = -238.7$ kJ/mol) with 2-methylpropene according to the equation

$$CH_3-\underset{\underset{CH_3}{|}}{C}=CH_2 + CH_3OH \rightarrow CH_3-\underset{\underset{CH_3}{|}}{\overset{\overset{CH_3}{|}}{C}}-O-CH_3$$

2-Methylpropene   Methyl *tert*-butyl ether

If $\Delta H^\circ = -57.8$ kJ for the reaction, calculate ΔH_f° in kJ/mol for 2-methylpropene.

8.56 One possible use for all the cooking fat left over after making french fries is to burn it as a fuel. Write a balanced equation and then use the following data to calculate the amount of energy released in kJ/mL from the combustion of cooking fat.

> Cooking fat: formula = $C_{51}H_{88}O_6$
> density = 0.94 g/mL
> $\Delta H_f^\circ = -1310$ kJ/mol

BOND DISSOCIATION ENERGIES

8.57 Use the information in Table 7.1 on page 231 to calculate an approximate ΔH° in kJ for the reaction of ethylene with hydrogen to yield ethane. (The strength of the C=C bond is 720 kJ/mol.)

$$H_2C=CH_2(g) + H_2(g) \rightarrow CH_3CH_3(g)$$

8.58 Use the bond dissociation energies in Table 7.1 to calculate an approximate ΔH° in kJ for the industrial synthesis of isopropyl alcohol (rubbing alcohol) by reaction of water with propylene. (The strength of the C=C bond is 720 kJ/mol.)

$$CH_3CH=CH_2 + H_2O \rightarrow CH_3\underset{\underset{}{|}}{\overset{\overset{OH}{|}}{C}}HCH_3$$

8.59 Calculate an approximate heat of combustion for butane in kJ by using the bond dissociation energies in Table 7.1. (The strength of the O=O bond is 498 kJ/mol, and that of a C=O bond in CO_2 is 804 kJ/mol.)

Butane

8.60 Use the bond dissociation energies in Table 7.1 to calculate an approximate heat of reaction, $\Delta H°$, in kJ for the industrial reaction of ethanol with acetic acid to yield ethyl acetate (nail-polish remover).

Acetic acid   Ethanol

Ethyl acetate

FREE ENERGY AND ENTROPY

8.61 What does entropy measure?

8.62 What are the two terms that make up the free-energy change for a reaction, ΔG, and which of the two is usually more important?

8.63 How is it possible for a reaction to be spontaneous yet endothermic?

8.64 Is it possible for a reaction to be nonspontaneous yet exothermic? Explain.

8.65 Tell whether the entropy changes for the following processes are likely to be positive or negative:

(a) The conversion of liquid water to water vapor at 100°C

(b) The freezing of liquid water to ice at 0°C

(c) The shuffling of an ordered deck of cards to give a random arrangement

(d) The eroding of a mountain by a glacier

8.66 One of the steps in the cracking of petroleum into gasoline involves the thermal breakdown of large hydrocarbon molecules into smaller ones. For example, the following reaction might occur:

$$C_{11}H_{24} \rightarrow C_4H_{10} + C_4H_8 + C_3H_6$$

Is ΔS for this reaction likely to be positive or negative? Explain.

8.67 The commercial production of dichloroethane, a solvent used in dry cleaning, involves the reaction of ethylene with chlorine:

$$C_2H_4(g) + Cl_2(g) \rightarrow C_2H_4Cl_2(l)$$

Is ΔS for this reaction likely to be positive or negative? Explain.

8.68 Tell whether reactions with the following values of ΔH and ΔS are spontaneous or nonspontaneous and whether they are exothermic or endothermic.

(a) $\Delta H = -48$ kJ; $\Delta S = +135$ J/K at 400 K

(b) $\Delta H = -48$ kJ; $\Delta S = -135$ J/K at 400 K

(c) $\Delta H = +48$ kJ; $\Delta S = +135$ J/K at 400 K

(d) $\Delta H = +48$ kJ; $\Delta S = -135$ J/K at 400 K

8.69 Tell whether reactions with the following values of ΔH and ΔS are spontaneous or nonspontaneous and whether they are exothermic or endothermic.

(a) $\Delta H = -128$ kJ; $\Delta S = 35$ J/K at 500 K

(b) $\Delta H = +67$ kJ; $\Delta S = -140$ J/K at 250 K

(c) $\Delta H = +75$ kJ; $\Delta S = 95$ J/K at 800 K

8.70 Which of the reactions (a)–(d) in Problem 8.68 is spontaneous at all temperatures, which is nonspontaneous at all temperatures, and which has a crossover temperature?

8.71 Suppose that a reaction has $\Delta H = -33$ kJ and $\Delta S = -58$ J/K. At what temperature will it change from spontaneous to nonspontaneous?

8.72 Suppose that a reaction has $\Delta H = +41$ kJ and $\Delta S = -27$ J/K. At what temperature, if any, will it change between spontaneous and nonspontaneous?

8.73 Vinyl chloride ($H_2C=CHCl$), the starting material used in the industrial preparation of polyvinyl chloride, is prepared by a two-step process that begins with the reaction of Cl_2 with ethylene to yield ethylene dichloride:

$$Cl_2(g) + H_2C=CH_2(g) \rightarrow ClCH_2CH_2Cl(l)$$
$$\Delta H° = -217.5 \text{ kJ/mol}$$
$$\Delta S° = -233.9 \text{ J/(K} \cdot \text{mol)}$$

(a) Tell whether the reaction is favored by entropy, by enthalpy, by both, or by neither, and then calculate $\Delta G°$ at 298 K.

(b) Tell whether the reaction has a crossover temperature between spontaneous and nonspontaneous. If yes, calculate the crossover temperature.

GENERAL PROBLEMS

8.74 When 1.50 g of magnesium metal is allowed to react with 200 mL of 6.00 M aqueous HCl, the temperature rises from 25.0°C to 42.9°C. Calculate ΔH in kJ for the reaction, assuming that the heat capacity of the calorimeter is 776 J/°C, that the specific heat of the final solution is the same as that of water [(4.18 J/(g · °C)], and that the density of the solution is 1.00 g/mL.

8.75 Use the data in Appendix B to find standard enthalpies of reaction in kJ for the following processes.
(a) $C(s) + CO_2(g) \rightarrow 2\ CO(g)$
(b) $2\ H_2O_2(l) \rightarrow 2\ H_2O(l) + O_2(g)$
(c) $Fe_2O_3(s) + 3\ CO(g) \rightarrow 2\ Fe(s) + 3\ CO_2(g)$

8.76 Find $\Delta H°$ in kJ for the reaction of nitric oxide with oxygen

$$2\ NO(g) + O_2(g) \rightarrow N_2O_4(g)$$

given the following thermochemical data:
$N_2O_4(g) \rightarrow 2\ NO_2(g) \qquad \Delta H° = 57.20\ kJ$
$NO(g) + 1/2\ O_2(g) \rightarrow NO_2(g) \qquad \Delta H° = -57.0\ kJ$

8.77 The boiling point of a substance is defined as the temperature at which liquid and gas exist in equilibrium. Use the heat of vaporization (ΔH_{vap} = 30.91 kJ/mol) and entropy of vaporization [ΔS_{vap} = 93.2 J/(K · mol)] to calculate the boiling point of liquid bromine in °C.

8.78 What is the melting point of benzene in K if ΔH_{fusion} = 9.95 kJ/mol and ΔS_{fusion} = 35.7 J/(K · mol)?

8.79 Metallic mercury is obtained by heating the mineral cinnabar (HgS) in air:

$$HgS(s) + O_2(g) \rightarrow Hg(l) + SO_2(g)$$

(a) Use the data in Appendix B to calculate $\Delta H°$ in kJ for the reaction.
(b) The entropy change for the reaction is $\Delta S°$ = +36.7 J/K. Is the reaction spontaneous at 25°C?
(c) Under what conditions, if any, is the reaction nonspontaneous? Explain.

8.80 Use the average bond dissociation energies in Table 7.1 to calculate approximate reaction enthalpies in kJ for the following processes.
(a) $2\ CH_4(g) \rightarrow C_2H_6(g) + H_2(g)$
(b) $C_2H_6(g) + F_2(g) \rightarrow C_2H_5F(g) + HF(g)$
(c) $N_2(g) + 3\ H_2(g) \rightarrow 2\ NH_3(g)$

8.81 Methanol (CH_3OH) is made industrially in two steps from CO and H_2. So cheap is methanol to make that it is being considered for use as a precursor to hydrocarbon fuels such as methane (CH_4):

Step 1. $CO(g) + 2\ H_2(g) \rightarrow CH_3OH(l)$
$$\Delta S° = -332\ J/K$$

Step 2. $CH_3OH(l) \rightarrow CH_4(g) + 1/2\ O_2(g)$
$$\Delta S° = 162\ J/K$$

(a) Calculate $\Delta H°$ in kJ for step 1.
(b) Calculate $\Delta G°$ in kJ for step 1.
(c) Is step 1 spontaneous at 298 K?
(d) Which term is more important, $\Delta H°$ or $\Delta S°$?
(e) In what temperature range is reaction 1 spontaneous?
(f) Calculate $\Delta H°$ for step 2.
(g) Calculate $\Delta G°$ for step 2.
(h) Is reaction 2 spontaneous at 298 K?
(i) Which term is more important, $\Delta H°$ or $\Delta S°$?
(j) In what temperature range is step 2 spontaneous?
(k) Calculate an overall $\Delta G°$, $\Delta H°$, and $\Delta S°$ for the formation of CH_4 from CO and H_2.
(l) Is the overall reaction spontaneous?
(m) If you were designing a production facility, would you plan on carrying out the reactions in separate steps or together? Explain.

8.82 Isooctane, C_8H_{18}, is the component of gasoline from which the term *octane rating* derives.
(a) Write a balanced equation for the combustion of isooctane with O_2 to yield $CO_2(g)$ and $H_2O(g)$.
(b) The standard molar heat of combustion for isooctane is −5456.6 kJ/mol. Calculate $\Delta H_f°$ for isooctane.

8.83 We said in Section 8.2 that the potential energy of water at the top of a dam or waterfall is converted into heat when the water dashes against rocks at the bottom. The potential energy of the water at the top is equal to $E_P = m \cdot g \cdot h$, where m is the mass of the water, g is the acceleration of the falling water due to gravity (g = 9.81 m/s^2), and h is the height of the water. Assuming that all the energy is converted to heat, calculate the temperature rise of the water in °C after falling over California's Yosemite Falls, a distance of 739 m. The specific heat of water is 4.18 J/(g · K).

chapter 9 GASES: THEIR PROPERTIES AND BEHAVIOR

A quick look around tells us that matter takes different forms. Most of the things around us are *solids*, substances whose constituent atoms, molecules, or ions are held rigidly together in a definite way, giving the solid a definite volume and shape. Other substances are *liquids*, whose constituent atoms or molecules are held together only weakly, giving the liquid a definite volume but a changeable and indefinite shape. Still other substances are *gases*, whose constituent atoms or molecules have little attraction for one another and are therefore free to move about in whatever volume is available.

Though few in number—only about a hundred substances are gases at room temperature—early studies of gases were enormously important in the historical development of chemical theories. We'll look briefly at this historical development in the present chapter and will see how the behavior of gases can be described.

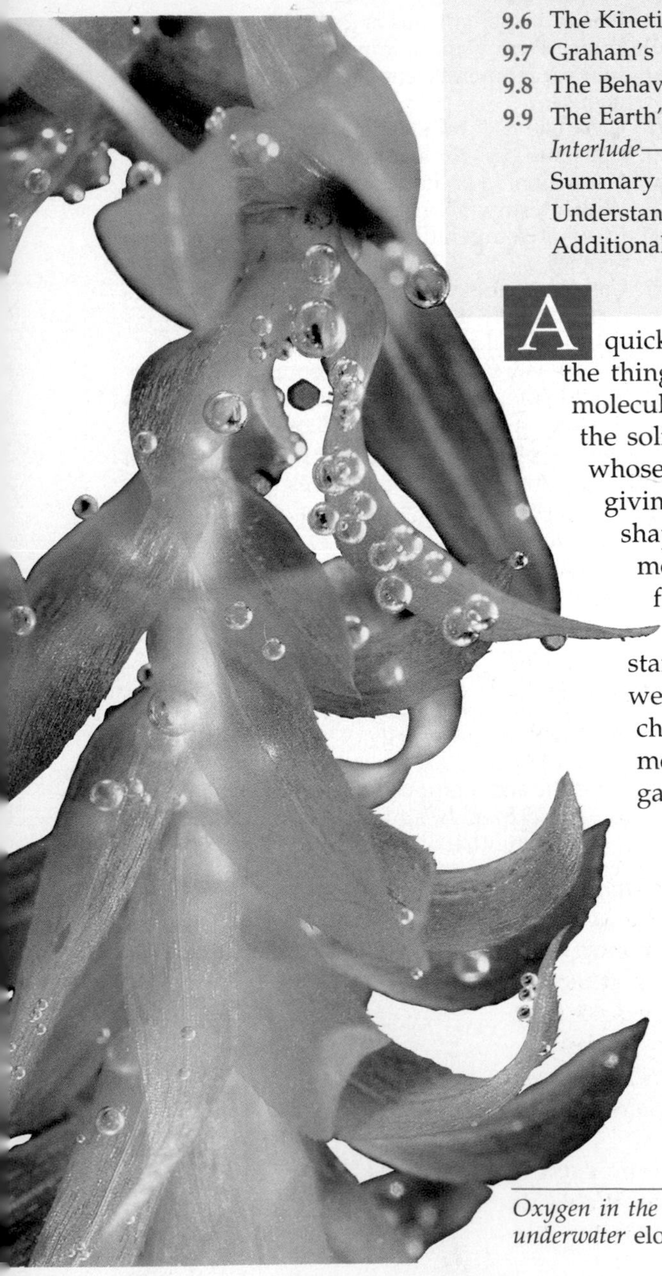

Oxygen in the earth's atmosphere is produced by photosynthesis in plants, such as in this underwater elodea.

9.1 ►GASES AND GAS PRESSURE

We live surrounded by a blanket of air, the mixture of gases that make up the earth's atmosphere. As shown in Table 9.1, nitrogen and oxygen account for more than 99% by volume of dry air. The remaining 1% is largely argon, although trace amounts of several other substances are also present. Carbon dioxide, about which there is so much current concern because of the so-called greenhouse effect, is present in air only to the extent of about 0.034%, or 340 parts per million (ppm). Though small, this value has risen in the past century from an estimated 290 ppm in 1850 as the burning of fossil fuels and the deforestation of tropical rain forests has increased.

TABLE 9.1   Composition of Dry Air at Sea Level

Constituent	% Volume	% Mass
N_2	78.08	75.52
O_2	20.95	23.14
Ar	0.93	1.29
CO_2	0.034	0.05
Ne	1.82×10^{-3}	1.27×10^{-3}
He	5.24×10^{-4}	7.24×10^{-5}
CH_4	1.7×10^{-4}	9.4×10^{-5}
Kr	1.14×10^{-4}	3.3×10^{-4}

Air is typical of gases in many respects, and its behavior illustrates several important points about gases. One such point is that gas mixtures are always *homogeneous*. Unlike liquids, which often fail to dissolve each other and which separate into distinct layers—oil and water, for example—gases always mix thoroughly. A second important point is that gases are *compressible*. When pressure is applied, the volume of a gas contracts proportionately. Liquids, however, are nearly incompressible, and even the application of great pressure changes their volume only slightly.

Homogeneous mixing and compressibility both result from the fact that individual molecules are so far apart in gases (Figure 9.1). Mixing occurs because individual gaseous molecules have little interaction with their neighbors, and, assuming that no reaction takes place, the chemical identities of those neighbors are irrelevant. In liquids, by contrast, molecules are packed closely together, where they are affected by various attractive and repulsive forces that can inhibit their mixing. Compressibility is possible in gases because only about 0.1% of the volume of a typical gas is taken up by the molecules under normal circumstances; the remaining 99.9% is empty space. By contract, approximately 70% of a liquid's volume is taken up by the molecules.

One of the most obvious characteristics of gases is that they exert a measurable *pressure* on the walls of their container (Figure 9.1b). We're all familiar with pumping up a tire or inflating a balloon and feeling the

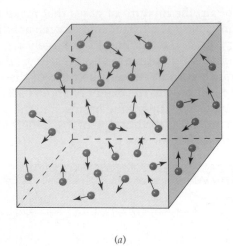
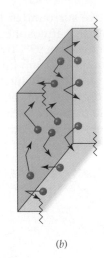

(a)   (b)

FIGURE 9.1   **(a)** A gas is a large collection of particles moving at random through a volume that is primarily empty space. **(b)** Collisions of randomly moving particles with the walls of the container exert a force per unit area that we perceive as gas pressure.

hardness that results from the "pressure" inside. In scientific terms, pressure is defined as a force F exerted per unit area A:

$$\text{Pressure } (P) = \frac{\text{force}}{\text{area}} = \frac{F}{A}$$

Force, in turn, is defined as mass (m) times acceleration (a), which, on earth, is usually the acceleration due to gravity, $a = 9.81 \text{ m/s}^2$.

$$F = \text{mass} \times \text{acceleration} = m \times a$$

To illustrate, if a 1.00 kg weight is resting on the front of your foot, with an area A of about 100 cm^2 (10^{-2} m^2), then the force exerted by the weight against your foot is 9.81 (kg · m)/s^2, or 9.81 *newtons* (N), and the pressure against your foot is 9.81×10^2 N/m^2, or 9.81×10^2 *pascals* (Pa):

Forcing more air into the tire increases the pressure and makes the tire feel "hard."

$$P = \frac{F}{A} = \frac{m \times a}{A}$$

$$P = \frac{1.00 \text{ kg} \times 9.81 \dfrac{\text{m}}{\text{s}^2}}{1.00 \times 10^{-2} \text{ m}^2} = 9.81 \times 10^2 \frac{\dfrac{\text{kg} \cdot \text{m}}{\text{s}^2}}{\text{m}^2} = 9.81 \times 10^2 \frac{\text{N}}{\text{m}^2}$$

$$= 9.81 \times 10^2 \text{ Pa}$$

As indicated, the SI unit for force is the newton, where 1 N = 1 (kg · m)/s^2, and the SI unit for pressure is the pascal, where 1 Pa = 1 N/m^2. Expressed in more familiar units, a pascal is actually a very small amount—the pressure exerted by a mass of 10.2 mg resting on an area of 1.00 cm^2:

$$\frac{(10.2 \text{ mg})\left(\dfrac{1 \text{ kg}}{10^6 \text{ mg}}\right)\left(9.81 \dfrac{\text{m}}{\text{s}^2}\right)}{(1.00 \text{ cm}^2)\left(\dfrac{1 \text{ m}^2}{10^4 \text{ cm}^2}\right)} = \frac{1.00 \times 10^{-4} \dfrac{\text{kg} \cdot \text{m}}{\text{s}^2}}{1.00 \times 10^{-4} \text{ m}^2} = 1.00 \text{ Pa}$$

In rough terms, a penny sitting on the tip of your finger exerts a pressure of about 250 Pa.

Just as a weight on your foot and a penny on your fingertip exert a pressure, the mass of the atmosphere pressing down on the earth's surface causes what we call *atmospheric pressure*. In fact, a 1.00 m² column of air extending from the earth's surface through the upper atmosphere has a mass of about 10,300 kg, producing an atmospheric pressure of approximately 101,000 Pa, or 101 kPa (Figure 9.2)

$$\frac{10{,}300 \text{ kg} \times 9.81 \frac{\text{m}}{\text{s}^2}}{1.00 \text{ m}^2} = 101{,}000 \text{ Pa} = 101 \text{ kPa}$$

FIGURE 9.2 A 1.00 m² column of air extending from the earth's surface through the upper atmosphere has a mass of about 10,300 kg, producing an atmospheric pressure of approximately 101,000 Pa.

As is frequently the case with SI units, which must serve many scientific disciplines, the pascal is an inconvenient size for most chemical measurements. Thus, the alternative pressure units *millimeters of mercury* (*mm Hg*) and *atmosphere* (*atm*) are more frequently used. The millimeter of mercury, also called a *torr* after the seventeenth-century Italian scientist Evangelista Torricelli, is based on atmospheric pressure measurements using a mercury *barometer*.

As shown in Figure 9.3, a barometer consists simply of a long, thin tube that is sealed at one end, filled with mercury, and then inverted into a dish of mercury. Some mercury runs from the tube into the dish until the downward pressure of the mercury in the column is exactly balanced by outside atmospheric pressure, which presses down on the mercury in the dish and pushes it up the column. The height of the mercury column varies slightly from day to day depending on the altitude and weather conditions, but standard atmospheric pressure at sea level is defined to be exactly 760 mm.

Knowing the density of mercury (13.5951 g/cm³ at 0°C) makes it possible to calculate the mass in the barometer column pressing down on a given area. An accurate value for the acceleration due to gravity (9.806 65

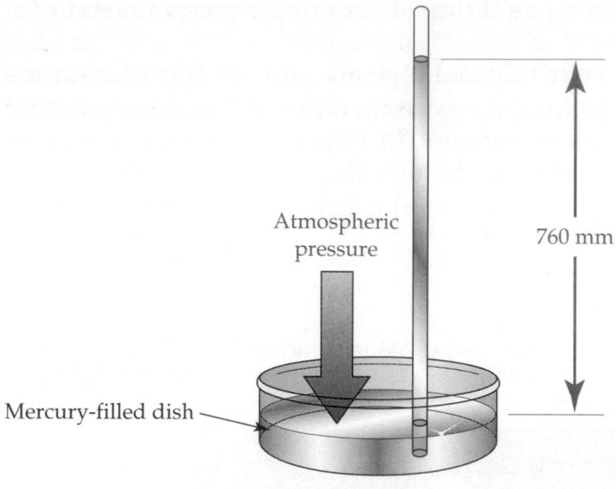

FIGURE 9.3 A mercury barometer is used to measure atmospheric pressure by determining the height of a mercury column supported in a sealed glass tube. The downward pressure of the mercury in the column is exactly balanced by outside atmospheric pressure that presses down on the mercury in the dish and pushes it up the column.

Atmospheric pressure

760 mm

Mercury-filled dish

m/s^2) then allows a conversion between mm Hg and pascal units. Thus, one standard atmosphere of pressure is equal to 760 mm Hg, or 101,325 Pa:

$$13.5951 \, \frac{\text{g}}{\text{cm}^3} \times 76 \, \text{cm} = 1033.227 \, \frac{\text{g}}{\text{cm}^2} = 1.033\,227 \times 10^4 \, \frac{\text{kg}}{\text{m}^2}$$

$$\left(1.033\,227 \times 10^4 \, \frac{\text{kg}}{\text{m}^2}\right)\left(9.806\,65 \, \frac{\text{m}}{\text{s}^2}\right) = 101,325 \, \text{Pa}$$

$$1 \, atm = 760 \, mm \, Hg = 101,325 \, Pa$$

Gas pressures in a container are often measured using an open-end *manometer*, a simple instrument similar in principle to the mercury barometer. As shown in Figure 9.4, an open-end manometer consists of a U-tube filled with mercury, with one end connected to the gas-filled container and the other end open to the atmosphere. The difference between the pressure of the gas and the pressure of the atmosphere is equal to the difference in

FIGURE 9.4 Open-end manometers for measuring pressure in a gas-filled bulb. In **(a)**, the pressure in the bulb is lower than atmospheric, so the mercury level is higher in the arm open to the bulb; in **(b)**, the pressure in the bulb is higher than atmospheric, so the mercury level is higher in the arm open to the atmosphere.

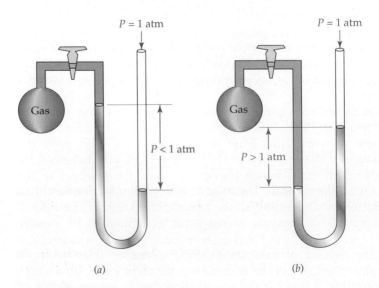

(a)   *(b)*

height between the mercury levels in the two arms. If the gas pressure inside the container is less than atmospheric, the mercury level is higher in the arm connected to the container (Figure 9.4a). If the gas pressure inside the container is greater than atmospheric, the mercury level is higher in the arm open to the atmosphere (Figure 9.4b).

EXAMPLE 9.1

Normal atmospheric pressure on top of Mt. Everest (29,028 ft) is 265 mm Hg. Convert this value to pascals and to atm.

BALLPARK SOLUTION   One atmosphere is defined as 760 mm Hg pressure. Since 265 mm Hg is about 1/3 of 760 mm Hg, the air pressure on Mt. Everest is about 1/3 of standard atmospheric pressure—approximately 30,000 Pa, or 0.3 atm.

DETAILED SOLUTION   Use the conversion factors 1 atm/760 mm Hg and 101,325 Pa/760 mm Hg to carry out the necessary calculations:

$$265 \text{ mm Hg} \times \frac{101,325 \text{ Pa}}{760 \text{ mm Hg}} = 3.53 \times 10^4 \text{ Pa}$$

$$265 \text{ mm Hg} \times \frac{1 \text{ atm}}{760 \text{ mm Hg}} = 0.349 \text{ atm}$$

Atmospheric pressure decreases as altitude increases.

EXAMPLE 9.2

Assume that you are using an open-end manometer (Figure 9.4) filled with mineral oil rather than mercury. What is the gas pressure in mm Hg if the level of mineral oil in the arm connected to the bulb is 237 mm higher than the level in the arm connected to the atmosphere and if atmospheric pressure is 746 mm Hg? The density of Hg is 13.6 g/mL, and the density of mineral oil is 0.822 g/mL.

SOLUTION   The pressure of the gas is less than atmospheric, since the liquid level is higher on the side connected to the sample. Because mercury is more dense than mineral oil by a factor of 13.6/0.822, or 16.5, a given pressure will hold a column of mineral oil 16.5 times higher than it will hold a column of mercury. Thus, a pressure of 237 mm mineral oil corresponds to a pressure of 14.3 mm Hg:

$$237 \text{ mm mineral oil} \times \frac{0.822 \text{ g/mL mineral oil}}{13.6 \text{ g/mL Hg}} = 14.3 \text{ mm Hg}$$

The total gas pressure is therefore equal to:

$$P_{\text{gas}} = 746 \text{ mm Hg} - 14.3 \text{ mm Hg} = 732 \text{ mm Hg}$$

PROBLEM 9.1 Pressures are often given in the familiar unit pounds per square inch (psi). How many psi correspond to one atmosphere? to 1.0 mm Hg?

PROBLEM 9.2 If the density of water is 1.00 g/mL and the density of mercury is 13.6 g/mL, how high a column of water in meters can be supported by standard atmospheric pressure?

PROBLEM 9.3 What is the pressure in atm in a container of gas connected to a mercury-filled, open-end manometer if the level in the arm connected to the container is 24.7 cm higher than in the arm open to the atmosphere and if atmospheric pressure is 0.975 atm?

9.2 ►THE GAS LAWS

Unlike solids and liquids, different gases show remarkably similar physical behavior, regardless of their chemical makeup. Helium and fluorine, for example, are vastly different in their chemical properties yet are almost identical in much of their physical behavior. Numerous observations made in the late 1600s showed that the physical condition of any gas can be defined by four variables: pressure (P), temperature (T), volume (V), and number of moles (n). The specific relationships among these variables are called the **gas laws**.

Boyle's Law: The Relationship Between Volume and Pressure

Imagine that you have a sample of gas inside a cylinder with a moveable piston at one end (Figure 9.5). What would happen if you were to decrease the volume of the gas by pushing the piston partway down? Common sense tells you that you would feel a resistance to pushing the piston because the pressure of the gas in the cylinder would increase. According to **Boyle's law**, the volume of a fixed amount of gas at a constant temperature varies inversely with its pressure. If the gas volume is halved, the gas pressure doubles; if the volume is doubled, the pressure is halved.

BOYLE'S LAW   The volume of a gas varies inversely with pressure. That is, P times V is constant when n and T are kept constant. (The symbol $\propto$ means "is proportional to," and k denotes a constant.)

$$V \propto 1/P \quad \text{or} \quad PV = k \text{ at constant } n, T$$

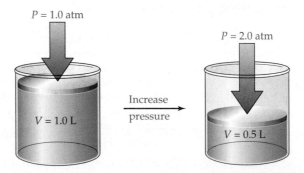

FIGURE 9.5 Boyle's law: At constant n and T, the volume of a gas decreases proportionately as its pressure increases. If the pressure is doubled, the volume is halved.

The validity of Boyle's law is easy to demonstrate by making a simple series of pressure/volume measurements on a gas sample (Table 9.2) and plotting them as in Figure 9.6. When V is plotted versus P as in Figure 9.6a, the result is a curve in the form of a hyperbola. When V is plotted versus $1/P$ as in Figure 9.6b, the result is a straight line. Such graphical behavior is characteristic of mathematical equations of the form $y = mx + b$. In the present instance, $y = V$, $m = $ the slope of the line (the constant k in

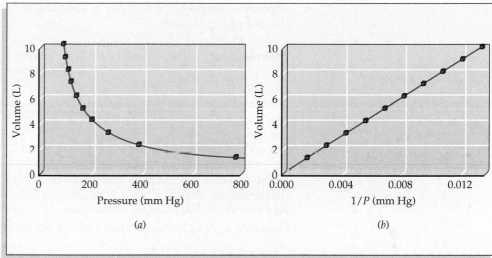

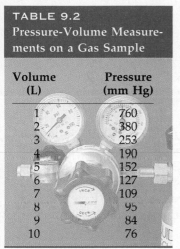

Volume (L)	Pressure (mm Hg)
1	760
2	380
3	253
4	190
5	152
6	127
7	109
8	95
9	84
10	76

TABLE 9.2
Pressure-Volume Measurements on a Gas Sample

FIGURE 9.6 Boyle's law plots based on measurements of volume versus pressure for a gas sample. **(a)** A plot of *V* versus *P* is a hyperbola, but **(b)** a plot of *V* versus 1/*P* is a straight line. Such a straight-line graph is characteristic of equations having the form *y* = *mx* + *b*.

the present instance), *x* = 1/*P*, and *b* = the *y*-intercept (a constant; 0 in the present instance). (See Appendix A.3 for a review of linear equations.)

$$V = k\left(\frac{1}{P}\right) + 0 \qquad (\text{or } PV = k)$$
$$\uparrow \quad \uparrow \; \uparrow \qquad \uparrow$$
$$y = m \; x \; + \; b$$

Charles' Law: The Relationship Between Volume and Temperature

Imagine again that you have a gas sample inside a cylinder with a moveable piston at one end (Figure 9.7). What would happen if you were to raise the temperature of the sample while letting the piston move freely to keep the pressure constant? Common sense tells you that the piston would move up because the volume of the gas in the cylinder would expand. According to **Charles' law**, the volume of a fixed amount of gas at a constant pressure varies directly with its absolute temperature. If the gas temperature in

FIGURE 9.7 Charles' law: At constant *n* and *P*, the volume of a gas increases proportionately as its absolute temperature increases. If the absolute temperature is doubled, the volume is doubled.

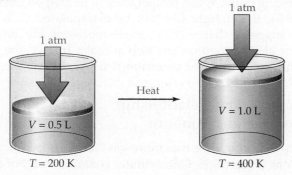

kelvins is doubled, the volume is doubled; if the gas temperature is halved, the volume is halved.

CHARLES' LAW   The volume of a gas varies directly with absolute temperature. That is, V divided by T is constant when n and P are held constant.

$$V \propto T \quad \text{or} \quad V/T = k \text{ at constant } n, P$$

The validity of Charles' law can be demonstrated by making a series of temperature/volume measurements on a gas sample, giving the results listed in Table 9.3. As was true of Boyle's law, Charles' law also takes the mathematical form $y = mx + b$, where $y = V$, m = the slope of the line, $x = T$, and b = the y-intercept (0 in the present instance). A plot of V versus T is therefore a straight line whose slope is equal to the constant k (Figure 9.8).

$$V = kT + 0 \quad \left(\text{or } \frac{V}{T} = k\right)$$
$$y = mx + b$$

TABLE 9.3 Temperature-Volume Measurements on a Gas Sample	
Volume (L)	Temperature (K)
0.45	123
0.63	173
0.82	223
1.00	273
1.18	323
1.37	373

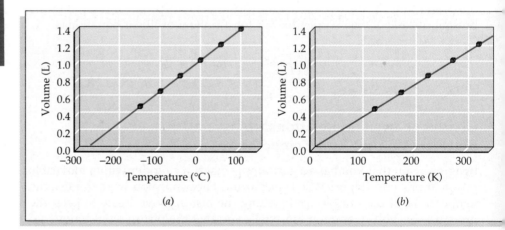

FIGURE 9.8   Charles'-law plots based on measurements of volume versus temperature for a gas sample. A plot of V versus T is a straight line that can be extrapolated to *absolute zero*.

The plots of volume versus temperature shown in Figure 9.8 demonstrate an interesting point. When temperature in plotted on the Celsius scale, as in Figure 9.8a, the straight line can be extrapolated to $V = 0$ at $T = -273°C$. This suggests that $-273°C$ must represent the lowest possible temperature, or *absolute zero* on the Kelvin scale. In fact, the approximate value of absolute zero was first determined using this simple method.

Avogadro's Law: The Relationship Between Volume and Amount

Imagine finally that you have two more gas samples inside cylinders with moveable pistons (Figure 9.9). One cylinder contains 1 mol of a gas and the

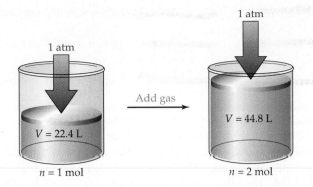

FIGURE 9.9 Avogadro's law: At constant T and P, the volume of a gas increases proportionately as its molar amount increases. If the molar amount is doubled, the volume is doubled.

other cylinder contains 2 mol of a gas at the same temperature and pressure as the first. Common sense tells you that the second cylinder will have twice the volume of the first cylinder because it contains twice as much gas. According to **Avogadro's law**, the volume of a gas at a fixed pressure and temperature depends on its molar amount. If the amount of the gas is halved, the gas volume is halved; if the amount is doubled, the volume is doubled.

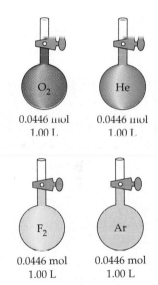

FIGURE 9.10 Avogadro's law: Each of these 1.00 L bulbs contains the same molar amount of gas: 1.00 L/(22.4 L/mol) = 0.0446 mol at 0°C and 1 atm pressure.

AVOGADRO'S LAW   The volume of a gas varies directly with its molar amount. That is, V divided by n is constant when T and P are held constant.

$$V \propto n \quad \text{or} \quad V/n = k \text{ at constant } T, P$$

Put another way, Avogadro's law also says that equal volumes of different gases at the same temperature and pressure contain the same molar amounts. A 1 L container of oxygen contains the same number of moles of gas as a 1 L sample of helium, fluorine, argon, or any other gas at the same T and P (Figure 9.10). Experiments show that 1.00 mol of a gas occupies a volume (the **standard molar volume**) of 22.4 L at 0°C and 1.00 atm pressure.

9.3 ►THE IDEAL-GAS LAW

All three of the gas laws discussed in the previous section can be combined into a single statement, the **ideal-gas law**, which describes how the volume of a gas is affected by changes in pressure, temperature, and amount. The constant R in the equation is called the **gas constant** and has the same value for all gases.

IDEAL-GAS LAW $$V = \frac{nRT}{P} \quad \text{or} \quad PV = nRT$$

Note how the ideal-gas law can be rearranged in different ways to take the form of Boyle's law, Charles' law, or Avogadro's law:

Boyle's law: $PV = nRT = k$ (when n and T are constant)

Charles' law: $\dfrac{V}{T} = \dfrac{nR}{P} = k$ (when n and P are constant)

Avogadro's law: $\dfrac{V}{n} = \dfrac{RT}{P} = k$ (when T and P are constant)

Note also that the value of the gas constant R can be calculated from a knowledge of the molar volume of a gas. Since 1.000 mole of a gas occupies a volume of 22.414 L at 0°C (273.15 K) and 1.000 atm pressure, the gas constant R is equal to 0.08206 (L · atm)/(K · mol), or 8.314 J/(K · mol) in SI units:

$$R = \frac{P \cdot V}{n \cdot T} = \frac{(1.000 \text{ atm})(22.414 \text{ L})}{(1.000 \text{ mol})(273.15 \text{ K})} = 0.08206 \; \frac{\text{L} \cdot \text{atm}}{\text{K} \cdot \text{mol}}$$

$$= 8.314 \text{ J/(K} \cdot \text{mol) (when } P \text{ is in pascals and } V \text{ is in m}^3)$$

The specific conditions used for the calculation, 1 atm and 0°C (273 K), are said to represent **standard temperature and pressure** (abbreviated **STP**). These standard conditions are generally used when reporting measurements on gases.

Standard temperature and pressure (STP) for gases $T = 273$ K; $P = 1$ atm

The name *ideal-gas* law implies that there must be some gases that have *nonideal* behavior. In fact, there's no such thing as an ideal gas that obeys the equation perfectly under all circumstances; all real gases deviate slightly from the behavior predicted by the law. As Table 9.4 shows, for example, the actual molar volume of a real gas often differs slightly from the 22.414 L ideal volume. Under normal conditions, though, the deviations from ideal behavior are so slight as to make little difference. We'll discuss circumstances in Section 9.8 where the deviations are greater.

TABLE 9.4   Standard Molar Volumes of Some Real Gases

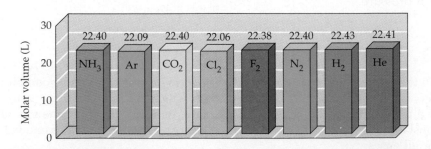

The importance of the ideal-gas law is that it allows us to calculate the value of any one of the variables P, V, n, or T, as long as the values of the other three are known. Examples 9.3 and 9.4 show how this can be done.

EXAMPLE 9.3

How many moles of air are there in the lungs of an average adult with a total lung capacity of 3.8 L? Assume that the person is at 1.0 atm pressure and has a normal body temperature of 37°C.

BALLPARK SOLUTION   A lung volume of 4 L is about 1/6 of 22.4 L, the standard molar volume of an ideal gas. Thus, the lungs should have a capacity of about 1/6 mol, or 0.17 mol.

DETAILED SOLUTION   This problem asks for a value of n when V, P, and T are given. The necessary form of the ideal-gas law is therefore $n = PV/RT$. Converting the temperature from Celsius to Kelvin and substituting the correct values into the equation gives

$$n = \frac{PV}{RT} = \frac{(1.0 \text{ atm})(3.8 \text{ L})}{\left(0.08206 \dfrac{\text{L} \cdot \text{atm}}{\text{K} \cdot \text{mol}}\right)(310 \text{ K})} = 0.15 \text{ mol}$$

The lungs of an average adult hold 0.15 mol of air.

EXAMPLE 9.4

In a typical automobile engine, the mixture of gasoline and air in a cylinder is compressed from 1.0 atm to 9.5 atm. If the uncompressed volume of the cylinder is 410 mL, what is the volume in mL when the mixture is fully compressed?

BALLPARK SOLUTION   Since the pressure in the cylinder increases about 10-fold, the volume must decrease about 10-fold according to Boyle's law, from 400 mL to 40 mL.

DETAILED SOLUTION   This is a Boyle's-law problem, since only P and V are changing while n and T remain fixed. We can therefore set up the following equality:

$$nRT = (PV)_{\text{initial}} = (PV)_{\text{final}}$$

Solving for V_{final} gives

$$V_{\text{final}} = \frac{(PV)_{\text{initial}}}{P_{\text{final}}} = \frac{1.0 \text{ atm} \times 410 \text{ mL}}{9.5 \text{ atm}} = 43 \text{ mL}$$

⌐ **PROBLEM 9.4** How many moles of methane gas, CH_4, are in a 100,000 L storage tank at STP? How many grams is this?

⌐ **PROBLEM 9.5** An aerosol spray can with a volume of 350 mL contains 3.2 g of propane gas (C_3H_8) as propellant. What is the pressure in atm of gas in the can at 20°C?

⌐ **PROBLEM 9.6** A helium gas cylinder of the sort used to fill balloons has a volume of 43.8 L and a pressure of 1.51×10^4 kPa at 25.0°C. How many moles of helium are in the tank?

⌐ **PROBLEM 9.7** What final temperature in °C is required for the pressure inside an automobile tire to increase from 2.15 atm at 0°C to 2.37 atm assuming the volume remains constant?

How many moles of methane are in these tanks?

9.4 ▶STOICHIOMETRIC RELATIONSHIPS WITH GASES

Many chemical reactions, including some of the most important processes in the chemical industry, involve gases. Tens of millions of tons of ammonia, for example, are manufactured each year by reaction of hydrogen with nitrogen according to the equation $3 H_2 + N_2 \rightarrow 2 NH_3$. Thus, it's necessary to be able to calculate amounts of gaseous reactants just as it's necessary to calculate amounts of solids, liquids, and solutions (Sections 3.4–3.9).

Most stoichiometric calculations are just applications of the ideal-gas law in which three of the variables P, V, T, and n are known, and the fourth variable must be calculated. For example, the reaction used in the deployment of automobile air bags is the high-temperature decomposition of sodium azide, NaN_3, to produce N_2 gas. (The sodium is then removed by a subsequent reaction.) How many liters of N_2 at 1.15 atm and 30°C are produced by decomposition of 135 g of NaN_3?

Automobile air bags are inflated with N_2 gas produced by decomposition of sodium azide.

$$2 NaN_3(s) \rightarrow 2 Na(s) + 3 N_2(g)$$

Values for P and T are given in this problem, the value of n can be calculated, and the ideal-gas law will then let us find V. To find n, the number of moles of N_2 gas produced, we first need to find how many moles of NaN_3 are in 135 g:

$$\text{Molar mass of } NaN_3 = 65.0 \text{ g/mol}$$

$$\text{Moles } NaN_3 = 135 \text{ g } NaN_3 \times \frac{1 \text{ mol } NaN_3}{65.0 \text{ g } NaN_3} = 2.08 \text{ mol } NaN_3$$

Next, find how many moles of N_2 are produced in the decomposition reaction. The balanced equation says that 2 mol of NaN_3 yield 3 mol of N_2. Thus, 2.08 mol of NaN_3 yields 3.12 mol of N_2:

$$\text{Moles } N_2 = 2.08 \text{ mol } NaN_3 \times \frac{3 \text{ mol } N_2}{2 \text{ mol } NaN_3} = 3.12 \text{ mol } N_2$$

Finally, use the ideal-gas law to calculate the volume of N_2. Remember to use the Kelvin temperature (303 K) rather than the Celsius temperature (30°C) in the calculation:

$$V = \frac{nRT}{P} = \frac{3.12 \text{ mol} \times 0.08206 \frac{L \cdot atm}{K \cdot mol} \times 303 \text{ K}}{1.15 \text{ atm}} = 67.5 \text{ L}$$

FIGURE 9.11 Apparatus for determining the density of an unknown gas. A bulb of known volume is evacuated, weighed when empty, filled with gas at a known pressure and temperature, and weighed again. Dividing the mass by the volume gives the density.

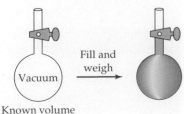

Vacuum

Fill and weigh

Known volume

Example 9.5 illustrates another gas stoichiometry calculation.

Still other applications of the ideal-gas law make it possible to calculate such properties as density and molar mass. Densities are calculated simply by weighing a known volume of a gas at a known temperature and pressure as shown in Figure 9.11. Using the ideal-gas law to find the volume at STP and then dividing the measured mass by the volume gives the density at STP. Example 9.6 gives a sample calculation.

Molar masses, and therefore molecular weights, are also calculated by using the ideal-gas law. Imagine, for example, that an unknown gas found bubbling up in a swamp is collected, placed in a sample bulb, and found to have a density of 0.714 g/L at STP. What is the molecular weight of the gas?

Let's assume that we have 1.00 L of sample, which has a mass of 0.714 g. Since we're told that the density is measured at STP, we know T, P, and V, and we need to find n:

$$n = \frac{PV}{RT} = \frac{1.00 \text{ atm} \times 1.00 \text{ L}}{0.08206 \frac{\text{L} \cdot \text{atm}}{\text{mol} \cdot \text{K}} \times 273 \text{ K}} = 0.0446 \text{ mol}$$

Dividing the mass of the sample by the number of moles then gives the molar mass:

$$\text{Molar mass} = \frac{0.714 \text{ g}}{0.0446 \text{ mol}} = 16.0 \text{ g/mol}$$

Thus, the molar mass of the unknown gas (actually methane, CH_4) is 16.0 g/mol, and the molecular weight is 16.0 amu.

It's often true in chemistry, particularly in gas-law calculations, that a problem can be solved in more than one way. As an alternative method of calculating the molar mass of the unknown swamp gas, you might recognize that 1 mol of an ideal gas has a volume of 22.41 L at STP. Since 1 L of the unknown gas has a mass of 0.714 g, 22.41 L of the gas (1 mol) has a mass of 16.0 g:

$$\text{Molar mass} = 0.714 \frac{\text{g}}{\text{L}} \times 22.41 \frac{\text{L}}{\text{mol}} = 16.0 \text{ g/mol}$$

Example 9.7 illustrates another calculation of the molar mass of an unknown gas.

EXAMPLE 9.5

A typical high-pressure tire on a racing bicycle might have a volume of 365 mL and a pressure of 7.80 atm at 25°C. Suppose the rider filled the tire with helium to minimize weight. What would be the mass of the helium in the tire?

SOLUTION We are given V, P, and T and need to use the ideal-gas law to calculate n, the number of moles of helium in the tire:

$$n = \frac{PV}{RT} = \frac{7.80 \text{ atm} \times 0.365 \text{ L}}{0.08206 \frac{\text{L} \cdot \text{atm}}{\text{mol} \cdot \text{K}} \times 298 \text{ K}} = 0.116 \text{ mol}$$

A mole-to-gram conversion using the molar mass of helium (4.00 g/mol) then gives the mass of the helium:

$$\text{Grams helium} = 0.116 \text{ mol He} \times \frac{4.00 \text{ g He}}{1 \text{ mol He}} = 0.464 \text{ g}$$

EXAMPLE 9.6

What is the density of ammonia, NH_3, in g/L at STP if the gas in a 1.000 L bulb weighs 0.672 g at 25°C and 733.4 mm Hg pressure?

SOLUTION   The density of any substance is mass divided by volume. For the ammonia sample, the mass is 0.672 g but the volume of the gas is given under nonstandard conditions and must first be converted to STP. Since the amount of sample n is constant, we can set the quantity PV/RT measured under nonstandard conditions equal to PV/RT at STP and then solve for V at STP:

$$n = \left(\frac{PV}{RT}\right)_{\text{measured}} = \left(\frac{PV}{RT}\right)_{\text{STP}} \quad \text{or} \quad V_{\text{STP}} = \left(\frac{PV}{RT}\right)_{\text{measured}} \times \left(\frac{RT}{P}\right)_{\text{STP}}$$

$$V_{\text{STP}} = \left(\frac{733.4 \text{ mm Hg} \times 1.000 \text{ L}}{298 \text{ K}}\right)\left(\frac{273 \text{ K}}{760 \text{ mm Hg}}\right) = 0.884 \text{ L}$$

Thus, the amount of gas in the 1.000 L bulb under the measured nonstandard conditions would have a volume of only 0.884 L at STP. Dividing this volume into mass gives the density of ammonia at STP:

$$\text{Density} = \frac{\text{mass}}{\text{volume}} = \frac{0.672 \text{ g}}{0.884 \text{ L}} = 0.760 \text{ g/L}$$

EXAMPLE 9.7

To identify the contents of an unlabeled cylinder of gas, a sample was collected and found to have a density of 5.380 g/L at 15°C and 736 mm Hg pressure. What is the molar mass of the gas?

SOLUTION   Let's assume we have a 1.000 L sample of the gas, which weighs 5.380 g. We know the temperature, volume, and pressure of the gas and can therefore use the ideal-gas law to find n, the number of moles in the sample:

$$PV = nRT \quad \text{or} \quad n = \frac{PV}{RT}$$

$$n = \frac{\left(736 \text{ mm Hg} \times \dfrac{1 \text{ atm}}{760 \text{ mm Hg}}\right)(1.000 \text{ L})}{0.08206 \dfrac{\text{L} \cdot \text{atm}}{\text{K} \cdot \text{mol}} \times 288 \text{ K}} = 0.0410 \text{ mol}$$

Dividing the number of grams in the sample by the number of moles gives the molar mass:

$$\frac{5.380 \text{ g}}{0.0410 \text{ mol}} = 131 \text{ g/mol}$$

The gas is probably xenon (atomic weight = 131.3 amu).

⌐ **PROBLEM 9.8** Carbonate-bearing rocks like limestone ($CaCO_3$) react with dilute acids such as HCl to produce carbon dioxide according to the equation $CaCO_3(s) + 2 \text{ HCl}(aq) \rightarrow CaCl_2(aq) + CO_2(g) + H_2O(l)$. How many grams of CO_2 would be formed by complete reaction of 33.7 g of limestone? What is the volume in liters of this CO_2 at STP?

PROBLEM 9.9 Propane gas (C_3H_8) is used as a fuel in rural areas. How many liters of CO_2 are formed at STP by complete combustion of the propane in a metal bottle with a volume of 15.0 L and a pressure of 4.5 atm at 25°C? The equation is

$$C_3H_8(g) + 5\,O_2(g) \rightarrow 3\,CO_2(g) + 4\,H_2O(l).$$

PROBLEM 9.10 A foul-smelling gas produced by reaction of HCl with Na_2S was collected, and a 1.00 L sample was found to have a mass of 1.52 g at STP. What is the molecular weight of the gas? What is its likely formula and name?

9.5 ➤PARTIAL PRESSURE AND DALTON'S LAW

Just as the gas laws apply to all pure gases, regardless of chemical identity, they also apply to *mixtures* of gases, such as air. The pressure, volume, temperature, and amount of a gas mixture are all related by the ideal-gas law.

What is responsible for the gas pressure of a mixture? Since the pressure of a pure gas at constant temperature and volume is proportional to its amount ($P = nRT/V$), the pressure contribution from each individual gas in a mixture must also be proportional to *its* amount in the mixture. In other words, the total pressure exerted by a mixture of gases in a container at constant V and T is equal to the sum of the pressures exerted by each individual gas in the container, a statement known as **Dalton's law of partial pressures**.

DALTON'S LAW OF PARTIAL PRESSURES $P_{total} = P_1 + P_2 + P_3 + \ldots$ at constant V, T, where P_1, P_2, $\ldots$ refer to the pressures of the individual gases in the mixture

The individual pressures of the various gases in the mixture, P_1, P_2, and so forth, are called *partial pressures* and refer to the pressure each individual gas would exert if it were alone in the container. That is,

$$P_1 = n_1\left(\frac{RT}{V}\right) \quad P_2 = n_2\left(\frac{RT}{V}\right) \quad P_3 = n_3\left(\frac{RT}{V}\right) \quad \ldots \text{ and so forth}$$

But since all the gases in the mixture have the same temperature and volume, we can rewrite Dalton's law to indicate that the total pressure depends on the total molar amount of gas present, not on the chemical identity of the individual gases.

$$P_{total} = (n_1 + n_2 + n_3 + \ldots)\left(\frac{RT}{V}\right)$$

The concentration of any individual component in a gas mixture is often expressed as a **mole fraction** (X), which is defined simply as the number of moles of the component divided by the total number of moles in the mixture:

$$\textbf{Mole fraction } (X) = \frac{\text{moles of component}}{\text{total moles in mixture}}$$

The mole fraction of component 1, for example, is

$$X_1 = \frac{n_1}{n_1 + n_2 + n_3 + \ldots} = \frac{n_1}{n_{\text{total}}}$$

But since $n = PV/RT$, we can also write

$$X_1 = \frac{P_1 \left(\dfrac{V}{RT} \right)}{P_{\text{total}} \left(\dfrac{V}{RT} \right)} = \frac{P_1}{P_{\text{total}}}$$

which can be rearranged to solve for P_1, the partial pressure of component 1.

$$P_1 = X_1 \cdot P_{\text{total}}$$

This equation says that the partial pressure exerted by any component in a gas mixture is equal to the mole fraction of that component times the total pressure. In air, for example, the mole fractions of N_2, O_2, Ar, and CO_2 are 0.7808, 0.2095, 0.0093, and 0.00034 respectively (Table 9.1), and the total pressure of the air is the sum of the individual partial pressures.

$$P_{\text{air}} = P_{N_2} + P_{O_2} + P_{Ar} + P_{CO_2} + \ldots$$

Thus, at a total air pressure of 1 atm (760 mm Hg), the partial pressures of the individual components are

$$
\begin{aligned}
P_{N_2} &= 0.7808 \text{ atm } N_2 &&= 593.4 \text{ mm Hg} \\
P_{O_2} &= 0.2095 \text{ atm } O_2 &&= 159.2 \text{ mm Hg} \\
P_{Ar} &= 0.0093 \text{ atm Ar} &&= 7.1 \text{ mm Hg} \\
P_{CO_2} &= 0.00034 \text{ am } CO_2 &&= 0.3 \text{ mm Hg} \\
\hline
P_{\text{air}} &= 1.0000 \text{ atm air} &&= 760.0 \text{ mm Hg}
\end{aligned}
$$

There are numerous practical applications of Dalton's law, ranging from the use of anesthetic agents in hospital operating rooms, where partial pressures of both oxygen and anesthetic in the patient's lungs must be constantly monitored, to the composition of diving gases used for underwater exploration. Example 9.8 gives an illustration.

EXAMPLE 9.8

At an underwater depth of 250 ft, the pressure is 8.38 atm. What should the mole percent of oxygen in the diving gas be for the partial pressure of oxygen in the mixture to be 0.21 atm, the same as it is in air at 1 atm?

SOLUTION   The partial pressure of a gas in a mixture is equal to the total pressure times the mole fraction of the gas. Rearranging this equation lets us solve for mole fraction:

Since
$$P_{O_2} = X_{O_2} \cdot P_{\text{total}}$$

then
$$X_{O_2} = \frac{P_{O_2}}{P_{\text{total}}} = \frac{0.21 \text{ atm}}{8.38 \text{ atm}} = 0.025$$

$$\text{Percent } O_2 = 0.025 \times 100\% = 2.5\% \ O_2$$

The diving gas should contain 2.5% O_2 for the partial pressure of O_2 to be the same at 8.38 atm as it is in normal air at 1 atm.

⌐ **PROBLEM 9.11** What are the mole fractions of each component in a mixture of 12.45 g of H_2, 60.67 g of N_2, and 2.38 g of NH_3?

⌐ **PROBLEM 9.12** What is the total pressure in atm and what is the partial pressure of each component if the gas mixture in Problem 9.11 is in a 10.00 L steel container at 90°C?

⌐ **PROBLEM 9.13** On a humid day in summer, the mole fraction of gaseous H_2O (water vapor) in the air at 25°C can be as high as 0.0287. Assuming a total pressure of 0.977 atm, what is the partial pressure in atm of H_2O in the air? ⌐

The partial pressure of oxygen in the scuba tanks must be the same underwater as in air at atmospheric pressure.

9.6 ➤THE KINETIC-MOLECULAR THEORY OF GASES

Thus far, we've concentrated on *describing* the behavior of gases rather than on understanding the reasons for that behavior. Actually, the reasons are straightforward and were explained more than a century ago using a model called the **kinetic-molecular theory**. The kinetic-molecular theory is based on the following assumptions:

1. A gas consists of tiny particles, either atoms or molecules, moving about at random.
2. The volume of the particles themselves is negligible compared with the total volume of the gas; most of the volume of a gas is empty space.
3. The gas particles act independently of one another; there are no attractive or repulsive forces between particles.
4. Collisions of the gas particles, either with other particles or with the walls of the container, are elastic; that is, the total kinetic energy of the gas particles is constant at constant T.
5. The average kinetic energy of the gas particles is proportional to the Kelvin temperature of the sample.

Beginning with these assumptions, it's possible not only to understand the behavior of gases but also to derive quantitatively the ideal-gas law (though we'll not do so here). For example, let's look at how the individual gas laws follow from the five postulates of kinetic-molecular theory:

1. **Boyle's law** ($P \propto 1/V$): Gas pressure is a measure of the number and forcefulness of collisions between gas particles and the walls of the container. The smaller the volume at constant n and T, the more crowded together the particles are and the greater the likelihood of collisions. Thus, pressure increases as volume decreases (Figure 9.12a)

2. **Charles' law** ($V \propto T$): Temperature is a measure of the average kinetic energy of the gas particles. The higher the temperature at constant n and P, the faster the gas particles move and the more room they require to move around in. Thus, volume increases as temperature increases (Figure 9.12b).

3. **Avogadro's law** ($V \propto n$): The more particles there are in a gas sample, the more volume the particles need at constant P and T. Thus, volume increases as amount increases (Figure 9.12c).

4. **Dalton's law** ($P_{total} = P_1 + P_2 + \ldots$): The chemical identity of the particles in a gas is irrelevant. Total pressure depends only on the total number of collisions with the walls of the container and does not change at constant n and T. The number of collisions undergone by a specific kind of particle depends on the mole percent of that kind of particle in the mixture (Figure 9.12d).

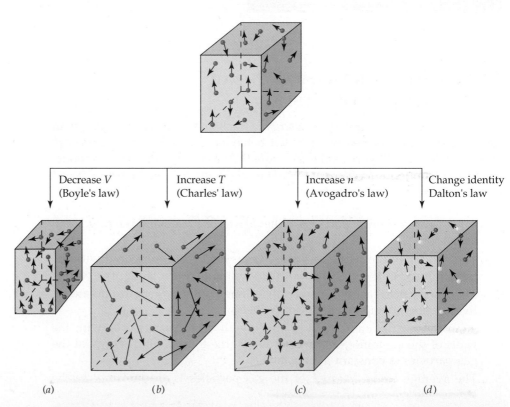

Decrease V
(Boyle's law)

Increase T
(Charles' law)

Increase n
(Avogadro's law)

Change identity
Dalton's law

(a)   *(b)*   *(c)*   *(d)*

FIGURE 9.12 **(a)** Decreasing the volume of the gas at constant n and T increases the likelihood of collisions with the container walls and therefore increases the pressure (Boyle's law). **(b)** Increasing the temperature (kinetic energy) at constant n and P increases the volume of the gas (Charles' law). **(c)** Increasing the amount of gas at constant T and P increases the volume (Avogadro's law). **(d)** Changing the identity of some gas molecules at constant T and V has no effect on the pressure (Dalton's law).

One of the more important conclusions from kinetic-molecular theory has to do with the relationship between temperature and E_K, the kinetic energy of molecular motion (assumption 5). It can be shown by a somewhat complex derivation that the total kinetic energy of a mole of gas particles is equal to (3/2) RT and that the average kinetic energy per particle is thus (3/2) RT/N_A, where N_A is Avogadro's number. Knowing this relationship makes it possible to calculate the average speed u of a gas particle.[1] To take a helium atom at room temperature (298 K), for example, we can write

$$E_K = \frac{3RT}{2N_A} = \frac{1}{2}mu^2$$

which can be rearranged to give $\qquad u^2 = \frac{3RT}{mN_A}$

or $\qquad u = \sqrt{\frac{3RT}{mN_A}} = \sqrt{\frac{3RT}{M}} \qquad$ where M is the molar mass

Substituting appropriate values for R [8.314 J/(K·mol)] and for M, the molar mass of helium (4.00×10^{-3} kg/mol), we have

$$u = \sqrt{\frac{3 \times 8.314 \frac{J}{K \cdot mol} \times 298 \text{ K}}{4.00 \times 10^{-3} \frac{kg}{mol}}} = \sqrt{1.86 \times 10^6 \frac{J}{kg}}$$

$$= \sqrt{1.86 \times 10^6 \frac{\frac{kg \cdot m^2}{s^2}}{kg}} = 1.36 \times 10^3 \text{ m/s}$$

Thus, the average speed of a helium atom at room temperature is more than 1.3 km/s, or about 3000 mi/h! Average speeds of some other molecules at 25°C are given in Table 9.5. The heavier the molecule, the slower the speed.

TABLE 9.5 Average Speeds (m/s) of Some Molecules at 25°C

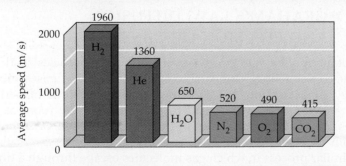

[1] The abbreviation u to represent the "average" speed of gas particles in a sample actually refers to the *root mean square* (rms) speed, which is defined as the square root of the average of the squares of the particle speeds.

Just because the *average* speed of helium atoms is 1.36 km/s at 298 K doesn't mean that all helium atoms are moving at that rate or that a specific atom will travel from New York to California in 1 h. As shown in Figure 9.13, there is a broad distribution of speeds among particles in a gas, a distribution that flattens out and moves higher as the temperature increases. Furthermore, an individual gas particle is likely to travel only a very short distance before it collides with another particle and caroms off in a different direction. Thus, the actual path followed by a gas particle is a random zigzag. For helium at room temperature and 1 atm pressure, the distance between collisions (the **mean free path**) is only about 2×10^{-7} m, or 1000 atomic diameters, and there are approximately 10^{10} collisions per second. For a larger O_2 molecule, the mean free path is about 6×10^{-8} m.

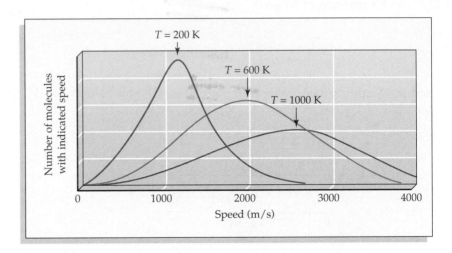

FIGURE 9.13   The distribution of speeds for helium atoms at different temperatures.

┌ **PROBLEM 9.14**   Calculate the average speed of a nitrogen molecule in m/s on a hot day in summer ($T = 37°C$) and on a cold day in winter ($T = -25°C$).   ┘

9.7 ➤ GRAHAM'S LAW: DIFFUSION AND EFFUSION OF GASES

The constant motion and high velocities of gas particles lead to some important consequences. One such consequence is that gases mix rapidly when they come in contact. Take the stopper off a bottle of perfume, for instance, and the odor will spread rapidly through the room as perfume molecules mix with the molecules in the air. This mixing of different gases by random molecular motion and with frequent collisions is called **diffusion**. A similar process in which gas molecules escape through a tiny hole in a membrane without collisions is called **effusion** (Figure 9.14).

According to **Graham's law**, a law formulated in the mid 1800s by the Scottish chemist Thomas Graham, the rate of effusion of a gas is inversely proportional to the square root of its molar mass. In other words, the lighter the molecule, the more rapidly it effuses.

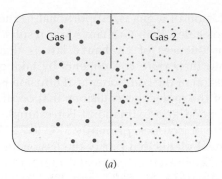
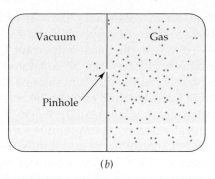

(a)   (b)

FIGURE 9.14   **(a)** Diffusion is the mixing of gas molecules by random motion under conditions where molecular collisions occur. **(b)** Effusion is the escape of a gas through a pinhole without molecular collisions.

GRAHAM'S LAW   The rate of effusion of a gas is inversely proportional to the square root of its molar mass, *M*.

$$\text{Rate} \propto \frac{1}{\sqrt{M}}$$

In comparing two gases at the same temperature and pressure, we can set up an equation saying that the ratio of the effusion rates of the two gases is inversely proportional to the ratio of the square roots of their molar masses:

$$\frac{\text{Rate}_1}{\text{Rate}_2} = \frac{\sqrt{M_2}}{\sqrt{M_1}} = \sqrt{\frac{M_2}{M_1}}$$

The inverse relationship between rate of effusion and square root of the molar mass follows directly from the connection between temperature and kinetic energy described in the previous section. Since temperature is a measure of average kinetic energy, different gases at the same temperature have the same average kinetic energy:

$$\text{Since} \quad \frac{1}{2}mu^2 = \frac{3RT}{2N_A} \quad \text{for any gas}$$

$$\text{then} \quad \left(\frac{1}{2}mu^2\right)_{\text{gas 1}} = \left(\frac{1}{2}mu^2\right)_{\text{gas 2}} \quad \text{at the same } T$$

Canceling the factor of 1/2 from both sides and rearranging, we find that the relative average speeds of the molecules in two gases vary as the inverse ratio of the square roots of their masses:

$$\text{Since} \quad \left(\frac{1}{2}mu^2\right)_{\text{gas 1}} = \left(\frac{1}{2}mu^2\right)_{\text{gas 2}}$$

$$\text{then} \quad (mu^2)_{\text{gas 1}} = (mu^2)_{\text{gas 2}} \quad \text{and} \quad \frac{u^2_{\text{gas 1}}}{u^2_{\text{gas 2}}} = \frac{m_2}{m_1}$$

$$\text{so} \quad \frac{u_{\text{gas 1}}}{u_{\text{gas 2}}} = \frac{\sqrt{m_2}}{\sqrt{m_1}} = \sqrt{\frac{m_2}{m_1}}$$

If, as seems reasonable, the rate of effusion of a gas is proportional to the average molecular speed of the gas molecules, then Graham's law results.

One of the most important consequences of Graham's law is that mixtures of gases can be separated into their pure components by taking advantage of the different rates of diffusion of the components. (Diffusion is more complex than effusion because of the molecular collisions that occur, but Graham's law usually works as a good approximation.) For example, naturally occurring uranium is a mixture of isotopes, primarily ^{235}U (0.72%) and ^{238}U (99.28%). In uranium enrichment plants that purify the fissionable uranium-235 (^{235}U) used for fuel in nuclear reactors, elemental uranium is converted into volatile uranium hexafluoride, bp 56°C, and UF_6 gas is allowed to diffuse from one chamber to another through a permeable membrane. The $^{235}UF_6$ and $^{238}UF_6$ molecules diffuse through the membrane at slightly different rates according to the square root of the ratio of their masses:

$$\text{For } ^{235}UF_6, \ m = 349.03 \text{ amu}$$

$$\text{For } ^{238}UF_6, \ m = 352.04 \text{ amu}$$

so
$$\frac{\text{Rate of } ^{235}U \text{ diffusion}}{\text{Rate of } ^{238}U \text{ diffusion}} = \sqrt{\frac{352.04 \text{ amu}}{349.03 \text{ amu}}} = 1.0043$$

The UF_6 gas that passes through the membrane is thus very slightly enriched in the lighter, faster-moving isotope. After repeating the process many thousands of times, a separation can be achieved. Approximately 85% of the Western world's nuclear fuel supply—some 5000 tons per year—is produced by this gas diffusion method.

The uranium-235 used as a fuel in nuclear reactors is obtained by gas diffusion of UF_6 in these cylinders.

EXAMPLE 9.9

Assume that you have a sample of hydrogen gas containing H_2, HD, and D_2 that you want to separate into pure components (H = ^{1}H, and D = ^{2}H). If Graham's law holds, what are the relative rates of diffusion of the three molecules?

SOLUTION   First, find the masses of the three molecules: for H_2, $m = 2.0$ amu; for HD, $m = 3.0$ amu; for D_2, $m = 4.0$ amu.

Next, apply Graham's law to different pairs of gas molecules. Since D_2 is the heaviest of the three molecules, it will diffuse most slowly, and we'll call its relative rate 1.00. We can then compare HD and H_2 with D_2:

Comparing HD with D_2, we have

$$\frac{\text{Rate of HD diffusion}}{\text{Rate of } D_2 \text{ diffusion}} = \sqrt{\frac{\text{mass of } D_2}{\text{mass of HD}}} = \sqrt{\frac{4.0 \text{ amu}}{3.0 \text{ amu}}} = 1.15$$

Comparing H_2 with D_2 we have

$$\frac{\text{Rate of } H_2 \text{ diffusion}}{\text{Rate of } D_2 \text{ diffusion}} = \sqrt{\frac{\text{mass of } D_2}{\text{mass of } H_2}} = \sqrt{\frac{4.0 \text{ amu}}{2.0 \text{ amu}}} = 1.41$$

Thus, the relative rates of diffusion are H_2 (1.41) > HD (1.15) > D_2 (1.00)

PROBLEM 9.15   Which gas in each of the following pairs diffuses more rapidly, and what are the relative rates of diffusion? **(a)** Kr or O_2 **(b)** N_2 or acetylene, C_2H_2

PROBLEM 9.16   What are the relative rates of effusion of the three naturally occurring isotopes of neon, ^{20}Ne, ^{21}Ne, and ^{22}Ne?

9.8 ►THE BEHAVIOR OF REAL GASES

Before ending our discussion of gases, it's worthwhile reiterating a point made earlier: The behavior of a real gas is often a bit different from that of an ideal gas. For instance, an ideal gas is assumed by kinetic-molecular theory to have no attractive forces between particles. At lower pressures, this assumption is a good one because the gas particles are far apart and the attractive forces between them are negligible. At higher pressures, however, the particles are much closer together and the attractive forces between them become more important. In general, intermolecular attractions become significant at a distance of about 10 molecular diameters and increase rapidly as the distance diminishes (Figure 9.15). As a result, the molecules draw together slightly, making the actual volume of most real gases at pressures up to 300 atm a bit smaller than that predicted by the ideal-gas law.

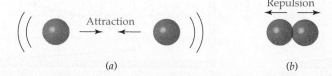

(a)   *(b)*

FIGURE 9.15 Molecules attract each other at distances up to about 10 molecular diameters **(a)** but repel each other when they become too close **(b)**. The result of the attraction is to decrease the actual volume of most real gases when compared with ideal gases at pressures up to 300 atm.

Another problem with real gases is the assumption that the volume of the gas particles themselves is negligible compared with the total gas volume. This assumption, too, is not valid at higher pressures. For instance, the volume taken up by N_2 molecules (approximately 13 cm^3/mol) is only about 0.06% of the total volume of the gas at STP but is about 20% of the total volume at 500 atm and 0°C (Figure 9.16). As a result, the actual volume of a gas at high pressure is larger than predicted by the ideal-gas law.

(a)   (b)

FIGURE 9.16 The volume taken up by the gas particles themselves is more important at higher pressure **(b)** than at lower pressure **(a)**. As a result, the actual volume of the gas at high pressure is somewhat larger than the ideal value.

Note that the effect of molecular volume (to increase V) is opposite that of intermolecular attractions (to decrease V). The two factors therefore tend to cancel at intermediate pressures, but the effect of molecular volume is dominant above about 350 atm. Both problems can be dealt with by modifications of the ideal-gas law (see Problem 9.79 at the end of the chapter), but we won't pursue the matter further at this point.

9.9 ➤ THE EARTH'S ATMOSPHERE

The mantle of gases surrounding the earth is far from the uniform mixture you might expect. Although atmospheric pressure decreases in a regular way at higher altitudes (Figure 9.17a), the profile of temperature versus altitude is much more complex (Figure 9.17b). Four regions of the atmosphere have been defined based on this temperature curve. The temperature in the *troposphere*, the region nearest the earth's surface, decreases regularly up to about 12 km altitude, where it reaches a minimum value and then

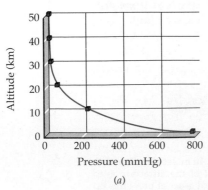

(a)

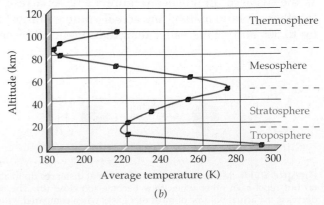

(b)

FIGURE 9.17 Variations of atmospheric pressure **(a)** and average temperature **(b)** with altitude. Four regions of the earth's atmosphere can be defined based on the temperature variations.

increases in the *stratosphere*, up to about 50 km. Above the stratosphere in the region from 50 to 85 km, the temperature again decreases in the *mesosphere* but then increases again in the *thermosphere*. To give you a feel for these altitudes, passenger jets normally fly near the top of the troposphere at altitudes of 10 to 12 km, and the world altitude record for aircraft is 37.65 km—roughly in the middle of the stratosphere.

Chemistry of the Troposphere

Not surprisingly, it's the troposphere that is the most easily disturbed by human activities and that has the greatest effect on the earth's surface conditions. Among those effects, three particularly important ones are air pollution, acid rain, and the greenhouse effect.

The earth and its atmosphere as seen from the moon.

   Air pollution has appeared in the last two centuries as an unwanted byproduct of our industrialized societies. Its causes are relatively straightforward; its control is not. The main causes of air pollution are the release of unburned hydrocarbon molecules and the production of nitric oxide, NO, during combustion of petroleum products in automobile engines. The NO is further oxidized by reaction with air to yield nitrogen dioxide, NO_2, which splits into NO plus free oxygen atoms in the presence of sunlight (symbolized by $h\nu$). Reaction of the oxygen atoms with O_2 molecules then yields ozone, O_3, a highly reactive substance that can further combine with unburned hydrocarbons in the air. The end result is the production of so-called photochemical smog, the hazy, brownish layer lying over so many cities.

$$NO_2(g) + h\nu \rightarrow NO(g) + O(g)$$

$$O(g) + O_2(g) \rightarrow O_3(g)$$

The photochemical smog over many cities is the end result of pollution from automobile exhausts.

Acid rain, a second major environmental problem, results primarily from the production of sulfur dioxide, SO_2, that accompanies the burning of sulfur-containing coal in power-generating plants. Sulfur dioxide is slowly converted to SO_3 by reaction with oxygen in air, and SO_3 dissolves in rainwater to yield dilute sulfuric acid, H_2SO_4.

$$S(\text{in coal}) + O_2(g) \rightarrow SO_2(g)$$

$$2\ SO_2(g) + O_2(g) \rightarrow 2\ SO_3(g)$$

$$SO_3(g) + H_2O(l) \rightarrow H_2SO_4(aq)$$

Among the many dramatic effects of acid rain are the extinction of fish from acidic lakes throughout parts of the northeastern United States, Canada, and Scandinavia, the damage to forests throughout much of central and eastern Europe, and the deterioration everywhere of marble buildings and statuary. Marble is a form of calcium carbonate, $CaCO_3$, and, like all metal

The details on this marble statue have been eaten away over the years by acid rain.

carbonates, reacts with acid to produce CO_2. The result is a slow eating away of the stone.

$$CaCO_3(s) + H_2SO_4(aq) \rightarrow CaSO_4(aq) + H_2O(l) + CO_2(g)$$

The third major atmospheric problem, the so-called greenhouse effect and the global warming that could result, is less well documented and less well understood than either air pollution or acid rain. The basis of concern about the greenhouse problem is the fear that human activities over the past century may have disturbed the earth's delicate thermal balance. One component of this balance is the radiant energy that the earth's surface receives from the sun, a certain amount of which is radiated back into space as infrared energy. Although much of this radiation passes out through the atmosphere, some of it is absorbed by atmospheric gases, particularly water vapor, carbon dioxide, and methane. This absorbed radiation warms the atmosphere and acts to maintain a relatively stable temperature at the earth's surface. Should increasing amounts of radiation be absorbed, however, increased atmospheric heating could result, and global temperatures could rise.

Will Alaska look like this in a few hundred years?

Careful measurements show that concentrations of atmospheric carbon dioxide have been rising in the last century, largely because of the increased use of fossil fuels, from an estimated 290 parts per million (ppm) in 1850 to a current level of 345 ppm (Figure 9.18). Thus, there is concern among many atmospheric scientists that increased absorption of radiation and widespread global warming might follow. Thus far, there have been no detectable consequences of a greenhouse effect, and current worries stem largely from the predictions of sophisticated computer models. These models predict a potential warming by as much as 3°C by the year 2050, an amount that would cause dramatic effects on climate.

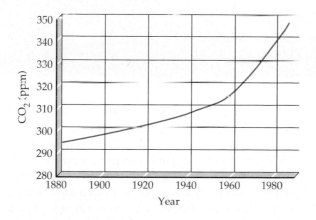

FIGURE 9.18 Concentrations of atmospheric CO_2 have increased dramatically in the last century as a result of increased fossil fuel use. Though no observable consequences have yet been noticed, atmospheric scientists worry that global atmospheric warming may soon occur.

Chemistry of the Upper Atmosphere

Relatively little of the atmosphere's mass is located above the troposphere, but the chemistry that occurs there is nonetheless crucial to maintaining life on earth. Particularly important is what takes place in the *ozone layer*, an atmospheric band stretching from about 20 to 40 km above the earth's surface. Ozone (O_3) is a severe pollutant at low altitudes but is critically important in the upper atmosphere because it absorbs intense ultraviolet radiation from the sun. Even though it is present in very small amounts in the stratosphere, ozone acts as a shield to prevent high-energy solar radiation from reaching the earth's surface, where it can cause such problems as eye cataracts and skin cancer.

Around 1976, a disturbing decrease in the amount of ozone present over the South Pole began showing up (Figure 9.19), and more recently a similar phenomenon has been found over the North Pole. Ozone levels drop to below 50% of normal in the polar spring before returning to near normal in the autumn.

FIGURE 9.19 The antarctic ozone hole from February 1979 to February 1992 as indicated by the Total Ozone Mapping Spectrometer (TOMS). Ozone values in the middle of the hole are up to 50% lower than normal.

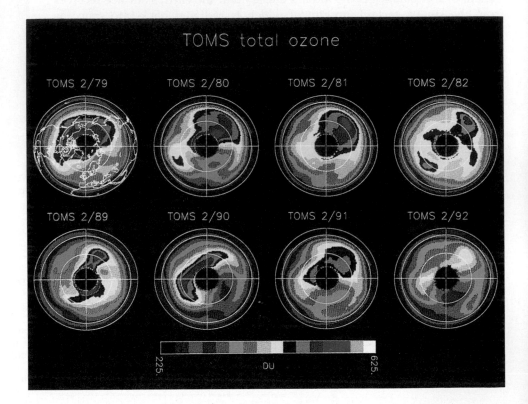

The principal cause of ozone depletion appears to be the presence in the stratosphere of *chlorofluorocarbons* (CFCs), such as CF_2Cl_2 and $CFCl_3$. Widely used as aerosol propellants and as refrigerants in air conditioners, CFCs have been released in large amounts in the past several decades and have ultimately found their way into the stratosphere. There are several different mechanisms of ozone destruction that predominate under different stratospheric conditions. All are multistep processes that begin when ultraviolet light ($h\nu$) strikes a CFC molecule, breaking a carbon–chlorine bond and generating a chlorine atom:

$$CFCl_3 + h\nu \rightarrow CFCl_2 + Cl$$

The resultant chlorine atom reacts with ozone to yield O_2 and ClO, and two ClO molecules then give Cl_2O_2. Further reaction occurs when Cl_2O_2 is struck by more ultraviolet light to generate O_2 and two more chlorine atoms.

$$
\begin{array}{ll}
(1) & 2\,[Cl + O_3 \rightarrow O_2 + ClO] \\
(2) & 2\,ClO \rightarrow Cl_2O_2 \\
(3) & \underline{Cl_2O_2 + h\nu \rightarrow 2\,Cl + O_2} \\
\text{Net:} & 2\,O_3 + h\nu \rightarrow 3\,O_2
\end{array}
$$

Look at the overall result of the above reaction sequence. Chlorine atoms are used in the first step but are regenerated in the third step, so they do not appear in the net equation. Thus, the net sequence is a *chain reaction*, in which the generation of just a few chlorine atoms from a few CFC molecules leads to the destruction of a great many ozone molecules

Recognition of the problem led the U.S. government in 1980 to ban the use of CFCs for aerosol propellants, though they are still widely used as refrigerants. Worldwide action to reduce CFC use began in September 1987, and an international agreement calling for a total ban on the release of CFCs by 1996 was reached in 1992. If nations comply with the agreement, the amount of CFCs in the stratosphere will peak around the year 2000 before slowly declining over the next century.

PROBLEM 9.17 The ozone layer is about 20 km thick, has an average total pressure of 10 mm Hg (1.3×10^{-2} atm), and has an average temperature of 230 K. The partial pressure of ozone in the layer is only about 1.2×10^{-6} mm Hg (1.6×10^{-9} atm). How many meters thick would the layer be if all the ozone contained in it were compressed into a thin layer of pure O_3 at STP?

interlude—INHALED ANESTHETICS

William Morton's demonstration in 1846 of ether-induced anesthesia during dental surgery ranks as one of the most important medical breakthroughs of all time. Before that date, all surgery had been carried out with the patient fully conscious. Use of chloroform as an anesthetic quickly followed Morton's work, made popular by Queen Victoria of England, who in 1853 gave birth to a child while anesthetized by chloroform.

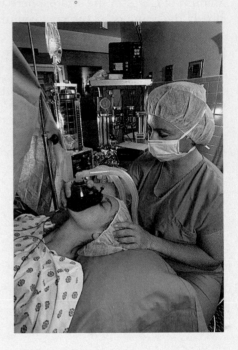

Inhaled anesthetic agents are used to prepare patients for surgery.

Literally hundreds of other substances have subsequently been shown to act as inhaled anesthetics, although halothane, enflurane, isoflurane, and methoxyflurane are at present the most commonly used agents in hospital operating rooms. All four of these substances are potent at relatively low doses, are nontoxic, and are nonflammable, an important safety feature.

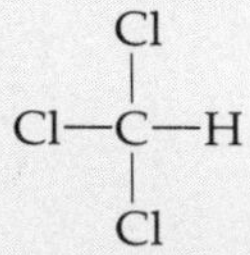

Chloroform

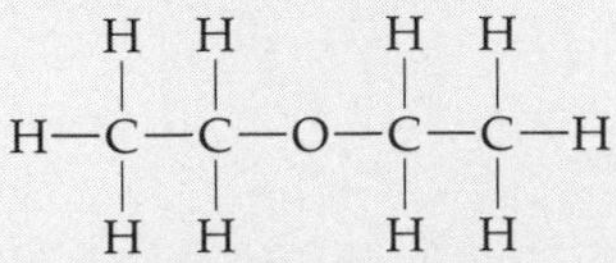

Diethyl ether ("Ether")

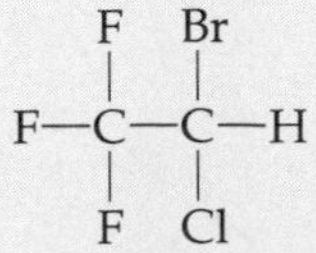

Halothane

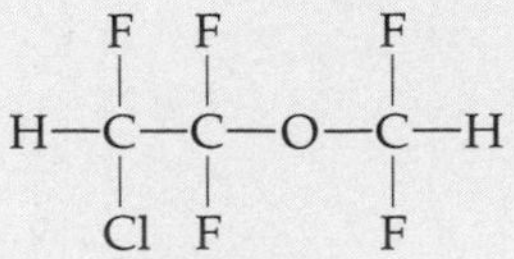

Enflurane

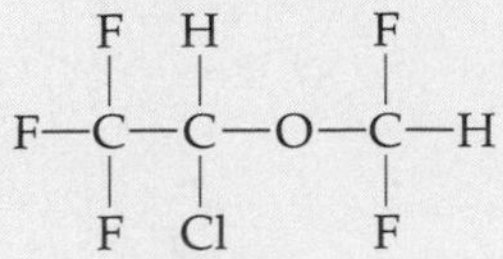

Isoflurane

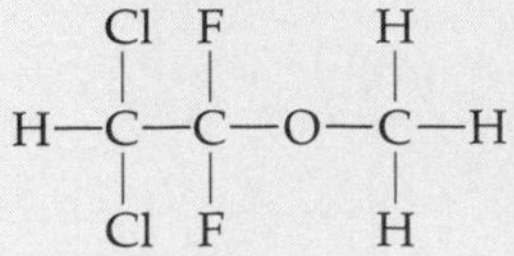

Methoxyflurane

Despite their great importance and utility, surprisingly little is known about how inhaled anesthetics work in the body. Even the definition of anesthesia as a behavioral state is imprecise, and the natures of the changes in brain function leading to anesthesia are unknown. Remarkably enough, the potency of different inhaled anesthetics correlates well with their solubility in olive oil: the more soluble in olive oil, the more potent as an anesthetic. This unusual observation has led many scientists to believe that anesthetics act by dissolving in the fatty membranes surrounding nerve cells. The resultant changes in the fluidity and shape of the membranes apparently decrease the ability of sodium ions to pass into the nerve cells, thereby blocking the firing of nerve impulses.

Depth of anesthesia is determined by the concentration of anesthetic agent that reaches the brain. Brain concentration, in turn, depends on the solubility and transport of the anesthetic agent in the bloodstream and on its partial pressure in inhaled air. Anesthetic potency is usually expressed as a *minimum alveolar concentration (MAC)*, defined as the concentration of anesthetic in inhaled air that results in anesthesia in 50% of patients. As shown in Table 9.6, nitrous oxide, N_2O, is the least potent of the common anesthetics. Fewer than 50% of patients are immobilized by breathing an 80/20 mix of nitrous oxide and oxygen. Methoxyflurane is the most potent agent; a partial pressure of only 1.2 mm Hg is sufficient to anesthetize 50% of patients, and a partial pressure of 1.4 mm Hg will anesthetize 95%.

TABLE 9.6   Relative Potency of Inhaled Anesthetics

Anesthetic	MAC (%)	MAC (partial pressure mm Hg)
Nitrous oxide		>760
Isoflurane	1.4	11
Enflurane	1.7	13
Halothane	0.75	5.7
Methoxyflurane	0.16	1.2

KEY WORDS

SUMMARY

A gas is a large collection of atoms or molecules moving independently through a volume that is largely empty space. Collisions of the randomly moving particles with the walls of the container exert a force per unit area that we perceive as gas pressure. The SI unit for pressure is the *pascal*, but the *atmosphere* and the *millimeter of mercury*, or *torr*, are more commonly used. The physical condition of any gas is defined by four variables: pressure (P), temperature (T), volume (V), and molar amount (n). The specific relationships among these variable are called the **gas laws**:

BOYLE'S LAW — The volume of a gas varies inversely with its pressure. That is, $V \propto 1/P$ or $PV = k$ at constant n, T.

CHARLES' LAW — The volume of a gas varies directly with Kelvin temperature. That is, $V \propto T$ or $V/T = k$ at constant n, P.

AVOGADRO'S LAW — The volume of a gas varies directly with its molar amount. That is, $V \propto n$ or $V/n = k$ at constant T, P.

The three individual gas laws can be combined into a single **ideal-gas law**. The constant R in the equation is called the **gas constant** and has the same value for all gases. At **standard temperature and pressure (STP**; 1 atm and 0°C), the **standard molar volume** of an ideal gas is 22.414 L.

$$\text{IDEAL-GAS LAW} \qquad V = \frac{nRT}{P} \quad \text{or} \quad PV = nRT$$

The value of the ideal gas law is that it simplifies stoichiometric calculations for reactions where one of the components is a gas. If P, V, and T are known, then n can be calculated.

The gas laws apply to mixtures of gases as well as to pure gases. According to **Dalton's law of partial pressures,** the total pressure exerted by a mixture of gases in a container is equal to the sum of the pressures exerted by each individual gas.

The behavior of gases can be accounted for using a model called the **kinetic-molecular theory,** a group of five postulates from which the ideal-gas law can be derived:

1. A gas consists of tiny particles moving about at random.
2. The volume of the particles is negligible compared with the total volume of the gas.
3. There are no forces between particles, either attractive or repulsive.
4. Collisions of the gas particles are elastic.
5. The average kinetic energy of the gas particles is proportional to the absolute temperature.

The connection between temperature and kinetic energy derived from the kinetic-molecular theory makes it possible to calculate the average speed of a gas particle at any temperature. The most important consequence of this relationship is **Graham's law**, which states that the rate of a gas's **effusion**, or spontaneous passage through a pinhole in a membrane, depends inversely on the square root of the gas's mass.

Real gases differ in their behavior from that predicted by the ideal-gas law, particularly at high pressure, where gas particles are forced close together and intermolecular attractions become significant.

1. Assume that you have a sample of gas in a cylinder with a movable piston, as in the following drawing.

Redraw the apparatus to represent the sample after **(a)** the temperature is increased from 300 K to 450 K at constant pressure; **(b)** the pressure is increased from 1 atm to 2 atm at constant temperature; **(c)** the temperature is decreased from 300 K to 200 K, and the pressure is decreased from 3 atm to 2 atm.

2. Assume that you have a sample of gas at 350 K in a sealed container, as represented in **(a)**. Which of the drawings **(b)–(d)** represents the gas after the temperature is lowered from 350 K to 150 K?

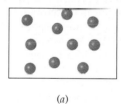

(a)

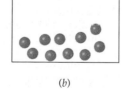

(b)

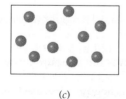

(c)

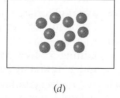

(d)

3. Assume that you have a mixture of He (at. wt = 4 amu) and Xe (at. wt = 131 amu) at 300 K. Which of the drawings best represents the mixture? (blue = He; green = Xe)

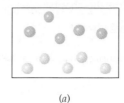

(a)

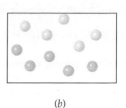

(b)

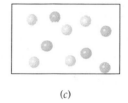

(c)

4. Three bulbs, two of which contain different gases and one of which is empty, are connected as shown in the following drawing.

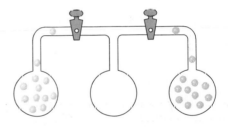

Redraw the apparatus to represent the gases after the stopcocks are opened and the system is allowed to come to equilibrium.

5. The apparatus shown is called a *closed-end* manometer because the arm not connected to the gas sample is closed to the atmosphere and is under vacuum. Explain how you can read the gas pressure in the bulb.

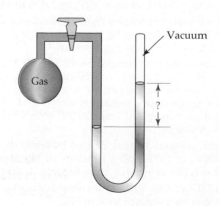

ADDITIONAL PROBLEMS

Problems 9.1–9.17 appear within the chapter.

GASES AND GAS PRESSURE

9.18 What physical phenomenon underlies what we measure as gas pressure?

9.19 What is temperature a measure of?

9.20 Why are gases so much more compressible than solids or liquids?

9.21 Why is atmospheric pressure less at the top of a mountain than at sea level?

9.22 Atmospheric pressure at the top of Pike's Peak in Colorado is approximately 480 mm Hg. Convert this value to atmospheres and to pascals.

9.23 Carry out the following conversions.
  (a) 4.81 atm to pascals
  (b) 1023 mm Hg to atm
  (c) 0.0023 atm to Pa

9.24 Carry out the following conversions.
  (a) 352 torr to kPa   **(b)** 0.255 atm to mm Hg
  (c) 0.0382 mm Hg to Pa

9.25 What is the pressure in mm Hg inside a container of gas connected to a mercury-filled, open-end manometer of the sort shown in Figure 9.4 when the level in the arm connected to the container is 17.6 cm lower than the level in the arm open to the atmosphere, and the atmospheric pressure reading outside the apparatus is 754.3 mm Hg?

9.26 What is the pressure in atm inside a container of gas connected to a mercury-filled, open-end manometer when the level in the arm connected to the container is 28.3 cm higher than the level in the arm open to the atmosphere, and the atmospheric pressure reading outside the apparatus is 1.021 atm?

9.27 Assume that you have an open-end manometer filled with ethyl alcohol (density = 0.7893 g/mL at 20°C) rather than mercury (density = 13.546 g/mL at 20°C). What is the pressure in pascals if the level in the arm open to the atmosphere is 55.1 cm higher than the level in the arm connected to the gas sample, and the atmospheric pressure reading is 752.3 mm Hg?

9.28 Calculate the average molecular weight of air from the data given in Table 9.1.

THE GAS LAWS

9.29 What conditions are defined as standard temperature and pressure?

9.30 Assume that you have a cylinder with a movable piston. What would happen to the gas pressure inside the cylinder if you were to **(a)** triple the Kelvin temperature while holding the volume constant; **(b)** reduce the amount of gas by 1/3 while holding the temperature and volume constant; **(c)** decrease the volume by 45% at constant T; **(d)** halve the Kelvin temperature and triple the volume?

9.31 Which sample contains the most molecules: 1.00 L of O_2 at STP, 1.00 L of air at STP, or 1.00 L of H_2 at STP?

9.32 Which sample contains more molecules: 2.50 L of air at 50°C and 750 mm Hg pressure or 2.16 L of CO_2 at −10°C and 765 mm Hg pressure?

9.33 Oxygen gas is commonly sold in 49.0 L steel containers at a pressure of 150 atm. What volume in liters would the gas occupy at a pressure of 1.02 atm if its temperature remained unchanged? if its temperature was raised from 20.0°C to 35°C at constant $P = 150$ atm?

9.34 A compressed air tank carried by scuba divers has a volume of 8.0 L and a pressure of 140 atm at 20°C. What is the volume of air in the tank in liters at STP?

9.35 If 15.0 g of CO_2 gas has a volume of 0.30 L at 300 K, what is its pressure in mm Hg?

9.36 If 20.0 g of N_2 gas has a volume of 0.40 L and a pressure of 6.0 atm, what is its temperature in K?

9.37 Gas pressure in interplanetary space is approximately 10^{-14} mm Hg at a temperature of approximately 1 K. If the gas is almost entirely hydrogen, what volume in liters is occupied by 1 mol of H_2 molecules? What is the density of H_2 gas in molecules per liter?

9.38 Many laboratory gases are sold in steel cylinders with a volume of 43.8 L. What mass in grams of argon is inside a cylinder whose pressure is 17,180 kPa at 20°C?

9.39 What volume in liters at STP would the argon gas in Problem 9.38 occupy?

9.40 Methane gas, CH_4, is sold in a 43.8 L cylinder containing 5.54 kg. What is the pressure inside the cylinder in kPa at 20°C?

9.41 A small cylinder of helium gas used for filling balloons has a volume of 2.30 L and a pressure of 13,800 kPa at 25°C. How many balloons can you fill if each one has a volume of 1.5 L and a pressure of 1.25 atm at 25°C?

GAS STOICHIOMETRY

9.42 Which sample contains more molecules: 15.0 L of steam (gaseous H_2O) at 123.0°C and 0.93 atm pressure or a 10.5 g ice cube at -5°C?

9.43 Imagine that you have two identical vessels, one containing hydrogen at STP and the other containing oxygen at STP. How can you tell which is which without opening them?

9.44 What is the total mass in grams of oxygen in a room measuring 4.0 m by 5.0 m by 2.5 m? Assume that the gas is at STP and that air contains 20.95% oxygen by volume.

9.45 The average oxygen content of arterial blood is approximately 0.25 g of O_2 per liter. Assuming a body temperature of 37°C, how many moles of oxygen are transported by each liter of arterial blood? How many mL?

9.46 One mole of any gas has a volume of 22.4 L at STP. What are the densities of the following gases in g/L at STP?
   (a) CH_4 **(b)** CO_2 **(c)** O_2 **(d)** UF_6

9.47 An unknown gas is placed in a 1.500 L bulb at a pressure of 356 mm Hg and a temperature of 22.5°C and found to weigh 0.9847 g. What is the molecular weight of the gas?

9.48 What is the molecular weight of a gas with each of the following densities:
   (a) 1.342 g/L at STP
   (b) 1.053 g/L at 25°C and 752 mm Hg

9.49 A mixture of Ar and N_2 gases has a density of 1.413 g/L at STP. What is the mole fraction of each gas?

9.50 What is the density in g/L of a gas mixture that contains 27.0% F_2 and 73.0% He by volume at 714 mm Hg and 27.5°C?

9.51 Pure oxygen gas was first prepared by heating mercuric oxide, HgO:

$$2\ HgO(s) \rightarrow 2\ Hg(l) + O_2(g)$$

What volume in liters of oxygen at STP is released by heating 10.57 g of HgO?

9.52 How many grams of HgO would you need to heat if you wanted to prepare 0.0155 mol of O_2 according to the equation in Problem 9.51?

9.53 Hydrogen gas can be prepared by reaction of zinc metal with aqueous HCl

$$Zn(s) + 2\ HCl(aq) \rightarrow ZnCl_2(aq) + H_2(g)$$

   (a) How many liters of H_2 would be formed at 742 mm Hg and 15°C if 25.5 g of zinc was allowed to react?
   (b) How many grams of zinc would you start with if you wanted to prepare 5.00 L of H_2 at 350 mm Hg and 30.0°C?

9.54 Ammonium nitrate can decompose explosively when heated according to the equation

$$2\ NH_4NO_3(s) \rightarrow 2\ N_2(g) + 4\ H_2O(g) + O_2(g)$$

How many liters of gas would be formed at 450°C and 1.00 atm pressure by explosion of 450 g of NH_4NO_3?

9.55 The reaction of sodium peroxide (Na_2O_2) with CO_2 is used in space vehicles to remove CO_2 from the air and generate O_2 for breathing.

$$2\ Na_2O_2(s) + 2\ CO_2(g) \rightarrow 2\ Na_2CO_3(s) + O_2(g)$$

   (a) Assuming that air is breathed at an average rate of 4.50 L/min (25°C; 735 mm Hg) and that the concentration of CO_2 in expelled air is 3.4% by volume, how many grams of CO_2 are produced in 24 h?
   (b) How many days would a 3.65 kg supply of Na_2O_2 last?

DALTON'S LAW

9.56 Use the information in Table 9.1 to calculate the partial pressure in atm of each gas in dry air at STP.

9.57 Natural gas is a mixture of many substances, primarily CH_4, C_2H_6, C_3H_8, and C_4H_{10}. Assuming that the total pressure of the gases is 1.48 atm and that their mole ratio is 94:4.0:1.5:0.50, calculate the partial pressure in atm of each gas.

9.58 A special gas mixture used in bacterial growth chambers contains 1.00% by weight CO_2 and 99.0% O_2. What is the partial pressure in atm of each gas at a total pressure of 0.977 atm?

9.59 A gas mixture sold commercially for use in certain lasers contains 5.00% by weight HCl, 1.00% H_2, and 94% Ne. What is the mole fraction of each gas in the mixture?

9.60 The HCl/H_2/Ne gas mixture in Problem 9.59 is sold in cylinders that have a volume of 49.0 L and a pressure of 13,800 kPa at 21.0°C. What is the partial pressure in kPa of each gas in the mixture?

9.61 Magnesium metal reacts with aqueous HCl to yield H_2 gas:

$$Mg(s) + 2\ HCl(aq) \rightarrow MgCl_2(aq) + H_2(g)$$

The gas that forms is found to have a volume of 3.557 L at 25°C and a pressure of 747 mm Hg. Assuming that the gas is saturated with water vapor at a partial pressure of 23.8 mm Hg, what is the partial pressure in mm Hg of the H_2? How many grams of magnesium metal were used in the reaction?

KINETIC MOLECULAR THEORY AND GRAHAM'S LAW

9.62 What are the basic assumptions of the kinetic-molecular theory?

9.63 What is the difference between effusion and diffusion?

9.64 What is the difference between heat and temperature?

9.65 Why does a helium-filled balloon lose pressure faster than an air-filled balloon?

9.66 The average temperature at an altitude of 20 km is 220 K. What is the average speed in m/s of an N_2 molecule at this altitude?

9.67 At what temperature in °C will xenon atoms have the same average speed as Br_2 molecules have at 20°C?

9.68 Which has a higher average speed, H_2 at 150 K or He at 375°C?

9.69 An unknown gas is found to diffuse through a porous membrane 2.92 times more slowly than H_2. What is the molecular weight of the gas?

9.70 Rank the following gases in order of their speed of diffusion through a membrane, and calculate the ratio of their diffusion rates: HCl, F_2, Ar.

9.71 Which will diffuse through a membrane more rapidly, CO or N_2? Assume that the samples contain only the most abundant isotopes of each element, ^{12}C, ^{16}O, and ^{14}N.

9.72 A big-league fastball travels at about 45 m/s. At what temperature in °C do helium atoms have this same average speed?

9.73 Traffic on the German autobahns reaches speeds of up to 230 km/h. At what temperature in °C do oxygen molecules have this same average speed?

GENERAL PROBLEMS

9.74 What would the atmospheric pressure be in mm Hg if our atmosphere were composed of pure CO_2 gas?

9.75 Assume that you take a soccer ball, deflate it to remove all the air, squash it flat, and find its mass to be 478.1 g. You then fill the soccer ball with argon to a pressure of 2.15 atm and reweigh it. What would the balance read in grams if the soccer ball has a volume of 7.35 L and the temperature is 20.0°C? You can assume that the volume of the squashed and deflated ball is zero.

9.76 The apparatus shown consists of three bulbs connected by stopcocks. What is the pressure inside the system when the stopcocks are opened? Assume that the lines connecting the bulbs have zero volume and that the temperature remains constant.

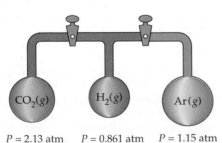

$P = 2.13$ atm $P = 0.861$ atm $P = 1.15$ atm
$V = 1.50$ L $V = 1.00$ L $V = 2.00$ L

9.77 The apparatus shown consists of three temperature-jacketed 1.000 L bulbs connected by stopcocks. Bulb **A** contains a mixture of $H_2O(g)$, $CO_2(g)$, and $N_2(g)$ at 25°C and a total pressure of 564 mm Hg. Bulb **B** is empty and is held at a temperature of −70°C. Bulb **C** is also empty and is held at a temperature of −190°C. The stopcocks are closed, and the volume of the lines connecting the bulbs is zero. CO_2 sublimes at −78°C, and N_2 boils at −196°C.

(a) The stopcock between **A** and **B** is opened, and the system is allowed to come to equilibrium. The pressure in **A** and **B** is now 219 mm Hg. What do bulbs **A** and **B** contain?

(b) How many moles of H_2O are in the system?

(c) Both stopcocks are opened, and the system is again allowed to come to equilibrium. The pressure throughout the system is 33.5 mm Hg. What do bulbs **A**, **B**, and **C** contain?

(d) How many moles of N_2 are in the system?

(e) How many moles of CO_2 are in the system?

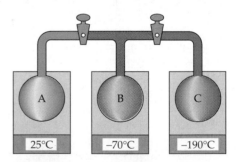

9.78 Assume that you have 1.00 g of nitroglycerin in a 500 mL steel container at 20.0°C and 1.00 atm pressure. An explosion occurs, raising the temperature of the container and its contents to 425°C. The balanced equation is

$$4\ C_3H_5N_3O_9(l) \rightarrow$$
$$12\ CO_2(g) + 10\ H_2O(g) + 6\ N_2(g) + O_2(g)$$

(a) How many moles of nitroglycerin and how many moles of gas (air) were in the container originally?

(b) How many moles of gas are in the container after the explosion?

(c) What is the pressure in atm inside the container after the explosion according to the ideal-gas law?

9.79 As noted in Section 9.8, the behavior of real gases at high pressure differs from that of ideal gases. One way to correct for the difference is to use the van der Waals equation instead of the ideal-gas law:

van der Waals equation: $\left(P + \dfrac{an^2}{V^2}\right)(V - nb) = nRT$

The van der Waals constants a and b are characteristic of each individual gas. Calculate the pressure in atm of 0.60 mol of N_2 in a 200 mL container at 20°C using both the ideal-gas law and the van der Waals equation. [For N_2, $a = 1.39$ (L² · atm)/mol² and $b = 0.0391$ L/mol.]

9.80 Use the van der Waals equation (Problem 9.79) to calculate the pressure in atm of 45.0 g of NH_3 gas in a 250 mL container at 0°C, at 50°C, and at 100°C. Plot your results on a graph with pressure on the horizontal axis and PV/RT on the vertical axis. [For NH_3, $a = 4.17$ (L² · atm)/mol² and $b = 0.0371$ L/mol.]

9.81 Isooctane, C_8H_{18}, is the component of gasoline from which the term *octane rating* derives.
 (a) Write a balanced equation for the combustion of isooctane to yield CO_2 and H_2O.
 (b) Assuming that gasoline is 100% isooctane, that isooctane burns to produce only CO_2 and H_2O, and that the density of isooctane is 0.792 g/mL, what mass of CO_2 in kg is produced each year by the U.S.'s annual gasoline consumption of 4.6×10^{10} L?
 (c) What is the volume in liters of this CO_2 at STP?

9.82 How many moles of air are necessary for the combustion of 1 mol of isooctane (Problem 9.81) assuming that air is 21.0% O_2 by volume? What is the volume in liters of this air at STP?

9.83 The Rankine temperature scale used in engineering is to the Fahrenheit scale as the Kelvin scale is to the Celsius scale. That is, one Rankine degree is the same size as one Fahrenheit degree, and 0°R = absolute zero. What temperature corresponds to the freezing point of water on the Rankine scale? What is the value of the gas constant R in (L · atm)/(°R · mol)?

9.84 When solid mercurous carbonate, Hg_2CO_3, is added to nitric acid, HNO_3, a reaction occurs to give mercuric nitrate, $Hg(NO_3)_2$, water, and two gases **A** and **B**:

$$Hg_2CO_3(s) + HNO_3(aq) \rightarrow Hg(NO_3)_2(aq)$$
$$+ H_2O(l) + \mathbf{A}(g) + \mathbf{B}(g)$$

 (a) When the gases are placed in a 500.0 mL bulb at 20°C, the pressure is 258 mm Hg. How many moles of gas are present?
 (b) When the gas mixture is passed over CaO(s), gas **A** reacts, forming $CaCO_3(s)$:

$$CaO(s) + \mathbf{A}(g) + \mathbf{B}(g) \rightarrow CaCO_3(s) + \mathbf{B}(g)$$

The remaining gas **B** is collected in a 250.0 mL volume at 20°C and found to have a pressure of 344 mm Hg. How many moles of **B** are present?

 (c) The mass of gas **B** collected in part (b) was found to be 0.218 g. What is the density of **B** in g/L?
 (d) What is the molecular weight of **B**, and what is its formula?
 (e) Write a balanced equation for the reaction of mercurous carbonate with nitric acid.

9.85 Chemical explosions are characterized by the instantaneous release of large quantities of hot gases, which set up a shock wave of enormous pressure (up to 700,000 atm) and velocity (up to 20,000 mi/h). For example, explosion of nitroglycerin ($C_3H_5N_3O_9$) releases four gases **A, B, C,** and **D:**

$$n\ C_3H_5N_3O_9(s) \rightarrow a\ \mathbf{A}(g) + b\ \mathbf{B}(g)$$
$$+ c\ \mathbf{C}(g) + d\ \mathbf{D}(g)$$

Assume that the explosion of 1 mol (227 g) of nitroglycerin releases gases with a temperature of 1950°C and a volume of 1323 L at 1.00 atm pressure.
 (a) How many moles of hot gas are released by explosion of 0.00400 mol of nitroglycerin?
 (b) When the products released by explosion of 0.00400 mol of nitroglycerin were placed in a 500.0 mL flask and the flask was cooled to −10°C, product **A** solidified, and the pressure inside the flask was 623 mm Hg. How many moles of **A** were present, and what is its likely identity?
 (c) When gases **B, C,** and **D** were passed through a tube of powdered Li_2O, gas **B** reacted to form Li_2CO_3. The remaining gases, **C** and **D**, were collected in another 500.0 mL flask and found to have a pressure of 260 mm Hg at 25°C. How many moles of **B** were present, and what is its likely identity?
 (d) When gases **C** and **D** were passed through a hot tube of powdered copper, gas **C** reacted to form CuO. The remaining gas, **D**, was collected in a third 500.0 mL flask and found to have a mass of 0.168 g and a pressure of 223 mm Hg at 25°C. How many moles each of **C** and **D** were present, and what are their likely identities?
 (e) Write a balanced equation for the explosion of nitroglycerin.

9.86 Chlorine occurs as a mixture of two isotopes, ^{35}Cl and ^{37}Cl. What is the ratio of the diffusion rates of the three species $(^{35}Cl)_2$, $^{35}Cl^{37}Cl$, and $(^{37}Cl)_2$?

9.87 The surface temperature of Venus is about 1050 K, and the pressure is about 75 earth atmospheres. Assuming that these conditions represent a Venusian "STP," what is the molar volume in liters of a gas on Venus?

9.88 Calcium carbide, CaC_2, reacts with water to give acetylene, C_2H_2, and $Ca(OH)_2$.
 (a) Write a balanced equation for the reaction.
 (b) How many liters of acetylene at 755 mm Hg pressure and 27°C are formed from the reaction of 25.0 g of CaC_2?

chapter 10

LIQUIDS, SOLIDS, AND CHANGES OF STATE

T he kinetic-molecular theory developed in the previous chapter accounted for the properties of gases by assuming that gas particles, whether atoms or molecules, act independently of one another. Because the attractive forces between them are extremely weak, the particles in gases are free to move about at random and occupy whatever space is available. The same is not true in liquids and solids, however. Liquids and solids are distinguished from gases by the presence of strong attractive forces between particles. In liquids, these attractive forces are strong enough to hold the particles in close contact while still letting them slip and slide over one another. In solids, the forces are so strong that they hold the particles rigidly in place and prevent their movement (Figure 10.1).

In this chapter, we'll examine the nature of the forces responsible for the properties of liquids and solids, paying particular attention to the ordering of particles

Water beads up on this lily pad because of surface tension, the tendency of a liquid to minimize surface area.

in solids and to the different kinds of solids that result. In addition, we'll look at what happens during transitions between solid, liquid, and gaseous states and at the effects of temperature and pressure on these transitions.

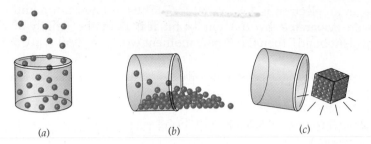

FIGURE 10.1 A molecular comparison of gases, liquids, and solids. **(a)** In gases, the particles feel little attraction for one another and are free to move about randomly. **(b)** In liquids, the particles are held close together by attractive forces but are free to move over each other. **(c)** In solids, the particles are rigidly held in an ordered arrangement.

10.1 ➤ POLAR COVALENT BONDS AND DIPOLE MOMENTS

Before looking at the forces between molecules, it's first necessary to develop the ideas of *bond dipoles* and *dipole moments*. We saw in Section 7.7 that polar covalent bonds form between atoms of different electronegativity. Chlorine is more electronegative than carbon, for example, and the chlorine atom in chloromethane (CH_3Cl) thus attracts the electrons in the C–Cl bond toward itself. The C–Cl bond is therefore polarized so that the chlorine atom is slightly electron rich (δ^-) and the carbon atom is slightly electron poor (δ^+).

Because the polar C–Cl bond in chloromethane has a positive end and a negative end, we describe it as being a bond **dipole,** and we often represent the dipole using an arrow with a cross at one end (↦) to indicate the direction of electron displacement. The point of the arrow represents the negative end of the dipole (δ^-), and the crossed end (which looks like a plus sign) represents the positive end (δ^+):

Cl^{δ^-}
|
C^{δ^+}
H H
H

Chloromethane, CH_3Cl

or

Chlorine is at negative end of bond dipole.

Carbon is at positive end of bond dipole.

Just as individual bonds in molecules are often polar, molecules as a whole are also often polar, because of the net sum of individual bond polarities and lone-pair contributions in the molecule. These *molecular dipoles* can be looked at in the following way: Assume that there is a "center of mass" of all positive charges (nuclei) in a molecule and a center of mass of all negative charges (electrons). If these two centers don't coincide, then the molecule has a net polarity.

The measure of net molecular polarity is a quantity called the **dipole moment** μ (Greek mu), which is defined as the magnitude of the charge Q at either end of the molecular dipole times the distance r between the charges,

$\mu = Q \times r$. Dipole moments are expressed in *debyes* (D), where 1 D = 3.336 × 10^{-30} coulomb meters (C · m) in SI units. To help calibrate your thinking, the charge on an electron is 1.60 × 10^{-19} C. Thus, if one proton and one electron were separated by 100 pm (a bit less than the length of an average covalent bond), then the dipole moment would be 1.60 × 10^{-29} C · m, or 4.80 D.

$$\mu = Q \times r$$

$$\mu = (1.60 \times 10^{-19} \text{ C})(100 \times 10^{-12} \text{ m})\left(\frac{1 \text{ D}}{3.336 \times 10^{-30} \text{ C} \cdot \text{m}}\right) = 4.80 \text{ D}$$

TABLE 10.1
Dipole Moments
of Some Compounds

Compound	Dipole Moment (D)
NaCl[a]	9.0
CH₃Cl	1.87
H₂O	1.85
NH₃	1.47
CO₂	0
CCl₄	0

[a] Measured in the gas phase

It is relatively easy to measure dipole moments experimentally, and values for some common substances are given in Table 10.1. Once the dipole moment is known, it's then possible to get an idea of the amount of charge separation in a molecule. In chloromethane, for example, the experimentally measured dipole moment is μ = 1.87 D. If we assume that the contributions of the nonpolar C–H bonds are small, then most of the chloromethane dipole moment is due to the C–Cl bond. Since the C–Cl bond distance is 178 pm, we can calculate that the dipole moment of chloromethane would be 1.78 × 4.80 D = 8.54 D if the C–Cl bond were ionic (that is, if a full negative charge on chlorine were separated from a full positive charge on carbon by a distance of 178 pm). But because the measured dipole moment of chloromethane is only 1.87 D, the C–Cl bond is only about (1.87/8.54)(100%) = 22% ionic. Thus, the chlorine atom in chloromethane has an excess of about 0.2 electron, and the carbon atom has a deficiency of about 0.2 electron.

Chloromethane (μ = 1.87 D)

Not surprisingly, the largest dipole moment listed in Table 10.1 belongs to the ionic compound NaCl. Water and ammonia also have substantial dipole moments because both oxygen and nitrogen are electronegative relative to hydrogen and because both have lone pairs of electrons:

Ammonia (μ = 1.47 D)   Water (μ = 1.85 D)

In contrast with water and ammonia, carbon dioxide and tetrachloromethane (CCl₄) have zero dipole moments. Molecules in each substance contain *individual* polar covalent bonds, but because of the symmetry of their structures the individual bond polarities in these molecules exactly cancel.

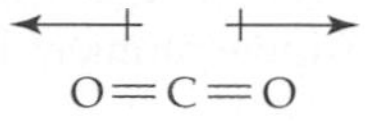

Carbon dioxide ($\mu = 0$) 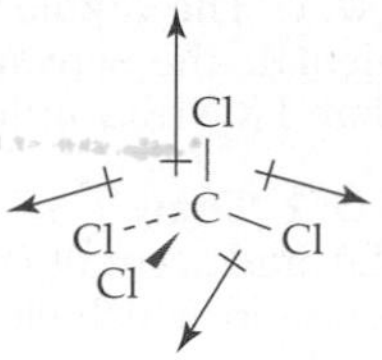   Tetrachloromethane ($\mu = 0$)

EXAMPLE 10.1

The dipole moment of HCl is 1.03 D, and the distance between atoms is 127 pm. Calculate the percent ionic character of the HCl bond.

SOLUTION   If HCl were 100% ionic, then a negative charge (Cl^-) would be separated from a positive charge (H^+) by 127 pm, and we could calculate an expected dipole moment of 6.09 D:

$$\mu = Q \times r$$

$$\mu = (1.60 \times 10^{-19}\ C)(127 \times 10^{-12}\ m)\left(\frac{1\ D}{3.336 \times 10^{-30}\ C \cdot m}\right) = 6.09\ D$$

The fact that the observed dipole moment of HCl is 1.03 D implies that the H–Cl bond is only about 17% ionic.

$$\frac{1.03\ D}{6.09\ D} \times 100\% = 17\%$$

EXAMPLE 10.2

Would you expect vinyl chloride ($H_2C{=}CHCl$), the starting material used for preparation of poly(vinyl chloride) polymer, to have a dipole moment? If yes, indicate the direction.

SOLUTION   First, use the VSEPR model described in Section 7.9 to predict a molecular shape for vinyl chloride. Since both carbon atoms have three charge clouds, each has trigonal planar geometry, and the molecule is planar:

Vinyl chloride

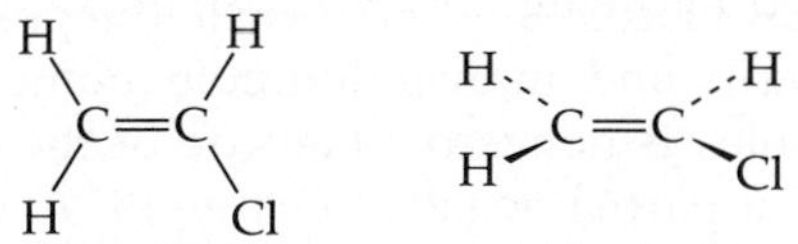

(top view)   (side view)

Next assign polarities to the individual bonds according to the differences in electronegativity of the bonded atoms (Figure 7.4), and then make a reasonable guess about the overall polarity. In vinyl chloride, only the C–Cl bond is polar, giving the molecule a net polarity:

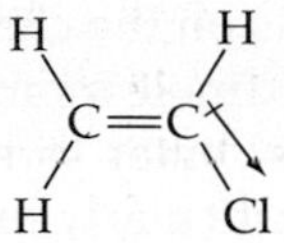

⌐ **PROBLEM 10.1** The dipole moment of HF is $\mu = 1.82$ D, and the bond length is 92 pm. Calculate the percent ionic character of the H–F bond. Is HF more ionic or less ionic than HCl (Example 10.1)?

⌐ **PROBLEM 10.2** Draw a rough three-dimensional structure for methanol (CH_3OH; $\mu = 1.70$ D), and account for its observed dipole moment by using arrows to indicate the direction in which electrons are displaced.

⌐ **PROBLEM 10.3** Tell which of the following compounds is likely to have a dipole moment, and show the direction of each.

(a) SF_6 **(b)** $H_2C=CH_2$ **(c)** $CHCl_3$ **(d)** CH_2Cl_2

10.2 ➤INTERMOLECULAR FORCES

Now that we know a bit about molecular polarities, let's see how they give rise to some of the forces that occur between molecules. The existence of such forces is easy to show. Take H_2O, for example. An individual H_2O molecule consists of two hydrogen atoms and one oxygen atom joined together in a specific way by the *intra*molecular forces that we call covalent bonds. But a visible sample of H_2O exists either as solid ice, liquid water, or gaseous steam depending on its temperature. Thus, there must also be some **inter**molecular forces between molecules that hold them together at certain temperatures. Although strictly speaking the term *intermolecular* refers only to molecular substances, we'll use it generally to refer to interactions among all kinds of particles, including molecules, ions, and atoms.

Intermolecular forces as a whole are usually called **van der Waals forces** after the Dutch scientist Johannes van der Waals (1837–1923). They are usually divided into three categories: *ion–dipole, dipole–dipole,* and so-called *London dispersion forces.* In addition, there is a fourth, more specific intermolecular force that affects only certain kinds of molecules: the *hydrogen bond.* All these intermolecular forces among particles are electrical—they result from the mutual attraction of unlike charges or the mutual repulsion of like charges. If the particles are ions, then full charges are present and the ion–ion attractions are so strong (on the order of 500–1000 kJ/mol) that they give rise to what we call ionic bonds (Section 6.4). If the particles are neutral, then at most partial charges are present, but even so the attractive forces can be substantial.

ION–DIPOLE FORCES   We saw in the previous section that a molecule has a net polarity and an overall dipole moment μ if the sum of its individual bond dipoles is nonzero. One side of the molecule has a net excess of electrons and a partial negative charge (δ^-), while the other side has a net deficiency of electrons and a partial positive charge (δ^+). An **ion–dipole force** is the result of electrical interactions between an ion and the partial charges on a polar molecule (Figure 10.2).

As you might expect, the favored orientation of a dipolar molecule in the presence of ions is one where the positive end of the dipole is near an anion and the negative end of the dipole is near a cation. The magnitude of the interaction energy E depends on the charge on the ion z, on the strength of the dipole as measured by its dipole moment μ, and on the inverse square of the distance r from the ion's center to the midpoint of the dipole: $E = z \cdot \mu/r^2$. Ion–dipole forces are particularly important in aqueous solutions of

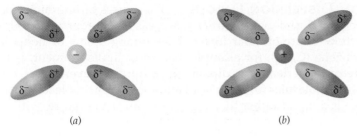

(a)   *(b)*

FIGURE 10.2 Dipolar molecules orient toward ions so that the positive end of the dipole is near an anion **(a)** and the negative end of the dipole is near a cation **(b)**.

ionic substances such as NaCl, in which dipolar water molecules surround the ions. We'll explore this point in more detail in the next chapter.

DIPOLE–DIPOLE FORCES Neutral but polar molecules experience **dipole–dipole forces** as the result of electrical interactions among dipoles on neighboring molecules. The forces can be either attractive or repulsive depending on the orientation of the molecules (Figure 10.3), and the net force in a large collection of molecules is a compromise among many interactions of both types. The forces are generally weak—on the order of 3–4 kJ/mol—and are significant only when molecules are in close contact.

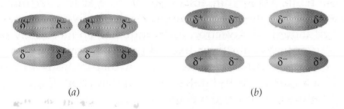

(a)   *(b)*

FIGURE 10.3 Dipolar molecules attract each other when they orient with unlike charges close together as in **(a)** but repel each other when they orient with like charges together as in **(b)**.

Not surprisingly, the strength of a given dipole–dipole interaction depends on the sizes of the dipole moments involved. The more polar the substance, the greater the strength of its dipole–dipole interactions. Table 10.2 lists several substances with approximately equal molecular weight but different dipole moments and shows that there is a reasonable correlation between dipole moment and boiling point. The higher the dipole moment, the stronger the intermolecular forces that must be overcome for a substance to boil, and the higher the boiling point.

TABLE 10.2	Comparison of Molecular Weights, Dipole Moments, and Boiling Points		
Substance	**Mol wt (amu)**	**Dipole Moment (D)**	**bp (K)**
$CH_3CH_2CH_3$	44.10	0.1	231
CH_3OCH_3	46.07	1.3	248
CH_3Cl	50.49	1.9	249
CH_3CN	41.05	3.9	355

LONDON DISPERSION FORCES   The causes of intermolecular forces among charged and polar particles are easy to understand, but it's less obvious how intermolecular forces arise among atoms or nonpolar molecules. Benzene (C_6H_6), for example, has zero dipole moment, is nonpolar, and therefore experiences no dipole–dipole forces. Nevertheless, there must be *some* intermolecular forces present among benzene molecules because the substance is a liquid rather than a gas at room temperature, with a melting point of 5.5°C and a boiling point of 80.1°C.

$\mu = 0$
mp = 5.5°C
bp = 80.1°C

Benzene

All particles, regardless of their polarity, experience **London dispersion forces**, which result from the motion of electrons around atoms. Take atoms of the noble gas helium, for example. Averaged over time, the electron distribution around a helium atom is spherically symmetrical. At any given *instant*, though, the electron distribution in an atom may be unsymmetrical, giving the atom a short-lived dipole moment. This instantaneous dipole on one atom can affect the electron distributions in neighboring atoms and *induce* temporary dipoles in those neighbors (Figure 10.4). As a result, weak attractive forces develop.

(a)   (b)

FIGURE 10.4 Averaged over time, the electron distribution in a helium atom is spherically symmetrical **(a)**. At any given instant, however, the electron distribution in an atom may be unsymmetrical, resulting in a temporary dipole moment in that atom and inducing a complementary attractive dipole in neighboring atoms **(b)**.

TABLE 10.3
Melting Points and Boiling Points of the Halogens

Halogen	mp (K)	bp (K)
F_2	53.5	85.0
Cl_2	172.2	238.6
Br_2	265.9	331.9
I_2	386.7	457.5

London forces are generally small—in the range 1–10 kJ/mol—and their exact magnitude depends on the ease with which a molecule's electron cloud can be distorted by a nearby electric field, a property referred to as **polarizability**. Smaller molecules and lighter atoms with fewer electrons are relatively nonpolarizable and have smaller dispersion forces, whereas larger molecules and heavier atoms with more electrons are more polarizable and have larger dispersion forces. Among the halogens, for example, the F_2 molecule is small and nonpolarizable, while I_2 is larger and more polarizable. As a result, F_2 has smaller dispersion forces and is a gas at room temperature, while I_2 has larger dispersion forces and is a solid (Table 10.3).

Shape is also important in determining the magnitude of the dispersion forces affecting a molecule. More spread-out shapes, which maximize molecular surface area, allow greater contact between molecules and give rise to higher dispersion forces than do more compact shapes, which minimize molecular contact. Pentane, for example, boils at 309.4 K, whereas 2,2-dimethylpropane boils at 282.7 K. Both substances have the same molecular formula, C_5H_{12}, but pentane is long and somewhat spread out, whereas 2,2-dimethylpropane is more compact (Figure 10.5).

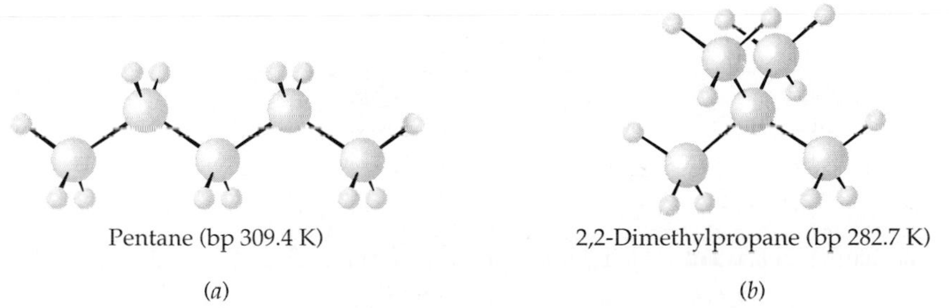

Pentane (bp 309.4 K)

(a)

2,2-Dimethylpropane (bp 282.7 K)

(b)

FIGURE 10.5 Longer, less compact molecules like pentane **(a)** feel stronger dispersion forces and have consequently higher boiling points than do more compact molecules like 2,2-dimethylpropane **(b)**.

HYDROGEN BONDS A **hydrogen bond** is an attractive interaction between a hydrogen atom bonded to an electronegative O, N, or F atom and an unshared electron pair on another nearby electronegative atom. For example, hydrogen bonds occur in both water and ammonia:

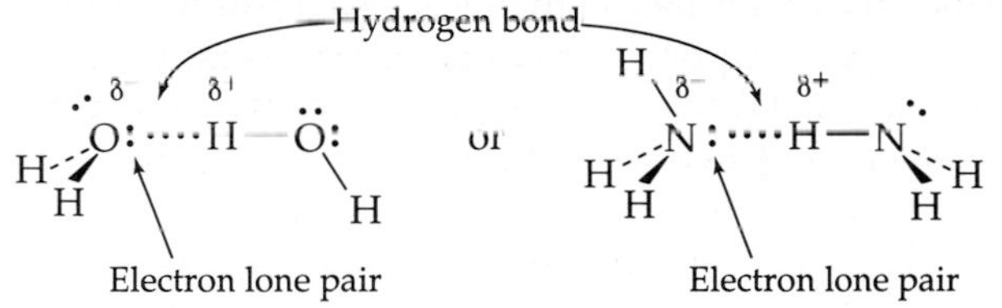

Hydrogen bonds can be quite strong—up to 40 kJ/mol—and are largely responsible for giving water its remarkable properties. To see one effect of hydrogen bonding, look at Figure 10.6, which plots the boiling points of the covalent binary hydrides for the group 4A–7A elements. As you might expect, the boiling points generally increase with molecular weight down a group of the periodic table as a result of increased London dispersion forces—$CH_4 < SiH_4 < GeH_4 < SnH_4$, for example. Three substances, however, are clearly anomalous: NH_3, H_2O, and HF. All three have higher boiling points than might be expected, because of the hydrogen bonds between molecules.

Hydrogen bonds are due to a combination of forces, the most important of which is the dipole–dipole interaction described previously. Bonds between hydrogen and nitrogen, oxygen, or fluorine are highly polar, with a partial positive charge on the hydrogen and a partial negative charge on the electronegative atom. In addition, the hydrogen atom has no core electrons

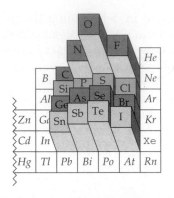

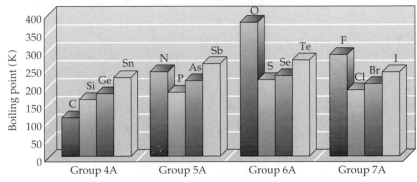

FIGURE 10.6 Boiling points of the covalent binary hydrides of groups 4A, 5A, 6A, and 7A. The boiling points generally increase with increasing molecular weight down a group of the periodic table, but the hydrides of nitrogen (NH₃), oxygen (H₂O), and fluorine (HF) have abnormally high boiling points.

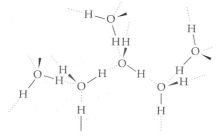

FIGURE 10.7 Liquid water contains a vast three-dimensional network of hydrogen bonds resulting from the attraction between positively polarized hydrogens and negatively polarized oxygens. Each oxygen can form two hydrogen bonds, represented by dotted lines.

to shield its nucleus, and it has a small size, so it can be approached closely. As a result, the dipole–dipole attractions are unusually strong, and hydrogen bonds can form. Water, in particular, is able to form a vast three-dimensional network of hydrogen bonds because each H_2O molecule has two hydrogens and two electron pairs (Figure 10.7).

A comparison of the four kinds of intermolecular forces is shown in Table 10.4.

TABLE 10.4   A Comparison of Intermolecular Forces

Force	Strength	Characteristics
Ion–dipole	Moderate (10–50 kJ/mol)	Occurs between ions and polar solvents
Dipole–dipole	Weak (3–4 kJ/mol)	Occurs between polar molecules
London dispersion	Weak (1–10 kJ/mol)	Occurs between all molecules; strength depends on size, polarizability
Hydrogen bond	Moderate (10–40 kJ/mol)	Occurs between molecules with O—H, N—H, and F—H bonds

EXAMPLE 10.3

Identify the likely kinds of intermolecular forces in the following substances.

(a) HCl   **(b)** CH_3CH_3 **(c)** CH_3NH_2 **(d)** Kr

SOLUTION   **(a)** HCl is a polar molecule but can't form hydrogen bonds. It has dipole–dipole forces and dispersion forces.
(b) CH_3CH_3 is a nonpolar molecule. It has only dispersion forces.
(c) CH_3NH_2 is a polar molecule that can form hydrogen bonds. In addition, it has dipole–dipole forces and dispersion forces.
(d) Kr is nonpolar and has only dispersion forces.

┌ **PROBLEM 10.4** Of the substances Ar, Cl_2, CCl_4, and HNO_3, which has the largest dipole–dipole forces? **(b)** The largest hydrogen-bond forces? **(c)** The smallest dispersion forces?

┌ **PROBLEM 10.5** Consider the kinds of intermolecular forces present in the following compounds, and then rank the substances in likely order of increasing boiling point: H_2S, CH_3OH, CBr_4, Ne

10.3 ➤SOME PROPERTIES OF LIQUIDS

Many familiar and observable properties of liquids can be explained by the intermolecular forces just discussed. We all know, for instance, that some liquids, such as gasoline, flow easily when poured whereas others, such as molasses, flow sluggishly (Figure 10.8). The measure of a liquid's resistance to flow is called the liquid's **viscosity** and is given in the SI unit $N \cdot s/m^2$. Not surprisingly, viscosity is related to the ease with which individual molecules move around in the liquid and thus to the intermolecular forces present. Substances with small, nonpolar molecules, like pentane or benzene, experience only weak intermolecular forces and have relatively low viscosities, whereas more polar substances, like ethanol or glycerol $[C_3H_5(OH)_3]$, experience stronger intermolecular forces and so have higher viscosities.

Another familiar property of liquids is *surface tension*. We're all familiar with seeing water bead-up on a newly waxed car and with the trick of "floating" a needle on water (Figure 10.9). **Surface tension** is the resistance

FIGURE 10.8 Viscous liquids like molasses flow sluggishly when poured. Less viscous liquids like water flow more freely.

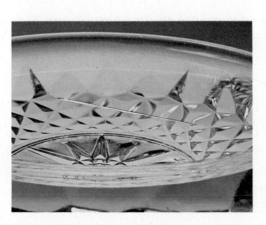

FIGURE 10.9 The ability of a metal needle to "float" on water despite the metal's greater density is due to surface tension, the pulling together of molecules on the surface of the liquid. Were the needle to depress the surface further, the area of the surface would have to increase, which would require energy.

of a liquid to spreading out and increasing its surface area. It is caused by the difference between intermolecular forces felt by molecules at the surface of the liquid and molecules in the interior. Molecules on the interior of a liquid are surrounded by other molecules and are attracted equally in all directions by intermolecular forces. Molecules at the surface, however, are not completely surrounded and are only attracted inward. The result of this uneven pull on the surface molecules is to cause the liquid to minimize its surface area and to make it more difficult for an object to penetrate the surface.

Surface tension, like viscosity, is generally higher in liquids that have stronger intermolecular forces. Both properties are also temperature dependent because molecules at higher temperatures have more kinetic energy to counteract the attractive forces holding them together. Data for some common substances are given in Table 10.5.

TABLE 10.5	Viscosities and Surface Tensions of Some Common Substances at 20°C		
Name	Formula	Viscosity $(N \cdot s/m^2)$	Surface Tension (J/m^2)
Pentane	C_5H_{12}	2.4×10^{-4}	1.61×10^{-2}
Benzene	C_6H_6	6.5×10^{-4}	2.89×10^{-2}
Water	H_2O	1.00×10^{-3}	7.29×10^{-2}
Ethanol	C_2H_5OH	1.20×10^{-3}	2.23×10^{-2}
Mercury	Hg	1.55×10^{-3}	4.6×10^{-1}
Glycerol	$C_3H_5(OH)_3$	1.49	6.34×10^{-2}

10.4 ➤CHANGES OF STATE

Solid ice melts to a liquid, liquid water freezes to a solid or evaporates to a gas, steam condenses to a liquid. Such processes, in which the physical form but not the chemical identity of a substance changes, are called **phase changes**, or **changes of state**. Matter in any one state, or **phase**, can change into either of the other two (Figure 10.10). Solids can even change directly into gases, as occurs when dry ice (solid CO_2) disappears, a process called *sublimation*.

FIGURE 10.10 Changes of state, their names, and their associated energy changes.

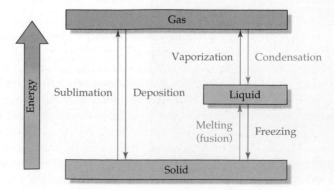

As with all naturally occurring processes, both chemical and physical, every phase change has associated with it a free-energy change, ΔG. This free-energy change ΔG is made up of two contributions, an enthalpy part (ΔH) and a temperature-dependent entropy part ($T\Delta S$), according to the usual equation, $\Delta G = \Delta H - T\Delta S$ (Section 8.13). The enthalpy part is the energy change associated with making or breaking the intermolecular attractions that hold liquids and solids together, while the entropy part is associated with the change in disorder between the various states. As you would expect, gases are more disordered and have more entropy than liquids, which in turn are more disordered and have more entropy than solids.

The melting of a solid to a liquid, the sublimation of a solid to a gas, and the vaporization of a liquid to a gas all involve a change from a less disordered to a more disordered state, and all absorb energy to overcome the intermolecular forces holding particles together. Thus, both ΔS and ΔH are positive for these phase changes. By contrast, the freezing of a liquid to a solid, the deposition of a gas to a solid, and the condensation of a gas to a liquid all involve a change from a more disordered state to a less disordered one, and all *release* energy as intermolecular attractions increase to hold particles more tightly together. Thus, both ΔS and ΔH have negative values for these phase changes. The situations are summarized in Figure 10.11.

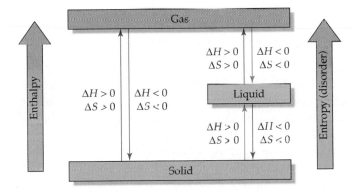

FIGURE 10.11 Phase changes from a less disordered to a more disordered state have positive values of ΔH and ΔS. Phase changes from a more disordered to a less disordered state have negative values of ΔH and ΔS.

Let's look at the ice–to–liquid-water and liquid-water–to–vapor transitions to see some examples of energy relationships during phase changes. For the melting (also called *fusion*) of ice to water, $\Delta H = +6.01$ kJ/mol and $\Delta S = +22.0$ J/(K · mol); for the vaporization of water to steam, $\Delta H = +40.67$ kJ/mol and $\Delta S = +109$ J/(K · mol). Notice that both ΔH and ΔS are larger for the liquid-to-vapor change than for the solid-to-liquid change because many more intermolecular attractions need to be overcome and much more disorder is gained in the liquid-to-vapor change. This highly endothermic conversion of liquid water to gaseous vapor is used by many organisms as a cooling mechanism. When our bodies perspire on a warm day, evaporation of the perspiration absorbs heat and leaves the skin feeling cooler.

Evaporation of perspiration carries away heat and cools the body.

Spraying the trees with water protects them from frost damage, since heat is released when the covering of water freezes.

For phase changes in the opposite direction, the numbers have the same absolute value but opposite signs. That is, $\Delta H = -6.01$ kJ/mol and $\Delta S = -22.0$ J/(K·mol) for the freezing of liquid water to ice, and $\Delta H = -40.67$ kJ/mol and $\Delta S = -109$ J/(K·mol) for the condensation of steam to liquid water. Citrus growers take advantage of the exothermic freezing of water when they spray their trees with water on cold nights to prevent frost damage. As water freezes on the leaves, it releases enough heat to protect the tree.

Knowing the values of ΔH and ΔS for a phase transition makes it possible to calculate the temperature at which the change occurs. Recall from Section 8.13 that ΔG is negative for a spontaneous process, positive for a nonspontaneous process, and zero for a process that is at equilibrium. Thus, by setting $\Delta G = 0$ and solving for T in the free-energy equation, we can calculate the temperature at which two phases are in equilibrium. For the solid-to-liquid phase change in water, for instance, we have

$$\Delta G = \Delta H - T\Delta S = 0$$

or $$T = \Delta H/\Delta S$$

where $\Delta H = +6.01$ kJ/mol and $\Delta S = +22.0$ J/(K·mol).

So $$T = \frac{6.01 \ \dfrac{\text{kJ}}{\text{mol}}}{0.0220 \ \dfrac{\text{kJ}}{\text{K·mol}}} = 273 \text{ K}$$

In other words, ice turns into liquid water, and liquid water turns into ice, at 273 K or 0°C—hardly a surprise. (In practice, the calculation is done in the opposite direction. That is, the temperature at which the phase change occurs is measured and is then used to calculate ΔS for the process: $\Delta S = \Delta H/T$.)

The results of continuously adding heat to a sample can be displayed on a **heating curve** like that in Figure 10.12. Beginning with ice at an arbitrary temperature of -25.0°C, addition of heat raises the ice's tempera-

FIGURE 10.12 A heating curve for H_2O, showing the temperature changes and phase transitions that occur when heat is added. The plateau at 0°C represents the melting of ice, and the plateau at 100°C represents the boiling of water.

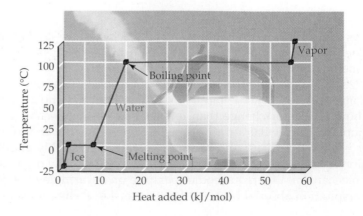

ture until it reaches 0°C. Since the molar heat capacity of ice (Section 8.7) is 36.57 J/(mol·°C), and since we need to raise the temperature 25.0°C, 914 J/mol is required.

$$\text{Energy to heat ice from } -25°C \text{ to } 0°C = 36.57 \frac{J}{mol \cdot °C} \times 25.0°C = 914 \text{ J/mol}$$

Once the temperature of the ice reaches 0°C, addition of further energy goes into disrupting hydrogen bonds and other intermolecular forces rather than into increasing the temperature, as indicated by the plateau at 0°C on the heating curve. At this temperature, the **melting point**, solid and liquid coexist in equilibrium as molecules break free from their positions in the ice crystals and enter the liquid phase. Not until the solid turns completely to liquid does heat again cause the temperature to rise. The amount of energy required for overcoming enough intermolecular forces to convert a solid into a liquid is called the substance's enthalpy of fusion, more commonly called the **heat of fusion**, ΔH_{fusion}. For ice, $\Delta H_{fusion} = +6.01$ kJ/mol.

Continued addition of heat to liquid water raises the temperature until it reaches 100°C. We can calculate from the molar heat capacity of water [75.4 J/(mol·°C)] that 7.54 kJ/mol is required:

$$\text{Energy to heat water from } 0°C \text{ to } 100°C = 75.4 \frac{J}{mol \cdot °C} \times 100°C = 7.54 \times 10^3 \text{ J/mol}$$

Once the temperature of the water reaches 100°C, addition of further energy again goes into disrupting various intermolecular forces rather than into increasing the temperature, as indicated by the second plateau at 100°C on the heating curve. At this temperature, the **boiling point**, liquid and vapor coexist in equilibrium as molecules break free from the surface of the liquid and enter the gas phase. The amount of energy necessary to convert a liquid into a gas is called the enthalpy of vaporization, or **heat of vaporization**, ΔH_{vap}. For water, $\Delta H_{vap} = +40.67$ kJ/mol. Only after the liquid has been completely vaporized does the temperature again rise as the vapor is heated.

Notice that the largest part (40.67 kJ/mol) of the 55.1 kJ/mol required to convert solid ice at −25°C to gaseous steam at 100°C is used for vaporization. The heat of vaporization for water is so large because so many hydrogen bonds must be broken before molecules can escape from the liquid. Table 10.6 gives further data on both heat of fusion and heat of vaporization

TABLE 10.6   Heats of Fusion and Heats of Vaporization for Some Common Compounds

Name	Formula	ΔH_{fusion} (kJ/mol)	ΔH_{vap} (kJ/mol)
Ammonia	NH_3	5.97	23.4
Benzene	C_6H_6	9.95	30.8
Ethanol	C_2H_5OH	5.02	38.6
Helium	He	0.02	0.10
Mercury	Hg	2.33	56.9
Water	H_2O	6.01	40.67

for some common compounds. Note that what is true for water is also true for other compounds: The heat of vaporization of a compound is always larger than its heat of fusion, because almost all intermolecular forces must be overcome before vaporization can occur, but relatively fewer intermolecular forces must be overcome for a solid to change to a liquid.

EXAMPLE 10.4

The boiling point of water is 100°C, and the enthalpy change for the conversion of water to steam is $\Delta H_{vap} = 40.67$ kJ/mol. What is the entropy change in J/(K · mol) for vaporization, ΔS_{vap}?

SOLUTION   At the temperature where a phase change occurs, the two phases coexist in equilibrium and ΔG, the free-energy difference between the phases, is zero: $\Delta G = \Delta H - T\Delta S = 0$.

Rearranging this equation gives $\Delta S = \Delta H/T$, where both ΔH and T are known. Remember that T must be expressed in kelvins:

$$\Delta S_{vap} = \frac{\Delta H_{vap}}{T} = \frac{40.67 \ \frac{kJ}{mol}}{373 \ K} = 0.109 \ kJ/(K \cdot mol) = 109 \ J/(K \cdot mol)$$

As you would expect, there is a large positive entropy change, corresponding to a large increase in disorder on converting water from a liquid to a gas.

⌐ **PROBLEM 10.6**   Which of the following processes would you expect to have a positive value of ΔS and which a negative value? **(a)** sublimation of dry ice **(b)** formation of dew on a cold morning **(c)** mixing of cigarette smoke with air in a closed room

⌐ **PROBLEM 10.7**   Chloroform has an enthalpy of vaporization of 29.2 kJ/mol and an entropy of vaporization of 87.5 J/(K · mol). What is the boiling point of chloroform in K?

10.5 ➤EVAPORATION, VAPOR PRESSURE, AND BOILING POINT

The conversion of a liquid to a vapor takes place in a visible way when the liquid boils, but it takes place under other conditions as well. Let's imagine the two experiments illustrated in Figure 10.13. In the first experiment (a),

FIGURE 10.13 Liquids after sitting for a length of time in an open container **(a)** and a closed container **(b)**. The liquid in the open container has evaporated, but the liquid in the closed container has brought about a rise in pressure.

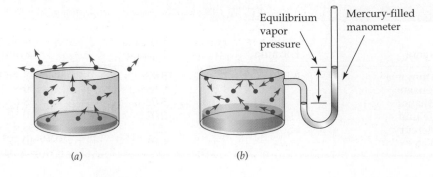

(a)   (b)

we place a liquid in an open container; in the second (b), we place the liquid in a closed container connected to a mercury manometer (Section 9.1). After a certain amount of time has passed, the liquid in the first container has evaporated, while the liquid in the second container remains but the pressure has risen. At equilibrium and at a constant temperature, the pressure has a constant value called the **vapor pressure** of the liquid.

Evaporation and vapor pressure are both explained on a molecular level by the kinetic-molecular theory developed in Section 9.6 to account for the behavior of gases. The molecules in a liquid are in constant motion but at a variety of speeds depending on the amount of kinetic energy each contains. In considering a large sample, molecular kinetic energies follow a distribution curve like that shown in Figure 10.14. The exact shape of the curve depends on the temperature. The higher the temperature and the lower the boiling point of the substance, the greater the fraction of molecules in the sample that have sufficient kinetic energy to break free from the surface of the liquid and escape into the vapor.

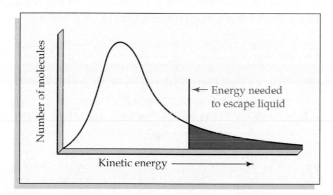

FIGURE 10.14 The distribution of molecular kinetic energies in a hypothetical liquid. Only the faster moving molecules have sufficient kinetic energy to escape from the liquid and enter the vapor.

Molecules that enter the vapor phase in an open container can escape the liquid completely and drift away until the liquid evaporates entirely, but molecules in a closed container are trapped. As more and more molecules pass from the liquid to the vapor, chances increase that random motion will cause some of them to return occasionally to the liquid. Ultimately, the number of molecules escaping the liquid and the number returning to it become equal and a dynamic equilibrium is reached. The total *numbers* of molecules in both liquid and vapor phases are steady, but *individual* molecules are constantly passing back and forth from one phase to the other.

The value of a liquid's vapor pressure depends on the magnitude of the intermolecular forces present and on the temperature. The smaller the intermolecular forces, the higher the vapor pressure because loosely held molecules escape more easily. The higher the temperature, the higher the vapor pressure because a larger fraction of molecules have sufficient kinetic energy to escape.

As indicated by the plots in Figure 10.15(a), the vapor pressure of a liquid rises with temperature in a nonlinear way. A linear relationship *is* found, however, when the logarithm of the vapor pressure, log (P_{vap}) is plotted against the inverse of the Kelvin temperature, $1/T$. Table 10.7 gives the appropriate data for water, and Figure 10.15(b) shows the plot. As noted in Section 9.2, a linear graph is characteristic of mathematical equations of the form $y = mx + b$. In the present instance, $y = \log(P_{vap})$, $x = 1/T$, m is the slope of the line ($-\Delta H_{vap}/2.303\ R$), and b is the y intercept (a constant, C). Thus, the data fit an expression known as the *Clausius-Clapeyron equation*:

Clausius-Clapeyron equation:

$$\underset{\underset{y}{\uparrow}}{\log(P_{vap})} = \underset{\underset{m}{\uparrow}}{\left(-\frac{\Delta H_{vap}}{2.303\ R}\right)} \underset{\underset{x}{\uparrow}}{\frac{1}{T}} + \underset{\underset{b}{\uparrow}}{C} = -\frac{\Delta H_{vap}}{2.303\ RT} + C$$

where ΔH_{vap} is the heat of vaporization of the liquid, R is the familiar gas constant (Section 9.3), and C is a constant characteristic of each specific substance. You might also recognize the value 2.303 as the conversion factor between common and natural logarithms; that is, $\ln x = 2.303 \log x$ (Appendix A.2).

TABLE 10.7	Vapor Pressure of Water at Various Temperatures						
Temp (K)	P_{vap} (mm Hg)	log (P_{vap})	1/T (1/K)	Temp (K)	P_{vap} (mm Hg)	log (P_{vap})	1/T (1/K)
273	4.58	0.661	0.00366	333	149.4	2.174	0.00300
283	9.21	0.964	0.00353	343	233.7	2.369	0.00292
293	17.5	1.243	0.00341	353	355.1	2.550	0.00283
303	31.8	1.502	0.00330	363	525.9	2.721	0.00275
313	55.3	1.743	0.00319	373	760.0	2.881	0.00268
323	92.5	1.966	0.00310	378	906.0	2.957	0.00265

FIGURE 10.15 (a) The vapor pressures of water, ethanol, and diethyl ether show a nonlinear rise when plotted as a function of temperature. (b) A plot of log (P_{vap}) versus $1/T$ (Kelvin) for water, prepared from the data in Table 10.7, shows a linear relationship.

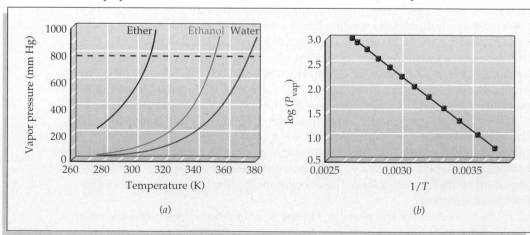

The Clausius-Clapeyron equation makes it possible to calculate the heat of vaporization of a liquid by measuring the vapor pressure at several temperatures and then plotting the results to obtain the slope of the line. Alternatively, once the heat of vaporization and the vapor pressure at one temperature are known, the vapor pressure of the liquid at any other temperature can be calculated, as shown in Example 10.5.

When the vapor pressure of a liquid rises to the point where it is equal to the external pressure pushing on the surface, all the liquid is able to change into the vapor phase, and the liquid boils. You might think of boiling as taking place in the following way: Imagine that a few molecules deep in the interior of the liquid momentarily break free from their neighbors and form a microscopic bubble. If the external pressure from the atmosphere is greater than the vapor pressure inside the bubble, the bubble is immediately crushed and no visible change occurs. At the temperature where the external pressure and the vapor pressure in the bubble are the same, however, the bubble is not crushed. Instead, it rises through the denser liquid, grows larger as more molecules join it, and appears as part of the vigorous action we associate with boiling.

The temperature at which boiling occurs when there is exactly 1 atm of external pressure is called the **normal boiling point** of the liquid. On the plots in Figure 10.15(a), the normal boiling points of the three liquids are reached when the curves cross the dashed line representing 760 mm Hg: for ether, 34.6°C (307.8 K); for ethanol, 78.3°C (351.5 K); and for water, 100.0°C (373.2 K).

If the external pressure on a liquid is *less* than 1 atm, then the vapor pressure necessary for boiling is reached earlier than it is at 1 atm and the liquid boils at a lower-than-normal temperature. On top of Mt. Everest, for example, where the pressure is only about 260 mm Hg, water boils at approximately 71°C. Conversely, if the external pressure on a liquid is *greater* than 1 atm, the vapor pressure necessary for boiling is reached later than it is at 1 atm and the liquid boils at a greater-than-normal temperature. Pressure cookers take advantage of this effect, causing water to boil at a higher temperature and allowing food to cook more rapidly.

At the boiling point, the vapor pressure of the liquid is equal to the external pressure.

It takes a long time to cook at high altitudes because water boils at a lower temperature.

EXAMPLE 10.5

The vapor pressure of ethanol at 34.7°C is 100.0 mm Hg, and the heat of vaporization of ethanol is 38.6 kJ/mol. What is the vapor pressure of ethanol in mm Hg at 65.0°C?

SOLUTION   There are several ways to do this problem. One way is to use the vapor pressure at $T = 34.7$°C to find a value for C in the Clausius-Clapeyron equation

$$C = \log P_{vap} + \frac{\Delta H_{vap}}{2.303\ RT}$$

and then use that value to solve for $\log P_{vap}$ at $T = 65.0$°C.
   Alternatively, because C is a constant, you might recognize that

$$C = \log P_1 + \frac{\Delta H_{vap}}{2.303\ RT_1} = \log P_2 + \frac{\Delta H_{vap}}{2.303\ RT_2}$$

which can be rearranged to solve for the desired quantity, $\log P_2$:

$$\log P_2 = \log P_1 + \frac{\Delta H_{vap}}{2.303\ R}\left(\frac{1}{T_1} - \frac{1}{T_2}\right)$$

where $P_1 = 100.0$ mm Hg and $\log P_1 = 2.0000$, $\Delta H_{vap} = 38.6$ kJ/mol, $R = 8.3145$ J/(K·mol), $T_2 = 65.0$°C (338.2 K), and $T_1 = 34.7$°C (307.9 K).
   Thus,

$$\log P_2 = 2.0000 + \left(\frac{38{,}600\ \dfrac{J}{mol}}{(2.303)\left(8.3145\ \dfrac{J}{K\cdot mol}\right)}\right)\left(\frac{1}{307.9\ K} - \frac{1}{338.2\ K}\right)$$

$$\log P_2 = 2.0000 + 0.5866 = 2.5866$$

$$P_2 = antilog(2.5866) = 386\ mm\ Hg$$

Antilogarithms are reviewed in Appendix A.2.

EXAMPLE 10.6

Ether has $P_{vap} = 400$ mm Hg at 17.9°C and a normal boiling point of 34.6°C. What is the heat of vaporization, ΔH_{vap}, for ether in kJ/mol?

SOLUTION   The heat of vaporization, ΔH_{vap}, of a liquid can be obtained either graphically by taking the slope of a plot of $\log P_{vap}$ versus $1/T$, or algebraically from the Clausius-Clapeyron equation. As derived in Example 10.5,

$$\log P_2 = \log P_1 + \frac{\Delta H_{vap}}{2.303\ R}\left(\frac{1}{T_1} - \frac{1}{T_2}\right)$$

which can be solved for ΔH_{vap}:

$$\Delta H_{vap} = \frac{(\log P_2 - \log P_1)(2.303\ R)}{\left(\dfrac{1}{T_1} - \dfrac{1}{T_2}\right)}$$

where P_1 = 400 mm Hg and log P_1 = 2.602, P_2 = 760 mm Hg at the normal boiling point and log P_2 = 2.881, R = 8.3145 J/(K · mol), T_1 = 17.9°C (291.1 K), and T_2 = 34.6°C (307.8 K).

Thus,

$$\Delta H_{vap} = \frac{(2.881 - 2.602)(2.303)\left(8.3145 \frac{J}{K \cdot mol}\right)}{\left(\frac{1}{291.1 \text{ K}} - \frac{1}{307.8 \text{ K}}\right)} = 28{,}660 \text{ J/mol} = 28.7 \text{ kJ/mol}$$

◸ **PROBLEM 10.8**   The normal boiling point of benzene is 80.1°C, and the heat of vaporization is ΔH_{vap} = 30.8 kJ/mol. What is the boiling point of benzene in °C on top of Mt. Everest, where P = 260 mm Hg?

◸ **PROBLEM 10.9**   Bromine has P_{vap} = 400 mm at 41.0°C and a normal boiling point of 331.9 K. What is the heat of vaporization, ΔH_{vap}, of bromine in kJ/mol? ◿

10.6 ►KINDS OF SOLIDS

It's obvious from a brief look around that most substances are solids rather than liquids or gases. That same brief look also makes it obvious that there are many different kinds of solids. Some solids, such as iron and aluminum, are hard and metallic; others, such as sugar and table salt, are crystalline and easily broken; and still others, such as rubber and many plastics, are soft and amorphous.

The most fundamental distinction between solids is that some are crystalline and some are amorphous. **Crystalline solids** are those whose atoms, ions, or molecules have an ordered arrangement extending over a long range. This order on the atomic level is also seen on the visible level, because crystalline solids usually have flat faces and sharp angles (Figure 10.16). **Amorphous solids**, by contrast, are those whose constituent particles are randomly arranged and have no ordered long-range structure.

Crystalline solids can be further categorized as ionic, molecular, covalent network, or metallic. **Ionic solids** are those like sodium chloride, whose constituent particles are ions. A crystal of sodium chloride is composed of alternating Na^+ and Cl^- ions ordered into a regular three-dimensional

FIGURE 10.16 Crystalline solids, such as the minerals shown here, have flat faces and sharply defined angles. These regular macroscopic features reflect a similarly ordered arrangement at the molecular level.

arrangement and held together by ionic bonds (Section 6.4). **Molecular solids** are those like sucrose or ice, whose constituent particles are molecules held together by the intermolecular forces discussed in Section 10.2. A crystal of ice, for example, is composed of H_2O molecules held together in a regular way by hydrogen bonding (Figure 10.17a). **Covalent network solids** are those like diamond (Figure 10.17b) or quartz (SiO_2), whose atoms are linked together by covalent bonds into a giant three-dimensional array. In effect, a covalent network solid is one *very* large molecule. **Metallic solids** such as silver or iron are like covalent network solids, but they consist of metal atoms, and their crystals have metallic properties such as electrical conductivity.

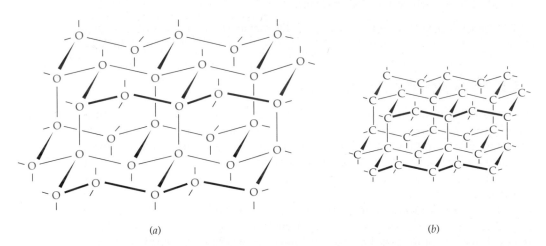

(a)   (b)

FIGURE 10.17 Crystal structures of **(a)** ice, a molecular solid, and **(b)** diamond, a covalent network solid. Ice consists of individual H_2O molecules held together in a regular manner by hydrogen bonds. (Only the oxygen atoms are shown; there is a hydrogen atom near the midpoint of each line between neighboring oxygens.) Diamond is essentially one very large molecule of carbon atoms linked by covalent bonds.

A summary of the different types of crystalline solids and their characteristics is shown in Table 10.8.

TABLE 10.8   Types of Crystalline Solids and Their Characteristics

Type of Solid	Intermolecular Forces	Properties	Examples
Ionic	Ion–ion forces	Brittle, hard, high-melting	NaCl, KBr, $MgCl_2$
Molecular	Dispersion forces, dipole–dipole forces, hydrogen bonds	Soft, low-melting, nonconducting	H_2O, Br_2, CO_2, CH_4
Covalent network	Covalent bonds	Hard, high-melting, nonconducting	C (diamond), SiO_2
Metallic	Metallic bonds	Variable hardness and melting point, conducting	Na, Zn, Cu, Fe

10.7 ►PROBING THE STRUCTURE OF SOLIDS: X-RAY CRYSTALLOGRAPHY

A fundamental principle of optics says that the wavelength of the light used to observe an object must be smaller than the object itself. Since atoms have diameters of around 2×10^{-10} m and the visible light detected by our eyes has wavelengths of $4–7 \times 10^{-7}$ m, it is impossible to see atoms using even the finest optical microscope. To "see" atoms, we must use "light" with a wavelength of less than 10^{-10} m, which is in the X-ray region of the electromagnetic spectrum (Section 5.2).

The origins of X-ray crystallography go back to the work of Max von Laue in 1912. On passing X rays through a crystal of sodium chloride and letting them strike a photographic plate, Laue noticed that a pattern of spots was produced on the plate, indicating that the X rays were being *diffracted* by the atoms in the crystal. A typical modern diffraction pattern is shown in Figure 10.18.

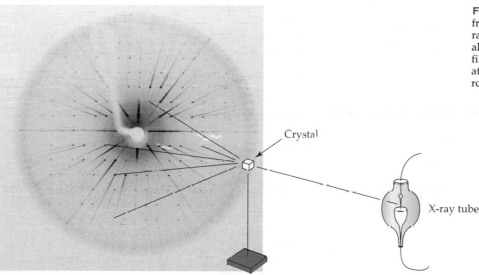

Crystal

X-ray tube

FIGURE 10.18 An X-ray diffraction experiment. A beam of X rays is passed through a crystal and allowed to strike a photographic film. The rays are diffracted by atoms in the crystal, giving rise to a regular pattern of spots on the film.

Diffraction of electromagnetic radiation occurs when a beam is scattered by an object containing regularly spaced lines (such as those in a diffraction grating) or points (such as the atoms in a crystal). This scattering can happen only if the spacing between the lines or points is comparable to the wavelength of the radiation. As shown schematically in Figure 10.19, diffraction is due to *interference* between two overlapping waves. If the waves are in phase, peak to peak and trough to trough, the interference is

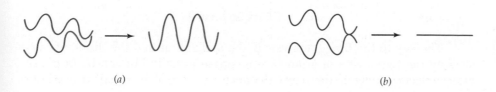

(a)   (b)

FIGURE 10.19 Interference of electromagnetic waves. **(a)** Constructive interference occurs if the waves are in phase and produces a wave with increased intensity. **(b)** Destructive interference occurs if the waves are out of phase and results in cancellation.

constructive and the combined wave is increased in intensity. If the waves are out of phase, however, the interference is destructive and the wave is canceled. Constructive interference gives rise to the intense spots observed on Laue's photographic plate, while destructive interference causes the surrounding light areas.

How does the diffraction of X rays by atoms in a crystal give rise to the observed pattern of spots on a photographic plate? According to an explanation advanced in 1913 by the English physicist William H. Bragg and his 22 year old son William L. Bragg, the X rays are diffracted by different layers of atoms in the crystal, leading to constructive interference in some instances but to destructive interference in others.

To understand the Bragg analysis, imagine that incoming X rays with wavelength λ strike a crystal face at an angle θ and then bounce off at the same angle, just as light bounces off a mirror (Figure 10.20). Those rays that strike an atom in the top layer are reflected off at the same angle θ, and those rays that strike an atom in the second layer are also reflected at angle θ. Because the second layer of atoms is farther from the X-ray source, though, the distance that the X rays have to travel to reach the second layer is farther than the distance they have to travel to reach the first layer by an amount indicated as BC in Figure 10.20. Using trigonometry, it's easy to show that the extra distance BC is equal to the distance between atomic layers d $(= AC)$ times the sine of the angle θ.

$$\sin \theta = \frac{BC}{d}$$

so $$BC = d \times \sin \theta$$

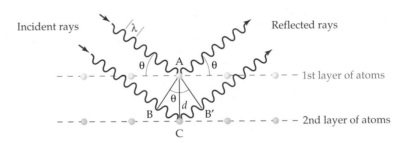

Incident rays   Reflected rays

1st layer of atoms

2nd layer of atoms

FIGURE 10.20 Diffraction of X rays of wavelength λ from atoms in the top two layers of a crystal. Rays striking atoms in the second layer travel a distance equal to $BC + CB'$ farther than do rays striking atoms in the first layer. If this distance is a whole number of wavelengths, the reflected rays are in phase and interfere constructively. Knowing the angle θ then makes it possible to calculate the distance d between the layers.

Of course the extra distance $BC = CB'$ must also be traveled again by the *reflected* rays as they exit the crystal, making the total extra distance traveled equal to $2d \times \sin \theta$.

$$BC + CB' = 2d \times \sin \theta$$

The key to the Bragg analysis is the realization that the different rays striking the two layers of atoms are in phase initially but can be in phase after reflection only if the extra distance $BC + CB'$ is equal to a whole number of wavelengths $n\lambda$, where n is an integer (1, 2, 3 . . .). If the extra distance is not a whole number of wavelengths, then the reflected rays will be out of phase and will cancel. Imposing this requirement on the previously derived relationship between distance and angle θ gives the Bragg equation:

$$\textit{Bragg equation} \qquad n\lambda = 2d \times \sin\theta$$

$$d = \frac{n\lambda}{2\sin\theta}$$

Of the variables in the Bragg equation, the value of the wavelength λ is known, the value of $\sin\theta$ can be measured, and the value of n is an integer generally assumed to be 1. Thus, the distance d between atomic layers in a crystal can be calculated. For their work, the Braggs shared the 1915 Nobel Prize in physics; the younger Bragg was 25 at the time.

Computer-controlled *X-ray diffractometers* are now available that automatically rotate a crystal and measure the diffraction from all planes. Analysis of the X-ray diffraction pattern then makes it possible to measure the interatomic distance between any two atoms in a crystal. For molecular substances, this knowledge of interatomic distances indicates which atoms are close enough to form a bond. X-ray analysis thus provides a means for determining the structures of molecules (Figure 10.21).

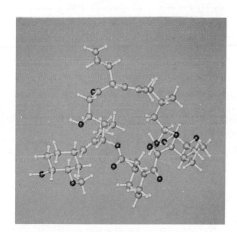

FIGURE 10.21 A computer-generated structure of the experimental drug FK506 as determined by X-ray crystallography. The drug is undergoing current testing for use during organ transplants because of its ability to suppress the human immune system.

10.8 ▸UNIT CELLS IN CRYSTALLINE SOLIDS

How do particles—atoms, ions, or molecules—pack together in crystals? Every crystal is made up of small repeating units called **unit cells,** which are stacked together in three dimensions like the bricks in a wall or the oranges in a grocery store display. Also like bricks and oranges, unit cells have symmetrical geometries that allow them to stack so that minimum space is wasted.

There are 14 different unit-cell geometries that occur in crystalline solids. All unit cells are parallelepipeds (six-sided geometric solids whose faces are parallelograms) and differ only in the lengths of the cell edges and the angles between the edges. We'll be concerned here only with those unit cells that have cubic symmetry, in which all edges are equal in length and all angles are 90°, and we'll be concerned only with those crystals whose

The wall is made of individual repeating units stacked together in a regular way like that of unit cells in crystals.

particles are spheres (atoms or monatomic ions). The situation is similar but a bit more complex for molecular substances.

There are three kinds of cubic unit cells: primitive cubic, body-centered cubic, and face-centered cubic. As shown in Figure 10.22, a **primitive-cubic** unit cell has an atom at each of its eight corners, where it is shared with seven other neighboring cubes that come together at the same point. As a result, we can say that 1/8 of each corner atom "belongs to" a given cubic unit. A **body-centered cubic** unit cell has an additional atom in the center of the cube, and a **face-centered cubic** unit cell has additional atoms on each of its six faces, where each is shared with one other neighboring cube. Thus, 1/2 of each face atom belongs to a given cubic unit.

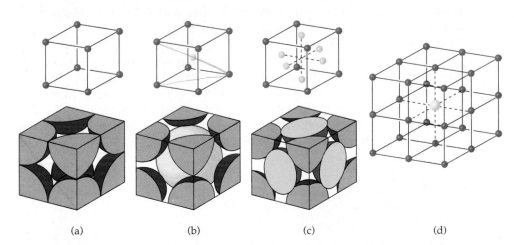

(a)   (b)   (c)   (d)

FIGURE 10.22 Geometries of the three cubic unit cells in both a skeletal view (top) in which each atom is represented by a point, and a space-filling view (bottom): **(a)** primitive cubic, **(b)** body-centered cubic, and **(c)** face-centered cubic. Structure **(d)** shows how eight primitive-cubic unit cells stack together to share a common corner.

EXAMPLE 10.7

How many atoms are in a primitive-cubic unit cell of a metal?

SOLUTION   As shown in Figure 10.22, there is an atom at each corner of the primitive-cubic unit cell, for a total of eight. When unit cells are stacked together, however, each corner atom is shared by eight cubes so that only 1/8 of each atom "belongs" to each unit cell. Thus there is $1/8 \times 8 = 1$ atom per unit cell.

PROBLEM 10.10 When cubic unit cells stack together, how many cubes share a common face? How many cubes share a common edge?

PROBLEM 10.11 How many atoms are in **(a)** a body-centered cubic unit cell and **(b)** a face-centered cubic unit cell of a metal?

10.9 ▸PACKING OF SPHERES
AND THE STRUCTURES OF METALS

Metals represent the simplest examples of crystal packing because individual atoms can be treated as uniform spherical objects. Not surprisingly, metal atoms (and other kinds of particles as well) generally pack together in crystals so that they can be as close together as possible to maximize intermolecular attractions.

If you were to take a large number of uniformly sized marbles and arrange them in a box in some orderly way, there are four possibilities you might come up with. The easiest way to arrange the marbles is in orderly rows and stacks, with the spheres in one layer sitting directly on top of those in the previous layer so that all layers are identical (Figure 10.23a). Called **simple cubic packing**, this arrangement has a primitive-cubic unit cell. Each sphere is touched by six neighbors—four in its own layer, one above, and one below—and is thus said to have a **coordination number** of six. Only 52% of the available volume is occupied by the spheres in simple cubic packing, making very inefficient use of space and thereby minimizing attractive forces. Of all the metals in the periodic table, only polonium crystallizes with this unit cell.

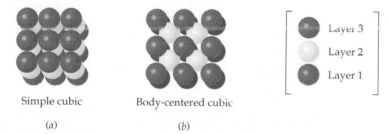

Simple cubic

(a)

Body-centered cubic

(b)

Layer 3
Layer 2
Layer 1

FIGURE 10.23 In simple cubic packing of spheres **(a)**, all the layers are identical, and all atoms are lined up in stacks and rows. Each sphere is touched by six neighbors, four in the same layer, one directly above, and one directly below. In body-centered cubic packing of spheres **(b)**, the second layer is offset from the first so that the atoms in layer 2 fit into the small depressions between atoms in layer 1. The third layer is a repeat of the first. Each sphere is touched by eight neighbors, four in the layer below and four in the layer above.

Alternatively, you might recognize that space could be used more efficiently. Instead of stacking the spheres directly on top of each other, you could offset alternate layers in an *a-b-a-b* arrangement so that the spheres in the *b* layers fit into the small depressions between spheres in the neighboring *a* layers, and vice versa (Figure 10.23b). Called **body-centered cubic**

These spheres have simple cubic packing.

These spheres have body-centered cubic packing.

These spheres have hexagonal closest packing.

These spheres have cubic closest packing.

packing, this arrangement has a body-centered cubic unit cell. Each sphere has a coordination number of eight—four neighbors above and four below—and space is used quite efficiently: 68% of the available volume is occupied. Iron, sodium, and 14 other metals crystallize in this way.

The remaining two packing arrangements of spheres are both called *closest packed*. The **hexagonal closest-packed** arrangement has a noncubic unit cell and has two alternating layers, *a-b-a-b*. Each layer has a hexagonal arrangement of touching spheres, which are offset so that spheres in a *b* layer fit into the small triangular depressions between spheres in an *a* layer (Figure 10.24). Zinc, magnesium, and 19 other metals crystallize in this way.

FIGURE 10.24 Hexagonal closest packing of spheres. There are two alternating hexagonal layers offset from each other so that the spheres in one layer sit in the small triangular depressions of neighboring layers. Layer 3 is a repeat of layer 1. Each sphere is touched by 12 neighbors, 6 in the same layer, 3 in the layer above, and 3 in the layer below.

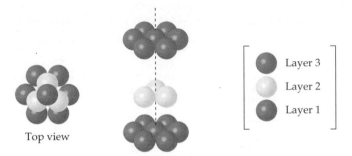

The **cubic closest-packed** arrangement has a face-centered cubic unit cell and has *three* alternating layers, *a-b-c-a-b-c*. The *a-b* layers are identical to those in the hexagonal closest-packed arrangement, but the third layer is offset from both *a* and *b* layers (Figure 10.25). In both kinds of closest-packed arrangements, each sphere has a coordination number of 12—6 neighbors in the same layer, 3 above, and 3 below—and 74% of the available volume is filled. Silver, copper, and 16 other metals crystallize with this arrangement. (The next time you're in a grocery store, look to see how the oranges are stacked in their display box. Chances are good they'll have a closest packed arrangement.)

FIGURE 10.25 Cubic closest packing of spheres. There are three alternating hexagonal layers offset from each other so that the spheres in one layer sit in the small triangular depressions of neighboring layers. Layer 4 is a repeat of layer 1. Each sphere is touched by 12 neighbors, 6 in the same layer, 3 in the layer above, and 3 in the layer below.

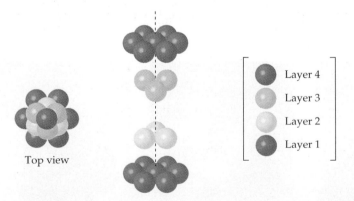

A summary of stacking patterns, coordination numbers, amount of space used, and unit cell for all four kinds of spherical packing is given in Table 10.9.

TABLE 10.9 Summary of the Four Kinds of Spherical Packing				
Structure	Stacking Pattern	Coordination Number	Space Used (%)	Unit Cell
Simple cubic	a-a-a-a-	6	52	Primitive cubic
Body-centered cubic	a-b-a-b-	8	68	Body-centered cubic
Hexagonal closest-packed	a-b-a-b-	12	74	(Noncubic)
Cubic closest-packed	a-b-c-a-b-c-	12	74	Face-centered cubic

EXAMPLE 10.8

Silver metal crystallizes in a cubic closest-packed arrangement with the edge of the unit cell having a length $d = 407$ pm. What is the radius in pm of a silver atom?

SOLUTION   Cubic closest packing uses a face-centered cubic unit cell. Looking at any one face of the cube head-on shows that the face atoms touch the corner atoms along the diagonal of the face but not along the edges. Each diagonal is therefore equal to four atomic radii, $4r$.

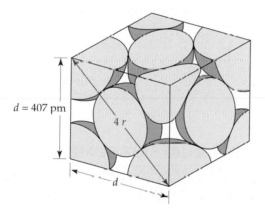

Since the diagonal and two edges of the cube form a right triangle, we can use the Pythagorean theorem to set the sum of the squares of the two edges equal to the square of the diagonal, $d^2 + d^2 = (4r)^2$, and then solve for r, the radius of one atom.

$$d^2 + d^2 = (4r)^2$$

$$2d^2 = 16r^2 \quad \text{and} \quad r^2 = \frac{d^2}{8}$$

Thus, $\qquad r = \sqrt{\frac{d^2}{8}} = \sqrt{\frac{(407 \text{ pm})^2}{8}} = 144 \text{ pm}$

The radius of a silver atom is 144 pm, or 1.44 Å.

EXAMPLE 10.9

Nickel has a face-centered cubic unit cell with a length of 352.4 pm along an edge. What is the density of nickel in g/cm³?

SOLUTION Density is equal to mass divided by volume. The volume of a single cubic unit cell with edge d is simply $d^3 = (3.524 \times 10^{-8} \text{ cm})^3 = 4.376 \times 10^{-23} \text{ cm}^3$. The mass of a single unit cell can be calculated by counting the number of atoms in the cell and multiplying by the mass of a single atom. Each of the eight corner atoms in a face-centered cubic unit cell is shared by eight unit cells so that only $1/8 \times 8 = 1$ atom belongs to a single cell. In addition, each of the six face atoms is shared by two unit cells, so that $1/2 \times 6 = 3$ atoms belong to a single cell. Thus, a single cell has 1 corner atom and 3 face atoms for a total of 4, and each atom has a mass equal to the molar mass of nickel (58.69 g/mol) divided by Avogadro's number (6.022×10^{23} atoms/mol). We can now calculate density:

$$\text{Density} = \frac{\text{mass}}{\text{volume}} = \frac{(4 \text{ atoms})\left(\dfrac{58.69 \dfrac{\text{g}}{\text{mol}}}{6.022 \times 10^{23} \dfrac{\text{atoms}}{\text{mol}}}\right)}{(4.376 \times 10^{-23} \text{ cm}^3)} = 8.91 \text{ g/cm}^3$$

The calculated density of nickel is 8.91 g/cm³. (The actual value is 8.90 g/cm³.)

┌ **PROBLEM 10.12** Polonium metal crystallizes in a simple cubic arrangement with the edge of a unit cell having a length $d = 334$ pm. What is the radius of a polonium atom in pm?

┌ **PROBLEM 10.13** What is the density of polonium (Problem 10.12) in g/cm³?

┌ **PROBLEM 10.14** What is the atomic radius in pm of an argon atom if solid argon has a density of 1.623 g/cm³ and crystallizes at low temperature in a face-centered cubic unit cell? ┘

10.10 ▸STRUCTURES OF SOME IONIC SOLIDS

Simple ionic solids such as NaCl and KBr are similar to metals in that the individual ions are spheres that pack together in a regular way. They differ from metals, however, in that the spheres are not all the same size—anions are generally larger than cations (Section 5.16). As a result, ionic solids adopt a variety of different unit cells depending on the size and charge of the particles. NaCl, KCl, and a number of other salts have a face-centered cubic unit cell in which the larger Cl^- anions occupy corners and face centers, while the smaller Na^+ cations fit into the holes between adjacent anions (Figure 10.26).

FIGURE 10.26 The unit cell of NaCl in both skeletal **(a)** and space-filling **(b)** views. The larger chloride anions adopt a face-centered cubic unit cell with the smaller sodium cations fitting into the holes between adjacent anions.

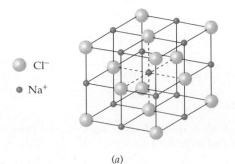

○ Cl⁻

● Na⁺

(a)

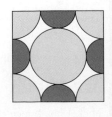

(b)

It's necessary, of course, that the unit cell of an ionic substance have equal numbers of positive and negative charges. In the NaCl unit cell, for instance, there are four Cl^- anions ($1/8 \times 8 = 1$ corner atom, plus $1/2 \times 6 = 3$ face atoms) and also four Na^+ cations ($1/4 \times 12 = 3$ edge atoms, plus 1 center atom). (Remember that each corner atom in a cubic unit cell is shared by eight cells, each face atom is shared by two cells, and each edge atom is shared by four cells.)

Several other common ionic unit cells are shown in Figure 10.27. Copper(I) chloride has a face-centered cubic arrangement of the larger Cl^- anions, with the smaller Cu^+ cations in holes so that each is surrounded by a tetrahedron of four anions. Barium chloride, by contrast, has a face-centered cubic arrangement of the smaller Ba^{2+} *cations*, with the larger Cl^- anions surrounded tetrahedrally. As required for charge neutrality, there are twice as many Cl^- anions as there are Ba^{2+} cations.

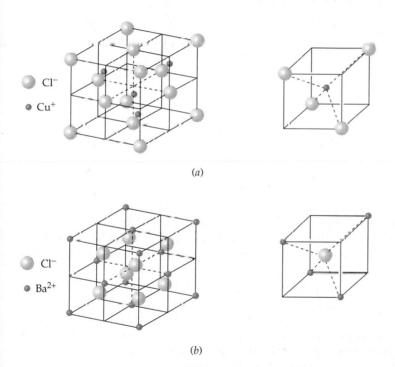

Cl^-

Cu^+

(a)

Cl^-

Ba^{2+}

(b)

FIGURE 10.27 Unit cells of **(a)** CuCl and **(b)** $BaCl_2$. Both are based on a face-centered cubic arrangement of one ion, with the other ion tetrahedrally surrounded in holes.

PROBLEM 10.15 Count the numbers of + and − charges in the CuCl and $BaCl_2$ unit cells (Figure 10.27), and show that both cells are electrically neutral.

10.11 ▶STRUCTURES OF SOME COVALENT NETWORK SOLIDS

Carbon

We saw in Figure 10.17b that the diamond form of elemental carbon is a covalent network solid. Each carbon atom is sp^3 hybridized and covalently bonded with tetrahedral geometry to four other atoms. In addition to diamond, there are more than 40 other known structural forms, or **allotropes**, of carbon, two of which are crystalline and the rest of which are amorphous. Graphite, the most common allotrope of carbon and the most stable under normal conditions, consists of two-dimensional sheets of fused six-membered rings (Figure 10.28a). Each carbon atom is sp^2 hybridized and is connected to three other carbons. Fullerene, a carbon allotrope discovered in 1990 as a constituent of soot, consists of spherical C_{60} molecules with the extraordinary shape of a soccer ball. The C_{60} ball has 12 pentagonal and 20 hexagonal facets, with each atom sp^2 hybridized and bonded to three other atoms (Figure 10.28b).

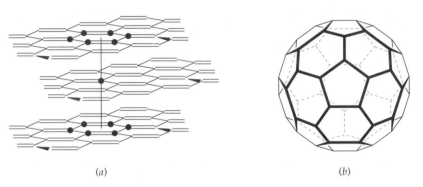

(a)   (b)

FIGURE 10.28 Two allotropes of carbon: **(a)** Graphite is a covalent network solid consisting of two-dimensional sheets of six-membered rings. The atoms in each sheet are offset slightly from the atoms in the neighboring sheets. **(b)** Fullerene, C_{60}, is a molecular solid whose molecules have the shape of a soccer ball. The ball has 12 pentagonal and 20 hexagonal faces, and each carbon atom is sp^2 hybridized.

The different structures of the carbon allotropes lead to widely different properties. Because of its three-dimensional network of single bonds, diamond is the hardest known substance. In addition to its use in jewelry, diamond is widely used industrially for the tips of saw blades and drilling bits. It is an electrical insulator and has a melting point of over 3550°C. Clear, colorless, and highly crystalline, diamonds are very rare and are found in only a few places in the world, primarily in South Africa.

Graphite is the black, slippery substance used as the "lead" in pencils, as an electrode material in batteries, and as a lubricant in locks. All these properties result from its sheetlike structure. Since the layers are held together only by London dispersion forces, they can slide over one another, giving graphite its greasy feeling and lubricating properties. Graphite is more stable than diamond at normal pressures but can be converted into diamond at very high pressure. Some industrial diamonds are now made from graphite by applying 150,000 atm pressure at high temperature.

Fullerene, black and shiny like graphite, is yet too newly discovered to have any useful properties, though research is currently going on at a furious pace.

Diamonds are prized gemstones and an industrial abrasive. Graphite is a soft shiny material used as an electrode material in batteries and as a lubricant.

Silica

Just as living organisms are based on carbon compounds, most rocks and minerals are based on silicon compounds. Quartz and many sands, for instance, are nearly pure *silica*, SiO_2; silicon and oxygen together make up

The sand on this beach is nearly pure SiO_2.

nearly 75% of the mass of the earth's crust. Considering that silicon and carbon are both in group 4A of the periodic table, you might expect SiO_2 to be similar in its properties to CO_2. In fact, though, CO_2 is a molecular substance and a gas at room temperature, whereas SiO_2 is a covalent network solid with a melting point over 1600°C.

The dramatic difference in properties between CO_2 and SiO_2 is due primarily to the difference in electronic structure between carbon and silicon. The π part of a *carbon–oxygen* double bond is formed by sideways overlap of a carbon $2p$ orbital with an oxygen $2p$ orbital. Were a similar *silicon–oxygen* double bond to form, however, it would require overlap of an oxygen $2p$ orbital and a silicon $3p$ orbital. A $3p$ orbital is larger than a $2p$ orbital, however, and overlap between the two is not as effective. As a result, silicon forms four single bonds to four oxygens in a covalent network structure rather than two double bonds to two oxygens in a molecular structure.

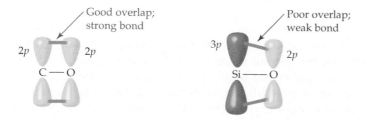

Heating silica above about 1600°C breaks many of its Si–O bonds and turns it from a crystalline solid into a viscous liquid. When this fluid is cooled, some of the Si–O bonds re-form in a random arrangement, and a noncrystalline, amorphous solid called *quartz glass* is formed. If additives are mixed in before cooling, a wide variety of glasses can be prepared. Common window glass, for instance, is prepared by adding $CaCO_3$ and Na_2CO_3. Addition of various transition-metal ions results in the preparation of

Borosilicate glass is used in cooking utensils because of its resistance to thermal shock.

colored glasses, and addition of B_2O_3 produces a high-melting *borosilicate glass* that is sold under the trade name Pyrex. Borosilicate glass is particularly useful for cooking utensils and laboratory ware because it expands very little when heated and is thus highly resistant to thermal shock.

10.12 ➤PHASE DIAGRAMS

Now that we've looked at the three states of matter individually, let's take an overall view. As noted previously, any one state of matter can change spontaneously into either of the other two depending on the temperature and pressure. A particularly convenient way to picture these pressure and temperature dependencies of a pure substance in a closed system is to use what are called **phase diagrams**. As illustrated in Figure 10.29 for water, a typical phase diagram shows which state is stable at different combinations of pressure and temperature, and shows the boundaries between phases. When a boundary line is crossed by changing either the temperature or the pressure, a phase change occurs.

The simplest way to understand a phase diagram is to begin at the origin in the lower left corner of Figure 10.29 and travel up and right along the boundary line between solid on the left and gas on the right. Points on this line represent pressure/temperature combinations at which the two phases are in equilibrium and a phase transition between solid ice and gaseous water vapor (sublimation) occurs. At some point along the solid/gas line, an intersection is reached where two lines diverge to form the bounds of the liquid region. The solid/liquid boundary for H_2O goes up and slightly left, while the liquid/gas boundary curves up and to the right. Called the **triple point**, this three-way intersection represents a unique combination of pressure and temperature at which all three phases coexist in equilibrium. For water, the triple-point temperature T_t is 0.0098°C, and the triple-point pressure P_t is 6.0×10^{-3} atm.

Continuing from the triple point, the solid/liquid boundary line represents the melting point of solid ice (or the freezing point of liquid water) at

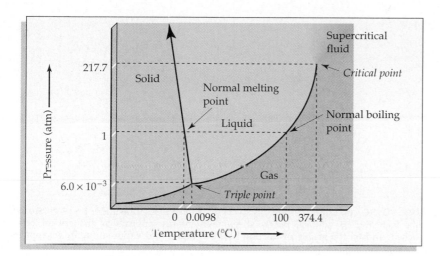

FIGURE 10.29 A phase diagram for H_2O, showing a negative slope for the solid/liquid boundary. Various features of the diagram are discussed in the text. Note that the pressure axis is not to scale.

various pressures. When the pressure is 1 atm, of course, the melting point (called the **normal melting point**) is exactly 0°C. There is a slight negative slope to the line, indicating that the melting point of ice decreases as pressure increases. Water is unusual in this respect, since most substances have a positive slope to their solid/liquid line, indicating that their melting points *increase* with pressure. We'll say more about this behavior shortly.

Starting also from the triple point, the liquid/gas line represents those pressure/temperature combinations at which liquid and gas coexist and water boils (or steam condenses). In fact, the part of the curve up to 1 atm pressure is identical to the vapor pressure curve we saw previously in Section 10.5. When the pressure is 1 atm, water is at its normal boiling point of 100°C. Continuing along the liquid/gas line, we suddenly reach the **critical point**, where the line ends abruptly. The critical temperature T_c is the temperature beyond which a gas cannot be liquefied, no matter how great the pressure; the critical pressure P_c is the pressure required to liquefy a gas at the critical point.

We're all used to seeing solid/liquid and liquid/gas phase transitions, but behavior at the critical point lies so far outside our normal experiences that it's hard to imagine. A gas at the critical point is under such high pressure, and its molecules are so close together, that it becomes indistinguishable from a liquid. A liquid at the critical point is at such a high temperature that it becomes indistinguishable from a gas. The two phases simply blend into one another to form a **supercritical fluid** that is neither liquid nor gas. No distinct physical phase change occurs on going beyond the critical point. Rather, a whitish, pearly sheen momentarily appears, and the visible boundary between liquid and gas then vanishes (Figure 10.30). Frankly, you have to see it to believe it.

The phase diagram of CO_2 shown in Figure 10.31 has many of the same features as that of water but differs in several interesting respects. First, the triple point has $P_t = 5.11$ atm, meaning that CO_2 can't be a liquid below this pressure, no matter what the temperature. At 1 atm pressure, CO_2 is a solid

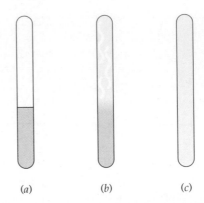

FIGURE 10.30 **(a)** Liquid carbon dioxide is sealed in a glass tube under pressure. **(b)** When the tube is heated, a whitish, pearly sheen momentarily appears as the critical point is approached, and **(c)** the visible distinction between liquid and vapor suddenly disappears at the critical point.

below −78.5°C but a gas above this temperature. Second, the slope of the solid/liquid boundary is positive, meaning that the solid phase is favored as the pressure rises and that the melting point of solid CO_2 therefore increases with pressure.

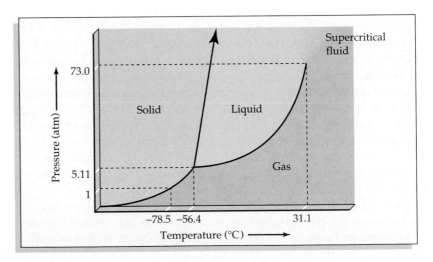

FIGURE 10.31 A phase diagram for CO_2, showing a positive slope for the solid/liquid boundary. The pressure axis is not to scale.

The effect of pressure on the slope of the solid/liquid boundary line—negative for H_2O but positive for CO_2 and most other substances—depends on the relative densities of the solid and liquid phases. For CO_2 and most other substances, the solid phase is denser than the liquid because particles are held closer together in the solid. Increasing the pressure pushes the molecules even closer together, thereby favoring the solid phase even more and giving the solid/liquid boundary line a positive slope. Water, however, reaches its maximum density as a liquid at 4°C (Section 1.13). Water actually becomes *less* dense when it freezes to a solid because large empty spaces are left between molecules due to the ordered three-dimensional network of hydrogen bonds in ice (Figure 10.17a). As a result, increasing the pressure favors the liquid phase, and the solid/liquid boundary has a negative slope.

Figure 10.32 shows a simple demonstration of the effect of pressure on melting point. If a thin wire with heavy weights at each end is draped over a block of ice, the wire rapidly cuts through the block because the increased pressure lowers the melting point of the ice under the wire, causing the ice to liquefy. If the same experiment is tried with a block of dry ice (solid CO_2), however, nothing happens: The dry ice is unaffected because increased pressure makes melting harder rather than easier.

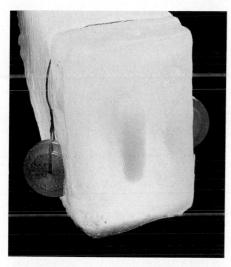

FIGURE 10.32 The weighted wire cuts easily through this block of ice as the increased pressure causes the ice under the wire to melt.

EXAMPLE 10.10

Freeze-dried foods are prepared by freezing the food and removing water by sublimation of ice at low pressure. Look at the phase diagram of water in Figure 10.29, and tell the maximum pressure (in mm Hg) at which ice and water vapor are in equilibrium.

SOLUTION   Ice and water vapor can be in equilibrium only below the triple-point pressure, $P_t = 6.0 \times 10^{-3}$ atm. Converting to mm Hg gives

$$6.0 \times 10^{-3} \text{ atm} \times \frac{760 \text{ mm Hg}}{1 \text{ atm}} = 4.6 \text{ mm Hg}$$

PROBLEM 10.16 Look at the phase diagram of CO_2 in Figure 10.31, and describe what happens to a CO_2 sample when the following changes are made:
(a) The temperature is increased from −100°C to 0°C at a constant pressure of 2 atm.
(b) The pressure is reduced from 72 atm to 5.0 atm at a constant temperature of 30°C.
(c) The pressure is first increased from 3.5 atm to 76 atm, and the temperature is then increased from −10°C to 45°C.

interlude—LIQUID CRYSTALS

The world is rarely as orderly as textbooks make it appear. In this chapter, for instance, we've made it seem that the distinction between liquids and solids is clear-cut and that the phase transition between states is always sharply defined. The truth, though, is more complex. At certain temperatures, many substances (about 0.5% of compounds to be exact) exist in a state that is neither fully liquid nor fully solid. The molecules in these **liquid crystals** can move around as in viscous liquids, but have a restricted range of motion as in solids.

The molecules in most liquid crystals have a rigid, rodlike shape with a length four to eight times greater than the diameter. When packed together, the molecules tend to orient with their long axes roughly parallel, like logs in a stack of firewood. Individual molecules can migrate through the fluid and can spin around their long axis, but they can't rotate end over end. Several different liquid crystalline phases exist depending on the amount of ordering. Two of the most common are the *nematic* phase, in which the ends of the molecules are randomly arranged, and the *smectic* phase, in which the molecules are arranged in layers.

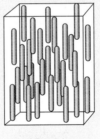
Nematic

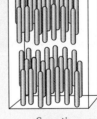
Smectic

The widespread use of liquid crystals for displays in digital watches, pocket calculators, and lap-top computer screens hinges on the fact that the orientation of liquid-crystal molecules is extremely sensitive to the presence of small electric fields and to the nature of nearby surfaces. As shown schematically in Figure 10.33, a typical liquid-crystal display (LCD) contains a thin layer of nematic liquid-crystal molecules sandwiched between two glass sheets that have been rubbed in different directions with a thin nylon brush and then layered with tiny transparent electrode strips made of indium/tin oxide. The outside of each glass sheet is coated with a *polarizer* oriented parallel to the rubbing direction, and the glass sheet that will be on the inside of the display is further coated with a reflecting mirror. Because the molecules in the liquid crystal align parallel to the direction of rubbing, and because the two glass sheets are rubbed at 90° angles to one another, the molecules undergo a gradual 90° twist in orientation between the two surfaces.

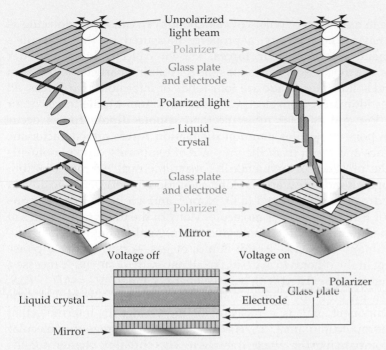

Unpolarized
light beam

Polarizer

Glass plate
and electrode

Polarized light

Liquid
crystal

Glass plate
and electrode

Polarizer

Mirror

Voltage off   Voltage on

Polarizer
Glass plate
Electrode

Liquid crystal →

Mirror →

FIGURE 10.33 Operation of a liquid-crystal display. In the off state, ordinary unpolarized light passes through a polarizer, and the plane of the resultant polarized light is then twisted 90° by interaction with the ordered molecules in the liquid crystal. The polarized light passes through a second polarizer oriented 90° to the first one, strikes a mirror, and is reflected back, where it is seen as white. When a voltage is applied to a pattern of tiny electrodes, the liquid-crystal molecules reorient parallel to the electric field, so that they can no longer rotate the plane of polarized light. The light is thus unable to pass through the second polarizer, can't strike the mirror, and is not reflected back. The pattern is therefore seen as dark.

An LCD works in the following way: A beam of ordinary light consists of electromagnetic waves vibrating in various planes perpendicular to the direction of travel. When light passes through a polarizer, such as the lenses of Polaroid sunglasses, only those waves vibrating in a single plane pass through, resulting in a beam of *plane-polarized light*. The light then passes through a layer of liquid crystal that twists the plane of polarization by 90°, thereby allowing the beam to pass through the second polarizer, strike the mirror, and reflect back to be seen as a white background. To form an LCD image, a voltage is applied to an appropriate pattern of tiny electrodes, causing the molecules in the liquid crystal to reorient parallel to the electric field. As a result, the molecules are no longer able to twist the plane of the polarized light, and the light can no longer pass through the second polarizer to reflect off the mirror. The areas under the electrodes are therefore seen as dark.

The screen in this lap-top computer uses liquid crystals for its display.

The presence in molecules of polar covalent bonds can cause the molecule as a whole to have a net polarity, a property measured by the molecule's **dipole moment**. **Intermolecular forces**, known collectively as **van der Waals forces**, are the attractions responsible for holding particles together in the liquid and solid states. There are four kinds of intermolecular forces, all of which arise ultimately from electrical attractions. **Ion–dipole forces** occur between an ion and a polar molecule, and **dipole–dipole forces** occur between two polar molecules. **London dispersion forces** are characteristic of all molecules and result from the presence of temporary dipole moments caused by the presence of momentarily unsymmetrical electron distributions. **Hydrogen bonds** are a special kind of attraction between a positively polarized hydrogen atom bonded to O, N, or F and a lone-pair of electrons on an O, N, or F atom of another molecule. Many of the unique properties of water are due to extensive intermolecular hydrogen bonding.

Matter in any one state—solid, liquid, or gas—can undergo a **phase change** to any of the other two states. Like all naturally occurring processes, a phase change has associated with it a free-energy change, $\Delta G = \Delta H - T\Delta S$, that has both an enthalpy component and an entropy component. The enthalpy component, ΔH, is a measure of the change in intermolecular forces; the entropy component, ΔS, is a measure of the change in molecular disorder accompanying the phase transition. The enthalpy change for the solid–liquid transition is called the **heat of fusion**, and the enthalpy change for the liquid–vapor transition is the **heat of vaporization**.

The effect of temperature and pressure on phase changes can be displayed graphically on **phase diagrams**. A typical phase diagram has three regions—solid, liquid, and gas—separated by three boundary lines—solid/gas, solid/liquid, and liquid/gas. The boundary lines represent pressure/temperature combinations at which two phases are in equilibrium and phase changes occur. At exactly 1 atm pressure, the temperature at the solid/liquid boundary corresponds to the **normal melting point** of the substance, and the temperature at the liquid/gas boundary corresponds to the **normal boiling point**. The three lines meet at the **triple point**, a unique combination of temperature and pressure at which all three phases coexist in equilibrium. The liquid/gas line runs from the triple point to the **critical point**, a pressure/temperature combination beyond which liquid and gas phases blend together to form a **supercritical fluid** that is neither liquid nor gas.

Solids can be characterized as **amorphous** if their particles are randomly arranged or **crystalline** if their particles are ordered. Crystalline solids can be further characterized as **ionic solids** if their particles are ions, **molecular solids** if their particles are molecules, **covalent network solids** if they consist of a covalently bonded array of atoms without discrete molecules, or **metallic solids** if their particles are metal atoms.

The regular three-dimensional network of particles in a crystal is made up of small repeating units called **unit cells**. There are 14 kinds of unit cells, three of which have cubic symmetry and are commonly used by metals and simple ionic compounds. **Simple cubic packing** uses a primitive-cubic unit cell, with an atom at each corner of the cube. **Body-centered cubic packing** uses a body-centered cubic unit cell, with an atom at the center and at each corner of the cube. **Cubic closest packing** uses a face-centered cubic unit cell, with an atom at the center of each face and at each corner of the cube. A fourth kind of packing, called **hexagonal closest packing**, uses a noncubic unit cell.

UNDERSTANDING KEY CONCEPTS

1. What are the most important kinds of intermolecular forces present in each of the following substances?
(a) toluene, $C_7H_8(l)$ (b) mercury, $Hg(l)$ (c) $NH_3(l)$

2. Draw a picture that shows how hydrogen bonding takes place between two molecules of methanol, CH_3OH.

3. Identify each of the following kinds of packing.

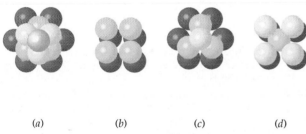

(a)   (b)   (c)   (d)

4. Sketch face-on views of a primitive-cubic unit cell, a face-centered cubic unit cell, and a body-centered cubic unit cell.

5. Zinc sulfide, or sphalerite, crystallizes in the following cubic unit cell.

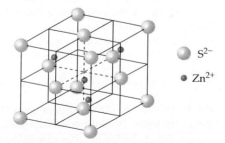

S^{2-}

Zn^{2+}

(a) What kind of packing do the sulfide ions adopt?
(b) How many S^{2-} ions and how many Zn^{2+} ions are in the unit cell?

6. The phase diagram of a substance is shown below.
(a) Approximately what is the normal boiling point and what is the normal melting point of the substance?
(b) Does the density of the solid increase or decrease as the temperature rises?
(c) What is the physical state of the substance under the following conditions. (i) $T = 150$ K, $P = 0.5$ atm (ii) $T = 325$ K, $P = 0.9$ atm (iii) $T = 450$ K, $P = 165$ atm

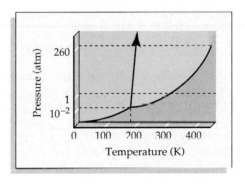

ADDITIONAL PROBLEMS

Problems 10.1–10.16 apear within the chapter.

DIPOLE MOMENTS AND INTERMOLECULAR FORCES

10.17 What is a molecule's dipole moment?

10.18 Why do some molecules have dipole moments while others don't?

10.19 What are the four kinds of intermolecular forces?

10.20 What fundamental interaction gives rise to all intermolecular forces?

10.21 What is the difference between London dispersion forces and dipole-dipole forces?

10.22 What are the most important kinds of intermolecular forces present in each of the following substances?

    (a) chloroform, $CHCl_3$
    (b) oxygen, O_2
    (c) polyethylene, C_nH_{2n+2}
    (d) methanol, CH_3OH

10.23 Methanol (CH_3OH; bp = 65°C) boils nearly 230° higher than methane (CH_4; bp = −164°C), but 1-decanol ($C_{10}H_{21}OH$; bp = 229°C) boils only 55° higher than decane ($C_{10}H_{22}$; bp = 174°C). Explain.

10.24 Which of the following substances would you expect to have a nonzero dipole moment? Explain, and show the direction of each.

    (a) Cl_2O
    (b) XeF_4
    (c) chloroethane, CH_3CH_2Cl
    (d) BF_3

10.25 Which of the following substances would you expect to have a nonzero dipole moment? Explain, and show the direction of each.

    (a) NF_3 **(b)** CH_3NH_2 **(c)** XeF_2 **(d)** PCl_5

10.26 How can you account for the fact that the dipole moment of SO_2 is 1.63 D, but that of CO_2 is zero?

10.27 Of the substances F_2, Kr, CH_3Cl, HF, which has

    (a) the smallest dipole–dipole forces?
    (b) the largest hydrogen-bond forces?
    (c) the largest dispersion forces?

10.28 Which substance in each of the following pairs would you expect to have larger dispersion forces:

    (a) ethane, C_2H_6, or octane, C_8H_{18}
    (b) HCl or HI
    (c) H_2O or H_2Se

10.29 Draw a picture that shows how hydrogen bonding takes place between two ammonia molecules.

VAPOR PRESSURE AND CHANGES OF STATE

10.30 What is the difference between ΔH_{vap} and ΔH_{fusion}? What is the sign of each?

10.31 Why is ΔH_{vap} usually larger than ΔH_{fusion}?

10.32 Why are both ΔH_{vap} and ΔH_{fusion} usually larger than the molar heat capacity of a substance?

10.33 Mercury has mp = −38.9°C and bp = 356.6°C. What, if any, changes of state take place under the following conditions?

    (a) The temperature of a sample is raised from −30°C to 365°C.
    (b) The temperature of a sample is lowered from 291 K to 238 K.
    (c) The temperature of a sample is lowered from 638 K to 231 K.

10.34 Water at room temperature is placed in a flask connected by rubber tubing to a vacuum pump, and the pump is turned on. After several minutes, the volume of the water has decreased and what remains has turned to ice. Explain.

10.35 Ether at room temperature is placed in a flask connected by a rubber tube to a vacuum pump, the pump is turned on, and the ether begins boiling. Explain.

10.36 How much energy (in kJ) is required to heat 5.00 g of ice from −10.0°C to 30.0°C? The heat of fusion of water is 6.01 kJ/mol, and the molar heat capacity of both ice and water is 75.3 J/(K · mol).

10.37 How much energy in kJ is released when 7.55 g of water at 33.5°C is cooled to −10.0°C? (See Problem 10.36.)

10.38 How much energy in kJ is released when 15.3 g of steam at 115.0°C is condensed to give water at 75.0°C? The heat of vaporization of water is 40.67 kJ/mol, and the molar heat capacities of both steam and water are 75.3 J/(K · mol).

10.39 Draw a molar heating curve for ethanol, C_2H_5OH, similar to that shown for water in Figure 10.12. Begin with solid ethanol at its melting point, and raise the temperature to 100°C. The data are: mp = −117.3°C, bp = 78.5°C, ΔH_{fusion} = 4.60 kJ/mol, ΔH_{vap} = 38.6 kJ/mol, heat capacity of liquid = 0.113 kJ/(K · mol), and heat capacity of vapor = 0.0657 kJ/(K · mol).

10.40 Would you expect entropy to increase or decrease in each of the following processes? What is the sign of ΔS in each case?

    (a) Air is suddenly released from a tire in a blow-out.
    (b) Mothballs slowly disappear over time.
    (c) A chemical reaction is carried out:
        $2\ Al(s) + 3\ Br_2(l) \longrightarrow 2\ AlBr_3(s)$

10.41 Benzene has ΔH_{vap} = 30.8 kJ/mol and ΔS_{vap} = +87.2 J/(K · mol). What is the normal boiling point of benzene in °C?

10.42 Naphthalene, better known as "mothballs," has bp = 218°C and ΔH_{vap} = 43.3 kJ/mol. What is the entropy of vaporization, ΔS_{vap}, in J/(K · mol) for naphthalene?

10.43 Sodium metal has mp = 97.8°C and ΔH_{fusion} = 2.64 kJ/mol. What is the entropy of fusion, ΔS_{fusion}, in J/(K · mol) for sodium?

10.44 Carbon disulfide, CS_2, has P_{vap} = 100 mm at −5.1°C and has a normal boiling point of 46.5°C. What is ΔH_{vap} for carbon disulfide in kJ/mol?

10.45 What is the vapor pressure of CS_2 in mm Hg at 20.0°C? (See Problem 10.44.)

10.46 The vapor pressure of $SiCl_4$ is 100 mm Hg at 5.4°C, and the normal boiling point is 56.8°C. What is ΔH_{vap} for $SiCl_4$ in kJ/mol?

10.47 What is the vapor pressure of $SiCl_4$ in mm Hg at 30.0°C? (See Problem 10.46.)

10.48 Dichloromethane, CH_2Cl_2, is an organic solvent used for removing caffeine from coffee beans. The following table gives the vapor pressure of dichloromethane at various temperatures. Fill in the rest of the table, and use the data to plot curves of P_{vap} versus T and log (P_{vap}) versus $1/T$.

Temp (K)	P_{vap} (mm Hg)	log (P_{vap})	1/T
263	80.1	?	?
273	133.6	?	?
283	213.3	?	?
293	329.6	?	?
303	495.4	?	?
313	724.4	?	?

10.49 Use the plot you made in Problem 10.48 to find a value in kJ/mol for ΔH_{vap} for dichloromethane and a value for the constant C in the Clausius-Clapeyron equation.

10.50 Choose any two temperatures and corresponding vapor pressures in the table given in Problem 10.48, and use those values to calculate ΔH_{vap} for dichloromethane in kJ/mol. How does the value you calculated compare to that read from your plot in Problem 10.49?

STRUCTURES OF SOLIDS

10.51 List the four main classes of crystalline solids, and give a specific example of each.

10.52 What kinds of particles are present in each of the four main classes of crystalline solids?

10.53 What is a unit cell?

10.54 Which of the four kinds of packing used by metals makes the most efficient use of space, and which makes the least efficient use of space?

10.55 Copper crystallizes in a face-centered cubic unit cell with an edge length of 362 pm. What is the radius of a copper atom in pm?

10.56 What is the molar volume of copper in cm³ (Problem 10.55)? What is the density of copper in g/cm³?

10.57 Lead crystallizes in a face-centered cubic unit cell with an edge length of 495 pm. What is the radius of a lead atom in pm? What is the density of lead in g/cm³?

10.58 Aluminum has a density of 2.699 g/cm³ and crystallizes with a face-centered cubic unit cell. What is the edge length of a unit cell in pm?

10.59 Tungsten crystallizes in a body-centered cubic unit cell with an edge length of 317 pm. What is the length in pm of a unit-cell diagonal that passes through the center atom?

10.60 In light of your answer to Problem 10.59, what is the radius in pm of a tungsten atom?

10.61 Sodium has a density of 0.971 g/cm³ and crystallizes with a body-centered cubic unit cell. What is the radius in pm of a sodium atom, and what is the edge length in pm of the cell?

10.62 Titanium metal has a density of 4.54 g/cm³ and an atomic radius of 144.8 pm. In what cubic unit cell does titanium crystallize?

10.63 Calcium metal has a density of 1.55 g/cm³ and crystallizes in a cubic unit cell with an edge length of 558.2 pm.
  (a) How many Ca atoms are in one unit cell?
  (b) In which of the three cubic unit cells does calcium crystallize?

10.64 Sodium hydride, NaH, crystallizes in a face-centered cubic unit cell similar to that of NaCl (Figure 10.26). How many Na^+ ions touch each H^- ion, and how many H^- ions touch each Na^+ ion?

10.65 If the edge length of an NaH unit cell is 488 pm, what is the length in pm of an Na–H bond? (See Problem 10.64.)

10.66 Cesium chloride crystallizes in a cubic unit cell with Cl^- ions at the corners and a Cs^+ ion in the center. Count the numbers of + and − charges, and show that the unit cell is electrically neutral.

10.67 The edge length of a CsCl unit cell (Problem 10.66) is 412.3 pm. What is the length in pm of the Cs–Cl bond?

10.68 Use the information in Problem 10.67 and the fact that the ionic radius of a Cl^- ion is 181 pm to calculate the ionic radius in pm of a Cs^+ ion.

10.69 Boron nitride, BN, is a covalent network solid with a structure similar to that of graphite. Sketch a small portion of the boron nitride structure.

PHASE DIAGRAMS

10.70 What two variables are plotted in a phase diagram?

10.71 Look at the phase diagram of CO_2 in Figure 10.31, and tell what phases are present under the following conditions.
  (a) $T = -60°C$, $P = 0.75$ atm
  (b) $T = -35°C$, $P = 18.6$ atm
  (c) $T = -80°C$, $P = 5.42$ atm

10.72 Look at the phase diagram of H_2O in Figure 10.29, and tell what happens to an H_2O sample when the following changes are made:
  (a) The temperature is reduced from 48°C to $-4.4°C$ at a constant pressure of 6.5 atm.
  (b) The pressure is increased from 85 atm to 226 atm at a constant temperature of 380°C.

10.73 Bromine has $T_t = -7.3°C$, $P_t = 44$ mm Hg, $T_c = 315°C$, and $P_c = 102$ atm. The density of the liquid is 3.1 g/cm^3, and the density of the solid is 3.4 g/cm^3. Sketch a rough phase diagram of bromine, and label all points of interest.

10.74 Refer to the bromine phase diagram you sketched in Problem 10.73, and tell what phases are present under the following conditions.
  (a) $T = -10°C$, $P = 0.0075$ atm
  (b) $T = 25°C$, $P = 16$ atm

10.75 Oxygen has $T_t = 54.3$ K, $P_t = 1.14$ mm Hg, $T_c = 154.6$ K, and $P_c = 49.77$ atm. The density of the liquid is 1.14 g/cm^3, and the density of the solid is 1.33 g/cm^3. Sketch a rough phase diagram for oxygen, and label all points of interest.

10.76 Does solid oxygen (Problem 10.75) melt when pressure is applied as water does? Explain.

10.77 Benzene has a melting point of 5.53°C and a boiling point of 80.09°C at atmospheric pressure. Its density is 0.8787 g/cm^3 when liquid and 0.899 g/cm^3 when solid; it has $T_c = 289.01°C$, $P_c = 48.34$ atm, $T_t = 5.52°C$, and $P_t = 0.0473$ atm. Sketch a rough phase diagram for benzene, identifying the triple point, the critical point, the point corresponding to STP, and the solid, liquid, and vapor regions.

10.78 Starting from a point at 200 K and 66.5 atm on the benzene phase diagram you drew in Problem 10.77, trace the following path on your diagram:
  (1) First, increase T to 585 K while keeping P constant.
  (2) Next, decrease P to 38.5 atm while keeping T constant.
  (3) Then, decrease T to 278.66 K while keeping P constant.
  (4) Finally, decrease P to 0.0025 atm while keeping T constant.
  What is your starting phase and what is your final phase?

10.79 How many phase transitions did you pass through in Problem 10.78, and what are they?

10.80 Assume that you have samples of the following three gases at 25°C. Which of the three can be liquefied by applying pressure and which cannot? Explain.
  ammonia: $T_c = 132.5°C$, and $P_c = 112.5$ atm
  methane: $T_c = -82.1°C$, and $P_c = 45.8$ atm
  sulfur dioxide: $T_c = 157.8°C$, and $P_c = 77.7$ atm

GENERAL PROBLEMS

10.81 What are the most important kinds of intermolecular forces present in each of the following?
  (a) octane, C_8H_{18}
  (b) ethylamine, $C_2H_5NH_2$
  (c) glucose, $C_6H_6(OH)_6$
  (d) a solution of NaOH in water

10.82 Mercury has a melting point of $-38.9°C$, a heat capacity of 28.3 J/mol, and a heat of fusion of 2.33 kJ/mol. Assuming that the heat capacities of both solid and liquid are the same and do not change with temperature, how much energy in kJ is needed to heat 7.50 g of Hg from a temperature of $-50.0°C$ to $+50.0°C$?

10.83 Silicon carbide, SiC, is a covalent network solid with a structure similar to that of diamond. Sketch a small portion of the SiC structure.

10.84 In Denver, the Mile-High City, water boils at 95°C. What is atmospheric pressure in atm in Denver? ΔH_{vap} for H_2O is 40.67 kJ/mol.

10.85 Acetic acid, the principal nonaqueous constituent of vinegar, exists as a dimer in the liquid state, with two acetic acid molecules joined together by two hydrogen bonds. Sketch the structure you would expect this dimer to have.

Acetic acid

10.86 How much heat in kJ is released when 15.0 g of gaseous ethanol at 90°C condenses to a liquid at 25°C? The necessary data are given in Problem 10.39.

10.87 Magnesium metal has $\Delta H_{fusion} = 9.037$ kJ/mol and $\Delta S_{fusion} = 9.79$ J/(K · mol). What is the melting point in °C of magnesium?

10.88 Titanium tetrachloride, $TiCl_4$, has a melting point of $-23.2°C$ and has $\Delta H_{fusion} = 9.37$ kJ/mol. What is the entropy of fusion, ΔS_{fusion}, in J/(K·mol) for $TiCl_4$?

10.89 Dichlorodifluoromethane, CCl_2F_2, one of the chlorofluorocarbon refrigerants responsible for destroying part of the earth's ozone layer, has $P_{vap} = 40.0$ mm Hg at $-81.6°C$ and has $P_{vap} = 400$ mm Hg at $-43.9°C$. What is the normal boiling point of CCl_2F_2 in °C?

10.90 Trichlorofluoromethane, CCl_3F, another chlorofluorocarbon refrigerant, has $P_{vap} = 100.0$ mm Hg at $-23°C$ and has $\Delta H_{vap} = 24.77$ kJ/mol.
   (a) What is the normal boiling point of trichlorofluoromethane in °C?
   (b) What is ΔS_{vap} for trichlorofluoromethane?

10.91 Nitrous oxide, N_2O, occasionally used by dentists under the name "laughing gas," has $P_{vap} = 100$ mm Hg at $-110.3°C$ and has a normal boiling point of $-88.5°C$. What is the heat of vaporization of nitrous oxide in kJ/mol?

10.92 Acetone, a common laboratory solvent, has $\Delta H_{vap} = 29.1$ kJ/mol and has a normal boiling point of $56.2°C$. At what temperature in °C does acetone have $P_{vap} = 105$ mm Hg?

10.93 Use the following data to sketch a rough phase diagram for krypton: $T_t = -169°C$, $P_t = 133$ mm Hg, $T_c = -63°C$, $P_c = 54$ atm, mp $= -156.6°C$,

bp $= -152.3°C$. The density of solid krypton is 2.8 g/cm³, and the density of the liquid is 2.4 g/cm³. Can a sample of gaseous krypton at room temperature be liquefied by raising the pressure?

10.94 What is the physical state of krypton (Problem 10.93) under the following conditions:
   (a) $P = 5.3$ atm; $T = -153°C$
   (b) $P = 65$ atm; $T = 250$ K

10.95 Calculate the percent volume occupied by the spheres in a body-centered cubic unit cell.

10.96 Iron crystallizes in a body-centered cubic unit cell with an edge length of 287 pm. What is the radius of an iron atom in pm?

10.97 Iron metal has a density of 7.86 g/cm³ and a molar mass of 55.85 g. Use this information together with the data in Problem 10.96 to calculate a value for Avogadro's number.

10.98 Silver metal crystallizes in a face-centered cubic unit cell, with an edge length of 408 pm. The molar mass of silver is 107.9 g/mol, and its density is 10.50 g/cm³. Use these data to calculate a value for Avogadro's number.

10.99 What is the edge length in pm of the NaCl unit cell (Figure 10.26)? The ionic radius of Na^+ is 97 pm, and the ionic radius of Cl^- is 181 pm.

10.100 Use the information you calculated in Problem 10.99 to determine the density of NaCl in g/cm³.

chapter 11 SOLUTIONS AND THEIR PROPERTIES

U p to this point, we've been concerned primarily with pure substances, both elements and compounds. If you look around, though, it's clear that most of the substances you see in day-to-day life are *mixtures*. Air is a gaseous mixture of (primarily) oxygen and nitrogen, blood is a liquid mixture of many different components, and rocks are solid mixtures of different minerals.

We saw in Section 2.7 that a mixture is any physical combination of two or more pure substances and that mixtures are classified as either *heterogeneous* or *homogeneous*, depending on their appearance. Heterogeneous mixtures are those in which the mixing of components is visually nonuniform; homogeneous mixtures are those in which the mixing *is* uniform, at least to the naked eye. We'll explore the properties of some homogeneous mixtures in this chapter, with particular emphasis on the mixtures we call *solutions*.

These tufa towers in Mono Lake, California, are formed by precipitation of minerals from the highly saline water.

11.1 ►SOLUTIONS

Homogeneous mixtures can be classified according to the size of their constituent particles as suspensions, colloids, or solutions. The divisions among the three groups are not always clear-cut, but the concepts are useful nonetheless. **Suspensions**, such as blood, paint, and aerosol sprays, contain particles that are greater than about 1000 nm in diameter and are visible with a low-power microscope. **Colloids**, such as milk and fog, contain particles with diameters in the range 2–1000 nm. **Solutions**, the most important class of homogeneous mixtures, contain particles the size of a typical ion or covalent molecule—0.2–2 nm in diameter. Table 11.1 summarizes some characteristics of the three classes.

TABLE 11.1	Some Characteristics of Solutions, Colloids, and Suspensions		
Kind of Mixture	**Particle Size**	**Examples**	**Characteristics**
Solution	0.2–2.0 nm	Air, sea water, gasoline, wine	Transparent to light, does not separate on standing, nonfilterable
Colloid	2.0–1000 nm	Butter, milk, fog, pearl	Often murky or opaque to light, does not separate on standing, nonfilterable
Suspension	>1000 nm	Blood, paint, aerosol sprays	Murky or opaque to light, separates on standing, filterable

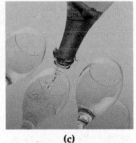

(a) (b) (c)

(a) An aerosol spray is a suspension of small particles that are visible to the naked eye. With time, the particles slowly fall to the ground. **(b)** Milk, a colloid, contains particles so fine that they do not separate out on standing. **(c)** Wine, a solution, contains dissolved molecules.

Although we usually think of a solution as a solid dissolved in a liquid or as a mixture of liquids, there are many other kinds of solutions as well. In fact, any one state of matter can form a solution with any other state, and seven different kinds of solutions are known (Table 11.2). Even solutions of one solid with another and solutions of a gas in a solid are well known. Metal alloys such as stainless steel (4% chromium in iron) and brass (10–40% zinc in copper) are examples of solid/solid solutions; hydrogen in palladium is an example of a gas/solid solution. Metallic palladium, in fact, is able to absorb up to 935 times its own volume of H_2 gas.

TABLE 11.2 Some Different Kinds of Solutions	
Kind of Solution	**Example**
Gas in gas	Air (O_2, N_2, Ar, and other gases)
Gas in liquid	Carbonated water (CO_2 in water)
Liquid in liquid	Gasoline (mixture of hydrocarbons)
Solid in liquid	Sea water (NaCl and other salts in water)
Gas in solid	H_2 in palladium metal
Liquid in solid	Dental amalgam (mercury in silver)
Solid in solid	Metal alloys such as 14-karat gold (Au and Ag)

For solutions in which a gas or solid is dissolved in a liquid, the dissolved substance is called the **solute**, and the liquid is called the **solvent**. When one liquid is dissolved in another, the distinction between solute and solvent is less clear. Usually, though, the minor component is considered the solute, and the major component is the solvent. Thus, ethyl alcohol is the solute and water the solvent in a mixture of 10% ethyl alcohol and 90% water, but water is the solute and ethyl alcohol the solvent in a mixture of 90% ethyl alcohol and 10% water.

Carbonated water is a solution of gaseous CO_2 in water; the bicycle frames are made of a steel alloy of Cr, Mo, and Fe.

11.2 ►ENERGY CHANGES AND THE SOLUTION PROCESS

With the exception of gas/gas mixtures, such as air, the different kinds of solutions listed in Table 11.2 involve *condensed phases*, either liquid or solid. Thus, all the intermolecular forces described in Chapter 10 and used for explaining the properties of pure liquids and solids are also important for explaining the properties of solutions. The situation is more complex for solutions than for pure substances, however, because there are three types of interactions among particles to take into account: solvent–solvent interactions, solute–solute interactions, and solvent–solute interactions.

A good rule of thumb, often summarized in the phrase "like dissolves like," is that solutions can form when the three types of interactions are similar in kind and in magnitude. Thus, ionic solids like NaCl dissolve in polar solvents like water because the strong ion–dipole attractions between

Na$^+$ and Cl$^-$ ions and polar H$_2$O molecules are similar in magnitude to the strong dipole–dipole attractions between water molecules and to the strong ion–ion attractions between Na$^+$ and Cl$^-$ ions. In the same way, nonpolar organic substances like cholesterol, C$_{27}$H$_{46}$O, dissolve in nonpolar organic solvents like benzene, C$_6$H$_6$, because of the similar London dispersion forces present among these similar kinds of molecules. Oil, however, does not dissolve in water because the two liquids have different kinds of intermolecular forces.

The dissolution of a solid in a liquid can be visualized as shown in Figure 11.1 for NaCl. When placed in water, those ions that are less tightly held because of their position at a corner or an edge of the crystal are exposed to water molecules, which collide with them until an ion happens to break free. More water molecules then orient around the ion, stabilizing it by means of ion–dipole attractions. A new edge or corner is thereby exposed on the crystal, and the process continues until the entire crystal has dissolved. The ions in solution are said to be *solvated*—more specifically, *hydrated* when water is the solvent—meaning that they are surrounded and stabilized by a shell of solvent molecules.

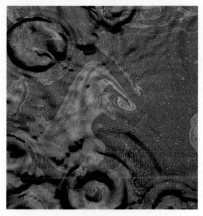

Oil doesn't dissolve in water because the two substances have different kinds of intermolecular forces.

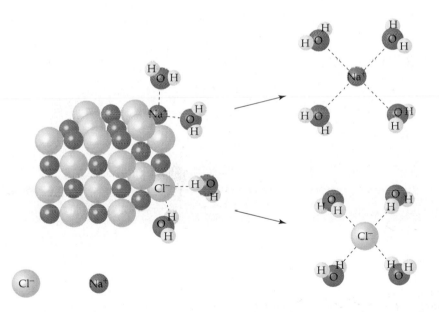

FIGURE 11.1 Dissolution of an NaCl crystal in water. Water molecules surround an accessible edge or corner ion of the crystal and collide with it until the ion breaks free. Additional water molecules then surround the ion and stabilize it by means of ion–dipole attractions.

These sodium chloride crystals dissolve in water because the attractive forces between ions and polar water molecules in the solution outweigh the attractive forces between ions in the crystal.

As with all chemical and physical processes, the dissolution of a substance in a solvent has associated with it a free-energy change, ΔG (Section 8.13). If ΔG is negative, the process is spontaneous and the substance dissolves; if ΔG is positive, the process is nonspontaneous and the substance does not dissolve. The free-energy change has two terms: $\Delta G = \Delta H - T\Delta S$. The enthalpy term ΔH measures the heat flow into or out of the system during dissolution, and the temperature-dependent entropy term $T\Delta S$ measures the change in the amount of molecular disorder or randomness in the system. The enthalpy change, ΔH, is called the enthalpy of solution or **heat of solution** (ΔH_{soln}), and the entropy change, ΔS, is called the **entropy of solution** (ΔS_{soln}).

What range of values might we expect for ΔH_{soln} and ΔS_{soln}? Let's take the entropy change first. Entropies of solution are usually positive because molecular randomness usually increases during dissolution: $+43.4$ J/(K·mol) for NaCl in water, for example. When a solid dissolves in a liquid, randomness increases on going from a well-ordered crystal to a less-ordered state in which solvated ions or molecules are able to move freely in solution. When one liquid dissolves in another, randomness increases as the different molecules intermingle. Table 11.3 lists values of ΔS_{soln} for some common substances.

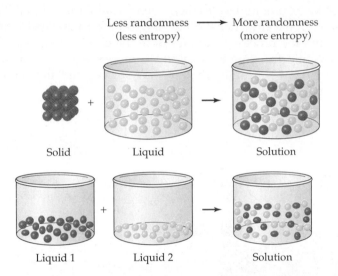

TABLE 11.3 Some Heats of Solution and Entropies of Solution in Water at 25°C

Substance	ΔH_{soln} (kJ/mol)	ΔS_{soln} [J/(K · mol)]
LiCl	−37.0	10.5
NaCl	3.9	43.4
KCl	17.2	75.0
LiBr	−48.8	21.5
NaBr	−0.6	54.6
KBr	19.9	89.0
NaOH	−44.5	−16.2
KOH	−57.6	12.9

Values for the enthalpy of solution, ΔH_{soln}, are difficult to predict (Table 11.3). Some solids dissolve exothermically and have negative values of ΔH_{soln} (−37.0 kJ/mol for LiCl in water), but others dissolve endothermically and have positive values of ΔH_{soln} (+17.2 kJ/mol for KCl in water). Athletes take advantage of both situations when they use instant hot packs or cold packs to treat injuries (Figure 11.2). Both kinds of instant packs consist of a pouch of water and a dry chemical, either $CaCl_2$ or $MgSO_4$ for hot packs, and NH_4NO_3 for cold packs. When the pack is squeezed, the pouch breaks and the solid dissolves, either raising or lowering the temperature.

Hot packs: $CaCl_2(s)$ $\Delta H_{soln} = -81.3$ kJ/mol

$MgSO_4(s)$ $\Delta H_{soln} = -91.2$ kJ/mol

Cold packs: $NH_4NO_3(s)$ $\Delta H_{soln} = +25.7$ kJ/mol

(a)   (b)

FIGURE 11.2 **(a)** Dissolution of $CaCl_2$ in water is exothermic, which causes the temperature of the water to rise. **(b)** Dissolution of NH_4NO_3 is endothermic, which causes the temperature of the water to drop.

The variations in heats of solution for different substances result from an interplay of the three kinds of interactions mentioned earlier:

Solvent–solvent interactions—Energy is required (positive ΔH) to overcome intermolecular forces between solvent molecules because the molecules must be separated and pushed apart to make room for solute particles.

Solute–solute interactions—Energy is required (positive ΔH) to overcome intermolecular forces holding solute particles together in the crystal. For an ionic solid, this is the lattice energy (Section 6.4). Substances with higher lattice energies therefore tend to be less soluble than substances with lower lattice energies.

Solvent–solute interactions—Energy is released (negative ΔH) when solvent molecules cluster around solute particles and solvate them. For ionic substances in water, the amount of hydration energy released is generally greater for smaller cations than for larger ones because water molecules can approach the positive nuclei of smaller ions more closely and thus bind more tightly. Similarly, hydration energy generally increases as the charge on the ion increases.

The first two kinds of interactions are endothermic, requiring an input of energy to spread apart solvent molecules and to break apart crystals. Only the third interaction is exothermic, as attractive intermolecular forces develop between solvent and solute. The sum of the three interactions determines whether ΔH_{soln} is endothermic or exothermic. For some substances, the one exothermic interaction is sufficiently large to outweigh the two endothermic interactions, but for other substances, the reverse is true (Figure 11.3).

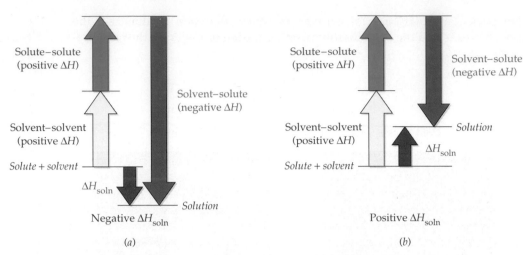

FIGURE 11.3 The value of ΔH_{soln} is the sum of three terms: solute–solute, solvent–solute, and solvent–solvent. It can be either negative as in **(a)** or positive as in **(b)**.

EXAMPLE 11.1

Pentane (C_5H_{12}) and 1-butanol (C_4H_9OH) are organic liquids with similar molecular weights, but their solubility properties differ substantially. Which of the two would you expect to be more soluble in water? Explain.

SOLUTION Pentane is a nonpolar molecule and is unlikely to form strong intermolecular interactions with polar water molecules. 1-Butanol, however, has an –OH group just as water does, and is therefore a polar molecule that can form hydrogen bonds with water. 1-Butanol is therefore more soluble in water.

⌐ **PROBLEM 11.1** Arrange the following compounds in order of their expected increasing solubility in water: Br_2, KBr, toluene (C_7H_8, a constituent of gasoline).

⌐ **PROBLEM 11.2** Which would you expect to have the larger (more negative) hydration energy, Na^+ or Cs^+? K^+ or Ba^{2+}?

11.3 ➤UNITS OF CONCENTRATION

In daily life, it's often sufficient to describe a solution qualitatively as either *dilute* or *concentrated*. In scientific work, though, it's usually necessary to know the exact **concentration** of a solution—that is, to know the exact amount of solute dissolved in a given amount of solvent. There are many ways of expressing concentration, each of which has its own advantages and disadvantages. We'll look briefly at four of the most common methods: *molarity, mole fraction, weight percent,* and *molality*.

Molarity (M)

The most common way of expressing concentration in a chemistry laboratory is to use *molarity*. As discussed in Section 3.7, a solution's molarity is given by the number of moles of solute per liter of solution (mol/L, abbreviated M). If, for example, you dissolve 0.500 mol (20.0 g) of NaOH in enough water to give 1.000 L of solution, then the solution has a concentration of 0.500 M.

$$\text{Molarity (M)} = \frac{\text{moles of solute}}{\text{liters of solution}}$$

The advantages of using molarity are twofold: (1) Stoichiometry calculations are simplified because numbers of moles are used rather than mass, and (2) amounts of solution (and therefore of solute) are measured by volume rather than by mass. As a result, titrations are particularly easy. The disadvantages of using molarity are also twofold: (1) The exact concentration depends on the temperature, because the volume of a solution expands or contracts as the temperature changes, and (2) the exact amount of solvent in a given volume can't be determined unless the density of the solution is known. (Remember from Section 3.7 that solutions of a given molarity are prepared by dissolving a solute in a small amount of solvent and then diluting with solvent to the desired volume. The solution is not made by dissolving the solute *in* the desired volume of solvent.)

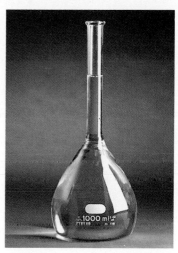

The solution in this flask was made by dissolving 0.500 mol of NaOH in water and diluting to a final volume of 1.000 L.

Mole Fraction (X)

As discussed in Section 9.5, the mole fraction (X) of any component in a solution is given by the number of moles of the component divided by the total number of moles making up the solution (including solvent):

$$\text{Mole fraction (X)} = \frac{\text{moles of component}}{\text{total moles making up the solution}}$$

For example, a solution prepared by dissolving 1.00 mol (32.0 g) of methyl alcohol (CH_3OH) in 5.00 mol (90.0 g) of water has a methyl alcohol concentration $X = 1.00 \text{ mol}/(1.00 \text{ mol} + 5.00 \text{ mol}) = 0.167$. Note that mole fractions are dimensionless, since the units cancel.

Mole fractions are independent of temperature and are particularly useful for calculations involving gas mixtures. Except in special situations, mole fractions are not often used for liquid solutions because other methods are generally more convenient.

Weight Percent (wt %)

As the name suggests, the weight percent of any component in a solution is the mass of that component divided by the total mass of the solution times 100%:

$$\text{Wt \%} = \frac{\text{mass of component}}{\text{total mass of solution}} \times 100\%$$

For example, a solution prepared by dissolving 10.0 g of glucose in 100.0 g of water has a glucose concentration of 9.09 wt %.

$$\text{Wt \% glucose} = \frac{10.0 \text{ g}}{10.0 \text{ g} + 100.0 \text{ g}} \times 100\% = 9.09 \text{ wt \%}$$

Closely related to wt %, and particularly useful for very dilute solutions, are the concentration units parts per million (ppm) and parts per billion (ppb):

$$\text{parts per million (ppm)} = \frac{\text{mass of component}}{\text{total mass of solution}} \times 10^6$$

$$\text{parts per billion (ppb)} = \frac{\text{mass of component}}{\text{total mass of solution}} \times 10^9$$

A concentration of 1 ppm for a substance means that each kilogram of solution contains 1 mg of the solute or, for dilute aqueous solutions near room temperature where 1 kg = 1 L, that each liter of solution contains 1 mg of solute. Similarly, a concentration of 1 ppb means that each liter of an aqueous solution contains 0.001 mg of solute. Values in ppm and ppb are frequently used for expressing the concentrations of trace amounts of impurities in air or water. For example, the maximum allowable concentration of lead in drinking water is 50 ppb, or about 1 g per 20,000 L.

The advantage of using wt % (or ppm) for expressing concentration is that values of wt % are independent of temperature because masses don't change when substances are heated or cooled. The disadvantage of using wt % is that it is generally more difficult when working with liquid solutions to measure amounts by mass rather than by volume. Furthermore, the density of a solution must be known before a concentration in wt % can be converted into molarity. Example 11.3 shows how to convert from wt % into molarity.

EXAMPLE 11.2

Assume that you have a 5.75 wt % solution of LiCl in water. What mass of solution in grams contains 1.60 g of LiCl?

SOLUTION Describing a solution as 5.75 wt % LiCl in water means that 100.0 g of solution contains 5.75 g of LiCl (and 94.25 g of water). The mass of solution per 1.60 g of LiCl is therefore

$$\text{Mass of soln} = 1.60 \text{ g LiCl} \times \frac{100 \text{ g soln}}{5.75 \text{ g LiCl}} = 27.8 \text{ g soln}$$

Thus, 27.8 g of solution contains 1.60 g of LiCl.

EXAMPLE 11.3

The density of a 25.0 wt % solution of sulfuric acid (H_2SO_4) in water is 1.1783 g/mL at 25.0°C. What is the molarity of the solution?

SOLUTION Describing a solution as 25.0 wt % sulfuric acid in water means that 100 g of solution contains 25.0 g of H_2SO_4 and 75.0 g of water. Since we want to calculate the concentration in molarity, we need to find the number of moles of sulfuric acid dissolved in a specific volume of solution. First, convert the 25.0 g of H_2SO_4 into moles:

$$\frac{\text{moles of } H_2SO_4}{100 \text{ g solution}} = \frac{25.0 \text{ g } H_2SO_4}{100 \text{ g solution}} \times \frac{1 \text{ mol } H_2SO_4}{98.1 \text{ g } H_2SO_4} = \frac{0.255 \text{ mol } H_2SO_4}{100 \text{ g solution}}$$

Next, find the volume of 100.0 g of solution using density as the conversion factor:

$$\text{Volume} = 100.0 \text{ g soln} \times \frac{1 \text{ mL}}{1.1783 \text{ g soln}} = 84.87 \text{ mL} = 0.08487 \text{ L}$$

Finally, calculate the molarity of the solution:

$$\text{Molarity} = \frac{\text{moles } H_2SO_4}{\text{liters of solution}} = \frac{0.255 \text{ mol } H_2SO_4}{0.08487 \text{ L}} = 3.00 \text{ M}$$

The molarity of the 25.0 wt % sulfuric acid solution is 3.00 M.

⌐ PROBLEM 11.3 What is the wt % concentration of a saline solution prepared by dissolving 1.00 mole of NaCl in 1.00 L of water?

⌐ PROBLEM 11.4 The legal limit for human exposure to carbon monoxide in the workplace is 35 ppm. Assuming that the density of air is 1.3 g/L, how many grams of carbon monoxide are in 1.0 L of air at the maximum allowable concentration?

⌐ PROBLEM 11.5 Assuming that sea water is an aqueous solution of NaCl, what is its molarity? The density of sea water is 1.025 g/mL at 20°C, and the NaCl concentration is 3.50 wt %.

Molality (*m*)

The *molality* of a solution is defined as the number of moles of solute per kilogram of solvent (mol/kg):

$$\textbf{Molality (\textit{m})} = \frac{\text{moles of solute}}{\text{mass of solvent (kg)}}$$

To prepare a 1.000 *m* solution of KBr in water, for example, you would dissolve 1.000 mol of KBr (119.0 g) in 1.000 kg (1000 mL) of water. You can't say for sure what the final volume of the solution will be, although it will almost certainly be a bit larger than 1000 mL. Although the names sound similar, note the differences between molarity and molality. *Molarity* is the number of moles of solute per *volume* (liter) of *solution*, whereas *molality* is the number of moles of solute per *mass* (kilogram) of *solvent*.

The advantages of using molality are that it is temperature independent, since masses don't change when substances are heated or cooled, and it is well suited for calculating certain properties of solutions that we'll discuss later in this chapter. The disadvantages are that amounts of solution are measured by mass rather than by volume and that the density of the solution must be known to convert molality into molarity (Example 11.5).

A summary of the four methods of expressing concentration, together with a comparison of their relative advantages and disadvantages, is given in Table 11.4.

TABLE 11.4	A Comparison of Various Concentration Units		
Name	**Units**	**Advantages**	**Disadvantages**
Molarity (M)	$\dfrac{\text{mol solute}}{\text{L solution}}$	Useful in stoichiometry; measure by volume	Temperature dependent; must know density to find solvent mass
Mole fraction (X)	none	Temperature independent; useful in special applications	Measure by mass; must know density to convert to molarity
Weight %	%	Temperature independent; useful for small amounts	Measure by mass; must know density to convert to molarity
Molality (m)	$\dfrac{\text{mol solute}}{\text{kg solvent}}$	Temperature independent; useful in special applications	Measure by mass; must know density to convert to molarity

EXAMPLE 11.4

What is the molality of a solution made by dissolving 1.45 g of table sugar (sucrose, $C_{12}H_{22}O_{11}$) in 30.0 mL of water?

SOLUTION The molar mass of sucrose, $C_{12}H_{22}O_{11}$, is 342.3 g/mol, so 1.45 g of sucrose is 4.24×10^{-3} mol:

$$1.45 \text{ g sucrose} \times \frac{1 \text{ mol sucrose}}{342.3 \text{ g sucrose}} = 4.24 \times 10^{-3} \text{ mol sucrose}$$

The molality of the solution is obtained by dividing the number of moles of sucrose by the mass of the solvent in kilograms. Since the density of water is 1.00 g/mL, 30.0 mL of water has a mass of 30.0 g, or 0.0300 kg. Thus, the molality of the solution is

$$\text{Molality} = \frac{4.24 \times 10^{-3} \text{ mol}}{0.0300 \text{ kg}} = 0.141 \ m$$

EXAMPLE 11.5

Ethylene glycol, $C_2H_4(OH)_2$, is a colorless liquid used as automobile antifreeze. If the density at 20°C of a 4.028 m solution of ethylene glycol in water is 1.0241 g/mL, what is the molarity of the solution? The molar mass of ethylene glycol is 62.07 g/mol.

SOLUTION A 4.028 m solution of ethylene glycol in water contains 4.028 mol of ethylene glycol per kilogram of water. To find the solution's molarity, we need to find the number of moles of solute per volume of solution. The volume, in turn, can be found from the mass of the solution by using density as a conversion factor. Thus, we first need to find the mass of the solution.

The mass of the solution is the sum of the masses of solute and solvent. Assuming that 1.000 kg of solvent is used to dissolve 4.028 mol of ethylene glycol, we need to find the mass of the ethylene glycol:

$$\text{Mass of ethylene glycol} = 4.028 \text{ mol} \times 62.07 \frac{g}{mol} = 250.0 \text{ g}$$

Dissolving this 250.0 g of ethylene glycol in 1.000 kg (or 1000 g) of water gives a total mass of the solution of 1250 g:

$$\text{Mass of solution} = 250.0 \text{ g} + 1000 \text{ g} = 1250 \text{ g}$$

The volume of the solution is obtained from its mass by using its density as a conversion factor:

$$\text{Volume of solution} = 1250 \text{ g} \times \frac{1 \text{ mL}}{1.0241 \text{ g}} = 1221 \text{ mL} = 1.221 \text{ L}$$

The molarity of the solution is the number of moles of solute divided by the volume of solution, or 3.299 M:

$$\text{Molarity of solution} = \frac{4.028 \text{ mol}}{1.221 \text{ L}} = 3.299 \text{ M}$$

EXAMPLE 11.6

A 0.750 M solution of H_2SO_4 in water has a density of 1.046 g/mL at 20°C. What is the concentration of this solution in **(a)** mole fraction, **(b)** weight percent, and **(c)** molality? The molar mass of H_2SO_4 is 98.1 g/mol.

SOLUTION   **(a)** Let's pick an arbitrary amount of the solution that will make the calculations easy, say 1.00 L. Since the concentration of the solution is 0.750 mol/L and the density is 1.046 g/mL (or 1.046 kg/L), 1.00 L of the solution contains 0.750 mol (73.6 g) of H_2SO_4 and has a mass of 1.046 kg.

$$\text{Moles of } H_2SO_4 \text{ in 1.00 L soln} = 0.750 \frac{mol}{L} \times 1.00 \text{ L} = 0.750 \text{ mol}$$

$$\text{Mass of } H_2SO_4 \text{ in 1.00 L soln} = 0.750 \text{ mol} \times 98.1 \frac{g}{mol} = 73.6 \text{ g}$$

$$\text{Mass of 1.00 L soln} = 1.00 \text{ L} \times 1.046 \frac{kg}{L} = 1.046 \text{ kg}$$

Subtracting the mass of H_2SO_4 from the total mass of the solution gives 0.972 kg of water, or 54.0 mol in 1.00 L of solution.

$$\text{Mass of } H_2O \text{ in 1.00 L soln} = (1.046 \text{ kg}) - (0.0736 \text{ kg}) = 0.972 \text{ kg } H_2O$$

$$\text{Moles of } H_2O \text{ in 1.00 L of soln} = 972 \text{ g} \times \frac{1 \text{ mol}}{18.0 \text{ g}} = 54.0 \text{ mol } H_2O$$

Thus, the mole fraction of H_2SO_4 is

$$X_{H_2SO_4} = \frac{0.750 \text{ mol } H_2SO_4}{0.750 \text{ mol } H_2SO_4 + 54.0 \text{ mol } H_2O} = 0.0137$$

(b) The wt % concentration can be determined from the calculations in (a):

$$\text{Wt \% of } H_2SO_4 = \frac{0.0736 \text{ kg } H_2SO_4}{1.046 \text{ kg total}} \times 100\% = 7.04\%$$

(c) The molality of the solution can also be determined from the calculations in (a). Since 0.972 kg of water has 0.750 mol of H_2SO_4 dissolved in it, 1.00 kg of water would have 0.772 mol of H_2SO_4 dissolved in it.

$$\frac{1.00 \text{ kg } H_2O}{0.972 \text{ kg } H_2O} \times 0.750 \text{ mol } H_2SO_4 = 0.772 \text{ mol } H_2SO_4$$

Thus, the molality of the sulfuric acid solution is 0.772 *m*.

What is the molality of NaCl in sea water?

PROBLEM 11.6 What is the molality of a solution prepared by dissolving 0.385 g of cholesterol, $C_{27}H_{46}O$, in 40.0 g of chloroform, $CHCl_3$? What is the mole fraction of cholesterol in the solution?

PROBLEM 11.7 What mass in grams of a 0.500 *m* solution of sodium acetate, CH_3CO_2Na, in water would you use to obtain 0.150 mol of sodium acetate?

PROBLEM 11.8 The density at 20°C of a 0.258 *m* solution of glucose in water is 1.0173 g/mL, and the molar mass of glucose is 180.2 g. What is the molarity of the solution?

PROBLEM 11.9 The density at 20°C of a 0.500 M solution of acetic acid in water (vinegar) is 1.0042 g/mL. What is the concentration of this solution in molality? The molar mass of acetic acid, CH_3COOH, is 60.05 g.

PROBLEM 11.10 Assuming that sea water is a 3.50 wt % aqueous solution of NaCl, what is the molality of sea water?

11.4 ▸SOME FACTORS AFFECTING SOLUBILITY

FIGURE 11.4 A supersaturated solution of sodium acetate in water. When a tiny seed crystal is added, larger crystals begin to grow and precipitate from the solution until equilibrium is reached.

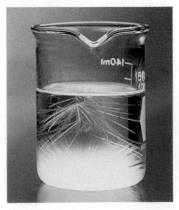

If you take solid NaCl and add it to water, dissolution occurs rapidly at first but then slows down as more and more NaCl is added. Eventually the dissolution stops because a dynamic equilibrium is reached, where the number of Na^+ and Cl^- ions leaving a crystal to go into solution is equal to the number of ions returning from solution to the crystal. At this point, the solution is said to be **saturated** in that solute.

$$\text{Solute + solvent} \underset{\text{crystallize}}{\overset{\text{dissolve}}{\rightleftarrows}} \text{solution}$$

Note that the preceding definition requires a saturated solution to be at *equilibrium* with undissolved solid. Substances that are more soluble at high temperature than at low temperature can sometimes form what are called **supersaturated** solutions, which contain a greater-than-equilibrium amount of solute. For example, when a saturated solution of sodium acetate is prepared at high temperature and then slowly cooled, a supersaturated solution results, as shown in Figure 11.4. Such a solution is unstable, however, and precipitation occurs when a tiny seed crystal of sodium acetate is added to initiate crystallization.

Effect of Temperature on Solubility

The amount of solute per unit of solvent needed to form a saturated solution is called the solute's **solubility**. Like melting point and boiling point, solubility is a physical property characteristic of a particular substance. Different substances can have greatly different solubilities, as shown in Figure 11.5. Sodium chloride, for instance, has a solubility of 36.0 g/100 mL of water at 20°C, and sodium nitrate has a solubility of 87.6 g/100 mL of water at 20°C. Sometimes, particularly when two liquids are involved, the solvent and solute are **miscible**—that is, they are mutually soluble in all proportions. Ethyl alcohol and water are an example.

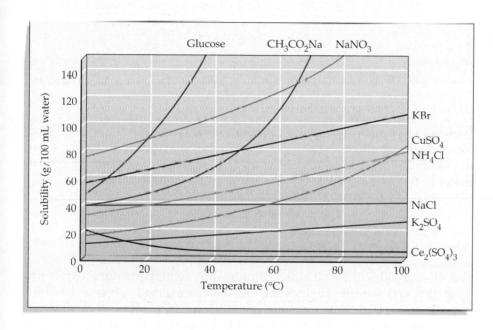

FIGURE 11.5 Solubilities of some common solids in water as a function of temperature. Most substances become more soluble as temperature rises, although the exact relationship is usually complex.

Solubilities are temperature dependent, and the temperature at which a specific measurement is made must be reported. As Figure 11.5 shows, there is no obvious correlation between structure and solubility or between solubility and temperature. The solubilities of most molecular and ionic solids increase with increasing temperature, though the solubilities of some (NaCl) are almost unchanged, and the solubilities of others [$Ce_2(SO_4)_3$] decrease.

The effect of temperature on the solubility of gases is more predictable than its effect on the solubility of solids. All gases become less soluble in water as the temperature increases (Figure 11.6). One trivial consequence of this decreased solubility is that carbonated drinks bubble continuously as they warm up to room temperature after being refrigerated. After a while, they lose so much dissolved CO_2 that they taste flat. A much more important consequence is the damage to aquatic life that can result from the decrease in dissolved oxygen when hot water is discharged from power stations into lakes and rivers, an effect known as thermal pollution.

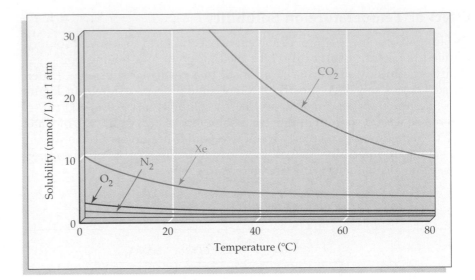

FIGURE 11.6 Solubilities of some gases in water as a function of temperature. All gases become less soluble in water as the temperature rises. (The concentration units are millimoles per liter (mmol/L) at a gas pressure of 1 atm.)

Effect of Pressure on Solubility

Pressure has practically no effect on the solubility of liquids and solids, but has a profound effect on the solubility of gases. According to **Henry's law**, the solubility of a gas in a liquid at a given temperature is directly proportional to the partial pressure of the gas over the solution. That is,

$$\text{Solubility} = k \cdot P$$

where k is the Henry's law constant characteristic of each specific gas, and P is the partial pressure of the gas over the solution. Doubling the partial pressure doubles the solubility, tripling the partial pressure triples the solubility, and so forth. Henry's law constants are usually given in the units mol/(L·atm), and measurements are made at 25°C. Note that, at a gas partial pressure P of 1 atm, the Henry's law constant k is numerically equal to the solubility of the gas in mol/L.

The most common example of Henry's law behavior occurs when you open a can of soda or other carbonated drink. Bubbles of gas come fizzing out of solution because the pressure of CO_2 in the can drops and CO_2 suddenly becomes less soluble. A more serious example of Henry's law behavior occurs when a deep-sea diver surfaces too quickly and develops

Divers must ascend slowly to prevent the bends caused by formation of nitrogen bubbles in the blood.

the "bends." Bends occur because large amounts of nitrogen dissolve in the blood at high underwater pressures. When the diver ascends and pressure decreases too rapidly, nitrogen bubbles form in the blood, blocking capillaries and inhibiting blood flow. The condition can be prevented by using an oxygen/helium mixture for breathing rather than air (oxygen/nitrogen), because helium has a much lower solubility in blood than nitrogen has.

On a molecular level, the increase in gas solubility with increasing gas pressure occurs because of a change in the position of the equilibrium between dissolved and undissolved gas. At a given pressure, an equilibrium is established in which equal numbers of gas particles enter and leave the solution (Figure 11.7a). When the pressure is increased, however, more gas particles are forced into the solution than leave it, and gas solubility therefore increases until a new equilibrium is established (Figure 11.7b).

(a)

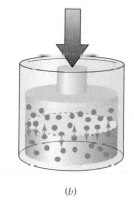
(b)

FIGURE 11.7 A molecular view of Henry's law. **(a)** At a given pressure, an equilibrium exists in which equal numbers of gas particles enter and leave the solution. **(b)** When pressure is increased by pushing on the piston, more gas particles are temporarily forced into solution than are able to leave, and solubility increases until a new equilibrium is reached.

EXAMPLE 11.7

The Henry's law constant of methyl bromide (CH_3Br), a gas used as a fumigating agent against termites, is $k = 0.159$ mol/(L · atm) at 25°C. What is the solubility of methyl bromide in water in mol/L at 25°C and a partial pressure of 125 mm Hg?

SOLUTION   Henry's law says that the solubility of a gas in water is equal to $k \cdot P$. In the present instance,

$$k = 0.159 \text{ mol/(L} \cdot \text{atm) for methyl bromide}$$

$$P = 125 \text{ mm Hg} \times \frac{1 \text{ atm}}{760 \text{ mm Hg}} = 0.164 \text{ atm}$$

$$\text{Solubility} = k \cdot P = 0.159 \frac{\text{mol}}{\text{L} \cdot \text{atm}} \times 0.164 \text{ atm} = 0.0261 \text{ M}$$

The solubility of methyl bromide in water at a partial pressure of 125 mm Hg is 0.0261 M.

⌐ **PROBLEM 11.11**   The solubility of CO_2 in water is 3.2×10^{-2} M at 25°C and 1 atm pressure. What is the Henry's law constant for CO_2 in mol/(L · atm)?

⌐ **PROBLEM 11.12**   The partial pressure of CO_2 in air is approximately 4.0×10^{-4} atm. Use the Henry's law constant you calculated in Problem 11.11 to find the concentration of CO_2 in **(a)** a can of soda under a CO_2 pressure of 2.5 atm at 25°C, and **(b)** a can of soda open to the atmosphere at 25°C.

The fizz in this glass of ginger ale is an example of Henry's law behavior. When the soda bottle is opened, the pressure of CO_2 drops and CO_2 suddenly becomes less soluble.

11.5 ►PHYSICAL BEHAVIOR OF SOLUTIONS: COLLIGATIVE PROPERTIES

The behavior of solutions is qualitatively similar to that of pure solvents but is quantitatively different. Pure water boils at 100°C and freezes at 0°C, for example, but a 1.00 *m* (molal) solution of NaCl in water boils at 101.02°C and freezes at −3.72°C.

The elevation of boiling point and the lowering of freezing point that are observed on comparing a pure solvent with a solution are examples of **colligative properties**, properties that depend on the *amount* of dissolved solute but not on the chemical identity of the solute. (The word *colligative* means "bound together in a collection" and is used because a "collection" of solute particles is responsible for the observed effects.) Other colligative properties are a decrease in vapor pressure of a solution compared with the pure solvent and osmosis, the migration of solvent and other small molecules through a semipermeable membrane.

In comparing the properties of a pure solvent
with those of a solution. . .

Colligative
properties
{
Vapor pressure of solution is lower.

Boiling point of solution is higher.

Freezing point of solution is lower.

Osmosis, the migration of solvent molecules
  through a semipermeable membrane, occurs
  when solvent and solution are separated by
  the membrane.
}

Let's look at each of the four colligative properties in more detail.

11.6 ►VAPOR-PRESSURE LOWERING OF SOLUTIONS: RAOULT'S LAW

We said in Section 10.5 that a liquid in a closed container is in equilibrium with its vapor and that the amount of pressure exerted by the vapor is called the *vapor pressure*. When you compare the vapor pressure of a pure solvent with that of a solution at the same temperature, however, you find that the two values are different. If the solute is nonvolatile and has no appreciable vapor pressure of its own, then the vapor pressure of the solution is always lower than that of the pure solvent. If the solute *is* volatile and has a significant vapor pressure of its own, as usually occurs in a mixture of two liquids, then the combined vapor pressure of the two liquids is always intermediate between the vapor pressures of the two pure substances.

Solutions with a Nonvolatile Solute

It's easy to demonstrate with manometers that a solution of a nonvolatile solute has a lower vapor pressure than a pure solvent has (Figure 11.8). Alternatively, you can show the same effect by comparing the rate of

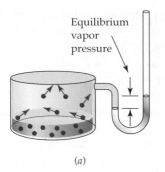

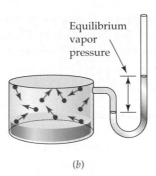

(a)   (b)

FIGURE 11.8 The equilibrium vapor pressure of a solution with a nonvolatile solute **(a)** is always lower than that of the pure solvent **(b)** by an amount that depends on the mole fraction of the solvent.

evaporation of a sample of pure solvent with the rate of evaporation of a solution. A solution always evaporates more slowly than a pure solvent does because its vapor pressure is lower and its molecules therefore escape less readily.

According to **Raoult's law**, the vapor pressure of a solution containing a nonvolatile solute is equal to the vapor pressure of pure solvent times the mole fraction of the solvent. That is:

RAOULT'S LAW $P_{soln} = P_{solv} \times X_{solv}$

where P_{soln} is the vapor pressure of the solution, P_{solv} is the vapor pressure of pure solvent at the same temperature, and X_{solv} is the mole fraction of the solvent in the solution. Take a solution of 1.00 mol of glucose in 15.0 mol of water at 25°C, for instance. The vapor pressure of pure water at 25°C is 23.76 mm Hg, and the mole fraction of water in the solution is 15.0 mol/(1.0 mol + 15.0 mol) = 0.938. Thus, Raoult's law predicts a vapor pressure for the solution of 23.76 mm Hg × 0.938 = 22.3 mm Hg, which corresponds to a vapor-pressure lowering, ΔP_{soln}, of 1.5 mm Hg.

$$P_{soln} = P_{solv} \times X_{solv} = 23.76 \text{ mm Hg} \times \frac{15.0 \text{ mol}}{1.00 \text{ mol} + 15.0 \text{ mol}} = 22.3 \text{ mm Hg}$$

$$\Delta P_{soln} = P_{solv} - P_{soln} = 23.76 \text{ mm Hg} - 22.3 \text{ mm Hg} = 1.5 \text{ mm Hg}$$

Note that the amount of vapor-pressure lowering can also be calculated by multiplying the mole fraction of the *solute* times the vapor pressure of the pure solvent. That is,

$$\Delta P_{soln} = P_{solv} \times X_{solute} = 23.76 \text{ mm Hg} \times \frac{1.00 \text{ mol}}{1.00 \text{ mol} + 15.0 \text{ mol}} = 1.5 \text{ mm Hg}$$

If an ionic substance such as NaCl is the solute, we have to calculate mole fractions based on the total number of solute *particles* (ions) rather than on the number of NaCl formula units. A solution of 1.00 mol NaCl in 15.0 mol water at 25°C, for example, results in a mole fraction for water of 0.882:

$$X_{water} = \frac{15.0 \text{ mol } H_2O}{1.00 \text{ mol Na}^+ + 1.00 \text{ mol Cl}^- + 15.0 \text{ mol } H_2O} = 0.882$$

Since the mole fraction of water is smaller in the 1.00 mol NaCl solution than it is in the 1.00 mol glucose solution, the vapor pressure of the NaCl solution is lower: 21.0 mm Hg for NaCl versus 22.3 mm Hg for glucose at 25°C.

$$P_{soln} = P_{solv} \times X_{solv} = 23.76 \text{ mm Hg} \times 0.882 = 21.0 \text{ mm Hg}$$

What accounts for the lowering of vapor pressure when a nonvolatile solute is dissolved in a solvent? As we've noted on several occasions, a physical process such as the vaporization of a liquid to a gas is accompanied by a free-energy change, $\Delta G_{vap} = \Delta H_{vap} - T\Delta S_{vap}$. The more negative the value of ΔG_{vap}, the easier the vaporization process. Thus, if we want to compare the ease of vaporization of a pure solvent with that of the solvent in a solution, we have to compare the signs and relative magnitudes of the ΔH_{vap} and ΔS_{vap} terms in the two cases.

The vaporization of a liquid to a gas is always *disfavored* by enthalpy (positive ΔH_{vap}) because energy is required to overcome intermolecular attractions between liquid molecules. At the same time, vaporization is always *favored* by entropy (positive ΔS_{vap}) because molecular disorder increases when molecules go from a semi-ordered liquid state to a much more disordered gaseous state.

The heats of vaporization for a pure solvent and for a solution are similar because similar intermolecular forces must be overcome in both cases for solvent molecules to escape from the liquid. Thus, the difference in ease of vaporization can't be due to a difference in ΔH_{vap}. The entropies of vaporization for a pure solvent and for a solution are *not* similar, however. Because a solution has more molecular disorder and higher entropy than a pure solvent does, the entropy *change* on going from liquid to vapor is smaller for the solution than it is for the pure solvent. Subtracting a *smaller* $T\Delta S_{vap}$ from ΔH_{vap} thus results in a *larger* (more positive) ΔG_{vap} for the solution. As a result, vaporization is more difficult for the solution, and the vapor pressure of the solution at equilibrium is lower (Figure 11.9).

FIGURE 11.9 The lower vapor pressure of a solution (a) compared with that of a pure solvent (b) is caused by differences in the entropies of vaporization, ΔS_{vap}. ΔS_{vap} is smaller for the solution than for the pure solvent because the entropy of the solution is higher to begin with (c). As a result, vaporization of the solution is more difficult (larger ΔG_{vap}), and the vapor pressure of the solution is lower.

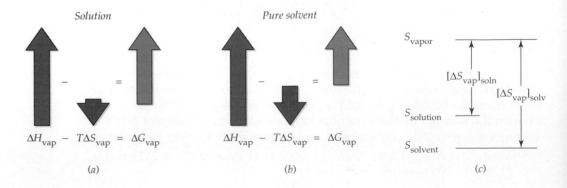

It should be pointed out that, just as the ideal-gas law discussed in Section 9.3 applies only to "ideal" gases, Raoult's law applies only to ideal solutions. Raoult's law approximates the behavior of most real solutions but significant deviations from ideality occur as solute concentration increases. The law works best when solute concentrations are low and when solute and solvent particles have similar intermolecular forces.

If the intermolecular forces between solute particles and solvent molecules are weaker than the forces between solvent molecules alone, then the solvent molecules are less tightly held in the solution, and the vapor pressure is higher than Raoult's law predicts. Conversely, if the intermolecular forces between solute and solvent molecules are stronger than the forces between solvent molecules alone, then the solvent molecules are more tightly held in the solution, and the vapor pressure is lower than predicted. Solutions of ionic substances, in particular, often have a vapor pressure significantly lower than predicted because the ion–dipole forces between dissolved ions and polar water molecules are so strong.

EXAMPLE 11.8

What is the vapor pressure in mm Hg of a solution made by dissolving 18.3 g of NaCl in 500.0 g of H_2O at 70°C? The vapor pressure of pure water at 70°C is 233.7 mm Hg.

SOLUTION   Raoult's law says that the vapor pressure of the solution is equal to the vapor pressure of pure solvent times the mole fraction of the solvent in the solution. Thus, we have to find a value for the mole fraction of solvent. First, calculate the number of moles of NaCl and H_2O. The molar mass of NaCl is 58.44 g/mol, and the molar mass of water is 18.02 g/mol.

$$\text{Moles NaCl} = 18.3 \text{ g NaCl} \times \frac{1 \text{ mol NaCl}}{58.44 \text{ g NaCl}} = 0.313 \text{ mol NaCl}$$

$$\text{Moles } H_2O = 500.0 \text{ g } H_2O \times \frac{1 \text{ mol } H_2O}{18.02 \text{ g } H_2O} = 27.75 \text{ mol } H_2O$$

Next, calculate the mole fraction of water in the solution. Since NaCl is an ionic substance that yields two particles when dissolved in water, the solution contains 0.626 mol of dissolved particles—0.313 mol of Na^+ ions and 0.313 mol of Cl^- ions. Thus, the mole fraction of water is

$$\text{Mole fraction } H_2O = \frac{27.75 \text{ mol}}{0.626 \text{ mol} + 27.75 \text{ mol}} = 0.9779$$

From Raoult's law, the vapor pressure of the solution is

$$P_{\text{soln}} = P_{\text{solv}} \times X_{\text{solv}} = 233.7 \text{ mm Hg} \times 0.9779 = 228.5 \text{ mm Hg}$$

EXAMPLE 11.9

How many grams of sucrose must be added to 320 g of water to lower the vapor pressure by 1.5 mm Hg at 25°C? The vapor pressure of water at 25°C is 23.8 mm Hg, and the molar mass of sucrose is 342.3 g/mol.

SOLUTION   The vapor pressure of the solution, P_{soln}, is equal to the vapor pressure of the pure solvent, P_{solv}, minus the amount of vapor-pressure lowering:

$$P_{soln} = 23.8 \text{ mm Hg} - 1.5 \text{ mm Hg} = 22.3 \text{ mm Hg}$$

To find the amount of sucrose that must be added, we must find the total number of moles of sucrose plus water and then find the number of moles of sucrose. According to Raoult's law, the vapor pressure of the solution is equal to the vapor pressure of pure solvent times the mole fraction of the solvent, X_{solv}. Thus, X_{solv} is equal to P_{soln} divided by P_{solv}:

$$\text{Since} \quad P_{soln} = P_{solv} \times X_{solv}$$

$$\text{then} \quad X_{solv} = \frac{P_{soln}}{P_{solv}} = \frac{23.8 \text{ mm Hg} - 1.5 \text{ mm Hg}}{23.8 \text{ mm Hg}} = 0.937$$

The mole fraction of water, X_{solv}, just calculated is the number of moles of water divided by the total number of moles of sucrose plus water. Since there are 17.8 mol of water, there must be a total of 20.1 mol of sucrose plus water.

$$\text{Moles of water} = 320 \text{ g} \times \frac{1 \text{ mol}}{18.0 \text{ g}} = 17.8 \text{ mol}$$

$$\text{Total moles} = \frac{\text{mol water}}{X_{solv}} = \frac{17.8 \text{ mol}}{0.937} = 19.0 \text{ mol}$$

Subtracting the number of moles of water from the total number of moles gives the number of moles of sucrose we need to add:

$$\text{Moles sucrose} = 19.0 \text{ mol} - 17.8 \text{ mol} = 1.2 \text{ mol}$$

Converting into grams then gives the mass of sucrose needed:

$$\text{Grams of sucrose} = 1.2 \text{ mol} \times 342.3 \frac{\text{g}}{\text{mol}} = 4.1 \times 10^2 \text{ g}$$

PROBLEM 11.13 What is the vapor pressure in mm Hg of a solution prepared by dissolving 5.00 g of benzoic acid ($C_7H_6O_2$) in 100 g of ethyl alcohol (C_2H_6O) at 35°C? The vapor pressure of pure ethyl alcohol at 35°C is 100.5 mm Hg.

PROBLEM 11.14 How many grams of NaBr must be added to 250 g of water to lower the vapor pressure by 1.30 mm Hg at 40°C? The vapor pressure of water at 40°C is 55.3 mm Hg.

Solutions with a Volatile Solute

As you might expect from Dalton's law (Section 9.5), the overall vapor pressure P_{total} of a mixture of two volatile liquids A and B is the sum of the vapor-pressure contributions of the individual components, P_A and P_B.

$$P_{total} = P_A + P_B$$

The individual vapor pressures P_A and P_B are calculated by Raoult's law. That is, the vapor pressure of A is equal to the mole fraction of A (X_A) times the vapor pressure of pure A (P_A°), and the vapor pressure of B is equal to the mole fraction of B (X_B) times the vapor pressure of pure B (P_B°). Thus, the total vapor pressure of the solution is

$$P_{total} = P_A + P_B = (P_A^\circ \, X_A) + (P_B^\circ \, X_B)$$

Take a mixture of the two similar organic liquids benzene (C_6H_6) and toluene (C_7H_8) for example. Pure benzene has a vapor pressure $P^\circ = 96.0$ mm Hg at 25°C, and pure toluene has $P^\circ = 30.3$ mm Hg at the same temperature. In a 1:1 molar mixture of the two, where the mole fraction of each is $X = 0.500$, the vapor pressure of the solution is 63.2 mm Hg.

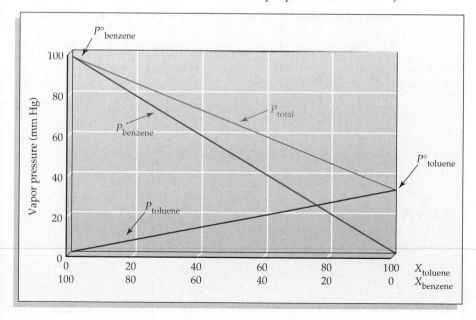

$$P_{total} = (P_{benzene}^\circ \cdot X_{benzene}) + (P_{toluene}^\circ \cdot X_{toluene})$$

$$= (96.0 \text{ mm Hg} \times 0.500) + (30.3 \text{ mm Hg} \times 0.500) = 63.2 \text{ mm Hg}$$

Not surprisingly, the vapor pressure of the mixture has a value intermediate between the vapor pressures of the two pure liquids (Figure 11.10).

FIGURE 11.10 The vapor pressure of a solution of the two volatile liquids benzene and toluene at 25°C is the sum of the two individual vapor pressures, calculated by Raoult's law.

As with nonvolatile solutes, the Raoult's-law behavior shown in Figure 11.10 for a mixture of benzene and toluene applies only to ideal solutions. Most real solutions show behavior that deviates slightly from the ideal in either a positive or negative way, depending on the kinds and strengths of intermolecular forces present in the solution.

┌ **PROBLEM 11.15** What is the vapor pressure in mm Hg of a solution prepared by dissolving 25.0 g of ethyl alcohol (C_2H_6O) in 100.0 g of water at 25°C? The vapor pressure of pure water is 23.8 mm Hg, and the vapor pressure of ethyl alcohol is 61.2 mm Hg at 25°C. What is the vapor pressure of the solution if 25.0 g of water is dissolved in 100 g of ethyl alcohol at 25°C?

11.7 ➤ BOILING-POINT ELEVATION AND FREEZING-POINT DEPRESSION OF SOLUTIONS

We said in Section 10.5 that the vapor pressure of a liquid rises with increasing temperature and that the liquid boils when its vapor pressure equals atmospheric pressure. Since a solution of a nonvolatile solute has a lower vapor pressure than a pure solvent has at a given temperature, it follows that a solution must be heated to a higher temperature than a pure solvent in order to boil. Similarly, the solution must be cooled to a lower temperature in order to freeze. Both situations are represented on a phase diagram as shown in Figure 11.11.

The boiling-point elevation of a solution relative to that of a pure solvent depends on the number of solute particles, just as vapor-pressure lowering does. Thus, a 1.00 m solution of glucose in water boils at 100.51°C

FIGURE 11.11 Phase diagrams for a pure solvent (red line) and a solution (green line). Since the vapor pressure of the solution is lower than that of the pure solvent, the temperature at which the vapor pressure of the solution reaches atmospheric pressure is higher than that of the solvent. Thus, the boiling point of the solution is higher by an amount ΔT_b. Similarly, the freezing point of the solution is lower than that of the pure solvent by an amount ΔT_f.

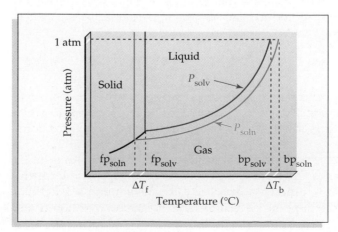

at 1 atm pressure (0.51°C above normal), but a 1.00 m solution of NaCl in water boils at 101.02°C (1.02°C above normal) because there are twice as many particles (ions) dissolved in the NaCl solution as there are in the glucose solution.

The increase in boiling point ΔT_b for a solution is

$$\Delta T_b = K_b \cdot m$$

where m is the *molal* (not molar) concentration of solute particles and K_b is the **molal boiling-point-elevation constant** characteristic of each liquid. The concentration must be expressed in molality—number of moles of solute particles per kilogram of solvent—rather than molarity so that the numbers of solute and solvent particles are independent of temperature. Molal boiling-point-elevation constants for some common substances are given in Table 11.5.

TABLE 11.5	Molal Boiling-Point-Elevation Constants (K_b) and Molal Freezing-Point-Depression Constants (K_f) for Some Common Substances	
Substance	K_b [(°C · kg)/mol]	K_f [(°C · kg)/mol]
Benzene (C_6H_6)	2.53	5.12
Camphor ($C_{10}H_{16}O$)	5.95	37.7
Chloroform ($CHCl_3$)	3.63	4.70
Diethyl ether ($C_4H_{10}O$)	2.02	1.79
Ethyl alcohol (C_2H_6O)	1.22	1.99
Water (H_2O)	0.51	1.86

The freezing-point depression of a solution relative to that of a pure solvent depends on the number of solute particles, just as boiling-point elevation does. For example, a 1.00 m solution of glucose in water freezes at −1.86°C, and a 1.00 m solution of NaCl in water freezes at −3.72°C. The decrease in freezing point ΔT_f for a solution is

$$\Delta T_f = K_f \cdot m$$

where m is the molal concentration of solute particles and K_f is the **molal freezing-point-depression constant** characteristic of each solvent. Molal freezing-point-depression constants are also given in Table 11.5 for some common substances.

The fundamental cause of boiling-point elevation and freezing-point depression in solutions is the same as the cause of vapor-pressure lowering (Section 11.6): an entropy difference between pure solvent and solvent in a solution. Let's take boiling-point elevations first. We know that liquid and vapor phases are in equilibrium at the boiling point (T_b) and that the free-energy difference between the two phases (ΔG_{vap}) is therefore zero (Section 8.13). Thus, $\Delta H_{vap} = T_b \Delta S_{vap}$, and $T_b = \Delta H_{vap}/\Delta S_{vap}$.

Since
$$\Delta G_{vap} = \Delta H_{vap} - T_b \Delta S_{vap} = 0$$

then
$$\Delta H_{vap} = T_b \Delta S_{vap}$$

and
$$T_b = \frac{\Delta H_{vap}}{\Delta S_{vap}}$$

When we compare the heats of vaporization (ΔH_{vap}) for a pure solvent and for a solution, we find that the two values are similar because similar intermolecular forces between solvent molecules must be overcome in both cases. In comparing the entropies of vaporization, however, we find that the two values are not similar. Because a solution has more molecular randomness than a pure solvent has, the entropy change between solution and vapor is smaller than the entropy change between pure solvent and vapor. But if ΔS_{vap} is *smaller* for the solution, then T_b must be correspondingly *larger*. In other words, the boiling point of the solution (T_b) is elevated over that of the pure solvent (Figure 11.12).

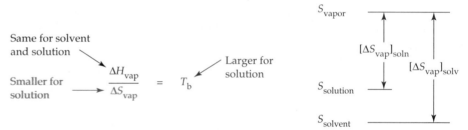

FIGURE 11.12 Relative entropy levels of a pure solvent, a solution, and a vapor. The entropy of vaporization ΔS_{vap} is smaller for the solution than for the pure solvent because the solution has a higher entropy level to begin with. The boiling point of the solution T_b must therefore be higher than that of the pure solvent.

An explanation similar to that just developed for boiling-point elevation also explains freezing-point depression. We know that liquid and solid phases are in equilibrium at the freezing point, that the free-energy difference between the phases (ΔG_{fusion}) is zero, and that T_f must therefore equal $\Delta H_{fusion}/\Delta S_{fusion}$.

Since
$$\Delta G_{fusion} = \Delta H_{fusion} - T_f \Delta S_{fusion} = 0$$

then
$$\Delta H_{fusion} = T_f \Delta S_{fusion}$$

and
$$T_f = \frac{\Delta H_{fusion}}{\Delta S_{fusion}}$$

In comparing a solution with a pure solvent, we find that the heats of fusion (ΔH_{fusion}) are similar because similar intermolecular forces between solvent molecules are involved. The entropies of fusion (ΔS_{fusion}) are not similar, however. Because a solution has more molecular randomness than a pure solvent has, the entropy change between solution and solid is larger than the entropy change between pure solvent and solid. With ΔS_{fusion} *larger* for the solution, T_f must be correspondingly *smaller*, and the freezing point of the solution (T_f) must be depressed relative to that of the pure solvent (Figure 11.13).

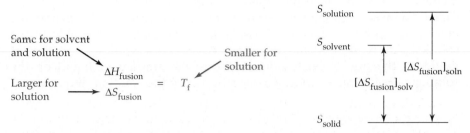

FIGURE 11.13   Relative entropy levels of a pure solvent, a solution, and a solid. ΔS_{fusion} is higher for the solution because the solution has a higher entropy level to begin with. As a result, the freezing point of the solution T_f is lower.

EXAMPLE 11.10

What is the molality of an aqueous glucose solution if the boiling point of the solution at 1 atm pressure is 101.27°C? The molal boiling-point-elevation constant for water is given in Table 11.5.

SOLUTION   Rearrange the equation for molal boiling-point elevation to solve for *m*.

$$\Delta T_b = K_b \cdot m$$

so

$$m = \frac{\Delta T_b}{K_b}$$

where $K_b = 0.51$ (°C · kg)/mol and $\Delta T_b = 101.27°C - 100.00°C = 1.27°C$

$$m = \frac{1.27 \ °C}{0.51 \dfrac{°C \cdot kg}{mol}} = 2.5 \frac{mol}{kg} = 2.5 \ m$$

The molality of the solution is 2.5 *m*.

⌐ **PROBLEM 11.16**   What is the normal boiling point in °C of a solution prepared by dissolving 1.50 g of aspirin (acetylsalicylic acid, $C_9H_8O_4$) in 75.00 g of chloroform ($CHCl_3$)? The normal boiling point of chloroform is 61.7°C, and K_b for chloroform is given in Table 11.5.

⌐ **PROBLEM 11.17**   What is the freezing point in °C of a solution prepared by dissolving 7.40 g of K_2SO_4 in 110 g of water? The value of K_f for water is given in Table 11.5.

⌐ **PROBLEM 11.18**   What is the molality of an aqueous solution of KBr whose freezing point is −2.95°C? The molal freezing-point-depression constant of water is given in Table 11.5.

11.8 ►OSMOSIS AND OSMOTIC PRESSURE

Certain materials, including those that make up the membranes around living cells, are **semipermeable**. That is, they allow water or other small molecules to pass through, but they block the passage of large solute molecules or ions. When a solution and a pure solvent (or two solutions of different concentration) are separated by the right kind of semipermeable membrane, solvent molecules pass through the membrane in a process called **osmosis**. Although the passage of solvent through the membrane takes place in both directions, passage from the pure solvent side to the solution side is more likely and occurs faster. As a result, the amount of liquid on the pure solvent side decreases, the amount of liquid on the solution side increases, and the concentration of the solution decreases.

A cucumber shrivels into a pickle when immersed in salt water because osmotic pressure drives water from the cucumber's cells.

Osmosis can be demonstrated with the experimental setup shown in Figure 11.14, in which a solution in the bulb is separated by a semipermeable membrane from pure solvent in the beaker. Solvent passes through the membrane from the beaker to the bulb, causing the liquid in the attached tube to rise. This rise in liquid level exerts a pressure that effectively pushes solvent back through the membrane until the rates of forward and reverse passage become equal and the liquid level stops rising. The amount of pressure necessary to achieve this equilibrium is called the **osmotic pressure** of the solution. Osmotic pressures can be extremely high, even for relatively dilute solutions. The osmotic pressure of a 0.15 M NaCl solution at 25°C, for example, is 7.3 atm, a value that will support a difference in water level of approximately 250 ft!

FIGURE 11.14
The phenomenon of osmosis. A solution inside the bulb is separated from pure solvent in the beaker by a semipermeable membrane. Net passage of solvent from the beaker through the membrane occurs, and the liquid in the tube rises until an equilibrium is reached. At equilibrium, the osmotic pressure exerted by the column of liquid in the tube is sufficient to prevent further net passage of solvent.

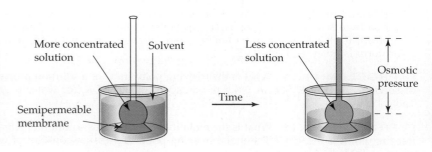

More concentrated solution   Solvent

Semipermeable membrane

Time

Less concentrated solution

Osmotic pressure

The amount of osmotic pressure Π (Greek capital pi) between solution and pure solvent depends on the concentration of solute particles in the solution according to the equation

$$\Pi = MRT$$

where M is the molar concentration of solute particles, R is the gas constant $[0.08206\ (L \cdot atm)/(K \cdot mol)]$, and T is the temperature in kelvins. For example, a 1.00 M solution of glucose in water at 300 K has an osmotic pressure of 24.6 atm.

$$\Pi = MRT = 1.00\frac{mol}{L} \times 0.08206\frac{L \cdot atm}{K \cdot mol} \times 300\ K = 24.6\ atm$$

Note that the solute concentration is given in *molarity* when calculating osmotic pressure rather than in molality as with other colligative properties. Because osmotic-pressure measurements are made at the specific temperature given in the equation $\Pi = MRT$, it's not necessary to express concentration in a temperature-independent unit such as molality.

Osmosis, like all colligative properties, results from an increase in entropy when pure solvent passes through the membrane and mixes with the solution. Perhaps the simplest explanation for why osmosis occurs is to draw an analogy with vapor-pressure lowering. The vapor pressure of a solution is, in effect, a measure of a molecule's tendency to escape the solution. The lower the vapor pressure, the less tendency a molecule has to leave. Since the molecules in a solution have a lower tendency to escape than do the molecules in a pure solvent, the molecules on the solution side of the semipermeable membrane have a lower tendency to pass to the pure solvent side than molecules on the pure solvent side have to pass to the solution side. As a result, osmotic pressure builds up.

EXAMPLE 11.11

The total concentration of dissolved particles inside red blood cells is approximately 0.30 M, and the membrane surrounding the cells is semipermeable. What would the osmotic pressure (in atm) inside the cells become if the cells were removed from blood plasma and placed in pure water at 298 K?

SOLUTION   If red blood cells were removed from the body and placed in pure water, water would pass through the cell membrane, causing an increase in pressure inside the cells. The amount of this pressure would be

$$\Pi = MRT$$

where M = 0.30 mol/L, R = 0.08206 $(L \cdot atm)/(K \cdot mol)$, T = 298 K

$$\Pi = 0.30\frac{mol}{L} \times 0.08206\frac{L \cdot atm}{K \cdot mol} \times 298\ K = 7.3\ atm$$

The buildup of internal pressure would cause the blood cells to burst.

EXAMPLE 11.12

A solution of an unknown substance in water at 293 K gives rise to an osmotic pressure of 5.66 atm. What is the molarity of the solution?

SOLUTION   We are given values for Π and T, and we need to solve for M in the equation $\Pi = MRT$. Rearranging this equation, we get

$$M = \frac{\Pi}{RT}$$

where $\Pi = 5.66$ atm, $R = 0.08206$ (L · atm)/(K · mol), and $T = 293$ K

Thus, $$M = \frac{5.66 \text{ atm}}{0.08206 \dfrac{\text{L} \cdot \text{atm}}{\text{K} \cdot \text{mol}} \times 293 \text{ K}} = 0.235 \text{ M}$$

┌ **PROBLEM 11.19**   What osmotic pressure in atm would you expect for a solution of 0.125 M $CaCl_2$ that is separated from pure water by a semipermeable membrane at 310 K?

┌ **PROBLEM 11.20**   A solution of an unknown substance in water at 300 K gives rise to an osmotic pressure of 3.85 atm. What is the molarity of the solution?┘

11.9 ➤ SOME USES OF COLLIGATIVE PROPERTIES

We use colligative properties in many ways, both in the chemical laboratory and in day-to-day life. Motorists in winter, for example, take advantage of freezing-point lowering when they drive on streets where the snow has been melted by a sprinkling of salt. The antifreeze added to automobile radiators and the de-icer solution sprayed on airplane wings also work by lowering the freezing point of water. That same antifreeze raises the boiling point of water in automobile radiators and keeps it from boiling over in summer.

One of the more interesting and practical uses of colligative properties is the desalination of sea water by **reverse osmosis**. When pure water and sea water are separated by a suitable membrane, the passage of water

A sprinkling of salt melts the snow by lowering the temperature at which water and ice are in equilibrium.

Fresh water is obtained from sea water by desalination using reverse osmosis.

molecules from the pure side to the solution side is faster than passage in the reverse direction. As osmotic pressure builds up, though, the rates of forward and reverse water passage converge and eventually become equal at an osmotic pressure of about 30 atm at 25°C. If a pressure even *greater* than 30 atm is now applied to the solution side, then the reverse passage of water becomes favored. As a result, pure water can be obtained from sea water (Figure 11.15).

FIGURE 11.15 A schematic for the desalination of sea water by reverse osmosis. By applying a pressure on the sea water that is greater than osmotic pressure, water flows through the osmotic membrane from the sea water side to the pure water side.

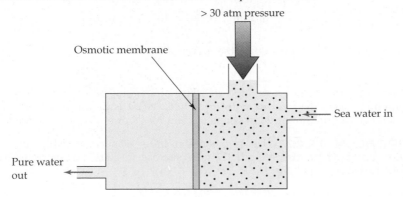

The most important use of colligative properties in the laboratory is for determining the molar mass (hence, molecular weight) of an unknown substance. Any of the four colligative properties we've discussed can be used, but the most accurate values are obtained from osmotic pressure measurements because the magnitude of the osmosis effect is so great. For example, a solution of 0.0200 M glucose in water at 300 K will give an osmotic pressure reading of 374.2 mm Hg, a value that can easily be read to four significant figures. The same solution, however, will lower the freezing point by only 0.04°C, a value that can be read easily to only one significant figure. Example 11.13 shows how osmosis can be used to determine molar mass.

EXAMPLE 11.13

A solution prepared by dissolving 20.0 mg of insulin in water and diluting to a volume of 5.00 mL gives an osmotic pressure of 12.5 mm Hg at 300 K. What is the molecular weight of insulin?

SOLUTION   To determine molecular weight, we need to know the number of moles of insulin represented by the 20.0 mg sample. We can do this by first rearranging the equation for osmotic pressure to find the molar concentration of the insulin solution:

$$\text{Since } \Pi = MRT, \quad \text{then } M = \frac{\Pi}{RT}$$

$$M = \frac{12.5 \text{ mm Hg} \times \dfrac{1 \text{ atm}}{760 \text{ mm Hg}}}{0.08206 \dfrac{\text{L} \cdot \text{atm}}{\text{K} \cdot \text{mol}} \times 300 \text{ K}} = 6.68 \times 10^{-4} \text{ M}$$

Since the volume of the solution is 5.00 mL, we can calculate the number of moles of insulin:

$$\text{Moles insulin} = 6.68 \times 10^{-4} \frac{\text{mol}}{\text{L}} \times \frac{1 \text{ L}}{1000 \text{ mL}} \times 5.00 \text{ mL} = 3.34 \times 10^{-6} \text{ mol}$$

Knowing both the mass and the number of moles of insulin, we can calculate the molar mass:

$$\text{Molar mass} = \frac{\text{mass of insulin}}{\text{moles of insulin}} = \frac{0.0200 \text{ g insulin}}{3.34 \times 10^{-6} \text{ mol insulin}} = 5990 \text{ g/mol}$$

The molecular weight of insulin is therefore 5990 amu.

⌐ **PROBLEM 11.21**   A solution of 0.250 g of naphthalene (mothballs) in 35.00 g of camphor lowers the freezing-point by 2.10°C. What is the molar mass of naphthalene? The freezing-point-depression constant for camphor is given in Table 11.5.

⌐ **PROBLEM 11.22**   What is the molar mass of sucrose (table sugar) if a solution prepared by dissolving 0.822 g of sucrose in 300.0 mL of water has an osmotic pressure of 149 mm Hg at 298 K?

11.10 ►FRACTIONAL DISTILLATION OF LIQUID MIXTURES

Petroleum refineries such as that shown in Figure 11.16 appear as a vast maze of pipes, towers, and tanks. The pipes, though, are just for transferring the petroleum or its products, and the tanks are just for storage. It's in the towers that the important separation of crude petroleum into usable fractions takes place. As we saw in Section 8.11, petroleum is a complex mixture of hydrocarbon molecules that is refined by distillation into different fractions: straight-run gasoline (bp 30–200°C), kerosene (bp 175–300°C), and gas oil (bp 275 400°C).

FIGURE 11.16 A typical petroleum refinery. Distillation of petroleum into fractions is carried out in the large towers.

Called **fractional distillation**, this separation of volatile liquids is yet another practical application of colligative properties. In essence, a mixture of volatile liquids is boiled, and the vapors are condensed. Because the vapor is enriched in the more volatile component with the higher vapor pressure according to Raoult's law (Section 11.6), the condensed vapors are also enriched in that component, and a partial purification has been effected. If the boil/condense cycle is then repeated a large number of times, complete purification of the more volatile liquid component can be achieved.

Let's look at a 1:1 molar mixture of benzene and toluene to see how they can be separated by fractional distillation. If we begin by heating the mixture, boiling occurs when the sum of the vapor pressures equals atmospheric pressure—that is, when $X \cdot P°_{benzene} + X \cdot P°_{toluene} = 760$ mm Hg, according to Raoult's law. Reading from the vapor-pressure curves in Figure 11.17 (or calculating values with the Clausius-Clapeyron equation as discussed in Section 10.5), we find that boiling occurs at 365.3 K (92.2°C), where $P°_{benzene} = 1084$ mm Hg and $P°_{toluene} = 436$ mm Hg:

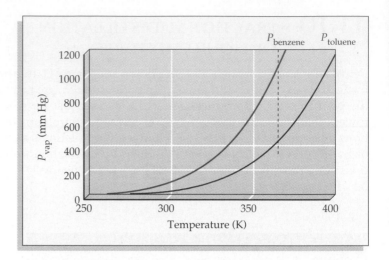

FIGURE 11.17 Vapor-pressure curves for benzene and toluene. A 1:1 mixture of the two liquids boils at 92.2°C (365.3 K).

$$P_{\text{mixt}} = X \cdot P^{\circ}_{\text{benzene}} + X \cdot P^{\circ}_{\text{toluene}}$$
$$= (0.500)(1084 \text{ mm Hg}) + (0.500)(436 \text{ mm Hg})$$
$$= 542 \text{ mm Hg} + 218 \text{ mm Hg}$$
$$= 760 \text{ mm Hg}$$

Although the starting *liquid* mixture of benzene and toluene has a 1:1 molar composition, the composition of the *vapor* is not 1:1. Of the 760 mm Hg total vapor pressure for the mixture, 542/760 = 71.3% is due to benzene and 218/760 = 28.7% is due to toluene. If we now condense the vapor, the liquid we get has this 71.3:28.7 composition. On boiling this new liquid mixture, the composition of the vapor now becomes 86.4% benzene and 13.6% toluene. A third condense/boil cycle brings the composition of the vapor to 94.4% benzene/5.6% toluene, and so on through further cycles until the desired level of purity is reached.

Fractional distillation can be represented on a liquid/vapor phase diagram by plotting temperature versus composition, as shown in Figure 11.18. The lower region of the diagram represents the liquid phase, and the upper region represents the vapor phase. Between the two is a thin equilibrium region where liquid and vapor coexist. To understand how the diagram works, let's imagine starting with our 50:50 benzene/toluene mixture and heating it to its boiling point (92.2°C on the diagram). The lower curve represents the liquid composition (50:50), but the upper curve represents the vapor composition (approximately 71:29). The two points are connected by a short horizontal line called a *tie line* to indicate that the temperature is the same at both points. Condensing the 71:29 vapor mixture by lowering the temperature gives a 71:29 liquid mixture that, when heated to *its* boiling point (86.6°C), has an 86:14 vapor composition, as represented by another tie line. In essence, fractional distillation is simply a walk across successive tie lines in whatever number of steps is necessary to reach the desired purity.

In practice, the successive boil/condense cycles occur naturally in the distillation column, and there is no need to isolate liquid mixtures at intermediate stages of purification. Fractional distillation is therefore relatively simple to carry out and is routinely used on a daily basis in chemical plants and laboratories throughout the world (Figure 11.19).

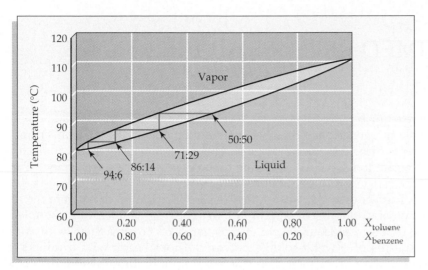

FIGURE 11.18 A phase diagram of temperature versus composition (mole fraction) for a mixture of benzene and toluene. Liquid composition is given by the lower curve, and vapor composition is given by the top curve. The thin region in the middle represents an equilibrium between phases. Liquid and vapor compositions at any given temperature are connected by a horizontal tie line.

FIGURE 11.19 A simple fractional distillation column used in a chemistry laboratory. The vapors from a boiling mixture of liquids rise inside the vacuum-jacketed distillation column, where they condense on contact with the cool column walls, drip back, and are reboiled by contact with more hot vapor. Numerous boil/condense cycles occur until vapors finally pass out the top of the column, reach the water-cooled condenser, and drip into the receiver.

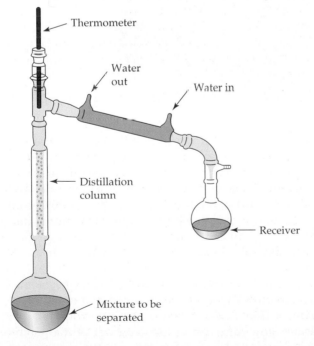

PROBLEM 11.23 Assume that you have a mixture of toluene and benzene where $X_{toluene} = 0.80$ and $X_{benzene} = 0.20$. Look at Figure 11.18, and predict **(a)** the approximate boiling point in °C of the mixture and **(b)** the composition of the vapor when the mixture boils.

interlude—TIMED-RELEASE MEDICATIONS

There's much more in most medications than medicine. Even something as simple as a generic aspirin tablet contains a binder to keep it from crumbing, a filler to bring it to the right size and help it disintegrate in the stomach, and a lubricant to keep it from sticking to the manufacturing equipment. Timed-release medications are more complex still.

The widespread use of timed-release medication dates from the introduction of Contac decongestant in 1961. The original idea was a simple one: Tiny beads of medicine were encapsulated by coating them with varying thicknesses of a water-soluble polymer. Those beads with a thinner coat dissolve and release their medicine more rapidly; those with a thicker coat dissolve more slowly. Combining the right number of beads with the right thicknesses into a single capsule makes possible the gradual release of medication over a predictable time.

The small beads of medicine are coated with different thicknesses of a water-soluble polymer so they dissolve and release medicine at different times.

The technology of timed-release medications has become much more sophisticated in recent years, and the kinds of medications that can be delivered have become more numerous. Some medicines, for instance, either damage the stomach lining or are destroyed by the highly acidic environment in the stomach but can be delivered safely if given an *enteric coating*. The enteric coating is simply a polymeric material (usually cellulose acetate phthalate) formulated so that it is stable under acidic conditions but reacts and is destroyed when it passes into the more basic environment of the intestines.

Still other kinds of timed-release drug delivery systems rely either on patches placed on the skin or on inserts implanted in the body. The medication is placed in a thin, disk-shaped reservoir surrounded by a soft, permeable membrane. Slow diffusion of the drug out of the patch or insert then allows the drug either to be absorbed through the skin or absorbed directly in the body. The Ocusert system for the treatment of glaucoma by inserting a

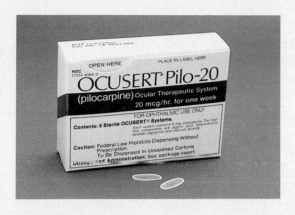

The Ocusert eye insert placed under the eyelid slowly releases a steady measured amount of pilocarpine, a drug used to treat glaucoma.

small medicated disk under the eyelid was the first such device. Long term birth-control pellets implanted in fatty tissue, and nicotine patches worn by smokers on the upper arm, operate similarly.

One clever new device now being developed for timed release of medication through the skin uses the osmosis effect to force a drug from its reservoir. Useful only for drugs that don't dissolve in water, the device is divided into two compartments, one containing medication covered by a perforated membrane and the other containing a water-soluble polymer covered by a semipermeable membrane. As moisture from the air diffuses through the membrane into the polymer-containing compartment, the buildup of osmotic pressure squeezes the medication out of the other compartment through tiny holes.

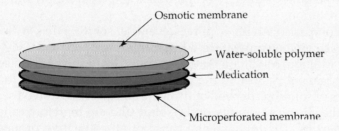

Osmotic membrane
Water-soluble polymer
Medication
Microperforated membrane

Solutions are homogeneous mixtures that contain particles the size of a typical ion or small covalent molecule. Any one state of matter can mix with any other state, leading to seven possible kinds of solutions. For solutions in which a gas or solid is dissolved in a liquid, the dissolved substance is called the **solute**, and the liquid is called the **solvent**.

The dissolution of a solute in a solvent has associated with it a free-energy change, ΔG, which has two terms, an enthalpy term ΔH and a temperature-dependent entropy term $T\Delta S$. The enthalpy change, ΔH, is the **heat of solution** (ΔH_{soln}), and the entropy change, ΔS, is the **entropy of solution** (ΔS_{soln}). Heats of solution can be either positive or negative depending on the relative strengths of solvent–solvent, solute–solute, and solvent–solute intermolecular forces. Entropies of solution are usually positive, since disorder increases when a pure solute dissolves in a pure solvent.

The concentration of a solution can be expressed in many ways, including **molarity** (moles of solute per liter of solution), **mole fraction** (moles of solute per moles of solution), **weight percent** (mass of solute per mass of solution), and **molality** (moles of solute per kilogram of solvent). When equilibrium is reached and no further solute dissolves in a given amount of solvent, a solution is said to be **saturated**. The concentration at this point represents the **solubility** of the solute. Solubilities are usually temperature dependent, though often not in a simple linear way. Gas solubilities always decrease with increasing temperature, but the solubilities of solids in liquids can either increase or decrease. The solubility of gases also depends on pressure. According to **Henry's law**, the solubility of a gas in a liquid at a given temperature is directly proportional to the partial pressure of the gas over the solution.

In comparison with a pure solvent, a solution has a lower vapor pressure at a given temperature, a lower freezing point, and a higher boiling point. In addition, a solution that is separated from solvent by a semipermeable membrane gives rise to the phenomenon of **osmosis**. All four of these properties of solutions depend only on the amount of dissolved solute rather than on the chemical identity of the solute and are therefore called **colligative properties**. The fundamental cause of all colligative properties is the same: a rise in entropy on going from pure solvent to a solution.

Colligative properties have many practical uses, including the melting of snow by salt, the desalinating of sea water by reverse osmosis, the separation and purification of volatile liquids by fractional distillation, and the determination of molecular weight by osmotic pressure measurement.

1. Rank the situations represented by the following drawings according to increasing entropy.

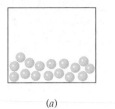 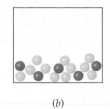

(a)   (b)

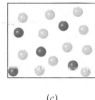

(c)

2. If a single 5 g block of NaCl is placed in water, it dissolves slowly, but if 5 g of powdered NaCl is placed in water, it dissolves rapidly. Explain.

3. Assume that two liquids are separated by a semipermeable membrane. Make a drawing that shows the situation after equilibrium is reached.

 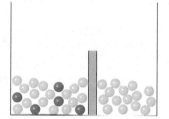

Before equilibrium

4. When 1 mol of NaCl is added to 1 L of water, the boiling point increases. When 1 mol of methyl alcohol is added to 1 L of water, the boiling point decreases. Explain.

5. When salt is spread on snow-covered roads at $-2°C$, the snow melts. When salt is spread on snow-covered roads at $-30°C$, nothing happens. Explain.

6. A phase diagram of temperature versus composition for a mixture of the two volatile liquids octane (bp = 69°C) and decane (bp = 126°C) is shown. Assume that you begin with a mixture containing 0.60 mol of decane and 0.40 mol of octane.

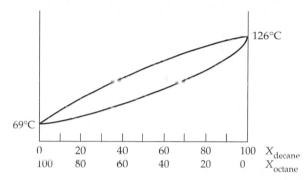

(a) What region on the diagram corresponds to vapor, and what region corresponds to liquid?
(b) At what approximate temperature will the mixture begin to boil? Mark as point a on the diagram the liquid composition at the boiling point, and mark as point b the vapor composition at the boiling point.
(c) Assume that the vapor at point b condenses and is reboiled. Mark as point c on the diagram the liquid composition of the condensed vapor and as point d on the diagram the vapor composition of the reboiled material.

ADDITIONAL PROBLEMS

Problems 11.1–11.23 appear within the chapter.

SOLUTIONS AND ENERGY CHANGES

11.24 What are the differences among colloids, suspensions, and solutions?

11.25 Give an example of each of the following kinds of solutions.

   (a) a gas in a liquid
   (b) a solid in a solid
   (c) a liquid in a solid

11.26 Explain the solubility rule of thumb "like dissolves like" in terms of the intermolecular forces that occur in solutions.

11.27 Why do ionic substances with higher lattice ener-

gies tend to be less soluble than substances with lower lattice energies?

11.28 Which would you expect to have the larger hydration energy, SO_4^{2-} or ClO_4^-? Explain.

11.29 Br_2 is much more soluble in tetrachloromethane, CCl_4, than it is in water. Explain.

11.30 Suppose you had a mixture of an ionic solid such as KBr and a molecular solid such as cholesterol. How could you take advantage of the differences in their solubility behavior to separate the mixture into its two pure components?

11.31 Ethyl alcohol, CH_3CH_2OH, is miscible with water at 20°C, but pentyl alcohol, $CH_3CH_2CH_2CH_2CH_2OH$, is soluble in water only to the extent of 2.7 g/100 mL. Explain.

11.32 Pentyl alcohol (Problem 11.31) is miscible with octane, C_8H_{18}, but methyl alcohol, CH_3OH, is insoluble in octane. Explain.

11.33 The enthalpy of solution (ΔH_{soln}) for HBr(*g*) in water is −85.1 kJ/mol and that for AgNO₃(*s*) is +22.6 kJ/mol. Assuming that you begin at room temperature and make a 0.10 M solution of each, which solution will be warm to the touch and which will be cool?

11.34 The dissolution of $CaCl_2(s)$ in water is exothermic, with ΔH_{soln} = −81.3 kJ/mol. If you were to pre-

pare a 1.00 *m* solution of $CaCl_2$ beginning with water at 25.0°C, what would the final temperature of the solution be in °C? Assume that the specific heats of both pure H_2O and the solution are the same, 4.18 J/(K · g).

11.35 The dissolution of $NH_4ClO_4(s)$ in water is endothermic, with ΔH_{soln} = +33.5 kJ/mol. If you prepare a 1.00 *m* solution of NH_4ClO_4 beginning with water at 25.0°C, what is the final temperature of the solution in °C? Assume that the specific heats of both pure H_2O and the solution are the same, 4.18 J/(K · g).

11.36 Sodium acetate dissolves exothermically (ΔH_{soln} = −17.3 kJ/mol) and is highly soluble in water. What can you conclude about the entropy of solution (ΔS_{soln}) for CH_3CO_2Na?

UNITS OF CONCENTRATION

11.37 What is the difference between molarity and molality?

11.38 What is the difference between a saturated and a supersaturated solution?

11.39 Describe how you would prepare each of the following solutions.
  (a) a 0.150 M solution of glucose in water
  (b) a 1.135 *m* solution of KBr in water
  (c) a solution of methyl alcohol and water in which X_{CH_3OH} = 0.15 and X_{H_2O} = 0.85

11.40 How would you prepare 165 mL of a 0.0268 M solution of benzoic acid ($C_7H_6O_2$) in chloroform ($CHCl_3$)?

11.41 How would you prepare 165 mL of a 0.0268 *m* solution of benzoic acid ($C_7H_6O_2$) in chloroform ($CHCl_3$)?

11.42 Which of the following solutions is more concentrated?
  (a) 0.500 M KCl or 0.500 wt % KCl in water
  (b) 1.75 M glucose or 1.75 *m* glucose in water

11.43 What is the wt % concentration of solutions made in the following way?
  (a) dissolve 0.655 mol of citric acid, $C_6H_8O_7$, in 1.00 kg of water
  (b) dissolve 0.135 mg of KBr in 5.00 mL of water
  (c) dissolve 5.50 g of aspirin, $C_9H_8O_4$, in 145 g of dichloromethane, CH_2Cl_2.

11.44 What is the molality of each solution prepared in Problem 11.43?

11.45 The threshold limit for continuous human exposure to ozone, O_3, in air is 0.1 ppm. Assuming an average air density of 1.3 g/L, how many grams of ozone are allowable in a room that measures 3 m wide, 5 m long, and 2.5 m high?

11.46 The so-called ozone layer in the earth's stratosphere has an average total pressure of 10 mm Hg (1.3×10^{-2} atm). The partial pressure of ozone in the layer is about 1.2×10^{-6} mm Hg (1.6×10^{-9} atm). What is the concentration of ozone in ppm, assuming that the average molar mass of air is 29 g/mol?

11.47 What is the concentration of each of the following solutions?
  (a) the molality of a solution prepared by dissolving 25.0 g of H_2SO_4 in 1.30 L of water
  (b) the mole fraction of each component of a solution prepared by dissolving 2.25 g of nicotine, $C_{10}H_{14}N_2$, in 80.0 g of CH_2Cl_2

11.48 How many grams of water should you add to 32.5 g of sucrose, $C_{12}H_{22}O_{11}$, to get a 0.85 *m* solution?

11.49 Household bleach is a 5.0 wt % aqueous solution of sodium hypochlorite, NaOCl. What is the molality of the bleach? What is the mole fraction of NaOCl in the bleach?

11.50 The density of a 16.0 wt % solution of sulfuric acid in water is 1.1094 g/mL at 25.0°C What is the molarity of the solution?

11.51 Ethylene glycol, $C_2H_6O_2$, is the principal constituent of automobile antifreeze. If the density of a 40.0 wt % solution of ethylene glycol in water is 1.0514 g/mL at 20°C, what is the molarity?

11.52 What is the molality of the 40.0 wt % ethylene glycol solution used for automobile antifreeze (Problem 11.51)?

11.53 Nalorphine ($C_{19}H_{21}NO_3$), a relative of morphine, is used to combat withdrawal symptoms in narcotics users. How many grams of a 1.3×10^{-3} *m* aqueous solution of nalorphine are needed to obtain a dose of 1.5 mg?

11.54 A 0.944 M solution of glucose, $C_6H_{12}O_6$, in water has density of 1.0624 g/mL at 20°C. What is the concentration of this solution in
  (a) mole fraction
  (b) weight percent
  (c) molality

11.55 Lactose, $C_{12}H_{22}O_{11}$, is a naturally occurring sugar found in mammalian milk. A 0.335 M solution of lactose in water has a density of 1.0432 g/mL at 20°C. What is the concentration of this solution in
  (a) mole fraction
  (b) weight percent
  (c) molality

SOLUBILITY AND HENRY'S LAW

11.56 Vinyl chloride (CH_2=CHCl), the starting material from which PVC polymer is made, has a Henry's law constant of 0.091 mol/(L · atm) at 25°C. What is the solubility of vinyl chloride in water in mol/L at 25°C and a pressure of 0.75 atm?

11.57 Hydrogen sulfide, H_2S, is a toxic gas responsible for the odor of rotten eggs. The solubility of $H_2S(g)$ in water at STP is 0.195 M. What is the Henry's law constant of H_2S at 0°C?

11.58 What is the solubility of H_2S (Problem 11.57) in water at 0°C and a partial pressure of 25.5 mm Hg?

11.59 Look at the solubility graph in Figure 11.6, and estimate an approximate Henry's law constant for xenon at STP.

11.60 Fish generally need an O_2 concentration in water of at least 4 mg/L for survival. What partial pressure of oxygen above the water (in atm at 0°C) is needed to obtain this concentration? The solubility of O_2 in water at 0°C and 1 atm partial pressure is 2.21 mmol/L.

11.61 Ammonia, NH_3, is one of the few gases that does not obey Henry's law. Suggest a reason.

COLLIGATIVE PROPERTIES

11.62 What single factor is responsible for all colligative properties?

11.63 What are the four colligative properties of solutions?

11.64 What is osmotic pressure?

11.65 Order the phases solid, liquid, vapor, and solution according to increasing entropy.

11.66 Draw a phase diagram showing how the phase boundaries differ for a pure solvent compared with a solution.

11.67 Why must a solution concentration be expressed in molality when considering boiling-point elevation or freezing-point depression but can be expressed in molarity when considering osmotic pressure?

11.68 What is the vapor pressure in mm Hg of the following solutions, each of which contains a non-volatile solute? The vapor pressure of pure water at 45.0°C is 71.93 mm Hg.
  (a) a solution of 10.0 g of urea, CH_4N_2O, in 150.0 g of water at 45.0°C
  (b) a solution of 10.0 g of LiCl in 150.0 g of water at 45.0°C

11.69 What is the boiling point in °C of each of the solutions in Problem 11.68? The molal boiling-point-elevation constant for water is given in Table 11.5.

11.70 What is the freezing point in °C of each of the solutions in Problem 11.68? The molal freezing-point-depression constant for water is given in Table 11.5.

11.71 Bromine is sometimes used as a solution in tetrachloromethane, CCl_4. What is the vapor pressure in mm Hg of a solution of 1.50 g Br_2 in 145.0 g CCl_4 at 300 K? The vapor pressure of pure bromine at 300 K is 30.5 kPa, and the vapor pressure of CCl_4 is 16.5 kPa.

11.72 Acetone, C_3H_6O, and ethyl acetate, $C_4H_8O_2$, are organic liquids often used as solvents. At 30°C, the vapor pressure of acetone is 285 mm Hg and the vapor pressure of ethyl acetate is 118 mm Hg. What is the vapor pressure (in mm Hg) at 30°C of a solution prepared by dissolving 25.0 g of acetone in 25.0 g of ethyl acetate?

11.73 What is the mole fraction of each component in the liquid mixture in Problem 11.72, and what is the mole fraction of each component in the vapor at 30°C?

11.74 The industrial solvents chloroform, $CHCl_3$, and dichloromethane, CH_2Cl_2, are prepared commercially by reaction of methane with chlorine, followed by fractional distillation of the product mixture. At 25°C, the vapor pressure of $CHCl_3$ is 205 mm Hg, and the vapor pressure of CH_2Cl_2 is 415 mm Hg. What is the vapor pressure (in mm Hg) at 25°C of a mixture of 15.0 g of $CHCl_3$ and 37.5 g of CH_2Cl_2?

11.75 What is the mole fraction of each component in the liquid mixture in Problem 11.74, and what is the mole fraction of each component in the vapor at 25°C?

11.76 What is the normal boiling point in °C of ethyl alcohol if a solution prepared by dissolving 26.0 g of glucose ($C_6H_{12}O_6$) in 285 g of ethyl alcohol has a boiling point of 79.1°C? See Table 11.5 to find K_b for ethyl alcohol.

11.77 A solution prepared by dissolving 5.00 g of aspirin, $C_9H_8O_4$, in 215 g of chloroform has a normal boiling point that is elevated by $\Delta T = 0.47°C$ over that of pure chloroform. What is the value of the molal boiling-point-elevation constant for chloroform?

11.78 A solution prepared by dissolving 3.00 g of ascorbic acid (vitamin C, $C_6H_8O_6$) in 50.0 g of acetic acid has a freezing point that is depressed by $\Delta T = 1.33°C$ below that of pure acetic acid. What is the value of the molal freezing-point-depression constant for acetic acid?

11.79 A solution of citric acid, $C_6H_8O_7$, in 50.0 g of acetic acid has a boiling point elevation of $\Delta T = 1.76°C$. What is the molality of the solution if the molal boiling-point-elevation constant for acetic acid is $K_b = 3.07$ (°C · kg)/mol.

11.80 What osmotic pressure in atm would you expect for each of the following solutions?
(a) 5.00 g of NaCl in 350.0 mL water at 50°C
(b) 6.33 g of sodium acetate, CH_3CO_2Na, in 55.0 mL of water at 10°C

11.81 What osmotic pressure in mm Hg would you expect for a solution of 11.5 mg of insulin, MW = 5990 amu, in 6.60 mL of water at 298 K?

11.82 What would the height of the water column be in meters in Problem 11.81? The density of mercury is 13.534 g/mL at 298 K.

11.83 A solution of an unknown molecule in water at 300 K gives rise to an osmotic pressure of 4.85 atm. What is the molarity of the solution?

11.84 Human blood gives rise to an osmotic pressure of approximately 7.7 atm at body temperature, 37.0°C. What must the molarity of an intravenous glucose solution be to have the same osmotic pressure as blood?

USES OF COLLIGATIVE PROPERTIES

11.85 Which of the four colligative properties is most often used for molecular weight determination, and why?

11.86 If cost per gram were not a concern, which of the following substances would be the most efficient per unit mass for melting snow from sidewalks and roads: glucose ($C_6H_{12}O_6$), LiCl, NaCl, $CaCl_2$? Explain.

11.87 Cellobiose is a sugar obtained by degradation of cellulose. If a solution of 1.500 g of cellobiose in 200.0 mL of water at 25.0°C gives rise to an osmotic pressure of 407.2 mm Hg, what is the molecular weight of cellobiose?

11.88 Met-enkephalin is one of the so-called endorphins, a class of naturally occurring morphinelike chemicals in the brain. What is the molecular weight of met-enkephalin if a solution of 15.0 mg met-enkephalin in 20.0 mL water at 298 K supports a column of water 32.9 cm high? The density of mercury at 298 K is 13.534 g/mL.

11.89 The freezing point of a solution prepared by dissolving 1.00 mol of hydrogen fluoride, HF, in 500 g water is −3.8°C, but the freezing point of a solution prepared by dissolving 1.00 mol of hydrogen chloride, HCl, in 500 g water is −7.4°C. Explain.

11.90 Elemental analysis of β-carotene, a dietary source of vitamin A, shows that it contains 10.51% H and 89.49% C. Dissolving 0.0250 g of β-carotene in 1.50 g of camphor gives a freezing-point depression of 1.17°C. What are the molecular weight and formula of β-carotene? [K_f for camphor is 37.7 (°C · kg)/mol.]

11.91 Lysine, one of the amino acid building blocks found in proteins, contains 49.29% C, 9.65% H, 19.16% N, and 21.89% O by elemental analysis. A solution prepared by dissolving 30.0 mg of lysine in 1.200 g of the organic solvent biphenyl, gives a freezing-point depression of 1.37°C. What are the molecular weight and formula of lysine? [K_f for biphenyl is 8.00 (°C · kg)/mol.]

11.92 The following graph is a phase diagram of temperature versus composition for mixtures of the two liquids chloroform and dichloromethane. Assuming that you begin with a mixture of 70% chloroform and 30% dichloromethane, answer the following questions.

(a) What region on the diagram corresponds to vapor, and what region corresponds to liquid?
(b) At what approximate temperature will the mixture begin to boil? Mark as point *a* on the diagram the liquid composition at the boiling point, and mark as point *b* the vapor composition at the boiling point.
(c) Assume that the vapor at point *b* condenses and is reboiled. Mark as point *c* on the diagram the liquid composition of the condensed vapor and as point *d* on the diagram the vapor composition of the reboiled material.

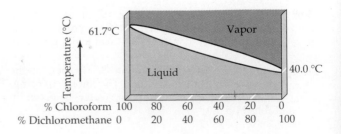

GENERAL PROBLEMS

11.93 How many grams of KBr dissolved in 125 g of water is needed to raise the boiling point of water to 103.2°C?

11.94 How many grams of ethylene glycol (automobile antifreeze, $C_2H_6O_2$) dissolved in 3.55 kg of water is needed to lower the freezing point of water in an automobile radiator to −22.0°C?

11.95 Which solution has the higher boiling point, 0.500 *m* glucose or 0.300 *m* KCl? Explain.

11.96 Which solution has the higher freezing point, 0.665 *m* KBr or 0.400 *m* $CaCl_2$? Explain.

11.97 When 1 mL of toluene is added to 100 mL of benzene (bp 80.1°C) the boiling point of the benzene solution rises, but when 1 mL of benzene is added to 100 mL of toluene (bp 110.6°C), the boiling point of the toluene solution falls. Explain.

11.98 When solid $CaCl_2$ is added to liquid water, the temperature rises. When solid $CaCl_2$ is added to ice at 0°C, the temperature falls. Explain.

11.99 Silver chloride has a solubility of 0.007 mg/mL in water at 5°C. What osmotic pressure in atm will a saturated solution of AgCl give rise to?

11.100 How many grams of naphthalene, $C_{10}H_8$, commonly used as household mothballs, should be added to 150.0 g of benzene to depress its freezing point by 0.35°C? See Table 11.5 to find K_f for benzene.

11.101 Assuming that sea water is a 3.5 wt % solution of NaCl and that its density is 1.00 g/mL, calculate both its boiling point and its freezing point in °C.

11.102 There's actually much more in sea water than just dissolved NaCl. Major ions present include 19,000 ppm Cl^-, 10,500 ppm Na^+, 2650 ppm SO_4^{2-}, 1350 ppm Mg^{2+}, 400 ppm Ca^{2+}, 380 ppm K^+, 140 ppm HCO_3^-, and 65 ppm Br^-.

(a) What is the total molality of all ions present in sea water?

(b) Assuming molality and molarity to be equal, what amount of osmotic pressure in atm would sea water give rise to at 300 K?

11.103 Rubbing alcohol is a 90 wt % solution of isopropyl alcohol, C_3H_8O, in water. How many grams of rubbing alcohol contains 10.5 g of isopropyl alcohol?

11.104 How many moles of isopropyl alcohol are in 50.0 g of rubbing alcohol (Problem 11.103)?

11.105 A 0.50 wt % solution of boric acid, $B(OH)_3$, is often used as an eyewash. What is the molality of the solution?

11.106 Although inconvenient, it's possible to use osmotic pressure to measure temperature. What is the temperature in kelvins if a solution prepared by dissolving 17.5 mg of glucose ($C_6H_{12}O_6$) in 50.0 mL of water gives rise to an osmotic pressure of 37.8 mm Hg?

chapter 12 CHEMICAL KINETICS

Chemists have three fundamental questions in mind when they study chemical reactions: What happens? How fast does it happen? To what extent does it happen? The answer to the first question is given by the balanced chemical equation, which identifies the nature of the reactants and products and the stoichiometry of the reaction. The answer to the third question will be addressed in the next chapter, which deals with chemical equilibrium. In this chapter, we'll look at the answer to the second question—the speeds, or rates, at which chemical reactions occur. The area of chemistry concerned with reaction rates and the sequence of steps by which reactions occur is called **chemical kinetics**.

Chemical kinetics is a subject of crucial environmental and economic importance. In the upper atmosphere, for example, maintenance or depletion of the ozone layer, which protects us from the sun's harmful ultraviolet radiation, depends on the relative rates of reactions that produce and destroy O_3 molecules. In the chemical industry, the profitability of the process for synthesis of ammonia, which is used as a fertilizer, depends on the rate at which gaseous N_2 and H_2 can be converted to NH_3.

The speed of these race cars at the Riverside Can-Am is defined as the change in location per unit time (mi/hr or m/s). Similarly, the speed, or rate, of a chemical reaction is defined as a change in concentration per unit time (M/s).

An industrial plant for synthesis of ammonia.

In this chapter, we'll learn how to describe reaction rates and will examine how they are affected by variables such as reactant concentrations and temperature. We'll also see how chemists use rate data to propose a mechanism, or pathway, by which a reaction takes place. By understanding reaction mechanisms, we can control known reactions and predict new ones.

12.1 ►REACTION RATES

The rates of chemical reactions are highly variable. Some reactions, such as the explosion of a gaseous mixture of H_2 and O_2, occur in a fraction of a second. Other reactions, such as the weathering of minerals, are imperceptibly slow. To describe the rate of a reaction in quantitative terms, we must

The explosion of gaseous H_2 and O_2 is a very fast reaction **(a)**, whereas the weathering of minerals is a very slow reaction **(b)**. The eroded rock is Medicine Hat Rock, Utah.

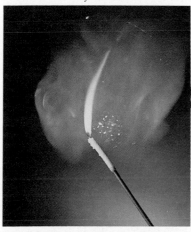

(a)

(b)

specify how fast the concentration of a reactant or a product changes per unit time.

$$\text{Rate} = \frac{\text{concentration change}}{\text{time change}}$$

As an example of a reaction that has been studied in detail, consider the thermal decomposition of dinitrogen pentoxide, a colorless ionic solid, $[NO_2]^+[NO_3]^-$, that gives gaseous N_2O_5 molecules when it's heated to 32°C at 1 atm pressure. In the gas phase, N_2O_5 decomposes to the brown gas nitrogen dioxide and molecular oxygen:

$$[NO_2]^+[NO_3]^-(s) \rightarrow N_2O_5(g)$$

$$2\,N_2O_5(g) \rightarrow 4\,NO_2(g) + O_2(g)$$

If we measure the reactant and product concentrations as a function of time at 55°C, we might obtain the data in Table 12.1. Note that the concentrations of NO_2 and O_2 increase as the concentration of N_2O_5 decreases.

TABLE 12.1 Concentrations as a Function of Time at 55°C for the Reaction $2\,N_2O_5(g) \rightarrow 4\,NO_2(g) + O_2(g)$

Time (s)	Concentration (M)		
	N_2O_5	NO_2	O_2
0	0.0200	0	0
100	0.0169	0.0063	0.0016
200	0.0142	0.0115	0.0029
300	0.0120	0.0160	0.0040
400	0.0101	0.0197	0.0049
500	0.0086	0.0229	0.0057
600	0.0072	0.0256	0.0064
700	0.0061	0.0278	0.0070

We define the **reaction rate** either as the increase in the concentration of a product per unit time or as the decrease in the concentration of a reactant per unit time (the two definitions are equivalent). Let's look first at product formation. In the decomposition of N_2O_5, the rate of formation of O_2 is given by the equation

$$\text{Rate of formation of } O_2 = \frac{\Delta[O_2]}{\Delta t} = \frac{\text{conc of } O_2 \text{ at time } t_2 - \text{conc of } O_2 \text{ at time } t_1}{t_2 - t_1}$$

where the square brackets surrounding O_2 denote the molar concentration of the substance within the brackets. As usual, the Greek letter Δ designates the change in the quantity that follows the Δ; thus, $\Delta[O_2]$ is the change in the molar concentration of O_2, Δt is the change in time, and $\Delta[O_2]/\Delta t$ is the change in the molar concentration of O_2 during the time interval from time

t_1 to t_2. During the time period 300 to 400 s, for example, the **average rate** of formation of O_2 is 9×10^{-6} M/s:

$$\text{Rate of formation of } O_2 = \frac{\Delta[O_2]}{\Delta t} = \frac{0.0049 \text{ M} - 0.0040 \text{ M}}{400 \text{ s} - 300 \text{ s}} = 9 \times 10^{-6} \text{ M/s}$$

The units of reaction rate are M/s (molar per second) or, equivalently, mol/(L · s). We define reaction rate in terms of concentration (mol/L) rather than amounts (moles) because we want the rate to be independent of the scale of the reaction. When twice as much 0.0200 M N_2O_5 decomposes in a vessel of twice the volume, twice the number of moles of O_2 form per second but the number of moles of O_2 *per liter* that form per second is unchanged.

We can gain additional insight into the concept of reaction rate if we plot the data of Table 12.1 to give the three lines that appear in Figure 12.1. Looking at the time period 300 to 400 s on the O_2 line, $\Delta[O_2]$ and Δt are represented, respectively, by the vertical and horizontal sides of a right triangle. The slope of the third side, the hypotenuse of the triangle, is $\Delta[O_2]/\Delta t$, the average rate of O_2 formation during that time period. The steeper the slope of the hypotenuse, the faster the rate. Look, for example, at the triangle defined by $\Delta[NO_2]$ and Δt. The average rate of formation of NO_2

FIGURE 12.1 Concentrations measured as a function of time when gaseous N_2O_5 at an initial concentration of 0.0200 M decomposes to gaseous NO_2 and O_2 at 55°C. Note that the concentrations of O_2 and NO_2 go up as the concentration of N_2O_5 goes down. The slope of the hypotenuse of the triangles gives the average rate of change of a product or a reactant concentration during the indicated time interval.

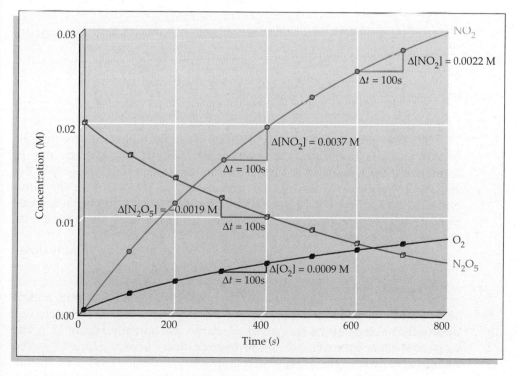

during the time period 300 to 400 s is 3.7×10^{-5} M/s, which is four times the rate of formation of O_2, in accord with the stoichiometry of the chemical equation for decomposition of N_2O_5:

$$\text{Rate of formation of NO}_2 = \frac{\Delta[NO_2]}{\Delta t} = \frac{0.0197 \text{ M} - 0.0160 \text{ M}}{400 \text{ s} - 300 \text{ s}} = 3.7 \times 10^{-5} \text{ M/s}$$

As O_2 and NO_2 form, N_2O_5 disappears. Consequently, $\Delta[N_2O_5]/\Delta t$ is negative, in accord with the negative slope of the hypotenuse of the triangle defined by $\Delta[N_2O_5]$ and Δt in Figure 12.1. Because *reaction rate is defined as a positive quantity*, we must always introduce a minus sign in calculating the rate of disappearance of a reactant. During the time period 300 to 400 s, for example, the average rate of decomposition of N_2O_5 is 1.9×10^{-5} M/s:

$$\text{Rate of decomposition of N}_2O_5 = \frac{-\Delta[N_2O_5]}{\Delta t}$$

$$= \frac{-(0.0101 \text{ M} - 0.0120 \text{ M})}{400 \text{ s} - 300 \text{ s}} = 1.9 \times 10^{-5} \text{ M/s}$$

It's important to specify the reactant or product when we quote a rate because the relative rates of product formation and reactant consumption depend on the coefficients in the balanced equation. For the decomposition of N_2O_5, 4 mol of NO_2 form and 2 mol of N_2O_5 disappear for each mole of O_2 that forms. Therefore, the rate of formation of O_2 is one-fourth the rate of formation of NO_2 and one-half the rate of decomposition of N_2O_5:

$$\begin{bmatrix} \text{Rate of formation} \\ \text{of O}_2 \end{bmatrix} = \frac{1}{4} \begin{bmatrix} \text{rate of formation} \\ \text{of NO}_2 \end{bmatrix} = \frac{1}{2} \begin{bmatrix} \text{rate of decomposition} \\ \text{of N}_2O_5 \end{bmatrix}$$

or

$$\frac{\Delta[O_2]}{\Delta t} = \frac{1}{4}\frac{\Delta[NO_2]}{\Delta t} = -\frac{1}{2}\frac{\Delta[N_2O_5]}{\Delta t}$$

It's also important to specify the time when quoting a rate because the rate changes as the reaction proceeds. For example, the average rate of formation of NO_2 is 3.7×10^{-5} M/s during the time period 300 to 400 s but is considerably slower during the period 600 to 700 s. During the latter period, $\Delta[NO_2]/\Delta t = 0.0022$ M/100 s $= 2.2 \times 10^{-5}$ M/s. Ordinarily, reaction rates decrease as the reaction mixture runs out of reactants, as indicated by the decreasing slope of the curves in Figure 12.1 as time passes.

Often, chemists want to know the rate of a reaction at a specific time t rather than the rate averaged over a time interval Δt. For example, what is the rate of formation of NO_2 at time $t = 350$ s? If we make our measurements at shorter and shorter time intervals, the triangle defined by $\Delta[NO_2]$ and Δt will shrink to a point, and the slope of the hypotenuse of the triangle will approach the slope of the tangent to the curve, as shown in Figure 12.2. The slope of the tangent to a concentration-versus-time curve at a time t is called the **instantaneous rate** of the reaction at that particular time. The instantaneous rate at the beginning of a reaction ($t = 0$) is called the **initial rate**.

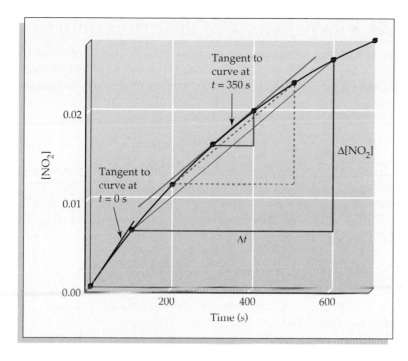

FIGURE 12.2 Concentration of NO_2 versus time when N_2O_5 decomposes at 55°C. The average rate of formation of NO_2 during a time interval Δt equals the slope of the hypotenuse of the triangle defined by $\Delta[NO_2]$ and Δt. As the time interval about the time t gets smaller, the triangle shrinks to a point, and the slope of the hypotenuse approaches the slope of the tangent to the curve at time t. The slope of the tangent at time t is defined as the instantaneous rate of the reaction at that particular time—in this figure, $t = 350$ s. The initial rate is the slope of the tangent to the curve at $t = 0$.

EXAMPLE 12.1

Ethanol (C_2H_5OH), the active ingredient in alcoholic beverages and an octane booster in gasoline, is produced by fermentation of glucose. The balanced equation is

$$C_6H_{12}O_6(aq) \rightarrow 2\ C_2H_5OH(aq) + 2\ CO_2(g)$$

(a) How is the rate of formation of ethanol related to the rate of consumption of glucose? **(b)** Write this relation in terms of $\Delta[C_2H_5OH]/\Delta t$ and $\Delta[C_6H_{12}O_6]/\Delta t$.

SOLUTION   **(a)** According to the balanced equation, 2 mol of ethanol are produced for each mole of glucose that reacts. Therefore, the rate of formation of ethanol is twice the rate of consumption of glucose. **(b)** Since the rate of formation of ethanol is $\Delta[C_2H_5OH]/\Delta t$ and the rate of consumption of glucose is $-\Delta[C_6H_{12}O_6]/\Delta t$ (note the minus sign), we can write

$$\frac{\Delta[C_2H_5OH]}{\Delta t} = -2\frac{\Delta[C_6H_{12}O_6]}{\Delta t}$$

⌐ PROBLEM 12.1 Use the data in Table 12.1 to calculate the average rate of decomposition of N_2O_5 and the average rate of formation of O_2 during the time interval 200 to 300 s.

⌐ PROBLEM 12.2 The oxidation of iodide ion by arsenic acid, H_3AsO_4, is described by the balanced equation

$$3\ I^-(aq) + H_3AsO_4(aq) + 2\ H^+(aq) \rightarrow I_3^-(aq) + H_3AsO_3(aq) + H_2O(l)$$

(a) If $-\Delta[I^-]/\Delta t = 4.8 \times 10^{-4}$ M/s, what is the value of $\Delta[I_3^-]/\Delta t$ during the same time interval? **(b)** What is the average rate of consumption of H^+ during that time interval?

The fermentation of sugar in cider produces ethanol and bubbles of CO_2 gas.

12.2 ►RATE LAWS AND REACTION ORDER

We noted in the previous section that the rate of decomposition of N_2O_5 depends on its concentration: The rate slows down as the N_2O_5 concentration decreases. In general, the rate of a chemical reaction depends on the concentrations of at least some of the reactants.

Let's consider the general reaction

$$a\,A + b\,B \rightarrow \text{products}$$

where A and B are chemical formulas for the reactants, and a and b are stoichiometric coefficients in the balanced chemical equation. The dependence of the reaction rate on concentration is given by the **rate law**, an equation that tells how the rate depends on the concentration of each reactant. The rate law is usually written in the form

$$\text{Rate} = -\frac{\Delta[A]}{\Delta t} = k[A]^m[B]^n$$

where k is a proportionality constant called the **rate constant**. We have arbitrarily expressed the rate as the rate of disappearance of A ($-\Delta[A]/\Delta t$), but we could equally well have written it in terms of the rate of disappearance of any reactant—say, $-\Delta[B]/\Delta t$—or in terms of the rate of appearance of any product.

The exponents m and n in the rate law indicate how sensitive the rate is to changes in [A] and [B] and are usually unrelated to the coefficients a and b in the balanced equation. For the simple reactions discussed in this book, the exponents are usually small positive integers. For more complex reactions, however, the exponents can be negative, zero, or even fractions. An exponent of 1 means that the rate depends linearly on the concentration of the corresponding reactant. For example, if $m = 1$ and [A] is doubled, the rate doubles. If $m = 2$ and [A] is doubled, $[A]^2$ quadruples and the rate increases by a factor of 4. Figure 12.3 will give you a feel for how the rate changes

FIGURE 12.3 Change in reaction rate when the concentration of a reactant, A, is doubled for different values of the exponent m in the rate law, rate = $k[A]^m[B]^n$. Note that the rate change increases as m increases.

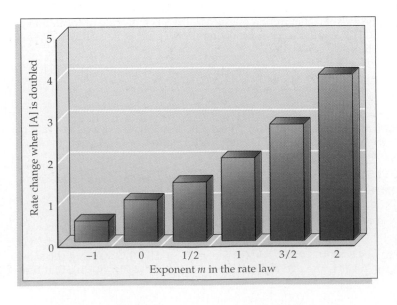

depend on the value of the exponent. When m is zero, the rate is independent of the concentration of A because any number raised to the zeroth power is unity ($[A]^0 = 1$). When m is negative, the rate *decreases* as $[A]$ increases. For example, if $m = -1$ and $[A]$ is doubled, $[A]^{-1}$ is halved and the rate decreases by a factor of 2.

The values of the exponents m and n determine the **reaction order** with respect to A and B, respectively, and the sum of the exponents $(m + n)$ defines the *overall* reaction order. Thus, if the rate law is

$$\text{Rate} = k[A]^2[B] \quad (m = 2; n = 1; \text{ and } m + n = 3)$$

we say that the reaction is second order in A, first order in B, and third order overall.

The values of the exponents in a rate law must be determined by experiment; they can't be deduced from the stoichiometry of the reaction. As Table 12.2 shows, there is no general relation between the stoichiometric coefficients in the balanced chemical equation and the exponents in the rate law. In the first reaction, for example, the coefficients of $(CH_3)_3CBr$ and H_2O in the balanced equation are both 1, but the exponents in the rate law are 1 for $(CH_3)_3CBr$ and 0 for H_2O: Rate $= k[(CH_3)_3CBr]^1[H_2O]^0 = k[(CH_3)_3CBr]$. In Section 12.8, we'll see that the exponents in the rate law depend on the reaction mechanism.

TABLE 12.2   Balanced Chemical Equations and Experimentally Determined Rate Laws for Some Typical Reactions

Reaction	Rate Law
$(CH_3)_3CBr(aq) + H_2O(aq) \rightarrow (CH_3)_3COH(aq) + H^+(aq) + Br^-(aq)$	Rate $= k[(CH_3)_3CBr]$
$HCOOH(aq) + Br_2(aq) \rightarrow 2\ H^+(aq) + 2\ Br^-(aq) + CO_2(g)$	Rate $= k[Br_2]$
$BrO_3^-(aq) + 5\ Br^-(aq) + 6\ H^+(aq) \rightarrow 3\ Br_2(aq) + 3\ H_2O(l)$	Rate $= k[BrO_3^-][Br^-][H^+]^2$
$H_2(g) + I_2(g) \rightarrow 2\ HI(g)$	Rate $= k[H_2][I_2]$
$CH_3CHO(g) \rightarrow CH_4(g) + CO(g)$	Rate $= k[CH_3CHO]^{3/2}$

EXAMPLE 12.2

Consider the second reaction in Table 12.2, shown in progress in Figure 12.4. What is the order of the reaction with respect to each of the reactants, and what is the overall reaction order?

FIGURE 12.4 The reaction of formic acid (HCOOH) and bromine (Br_2). As time passes (left to right), the red color of bromine disappears as Br_2 is consumed.

SOLUTION   To find the reaction order, look at the exponents on the concentration terms in the rate law:

$$\text{Rate} = k[Br_2]$$

Because HCOOH (formic acid) does not appear in the rate law, the rate must be independent of the concentration of HCOOH. Therefore, the reaction is zero order in HCOOH. Because the exponent on $[Br_2]$ (understood) is 1, the reaction is first order in Br_2. Therefore, the rate law is

$$\text{Rate} = k[HCOOH]^m[Br_2]^n = k[HCOOH]^0[Br_2]^1 \quad \text{or} \quad \text{Rate} = k[Br_2]$$

The reaction is first order overall because the sum of the exponents $(m + n)$ is 1.

◤ **PROBLEM 12.3**   Consider the last three reactions in Table 12.2. What is the order of each reaction in the various reactants, and what is the overall reaction order? ◢

12.3 ►EXPERIMENTAL DETERMINATION OF A RATE LAW

One method of determining the values of the exponents in a rate law is to carry out a series of experiments in which the initial rate of a reaction is measured as a function of different sets of initial concentrations. Consider, for example, the air oxidation of nitric oxide, one of the reactions that contributes to the formation of acid rain:

$$2\,NO(g) + O_2(g) \rightarrow 2\,NO_2(g)$$

Look at the initial rate data in Table 12.3.

TABLE 12.3	Initial Concentration and Rate Data for the Reaction $2\,NO(g) + O_2(g) \rightarrow 2\,NO_2(g)$		
Experiment	Initial [NO]	Initial [O$_2$]	Initial Rate (M/s)
1	0.015	0.015	0.024
2	0.030	0.015	0.096
3	0.015	0.030	0.048
4	0.030	0.030	0.192

Note that pairs of experiments are designed to investigate the effect on the initial rate of a change in the initial concentration of a single reactant. In the first two experiments, for example, the concentration of NO is doubled from 0.015 M to 0.030 M while the initial concentration of O_2 is held constant. The initial rate increases by a factor of four, from 0.024 M/s to 0.096 M/s, indicating that the rate depends on the concentration of NO squared, $[NO]^2$. When $[NO]$ is held constant and $[O_2]$ is doubled (experiments 1 and 3), the initial rate doubles from 0.024 M/s to 0.048 M/s, showing that the rate depends on the concentration of O_2 to the first power, $[O_2]^1$. Therefore, the rate law is

$$Rate = k[NO]^2[O_2]$$

In accord with this rate law, the initial rate increases by a factor of eight when the concentrations of both NO and O_2 are doubled (experiments 1 and 4).

The preceding method uses initial rates rather than rates at a later stage of the reaction because we want to avoid complications from the reverse reaction: reactants ⟵ products. As the product concentrations build up, the rate of the reverse reaction increases. If the reverse rate becomes comparable to the forward rate, the measured reaction rate will depend on the concentrations of both reactants and products. At the beginning of the reaction, however, the product concentrations are zero, and therefore the products can't affect the measured rate. When we measure an initial rate, we are measuring the rate of only the forward reaction, and only reactants (and catalysts—see Section 12.11) can appear in the rate law.

One aspect of determining a rate law is to establish the reaction order. Another is to evaluate the numerical value of the rate constant k. Each reaction has its own characteristic value of the rate constant, which depends on temperature but does not depend on concentrations. To evaluate k for the oxidation of nitric oxide, for instance, we can use the data from any one of the experiments in Table 12.3. Solving the rate law for k and substituting in the initial rate and concentrations from the first experiment, we obtain

$$k = \frac{rate}{[NO_2]^2[O_2]} = \frac{0.024\,\frac{M}{s}}{(0.015\ M)^2(0.015\ M)} = 7.1 \times 10^3/(M^2 \cdot s)$$

Try repeating the calculation for experiments 2–4, and show that you get the same value of k. Note that the units of k are $1/(M^2 \cdot s)$ (read "one over molar squared second") or, equivalently, $L^2/(mol^2 \cdot s)$. Because rates have units of M/s, the units of k depend on the number of concentration terms in the rate law and on the values of the exponents. Units for some common cases are given below:

Rate Law	Overall Reaction Order	Units for k
Rate = $k[A]$	First order	$1/s$
Rate = $k[A][B]$	Second order	$1/(M \cdot s)$
Rate = $k[A][B]^2$	Third order	$1/(M^2 \cdot s)$

Example 12.3 gives another instance of how a rate law can be determined from initial rates.

EXAMPLE 12.3

Given the following initial rate data for decomposition of gaseous N_2O_5 at 55°C, **(a)** determine the rate law and **(b)** calculate the value of the rate constant.

Experiment	Initial [N_2O_5]	Initial Rate of Decomposition of N_2O_5 (M/s)
1	0.020	3.4×10^{-5}
2	0.050	8.5×10^{-5}

SOLUTION   **(a)** The rate law for decomposition of N_2O_5 can be written as

$$\text{Rate} = -\frac{\Delta[N_2O_5]}{\Delta t} = k[N_2O_5]^m$$

where m is both the order of the reaction in N_2O_5 and the overall reaction order. Comparing experiments 1 and 2 shows that an increase in the initial concentration of N_2O_5 by a factor of 2.5 increases the initial rate by a factor of 2.5.

$$\frac{[N_2O_5]_2}{[N_2O_5]_1} = \frac{0.050 \text{ M}}{0.020 \text{ M}} = 2.5 \qquad \frac{(\text{Rate})_2}{(\text{Rate})_1} = \frac{8.5 \times 10^{-5} \text{ M/s}}{3.4 \times 10^{-5} \text{ M/s}} = 2.5$$

The rate depends linearly on the concentration of N_2O_5, and therefore the rate law is

$$\text{Rate} = -\frac{\Delta[N_2O_5]}{\Delta t} = k[N_2O_5]$$

Note that the reaction is first order in N_2O_5. If the rate had increased by a factor of $(2.5)^2 = 6.25$, the reaction would have been second order in N_2O_5. If the rate had increased by a factor of $(2.5)^3 = 15.6$, the reaction would have been third order in N_2O_5, and so on.

A more formal way to approach this problem is to write the rate law for each experiment:

$$(\text{Rate})_1 = k[N_2O_5]_1{}^m = k(0.020 \text{ M})^m \qquad (\text{Rate})_2 = k[N_2O_5]_2{}^m = k(0.050 \text{ M})^m$$

If we then divide the second equation by the first, we obtain

$$\frac{(\text{Rate})_2}{(\text{Rate})_1} = \frac{k(0.050 \text{ M})^m}{k(0.020 \text{ M})^m} = (2.5)^m$$

Comparing this ratio to the ratio of the experimental rates

$$\frac{(\text{Rate})_2}{(\text{Rate})_1} = \frac{8.5 \times 10^{-5} \text{ M/s}}{3.4 \times 10^{-5} \text{ M/s}} = 2.5$$

shows that the exponent m must have a value of 1. Therefore, the rate law is

$$\text{Rate} = -\frac{\Delta[N_2O_5]}{\Delta t} = k[N_2O_5]$$

(b) The value of the rate constant k can be found by solving the rate law for k and then substituting in the data from either experiment. Using the data from the first experiment, we obtain

$$k = \frac{\text{Rate}}{[N_2O_5]} = \frac{3.4 \times 10^{-5}\,\dfrac{M}{s}}{0.020\,M} = 1.7 \times 10^{-3}/s$$

Note that the units of k, 1/s, are given by the ratio of the units for the rate and the concentration. These are the expected units for a first-order reaction.

EXAMPLE 12.4

What is the rate of decomposition of N_2O_5 at 55°C when its concentration is 0.030 M?

SOLUTION   Calculate the rate by substituting the rate constant ($1.7 \times 10^{-3}/s$) and concentration (0.030 M) into the rate law:

$$\text{Rate} = -\frac{\Delta[N_2O_5]}{\Delta t} = k[N_2O_5] = \left(\frac{1.7 \times 10^{-3}}{s}\right)(0.030\,M) = 5.1 \times 10^{-5}\,M/s$$

Be careful not to confuse the rate of a reaction and the rate constant. The rate depends on concentrations, whereas the rate *constant* does not (it is a constant). The rate has units of M/s, whereas the units of the rate constant depend on the overall reaction order.

PROBLEM 12.4 The oxidation of iodide ion by hydrogen peroxide in an acidic solution is described by the balanced equation

$$H_2O_2(aq) + 3\,I^-(aq) + 2\,H^+(aq) \rightarrow I_3^-(aq) + 2\,H_2O(l)$$

The rate of formation of the red-colored triiodide ion, $\Delta[I_3^-]/\Delta t$, can be determined by measuring the rate of appearance of the color (Figure 12.5). Given the following initial rate data at 25°C, **(a)** determine the rate law for formation of I_3^-; **(b)** calculate the value of the rate constant; **(c)** calculate the rate of formation of I_3^- when the concentrations are $[H_2O_2] = 0.300$ M and $[I^-] = 0.400$ M.

Experiment	Initial [H$_2$O$_2$]	Initial [I$^-$]	Initial Rate (M/s)
1	0.100	0.100	1.15×10^{-4}
2	0.100	0.200	2.30×10^{-4}
3	0.200	0.100	2.30×10^{-4}
4	0.200	0.200	4.60×10^{-4}

FIGURE 12.5 A sequence of photographs showing the progress of the reaction of hydrogen peroxide (H_2O_2) and iodide ion (I^-). As time passes (left to right), the red color due to triiodide ion (I_3^-) increases in intensity.

┌─ **PROBLEM 12.5** What are the units of the rate constant for each of the
reactions in Table 12.2? ─┘

12.4 ➤INTEGRATED RATE LAW
FOR A FIRST-ORDER REACTION

Thus far we've focused on the rate law, an equation that tells how a reaction rate depends on reactant concentrations. But we're also interested in how reactant and product concentrations vary with time. For example, it's important to know if the atmospheric ozone layer is being destroyed at a certain rate, but we also want to know what the ozone concentration will be 50 years from now and how long it will take for the concentration to be reduced by a given amount, say, 10%.

Because the kinetics of the pollution-induced decomposition of ozone is a very complicated problem, let's consider instead a simple, general first-order reaction

$$a\,A \rightarrow \text{products}$$

having the rate law $\quad\text{Rate} = -\dfrac{\Delta[A]}{\Delta t} = k[A]$

An example is the decomposition of hydrogen peroxide in basic solution, discussed in Example 12.5:

$$2\,H_2O_2(aq) \rightarrow 2\,H_2O(l) + O_2(g)$$

Using mathematics, it's possible to convert the rate law to another form, called the **integrated rate law**:

$$\ln\frac{[A]_t}{[A]_0} = -kt \quad\text{or}\quad \log\frac{[A]_t}{[A]_0} = \frac{-kt}{2.303}$$

The integrated rate law can be written in terms of the natural logarithm (base $e = 2.71828\ldots$), symbolized by ln, or in terms of the common logarithm (base 10), denoted by log. The two logarithms are related by the equation $\ln x = 2.303(\log x)$. (See Appendix A.2 for a review of logarithms.) The symbol $[A]_0$ designates the concentration of A at some initial time, arbitrarily considered to be $t = 0$, and $[A]_t$ is the concentration of A at any time t thereafter. The ratio $[A]_t/[[A]_0$ is the fraction of A that remains at time t. Thus, the integrated rate law is a *concentration-time equation* that allows us to calculate the concentration of A at any time t or the fraction of A that remains at any time t. The integrated rate law can also be used to calculate the time required for the initial concentration of A to drop to any particular value or to any particular fraction of its initial concentration. Example 12.5 shows how to use the integrated rate law.

Since $\log\{[A]_t/[A]_0\} = \log[A]_t - \log[A]_0$, we can rewrite the integrated rate law as

$$\log[A]_t = \frac{-kt}{2.303} + \log[A]_0$$

This equation is of the form $y = mx + b$, which says that log $[A]_t$ is a linear function of time.

$$\log [A]_t = \left(\frac{-k}{2.303}\right)t + \log [A]_0$$

$$\underset{y}{\uparrow} \qquad \underset{m}{\uparrow}\ \underset{x}{\uparrow} \qquad \underset{b}{\uparrow}$$

If we graph log [A] versus time, we obtain a straight line having a slope $m = -k/2.303$ and an intercept $b = \log [A]_0$ (Figure 12.6). The value of the rate constant is readily calculated from the slope of the straight line:

$$k = -2.303(\text{slope})$$

This graphical method of determining a rate constant, illustrated in Example 12.6, is an alternative to the method of initial rates used in Example 12.3. It should be emphasized, however, that a plot of log [A] versus time will give a straight line only if the reaction is first order in A. Indeed, a good way of testing whether a reaction is first order is to examine the appearance of a such a plot.

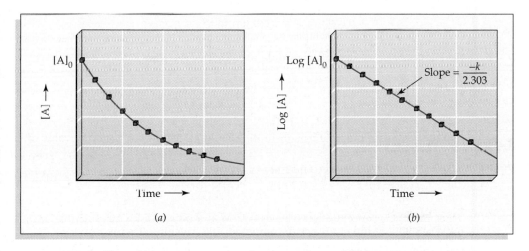

FIGURE 12.6 Plots of **(a)** reactant concentration versus time and **(b)** logarithm of reactant concentration versus time for a first-order reaction. A first-order reaction exhibits an exponential decay of the reactant concentration **(a)** and a linear decay of the logarithm of the reactant concentration **(b)**. The slope of the log [A] versus time plot gives the rate constant.

EXAMPLE 12.5

The decomposition of hydrogen peroxide in dilute sodium hydroxide solution is described by the equation

$$2\ H_2O_2(aq) \rightarrow 2\ H_2O(l) + O_2(g)$$

The reaction is first order in H_2O_2, and the rate constant at 20°C is 1.8×10^{-5}/s. If the initial concentration of H_2O_2 is 0.30 M, **(a)** what is the concentration of H_2O_2 after 4 h, **(b)** how long will it take for the H_2O_2 concentration to drop to 0.12 M, and **(c)** how long will it take for 90% of the H_2O_2 to decompose?

SOLUTION   (a) Since this reaction has a first-order rate law, $-\Delta[H_2O_2]/\Delta t = k[H_2O_2]$, we can use the corresponding concentration-time equation for a first-order reaction:

$$\log \frac{[H_2O_2]_t}{[H_2O_2]_0} = \frac{-kt}{2.303}$$

Because k has units of $1/s$, we must first convert the time in hours to units of seconds:

$$t = (4 \text{ h})\left(\frac{60 \text{ min}}{h}\right)\left(\frac{60 \text{ s}}{\min}\right) = 14{,}400 \text{ s}$$

Then substitute the values of $[H_2O_2]_0$, k, and t into the concentration-time equation:

$$\log \frac{[H_2O_2]_t}{0.30 \text{ M}} = \frac{-(1.8 \times 10^{-5} \text{ s}^{-1})(1.44 \times 10^4 \text{ s})}{2.303} = -0.112$$

Taking the antilog of both sides gives

$$\frac{[H_2O_2]_t}{0.30 \text{ M}} = 10^{-0.112} = 0.773$$

$$[H_2O_2]_t = (0.773)(0.30 \text{ M}) = 0.23 \text{ M}$$

(b) First, solve the concentration-time equation for the time

$$t = \left(\frac{-2.303}{k}\right) \log \frac{[H_2O_2]_t}{[H_2O_2]_0}$$

and then evaluate the time by substituting in the concentrations and the value of k:

$$t = \left(\frac{-2.303}{1.8 \times 10^{-5} \text{ s}^{-1}}\right) \log \frac{0.12 \text{ M}}{0.30 \text{ M}} = \left(\frac{-2.303}{1.8 \times 10^{-5} \text{ s}^{-1}}\right)(-0.398) = 5.1 \times 10^4 \text{ s}$$

Thus, the H_2O_2 concentration reaches 0.12 M at a time of 5.1×10^4 s (14 h).
(c) When 90% of the H_2O_2 has decomposed, 10% remains. Therefore,

$$\frac{[H_2O_2]_t}{[H_2O_2]_0} = \frac{(0.10)(0.30 \text{ M})}{0.30 \text{ M}} = 0.10$$

The time required for 90% decomposition is

$$t = \left(\frac{-2.303}{1.8 \times 10^{-5} \text{ s}^{-1}}\right) \log 0.10 = \left(\frac{-2.303}{1.8 \times 10^{-5} \text{ s}^{-1}}\right)(-1.0) = 1.3 \times 10^5 \text{ s} = 36 \text{ h}$$

EXAMPLE 12.6

Experimental concentration-versus-time data for the decomposition of gaseous N_2O_5 at 55°C are listed in Table 12.1 and are plotted in Figure 12.1. Use these data to confirm that the decomposition of N_2O_5 is a first-order reaction. What is the value of the rate constant?

SOLUTION To confirm that the reaction is first order, check to see if a plot of log [N_2O_5] versus time gives a straight line. Values of log [N_2O_5] are listed in the following table and are plotted versus time in the graph.

Time (s)	[N_2O_5]	Log [N_2O_5]
0	0.0200	−1.699
100	0.0169	−1.772
200	0.0142	−1.848
300	0.0120	−1.921
400	0.0101	−1.996
500	0.0086	−2.066
600	0.0072	−2.143
700	0.0061	−2.215

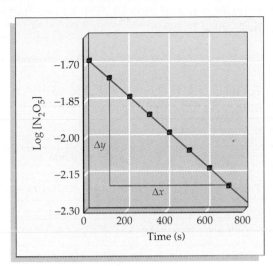

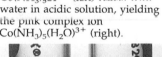

The purple complex ion $Co(NH_3)_5Br^{2+}$ (left) reacts with water in acidic solution, yielding the pink complex ion $Co(NH_3)_5(H_2O)^{3+}$ (right).

Because the data points lie on a straight line, the reaction is first order in N_2O_5. The slope of the line can be determined from the coordinates of any two widely separated points on the line, and the rate constant k can be calculated from the slope:

$$\text{Slope} = \frac{\Delta y}{\Delta x} = \frac{(-2.22) - (-1.77)}{700 \text{ s} - 100 \text{ s}} = \frac{-0.45}{600 \text{ s}} = -7.5 \times 10^{-4}/\text{s}$$

$$k = -2.303(\text{slope}) = -2.303(-7.5 \times 10^{-4}/\text{s}) = 1.7 \times 10^{-3}/\text{s}$$

Note that the slope is negative, k is positive, and the value of k agrees with the value obtained earlier in Example 12.3 by the method of initial rates.

▶ **PROBLEM 12.6** In acidic aqueous solution, the complex ion $Co(NH_3)_5Br^{2+}$ undergoes a slow reaction in which the bromide ion is replaced by a water molecule:

$$Co(NH_3)_5Br^{2+}(aq) + H_2O(l) \rightarrow Co(NH_3)_5(H_2O)^{3+}(aq) + Br^-(aq)$$

The reaction is first order in $Co(NH_3)_5Br^{2+}$, and the rate constant at 25°C is 6.3×10^{-6}/s. If the initial concentration of $Co(NH_3)_5Br^{2+}$ is 0.100 M, **(a)** what is its molarity after 10 h reaction time, and **(b)** how many hours are required for 75% of the $Co(NH_3)_5Br^{2+}$ to react?

These toy trucks are made of polypropylene.

▶ **PROBLEM 12.7** At elevated temperatures, cyclopropane is converted to propene, the material from which polypropylene is made:

$$\underset{\text{Cyclopropane}}{CH_2 - CH_2 \overset{\displaystyle CH_2}{\diagup \diagdown}} \rightarrow \underset{\text{Propene}}{CH_3 - CH = CH_2}$$

Given the following concentration data, test whether the reaction is first order, and calculate the value of the rate constant:

Time (min)	0	5.0	10.0	15.0	20.0
[Cyclopropane]	0.098	0.080	0.066	0.054	0.044

12.5 ➤ HALF-LIFE OF A FIRST-ORDER REACTION

The **half-life** of a reaction, symbolized by $t_{1/2}$, is the time required for the reactant concentration to drop to one-half of its initial value. Consider the first-order reaction

$$a\,\text{A} \rightarrow \text{products}$$

To relate the reaction's half-life to the rate constant, let's begin with the integrated rate law:

$$\log \frac{[\text{A}]_t}{[\text{A}]_0} = \frac{-kt}{2.303} \quad \text{or} \quad \ln \frac{[\text{A}]_t}{[\text{A}]_0} = -kt$$

When $t = t_{1/2}$, the fraction of A that remains, $[\text{A}]_t/[\text{A}]_0$, is $1/2$. Therefore,

$$\ln \frac{1}{2} = -kt_{1/2}$$

so

$$t_{1/2} = \frac{-\ln 1/2}{k} = \frac{\ln 2}{k}$$

or

$$t_{1/2} = \frac{0.693}{k}$$

Thus, the half-life of a reaction is readily calculated from the rate constant, and vice versa.

It's evident from the equation for $t_{1/2}$ that the half-life of a first-order reaction is a constant because it depends only on the rate constant and not on the reactant concentration. This point is worth noting because reactions of other orders have half-lives that *do* depend on concentration. In reactions that are not first order, the amount of time in one half-life changes as the reactant concentration changes.

The constancy of the half-life for a first-order reaction is illustrated in Figure 12.7. Each successive half-life is an equal period of time in which the

FIGURE 12.7 Concentration of a reactant A as a function of time for a first-order reaction. The concentration falls from its initial value, $[\text{A}]_0$, to $[\text{A}]_0/2$ after one half-life, to $[\text{A}]_0/4$ after a second half-life, to $[\text{A}]_0/8$ after a third half-life, and so on. For a first-order reaction, each half-life represents an equal amount of time.

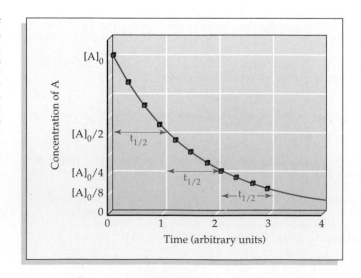

concentration decreases by a factor of 2. We'll see in Chapter 22 that half-lives are widely used in describing radioactive decay rates.

EXAMPLE 12.7

(a) Estimate the half-life for the decomposition of gaseous N_2O_5 at 55°C from the concentration-versus-time plot in Figure 12.1. (b) Calculate the half-life from the rate constant (1.7×10^{-3}/s). (c) What is the concentration of N_2O_5 after five half-lives? (d) How long will it take for the N_2O_5 concentration to fall to 12.5% of its initial value?

SOLUTION (a) Figure 12.1 shows that the concentration of N_2O_5 falls from 0.020 M to 0.010 M during a time period of approximately 400 s. Therefore, $t_{1/2} \approx$ 400 s. At the end of the second half-life, $2t_{1/2} = 800$ s, $[N_2O_5]$ has decreased by another factor of 2, to 0.050 M.
(b) Based on the value of the rate constant,

$$t_{1/2} = \frac{0.693}{k} = \frac{0.693}{1.7 \times 10^{-3} \text{ s}^{-1}} = 4.1 \times 10^2 \text{ s (6.8 min)}$$

(c) At $5t_{1/2}$, $[N_2O_5]$ will be $(1/2)^5 = 1/32$ of its initial value. Therefore,

$$[N_2O_5] = \frac{0.020 \text{ M}}{32} = 0.00062 \text{ M}$$

(d) Since 12.5% of the initial concentration corresponds to 1/8 or $(1/2)^3$ of the initial concentration, the time required is three half-lives:

$$t = 3t_{1/2} = 3(4.1 \times 10^2 \text{ s}) = 1.2 \times 10^3 \text{ s (20 min)}$$

PROBLEM 12.8 Consider the first-order decomposition of H_2O_2 in Example 12.5. (a) What is the half-life of the reaction in hours at 20°C? (b) If the initial concentration of H_2O_2 is 0.30 M, what is the molarity of H_2O_2 after four half-lives? (c) How many hours will it take for the concentration to drop to 25% of its initial value?

12.6 ▶SECOND-ORDER REACTIONS

If a reaction, a A $\longrightarrow$ products, is second order, its rate law is

$$\text{Rate} = -\frac{\Delta[A]}{\Delta t} = k[A]^2$$

An example is the thermal decomposition of nitrogen dioxide to yield NO and O_2:

$$2 \text{ NO}_2(g) \rightarrow 2 \text{ NO}(g) + O_2(g)$$

Using mathematics, it's possible to convert the rate law to the integrated rate law

$$\frac{1}{[A]_t} = kt + \frac{1}{[A]_0}$$

This integrated rate law then allows us to calculate the concentration of A at any time t if the initial concentration $[A]_0$ is known.

Since the integrated rate law has the form $y = mx + b$, a graph $1/[A]$ versus time is a straight line if the reaction is second order.

$$\underset{y}{\frac{1}{[A]_t}} = \underset{m}{k}\underset{x}{t} + \underset{b}{\frac{1}{[A]_0}}$$

The slope of the straight line is $m = k$, and the intercept is $b = 1/[A]_0$. Thus, by plotting $1/[A]$ versus time we can test whether the reaction is second order and can determine the value of the rate constant (Example 12.8).

We can obtain an expression for the half-life of a second-order reaction by substituting $[A]_t = [A]_0/2$ and $t = t_{1/2}$ into the integrated rate law:

$$t_{1/2} = \frac{1}{k[A]_0}$$

In contrast with a first-order reaction, the time required for the concentration of A to drop to one-half of its initial value depends on both the rate constant and the initial concentration. Thus, the value of $t_{1/2}$ increases as the reaction proceeds because the value of $[A]_0$ at the beginning of each successive half-life is smaller by a factor of 2. Consequently, each half-life for a second-order reaction is twice as long as the preceding one (Figure 12.8).

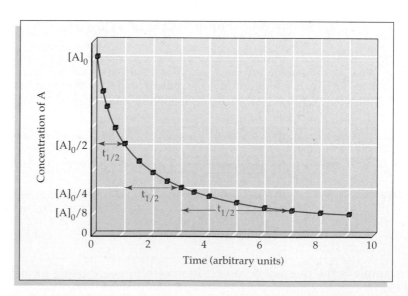

FIGURE 12.8 Concentration of a reactant A as a function of time for a second-order reaction. Note that each half-life is twice as long as the preceding one because $t_{1/2} = 1/k[A]_0$ and because the concentration of A at the beginning of each successive half-life is smaller by a factor of 2.

Table 12.4 summarizes some important differences between first-order and second-order reactions of the type $a\ A \longrightarrow$ products.

TABLE 12.4	Characteristics of First- and Second-Order Reactions of the Type $a\,A \rightarrow$ Products	

	First-Order	**Second-Order**
Rate law	$-\dfrac{\Delta[A]}{\Delta t} = k[A]$	$-\dfrac{\Delta[A]}{\Delta t} = k[A]^2$
Concentration-time equation	$\log [A]_t = \dfrac{-kt}{2.303} + \log [A]_0$	$\dfrac{1}{[A]_t} = kt + \dfrac{1}{[A]_0}$
Linear graph	$\log [A]$ versus t	$\dfrac{1}{[A]}$ versus t
Graphical determination of k	$k = -2.303(\text{slope})$	$k = \text{slope}$
Half-life	$t_{1/2} = \dfrac{0.693}{k}$	$t_{1/2} = \dfrac{1}{k[A]_0}$

EXAMPLE 12.8

At elevated temperatures, nitrogen dioxide decomposes to nitric oxide and molecular oxygen:

$$2\,NO_2(g) \rightarrow 2\,NO(g) + O_2(g)$$

Concentration-time data for this reaction at 300°C follow:

Time (s)	$[NO_2]$	Time (s)	$[NO_2]$
0	8.00×10^{-3}	200	4.29×10^{-3}
50	6.58×10^{-3}	300	3.48×10^{-3}
100	5.59×10^{-3}	400	2.93×10^{-3}
150	4.85×10^{-3}	500	2.53×10^{-3}

(a) Use these data to determine if the reaction is first order or second order. **(b)** Calculate the value of the rate constant. **(c)** What is the concentration of NO_2 at $t = 20$ min? **(d)** What is the half-life of the reaction when the initial concentration of NO_2 is 6.00×10^{-3} M? **(e)** What is $t_{1/2}$ when $[NO_2]_0$ is 3.00×10^{-3} M?

SOLUTION   **(a)** To determine whether the reaction is first order or second order, calculate values of $\log [NO_2]$ and $1/[NO_2]$ and then graph these values versus time.

Time (s)	$[NO_2]$	Log $[NO_2]$	$1/[NO_2]$
0	8.00×10^{-3}	−2.097	125
50	6.58×10^{-3}	−2.182	152
100	5.59×10^{-3}	−2.253	179
150	4.85×10^{-3}	−2.313	206
200	4.29×10^{-3}	−2.367	233
300	3.48×10^{-3}	−2.458	287
400	2.93×10^{-3}	−2.533	341
500	2.53×10^{-3}	−2.597	395

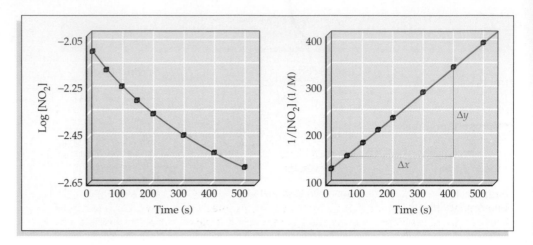

The plot of log $[NO_2]$ versus time is curved, but the plot of $1/[NO_2]$ versus time is a straight line. The reaction is therefore second order in NO_2.

(b) The rate constant is equal to the slope of the line in the plot of $1/[NO_2]$ versus time, which we can estimate from the coordinates of two widely separated points on the graph:

$$k = \text{slope} = \frac{\Delta y}{\Delta x} = \frac{340 \text{ M}^{-1} - 150 \text{ M}^{-1}}{400 \text{ s} - 50 \text{ s}} = \frac{190 \text{ M}^{-1}}{350 \text{ s}} = 0.54/(\text{M} \cdot \text{s})$$

(c) The concentration of NO_2 at $t = 20$ min (1200 s) can be calculated using the integrated rate law:

$$\frac{1}{[NO_2]_t} = kt + \frac{1}{[NO_2]_0}$$

Substituting in the values of k, t, and $[NO_2]_0$ gives

$$\frac{1}{[NO_2]_t} = \left(\frac{0.54}{\text{M} \cdot \text{s}}\right)(1200 \text{ s}) + \frac{1}{8.00 \times 10^{-3} \text{ M}}$$

$$= \frac{648}{\text{M}} + \frac{125}{\text{M}} = \frac{773}{\text{M}}$$

$$[NO_2]_t = 1.29 \times 10^{-3} \text{ M}$$

(d) The half-life of a second-order reaction can be calculated from the rate constant and the initial concentration of NO_2 (6.00×10^{-3} M):

$$t_{1/2} = \frac{1}{k[NO_2]_0} = \frac{1}{\left(\dfrac{0.54}{\text{M} \cdot \text{s}}\right)(6.00 \times 10^{-3} \text{ M})} = 3.1 \times 10^2 \text{ s}$$

(e) When $[NO_2]_0$ is 3.00×10^{-3}M, $t_{1/2} = 6.2 \times 10^2$ s (twice as long because $[NO_2]_0$ is now smaller by a factor of 2).

⌐ **PROBLEM 12.9** Hydrogen iodide gas decomposes at 410°C:

$$2 \, HI(g) \rightarrow H_2(g) + I_2(g)$$

The following data describe this decomposition:

Time (min)	0	20	40	60	80
[HI]	0.500	0.382	0.310	0.260	0.224

(a) Determine whether the reaction is first order or second order. **(b)** Calculate the value of the rate constant. **(c)** At what time in minutes does the HI concentration reach 0.100 M? **(d)** How many minutes does it take for the HI concentration to drop from 0.400 M to 0.200 M?

12.7 ►REACTION MECHANISMS

Thus far, our discussion of chemical kinetics has centered on reaction rates. We've seen that the rate of a reaction depends on both reactant concentrations and the value of the rate constant. An equally important issue in chemical kinetics is the **reaction mechanism**, the sequence of molecular events, or reaction steps, that defines the pathway from reactants to products. The reaction steps involve the breaking of chemical bonds and/or the making of new bonds. Chemists want to know the sequence in which the various reaction steps take place so they can better control known reactions and predict new ones.

A single step in a reaction mechanism is called an **elementary reaction** or an **elementary step**. To clarify the crucial distinction between an elementary reaction and an overall reaction, let's consider the gas-phase reaction of nitrogen dioxide and carbon monoxide to give nitric oxide and carbon dioxide:

$$NO_2(g) + CO(g) \rightarrow NO(g) + CO_2(g) \quad \text{(overall reaction)}$$

Experimental evidence suggests that this reaction takes place by a two-step mechanism:

Step 1. $NO_2(g) + NO_2(g) \rightarrow NO(g) + NO_3(g)$ (elementary reaction)

Step 2. $NO_3(g) + CO(g) \rightarrow NO_2(g) + CO_2(g)$ (elementary reaction)

In the first elementary step, two NO_2 molecules collide with enough energy to break one N–O bond and form another, resulting in the transfer of an oxygen atom from one NO_2 molecule to the other. In the second step, the NO_3 molecule formed in the first step collides with a CO molecule, and the transfer of an oxygen atom from NO_3 to CO yields an NO_2 molecule and a CO_2 molecule (Figure 12.9).

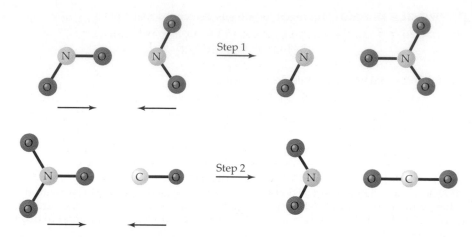

FIGURE 12.9 Computer-generated models of the elementary steps in the reaction of NO_2 with CO.

The chemical equation for an elementary reaction is a description of an individual molecular event. By contrast, the balanced equation for an *overall* reaction provides no information about how the reaction occurs. The equation for the reaction of NO_2 with CO, for example, does not tell us that the reaction occurs by direct transfer of an oxygen atom from an NO_2 molecule to a CO molecule; it describes only the stoichiometry of the overall process. To keep straight the distinction between an elementary reaction and an overall reaction, just remember that elementary reactions describe the reaction mechanism (think of collisions of individual molecules), whereas the overall reaction describes reaction stoichiometry (think of moles of reactants and products).

ELEMENTARY REACTION   describes behavior of individual molecules

OVERALL REACTION   describes reaction stoichiometry

Of course, the elementary steps in a proposed reaction mechanism must sum to give the overall reaction. When we sum the elementary steps in the reaction of NO_2 with CO and then cancel the molecules that appear on both sides of the resulting equation, we obtain the overall reaction:

$$Step\ 1.\quad NO_2(g) + NO_2(g) \rightarrow NO(g) + NO_3(g)\quad \text{(elementary reaction)}$$
$$Step\ 2.\quad NO_3(g) + CO(g) \rightarrow NO_2(g) + CO_2(g)\quad \text{(elementary reaction)}$$

$$NO_2(g) + \cancel{NO_2(g)} + \cancel{NO_3(g)} + CO(g) \rightarrow NO(g) + \cancel{NO_3(g)} + \cancel{NO_2(g)} + CO_2(g)$$
$$NO_2(g) + CO(g) \rightarrow NO(g) + CO_2(g)\quad \text{(overall reaction)}$$

A species that is formed in one step of a reaction mechanism and consumed in a subsequent step, such as NO_3 in our example, is called a **reaction intermediate**. Note that reaction intermediates do not appear in the net equation for the overall reaction, and it's only by looking at the elementary steps that their presence is noticed.

Elementary reactions are classified on the basis of their **molecularity**, the number of molecules (or atoms) on the reactant side of the chemical equation for the elementary reaction. A **unimolecular reaction** is an elementary reaction that involves a single reactant molecule—for example, the unimolecular decomposition of ozone in the upper atmosphere:

$$O_3^*(g) \rightarrow O_2(g) + O(g)$$

(The asterisk on O_3 indicates that the ozone molecule is in an energetically excited state as a result of previous absorption of ultraviolet light from the sun. The absorbed energy causes one of the two O–O bonds to break, with loss of an oxygen atom.)

Bimolecular reactions are elementary reactions that result from energetic collisions between two reactant molecules. An example taken from the chemistry of the upper atmosphere is the reaction of an ozone molecule with an oxygen atom to yield two O_2 molecules:

$$O_3(g) + O(g) \rightarrow 2\ O_2(g)$$

Both unimolecular and bimolecular reactions are common, but **termolecular reactions**, which involve three atoms or molecules, are rare. As any pool player knows, three-body collisions are much less probable than two-body collisions. There are some reactions, however, that require a three-body collision, notably the combination of two atoms to form a diatomic molecule. For example, oxygen atoms in the upper atmosphere combine as a result of collisions involving some third molecule M:

$$O(g) + O(g) + M(g) \rightarrow O_2(g) + M(g)$$

In the atmosphere, M is most likely N_2, but in principle it could be any atom or molecule. The role of M is to carry away the energy that is released when the O–O bond is formed. If M were not involved in the collision, the two oxygen atoms would simply bounce off each other, and no reaction would occur.

The northern lights, the beautiful auroras often observed in the Northern Hemisphere at high latitudes. The light is produced in part by excited O atoms in the upper atmosphere.

EXAMPLE 12.9

The following two-step mechanism has been proposed for the gas-phase decomposition of nitrous oxide (N_2O):

$$N_2O(g) \rightarrow N_2(g) + O(g)$$
$$N_2O(g) + O(g) \rightarrow N_2(g) + O_2(g)$$

(a) Write the chemical equation for the overall reaction. **(b)** Identify any reaction intermediates. **(c)** What is the molecularity of each of the elementary reactions? **(d)** What is the molecularity of the overall reaction?

SOLUTION **(a)** The overall reaction is the sum of the two elementary steps:

$$2\ N_2O(g) + \cancel{O(g)} \rightarrow 2\ N_2(g) + \cancel{O(g)} + O_2(g)$$

Cancellation of $O(g)$, which appears on both sides, gives the net equation for the overall reaction:

$$2\ N_2O(g) \rightarrow 2\ N_2(g) + O_2(g)$$

(b) The oxygen atom is a reaction intermediate, since it's formed in the first elementary step and is consumed in the second step.
(c) The first elementary reaction is unimolecular because it involves a single reactant molecule. The second elementary step is bimolecular because it involves two reactant molecules.
(d) It's inappropriate to use the word *molecularity* in connection with the overall reaction because the overall reaction does not describe an individual molecular event. Only an elementary reaction can have a molecularity.

┌ **PROBLEM 12.10** A suggested mechanism for the reaction of nitrogen dioxide and molecular fluorine is

$$NO_2(g) + F_2(g) \rightarrow NO_2F(g) + F(g)$$

$$F(g) + NO_2(g) \rightarrow NO_2F(g)$$

(a) Give the chemical equation for the overall reaction, and identify any reaction intermediates. (b) Specify the molecularity of each of the elementary reactions. ◢

12.8 ➤ RATE LAWS AND REACTION MECHANISMS

We emphasized in Section 12.2 that the rate law for a chemical reaction can't be deduced from the stoichiometric coefficients in the balanced equation for the overall reaction. The rate law for an overall reaction must be determined by experiment. By contrast, the rate law for an elementary reaction follows directly from its molecularity. The concentration of each reactant in an elementary reaction appears in the rate law, with an exponent equal to its coefficient in the chemical equation for the elementary reaction. Consider, for example, the unimolecular decomposition of ozone:

$$O_3(g) \rightarrow O_2(g) + O(g)$$

The number of moles of O_3 per liter that decompose per unit time is directly proportional to the molar concentration of O_3:

$$\text{Rate} = -\frac{\Delta[O_3]}{\Delta t} = k[O_3]$$

In general, the rate of a unimolecular reaction is first order in the concentration of the reactant molecule.

For a bimolecular elementary reaction of the type $A + B \longrightarrow$ products, the reaction rate depends on the frequency of collisions between A and B molecules. The frequency of AB collisions involving any *particular* A molecule is proportional to the molar concentration of B, and the total frequency of AB collisions involving *all* A molecules is proportional to the molar concentration of A times the molar concentration of B. Therefore the reaction obeys the second-order rate law

$$\text{Rate} = -\frac{\Delta[A]}{\Delta t} = -\frac{\Delta[B]}{\Delta t} = k[A][B]$$

A similar line of reasoning indicates that a bimolecular reaction of the type

$$A + A \rightarrow products \qquad (2\ A \rightarrow products)$$

has the second-order rate law

$$Rate = -\frac{\Delta[A]}{\Delta t} = k[A][A] = k[A]^2$$

Rate laws for elementary reactions are summarized in Table 12.5. Note that the overall reaction order for an elementary reaction is equal to its molecularity.

TABLE 12.5   Rate Laws for Elementary Reactions

Elementary Reaction	Molecularity	Rate Law
A → products	Unimolecular	Rate = $k[A]$
A + A → products (2 A → products)	Bimolecular	Rate = $k[A]^2$
A + B → products	Bimolecular	Rate = $k[A][B]$
A + A + B → products (2 A + B → products)	Termolecular	Rate = $k[A]^2[B]$
A + B + C → products	Termolecular	Rate = $k[A][B][C]$

The experimentally observed rate law for an overall reaction depends on the reaction mechanism—that is, on the sequence of elementary steps and their relative rates. Of course, if the overall reaction occurs in a single elementary step, the experimental rate law is the same as the rate law for the elementary reaction. An example is the conversion of bromomethane to methanol in basic solution:

$$\underset{\text{Bromomethane}}{CH_3Br(aq)} + OH^-(aq) \rightarrow \underset{\text{Methanol}}{CH_3OH(aq)} + Br^-(aq)$$

This reaction occurs in a single bimolecular step, and the experimental rate law is

$$Rate = -\frac{\Delta[CH_3Br]}{\Delta t} = k[CH_3Br][OH^-]$$

When an overall reaction occurs in two or more elementary steps, one of the steps is often much slower than the others. The slowest step in a reaction mechanism is called the **rate-determining step** because that step acts as a bottleneck, limiting the rate at which reactants can be converted to products. The overall reaction can occur no faster than the speed of the rate-determining step. In the reaction of nitrogen dioxide with carbon monoxide, for example, the first step in the mechanism is slower and rate determining, whereas the second step occurs more rapidly:

$$NO_2(g) + NO_2(g) \xrightarrow{k_1} NO(g) + NO_3(g) \quad \text{(slower, rate determining)}$$

$$\underline{NO_3(g) + CO(g) \xrightarrow{k_2} NO_2(g) + CO_2(g)} \quad \text{(faster)}$$

$$NO_2(g) + CO(g) \longrightarrow NO(g) + CO_2(g) \quad \text{(overall reaction)}$$

Consequently, $k_1 \ll k_2$, where k_1 and k_2, written above the arrows in the preceding equations, are the rate constants for the elementary reactions. The rate of the overall reaction is determined by the rate of the first, slower step. In the second step, the unstable intermediate (NO_3) reacts as soon as it is formed in the first step.

A chemical reaction is somewhat like the cafeteria line in a dining hall. The rate at which the line moves is determined not by the faster steps, perhaps picking up a salad or a beverage, but by the slowest step, perhaps waiting for a well-done hamburger.

An acceptable reaction mechanism must meet two criteria: (1) The elementary steps must sum to give the overall reaction, and (2) the mechanism must be consistent with the observed rate law for the overall reaction. In the reaction of NO_2 with CO, for example, the experimental rate law is

$$\text{Rate} = -\frac{\Delta[NO_2]}{\Delta t} = k[NO_2]^2$$

The rate at which this cafeteria line moves is determined by the rate of the slowest step.

The rate law predicted by the proposed mechanism is that for the rate-determining step, where k_1 is the rate constant for that step:

$$\text{Rate} = -\frac{\Delta[NO_2]}{\Delta t} = k_1[NO_2]^2$$

(Note that the predicted rate law follows directly from the molecularity of the rate-determining step.) Because the observed and predicted rate laws have the same form (second-order dependence on $[NO_2]$), the proposed mechanism is consistent with the experimental rate law. The observed rate constant k is the same as k_1, the rate constant for the first elementary step.

The procedure that chemists use in establishing a reaction mechanism goes something like this: First, the rate law is determined by experiment. Then, a series of elementary steps is devised, and the rate law predicted by the proposed mechanism is worked out. If the observed and predicted rate laws don't agree, the mechanism must be discarded and another one must be devised. If the observed and predicted rate laws *do* agree, the proposed mechanism is a plausible pathway for the reaction. Figure 12.10 summarizes the steps.

The case for a particular mechanism is strengthened considerably if a reaction intermediate can be isolated or if an unstable intermediate (such as NO_3) can be detected. It's easy to disprove a mechanism, but it's never possible to finally "prove" a mechanism because there may be an alternative reaction pathway, not yet imagined, that also fits the experimental facts. The

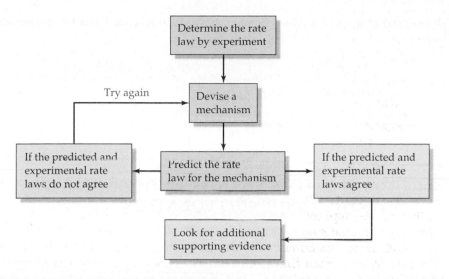

FIGURE 12.10 Flow chart illustrating the logic used in studies of reaction mechanisms

best that we can do to establish a mechanism is to accumulate a convincing body of experimental evidence that supports it. Proving a reaction mechanism is more like proving a case in a law court than like proving a theorem in mathematics.

EXAMPLE 12.10

The following reaction has a second-order rate law:

$$2\ NO_2(g) + F_2(g) \rightarrow 2\ NO_2F(g) \qquad Rate = k[NO_2][F_2]$$

Devise a possible reaction mechanism.

SOLUTION The reaction doesn't occur in a single elementary step because, if it did, the rate law would be third order: rate = $k[NO_2]^2[F_2]$. The observed rate law will be obtained if the rate-determining step involves the bimolecular reaction of NO_2 and F_2. A plausible sequence of elementary steps is

$NO_2(g) + F_2(g) \xrightarrow{k_1} NO_2F(g) + F(g)$	(slower, rate-determining step)
$F(g) + NO_2(g) \xrightarrow{k_2} NO_2F(g)$	(faster)
$\overline{2\ NO_2(g) + F_2(g) \longrightarrow 2\ NO_2F(g)}$	(overall reaction)

The rate law predicted by this mechanism, rate = $k_1[NO_2][F_2]$, is in accord with the observed rate law.

⌐ **PROBLEM 12.11** Write the rate law for the following elementary reactions:
(a) $O_3(g) + O(g) \rightarrow 2\ O_2(g)$
(b) $Br(g) + Br(g) + Ar(g) \rightarrow Br_2(g) + Ar(g)$
(c) $Co(CN)_5(H_2O)^{2-}(aq) \rightarrow Co(CN)_5^{2-}(aq) + H_2O(l)$

⌐ **PROBLEM 12.12** The following substitution reaction has a first-order rate law:

$$Co(CN)_5(H_2O)^{2-}(aq) + I^-(aq) \rightarrow Co(CN)_5I^{3-}(aq) + H_2O(l)$$

$$Rate = k[Co(CN)_5(H_2O)^{2-}]$$

Suggest a possible reaction mechanism, and show that your mechanism is in accord with the observed rate law. ⌐

12.9 ➤ REACTION RATES AND TEMPERATURE: THE ARRHENIUS EQUATION

Everyday experience tells us that the rates of chemical reactions increase with increasing temperature. Familiar fuels such as gas, oil, and coal are relatively inert at room temperature but burn rapidly at elevated temperatures. Many foods spoil at room temperature but resist decay when stored in a freezer. Metallic magnesium reacts with hot water but hardly at all with cold water (Figure 12.11). As a rule of thumb, reaction rates tend to double when the temperature is increased by 10°C (Figure 12.12).

FIGURE   12.11 Magnesium reacts with hot water (left), but its reaction with cold water (right) is too slow to be observed. Evidence of reaction is the pink color of phenolphthalein, which indicates formation of an alkaline solution. The reaction is $Mg(s) + 2 H_2O(l) \rightarrow Mg^{2+}(aq) + 2 OH^-(aq) + H_2(g)$.

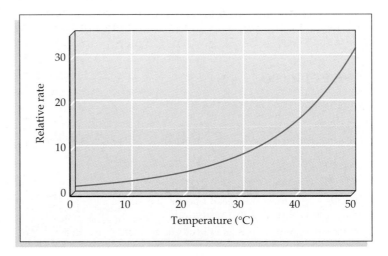

FIGURE 12.12 Exponential increase in the rate of a reaction whose rate doubles when the temperature is raised by 10°C. The factor by which the rate increases for a 10°C temperature increase depends on the particular reaction and on the temperature range, but it generally lies in the range of 1.5 to 4.

To understand why reaction rates depend on temperature, we need a picture of how reactions take place. According to the **collision theory** model, a bimolecular reaction occurs when two properly oriented reactant molecules come together in a sufficiently energetic collision. To be specific, let's consider one of the simplest possible reactions, the reaction of an atom A with a diatomic molecule BC to give a diatomic molecule AB and an atom C:

$$A + BC \rightarrow AB + C$$

An example from atmospheric chemistry is the reaction of an oxygen atom with an HCl molecule to give an OH molecule and a chlorine atom:

$$O(g) + HCl(g) \rightarrow OH(g) + Cl(g)$$

If the reaction occurs in a single step, the electron distribution about the three nuclei must change in the course of the collision such that a new bond, A–B, develops at the same time as the old bond, B–C, breaks. Between the reactant and product stages, the nuclei pass through a configuration in which all three atoms are weakly linked together. We can picture the progress of the reaction as

$$A + B—C \rightarrow A\text{-}{-}\text{-}B\text{-}{-}\text{-}C \rightarrow A—B + C$$

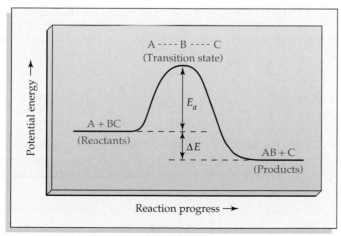

If A and BC have filled shells of electrons (no unpaired electrons or vacant, low-energy orbitals), they will repel each other. To achieve the configuration A---B---C, then, the atoms require energy to overcome this repulsion. The energy comes from the kinetic energy of the colliding particles and is stored in A---B---C as potential energy. In fact, A---B---C has more potential energy than either the reactants or the products. Thus, there is a potential energy barrier that must be surmounted before reactants can be converted to products, as depicted graphically on the potential energy profile in Figure 12.13.

The height of the barrier is called the **activation energy**, E_a, and the configuration of atoms at the maximum in the potential energy profile is

FIGURE 12.13 Potential energy profile for the reaction A + BC → AB + C, showing the activation energy barrier between the reactants and products. As the reaction progresses, kinetic energy of the reactants is first converted into potential energy of the transition state and is then transformed into kinetic energy of the products. At each point along the profile, the *total* energy is conserved. The profile is drawn for an exothermic reaction (ΔE, the energy of reaction, is negative).

called the **transition state**, or the **activated complex**. Since energy is conserved in the collision, all the energy needed to climb the potential energy hill must come from the kinetic energy of the colliding molecules. If the collision energy is less than E_a, the reactant molecules can't surmount the barrier and so they simply bounce apart. If the collision energy is at least as great as E_a, however, the reactants can climb over the barrier and be converted to products.

Experimental evidence for the notion of an activation energy barrier comes from a comparison of collision rates and reaction rates. Collision rates in gases can be calculated from kinetic-molecular theory (Section 9.6). For a gas at room temperature (298 K) and 1 atm pressure, each molecule undergoes approximately 10^9 collisions per second, or one collision every 10^{-9} s. Thus, if every collision resulted in reaction, every gas-phase reaction would be complete in about 10^{-9} s. By contrast, observed reactions often have half-lives of minutes or hours, so it's clear that only a tiny fraction of the collisions can lead to reaction.

Very few collisions are productive because very few occur with a kinetic energy as large as the activation energy. The fraction of the collisions with an energy equal to or greater than the activation energy E_a is represented in Figure 12.14 at two different temperatures by the area under the curve to the right of E_a. This fraction f is given quantitatively by the equation

$$f = e^{-E_a/RT}$$

where $e = 2.71828 \ldots$ is the base of natural logarithms, R is the gas constant [8.314 J/(K $\cdot$ mol)], and T is the absolute temperature in kelvins. Note that f is a very small number. For example, for a reaction having an activation energy of 75 kJ/mol, the value of f at 298 K is 7×10^{-14}:

$$f = \exp\left\{ \frac{-75{,}000\, \dfrac{\text{J}}{\text{mol}}}{\left(8.314\, \dfrac{\text{J}}{\text{K} \cdot \text{mol}}\right)(298\ \text{K})} \right\} = e^{-30.3} = 7 \times 10^{-14}$$

FIGURE 12.14 Plots at two different temperatures of the fraction of the collisions with a particular kinetic energy. For each plot, the total area under the curve is unity, and the area to the right of E_a is the fraction f of the collisions with an energy greater than or equal to E_a. The fraction of the collisions that are sufficiently energetic to result in reaction increases exponentially with increasing temperature.

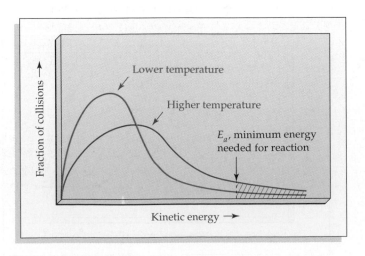

Lower temperature

Higher temperature

E_a, minimum energy needed for reaction

Fraction of collisions →

Kinetic energy →

Thus, only *7 collisions in 100 trillion* are sufficiently energetic to convert reactants to products.

As the temperature increases, the distribution of collision energies broadens and shifts to higher energies (Figure 12.14), resulting in an exponential increase in the fraction of the collisions that lead to products. At 308 K, for example, the calculated value of f is 2×10^{-13}, about 3 times larger than the value of f at 298 K. Collision theory nicely accounts for the exponential dependence of reaction rates on temperature and also explains why reaction rates are so much lower than collision rates. (Collision rates also increase with increasing temperature, but only by a small amount—less than 2% on going from 298 K to 308 K.)

The fraction of the collisions that lead to products is further reduced by an orientation requirement. Even if the reactants collide with sufficient energy, they won't react unless the orientation of the reaction partners is correct for formation of the transition state. For example, a collision of A with the C end of the molecule BC can't result in formation of AB:

$$A + C-B \;\rightarrow\; A\text{---}C\text{---}B \;\nrightarrow\; A-B + C$$

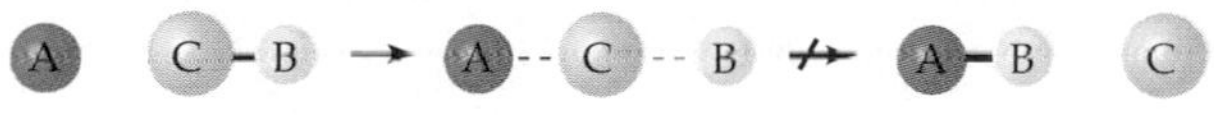

The reactant molecules would simply collide and then separate without reaction:

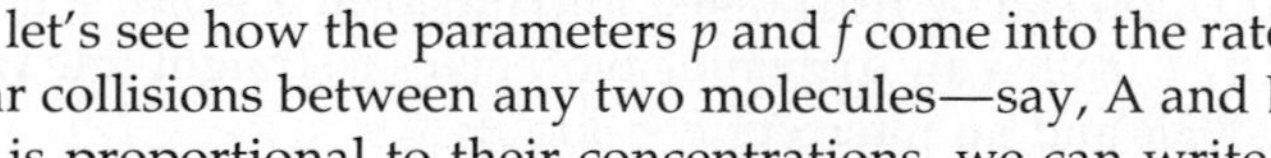

$$A + C-B \rightarrow A\text{---}C\text{---}B \rightarrow A + C-B$$

The fraction of the collisions having proper orientation for conversion of reactants to products is called the **steric factor**, p. For the reaction, $2\,HI(g) \longrightarrow H_2(g) + I_2(g)$, the value of p is about 0.5, but for reactions of complex molecules, p is a fraction considerably less than 0.5.

Now let's see how the parameters p and f come into the rate law. Since bimolecular collisions between any two molecules—say, A and B—occur at a rate that is proportional to their concentrations, we can write

$$\text{Collision rate} = Z[A][B]$$

where Z is a constant related to the collision frequency and has units of a second-order rate constant, $1/(M \cdot s)$. The reaction rate is lower than the collision rate by a factor $p \times f$ because only a fraction of the colliding molecules have the correct orientation and the minimum energy needed for reaction:

$$\text{Reaction rate} = p \times f \times \text{collision rate} = pfZ[A][B]$$

Since the rate law is

$$\text{Reaction rate} = k[\text{A}][\text{B}]$$

the rate constant predicted by collision theory is $k = pfZ$, or

$$k = pZe^{-E_a/RT}$$

This expression is usually written in a form called the **Arrhenius equation** after Svante Arrhenius, the Swedish chemist who proposed it in 1889 on the basis of experimental studies of reaction rates:

$$\textit{Arrhenius equation} \qquad k = Ae^{-E_a/RT}$$

The parameter $A \ (= pZ)$ is called the **frequency factor** (or pre-exponential factor).

12.10 ► USING THE ARRHENIUS EQUATION

The Arrhenius equation is extremely useful because it allows us to determine the activation energy E_a for a reaction if values of the rate constant are known at different temperatures. Taking the natural logarithm of both sides of the Arrhenius equation, we obtain the logarithmic form

$$\ln k = \ln A - \frac{E_a}{RT}$$

or, in terms of base-10 logarithms,

$$\log k = \log A - \frac{E_a}{2.303RT}$$

Since this equation can be rearranged into the form $y = mx + b$, a graph of $\log k$ versus $1/T$ (called an Arrhenius plot) gives a straight line with a slope $m = -E_a/2.303R$ and an intercept $b = \log A$:

$$\log k = \underset{\underset{y}{\uparrow}}{} \left(\underset{\underset{m}{\uparrow}}{\frac{-E_a}{2.303R}}\right)\left(\underset{\underset{x}{\uparrow}}{\frac{1}{T}}\right) + \underset{\underset{b}{\uparrow}}{\log A}$$

As Example 12.11 shows, the experimental value of the activation energy is easily determined from the slope of the straight line:

$$E_a = -2.303R(\text{slope})$$

Still another form of the Arrhenius equation can be derived that allows us to estimate the activation energy from rate constants at just two tempera-tures. At temperature T_1

$$\log k_1 = \left(\frac{-E_a}{2.303R}\right)\left(\frac{1}{T_1}\right) + \log A$$

and at temperature T_2

$$\log k_2 = \left(\frac{-E_a}{2.303R}\right)\left(\frac{1}{T_2}\right) + \log A$$

Subtracting the first equation from the second, and remembering that $\log k_2 - \log k_1 = \log (k_2/k_1)$, we obtain

$$\log \left(\frac{k_2}{k_1}\right) = \left(\frac{-E_a}{2.303R}\right)\left(\frac{1}{T_2} - \frac{1}{T_1}\right)$$

This equation can be used to calculate E_a from rate constants k_1 and k_2 at temperatures T_1 and T_2. By the same token, if we know E_a and the rate constant k_1 at one temperature T_1, we can calculate the rate constant k_2 at another temperature T_2. Example 12.11 shows how this is done.

EXAMPLE 12.11

Rate constants for the gas phase decomposition of hydrogen iodide, $2\ HI(g) \longrightarrow H_2(g) + I_2(g)$, are listed in the following table:

Temperature (°C)	k [1/(M · s)]	Temperature (°C)	k [1/(M · s)]
283	3.52×10^{-7}	427	1.16×10^{-3}
356	3.02×10^{-5}	508	3.95×10^{-2}
393	2.19×10^{-4}		

(a) Find the activation energy in kJ/mol using all five data points. **(b)** Calculate E_a from the rate constants at 283 and 508°C. **(c)** Given the rate constant at 283°C and the value of E_a obtained in part (b), calculate the rate constant at 293°C.

SOLUTION   **(a)** The activation energy E_a can be determined from the slope of a linear plot of $\log k$ versus $1/T$. Since the temperature in the Arrhenius equation is expressed in kelvins, we must first convert the Celsius temperatures to absolute temperatures. Then calculate values of $1/T$ and $\log k$, and plot $\log k$ versus $1/T$. The results are shown in the following table and graph.

t (°C)	T (K)	k [1/(M · s)]	$1/T$ (1/K)	Log k
283	556	3.52×10^{-7}	0.00180	−6.453
356	629	3.02×10^{-5}	0.00159	−4.520
393	666	2.19×10^{-4}	0.00150	−3.660
427	700	1.16×10^{-3}	0.00143	−2.936
508	781	3.95×10^{-2}	0.00128	−1.403

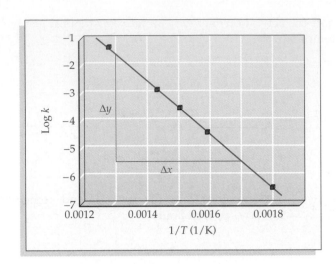

The slope of the straight-line plot can be determined from the coordinates of two widely separated points:

$$\text{Slope} = \frac{\Delta y}{\Delta x} = \frac{(-5.5) - (-1.6)}{(0.00170 \text{ K}^{-1}) - (0.00130 \text{ K}^{-1})} = \frac{-3.9}{0.00040 \text{ K}^{-1}} = -9.8 \times 10^3 \text{ K}$$

Finally, calculate the activation energy from the slope:

$$E_a = -2.303R(\text{slope}) = -2.303\left(8.314 \frac{\text{J}}{\text{K} \cdot \text{mol}}\right)(-9.8 \times 10^3 \text{ K})$$

$$= 1.9 \times 10^5 \text{ J/mol} = 190 \text{ kJ/mol}$$

Note that the slope of the Arrhenius plot is negative and the activation energy is positive. The greater the activation energy for a particular reaction, the steeper the slope of the log k versus $1/T$ plot and the greater the increase in the rate constant for a given increase in temperature.

(b) To calculate E_a from values of the rate constant at two temperatures, use the equation

$$\log\left(\frac{k_2}{k_1}\right) = \left(\frac{-E_a}{2.303R}\right)\left(\frac{1}{T_2} - \frac{1}{T_1}\right)$$

Substituting in the values of $k_1 = 3.52 \times 10^{-7}/(\text{M} \cdot \text{s})$ at $T_1 = 556$ K (283°C) and $k_2 = 3.95 \times 10^{-2}/(\text{M} \cdot \text{s})$ at $T_2 = 781$ K (508°C) gives

$$\log\left[\frac{\left(\frac{3.95 \times 10^{-2}}{\text{M} \cdot \text{s}}\right)}{\left(\frac{3.52 \times 10^{-7}}{\text{M} \cdot \text{s}}\right)}\right] = \left[\frac{-E_a}{(2.303)\left(8.314 \frac{\text{J}}{\text{K} \cdot \text{mol}}\right)}\right]\left(\frac{1}{781 \text{ K}} - \frac{1}{556 \text{ K}}\right)$$

Simplifying this equation gives

$$5.050 = \left[\frac{-E_a}{\left(19.15\,\frac{J}{K \cdot mol}\right)}\right]\left(\frac{-5.18 \times 10^{-4}}{K}\right)$$

$$E_a = 1.87 \times 10^5 \text{ J/mol} = 187 \text{ kJ/mol}$$

(c) Use the same equation as in part b, but now $k_1 = 3.52 \times 10^{-7}/(M \cdot s)$ at $T_1 = 556$ K (283°C) and $E_a = 1.87 \times 10^5$ J/mol are known, and k_2 at $T_2 = 566$ K (293°C) is the unknown:

$$\log\left[\frac{k_2}{\left(\frac{3.52 \times 10^{-7}}{M \cdot s}\right)}\right] = \left|\frac{\left(-1.87 \times 10^5\,\frac{J}{mol}\right)}{(2.303)\left(8.314\,\frac{J}{K \cdot mol}\right)}\right|\left(\frac{1}{566\text{ K}} - \frac{1}{556\text{ K}}\right) = 0.310$$

Taking the antilog of both sides gives

$$\frac{k_2}{\left(\frac{3.52 \times 10^{-7}}{M \cdot s}\right)} = 10^{0.010} = 2.04$$

so $k_2 = 7.18 \times 10^{-7}/(M \cdot s)$

In this temperature range, a 10 K rise in temperature doubles the reaction rate.

┌ **PROBLEM 12.13** Rate constants for decomposition of gaseous dinitrogen pentoxide are 3.7×10^{-5}/s at 25°C and 1.7×10^{-3}/s at 55°C.

$$2 \text{ N}_2\text{O}_5(g) \rightarrow 4 \text{ NO}_2(g) + \text{O}_2(g)$$

(a) Determine the activation energy for this reaction in kJ/mol. (b) Calculate the rate constant at 35°C.

12.11 ▸CATALYSIS

A **catalyst** is a substance that increases the rate of a reaction without being consumed in the reaction. An example is manganese dioxide, a black powder that speeds up the thermal decomposition of potassium chlorate:

$$2 \text{ KClO}_3(s) \xrightarrow[\text{heat}]{\text{MnO}_2 \text{ catalyst}} 2 \text{ KCl}(s) + 3 \text{ O}_2(g)$$

In the absence of a catalyst, $KClO_3$ decomposes very slowly, even when heated vigorously, but when a small amount of MnO_2 is mixed with the $KClO_3$ before heating, rapid evolution of oxygen ensues. The MnO_2 can be recovered unchanged after the reaction is complete.

Catalysts are of enormous importance, both in the chemical industry and in living organisms. Nearly all processes for the manufacture of essential chemicals use catalysts to favor formation of specific products and to lower reaction temperatures, thus reducing energy costs. In environmental chemistry, catalysts such as nitric oxide play a role in the formation of air pollutants, while other catalysts, such as platinum in automobile catalytic converters, are potent weapons in the battle to control air pollution.

In living organisms chemical reactions are catalyzed by large protein molecules called *enzymes*, which facilitate specific reactions of crucial biological importance (Section 24.6). For example, nitrogenase, an enzyme present in bacteria on the root nodules of leguminous plants such as peas and beans, catalyzes the conversion of atmospheric nitrogen to ammonia. The ammonia serves as a nitrogen fertilizer for plant growth. In the human body, the enzyme carbonic anhydrase catalyzes the reaction of carbon dioxide with water:

$$CO_2(aq) + H_2O(l) \rightleftarrows H^+(aq) + HCO_3^-(aq)$$

The forward reaction occurs when the blood takes up CO_2 in the tissues, and the reverse reaction occurs when the blood releases CO_2 in the lungs. Remarkably, carbonic anhydrase increases the rate of these reactions by a factor of about 10^6.

How does a catalyst work? A catalyst accelerates the rate of a reaction by making available a new and more efficient pathway for conversion of reactants to products. Let's consider the decomposition of hydrogen peroxide in a basic, aqueous solution:

$$2\ H_2O_2(aq) \rightarrow 2\ H_2O(l) + O_2(g)$$

Although unstable with respect to water and oxygen, hydrogen peroxide nevertheless decomposes slowly because the reaction has a high activation energy (76 kJ/mol). In the presence of iodide ion, however, the reaction is appreciably faster (Figure 12.15) because it can proceed by a different, lower-energy pathway:

Step 1. $H_2O_2(aq) + I^-(aq) \rightarrow H_2O(l) + IO^-(aq)$ (slower, rate determining)

Step 2. $\underline{H_2O_2(aq) + IO^-(aq) \rightarrow H_2O(l) + O_2(g) + I^-(aq)}$ (faster)

        $2\ H_2O_2(aq) \rightarrow 2\ H_2O(l) + O_2(g)$ (overall reaction)

FIGURE 12.15 The rate of decomposition of aqueous hydrogen peroxide can be monitored qualitatively by collecting the evolved oxygen gas in a balloon. **(a)** In the absence of a catalyst, little O_2 is produced. **(b)** After addition of aqueous sodium iodide, the balloon rapidly inflates with O_2.

(a)   (b)

The H_2O_2 first oxidizes the catalyst (I^-) to hypoiodite ion (IO^-) and then reduces the intermediate IO^- back to I^-. Note that the catalyst does not appear in the overall reaction because it is consumed in one reaction step and regenerated in a later step. The catalyst is, however, intimately involved in the reaction, as is evident from the presence of I^- in the observed rate law:

$$\text{Rate} = k[H_2O_2][I^-]$$

The rate law is consistent with the reaction of H_2O_2 and I^- as the rate-determining step.

A catalyst can speed up a reaction in two ways: either by increasing the frequency factor A or by decreasing the activation energy E_a in the Arrhenius equation. Because the rate constant is more sensitive to the value of E_a, a catalyst usually functions by lowering the activation energy (Figure 12.16). In the decomposition of hydrogen peroxide, for example, catalysis by I^- lowers E_a for the overall reaction by 19 kJ/mol.

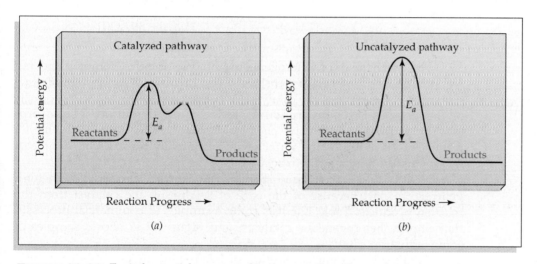

FIGURE 12.16 Typical potential energy profiles for a reaction whose activation energy is lowered by the presence of a catalyst. The shape of the barrier for the catalyzed pathway **(a)** describes the decomposition of H_2O_2: The first of the two maxima is higher, because the first elementary step is rate-determining. Maxima for both steps, though, are lower than the top of the barrier for the uncatalyzed reaction **(b)**. Note that a catalyst does not affect the energy of the reactants and products.

12.12 ➤HOMOGENEOUS AND HETEROGENEOUS CATALYSTS

Catalysts are commonly classified as either *homogeneous* or *heterogeneous*. A **homogeneous catalyst** is one that exists in the same phase as the reactants. For example, iodide ion is a homogeneous catalyst for the decomposition of aqueous hydrogen peroxide because both I^- and H_2O_2 are present in the same aqueous solution phase.

In the atmosphere, nitric oxide is a gas-phase homogeneous catalyst for the conversion of molecular oxygen to ozone, a process described by the following series of reactions:

$$1/2\ O_2(g) + NO(g) \rightarrow NO_2(g)$$

$$NO_2(g) \xrightarrow{\text{sunlight}} NO(g) + O(g)$$

$$O(g) + O_2(g) \rightarrow O_3(g)$$

$$3/2\ O_2(g) \rightarrow O_3(g) \qquad \text{(overall reaction)}$$

Nitric oxide first reacts with atmospheric O_2 to give nitrogen dioxide, a poisonous brown gas. Subsequent absorption of sunlight by NO_2 results in dissociation of an oxygen atom, which then reacts with O_2 to form ozone. As usual, the catalyst (NO) and the intermediates (NO_2 and O) do not appear in the chemical equation for the overall reaction.

A **heterogeneous catalyst** is one that exists in a different phase from the reactants. Ordinarily, the catalyst is a solid, and the reactants are either gases or liquids. For example, in the Fischer-Tropsch process for manufacturing synthetic gasoline, tiny particles of a metal, such as iron or cobalt, supported on alumina (Al_2O_3) catalyze conversion of gaseous carbon monoxide and hydrogen to hydrocarbons such as octane (C_8H_{18}):

$$8\ CO(g) + 17\ H_2(g) \xrightarrow[\text{catalyst}]{Co/Al_2O_3} C_8H_{18}(l) + 8\ H_2O(l)$$

The manufacture of these products involves reactions carried out using heterogeneous catalysts.

The mechanism of heterogeneous catalysis is often complex and not well understood. Important steps, however, involve (1) adsorption of reactants onto the surface of the catalyst, (2) conversion of reactants to products on the surface, and (3) desorption of products from the surface. The adsorption step is thought to involve chemical bonding of reactants to the highly reactive metal atoms on the surface with accompanying breaking, or at least weakening, of bonds in the reactants.

Most of the catalysts used in industrial chemical processes are heterogeneous, in part because of the ease of separating the catalyst from the reaction products. Table 12.6 lists some examples of commercial processes that employ heterogeneous catalysts, and Figure 12.17 shows samples of some industrial catalysts.

Another important application of heterogeneous catalysts is in automobile catalytic converters. Despite much work on engine design and fuel composition, automotive exhaust emissions contain air pollutants such as unburned hydrocarbons (C_xH_y), carbon monoxide, and nitric oxide. Carbon monoxide results from incomplete combustion of hydrocarbon fuels, and nitric oxide is produced when atmospheric nitrogen and oxygen combine at

FIGURE 12.17 Samples of some heterogeneous catalysts used in industry.

TABLE 12.6	Some Heterogeneous Catalysts Used in Commercially Important Reactions		

Reaction	Catalyst	Commercial Process	End Product: Commercial Uses
$2\ SO_2 + O_2 \rightarrow 2\ SO_3$	Pt or V_2O_5	Intermediate step in the contact process for synthesis of sulfuric acid	H_2SO_4: Manufacture of fertilizers, chemicals; oil refining
$4\ NH_3 + 5\ O_2 \rightarrow 4\ NO + 6\ H_2O$	Pt and Rh	First step in the Ostwald process for synthesis of nitric acid	HNO_3: Manufacture of explosives, fertilizers, plastics, dyes, lacquers
$N_2 + 3\ H_2 \rightarrow 2\ NH_3$	Fe, K_2O, and Al_2O_3	Haber process for synthesis of ammonia	NH_3: Manufacture of fertilizers, nitric acid
$H_2O + CH_4 \rightarrow CO + 3\ H_2$	Ni	Steam–hydrocarbon re-forming process for synthesis of hydrogen	H_2: Manufacture of ammonia, methanol
$CO + H_2O \rightarrow CO_2 + H_2$	ZnO and CuO	Gas shift reaction to improve yield in the synthesis of H_2	
$CO + 2\ H_2 \rightarrow CH_3OH$	ZnO and Cr_2O_3	Industrial synthesis of methanol	CH_3OH: Manufacture of plastics, adhesives, gasoline additives; industrial solvent
$C_2H_4 + H_2 \rightarrow C_2H_6$	Ni, Pd, or Pt	Catalytic hydrogenation of compounds with C=C bonds, as in conversion of unsaturated vegetable oils to solid fats	Food products: margarine, shortening

the high temperatures present in an automobile engine. Catalytic converters promote conversion of the offending pollutants to carbon dioxide, water, nitrogen and oxygen (Figure 12.18):

$$C_xH_y(g) + (x + y/4)\ O_2(g) \rightarrow x\ CO_2(g) + (y/2)\ H_2O(g)$$

$$2\ CO(g) + O_2(g) \rightarrow 2\ CO_2(g)$$

$$2\ NO(g) \rightarrow N_2(g) + O_2(g)$$

Typical catalysts for these reactions are the so-called noble metals, such as Pt and Pd, and transition-metal oxides, such as V_2O_5, Cr_2O_3, and CuO. Because the surface of the catalyst is rendered ineffective ("poisoned") by adsorption of lead, automobiles with catalytic converters must use unleaded gasolines.

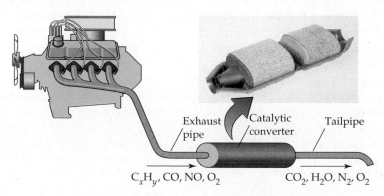

Exhaust pipe / Catalytic converter / Tailpipe

C_xH_y, CO, NO, O_2 CO_2, H_2O, N_2, O_2

Figure 12.18 The gases exhausted from an automobile engine pass through a catalytic converter where air pollutants such as unburned hydrocarbons (C_xH_y), CO, and NO are converted to CO_2, H_2O, N_2, and O_2. The photo shows a cross-sectional view of the catalytic converter. The beads are impregnated with the heterogeneous catalyst.

interlude—EXPLOSIVES

Some chemical reactions are slow and some are fast, but *explosions* are in a class by themselves. Chemical explosions are characterized by the nearly instantaneous release of large quantities of hot gases. The rapidly expanding gases set up a shock wave of enormous pressure—up to 700,000 atm—that propagates through the surroundings, causing the physical devastation we're all familiar with seeing in pictures.

The high-pressure shock wave caused by a chemical explosion is powerful enough to demolish this building.

Many substances undergo rapid reaction with evolution of heat—the burning of hydrogen in oxygen, for example—but the rate at which a bimolecular combustion reaction occurs is limited by the rate at which oxygen and fuel molecules encounter one another. True explosions, by contrast, are unimolecular *detonations* in which a sudden, high-pressure wave causes molecules to break down spontaneously into fragments, which then recombine to give the final products—usually stable gases such as N_2, H_2O, and CO_2. The shock wave can travel at speeds of up to 9000 m/s (approximately 20,000 mi/h), causing the explosion to occur at a rate far faster than that of other chemical reactions.

Explosives are generally categorized as either *primary* or *secondary*, depending on their sensitivity to shock. Primary explosives are the most sensitive to heat and shock. They are generally used in detonators, blasting caps, and military fuses to initiate the explosion of the less-sensitive secondary explosive. Mercuric fulminate, $Hg(ONC)_2$, was the first initiator to be used commercially, but it has been largely replaced by lead azide, $Pb(N_3)_2$, which is more stable when stored under hot conditions.

Secondary explosives, or *high explosives*, are generally less sensitive to heat and shock than are primary explosives and are therefore safer to manufacture, transport, and handle. Most secondary explosives will simply burn rather than explode when ignited in air, and most can be detonated only by the nearby explosion of a primary initiator. Among the most common secondary explosives are nitroglycerin, trinitrotoluene (TNT), RDX, and HMX.

488

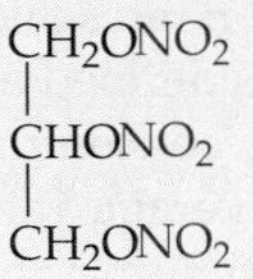

Nitroglycerin

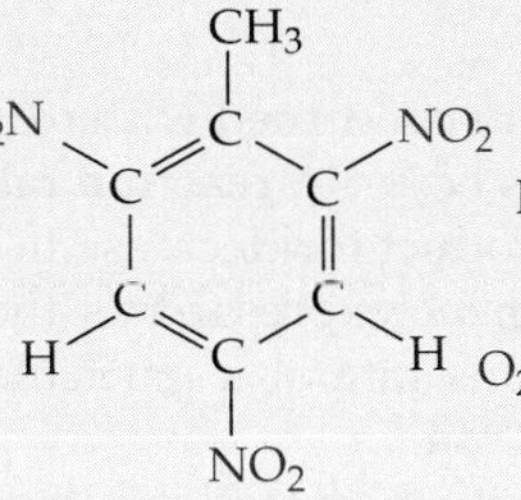

Trinitrotoluene
(TNT)

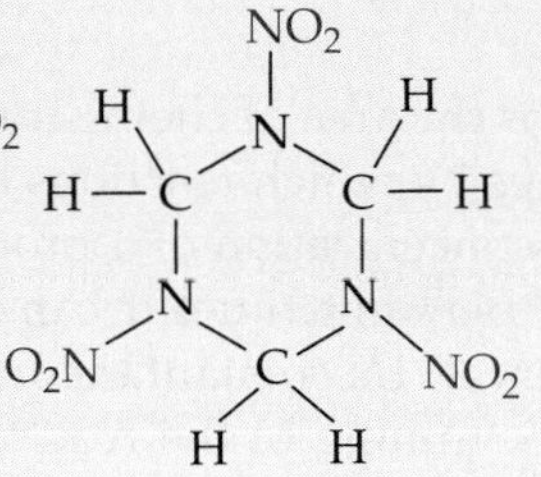

RDX

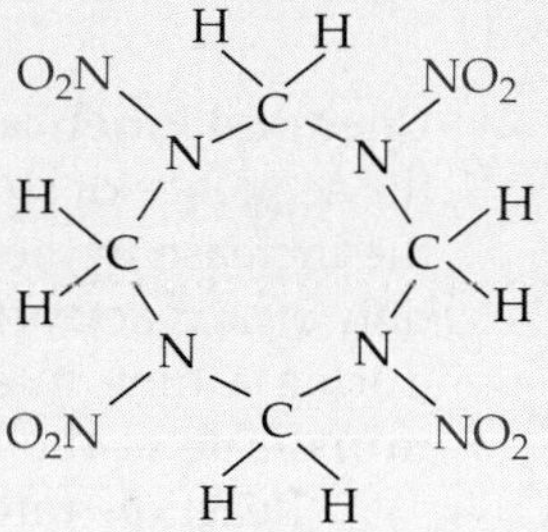

HMX

As is clear from looking at their structures, most explosives are rich in oxygen and nitrogen, and most contain *nitro groups*, $-NO_2$. Because the chemical bonds in nitro groups are relatively weak (about 200 kJ/mol), and because the explosion products (CO_2, N_2, H_2O, and others) are extremely stable, a great deal of energy is released within a few microseconds during an explosion. One mole (227 g) of nitroglycerin, for example, releases 1427 kJ of energy when it explodes. The actual mix of reaction products is complex, but the reaction can be approximated by the balanced equation

$$4\ C_3H_5N_3O_9(l) \rightarrow 12\ CO_2(g) + 10\ H_2O(g) + 6\ N_2(g) + O_2(g)$$

Note that 29 mol of gaseous products with a volume of 650 L at STP (Section 9.3) come from just 4 mol of liquid nitroglycerin.

The first commercially important high explosive was nitroglycerin, prepared in 1847 by reaction of glycerin with nitric acid in the presence of sulfuric acid.

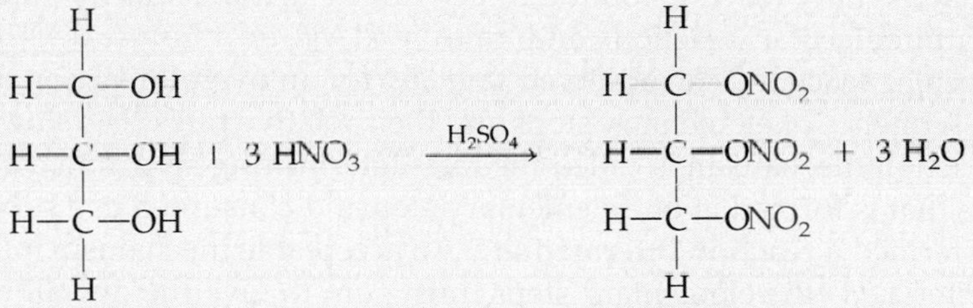

Glycerin

Nitroglycerin

As you might expect, the reaction is extremely hazardous to carry out. In fact, it wasn't until 1865 that the Swedish chemist Alfred Nobel succeeded in finding a relatively safe method of producing nitroglycerin and of incorporating it into a reliable commercial blasting product known as *dynamite*. (Nobel's younger brother was killed in an accidental explosion during the developmental work.) Modern industrial dynamite used for quarrying stone and blasting roadbeds is a mixture of ammonium nitrate and nitroglycerin absorbed onto diatomaceous earth.

Military explosives are generally used as fillings for bombs or shells and must therefore have a very low sensitivity to impact shock on firing. In addition, they must have good stability for long-term storage under adverse conditions, and they should have a low mass-to-energy ratio. TNT, RDX, and HMX are the most commonly used military high explosives. RDX is also often compounded with waxes or synthetic polymers to make the so-called plastic explosives favored by terrorist groups.

489

Chemical kinetics is the area of chemistry concerned with reaction rates and the sequence of steps by which reactions occur. A **reaction rate** is defined as the increase in the concentration of a product (or decrease in the concentration of a reactant) per unit time. It can be expressed as the **average rate** during a time interval, the **instantaneous rate** at a particular time, or the **initial rate** at the beginning of the reaction.

Reaction rates depend on reactant concentrations, temperature, and the presence of catalysts. The concentration dependence is given by the **rate law**, rate $= k[A]^m[B]^n$, where k is the **rate constant**, m and n specify the reaction order with respect to reactants A and B, and $m + n$ is the overall reaction order. The values of m and n must be determined by experiment; they can't be deduced from the stoichiometry of the overall reaction.

The **integrated rate law** is a concentration-time equation that allows us to calculate concentrations at any time t and the time required for an initial concentration to reach any particular value. For a first-order reaction, the integrated rate law is $\log [A]_t = -kt/2.303 + \log [A]_0$. A graph of $\log [A]$ versus time is a straight line with a slope equal to $-k/2.303$. For a second-order reaction, the integrated rate law is $1/[A]_t = kt + 1/[A]_0$. A graph of $1/[A]$ versus time is linear with a slope equal to k. The **half-life** ($t_{1/2}$) of a reaction is the time required for the reactant concentration to drop to one-half of its initial value.

A **reaction mechanism** is the sequence of **elementary reactions (elementary steps)** that defines the pathway from reactants to products. Elementary reactions are classified as **unimolecular**, **bimolecular**, or **termolecular**, depending on the number of reactant molecules. The rate law for an elementary reaction follows directly from its **molecularity**: rate $= k[A]$ for a unimolecular reaction, and rate $= k[A]^2$ or rate $= k[A][B]$ for a bimolecular reaction. The observed rate law for an overall reaction depends on the sequence of elementary steps and their relative rates. The slowest step in a reaction mechanism is called the **rate-determining step**, and a chemical species that is formed in one elementary step and consumed in a subsequent step is termed a **reaction intermediate**. An acceptable mechanism must meet two criteria: (1) the elementary steps must sum to give the overall reaction, and (2) the mechanism must be consistent with the observed rate law.

The exponential temperature dependence of rate constants is described by the **Arrhenius equation**, $k = Ae^{-E_a/RT}$, where A is the **frequency factor** and E_a is the **activation energy**. The value of E_a can be determined from the slope of a linear plot of $\log k$ versus $1/T$, and it can be interpreted as the height of the potential energy barrier between reactants and products. The configuration of atoms at the top of the barrier is called the **transition state**. According to **collision theory**, the rate constant is given by $k = pZe^{-E_a/RT}$, where p is a **steric factor** (the fraction of the collisions in which the molecules have the proper orientation for reaction), Z is a constant related to the collision frequency, and $e^{-E_a/RT}$ is the fraction of the collisions with energy equal to E_a or greater.

A **catalyst** is a substance that increases the rate of a reaction without being consumed in the reaction. It functions by making available an alternative reaction pathway that has a lower activation energy. A **homogeneous catalyst** is present in the same phase as the reactants, whereas a **heterogeneous catalyst** is present in a different phase.

1. The following reaction is first order in A and first order in B:

$$A + B \longrightarrow products \qquad Rate = k[A][B]$$

What are the relative rates of this reaction in vessels **(a)–(d)**? Each vessel has the same volume. Red spheres represent A molecules, and blue spheres represent B molecules.

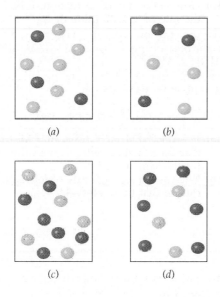

(a)   (b)

(c)   (d)

2. What are the relative values of the rate constant k for cases **(a)–(d)** in Problem 1?

3. Consider the first-order reaction A $\longrightarrow$ B in which A molecules (red spheres) rearrange to give B molecules (blue spheres). **(a)** Given the following pictures at $t = 0$ min and $t = 1$ min, draw pictures that show the number of A and B molecules present at $t = 2$ min and $t = 3$ min. **(b)** What is the half-life of the reaction?

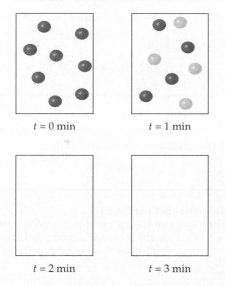

$t = 0$ min $t = 1$ min

$t = 2$ min $t = 3$ min

4. The following pictures represent the progress of a reaction in which two A molecules combine to give a more complex molecule A_2, 2 A $\longrightarrow$ A_2. **(a)** Is the reaction first order or second order in A? **(b)** Write the rate law. **(c)** Draw an appropriate picture in the last box, and specify the time.

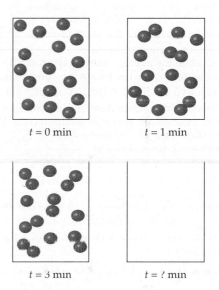

$t = 0$ min $t = 1$ min

$t = 3$ min $t = ?$ min

5. What is the molecularity of each of the following elementary reactions? Each reaction is written in the format

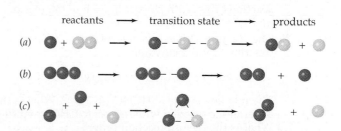

reactants $\longrightarrow$ transition state $\longrightarrow$ products

(a)

(b)

(c)

6. The rate of the reaction A + B_2 $\longrightarrow$ AB + B is directly proportional to the concentration of B_2, independent of the concentration of A, and directly proportional to the concentration of a substance C.

(a) Write the rate law.
(b) Write a mechanism that accords with the experimental facts.
(c) What is the role of C in this reaction, and why doesn't C appear in the chemical equation for the overall reaction?

ADDITIONAL PROBLEMS

Problems 12.1–12.13 appear within the chapter.

REACTION RATES

12.14 What are the usual units of reaction rate?

12.15 Concentrations of trace constituents of the atmosphere are sometimes expressed in terms of molecules/cm^3. If those units are used for concentrations, what are the units of reaction rate?

12.16 Given the concentration-time data in Problem 12.7, calculate the average rate of decomposition of cyclopropane during the following time intervals.
(a) 0 to 5.0 min   **(b)** 15.0 to 20.0 min

12.17 Concentration-time data for decomposition of nitrogen dioxide are given in Example 12.8. What is the average rate of decomposition of NO_2 during the following time periods?
(a) 50 to 100 s   **(b)** 100 to 150 s

12.18 From a plot of the concentration-time data in Example 12.8, estimate **(a)** the instantaneous rate of decomposition of NO_2 at $t = 100$ s and **(b)** the initial rate of decomposition of NO_2.

12.19 Ammonia is manufactured in large amounts by the reaction

$$N_2(g) + 3\,H_2(g) \rightarrow 2\,NH_3(g)$$

(a) How is the rate of consumption of H_2 related to the rate of consumption of N_2?
(b) How is the rate of formation of NH_3 related to the rate of consumption of N_2?

12.20 For the reaction $N_2(g) + 3\,H_2(g) \rightarrow 2\,NH_3(g)$, what is the relationship between $\Delta[N_2]/\Delta t$, $\Delta[H_2]/\Delta t$, and $\Delta[NH_3]/\Delta t$?

12.21 The oxidation of iodide ion by peroxydisulfate ion is described by the equation

$$3\,I^-(aq) + S_2O_8{}^{2-}(aq) \rightarrow I_3{}^-(aq) + 2\,SO_4{}^{2-}(aq)$$

(a) If $-\Delta[S_2O_8{}^{2-}]/\Delta t = 1.5 \times 10^{-3}$ M/s for a particular time interval, what is the value of $\Delta[I^-]/\Delta t$ for the same time interval?
(b) What is the average rate of formation of $SO_4{}^{2-}$ during that time interval?

RATE LAWS

12.22 The gas-phase reaction of nitric oxide and bromine yields nitrosyl bromide:

$$2\,NO(g) + Br_2(g) \rightarrow 2\,NOBr(g)$$

The rate law is rate = $k[NO]^2[Br_2]$. What is the reaction order with respect to each of the reactants, and what is the overall reaction order?

12.23 The reaction of gaseous chloroform and chlorine is described by the equation

$$CHCl_3(g) + Cl_2(g) \rightarrow CCl_4(g) + HCl(g)$$

The rate law is rate = $k[CHCl_3][Cl_2]^{1/2}$. What is the order of the reaction with respect to $CHCl_3$ and Cl_2? What is the overall reaction order?

12.24 The gas-phase reaction of hydrogen and iodine monochloride,

$$H_2(g) + 2\,ICl(g) \rightarrow 2\,HCl(g) + I_2(g)$$

is first order in H_2 and first order in ICl. What is the rate law, and what are the units of the rate constant?

12.25 The reaction $2\,NO(g) + 2\,H_2(g) \rightarrow N_2(g) + 2\,H_2O(g)$ is first order in H_2 and second order in NO. Write the rate law, and specify the units of the rate constant.

12.26 At 600°C, acetone (CH_3COCH_3) decomposes to ketene ($CH_2{=}C{=}O$) and various hydrocarbons. Given the following initial rate data, **(a)** determine the rate law, **(b)** calculate the rate constant, and **(c)** calculate the rate of decomposition when the acetone concentration is 1.8×10^{-3} M.

Experiment	Initial [CH₃COCH₃]	Initial Rate of Decomp. of CH₃COCH₃ (M/s)
1	6.0×10^{-3}	5.2×10^{-5}
2	9.0×10^{-3}	7.8×10^{-5}

12.27 The following initial rates were measured for the thermal decomposition of azomethane (CH_3NNCH_3):

$$CH_3NNCH_3(g) \rightarrow C_2H_6(g) + N_2(g)$$

Experiment	Initial [CH₃NNCH₃]	Initial Rate of Decomp. of CH₃NNCH₃ (M/s)
1	2.4×10^{-2}	6.0×10^{-6}
2	8.0×10^{-3}	2.0×10^{-6}

(a) What is the rate law?
(b) Calculate the value of the rate constant.
(c) What is the rate of decomposition when the concentration of azomethane is 0.020 M?

12.28 Initial rate data at 25°C are listed for the reaction

$$NH_4^+(aq) + NO_2^-(aq) \rightarrow N_2(g) + H_2O(l)$$

Experiment	Initial $[NH_4^+]$	Initial $[NO_2^-]$	Initial Rate (M/s)
1	0.24	0.10	7.2×10^{-6}
2	0.12	0.10	3.6×10^{-6}
3	0.12	0.15	5.4×10^{-6}

(a) What is the rate law?
(b) What is the value of the rate constant?
(c) What is the reaction rate when the concentrations are $[NH_4^+] = 0.39$ M and $[NO_2^-] = 0.052$ M?

12.29 The following initial rates were measured for the reaction

$$2 NO(g) + Cl_2(g) \rightarrow 2 NOCl(g)$$

Experiment	Initial $[NO]$	Initial $[Cl_2]$	Initial Rate (M/s)
1	0.13	0.20	1.0×10^{-2}
2	0.26	0.20	4.0×10^{-2}
3	0.13	0.10	5.0×10^{-3}

(a) Write the rate law.
(b) Calculate the value of the rate constant.
(c) What is the reaction rate when both reactant concentrations are 0.12 M?

INTEGRATED RATE LAW; HALF-LIFE

12.30 At 500°C, cyclopropane (C_3H_6) rearranges to propene (CH_3–CH=CH_2). The reaction is first order, and the rate constant is 6.7×10^{-4}/s. If the initial concentration of C_3H_6 is 0.0500 M, (a) what is the molarity of C_3H_6 after 30 min, (b) how many minutes does it take for the C_3H_6 concentration to drop to 0.0100 M, and (c) how many minutes does it take for 25% of the C_3H_6 to react?

12.31 What is the half-life in minutes of the reaction in Problem 12.30? How many minutes will it take for the concentration of cyclopropane to drop to 6.25% of its initial value?

12.32 Butadiene (C_4H_6) reacts with itself to form a dimer with the formula C_8H_{12}. The reaction is second order in C_4H_6. If the rate constant at a particular temperature is 4.0×10^{-2}/(M·s) and the initial concentration of C_4H_6 is 0.0200 M, (a) what is its molarity after a reaction time of 1.00 h, and (b) what is the time in hours when the C_4H_6 concentration reaches a value of 0.0020 M?

12.33 What is the half-life in minutes of the reaction in Problem 12.32 when the initial C_4H_6 concentration is 0.0200 M? How many minutes does it take for the concentration of C_4H_6 to drop from 0.0100 M to 0.0050 M?

12.34 At elevated temperatures, nitrous oxide decomposes according to the equation

$$2 N_2O(g) \rightarrow 2 N_2(g) + O_2(g)$$

Given the following data, make the appropriate graphs to determine whether the reaction is first order or second order. What is the value of the rate constant?

Time (min)	0	60	90	120	180
$[N_2O]$	0.250	0.218	0.204	0.190	0.166

12.35 Nitrosyl bromide decomposes at 10°C:

$$2 NOBr(g) \rightarrow 2 NO(g) + Br_2(g)$$

Use the following kinetic data to determine the order of the reaction and the value of the rate constant.

Time (s)	0	10	20	30	40
$[NOBr]$	0.0400	0.0303	0.0244	0.0204	0.0175

12.36 Decomposition of N_2O_5 is a first-order reaction. At 25°C, it takes 5.2 h for the concentration to drop from 0.120 M to 0.060 M. How many hours does it take for the concentration to drop from 0.030 M to 0.015 M? From 0.480 M to 0.015 M?

12.37 At 25°C, the half-life of a certain first-order reaction is 248 s. What is the value of the rate constant at this temperature?

REACTION MECHANISMS

12.38 Tell what is meant by each of the following terms.

(a) reaction mechanism   (b) elementary reaction
(c) molecularity   (d) reaction intermediate

12.39 What distinguishes the rate-determining step from the other steps in a reaction mechanism? How does the rate-determining step affect the observed rate law?

12.40 Consider the following mechanism for reaction of hydrogen and iodine monochloride:

Step 1. $H_2(g) + ICl(g) \rightarrow HI(g) + HCl(g)$

Step 2. $HI(g) + ICl(g) \rightarrow I_2(g) + HCl(g)$

(a) Write the equation for the overall reaction.
(b) Identify any reaction intermediates.
(c) What is the molecularity of each elementary step?

12.41 The following mechanism has been proposed for the reaction of nitric oxide and chlorine:

Step 1. $NO(g) + Cl_2(g) \rightarrow NOCl_2(g)$

Step 2. $NOCl_2(g) + NO(g) \rightarrow 2 NOCl(g)$

(a) What is the overall reaction?
(b) Identify any reaction intermediates.
(c) Identify the molecularity of each elementary step.

12.42 Give the molecularity and the rate law for each of the following elementary reactions.

(a) $O_3(g) + Cl(g) \rightarrow O_2(g) + ClO(g)$
(b) $NO_2(g) \rightarrow NO(g) + O(g)$
(c) $ClO(g) + O(g) \rightarrow Cl(g) + O_2(g)$
(d) $Cl(g) + Cl(g) + N_2(g) \rightarrow Cl_2(g) + N_2(g)$

12.43 Identify the molecularity and write the rate law for each of the following elementary reactions.

(a) $I_2(g) \rightarrow 2 I(g)$
(b) $2 NO(g) + Br_2(g) \rightarrow 2 NOBr(g)$
(c) $CH_3Br(aq) + OH^-(aq) \rightarrow CH_3OH(aq) + Br^-(aq)$
(d) $N_2O_5(g) \rightarrow NO_2(g) + NO_3(g)$

12.44 The thermal decomposition of nitryl chloride is believed to occur by the following mechanism:

$$NO_2Cl(g) \xrightarrow{k_1} NO_2(g) + Cl(g)$$

$$Cl(g) + NO_2Cl(g) \xrightarrow{k_2} NO_2(g) + Cl_2(g)$$

(a) What is the overall reaction?
(b) What rate law is predicted by this mechanism if the first step is the rate-determining step?

12.45 The reaction $H_2(g) + 2 ICl(g) \longrightarrow I_2(g) + 2 HCl(g)$ has a second-order rate law, rate = $k[H_2][ICl]$. Suggest a mechanism that is consistent with this rate law.

12.46 Consider the hypothetical reaction $A + 2 B \longrightarrow AB_2$. Propose a mechanism that is consistent with the rate law: rate = $k[A][B]$.

THE ARRHENIUS EQUATION

12.47 Rate constants for the reaction $2 N_2O_5(g) \longrightarrow 4 NO_2(g) + O_2(g)$ exhibit the following temperature dependence:

Temperature (°C)	k (1/s)	Temperature (°C)	k (1/s)
25	3.7×10^{-5}	55	1.7×10^{-3}
45	5.1×10^{-4}	65	5.2×10^{-3}

From an appropriate graph of the data, determine the activation energy for this reaction in kJ/mol.

12.48 The following rate constants describe the thermal decomposition of nitrogen dioxide.

$$2 NO_2(g) \rightarrow 2 NO(g) + O_2(g)$$

Temperature (°C)	k [1/(M · s)]	Temperature (°C)	k [1/(M · s)]
330	0.77	378	4.1
354	1.8	383	4.7

Make an appropriate graph of the data, and calculate the value of E_a for this reaction in kJ/mol.

12.49 Rate constants for the reaction $NO_2(g) + CO(g) \longrightarrow NO(g) + CO_2(g)$ are 1.3/(M · s) at 700 K and 23.0/(M · s) at 800 K.

(a) Calculate the value of the activation energy in kJ/mol.
(b) Calculate the rate constant at 750 K.

12.50 A certain first-order reaction has a rate constant of 1.0×10^{-3} at 25°C.

(a) If the reaction rate doubles when the temperature is increased to 35°C, what is the activation energy for this reaction in kJ/mol?
(b) What is E_a in kJ/mol if the same temperature change causes the rate to triple?

12.51 Why don't all collisions between reactant molecules lead to chemical reaction?

12.52 Tell what is meant by each of the following terms.

(a) frequency factor   (b) steric factor
(c) activation energy   (d) transition state

12.53 Values of $E_a = 183$ kJ/mol and $\Delta E = 12$ kJ/mol have been measured for the reaction

$$2 HI(g) \rightarrow H_2(g) + I_2(g)$$

(a) Sketch a potential energy profile for this reaction that shows the potential energy of reactants, products, and the transition state. Include labels that define E_a and ΔE.
(b) Suggest a plausible structure for the transition state.

CATALYSIS

12.54 What effect does a catalyst have on the rate, mechanism, and activation energy of a chemical reaction?

12.55 Why doesn't a catalyst appear in the chemical equation for a reaction?

12.56 Distinguish between a homogeneous and a heterogeneous catalyst. Give an example of each, and identify a reaction that is accelerated by the catalyst.

12.57 In the upper atmosphere, chlorofluorocarbons absorb sunlight, and subsequent fragmentation produces Cl atoms. The Cl atoms participate in the following mechanism for destruction of ozone:

$$Cl(g) + O_3(g) \rightarrow ClO(g) + O_2(g)$$

$$ClO(g) + O(g) \rightarrow Cl(g) + O_2(g)$$

(a) Write the equation for the overall reaction.
(b) What is the role of the Cl atoms in this reaction?
(c) Is ClO a catalyst or a reaction intermediate?
(d) What distinguishes a catalyst from an intermediate?

12.58 In a formerly important process for the manufacture of sulfuric acid, sulfur dioxide was oxidized to sulfur trioxide in the following sequence of reactions:

$$2 SO_2(g) + 2 NO_2(g) \rightarrow 2 SO_3(g) + 2 NO(g)$$

$$2 NO(g) + O_2(g) \rightarrow 2 NO_2(g)$$

(a) Write the equation for the overall reaction.
(b) Identify any molecule that acts as a catalyst or a reaction intermediate. Briefly justify your answer.

GENERAL PROBLEMS

12.59 List three factors that affect the rate of a chemical reaction. How do these factors affect the rate law?

12.60 Consider the reaction $H_2(g) + I_2(g) \longrightarrow 2 HI(g)$. The reaction of a fixed amount of H_2 and I_2 is studied in a cylinder with a moveable piston. Indicate the effect of each of the following changes on the reaction rate.

(a) an increase in temperature at constant volume
(b) an increase in volume at constant temperature
(c) addition of a catalyst
(d) addition of argon (inert) at constant volume

12.61 Listed below are concentration-time data for the conversion of A and B to D.

Experiment	Time (s)	[A]	[B]	[C]	[D]
1	0	5.00	2.00	1.00	0.00
	60	4.80	1.90	1.00	0.10
2	0	10.00	2.00	1.00	0.00
	60	9.60	1.80	1.00	0.20
3	0	5.00	4.00	1.00	0.00
	60	4.80	3.90	1.00	0.10
4	0	5.00	2.00	2.00	0.00
	60	4.60	1.80	2.00	0.20

(a) Write a balanced equation for the reaction.
(b) What is the reaction order with respect to A, B, and C, and what is the overall reaction order?
(c) Write the rate law.
(d) Is a catalyst involved in this reaction? Briefly justify your answer.
(e) Suggest a mechanism consistent with the data.
(f) Calculate the rate constant for formation of D.

12.62 When the temperature of a gas is raised by 10°C, the collision frequency increases by only ~ 2%, but the reaction rate increases by 100% or more. Explain.

12.63 What fraction of the molecules in a gas at 300 K collide with an energy equal to E_a or greater when E_a is equal to 50 kJ/mol? What is the value of this fraction when E_a is 100 kJ/mol?

12.64 How does the height of the barrier in a potential energy profile affect the rate of the reaction?

12.65 Radioactive decay exhibits a first-order rate law—rate = $-\Delta N/\Delta t = kN$—where N denotes the number of radioactive nuclei present at time t. The half-life of strontium 90, a dangerous nuclear fission product, is 29 y.

(a) What fraction of the strontium-90 remains after three half-lives?
(b) What is the value of the rate constant for the decay of strontium-90?
(c) How many years are required for 99% of the strontium-90 to disappear?

12.66 If the rate of a reaction increases by a factor of 2.5 for a temperature rise from 20°C to 30°C, what is the value of the activation energy in kJ/mol? By what factor does the rate of this reaction increase when the temperature is raised from 120°C to 130°C?

12.67 A two-step mechanism has been suggested for the reaction of nitric oxide and bromine:

$$NO(g) + Br_2(g) \xrightarrow{k_1} NOBr_2(g)$$

$$NOBr_2(g) + NO(g) \xrightarrow{k_2} 2 NOBr(g)$$

(a) What is the overall reaction?
(b) What is the role of $NOBr_2$ in this reaction?
(c) What is the predicted rate law if the first step is much slower than the second step?
(d) The observed rate law is: rate = $k[NO]^2[Br_2]$. What can you conclude about the rate-determining step?

chapter *13* CHEMICAL EQUILIBRIUM

I n previous chapters, we've generally assumed that chemical reactions result in complete conversion of reactants to products. Stoichiometric amounts of gaseous H_2 and O_2, for example, react to give water vapor, with only negligible amounts of H_2 and O_2 remaining after reaction. The reverse reaction, decomposition of water vapor to gaseous H_2 and O_2, does not occur to an appreciable extent at ordinary temperatures, and we therefore say that the forward reaction is *irreversible*.

$$2\,H_2(g) + O_2(g) \rightarrow 2\,H_2O(g) \qquad \textit{reaction occurs only in forward direction}$$

In contrast to the reaction of H_2 with O_2, many other chemical reactions do not completely convert reactants to products. When the colorless gas dinitrogen tetroxide (N_2O_4) decomposes to the dark brown gas nitrogen dioxide (NO_2), for example, the concentration of N_2O_4 decreases and the concentration of NO_2 increases over time until both concentrations level off at constant values that persist indefinitely. The state that's reached when the concentrations of reactants and products remain constant in time is called the state of **chemical equilibrium**. A mixture of reactants and products in the equilibrium state is an **equilibrium mixture**.

$$N_2O_4(g) \rightleftarrows 2\,NO_2(g) \qquad \textit{reaction occurs in both directions}$$

When the rate at which people move from the first floor to the second equals the rate at which people move from the second floor to the first, the number of people on each floor remains constant, and the two populations are in equilibrium.

In this chapter, we'll address a number of important questions about the composition of equilibrium mixtures. What is the relationship between the concentrations of reactants and products in an equilibrium mixture? How can we determine equilibrium concentrations from initial concentrations? What factors can be exploited to alter the composition of an equilibrium mixture? This last question is particularly important when choosing conditions for the industrial synthesis of many chemicals—for example, hydrogen, ammonia, and lime (CaO).

13.1 ➤THE EQUILIBRIUM STATE

Let's look more carefully at the decomposition reaction of N_2O_4 to NO_2. This reaction is said to be *reversible* because the product NO_2 molecules can recombine to form reactant N_2O_4 molecules. To indicate that the reaction can proceed in both forward and reverse directions, we write the balanced equation with two arrows, one pointing from reactants to products, and the other pointing from products to reactants[1]:

$$N_2O_4(g) \rightleftarrows 2\, NO_2(g)$$

Figure 13.1 shows the results of two experiments that illustrate the reversible interconversion of N_2O_4 and NO_2. In the first experiment, imagine that 0.0400 mol of N_2O_4 is placed in a 1.000 L flask at 25°C to give an initial N_2O_4 concentration of 0.0400 M. The formation of NO_2 is indicated by the appearance of a brown color, and its concentration can be monitored by measuring the intensity of the color. According to the balanced equation, 0.5 mol of N_2O_4 disappears for each mol of NO_2 that forms, so the concentra-

FIGURE 13.1 Change in the concentrations of N_2O_4 and NO_2 with time in two experiments: **(a)** only N_2O_4 is present initially; **(b)** only NO_2 is present initially. In experiment **(a)**, $[NO_2]$ increases as $[N_2O_4]$ decreases. In experiment **(b)**, $[N_2O_4]$ increases as $[NO_2]$ decreases. In both experiments, a state of chemical equilibrium is reached when the concentrations level off at constant values ($[N_2O_4] = 0.0337$ M; $[NO_2] = 0.0125$ M).

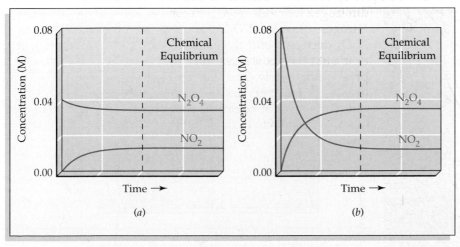

[1] The terms "reactants" and "products" could be confusing in this context because the products of the forward reaction are reactants in the reverse reaction. To avoid confusion, we'll restrict the term *reactants* to the substances on the left side of the chemical equation and the term *products* to the substances on the right side of the equation.

tion of N_2O_4 at any time is equal to the initial concentration of N_2O_4 minus half the concentration of NO_2. As time passes, the concentration of N_2O_4 decreases and the concentration of NO_2 increases until both concentrations level off at constant, equilibrium values: $[N_2O_4]$ = 0.0337 M; $[NO_2]$ = 0.0125 M.

In the second experiment shown in Figure 13.1, imagine that we begin with NO_2, rather than N_2O_4, at a concentration of 0.0800 M. The conversion of NO_2 to N_2O_4 proceeds, but it is incomplete, and the concentrations level off at the same values as obtained in the first experiment. Taken together, the two experiments demonstrate that the interconversion of N_2O_4 and NO_2 is reversible and that the same equilibrium state is reached starting from either substance.

Why do the reactions of N_2O_4 and NO_2 appear to "stop" after the concentrations reach their equilibrium values? We'll explore that question in more detail in Section 13.11, but we'll note for now that the concentrations reach constant values not because the reactions stop, but because the *rates* of the forward and reverse reactions become equal. Take, for example, the experiment in which N_2O_4 is converted to an equilibrium mixture of NO_2 and N_2O_4. Because reaction rates depend on concentrations (Section 12.2), the rate of the forward reaction ($N_2O_4 \longrightarrow 2\ NO_2$) decreases as the concentration of N_2O_4 decreases, while the rate of the reverse reaction ($N_2O_4 \longleftarrow 2\ NO_2$) increases as the concentration of NO_2 increases. Eventually, the decreasing rate of the forward reaction and the increasing rate of the reverse reaction become equal. At that point, there are no further changes in concentrations because N_2O_4 and NO_2 both disappear as fast as they're formed. Thus, chemical equilibrium is a *dynamic state* in which forward and reverse reactions continue at equal rates, and there is no net conversion of reactants to products (Figure 13.2).

In the environment, water is involved in a dynamic equilibrium. The amount of surface water and atmospheric water vapor is roughly constant because evaporation and rain occur at the same rates on a global level.

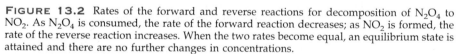

FIGURE 13.2 Rates of the forward and reverse reactions for decomposition of N_2O_4 to NO_2. As N_2O_4 is consumed, the rate of the forward reaction decreases; as NO_2 is formed, the rate of the reverse reaction increases. When the two rates become equal, an equilibrium state is attained and there are no further changes in concentrations.

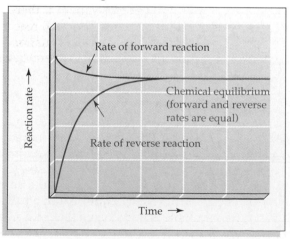

Table 13.1 lists concentration data for the preceding experiments along with data for three additional experiments. In experiments 1 and 2, the equilibrium mixtures have identical compositions because the initial concentration of N_2O_4 in experiment 1 is half the initial concentration of NO_2 in experiment 2 (that is, the total number of N and O atoms is the same in both experiments). In experiments 3–5, different initial concentrations of N_2O_4 and/or NO_2 give different equilibrium concentrations. In all the experiments, however, the equilibrium concentrations are related. The last column of Table 13.1 shows that, at equilibrium, the expression $[NO_2]^2/[N_2O_4]$ has a constant value of 4.64×10^{-3} M.

⌐ **PROBLEM 13.1** From measurements of the intensity of its brown color, the concentration of NO_2 in an equilibrium mixture of NO_2 and N_2O_4 was found to be 0.0200 M. Use the fact that $[NO_2]^2/[N_2O_4] = 4.64 \times 10^{-3}$ M to calculate the molar concentration of N_2O_4.

TABLE 13.1 Concentration Data at 25°C for the Reaction
$N_2O_4(g) \rightleftarrows 2\ NO_2(g)$

Experiment Number	Initial Concentrations (M)		Equilibrium Concentrations (M)		
	$[N_2O_4]$	$[NO_2]$	$[N_2O_4]$	$[NO_2]$	$[NO_2]^2/[N_2O_4]$
1	0.0400	0.0000	0.0337	0.0125	4.64×10^{-3}
2	0.0000	0.0800	0.0337	0.0125	4.64×10^{-3}
3	0.0600	0.0000	0.0522	0.0156	4.66×10^{-3}
4	0.0000	0.0600	0.0246	0.0107	4.65×10^{-3}
5	0.0200	0.0600	0.0429	0.0141	4.63×10^{-3}

13.2 ➤THE EQUILIBRIUM CONSTANT K_C

Confronted with a constant value of the expression $[NO_2]^2/[N_2O_4]$ for the decomposition of N_2O_4 to NO_2, you might wonder if there is an analogous expression having a constant value for every chemical reaction. If so, how is the form of that expression related to the balanced equation for the reaction? To answer those questions, let's consider a general reversible reaction:

$$a\ A + b\ B \rightleftarrows c\ C + d\ D$$

where A and B are chemical formulas for the reactants, C and D are formulas for the products, and a, b, c, and d are stoichiometric coefficients in the balanced chemical equation. On the basis of experimental studies of many reversible reactions, the Norwegian chemists Cato Maximilian Guldberg and Peter Waage proposed in 1864 that the composition of an equilibrium mixture obeys the following **equilibrium equation**, where K_c is the **equilibrium constant** and the expression on the right side of the equilibrium equation is called the **equilibrium constant expression**.

Equilibrium equation $K_c = \dfrac{[C]^c[D]^d}{[A]^a[B]^b}$

Equilibrium constant   *Equilibrium constant expression*

As usual, square brackets indicate the molar concentration of the substance within the brackets (hence the subscript c for "concentration" in K_c). The equilibrium equation is also known as the *law of mass action* because in the early days of chemistry, concentration was called "active mass."

The equilibrium constant K_c is the number obtained by multiplying the equilibrium concentrations of the reaction products and dividing by the equilibrium concentrations of the reactants, with the concentration of each substance raised to the power of its coefficient in the balanced chemical equation. No matter what the individual equilibrium concentrations may be in a particular experiment, the equilibrium constant expression for a reaction at a particular temperature always has the same value. Thus, the equilibrium equation for the decomposition reaction of N_2O_4 to give 2 NO_2 is

$$K_c = \frac{[NO_2]^2}{[N_2O_4]} = 4.64 \times 10^{-3} \quad \text{at } 25°C$$

The equilibrium constant expression is $[NO_2]^2/[N_2O_4]$, and the equilibrium constant K_c has a value of 4.64×10^{-3} at 25°C (Table 13.1).

The units of K_c depend on the particular equilibrium equation. For this example, K_c has units of $(M)^2/(M) = M$. It's customary, however, to omit the units when quoting values of equilibrium constants. The value of an equilibrium constant depends on temperature, which must be given when citing a value of K_c. For example, K_c for the decomposition of N_2O_4 increases from 4.64×10^{-3} at 25°C to 1.53 at 127°C.

The form of the equilibrium constant expression and the numerical value of the equilibrium constant depend on the form of the balanced chemical equation. Consider again the chemical equation and the equilibrium equation for a general reaction:

$$a \text{ A} + b \text{ B} \rightleftharpoons c \text{ C} + d \text{ D} \qquad K_c = \frac{[C]^c[D]^d}{[A]^a[B]^b}$$

If we write the chemical equation in the reverse direction, the new equilibrium constant expression is the reciprocal of the original expression, and the new equilibrium constant K_c' is the reciprocal of the original equilibrium constant K_c:

$$c \text{ C} + d \text{ D} \rightleftharpoons a \text{ A} + b \text{ B} \qquad K_c' = \frac{[A]^a[B]^b}{[C]^c[D]^d} = \frac{1}{K_c}$$

(The "prime" superscript differentiates K_c' from K_c.) Because the equilibrium constants K_c and K_c' have different numerical values, it's important to specify the form of the balanced chemical equation when quoting the value of an equilibrium constant.

EXAMPLE 13.1

Write the equilibrium equation for each of the following reactions.
(a) $N_2(g) + 3 H_2(g) \rightleftarrows 2 NH_3(g)$
(b) $2 NH_3(g) \rightleftarrows N_2(g) + 3 H_2(g)$

SOLUTION   (a) Put the concentration of the reaction product, NH_3, in the numerator of the equilibrium constant expression and the concentrations of the reactants, N_2 and H_2, in the denominator. Then raise the concentration of each substance to the power of its coefficient in the balanced chemical equation.

$$K_c = \frac{[NH_3]^2}{[N_2][H_2]^3}$$

(b) Because the balanced equation is the reverse of that in (a), the equilibrium constant expression is the reciprocal of that in (a) and the equilibrium constant K_c' is the reciprocal of the equilibrium constant in (a).

$$K_c' = \frac{[N_2][H_2]^3}{[NH_3]^2} \qquad K_c' = \frac{1}{K_c}$$

EXAMPLE 13.2

The following concentrations were measured for an equilibrium mixture at 500 K: $[N_2] = 3.0 \times 10^{-2}$ M; $[H_2] = 3.7 \times 10^{-2}$ M; $[NH_3] = 1.6 \times 10^{-2}$ M. Calculate the equilibrium constant at 500 K for each of the reactions in Example 13.1.

SOLUTION   (a) Calculate the value of K_c by substituting the equilibrium concentrations into the equilibrium equation:

$$K_c = \frac{[NH_3]^2}{[N_2][H_2]^3} = \frac{(1.6 \times 10^{-2})^2}{(3.0 \times 10^{-2})(3.7 \times 10^{-2})^3} = 1.7 \times 10^2$$

(b) The equilibrium constant K_c' is

$$K_c' = \frac{[N_2][H_2]^3}{[NH_3]^2} = \frac{(3.0 \times 10^{-2})(3.7 \times 10^{-2})^3}{(1.6 \times 10^{-2})^2} = 5.9 \times 10^{-3}$$

Note that K_c' is the reciprocal of K_c. That is,

$$5.9 \times 10^{-3} = \frac{1}{1.7 \times 10^2}$$

PROBLEM 13.2 The oxidation of sulfur dioxide to give sulfur trioxide is an important step in the industrial process for synthesis of sulfuric acid. Write the equilibrium equation for each of the following reactions.
(a) $2 SO_2(g) + O_2(g) \rightleftarrows 2 SO_3(g)$
(b) $2 SO_3(g) \rightleftarrows 2 SO_2(g) + O_2(g)$

PROBLEM 13.3 The following equilibrium concentrations were measured at 800 K: $[SO_2] = 3.0 \times 10^{-3}$ M; $[O_2] = 3.5 \times 10^{-3}$ M; $[SO_3] = 5.0 \times 10^{-2}$ M. Calculate the equilibrium constant at 800 K for each of the reactions in Problem 13.2.

13.3 ▶THE EQUILIBRIUM CONSTANT K_P

Because gas pressures are easily measured, equilibrium equations for gas-phase reactions are often written using partial pressures rather than molar concentrations. For example, the equilibrium equation for the decomposition of N_2O_4 can be written as

$$K_p = \frac{(P_{NO_2})^2}{P_{N_2O_4}} \qquad \text{for the reaction } N_2O_4(g) \rightleftarrows 2\,NO_2(g)$$

$P_{N_2O_4}$ and P_{NO_2} are the partial pressures in atmospheres of reactants and products at equilibrium, and the subscript p on K reminds us that the **equilibrium constant K_p** is defined using partial pressures. In this particular case, the units of K_p are $(atm)^2/(atm) = atm$, but, as for K_c, it's common practice to omit the units. Note that the equilibrium equations for K_p and K_c have the same form except that the expression for K_p contains partial pressures instead of molar concentrations.

It can be shown that the values of K_p and K_c for a general gas-phase reaction are related by the equation

$$K_p = K_c(RT)^{\Delta n} \qquad \text{for the reaction } a\,A + b\,B \rightleftarrows c\,C + d\,D$$

Here, R is the gas constant, $0.0821 \text{ L} \cdot \text{atm}/(\text{K} \cdot \text{mol})$, T is the absolute temperature in kelvins, and $\Delta n = (c + d) - (a + b)$ is the sum of the coefficients of the gaseous products minus the sum of the coefficients of the gaseous reactants.

For the decomposition of 1 mol of N_2O_4 to 2 mol of NO_2, $\Delta n = 2 - 1 = 1$, and $K_p = K_c(RT)$.

$$N_2O_4(g) \rightleftarrows 2\,NO_2(g) \qquad K_p = K_c(RT)$$

For the reaction of 1 mol of hydrogen with 1 mol of iodine to give 2 mol of hydrogen iodide, $\Delta n = 2 - (1 + 1) = 0$, and $K_p = K_c(RT)^0 = K_c$.

$$H_2(g) + I_2(g) \rightleftarrows 2\,HI(g) \qquad K_p = K_c$$

In general, K_p is equal to K_c only if the same number of moles of gases appear on both sides of the balanced chemical equation.

EXAMPLE 13.3

Methane (CH_4) reacts with hydrogen sulfide to yield H_2 gas and carbon disulfide, a solvent used in manufacturing rayon and cellophane:

$$CH_4(g) + 2\,H_2S(g) \rightleftarrows CS_2(g) + 4\,H_2(g)$$

What is the value of K_p at 1000 K if the partial pressures in an equilibrium mixture at 1000 K are 0.20 atm of CH_4, 0.25 atm of H_2S, 0.52 atm of CS_2, and 0.10 atm of H_2?

SOLUTION   Write the equilibrium equation by setting K_p equal to the equilibrium constant expression in terms of partial pressures:

$$K_p = \frac{(P_{CS_2})(P_{H_2})^4}{(P_{CH_4})(P_{H_2S})^2}$$

The partial pressures of products are in the numerator and the partial pressures of reactants are in the denominator, with the pressure of each substance raised to the power of its coefficient in the balanced chemical equation. Then substitute the partial pressures into the equilibrium equation and solve for K_p :

$$K_p = \frac{(P_{CS_2})(P_{H_2})^4}{(P_{CH_4})(P_{H_2S})^2} = \frac{(0.52)(0.10)^4}{(0.20)(0.25)^2} = 4.2 \times 10^{-3}$$

EXAMPLE 13.4

Hydrogen is produced industrially by the steam-hydrocarbon reforming process. The reaction that takes place in the first step of this process is

$$H_2O(g) + CH_4(g) \rightleftarrows CO(g) + 3 H_2(g)$$

(a) If $K_c = 3.8 \times 10^{-3}$ at 1000 K, what is the value of K_p at the same temperature?
(b) If $K_p = 6.2 \times 10^4$ at 1400 K, what is the value of K_c at 1400 K?

SOLUTION   **(a)** For this reaction, $\Delta n = (1 + 3) - (1 + 1) = 2$. Therefore,

$$K_p = K_c(RT)^{\Delta n} = K_c(RT)^2 = (3.8 \times 10^{-3})[(0.0821)(1000)]^2 = 26$$

(b) Since $K_p = K_c(RT)^2$,

$$K_c = \frac{K_p}{(RT)^2} = \frac{6.2 \times 10^4}{[(0.0821)(1400)]^2} = 4.7$$

PROBLEM 13.4   In the industrial synthesis of hydrogen, mixtures of CO and H_2 are enriched in H_2 by reacting the CO with steam. The chemical equation for this so-called *gas shift reaction* is

$$CO(g) + H_2O(g) \rightleftarrows CO_2(g) + H_2(g)$$

What is the value of K_p at 700 K if the partial pressures in an equilibrium mixture at 700 K are 1.31 atm of CO, 10.0 atm of H_2O, 6.12 atm of CO_2, and 20.3 atm of H_2?

PROBLEM 13.5   Nitric oxide reacts with oxygen to give nitrogen dioxide, an important reaction in the Ostwald process for the industrial synthesis of nitric acid:

$$2 NO(g) + O_2(g) \rightleftarrows 2 NO_2(g)$$

If $K_c = 6.9 \times 10^5$ at 500 K, what is the value of K_p at this temperature? If $K_p = 1.3 \times 10^{-2}$ at 1000 K, what is the value of K_c at 1000 K?

13.4 ►HETEROGENEOUS EQUILIBRIA

Thus far we've been discussing **homogeneous equilibria**, in which all reactants and products are in a single phase, usually either gaseous or solution. **Heterogeneous equilibria**, by contrast, are those in which reactants and products are present in more than one phase. Take, for example, the thermal decomposition of solid calcium carbonate, a reaction used in manufacturing cement:

$$CaCO_3(s) \rightleftarrows CaO(s) + CO_2(g)$$

The manufacture of cement begins by thermal decomposition of limestone, $CaCO_3$, in large kilns.

When the reaction is carried out in a closed container, three phases are present at equilibrium: solid calcium carbonate, solid calcium oxide, and gaseous carbon dioxide. If we were to write the usual equilibrium equation for the reaction, including all the reactants and products, we would have

$$"K_c" = \frac{[CaO][CO_2]}{[CaCO_3]}$$

But because both CaO and $CaCO_3$ are solids, their molar "concentrations" are constants that can be calculated from their densities and molar masses:

$$[CaO] = \frac{3.25 \text{ g}}{\text{cm}^3} \times \frac{1 \text{ mol}}{56.1 \text{ g}} \times \frac{1000 \text{ cm}^3}{L} = 57.9 \text{ mol/L}$$

$$[CaCO_3] = \frac{2.71 \text{ g}}{\text{cm}^3} \times \frac{1 \text{ mol}}{100 \text{ g}} \times \frac{1000 \text{ cm}^3}{L} = 27.1 \text{ mol/L}$$

In general, the concentration of any pure solid is independent of its amount because concentration is the *ratio* of amount (in moles) to volume (in liters). If, for example, you double the amount of $CaCO_3$, you also double its volume, and the ratio of the two (the concentration) remains constant.

Rearranging the equilibrium equation for the decomposition of $CaCO_3$ to combine the constants $[CaCO_3]$, $[CaO]$, and "K_c", we obtain

$$"K_c" \times \frac{[CaCO_3]}{[CaO]} = K_c = [CO_2]$$

where the new equilibrium constant is $K_c = "K_c"[CaCO_3]/[CaO]$. The analogous equilibrium equation in terms of pressure is $K_p = P_{CO_2}$, where P_{CO_2} is the equilibrium pressure of CO_2 in atmospheres.

$$K_c = [CO_2] \qquad\qquad K_p = P_{CO_2}$$

As a general rule, the concentrations of pure solids or pure liquids are not included when writing the equilibrium equation for any heterogeneous equilibrium. Only the concentrations of gases and solutes in liquids are included.

To establish equilibrium between solid $CaCO_3$, solid CaO, and gaseous CO_2, all three components must be present. It follows from the equation $K_p = P_{CO_2}$, however, that the concentration and pressure of CO_2 at equilibrium are constant, independent of how much solid CaO and $CaCO_3$ is present (Figure 13.3). Of course, if the temperature is changed, the concentration and pressure of CO_2 will also change because the values of K_c and K_p depend on temperature.

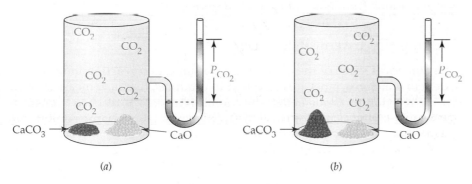

(a)   (b)

FIGURE 13.3   At the same temperature, the equilibrium pressure of CO_2 (measured with a closed-end manometer) is the same in **(a)** and **(b)**, independent of how much solid $CaCO_3$ and CaO is present.

EXAMPLE 13.5

Write the equilibrium equation for each of the following reactions
(a) $CO_2(g) + C(s) \rightleftarrows 2\,CO(g)$
(b) $Hg(l) + Hg^{2+}(aq) \rightleftarrows Hg_2^{2+}(aq)$

SOLUTION   **(a)** Because carbon is a pure solid, its molar concentration is a constant that is incorporated into the equilibrium constant K_c. Therefore,

$$K_c = \frac{[CO]^2}{[CO_2]}$$

When liquid mercury is in contact with a solution containing Hg_2^{2+} and Hg^{2+}, the concentration ratio $[Hg_2^{2+}]/[Hg^{2+}]$ has a constant value.

Alternatively, because CO and CO_2 are gases, the equilibrium equation can be written using partial pressures:

$$K_P = \frac{(P_{CO})^2}{P_{CO_2}}$$

The relationship between K_P and K_c is $K_P = K_c(RT)^{\Delta n}$, where $\Delta n = 2 - 1 = 1$.
(b) The concentrations of mercury(I) and mercury(II) ions appear in the equilibrium equation, but the concentration of mercury metal is omitted because, as a pure liquid, its concentration is a constant. Therefore,

$$K_c = \frac{[Hg_2^{2+}]}{[Hg^{2+}]}$$

┌ **PROBLEM 13.6** Write the equilibrium constant expressions for K_c and K_p for each of the following reactions.
(a) $2\ Fe(s) + 3\ H_2O(g) \rightleftarrows Fe_2O_3(s) + 3\ H_2(g)$
(b) $2\ H_2O(l) \rightleftarrows 2\ H_2(g) + O_2(g)$

13.5 ➤USING THE EQUILIBRIUM CONSTANT

Knowing the value of the equilibrium constant for a chemical reaction is important in many ways. For example, it lets us judge the extent of the reaction, predict the direction of the reaction, and calculate equilibrium concentrations from any initial concentrations.

Judging the Extent of Reaction

The numerical value of the equilibrium constant for a reaction indicates the extent to which reactants are converted to products; that is, it measures how far the reaction proceeds before the equilibrium state is reached. Consider, for example, the reaction of H_2 with O_2, which has a very large equilibrium constant ($K_c = 2.4 \times 10^{47}$ at 500 K):

$$2\ H_2(g) + O_2(g) \rightleftarrows 2\ H_2O(g)$$

$$K_c = \frac{[H_2O]^2}{[H_2]^2[O_2]} = 2.4 \times 10^{47} \quad \text{at 500 K}$$

Because reaction products appear in the numerator of the equilibrium constant expression and reactants are in the denominator, a very large value of K_c means that the equilibrium ratio of products to reactants is very large. In other words, the reaction proceeds almost all the way to completion. For example, if stoichiometric amounts of H_2 and O_2 are allowed to react and $[H_2O] = 1.0$ M at equilibrium, then the concentrations of H_2 and O_2 that remain at equilibrium are negligibly small ($[H_2] = 2.0 \times 10^{-16}$ M and $[O_2] = 1.0 \times 10^{-16}$ M). (Try substituting these concentrations into the equilibrium equation to show that they satisfy the equation.)

By contrast, if a reaction has a very small value of K_c, the equilibrium ratio of products to reactants is very small and the reaction proceeds hardly at all before equilibrium is reached. For example, the reverse of the reaction of H_2 with O_2 gives the same equilibrium mixture as obtained from the forward reaction ($[H_2] = 2.0 \times 10^{-16}$ M, $[O_2] = 1.0 \times 10^{-16}$ M, $[H_2O] = 1.0$ M). The reverse reaction does not occur to any appreciable extent, however, because its equilibrium constant is so small: $K_c' = 1/K_c = 1/(2.4 \times 10^{47}) = 4.1 \times 10^{-48}$.

$$2 H_2O(g) \rightleftarrows 2 H_2(g) + O_2(g)$$

$$K_c' = \frac{[H_2]^2[O_2]}{[H_2O]^2} = 4.1 \times 10^{-48} \quad \text{at 500 K}$$

If a reaction has a value of K_c that is neither large nor small, appreciable concentrations of both reactants and products are present in the equilibrium mixture. As a rule of thumb, K_c is neither large nor small if it has a value in the range of 10^3 to 10^{-3}. Take the reaction of hydrogen with iodine, which has $K_c = 57.0$ at 700 K:

$$H_2(g) + I_2(g) \rightleftarrows 2 HI(g)$$

$$K_c = \frac{[HI]^2}{[H_2][I_2]} = 57.0 \quad \text{at 700 K}$$

If the equilibrium H_2 and I_2 concentrations are 0.010 M, then the concentration of HI at equilibrium is 0.075 M:

$$[HI]^2 = K_c[H_2][I_2]$$

$$[HI] = \sqrt{K_c[H_2][I_2]} = \sqrt{(57.0)(0.010)(0.010)} = 0.075 \text{ M}$$

Thus, the concentrations of both reactants and products—0.010 M and 0.075 M—are appreciable.

The gas-phase decomposition of N_2O_4 to NO_2 is another example of a reaction with a value of K_c that is neither large nor small: $K_c = 4.64 \times 10^{-3}$ at 25°C. Accordingly, equilibrium mixtures contain appreciable concentrations of both N_2O_4 and NO_2, as shown previously in Table 13.1.

We can make the following generalizations concerning the composition of equilibrium mixtures:

1. If $K_c > 10^3$, products predominate over reactants. If K_c is very large, the reaction proceeds almost all the way to completion.
2. If $K_c < 10^{-3}$, reactants predominate over products. If K_c is very small, the reaction proceeds hardly at all toward completion.
3. If K_c is in the range 10^{-3} to 10^3, appreciable concentrations of both reactants and products are present.

These points are illustrated in Figure 13.4.

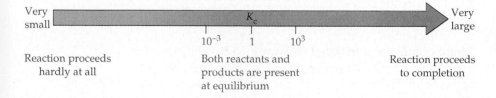

FIGURE 13.4 Judging the extent of a reaction. The larger the value of the equilibrium constant K_c, the farther the reaction proceeds to the right before reaching the equilibrium state.

┌ **PROBLEM 13.7** The value of K_c for the dissociation reaction $H_2(g) \rightleftarrows 2\,H(g)$ is 1.2×10^{-42} at 500 K. Does the equilibrium mixture contain mainly H_2 molecules or H atoms? ┘

Predicting the Direction of Reaction

Consider again the gaseous reaction of hydrogen with iodine:

$$H_2(g) + I_2(g) \rightleftarrows 2\,HI(g) \qquad K_c = 57.0 \text{ at } 700 \text{ K}$$

Suppose that we have a mixture of $H_2(g)$, $I_2(g)$, and $HI(g)$ at 700 K and that the concentrations are $[H_2]_t = 0.10$ M, $[I_2]_t = 0.20$ M, and $[HI]_t = 0.40$ M. (The subscript t on the concentration symbols means that the concentrations were measured at some arbitrary time t, not necessarily at equilibrium.) If we substitute these concentrations into the equilibrium constant expression, we obtain a value called the **reaction quotient Q_c**:

$$\textit{Reaction quotient} \qquad Q_c = \frac{[HI]_t^2}{[H_2]_t[I_2]_t} = \frac{(0.40)^2}{(0.10)(0.20)} = 8.0$$

The reaction quotient Q_c is defined in the same way as the equilibrium constant K_c except that the concentrations in the equilibrium constant expression are not necessarily equilibrium values.

For the case at hand, the numerical value of Q_c (8.0) is not equal to K_c (57.0), which means that the mixture of $H_2(g)$, $I_2(g)$, and $HI(g)$ is not at equilibrium—hardly surprising since the concentrations were chosen at random. As time passes, though, reaction will occur, changing the concentrations and changing the value of Q_c in the direction of K_c. After a sufficiently long time, an equilibrium state is reached, and $Q_c = K_c$.

The reaction quotient Q_c lets us predict the direction of reaction by comparing the values of Q_c and K_c. When Q_c is less than K_c, movement toward equilibrium increases Q_c by converting reactants to products, meaning that the reaction proceeds from left to right. When Q_c is greater than K_c, movement toward equilibrium decreases Q_c by converting products to reactants, meaning that the reaction proceeds from right to left.

To predict the direction of any chemical reaction, we simply calculate the value of Q_c by substituting the concentrations of reactants and products into the equilibrium constant expression. Then we compare the values of Q_c and K_c.

1. If $Q_c < K_c$, the reaction goes from left to right.
2. If $Q_c > K_c$, the reaction goes from right to left.
3. If $Q_c = K_c$, the reaction mixture is already at equilibrium, and no net reaction occurs.

These points are illustrated in Figure 13.5.

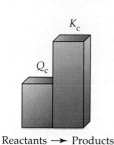
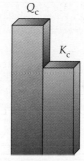

Reactants → Products

Reactants and products are at equilibrium

Reactants ← Products

FIGURE 13.5 Predicting the direction of reaction. The direction of reaction depends on the relative values of Q_c and K_c.

EXAMPLE 13.6

A mixture of 4.2 mol of N_2, 2.0 mol of H_2, and 10.0 mol of NH_3 is introduced into a 20 L reaction vessel at 500 K. At this temperature, the equilibrium constant K_c for the reaction $N_2(g) + 3 H_2(g) \rightleftharpoons 2 NH_3(g)$ is 1.7×10^2. Is the reaction mixture at equilibrium? If not, what is the direction of the reaction?

SOLUTION   The initial concentrations are calculated by dividing the number of moles of each substance by the volume; $[N_2] = (4.2 \text{ mol})/(20 \text{ L}) = 0.21$ M. Similarly, $[H_2] = 0.10$ M and $[NH_3] = 0.50$ M. The value of the reaction quotient is obtained by substituting these concentrations into the equilibrium constant expression:

$$Q_c = \frac{[NH_3]_t^2}{[N_2]_t[H_2]_t^3} = \frac{(0.50)^2}{(0.21)(0.10)^3} = 1.2 \times 10^3$$

Since Q_c is not equal to K_c, the reaction mixture is not at equilibrium. Because Q_c is greater than K_c, the reaction will proceed from right to left, decreasing the concentration of NH_3 and increasing the concentrations of N_2 and H_2 until $Q_c = K_c = 1.7 \times 10^2$.

PROBLEM 13.8 The equilibrium constant K_c for the reaction $2 NO(g) + O_2(g) \rightleftharpoons 2 NO_2(g)$ is 6.9×10^5 at 500 K. A 5.0 L reaction vessel at this temperature was filled with 0.060 mol of NO, 1.0 mol of O_2, and 0.80 mol of NO_2.

(a) Is the reaction mixture at equilibrium? If not, in which direction does the reaction proceed?

(b) What is the direction of reaction if the initial amounts are 5.0×10^{-3} mol of NO, 0.20 mol of O_2, and 4.0 mol of NO_2?

Calculating Equilibrium Concentrations

If the equilibrium constant and all the equilibrium concentrations but one are known, then the unknown concentration can be calculated directly from the equilibrium equation, as shown in Example 13.7. Often, however, we know the initial concentrations but don't know any of the equilibrium concentrations. Examples 13.8 and 13.9 illustrate how to use the equilibrium constant to calculate equilibrium concentrations from initial concentrations.

The series of steps followed in all calculations of equilibrium concentrations from initial concentrations is summarized in Figure 13.6.

Step 1. Write the balanced equation for the reaction.

Step 2. Under the balanced equation, make a table that lists for each substance involved in the reaction
    (a) the initial concentration
    (b) the change in concentration on going to equilibrium
    (c) the equilibrium concentration
In constructing the table, define *x* as the concentration (mol/L) of one of the substances that reacts on going to equilibrium.

Step 3. Substitute the equilibrium concentrations into the equilibrium equation for the reaction and solve for *x*. If you must solve a quadratic equation, choose the mathematical solution that makes chemical sense.

Step 4. Calculate the equilibrium concentrations from the calculated value of *x*.

FIGURE 13.6 Steps to follow in calculating equilibrium concentrations from initial concentrations.

Step 5. Check your results by substituting them into the equilibrium equation.

EXAMPLE 13.7

An equilibrium mixture of gaseous O_2, NO, and NO_2 at 500 K contains 1.0×10^{-3} M O_2 and 5.0×10^{-2} M NO_2. At this temperature, the equilibrium constant K_c for the reaction $2\,NO(g) + O_2(g) \rightleftarrows 2\,NO_2(g)$ is 6.9×10^5. What is the concentration of NO?

SOLUTION In this problem, K_c and all of the equilibrium concentrations but one are known. We're asked to calculate the unknown equilibrium concentration. First, write the equilibrium equation for the reaction:

$$K_c = \frac{[NO_2]^2}{[NO]^2[O_2]}$$

Into this expression, substitute the known values of K_c, $[O_2]$, and $[NO_2]$:

$$6.9 \times 10^5 = \frac{(5.0 \times 10^{-2})^2}{[NO]^2(1.0 \times 10^{-3})}$$

Solving for [NO] gives the equilibrium concentration of NO:[2]

$$[NO] = \sqrt{\frac{(5.0 \times 10^{-2})^2}{(1.0 \times 10^{-3})(6.9 \times 10^5)}} = \sqrt{3.6 \times 10^{-6}} = \pm 1.9 \times 10^{-3} \text{ M}$$

Of the two roots, choose the positive one ([NO] = 1.9×10^{-3} M) because the concentration of a chemical substance is always a positive quantity.

[2]When you press the $\sqrt{x}$ button on your calculator, you get a positive number. Remember, though, that the square root of a positive number can be positive or negative.

To be sure that we haven't made any errors, it's a good idea to check the result by substituting it into the equilibrium equation:

$$K_c = 6.9 \times 10^5 = \frac{[NO_2]^2}{[NO]^2[O_2]} = \frac{(5.0 \times 10^{-2})^2}{(1.9 \times 10^{-3})^2(1.0 \times 10^{-3})} = 6.9 \times 10^5$$

EXAMPLE 13.8

The equilibrium constant K_c for the reaction of H_2 with I_2 is 57.0 at 700 K:

$$H_2(g) + I_2(g) \rightleftharpoons 2\ HI(g) \qquad K_c = 57.0 \text{ at } 700\text{ K}$$

If 1.00 mol of H_2 is allowed to react with 1.00 mol of I_2 in a 10.0 L reaction vessel at 700 K, what are the concentrations of H_2, I_2, and HI at equilibrium? What is the composition of the equilibrium mixture in moles?

SOLUTION   This problem involves calculation of equilibrium concentrations from *initial* concentrations. Therefore, use the method outlined in Figure 13.6.

Step 1. The balanced equation is given: $H_2(g) + I_2(g) \rightleftharpoons 2\ HI(g)$

Step 2. The initial concentrations are $[H_2] = [I_2] = (1.00 \text{ mol})/(10.0 \text{ L}) = 0.100$ M. For convenience, define an unknown, x, as the concentration (mol/L) of H_2 that reacts on going to the equilibrium state. According to the balanced equation for the reaction, x mol/L of H_2 reacts with x mol/L of I_2 to give $2x$ mol/L of HI. This reduces the initial concentrations of H_2 and I_2 from 0.100 mol/L to $(0.100 - x)$ mol/L at equilibrium. Let's summarize these results in a table under the balanced equation:

	$H_2(g)$	+	$I_2(g)$	$\rightleftharpoons$	$2\ HI(g)$
Initial concentration (M)	0.100		0.100		0
Change (M)	$-x$		$-x$		$+2x$
Equilibrium concentration (M)	$(0.100 - x)$		$(0.100 - x)$		$2x$

Step 3. Substitute the equilibrium concentrations into the equilibrium equation for the reaction:

$$K_c = 57.0 = \frac{[HI]^2}{[H_2][I_2]} = \frac{(2x)^2}{(0.100 - x)(0.100 - x)} = \left(\frac{2x}{0.100 - x}\right)^2$$

Because the right side of this equation is a perfect square, we can take the square root of both sides:

$$\sqrt{57.0} = \pm 7.55 = \frac{2x}{0.100 - x}$$

Solving for x, we obtain two solutions. The equation with the positive square root of 57.0 gives

$$+7.55(0.100 - x) = 2x$$

$$0.755 = 2x + 7.55x$$

$$x = \frac{0.755}{9.55} = 0.0791 \text{ M}$$

The equation with the negative square root of 57.0 gives

$$-7.55(0.100 - x) = 2x$$

$$-0.755 = 2x - 7.55x$$

$$x = \frac{-0.755}{-5.55} = 0.136 \text{ M}$$

Because the initial concentrations of H_2 and I_2 are 0.100 M, x can't exceed 0.100 M. Therefore, discard $x = 0.136$ M as chemically unreasonable and choose the first solution, $x = 0.0791$ M.

Step 4. Calculate the equilibrium concentrations from the calculated value of x:

$$[H_2] = [I_2] = 0.100 - x = 0.100 - 0.0791 = 0.021 \text{ M}$$

$$[HI] = 2x = (2)(0.0791) = 0.158 \text{ M}$$

Step 5. Check the results by substituting them into the equilibrium equation:

$$K_c = 57.0 = \frac{[HI]^2}{[H_2][I_2]} = \frac{(0.158)^2}{(0.021)(0.021)} = 57$$

The number of moles of each substance in the equilibrium mixture can be obtained by multiplying each concentration by the volume of the reaction vessel:

$$\text{Moles of } H_2 = \text{Moles of } I_2 = (0.021 \text{ mol/L})(10.0 \text{ L}) = 0.21 \text{ mol}$$

$$\text{Moles of HI} = (0.158 \text{ mol/L})(10.0 \text{ L}) = 1.58 \text{ mol}$$

EXAMPLE 13.9

Calculate the equilibrium concentrations of H_2, I_2, and HI at 700 K if the initial concentrations are $[H_2] = 0.100$ M and $[I_2] = 0.200$ M. K_c for the reaction $H_2(g) + I_2(g) \rightleftarrows 2 \text{ HI}(g)$ is 57.0 at 700 K.

SOLUTION   This problem is similar to Example 13.8 except that the initial concentrations of H_2 and I_2 are unequal.

Step 1. Again, the balanced equation is $H_2(g) + I_2(g) \rightleftarrows 2 \text{ HI}(g)$

Step 2. Set up a table of concentrations under the balanced equation:

	$H_2(g)$	+	$I_2(g)$	$\rightleftarrows$	$2 \text{ HI}(g)$
Initial concentration (M)	0.100		0.200		0
Change (M)	$-x$		$-x$		$+2x$
Equilibrium concentration (M)	$(0.100 - x)$		$(0.200 - x)$		$2x$

Step 3. Substitute the equilibrium concentrations into the equilibrium equation:

$$K_c = 57.0 = \frac{[HI]^2}{[H_2][I_2]} = \frac{(2x)^2}{(0.100 - x)(0.200 - x)}$$

Because the right side of this equation is not a perfect square, we must put the equation into the standard quadratic form, $ax^2 + bx + c = 0$, and then solve for x using the quadratic formula (Appendix A.4):

$$x = \frac{-b \pm \sqrt{b^2 - 4\,ac}}{2a}$$

Rearranging the equilibrium equation gives

$$(57.0)(0.0200 - 0.300x + x^2) = 4x^2$$

or

$$53.0x^2 - 17.1x + 1.14 = 0$$

Substituting the values of a, b, and c into the quadratic formula gives two solutions:

$$x = \frac{17.1 \pm \sqrt{(17.1)^2 - 4(53.0)(1.14)}}{2(53.0)} = \frac{17.1 \pm 7.1}{106} = 0.228 \text{ and } 0.0943$$

Discard the solution that uses the positive square root ($x = 0.228$) because the H_2 concentration can't change by more than its initial value (0.100 M). Therefore, choose the solution that uses the negative square root ($x = 0.0943$).

Step 4. Calculate the equilibrium concentrations from the calculated value of x:

$$[H_2] = 0.100 - x = 0.100 - 0.0943 = 0.006 \text{ M}$$

$$[I_2] = 0.200 - x = 0.200 - 0.0943 = 0.106 \text{ M}$$

$$[HI] = 2x = (2)(0.0943) = 0.189 \text{ M}$$

Step 5. Check the results by substituting them into the equilibrium equation:

$$K_c = 57.0 = \frac{[HI]^2}{[H_2][I_2]} = \frac{(0.189)^2}{(0.006)(0.106)} = 6 \times 10^1$$

The calculated value of K_c (6×10^1), which must be rounded to one significant figure, agrees with the value given in the problem (57.0).

⌐ **PROBLEM 13.9** In Problem 13.7, we found that an equilibrium mixture of H_2 molecules and H atoms at 500 K contains mainly H_2 molecules because the equilibrium constant for the dissociation reaction $H_2(g) \rightleftarrows 2\,H(g)$ is very small ($K_c = 1.2 \times 10^{-42}$).

(a) What is the molar concentration of H atoms if $[H_2] = 0.10$ M?
(b) How many H atoms and H_2 molecules are present in 1.0 L of 0.10 M H_2 at 500 K?

⌐ **PROBLEM 13.10** The H_2/CO ratio in mixtures of carbon monoxide and hydrogen (synthesis gas) is increased by the gas shift reaction $CO(g) + H_2O(g) \rightleftarrows CO_2(g) + H_2(g)$, which has an equilibrium constant $K_c = 4.24$ at 800 K. Calculate the equilibrium concentrations of CO_2, H_2, CO, and H_2O at 800 K if only CO and H_2O are present initially at concentrations of 0.150 M.

┌ **PROBLEM 13.11** Calculate the equilibrium concentrations of N_2O_4 and NO_2 at 25°C in a vessel that contains an initial N_2O_4 concentration of 0.0500 M. The equilibrium constant K_c for the reaction $N_2O_4(g) \rightleftarrows 2\,NO_2(g)$ is 4.64×10^{-3} at 25°C.

13.6 ►FACTORS THAT ALTER THE COMPOSITION OF AN EQUILIBRIUM MIXTURE

Liquid ammonia is used as a nitrogen fertilizer.

One of the principal goals of chemical synthesis is to effect a maximum conversion of reactants to products with a minimum expenditure of energy. This objective is achieved easily if the reaction goes nearly to completion at mild temperature and pressure. But if the reaction gives an equilibrium mixture of reactants and products, then the experimental conditions must be adjusted. For example, in the Haber process for the synthesis of ammonia from N_2 and H_2 (Figure 13.7), the choice of experimental conditions is of real economic importance. Annual U.S. production of ammonia is about 17 million tons, more than three-fourths of which is used in the fertilizer industry.

Several factors can be exploited to alter the composition of an equilibrium mixture:

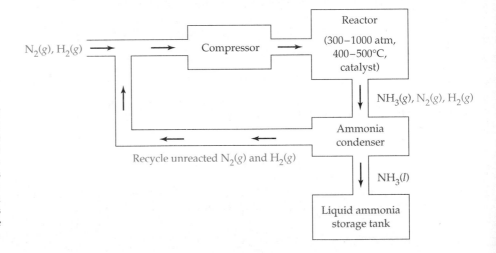

FIGURE 13.7 Representation of the Haber process for the industrial production of ammonia. A mixture of gaseous N_2 and H_2 at 300–1000 atm pressure is passed over a catalyst at 400–500°C, and ammonia is produced by the reaction $N_2(g) + 3\,H_2(g) \rightleftarrows 2\,NH_3(g)$. The NH_3 in the gaseous mixture of reactants and products is liquefied, and the unreacted N_2 and H_2 are recycled.

1. The concentration of reactants or products can be changed.
2. The pressure and volume can be changed.
3. The temperature can be changed.

(A possible fourth factor, addition of a catalyst, increases only the rate at which equilibrium is reached; it doesn't affect the equilibrium concentrations.)

The qualitative effect of the listed changes on the composition of an equilibrium mixture can be predicted using a principle first described by the French chemist Henri-Louis Le Châtelier.

| LE CHÂTELIER'S PRINCIPLE | If a stress is applied to a reaction mixture at equilibrium, reaction occurs in the direction that relieves the stress. |

The word "stress" in this context means a change in concentration, pressure, volume, or temperature that disturbs the original equilibrium. Reaction then occurs to change the composition of the mixture until a new state of equilibrium is reached. The direction that the reaction takes (reactants to products, or vice versa) is the one that reduces the stress. In the next three sections, we'll look at the different kinds of stress that can change the composition of an equilibrium mixture.

13.7 ►CHANGES IN CONCENTRATIONS

Let's consider the equilibrium that occurs in the Haber process for synthesis of ammonia:

$$N_2(g) + 3 H_2(g) \rightleftharpoons 2 NH_3(g) \qquad K_c = 0.291 \text{ at } 700 \text{ K}$$

Suppose that we have an equilibrium mixture of 0.50 M N_2, 3.00 M H_2, and 1.98 M NH_3 at 700 K and that we disturb the equilibrium by increasing the N_2 concentration to 1.50 M. Le Châtelier's principle tells us that reaction will occur to relieve the stress of the increased concentration of N_2. Stress will be relieved if reaction occurs from left to right, thus converting some of the N_2 to NH_3. Of course, as the N_2 concentration decreases, the H_2 concentration must also decrease and the NH_3 concentration must increase in accord with the stoichiometry of the balanced equation. These changes are illustrated in Figure 13.8.

FIGURE 13.8 Changes in concentrations when N_2 is added to an equilibrium mixture of N_2, H_2, and NH_3. Net conversion of N_2 and H_2 to NH_3 occurs until a new equilibrium is established.

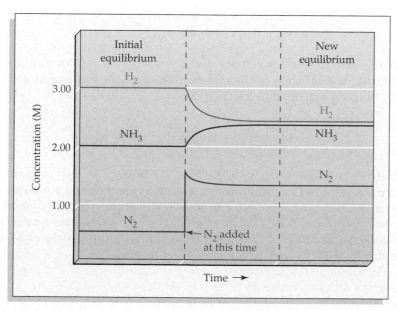

In general, when an equilibrium is disturbed by the addition or removal of any reactant or product, Le Châtelier's principle predicts that

1. The concentration stress of an *added* reactant or product is relieved by reaction in the direction that *consumes* the added substance.
2. The concentration stress of a *removed* reactant or product is relieved by reaction in the direction that *replenishes* the removed substance.

Application of these rules to the equilibrium $N_2(g) + 3\ H_2(g) \rightleftarrows 2\ NH_3(g)$ indicates that the yield of ammonia is increased by an increase in the N_2 or H_2 concentration or by a decrease in the NH_3 concentration (Table 13.2). In the industrial production of ammonia, the concentration of gaseous NH_3 is decreased by liquefying the ammonia (bp $-33\ °C$) as it's formed, and so more ammonia is produced.

TABLE 13.2	Effect of Concentration Changes on the Equilibrium, $N_2(g) + 3\ H_2(g) \rightleftarrows 2\ NH_3(g)$
Concentration Stress	**Direction of Net Reaction**
Increase in $[N_2]$ or $[H_2]$	Left to right
Increase in $[NH_3]$	Right to left
Decrease in $[N_2]$ or $[H_2]$	Right to left
Decrease in $[NH_3]$	Left to Right

Le Châtelier's principle is a handy rule for predicting changes in the composition of an equilibrium mixture, but it doesn't explain *why* those changes occur. To see why Le Châtelier's principle "works," let's look again at the reaction quotient Q_c. For the initial equilibrium mixture of 0.50 M N_2, 3.00 M H_2, and 1.98 M NH_3, Q_c is equal to the equilibrium constant K_c (0.291) because the system is at equilibrium:

$$Q_c = \frac{[NH_3]_t^2}{[N_2]_t[H_2]_t^3} = \frac{(1.98)^2}{(0.50)(3.00)^3} = 0.29 = K_c$$

When we disturb the equilibrium by increasing the N_2 concentration to 1.50 M, the denominator of the equilibrium constant expression increases and Q_c decreases to a value less than K_c:

$$Q_c = \frac{[NH_3]_t^2}{[N_2]_t[H_2]_t^3} = \frac{(1.98)^2}{(1.50)(3.00)^3} = 0.0968 < K_c$$

For the system to move to a new state of equilibrium, Q_c must increase; that is, the numerator of the equilibrium constant expression must increase and the denominator must decrease. This implies net conversion of N_2 and H_2 to NH_3, just as predicted by Le Châtelier's principle. When the new equilibrium is established (Figure 13.8), the concentrations are 1.31 M N_2, 2.43 M H_2, and 2.36 M NH_3, and Q_c is again equal to K_c:

$$Q_c = \frac{[NH_3]_t^2}{[N_2]_t[H_2]_t^3} = \frac{(2.36)^2}{(1.31)(2.43)^3} = 0.296 = K_c$$

As another example of how a change in concentration affects an equilibrium, let's consider the reaction in aqueous solution of iron(III) and thiocyanate (SCN^-) ions to give an equilibrium mixture that contains the Fe—N bonded red complex ion $FeNCS^{2+}$:

$$Fe^{3+}(aq) + SCN^-(aq) \rightleftarrows FeNCS^{2+}(aq)$$
$$\text{pale yellow} \qquad \text{colorless} \qquad \text{red}$$

Shifts in the position of this equilibrium can be detected by observing how the color of the solution changes when we add various reagents (Figure 13.9). If we add aqueous $FeCl_3$, the red color gets darker, as predicted by Le Châtelier's principle. The concentration stress of added Fe^{3+} is relieved by reaction from left to right, which consumes some of the Fe^{3+} and increases the concentration of $FeNCS^{2+}$. (Note that the Cl^- ions are not involved in the reaction.) Similarly, if we add aqueous KSCN, the stress of added SCN^- shifts the equilibrium from left to right, and again the red color gets darker.

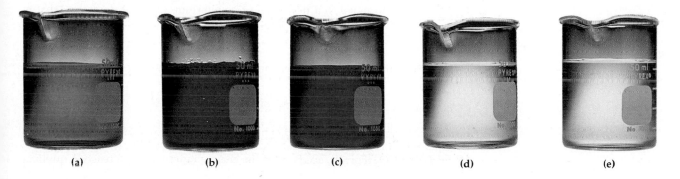

(a)	(b)	(c)	(d)	(e)

FIGURE 13.9 Color changes produced by adding various reagents to an equilibrium mixture of Fe^{3+} (pale yellow), SCN^- (colorless), and $FeNCS^{2+}$ (red): **(a)** The original solution. **(b)** After adding $FeCl_3$ to the original solution, the red color is darker because of an increase in $[FeNCS^{2+}]$. **(c)** After adding KSCN to the original solution, the red color again deepens. **(d)** After adding $H_2C_2O_4$ to the original solution, the red color disappears because of a decrease in $[FeNCS^{2+}]$; the yellow color is due to $Fe(C_2O_4)_3^{3-}$. **(e)** After adding $HgCl_2$ to the original solution, the red color again vanishes.

The equilibrium can be shifted in the opposite direction by adding reagents that remove Fe^{3+} or SCN^- ions. For example, oxalic acid ($H_2C_2O_4$), a poisonous substance present in plants such as rhubarb, reacts with Fe^{3+} to form the stable complex ion $Fe(C_2O_4)_3^{3-}$, thus decreasing the concentration of free $Fe^{3+}(aq)$. In accord with Le Châtelier's principle, the concentration stress of removed Fe^{3+} is relieved by dissociation of $FeNCS^{2+}$ to replenish the Fe^{3+} ions. Because the concentration of $FeNCS^{2+}$ decreases, the red color disappears.

Addition of aqueous $HgCl_2$ also eliminates the red color because Hg^{2+} reacts with SCN^- ions to form the stable Hg—S bonded complex ion $Hg(SCN)_4^{2-}$. Removal of free $SCN^-(aq)$ shifts the equilibrium $Fe^{3+}(aq) + SCN^-(aq) \rightleftarrows FeNCS^{2+}(aq)$ from right to left to replenish the SCN^- ions.

EXAMPLE 13.10

The reaction of iron(III) oxide with carbon monoxide occurs in a blast furnace when iron ore is reduced to iron metal:

$$Fe_2O_3(s) + 3\ CO(g) \rightleftarrows 2\ Fe(l) + 3\ CO_2(g)$$

Use Le Châtelier's principle to predict the direction of reaction when an equilibrium mixture is disturbed by **(a)** adding Fe_2O_3; **(b)** removing CO_2; **(c)** removing CO. For **(c)**, account for the change in terms of the reaction quotient Q_c.

SOLUTION   **(a)** Because Fe_2O_3 is a solid, its "concentration" doesn't change when more Fe_2O_3 is added. Therefore, there is no concentration stress, and the original equilibrium is undisturbed.
(b) Le Châtelier's principle predicts that the concentration stress of removed CO_2 will be relieved by reaction from left to right to replenish the CO_2.
(c) Le Châtelier's principle predicts that the concentration stress of removed CO will be relieved by reaction from right to left to replenish the CO.
   The reaction quotient is

$$Q_c = \frac{[CO_2]_t^3}{[CO]_t^3}$$

When the equilibrium is disturbed by reducing [CO], Q_c increases, so that $Q_c > K_c$. For the system to move to a new state of equilibrium, Q_c must decrease—that is, $[CO_2]$ must decrease and [CO] must increase. Therefore the reaction goes from right to left.

┌ **PROBLEM 13.12**   Consider the equilibrium for the gas shift reaction:

$$CO(g) + H_2O(g) \rightleftarrows CO_2(g) + H_2(g)$$

Use Le Châtelier's principle to predict how the concentration of H_2 will change when the equilibrium is disturbed by: **(a)** adding CO; **(b)** adding CO_2; **(c)** removing H_2O; **(d)** removing CO_2. For **(d)**, account for the change in terms of the reaction quotient Q_c.

13.8 ►CHANGES IN PRESSURE AND VOLUME

If the number of moles of gaseous reactants in the balanced equation is different from the number of moles of gaseous products, then the composition of the equilibrium mixture changes when the pressure is changed by changing the volume. To illustrate, let's return to the Haber synthesis of ammonia. The balanced equation for the reaction has 4 mol of gas on the left side of the equation and 2 mol on the right side:

$$N_2(g) + 3\ H_2(g) \rightleftarrows 2\ NH_3(g) \qquad K_c = 0.291 \text{ at } 700 \text{ K}$$

   What happens to the composition of the equilibrium mixture if we double the pressure by decreasing the volume? (Recall from Section 9.2 that the pressure of an ideal gas is inversely proportional to the volume at constant temperature and constant number of moles of gas; $P = nRT/V$.) According to Le Châtelier's principle, reaction will occur in the direction that relieves the stress of the increased pressure. At constant temperature and constant volume, the pressure can be reduced only by reducing the number of moles of gas. Therefore, we predict that the reaction will proceed from left to right because the forward reaction converts 4 mol of gaseous reactants to 2 mol of gaseous products.

In general, Le Châtelier's principle predicts that

1. An *increase* in pressure by reducing the volume will bring about net reaction in the direction that *decreases* the number of moles of gas.
2. A *decrease* in pressure by enlarging the volume will bring about net reaction in the direction that *increases* the number of moles of gas.

To see why Le Châtelier's principle works for pressure (volume) changes, let's look again at the reaction quotient for the equilibrium mixture of 0.50 M N_2, 3.00 M H_2, and 1.98 M NH_3:

$$Q_c = \frac{[NH_3]_t^2}{[N_2]_t[H_2]_t^3} = \frac{(1.98)^2}{(0.50)(3.00)^3} = 0.29 = K_c$$

When we disturb the equilibrium by reducing the volume by a factor of 2, we not only double the total pressure, we also double the molar concentration of each reactant and product (because molarity $= n/V$). Because the balanced equation has more moles of gaseous reactants than gaseous products, the increase in the denominator of the equilibrium constant expression is greater than the increase in the numerator, and the new value of Q_c is less than the equilibrium constant K_c:

$$Q_c = \frac{[NH_3]_t^2}{[N_2]_t[H_2]_t^3} = \frac{(3.96)^2}{(1.00)(6.00)^3} = 0.0726 < K_c$$

For the system to move to a new state of equilibrium, Q_c must increase, which means that reaction must go from left to right, as predicted by Le Châtelier's principle. In practice, the yield of ammonia in the Haber process is increased by running the reaction at high pressure, typically 300–1000 atm.

Of course, the composition of an equilibrium mixture is unaffected by a change in pressure if the reaction involves no change in the number of moles of gas. The reaction of hydrogen with iodine, for example, has 2 mol of gas on both sides of the balanced equation:

$$H_2(g) + I_2(g) \rightleftharpoons 2\,HI(g)$$

If we double the pressure by halving the volume, the numerator and denominator of the reaction quotient change by the same factor, and Q_c remains unchanged.

$$Q_c = \frac{[HI]_t^2}{[H_2]_t[I_2]_t}$$

In applying Le Châtelier's principle to a heterogeneous equilibrium, the effect of pressure changes on solids and liquids can be ignored because the volume (and concentration) of a solid or a liquid is nearly independent of pressure. Consider, for example, the high-temperature reaction of carbon with steam, the first step in converting coal to gaseous fuels:

$$C(s) + H_2O(g) \rightleftharpoons CO(g) + H_2(g)$$

Ignoring the carbon because it's a solid, we predict that a decrease in volume (increase in pressure) will shift the equilibrium from right to left because the reverse reaction decreases the amount of gas from 2 mol to 1 mol.

Throughout this section, we've been careful to limit the application of Le Châtelier's principle to pressure changes that result from a change *in volume*. What happens, though, if we keep the volume constant but increase the total pressure by adding a gas that is not involved in the reaction—say, an inert gas such as argon? In that case, the equilibrium remains undisturbed because adding an inert gas does not change the molar concentrations of reactants or products. Only if the added gas is a reactant or product does the reaction quotient change.

EXAMPLE 13.11

Does the number of moles of reaction products increase, decrease, or remain the same when each of the following equilibria is subjected to a decrease in pressure by increasing the volume?

(a) $PCl_5(g) \rightleftarrows PCl_3(g) + Cl_2(g)$
(b) $CaO(s) + CO_2(g) \rightleftarrows CaCO_3(s)$
(c) $3\ Fe(s) + 4\ H_2O(g) \rightleftarrows Fe_3O_4(s) + 4\ H_2(g)$

SOLUTION   **(a)** According to Le Châtelier's principle, the stress of a decrease in pressure is relieved by net reaction in the direction that increases the number of moles of gas. Since the forward reaction converts 1 mol of gas to 2 mol of gas, the reaction will go from left to right, thus increasing the number of moles of PCl_3 and Cl_2.
(b) Because there is 1 mol of gas on the left side of the balanced equation and none on the right side, the stress of a decrease in pressure is relieved by reaction from right to left. The number of moles of $CaCO_3$ therefore decreases.
(c) Because there are 4 mol of gas on both sides of the balanced equation, the composition of the equilibrium mixture is unaffected by a change in pressure. The number of moles of Fe_3O_4 and H_2 remains the same.

⌐ PROBLEM 13.13 Does the number of moles of reaction products increase, decrease, or remain the same when each of the following equilibria is subjected to an increase in pressure by decreasing the volume?

(a) $CO(g) + H_2O(g) \rightleftarrows CO_2(g) + H_2(g)$
(b) $2\ CO(g) \rightleftarrows C(s) + CO_2(g)$
(c) $N_2O_4(g) \rightleftarrows 2\ NO_2(g)$

13.9 ➤ CHANGES IN TEMPERATURE

When an equilibrium is disturbed by a change in concentration, pressure, or volume, the composition of the equilibrium mixture changes because the reaction quotient Q_c after the disturbance is no longer equal to the equilibrium constant K_c. Concentration, pressure, or volume changes, however, don't change the value of the equilibrium constant, so long as the temperature remains constant.

By contrast, a change in temperature nearly always changes the value of the equilibrium constant. For the Haber synthesis of ammonia, an exothermic reaction, the equilibrium constant K_c decreases by a factor of 10^{11} over the temperature range 300–1000 K (Figure 13.10).

$$N_2(g) + 3\ H_2(g) \rightleftarrows 2\ NH_3(g) + 92.2\ kJ \qquad (\Delta H° = -92.2\ kJ)$$

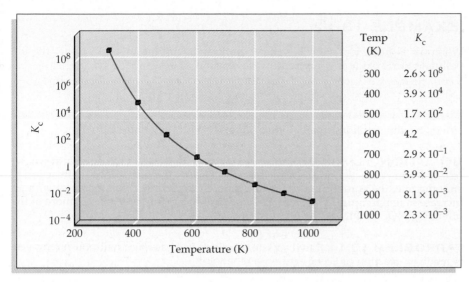

Temp (K)	K_c
300	2.6×10^8
400	3.9×10^4
500	1.7×10^2
600	4.2
700	2.9×10^{-1}
800	3.9×10^{-2}
900	8.1×10^{-3}
1000	2.3×10^{-3}

FIGURE 13.10 Temperature dependence of the equilibrium constant for the reaction $N_2(g)$ + $3 H_2(g) \rightleftarrows 2 NH_3(g)$. Note that K_c is plotted on a logarithmic scale and decreases by a factor of 10^{11} on raising the temperature from 300 K to 1000 K.

At low temperatures, the equilibrium mixture is rich in NH_3 because K_c is large. At high temperatures, the equilibrium shifts in the direction of N_2 and H_2.

In general, the temperature dependence of the equilibrium constant depends on the sign of $\Delta H°$ for the reaction:

1. The equilibrium constant for an exothermic reaction (negative $\Delta H°$) decreases as the temperature increases.

2. The equilibrium constant for an endothermic reaction (positive $\Delta H°$) increases as the temperature increases.

If you forget the way in which K_c depends on temperature, you can predict it using Le Châtelier's principle. Take the endothermic decomposition of N_2O_4, for example:

$$N_2O_4(g) + 57.2 \text{ kJ} \rightleftarrows 2 NO_2(g) \qquad (\Delta H° = +57.2 \text{ kJ})$$
$$\text{colorless} \qquad\qquad\qquad \text{brown}$$

Le Châtelier's principle says that if heat is added to an equilibrium mixture, thus increasing its temperature, net reaction occurs in the direction that relieves the stress of the added heat. For an endothermic reaction, such as the decomposition of N_2O_4, heat is absorbed by reaction in the forward direction, and therefore K_c increases with increasing temperature. Because N_2O_4 is colorless and NO_2 has a brown color, the effect of temperature on the N_2O_4–NO_2 equilibrium is readily apparent from the color of the gas (Figure 13.11). For an exothermic reaction, such as the Haber synthesis of NH_3, heat is absorbed by reaction in the reverse direction, and consequently K_c decreases with increasing temperature.

FIGURE 13.11 Sample tubes containing an equilibrium mixture of N_2O_4 and NO_2 immersed in ice water (left) and hot water (right). The darker brown color of the sample at the higher temperature indicates that the equilibrium $N_2O_4(g) \rightleftarrows NO_2(g)$ shifts from left to right (K_c increases) with increasing temperature.

EXAMPLE 13.12

When air is heated at very high temperatures in an automobile engine, the air pollutant nitric oxide is produced by the reaction

$$N_2(g) + O_2(g) \rightleftarrows 2\ NO(g) \qquad \Delta H° = 180.5\ kJ$$

How does the equilibrium amount of NO vary with the temperature?

SOLUTION   Le Châtelier's principle predicts that a stress of added heat will be relieved by net reaction in the direction that absorbs the heat. Since the reaction is endothermic, the equilibrium will shift from left to right (K_c will increase) with an increase in temperature. Therefore, the equilibrium mixture will contain more of the offending NO, the higher the temperature.

PROBLEM 13.14   Ethyl acetate, a solvent used as fingernail-polish remover, is made by reaction of acetic acid with ethanol:

$$\underset{\text{Acetic acid}}{CH_3COOH(l)} + \underset{\text{Ethanol}}{C_2H_5OH(l)} \rightleftarrows \underset{\text{Ethyl acetate}}{CH_3COOC_2H_5(l)} + H_2O(l) \qquad \Delta H° = -2.9\ kJ$$

Does the amount of ethyl acetate in an equilibrium mixture increase or decrease when the temperature is increased? How does K_c change when the temperature is decreased? Justify your answers using Le Châtelier's principle.

13.10 ▸ THE EFFECT OF A CATALYST

A catalyst increases the rate of a chemical reaction by making available a new, lower-energy pathway for conversion of reactants to products (Section 12.11). Because the forward and reverse reactions pass through the same transition state, a catalyst lowers the activation energy for the forward and reverse reactions by exactly the same amount (Figure 13.12). As a result, the rates of the forward and reverse reactions increase by the same factor.

FIGURE 13.12 Potential energy profiles for a reaction whose activation energy is lowered by the presence of a catalyst. The activation energy for the catalyzed pathway (red line) is lower than that for the uncatalyzed pathway (blue line) by an amount ΔE_a. The catalyst lowers the activation energy barrier for the forward and reverse reactions by exactly the same amount (ΔE_a). The catalyst therefore accelerates the forward and reverse reactions by the same factor, and the composition of the equilibrium mixture is unchanged.

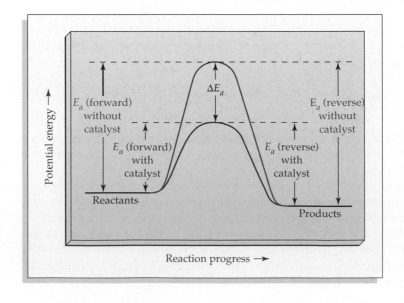

If a reaction mixture is at equilibrium in the absence of a catalyst (the forward and reverse rates are equal) it will still be at equilibrium after addition of a catalyst because the forward and reverse rates, though faster, remain equal. If a reaction mixture is not at equilibrium, a catalyst accelerates the rate at which equilibrium is reached, but it does not affect the composition of the equilibrium mixture. A catalyst can't alter the equilibrium constant or the equilibrium concentrations because it doesn't appear in the balanced chemical equation or in the equilibrium constant expression.

Even though it doesn't change the position of an equilibrium, a catalyst can nevertheless have an important influence on the choice of optimum conditions for a reaction. Consider again the Haber synthesis of ammonia. Because the reaction $N_2(g) + 3 H_2(g) \rightleftarrows 2 NH_3(g)$ is exothermic, its equilibrium constant decreases with increasing temperature, and optimum yields of NH_3 are obtained at low temperatures. At those low temperatures, however, the rate at which equilibrium is reached is too slow for the reaction to be practical. We thus have what appears to be a no-win situation: low temperatures give good yields but slow rates, whereas high temperatures give satisfactory rates but poor yields. The answer to the dilemma is to find a catalyst. In the early 1900s, the German chemist Fritz Haber discovered that a catalyst consisting of iron mixed with certain metal oxides causes the reaction to occur at a satisfactory rate at temperatures where the equilibrium concentration of NH_3 is reasonably favorable (Figure 13.13).[3] The yield of NH_3 can be further improved by running the reaction at high pressures. Typical reaction conditions for the industrial synthesis of ammonia are 400–500°C and 300–1000 atm.

FIGURE 13.13 An industrial catalyst for synthesis of ammonia.

[3]Fritz Haber (1868–1934) was awarded the 1918 Nobel Prize in chemistry for the synthesis of ammonia from its elements. It's interesting to note that Haber's work was of great help to German military efforts in World War I. Ammonia was the starting material for synthesis of nitric acid, which was used to make explosives such as nitroglycerin and trinitrotoluene (TNT).

⌐ **PROBLEM 13.15** A platinum catalyst is used in automobile catalytic converters to hasten the oxidation of carbon monoxide:

$$2\ CO(g) + O_2(g) \rightleftarrows 2\ CO_2(g) \qquad \Delta H° = -566\ kJ$$

Suppose that you have a reaction vessel containing an equilibrium mixture of $CO(g)$, $O_2(g)$, and $CO_2(g)$. Will the amount of CO increase, decrease, or remain the same when **(a)** a platinum catalyst is added; **(b)** the temperature is increased; **(c)** the pressure is increased by decreasing the volume; **(d)** the pressure is increased by adding argon gas; **(e)** the pressure is increased by adding O_2 gas? ◢

13.11 ➤THE LINK BETWEEN CHEMICAL EQUILIBRIUM AND CHEMICAL KINETICS

We've emphasized on numerous occasions that the equilibrium state is a dynamic one in which reactant and product concentrations remain constant, not because the reaction stops, but because the rates of the forward and reverse reactions are equal. To explore this idea further, let's consider the general, reversible reaction

$$A + B \rightleftarrows C + D$$

Let's assume that the forward and reverse reactions occur in a single bimolecular step; that is, they are elementary reactions (Section 12.7). We can then write the following rate laws:

$$\text{Rate of forward reaction} = k_f\,[A][B]$$

$$\text{Rate of reverse reaction} = k_r\,[C][D]$$

If we begin with a mixture that contains all reactants and no products, the initial rate of the reverse reaction is zero because $[C] = [D] = 0$. As A and B are converted to C and D by the forward reaction, the rate of the forward reaction decreases because [A] and [B] are getting smaller, while the rate of the reverse reaction increases because [C] and [D] are getting larger. Eventually, the rates of the forward and reverse reactions become equal, and thereafter the concentrations remain constant; the system is at chemical equilibrium.

Because the forward and reverse rates are equal at equilibrium, we can write

$$k_f\,[A][B] = k_r\,[C][D]$$

which can be rearranged to give

$$\frac{k_f}{k_r} = \frac{[C][D]}{[A][B]}$$

The right side of this equation is the equilibrium constant expression for the forward reaction, which is equal to the equilibrium constant K_c since the reaction mixture is at equilibrium. Therefore, the equilibrium constant is simply equal to the ratio of the rate constants for the forward and reverse reactions:[4]

$$K_c = \frac{k_f}{k_r} = \frac{[C][D]}{[A][B]}$$

The equation relating K_c to k_f and k_r provides a fundamental link between chemical equilibrium and chemical kinetics: The relative values of the rate constants for the forward and reverse reactions determine the composition of the equilibrium mixture. When k_f is much larger than k_r, K_c is very large and the reaction goes almost to completion. Such a reaction is irreversible because the reverse reaction is often too slow to be detected. When k_f and k_r have comparable values, K_c has a value near unity, and both reactants and products are present at equilibrium. This is the usual situation for a reversible reaction.

When a catalyst is added to a reaction mixture, both rate constants k_f and k_r increase because the reaction takes place by a different, lower-energy mechanism. Because k_f and k_r increase by the same factor, though, the ratio k_f/k_r is unaffected, and the value of the equilibrium constant $K_c = k_f/k_r$ remains unchanged. Thus, addition of a catalyst does not alter the composition of an equilibrium mixture.

EXAMPLE 13.12

The equilibrium constant K_c for the reaction of hydrogen with iodine is 57.0 at 700 K:

$$H_2(g) + I_2(g) \underset{k_r}{\overset{k_f}{\rightleftharpoons}} 2\,HI(g) \qquad K_c = 57.0 \text{ at } 700 \text{ K}$$

(a) Is the rate constant k_f for the formation of HI larger or smaller than the rate constant k_r for the decomposition of HI?
(b) The value of k_r at 700 K is $1.16 \times 10^{-3}/(M \cdot s)$. What is the value of k_f at the same temperature?
(c) How are the values of k_f, k_r, and K_c affected by the addition of a catalyst?

SOLUTION   **(a)** Because $K_c = k_f/k_r = 57.0$, the rate constant for the formation of HI (forward reaction) is larger than the rate constant for the decomposition of HI (reverse reaction) by a factor of 57.0.
(b) Since $K_c = k_f/k_r$,

$$k_f = (K_c)(k_r) = (57.0)\left(\frac{1.16 \times 10^{-3}}{M \cdot s}\right) = 6.61 \times 10^{-2}/(M \cdot s)$$

(c) A catalyst lowers the activation energy barrier for the forward and reverse reactions by the same amount, thus increasing the rate constants k_f and k_r by the same factor. Because the equilibrium constant K_c equals the ratio of k_f to k_r, the value of K_c is unaffected by the addition of a catalyst.

PROBLEM 13.16   At 575 K, the forward and reverse rate constants for the reaction of H_2 with I_2 (Example 13.12) have values of $k_f = 1.32 \times 10^{-4}/(M \cdot s)$ and $k_r = 1.22 \times 10^{-6}/(M \cdot s)$.

(a) If you begin with equal concentrations of H_2 and I_2, will the equilibrium mixture have larger concentrations of reactants or products?
(b) Calculate the equilibrium constant K_c at 575 K.

[4] In deriving this equation for K_c, we have assumed a single-step mechanism. For a multistep mechanism, each step has a characteristic rate constant ratio, k_f/k_r, and K_c is equal to the product of the rate constant ratios for the various steps.

interlude—BREATHING AND OXYGEN TRANSPORT

Humans, like all animals, need oxygen. The oxygen comes, of course, from breathing: About 500 mL of air is drawn into the lungs of an average person with each breath. As the freshly inspired air travels through the bronchial passages and enters the approximately 150 million alveolar sacs of the lungs, it picks up moisture and mixes with air remaining from the previous breath. As it mixes, the concentrations of both water vapor and carbon dioxide increase. These gas concentrations are measured by their *partial pressures* (Section 9.5), with the partial pressure of oxygen in the lungs usually around 100 mm Hg (Table 13.3). Oxygen then diffuses through the delicate walls of the lung alveoli and into arterial blood, which transports it to all body tissues.

TABLE 13.3 Partial Pressures of Oxygen in the Lungs and Blood at Sea Level

Source	P_{O_2} (mm Hg)
Dry air	159
Alveolar air	100
Arterial blood	95
Venous blood	40

Only about 3% of the oxygen in blood is dissolved; the rest is chemically bound to *hemoglobin* molecules (Hb), large proteins that contain *heme* groups embedded in them. Each hemoglobin molecule contains four heme groups, and each heme group contains an iron atom that is able to bind to one O_2 molecule. Thus, a single hemoglobin molecule can bind four molecules of oxygen.

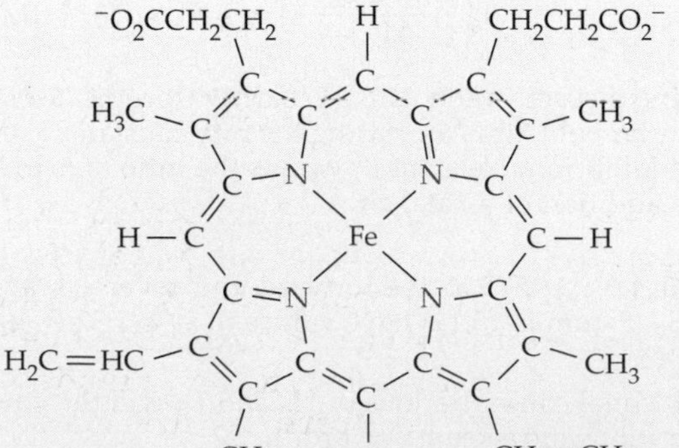

Heme — an O_2 molecule binds to the central iron atom.

The entire system of oxygen transport and delivery in the body depends on the pickup and release of O_2 by hemoglobin (Hb) according to the following series of equilibria:

$$Hb + O_2 \rightleftharpoons Hb(O_2)$$

$$Hb(O_2) + O_2 \rightleftharpoons Hb(O_2)_2$$

$$Hb(O_2)_2 + O_2 \rightleftharpoons Hb(O_2)_3$$

$$Hb(O_2)_3 + O_2 \rightleftharpoons Hb(O_2)_4$$

The positions of the different equilibria depend on the partial pressures of O_2 (P_{O_2}) in the various tissues. In hard-working, oxygen-starved muscles, where P_{O_2} is low, oxygen is released from hemoglobin as the equilibria shift toward the left, according to Le Châtelier's principle. In the lung, where P_{O_2} is high, oxygen is absorbed by hemoglobin as the equilibria shift toward the right.

The amount of oxygen carried by hemoglobin at any given value of P_{O_2} is usually expressed as a percent saturation and can be found from the curve shown in Figure 13.14. The saturation is 97.5% in the lungs, where $P_{O_2} = 100$ mm Hg, meaning that each hemoglobin is carrying close to its maximum possible amount of four O_2 molecules. When $P_{O_2} = 26$ mm Hg, however, the saturation drops to 50%.

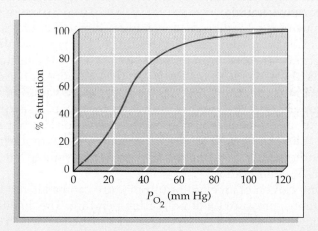

FIGURE 13.14 An oxygen-carrying curve for hemoglobin. The percent saturation of the oxygen binding sites on hemoglobin depends on the partial pressure of oxygen (P_{O_2}).

What about people who live at high altitudes? In Leadville, Colorado, for example, where the altitude is 10,000 ft, the partial pressure of O_2 in the lungs is only about 68 mm Hg. Hemoglobin is only 90% saturated with O_2 at this pressure, meaning that less oxygen is available for delivery to tissues. People who climb suddenly from sea level to high altitude thus experience a feeling of oxygen deprivation, or *hypoxia*, as their bodies are unable to supply enough oxygen to the tissues. The body soon copes with the situation, though, by producing more hemoglobin molecules, which both provide more capacity for O_2 transport and also drive the Hb + O_2 equilibria to the right, making each hemoglobin molecule a more efficient carrier.

Chemical equilibrium is a dynamic state in which the concentrations of reactants and products remain constant because the rates of the forward and reverse reactions are equal. For the general reaction $a\,A + b\,B \rightleftarrows c\,C + d\,D$, concentrations in the equilibrium mixture are related by the **equilibrium equation**:

$$K_c = \frac{[C]^c[D]^d}{[A]^a[B]^b}$$

The quotient on the right side of the equation is called the **equilibrium constant expression**. The **equilibrium constant K_c** is the number obtained when equilibrium concentrations (in mol/L) are substituted into the equilibrium constant expression. The value of K_c varies with temperature and depends on the form of the balanced chemical equation.

The **equilibrium constant K_p** can be used for gas-phase reactions. It is defined in the same way as K_c except that the equilibrium constant expression contains partial pressures in atmospheres instead of molar concentrations. K_p and K_c are related by the equation $K_p = K_c(RT)^{\Delta n}$, where $\Delta n = (c + d) - (a + b)$.

Homogeneous equilibria are those in which all reactants and products are in a single phase; **heterogeneous equilibria** are those in which reactants and products are present in more than one phase. The equilibrium expression for a heterogeneous equilibrium does not include concentrations of pure solids or pure liquids.

The value of the equilibrium constant for a reaction makes it possible to judge the extent of reaction, predict the direction of reaction, and calculate equilibrium concentrations from initial concentrations. The farther the reaction proceeds toward completion, the larger the value of K_c. The direction of a reaction not at equilibrium depends on the relative values of K_c and the **reaction quotient Q_c.** Q_c is defined in the same way as K_c except that the concentrations in the equilibrium constant expression are not necessarily equilibrium concentrations. If $Q_c < K_c$, the reaction goes from left to right to attain equilibrium; if $Q_c > K_c$, the reaction goes from right to left; if $Q_c = K_c$, the system is at equilibrium.

The composition of an equilibrium mixture can be altered by changes in concentrations, pressure (volume), or temperature. The qualitative effect of these changes is predicted by **Le Châtelier's principle**, which says that if a stress is applied to a reaction mixture at equilibrium, reaction occurs in the direction that relieves the stress. Temperature changes affect equilibrium concentrations because K_c is temperature dependent. As the temperature increases, K_c for an exothermic reaction decreases, and K_c for an endothermic reaction increases.

A catalyst increases the rate at which chemical equilibrium is reached, but it does not affect the equilibrium constant or the equilibrium concentrations. The equilibrium constant for a reaction is equal to the ratio of the rate constants for the forward and reverse reactions; $K_c = k_f/k_r$.

1. Consider the interconversion of A molecules (red spheres) and B molecules (blue spheres) according to the reaction A $\rightleftarrows$ B. Each of the following series of pictures represents a separate experiment in which time increases from left to right.

—————————— Increasing time ——————————→

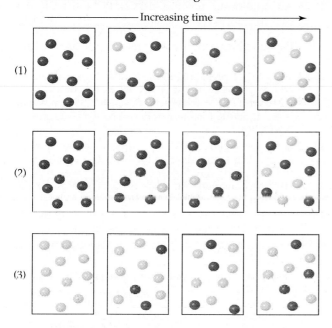

(1)

(2)

(3)

(a) Which of the experiments has resulted in an equilibrium state?

(b) What is the value of the equilibrium constant K_c for the reaction A $\rightleftarrows$ B?

(c) Explain why you can calculate K_c without knowing the volume of the reaction vessel.

2. The following pictures represent the equilibrium state for three different reactions of the type $A_2 + X_2 \rightleftarrows$ 2 AX (X = B, C, or D).

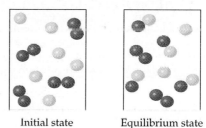

A$_2$ + B$_2$ $\rightleftarrows$ AB   A$_2$ + C$_2$ $\rightleftarrows$ AC   A$_2$ + D$_2$ $\rightleftarrows$ AD

(a) Which reaction has the largest equilibrium constant?

(b) Which reaction has the smallest equilibrium constant?

3. The reaction $A_2 + B_2 \rightleftarrows 2$ AB has an equilibrium constant $K_c = 4$. The following pictures represent reaction mixtures that contain A$_2$ molecules (red), B$_2$ molecules (blue), and AB molecules.

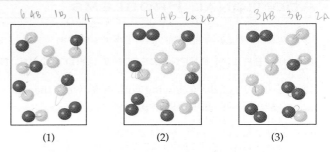

6 AB   1B₁ 1A   4 AB 2a₂2B   3AB 3B 2A

(1)   (2)   (3)

(a) Which reaction mixture is at equilibrium?

(b) For those reaction mixtures that are not at equilibrium, will the reaction go in the forward or the reverse direction to reach equilibrium?

4. The following pictures represent the initial state and the equilibrium state for the reaction of A$_2$ molecules (red) with B atoms (blue) to give AB molecules.

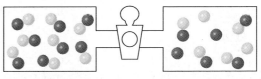

Initial state   Equilibrium state

(a) Write a balanced equation for the reaction.

(b) If the volume of the equilibrium mixture is decreased, will the number of AB molecules increase, decrease, or remain the same? Explain.

5. Consider the reaction A + B $\rightleftarrows$ AB. The vessel on the right contains an equilibrium mixture of A molecules (red spheres), B molecules (blue spheres), and AB molecules. If the stopcock is opened and the contents of the two vessels are allowed to mix, will the reaction go in the forward or the reverse direction? Explain.

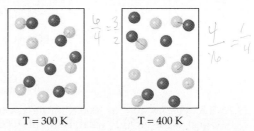

6. The following pictures represent the equilibrium mixture for the reaction A + B $\rightleftarrows$ AB at 300 K and 400 K.

$\frac{6}{4} = \frac{3}{2}$ $\frac{4}{16} = \frac{1}{4}$

T = 300 K   T = 400 K

Is the reaction exothermic or endothermic? Explain in terms of Le Châtelier's principle.

529

ADDITIONAL PROBLEMS

Problems 13.1–13.16 appear within the chapter.

EQUILIBRIUM EXPRESSIONS AND EQUILIBRIUM CONSTANTS

13.17 For each of the following equilibria, write the equilibrium constant expression for K_c.
  (a) $PCl_5(g) \rightleftarrows PCl_3(g) + Cl_2(g)$
  (b) $2\,NO(g) + Cl_2(g) \rightleftarrows 2\,NOCl(g)$
  (c) $CH_4(g) + 2\,H_2S(g) \rightleftarrows CS_2(g) + 4\,H_2(g)$

13.18 For each of the equilibria in Problem 13.17, write the equilibrium constant expression for K_p, and give the equation that relates K_p and K_c.

13.19 Diethyl ether, used as an anesthetic, is synthesized by heating ethanol with concentrated sulfuric acid. Write the equilibrium constant expression for K_c.

$$2\,C_2H_5OH(l) \rightleftarrows C_2H_5OC_2H_5(l) + H_2O(l)$$
$$\underset{\text{Ethanol}}{} \qquad \underset{\text{Diethyl ether}}{}$$

13.20 Ethylene glycol, used as antifreeze in automobile radiators, is manufactured by hydration of ethylene oxide. Write the equilibrium constant expression for K_c.

$$\underset{\text{Ethylene oxide}}{CH_2 \!-\! CH_2(l)} + H_2O(l) \rightleftarrows \underset{\text{Ethylene glycol}}{HOCH_2CH_2OH(l)}$$

13.21 Write the equilibrium constant expression for K_c for the following metabolic reactions.

(a) $\underset{\text{Fumaric acid}}{HOOCCH\!=\!CHCOOH(aq)} + H_2O(l) \rightleftarrows$

$$\underset{\text{Malic acid}}{HOOCCHCH_2COOH(aq)}$$
with OH group shown above.

(b) $\underset{\text{Acetic acid}}{CH_3COOH(aq)} + \underset{\text{Oxaloacetic acid}}{HOOCCCH_2COOH(aq)} \rightleftarrows$

$$\underset{\text{Citric acid}}{HO\!-\!C\!-\!COOH(aq)}$$
with CH_2COOH groups above and below.

13.22 If $K_c = 7.5 \times 10^{-9}$ at 1000 K for the reaction $N_2(g) + O_2(g) \rightleftarrows 2\,NO(g)$, what is K_c at 1000 K for the reaction $2\,NO(g) \rightleftarrows N_2(g) + O_2(g)$?

13.23 An equilibrium mixture of PCl_5, PCl_3, and Cl_2 at a certain temperature contains 8.3×10^{-3} M PCl_5, 1.5×10^{-2} M PCl_3, and 3.2×10^{-2} M Cl_2. Calculate the equilibrium constant K_c for the reaction $PCl_5(g) \rightleftarrows PCl_3(g) + Cl_2(g)$.

13.24 The partial pressures in an equilibrium mixture of NO, Cl_2, and NOCl at 500 K are as follows: $P_{NO} = 0.240$ atm; $P_{Cl_2} = 0.608$ atm; $P_{NOCl} = 1.35$ atm. What is K_p at 500 K for the reaction $2\,NO(g) + Cl_2(g) \rightleftarrows 2\,NOCl(g)$?

13.25 A sample of HI (9.30×10^{-3} mol) was placed in an empty 2.00 L container at 1000 K. After equilibrium was reached, the concentration of I_2 was 6.29×10^{-4} M. Calculate the value of K_c at 1000 K for the reaction $H_2(g) + I_2(g) \rightleftarrows 2\,HI(g)$.

13.26 Vinegar contains acetic acid, a weak acid that is partially dissociated in aqueous solution:

$$CH_3COOH(aq) \rightleftarrows H^+(aq) + CH_3CO_2^-(aq)$$

  (a) Write the equilibrium constant expression for K_c.
  (b) What is the value of K_c if the extent of dissociation in 1.0 M CH_3COOH is 0.42%?

13.27 The industrial solvent ethyl acetate is produced by reaction of acetic acid with ethanol:

$$\underset{\text{Acetic acid}}{CH_3COOH(l)} + \underset{\text{Ethanol}}{C_2H_5OH(l)} \rightleftarrows$$
$$\underset{\text{Ethyl acetate}}{CH_3COOC_2H_5(l)} + H_2O(l)$$

  (a) Write the equilibrium constant expression for K_c.
  (b) A solution prepared by mixing 1.00 mol of acetic acid and 1.00 mol of ethanol contains 0.65 mol of ethyl acetate at equilibrium. Calculate the value of K_c. Explain why you can calculate K_c without knowing the volume of the solution.

13.28 A characteristic reaction of ethyl acetate is hydrolysis, the reverse of the reaction in Problem 13.27. Write the equilibrium equation for hydrolysis of ethyl acetate, and use the data in Problem 13.27 to calculate K_c for the hydrolysis reaction.

13.29 At 800 K, $K_c = 4.24$ for the reaction $CO(g) + H_2O(g) \rightleftarrows CO_2(g) + H_2(g)$. What is K_p at the same temperature?

13.30 One step in the manufacture of sulfuric acid involves the oxidation of sulfur dioxide, $2\,SO_2(g) + O_2(g) \rightleftarrows 2\,SO_3(g)$. If K_p for this reaction is 3.30 at 1000 K, what is the value of K_c at the same temperature?

13.31 The vapor pressure of water at 25°C is 0.0313 atm. Calculate the values of K_p and K_c at 25°C for the equilibrium $H_2O(l) \rightleftharpoons H_2O(g)$.

13.32 Naphthalene, a white solid used to make mothballs, has a vapor pressure of 0.10 mm Hg at 27°C. Calculate the values of K_p and K_c at 27°C for the equilibrium $C_{10}H_8(s) \rightleftharpoons C_{10}H_8(g)$.

13.33 For each of the following equilibria, write the equilibrium constant expression for K_c. Where appropriate, also write the equilibrium constant expression for K_p.
(a) $Fe_2O_3(s) + 3\ CO(g) \rightleftharpoons 2\ Fe(l) + 3\ CO_2(g)$
(b) $4\ Fe(s) + 3\ O_2(g) \rightleftharpoons 2\ Fe_2O_3(s)$
(c) $BaSO_4(s) \rightleftharpoons BaO(s) + SO_3(g)$
(d) $BaSO_4(s) \rightleftharpoons Ba^{2+}(aq) + SO_4^{2-}(aq)$

USING THE EQUILIBRIUM CONSTANT

13.34 When the following reactions come to equilibrium, does the equilibrium mixture contain mostly reactants or mostly products?
(a) $2\ SO_2(g) + O_2(g) \rightleftharpoons 2\ SO_3(g)$ $K_c = 1.2 \times 10^9$
(b) $2\ HCl(g) \rightleftharpoons H_2(g) + Cl_2(g)$ $K_c = 2.0 \times 10^{-17}$

13.35 Which of the following reactions goes almost all the way to completion, and which proceeds hardly at all toward completion?
(a) $N_2(g) + O_2(g) \rightleftharpoons 2\ NO(g)$ $K_c = 2.7 \times 10^{-18}$
(b) $2\ NO(g) + O_2(g) \rightleftharpoons 2\ NO_2(g)$ $K_c = 6.0 \times 10^{13}$

13.36 For which of the following reactions will the equilibrium mixture contain an appreciable concentration of both reactants and products?
(a) $Cl_2(g) \rightleftharpoons 2\ Cl(g)$ $K_c = 6.4 \times 10^{-39}$
(b) $Cl_2(g) + 2\ NO(g) \rightleftharpoons 2\ NOCl(g)$ $K_c = 3.7 \times 10^8$
(c) $Cl_2(g) + 2\ NO_2(g) \rightleftharpoons 2\ NO_2Cl(g)$ $K_c = 1.8$

13.37 When wine spoils, ethanol is oxidized to acetic acid as O_2 from the air dissolves in the wine:

$$\underset{\text{Ethanol}}{C_2H_5OH(aq)} + O_2(aq) \rightleftharpoons \underset{\text{Acetic acid}}{CH_3COOH(aq)} + H_2O(l)$$

The value of K_c for this reaction at 25°C is 1.2×10^{82}. Will much ethanol remain when the reaction has reached equilibrium? Explain.

13.38 The value of K_c for the reaction $3\ O_2(g) \rightleftharpoons 2\ O_3(g)$ is 1.7×10^{-56} at 25°C. Do you expect pure air at 25°C to contain much O_3 (ozone) when O_2 and O_3 are in equilibrium? If the concentration of O_2 in air at 25°C is 8×10^{-3} M, what is the equilibrium concentration of O_3?

13.39 At a particular temperature, the equilibrium constant has a single, unique value, but the reaction quotient can have an infinite number of values. Explain.

13.40 At 1400 K, $K_c = 2.5 \times 10^{-3}$ for the reaction $CH_4(g) + 2\ H_2S(g) \rightleftharpoons CS_2(g) + 4\ H_2(g)$. A 10 L reaction vessel at 1400 K contains 2.0 mol of CH_4, 3.0 mol of CS_2, 3.0 mol of H_2, and 4.0 mol of H_2S. Is the reaction mixture at equilibrium? If not, in which direction does the reaction proceed to reach equilibrium?

13.41 The first step in the industrial synthesis of hydrogen is the reaction of steam and methane to give synthesis gas, a mixture of carbon monoxide and hydrogen:

$$H_2O(g) + CH_4(g) \rightleftharpoons CO(g) + 3\ H_2(g)$$

The equilibrium constant K_c is 4.7 at 1400 K. A mixture of reactants and products at 1400 K contains 0.035 M H_2O, 0.050 M CH_4, 0.15 M CO, and 0.20 M H_2. In which direction does the reaction proceed to reach equilibrium?

13.42 An equilibrium mixture of N_2, H_2, and NH_3 at 500 K contains 0.020 M N_2 and 0.18 M H_2. At this temperature, K_c for the reaction $N_2(g) + 3\ H_2(g) \rightleftharpoons 2\ NH_3(g)$ is 1.7×10^2. What is the concentration of NH_3?

13.43 An equilibrium mixture of O_2, SO_2, and SO_3 contains equal concentrations of SO_2 and SO_3. Calculate the concentration of O_2 if $K_c = 2.7 \times 10^2$ for the reaction $2\ SO_2(g) + O_2(g) \rightleftharpoons 2\ SO_3(g)$.

13.44 The air pollutant NO is produced in automobile engines because of the high-temperature reaction $N_2(g) + O_2(g) \rightleftharpoons 2\ NO(g)$ ($K_c = 1.7 \times 10^{-3}$ at 2300 K). If the initial concentrations of N_2 and O_2 at 2300 K are both 1.40 M, what are the concentrations of NO, N_2, and O_2 when the reaction mixture reaches equilibrium?

13.45 Recalculate the equilibrium concentrations in Problem 13.44 if the initial concentrations are 2.24 M N_2 and 0.56 M O_2. (This N_2/O_2 concentration ratio is the ratio found in air.)

13.46 At a certain temperature, the reaction $PCl_5(g) \rightleftharpoons PCl_3(g) + Cl_2(g)$ has an equilibrium constant $K_c = 5.8 \times 10^{-2}$. Calculate the equilibrium concentrations of PCl_5, PCl_3, and Cl_2 if only PCl_5 is present initially at a concentration of 0.160 M.

13.47 The value of K_c for the reaction of acetic acid with ethanol is 3.4 at 25°C:

$$CH_3COOH(l) + C_2H_5OH(l) \rightleftarrows$$
 Acetic acid   Ethanol
$$CH_3COOC_2H_5(l) + H_2O(l) \qquad K_c = 3.4$$
   Ethyl acetate

(a) How many moles of ethyl acetate are present in an equilibrium mixture that contains 4.0 mol of acetic acid, 6.0 mol of ethanol, and 12.0 mol of water at 25°C?

(b) Calculate the number of moles of all reactants and products in an equilibrium mixture prepared by mixing 1.00 mol of acetic acid and 10.00 mol of ethanol.

13.48 In a basic aqueous solution, chloromethane undergoes a substitution reaction in which Cl^- is replaced by OH^-:

$$CH_3Cl + OH^- \rightleftarrows CH_3OH + Cl^-$$
 Chloromethane   Methanol

The equilibrium constant K_c is 10^{16}. Calculate the equilibrium concentrations of CH_3Cl, CH_3OH, OH^-, and Cl^- in a solution prepared by mixing equal volumes of 0.1 M CH_3Cl and 0.2 M NaOH. (*Hint:* In defining x, assume that the reaction goes 100% to completion, and then take account of a small amount of the reverse reaction.)

LE CHÂTELIER'S PRINCIPLE

13.49 Consider the following equilibrium: $Ag^+(aq) + Cl^-(aq) \rightleftarrows AgCl(s)$. Use Le Châtelier's principle to predict how the amount of solid silver chloride will change when the equilibrium is disturbed by
(a) adding NaCl
(b) adding $AgNO_3$
(c) removing Cl^-
(d) adding NH_3, which reacts with Ag^+ to form the complex ion, $Ag(NH_3)_2^+$

13.50 Will the concentration of NO_2 increase, decrease, or remain the same when the equilibrium $ClNO_2(g) + NO(g) \rightleftarrows ClNO(g) + NO_2(g)$ is disturbed by
(a) adding $ClNO_2$
(b) adding $ClNO$
(c) adding NO
(d) removing NO
For (a), account for the change in terms of the reaction quotient Q_c.

13.51 When each of the following equilibria is disturbed by increasing the pressure as a result of decreasing the volume, does the number of moles of reaction products increase, decrease, or remain the same?
(a) $2 CO_2(g) \rightleftarrows 2 CO(g) + O_2(g)$
(b) $N_2(g) + O_2(g) \rightleftarrows 2 NO(g)$
(c) $Si(s) + 2 Cl_2(g) \rightleftarrows SiCl_4(g)$

13.52 For each of the following equilibria, use Le Châtelier's principle to predict the direction of reaction when the volume is increased.
(a) $C(s) + H_2O(g) \rightleftarrows CO(g) + H_2(g)$
(b) $2 H_2(g) + O_2(g) \rightleftarrows 2 H_2O(g)$
(c) $2 Fe(s) + 3 H_2O(g) \rightleftarrows Fe_2O_3(s) + 3 H_2(g)$

13.53 For the gas shift reaction $CO(g) + H_2O(g) \rightleftarrows CO_2(g) + H_2(g)$, $\Delta H° = -41.2$ kJ. Does the amount of H_2 in an equilibrium mixture increase or decrease when the temperature is increased? How does K_c change when the temperature is decreased? Justify your answers using Le Châtelier's principle.

13.54 The value of $\Delta H°$ for the reaction $3 O_2(g) \rightleftarrows 2 O_3(g)$ is 285 kJ. Does the equilibrium constant for this reaction increase or decrease when the temperature increases? Justify your answer using Le Châtelier's principle.

13.55 For the reaction $H_2(g) + I_2(g) \rightleftarrows 2 HI(g)$, $\Delta H° = -9.4$ kJ. Will the equilibrium concentration of HI increase or decrease when the temperature increases?

13.56 Consider the endothermic reaction $Fe^{3+}(aq) + Cl^-(aq) \rightleftarrows FeCl^{2+}(aq)$. Use Le Châtelier's principle to predict how the equilibrium concentration of the complex ion $FeCl^{2+}$ will change when:
(a) $Fe(NO_3)_3$ is added
(b) Cl^- is precipitated as AgCl by addition of $AgNO_3$
(c) the temperature is increased
(d) a catalyst is added

13.57 Methanol (CH_3OH) is manufactured by reaction of carbon monoxide and hydrogen in the presence of a ZnO/Cr_2O_3 catalyst:

$$CO(g) + 2 H_2(g) \rightleftarrows CH_3OH(g) \qquad \Delta H° = -91 \text{ kJ}$$

Does the amount of methanol increase, decrease, or remain the same when an equilibrium mixture of reactants and products is subjected to the following changes?
(a) the temperature is increased
(b) the volume is decreased
(c) helium is added
(d) CO is added
(e) the catalyst is removed

13.58 In the gas phase at 179°C, isopropanol (rubbing alcohol) decomposes to acetone, an important industrial solvent:

$(CH_3)_2CHOH(g) \rightleftarrows (CH_3)_2CO(g) + H_2(g)$
   Isopropanol   Acetone

$$\Delta H° = +57.3 \text{ kJ}$$

Does the amount of acetone increase, decrease, or remain the same when an equilibrium mixture of reactants and products is subjected to the following changes?

(a) the temperature is increased
(b) the volume is increased
(c) argon is added
(d) H_2 is added
(e) a catalyst is added

CHEMICAL EQUILIBRIUM AND CHEMICAL KINETICS

13.59 Consider a general, single-step reaction of the type $A + B \rightleftarrows C$. Show that the equilibrium constant is equal to the ratio of the rate constants for the forward and reverse reactions, $K_c = k_f/k_r$.

13.60 Which of the following relative values of k_f and k_r results in an equilibrium mixture that contains large amounts of reactants and small amounts of products?

(a) $k_f > k_r$ (b) $k_f = k_r$ (c) $k_f < k_r$

13.61 Consider the gas-phase hydration of hexafluoroacetone, $(CF_3)_2CO$:

$$(CF_3)_2CO(g) + H_2O(g) \underset{k_r}{\overset{k_f}{\rightleftarrows}} (CF_3)_2C(OH)_2(g)$$

At 76°C, the forward and reverse rate constants are $k_f = 0.13/(M \cdot s)$ and $k_r = 6.2 \times 10^{-4}/s$. What is the value of the equilibrium constant K_c?

13.62 Consider the reaction of chloromethane with OH^- in aqueous solution:

$$CH_3Cl(aq) + OH^-(aq) \underset{k_r}{\overset{k_f}{\rightleftarrows}} CH_3OH(aq) + Cl^-(aq)$$

At 25°C, the rate constant for the forward reaction is $6 \times 10^{-6}/(M \cdot s)$, and the equilibrium constant K_c is 10^{16}. Calculate the rate constant for the reverse reaction at 25°C.

GENERAL PROBLEMS

13.63 List three factors that can alter the composition of an equilibrium mixture. Which of these factors affects the value of the equilibrium constant?

13.64 Explain why a catalyst does not affect the composition of an equilibrium mixture.

13.65 Given the Arrhenius equation, $k = Ae^{-E_a/RT}$, and given the relation between the equilibrium constant and the forward and reverse rate constants, $K_c = k_f/k_r$, explain why K_c for an exothermic reaction decreases with increasing temperature.

13.66 The affinity of hemoglobin (Hb) for CO is greater than its affinity for O_2. Use Le Châtelier's principle to predict how CO affects the equilibrium $Hb + O_2 \rightleftarrows Hb(O_2)$. Suggest a reason for the toxicity of CO.

13.67 Consider the reaction $C(s) + CO_2(g) \rightleftarrows 2 CO(g)$. When 1.50 mol of CO_2 and an excess of solid carbon are heated in a 20.0 L container at 1100 K, the equilibrium concentration of CO is 7.00×10^{-2} M. Calculate

(a) the equilibrium concentration of CO_2
(b) the value of the equilibrium constant K_c at 1100 K.

13.68 When 0.500 mol of N_2O_4 is placed in a 4.00 L reaction vessel and heated at 400 K, 79.3% of the N_2O_4 decomposes to NO_2. Calculate K_c and K_p at 400 K for the reaction $N_2O_4(g) \rightleftarrows 2 NO_2(g)$.

13.69 At 25°C, $K_c = 1.6 \times 10^{24}$ for the reaction $CaO(s) + CO_2(g) \rightleftarrows CaCO_3(s)$. Use the value of K_c to make a qualitative estimate of the extent of this reaction. Verify your estimate by calculating the concentration of CO_2 gas that is in equilibrium with solid CaO and solid $CaCO_3$ at 25°C.

13.70 What concentration of NH_3 is in equilibrium with 1.0×10^{-3} M N_2 and 2.0×10^{-3} M H_2 at 700 K? At this temperature, $K_c = 0.291$ for the reaction $N_2(g) + 3 H_2(g) \rightleftarrows 2 NH_3(g)$.

13.71 The F–F bond in F_2 is relatively weak; $K_p = 7.83$ at 1500 K for the reaction $F_2(g) \rightleftarrows 2 F(g)$. If the equilibrium partial pressure of F_2 molecules at 1500 K is 0.200 atm, what is the equilibrium partial pressure of F atoms in atm? What fraction of the F_2 molecules dissociates at 1500 K?

13.72 The equilibrium concentrations in a gas mixture at a particular temperature are 0.13 M H_2, 0.70 M I_2, and 2.1 M HI. What equilibrium concentrations are obtained at the same temperature when 0.20 mol of HI is injected into an empty 500 mL container?

13.73 A 5.00 L reaction vessel is filled with 1.00 mol of H_2, 1.00 mol of I_2, and 2.50 mol of HI. Calculate the equilibrium concentrations of H_2, I_2, and HI at 500 K. The equilibrium constant K_c at 500 K for the reaction $H_2(g) + I_2(g) \rightleftarrows 2\,HI(g)$ is 129.

13.74 At 1000 K, the value of K_c for the reaction $C(s) + H_2O(g) \rightleftarrows CO_2(g) + H_2(g)$ is 3.0×10^{-2}. Calculate the equilibrium concentrations of H_2O, CO_2, and H_2 in a reaction mixture obtained by heating 6.00 mol of steam and an excess of solid carbon in a 5.00 L container. What is the molar composition of the equilibrium mixture?

13.75 The equilibrium constant K_c for the reaction $H_2(g) + I_2(g) \rightleftarrows 2\,HI(g)$ is 80.9 at 600 K and 57.0 at 700 K.
  (a) Is the reaction endothermic or exothermic?
  (b) How are the equilibrium concentrations affected by (i) an increase in volume, (ii) addition of an inert gas, (iii) addition of a catalyst?

13.76 A gaseous reaction mixture at equilibrium is subjected to the following changes:
  (a) a decrease in volume
  (b) an increase in temperature
  (c) addition of reactants
  (d) addition of a catalyst
  (e) addition of an inert gas
  Which of these changes affect the value of the reaction quotient, Q_c, but leave the value of the equilibrium constant, K_c, unchanged? Which of the changes affect the value of K_c? Which affect neither Q_c nor K_c?

13.77 Baking soda (sodium bicarbonate) decomposes when it is heated:

$$2\,NaHCO_3(s) \rightleftarrows NaCO_3(s) + CO_2(g) + H_2O(g)$$
$$\Delta H° = 136\ kJ$$

Consider an equilibrium mixture of reactants and products in a closed container. How does the number of moles of CO_2 change when the mixture is disturbed by
  (a) adding solid $NaHCO_3$
  (b) adding water vapor
  (c) decreasing the volume of the container
  (d) increasing the temperature

13.78 Acetic acid tends to form dimers, $(CH_3COOH)_2$, because of hydrogen bonding:

Monomer   Dimer

The equilibrium constant K_c for this reaction is 1.51 $\times$ 10^2 in benzene solution, but only 3.7×10^{-2} in water solution.
  (a) Calculate the ratio of dimers to monomers for 0.100 M acetic acid in benzene.
  (b) Calculate the ratio of dimers to monomers for 0.100 M acetic acid in water.
  (c) Why is K_c for the water solution so much smaller than K_c for the benzene solution?

13.79 Refining petroleum involves cracking large hydrocarbon molecules into smaller, more volatile pieces. A simple example of hydrocarbon cracking is the gas-phase thermal decomposition of butane to ethane and ethylene:

Butane, C_4H_{10}

Ethane, C_2H_6 Ethylene, C_2H_4

  (a) Write the equilibrium constant expressions for K_p and K_c.
  (b) The value of K_p at 500°C is 12. What is the value of K_c?
  (c) A sample of butane having a pressure of 50 atm is heated at 500°C in a closed container at constant volume. When equilibrium is reached, what percentage of the butane has been converted to ethane and ethylene? What is the total pressure at equilibrium?
  (d) How would the percent conversion in **(c)** be affected by a decrease in volume?

13.80 The equilibrium constant K_c for the gas-phase thermal decomposition of cyclopropane to propene is 1.0×10^5 at 500 K:

Cyclopropane   Propene

  (a) What is the value of K_p at 500 K?
  (b) What is the equilibrium partial pressure of cyclopropane at 500 K when the partial pressure of propene is 5.0 atm?
  (c) Can you alter the ratio of the two concentrations at equilibrium by adding cyclopropane or by decreasing the volume of the container? Explain.
  (d) Which has the larger rate constant, the forward reaction or the reverse reaction?

13.81 The equilibrium constant K_p for the gas-phase thermal decomposition of *tert*-butyl chloride is 3.45 at 500 K:

$$(CH_3)_3CCl(g) \rightleftarrows (CH_3)_2C{=}CH_2(g) + HCl(g)$$
tert-Butyl chloride   Isobutylene

(a) Calculate the value of K_c.

(b) Calculate the molar concentrations of reactants and products in an equilibrium mixture obtained by heating 1.00 mol of *tert*-butyl chloride in a 5.00 L vessel at 500 K.

(c) A mixture of isobutylene (0.400 atm partial pressure) and HCl (0.600 atm partial pressure) is heated at 500 K. What are the partial pressures of *tert*-butyl chloride, isobutylene, and HCl when equilibrium is reached?

13.82 Derive the equation $K_p = K_c(RT)^{\Delta n}$ for the general reaction $a\,A + b\,B \rightleftarrows c\,C + d\,D$, where $\Delta n = (c + d) - (a + b)$. *Hint:* The partial pressure of an ideal gas, P_A, is related to its molarity, $n_A/V = [A]$, by the equation $P_A = (n_A/V)RT = [A]RT$. Using analogous equations for P_B, P_C, and P_D, write the equilibrium constant expression for K_p in terms of molarity and then compare it with the equilibrium constant expression for K_c.

chapter 14

HYDROGEN, OXYGEN, AND WATER

Hydrogen, the most abundant element in the universe, and oxygen, the most abundant element on the earth's surface, are two of the most familiar and important of all the elements. Hydrogen forms more compounds than any other element, and large amounts of elemental hydrogen are produced in the chemical industry for use in the synthesis of such chemicals as ammonia and methanol. Oxygen is essential for respiration and is the oxidizing agent in the combustion processes that provide energy for maintaining our industrialized civilization. More than 23 million tons of oxygen are produced annually in the United States, largely for use in making steel, by fractional distillation of liquefied air.

In this chapter, we'll take a detailed look at the chemistry of hydrogen and oxygen, and we'll discuss some of the properties of water, the most important of all chemical compounds and the solvent for all the reactions to be discussed in Chapters 15 and 16.

A "water crown," produced when a drop of water strikes the surface of a pool of water.

14.1 ▸HYDROGEN

Henry Cavendish (1731–1810), an English chemist and physicist, is generally credited with being the first person to isolate hydrogen in pure form. Cavendish showed that the action of acids on metals like zinc, iron, and tin produces a flammable gas that can be distinguished from other gases by its unusually low density:

$$2\ H^+(aq) + Zn(s) \rightarrow H_2(g) + Zn^{2+}(aq)$$

The French chemist Lavoisier called the gas "hydrogen," meaning "water former," because it combines with oxygen to produce water.

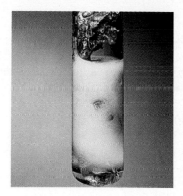

Pure hydrogen was first prepared by the reaction of aqueous acid with zinc metal.

At ordinary temperatures and pressures, hydrogen is a colorless, odorless, and tasteless gas comprised of diatomic molecules, H_2. Because H_2 is a nonpolar molecule that contains only two electrons, intermolecular forces are extremely weak (Section 10.2). As a result, hydrogen has a very low melting point (−259.2°C) and a very low boiling point (−252.8°C). The bonding forces within the H_2 molecule are exceptionally strong, however: The H–H bond dissociation energy is 436 kJ/mol, greater than that for any other single bond between two atoms of the same element (Section 7.2):

$$H_2(g) \rightarrow 2\ H(g) \qquad D = 436\ \text{kJ}$$

By comparison, the bond dissociation energies of the halogens range from 151 kJ/mol for I_2 to 243 kJ/mol for Cl_2. Because of the strong H–H bond, H_2 is thermally stable. Even at 2000 K, only one of every 2500 H_2 molecules is dissociated into H atoms.

Hydrogen is thought to account for approximately 75% of the mass of the universe. Our sun and other stars, for instance, are composed mainly of hydrogen, which serves as their nuclear fuel. On earth, though, hydrogen is rarely found in uncombined form because the earth's gravity is too weak to hold molecules as light as H_2. Nearly all the H_2 originally present in the earth's atmosphere has been lost to space, and the atmospheric abundance of H_2 is only 0.53 ppm by volume. In the earth's crust and oceans, hydrogen is the ninth most abundant element on a mass basis (0.9 mass %) and the third most abundant on an atom basis (15.4 atom %). Hydrogen is found in water, petroleum, proteins, carbohydrates, fats, and literally millions of other compounds.

Interstellar gas clouds consist largely of hydrogen, the most abundant element in the universe.

Approximately 95% of the H_2 produced in industry is synthesized and consumed in plants that manufacture other chemicals. The largest single consumer of hydrogen is the Haber process for synthesizing ammonia (Sections 13.6–13.10):

$$N_2(g) + 3\ H_2(g) \rightleftarrows 2\ NH_3(g)$$

Large amounts of hydrogen are also used for the synthesis of methanol, CH_3OH, from carbon monoxide:

$$CO(g) + 2\ H_2(g) \xrightarrow[\text{catalyst}]{\text{cobalt}} CH_3OH(l)$$

Methanol is an industrial solvent, a precursor to additives in unleaded gasolines, and a starting material for the manufacture of formaldehyde, CH_2O, used in making plastics. Annual U.S. production of methanol, about 1.7 billion gallons, consumes about 700,000 tons of hydrogen.

PROBLEM 14.1 Hydrogen is used to inflate weather balloons because it is much less dense than air. Calculate the density of gaseous H_2 at 25°C and 1 atm pressure. Compare your result with the density of dry air under the same conditions (1.185×10^{-3} g/cm^3).

14.2 ➤ISOTOPES OF HYDROGEN

As mentioned in Section 2.5, there are three isotopes of hydrogen: *protium*, or ordinary hydrogen (^1_1H), *deuterium*, or heavy hydrogen (^2_1H or D), and *tritium* (^3_1H or T). Nearly all (99.9844%) the atoms in naturally occurring hydrogen are protium. The terrestrial abundance of deuterium is only 0.0156 atom %, and tritium is present only in trace amounts ($\sim10^{-16}$ atom %).

The properties of protium, deuterium, and tritium are similar (Table 14.1) because chemical behavior is determined primarily by electronic structure and all three isotopes have the same electronic configuration ($1s^1$). There are, however, quantitative differences in properties, known as **isotope effects**, that arise from the differences in the mass of the isotopes. For example, D_2 has a higher melting point, a higher boiling point, and a greater

TABLE 14.1   Properties of Hydrogen Isotopes

Property	Protium	Deuterium	Tritium
Atomic hydrogen (H)			
Mass, amu	1.0078	2.0141	3.0160
Ionization energy, kJ/mol	1311.7	1312.2	
Nuclear stability	Stable	Stable	Radioactive
Molecular hydrogen (H$_2$)			
Melting point, K	13.96	18.73	20.62
Boiling point, K	20.39	23.67	25.04
Heat of dissociation, kJ/mol	435.9	443.4	446.9
Water (H$_2$O)			
Melting point, °C	0.00	3.81	4.48
Boiling point, °C	100.00	101.42	101.51
Density at 25°C, g/mL	0.997	1.104	1.214
Dissociation constant at 25°C	1.01×10^{-14}	0.195×10^{-14}	$\sim 0.06 \times 10^{-14}$

heat of dissociation than does H$_2$. Similarly, D$_2$O has a higher melting point and a higher boiling point than does H$_2$O, and the equilibrium constant for dissociation of D$_2$O is about five times smaller than that for H$_2$O:

$$H_2O(l) \rightleftharpoons H^+(aq) + OH^-(aq) \qquad K = 1.01 \times 10^{-14}$$

$$D_2O(l) \rightleftharpoons D^+(aq) + OD^-(aq) \qquad K = 0.195 \times 10^{-14}$$

Isotope effects are much greater for hydrogen than for any other element because the percentage differences between the masses of the various isotopes are considerably larger for hydrogen than for heavier elements.

Deuterium can be separated from protium by passing an electric current through a solution of an inert electrolyte in ordinary water. (*Electrolysis*, the process of using an electric current to bring about chemical change, will be discussed in Section 18.11.) Because D$_2$ evolves from D$_2$O more slowly than H$_2$ evolves from H$_2$O, the water is enriched in D$_2$O as the electrolysis proceeds.

$$2 H_2O(l) \xrightarrow{\text{electrolysis}} 2 H_2(g) + O_2(g) \quad \text{(faster)}$$

$$2 D_2O(l) \xrightarrow{\text{electrolysis}} 2 D_2(g) + O_2(g) \quad \text{(slower)}$$

In a typical experiment, reduction of the water's volume from 2400 L to 83 mL yields 99 percent pure D$_2$O. Large amounts of D$_2$O (about 160 tons per year in the United States) are manufactured by this method for use as a coolant and a moderator in nuclear reactors.

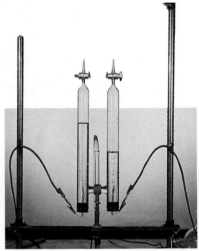

Electrolysis of water gives H$_2$ gas at one electrode and O$_2$ gas at the other electrode.

┌ **PROBLEM 14.2** The most abundant elements (by mass) in the body of a healthy human adult are oxygen (61.4%), carbon (22.9%), hydrogen (10.0%), and nitrogen (2.6%).

(a) Calculate the percent H if all the hydrogen atoms were deuterium atoms.
(b) Calculate the percent C if all the carbon atoms were atoms of the isotope having a mass of 13 amu ($^{13}_{6}$C).
(c) Are isotope effects larger for hydrogen or for carbon?

14.3 ►SYNTHESIS OF HYDROGEN

The purest hydrogen (>99.95% pure) is made by electrolysis of water. This process requires a large amount of energy, however—286 kJ per mole of H_2 produced—and the process is not economical for large-scale production.

$$2\ H_2O(l) \rightarrow 2\ H_2(g) + O_2(g) \qquad \Delta H° = +572\ kJ$$

Small amounts of hydrogen are conveniently prepared in the lab by reaction of dilute acid with an electropositive metal such as zinc:

$$Zn(s) + 2\ H^+(aq) \rightarrow H_2(g) + Zn^{2+}(aq)$$

Oxidation	↑	↑	↑	↑
numbers	0	+1	0	+2

This is a redox reaction (Section 4.6) in which hydrogen in H^+ is reduced from the +1 to the 0 oxidation state while zinc is oxidized from the 0 to the +2 oxidation state. A typical lab apparatus for generating and collecting hydrogen is shown in Figure 14.1.

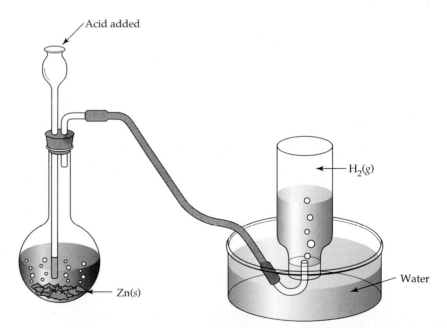

FIGURE 14.1 Preparation of hydrogen by reaction of zinc metal with dilute acid. The H_2 gas, which is nearly insoluble in water, displaces the water in the collection vessel.

Since water is the cheapest and most readily available source of hydrogen, all large-scale, industrial methods for producing hydrogen use a reducing agent such as hot iron, carbon, or methane (natural gas) to extract the oxygen from steam:

$$4\ H_2O(g) + 3\ Fe(s) \xrightarrow{\text{heat}} Fe_3O_4(s) + 4\ H_2(g) \qquad \Delta H° = -151\ kJ$$

$$H_2O(g) + C(s) \xrightarrow{1000°C} CO(g) + H_2(g) \qquad \Delta H° = +131\ kJ$$

Synthesis gas is formed from coal and steam in the Great Plains Synfuels Plant, Beulah, North Dakota.

At present, the most important industrial method for producing hydrogen is the three-step, **steam–hydrocarbon reforming process**. The first step in the process is the conversion of steam and methane to a mixture of carbon monoxide and hydrogen known as *synthesis gas* (so called because it can be used as the starting material for the synthesis of liquid fuels). The reaction requires high temperature, moderately high pressure, and a nickel catalyst:

$$H_2O(g) + CH_4(g) \xrightarrow[\text{Ni catalyst}]{1100°} CO(g) + 3\ H_2(g) \qquad \Delta H° = +206\ \text{kJ}$$

In the second step, the synthesis gas and additional steam are passed over a metal oxide catalyst at about 400°C. Under these conditions, the carbon monoxide component of the synthesis gas and the steam are converted to carbon dioxide and more hydrogen. This reaction of CO with H_2O is called the **gas shift reaction** because it shifts the composition of synthesis gas by removing the toxic carbon monoxide and producing more of the economically important hydrogen:

$$CO(g) + H_2O(g) \xrightarrow[\text{catalyst}]{400°C} CO_2(g) + H_2(g) \qquad \Delta H° = -41\ \text{kJ}$$

Finally, the unwanted carbon dioxide is removed from the hydrogen in a third step by passing the H_2/CO_2 mixture through a basic aqueous solution. This treatment converts the carbon dioxide to carbonate ion, which remains in the aqueous phase:

$$CO_2(g) + 2\ OH^-(aq) \rightarrow CO_3^{2-}(aq) + H_2O(l)$$

EXAMPLE 14.1

Write a balanced net ionic equation for the synthesis of hydrogen by reaction of magnesium metal with dilute acid.

SOLUTION   Magnesium is an active group 2A metal that reduces $H^+(aq)$ to $H_2(g)$. In turn, the magnesium is oxidized to Mg^{2+}. Balance the equation by using the oxidation number method (Section 4.8).

$$Mg(s) + 2\ H^+(aq) \rightarrow H_2(g) + Mg^{2+}(aq)$$

Oxidation   ↑   ↑   ↑   ↑
numbers   0   +1   0   +2

PROBLEM 14.3 Write a balanced net ionic equation for the reaction of gallium metal with dilute acid.

14.4 ▸REACTIVITY OF HYDROGEN

The hydrogen atom is the simplest of all atoms, containing only a single $1s$ electron and a single proton. In most versions of the periodic table, hydrogen is located above the alkali metals because they too have just one valence electron. Alternatively, hydrogen could be placed above the halogens because, like the halogens, hydrogen is just one electron short of a noble-gas configuration (He). Thus, hydrogen has properties similar to those of both alkali metals and halogens. A hydrogen atom can lose an electron to form a hydrogen cation, H^+, or it can gain an electron to yield a hydride anion, H^-:

$$\text{Alkali-metal-like reaction: } H(g) \rightarrow H^+(g) + e^- \qquad E_i = 1312 \text{ kJ/mol}$$

$$\text{Halogen-like reaction: } \qquad H(g) + e^- \rightarrow H^-(g) \qquad E_{ea} = -73 \text{ kJ/mol}$$

Because the amount of energy needed to ionize a hydrogen atom is so large ($E_i = 1312$ kJ/mol), hydrogen doesn't completely transfer its valence electron in ordinary chemical reactions. Instead, hydrogen shares this electron with a nonmetallic element to give a covalent compound such as CH_4, NH_3, H_2O, or HF. In this regard, hydrogen differs markedly from the alkali metals, which have much smaller ionization energies (ranging from 520 kJ/mol for Li to 376 kJ/mol for Cs) and which form ionic compounds with nonmetals.

Complete ionization of a hydrogen atom is possible in the gas phase, where the hydrogen ion is present as a bare proton of radius $\sim 1.5 \times 10^{-3}$ pm, about 100,000 times smaller than the radius of a hydrogen atom. In liquids and solids, however, the bare proton is too reactive to exist by itself. Instead, the proton bonds to a molecule that has a lone pair of electrons. In water, for example, the proton bonds to an H_2O molecule to give a hydronium ion, $H_3O^+(aq)$.

Although adding an electron to hydrogen ($E_{ea} = -73$ kJ/mol) releases less energy than adding an electron to one of the halogens ($E_{ea} = -295$ to

Explosive burning of the hydrogen-filled dirigible *Hindenburg* during landing at Lakehurst, New Jersey, on May 6, 1937.

−349 kJ/mol), hydrogen will accept an electron from an active metal to give an ionic hydride, such as NaH or CaH_2. In this regard, the behavior of hydrogen parallels that of the halogens.

At room temperature, H_2 is relatively unreactive because of its strong H–H bond, although it does react with F_2 to give HF and with Cl_2 to give HCl. Reactions of H_2 with O_2, N_2, or C, however, require high temperatures, the presence of a catalyst, or both. Catalysts such as metallic iron, nickel, palladium, or platinum facilitate the dissociation of H_2 into highly reactive H atoms. The reaction of hydrogen and oxygen is highly exothermic, and gas mixtures that contain as little as 4 volume % hydrogen in oxygen or in air are highly flammable and potentially explosive.

$$2\ H_2(g) + O_2(g) \rightarrow 2\ H_2O(l) \qquad \Delta H° = -572\ kJ$$

14.5 ►BINARY HYDRIDES

The **binary hydrides** are compounds that contain hydrogen and just one other element. Formulas and melting points of the simplest hydrides of the main-group elements are listed in Figure 14.2. Binary hydrides can be classified into three categories: ionic, covalent, and metallic.

(1) 1A	(2) 2A			(13) 3A	(14) 4A	(15) 5A	(16) 6A	(17) 7A	(18) 8A
LiH 692	BeH_2 d 250			B_2H_6 −165	CH_4 −182	NH_3 −78	H_2O 0	HF −83	
NaH d 800	MgH_2 d 280			AlH_3 d 150	SiH_4 −185	PH_3 −134	H_2S −86	HCl −115	
KH d	CaH_2 816			GaH_3 −15	GeH_4 −165	AsH_3 −116	H_2Se −66	HBr −88	
RbH d	SrH_2 d 675			InH_3 (?)	SnH_4 −146	SbH_3 −88	H_2Te −51	HI −51	
CsH d	BaH_2 d 675			TlH_3 (?)	PbH_4	BiH_3	H_2Po	HAt	

FIGURE 14.2 Formulas and melting points (°C) of the simplest hydrides of the main-group elements. The group 1A and the heavier group 2A hydrides, shown in blue, are ionic, while the other main-group hydrides, shown in red, are covalent. The change in bond type, however, is gradual and continuous. Transition-metal hydrides (not shown) are classified as metallic. The letter *d* indicates decomposition (rather than melting) on heating to the indicated temperature. The existence of InH_3 and TlH_3 is uncertain.

Ionic Hydrides

Ionic hydrides are saltlike, high-melting, white, crystalline compounds formed by the alkali metals and the heavier alkaline-earth metals Ca, Sr, and Ba. They can be prepared by direct reaction of the elements at about 400°C:

$$2\ Na(l) + H_2(g) \rightarrow 2\ NaH(s)$$

$$Ca(s) + H_2(g) \rightarrow CaH_2(s)$$

The alkali-metal hydrides contain alkali-metal cations and H^- anions in a face-centered cubic crystal structure like that of sodium chloride (Section 10.10). Alkali-metal hydrides are also ionic in the liquid state, as shown by the fact that the molten compounds conduct electricity.

The H^- anion is a good proton acceptor, and ionic hydrides therefore react with water to give H_2 gas in an acid–base reaction (Section 15.1):

$$CaH_2(s) + 2\ H_2O(l) \rightarrow 2\ H_2(g) + Ca^{2+}(aq) + 2\ OH^-(aq)$$

Calcium hydride reacts with water to give bubbles of H_2 gas and OH^- ions. The red color is due to added phenolphthalein, which turns from colorless to red in the presence of base.

This reaction of an ionic hydride with water can be considered a redox reaction as well as an acid–base reaction because the hydride reduces the water (+1 oxidation state for H) to H_2 (0 oxidation state). In turn, the hydride (−1 oxidation state for H) is oxidized to H_2. In general, ionic hydrides are good reducing agents. Some of them, such as potassium hydride, catch fire in air because of a rapid redox reaction with oxygen:

$$2\ KH(s) + O_2(g) \rightarrow H_2O(g) + K_2O(s)$$

Covalent Hydrides

Covalent hydrides, as their name implies, are compounds in which hydrogen is attached to another element by a covalent bond. The most common examples are the hydrides of the nonmetallic elements, such as diborane (B_2H_6), methane (CH_4), ammonia (NH_3), water (H_2O), and the hydrogen halides (HX). Only the simplest covalent hydrides are listed in Figure 14.2, though more complex examples, such as hydrogen peroxide (H_2O_2) and hydrazine (N_2H_4), are also known. Since most covalent hydrides consist of discrete, small molecules that have relatively weak intermolecular forces, they are gases or volatile liquids at ordinary temperatures.

Metallic Hydrides

Metallic hydrides are formed by reaction of the lanthanide and actinide metals and certain of the *d*-block transition metals with variable amounts of hydrogen. These hydrides have the general formula MH_x, where the x subscript represents the number of H atoms in the simplest formula. They are often called **interstitial hydrides** because they are thought to consist of a crystal lattice of metal atoms with the smaller hydrogen atoms occupying holes, or interstices, between the larger metal atoms (Figure 14.3).

FIGURE 14.3 Structure of an interstitial metallic hydride. The metal atoms (larger spheres) have a face-centered-cubic structure, and the hydrogen atoms (smaller spheres) occupy interstices (holes) between the metal atoms.

The nature of the bonding in metallic hydrides is not well understood, and it's not known whether the hydrogens are present as neutral H atoms, H^+ cations, or H^- anions. Because the hydrogen atoms can fill a variable number of interstices, many metallic hydrides are **nonstoichiometric compounds**, meaning that their atomic composition can't be expressed as a ratio of small whole numbers. Examples are $TiH_{1.7}$, $ZrH_{1.9}$, and PdH_x ($x < 1$). Other metallic hydrides, however, are stoichiometric compounds—for example, TiH_2 and UH_3.

The properties of metallic hydrides depend on their composition, which is a function of the partial pressure of H_2 gas in the surroundings. For example, PdH_x behaves as a metallic conductor for small values of x but becomes a semiconductor when x reaches about 0.5. (Semiconductors are discussed in Section 21.5.) The H atoms in PdH_x are highly mobile, and H_2 can pass through a membrane of palladium metal. The process probably involves dissociation of H_2 into H atoms on one surface of the membrane, diffusion of H atoms through the membrane as they jump from one interstice to another, and recombination to form H_2 on the opposite surface of the membrane. Because other gases don't penetrate palladium, this process can be used to separate H_2 or D_2 from other components of gas mixtures.

Interstitial hydrides are of current interest as potential hydrogen-storage devices because they can contain a remarkably large amount of hydrogen. Palladium, for example, absorbs up to 935 times its own volume of H_2. This amount corresponds to a density of hydrogen comparable to that in liquid hydrogen. For use as a fuel, hydrogen could be stored as PdH_x and then liberated when needed simply by heating the PdH_x.

Favored at higher temperature → $Pd(s) + x/2\ H_2(g) \rightleftarrows PdH_x(s)$ ← Favored at lower temperature

EXAMPLE 14.2

(a) Write a balanced net ionic equation for the reaction of barium hydride with water.
(b) How many grams of barium hydride must be treated with water to obtain 4.36 L of hydrogen at 20°C and 0.975 atm pressure?

SOLUTION   (a) Because barium is a group 2A metal, barium hydride, BaH_2, is an ionic hydride. The H^- ion reduces water to H_2 gas, and in the process H^- is oxidized to H_2 gas. Balance the equation either by inspection or by using the method of oxidation numbers.

$$BaH_2(s) + 2\ H_2O(l) \rightarrow 2\ H_2(g) + Ba^{2+}(aq) + 2\ OH^-(aq)$$

Oxidation numbers: -1 (BaH_2), $+1$ (H_2O), 0 (H_2)

(b) Follow a three-step procedure to find the mass of barium hydride needed. First, convert the volume of gaseous H_2 to moles by using the ideal-gas law:

$$n = \frac{PV}{RT} = \frac{(0.975\ \text{atm})(4.36\ \text{L})}{\left(0.08206\ \dfrac{\text{L} \cdot \text{atm}}{\text{mol} \cdot \text{K}}\right)(293\ \text{K})} = 0.177\ \text{mol}\ H_2$$

Next, use the balanced equation to calculate the number of moles of BaH_2 required to produce 0.177 moles of H_2:

$$\text{Moles of } BaH_2 = 0.177\ \text{mol}\ H_2 \times \frac{1\ \text{mol}\ BaH_2}{2\ \text{mol}\ H_2} = 0.0885\ \text{mol}\ BaH_2$$

Finally, calculate the molar mass of BaH_2, and use it to convert moles of BaH_2 to grams of BaH_2:

$$\text{Molar mass of } BaH_2 = 137.3 + 2(1.0) = 139.3\ \text{g/mol}$$

$$\text{Grams of } BaH_2 = 0.0885\ \text{mol}\ BaH_2 \times \frac{139.3\ \text{g}\ BaH_2}{1\ \text{mol}\ BaH_2} = 12.3\ \text{g}\ BaH_2$$

⌐ PROBLEM 14.4   Calcium hydride is a convenient, portable source of hydrogen used, for example, to inflate weather balloons. If the reaction of CaH_2 with water is used to inflate a balloon with 2.0×10^5 L of H_2 gas at 25°C and 1.00 atm pressure, how many kg of CaH_2 is needed?

⌐ PROBLEM 14.5   If palladium metal (density, $12.0\ \text{g/cm}^3$) dissolves 935 times its own volume of H_2 at STP, what is the value of x in the formula PdH_x? What is the density of hydrogen in PdH_x? What is the molarity of H atoms in PdH_x? Assume that the volume of palladium is unchanged when the H atoms go into the interstices. ⌐

14.6 ▸OXYGEN

Scuba divers require a supply of oxygen.

Oxygen, the most abundant element on the surface of our planet, is crucial to human life. It's in the air we breathe, the water we drink, and the food we eat. It's the oxidizing agent in the metabolic "burning" of foods, and it's an important component of biological molecules: Approximately one-fourth of the atoms in living organisms are oxygen. Moreover, oxygen is the oxidizing agent in the combustion processes that provide thermal and electrical energy for maintaining our industrialized civilization.

On a mass basis, oxygen constitutes 23% of the atmosphere (21% by volume), 46% of the lithosphere (the earth's crust), and more than 85% of the hydrosphere. In the atmosphere, oxygen is found primarily as O_2, sometimes called *dioxygen*. The oxygen in the hydrosphere is, of course, in the form of H_2O, but enough dissolved O_2 is present to maintain aquatic life. In the lithosphere, oxygen is combined with other elements in crustal rocks composed of silicates, carbonates, oxides, and other oxygen-containing minerals.

The amount of oxygen in the atmosphere remains fairly constant at about 1.18×10^{18} kg because the combustion and respiration processes that remove O_2 are balanced by photosynthesis, the complex process in which green plants use solar energy to produce O_2 and glucose from carbon dioxide and water:

$$6\ CO_2 + 6\ H_2O \rightarrow 6\ O_2 + C_6H_{12}O_6$$
$$\text{Glucose}$$

The metabolism of carbohydrates in our bodies to give carbon dioxide and water is essentially the reverse of the photosynthesis reaction. The energy from the sun that is absorbed in the endothermic photosynthetic process is released when organic matter is burned to carbon dioxide and water. This cycling of oxygen between the atmosphere and the biosphere serves as the mechanism for converting solar energy to heat and to the chemical energy needed for metabolic processes. Ultimately, nearly all our energy comes from the sun.

14.7 ➤PREPARATION AND USES OF OXYGEN

Oxygen was first isolated and characterized in the period 1771–1774 by the English chemist Joseph Priestley and the Swedish chemist Karl Wilhelm Scheele. Priestley and Scheele found that heating certain compounds such as mercury(II) oxide generates a colorless, odorless, tasteless gas that supports combustion better than air does:

$$2\ HgO(s) \xrightarrow{\text{heat}} 2\ Hg(l) + O_2(g)$$

Priestley called the gas "dephlogisticated air," but Lavoisier soon recognized it as an element and named it oxygen.

Small amounts of O_2 can be prepared in the laboratory by electrolysis of water, by decomposition of aqueous hydrogen peroxide in the presence of a catalyst such as Fe^{3+}, or by thermal decomposition of an oxoacid salt, such as potassium chlorate, $KClO_3$:

$$2\ H_2O(l) \xrightarrow{\text{electrolysis}} 2\ H_2(g) + O_2(g)$$

$$2\ H_2O_2(aq) \xrightarrow{\text{catalyst}} 2\ H_2O(l) + O_2(g)$$

$$2\ KClO_3(s) \xrightarrow[\text{MnO}_2\ \text{catalyst}]{\text{heat}} 2\ KCl(s) + 3\ O_2(g)$$

On heating, red mercury(II) oxide decomposes to give silvery droplets of mercury and colorless O_2 gas.

Because oxygen is relatively insoluble in water, it can be collected by water displacement using the same apparatus employed to collect hydrogen (Figure 14.1). Nowadays, though, oxygen is seldom prepared in the laboratory because it's commercially available as a compressed gas in high-pressure steel cylinders.

Oxygen is produced on an industrial scale, along with nitrogen and argon, by fractional distillation of liquefied air. When liquid air warms in

Crude iron is converted to steel by oxidizing impurities with O_2 gas.

a suitable distilling column, the more volatile components—nitrogen (bp $-196°C$) and argon (bp $-186°C$)—can be removed as gases from the top of the column. The less volatile oxygen (bp $-183°C$) remains as a liquid at the bottom. Oxygen (23 million tons) is the third-ranking industrial chemical produced in the United States; only sulfuric acid (40 million tons) and nitrogen (33 million tons) are produced in greater quantities.

More than two-thirds of the oxygen produced industrially is used in making steel. The crude iron obtained from a blast furnace contains impurities, such as carbon, silicon, and phosphorus, which adversely affect the mechanical properties of the metal. In refining crude iron to strong steels, the impurity levels are lowered by treating the iron with a controlled amount of O_2, thus converting the impurities to the corresponding oxides (Section 21.3).

Oxygen is used in sewage treatment to destroy malodorous compounds by oxidation and in paper bleaching to oxidize compounds that impart unwanted colors. In the oxyacetylene torch, the highly exothermic reaction between O_2 and acetylene provides the high temperatures ($>3000°C$) needed for cutting and welding metals:

$$2 \ HC{\equiv}CH(g) + 5 \ O_2(g) \rightarrow 4 \ CO_2(g) + 2 \ H_2O(g) \qquad \Delta H° = -2511 \ kJ$$
Acetylene

In all its applications, O_2 serves as an inexpensive and readily available oxidizing agent.

The oxyacetylene torch uses the highly exothermic reaction of acetylene with O_2 gas.

EXAMPLE 14.3

How many mL of O_2 gas at 25°C and 1.00 atm pressure can be obtained by thermal decomposition of 0.20 g of $KClO_3$?

$$2 \ KClO_3(s) \rightarrow 2 \ KCl(s) + 3 \ O_2(g)$$

SOLUTION   Follow a three-step procedure. First, determine the molar mass of $KClO_3$, and use it to calculate the number of moles of $KClO_3$ available:

Molar mass of $KClO_3 = 39.1 + 35.5 + 3(16.0) = 122.6 \ g/mol$

Moles of $KClO_3 = 0.20 \ g \ KClO_3 \times \dfrac{1 \ mol \ KClO_3}{122.6 \ g \ KClO_3} = 1.63 \times 10^{-3} \ mol \ KClO_3$

Next, use the balanced equation and the number of moles of $KClO_3$ to calculate the number of moles of O_2 produced:

$$\text{Moles of } O_2 = \frac{3 \text{ mol } O_2}{2 \text{ mol } KClO_3} \times (1.63 \times 10^{-3} \text{ mol } KClO_3) = 2.44 \times 10^{-3} \text{ mol } O_2$$

Finally, use the ideal-gas law and the number of moles of O_2 to calculate the volume of O_2:

$$V = \frac{nRT}{P} = \frac{(2.44 \times 10^{-3} \text{ mol})\left(0.08206 \dfrac{L \cdot atm}{K \cdot mol}\right)(298 \text{ K})}{1.00 \text{ atm}}$$

$$= 5.97 \times 10^{-2} \text{ L} = 6.0 \times 10^{-2} \text{ L or } 60 \text{ mL}$$

PROBLEM 14.6 How many mL of O_2 gas at 25°C and 1.00 atm pressure are obtained by thermal decomposition of 0.20 g of $KMnO_4$? The balanced equation for the reaction is $2 \, KMnO_4(s) \rightarrow K_2MnO_4(s) + MnO_2(s) + O_2(g)$.

14.8 ➤ PROPERTIES OF OXYGEN

Gaseous O_2 condenses at −183°C to form a pale blue liquid and freezes at −219°C to give a pale blue solid. In all three phases—gas, liquid, and solid— O_2 is paramagnetic, as illustrated previously in Figure 7.16. The oxygen–oxygen bond length in O_2 is 121 pm, appreciably shorter than the O–O single bond in H_2O_2 (146 pm), and the bond dissociation energy of O_2 (498 kJ) is intermediate between that for the single bond in F_2 (158 kJ) and the triple bond in N_2 (945 kJ). These properties are consistent with the presence of a double bond in O_2 (Section 7.14).

Liquid oxygen has a blue color.

We can anticipate the reactivity of oxygen on the basis of the electron configuration of an oxygen atom ($1s^2 \, 2s^2 \, 2p^4$) and its high electronegativity (3.5). With six valence electrons, oxygen is just two electrons short of the stable octet configuration of neon, the next noble gas. Oxygen can achieve an octet configuration either by accepting two electrons from an active metal or by gaining a share in two additional electrons through covalent bonding. Thus, oxygen reacts with active metals, such as lithium and magnesium, to give *ionic oxides*:

$$4\,Li(s) + O_2(g) \rightarrow 2\,Li_2O(s)$$

$$2\,Mg(s) + O_2(g) \rightarrow 2\,MgO(s)$$

With nonmetals, such as hydrogen, carbon, sulfur, and phosphorus, oxygen forms *covalent oxides*:

$$2\,H_2(g) + O_2(g) \rightarrow H_2O(l)$$

$$C(s) + O_2(g) \rightarrow CO_2(g)$$

$$S_8(s) + 8\,O_2(g) \rightarrow 8\,SO_2(g)$$

$$P_4(s) + 5\,O_2(g) \rightarrow P_4O_{10}(s)$$

In covalent compounds, oxygen generally achieves an octet configuration either by forming two single bonds, as in H_2O, or one double bond, as in CO_2. Double bonds often form between oxygen and other small atoms because there is good overlap between an oxygen $p\pi$ orbital and the $p\pi$ orbitals of small atoms, such as carbon and nitrogen. With larger atoms, such as silicon, however, there is less efficient π overlap, and double bond formation is therefore less common (Figure 14.4).

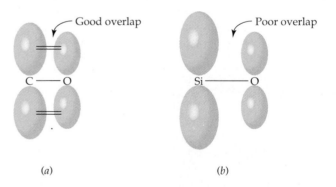

(a)   (b)

FIGURE 14.4 **(a)** With a small atom, such as carbon, oxygen forms a strong π bond. **(b)** With a larger atom, such as silicon, oxygen tends not to form π bonds because the longer Si–O distance and the more diffuse silicon orbitals result in poor π overlap.

Oxygen reacts directly with all the elements in the periodic table except the noble gases and a few inactive metals, such as platinum and gold. Fortunately, most of these reactions are slow at room temperature. Otherwise cars would rust faster, and many elements would spontaneously inflame in air. At higher temperatures, however, oxygen is extremely reactive. Figure 14.5 illustrates reactions of oxygen with magnesium, sulfur, and white phosphorus. To initiate burning of magnesium and sulfur requires heat, but white phosphorus spontaneously inflames in oxygen at room temperature. The brilliant white light emitted by burning magnesium makes this element particularly useful in fireworks.

(a)

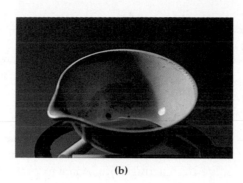

(b)

(c)

FIGURE 14.5 Some reactions of oxygen: **(a)** A piece of hot magnesium metal burns in air with a bright white flame, producing a white smoke of solid magnesium oxide, MgO. **(b)** Hot, molten sulfur burns in air with a blue flame, forming gaseous sulfur dioxide, SO_2. **(c)** White phosphorus spontaneously inflames in oxygen, yielding an incandescent white smoke of solid P_4O_{10}.

14.9 ➤OXIDES

It's convenient to classify binary oxygen-containing compounds on the basis of the oxygen's oxidation state. Binary compounds with oxygen in the -2 oxidation state are called **oxides**, compounds with oxygen in the -1 oxidation state are **peroxides**, and compounds with oxygen in the $-1/2$ oxidation state are **superoxides**. We'll look first at oxides and then consider peroxides and superoxides in the next section.

We can categorize oxides in terms of their acid–base character as basic, acidic, or amphoteric, as shown in Figure 14.6.

FIGURE 14.6 Formulas, acid–base properties, and covalent–ionic character of the oxides of main-group elements in their highest oxidation states. Basic oxides are shown in blue, acidic oxides are shown in red, and amphoteric oxides are shown in violet.

Increasing acidic character →

Increasing covalent character →

(1) 1A	(2) 2A		(13) 3A	(14) 4A	(15) 5A	(16) 6A	(17) 7A	(18) 8A
Li_2O	BeO		B_2O_3	CO_2	N_2O_5			
Na_2O	MgO		Al_2O_3	SiO_2	P_4O_{10}	SO_3	Cl_2O_7	
K_2O	CaO		Ga_2O_3	GeO_2	As_2O_5	SeO_3	BrO_2	
Rb_2O	SrO		In_2O_3	SnO_2	Sb_2O_5	TeO_3	I_2O_5	
Cs_2O	BaO		Tl_2O_3	PbO_2	Bi_2O_5			

Increasing basic character ↓

Increasing ionic character

Basic oxides are ionic and are formed by the metals on the left side of the periodic table. Water-soluble basic oxides, such as Na_2O, dissolve by reacting with water to produce OH^- ions:

$$Na_2O(s) + H_2O(l) \rightarrow 2\,Na^+(aq) + 2\,OH^-(aq)$$

Water-insoluble basic oxides, such as MgO, dissolve in strong acids because H^+ ions from the acid combine with the O^{2-} ion to produce water:

$$MgO(s) + 2\,H^+(aq) \rightarrow Mg^{2+}(aq) + H_2O(l)$$

Acidic oxides are covalent and are formed by the nonmetals on the right side of the periodic table. Water-soluble acidic oxides, such as Cl_2O_7, dissolve by reacting with water to produce aqueous H^+ ions:

$$Cl_2O_7(l) + H_2O(l) \rightarrow 2\,H^+(aq) + 2\,ClO_4^-(aq)$$

Water-insoluble acidic oxides, such as SiO_2, dissolve in strong bases:

$$SiO_2(s) + 2\,OH^-(aq) \rightarrow SiO_3^{2-}(aq) + H_2O(l)$$

Amphoteric oxides exhibit both acidic and basic properties. (The term **amphoteric** [am-fo-**tare**-ic] comes from the Greek word *amphoteros*, meaning "in both ways.") For example, Al_2O_3 is insoluble in water, but it dissolves both in strong acids and in strong bases. Al_2O_3 behaves as a base when it reacts with acids, giving the Al^{3+} ion, but it behaves as an acid when it reacts with bases, yielding the aluminate ion, $Al(OH)_4^-$.

Basic behavior $Al_2O_3(s) + 6\,H^+(aq) \rightarrow 2\,Al^{3+}(aq) + 3\,H_2O(l)$

Acidic behavior $Al_2O_3(s) + 2\,OH^-(aq) + 3\,H_2O(l) \rightarrow 2\,Al(OH)_4^-(aq)$

The elements that form amphoteric oxides have intermediate electronegativity, and their oxides have strongly polar covalent character.

The acid–base properties and the ionic–covalent character of an oxide of an element depend on the element's position in the periodic table and on its oxidation state. As Figure 14.6 shows, both the acidic character and the covalent character of an oxide increase across the periodic table from the active metals on the left to the electronegative nonmetals on the right. In the third row, for example, Na_2O and MgO are basic, Al_2O_3 is amphoteric, and SiO_2, P_4O_{10}, SO_3, and Cl_2O_7 are acidic. Within a group in the periodic table, both the basic character and the ionic character of an oxide increase from the more electronegative elements at the top to the less electronegative ones at the bottom. In group 3A, for example, B_2O_3 is acidic, Al_2O_3 and Ga_2O_3 are amphoteric, and In_2O_3 and Tl_2O_3 are basic. Combining the horizontal and vertical trends in acidity, we find the most acidic oxides in the upper right of the periodic table, the most basic oxides in the lower left, and the amphoteric oxides in a roughly diagonal band stretching across the middle.

Both the acidic character and the covalent character of different oxides of the same element increase with increasing oxidation state of the element.

Thus, sulfur(VI) oxide (sulfur trioxide, SO_3) is more acidic than sulfur(IV) oxide (sulfur dioxide, SO_2). Reaction of SO_3 with water gives a strong acid (sulfuric acid, H_2SO_4), whereas reaction of SO_2 with water yields a weak acid (sulfurous acid, H_2SO_3). The oxides of chromium exhibit the same trend. Chromium(VI) oxide (CrO_3) is acidic, chromium(III) oxide (Cr_2O_3) is amphoteric, and chromium(II) oxide (CrO) is basic.

As the bonding in oxides gradually changes from ionic to covalent, there is a corresponding change in structure and physical properties, as illustrated by the melting points in Figure 14.7. MgO, an ionic oxide with the face-centered-cubic NaCl crystal structure, has a very high melting point because of its high lattice energy (Section 6.4). Al_2O_3 and SiO_2 also have extended three-dimensional structures (Section 10.11), but they exhibit increasingly covalent character and have increasingly lower melting points. The trend continues to P_4O_{10}, SO_3, and Cl_2O_7, which are molecular substances with strong covalent bonds within each molecule but only weak intermolecular forces between molecules. Thus, P_4O_{10} is a volatile, low-melting solid, and SO_3 and Cl_2O_7 are volatile liquids.

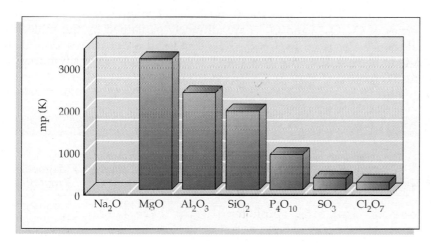

FIGURE 14.7 Melting points in kelvins of oxides of the third-period elements in their highest oxidation states. (No melting point is given for Na_2O because it sublimes at 1548 K.)

The uses of oxides are, of course, determined by their properties. Because of their high thermal stability, high strength, and high electrical resistance, MgO and Al_2O_3 are used as high-temperature electrical insulators in products such as electrical heaters and spark plugs. SiO_2 is the main component of the optical fibers used for communications. The acidic oxides of the nonmetals are important as precursors to industrial acids, such as H_3PO_4 and H_2SO_4.

⌈ **PROBLEM 14.7** Write balanced net ionic equations for the following reactions.

(a) dissolution of solid Li_2O in water
(b) dissolution of SO_3 in water
(c) dissolution of the amphoteric oxide Cr_2O_3 in strong acid
(d) dissolution of Cr_2O_3 in strong base to give $Cr(OH)_4^-$ ions

14.10 ▶ PEROXIDES AND SUPEROXIDES

When the heavier group 1A and 2A metals are heated in an excess of air, they form either *peroxides*, such as Na_2O_2 and BaO_2, or *superoxides*, such as KO_2, RbO_2, and CsO_2. Under the same conditions, however, the lighter group 1A and 2A metals form normal oxides, such as Li_2O, MgO, CaO, and SrO. All of these compounds are ionic solids, and the nature of the product obtained is determined by ion sizes and the lattice energies of the various solids.

The metal peroxides contain the peroxide ion, O_2^{2-}, and the metal superoxides contain the superoxide ion, O_2^-:

$$\left[:\ddot{O}:\ddot{O}:\right]^{2-} \qquad \left[:\ddot{O}:\dot{O}:\right]^- \longleftrightarrow \left[:\dot{O}:\ddot{O}:\right]^-$$

Peroxide ion   Superoxide ion

The peroxide ion is diamagnetic and has an O–O single bond length of 149 pm. The superoxide ion, with one unpaired electron, is paramagnetic and has an O–O bond length of 133 pm, intermediate between the peroxide ion's bond length and the O=O double bond length of 121 pm in O_2. This suggests a bond order of 1.5 for O_2^-. The trend in bond lengths and the magnetic properties of O_2, O_2^-, and O_2^{2-} are all nicely explained by the molecular orbital theory discussed in Section 7.14. The bond order decreases and the bond length increases with increasing population of the antibonding π^* orbitals (Table 14.2).

TABLE 14.2 Bond and Magnetic Properties of Diatomic Oxygen Species

Species	Number of π^* Electrons	Number of Unpaired Electrons	Bond Order	Bond Length (pm)	Magnetic Properties
O_2	2	2	2	121	Paramagnetic
O_2^-	3	1	1.5	133	Paramagnetic
O_2^{2-}	4	0	1	149	Diamagnetic

Dissolution of sodium peroxide in water that contains phenolphthalein gives a red color due to formation of OH^- ions.

Like the oxide ion, the peroxide ion is a base. When a metal peroxide dissolves in water, the O_2^{2-} ion picks up a proton from water, forming HO_2^- and OH^- ions. The dissolution of Na_2O_2 is typical:

$$Na_2O_2(s) + H_2O(l) \rightarrow 2\,Na^+(aq) + HO_2^-(aq) + OH^-(aq)$$

In the presence of strong acids, the O_2^{2-} ion combines with two protons, yielding hydrogen peroxide, H_2O_2. For example, barium peroxide reacts with a stoichiometric amount of sulfuric acid to give insoluble barium sulfate and an aqueous solution of hydrogen peroxide:

$$BaO_2(s) + H_2SO_4(aq) \rightarrow BaSO_4(s) + H_2O_2(aq)$$

When metal superoxides, such as KO_2, dissolve in water, they decompose with evolution of oxygen:

$$2\ KO_2(s) + H_2O(l) \rightarrow O_2(g) + 2\ K^+(aq) + HO_2^-(aq) + OH^-(aq)$$

Oxidation
numbers $-1/2$ 0 -1

This decomposition is a redox reaction in which the oxygen in KO_2 is simultaneously oxidized from the $-1/2$ oxidation state in O_2^- to the 0 oxidation state in O_2 and reduced from the $-1/2$ oxidation state in O_2^- to the -1 oxidation state in HO_2^-. Such a reaction, in which a substance is both oxidized and reduced, is called a **disproportionation** reaction. One useful consequence of the KO_2 disproportionation reaction in water is that potassium superoxide can serve as a convenient source of oxygen in masks worn by fire fighters. The oxygen results from reaction of KO_2 with exhaled water vapor.

A self-contained breathing apparatus used by fire fighters.

EXAMPLE 14.4

What is the oxidation state of oxygen in each of the following compounds? Is each compound an oxide, a peroxide, or a superoxide?

(a) KO_2 **(b)** BaO_2 **(c)** SnO_2

SOLUTION   First determine the oxidation number of the metal, which is generally equal to the periodic group number (Section 2.10). Then assign an oxidation number to oxygen so that the sum of the oxidation numbers is zero.

(a) Because K in group 1A has an oxidation number of $+1$, the oxidation state of oxygen in KO_2 must be $-1/2$. Thus, KO_2 is a superoxide.
(b) Because Ba in group 2A has an oxidation number of $+2$, the oxidation state of oxygen in BaO_2 must be -1. Thus, BaO_2 is a peroxide.
(c) Because Sn in group 4A has an oxidation number of $+4$, the oxidation state of oxygen in SnO_2 must be -2. Thus, SnO_2 is an oxide. (The alternative formulation as a peroxide of Sn^{2+} can be ruled out on the basis of the chemical properties of SnO_2.)

⌐ **PROBLEM 14.8**   What is the oxidation state of oxygen in each of the following compounds? Is each compound an oxide, a peroxide, or a superoxide?

(a) Rb_2O_2 **(b)** CaO **(c)** CsO_2 **(d)** SrO_2 **(e)** CO_2

⌐ **PROBLEM 14.9**   Write a balanced net ionic equation for the reaction of water with each of the oxygen compounds listed in Problem 14.8.   ⌐

14.11 ►HYDROGEN PEROXIDE

Hydrogen peroxide (HOOH) is sold in drugstores as a 3% aqueous solution for domestic use and is marketed as a 30% aqueous solution for industrial and laboratory use. Because of its oxidizing properties, hydrogen peroxide is used as a mild antiseptic and as a bleach for textiles, paper pulp, and hair. In the chemical industry, hydrogen peroxide is a starting material for synthesis of other compounds that contain the peroxide (–O–O–) linkage, some of which are used in the manufacture of plastics.

Pure hydrogen peroxide is an almost colorless, syrupy liquid that freezes at $-0.4°C$ and boils at an estimated $150°C$. The exact boiling point is not known with certainty, however, because pure H_2O_2 explodes when heated. The relatively high estimated boiling point indicates strong hydrogen bonding between H_2O_2 molecules in the liquid (Section 10.2). In aqueous solutions, hydrogen peroxide behaves as a weak acid, partially dissociating to give H^+ and HO_2^- ions.

Hydrogen peroxide is both a strong oxidizing agent and a reducing agent. When hydrogen peroxide acts as an oxidizing agent, oxygen is reduced from the -1 oxidation state in H_2O_2 to the -2 oxidation state in H_2O (or OH^-). For example, hydrogen peroxide oxidizes Br^- to Br_2 in acidic solution:

$$H_2O_2(aq) + 2\,H^+(aq) + 2\,Br^-(aq) \rightarrow 2\,H_2O(l) + Br_2(aq)$$

Oxidation numbers: -1 ; -1 ; -2 ; 0

When hydrogen peroxide acts as a reducing agent, oxygen is oxidized from the -1 oxidation state in H_2O_2 to the 0 oxidation state in O_2. A typical example is the reaction of hydrogen peroxide with permanganate ion, MnO_4^-, in acidic solution:

$$5\,H_2O_2(aq) + 2\,MnO_4^-(aq) + 6\,H^+(aq) \rightarrow 5\,O_2(g) + 2\,Mn^{2+}(aq) + 8\,H_2O(l)$$

Oxidation numbers: -1 ; $+7$; 0 ; $+2$

The reduction of manganese in this reaction from the $+7$ oxidation state in MnO_4^- to the $+2$ oxidation state in the manganese(II) ion, Mn^{2+}, is accompanied by a beautiful color change (Figure 14.8).

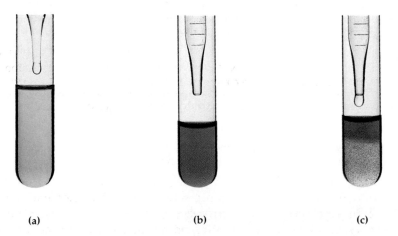

FIGURE 14.8 Some reactions of hydrogen peroxide: **(a)** Addition of a colorless aqueous solution of H_2O_2 to a colorless acidic solution of NaBr produces a yellow-orange color due to bromine, Br_2. **(b)** Addition of aqueous H_2O_2 to a violet acidic solution of $KMnO_4$ produces **(c)** bubbles of O_2 gas and decolorizes the solution as the violet MnO_4^- ion is converted to the nearly colorless Mn^{2+} ion.

(a)   (b)   (c)

Because hydrogen peroxide is both an oxidizing agent and a reducing agent, it can oxidize and reduce itself. Thus, hydrogen peroxide is unstable with respect to disproportionation to water and oxygen:

$$2\,H_2O_2(l) \rightarrow 2\,H_2O(l) + O_2(g) \qquad \Delta H° = -196\text{ kJ}$$

Oxidation numbers: -1 ; -2 ; 0

In the absence of catalysts, the disproportionation is too slow to be observed at room temperature. Rapid, exothermic, and potentially explosive decomposition of hydrogen peroxide is initiated, however, by heat and by a broad range of catalysts, including transition-metal ions, certain anions, such as I^-, metal surfaces, blood (Figure 14.9), and even tiny particles of dust. Because decomposition is accelerated by light, hydrogen peroxide is stored in dark-colored bottles. Hydrogen peroxide is best handled in dilute aqueous solutions; concentrated solutions and the pure liquid are extremely hazardous materials.

FIGURE 14.9 (a) When a few drops of blood are added to aqueous hydrogen peroxide, (b) the hydrogen peroxide decomposes rapidly, evolving bubbles of oxygen that produce a thick foam.

PROBLEM 14.10 Draw a Lewis electron dot structure for H_2O_2. Is your structure consistent with an O–O bond length of 148 pm? (Look at Table 14.2 to see how bond length and bond order are related.)

PROBLEM 14.11 The discoloration and restoration of old oil paintings involves some interesting chemistry. On exposure to polluted air containing H_2S, white lead carbonate pigments are converted to PbS, a black solid. Hydrogen peroxide has been used to restore the original white color. Write a balanced equation for the reaction, which involves oxidation of black PbS to white $PbSO_4$.

14.12 ▸OZONE

Oxygen exists in two allotropes: ordinary dioxygen, O_2, and ozone, O_3. (Recall from Section 10.11 that allotropes are forms of an element in which the atoms are connected in different ways.) Ozone is a toxic, pale blue gas with a characteristic sharp, penetrating odor that you can detect at concentrations as low as 0.01 ppm. You've probably noticed the odor of ozone around sparking electric motors or after a severe electrical storm. Ozone is produced when an electric discharge passes through O_2, providing the energy needed to bring about the endothermic reaction:

$$3\ O_2(g) \xrightarrow[\text{discharge}]{\text{electric}} 2\ O_3(g) \qquad \Delta H^\circ = 285 \text{ kJ}$$

Although ozone can be prepared in the laboratory by passing O_2 through an electrical device like that shown in Figure 14.10, it is unstable and decomposes exothermically to O_2. Decomposition of the dilute gas is slow, but the concentrated gas, liquid ozone (bp $-112°C$), or solid ozone (mp $-192°C$) can decompose explosively.

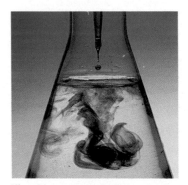

FIGURE 14.10 Generator for preparing ozone. A stream of O_2 gas flows through the ozonizer tube and is partially converted to O_3 when an electric discharge passes through the gas.

O$_2$ →

→ O_3 and unreacted O_2

Metal foil on inside of inner glass tube

Metal foil on outside of outer glass tube

High-voltage source

Two resonance structures are required for ozone because the two O–O bonds have equal length:

O   128 pm
O   O
117°

The bent structure is in accord with the VSEPR model described in Section 7.9, which predicts a bond angle near 120° for a triatomic molecule in which the central atom is surrounded by two σ-bond pairs and one lone pair. The π bond is delocalized over the three oxygen atoms, giving a net bond order of 1.5 between each pair of oxygen atoms (1σ bond + 0.5π bond; Section 7.15). This bonding description is in agreement with the O–O bond length of 128 pm, intermediate between the lengths of an O–O single bond and an O=O double bond.

Ozone is an extremely powerful oxidizing agent; of the common oxidizing agents, only F_2 is more potent. A standard method for detecting ozone in polluted air is to pass the air through a basic solution of potassium iodide that contains a starch indicator. The ozone oxidizes iodide ion to iodine, I_2, which combines with the starch to give the deep blue starch–iodine complex:

The blue starch–iodine complex forms when I^- is oxidized to I_2 in the presence of starch.

$$O_3(g) + 2\,I^-(aq) + H_2O(l) \rightarrow O_2(g) + I_2(aq) + 2\,OH^-(aq)$$

Oxidation numbers: 0   −1   0   −2

Because of its oxidizing properties, ozone is sometimes used to kill bacteria in drinking water.

14.13 ➤ WATER

Water is the most familiar and most abundant compound on earth. Nearly three-fourths of the earth's surface is covered with water; an estimated 1.35×10^{18} m³ of water is present in the oceans.[1] Water accounts for nearly two-thirds of the mass of the adult human body and 93% of the mass of the human embryo in the first month.

This photograph from space shows the Pacific Ocean off the west coast of the United States and Mexico.

[1] It's interesting to note that the volume of the oceans in cubic centimeters (1.35×10^{24} cm³) is roughly twice Avogadro's number.

Water reacts with the alkali metals, the heavier alkaline-earth metals (Ca, Sr, Ba, and Ra), and the halogens. With most other elements, water is unreactive at room temperature. Water is reduced to hydrogen by the alkali and alkaline-earth metals, which are oxidized to aqueous metal hydroxides:

$$2\ Na(s) + 2\ H_2O(l) \rightarrow H_2(g) + 2\ Na^+(aq) + 2\ OH^-(aq)$$

$$Ca(s) + 2\ H_2O(l) \rightarrow H_2(g) + Ca^{2+}(aq) + 2\ OH^-(aq)$$

Oxidation numbers

Ca	H$_2$O	H$_2$	Ca^{2+}
0	+1	0	+2

Only fluorine is more electronegative than oxygen, and fluorine is the only element able to oxidize water to oxygen. In the process, fluorine is reduced to hydrofluoric acid:

$$2\ F_2(g) + 2\ H_2O(l) \rightarrow O_2(g) + 4\ HF(aq)$$

Oxidation numbers

F$_2$	H$_2$O	O$_2$	HF
0	-2	0	-1

Chlorine doesn't oxidize water, but instead disproportionates to a limited extent. The products are hypochlorous acid, a weak acid in which chlorine is in the +1 oxidation state, and hydrochloric acid, a strong acid in which chlorine is in the −1 oxidation state:

$$Cl_2(g) + H_2O(l) \rightleftarrows HOCl(aq) + H^+(aq) + Cl^-(aq)$$

Oxidation numbers

Cl$_2$	HOCl	Cl$^-$
0	+1	-1

Bromine and iodine behave similarly, but the extent of disproportionation decreases markedly in the series Cl > Br > I.

▭ **PROBLEM 14.12** Write a balanced net ionic equation for the reaction of water with each of the following elements: **(a)** Li **(b)** Sr **(c)** Br$_2$

Calcium reacts with water to give bubbles of H$_2$ gas and an aqueous solution of calcium hydroxide.

14.14 ►HYDRATES

Solid compounds that contain water molecules are called **hydrates**. Examples are hydrated salts, such as magnesium perchlorate hexahydrate, $Mg(ClO_4)_2 \cdot 6\ H_2O$, and aluminum chloride hexahydrate, $AlCl_3 \cdot 6\ H_2O$. Because the structures of hydrates are sometimes complex or unknown, a dot is used in the formula of a hydrate to specify the composition without indicating how the water is bound. If the structure is known, a more informative formula can be given. The formulas $[Mg(H_2O)_6](ClO_4)_2$ and $[Al(H_2O)_6]Cl_3$, for instance, indicate that the six water molecules in each compound are attached to the metal ion. As shown in Figure 14.11, the negative (oxygen) end of each dipolar water molecule bonds to the positive metal cation, and the six water molecules are located at the vertices of an octahedron. Because bonding interactions between water and a metal cation increase with increasing charge on the cation, hydrate formation is common in the case of salts that contain +2 and +3 cations.

FIGURE 14.11 **(a)** A regular octahedron is a polyhedron that has eight equilateral triangular faces and six vertices. **(b)** Octahedral structure of the hydrated metal cation in $[Mg(H_2O)_6](ClO_4)_2$. The six Mg–O bonds point toward the six vertices of a regular octahedron. **(c)** A view of the $[Mg(H_2O)_6]^{2+}$ cation showing only the location of the atoms and the octahedral arrangement of the bonds to the six H_2O molecules.

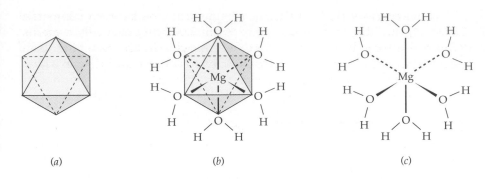

(a)   (b)   (c)

When hydrates are heated, the water is driven off. If you heat blue crystals of $CuSO_4 \cdot 5H_2O$ above 350°C, for example, you'll obtain anhydrous copper sulfate, $CuSO_4$. This transformation is readily observed, since the anhydrous compound is white (Figure 14.12).

FIGURE 14.12 Blue crystals of $CuSO_4 \cdot 5H_2O$ are converted to white anhydrous $CuSO_4$ when the sample is heated with a Bunsen flame.

Approximately 2.0% of the earth's water is in the form of ice.

Some anhydrous compounds have such a great tendency to form hydrates that they absorb water from the atmosphere. Anhydrous $Mg(ClO_4)_2$, for example, picks up water from air, yielding $Mg(ClO_4)_2 \cdot 6H_2O$. Compounds that absorb water from the air are said to be **hygroscopic** and are often useful as drying agents.

PROBLEM 14.13 A 5.62 g sample of a hydrate of nickel sulfate is heated until all the water is driven off. If 3.10 g of anhydrous $NiSO_4$ is obtained, what is the formula of the hydrate?

14.15 ►NATURAL WATERS

Approximately 97.3% of the world's vast supply of water (1.38×10^{18} m³) is in the oceans. Most of the rest is in the form of polar ice caps and glaciers (2.0%) and underground fresh water (0.6%). Freshwater lakes and rivers account for less than 0.01% of the total water supply, yet they nevertheless contain an enormous amount of water (1.26×10^{14} m³).

Sea water is unfit for drinking or agriculture because each kilogram contains about 35 g of dissolved salts. The major constituent of sea water is sodium chloride, but more than 60 different elements are present in small amounts. Table 14.3 lists the ions that account for more than 99% of the mass of the dissolved salts. Although the oceans represent an almost unlimited source of chemicals, ion concentrations are low and consequently recovery costs are high. Only three substances are obtained from sea water commercially: sodium chloride, magnesium, and bromine.

Water for use in homes, agriculture, and industry is generally obtained from freshwater lakes, rivers, or underground sources. The water you drink must be purified to remove solid particles, colloidal material, bacteria, and other harmful impurities. Important steps in a typical purification process include preliminary filtration, sedimentation, sand filtration, aeration, and sterilization (Figure 14.13).

TABLE 14.3 Major Ionic Constituents of Sea Water	
Ion	**g/kg of sea water**
Cl^-	19.0
Na^+	10.5
SO_4^{2-}	2.65
Mg^{2+}	1.35
Ca^{2+}	0.40
K^+	0.38
HCO_3^-	0.14
Br^-	0.065

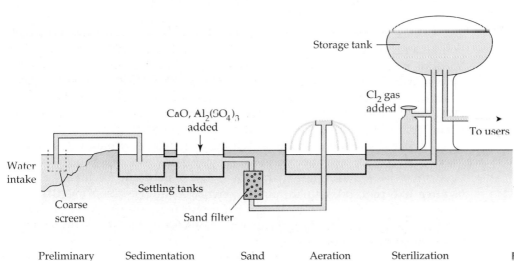

Preliminary filtration	Sedimentation	Sand filtration	Aeration	Sterilization

FIGURE 14.13 Purification of drinking water.

The sedimentation, or settling, of suspended matter takes place in large tanks and is accelerated by the addition of lime, CaO, and aluminum sulfate, $Al_2(SO_4)_3$. The lime makes the water slightly basic, which precipitates Al^{3+} ions from the added aluminum sulfate as aluminum hydroxide:

$$CaO(s) + H_2O(l) \rightarrow Ca^{2+}(aq) + 2\ OH^-(aq)$$

$$Al^{3+}(aq) + 3\ OH^-(aq) \rightarrow Al(OH)_3(s)$$

As the gelatinous precipitate of aluminum hydroxide slowly settles, it carries with it suspended solid and colloidal material and most of the bacteria as well. The water is then filtered through a bed of sand and subsequently sprayed into the air to oxidize dissolved organic impurities. Finally, the water is sterilized by adding chlorine or ozone, which kills the remaining bacteria. The water still contains up to 0.5 g/L of inorganic ions such as Na^+, K^+, Mg^{2+}, Ca^{2+}, Cl^-, F^-, SO_4^{2-}, and HCO_3^-, but in such low concentrations that these ions are not harmful.

interlude—A "HYDROGEN ECONOMY"

Hydrogen is an enormously attractive fuel because it's environmentally clean, giving only water as a combustion product. If hydrogen is burned in air, small amounts of nitrogen oxides can be produced because of the high-temperature combination of nitrogen and oxygen, but the combustion products are free of CO, CO_2, SO_2, unburned hydrocarbons, and other environmental pollutants that result from the combustion of petroleum fuels. The amount of heat liberated when hydrogen burns is 242 kJ/mol, more than twice that of gasoline, oil, or natural gas on a mass basis (121 kJ/g).

$$H_2(g) + 1/2\ O_2(g) \rightarrow H_2O(g) \qquad \Delta H° = -242\ kJ$$

Some people envision what they call a "hydrogen economy" in which our energy needs are met by gaseous, liquid, and solid hydrogen. For heating homes, gaseous hydrogen could be conveyed through underground pipes, while liquid hydrogen could be shipped by truck or by rail in large vacuum-insulated tanks. Automobiles might be powered by "solid hydrogen" in the form of solid interstitial hydrides. Prototype cars have already been built with their engines modified to run on hydrogen.

What's keeping us from reaching a hydrogen economy? Before a hydrogen economy can become a reality, cheaper ways of producing hydrogen must be found. Since hydrogen is not a naturally occurring energy source like coal, oil, or natural gas, energy from elsewhere must first be expended to produce the hydrogen before it can be used. Current research therefore focuses on finding cheaper methods for extracting hydrogen from its compounds.

One approach for producing hydrogen is to use solar energy to "split" water into hydrogen and oxygen. The feasibility of the scheme depends on the development of catalysts that absorb sunlight and then use the energy to reduce water to hydrogen. Another strategy employs thermal energy to effect a series of reactions that bring about the net conversion of water to hydrogen and oxygen. One such reaction series uses the following high-temperature reactions with iron compounds in which an iron atom is shuttled between different oxidation states:

$$3\ FeCl_2(s) + 4\ H_2O(g) \xrightarrow{500°C} Fe_3O_4(s) + 6\ HCl(g) + H_2(g)$$

$$Fe_3O_4(s) + 3/2\ Cl_2(g) + 6\ HCl(g) \xrightarrow{100°C} 3\ FeCl_3(s) + 3\ H_2O(g) + 1/2\ O_2(g)$$

$$3\ FeCl_3(s) \xrightarrow{300°C} 3\ FeCl_2(s) + 3/2\ Cl_2(g)$$

Net: $H_2O(g) \longrightarrow H_2(g) + 1/2\ O_2(g)$

The use of liquid hydrogen as a fuel in the U.S. space program is well known. Hydrogen powered the *Saturn V* rocket that carried the first astronauts to the moon, and it fuels the rocket engines of the space shuttle (Figure 14.14a). Although liquid hydrogen has been handled safely for many years, it is an extremely dangerous substance. The disastrous breakup of the *Challenger* space shuttle (Figure 14.14b), which took the lives of seven astronauts in 1986, resulted from the accidental release and explosive burning of massive amounts of hydrogen and oxygen. Before liquid hydrogen can come into more general use as a fuel, the hazards of storing and distributing this flammable and explosive material must be addressed.

(a)

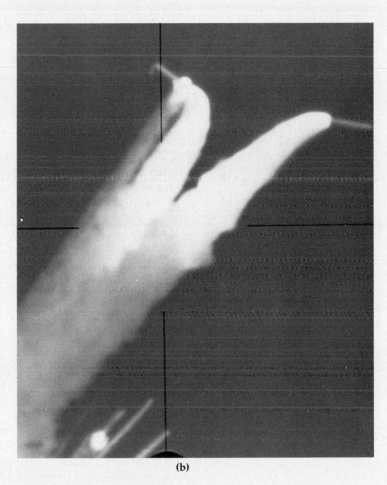

(b)

FIGURE 14.14 **(a)** The space shuttle, consisting of the orbiter, two solid rocket boosters, and the huge external fuel tank. The fuel tank, which is 47.0 m high and 8.4 m in diameter, contains 1.45×10^6 L of liquid hydrogen and 5.41×10^5 L of liquid oxygen at liftoff. **(b)** Breakup of the *Challenger* space shuttle. The rocket boosters are at the top of the photo, and the orbiter is at the bottom left, with its rocket engines still firing. Wreckage of the external fuel tank is obscured by vapor and smoke.

Hydrogen, the most abundant element in the universe, has three isotopes: protium (^1_1H), deuterium (^2_1H), and tritium (^3_1H). The isotopes of hydrogen exhibit small differences in properties known as **isotope effects**. A hydrogen atom, which has the electronic configuration $1s^1$, can lose its electron, forming a hydrogen cation (H^+), or it can gain an electron, yielding a hydride anion (H^-). At ordinary temperatures, hydrogen exists as diatomic molecules (H_2), which are thermally stable and unreactive because of the strong H–H bond.

Hydrogen is produced industrially by the **steam–hydrocarbon reforming process** and is used in the synthesis of ammonia and methanol. In the laboratory, hydrogen is prepared by reaction of dilute acid with an active metal, such as zinc.

Hydrogen forms three types of **binary hydrides**. Active metals give *ionic hydrides*, such as LiH and CaH_2; nonmetals give *covalent hydrides*, such as NH_3, H_2O, and HF; and transition metals give **metallic**, or **interstitial**, **hydrides**, such as PdH_x. Interstitial hydrides are often **nonstoichiometric compounds**.

Oxygen is the most abundant element in the earth's crust. Dioxygen (O_2) can be prepared in the laboratory by electrolysis of water, by catalytic decomposition of hydrogen peroxide, or by thermal decomposition of $KClO_3$. O_2 is manufactured by fractional distillation of liquefied air and is used in making steel. The O_2 molecule is paramagnetic and has an O=O double bond. Ozone (O_3), an allotrope of oxygen, is a powerful oxidizing agent.

Oxygen forms ionic oxides, such as Li_2O and MgO, with active metals, and covalent oxides, such as P_4O_{10} and SO_3, with nonmetals. Oxides can also be classified in terms of their acid–base properties. Basic oxides are ionic, and acidic oxides are covalent. **Amphoteric** oxides, such as Al_2O_3, exhibit both acidic and basic properties.

Metal **peroxides**, such as Na_2O_2, are ionic compounds that contain O_2^{2-} and have oxygen in the -1 oxidation state. Metal **superoxides**, such as KO_2, contain O_2^- and have oxygen in the $-1/2$ oxidation state. Hydrogen peroxide (H_2O_2), a strong oxidizing agent and also a reducing agent, is unstable with respect to **disproportionation** to H_2O and O_2. A disproportionation reaction is one in which a substance is simultaneously oxidized and reduced.

Water is the most abundant compound on earth. It is reduced to H_2 by the alkali and heavier alkaline-earth metals and is oxidized to O_2 by fluorine. Solid compounds that contain water are known as **hydrates**. Sea water, which accounts for 97.3% of the world's water supply, contains 3.5 mass percent of dissolved salts. Purification of drinking water involves preliminary filtration, sedimentation, sand filtration, aeration, and sterilization.

1. In the following pictures of binary hydrides, green spheres represent H atoms or ions, and brown spheres represent atoms or ions of the other element.

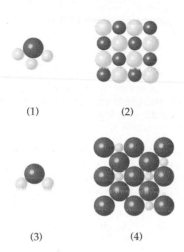

(1)   (2)

(3)   (4)

(a) Identify each binary hydride as ionic, covalent, or interstitial.
(b) What is the oxidation state of hydrogen in compounds (1), (2), and (3)? What is the oxidation state of the other element?

2. Look at the location of elements A, B, C, and D in the following periodic table.

(a) Write the formula of the simplest binary hydride of each element.
(b) Classify each binary hydride as ionic, covalent, or interstitial.
(c) Which of these hydrides are molecular? Which are solids with an infinitely extended three-dimensional crystal structure?
(d) What are the oxidation states of hydrogen and the other element in the hydrides of A, C, and D?

3. To prepare hydrogen gas from water, would you react water with an oxidizing agent or a reducing agent? Which of the following metals could be used in the reaction?
(a) Ag   (b) Al   (c) Au   (d) Ca

4. There are three isotopes of oxygen (^{16}O, ^{17}O, and ^{18}O).
(a) How many kinds of dioxygen (O_2) are there? Draw their structures.
(b) How many kinds of ozone (O_3) are there? Draw their structures.

5. Look at the location of elements A, B, C, and D in the following periodic table.

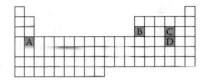

(a) Write the formula of the oxide that has each of these elements in its highest oxidation state.
(b) Classify each oxide as basic, acidic, or amphoteric.
(c) Which oxide is the most ionic? Which is the most covalent?
(d) Which of these oxides are molecular? Which are solids with an infinitely extended three-dimensional crystal structure?
(e) Which of these oxides has the highest melting point? Which has the lowest melting point?

6. Can water undergo a disproportionation reaction? Explain.

7. Which of the following elements are oxidized by water? Which are reduced by water? Which undergo a disproportionation reaction when treated with water?
(a) Cl_2 (b) F_2 (c) K   (d) Br_2

ADDITIONAL PROBLEMS

Problems 14.1–14.13 appear within the chapter.

ISOTOPES OF HYDROGEN

14.14 Name the three isotopes of hydrogen. For each isotope, give the composition of the nucleus, and indicate the approximate natural abundance in atom %.

14.15 How can protium and deuterium be separated?

14.16 Explain what is meant by an isotope effect, and give two examples.

14.17 List three properties that distinguish H_2O from D_2O.

14.18 Calculate the percentage mass difference between:
(a) ^1H and ^2H (b) ^2H and ^3H.

Would you expect the differences in properties for H_2O and D_2O to be larger or smaller than the differences in properties for D_2O and T_2O? Do the data in Table 14.1 support your prediction?

14.19 (a) If the volume of the oceans is 1.35×10^{18} m^3 and the abundance of deuterium is 0.0156 atom %, how many kg of deuterium are present in the oceans? You may neglect the presence of dissolved substances and assume that the density of water is 1.00 g/cm^3.
(b) Do the same calculation for tritium, assuming that the abundance of tritium is approximately 10^{-16} atom %.

14.20 There are three isotopes of hydrogen and three isotopes of oxygen (^{16}O, ^{17}O, and ^{18}O). How many kinds of water are there? Draw their structures.

14.21 There are three isotopes of hydrogen and just one isotope of phosphorus (^{31}P). How many kinds of phosphine (PH_3) are there? Draw their structures.

SYNTHESIS OF HYDROGEN

14.22 Write a balanced equation for the synthesis of hydrogen using each of the following starting materials.
(a) Zn   (b) C   (c) CH_4 (d) H_2O

14.23 Complete and balance the equation for each of the following reactions:
(a) $Fe(s) + H^+(aq) \longrightarrow$
(b) $Ca(s) + H_2O(l) \longrightarrow$
(c) $Al(s) + H^+(aq) \longrightarrow$
(d) $C_2H_6(g) + H_2O(g) \xrightarrow[\text{catalyst}]{\text{heat}}$

14.24 What is the most important method for the industrial production of hydrogen? Write balanced equations for the reactions involved.

14.25 Write a balanced equation for each of the following reactions:
(a) reduction of steam by hot iron
(b) production of synthesis gas from propane, C_3H_8
(c) the gas shift reaction.

14.26 Ionic metal hydrides react with water to give hydrogen gas and an aqueous solution of the metal hydroxide.
(a) On reaction of equal masses of LiH and CaH_2 with water, which compound gives more hydrogen?
(b) How many kg of CaH_2 is needed to fill a 100 liter tank with compressed H_2 gas at 150 atm pressure and 25°C?

14.27 The hydrogen-filled dirigible *Hindenburg* had a volume of 1.99×10^8 L. If the hydrogen used was produced by reaction of carbon with steam, how many kg of carbon would have been needed to produce enough hydrogen to fill the dirigible at 20°C and 740 mm pressure?

$$C(s) + H_2O(g) \rightarrow CO(g) + H_2(g)$$

REACTIVITY OF HYDROGEN

14.28 Why is hydrogen located in group 1A of the periodic table? Why, then, is the chemical behavior of hydrogen so different from that of the alkali metals?

14.29 Hydrogen could be located in group 7A of the periodic table instead of in group 1A. Cite some reasons for putting hydrogen in group 7A.

14.30 In the following compounds, is hydrogen present as H^+, H^-, or as a covalently bound H atom?
(a) HCl   (b) CaH_2 (c) SiH_4 (d) RbH

14.31 In the following compounds, is hydrogen present as H^+, H^-, or as a covalently bound H atom?
(a) KH   (b) PH_3 (c) H_2S (d) BaH_2

BINARY HYDRIDES

14.32 List the three classes of binary hydrides, and give three examples of each.

14.33 Identify the following hydrides as ionic, covalent, or interstitial.
(a) H_2S (b) NaH (c) PdH_x (d) SrH_2 (e) CH_4

14.34 Identify the oxidation state of hydrogen in
(a) HF (b) LiH (c) SnH_4 (d) MgH_2

14.35 Identify the oxidation state of hydrogen in
(a) SrH_2 (b) H_2Se (c) HBr (d) NaH

14.36 Qualitatively compare some of the physical properties of H_2S, NaH, and PdH_x.

14.37 Describe the bonding in
(a) KH (b) PH_3

14.38 Predict the molecular structure of
(a) H_2Se (b) AsH_3 (c) SiH_4

14.39 Describe the molecular geometry of
(a) GeH_4 (b) H_2S (c) NH_3

14.40 Write a balanced net ionic equation for each of the following reactions.

(a) preparation of potassium hydride
(b) reaction of potassium hydride with water
(c) reaction of calcium hydride with hydrochloric acid
(d) preparation of strontium hydride

14.41 Complete and balance the net ionic equation for each of the following reactions.
(a) $MgH_2(s) + H^+(aq) \longrightarrow$
(b) $NaH(s) + H_2O(l) \longrightarrow$
(c) $Li(l) + H_2(g) \longrightarrow$
(d) $BaH_2(s) + H^+(aq) + HSO_4^-(aq) \longrightarrow$

14.42 What is a nonstoichiometric compound? Give an example, and account for its lack of stoichiometry in terms of structure.

14.43 TiH_2 has a density of 3.9 g/cm³. Calculate the density of hydrogen in TiH_2, and compare it with that in liquid H_2 (0.070 g/cm³). How many cm³ of H_2 at STP are absorbed in making 1.00 cm³ of TiH_2?

PREPARATION AND USES OF OXYGEN

14.44 How is O_2 prepared (a) in industry and (b) in the laboratory? Write balanced equations for the reactions involved.

14.45 List three uses of oxygen.

14.46 How many liters of O_2 gas at 25°C and 1.00 atm

pressure can be obtained by catalytic decomposition of 20.4 g of hydrogen peroxide?

14.47 In the oxyacetylene torch, how many grams of acetylene and how many liters of O_2 at STP are needed to generate 1000 kJ of heat?

PROPERTIES OF OXYGEN

14.48 List four physical properties of O_2.

14.49 List three chemical properties of O_2.

14.50 Try drawing some electron-dot structures for O_2, and explain why they are inconsistent with the paramagnetism of O_2 and its O=O double bond.

14.51 Account for the paramagnetism of O_2 and its O=O double bond in terms of the molecular orbital theory (Section 7.14).

14.52 Write a balanced equation for the reaction of an excess of O_2 with
(a) P (b) Li (c) Al (d) Si

14.53 Write a balanced equation for the reaction of an excess of O_2 with
(a) Ca (b) C (c) As (d) B

OXIDES

14.54 Distinguish between acidic, basic, and amphoteric oxides, and give two examples of each.

14.55 Arrange the following oxides in order of increasing covalent character: B_2O_3, BeO, CO_2, Li_2O, N_2O_5

14.56 Arrange the following oxides in order of increasing ionic character: SiO_2, K_2O, P_4O_{10}, Ga_2O_3, GeO_2

14.57 Arrange the following oxides in order of increasing basic character: Al_2O_3, Cl_2O_7, Cs_2O, Na_2O

14.58 Arrange the following oxides in order of increasing acidic character: CO_2, BaO, N_2O_5, SnO_2

14.59 Which is more acidic?
(a) Cr_2O_3 or CrO_3 (b) N_2O_5 or N_2O_3
(c) SO_2 or SO_3

14.60 Which is more basic?
(a) CrO or Cr_2O_3 (b) SnO_2 or SnO
(c) As_2O_3 or As_2O_5

14.61 Write a balanced net ionic equation for reaction of each of the following oxides with water.
(a) Cl_2O_7 (b) K_2O (c) SO_3

14.62 Write a balanced net ionic equation for reaction of each of the following oxides with water.

(a) BaO   (b) Cs_2O (c) N_2O_5

14.63 Write a balanced net ionic equation for reaction of the amphoteric oxide ZnO with (a) hydrochloric acid and (b) aqueous sodium hydroxide. The product in (b) is $Zn(OH)_4{}^{2-}$.

14.64 Write a balanced net ionic equation for reaction of the amphoteric oxide Ga_2O_3 with (a) sulfuric acid and (b) aqueous potassium hydroxide. The product in (b) is $Ga(OH)_4{}^-$.

PEROXIDES AND SUPEROXIDES; HYDROGEN PEROXIDE

14.65 Give an example of a peroxide and a superoxide. For each example, list one physical property and one chemical property.

14.66 List three properties and two uses of hydrogen peroxide.

14.67 What products are formed when the following metals are burned in an excess of air?

(a) Ba   (b) Ca   (c) Cs   (d) Li   (e) Na

14.68 Write balanced net ionic equations for the reaction of water with (a) BaO_2 and (b) RbO_2.

14.69 Draw molecular orbital (MO) energy level diagrams for O_2, $O_2{}^-$, and $O_2{}^{2-}$, including only MOs derived from the oxygen $2p$ atomic orbitals. Show the electronic population of the MOs.

(a) Why does the O–O bond length increase in the series O_2, $O_2{}^-$, $O_2{}^{2-}$?

(b) Why is $O_2{}^-$ paramagnetic, whereas $O_2{}^{2-}$ is diamagnetic?

14.70 Draw a molecular orbital (MO) energy level diagram for $O_2{}^+$, including only MOs derived from the oxygen $2p$ atomic orbitals. Show the electronic population of the MOs.

(a) Predict the bond order in $O_2{}^+$, and state whether the O–O bond should be longer or shorter than that in O_2.

(b) Should $O_2{}^+$ be paramagnetic or diamagnetic?

14.71 Tell what is meant by a disproportionation reaction, and illustrate with an example.

14.72 Write a balanced equation for the catalyzed decomposition of hydrogen peroxide.

14.73 Write a balanced net ionic equation for each of the following reactions:

(a) oxidation by H_2O_2 of I^- to I_2 in acidic solution

(b) reduction by H_2O_2 of $Cr_2O_7{}^{2-}$ to Cr^{3+} in acidic solution

14.74 Write a balanced net ionic equation for each of the following reactions:

(a) oxidation by H_2O_2 of Fe^{2+} to Fe^{3+} in acidic solution

(b) reduction by H_2O_2 of $IO_4{}^-$ to $IO_3{}^-$ in basic solution

OZONE

14.75 Name the allotropes of oxygen, and draw electron-dot structures for each.

14.76 Give a description of the electronic structure of ozone that is consistent with the fact that the two O–O bond lengths are equal.

14.77 How is ozone made in the laboratory?

14.78 How many kJ of energy must be supplied to convert 10.0 g of O_2 to O_3?

14.79 List two physical properties and two chemical properties of ozone.

14.80 What experiment could you perform to distinguish O_3 from O_2?

WATER AND HYDRATES

14.81 Write a balanced net ionic equation for the reaction of water with each of the following.

(a) F_2 (b) I_2 (c) K

14.82 Write a balanced net ionic equation for the reaction of water with each of the following.

(a) Na   (b) Ba   (c) Cl_2

14.83 Give an example of a hydrate, and indicate how the water is bound.

14.84 What is the mass percent water in plaster of paris, $CaSO_4 \cdot H_2O$?

14.85 When 3.44 g of the mineral gypsum, $CaSO_4 \cdot xH_2O$, is heated to 128°C, 2.90 g of $CaSO_4 \cdot \frac{1}{2}H_2O$ is obtained. What is the value of x in the formula of gypsum?

14.86 Anhydrous, hygroscopic, blue $CoCl_2$ forms red-violet $CoCl_2 \cdot xH_2O$ on exposure to moist air. If the color change is accompanied by an 83.0% increase in mass, what is the formula of the hydrate?

NATURAL WATERS

14.87 How are the following chemicals used in the purification of drinking water? Cl_2, CaO, $Al_2(SO_4)_3$.

14.88 If sea water contains 3.5 mass percent of dissolved salts, how many kg of salts are present in 1.0 mi^3 of sea water? (1 mi = 1609 m; density of sea water = 1.025 g/cm^3).

14.89 How many kg of magnesium are present in a cubic meter of sea water? Assume the Mg^{2+} ion concentration listed in Table 14.3 and a density for sea water of 1.025 g/cm^3.

14.90 How many liters of sea water (density 1.025 g/cm^3) must be processed to obtain 2.0 million kg of bromine? Assume the Br^- ion concentration listed in Table 14.3 and a recovery rate of 20%.

GENERAL PROBLEMS

14.91 List three physical properties of hydrogen.

14.92 List three chemical properties of hydrogen.

14.93 List two industrial uses of hydrogen.

14.94 What are the advantages and disadvantages of hydrogen as a fuel?

14.95 How many tons of hydrogen are required for the annual U.S. production of ammonia (17 million tons)?

14.96 How many liters of H_2 at STP are required for the hydrogenation of 2.7 kg of butadiene?

$$H_2C=CH-CH=CH_2(g) + 2\ H_2(g) \rightarrow$$
Butadiene

$$CH_3-CH_2-CH_2-CH_3(g)$$
Butane

14.97 What is the oxidation state of hydrogen in each of the following substances?
(a) NH_3 (b) BaH_2 (c) $Ca(OH)_2$
(d) HBr (e) H_3PO_4

14.98 In what form is oxygen found in nature?

14.99 Give the formula and the name of a compound that has oxygen in each of the following oxidation states: $-1/2$, -1, -2.

14.100 Name each of the following compounds.
(a) B_2O_3 (b) H_2O_2 (c) SrH_2
(d) CsO_2 (e) $HClO_4$ (f) BaO_2

14.101 Give the chemical formula for each of the following compounds.
(a) calcium hydroxide
(b) chromium(III) oxide
(c) rubidium superoxide
(d) sodium peroxide
(e) barium hydride
(f) hydrogen selenide

14.102 Write a balanced equation for a reaction in which each of the following acts as an oxidizing agent.
(a) O_2 (b) O_3 (c) H_2O_2 (d) H_2 (e) H_2O

14.103 Write a balanced equation for a reaction in which each of the following acts as a reducing agent.
(a) H_2 (b) H_2O_2 (c) H_2O

14.104 Use the standard heats of formation in Appendix B to calculate $\Delta H°$ in kJ for each of these reactions:
(a) $CO(g) + 2\ H_2(g) \longrightarrow CH_3OH(l)$
(b) $CO(g) + H_2O(g) \longrightarrow CO_2(g) + H_2(g)$
(c) $2\ KClO_3(s) \longrightarrow 2\ KCl(s) + 3\ O_2(g)$
(d) $6\ CO_2(g) + 6\ H_2O(l) \longrightarrow 6\ O_2(g) + C_6H_{12}O_6(s)$

chapter 15 AQUEOUS EQUILIBRIA: ACIDS AND BASES

A cids and bases are among the most familiar and important of all chemical compounds. Acetic acid in vinegar, citric acid in lemons and other citrus fruits, magnesium hydroxide (milk of magnesia) in commercial antacids, and ammonia in household cleaning products are among the acids and bases that you encounter in everyday life. Sulfuric acid, which is used in the production of fertilizers, petroleum, plastics, and numerous other materials, is the world's most important industrial chemical. In the United States alone, annual production of sulfuric acid is 40 million tons, an amount far greater than that of any other chemical.

The characteristic properties of acids and bases have been known for centuries. Acids have a sour taste,[1] they dissolve metals such as iron and zinc, and they change the color of the plant dye litmus from blue to red. By contrast, bases have a bitter taste and slippery feel, and they change the color of litmus from red to blue. When acids and bases are mixed in the right

---

[1]Although many early chemists tasted the substances they worked with and survived, you should never taste any laboratory chemical.

Bubbles of CO_2 gas, a weak acid, emerging from an Alka Seltzer tablet.

Some familiar acids.

proportion, the characteristic acidic and basic properties disappear, and new substances known as *salts* are obtained.

What is it that makes an acid an acid and a base a base? We first raised those questions in Section 4.5, and we'll now take a closer look at some of the concepts that chemists have developed to describe the chemical behavior of acids and bases. We'll also apply the principles of chemical equilibrium discussed in Chapter 13 to determine the concentrations of the substances present in aqueous solutions of acids and bases. An enormous amount of chemistry can be understood in terms of acid–base reactions, perhaps the most important reaction type in all of chemistry.

15.1 ►ACID–BASE CONCEPTS: THE BRØNSTED-LOWRY THEORY

Up to this point we've been using the Arrhenius theory of acids and bases (Section 4.5). According to Arrhenius, acids are substances that dissociate in water to produce hydrogen ions (H^+), and bases are substances that dissociate in water to yield hydroxide ions (OH^-). Thus, HCl and H_2SO_4 are acids, and NaOH and $Ba(OH)_2$ are bases.

A generalized Arrhenius acid $HA(aq) \rightarrow H^+(aq) + A^-(aq)$

A generalized Arrhenius base $MOH(aq) \rightarrow M^+(aq) + OH^-(aq)$

The Arrhenius theory accounts for the properties of many common acids and bases, but it has important limitations. For one thing, the Arrhenius theory is restricted to aqueous solutions; for another, it doesn't account for the basicity of substances like ammonia (NH_3) that don't contain OH groups. In 1923, a more general theory of acids and bases was proposed independently by the Danish chemist Johannes Brønsted and the English chemist Thomas Lowry. According to the **Brønsted-Lowry theory**, an acid is any substance (molecule or ion) that can transfer a proton (H^+ ion) to another substance, and a base is any substance that can accept a proton. In short, acids are proton donors, bases are proton acceptors, and acid–base reactions are proton-transfer reactions:

Brønsted-Lowry acid   A substance that can transfer H^+

Brønsted-Lowry base   A substance that can accept H^+

$$\underset{\substack{H^+ \text{ donor} \\ \text{acid}}}{HA} \; + \; \underset{\substack{H^+ \text{ acceptor} \\ \text{base}}}{B} \; \rightleftarrows \; \underset{\substack{H^+ \text{ donor} \\ \text{acid}}}{BH^+} \; + \; \underset{\substack{H^+ \text{ acceptor} \\ \text{base}}}{A^-}$$

conjugate acid–base pairs

It's clear from the preceding equation that the products of a Brønsted-Lowry acid–base reaction, BH^+ and A^-, are themselves acids and bases. The acid HA donates a proton to the base B, leaving A^-, but A^- itself can accept a proton back, and so it is a Brønsted-Lowry base. Similarly, the species BH^+ produced when the base B accepts a proton can donate a proton back and is therefore a Brønsted-Lowry acid. Chemical species whose formulas differ only by one proton are said to be **conjugate acid- –base pairs**. Thus, A^- is the **conjugate base** of the acid HA, and HA is the **conjugate acid** of the base A^-. Similarly, B is the conjugate base of the acid BH^+, and BH^+ is the conjugate acid of the base B.

To see what's going on in an acid–base reaction, keep your eye on the proton. For example, when a Brønsted-Lowry acid HA is placed in water, it reacts reversibly with water in an *acid-dissociation equilibrium*. The acid transfers a proton to the solvent, which acts as a base (a proton acceptor). The products are the hydronium ion, H_3O^+ (the conjugate acid of H_2O), and A^- (the conjugate base of HA):

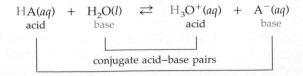

$$\underset{\text{acid}}{HA(aq)} \; + \; \underset{\text{base}}{H_2O(l)} \; \rightleftarrows \; \underset{\text{acid}}{H_3O^+(aq)} \; + \; \underset{\text{base}}{A^-(aq)}$$

conjugate acid–base pairs

In the reverse reaction, H_3O^+ acts as the proton donor (acid) and A^- acts as the proton acceptor (base). Typical examples of Brønsted-Lowry acids include not only electrically neutral molecules, such as HCl, HNO_3, and HF, but also cations and anions of salts that contain transferable protons, such as NH_4^+, HSO_4^-, and HCO_3^-.

When a Brønsted-Lowry base such as NH_3 dissolves in water, it accepts a proton from the solvent, which acts as an acid. The products are the hydroxide ion, OH^- (the conjugate base of water), and the ammonium ion, NH_4^+ (the conjugate acid of NH_3). In the reverse reaction, NH_4^+ acts as the proton donor (acid), and OH^- acts as the proton acceptor (base):

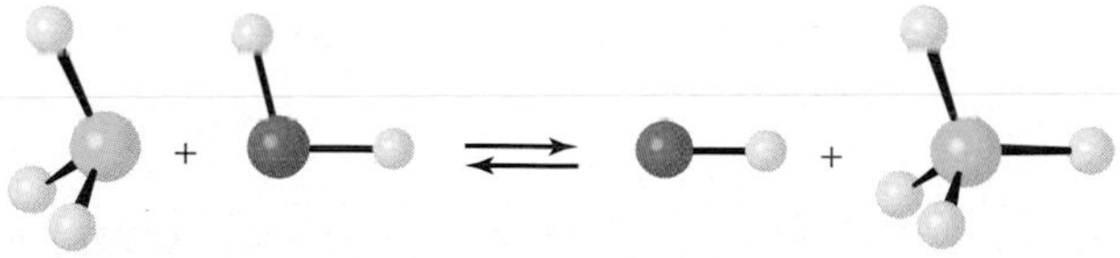

$$NH_3(aq) \;+\; H_2O(l) \rightleftharpoons OH^-(aq) \;+\; NH_4^+(aq)$$

base   acid   base   acid

conjugate acid–base pairs

For a molecule or ion to accept a proton, it must have at least one unshared pair of electrons that it can use for bonding to the proton. As shown by the following Lewis electron-dot structures, all Brønsted-Lowry bases have one or more lone pairs of electrons:

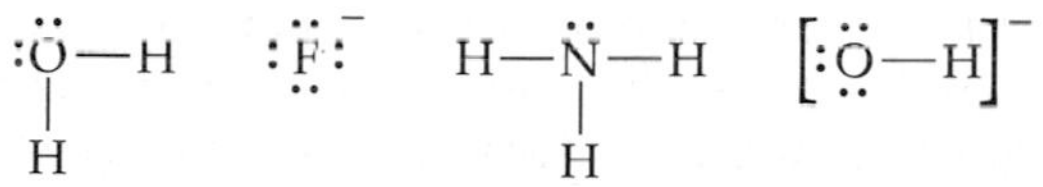

Some Brønsted-Lowry bases

EXAMPLE 15.1

Account for the acidic properties of nitrous acid (HNO_2) in terms of the Arrhenius theory and the Brønsted-Lowry theory, and identify the conjugate base of HNO_2.

SOLUTION HNO_2 is an Arrhenius acid because it undergoes dissociation in water to produce H^+ ions:

$$HNO_2(aq) \rightleftharpoons H^+(aq) + NO_2^-(aq)$$

Nitrous acid is a Brønsted-Lowry acid because it acts as a proton donor when it dissociates, transferring a proton to water to give the hydronium ion:

$$HNO_2(aq) \;+\; H_2O(l) \rightleftharpoons H_3O^+(aq) \;+\; NO_2^-(aq)$$

acid   base   acid   base

conjugate acid–base pairs

The conjugate base of HNO_2 is NO_2^-, the fragment that remains after HNO_2 has lost a proton.

⌐ **PROBLEM 15.1**   Write a balanced equation for the dissociation of each of the following Brønsted-Lowry acids in water.

(a) H_2SO_4 **(b)** HSO_4^- **(c)** H_3O^+

What is the conjugate base of the acid in each case?

⌐ **PROBLEM 15.2**   What is the conjugate acid of each of the following Brønsted-Lowry bases?

(a) HCO_3^- **(b)** CO_3^{2-} **(c)** OH^-

15.2 ➤ACID STRENGTH AND BASE STRENGTH

A helpful way of viewing an acid-dissociation equilibrium is to realize that the two bases, H_2O and A^-, are competing for protons:

$$\underset{\text{acid}}{HA(aq)} \quad + \quad \underset{\text{base}}{H_2O(l)} \quad \rightleftharpoons \quad \underset{\text{acid}}{H_3O^+(aq)} \quad + \quad \underset{\text{base}}{A^-(aq)}$$

If H_2O is a stronger base than A^-, the H_2O molecules will get the protons, and the solution will contain mainly H_3O^+ and A^-. If A^- is a stronger base than H_2O, the A^- ions will get the protons, and the solution will contain mainly HA and H_2O. *In every acid-base reaction, the proton is transferred to the stronger base.*

Different acids differ in their ability to donate protons. A **strong acid** is one that is almost completely dissociated in aqueous solution (Section 4.5). Thus, the acid-dissociation equilibrium of a strong acid lies nearly 100% to the right, and the solution contains almost entirely H_3O^+ and A^- ions with only a negligible amount of undissociated HA molecules. Typical examples of strong acids are perchloric acid ($HClO_4$), hydrochloric acid (HCl), hydrobromic acid (HBr), hydroiodic acid (HI), sulfuric acid (H_2SO_4), and nitric acid (HNO_3). It follows from this definition that *strong acids have weak conjugate bases.* The ClO_4^-, Cl^-, Br^-, I^-, HSO_4^-, and NO_3^- ions have only a negligible tendency to be protonated (to combine with a proton) in aqueous solution, and they are therefore much weaker bases than H_2O.

A **weak acid** is one that is only partially dissociated in aqueous solution. Proton transfer from a weak acid to water is incomplete, and the solution contains a mixture of undissociated acid (HA), its conjugate base (A^-), and H_3O^+ ions. Typical examples of weak acids are nitrous acid (HNO_2), hydrofluoric acid (HF), and acetic acid (CH_3COOH). In the case of very weak acids such as NH_3, OH^-, and H_2, the acid has practically no tendency to transfer a proton to water, and the acid-dissociation equilibrium lies essentially 100% to the left. It follows from this definition that *very weak acids have strong conjugate bases.* For example, the NH_2^-, O^{2-}, and H^- ions are essentially 100% protonated in aqueous solution and are much stronger bases than H_2O.

The equilibrium concentrations of HA, H_3O^+, and A^- for strong acids, weak acids, and very weak acids are represented graphically in Figure 15.1. The inverse relationship between the strength of an acid and the strength of its conjugate base is illustrated in Table 15.1.

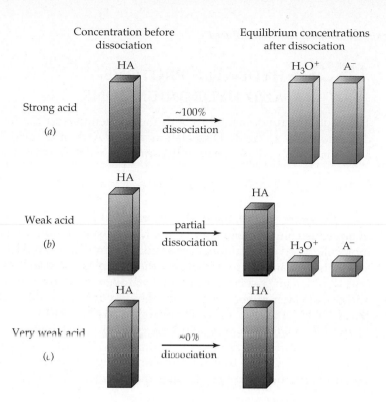

Concentration before dissociation

Equilibrium concentrations after dissociation

Strong acid
(a)

HA

$\xrightarrow{\sim 100\%}$ dissociation

H_3O^+ A^-

Weak acid
(b)

HA

$\xrightarrow{\text{partial}}$ dissociation

HA

H_3O^+ A^-

Very weak acid
(c)

HA

$\xrightarrow{\approx 0\%}$ dissociation

HA

FIGURE 15.1 Equilibrium concentrations of HA, H_3O^+, and A^- for **(a)** a strong acid, **(b)** a weak acid, and **(c)** a very weak acid. Dissociation of HA involves H^+ transfer to H_2O, yielding H_3O^+ and A^-. The extent of dissociation is nearly 100% for a strong acid, less than 100% for a weak acid, and nearly 0% for a very weak acid.

TABLE 15.1   Relative Strengths of Conjugate Acid–Base Pairs

	Acid, HA		Base, A⁻		
Stronger acid	$HClO_4$ HCl H_2SO_4 HNO_3	Strong acids. 100% dissociated in aqueous solution.	ClO_4^- Cl^- HSO_4^- NO_3^-	Very weak bases. Negligible tendency to be protonated in aqueous solution.	Weaker base
	H_3O^+		H_2O		
	HSO_4^- H_3PO_4 HNO_2 HF CH_3COOH H_2CO_3 H_2S NH_4^+ HCN HCO_3^-	Weak acids. Exist in solution as a mixture of HA, A^-, and H_3O^+.	SO_4^{2-} $H_2PO_4^-$ NO_2^- F^- $CH_3CO_2^-$ HCO_3^- HS^- NH_3 CN^- CO_3^{2-}	Weak bases. Moderate tendency to be protonated in aqueous solution.	
	H_2O		OH^-		
Weaker acid	NH_3 OH^- H_2	Very weak acids. Negligible tendency to dissociate.	NH_2^- O^{2-} H^-	Strong bases. 100% protonated in aqueous solution.	Stronger base

15.3 ►HYDRATED PROTONS AND HYDRONIUM IONS

The proton is fundamental to both the Arrhenius and the Brønsted-Lowry definitions of an acid. Dissociation of an Arrhenius acid HA gives an aqueous hydrogen ion, or hydrated proton, written as $H^+(aq)$:

$$HA(aq) \rightarrow H^+(aq) + A^-(aq)$$

As a bare proton, the positively charged H^+ ion is too reactive to exist in aqueous solution, and so it bonds to the oxygen atom of a solvent water molecule to give the trigonal pyramidal **hydronium ion, H_3O^+.** The H_3O^+ ion, which can be regarded as the simplest hydrate of the proton, $[H(H_2O)]^+$, can associate through hydrogen bonding (Section 10.2) with additional water molecules to give higher hydrates with the general formula $[H(H_2O)_n]^+$ (n = 2, 3, or 4), for example, $H_5O_2^+$, $H_7O_3^+$, and $H_9O_4^+$. It's likely that acidic aqueous solutions contain a distribution of $[H(H_2O)_n]^+$ ions having different values of n. In this book, though, we'll use the symbols $H^+(aq)$ and $H_3O^+(aq)$ to mean the same thing—namely, a proton hydrated by an unspecified number of water molecules.

H_3O^+ – the hydronium ion, or hydrated H^+

15.4 ►DISSOCIATION OF WATER

One of the most important properties of water is its ability to act both as an acid (a proton donor) and as a base (a proton acceptor). In the presence of an acid, water acts as a base, whereas in the presence of a base, water acts as an acid. It's not surprising, therefore, that in pure water one water molecule can donate a proton to another water molecule in a reaction in which water acts both as an acid and as a base *at the same time*:

$$H-\overset{..}{\underset{|}{O}}: + :\overset{..}{\underset{|}{O}}-H \rightleftarrows \left[H-\overset{..}{\underset{|}{O}}-H\right]^+ + \left[H-\overset{..}{\underset{..}{O}}:\right]^-$$

acid   base   acid   base

conjugate acid–base pairs

Called the **dissociation of water**, this reaction is characterized by the usual kind of equilibrium equation:

For the reaction $2 H_2O(l) \rightleftarrows H_3O^+(aq) + OH^-(aq)$

$$K = \frac{[H_3O^+][OH^-]}{[H_2O]^2}$$

Because the molar concentration of pure water is constant, we can combine $[H_2O]^2$ and the equilibrium constant K into a single constant, denoted by K_w and called the **ion-product constant for water**:

$$K_w = K[H_2O]^2 = [H_3O^+][OH^-]$$

There are two important aspects of the dynamic equilibrium in the dissociation of water. First, the forward and reverse reactions are rapid; H_2O molecules, H_3O^+ ions, and OH^- ions continually interconvert as protons transfer quickly from one species to another. Second, the position of the equilibrium lies far to the left; at any one instant, only a very few molecules are dissociated into H_3O^+ and OH^- ions. The vast majority of water molecules are undissociated.

We can calculate the extent of dissociation of water molecules starting from experimental measurements that show the H_3O^+ concentration in pure water to be 1.0×10^{-7} M at 25°C:

$$[H_3O^+] = 1.0 \times 10^{-7} \text{ M} \qquad \text{at } 25°\text{C}$$

Since the dissociation reaction of water produces equal concentrations of H_3O^+ and OH^- ions, the OH^- concentration in pure water is also 1.0×10^{-7} M at 25°C:

$$[H_3O^+] = [OH^-] = 1.0 \times 10^{-7} \text{ M} \qquad \text{at } 25°\text{C}$$

Furthermore, we know that the molar concentration of pure water, calculated from its density and molar mass, is 55.4 M at 25°C:

$$[H_2O] = \left(\frac{997 \text{ g}}{1 \text{ }}\right)\left(\frac{1 \text{ mol}}{18.0 \text{ g}}\right) = 55.4 \text{ mol/L} \qquad \text{at } 25°\text{C}$$

From these facts, we can conclude that the ratio of dissociated to undissociated water molecules is about 2 in 10^9, a very small number indeed:

$$\frac{1.0 \times 10^{-7} \text{ M}}{55.4 \text{ M}} = 1.8 \times 10^{-9} \qquad \text{(about 2 in } 10^9\text{)}$$

In addition, we can calculate that the numerical value of K_w at 25°C is 1.0×10^{-14}:

$$K_w = [H_3O^+][OH^-] = (1.0 \times 10^{-7})(1.0 \times 10^{-7})$$

$$= 1.0 \times 10^{-14} \qquad \text{at } 25°\text{C}$$

As is common practice for equilibrium constants, the units of K_w (mol²/L²) are omitted. In very dilute solutions, the product of the H_3O^+ and OH^- concentrations is not affected by the presence of solutes. This is not true in more concentrated solutions, but we'll neglect that complication and assume that *the product of the H_3O^+ and OH^- concentrations is always 1.0×10^{-14} at 25°C in any aqueous solution.*

We can distinguish neutral, acidic, and basic aqueous solutions by the relative values of the H_3O^+ and OH^- concentrations:

Neutral: $[H_3O^+] = [OH^-]$ $[H_3O^+] = 1.0 \times 10^{-7}$ M at 25°C

Acidic: $[H_3O^+] > [OH^-]$ $[H_3O^+] > 1.0 \times 10^{-7}$ M at 25°C

Basic: $[H_3O^+] < [OH^-]$ $[H_3O^+] < 1.0 \times 10^{-7}$ M at 25°C

If one of the concentrations, $[H_3O^+]$ or $[OH^-]$, is known, the other is readily calculated:

Since $\qquad [H_3O^+][OH^-] = K_w = 1.0 \times 10^{-14}$

then $\quad [H_3O^+] = \dfrac{1.0 \times 10^{-14}}{[OH^-]}$ and $[OH^-] = \dfrac{1.0 \times 10^{-14}}{[H_3O^+]}$

In the previous discussion, we were careful to emphasize that the value of $K_w = 1.0 \times 10^{-14}$ applies at 25°C. Because K_w increases with increasing temperature, the H_3O^+ and OH^- concentrations in neutral aqueous solutions at temperatures other than 25°C deviate from 1.0×10^{-7} M (Problem 15.4). Unless otherwise indicated, we'll always assume a temperature of 25°C.

EXAMPLE 15.2

The concentration of H_3O^+ ions in a sample of lemon juice is 2.5×10^{-3} M. Calculate the concentration of OH^- ions, and classify the solution as neutral, acidic, or basic.

BALLPARK SOLUTION   Since the product of the H_3O^+ and OH^- concentrations must equal 10^{-14} M^2, and since the H_3O^+ concentration is in the range 10^{-3} to 10^{-2} M, the OH^- concentration must be in the range 10^{-11} to 10^{-12} M. Because $[H_3O^+] > [OH^-]$, the solution is acidic.

DETAILED SOLUTION   When $[H_3O^+]$ is known, the OH^- concentration can be found from the expression

$$[OH^-] = \frac{K_w}{[H_3O^+]} = \frac{1.0 \times 10^{-14}}{2.5 \times 10^{-3}} = 4.0 \times 10^{-12} \text{ M}$$

⌐ **PROBLEM 15.3**   The concentration of OH^- in a sample of sea water is 5.0×10^{-6} M. Calculate the concentration of H_3O^+ ions, and classify the solution as neutral, acidic, or basic.

⌐ **PROBLEM 15.4**   At 50°C the value of K_w is 5.5×10^{-14}. What are the concentrations of H_3O^+ and OH^- in a neutral solution at 50°C?

15.5 ►THE pH SCALE

Rather than write hydronium ion concentrations in molarity, it's more convenient to express them on a logarithmic scale known as the *pH scale*. The term **pH** is derived from the French *puissance d'hydrogène* ("power of hydrogen") and refers to the power of 10 (the exponent) used to express the molar

H_3O^+ concentration. The pH of a solution is defined as the negative base-10 logarithm (log) of the molar hydronium ion concentration[2]:

$$pH = -\log[H_3O^+] \quad \text{or} \quad [H_3O^+] = \text{antilog}(-pH) = 10^{-pH}$$

Thus, an acidic solution having $[H_3O^+] = 10^{-2}$ M has a pH of 2, a basic solution having $[OH^-] = 10^{-2}$ M and $[H_3O^+] = 10^{-12}$ M has a pH of 12, and a neutral solution having $[H_3O^+] = 10^{-7}$ M has a pH of 7. Note that although we express $[H_3O^+]$ in mol/L, we take the log of the number only, not the units.

If you use a calculator to find the pH from the H_3O^+ concentration, your answer will have more decimal places than the proper number of significant figures. For example, the pH of the lemon juice in Example 15.2 ($[H_3O^+] = 2.5 \times 10^{-3}$ M) is found on a calculator to be

$$pH = -\log(2.5 \times 10^{-3}) = 2.60206$$

This result should be rounded to pH 2.60 (two significant figures) because $[H_3O^+]$ has only two significant figures. Note that the only significant figures in a logarithm are the digits to the right of the decimal point; the number to the left of the decimal point is an exact number related to the integral power of 10 in the exponential expression for $[H_3O^+]$.

$$pH = -\log(\underset{\substack{\text{2 significant} \\ \text{figures (2 SF's)}}}{2.5} \times 10^{-3}) = -\log \underset{\substack{\text{exact} \\ \text{number}}}{10^{-3}} - \log \underset{\substack{\text{exact} \\ \text{number}}}{2.5} = \underset{\substack{\text{2 SF's}}}{3} - \underset{\substack{\text{exact} \\ \text{number}}}{0.40} = \underset{\substack{\text{2 SF's}}}{2.60}$$

Because the pH scale is logarithmic, the pH changes by 1 unit when $[H_3O^+]$ changes by a factor of 10, by 2 units when $[H_3O^+]$ changes by a factor of 100, and by 6 units when $[H_3O^+]$ changes by a factor of 1,000,000. To appreciate the extent to which the pH scale is a compression of the $[H_3O^+]$ scale, compare the amounts of 12 M HCl required to change the pH of the water in a backyard swimming pool. Only about 100 mL of 12 M HCl is needed to change the pH from 7 to 6, but a 10,000 L truckload of 12 M HCl is needed to change the pH from 7 to 1.

The pH scale and pH values for some common substances are shown in Figure 15.2. Because the pH is the *negative* log of $[H_3O^+]$, the pH decreases as $[H_3O^+]$ increases. Thus, when $[H_3O^+]$ increases from 10^{-7} M to 10^{-6} M, the pH decreases from 7 to 6. As a result, acidic solutions have pH less than 7.00, and basic solutions have pH greater than 7.00.

Acidic solution:	pH < 7
Neutral solution:	pH = 7
Basic solution:	pH > 7

[2] The pOH can be defined in the same way as the pH. Just as $pH = -\log[H_3O^+]$, so $pOH = -\log[OH^-]$. It follows from the equation $[H_3O^+][OH^-] = 1.0 \times 10^{-14}$ that $pH + pOH = 14.00$.

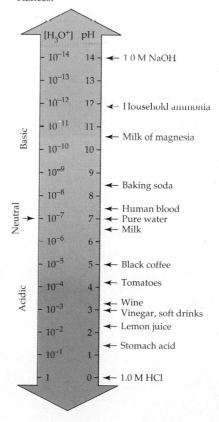

FIGURE 15.2 The pH scale and pH values for some common substances.

Basic

Neutral

Acidic

$[H_3O^+]$	pH	
10^{-14}	14	← 1.0 M NaOH
10^{-13}	13	
10^{-12}	12	← Household ammonia
10^{-11}	11	← Milk of magnesia
10^{-10}	10	
10^{-9}	9	← Baking soda
10^{-8}	8	← Human blood
10^{-7}	7	← Pure water / ← Milk
10^{-6}	6	
10^{-5}	5	← Black coffee
10^{-4}	4	← Tomatoes
10^{-3}	3	← Wine / ← Vinegar, soft drinks
10^{-2}	2	← Lemon juice
10^{-1}	1	← Stomach acid
1	0	← 1.0 M HCl

EXAMPLE 15.3

Calculate the pH of an aqueous ammonia solution that has an OH^- concentration of 1.9×10^{-3} M.

SOLUTION   First, calculate the H_3O^+ concentration from the OH^- concentration, and then take the negative of the logarithm of $[H_3O^+]$ to convert to pH:

$$[H_3O^+] = \frac{K_w}{[OH^-]} = \frac{1.0 \times 10^{-14}}{1.9 \times 10^{-3}} = 5.3 \times 10^{-12} \text{ M}$$

$$pH = -\log[H_3O^+] = -\log(5.3 \times 10^{-12}) = 11.28$$

Note that the pH is quoted to two significant figures (.28) because $[H_3O^+]$ is known to two significant figures (5.3).

EXAMPLE 15.4

Acid rain is a matter of serious concern because most species of fish die in waters having a pH lower than 4.5–5.0. Calculate the H_3O^+ concentration in a lake that has a pH of 4.5.

BALLPARK SOLUTION   A pH of 4.5 is between pH 4 and pH 5. Therefore $[H_3O^+]$ is between 10^{-4} M and 10^{-5} M.

DETAILED SOLUTION   Calculate the H_3O^+ concentration by taking the antilogarithm of the negative of the pH:

$$[H_3O^+] = \text{antilog}(-pH) = 10^{-pH} = 10^{-4.5} = 3 \times 10^{-5} \text{ M}$$

$[H_3O^+]$ is reported to only one significant figure because the pH has only one digit beyond the decimal point. (If you need help in finding the antilog of a number, see Appendix A.2.)

⌐ **PROBLEM 15.5**   Calculate the pH of each of the following solutions.
(a) a sample of sea water that has an OH^- concentration of 1.58×10^{-6} M
(b) a sample of acid rain that has an H_3O^+ concentration of 6.0×10^{-5} M

⌐ **PROBLEM 15.6**   Human blood has a pH of 7.40. Calculate the concentrations of H_3O^+ and OH^-.

15.6 ➤MEASURING pH

The approximate pH of a solution can be determined easily by using an **acid–base indicator**, a substance that changes color in a specific pH range (Figure 15.3). Indicators (abbreviated HIn) exhibit pH-dependent color changes because they are weak acids and happen to have different colors in their acid (HIn) and conjugate base (In^-) forms:

$$HIn(aq) + H_2O(l) \rightleftarrows H_3O^+(aq) + In^-(aq)$$
$$\text{color A} \qquad\qquad\qquad\qquad \text{color B}$$

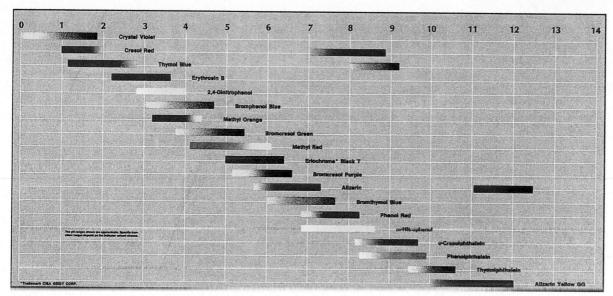

FIGURE 15.3 Some common acid–base indicators and their color changes. Note that the color of an indicator changes over a range of about 2 pH units.

Bromthymol blue, for example, changes color in the pH range 6.0–7.6 from yellow in its acid form to blue in its base form. Phenolphthalein changes from colorless in its acid form to pink in its base form in the pH range 8.2–9.8.

Because indicators change color over a range of about 2 pH units, it's possible to determine the pH of a solution to within approximately ±1 pH unit simply by adding to the solution a few drops of an indicator that changes color in the appropriate pH range. To make the determination particularly easy, a mixture of pH indicators known as *universal indicator* is commercially available for making approximate pH measurements in the range pH 3 to pH 10 (Figure 15.4). More accurate pH values can be determined with an electronic instrument called a *pH meter* (Figure 15.5), a device that measures a pH-dependent electrical potential of the test solution.

FIGURE 15.4 (a) The color of universal indicator in solutions of known pH from 1 to 12. (b) The color of universal indicator in some familiar products gives the following approximate pH values: vinegar, pH 3; club soda, pH 4–5; household ammonia, pH ≥ 10.

FIGURE 15.5 A pH meter with its electrical probe dipping into a grapefruit. An accurate value of the pH (3.7) is shown on the meter.

15.7 ▸THE pH OF STRONG ACIDS AND STRONG BASES

The commonly encountered strong acids listed in Table 15.1 include three **monoprotic acids** ($HClO_4$, HCl, and HNO_3), which contain a single dissociable proton, and one **diprotic acid** (H_2SO_4), which has two dissociable protons. Because strong acids are 100% dissociated in aqueous solution, the H_3O^+ and A^- concentrations are equal to the initial concentration of the acid, and the concentration of undissociated HA molecules is essentially zero.

$$HA(aq) + H_2O(l) \xrightarrow{100\%} H_3O^+(aq) + A^-(aq)$$

The pH of a solution of a strong monoprotic acid is easily calculated from the H_3O^+ concentration, as illustrated in Example 15.5. Calculation of the pH of an H_2SO_4 solution is more complicated because the second of the two protons is only partially dissociated.

The most familiar examples of strong bases are alkali metal hydroxides MOH, such as NaOH (*caustic soda*) and KOH (*caustic potash*). These compounds are water-soluble ionic solids that exist in aqueous solution as alkali-metal cations (M^+) and OH^- anions:

$$MOH(s) \xrightarrow{H_2O} MOH(aq) \xrightarrow{100\%} M^+(aq) + OH^-(aq)$$

Thus, 0.10 M NaOH contains 0.10 M Na^+ and 0.10 M OH^-, and the pH is readily calculated from the OH^- concentration, as shown in Example 15.6.

The alkaline-earth metal hydroxides $M(OH)_2$ (M = Mg, Ca, Sr, or Ba) are also strong bases (~100% dissociated), but they give lower OH^- concentrations because they are less soluble. Their solubility at room temperature varies from 38 g/L for the relatively soluble $Ba(OH)_2$ to ~10^{-2} g/L for the relatively insoluble $Mg(OH)_2$. Aqueous suspensions of $Mg(OH)_2$, called milk of magnesia, are used as an antacid. The most common and least expensive alkaline earth hydroxide is $Ca(OH)_2$, which is known as *slaked lime* and is made by mixing CaO (*lime*) and water:

$$CaO(s) + H_2O(l) \rightarrow Ca(OH)_2(s)$$

Slaked lime is used in making mortars and cements. Aqueous solutions of the slightly soluble $Ca(OH)_2$ are called limewater.

The oxide ion (O^{2-}) is an even stronger base than OH^- (Table 15.1), yet it can't exist in aqueous solutions because it is immediately and completely protonated by water, yielding OH^- ions:

$$O^{2-}(aq) + H_2O(l) \xrightarrow{100\%} OH^-(aq) + OH^-(aq)$$

Thus, dissolving 1 mol of lime in water gives 1 mol of Ca^{2+} and 2 mol of OH^-:

$$CaO(s) + H_2O(l) \rightarrow Ca^{2+}(aq) + 2\ OH^-(aq)$$

Lime, the world's most important strong base, is made by decomposition of limestone, $CaCO_3$, at temperatures of 800 to 1000°C:

$$CaCO_3(s) \xrightarrow{\text{heat}} CaO(s) + CO_2(g)$$

Lime is produced in enormous quantities (18 million tons per year in the United States) for use in steelmaking, water purification, and chemical manufacture.

EXAMPLE 15.5

Calculate the pH of a 0.025 M HNO_3 solution.

SOLUTION   Since nitric acid is a strong acid, it is almost completely dissociated in aqueous solution. Therefore, $[H_3O^+] = 0.025$ M, and pH = 1.60.

$$HNO_3(aq) + H_2O(l) \xrightarrow{100\%} H_3O^+(aq) + NO_3^-(aq)$$

$$pH = -\log[H_3O^+] = -\log(2.5 \times 10^{-2}) = 1.60$$

EXAMPLE 15.6

Calculate the pH of each of the following solutions.
(a) a 0.10 M solution of NaOH
(b) a 0.0050 M solution of slaked lime [$Ca(OH)_2$]
(c) a solution prepared by dissolving 0.28 g of lime (CaO) in enough water to make 1.00 L of limewater [$Ca(OH)_2(aq)$]

SOLUTION   **(a)** Because NaOH is a strong base, it is 100% dissociated. Therefore, $[OH^-] = 0.10$ M, $[H_3O^+] = 1.0 \times 10^{-13}$ M, and pH = 13.00.

$$[H_3O^+] = \frac{K_w}{[OH^-]} = \frac{1.0 \times 10^{-14}}{0.10} = 1.0 \times 10^{-13} \text{ M}$$

$$pH = -\log(1.0 \times 10^{-13}) = 13.00$$

(b) Because slaked lime is a strong base, it is 100% dissociated, providing 2 OH^- per $Ca(OH)_2$ formula unit. Therefore, $[OH^-] = 2(0.0050 \text{ M}) = 0.010$ M, $[H_3O^+] = 1.0 \times 10^{-12}$ M, and pH = 12.00.

$$[H_3O^+] = \frac{K_w}{[OH^-]} = \frac{1.0 \times 10^{-14}}{0.010} = 1.0 \times 10^{-12} \text{ M}$$

$$pH = -\log(1.0 \times 10^{-12}) = 12.00$$

(c) First calculate the number of moles of CaO dissolved from the mass of CaO and its molar mass (56 g/mol):

$$\text{Moles of CaO} = 0.28 \text{ g CaO} \times \frac{1 \text{ mol CaO}}{56 \text{ g CaO}} = 0.0050 \text{ mol CaO}$$

Protonation of the O^{2-} ion produces 2 mol of OH^- per mole of CaO dissolved:

$$CaO(s) + H_2O(l) \rightarrow Ca^{2+}(aq) + 2\ OH^-(aq)$$

Moles of OH^- produced = 2(0.0050 mol) = 0.010 mol

Since the solution volume is 1.00 L,

$$[OH^-] = \frac{0.010\ \text{mol}}{1.00\ \text{L}} = 0.010\ M$$

The $[OH^-]$ is identical to that in part **(b)**. Therefore, pH = 12.00.

▶ **PROBLEM 15.7** Calculate the pH of
(a) 0.050 M $HClO_4$ **(b)** 1.0 M HCl   **(c)** 6.0 M HCl

▶ **PROBLEM 15.8** Calculate the pH of
(a) 0.020 M KOH   **(b)** 0.010 M $Ba(OH)_2$

▶ **PROBLEM 15.9** Calculate the pH of a solution prepared by dissolving 0.25 g of BaO in enough water to make 500 mL of solution.

▶ **PROBLEM 15.10** H_3O^+ is the strongest acid that can exist in aqueous solution because stronger acids dissociate by transferring a proton to water. What is the strongest base that can exist in aqueous solution?

15.8 ▶EQUILIBRIA IN SOLUTIONS OF WEAK ACIDS

A weak acid is not the same thing as a dilute solution of a strong acid. Whereas a strong acid is 100% dissociated in aqueous solution, a weak acid is only partially dissociated. The position of any acid-dissociation equilibrium is characterized by an equilibrium equation, where the equilibrium constant K_a is called the **acid-dissociation constant**:

For the reaction $HA(aq) + H_2O(l) \rightleftarrows H_3O^+(aq) + A^-(aq)$

$$K = \frac{[H_3O^+][A^-]}{[HA][H_2O]} \qquad K_a = K[H_2O] = \frac{[H_3O^+][A^-]}{[HA]}$$

$$K_a = \frac{[H_3O^+][A^-]}{[HA]}$$

Note that $[H_2O]$, which is essentially constant in dilute aqueous solutions, has been incorporated into the equilibrium constant K_a and is therefore omitted from the equilibrium constant expression for K_a.

Values of K_a for some typical weak acids are listed in Table 15.2. As indicated by the equilibrium equation, the larger the value of K_a, the stronger the acid. Thus, hydrocyanic acid ($K_a = 4.9 \times 10^{-10}$) is the weakest of the acids listed in Table 15.2, and nitrous acid ($K_a = 4.6 \times 10^{-4}$) is the strongest.

For comparison, the K_a values of strong acids are much greater than 1; for example, HCl has $K_a = 2 \times 10^6$. K_a values for other weak acids are given in Appendix C.

Numerical values of acid-dissociation constants are determined from pH measurements, as shown in Example 15.7.

TABLE 15.2   Acid-Dissociation Constants at 25°C

	Acid	Molecular Formula	Structural Formula[a]	K_a
Stronger acid	Hydrochloric	HCl	H—Cl	2×10^6
	Nitrous	HNO_2	H—O—N=O	4.5×10^{-4}
	Hydrofluoric	HF	H—F	3.5×10^{-4}
	Acetylsalicylic (aspirin)	$C_9H_8O_4$		3.0×10^{-4}
	Formic	HCOOH		1.8×10^{-4}
	Ascorbic (vitamin C)	$C_6H_8O_6$		8.0×10^{-5}
	Benzoic	C_6H_5COOH		6.5×10^{-5}
	Acetic	CH_3COOH		1.8×10^{-5}
	Hypochlorous	HOCl	H—O—Cl	3.5×10^{-8}
Weaker acid	Hydrocyanic	HCN	H—C≡N	4.9×10^{-10}

[a] The proton that is transferred to water when the acid dissociates is shown in color.

EXAMPLE 15.7

The pH of 0.250 M HF is 2.036. What is the value of K_a for hydrofluoric acid?

SOLUTION   First, write the balanced equation for the dissociation equilibrium and the equilibrium equation that defines K_a.

$$HF(aq) + H_2O(l) \rightleftarrows H_3O^+(aq) + F^-(aq)$$

$$K_a = \frac{[H_3O^+][F^-]}{[HF]}$$

To obtain a value for K_a, we need to calculate the concentrations of the individual components in the equilibrium mixture. We can begin by calculating the H_3O^+ concentration from the pH:

$$[H_3O^+] = \text{antilog}(-pH) = 10^{-pH} = 10^{-2.036} = 9.20 \times 10^{-3}\ M$$

Since dissociation of one HF molecule gives one H_3O^+ ion and one F^- ion, the H_3O^+ and F^- concentrations are equal.

$$[F^-] = [H_3O^+] = 9.20 \times 10^{-3}\ M$$

Furthermore, the HF concentration at equilibrium is equal to the initial concentration (0.250 M) minus whatever dissociates (9.20×10^{-3} M):

$$[HF] = 0.250 - 0.00920 = 0.241\ M$$

Substituting the equilibrium concentrations into the equilibrium equation then gives the value of K_a:

$$K_a = \frac{[H_3O^+][F^-]}{[HF]} = \frac{(9.20 \times 10^{-3})(9.20 \times 10^{-3})}{0.241} = 3.52 \times 10^{-4}$$

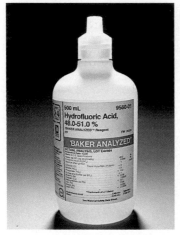

Of the hydrohalic acids HF, HCl, HBr, and HI, hydrofluoric acid is the only weak acid.

⌐ **PROBLEM 15.11**   The pH of 0.10 M HOCl is 4.23. Calculate K_a for hypochlorous acid, and check your answer against the value given in Table 15.2.   ⌐

15.9 ➤CALCULATING EQUILIBRIUM CONCENTRATIONS IN WEAK-ACID SOLUTIONS

Once the K_a value for a weak acid has been measured, it can be used to calculate equilibrium concentrations and the pH in a solution of the acid. Let's illustrate the approach to such a problem by calculating the concentrations of all species present (H_3O^+, CN^-, HCN, and OH^-) and the pH in a 0.10 M HCN solution. The approach we'll use is quite general and will be useful on numerous later occasions.

The key to solving acid–base equilibrium problems is to think about the chemistry—that is, to consider the possible proton-transfer reactions that can take place between Brønsted-Lowry acids and bases. Let's begin by listing the species present initially before any dissociation and identifying them as acids or bases. We include in our list the acid HCN and the solvent H_2O, but do not include the species produced only in small concentrations as a result of dissociation reactions (H_3O^+, CN^-, and OH^-). Since water can behave either as an acid or as a base, our list of species present initially is

$$HCN \qquad\qquad\qquad H_2O$$
$$\text{acid} \qquad\qquad\qquad \text{acid or base}$$

Because we have two acids (HCN and H_2O) and just one base (H_2O), there are two possible proton transfer reactions:

$$HCN(aq) + H_2O(l) \rightleftarrows H_3O^+(aq) + CN^-(aq) \qquad K_a = 4.9 \times 10^{-10}$$
$$H_2O(l) + H_2O(l) \rightleftarrows H_3O^+(aq) + OH^-(aq) \qquad K_w = 1.0 \times 10^{-14}$$

The proton-transfer reaction that proceeds farther to the right—the one that has the larger equilibrium constant—is called the **principal reaction**. Any other proton-transfer reactions are termed **subsidiary reactions**. Since K_a for HCN is more than 1000 times greater than K_w, the principal reaction in this case is dissociation of HCN. Dissociation of water is a subsidiary reaction. Although the principal reaction and the subsidiary reaction both produce H_3O^+ ions, there is only one H_3O^+ concentration in the solution, which must satisfy simultaneously the equilibrium equations for both the principal and subsidiary reactions. To make life simple, we'll assume that essentially all the H_3O^+ comes from the principal reaction:

$$[H_3O^+] \text{ (total)} = [H_3O^+] \text{ (from principal reaction)}$$
$$+ [H_3O^+] \text{ (from subsidiary reaction)}$$
$$\approx [H_3O^+] \text{ (from principal reaction)}$$

In other words, we'll assume that the equilibrium concentration of H_3O^+ is established by the dissociation of the stronger acid HCN, while dissociation of the weaker acid H_2O makes a negligible contribution.

$$[H_3O^+] \text{ (total)} \approx [H_3O^+] \text{ (from HCN)}$$

Next, we express the concentrations of the species involved in the principal reaction in terms of the concentration of HCN that dissociates—say, x moles per liter. According to the balanced equation for the dissociation of HCN, if x mol/L of HCN dissociates, then x mol/L of H_3O^+ and x mol/L of CN^- are formed, and the initial concentration of HCN before dissociation (0.10 mol/L) is reduced to $(0.10 - x)$ mol/L at equilibrium. Let's summarize these considerations in a table under the principal reaction:

Principal reaction:	$HCN(aq)$	$+ H_2O(l) \rightleftarrows$	$H_3O^+(aq)$	$+ CN^-(aq)$
Initial conc (M)	0.10		~0	0
Change (M)	$-x$		$+x$	$+x$
Equilibrium conc (M)	$(0.10 - x)$		x	x

Substituting the equilibrium concentrations into the equilibrium equation for the principal reaction gives a quadratic equation in x:

$$K_a = 4.9 \times 10^{-10} = \frac{[H_3O^+][CN^-]}{[HCN]} = \frac{(x)(x)}{(0.10 - x)}$$

Since K_a is very small, the principal reaction will not proceed very far to the right, and x will be negligibly small in comparison with 0.10. Therefore, we can make the approximation that $(0.10 - x) \approx 0.10$, which greatly simplifies the solution:

$$4.9 \times 10^{-10} = \frac{(x)(x)}{(0.10 - x)} \approx \frac{x^2}{0.10}$$

$$x^2 = 4.9 \times 10^{-11}$$

$$x = 7.0 \times 10^{-6}$$

Next, we use the calculated value of x to obtain the equilibrium concentration of all species involved in the principal reaction:

$$[H_3O^+] = [CN^-] = x = 7.0 \times 10^{-6} \text{ M}$$

$$[HCN] = 0.10 - x = 0.10 - (7.0 \times 10^{-6}) = 0.10 \text{ M}$$

Our simplifying approximation, $0.10 - x \approx 0.10$, is valid because x is only 7.0×10^{-6} and [HCN] is 0.10. *It's important to check the validity of the simplifying approximation in every problem* because x is not always negligible in comparison with the initial concentration of the acid. Example 15.8 illustrates such a case.

The concentrations of the species involved in the principal reaction are the "big" concentrations. The species involved in the subsidiary reaction(s) are present in smaller concentrations that can be calculated from equilibrium equations for the subsidiary reaction(s) and the big concentrations already determined. In the present problem, only the OH^- concentration remains to be calculated. It is determined from the subsidiary equilibrium equation, $[H_3O^+][OH^-] = K_w$, and the H_3O^+ concentration (7.0×10^{-6} M) already calculated from the principal reaction:

$$[OH^-] = \frac{K_w}{[H_3O^+]} = \frac{1.0 \times 10^{-14}}{7.0 \times 10^{-6}} = 1.4 \times 10^{-9} \text{ M}$$

Note that $[OH^-]$ is 5000 times smaller than $[H_3O^+]$.

At this point we can check our initial assumption that essentially all the H_3O^+ comes from the principal reaction. The $[H_3O^+]$ *from the dissociation of water* is equal to $[OH^-]$, which we just calculated to be 1.4×10^{-9} M. This value is negligible compared with $[H_3O^+]$ from the dissociation of HCN (7.0×10^{-6} M).

Finally, we can calculate the pH:

$$\text{pH} = -\log(\text{total } [H_3O^+]) = -\log(7.0 \times 10^{-6}) = 5.15$$

Figure 15.6 summarizes the steps that we've followed in solving this problem. We'll apply the same systematic approach to other aqueous equilibrium problems.

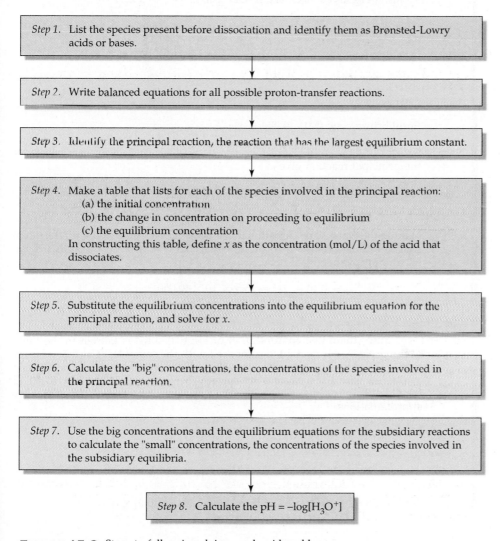

Step 1. List the species present before dissociation and identify them as Brønsted-Lowry acids or bases.

Step 2. Write balanced equations for all possible proton-transfer reactions.

Step 3. Identify the principal reaction, the reaction that has the largest equilibrium constant.

Step 4. Make a table that lists for each of the species involved in the principal reaction:
    (a) the initial concentration
    (b) the change in concentration on proceeding to equilibrium
    (c) the equilibrium concentration
In constructing this table, define x as the concentration (mol/L) of the acid that dissociates.

Step 5. Substitute the equilibrium concentrations into the equilibrium equation for the principal reaction, and solve for x.

Step 6. Calculate the "big" concentrations, the concentrations of the species involved in the principal reaction.

Step 7. Use the big concentrations and the equilibrium equations for the subsidiary reactions to calculate the "small" concentrations, the concentrations of the species involved in the subsidiary equilibria.

Step 8. Calculate the pH = $-\log[H_3O^+]$

FIGURE 15.6 Steps to follow in solving weak-acid problems.

EXAMPLE 15.8

Calculate the concentration of all species present (H_3O^+, F^-, HF, and OH^-) and the pH in 0.050 M HF.

SOLUTION Follow the 8-step sequence outlined in Figure 15.6.

Step 1. The species present initially are

$$HF \qquad\qquad\qquad H_2O$$
$$\text{acid} \qquad\qquad\qquad \text{acid or base}$$

Step 2. The possible proton-transfer reactions are

$$HF(aq) + H_2O(l) \rightleftharpoons H_3O^+(aq) + F^-(aq) \qquad K_a = 3.5 \times 10^{-4}$$

$$H_2O(l) + H_2O(l) \rightleftharpoons H_3O^+(aq) + OH^-(aq) \qquad K_w = 1.0 \times 10^{-14}$$

Step 3. Since $K_a \gg K_w$, the principal reaction is dissociation of HF.

Step 4. Principal reaction: $HF(aq) + H_2O(l) \rightleftharpoons H_3O^+(aq) + F^-(aq)$

Initial conc (M)	0.050	~0	0
Change (M)	$-x$	$+x$	$+x$
Equilibrium conc (M)	$(0.050 - x)$	x	x

Step 5. Substituting the equilibrium concentrations into the equilibrium equation for the principal reaction gives

$$K_a = 3.5 \times 10^{-4} = \frac{[H_3O^+][F^-]}{[HF]} = \frac{(x)(x)}{(0.050 - x)}$$

Making the usual approximation that x is negligible compared with the initial concentration of the acid, we assume that $(0.050 - x) \approx 0.050$ and then solve for an approximate value of x:

$$x^2 \approx (3.5 \times 10^{-4})(0.050)$$

$$x \approx 4.2 \times 10^{-3}$$

Since the initial concentration of HF (0.050 M) is known to the third decimal place, x is negligible compared with the initial [HF] only if x is less than 0.001 M. Our approximate value of x (0.0042 M) is not negligible compared with 0.050 M, and so our approximation, $0.050 - x \approx 0.050$, is invalid. We must therefore solve the quadratic equation without making approximations:

$$3.5 \times 10^{-4} = \frac{x^2}{(0.050 - x)}$$

$$x^2 + (3.5 \times 10^{-4})x - (1.75 \times 10^{-5}) = 0$$

We employ the standard quadratic formula (Appendix A.4)

$$x = \frac{-b \pm \sqrt{b^2 - 4ac}}{2a}$$

$$x = \frac{-(3.5 \times 10^{-4}) \pm \sqrt{(3.5 \times 10^{-4})^2 - 4(-1.75 \times 10^{-5})}}{2}$$

$$= \frac{-(3.5 \times 10^{-4}) \pm (8.37 \times 10^{-3})}{2}$$

$$= +4.0 \times 10^{-3} \quad \text{or} \quad -4.4 \times 10^{-3}$$

Of the two solutions for x, only the positive value of x has physical meaning, since x is the H_3O^+ concentration. Therefore,

$$x = 4.0 \times 10^{-3}$$

Step 6. The big concentrations are

$$[H_3O^+] = [F^-] = x = 4.0 \times 10^{-3} \text{ M}$$

$$[HF] = (0.050 - x) = (0.050 - 0.0040) = 0.046 \text{ M}$$

Step 7. The small concentrations are obtained from the subsidiary equilibrium, the dissociation of water:

$$[OH^-] = \frac{K_w}{[H_3O^+]} = \frac{1.0 \times 10^{-14}}{4.0 \times 10^{-3}} = 2.5 \times 10^{-12} \text{ M}$$

Step 8. $pH = -\log[H_3O^+] = -\log(4.0 \times 10^{-3}) = 2.40$

⌐ **PROBLEM 15.12** Acetic acid, CH_3COOH, is the solute that gives vinegar its characteristic odor and sour taste. Calculate the concentration of all species present (H_3O^+, $CH_3CO_2^-$, CH_3COOH, and OH^-) and the pH in **(a)** 1.00 M CH_3COOH and **(b)** 0.0100 M CH_3COOH. ⌐

15.10 ►PERCENT DISSOCIATION IN WEAK-ACID SOLUTIONS

In addition to K_a, another useful measure of weak-acid strength is the **percent dissociation**, defined as the concentration of the acid that dissociates divided by the initial concentration of the acid times 100%:

$$\text{Percent dissociation} = \frac{[HA] \text{ dissociated}}{[HA] \text{ initial}} \times 100\%$$

Take, for example, the 1.00 M acetic acid solution in Problem 15.12. If you solved that problem correctly, you found that 1.00 M CH_3COOH has an H_3O^+ concentration of 4.2×10^{-3} M. Because $[H_3O^+]$ equals the concentration of CH_3COOH that dissociates, the percent dissociation in 1.00 M CH_3COOH is 0.42%:

$$\text{Percent dissociation} = \frac{[CH_3COOH] \text{ dissociated}}{[CH_3COOH] \text{ initial}} = \frac{4.2 \times 10^{-3} \text{ M}}{1.00 \text{ M}} \times 100\% = 0.42\%$$

In general, the percent dissociation depends on the acid and increases with increasing value of K_a.

For a given weak acid, the percent dissociation increases with increasing dilution, as shown in Figure 15.7. The 0.0100 M CH_3COOH solution in Problem 15.12, for example, has $[H_3O^+] = 4.2 \times 10^{-4}$ M, and the percent dissociation is 4.2%:

$$\text{Percent dissociation} = \frac{[CH_3COOH] \text{ dissociated}}{[CH_3COOH] \text{ initial}} = \frac{4.2 \times 10^{-4} \text{ M}}{0.0100 \text{ M}} \times 100\% = 4.2\%$$

⌐ **PROBLEM 15.13** Calculate the percent dissociation of HF ($K_a = 3.5 \times 10^{-4}$) in **(a)** 0.050 M HF and **(b)** 0.50 M HF. ⌐

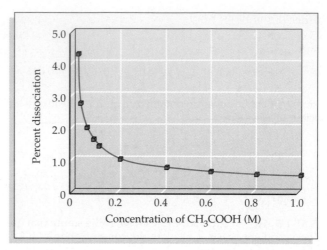

FIGURE 15.7   The percent dissociation of acetic acid increases as the concentration of the acid decreases. A 100-fold decrease in [CH$_3$COOH] results in a 10-fold *increase* in the percent dissociation.

15.11 ➤POLYPROTIC ACIDS

Acids that contain more than one dissociable proton are called **polyprotic acids.** Carbonic acid (H$_2$CO$_3$), for example, the diprotic acid that forms when gaseous carbon dioxide dissolves in water, is important in maintaining a constant pH in human blood. Polyprotic acids dissociate in a stepwise manner, and each dissociation step is characterized by its own acid-dissociation constant, K_{a1}, K_{a2}, and so forth. Carbonic acid, for example, undergoes the following dissociation reactions[3].

$$H_2CO_3(aq) + H_2O(l) \rightleftarrows H_3O^+(aq) + HCO_3^-(aq)$$

$$K_{a1} = \frac{[H_3O^+][HCO_3^-]}{[H_2CO_3]} = 4.3 \times 10^{-7}$$

$$HCO_3^-(aq) + H_2O(l) \rightleftarrows H_3O^+(aq) + CO_3^{2-}(aq)$$

$$K_{a2} = \frac{[H_3O^+][CO_3^{2-}]}{[HCO_3^-]} = 5.6 \times 10^{-11}$$

[3] Carbonic acid is in equilibrium with dissolved carbon dioxide, CO$_2$(aq), and the equilibrium lies far to the left:

$$CO_2(aq) + H_2O(l) \rightleftarrows H_2CO_3(aq)$$

Thus, the first dissociation step could be written as

$$\{CO_2(aq) + H_2O(l)\} + H_2O(l) \rightleftarrows H_3O^+(aq) + HCO_3^-(aq) \qquad K_{a1} = 4.3 \times 10^{-7}$$

It's common practice, however, to treat all diprotic acids as having the formula H$_2$A, and we therefore write the first dissociation step using H$_2$CO$_3$(aq) rather than {CO$_2$(aq) + H$_2$O(l)}.

The values of stepwise dissociation constants of polyprotic acids given in Table 15.3 decrease, typically by a factor of 10^4 to 10^6, in the order $K_{a1} > K_{a2} > K_{a3}$. Because of electrostatic forces, it's more difficult to remove a positively charged proton from a negative ion such as HCO_3^- than from an uncharged molecule such as H_2CO_3, so $K_{a2} < K_{a1}$. In the case of triprotic acids (such as H_3PO_4), it's more difficult to remove H^+ from an anion with a double negative charge (such as HPO_4^{2-}), than from an anion with a single negative charge (such as $H_2PO_4^-$), so $K_{a3} < K_{a2}$.

TABLE 15.3   Stepwise Dissociation Constants for Polyprotic Acids at 25°C

Name	Formula	K_{a1}	K_{a2}	K_{a3}
Carbonic acid	H_2CO_3	4.3×10^{-7}	5.6×10^{-11}	
Hydrogen sulfide[a]	H_2S	1.0×10^{-7}	$\sim 10^{-19}$	
Oxalic acid	$H_2C_2O_4$	5.9×10^{-2}	6.4×10^{-5}	
Phosphoric acid	H_3PO_4	7.5×10^{-3}	6.2×10^{-8}	4.8×10^{-13}
Sulfuric acid	H_2SO_4	Very large	1.2×10^{-2}	
Sulfurous acid	H_2SO_3	1.5×10^{-2}	6.3×10^{-8}	

[a] Because of its very small size, K_{a2} for H_2S is difficult to measure. Its value is therefore quite uncertain.

Polyprotic acid solutions contain a mixture of acids—H_2A, HA^-, and H_2O in the case of a diprotic acid. Because H_2A is by far the stronger acid, the principal reaction is dissociation of H_2A, and essentially all the H_3O^+ in the solution comes from the first dissociation step. Example 15.9 shows how calculations are done.

EXAMPLE 15.9

Calculate the concentration of all species present (H_2CO_3, HCO_3^-, CO_3^{2-}, H_3O^+, and OH^-) and the pH in a 0.040 M carbonic acid solution.

SOLUTION   Use the 8-step procedure summarized in Figure 15.6.

Steps 1–3. The species present initially are H_2CO_3 (acid) and H_2O (acid or base). Because $K_{a1} \gg K_w$, the principal reaction is dissociation of H_2CO_3.

Step 4. Principal reaction: $H_2CO_3(aq) + H_2O(l) \rightleftarrows H_3O^+(aq) + HCO_3^-(aq)$

Initial conc (M)	0.040	~ 0	0
Change (M)	$-x$	$+x$	$+x$
Equilibrium conc (M)	$(0.040 - x)$	x	x

Step 5. Substituting the equilibrium concentrations into the equilibrium equation for the principal reaction gives

$$K_{a1} = 4.3 \times 10^{-7} = \frac{[H_3O^+][HCO_3^-]}{[H_2CO_3]} = \frac{(x)(x)}{(0.040 - x)}$$

Assuming that $(0.040 - x) \approx 0.040$,

$$x^2 = (4.3 \times 10^{-7})(0.040)$$

$$x = 1.3 \times 10^{-4} \text{ (approximation, } 0.040 - x \approx 0.040 \text{, is justified)}$$

Step 6. The big concentrations are:

$$[H_3O^+] = [HCO_3^-] = x = 1.3 \times 10^{-4} \text{ M}$$

$$[H_2CO_3] = 0.040 - x = 0.040 - 0.00013 = 0.040 \text{ M}$$

Step 7. The small concentrations are obtained from the subsidiary equilibria—(1) dissociation of HCO_3^- and (2) dissociation of water—and from the big concentrations already determined:

(1) $$HCO_3^-(aq) + H_2O(l) \rightleftarrows H_3O^+(aq) + CO_3^{2-}(aq)$$

$$K_{a2} = 5.6 \times 10^{-11} = \frac{[H_3O^+][CO_3^{2-}]}{[HCO_3^-]} = \frac{(1.3 \times 10^{-4})[CO_3^{2-}]}{(1.3 \times 10^{-4})}$$

$$[CO_3^{2-}] = K_{a2} = 5.6 \times 10^{-11} \text{ M}$$

(2) $$[OH^-] = \frac{K_w}{[H_3O^+]} = \frac{1.0 \times 10^{-14}}{1.3 \times 10^{-4}} = 7.7 \times 10^{-11} \text{ M}$$

The second dissociation of H_2CO_3 produces a negligible amount of H_3O^+ compared with the 1.3×10^{-4} mol/L obtained from the first dissociation. Of the 1.3×10^{-4} mol/L of HCO_3^- produced by the first dissociation, only 5.6×10^{-11} mol/L dissociates to form H_3O^+ and CO_3^{2-}.

Step 8. $pH = -\log[H_3O^+] = -\log(1.3 \times 10^{-4}) = 3.89$

⌐ **PROBLEM 15.14** Calculate the concentration of all species present and the pH in 0.10 M H_2SO_3. Values of K_a are in Table 15.3.

⌐ **PROBLEM 15.15** Calculate the concentration of all species present and the pH in 0.50 M H_2SO_4. *Hint:* Dissociation of the first proton of H_2SO_4 is complete, but dissociation of the second proton is partial and takes place in the presence of H_3O^+ from the first dissociation. Values of K_a are given in Table 15.3.

15.12 ►EQUILIBRIA IN SOLUTIONS OF WEAK BASES

Weak bases, such as ammonia, accept a proton from water to give the conjugate acid of the base and OH^- ions:

$$NH_3(aq) + H_2O(l) \rightleftarrows NH_4^+(aq) + OH^-(aq)$$

The position of any base-dissociation equilibrium is characterized by an equilibrium constant K_b, called the **base-dissociation constant**:

For the reaction $B(aq) + H_2O(l) \rightleftarrows BH^+(aq) + OH^-(aq)$

$$K_b = \frac{[BH^+][OH^-]}{[B]}$$

As usual, $[H_2O]$ is omitted from the equilibrium constant expression.

Table 15.4 lists some typical weak bases and gives their K_b values. Many weak bases are organic compounds called **amines,** derivatives of ammonia in which one or more hydrogen atoms are replaced by another group. Methylamine, for example, is an organic amine responsible for the odor of rotting fish.

$$CH_3\overset{\overset{\displaystyle\cdot\cdot}{}}{-N}-H \qquad CH_3\overset{\overset{\displaystyle\cdot\cdot}{}}{-N}-CH_3 \qquad H\overset{\overset{\displaystyle\cdot\cdot}{}}{-N}-OH$$
$$\underset{\text{Methylamine}}{\overset{|}{H}} \qquad \underset{\text{Dimethylamine}}{\overset{|}{II}} \qquad \underset{\text{Hydroxylamine}}{\overset{|}{H}}$$

The basicity of an amine is due to the lone pair of electrons on the nitrogen atom, which can be used for bonding to a proton.

Equilibria in a solution of a weak base are treated by the same procedure used for solving weak acid problems. Example 15.10 illustrates the procedure.

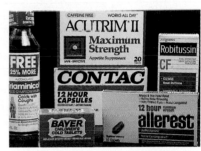

Many over-the-counter drugs contain salts formed from amines and hydrochloric acid.

TABLE 15.4	K_b Values for Some Weak Bases and K_a Values for the Corresponding Conjugate Acids				
Base	**Formula, B**	K_b	**Conjugate Acid, BH^+**	K_a
Ammonia	NH_3	1.8×10^{-5}	NH_4^+	5.6×10^{-10}
Aniline	$C_6H_5NH_2$	4.3×10^{-10}	$C_6H_5NH_3^+$	2.3×10^{-5}
Dimethylamine	$(CH_3)_2NH$	5.4×10^{-4}	$(CH_3)_2NH_2^+$	1.9×10^{-11}
Hydrazine	N_2H_4	8.9×10^{-7}	$N_2H_5^+$	1.1×10^{-8}
Hydroxylamine	NH_2OH	9.1×10^{-9}	NH_3OH^+	1.1×10^{-6}
Methylamine	CH_3NH_2	3.7×10^{-4}	$CH_3NH_3^+$	2.7×10^{-11}

EXAMPLE 15.10

Codeine ($C_{18}H_{21}NO_3$), a drug used in pain killers and cough medicines, is a naturally occurring amine that has $K_b = 1.6 \times 10^{-6}$. Calculate the concentration of all species present and the pH in a 0.0012 M solution of codeine.

SOLUTION   Use the procedure outlined in Figure 15.6.

Step 1. Let's use Cod as an abbreviation for codeine. The species present initially are Cod (base) and H_2O (acid or base).

Step 2. There are two possible proton-transfer reactions:

$$\text{Cod}(aq) + \text{H}_2\text{O}(l) \rightleftarrows \text{CodH}^+(aq) + \text{OH}^-(aq) \qquad K_b = 1.6 \times 10^{-6}$$

$$\text{H}_2\text{O}(l) + \text{H}_2\text{O}(l) \rightleftarrows \text{H}_3\text{O}^+(aq) + \text{OH}^-(aq) \qquad K_w = 1.0 \times 10^{-14}$$

Step 3. Since Cod is a stronger base than H_2O ($K_b \gg K_w$), the principal reaction involves protonation of codeine.

Step 4. Principal reaction: $\text{Cod}(aq) + \text{H}_2\text{O}(l) \rightleftarrows \text{CodH}^+(aq) + \text{OH}^-(aq)$

Initial conc (M)	0.0012	0	~0
Change (M)	$-x$	$+x$	$+x$
Equilibrium conc (M)	$0.0012 - x$	x	x

Step 5. The value of x is obtained from the equilibrium equation:

$$K_b = 1.6 \times 10^{-6} = \frac{[\text{CodH}^+][\text{OH}^-]}{[\text{Cod}]} = \frac{(x)(x)}{(0.0012 - x)}$$

Assuming that $(0.0012 - x) \approx 0.0012$,

$$x^2 = (1.6 \times 10^{-6})(0.0012)$$

$$x = 4.4 \times 10^{-5} \text{ (approximation } 0.0012 - x \approx 0.0012 \text{ is justified)}$$

Step 6. The big concentrations are

$$[\text{CodH}^+] = [\text{OH}^-] = x = 4.4 \times 10^{-5} \text{ M}$$

$$[\text{Cod}] = 0.0012 - x = 0.0012 - 0.000044 = 0.0012 \text{ M}$$

Step 7. The small concentration is obtained from the subsidiary equilibrium, the dissociation of water:

$$[\text{H}_3\text{O}^+] = \frac{K_w}{[\text{OH}^-]} = \frac{1.0 \times 10^{-14}}{4.4 \times 10^{-5}} = 2.3 \times 10^{-10} \text{ M}$$

Step 8. $\text{pH} = -\log[\text{H}_3\text{O}^+] = -\log(2.3 \times 10^{-10}) = 9.64$

Note that the pH is greater than 7.00, as expected for a weak base solution.

┌ **PROBLEM 15.16** Calculate the concentration of all species present and the pH in 0.40 M NH_3 ($K_b = 1.8 \times 10^{-5}$).

15.13 ►RELATION BETWEEN K_a AND K_b

We've seen in previous sections that the strength of an acid can be expressed by K_a, and the strength of a base can be expressed by K_b. For a conjugate acid–base pair, the two equilibrium constants are related in a simple way that makes it possible to calculate either one of the constants from the other.

Let's consider the conjugate acid–base pair NH_4^+ and NH_3, for example. K_a refers to proton transfer from the acid NH_4^+ to water, and K_b refers to proton transfer from water to the base NH_3. The sum of these two reactions is simply the dissociation of water:

$$NH_4^+(aq) + H_2O(l) \rightleftharpoons H_3O^+(aq) + NH_3(aq) \qquad K_a = \frac{[H_3O^+][NH_3]}{[NH_4^+]} = 5.6 \times 10^{-10}$$

$$NH_3(aq) + H_2O(l) \rightleftharpoons NH_4^+(aq) + OH^-(aq) \qquad K_b = \frac{[NH_4^+][OH^-]}{[NH_3]} = 1.8 \times 10^{-5}$$

$$\text{Net: } 2\,H_2O(l) \rightleftharpoons H_3O^+(aq) + OH^-(aq) \qquad K_w = [H_3O^+][OH^-] = 1.0 \times 10^{-14}$$

Note that the equilibrium constant for the net reaction is equal to the *product* of the equilibrium constants for the reactions added:

$$K_a \times K_b = \frac{[H_3O^+][NH_3]}{[NH_4^+]} \times \frac{[NH_4^+][OH^-]}{[NH_3]} = [H_3O^+][OH^-] = K_w$$

$$= (5.6 \times 10^{-10})(1.8 \times 10^{-5}) = 1.0 \times 10^{-14}$$

What we've shown in this particular case is true in general. *Whenever chemical equations for two (or more) reactions are added to get the chemical equation for a net reaction, the equilibrium constant for the net reaction is equal to the product of the equilibrium constants for the individual reactions:*

$$K_{net} = K_1 \times K_2 \times \cdots$$

For any conjugate acid–base pair, the product of the acid-dissociation constant for the acid and the base-dissociation constant for the base is always equal to the ion-product constant for water:

$$K_a \times K_b = K_w$$

This relation holds for all the conjugate acid–base pairs in Table 15.4. As the strength of an acid increases (larger K_a), the strength of its conjugate base decreases (smaller K_b) because the product $K_a \times K_b$ must remain constant at 1.0×10^{-14}.

Compilations of equilibrium constants, such as Appendix C, generally list either K_a or K_b, but not both because K_a is easily calculated from K_b and vice versa:

$$K_a = \frac{K_w}{K_b} \quad \text{and} \quad K_b = \frac{K_w}{K_a}$$

EXAMPLE 15.11

(a) K_b for trimethylamine is 6.5×10^{-5}. Calculate K_a for the trimethylammonium ion, $(CH_3)_3NH^+$.
(b) K_a for HCN is 4.9×10^{-10}. Calculate K_b for CN^-.

SOLUTION

(a) Since $K_b = 6.5 \times 10^{-5}$ for $(CH_3)_3N$, we can find $K_a = K_w/K_b$:

$$K_a = \frac{K_w}{K_b} = \frac{1.0 \times 10^{-14}}{6.5 \times 10^{-5}} = 1.5 \times 10^{-10}$$

K_a is the equilibrium constant for the acid-dissociation reaction

$$(CH_3)_3NH^+(aq) + H_2O(l) \rightleftarrows H_3O^+(aq) + (CH_3)_3N(aq)$$

(b) Since $K_a = 4.9 \times 10^{-10}$ for HCN, we can find $K_b = K_w/K_a$:

$$K_b = \frac{K_w}{K_a} = \frac{1.0 \times 10^{-14}}{4.9 \times 10^{-10}} = 2.0 \times 10^{-5}$$

K_b is the equilibrium constant for the base-dissociation reaction

$$CN^-(aq) + H_2O(l) \rightleftarrows HCN(aq) + OH^-(aq)$$

PROBLEM 15.17 **(a)** Piperidine $(C_5H_{11}N)$ is a base found in black pepper. Find K_b for piperidine in Appendix C, and then calculate K_a for the $C_5H_{11}NH^+$ cation. **(b)** Find K_a for HOCl in Appendix C, and then calculate K_b for OCl^-.

15.14 ➤ ACID–BASE PROPERTIES OF SALTS

Salt solutions can be neutral, acidic, or basic, depending on the acid–base properties of the constituent cations and anions (Figure 15.8). As a general rule, salts formed by reaction of a strong acid and a strong base are neutral, salts formed by reaction of a strong acid with a weak base are acidic, and salts formed by reaction of a weak acid with a strong base are basic. It's as if, in an acid–base reaction, the influence of the stronger partner is dominant:

Strong acid + **strong base** → neutral solution

Strong acid + weak base → **acidic** solution

Weak acid + **strong base** → **basic** solution

FIGURE 15.8 0.10 M aqueous salt solutions (left to right): NaCl, NH_4Cl, $AlCl_3$, NaCN, and $(NH_4)_2CO_3$. A few drops of universal indicator have been added to each solution. The color of the indicator shows that the NaCl solution is neutral, the NH_4Cl and $AlCl_3$ solutions are acidic, and the NaCN and $(NH_4)_2CO_3$ solutions are basic.

Salts that Yield Neutral Solutions

Salts such as NaCl that are derived from a strong base (NaOH) and a strong acid (HCl) yield neutral solutions because neither the cation nor the anion reacts with water to produce H_3O^+ or OH^- ions. As the conjugate base of a strong acid, Cl^- has no tendency to make the solution basic by picking up a proton from water. As the cation of a strong base, the hydrated Na^+ ion has no tendency to make the solution acidic by transfering a proton to a solvent water molecule. The following ions do *not* react with water to produce either H_3O^+ or OH^- ions:

1. Cations from strong bases.
    alkali-metal cations of group 1A (Li^+, Na^+, K^+)
    alkaline-earth cations of group 2A (Ca^{2+}, Sr^{2+}, Ba^{2+}) except for Be^{2+}
2. Anions from strong acids:
    Cl^-, Br^-, I^-, NO_3^-, and ClO_4^-

Salts that contain only these ions give neutral solutions in pure water (pH = 7).[4]

Salts that Yield Acidic Solutions

Salts such as NH_4Cl that are derived from a weak base (NH_3) and a strong acid (HCl) produce acidic solutions. In this case, the anion is inert (neither an acid nor a base), but the cation is a weak acid.[5]

$$NH_4^+(aq) + H_2O(l) \rightleftharpoons H_3O^+(aq) + NH_3(aq)$$

The pH of a solution that contains an acidic cation can be calculated by the standard procedure outlined in Figure 15.6. For a 0.10 M NH_4Cl solution, the pH is 5.12.

A second class of acidic cations is that of small, highly charged cations, such as Al^{3+}. In aqueous solution, the Al^{3+} ion is bonded to six water molecules to give the hydrated cation $Al(H_2O)_6^{3+}$. Because of the +3 charge on the Al^{3+} ion, electrons within the bound water molecules are attracted toward the Al^{3+} ion, thus decreasing the electron density in the O–H bonds. As a result, the O–H bonds are weakened, easing transfer of a proton to a solvent water molecule:

$$Al(H_2O)_6^{3+}(aq) + H_2O(l) \rightleftharpoons H_3O^+(aq) + Al(H_2O)_5(OH)^{2+}(aq)$$

The acid-dissociation constant for $Al(H_2O)_6^{3+}$, $K_a = 1.4 \times 10^{-5}$, is much larger than $K_w = 1.0 \times 10^{-14}$, which means that the water molecules in the hydrated cation are much stronger proton donors than are the free solvent water molecules. Other cations that give acidic solutions are Be^{2+} and transition-metal cations, such as Zn^{2+}, Cr^{3+}, and Fe^{3+}.

---

[4] Water usually contains dissolved atmospheric CO_2 and consequently is acidic. To prepare a neutral salt solution, you would have to remove the dissolved CO_2.

[5] The reaction of a cation or anion of a salt with water to produce H_3O^+ or OH^- ions is sometimes called a *salt hydrolysis reaction*. There is no fundamental difference, however, between a salt hydrolysis reaction and any other Brønsted-Lowry acid–base reaction.

EXAMPLE 15.12

Calculate the pH of a 0.10 M solution of $AlCl_3$. K_a for $Al(H_2O)_6^{3+}$ is 1.4×10^{-5}.

SOLUTION   Because this problem is similar to others done earlier, we'll abbreviate the procedure in Figure 15.6.

Steps 1–4. The species present initially are $Al(H_2O)_6^{3+}$ (acid), Cl^- (inert), and H_2O (acid or base). Because $Al(H_2O)_6^{3+}$ is a stronger acid than water ($K_a > K_w$), the principal reaction is dissociation of $Al(H_2O)_6^{3+}$:

$$Al(H_2O)_6^{3+}(aq) + H_2O(l) \rightleftarrows H_3O^+(aq) + Al(H_2O)_5(OH)^{2+}(aq)$$

Equilibrium
conc (M) $0.10 - x$ x x

Step 5. The value of x is obtained from the equilibrium equation:

$$K_a = 1.4 \times 10^{-5} = \frac{[H_3O^+][Al(H_2O)_5(OH)^{2+}]}{[Al(H_2O)_6^{3+}]} = \frac{(x)(x)}{0.10 - x} \approx \frac{x^2}{0.10}$$

$$x = [H_3O^+] = 1.2 \times 10^{-3} \text{ M}$$

Step 8. pH $= -\log(1.2 \times 10^{-3}) = 2.92$

Note that $Al(H_2O)_6^{3+}$ is a much stronger acid than NH_4^+, in accord with the colors of the indicator in Figure 15.8.

┌ **PROBLEM 15.18** Predict whether the following salt solutions are neutral, acidic, or basic, and calculate the pH: **(a)** 0.25 M NH_4Br **(b)** 0.40 M $ZnCl_2$
K_a for $Zn(H_2O)_6^{2+}$ is 2.5×10^{-10}. ┘

Salts that Yield Basic Solutions

Salts such as NaCN that are derived from a strong base (NaOH) and a weak acid (HCN) yield basic solutions. In this case, the cation is neither an acid nor a base, but the anion is a weak base:

$$CN^-(aq) + H_2O(l) \rightleftarrows HCN(aq) + OH^-(aq)$$

Other anions that exhibit basic properties are listed in Table 15.1 and include NO_2^-, F^-, $CH_3CO_2^-$, and CO_3^{2-}. The pH of a basic salt solution can be calculated by the standard procedure, as shown in Example 15.13.

EXAMPLE 15.13

Calculate the pH of a 0.10 M solution of NaCN. K_a for HCN is 4.9×10^{-10}.

SOLUTION
Step 1. The species present initially are Na^+ (inert), CN^- (base), and H_2O (acid or base).

Step 2. There are two possible proton-transfer reactions:

$$CN^-(aq) + H_2O(l) \rightleftarrows HCN(aq) + OH^-(aq) \qquad K_b$$

$$H_2O(l) + H_2O(l) \rightleftarrows H_3O^+(aq) + OH^-(aq) \qquad K_w$$

Step 3. As shown in Example 15.11, $K_b = K_w/(K_a$ for HCN$) = 2.0 \times 10^{-5}$. Because $K_b > K_w$, CN^- is a stronger base than H_2O, and the principal reaction is proton transfer from H_2O to CN^-.

Step 4. Principal reaction: $\qquad CN^-(aq) + H_2O(l) \rightleftarrows HCN(aq) + OH^-(aq)$

Equilibrium conc (M) $\qquad 0.10 - x \qquad\qquad\qquad x \qquad\qquad x$

Step 5. The value of x is obtained from the equilibrium equation:

$$K_b = 2.0 \times 10^{-5} = \frac{[HCN][OH^-]}{[CN^-]} = \frac{(x)(x)}{(0.10 - x)} \approx \frac{x^2}{0.10}$$

$$x = [OH^-] = 1.4 \times 10^{-3} \text{ M}$$

Step 7. $[H_3O^+] = \dfrac{K_w}{[OH^-]} = \dfrac{1.0 \times 10^{-14}}{1.4 \times 10^{-3}} = 7.1 \times 10^{-12}$

Step 8. pH $= -\log(7.1 \times 10^{-12}) = 11.15$.

In accord with the color of the indicator in Figure 15.8, the solution is basic.

⌐ PROBLEM 15.19 Calculate the pH of 0.20 M $NaNO_2$. K_a for HNO_2 is 4.6 × 10^{-4}.

Salts that Contain Acidic Cations and Basic Anions

Finally, let's look at a salt, such as $(NH_4)_2CO_3$, for which both the cation and the anion can undergo proton-transfer reactions. Because NH_4^+ is a weak acid and CO_3^{2-} is a weak base, the pH of an $(NH_4)_2CO_3$ solution depends on the relative acid strength of the cation and base strength of the anion.

$$NH_4^+(aq) + H_2O(l) \rightleftarrows H_3O^+(aq) + NH_3(aq) \qquad K_a$$

$$CO_3^{2-}(aq) + H_2O(l) \rightleftarrows HCO_3^-(aq) + OH^-(aq) \qquad K_b$$

We can distinguish three possible cases:

(a) $K_a > K_b$. If K_a for the cation is greater than K_b for the anion, the solution will contain an excess of H_3O^+ ions (pH < 7).

(b) $K_a < K_b$. If K_a for the cation is less than K_b for the anion, the solution will contain an excess of OH^- ions (pH > 7).

(c) $K_a \approx K_b$. If K_a for the cation and K_b for the anion are comparable, the solution will contain approximately equal concentrations of H_3O^+ and OH^- ions (pH ≈ 7).

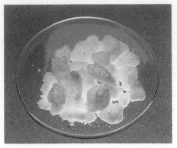

Ammonium carbonate has a strong odor of ammonia and has been used as a smelling salt: $NH_4^+ + CO_3^{2-} \rightleftarrows HCO_3^- + NH_3$.

To determine whether an $(NH_4)_2CO_3$ solution is acidic, basic, or neutral, let's work out the values of K_a for NH_4^+ and K_b for CO_3^{2-}:

$$K_a \text{ for } NH_4^+ = \frac{K_w}{K_b \text{ for } NH_3} = \frac{1.0 \times 10^{-14}}{1.8 \times 10^{-5}} = 5.6 \times 10^{-10}$$

$$K_b \text{ for } CO_3^{2-} = \frac{K_w}{K_a \text{ for } HCO_3^-} = \frac{K_w}{K_{a2} \text{ for } H_2CO_3} = \frac{1.0 \times 10^{-14}}{5.6 \times 10^{-11}} = 1.8 \times 10^{-4}$$

Because $K_a < K_b$, the solution is basic (pH > 7), in accord with the color of the indicator in Figure 15.8.

A summary of the acid–base properties of salts is given in Table 15.5.

TABLE 15.5   Acid–Base Properties of Salts

Type of Salt	Examples	Ions that React with Water	pH of Solution
Cation from strong base; anion from strong acid	$NaCl$, KNO_3 BaI_2	None	~7
Cation from weak base; anion from strong acid	NH_4Cl, NH_4NO_3 $[(CH_3)_3NH]Cl$	Cation	<7
Small, highly charged cation; anion from strong acid	$AlCl_3$, $Cr(NO_3)_3$ $Fe(ClO_4)_3$	Hydrated cation	<7
Cation from strong base; anion from weak acid	$NaCN$, KF, Na_2CO_3	Anion	>7
Cation from weak base; anion from weak acid	NH_4CN, NH_4F, $(NH_4)_2CO_3$	Cation and anion	<7 if $K_a > K_b$ >7 if $K_a < K_b$ ~7 if $K_a \approx K_b$

┌ **PROBLEM 15.20** Calculate K_a for the cation and K_b for the anion in an aqueous NH_4CN solution. Is the solution acidic, basic, or neutral?

┌ **PROBLEM 15.21** Classify each of the following salt solutions as acidic, basic, or neutral.
(a) KBr   **(b)** $NaNO_2$ **(c)** NH_4Br **(d)** $ZnCl_2$ **(e)** NH_4F ◢

15.15 ➤FACTORS THAT AFFECT ACID STRENGTH

Why is one acid stronger than another? Although a complete analysis of the factors that determine the strength of an acid is a complex business, the extent of dissociation of an acid HA is often determined by the strength and polarity of the H–A bond. The strength of the H–A bond is the enthalpy required to dissociate HA into an H atom and an A atom (Section 8.10). The polarity of the H–A bond increases with an increase in the electronegativity of A and is related to the ease of electron transfer from an H atom to an A atom to give an H^+ ion and an A^- ion. In general, the weaker the H–A bond, the stronger the acid, and the more polar the H–A bond, the stronger the acid.

Let's look first at the relative acidity of some binary acids, which contain hydrogen and just one other element. The acidity of the hydrohalic acids HA (A = F, Cl, Br, or I) increases from HF to HI, a trend that parallels a decrease in the H–A bond strength:

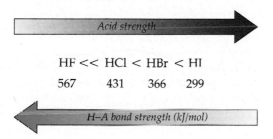

$$HF \ll HCl < HBr < HI$$
$$567 \quad\ 431 \quad\ 366 \quad\ 299$$

HF is a weak acid ($K_a = 3.5 \times 10^{-4}$), whereas HCl, HBr, and HI are strong acids. In general, the H–A bond strength decreases markedly with increasing size of A down a column of the periodic table. For binary acids of elements in the same column, the H–A bond strength is the most important determinant of acid strength. As a further example of this effect, H_2S ($K_{a1} = 1.0 \times 10^{-7}$) is a stronger acid than H_2O.

For binary acids of elements in the same *row* of the periodic table, changes in the H–A bond strength are smaller, and the polarity of the H–A bond is the most important determinant of acid strength. The strengths of binary acids of the second row elements, for example, increase with increasing electronegativity of A:

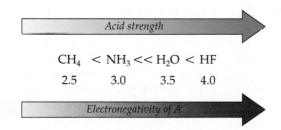

$$CH_4 \ < NH_3 \ll H_2O < HF$$
$$2.5 \quad\ 3.0 \quad\ 3.5 \quad\ 4.0$$

The C–H bond is nonpolar, and methane exhibits no tendency to dissociate in water into H_3O^+ and CH_3^- ions. The N–H bond is more polar, but dissociation of NH_3 into H_3O^+ and NH_2^- ions is still negligibly small. Water and hydrofluoric acid, however, are increasingly stronger acids.

Another familiar class of acids consists of oxoacids, such as H_2CO_3, HNO_3, H_2SO_4, and $HClO$. These compounds have the general formula H_nYO_m, where Y is a nonmetallic atom, such as C, N, S, or Cl, and n and m are integers. The atom Y is always bonded to one or more hydroxyl (OH) groups and can be bonded, in addition, to one or more oxygen atoms:

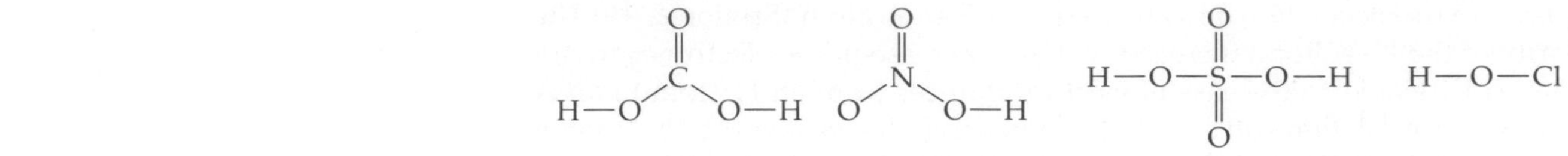

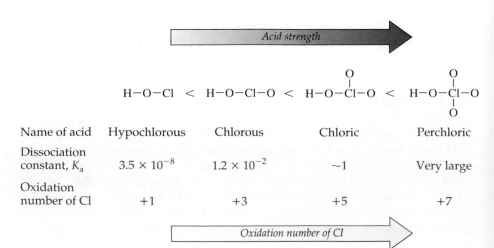

| Carbonic acid | Nitric acid | Sulfuric acid | Hypochlorous acid |

Because dissociation of an oxoacid involves breaking an O–H bond, any factor that weakens the O–H bond or increases its polarity increases the strength of the acid. Two such factors are the electronegativity of Y and the oxidation number of Y in the general reaction

$$\overset{|}{\underset{|}{-Y}}-O-H + H_2O \rightleftharpoons H_3O^+ + \overset{|}{\underset{|}{-Y}}-O^-$$

1. For oxoacids that contain the same number of OH groups and the same number of O atoms, acid strength increases with increasing electronegativity of Y. For example, the acid strength of the hypohalous acids HOY (Y = Cl, Br, or I) increases with increasing electronegativity of the halogen:

Acid strength →

	H–O–I <	H–O–Br <	H–O–Cl
Name of acid	Hypoiodous	Hypobromous	Hypochlorous
Dissociation constant, K_a	2.3×10^{-11}	2.0×10^{-9}	3.5×10^{-8}
Electronegativity	2.5	2.8	3.0

Electronegativity of Y →

2. For oxoacids that contain the same atom Y but different numbers of oxygen atoms, acid strength increases with increasing oxidation number of Y, which increases, in turn, with an increasing number of oxygen atoms. This effect is illustrated by the oxoacids of chlorine:

Acid strength →

$$H-O-Cl \;<\; H-O-Cl-O \;<\; H-O-\overset{O}{\underset{}{Cl}}-O \;<\; H-O-\overset{O}{\underset{O}{Cl}}-O$$

Name of acid	Hypochlorous	Chlorous	Chloric	Perchloric
Dissociation constant, K_a	3.5×10^{-8}	1.2×10^{-2}	~1	Very large
Oxidation number of Cl	+1	+3	+5	+7

Oxidation number of Cl →

Because oxygen is more electronegative than chlorine, adding more O atoms shifts electron density from the Cl atom to the O atoms, making the Cl atom more positive. In turn, electron density shifts toward the Cl atom from the OH group, thus weakening the O–H bond and increasing its polarity. Maximum weakening and polarization of the O–H bond occurs in $HClO_4$, the acid that contains the largest number of O atoms. This effect is further illustrated by the oxoacids of sulfur: H_2SO_4 is a stronger acid than H_2SO_3.

PROBLEM 15.22 Identify the stronger acid in each of the following pairs.
(a) H_2S or H_2Se **(b)** HI or H_2Te **(c)** HNO_2 or HNO_3 **(d)** H_2SO_3 or H_2SeO_3

15.16 ►LEWIS ACIDS AND BASES

In 1923, the same year in which Brønsted and Lowry defined acids and bases in terms of their proton donor/acceptor properties, the American chemist G. N. Lewis proposed an even more general concept of acids and bases. Lewis noticed that when a base accepts a proton, it shares a lone pair of electrons with the proton to form a new covalent bond. Using ammonia as an example, the reaction can be written in the following format, in which the curved arrow represents donation of the nitrogen lone pair to form a bond with H^+:

$$H^+ \ + \ \begin{matrix} & H & \\ & | & \\ :N & - & H \\ & | & \\ & H & \end{matrix} \ \rightarrow \ \begin{bmatrix} & H & \\ & | & \\ H - N & - & H \\ & | & \\ & H & \end{bmatrix}^+$$

In this reaction, the proton behaves as an electron-pair acceptor, and the ammonia molecule behaves as an electron-pair donor. Consequently, the Lewis definition of acids and bases states that a **Lewis acid** *is an electron-pair acceptor, and a **Lewis base** is an electron-pair donor.*

Since all proton acceptors have an unshared pair of electrons, and since all electron-pair donors can accept a proton, the Lewis and the Brønsted-Lowry definitions of a base are simply different ways of looking at the same property. All Lewis bases are Brønsted-Lowry bases, and all Brønsted-Lowry bases are Lewis bases. The Lewis definition of an acid, however, is considerably more general than the Brønsted-Lowry definition. Lewis acids include not only H^+ but also other cations and neutral molecules having vacant valence orbitals that can accept a share in a pair of electrons from a Lewis base.

Common examples of cationic Lewis acids are metal ions, such as Al^{3+} and Cu^{2+}. Hydration of the Al^{3+} ion, for example, is a Lewis acid–base reaction in which each of six H_2O molecules donates a pair of electrons to Al^{3+} to form the hydrated cation $Al(H_2O)_6{}^{3+}$:

$$Al^{3+} \ + \ 6 \begin{matrix} :\ddot{O} - H \\ | \\ H \end{matrix} \ \rightarrow \ Al \left(\begin{matrix} :\ddot{O} - H \\ | \\ H \end{matrix} \right)^{3+}_6$$

Lewis acid   Lewis base

Similarly, the reaction of Cu^{2+} ion with ammonia is a Lewis acid–base reaction in which each of four NH_3 molecules donates a pair of electrons to Cu^{2+} to form the deep blue complex ion $Cu(NH_3)_4^{2+}$ (Figure 15.9). (A *complex ion* is an ion that contains a metal cation bonded to one or more small molecules or ions, such as H_2O, NH_3, CN^-, or Cl^-.)

$$Cu^{2+} \;+\; 4\,:NH_3 \;\longrightarrow\; Cu(NH_3)_4^{2+}$$
$$\text{Lewis acid} \quad \text{Lewis base}$$

FIGURE 15.9 Addition of an excess of aqueous ammonia to a solution of the light blue $Cu^{2+}(aq)$ ion (left) gives a light blue precipitate of $Cu(OH)_2$ (center). Addition of excess ammonia yields the deep blue complex ion $Cu(NH_3)_4^{2+}$ (right).

Examples of neutral Lewis acids are halides of group 3A elements, such as BF_3. Boron trifluoride, a colorless gas, is an excellent Lewis acid because the boron atom in the trigonal planar BF_3 molecule is surrounded by only six valence electrons (Figure 15.10). The boron atom uses three sp^2 hybrid orbitals to bond to the three F atoms and has a vacant $2p$ valence orbital that can accept a share in a pair of electrons from a Lewis base, such as NH_3:

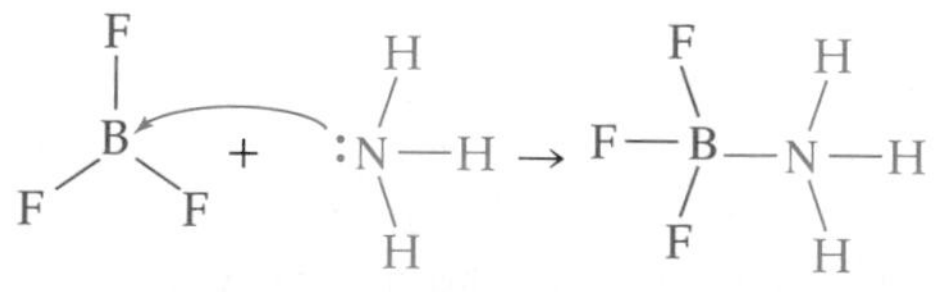

$$\text{Lewis acid} \qquad \text{Lewis base} \qquad \text{Acid–base adduct}$$

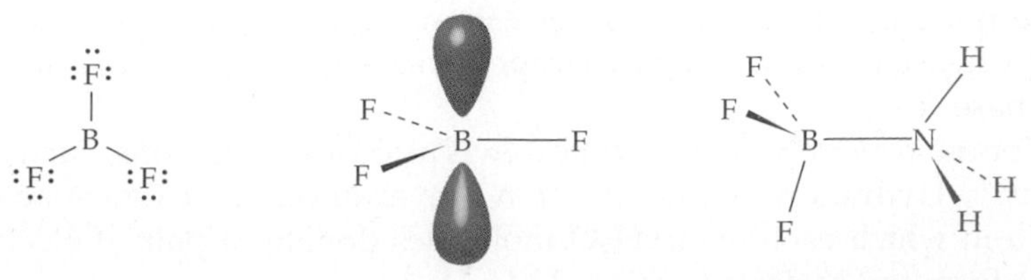

FIGURE 15.10 **(a)** Electron-dot structure of BF_3. **(b)** Trigonal planar structure of BF_3 and vacant $2p$ orbital perpendicular to the molecular plane. **(c)** Structure of the F_3B–NH_3 adduct. The geometry about boron changes from trigonal planar in BF_3 to tetrahedral in the adduct. Both boron and nitrogen use sp^3 hybrid orbitals in the adduct.

(a)   *(b)*   *(c)*

In the product, an addition compound called a Lewis acid–base adduct, the boron atom has acquired the usual stable octet of electrons.

Additional examples of neutral Lewis acids are oxides of nonmetals, such as CO_2, SO_2, and SO_3. The reaction of SO_3 with water, for example, can be viewed as a Lewis acid–base reaction in which SO_3 accepts a lone pair of electrons from a water molecule:

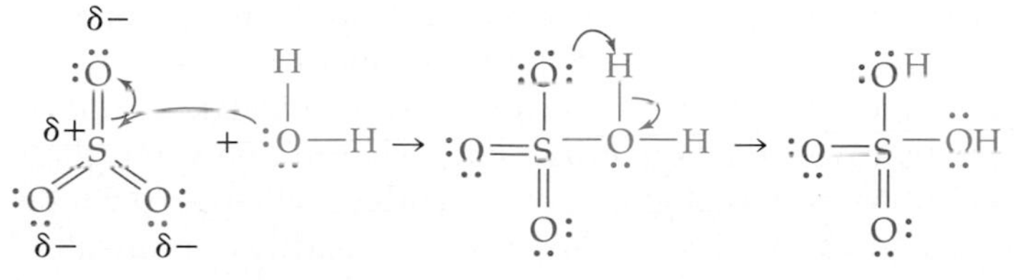

Because oxygen is more electronegative than sulfur, the S atom in SO_3 bears a partial positive charge (δ^+) and therefore attracts an electron pair from H_2O. Formation of a bond from the water O atom to the S atom in the first step is helped along by a shift of a shared pair of electrons to oxygen. In the second step, a proton shifts from one oxygen atom to another, yielding sulfuric acid (H_2SO_4).

EXAMPLE 15.14

For each of the following reactions, identify the Lewis acid and the Lewis base.
(a) $CO_2 + OH^- \longrightarrow HCO_3^-$
(b) $B(OH)_3 + OH^- \longrightarrow B(OH)_4^-$
(c) $Fe^{3+} + 6\ CN^- \longrightarrow Fe(CN)_6^{3-}$

SOLUTION   (a) The carbon atom of O=C–O bears a partial positive charge (δ^+) because oxygen is more electronegative than carbon. Therefore, the carbon atom attracts an electron pair from OH^-. Formation of a covalent bond from OH^- to CO_2 is helped along by a shift of a shared electron pair to oxygen.

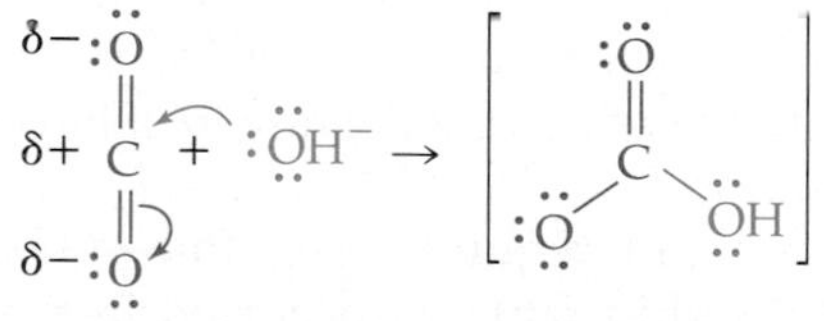

The Lewis acid (electron-pair acceptor) is CO_2; the Lewis base (electron-pair donor) is OH^-.
(b) The Lewis acid is boric acid, $B(OH)_3$, a weak acid and mild antiseptic used in eyewash. The boron atom completes its octet by accepting a pair of electrons from the Lewis base, OH^-.
(c) The Lewis acid is Fe^{3+}, and the Lewis base is CN^-.

⌐ **PROBLEM 15.23** For each of the following reactions, identify the Lewis acid and the Lewis base.
(a) $AlCl_3 + Cl^- \rightarrow AlCl_4^-$ (b) $2\ NH_3 + Ag^+ \rightarrow Ag(NH_3)_2^+$
(c) $SO_2 + OH^- \rightarrow HSO_3^-$ (d) $6\ H_2O + Cr^{3+} \rightarrow Cr(OH_2)_6^{3+}$

interlude—ACID RAIN

Plants that burn sulfur-containing coal and oil release large quantities of sulfur oxides into the atmosphere, ultimately leading to acid rain.

The problem of acid rain has emerged as one of the more important environmental issues of recent times. Both the causes and the effects of acid rain are well understood. The problem is what to do about it.

As the water that evaporates from oceans and lakes condenses into raindrops, it dissolves small quantities of gases from the atmosphere. Under normal conditions, rain is slightly acidic, with a pH close to 5.6 because of dissolved CO_2. In recent decades, however, the acidity of rainwater in many industrialized areas of the world has increased by a factor of over 100, to a pH between 3 and 3.5.

The primary cause of acid rain is industrial and automotive pollution. Each year in the United States and Canada, large power plants and smelters that burn sulfur-containing fossil fuels pour millions of tons of sulfur dioxide (SO_2) gas into the atmosphere, where some is oxidized by air to produce sulfur trioxide (SO_3). Sulfur oxides then dissolve in rain to form dilute sulfurous acid and sulfuric acid:

$$SO_2(g) + H_2O(l) \rightarrow H_2SO_3(aq) \qquad \text{Sulfurous acid}$$

$$SO_3(g) + H_2O(l) \rightarrow H_2SO_4(aq) \qquad \text{Sulfuric acid}$$

Nitrogen oxides produced by the high-temperature reaction of N_2 with O_2 in coal-burning plants and in automobile engines make a further contribution to the problem. Nitrogen dioxide (NO_2) dissolves in water to form dilute nitric acid (HNO_3) and nitric oxide (NO):

$$3 \ NO_2(g) + 2 \ H_2O(l) \rightarrow HNO_3(aq) + NO(g)$$

Oxides of both sulfur and nitrogen have always been present in the atmosphere, produced by such natural sources as volcanoes and lightning bolts, but their amounts have increased dramatically over the last century because of industrialization.

Marble statues are being slowly dissolved by reaction of calcium carbonate with acid rain.

Many processes in nature require such a fine pH balance that they are dramatically upset by the shift that has occurred in the pH of rain. Thousands of lakes in the Adirondack region of upper New York State and in southeastern Canada have become so acidic that all fish life has disappeared. Massive tree die-offs have occurred throughout Central Europe as acid rain has lowered the pH of the soil and has leached nutrients from leaves. Countless marble statues are being slowly dissolved away as their calcium carbonate is attacked by acid rain. What should be done about it?

$$CaCO_3(s) + 2 \ H^+(aq) \rightarrow Ca^{2+}(aq) + H_2O(l) + CO_2(g)$$

SUMMARY

According to the Arrhenius theory, acids are substances of general formula HA that dissociate in water to produce $H^+(aq)$. Bases are substances of general formula MOH that dissociate to yield $OH^-(aq)$. The more general **Brønsted-Lowry theory** defines an acid as a proton donor, a base as a proton acceptor, and an acid–base reaction as a proton-transfer reaction. Examples of Brønsted-Lowry acids are HCl, NH_4^+, and HSO_4^-; examples of Brønsted-Lowry bases are OH^-, F^-, and NH_3.

A **strong acid** HA is nearly 100% dissociated in aqueous solution, whereas a **weak acid** HA is only partially dissociated, existing as an equilibrium mixture of HA, A^-, and H_3O^+:

$$HA(aq) + H_2O(l) \rightleftarrows H_3O^+(aq) + A^-(aq)$$

The strength of an acid (HA) and the strength of its **conjugate base** (A^-) are inversely related. Strong acids, such as HCl, have very weak conjugate bases (Cl^-). Strong bases, such as OH^-, have very weak conjugate acids (H_2O). The H_3O^+ ion, a hydrated proton, is called the **hydronium ion**.

Water, which can act both as an acid and as a base, undergoes the **dissociation** reaction, $H_2O + H_2O \rightleftarrows H_3O^+ + OH^-$. In pure water at 25°C, $[H_3O^+] = [OH^-] = 1.0 \times 10^{-7}$ M. The **ion-product constant for water**, K_w, is given by $K_w = [H_3O^+][OH^-] = 1.0 \times 10^{-14}$. The acidity of an aqueous solution is expressed on the **pH** scale, where pH $= -\log[H_3O^+]$. Acidic solutions have pH < 7.00, basic solutions have pH > 7.00, and neutral solutions have pH $= 7.00$. The pH of a solution can be determined using an **acid–base indicator** or a pH meter. The extent of dissociation of a weak acid HA is measured by the **acid-dissociation constant**, K_a:

$$K_a = \frac{[H_3O^+][A^-]}{[HA]}$$

Polyprotic acids contain more than one dissociable proton and dissociate in a stepwise manner. Because the stepwise dissociation constants decrease in the order $K_{a1} \gg K_{a2} \gg K_{a3}$, nearly all the H_3O^+ in a polyprotic acid solution comes from the first dissociation step. Sulfuric acid is atypical in that the first step goes nearly 100% to completion.

The extent of dissociation of a weak base B is measured by the **base-dissociation constant**, K_b:

$$B(aq) + H_2O(l) \rightleftarrows BH^+(aq) + OH^-(aq) \qquad K_b = \frac{[BH^+][OH^-]}{[B]}$$

Examples of weak bases are NH_3 and derivatives of NH_3 called **amines**. For any **conjugate acid–base pair**, (K_a for the acid) $\times$ (K_b for the base) $= K_w$.

Aqueous solutions of salts can be neutral, acidic, or basic, depending on the acid–base properties of the constituent cations and anions. Cations of groups 1A and 2A (except Be^{2+}) and anions that are conjugate bases of strong acids, such as Cl^-, do not react with water to produce H_3O^+ or OH^- ions. Cations that are conjugate acids of weak bases, such as NH_4^+, and small, highly charged cations, such as Al^{3+}, yield acidic solutions, whereas anions that are conjugate bases of weak acids, such as CN^-, yield basic solutions.

The acid strength of a binary acid HA increases with decreasing strength and increasing polarity of the H–A bond. The acid strength of an oxoacid, H_nYO_m (Y = C, N, S, Cl), increases with increasing electronegativity and increasing oxidation number of the atom Y.

According to the **Lewis definition,** an acid is an electron-pair acceptor and a base is an electron-pair donor. The Lewis definition of an acid includes not only H^+ but also other cations and neutral molecules that can accept a share in a pair of electrons from a Lewis base. Examples of Lewis acids are Al^{3+}, Cu^{2+}, BF_3, SO_3, and CO_2.

UNDERSTANDING KEY CONCEPTS

1. For each of the following reactions, identify the Brønsted–Lowry acids and bases.

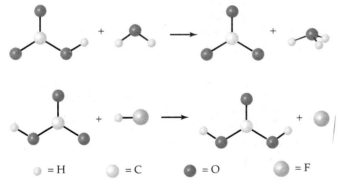

= H   = C   = O   = F

2. For the following Brønsted–Lowry acids, acid strength increases in the order $H_2X < HY < HZ$.
   (a) What is the conjugate base of each acid?
   (b) Arrange the conjugate bases in order of increasing base strength.

3. The following pictures represent aqueous solutions of three acids HA (A = X, Y, or Z). Water molecules have been omitted for clarity.

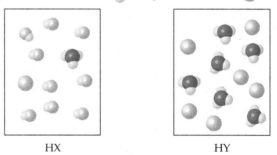

= HA   = H_3O^+ = A^-

HX   HY

HZ

   (a) Arrange the three acids in order of increasing acid strength.
   (b) Which acid, if any, is a strong acid?
   (c) Which acid has the smallest value of K_a?
   (d) What is the percent dissociation in the solution of HZ?

4. Which of the following pictures represents a solution of a weak diprotic acid H_2A? (Water molecules are omitted for clarity.) Which picture(s) represent(s) an impossible situation? Explain.

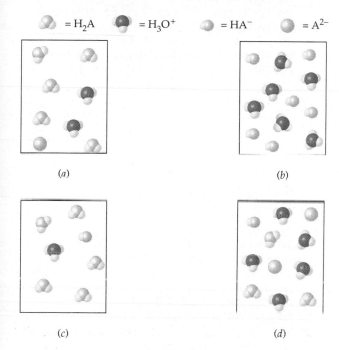

= H_2A = H_3O^+ = HA^- = A^{2-}

(a)   (b)

(c)   (d)

5. The hydrated cation $M(H_2O)_6^{3+}$ has $K_a = 10^{-4}$, and the acid HA has $K_a = 10^{-5}$. Identify the principal reaction in a solution of each of the following salts, and classify each solution as acidic, basic, or neutral.
   (a) NaA   (b) $M(NO_3)_3$ (c) $Na(NO_3)_3$ (d) MA_3

6. Locate sulfur, selenium, chlorine, and bromine in the periodic table.

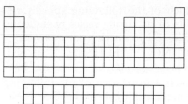

(a) Which binary acid (H_2S, H_2Se, HCl, or HBr) is the strongest? Which is the weakest? Explain.
(b) Which oxoacid (H_2SO_3, H_2SeO_3, $HClO_3$, or $HBrO_3$) is the strongest? Which is the weakest? Explain.

7. Look at the Lewis structures of the following molecules and ions.

$$Fe^{3+} \quad \begin{bmatrix} :\overset{..}{O}: \\ \| \\ :\overset{..}{O}-\overset{}{S}-\overset{..}{O}: \end{bmatrix}^{2-} \quad \begin{bmatrix} H \\ | \\ H-N-H \\ | \\ H \end{bmatrix}^{+} \quad \begin{matrix} :\overset{..}{Cl}: \\ | \\ .B. \\ :\overset{..}{Cl} \quad \overset{..}{Cl}: \end{matrix}$$

$$\begin{bmatrix} :\overset{..}{O}-\overset{..}{Cl}: \end{bmatrix}^{-} \quad \begin{bmatrix} :\overset{..}{O}-H \\ | \\ :\overset{..}{O}-P-\overset{..}{O}: \\ | \\ :\overset{..}{O}-H \end{bmatrix}^{-}$$

(a) Which of these molecules and ions can behave as a Brønsted–Lowry acid? Which can behave as a Brønsted–Lowry base?
(b) Which can behave as a Lewis acid? Which can behave as a Lewis base?

ADDITIONAL PROBLEMS

Problems 15.1–15.23 appear within the chapter.

ACID–BASE CONCEPTS

15.24 What is meant by each of the following terms? Illustrate each with two examples.
   (a) Arrhenius acid
   (b) Arrhenius base
   (c) Brønsted-Lowry acid
   (d) Brønsted-Lowry base

15.25 Give three examples of molecules or ions that are Brønsted-Lowry bases but not Arrhenius bases.

15.26 Write a balanced ionic equation for the reaction of equal volumes of the following solutions.
   (a) 1 M HNO_3 and 1 M KOH
   (b) 0.2 M HNO_3 and 0.1 M $Ba(OH)_2$
   Write balanced net ionic equations for these reactions.

15.27 Give the formula for the conjugate base of each of the following Brønsted-Lowry acids:

   (a) HOCl   (b) HPO_4^{2-} (c) H_2O
   (d) $CH_3NH_3^+$ (e) H_2CO_3 (f) H_2

15.28 Give the formula for the conjugate acid of each of the following Brønsted-Lowry bases:
   (a) F^- (b) H_2O (c) $(CH_3)_2NH$
   (d) HPO_4^{2-} (e) OH^- (f) SO_4^{2-}

15.29 For each of the following reactions, identify the Brønsted-Lowry acids and bases and the conjugate acid–base pairs.
   (a) $CH_3COOH(aq) + NH_3(aq) \rightleftarrows NH_4^+(aq) + CH_3CO_2^-(aq)$
   (b) $CO_3^{2-}(aq) + H_3O^+(aq) \rightleftarrows H_2O(l) + HCO_3^-(aq)$
   (c) $HSO_3^-(aq) + H_2O(l) \rightleftarrows H_3O^+(aq) + SO_3^{2-}(aq)$
   (d) $HSO_3^-(aq) + H_2O(l) \rightleftarrows H_2SO_3(aq) + OH^-(aq)$

15.30 Explain why a Brønsted-Lowry base must possess at least one lone pair of electrons.

15.31 Which of the following species behave as strong acids or as strong bases in aqueous solution? See Table 15.1 to check your answers.

(a) HNO_2 (b) HNO_3 (c) NH_4^+ (d) Cl^-
(e) H^- (f) O^{2-} (g) H_2SO_4

15.32 Which acid in each of the following pairs has the stronger conjugate base? See Table 15.1 for help with (c) and (d).

(a) H_2CO_3 or H_2SO_4 (b) HCl or HF
(c) HF or NH_4^+ (d) HCN or HSO_4^-

15.33 If you mix equal concentrations of reactants and products, which of the following reactions proceed to the right and which proceed to the left? Use the data in Table 15.1, and remember that the stronger base gets the proton.

(a) $H_2CO_3(aq) + HSO_4^-(aq) \rightleftarrows H_2SO_4(aq) + HCO_3^-(aq)$
(b) $HF(aq) + Cl^-(aq) \rightleftarrows HCl(aq) + F^-(aq)$
(c) $HF(aq) + NH_3(aq) \rightleftarrows NH_4^+(aq) + F^-(aq)$
(d) $HSO_4^-(aq) + CN^-(aq) \rightleftarrows HCN(aq) + SO_4^{2-}(aq)$

DISSOCIATION OF WATER; pH

15.34 Draw a Lewis electron-dot structure for water, and explain how water can act both as an acid and as a base.

15.35 Write the equilibrium expression for the dissociation of water, and explain why the concentration of water is omitted.

15.36 Classify each of the following solutions as acidic, basic, or neutral.

(a) $[H_3O^+] = 3.4 \times 10^{-9}$ M
(b) $[OH^-] = 0.010$ M
(c) $[OH^-] = 1.0 \times 10^{-10}$ M
(d) $[H_3O^+] = 1.0 \times 10^{-7}$ M
(e) $[H_3O^+] = 8.6 \times 10^{-5}$ M

15.37 For each of the following solutions, calculate $[OH^-]$ from $[H_3O^+]$, or $[H_3O^+]$ from $[OH^-]$.

(a) $[H_3O^+] = 2.5 \times 10^{-4}$ M
(b) $[H_3O^+] = 2.0$ M
(c) $[OH^-] = 5.6 \times 10^{-9}$ M
(d) $[OH^-] = 1.5 \times 10^{-3}$ M
(e) $[OH^-] = 1.0 \times 10^{-7}$ M

15.38 Calculate the pH to the correct number of significant figures for solutions having the following concentrations of H_3O^+ or OH^-.

(a) $[H_3O^+] = 2.0 \times 10^{-5}$ M
(b) $[OH^-] = 4 \times 10^{-3}$ M
(c) $[H_3O^+] = 10^{-3}$ M
(d) $[OH^-] = 1.0 \times 10^{-7}$ M
(e) $[H_3O^+] = 5.0$ M
(f) $[OH^-] = 12$ M

15.39 Calculate the H_3O^+ concentration to the correct number of significant figures for solutions having the following values of the pH.

(a) 4.1   (b) 10.82   (c) 0.00   (d) 14.25   (e) −1.0

15.40 What is the change in pH if $[H_3O^+]$ changes by each of the following factors?

(a) 1000   (b) 1.0×10^5 (c) 2.0

15.41 By what factor must $[H_3O^+]$ change to produce the following pH changes?

(a) 1.0 unit   (b) 10.00 units   (c) 0.10 unit

15.42 Given the following approximate concentrations of H_3O^+ or OH^- for various biological fluids, calculate the pH.

(a) gastric juice, $[H_3O^+] = 10^{-2}$ M
(b) spinal fluid, $[H_3O^+] = 4 \times 10^{-8}$ M
(c) bile, $[OH^-] = 8 \times 10^{-8}$ M
(d) urine, $[OH^-] = 6 \times 10^{-10}$ M to 2×10^{-6} M

STRONG ACIDS AND STRONG BASES

15.43 Calculate the pH of each of the following solutions.

(a) 0.20 M $HClO_4$
(b) 6.3×10^{-3} M NaOH
(c) 4.0×10^{-3} M $Ba(OH)_2$
(d) 1.5 M HBr
(e) 1.5 M KOH

15.44 Calculate the pH of each of the following.

(a) 4.8 g of lithium hydroxide in 250 mL of solution
(b) 0.93 g of hydrogen chloride in 0.40 L of solution
(c) 0.20 g of sodium oxide in 100 mL of solution

15.45 Calculate the pH of solutions prepared by

(a) diluting 50 mL of 0.10 M HCl to a volume of 1.00 L
(b) mixing 100 mL of 2.0×10^{-3} M HCl and 400 mL of 1.0×10^{-3} M $HClO_4$ (assume that volumes are additive)
(c) mixing equal volumes of 0.20 M HCl and 0.50 M HNO_3

15.46 Give the formula for the following industrial chemicals.

(a) caustic soda   (b) lime
(c) limestone   (d) slaked lime

WEAK ACIDS

15.47 Write a balanced net ionic equation and the corresponding equilibrium equation for dissociation of the following weak acids.
(a) chlorous acid, $HClO_2$
(b) hypobromous acid, $HOBr$
(c) formic acid, $HCOOH$

15.48 Using values of K_a in Appendix C, arrange the following acids in order of (a) increasing acid strength and (b) decreasing percent dissociation: C_6H_5OH, HNO_3, CH_3COOH, $HOCl$. Also, estimate $[H_3O^+]$ in a 1 M solution of each acid.

15.49 The pH of 0.040 M hypobromous acid (HOBr) is 5.05. Set up the equilibrium equation for dissociation of HOBr, and calculate the value of the acid-dissociation constant.

15.50 Lactic acid ($C_3H_6O_3$), which occurs in sour milk and foods such as sauerkraut, is a weak monoprotic acid. The pH of a 0.10 M solution of lactic acid is 2.43. What is the value of K_a for lactic acid?

15.51 Phenol (C_6H_5OH) is a weak acid used as a general disinfectant and in the manufacture of plastics. Calculate the concentration of all species present (H_3O^+, $C_6H_5O^-$, C_6H_5OH, and OH^-) and the pH in a 0.10 M solution of phenol ($K_a = 1.3 \times 10^{-10}$). Also calculate the percent dissociation.

15.52 Formic acid (HCOOH) is an organic acid secreted by ants and stinging nettles. Calculate the concentration of all species present (HCOOH, HCO_2^-, H_3O^+ and OH^-) and the pH in 0.20 M HCOOH ($K_a = 1.8 \times 10^{-4}$).

15.53 Calculate the pH and the percent dissociation in 1.5 M HNO_2 ($K_a = 4.5 \times 10^{-4}$).

15.54 A typical aspirin tablet contains 324 mg of aspirin (acetylsalicylic acid, $C_9H_8O_4$), a monoprotic acid having $K_a = 3.0 \times 10^{-4}$. If you dissolve two aspirin tablets in a 300 mL glass of water, what is the pH of the solution and the percent dissociation?

POLYPROTIC ACIDS

15.55 Write balanced net ionic equations and the corresponding equilibrium equations for stepwise dissociation of the triprotic acid H_3PO_4.

15.56 Calculate the concentration of all species present (H_2CO_3, HCO_3^-, CO_3^{2-}, H_3O^+, and OH^-) and the pH in 0.010 M H_2CO_3 ($K_{a1} = 4.3 \times 10^{-7}$; $K_{a2} = 5.6 \times 10^{-11}$).

15.57 Oxalic acid ($H_2C_2O_4$) is a diprotic acid that occurs in plants such as rhubarb. Calculate the pH and the

6.4×10^5

concentration of $C_2O_4^{2-}$ ions in 0.20 M $H_2C_2O_4$ ($K_{a1} = 5.9 \times 10^{-2}$; $K_{a2} = 6.4 \times 10^{-5}$).

15.58 Calculate the concentrations of H_3O^+, HSO_3^-, and SO_3^{2-} in 0.025 M H_2SO_3 ($K_{a1} = 1.5 \times 10^{-2}$; $K_{a2} = 6.3 \times 10^{-8}$).

15.59 Calculate the concentrations of H_3O^+ and SO_4^2 in a solution prepared by mixing equal volumes of 0.2 M HCl and 0.6 M H_2SO_4. K_{a2} for H_2SO_4 is 1.2×10^{-2}.

WEAK BASES; RELATION BETWEEN K_A AND K_B

15.60 Write a balanced net ionic equation and the corresponding equilibrium equation for reaction of the following weak bases with water.
(a) dimethylamine, $(CH_3)_2NH$
(b) aniline, $C_6H_5NH_2$
(c) pyridine, C_5H_5N
(d) cyanide ion, CN^-

15.61 Morphine ($C_{17}H_{19}NO_3$), a narcotic used in painkillers, is a weak organic base. If the pH of a 7.0×10^{-4} M solution of morphine is 9.5, what is the value of K_b?

15.62 Using the values of K_b in Appendix C, calculate $[OH^-]$ and the pH for each of the following solutions.
(a) 0.24 M methylamine
(b) 0.040 M pyridine
(c) 0.075 M hydroxylamine

15.63 Aniline ($C_6H_5NH_2$) is an organic base used in the manufacture of dyes. Calculate the concentration

of all species present ($C_6H_5NH_2$, $C_6H_5NH_3^+$, OH^-, and H_3O^+) and the pH in a 0.15 M solution of aniline ($K_b = 4.3 \times 10^{-10}$).

15.64 Use the conjugate acid–base pair HCN and CN^- to derive the relationship between K_a and K_b.

15.65 Using values of K_b in Appendix C, calculate values of K_a for each of the following ions.
(a) propylammonium ion, $C_3H_7NH_3^+$
(b) hydroxylammonium ion, NH_3OH^+
(c) anilinium ion, $C_6H_5NH_3^+$
(d) pyridinium ion, $C_5H_5NH^+$

15.66 Using values of K_a in Appendix C, calculate values of K_b for each of the following ions.
(a) fluoride ion, F^-
(b) hypobromite ion, OBr^-
(c) hydrogen sulfide ion, HS^-
(d) sulfide ion, S^{2-}

ACID–BASE PROPERTIES OF SALTS

15.67 Write a balanced net ionic equation for the reaction of each of the following ions with water. In each case, identify the Brønsted-Lowry acids and bases and the conjugate acid-base pairs.
(a) $CH_3NH_3^+$ (b) $Cr(H_2O)_6^{3+}$
(c) $CH_3CO_2^-$ (d) PO_4^{3-}

15.68 Classify each of the following ions according to whether they react with water to give a neutral, acidic, or basic solution.
(a) F^- (b) Br^- (c) NH_4^+
(d) $K(H_2O)_6^+$ (e) SO_3^{2-} (f) $Cr(H_2O)_6^{3+}$

15.69 Classify each of the following salt solutions as neutral, acidic, or basic. See Appendix C for values of equilibrium constants.
(a) $Fe(NO_3)_3$ (b) $Ba(NO_3)_2$ (c) NaOCl
(d) NH_4I (e) NH_4NO_2

15.70 Identify the principal reaction in solutions of each of the following salts.
(a) Na_2CO_3 (b) NH_4NO_3 (c) NaCl

15.71 Calculate the concentration of all species present and the pH in 0.10 M solutions of the following substances. See Appendix C for values of equilibrium constants.
(a) ethylammonium nitrate, $(C_2H_5NH_3)NO_3$
(b) sodium acetate, $Na(CH_3CO_2)$
(c) sodium nitrate, $NaNO_3$.

15.72 Calculate the pH and the percent dissociation of the hydrated cation in 0.020 M solutions of the following substances. See Appendix C for values of equilibrium constants.
(a) $Fe(NO_3)_2$ (b) $Fe(NO_3)_3$

FACTORS THAT AFFECT ACID STRENGTH

15.73 Arrange each group of compounds in order of increasing acid strength. Explain your reasoning.
(a) HCl, H_2S, PH_3
(b) NH_3, PH_3, AsH_3
(c) HBrO, $HBrO_2$, $HBrO_3$

15.74 Draw Lewis electron-dot structures for acetic acid (CH_3COOH) and trichloroacetic acid (CCl_3COOH), and explain why K_a is larger for CCl_3COOH.

15.75 Identify the stronger acid in each of the following pairs. Explain your reasoning.

(a) H_2Se or H_2Te
(b) H_3PO_4 or H_3AsO_4
(c) $H_2PO_4^-$ or HPO_4^{2-}
(d) CH_4 or NH_4^+

15.76 Identify the stronger base in each of the following pairs. Explain your reasoning.
(a) ClO_2^- or ClO_3^-
(b) HSO_4^- or $HSeO_4^-$
(c) HS^- or OH^-
(d) HS^- or Br^-

LEWIS ACIDS AND BASES

15.77 Give the Lewis definition of an acid and a base, and explain why this definition is more general than the Brønsted-Lowry definition.

15.78 Give three examples of Lewis acids that are not Brønsted-Lowry acids.

15.79 For each of the following reactions, identify the Lewis acid and the Lewis base:
(a) $SiF_4(g) + 2 F^-(aq) \rightarrow SiF_6^{2-}(aq)$
(b) $4 NH_3(aq) + Zn^{2+}(aq) \rightarrow Zn(NH_3)_4^{2+}(aq)$
(c) $Al(OH)_3(s) + OH^-(aq) \rightarrow Al(OH)_4^-(aq)$
(d) $AgCl(s) + Cl^-(aq) \rightarrow AgCl_2^-(aq)$

15.80 Classify each of the following as a Lewis acid or a Lewis base.
(a) CN^- (b) H^+ (c) H_2O
(d) Fe^{3+} (e) OH^- (f) CO_2
(g) $P(CH_3)_3$ (h) $B(CH_3)_3$

15.81 Which would you expect to be the stronger Lewis acid in each of the following pairs? Explain your answer.
(a) BF_3 or BH_3 (b) SO_2 or SO_3
(c) Sn^{2+} or Sn^{4+} (d) CH_3^+ or CH_4

GENERAL PROBLEMS

15.82 A vitamin C tablet containing 250 mg of ascorbic acid ($C_6H_8O_6$; $K_a = 8.0 \times 10^{-5}$) is dissolved in a 250 mL glass of water. What is the pH of the solution?

15.83 Strychnine ($C_{21}H_{22}N_2O_2$), a deadly poison used for killing rodents, is a weak base having $K_b = 1.8 \times 10^{-6}$. Calculate the pH of a saturated solution of strychnine (16 mg/100 mL).

15.84 Write a balanced net ionic equation for dissociation of each of the following Brønsted-Lowry acids in water, and identify the conjugate acid–base pairs.
(a) HBr   (b) NH_4^+ (c) HS^-

15.85 Write a balanced net ionic equation for a reaction in which HCO_3^- acts as **(a)** a Brønsted–Lowry acid and **(b)** a Brønsted–Lowry base.

15.86 Aqueous solutions of hydrogen sulfide contain H_2S, HS^-, S^{2-}, H_3O^+, OH^-, and H_2O in varying concentrations. Which of these species can act only as an acid? Which can act only as a base? Which can act both as an acid and as a base?

15.87 Sodium benzoate (C_6H_5COONa) is used as a food preservative. Calculate the concentration of all species present (Na^+, $C_6H_5CO_2^-$, C_6H_5COOH, H_3O^+, and OH^-) and the pH in 0.050 M sodium benzoate. K_a for benzoic acid (C_6H_5COOH) is 6.5×10^{-5}.

15.88 Given the following pH values for some common foods, calculate $[H_3O^+]$ and $[OH^-]$.
    (a) dill pickles, 3.2   **(b)** eggs, 7.8
    (c) apples, 3.1   **(d)** milk, 6.4
    (e) tomatoes, 4.2   **(f)** limes, 1.9

15.89 Draw a Lewis electron-dot structure for H_3O^+, and explain how H_3O^+ can form higher hydrates such as $H_5O_2^+$, $H_7O_3^+$, and $H_9O_4^+$.

15.90 Identify the strongest acid in each of the following sets.
    (a) H_2O, HF, or HCl
    (b) $HClO_2$, $HClO_3$, or $HBrO_3$
    (c) HBr, H_2S, or H_2Se

15.91 At 0°C, the density of water is 0.9987 g/mL and the value of K_w is 1.14×10^{-15}. What fraction of the water molecules are dissociated at 0°C? What is the percent dissociation at 0°C? What is the pH of a neutral solution at 0°C?

15.92 The acid–base indicator phenolphthalein changes from colorless to pink in the pH range 8.2–9.8. By what factor does $[H_3O^+]$ change over this pH range?

15.93 Calculate the concentration of all species present (H_3PO_4, $H_2PO_4^-$, HPO_4^{2-}, PO_4^{3-}, H_3O^+, and OH^-) and the pH in 0.10 M H_3PO_4. Values of equilibrium constants are listed in Appendix C.

15.94 Nicotine ($C_{10}H_{14}N_2$) can pick up two protons because it has two basic N atoms ($K_{b1} = 1.0 \times 10^{-6}$; $K_{b2} = 1.3 \times 10^{-11}$). Calculate the values of K_a for the conjugate acids $C_{10}H_{14}N_2H^+$ and $C_{10}H_{14}N_2H_2^+$.

15.95 Baking powder contains baking soda ($NaHCO_3$) and an acidic substance such as sodium alum ($NaAl(SO_4)_2 \cdot 12H_2O$). These components react in an aqueous medium to produce CO_2 gas, which "raises" the dough. Write a balanced net ionic equation for the reaction.

15.96 Arrange the following substances in order of increasing $[H_3O^+]$ for a 0.10 M solution of each.
    (a) $Zn(NO_3)_2$ **(b)** Na_2O **(c)** NaOCl
    (d) $NaClO_4$ **(e)** $HClO_4$

15.97 Arrange the following substances in order of increasing pH for a 0.10 M solution of each.
    (a) NaBr   **(b)** HBr   **(c)** NH_4Br
    (d) $(NH_4)_2CO_3$ **(e)** Na_2CO_3

15.98 Calculate the percent dissociation in each of the following solutions. What is the quantitative relationship between the percent dissociation and the concentration of the acid? What is the quantitative relationship between the percent dissociation and the value of K_a?
    (a) 2.0 M HOCl ($K_a = 3.5 \times 10^{-8}$)
    (b) 0.020 M HOCl
    (c) 2.0 M HF ($K_a = 3.5 \times 10^{-4}$)

15.99 Beginning with the equilibrium equation for the dissociation of a weak acid, show that the percent dissociation varies directly as the square root of K_a and inversely as the square root of the concentration of the acid.

15.100 Calculate the concentration of all species present (H_3O^+, F^-, HF, and OH^-) and the pH in a solution that contains 0.10 M HF ($K_a = 3.5 \times 10^{-4}$) and 0.10 M HCl.

15.101 For a solution of two weak acids with comparable values of K_a, there is no single principal reaction. The two acid-dissociation equilibrium equations must therefore be solved simultaneously. Calculate the pH in a solution that is 0.10 M in acetic acid (CH_3COOH, $K_a = 1.8 \times 10^{-5}$) and 0.10 M in benzoic acid (C_6H_5COOH, $K_a = 6.5 \times 10^{-5}$). *Hint:* Let $x = [CH_3COOH]$ that dissociates and $y = [C_6H_5COOH]$ that dissociates; then $[H_3O^+] = x + y$.

15.102 In the case of very weak acids, $[H_3O^+]$ from the dissociation of water is significant compared with $[H_3O^+]$ from dissociation of the weak acid. The sugar substitute saccharin ($C_7H_5NO_3S$), for example, is a very weak acid having $K_a = 2.1 \times 10^{-12}$ and a solubility in water of 0.019 mol/L. Calculate $[H_3O^+]$ in a saturated solution of saccharin. *Hint:* Equilibrium equations for the dissociation of saccharin and water must be solved simultaneously.

15.103 What is the pH and the principal source of H_3O^+ ions in 1.0×10^{-10} M HCl? (*Hint:* The pH of an acid solution can't exceed 7.00.) What is the pH of 1.0×10^{-7} M HCl?

15.104 Classify each of the following salt solutions as neutral, acidic, or basic. See Appendix C for values of equilibrium constants.
    (a) NH_4F **(b)** $NH_4(CH_3CO_2)$ **(c)** $(NH_4)_2SO_3$

15.105 Calculate the concentration of all species present and the pH in 0.25 M solutions of each of the salts in Problem 15.104. *Hint:* The principal reaction is proton transfer from the cation to the anion.

chapter 16

APPLICATIONS OF AQUEOUS EQUILIBRIA

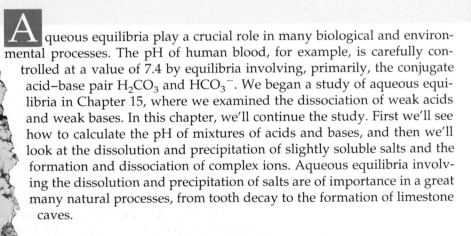

A queous equilibria play a crucial role in many biological and environmental processes. The pH of human blood, for example, is carefully controlled at a value of 7.4 by equilibria involving, primarily, the conjugate acid–base pair H_2CO_3 and HCO_3^-. We began a study of aqueous equilibria in Chapter 15, where we examined the dissociation of weak acids and weak bases. In this chapter, we'll continue the study. First we'll see how to calculate the pH of mixtures of acids and bases, and then we'll look at the dissolution and precipitation of slightly soluble salts and the formation and dissociation of complex ions. Aqueous equilibria involving the dissolution and precipitation of salts are of importance in a great many natural processes, from tooth decay to the formation of limestone caves.

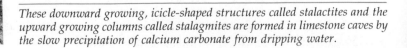

These downward growing, icicle-shaped structures called stalactites and the upward growing columns called stalagmites are formed in limestone caves by the slow precipitation of calcium carbonate from dripping water.

16.1 ►NEUTRALIZATION REACTIONS

We've seen on numerous occasions that the neutralization reaction of an acid with a base produces water and a salt. But to what extent does a neutralization reaction proceed to completion? Let's look at four types of neutralization reactions: (1) strong acid–strong base, (2) weak acid–strong base, (3) strong acid–weak base, and (4) weak acid–weak base.

Strong Acid–Strong Base

Let's consider the reaction of hydrochloric acid with aqueous sodium hydroxide to give water and an aqueous solution of sodium chloride:

$$HCl(aq) + NaOH(aq) \rightarrow H_2O(l) + NaCl(aq)$$

Because $HCl(aq)$, $NaOH(aq)$, and $NaCl(aq)$ are all 100% dissociated, the net ionic equation for the neutralization reaction is

$$H_3O^+(aq) + OH^-(aq) \rightarrow 2\, H_2O(l)$$

If we mix equal numbers of moles of $HCl(aq)$ and $NaOH(aq)$, the concentrations of H_3O^+ and OH^- remaining in the NaCl solution after neutralization will be the same as those in pure water, $[H_3O^+] = [OH^-] = 1.0 \times 10^{-7}$ M. In other words, the reaction of HCl with NaOH proceeds far to the right.

Another route to the same conclusion is to look at the equilibrium constant for the reaction. Because the neutralization reaction of any strong acid with a strong base is the reverse of the dissociation of water, its equilibrium constant, K_n, is just the reciprocal of the ion-product constant for water:

$$H_3O^+(aq) + OH^-(aq) \rightleftharpoons 2\, H_2O(l)$$

$$K_n = \frac{1}{[H_3O^+][OH^-]} = \frac{1}{K_w} = \frac{1}{1.0 \times 10^{-14}} = 1.0 \times 10^{14}$$

The value of K_n (1.0×10^{14}) is a large number, which means that the neutralization reaction proceeds essentially 100% to completion. After neutralization of equal molar amounts of acid and base, the solution contains a salt of a strong base and a strong acid. Because neither the cation nor the anion of the salt has acidic or basic properties, the pH is 7.00 (Section 15.14).

Weak Acid–Strong Base

Because a weak acid (HA) is largely undissociated, the net ionic equation for the neutralization reaction involves proton transfer from HA to the strong base, OH^-:

$$HA(aq) + OH^-(aq) \rightarrow H_2O(l) + A^-(aq)$$

Acetic acid (CH_3COOH), for example, reacts with aqueous NaOH to give water and aqueous sodium acetate (CH_3COONa):

$$CH_3COOH(aq) + OH^-(aq) \rightarrow H_2O(l) + CH_3CO_2^-(aq)$$

Note that Na^+ ions don't appear in the net ionic equation because both NaOH and CH_3COONa are completely dissociated.

To obtain the equilibrium constant, K_n, for the neutralization reaction, we can multiply known equilibrium constants for reactions that add to give the net ionic equation:

$$CH_3COOH(aq) + H_2O(l) \rightleftarrows H_3O^+(aq) + CH_3CO_2^-(aq) \qquad K_a = 1.8 \times 10^{-5}$$

$$\underline{H_3O^+(aq) + OH^-(aq) \rightleftarrows 2\,H_2O(l) \qquad\qquad 1/K_w = 1.0 \times 10^{14}}$$

$$\text{Net: } CH_3COOH(aq) + OH^-(aq) \rightleftarrows H_2O(l) + CH_3CO_2^-(aq) \qquad K_n = (K_a)(1/K_w) = (1.8 \times 10^{-5})(1.0 \times 10^{14})$$

$$= 1.8 \times 10^9$$

When NaOH(*aq*) is added to CH$_3$COOH(*aq*) containing the acid–base indicator phenolphthalein, the color of the indicator changes from colorless to pink in the pH range 8.2–9.8 due to neutralization of the acetic acid.

Remember that the equilibrium constant for a net reaction is always equal to the *product* of the equilibrium constants for the reactions added (Section 15.13). The large value of K_n (1.8×10^9) means that the neutralization reaction proceeds nearly 100% to completion.

We can generally assume that neutralization of any weak acid by a strong base goes 100% to completion because OH$^-$ has a great affinity for protons. After neutralization of equal molar amounts of CH$_3$COOH and NaOH, the solution contains Na$^+$, which has no acidic or basic properties, and CH$_3$CO$_2^-$, which is a weak base. Therefore, the pH is greater than 7.00 (Section 15.14).

Strong Acid–Weak Base

A strong acid (HA) is completely dissociated into H$_3$O$^+$ and A$^-$ ions, and the neutralization reaction therefore involves proton transfer from H$_3$O$^+$ to the weak base (B):

$$H_3O^+(aq) + B(aq) \rightarrow H_2O(l) + BH^+(aq)$$

For example, the net ionic equation for neutralization of hydrochloric acid with aqueous ammonia is

$$H_3O^+(aq) + NH_3(aq) \rightarrow H_2O(l) + NH_4^+(aq)$$

As in the weak acid–strong base case, we can obtain the equilibrium constant for the neutralization reaction by multiplying known equilibrium constants for reactions that add to give the net ionic equation:

$$NH_3(aq) + H_2O(l) \rightleftarrows NH_4^+(aq) + OH^-(aq) \qquad K_b = 1.8 \times 10^{-5}$$

$$\underline{H_3O^+(aq) + OH^-(aq) \rightleftarrows 2\,H_2O(l) \qquad\qquad 1/K_w = 1.0 \times 10^{14}}$$

$$\text{Net: } H_3O^+(aq) + NH_3(aq) \rightleftarrows H_2O(l) + NH_4^+(aq) \qquad K_n = (K_b)(1/K_w) = (1.8 \times 10^{-5})(1.0 \times 10^{14})$$

$$= 1.8 \times 10^9$$

Again, the neutralization reaction proceeds nearly 100% to the right because its equilibrium constant, K_n, is a very large number (1.8×10^9). (It's purely coincidence that the neutralization reactions of CH$_3$COOH with NaOH and of HCl with NH$_3$ have the same value of K_n. The acid-dissociation constant K_a for CH$_3$COOH happens to have the same value as the base-dissociation constant K_b for NH$_3$.)

We can generally assume that neutralization of any weak base with a strong acid goes 100% to completion because H_3O^+ is a powerful proton donor. After neutralization of equal molar amounts of NH_3 and HCl, the solution contains NH_4^+, which is a weak acid, and Cl^-, which has no acidic or basic properties. Therefore, the pH is less than 7.00 (Section 15.14).

Weak Acid–Weak Base

Both the acid (HA) and the base (B) are undissociated, and the neutralization reaction therefore involves proton transfer from the weak acid to the weak base. For example, the net ionic equation for neutralization of acetic acid with aqueous ammonia is

$$CH_3COOH(aq) + NH_3(aq) \rightleftharpoons NH_4^+(aq) + CH_3CO_2^-(aq) \qquad K_n = 3.2 \times 10^4$$

We can obtain the equilibrium constant, K_n, by adding equations for (1) the acid dissociation of acetic acid, (2) the base dissociation of ammonia, and (3) the reverse of the dissociation of water.

$$CH_3COOH(aq) + H_2O(l) \rightleftharpoons H_3O^+(aq) + CH_3CO_2^-(aq) \qquad K_a = 1.8 \times 10^{-5}$$

$$NH_3(aq) + H_2O(l) \rightleftharpoons NH_4^+(aq) + OH^-(aq) \qquad K_b = 1.8 \times 10^{-5}$$

$$H_3O^+(aq) + OH^-(aq) \rightleftharpoons 2\,H_2O(l) \qquad 1/K_w = 1.0 \times 10^{14}$$

Net: $CH_3COOH(aq) + NH_3(aq) \rightleftharpoons NH_4^+(aq) + CH_3CO_2^-(aq) \qquad K_n = (K_a)(K_b)(1/K_w)$

$$K_n = (K_a)(K_b)\left(\frac{1}{K_w}\right) = (1.8 \times 10^{-5})(1.8 \times 10^{-5})\left(\frac{1}{1.0 \times 10^{-14}}\right) = 3.2 \times 10^4$$

When hydrochloric acid is added to aqueous ammonia containing the acid–base indicator methyl red, the color of the indicator changes from yellow to red in the pH range 4.2–6.0 due to neutralization of the NH_3.

Note that the value of K_n in this case is smaller than it is for the preceding three cases, so the neutralization does not proceed as far toward completion.

In general, weak acid–weak base neutralizations have less tendency to proceed to completion than neutralizations involving strong acids or strong bases. The neutralization of HCN with aqueous ammonia, for example, has a value of K_n less than unity, which means that the reaction proceeds less than halfway to completion:

$$HCN(aq) + NH_3(aq) \rightleftharpoons NH_4^+(aq) + CN^-(aq) \qquad K_n = 0.88$$

EXAMPLE 16.1

Write a balanced net ionic equation for neutralization of equal molar amounts of nitric acid and methylamine (CH_3NH_2). Indicate whether the pH after neutralization is greater than, equal to, or less than 7.00.

SOLUTION   Since HNO_3 is a strong acid and CH_3NH_2 is a weak base, the net ionic equation is

$$H_3O^+(aq) + CH_3NH_2(aq) \rightarrow H_2O(l) + CH_3NH_3^+(aq)$$

After neutralization, the solution contains $CH_3NH_3^+$, a weak acid, and NO_3^-, which has no acidic or basic properties. Therefore, the pH is less than 7.00.

┌ **PROBLEM 16.1** Write a balanced net ionic equation for neutralization of equal molar amounts of the following reagents. Indicate whether the pH after neutralization is greater than, equal to, or less than 7.00. Values of K_a and K_b can be found in Appendix C.

(a) HNO_2 and KOH **(b)** HBr and NH_3 **(c)** $LiOH$ and $HClO_4$

┌ **PROBLEM 16.2** Write a balanced net ionic equation for neutralization of the following reagents, calculate the value of K_n for each neutralization reaction, and arrange the reactions in order of increasing tendency to proceed to completion. Values of K_a and K_b can be found in Appendix C.

(a) HF and $NaOH$ **(b)** HCl and KOH **(c)** HF and NH_3 ┘

16.2 ➤THE COMMON-ION EFFECT

Let's consider a solution of acetic acid and its conjugate base, acetate ion. What is the pH of a solution prepared by dissolving 0.10 mol of acetic acid and 0.10 mol of sodium acetate in water and then diluting the solution to a volume of 1.00 L? This problem, like those discussed in Chapter 15, can be solved by thinking about the chemistry. First, identify the acid–base properties of the various species in solution, and then consider the possible proton-transfer reactions these species can undergo. We'll follow the procedure outlined in Figure 15.6.

Since acetic acid is largely undissociated in aqueous solution, and since the salt sodium acetate is essentially 100% dissociated, the species present initially are

$$CH_3COOH \qquad Na^+ \qquad CH_3CO_2^- \qquad H_2O$$
$$\text{acid} \qquad\quad \text{inert} \qquad\quad \text{base} \qquad\qquad \text{acid or base}$$

Because we have two acids and two bases, there are four possible proton transfer reactions. We know, however, that acetic acid is a stronger acid than water and that the principal reaction therefore involves proton transfer from CH_3COOH to either $CH_3CO_2^-$ or H_2O:

$$CH_3COOH(aq) + CH_3CO_2^-(aq) \rightleftarrows CH_3CO_2^-(aq) + CH_3COOH(aq) \qquad K = 1$$

$$CH_3COOH(aq) + H_2O(l) \rightleftarrows H_3O^+(aq) + CH_3CO_2^-(aq) \qquad K_a = 1.8 \times 10^{-5}$$

Although the first of these reactions has the larger equilibrium constant, we can't consider it to be the principal reaction because reactants and products are identical. Proton transfer from acetic acid to its conjugate base is constantly occurring, but that reaction doesn't change any concentrations and therefore can't be used to calculate equilibrium concentrations. Consequently, the principal reaction is dissociation of acetic acid.

Let's set up a table of concentrations for the species involved in the principal reaction. As usual, we define x as the concentration of acid that dissociates—here, acetic acid—but we need to take account of the fact that the acetate ions come from two sources: 0.10 mol/L of acetate comes from the sodium acetate present initially, and x mol/L comes from dissociation of acetic acid.

Principal reaction: $CH_3COOH(aq) + H_2O(l) \rightleftarrows H_3O^+(aq) + CH_3CO_2{}^-(aq)$

Initial conc (M)	0.10	~0	0.10
Change (M)	$-x$	$+x$	$+x$
Equilibrium conc (M)	$(0.10 - x)$	x	$(0.10 + x)$

Substituting the equilibrium concentrations into the equilibrium equation for the principal reaction, we obtain

$$K_a = 1.8 \times 10^{-5} = \frac{[H_3O^+][CH_3CO_2{}^-]}{[CH_3COOH]} = \frac{(x)(0.10 + x)}{(0.10 - x)}$$

Because K_a is small, x is small compared with 0.10, and we can make the approximation that $(0.10 + x) \approx (0.10 - x) \approx 0.10$, which simplifies the solution of the equation to

$$1.8 \times 10^{-5} = \frac{(x)(0.10 + x)}{(0.10 - x)} \approx \frac{(x)(0.10)}{0.10}$$

$$x = [H_3O^+] = 1.8 \times 10^{-5} \text{ M}$$

$$pH = -\log(1.8 \times 10^{-5}) = 4.74$$

It's interesting to compare $[H_3O^+]$ in a 0.10 M solution of pure acetic acid with $[H_3O^+]$ in a solution that contains both 0.10 M acetic acid and 0.10 M sodium acetate. The principal reaction is the same in both cases, but in 0.10 M acetic acid all the acetate ions come from dissociation of acetic acid:

Principal reaction: $CH_3COOH(aq) + H_2O(l) \rightleftarrows H_3O^+(aq) + CH_3CO_2{}^-(aq)$

Equilibrium conc (M)	$(0.10 - x)$	x	x

The calculated $[H_3O^+]$ of 0.10 M acetic acid is 1.3×10^{-3} M (Section 15.9), and the pH is 2.89 versus a pH of 4.74 for the acetic acid–sodium acetate solution. The difference in pH is illustrated in Figure 16.1.

FIGURE 16.1 The 0.10 M acetic acid–0.10 M sodium acetate solution on the left has a lower H_3O^+ concentration ($[H_3O^+]$ = 1.8 × 10^{-5} M; pH 4.74) than the 0.10 M acetic acid on the right ($[H_3O^+]$ = 1.8 × 10^{-3} M; pH 2.89). The difference in pH is revealed by the color of the indicator methyl orange, which changes from red to yellow in the pH range 3.2–4.4.

The decrease in $[H_3O^+]$ on adding acetate ions to an acetic acid solution is an example of the **common-ion effect**, the shift in the position of an equilibrium on addition of a substance that provides an ion in common with one of the ions already involved in the equilibrium. Thus, added acetate ions shift the position of the acetic acid dissociation equilibrium to the left, as shown graphically in Figure 16.2.

$$CH_3COOH(aq) + H_2O(l) \rightleftarrows H_3O^+(aq) + CH_3CO_2^-(aq)$$

The common-ion effect is just another example of Le Châtelier's principle (Section 13.6), but it occurs often enough to merit a special name. Another case is found in Example 16.2.

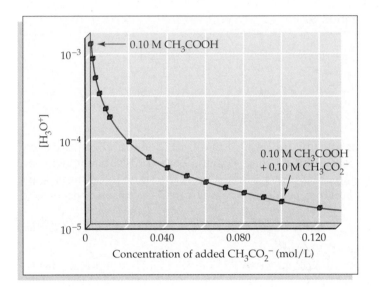

FIGURE 16.2 The common-ion effect. The concentration of H_3O^+ in a 0.10 M acetic acid solution decreases as the concentration of added sodium acetate increases because added acetate ions shift the acid-dissociation equilibrium to the left. Note that $[H_3O^+]$ is plotted on a logarithmic pH scale.

EXAMPLE 16.2

Calculate the concentration of all species present, the pH, and the percent dissociation of ammonia in a solution that is 0.15 M in NH_3 and 0.45 M in NH_4Cl.

SOLUTION   Because the salt NH_4Cl is 100% dissociated, the species present initially are NH_3 (a weak base), NH_4^+ (its conjugate acid), Cl^- (inert), and H_2O (acid or base). The strongest base present is NH_3; therefore, the principal reaction is proton transfer to NH_3 from H_2O. Since NH_4^+ ions come from the NH_4Cl present initially (0.45 M) and from reaction of NH_3 with H_2O, the concentrations of the species involved in the principal reaction are as follows:

Principal reaction:	$NH_3(aq)$ +	$H_2O(l)$ $\rightleftarrows$	$NH_4^+(aq)$ +	$OH^-(aq)$
Initial conc (M)	0.15		0.45	~0
Change (M)	$-x$		$+x$	$+x$
Equilibrium conc (M)	$(0.15 - x)$		$(0.45 + x)$	x

The common ion in this problem is NH_4^+. The equilibrium equation for the principal reaction is

$$K_b = 1.8 \times 10^{-5} = \frac{[NH_4^+][OH^-]}{[NH_3]} = \frac{(0.45 + x)(x)}{(0.15 - x)} \approx \frac{(0.45)(x)}{0.15}$$

where x is assumed to be negligible in comparison with 0.45 and 0.15 because the equilibrium constant K_b is small and because the equilibrium is shifted to the left by the common-ion effect. Therefore,

$$x = [OH^-] = \frac{(1.8 \times 10^{-5})(0.15)}{0.45} = 6.0 \times 10^{-6} \text{ M}$$

$$[NH_3] = 0.15 - x = 0.15 - (6.0 \times 10^{-6}) = 0.15 \text{ M}$$

$$[NH_4^+] = 0.45 + x = 0.45 + (6.0 \times 10^{-6}) = 0.45 \text{ M}$$

Note that the assumption concerning the size of x is justified. The H_3O^+ concentration and the pH are

$$[H_3O^+] = \frac{K_w}{[OH^-]} = \frac{1.0 \times 10^{-14}}{6.0 \times 10^{-6}} = 1.7 \times 10^{-9} \text{ M}$$

$$pH = -\log(1.7 \times 10^{-9}) = 8.77$$

The percent dissociation of ammonia is

$$\text{Percent dissociation} = \frac{[NH_3]_{dissociated}}{[NH_3]_{initial}} \times 100\% = \frac{6.0 \times 10^{-6}}{0.15} \times 100\% = 0.0040\%$$

By contrast, in 0.15 M NH_3, where there is no common-ion effect, the percent dissociation of NH_3 is 1.1%.

⌐ **PROBLEM 16.3** Calculate the concentration of all species present, the pH, and the percent dissociation of HCN ($K_a = 4.9 \times 10^{-10}$) in a solution that is 0.025 M in HCN and 0.010 M in NaCN.

⌐ **PROBLEM 16.4** Calculate the pH in a solution prepared by dissolving 0.10 mol of solid NH_4Cl in 500 mL of 0.40 M NH_3. Assume that there is no volume change.

16.3 ►BUFFER SOLUTIONS

Solutions like those discussed in the previous section, which contain a weak acid and its conjugate base, are called **buffer solutions** because they resist drastic changes in pH. If a small amount of base is added to a buffer solution, the pH increases, but not by much because the acid component of the buffer solution neutralizes the added base. If a small amount of acid is added to a buffer solution, the pH decreases, but again not by much because the base component of the buffer solution neutralizes the added acid.

Buffer solution
Weak acid + conjugate base
For example:
$\begin{cases} CH_3COOH + CH_3COO^- \\ HF + F^- \\ NH_4^+ + NH_3 \\ H_2PO_4^- + HPO_4^{2-} \end{cases}$

Buffer solutions are very important in biological systems. Blood, for example, is a buffer solution that can soak up the acids and bases produced in biological reactions with minimal change in pH. The pH of human blood is carefully controlled at a value very close to 7.4 by conjugate acid–base pairs, primarily H_2CO_3 and its conjugate base HCO_3^-. The oxygen-carrying ability of blood depends on control of the pH to better than 0.1 pH unit.

To see how a buffer solution works, let's return to the 0.10 M acetic acid–0.10 M sodium acetate solution discussed in the previous section. The principal reaction and the equilibrium concentrations for the solution are

Principal reaction: $CH_3COOH(aq) + H_2O(l) \rightleftarrows H_3O^+(aq) + CH_3CO_2^-(aq)$

Equilibrium conc (M) $(0.10 - x)$ x $(0.10 + x)$

If we solve the equilibrium equation for $[H_3O^+]$, we obtain

$$K_a = \frac{[H_3O^+][CH_3CO_2^-]}{[CH_3COOH]}$$

$$[H_3O^+] = K_a \frac{[CH_3COOH]}{[CH_3CO_2^-]}$$

Thus, the H_3O^+ concentration in a buffer solution has a value close to the value of K_a for the weak acid but differs by a factor equal to the concentration ratio [weak acid]/[conjugate base]. In the 0.10 M acetic acid–0.10 M sodium acetate solution, for example, the concentration ratio is unity, and $[H_3O^+]$ equals K_a:

$$[H_3O^+] = K_a \frac{(0.10 - x)}{(0.10 + x)} = K_a \left(\frac{0.10}{0.10}\right) = K_a = 1.8 \times 10^{-5}\ M$$

$$pH = -\log(1.8 \times 10^{-5}) = 4.74$$

Note that in calculating this result we have set the *equilibrium* concentrations, $(0.10 - x)$ and $(0.10 + x)$, equal to the *initial* concentrations, 0.10, because x is negligible compared with the initial concentrations. For commonly used buffer solutions, K_a is small, and the initial concentrations are relatively large. As a result, x is generally negligible compared with the initial concentrations, and we can use initial concentrations in the calculations.

Now let's consider what happens when we add acid or base to a buffer solution. Suppose that we add 0.01 mol of solid NaOH to 1.00 L of the 0.10 M acetic acid–0.10 M sodium acetate solution. The added strong base will neutralize some of the acetic acid, but as long as the concentration ratio $[CH_3COOH]/[CH_3CO_2^-]$ stays close to its original value, $[H_3O^+]$ won't change by very much.

Because neutralization reactions involving strong acids or strong bases go essentially 100% to completion, we must take account of neutralization before calculating $[H_3O^+]$. Initially, we have $(1.00\ L)(0.10\ mol/L) = 0.10$ mol of acetic acid and an equal amount of acetate ions. When we add 0.01 mol of NaOH, the neutralization reaction will alter the numbers of moles:

Neutralization reaction: $CH_3COOH(aq) + OH^-(aq) \xrightarrow{100\%} H_2O(l) + CH_3CO_2^-(aq)$

Before reaction (mol)	0.10	0.01	0.10
Change (mol)	−0.01	−0.01	+0.01
After reaction (mol)	0.09	~0	0.11

If we assume that the solution volume remains constant at 1.00 L, the concentrations of the buffer components after neutralization are

$$[CH_3COOH] = \frac{0.09 \text{ mol}}{1.00 \text{ L}} = 0.09 \text{ M}$$

$$[CH_3CO_2^-] = \frac{0.11 \text{ mol}}{1.00 \text{ L}} = 0.11 \text{ M}$$

Substituting these concentrations into the expression for $[H_3O^+]$, we can then calculate pH:

$$[H_3O^+] = K_a \frac{[CH_3COOH]}{[CH_3CO_2^-]}$$

$$= (1.8 \times 10^{-5})\left(\frac{0.09}{0.11}\right) = 1.5 \times 10^{-5} \text{ M}$$

$$pH = 4.82$$

Note that adding 0.01 mol of NaOH changes $[H_3O^+]$ by only a small amount because the concentration ratio [weak acid]/[conjugate base] changes by only a small amount, from unity to $\%_{11}$. The corresponding change in pH, from 4.74 to 4.82, is only 0.08 pH unit.

Now suppose that we add 0.01 mol of HCl to 1.00 L of the 0.10 M acetic acid–0.10 M sodium acetate buffer solution. The added strong acid will convert 0.01 mol of acetate ions to 0.01 mol of acetic acid because of the neutralization reaction

$$H_3O^+(aq) + CH_3CO_2^-(aq) \xrightarrow{100\%} H_2O(l) + CH_3COOH(aq)$$

The concentrations after neutralization will be $[CH_3COOH] = 0.11$ M and $[CH_3CO_2^-] = 0.09$ M, and the pH of the solution will be 4.66:

$$[H_3O^+] = K_a \frac{[CH_3COOH]}{[CH_3CO_2^-]}$$

$$= (1.8 \times 10^{-5})\left(\frac{0.11}{0.09}\right) = 2.2 \times 10^{-5} \text{ M}$$

$$pH = 4.66$$

Again, the change in pH, from 4.74 to 4.66, is small because the concentration ratio, [weak acid]/[conjugate base], remains close to its original value.

To appreciate the ability of a buffer solution to maintain a nearly constant pH, let's contrast the behavior of the 0.10 M acetic acid–0.10 M sodium acetate buffer with that of a 1.8×10^{-5} M HCl solution. The HCl solution has the same pH (4.74) as the buffer solution, but it doesn't have the capacity to soak up added acid or base. For example, if we add 0.01 mol of solid NaOH to 1.00 L of 1.8×10^{-5} M HCl, a negligible amount of OH⁻ (1.8×10^{-5} mol) is neutralized, and the concentration of OH⁻ after neutralization is 0.01 mol/1.00 L = 0.01 M. As a result, the pH rises from 4.74 to 12.0:

$$[H_3O^+] = \frac{K_w}{[OH^-]} = \frac{(1.0 \times 10^{-14})}{(0.01)} = 1 \times 10^{-12} \text{ M}$$

$$pH = 12.0$$

The abilities of the HCl solution and the buffer solution to absorb added base are contrasted in Figure 16.3.

FIGURE 16.3 The color of each solution is due to the presence of a few drops of methyl red, an acid–base indicator that is red at pH less than about 5.4 and yellow at pH greater than about 5.4. **(a)** 1.00 L of 1.8×10^{-5} M HCl (pH 4.74); **(b)** the solution in **(a)** turns yellow (pH > 5.4) after addition of only a few drops of 0.10 M NaOH; **(c)** 1.00 L of a 0.10 M acetic acid–0.10 M sodium acetate buffer solution (pH = 4.74); **(d)** the solution in **(c)** is still red (pH < 5.4) after addition of 100 mL of 0.10 M NaOH.

(a)   (b)   (c)   (d)

We sometimes talk about the buffering ability of a solution using the term **buffer capacity** as a measure of the amount of acid or base that the solution can absorb without a significant change in pH. Buffer capacity is also a measure of how little the pH changes for addition of a given amount of acid or base. Buffer capacity depends on how much weak acid and conjugate base is present. The more concentrated the solution, the greater the buffer capacity.

PROBLEM 16.5 Calculate the pH of 100 mL of a buffer solution that is 0.25 M in HF and 0.50 M in NaF. What is the change in pH on addition of **(a)** 0.002 mol of HNO₃ and **(b)** 0.004 mol of KOH?

PROBLEM 16.6 Calculate the change in pH when 0.002 mol of HNO₃ is added to 100 mL of a buffer solution that is 0.050 M in HF and 0.100 M in NaF. Does this solution have more or less buffer capacity than the one in Problem 16.5?

16.4 ➤ THE HENDERSON-HASSELBALCH EQUATION

We saw in the previous section that the H_3O^+ concentration in a buffer solution depends on the dissociation constant of the weak acid and on the concentration ratio [weak acid]/[conjugate base]:

$$[H_3O^+] = K_a \frac{[acid]}{[base]}$$

This equation can be rewritten in logarithmic form by taking the negative base-10 logarithm of both sides:

$$pH = -\log[H_3O^+] = -\log K_a - \log \frac{[acid]}{[base]}$$

If we then define $pK_a = -\log K_a$ by analogy with $pH = -\log[H_3O^+]$ and take account of the fact that

$$-\log \frac{[acid]}{[base]} = \log \frac{[base]}{[acid]}$$

we obtain an expression called the **Henderson-Hasselbalch equation:**

Henderson-Hasselbalch equation $\qquad pH = pK_a + \log \frac{[base]}{[acid]}$

The Henderson-Hasselbalch equation says that the pH of a buffer solution has a value close to the pK_a of the weak acid, differing only by the amount log [base]/[acid]. When [base]/[acid] = 1, then log [base]/[acid] = 0, and the pH equals the pK_a.

The real importance of the Henderson-Hasselbalch equation, particularly in biochemistry, is that it tells us how the pH affects the percent dissociation of a weak acid. Suppose, for example, that the pH of a solution is two pH units greater than the pK_a of the dissolved acid:

$$pH = pK_a + 2$$

Then log [base]/[acid] = 2, and [base]/[acid] = 10^2 = 100/1. Therefore, 100 of every 101 acid molecules are dissociated, which corresponds to 99% dissociation:

$$\log \frac{[base]}{[acid]} = pH - pK_a = 2$$

$$\frac{[base]}{[acid]} = 10^2 = 100/1 \quad \text{(99\% dissociation)}$$

The Henderson-Hasselbalch equation thus gives the following relationships:

At pH = pK_a + 2	$\dfrac{[\text{base}]}{[\text{acid}]} = 10^2 = 100/1$	99% dissociation
At pH = pK_a + 1	$\dfrac{[\text{base}]}{[\text{acid}]} = 10^1 = 10/1$	91% dissociation
At pH = pK_a + 0	$\dfrac{[\text{base}]}{[\text{acid}]} = 10^0 = 1/1$	50% dissociation
At pH = pK_a − 1	$\dfrac{[\text{base}]}{[\text{acid}]} = 10^{-1} = 1/10$	9% dissociation
At pH = pK_a − 2	$\dfrac{[\text{base}]}{[\text{acid}]} = 10^{-2} = 1/100$	1% dissociation

The Henderson-Hasselbalch equation also tells us how to prepare a buffer solution with a given pH. As detailed in Example 16.3, the general idea is to select a weak acid whose pK_a is close to the desired pH and then adjust the [base]/[acid] ratio to the value specified by the Henderson-Hasselbalch equation. For example, to prepare a buffer having its pH near 7, we might use the $H_2PO_4^-$–HPO_4^{2-} conjugate acid–base pair because pK_a for $H_2PO_4^-$ is $-\log(6.2 \times 10^{-8}) = 7.21$. Similarly, a mixture of NH_4Cl and NH_3 would be a good choice for a buffer having pH near 9 because pK_a for NH_4^+ is $-\log(5.6 \times 10^{-10}) = 9.25$. As a rule of thumb, the pK_a of the weak acid component of a buffer should be within ±1 pH units of the desired pH.

Because buffer solutions are widely used in the laboratory and in medicine, prepackaged buffers having a variety of precisely known pH values are commercially available (Figure 16.4). The manufacturer prepares these buffers by choosing a buffer system having an appropriate pK_a value and then adjusting the amounts of the ingredients so that the [base]/[acid] ratio has the proper value.

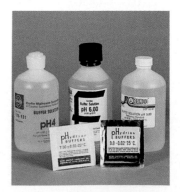

FIGURE 16.4 Prepackaged buffer solutions of known pH and solid ingredients for preparing buffer solutions of known pH.

It's important to realize that the pH of a buffer solution does not depend on the volume of the solution. Because a change in solution volume changes the concentrations of the acid and base by the same amount, the [base]/[acid] ratio and the pH remain unchanged. As a result, the volume of water used to prepare a buffer solution is not critical, and you can dilute a buffer without a change in pH. The pH depends only on pK_a and on the relative molar amounts of weak acid and conjugate base.

EXAMPLE 16.3

(a) Use the Henderson-Hasselbalch equation to calculate the pH of a buffer solution that is 0.45 M in NH_4Cl and 0.15 M in NH_3. **(b)** How would you prepare an NH_4Cl–NH_3 buffer that has a pH of 9.00?

SOLUTION **(a)** We've already solved this problem by another method in Example 16.2. Now that we've discussed equilibria in buffer solutions, though, you can use the Henderson-Hasselbalch equation as a shortcut. Since NH_4^+ is the weak acid in an NH_4^+–NH_3 buffer solution, you need pK_a for NH_4^+. Calculate it from the tabulated K_b value for NH_3 (Appendix C):

$$K_a = \frac{K_w}{K_b} = \frac{1.0 \times 10^{-14}}{1.8 \times 10^{-5}} = 5.6 \times 10^{-10}$$

$$pK_a = -\log K_a = -\log(5.6 \times 10^{-10}) = 9.25$$

Since [base] = $[NH_3]$ = 0.15 M and [acid] = $[NH_4^+]$ = 0.45 M,

$$pH = pK_a + \log \frac{[base]}{[acid]} = 9.25 + \log\left(\frac{0.15}{0.45}\right) = 9.25 - 0.48 = 8.77$$

The pH of the buffer solution is 8.77.

(b) Rearrange the Henderson-Hasselbalch equation to obtain an expression for the relative amounts of NH_3 and NH_4^+ in a solution having pH = 9.00:

$$\log \frac{[base]}{[acid]} = pH - pK_a = 9.00 - 9.25 = -0.25$$

Therefore,

$$\frac{[NH_3]}{[NH_4^+]} = \text{antilog}(-0.25) = 10^{-0.25} = 0.56$$

The solution must contain 0.56 mol of NH_3 for every 1.00 mol of NH_4Cl, but the volume of the solution isn't critical. One way of preparing the buffer would be to combine 1.00 mol of NH_4Cl (53.5 g) with 0.56 mol of NH_3 (say, 560 mL of 1.00 M NH_3).

EXAMPLE 16.4

What $[NH_3]/[NH_4^+]$ ratio is required for a buffer solution that has pH = 7.00? Why is a mixture of NH_3 and NH_4Cl a poor choice of reagents for a buffer having pH = 7.00?

SOLUTION We can calculate the required $[NH_3]/[NH_4^+]$ ratio from the Henderson-Hasselbalch equation:

$$\log \frac{[NH_3]}{[NH_4^+]} = pH - pK_a = 7.00 - 9.25 = -2.25$$

$$\frac{[NH_3]}{[NH_4^+]} = \text{antilog}(-2.25) = 10^{-2.25} = 5.6 \times 10^{-3}$$

For a typical value of $[NH_4^+]$—say, 1.0 M—$[NH_3]$ would have to be very small (0.0056 M). Such a solution is a poor buffer because it has little capacity to absorb added acid. Also, because the $[NH_3]/[NH_4^+]$ ratio is far from unity, addition of a small amount of H_3O^+ or OH^- will result in a large change in pH.

⌐ **PROBLEM 16.7** Use the Henderson-Hasselbalch equation to calculate the pH of a buffer solution prepared by mixing equal volumes of 0.20 M $NaHCO_3$ and 0.10 M Na_2CO_3. K_a values are given in Appendix C.

⌐ **PROBLEM 16.8** Give a recipe for preparing a $NaHCO_3$–Na_2CO_3 buffer solution that has pH = 10.40.

⌐ **PROBLEM 16.9** Suppose you are performing an experiment that requires a constant pH of 7.50. Suggest an appropriate buffer system based on the K_a values in Appendix C.

16.5 ➤pH TITRATION CURVES

In a typical acid–base titration (Section 3.10), a solution containing a known concentration of base (or acid) is added slowly from a buret to a second solution containing an unknown concentration of acid (or base). The progress of the titration is monitored, either by using a pH meter or by observing the color of a suitable acid–base indicator. With a pH meter, you can record data to produce a **pH titration curve**, a plot of the pH of the solution as a function of the volume of added titrant (Figure 16.5).

Why study titration curves? The shape of a pH titration curve makes it possible to identify the **equivalence point** in a titration, the point at which stoichiometrically equivalent quantities of acid and base have been mixed together. Knowing the shape of the titration curve is also useful in selecting a suitable indicator to signal the equivalence point. We'll explore both of these points later.

FIGURE 16.5 **(a)** In this pH titration, 0.100 M NaOH is being added slowly from a buret to an HCl solution of unknown concentration. The pH of the solution is measured with a pH meter and is recorded as a function of the volume of NaOH added. **(b)** pH titration curve for titration of 40.0 mL of 0.100 M HCl with 0.100 M NaOH. The pH increases gradually in the regions before and after the equivalence point, but increases rapidly in the region near the equivalence point. The equivalence point comes after addition of 40.0 mL of 0.100 M NaOH. The pH at the equivalence point is 7.00.

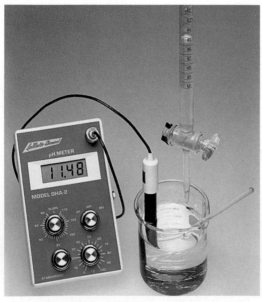

(a)

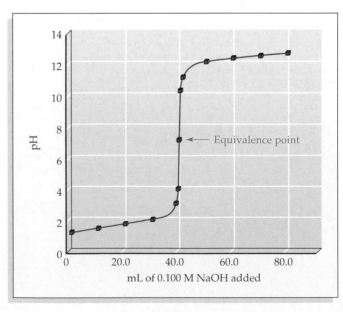

(b)

We can calculate pH titration curves from the principles of aqueous solution equilibria. To understand why titration curves have certain characteristic shapes, we'll calculate these curves for four important types of titration: (1) strong acid–strong base, (2) weak acid–strong base, (3) weak base–strong acid, and (4) polyprotic acid–strong base. For convenience, we'll express amounts of solute in millimoles (mmol) and solution volumes in mL. Molar concentration can thus be expressed in mmol/mL, a unit that is equivalent to mol/L:

$$\text{Molarity} = \frac{\text{mmol of solute}}{\text{mL of solution}} = \frac{10^{-3}\,\text{mol of solute}}{10^{-3}\,\text{L of solution}} = \frac{\text{mol of solute}}{\text{L of solution}}$$

16.6 ▸STRONG ACID–STRONG BASE TITRATIONS

As an example of a strong acid–strong base titration, let's consider the titration of 40.0 mL of 0.100 M HCl with 0.100 M NaOH. We'll calculate the pH at selected points in the course of the titration to illustrate the procedures that can be used to calculate the entire curve.

1. Before Addition of Any NaOH   Since HCl is a strong acid, the initial concentration of H_3O^+ is 0.100 M, and the pH is 1.00. (We've rounded the value of the pH to two significant figures.)

2. After Addition of 10.0 mL of 0.100 M NaOH   The added OH^- ions will decrease $[H_3O^+]$ because of the neutralization reaction

$$H_3O^+(aq) + OH^-(aq) \xrightarrow{100\%} 2\,H_2O(l)$$

The number of mmol of H_3O^+ present initially is the product of the initial volume of HCl and its molarity:

$$\text{mmol }H_3O^+\text{ initial} = (40.0\text{ mL})(0.100\text{ mmol/mL}) = 4.00\text{ mmol}$$

Similarly, the number of mmol of OH^- added is the product of the volume of NaOH added and its molarity:

$$\text{mmol }OH^-\text{ added} = (10.0\text{ mL})(0.100\text{ mmol/mL}) = 1.00\text{ mmol}$$

For each mmol of OH^- added, an equal amount of H_3O^+ will disappear because of the neutralization reaction. The number of mmol of H_3O^+ remaining after neutralization is therefore

$$\begin{aligned}\text{mmol }H_3O^+\text{ after neutralization} &= \text{mmol }H_3O^+\text{ initial} - \text{mmol }OH^-\text{ added}\\ &= 4.00\text{ mmol} - 1.00\text{ mmol} = 3.00\text{ mmol}\end{aligned}$$

We've carried out this calculation using *amounts* of acid and base (mmol) rather than *concentrations* (molarity) because the volume changes as the titration proceeds. If we divide the number of mmol of H_3O^+ after neutralization by the total volume (now $40.0 + 10.0 = 50.0$ mL), we obtain $[H_3O^+]$ after neutralization:

$$[H_3O^+]\text{ after neutralization} = \frac{3.00\text{ mmol}}{50.0\text{ mL}} = 6.00 \times 10^{-2}\text{ M}$$

$$pH = -\log(6.00 \times 10^{-2}) = 1.22$$

TABLE 16.1 Sample Results for pH Calculations at Various Points in the Titration of 40.0 mL of 0.100 M HCl with 0.100 M NaOH

mL NaOH added	mmol OH⁻ added	mmol H₃O⁺ after neutr.	Total volume (ml)		[H₃O⁺] after neutr.	pH
Before the equivalence point:						
0.0	0.0	4.00	40.0		1.00×10^{-1}	1.00
10.0	1.00	3.00	50.0		6.00×10^{-2}	1.22
20.0	2.00	2.00	60.0		3.33×10^{-2}	1.48
30.0	3.00	1.00	70.0		1.43×10^{-2}	1.84
39.0	3.90	0.10	79.0		1.27×10^{-3}	2.90
39.9	3.99	0.01	79.9		1.3×10^{-4}	3.9
At the equivalence point:						
40.0	4.00	0.00	80.0		1.0×10^{-7}	7.00
Beyond the equivalence point:		mmol OH⁻ after neutr.		[OH⁻] after neutr.		
40.1	4.01	0.01	80.1	1.2×10^{-4}	8.3×10^{-11}	10.1
41.0	4.10	0.10	81.0	1.2×10^{-3}	8.3×10^{-12}	11.08
50.0	5.00	1.00	90.0	1.11×10^{-2}	9.0×10^{-13}	12.05
60.0	6.00	2.00	100.0	2.00×10^{-2}	5.0×10^{-13}	12.30
70.0	7.00	3.00	110.0	2.73×10^{-2}	3.7×10^{-13}	12.43
80.0	8.00	4.00	120.0	3.33×10^{-2}	3.0×10^{-13}	12.52

This same procedure can be used to calculate the pH at other points prior to the equivalence point, giving the results summarized in Table 16.1.

3. After Addition of 40.0 mL of 0.100 M NaOH At this point, we have added (40.0 mL)(0.100 mmol/mL) = 4.00 mmol of NaOH, which is just enough OH⁻ to neutralize all the 4.00 mmol of HCl initially present. This is the equivalence point of the titration, and the pH is 7.00 because the solution contains only water and NaCl, a salt derived from a strong base and a strong acid.

4. After Addition of 60.0 mL of 0.100 M NaOH At this point, we have added (60.0 mL)(0.100 mmol/mL) = 6.00 mmol of NaOH, which is more than enough to neutralize the 4.00 mmol of HCl initially present. Consequently, an excess of OH⁻ (6.00 − 4.00 = 2.00 mmol) is present. Since the total volume is now 40.0 + 60.0 = 100.0 mL, the concentration of OH⁻ is

$$[OH^-] \text{ after neutralization} = \frac{2.00 \text{ mmol}}{100.0 \text{ mL}} = 2.00 \times 10^{-2} \text{ M}$$

The H₃O⁺ concentration and the pH are

$$[H_3O^+] = \frac{K_w}{[OH^-]} = \frac{1.0 \times 10^{-14}}{2.00 \times 10^{-2}} = 5.0 \times 10^{-13} \text{ M}$$

$$pH = -\log(5.0 \times 10^{-13}) = 12.30$$

Sample results for pH calculations at other points beyond the equivalence point are also included in Table 16.1.

If we plot the pH data in Table 16.1 as a function of the milliliters of NaOH added, we obtain the pH titration curve in Figure 16.5. This curve exhibits a gradual increase in pH in the regions before and after the equivalence point but a very sharp increase in pH in the region near the equivalence point. Thus, when the volume of added NaOH increases from 39.9 to 40.1 mL (0.2 mL is only about 4 drops from a buret), the pH increases from 3.9 to 10.1 (Table 16.1). The very sharp increase in pH in the region of the equivalence point is characteristic of the titration curve for any strong acid–strong base titration. It is this feature that allows us to identify the equivalence point when the concentration of the acid is unknown.

The pH titration curve for titration of a strong base with a strong acid is similar except that the initial pH is high and then decreases as acid is added.

PROBLEM 16.10 Calculate the pH of 40.0 mL of 0.100 M HCl after addition of **(a)** 35.0 mL and **(b)** 45.0 mL of 0.100 M NaOH. Are your results consistent with the pH data in Table 16.1?

PROBLEM 16.11 Calculate the pH of 40.0 mL of 0.100 M NaOH after addition of the following volumes of 0.0500 M HCl.
(a) 60.0 mL   **(b)** 80.2 mL   **(c)** 100.0 mL

16.7 ➤WEAK ACID–STRONG BASE TITRATIONS

As an example of a weak acid–strong base titration, let's consider the titration of 40.0 mL of 0.100 M acetic acid with 0.100 M NaOH. Calculation of the pH at selected points along the titration curve is straightforward because we've already met all the equilibrium problems that arise.

1. Before Addition of Any NaOH   The equilibrium problem at this point is the familiar one of calculating the pH of a solution of a weak acid. The calculated pH of 0.100 M acetic acid is 2.89.

2. After Addition of 20.0 mL of 0.100 M NaOH   Since acetic acid is largely undissociated and NaOH is completely dissociated, the neutralization reaction is

$$CH_3COOH(aq) + OH^-(aq) \xrightarrow{100\%} H_2O(l) + CH_3CO_2^-(aq)$$

At this point we have added (20.0 mL)(0.100 mmol/mL) = 2.00 mmol of NaOH, which is enough OH^- to neutralize exactly half the 4.00 mmol of CH_3COOH initially present. After neutralization, we have a buffer solution that contains 2.00 mmol of $CH_3CO_2^-$ and $4.00 - 2.00 = 2.00$ mmol of CH_3COOH. Consequently, the [base]/[acid] ratio is unity, and $pH = pK_a$:

$$pH = pK_a + \log \frac{[base]}{[acid]} = pK_a = 4.74$$

3. After Addition of 40.0 mL of 0.100 M NaOH   This is the equivalence point of the titration because we have added (40.0 mL)(0.100 mmol/mL) = 4.00 mmol of NaOH, which is just enough OH^- to neutralize all the 4.00 mmol of CH_3COOH initially present. After neutralization, the solution contains 0.0500 M CH_3COONa:

$$[Na^+] = [CH_3CO_2^-] = \frac{4.00 \text{ mmol}}{(40.0 \text{ mL} + 40.0 \text{ mL})} = 0.0500 \text{ M}$$

Because Na^+ is neither an acid nor a base and $CH_3CO_2^-$ is a weak base, we have a basic salt solution (Section 15.14), whose pH can be calculated as 8.72 by the method outlined in Example 15.13. For a weak monoprotic acid–strong base titration, the pH at the equivalence point is always greater than 7.00 because the anion of the weak acid is a base.

4. After Addition of 60.00 mL of 0.100 M NaOH At this point we have added $(60.0 \text{ mL})(0.100 \text{ mmol/mL}) = 6.00$ mmol of NaOH, which is more than enough OH^- to neutralize the 4.00 mmol of CH_3COOH initially present. The total volume is $40.0 + 60.0 = 100.0$ mL, and the concentrations after neutralization are

$$[CH_3CO_2^-] = \frac{4.00 \text{ mmol}}{100.0 \text{ mL}} = 0.0400 \text{ M}$$

$$[OH^-] = \frac{6.00 \text{ mmol} - 4.00 \text{ mmol}}{100.0 \text{ mL}} = 0.0200 \text{ M}$$

Because the only acid present is water, the principal reaction is the same as that at the equivalence point:

$$CH_3CO_2^-(aq) + H_2O(l) \rightleftarrows CH_3COOH(aq) + OH^-(aq)$$

In this case, however, $[OH^-]$ from base dissociation of $CH_3CO_2^-$ is negligible compared with $[OH^-]$ from the excess NaOH. $[H_3O^+]$ and the pH can therefore be calculated from $[OH^-]$:

$$[H_3O^+] = \frac{K_w}{[OH^-]} = \frac{1.0 \times 10^{-14}}{0.0200} = 5.0 \times 10^{-13} \text{ M}$$

$$pH = 12.30$$

In general, at points beyond the equivalence point, base dissociation of the anion of the weak acid is negligible, and the pH is determined by the concentration of OH^- from the excess NaOH.

The results of pH calculations for titration of 0.100 M CH_3COOH with 0.100 M NaOH are plotted in Figure 16.6. Comparison of the titration curves for the weak acid–strong base titration and the strong acid–strong base case shows several significant differences:

1. The initial rise in pH is greater for titration of the weak acid, but the curve then becomes more level in the region midway to the equivalence point. Both effects are due to the buffering action of the weak acid–conjugate base mixture. The curve has minimum slope exactly halfway to the equivalence point, where the buffering action is maximized and the pH equals pK_a of the weak acid.

2. The increase in pH in the region near the equivalence point is smaller than in the strong acid–strong base case. The weaker the acid, the smaller the increase in the pH, as illustrated in Figure 16.7.

3. The pH at the equivalence point is greater than 7.00 because the anion of a weak acid is a base.

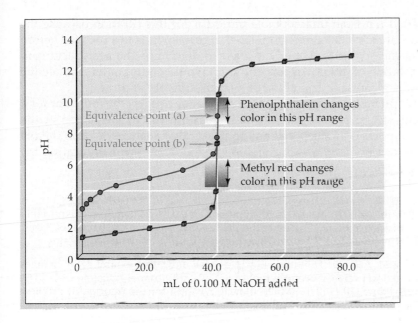

FIGURE 16.6 pH titration curves for titration of **(a)** 40.0 mL of 0.100 M CH_3COOH with 0.100 M NaOH (blue line) and **(b)** 40.0 mL of 0.100 M HCl with 0.100 M NaOH (red line). The pH ranges in which the acid–base indicators phenolphthalein and methyl red change color are indicated. Note that phenolphthalein is an excellent indicator for the weak acid–strong base titration because the equivalence point **(a)** is at pH 8.72; methyl red is an unsatisfactory indicator because it changes color well before the equivalence point. Either phenolphthalein or methyl red can be used for the strong acid–strong base titration because the curve rises very steeply in the region of the equivalence point **(b)** at pH 7.00.

Beyond the equivalence point, the curves for the weak acid–strong base and strong acid–strong base titrations are identical because the pH in both cases is determined by the concentration of OH^- from the excess NaOH.

Figure 16.6 shows how knowledge of the shape of a pH titration curve makes it possible to select a suitable acid–base indicator to signal the equivalence point of a titration. Phenolphthalein is an excellent indicator for the CH_3COOH–NaOH titration because the pH at the equivalence point (8.72) falls within the pH range (8.2–9.8) in which phenolphthalein changes

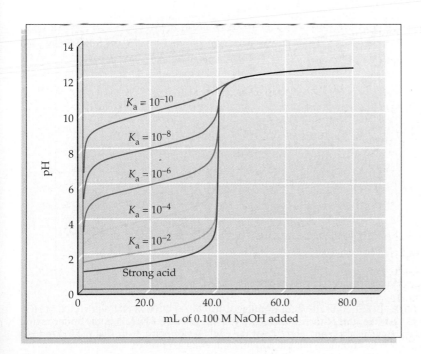

FIGURE 16.7 pH titration curves for titration of 40.0 mL of 0.100 M solutions of various weak acids with 0.100 M NaOH. In each case, the equivalence point comes after addition of 40.0 mL of 0.100 M NaOH, but the increase in pH at the equivalence point gets smaller and the equivalence point gets more difficult to detect as the K_a value of the weak acid decreases.

color. Methyl red is an unacceptable indicator for this titration because the pH range in which it changes color (4.2–6.0) corresponds to a pH well before the equivalence point. Anyone who tried to determine the acetic acid content in a solution of unknown concentration would badly underestimate the amount of acid present if methyl red was used as the indicator.

Either phenolphthalein or methyl red can be used as the indicator for a strong acid–strong base titration. The increase in pH in the region of the equivalence point is so steep that any indicator that changes color in the pH range 4–10 can be used without making a significant error in locating the equivalence point.

PROBLEM 16.12   Consider the titration of 100.0 mL of 0.016 M HOCl ($K_a = 3.5 \times 10^{-8}$) with 0.0400 M NaOH. How many mL of 0.0400 M NaOH are required to reach the equivalence point? Calculate the pH **(a)** after addition of 10.0 mL of 0.0400 M NaOH; **(b)** halfway to the equivalence point; **(c)** at the equivalence point.

PROBLEM 16.13   The following acid–base indicators exhibit color changes in the indicated pH ranges: bromthymol blue (6.0–7.6), thymolphthalein (9.4–10.6), and alizarin yellow (10.1–12.0). Which indicator is best for the titration in Problem 16.12? Which indicator is unacceptable? Explain.

16.8 ➤ WEAK BASE–STRONG ACID TITRATIONS

Figure 16.8 shows the pH titration curve for a typical weak base–strong acid titration—the titration of 40.0 mL of 0.100 M NH_3 with 0.100 M HCl. The pH calculations are simply outlined; you should verify the results yourself.

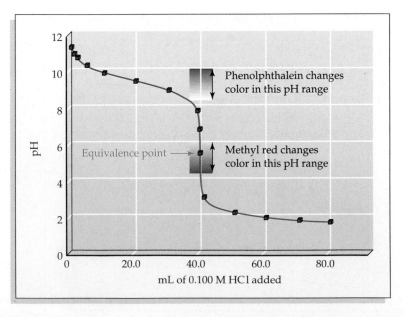

FIGURE 16.8   pH titration curve for titration of 40.0 mL of 0.100 M NH_3 with 0.100 M HCl. The pH is 11.12 at the start of the titration, 9.25 (the pK_a value for NH_4^+) in the buffer region halfway to the equivalence point, and 5.28 at the equivalence point. Note that methyl red is a good indicator for this titration, but phenolphthalein is unacceptable.

1. Before Addition of Any HCl The equilibrium problem at the start of the titration is the familiar one of calculating the pH of a solution of a weak base (Section 15.12). The principal reaction at this point is base dissociation of ammonia:

$$NH_3(aq) + H_2O(l) \rightleftharpoons NH_4^+(aq) + OH^-(aq) \qquad K_b = 1.8 \times 10^{-5}$$

The initial pH is 11.12.

2. Before the Equivalence Point As HCl is added to the NH_3 solution, NH_3 is converted to NH_4^+ because of the neutralization reaction

$$NH_3(aq) + H_3O^+(aq) \xrightarrow{\ 100\%\ } NH_4^+(aq) + H_2O(l)$$

The neutralization reaction goes to completion, but the amount of H_3O^+ added before the equivalence point is not sufficient to convert all the NH_3 to NH_4^+. We therefore have an NH_4^+–NH_3 buffer solution, which accounts for the leveling of the titration curve in the buffer region between the start of the titration and the equivalence point. The pH can be calculated from the Henderson-Hasselbalch equation. After addition of 20.0 mL of 0.100 M HCl (halfway to the equivalence point), $[NH_3] = [NH_4^+]$ and the pH is equal to pK_a for NH_4^+ (9.25).

3. At the Equivalence Point We reach the equivalence point after adding 40.0 mL of 0.100 M HCl (4.00 mmol). At this point the 4.00 mmol of NH_3 present initially has been converted to 4.00 mmol of NH_4^+; $[NH_4^+] = (4.00$ mmol$)/(80.0$ mL$) = 0.0500$ M. Since NH_4^+ is a weak acid and Cl^- is neither an acid nor a base, we have an acidic salt solution (Section 15.14). The principal reaction is

$$NH_4^+(aq) + H_2O(l) \rightleftharpoons H_3O^+(aq) + NH_3(aq) \qquad K_a = K_w/K_b = 5.6 \times 10^{-10}$$

The pH at the equivalence point is 5.28.

4. Beyond the Equivalence Point In this region, all the NH_3 has been converted to NH_4^+, and an excess of H_3O^+ is present from the excess HCl. Since acid dissociation of NH_4^+ produces a negligible $[H_3O^+]$ compared with $[H_3O^+]$ from the excess HCl, the pH can be calculated directly from the concentration of excess HCl. For example, after addition of 60.0 mL of 0.100 M HCl:

$$[H_3O^+] = \frac{6.00 \text{ mmol} - 4.00 \text{ mmol}}{100.0 \text{ mL}} = 0.0200 \text{ M}$$

$$pH = 1.70$$

16.9 ▸POLYPROTIC ACID–STRONG BASE TITRATIONS

As a final example of an acid–base titration, let's consider the gradual addition of NaOH to the protonated form of the amino acid alanine (H_2A^+), a substance that acts as a *diprotic acid*. Amino acids, about which we'll learn more in Chapter 24, are both acidic and basic and can be protonated by strong acids. The resultant protonated amino acid thus has two dissociable protons and can react with two molar amounts of OH$^-$ to give first the neutral form and then the anionic form:

$$\underset{\substack{\text{Protonated form (H}_2\text{A}^+) \\ \text{of alanine}}}{\overset{\overset{\displaystyle O}{\parallel}}{\text{H}_3\overset{+}{\text{N}}\text{CHCOH}}} \xrightarrow{\text{OH}^-} \underset{\substack{\text{Neutral form (HA)} \\ \text{of alanine}}}{\overset{\overset{\displaystyle O}{\parallel}}{\text{H}_3\overset{+}{\text{N}}\text{CHCO}^-}} \xrightarrow{\text{OH}^-} \underset{\substack{\text{Anionic form (A}^-) \\ \text{of alanine}}}{\overset{\overset{\displaystyle O}{\parallel}}{\text{H}_2\text{NCHCO}^-}}$$

$$\underset{\text{CH}_3}{\qquad} \qquad \underset{\text{CH}_3}{\qquad} \qquad \underset{\text{CH}_3}{\qquad}$$

The proton of the –COOH group is more acidic than the proton of the –NH$_3^+$ group and is neutralized first. The neutral form of alanine (HA) that results has a plus charge on the –NH$_3^+$ group and a minus charge on the –COO$^-$ group but is electrically neutral overall. In a second step, the proton of the –NH$_3^+$ group is neutralized, yielding the anionic form of alanine (A$^-$). The dissociation equilibria and their K_a values are

$$\text{H}_2\text{A}^+(aq) + \text{H}_2\text{O}(l) \;\rightleftarrows\; \text{H}_3\text{O}^+(aq) + \text{HA}(aq) \qquad K_{a1} = 4.6 \times 10^{-3};\, \text{p}K_{a1} = 2.34$$

$$\text{HA}(aq) + \text{H}_2\text{O}(l) \;\rightleftarrows\; \text{H}_3\text{O}^+(aq) + \text{A}^-(aq) \qquad K_{a2} = 2.0 \times 10^{-10};\, \text{p}K_{a2} = 9.69$$

Figure 16.9 shows the pH titration curve for addition of NaOH to 1.00 L of a 1.00 M solution of H$_2$A$^+$ (1.00 mol). To simplify the calculations, we've assumed that the added base is solid NaOH so that we can neglect volume changes in the course of the titration. Let's calculate the pH at some of the more important points.

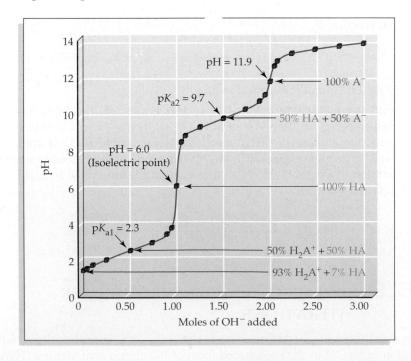

FIGURE 16.9 Change in the pH of 1.00 L of a 1.00 M solution of H$_2$A$^+$ on addition of solid NaOH. H$_2$A$^+$, the protonated form of alanine, is a diprotic acid. Note that the titration curve exhibits two equivalence points, at pH 6.0 and pH 11.9, and two buffer regions, near pH 2.3 and pH 9.7.

1. Before Addition of Any NaOH   The equilibrium problem at the start of the titration is the familiar one of calculating the pH of a diprotic acid (Section 15.11). The principal reaction is dissociation of H_2A^+, and $[H_3O^+]$ can be calculated from the equilibrium equation

$$K_{a1} = 4.6 \times 10^{-3} = \frac{[H_3O^+][HA]}{[H_2A^+]} = \frac{(x)(x)}{(1.00 - x)}$$

Solving the quadratic equation gives $[H_3O^+] = 0.066$ M and pH = 1.18.

2. Halfway to the First Equivalence Point   As NaOH is added, H_2A^+ is converted to HA because of the neutralization reaction

$$H_2A^+(aq) + OH^-(aq) \xrightarrow{100\%} H_2O(l) + HA(aq)$$

Halfway to the first equivalence point, we have an H_2A^+–HA buffer solution with $[H_2A^+] = [HA]$. The Henderson-Hasselbalch equation gives pH = $pK_{a1} = 2.34$.

3. At the First Equivalence Point   At this point, we have added just enough NaOH to convert all the H_2A^+ to HA. The principal reaction at the first equivalence point is proton transfer between HA molecules:

$$2\ HA(aq) \rightleftharpoons H_2A^+(aq) + A^-(aq) \qquad K = K_{a2}/K_{a1} = 4.3 \times 10^{-8}$$

The pH is equal to the average of the two pK_a values:

$$pH \text{ (at first equivalence point)} = \frac{pK_{a1} + pK_{a2}}{2}$$

Since H_2A^+ has $pK_{a1} = 2.34$ and $pK_{a2} = 9.69$, the pH at the first equivalence point is $(2.34 + 9.69)/2 = 6.02$. The same situation holds for most polyprotic acids: The pH at the first equivalence point is equal to the average of pK_{a1} and pK_{a2}.

For an amino acid, the pH value, $(pK_{a1} + pK_{a2})/2$, is called the *isoelectric point* (Figure 16.9). At that point the concentration of the neutral HA is at a maximum, and the very small concentrations of H_2A^+ and A^- are equal. Biochemists take advantage of the isoelectric point in separating mixtures of amino acids.

4. Halfway Between the First and Second Equivalence Points   At this point, half the HA has been converted to A^- because of the neutralization reaction

$$HA(aq) + OH^-(aq) \xrightarrow{100\%} H_2O(l) + A^-(aq)$$

We thus have an HA–A^- buffer solution with $[HA] = [A^-]$. Therefore, pH = $pK_{a2} = 9.69$.

5. At the Second Equivalence Point   At this point, we have added enough NaOH to convert all the HA to A^-, and we have a 1.00 M solution of a basic salt (Section 15.14). The principal reaction is

$$A^-(aq) + H_2O(l) \rightleftarrows HA(aq) + OH^-(aq) \qquad K_b = K_w/K_{a2} = 5.0 \times 10^{-5}$$

and its equilibrium constant, the base-dissociation constant of A^-, is

$$K_b = \frac{K_w}{K_a \text{ for HA}} = \frac{K_w}{K_{a2}} = \frac{1.0 \times 10^{-14}}{2.0 \times 10^{-10}} = 5.0 \times 10^{-5}$$

We can obtain $[OH^-]$ from the equilibrium equation for the principal reaction and then calculate $[H_3O^+]$ and pH in the usual way:

$$K_b = 5.0 \times 10^{-5} = \frac{[HA][OH^-]}{[A^-]} = \frac{(x)(x)}{1.00 - x}; x = [OH^-] = 7.1 \times 10^{-3} \text{ M}$$

$$[H_3O^+] = \frac{K_w}{[OH^-]} = \frac{1.0 \times 10^{-14}}{7.1 \times 10^{-3}} = 1.4 \times 10^{-12} \text{ M}$$

$$pH = 11.85$$

Since the initial solution of H_2A^+ contained 1.00 mol of H_2A^+, the amount of NaOH required to reach the second equivalence point is 2.00 mol. Beyond the second equivalence point, the pH is determined by $[OH^-]$ from the excess NaOH.

EXAMPLE 16.5

Consider the titration of 30.0 mL of 0.0600 M H_2CO_3 ($K_{a1} = 4.3 \times 10^{-7}$; $K_{a2} = 5.6 \times 10^{-11}$) with 0.0900 M NaOH. Calculate the pH after addition of 20.0 mL of base.

SOLUTION   The number of mmol of H_2CO_3 present initially and the number of mmol of NaOH added are

$$\text{mmol } H_2CO_3 \text{ initial} = (30.0 \text{ mL})(0.0600 \text{ mmol/mL}) = 1.80 \text{ mmol}$$

$$\text{mmol NaOH added} = (20.0 \text{ mL})(0.0900 \text{ mmol/mL}) = 1.80 \text{ mmol}$$

The added base is just enough to reach the first equivalence point, converting all the H_2CO_3 to HCO_3^- because of the neutralization reaction

$$H_2CO_3(aq) + OH^-(aq) \xrightarrow{100\%} H_2O(l) + HCO_3^-(aq)$$

Therefore, the pH is equal to the average of pK_{a1} and pK_{a2}:

$$pH = \frac{pK_{a1} + pK_{a2}}{2} = \frac{[-\log(4.3 \times 10^{-7})] + [-\log(5.6 \times 10^{-11})]}{2}$$

$$= \frac{6.37 + 10.25}{2} = 8.31$$

⌐ **PROBLEM 16.14**   Consider the titration of 40.0 mL of 0.0800 M H_2SO_3 ($K_{a1} = 1.5 \times 10^{-2}$; $K_{a2} = 6.3 \times 10^{-8}$) with 0.160 M NaOH. Calculate the pH after addition of the following volumes of 0.160 M NaOH.

(a) 20.0 mL   **(b)** 30.0 mL   **(c)** 35.0 mL

16.10 ►SOLUBILITY EQUILIBRIA

Many biological and environmental processes involve the dissolution or precipitation of a sparingly soluble ionic compound. Tooth decay, for example, begins when tooth enamel, composed of the mineral hydroxyapatite, $Ca_5(PO_4)_3OH$, dissolves in organic acids produced by bacterial decomposition of foods rich in sugar. Kidney stones form when moderately insoluble calcium salts, such as calcium oxalate, CaC_2O_4, precipitate slowly over a long period of time. To understand the quantitative aspects of such solubility and precipitation phenomena, we have to examine the principles of solubility equilibria.

Let's consider the solubility equilibrium in a saturated solution of calcium fluoride in contact with an excess of solid calcium fluoride. Like most ionic solutes, calcium fluoride is a strong electrolyte in water and exists in the aqueous phase as completely dissociated hydrated ions, $Ca^{2+}(aq)$ and $F^-(aq)$. At equilibrium, the ion concentrations remain constant because the rate at which solid CaF_2 dissolves to give $Ca^{2+}(aq)$ and $F^-(aq)$ exactly equals the rate at which these ions crystallize to form solid CaF_2:

$$CaF_2(s) \rightleftarrows Ca^{2+}(aq) + 2\,F^-(aq)$$

The equilibrium equation for the dissolution reaction is

$$K_{sp} = [Ca^{2+}][F^-]^2$$

where the equilibrium constant K_{sp} is called the **solubility product constant**, or simply the **solubility product**. As usual, the concentration of the solid, CaF_2, has been omitted from the equilibrium equation because it is constant (Section 13.4).

For the general solubility equilibrium

$$M_mX_x(s) \rightleftarrows m\,M^{n+}(aq) + x\,X^{y-}(aq)$$

the equilibrium constant expression for K_{sp} is

$$K_{sp} = [M^{n+}]^m\,[X^{y-}]^x$$

Thus, K_{sp} always equals the product of the equilibrium concentrations of the ions on the right side of the chemical equation, with the concentration of each ion raised to the power of its coefficient in the balanced equation.

EXAMPLE 16.6

Write the expression for the solubility product of silver chromate, Ag_2CrO_4.

SOLUTION   First write the balanced equation for the solubility equilibrium:

$$Ag_2CrO_4(s) \rightleftarrows 2\,Ag^+(aq) + CrO_4^{2-}(aq)$$

The exponents in the equilibrium constant expression for K_{sp} are the coefficients in the balanced equation. Therefore,

$$K_{sp} = [Ag^+]^2[CrO_4^{2-}]$$

Tooth decay begins when tooth enamel, $Ca_5(PO_4)_3OH$, dissolves in organic acids. Very tiny cavities can be detected by exposing the tooth to a beam of laser light.

Kidney stones consist of insoluble calcium salts, such as calcium oxalate.

A saturated solution of calcium fluoride in contact with solid CaF_2 contains constant, equilibrium concentrations of $Ca^{2+}(aq)$ and $F^-(aq)$ because at equilibrium the ions crystallize at the same rate as the solid dissolves.

⌐ **PROBLEM 16.15** Write the expression for K_{sp} of
(a) AgCl **(b)** PbI_2 **(c)** $Ca_3(PO_4)_2$ ⌐

16.11 ➤ MEASURING K_{sp} AND CALCULATING SOLUBILITY FROM K_{sp}

The solubility product K_{sp} is a number that is measured by experiment. We could, for example, determine K_{sp} for CaF_2 by adding an excess of solid CaF_2 to water, stirring the mixture to give a saturated solution of CaF_2, and then measuring the concentrations of Ca^{2+} and F^- in the saturated solution. To make sure that the concentrations had reached constant, equilibrium values, we would want to stir the mixture for an additional period of time and then repeat the measurements. Suppose that we found $[Ca^{2+}] = 3.3 \times 10^{-4}$ M and $[F^-] = 6.7 \times 10^{-4}$ M. (The value of $[F^-]$ is approximately twice the value of $[Ca^{2+}]$ because each mole of CaF_2 that dissolves yields 1 mol of Ca^{2+} ions and 2 mol of F^- ions.) We could then calculate K_{sp} for CaF_2:

$$K_{sp} = [Ca^{2+}][F^-]^2 = (3.3 \times 10^{-4})(6.7 \times 10^{-4})^2 = 1.5 \times 10^{-10}$$

The formation of a precipitate when aqueous NaF is added to aqueous $CaCl_2$ guarantees that the supernatant solution is saturated in CaF_2.

Another way to measure K_{sp} for CaF_2 is to approach the equilibrium from the opposite direction—that is, by mixing sources of Ca^{2+} and F^- ions to give a precipitate of solid CaF_2 and a saturated solution of CaF_2. Suppose, for example, that we mix solutions of $CaCl_2$ and NaF, allow time for equilibrium to be achieved, and then measure $[Ca^{2+}] = 1.5 \times 10^{-4}$ M and $[F^-] = 1.0 \times 10^{-3}$ M. These ion concentrations yield the same value of K_{sp}:[1]

$$K_{sp} = [Ca^{2+}][F^-]^2 = (1.5 \times 10^{-4})(1.0 \times 10^{-3})^2 = 1.5 \times 10^{-10}$$

If the saturated solution is prepared by a method other than dissolution of CaF_2 in pure water, there are no separate restrictions on $[Ca^{2+}]$ and $[F^-]$; the only restriction on the ion concentrations is that the equilibrium constant expression $[Ca^{2+}][F^-]^2$ must equal K_{sp}. That condition is satisfied by an infinite number of combinations of $[Ca^{2+}]$ and $[F^-]$, and therefore we can prepare many different solutions that are saturated with respect to CaF_2. For example, if $[F^-]$ is 1.0×10^{-2} M, then $[Ca^{2+}]$ must be 1.5×10^{-6} M:

$$[Ca^{2+}] = \frac{K_{sp}}{[F^-]^2} = \frac{1.5 \times 10^{-10}}{(1.0 \times 10^{-2})^2} = 1.5 \times 10^{-6} \text{ M}$$

Selected values of K_{sp} for various ionic compounds at 25°C are listed in Table 16.2, and additional values can be found in Appendix C. Like all equilibrium constants, values of K_{sp} depend on temperature (Section 13.9). Once the K_{sp} value for a compound has been measured, we can use it to calculate the *solubility* of the compound—the amount of compound that dissolves per unit volume of saturated solution.

---

[1] Values of K_{sp} are not affected by the presence of other ionic solutes, such as Na^+ from NaF and Cl^- from $CaCl_2$, as long as the solution is very dilute. As ion concentrations increase, K_{sp} values are somewhat modified because of electrostatic interactions between ions. We'll ignore this complication.

TABLE 16.2	K_{sp} Values at 25°C for Some Ionic Compounds	
Name	**Formula**	K_{sp}
Aluminum hydroxide	$Al(OH)_3$	1.9×10^{-33}
Barium carbonate	$BaCO_3$	2.6×10^{-9}
Calcium carbonate	$CaCO_3$	5.0×10^{-9}
Calcium fluoride	CaF_2	1.5×10^{-10}
Lead chloride	$PbCl_2$	1.2×10^{-5}
Lead chromate	$PbCrO_4$	2.8×10^{-13}
Silver chloride	$AgCl$	1.8×10^{-10}
Silver sulfate	Ag_2SO_4	1.2×10^{-5}

EXAMPLE 16.7

A particular saturated solution of silver chromate, Ag_2CrO_4, has $[Ag^+] = 5.0 \times 10^{-5}$ M and $[CrO_4^{2-}] = 4.4 \times 10^{-4}$ M. What is the value of K_{sp} for Ag_2CrO_4?

SOLUTION Substituting the equilibrium concentrations into the expression for K_{sp} of Ag_2CrO_4 (Example 16.6) gives the value of K_{sp}:

$$K_{sp} = [Ag^+]^2[CrO_4^{2-}] = (5.0 \times 10^{-5})^2(4.4 \times 10^{-4}) = 1.1 \times 10^{-12}$$

EXAMPLE 16.8

A saturated solution of Ag_2CrO_4 prepared by dissolving solid Ag_2CrO_4 in water has $[CrO_4^{2-}] = 6.5 \times 10^{-5}$ M. Calculate K_{sp} for Ag_2CrO_4.

SOLUTION Because the Ag^+ and CrO_4^{2-} ions come from dissolution of solid Ag_2CrO_4, $[Ag^+]$ must be twice $[CrO_4^{2-}]$. Therefore, $[Ag^+] = (2)(6.5 \times 10^{-5}) = 1.3 \times 10^{-4}$ M. The value of K_{sp} is

$$K_{sp} = [Ag^+]^2[CrO_4^{2-}] = (1.3 \times 10^{-4})^2(6.5 \times 10^{-5}) = 1.1 \times 10^{-12}$$

Addition of aqueous K_2CrO_4 to aqueous $AgNO_3$ gives a red precipitate of Ag_2CrO_4 and a saturated solution of Ag_2CrO_4.

EXAMPLE 16.9

Calculate the solubility of MgF_2 in water in units of **(a)** moles per liter and **(b)** grams per liter.

SOLUTION Write the balanced equation for the solubility equilibrium, and then look up the K_{sp} value for MgF_2 in Appendix C:

$$MgF_2(s) \rightleftharpoons Mg^{2+}(aq) + 2\ F^-(aq) \qquad K_{sp} = 7.4 \times 10^{-11}$$

(a) If we define x as the number of mol/L of MgF_2 that dissolves, then the saturated solution contains x mol/L of Mg^{2+} and $2x$ mol/L of F^-. Substituting these equilibrium concentrations into the expression for K_{sp} gives

$$K_{sp} = 7.4 \times 10^{-11} = [Mg^{2+}][F^-]^2 = (x)(2x)^2$$

$$4x^3 = 7.4 \times 10^{-11}$$

$$x^3 = 1.8 \times 10^{-11}$$

$$x = 2.6 \times 10^{-4} \text{ mol/L}$$

Thus, the molar solubility of MgF_2 in water at 25°C is 2.6×10^{-4} M.

Note that the number 2 appears twice in the expression $(x)(2x)^2$. The exponent 2 is required because of the equilibrium equation, $K_{sp} = [Mg^{2+}][F^-]^2$. The coefficient 2 in $2x$ is required because each mole of MgF_2 that dissolves gives 2 mol of $F^-(aq)$.

(b) To convert the solubility from units of moles per liter to units of grams per liter, multiply the molar solubility of MgF_2 by its molar mass (62.3 g/mol):

$$\text{Solubility (in g/L)} = \frac{2.6 \times 10^{-4} \text{ mol}}{L} \times \frac{62.3 \text{ g}}{\text{mol}} = 1.6 \times 10^{-2} \text{ g/L}$$

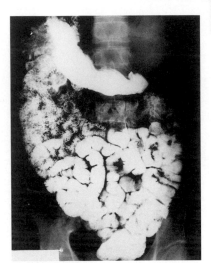

This X-ray photograph of the upper gastrointestinal tract was taken soon after the patient drank a barium sulfate "cocktail."

☐ **PROBLEM 16.16** A saturated solution of $Ca_3(PO_4)_2$ has $[Ca^{2+}] = 2.01 \times 10^{-8}$ M and $[PO_4{}^{3-}] = 1.6 \times 10^{-5}$ M. Calculate K_{sp} for $Ca_3(PO_4)_2$.

☐ **PROBLEM 16.17** Prior to having an X-ray exam of the upper gastrointestinal tract, a patient drinks an aqueous suspension of solid $BaSO_4$. (Scattering of X rays by barium greatly enhances the quality of the photograph.) Although Ba^{2+} is toxic, ingestion of $BaSO_4$ is safe because this salt is quite insoluble. If a saturated solution prepared by dissolving solid $BaSO_4$ in water has $[Ba^{2+}] = 1.05 \times 10^{-5}$ M, what is the value of K_{sp} for $BaSO_4$?

☐ **PROBLEM 16.18** Calculate **(a)** the molar solubility of AgCl in water and **(b)** the solubility of Ag_2SO_4 in water in units of grams per liter.

16.12 ►FACTORS THAT AFFECT SOLUBILITY

The Common-Ion Effect

We've already discussed the common-ion effect in connection with the dissociation of weak acids and bases (Section 16.2). To see how a common ion affects the position of a solubility equilibrium, let's look again at the solubility of MgF_2:

$$MgF_2(s) \rightleftarrows Mg^{2+}(aq) + 2 \text{ F}^-(aq)$$

In Example 16.9, we found that the molar solubility of MgF_2 in pure water is 2.6×10^{-4} M. Thus,

$$[Mg^{2+}] = 2.6 \times 10^{-4} \text{ M}$$

$$[F^-] = 5.2 \times 10^{-4} \text{ M}$$

When MgF_2 dissolves in a solution that contains a common ion from another source—say, F^- from NaF—the position of the solubility equilibrium is shifted to the left by the common-ion effect. If $[F^-]$ is larger than 5.2×10^{-4} M, then $[Mg^{2+}]$ must be correspondingly smaller than 2.6×10^{-4} M to maintain the equilibrium expression $[Mg^{2+}][F^-]^2$ at a constant value of $K_{sp} = 7.4 \times 10^{-11}$. A smaller value of $[Mg^{2+}]$ thus means that MgF_2 is less soluble in a sodium fluoride solution than it is in pure water. Similarly, the presence of Mg^{2+} from another source—say, $MgCl_2$—shifts the solubility equilibrium to the left and decreases the solubility of MgF_2.

In general, the solubility of a slightly soluble ionic compound is decreased by the presence of a common ion in the solution, as illustrated in Figure 16.10. The quantitative aspects of the common-ion effect are explored in Example 16.10.

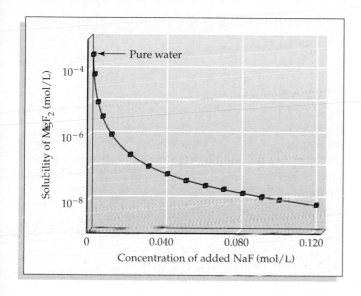

FIGURE 16.10 The common-ion effect. The solubility of MgF_2 decreases markedly on addition of F^- ions. Note that the solubility is plotted on a logarithmic scale.

EXAMPLE 16.10

Calculate the molar solubility of MgF_2 in 0.10 M NaF.

SOLUTION   Once again, define x as the molar solubility of MgF_2. Dissolution of x mol/L of MgF_2 provides x mol/L of Mg^{2+} and $2x$ mol/L of F^-, but the total concentration of F^- is $(0.10 + 2x)$ mol/L because the solution already contains 0.10 mol/L of F^- from the NaF. It's helpful to summarize the equilibrium concentrations under the balanced equation:

$$\text{Solubility equilibrium:}\quad MgF_2(s) \rightleftharpoons Mg^{2+}(aq) + 2\,F^-(aq)$$

$$\text{Equilibrium conc (M)}\qquad\qquad\qquad x\qquad\quad 0.10 + 2x$$

Substituting the equilibrium concentrations into the expression for K_{sp} gives

$$K_{sp} = 7.4 \times 10^{-11} = [Mg^{2+}][F^-]^2 = (x)(0.10 + 2x)^2$$

Because K_{sp} is small, $2x$ will be small compared with 0.10, and we can make the approximation that $(0.10 + 2x) \approx 0.10$. Therefore,

$$7.4 \times 10^{-11} = (x)(0.10 + 2x)^2 \approx (x)(0.10)^2$$

$$x = \frac{7.4 \times 10^{-11}}{(0.10)^2} = 7.4 \times 10^{-9}\ M$$

Recall Example 16.9, and note that MgF_2 is less soluble in 0.10 M NaF than in pure water by a factor of about 35,000!

PROBLEM 16.19 Calculate the molar solubility of MgF_2 in 0.10 M $MgCl_2$.

The pH of the Solution

The solubility of an ionic compound increases with decreasing pH of the solution if the compound contains a basic anion. The solubility of $CaCO_3$, for example, increases as the solution becomes more acidic (Figure 16.11) because the solubility equilibrium is shifted to the right by protonation of CO_3^{2-} ions:

$$CaCO_3(s) \rightleftarrows Ca^{2+}(aq) + CO_3^{2-}(aq)$$

$$\underline{H_3O^+(aq) + CO_3^{2-}(aq) \rightleftarrows HCO_3^-(aq) + H_2O(l)}$$

$$\text{Net: } CaCO_3(s) + H_3O^+(aq) \rightleftarrows Ca^{2+}(aq) + HCO_3^-(aq) + H_2O(l)$$

Other salts that contain basic anions, such as CN^-, PO_4^{3-}, S^{2-}, or F^-, behave similarly. By contrast, pH has no effect on the solubility of salts that contain anions of strong acids (Cl^-, Br^-, I^-, NO_3^-, and ClO_4^-) because these anions are not protonated by H_3O^+.

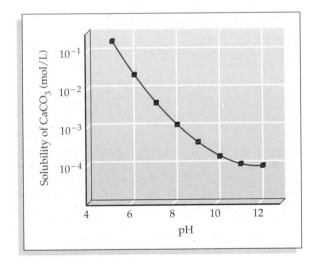

FIGURE 16.11 The solubility of $CaCO_3$ increases as the solution becomes more acidic because the CO_3^{2-} ions combine with protons, thus driving the solubility equilibrium to the right. Note that the solubility is plotted on a logarithmic scale.

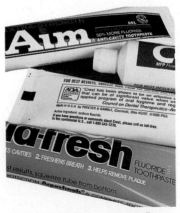

These toothpastes contain fluoride ions, which help to reduce tooth decay.

The effect of pH on the solubility of $CaCO_3$ has important environmental consequences. For instance, the formation of limestone caves, such as Mammoth Cave in Kentucky, is due to the slow dissolution of limestone ($CaCO_3$) in the slightly acidic natural water of underground streams. Marble, another form of $CaCO_3$, also dissolves in acid, which accounts for the deterioration of marble monuments on exposure to acid rain.

The effect of pH on solubility is also important in understanding how fluoride ion reduces tooth decay. When tooth enamel comes in contact with F^- ions in drinking water or fluoride-containing toothpaste, OH^- ions in hydroxyapatite, $Ca_5(PO_4)_3OH$, are replaced by F^- ions, giving the mineral fluorapatite, $Ca_5(PO_4)_3F$. Because F^- is a much weaker base than OH^-, $Ca_5(PO_4)_3F$ is much more resistant to dissolving in acids than is $Ca_5(PO_4)_3OH$.

⌐ PROBLEM 16.20 Which of the following compounds are more soluble in acidic solution than in pure water? **(a)** AgCN **(b)** PbI₂ **(c)** Al(OH)₃ **(d)** ZnS

Formation of Complex Ions

The solubility of an ionic compound increases dramatically if the solution contains a substance that can bond to the metal cation. Silver chloride, for example, is insoluble in water and in acid, but it dissolves in an excess of aqueous ammonia. Ammonia shifts the solubility equilibrium to the right by tying up the Ag^+ ion in the form of the complex ion $Ag(NH_3)_2^+$:

$$AgCl(s) \rightleftarrows Ag^+(aq) + Cl^-(aq)$$

$$Ag^+(aq) + 2\,NH_3(aq) \rightleftarrows Ag(NH_3)_2^+(aq)$$

Silver chloride is insoluble in water (left) but dissolves on addition of an excess of aqueous ammonia (right).

The stability of a complex ion is measured by its **formation constant K_f** (or **stability constant**), the equilibrium constant for formation of the complex ion from the hydrated metal cation. For example, K_f for $Ag(NH_3)_2^+$ is the equilibrium constant for the reaction of $Ag^+(aq)$ with $NH_3(aq)$:

$$Ag^+(aq) + 2\,NH_3(aq) \rightleftarrows Ag(NH_3)_2^+(aq)$$

$K_{sp} = [Ag][Cl]$
$$K_f = \frac{[Ag(NH_3)_2^+]}{[Ag^+][NH_3]^2} = 1.7 \times 10^7$$

The large value of K_f for $Ag(NH_3)_2^+$ means that the complex ion is quite stable, and nearly all the Ag^+ ion in an aqueous ammonia solution is therefore present in the form of the complex ion (Example 16.11).

The overall reaction for dissolution of AgCl in aqueous ammonia is the sum of the preceding equations for the dissolution of AgCl in water and the reaction of $Ag^+(aq)$ with $NH_3(aq)$ to give $Ag(NH_3)_2^+$:

$$AgCl(s) + 2\,NH_3(aq) \rightleftarrows Ag(NH_3)_2^+(aq) + Cl^-(aq)$$

Its equilibrium constant K is the product of the equilibrium constants for the reactions added:

$$K = \frac{[Ag(NH_3)_2^+][Cl^-]}{[NH_3]^2} = (K_{sp})(K_f) = (1.8 \times 10^{-10})(1.7 \times 10^7) = 3.1 \times 10^{-3}$$

Because K is much larger than K_{sp}, the solubility equilibrium in the presence of ammonia lies much farther to the right than it does in the absence of ammonia. The increase in the solubility of AgCl on addition of ammonia is shown graphically in Figure 16.12. In general, the solubility of an ionic compound increases when the metal cation is tied up in the form of a complex ion. The quantitative effect of complex formation on the solubility of AgCl is explored in Example 16.12.

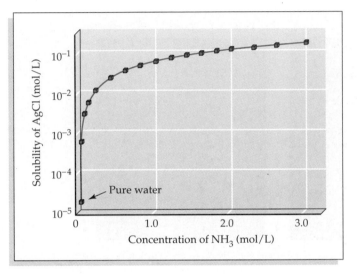

FIGURE 16.12 The solubility of AgCl in aqueous ammonia increases with increasing ammonia concentration owing to formation of the complex ion $Ag(NH_3)_2^+$. Note that the solubility is plotted on a logarithmic scale.

EXAMPLE 16.11

What are the concentrations of Ag^+ and $Ag(NH_3)_2^+$ in a solution prepared by adding 0.10 mol of $AgNO_3$ to 1.0 L of 3.0 M NH_3?

SOLUTION   Because K_f for $Ag(NH_3)_2^+$ is large (1.7×10^7), nearly all the Ag^+ from $AgNO_3$ will be converted to $Ag(NH_3)_2^+$:

$$Ag^+(aq) + 2\,NH_3(aq) \rightleftarrows Ag(NH_3)_2^+(aq)$$

To calculate the concentrations, it's convenient to introduce the fiction of 100% conversion of Ag^+ to $Ag(NH_3)_2^+$ followed by a tiny amount of back reaction (dissociation of $Ag(NH_3)_2^+$) to give a small equilibrium concentration of Ag^+. Conversion of 0.10 mol/L of Ag^+ to $Ag(NH_3)_2^+$ consumes 0.20 mol/L of NH_3. Assuming 100% conversion to $Ag(NH_3)_2^+$ gives the following concentrations:

$$[Ag^+] = 0\ M$$

$$[Ag(NH_3)_2^+] = 0.10\ M$$

$$[NH_3] = 3.0 - 0.20 = 2.8\ M$$

Dissociation of x mol/L of $Ag(NH_3)_2^+$ in the back reaction produces x mol/L of Ag^+ and $2x$ mol/L of NH_3. Therefore, the equilibrium concentrations (in mol/L) are

$$[Ag(NH_3)_2^+] = 0.10 - x$$

$$[Ag^+] = x$$

$$[NH_3] = 2.8 + 2x$$

Let's summarize our reasoning in a table under the balanced equation:

$$Ag^+(aq) + 2\,NH_3(aq) \rightleftharpoons Ag(NH_3)_2^+(aq)$$

Initial conc (M)	0.10	3.0	0
After 100% reaction (M)	0	2.8	0.10
Equilibrium conc (M)	x	$2.8 + 2x$	$0.10 - x$

Substituting the equilibrium concentrations into the expression for K_f, and making the approximation that x is negligible compared with 0.10 (and with 2.8) gives

$$K_f = 1.7 \times 10^7 = \frac{[Ag(NH_3)_2^+]}{[Ag^+][NH_3]^2} = \frac{0.10 - x}{(x)(2.8 + 2x)^2} \approx \frac{0.10}{(x)(2.8)^2}$$

$$[Ag^+] = x = \frac{0.10}{(1.7 \times 10^7)(2.8)^2} = 7.5 \times 10^{-10}\ M$$

$$[Ag(NH_3)_2^+] = 0.10 - x = 0.10 - (7.5 \times 10^{-10}) = 0.10\ M$$

Note that nearly all the Ag^+ is in the form of the complex ion.

EXAMPLE 16.12

Calculate the molar solubility of AgCl in (a) pure water and (b) 3.0 M NH_3.

SOLUTION   (a) In pure water, the solubility equilibrium is

$$AgCl(s) \rightleftharpoons Ag^+(aq) + Cl^-(aq)$$

Substituting the equilibrium concentrations (x mol/L) into the expression for K_{sp} gives

$$K_{sp} = 1.8 \times 10^{-10} = [Ag^+][Cl^-] = (x)(x)$$
$$x = \sqrt{1.8 \times 10^{-10}} = 1.3 \times 10^{-5}\ M$$

(b) The balanced equation for dissolution of AgCl in aqueous NH_3 is

$$AgCl(s) + 2\,NH_3(aq) \rightleftharpoons Ag(NH_3)_2^+(aq) + Cl^-(aq)$$

If we define x as the number of mol/L of AgCl that dissolves, then the saturated solution contains x mol/L of $Ag(NH_3)_2^+$, x mol/L of Cl^-, and $(3.0 - 2x)$ mol/L of

NH_3. (We're assuming that essentially all the Ag^+ is in the form of the complex ion, as proved in Example 16.11.) Substituting the equilibrium concentrations into the equilibrium equation gives

$$K = 3.1 \times 10^{-3} = \frac{[Ag(NH_3)_2{}^+][Cl^-]}{[NH_3]^2} = \frac{(x)(x)}{(3.0 - 2x)^2}$$

Taking the square root of both sides, we obtain:

$$5.6 \times 10^{-2} = \frac{x}{3.0 - 2x}$$

$$x = (5.6 \times 10^{-2})(3.0 - 2x) = 0.17 - 0.11x$$

$$x = \frac{0.17}{1.11} = 0.15 \text{ M}$$

Note that AgCl is much more soluble in $NH_3(aq)$ than in pure water, as shown in Figure 16.12.

┌ **PROBLEM 16.21** In an excess of $NH_3(aq)$, Cu^{2+} ion forms a deep blue complex ion, $Cu(NH_3)_4{}^{2+}$, which has a formation constant $K_f = 1.1 \times 10^{13}$. Calculate the concentration of Cu^{2+} in a solution prepared by adding 5.0×10^{-3} mol of $CuSO_4$ to 500 mL of 0.40 M NH_3.

┌ **PROBLEM 16.22** The "fixing" of photographic film involves dissolving unexposed silver bromide in a thiosulfate ($S_2O_3{}^{2-}$) solution:

$$AgBr(s) + 2\, S_2O_3{}^{2-}(aq) \rightleftarrows Ag(S_2O_3)_2{}^{3-}(aq) + Br^-(aq)$$

Using $K_{sp} = 5.4 \times 10^{-13}$ for AgBr and $K_f = 2.9 \times 10^{13}$ for $Ag(S_2O_3)_2{}^{3-}$, calculate the equilibrium constant K for the dissolution reaction, and calculate the molar solubility of AgBr in 0.10 M $Na_2S_2O_3$.

This photographic film was immersed in a thiosulfate solution to dissolve the unexposed silver bromide.

Amphoterism

We saw in Section 14.9 that amphoteric oxides, such as aluminum oxide, are soluble both in strongly acidic and in strongly basic solutions:

In acid: $Al_2O_3(s) + 6\, H_3O^+(aq) \rightleftarrows 2\, Al^{3+}(aq) + 9\, H_2O(l)$

In base: $Al_2O_3(s) + 2\, OH^-(aq) + 3\, H_2O(l) \rightleftarrows 2\, Al(OH)_4{}^-(aq)$

The corresponding hydroxides behave similarly (Figure 16.13):

In acid: $Al(OH)_3(s) + 3\, H_3O^+(aq) \rightleftarrows Al^{3+}(aq) + 6\, H_2O(l)$

In base: $Al(OH)_3(s) + OH^-(aq) \rightleftarrows Al(OH)_4{}^-(aq)$

Dissolution of $Al(OH)_3$ in excess base is just a special case of the effect of complex formation on solubility: $Al(OH)_3$ dissolves because excess OH^- ions convert it to the soluble complex ion $Al(OH)_4{}^-$ (aluminate ion). The effect of pH on the solubility of $Al(OH)_3$ is shown in Figure 16.14.

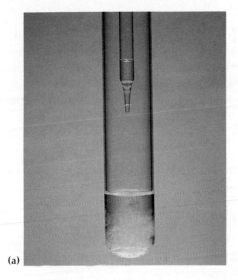
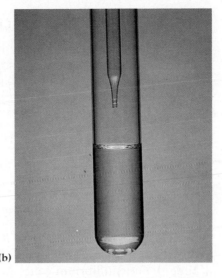

(a)   (b)

FIGURE 16.13 (a) Aluminum hydroxide, a gelatinous, white precipitate, forms on addition of aqueous NaOH to $Al^{3+}(aq)$. (b) The precipitate dissolves on addition of excess aqueous NaOH, yielding the colorless $Al(OH)_4^-$ ion. (The precipitate also dissolves in aqueous HCl, yielding the colorless Al^{3+} ion.)

Other examples of amphoteric hydroxides include $Zn(OH)_2$, $Cr(OH)_3$, and $Sn(OH)_2$, which react with excess OH^- ions to form the soluble complex ions $Zn(OH)_4^{2-}$ (zincate ion), $Cr(OH)_4^-$ (chromite ion), and $Sn(OH)_3^-$ (stannite ion), respectively. By contrast, basic hydroxides such as $Mn(OH)_2$, $Fe(OH)_2$, and $Fe(OH)_3$ dissolve in strong acid, but not in strong base.

The solution of Al_2O_3 in base is important in the production of aluminum metal from its ore. Aluminum is mined as bauxite ($Al_2O_3 \cdot xH_2O$), a hydrated oxide that is always contaminated with Fe_2O_3 and SiO_2. In the **Bayer process**, Al_2O_3 is purified by treating bauxite with hot aqueous NaOH. Al_2O_3 and SiO_2 dissolve, forming $Al(OH)_4^-$ and SiO_3^{2-} (silicate ion), but Fe_2O_3 remains undissolved. Subsequent filtration removes Fe_2O_3, and treatment of the filtrate with a weak acid (CO_2 in air) precipitates $Al(OH)_3$, leaving SiO_3^{2-} in solution. Removal of water from $Al(OH)_3$ by heating gives pure Al_2O_3, which is then converted to aluminum metal by electrolysis (Section 18.12).

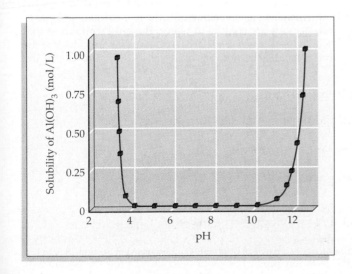

FIGURE 16.14 A plot of solubility versus pH shows that $Al(OH)_3$ is an amphoteric hydroxide. $Al(OH)_3$ is essentially insoluble between pH 4 and 10, but it dissolves both in strongly acidic and in strongly basic solution.

16.13 ►PRECIPITATION OF IONIC COMPOUNDS

A common problem in chemistry is to decide whether a precipitate of an ionic compound will form when solutions that contain the constituent ions are mixed. For example, will CaF_2 precipitate on mixing solutions of $CaCl_2$ and NaF? In other words, will the dissolution reaction proceed in the reverse direction, from right to left?

$$CaF_2(s) \rightleftarrows Ca^{2+}(aq) + 2\ F^-(aq)$$

The answer depends on the value of the **ion product (IP)**, a number defined by the expression

$$IP = [Ca^{2+}]_t[F^-]_t^2$$

IP is defined in the same way as K_{sp}, except that the concentrations in the expression for IP are initial concentrations (that is, arbitrary concentrations at time t), not equilibrium concentrations.[2]

If the value of IP is greater than K_{sp}, the solution is supersaturated with respect to CaF_2—a nonequilibrium situation. In that case, CaF_2 will precipitate, thus reducing the ion concentrations until IP equals K_{sp}. At that point, solubility equilibrium is reached, and the solution is saturated.

In general, we need only to calculate the value of IP and then compare it with K_{sp} to decide whether an ionic compound will precipitate. Three cases arise:

1. If $IP > K_{sp}$, the solution is supersaturated, and precipitation will occur.
2. If $IP = K_{sp}$, the solution is saturated, and equilibrium exists already.
3. If $IP < K_{sp}$, the solution is unsaturated, and precipitation will not occur.

EXAMPLE 16.13

Will a precipitate form when 0.150 L of 0.10 M $Pb(NO_3)_2$ and 0.100 L of 0.20 M NaCl are mixed?

SOLUTION   After the two solutions are mixed, the combined solution contains Pb^{2+}, NO_3^-, Na^+, and Cl^- ions, and it has a volume of 0.150 L + 0.100 L = 0.250 L. Because sodium salts and nitrate salts are soluble in water, the only compound that might precipitate is $PbCl_2$, which has $K_{sp} = 1.2 \times 10^{-5}$ (Appendix C). To calculate the value of the IP for $PbCl_2$, first calculate the number of moles of Pb^{2+} and Cl^- in the combined solution:

$$\text{Moles } Pb^{2+} = (0.150\ \text{L})(0.10\ \text{mol/L}) = 1.5 \times 10^{-2}\ \text{mol}$$

$$\text{Moles } Cl^- = (0.100\ \text{L})(0.20\ \text{mol/L}) = 2.0 \times 10^{-2}\ \text{mol}$$

Then convert moles to molar concentrations:

$$[Pb^{2+}] = \frac{1.5 \times 10^{-2}\ \text{mol}}{0.250\ \text{L}} = 6.0 \times 10^{-2}\ \text{M}$$

$$[Cl^-] = \frac{2.0 \times 10^{-2}\ \text{mol}}{0.250\ \text{L}} = 8.0 \times 10^{-2}\ \text{M}$$

---

[2] The ion product (IP) is actually a reaction quotient (Q_c), discussed in Section 13.5. In the present context, though, the term ion product is more descriptive, since the equilibrium constant expression isn't a quotient.

The ion product is

$$IP = [Pb^{2+}]_t[Cl^-]_t^2 = (6.0 \times 10^{-2})(8.0 \times 10^{-2})^2 = 3.8 \times 10^{-4}$$

Since $K_{sp} = 1.2 \times 10^{-5}$, the IP is greater than K_{sp}, and $PbCl_2$ will precipitate.

⌐ **PROBLEM 16.23** Will a precipitate form on mixing equal volumes of the following solutions?
(a) 3.0×10^{-3} M $BaCl_2$ and 2.0×10^{-3} M Na_2CO_3
(b) 1.0×10^{-5} M $Ba(NO_3)_2$ and 4.0×10^{-5} M Na_2CO_3

16.14 ➤SEPARATION OF IONS BY SELECTIVE PRECIPITATION

A convenient method for separating a mixture of ions in a solution is to add a reagent that will precipitate some of the ions but not others. The anions SO_4^{2-} and Cl^-, for example, can be separated by addition of a solution of $Ba(NO_3)_2$. Insoluble $BaSO_4$ precipitates, but Cl^- remains in solution because $BaCl_2$ is soluble. Similarly, the cations Ag^+ and Zn^{2+} can be separated by addition of dilute HCl. AgCl precipitates, but Zn^{2+} stays in solution because $ZnCl_2$ is soluble.

In the next section, we'll see how mixtures of metal cations M^{2+} can be separated into two groups by selective precipitation of metal sulfides, MS. For example, Pb^{2+}, Cu^{2+}, and Hg^{2+}, which form very insoluble sulfides, can be separated from Mn^{2+}, Fe^{2+}, Co^{2+}, Ni^{2+}, and Zn^{2+}, which form more soluble sulfides. The separation is carried out in an acidic solution and makes use of the following solubility equilibrium:

$$MS(s) + 2 H_3O^+(aq) \rightleftharpoons M^{2+}(aq) + H_2S(aq) + 2 H_2O(l)$$

The equilibrium constant for this reaction, called the *solubility product in acid*, is given the symbol K_{spa}:

$$K_{spa} = \frac{[M^{2+}][H_2S]}{[H_3O^+]^2}$$

The separation depends on adjusting the H_3O^+ concentration so that the reaction quotient Q_c exceeds K_{spa} for the very insoluble sulfides but not for the more soluble ones (Table 16.3).

In a typical experiment, the M^{2+} concentrations are about 0.01 M, and the H_3O^+ concentration is adjusted to about 0.3 M by adding HCl. The

When 0.150 L of 0.10 M $Pb(NO_3)_2$ and 0.100 L of 0.20 M NaCl are mixed, a white precipitate of $PbCl_2$ forms because the ion product is greater than K_{sp}.

TABLE 16.3	Solubility Products in Acid (K_{spa}) at 25°C for Metal Sulfides			
Metal Sulfide, MS	K_{spa}		**Metal Sulfide, MS**	K_{spa}
MnS	3×10^{10}		ZnS	3×10^{-2}
FeS	6×10^2		PbS	3×10^{-7}
CoS	3		CuS	6×10^{-16}
NiS	8×10^{-1}		HgS	2×10^{-32}

Adding H_2S to an acidic solution of Hg^{2+} and Ni^{2+} precipitates Hg^{2+} as black HgS but leaves green Ni^{2+} in solution.

solution is then saturated with H_2S gas, which gives an H_2S concentration of about 0.10 M. Substituting these concentrations into the equilibrium constant expression, we find that the reaction quotient Q_c is 1×10^{-2}:

$$Q_c = \frac{[M^{2+}]_t[H_2S]_t}{[H_3O^+]_t^2} = \frac{(0.01)(0.10)}{(0.3)^2} = 1 \times 10^{-2}$$

This value of Q_c exceeds K_{spa} for PbS, CuS, and HgS (Table 16.3) but does not exceed K_{spa} for MnS, FeS, CoS, NiS, or ZnS. As a result, PbS, CuS, and HgS precipitate under these acidic conditions, but Mn^{2+}, Fe^{2+}, Co^{2+}, Ni^{2+}, and Zn^{2+} remain in solution.

⌐ **PROBLEM 16.24** Show whether or not Cd^{2+} can be separated from Zn^{2+} by bubbling H_2S through a 0.3 M HCl solution that contains 0.005 M Cd^{2+} and 0.005 M Zn^{2+}. K_{spa} for CdS is 8×10^{-7}.

16.15 ▸QUALITATIVE ANALYSIS

Qualitative analysis is a procedure for identifying the ions present in an unknown solution. The ions are identified by specific chemical tests, but because one ion can interfere with the test for another, the ions must first be separated. In the traditional scheme of analysis for metal cations, some 20 cations are separated initially into five groups by selective precipitation (Figure 16.15).

FIGURE 16.15 Flow chart for separation of metal cations in qualitative analysis.

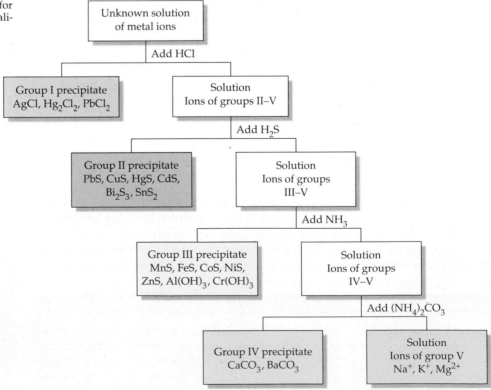

Group I—Ag⁺, Hg₂²⁺, and Pb²⁺   When aqueous HCl is added to the unknown solution, the cations of group I precipitate as insoluble chlorides—AgCl, Hg₂Cl₂, and PbCl₂. The cations of groups II–V, which form soluble chlorides, remain in solution.

Group II—Pb²⁺, Cu²⁺, Hg²⁺, Cd²⁺, Bi³⁺, and Sn⁴⁺   After the insoluble chlorides have been removed, the solution is treated with H_2S to precipitate the cations of group II as insoluble sulfides—PbS, CuS, HgS, CdS, Bi₂S₃, and SnS₂. Because the solution is strongly acidic at this point ($[H_3O^+] \approx 0.3$ M), only the most insoluble sulfides precipitate. The acid-insoluble sulfides are then removed from the solution.

Group III—Mn²⁺, Fe²⁺, Co²⁺, Ni²⁺, Zn²⁺, Al³⁺, and Cr³⁺   At this point aqueous NH_3 is added, neutralizing the acidic solution and giving an NH_4^+–NH_3 buffer that is slightly basic (pH ≈ 8). The decrease in $[H_3O^+]$ shifts the metal sulfide solubility equilibrium to the left, thus precipitating the 2+ cations of group III as insoluble sulfides—MnS, FeS, CoS, NiS, and ZnS. The 3+ cations precipitate from the basic solution, not as sulfides, but as insoluble hydroxides—$Al(OH)_3$ and $Cr(OH)_3$.

Group IV—Ca²⁺ and Ba²⁺   After the base-insoluble sulfides and the insoluble hydroxides have been removed, the solution is treated with $(NH_4)_2CO_3$ to precipitate the cations of group IV as insoluble carbonates—$CaCO_3$ and $BaCO_3$. $MgCO_3$ does not precipitate at this point because $[CO_3^{2-}]$ in the NH_4^+–NH_3 buffer is maintained at a low value.

Group V—Na⁺, K⁺, and Mg²⁺   The only ions remaining in solution are those whose chlorides, sulfides, and carbonates are soluble under conditions of the previous reactions. Magnesium ion is separated and identified by addition of a solution of $(NH_4)_2HPO_4$; if Mg^{2+} is present, a white precipitate of $Mg(NH_4)PO_4$ forms. The alkali metal ions are usually identified by the characteristic colors that they impart to a Bunsen flame (Figure 16.16).

(a)

(b)

FIGURE 16.16 Flame tests for **(a)** sodium (persistent yellow) and **(b)** potassium (fleeting violet).

When aqueous potassium chromate is added to a solution that contains Pb^{2+}, a yellow precipitate of $PbCrO_4$ forms.

Once the cations have been separated into groups, further separations and specific tests are carried out to determine the presence or absence of the ions in each group. In group I, for example, lead can be separated from silver and mercury by treating the precipitate with hot water. The more soluble PbCl₂ dissolves, but the less soluble AgCl and Hg₂Cl₂ do not. To test for lead, the solid chlorides are removed, and the solution is treated with a solution of K_2CrO_4. If Pb^{2+} is present, a yellow precipitate of $PbCrO_4$ forms.

Detailed procedures for separating and identifying all the ions can be found in general chemistry laboratory manuals. Although modern methods of metal-ion analysis employ sophisticated analytical instruments, qualitative analysis is still included in most general chemistry laboratory courses because it is an excellent vehicle for developing laboratory skills and for learning about acid–base, solubility, and complex-ion equilibria.

interlude—ANALYZING PROTEINS BY ELECTROPHORESIS

It has been estimated that there may be as many as 100,000 different proteins in the human body. Many of these proteins act as *enzymes*, biological catalysts that regulate the tens of thousands of different reactions going on in your body each minute. Other proteins protect against disease, act as transport agents to move ions and small molecules through the body, or provide mechanical support for muscle, skin, and bone. Regardless of the exact details, virtually all biological processes involve proteins in some way.

This muscle fiber is made of myosin, one of many proteins in the body.

As you might imagine, a biochemist faced with the need to purify one specific protein out of a mixture of thousands of substances with similar structures has a very difficult task. Using some knowledge of acid–base chemistry, however, makes the job easier. Proteins have various acidic and basic sites spread throughout their structures. Each of these acidic or basic sites in a given protein has its own pK_a or pK_b, and each site is therefore either charged or uncharged at a given pH. A protein with an abundance of acidic sites (AH), for example, will be uncharged at a low pH, where dissociation is suppressed, but will gain successively more negative charges as the pH is raised and dissociations begin to occur. Conversely, a protein with an abundance of basic sites (B) will have many positive charges at a low pH, where the basic sites are protonated, but will become successively less charged as the pH is raised and the basic sites are deprotonated. Different proteins have different numbers of acidic and basic sites and therefore have different overall charges at a given pH. By taking advantage of these different overall charges, biochemists can separate protein molecules from one another.

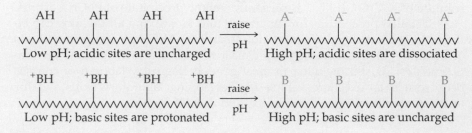

When a protein is placed in an electric field between two electrodes, a positively charged protein migrates toward the negative electrode, and a negatively charged protein migrates toward the positive electrode (Figure 16.17). The amount of this movement, called **electrophoresis**, varies with the size and shape of the protein, with the strength of the electric field, and with the overall charge on the protein. The overall charge, in turn, is determined by the number of acidic and basic sites on the protein and by the pH of the medium through which the protein moves.

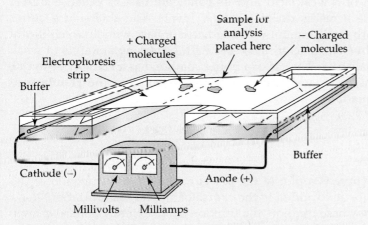

FIGURE 16.17 An electrophoresis apparatus for protein separation. A mixture of proteins is buffered to a given pH and placed in an electric field. Different proteins migrate toward one or the other of the electrodes at a rate dependent on the protein's overall charge.

Electrophoresis is routinely used both in research laboratories for isolating new proteins and in clinical laboratories for determining protein concentrations in blood serum. In the protein separation shown in Figure 16.18, blood serum is separated by electrophoresis into a pattern of five or six different protein fractions—albumin, two α-globulins, one or two β-globulins, and the γ-globulins. An abnormally high or low reading in each region indicates different clinical conditions.

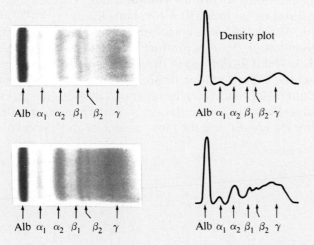

FIGURE 16.18 **(a)** A normal electrophoresis pattern of blood serum; **(b)** an abnormal pattern, showing elevated γ-globulin and indicating the possibility of liver disease, collagen disorder, or infection.

Neutralization reactions involving a strong acid and/or a strong base have very large equilibrium constants (K_n) and proceed nearly 100% to completion. Weak acid–weak base neutralizations tend not to go to completion.

The shift in the position of an equilibrium on addition of a substance that provides an ion in common with one of the ions already involved in the equilibrium is called the **common-ion effect**. An example is the decrease in percent dissociation of a weak acid on addition of its conjugate base.

A solution of a weak acid and its conjugate base is called a **buffer solution** because it resists drastic changes in pH. The ability of a buffer solution to absorb small amounts of added acid or base without a significant change in pH (**buffer capacity**) increases with increasing amounts of weak acid and conjugate base. The pH of a buffer solution has a value close to pK_a ($-\log K_a$) of the weak acid and can be calculated from the **Henderson-Hasselbalch equation**:

$$pH = pK_a + \log \frac{[\text{conjugate base}]}{[\text{weak acid}]}$$

A **pH titration curve** is a plot of the pH of a solution as a function of the volume of base (or acid) added in the course of an acid–base titration. For a strong acid–strong base titration, the titration curve exhibits a sharp change in pH in the region of the **equivalence point**, the point at which stoichiometrically equivalent amounts of acid and base have been mixed together. For weak acid–strong base and weak base–strong acid titrations, the titration curves display a relatively flat region midway to the equivalence point, a smaller change in pH in the region of the equivalence point, and a pH at the equivalence point that is not equal to 7.00.

The **solubility product**, K_{sp}, for an ionic compound is the equilibrium constant for dissolution of the compound in water. The solubility of the compound and K_{sp} are related by the equilibrium equation for the dissolution reaction. The solubility of an ionic compound is (1) suppressed by the presence of a common ion in the solution, (2) increased by decreasing the pH if the compound contains a basic anion, such as OH^-, S^{2-}, or CO_3^{2-}, and (3) increased by the presence of a Lewis base, such as NH_3, CN^-, or Cl^-, that can bond to the metal cation to form a **complex ion**. The stability of a complex ion is measured by its **formation constant**, K_f.

When solutions of soluble ionic compounds are mixed, an insoluble compound will precipitate if the **ion product (IP)** for the insoluble compound exceeds its K_{sp}. The IP is defined in the same way as K_{sp}, except that the concentrations in the expression for IP are not necessarily equilibrium concentrations. Certain metal cations can be separated by selective precipitation of metal sulfides. Selective precipitation is important in **qualitative analysis**, a procedure for identifying the ions present in an unknown solution.

1. The following pictures represent solutions that contain a weak acid HA and/or its sodium salt NaA. Na^+ ions and solvent water molecules have been omitted for clarity.

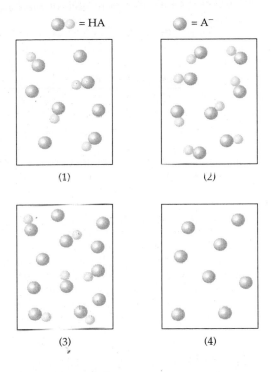

= HA   = A⁻

(1)   (2)

(3)   (4)

(a) Which of the solutions are buffer solutions?
(b) Which solution has the greatest buffer capacity?

2. The following pictures represent solutions that contain a weak acid HA ($pK_a - 6.0$) and its sodium salt NaA. Na^+ ions and solvent water molecules have been omitted for clarity.

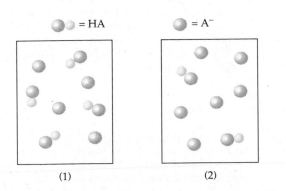

= HA   = A⁻

(1)   (2)

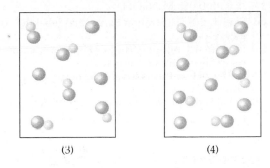

(3)   (4)

(a) Which solution has the highest pH? Which has the lowest pH?
(b) Draw a picture that represents the equilibrium state of solution (1) after addition of two H_3O^+ ions
(c) Draw a picture that represents the equilibrium state of solution (1) after addition of two OH^- ions.

3. The strong acid HA is mixed with an equal molar amount of aqueous NaOH. Which of the following pictures represents the equilibrium state of the solution? Na^+ ions and solvent water molecules have been omitted for clarity.

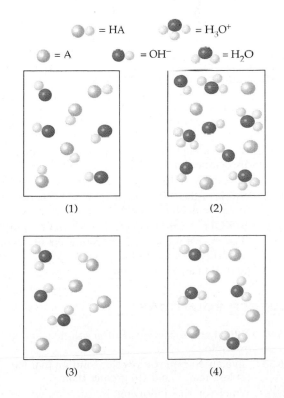

= HA   = H_3O^+
= A   = OH^- = H_2O

(1)   (2)

(3)   (4)

4. Consider the titration of 50.0 mL of 0.010 M HA ($K_a = 10^{-4}$) with 0.010 M NaOH.

    (a) Sketch the pH titration curve, and label the equivalence point.

    (b) How many mL of 0.010 M NaOH are required to reach the equivalence point?

    (c) Is the pH at the equivalence point greater than, equal to, or less than 7.00?

    (d) What is the pH exactly halfway to the equivalence point?

5. Consider saturated solutions of the slightly soluble salts AgBr and $BaCO_3$.

    (a) Is the solubility of AgBr increased, decreased, or unaffected by addition of each of the following reagents?
    (a) HBr   (b) HNO_3 (c) $AgNO_3$ (d) NH_3

    (b) Is the solubility of $BaCO_3$ increased, decreased, or unaffected by addition of each of the following reagents?
    (a) NaOH   (b) HNO_3 (c) $Ba(NO_3)_2$ (d) $NaCO_3$

ADDITIONAL PROBLEMS

Problems 16.1–16.24 appear within the chapter.

NEUTRALIZATION REACTIONS

16.25 List the four types of neutralization reactions, and give an example of each.

16.26 Write a balanced net ionic equation for neutralization of equal molar amounts of the following reagents, and indicate whether the pH after neutralization is greater than, equal to, or less than 7.00.

    (a) HI and LiOH

    (b) HOCl and $Ba(OH)_2$

    (c) HNO_3 and aniline ($C_6H_5NH_2$)

    (d) benzoic acid (C_6H_5COOH) and KOH

16.27 Calculate the equilibrium constant, K_n, for each neutralization reaction in Problem 16.26, and arrange the four reactions in order of increasing

tendency to proceed to completion. See Appendix C for values of K_a and K_b.

16.28 Which of the following mixtures has the highest pH?

    (a) 0.1 M HF and 0.1 M NaOH

    (b) 0.1 M HCl and 0.1 M NaOH

    (c) 0.1 M $HClO_4$ and 0.1 M NaOH

16.29 Phenol (C_6H_5OH, $K_a = 1.3 \times 10^{-10}$) is a weak acid used in mouthwashes, and pyridine (C_5H_5N, $K_b = 1.8 \times 10^{-9}$) is a weak base. Calculate the value of K_n for neutralization of phenol by pyridine. Does the neutralization reaction proceed very far toward completion?

THE COMMON-ION EFFECT

16.30 Tell what is meant by the common-ion effect, and give an example. Account for the common-ion effect in terms of Le Châtelier's principle.

16.31 Which of the following reagents affects the percent dissociation of HNO_2?

    (a) $NaNO_2$ (b) NaCl   (c) HCl   (d) $Ba(NO_2)_2$

16.32 Which of the following reagents affects the pH of an aqueous NH_3 solution?

    (a) KOH   (b) NH_4NO_3 (c) NH_4Br (d) KBr

16.33 Does the pH increase, decrease, or remain the same on addition of each of the following?

    (a) LiF to an HF solution

    (b) KI to an HI solution

    (c) $NaClO_4$ to an NaOH solution

    (d) NH_4Cl to an NH_3 solution

16.34 Calculate the pH of a solution that is 0.25 M in HF and 0.10 M in NaF.

16.35 Calculate the percent dissociation of 0.10 M hydrazoic acid (HN_3, $K_a = 1.9 \times 10^{-5}$). Recalculate the percent dissociation of 0.10 M HN_3 in the presence of 0.10 M HCl, and explain the change.

16.36 Calculate the pH of 100 mL of 0.30 M NH_3 after addition of 4.0 g of NH_4NO_3. Assume that the volume remains constant.

BUFFER SOLUTIONS

16.37 Tell what is meant by a buffer solution, and give an example.

16.38 Give an example of a buffer solution that has a pH (a) less than 7 and (b) greater than 7.

16.39 Which of the following gives a buffer solution when equal volumes of the two solutions are mixed?

weak acid + conjugate base

    (a) 0.1 M HF and 0.1 M NaF

    (b) 0.1 M HF and 0.1 M NaOH

    (c) 0.2 M HF and 0.1 M NaOH

    (d) 0.1 M HCl and 0.2 M NaF

16.40 Which of the following solutions has the greater buffer capacity: (a) 0.3 M HNO_2–0.3 M $NaNO_2$ or (b) 0.1 M HNO_2–0.1 M $NaNO_2$? Explain.

16.41 The following reaction is important in maintaining the pH of blood at a nearly constant value of about 7.4: $H_2CO_3 + H_2O \rightleftharpoons H_3O^+ + HCO_3^-$. What happens to the position of this equilibrium and the pH when blood absorbs added acid or base?

16.42 Explain how the $H_2PO_4^- - HPO_4^{2-}$ buffer system can help to maintain the pH of intracellular fluid at a value close to 7.4.

16.43 Calculate the pH of a buffer solution that is 0.20 M in HCN and 0.12 M in NaCN.

16.44 Calculate the pH of a buffer solution prepared by dissolving 4.2 g of $NaHCO_3$ and 5.3 g of Na_2CO_3 in 0.20 L of water. Will the pH change if the solution is diluted by a factor of 2? Explain.

16.45 Calculate the pH of 0.500 L of a 0.200 M NH_4Cl–0.200 M NH_3 buffer before and after addition of **(a)** 0.005 mol of NaOH and **(b)** 0.020 mol of HCl. Assume that the volume remains constant.

16.46 What is the pK_a value for each of the following acids? (K_a values are listed in Appendix C.)
(a) boric (H_3BO_3)
(b) formic (HCOOH)
(c) hypochlorous (HOCl)
How does pK_a vary as acid strength increases?

16.47 What is the value of K_a for an acid that has **(a)** pK_a = 5.00 and **(b)** pK_a = 8.70? Which of the two acids is weaker?

16.48 Use the Henderson-Hasselbalch equation to calculate the pH of a buffer solution that is 0.25 M in formic acid (HCOOH) and 0.50 M in sodium formate (HCOONa).

16.49 What is the ratio of H_2CO_3 to HCO_3^- in normal blood having a pH of 7.40?

16.50 In what ratio should you mix 1.0 M solutions of NH_4Cl and NH_3 to produce a buffer solution having pH 9.80?

16.51 Give a recipe for preparing a CH_3COOH–CH_3COONa buffer solution that has pH 4.44.

16.52 You need a buffer solution that has pH 7.00. Which of the following buffer systems should you choose? Explain.
(a) H_3PO_4–$H_2PO_4^-$
(b) $H_2PO_4^-$–HPO_4^{2-}
(c) HPO_4^{2-}–PO_4^{3-}

PH TITRATION CURVES

16.53 Consider the titration of 60.0 mL of 0.150 M HNO_3 with 0.450 M NaOH.
(a) How many mmol of HNO_3 are present at the start of the titration?
(b) How many mL of NaOH are required to reach the equivalence point?
(c) What is the pH at the equivalence point?
(d) Sketch the general shape of the pH titration curve.

16.54 A 60.0 mL sample of a monoprotic acid is titrated with 0.150 M NaOH. If 20.0 mL of base is required to reach the equivalence point, what is the concentration of the acid?

16.55 A 50.0 mL sample of 0.120 M HBr is titrated with 0.240 M NaOH. Calculate the pH after addition of the following volumes of base, and construct a plot of pH versus mL of NaOH added.
(a) 0.0 mL   **(b)** 20.0 mL   **(c)** 24.9 mL
(d) 25.0 mL   **(e)** 25.1 mL   **(f)** 40.0 mL

16.56 Consider the titration of 40.0 mL of 0.250 M HF with 0.200 M NaOH. How many mL of base are required to reach the equivalence point? Calculate the pH at each of the following points.
(a) after addition of 10.0 mL of base
(b) halfway to the equivalence point
(c) at the equivalence point
(d) after addition of 80.0 mL of base

16.57 On the same graph, sketch pH titration curves for titration of (1) a strong acid with a strong base and

(2) a weak acid with a strong base. How do the two curves differ with respect to
(a) the initial pH
(b) the pH in the region between the start of the titration and the equivalence point
(c) the pH at the equivalence point
(d) the pH beyond the equivalence point
(e) the volume of base required to reach the equivalence point

16.58 A 100.0 mL sample of 0.100 M methylamine (CH_3NH_2, $K_b = 3.7 \times 10^{-4}$) is titrated with 0.250 M HNO_3. Calculate the pH after addition of each of the following volumes of acid.
(a) 0.0 mL   **(b)** 20.0 mL
(c) 40.0 mL   **(d)** 60.0 mL

16.59 Consider the titration of 50.0 mL of a 0.100 M solution of the protonated form of alanine (H_2A^+; $K_{a1} = 4.6 \times 10^{-3}$, $K_{a2} = 2.0 \times 10^{-10}$) with 0.100 M NaOH. Calculate the pH after addition of each of the following volumes of base.
(a) 10.0 mL   **(b)** 25.0 mL
(c) 50.0 mL   **(d)** 75.0 mL

16.60 What is the pH at the equivalence point for titration of 0.10 M solutions of the following acids and bases, and which of the indicators in Figure 15.3 would be suitable for each titration?
(a) HNO_2 and NaOH
(b) HI and NaOH
(c) CH_3NH_2 (methylamine) and HCl
(d) $NaHSO_3$ and NaOH

16.61 Distinguish between the solubility of an ionic compound and its solubility product.

16.62 For each of the following compounds, write a balanced net ionic equation for dissolution of the compound in water, and write the equilibrium expression for K_{sp}.

(a) Ag_2CO_3 (b) $PbCrO_4$
(c) $Al(OH)_3$ (d) Hg_2Cl_2

16.63 For each of the following, write the equilibrium expression for K_{sp}.

(a) $Ca(OH)_2$ (b) Ag_3PO_4
(c) $BaCO_3$ (d) $Ca_5(PO_4)_3OH$

16.64 A particular saturated solution of PbI_2 has $[Pb^{2+}]$ = 5.0×10^{-3} M and $[I^-]$ = 1.3×10^{-3} M.

(a) What is the value of K_{sp} for PbI_2?
(b) What is $[I^-]$ in a saturated solution of PbI_2 that has $[Pb^{2+}]$ = 2.5×10^{-4} M?

(c) What is $[Pb^{2+}]$ in a saturated solution that has $[I^-]$ = 2.5×10^{-4} M?

16.65 If a saturated solution prepared by dissolving Ag_2CO_3 in water has $[Ag^+]$ = 2.56×10^{-4} M, what is the value of K_{sp} for Ag_2CO_3?

16.66 Use the following solubility data to calculate a value of K_{sp} for each compound.

(a) $CdCO_3$, 2.5×10^{-6} M
(b) $Ca(OH)_2$, 1.06×10^{-2} M
(c) $PbBr_2$, 4.34 g/L
(d) $BaCrO_4$, 2.8×10^{-3} g/L

16.67 Use the values of K_{sp} in Appendix C to calculate the molar solubility of the following compounds.

(a) $CuCO_3$ (b) Ag_2SO_4 (c) $Cr(OH)_3$

16.68 Use the values of K_{sp} in Appendix C to calculate the solubility of the following compounds in g/L.

(a) Ag_2CrO_4 (b) $CuBr$ (c) $Cu_3(PO_4)_2$

FACTORS THAT AFFECT SOLUBILITY

16.69 Use Le Châtelier's principle to explain the following changes in the solubility of Ag_2CO_3 in water.

(a) decrease on addition of $AgNO_3$
(b) increase on addition of HNO_3
(c) decrease on addition of Na_2CO_3
(d) increase on addition of NH_3

16.70 Calculate the molar solubility of AgBr in (a) pure water and (b) 0.050 M HBr.

16.71 Calculate the molar solubility of PbI_2 in (a) 0.20 M $Pb(NO_3)_2$ and (b) 0.20 M NaI.

16.72 What is the molar solubility of $Mg(OH)_2$ in a solution that has a pH of (a) 12.00 and (b) 9.00?

16.73 Which of the following compounds are more soluble in acidic solution than in pure water?

(a) AgBr   (b) $CaCO_3$ (c) $Ni(OH)_2$ (d) $Ca_3(PO_4)_2$

16.74 Write a balanced net ionic equation for dissolution of the following compounds in an acidic solution.

(a) MnS   (b) $Fe(OH)_3$ (c) AgCl   (d) $BaCO_3$

16.75 Silver ion reacts with excess CN^- to form a colorless complex ion, $Ag(CN)_2^-$, which has a formation constant $K_f = 1 \times 10^{21}$. Calculate the concentration of Ag^+ in a solution prepared by mixing equal volumes of 2.0×10^{-3} M $AgNO_3$ and 0.20 M NaCN.

16.76 Write a balanced net ionic equation for each of the following dissolution reactions, and use the appropriate K_{sp} and K_f values to calculate the equilibrium constant for each.

(a) AgI in aqueous NaCN to form $Ag(CN)_2^-$ ($K_f = 1 \times 10^{21}$)
(b) $Al(OH)_3$ in aqueous NaOH to form $Al(OH)_4^-$ ($K_f = 2.1 \times 10^{34}$)
(c) $Zn(OH)_2$ in aqueous NH_3 to form $Zn(NH_3)_4^{2+}$ ($K_f = 2.9 \times 10^9$).

16.77 Calculate the molar solubility of AgI in (a) pure water and (b) 0.10 M NaCN. K_f for $Ag(CN)_2^-$ is 1×10^{21}.

PRECIPITATION; QUALITATIVE ANALYSIS

16.78 Will a precipitate of $BaSO_4$ form when 100 mL of 4.0×10^{-3} M $BaCl_2$ and 300 mL of 6.0×10^{-4} M Na_2SO_4 are mixed? Explain.

16.79 Will a precipitate of $PbCl_2$ form on mixing equal volumes of 0.010 M $Pb(NO_3)_2$ and 0.010 M HCl? Explain. What minimum Cl^- concentration is required to begin precipitation of $PbCl_2$ from 5.0×10^{-3} M $Pb(NO_3)_2$?

16.80 What compound, if any, will precipitate when 80 mL of 1.0×10^{-5} M $Ba(OH)_2$ is added to 20 mL of 1.0×10^{-5} M $Fe_2(SO_4)_3$?

16.81 "Hard" water contains alkaline-earth cations such as Ca^{2+}, which reacts with CO_3^{2-} to form insoluble deposits of $CaCO_3$. Will a precipitate of $CaCO_3$ form if a 250 mL sample of hard water having $[Ca^{2+}]$ = 8.0×10^{-4} M is treated with (a) 0.10 mL

of 2.0×10^{-3} M Na_2CO_3 or (b) 10 mg of solid Na_2CO_3?

16.82 The pH of a sample of hard water having $[Mg^{2+}]$ = 2.5×10^{-4} M is adjusted to pH 10.80. Will $Mg(OH)_2$ precipitate?

16.83 Can Fe^{2+} be separated from Sn^{2+} by bubbling H_2S through a 0.3 M HCl solution that contains 0.01 M Fe^{2+} and 0.01 M Sn^{2+}? A saturated solution of H_2S has $[H_2S] \approx 0.10$ M. Values of K_{spa} are 6×10^2 for FeS and 1×10^{-5} for SnS.

16.84 Will CoS precipitate in a solution that is 0.10 M in $Co(NO_3)_2$, 0.5 M in HCl, and 0.10 M in H_2S? Will CoS precipitate if the pH of the solution is adjusted to pH 8 with an NH_4^+–NH_3 buffer? $K_{spa} = 3$ for CoS.

16.85 In qualitative analysis, Al^{3+} and Mg^{2+} are separated in an NH_4^+–NH_3 buffer having pH ≈ 8. Assuming cation concentrations of 0.01 M, show why $Al(OH)_3$ precipitates but $Mg(OH)_2$ doesn't.

16.86 Using the qualitative analysis flow chart in Figure 16.15, tell how you could separate the following pairs of ions.

(a) Ag^+ and Cu^{2+} (b) Na^+ and Ca^{2+}
(c) Mg^{2+} and Mn^{2+} (d) K^+ and Cr^{3+}

16.87 Give a method for separating the following pairs of ions by addition of no more than two reagents.

(a) Hg_2^{2+} and Co^{2+} (b) Na^+ and Mg^{2+}
(c) Fe^{2+} and Hg^{2+} (d) Ba^{2+} and Pb^{2+}

16.88 Assume that you have three white solids: NaCl, KCl, and $MgCl_2$. What tests could you do to tell which is which?

GENERAL PROBLEMS

16.89 Which of the following pairs of reagents, when mixed in any proportion you wish, can be used to prepare a buffer solution?

(a) NaCN and HCN
(b) NaCN and NaOH
(c) HCl and NaCN
(d) HCl and NaOH
(e) HCN and NaOH

16.90 Give an example of the common-ion effect on a solubility equilibrium. Does the common-ion effect change the solubility or the solubility product? Explain.

16.91 Write a balanced net ionic equation for each of the following reactions, and tell which reaction proceeds farthest to completion.

(a) $HNO_3(aq) + NH_3(aq) \longrightarrow$
(b) $HNO_3(aq) + CH_3NH_2(aq) \longrightarrow$
(c) $HNO_3(aq) + KOH(aq) \longrightarrow$

16.92 Does the pH increase, decrease, or remain unchanged when each of the following reagents is added to a solution of hypochlorous acid?

(a) KCl   (b) HCl   (c) NaOCl   (d) NaOH

16.93 Which of the following pairs gives a buffer solution when equal volumes of the two solutions are mixed?

(a) 0.1 M $NaHCO_3$ and 0.1 M H_2CO_3
(b) 0.1 M $NaHCO_3$ and 0.1 M Na_2CO_3
(c) 0.1 M $NaHCO_3$ and 0.1 M HCl
(d) 0.2 M $NaHCO_3$ and 0.1 M NaOH

16.94 Make a rough plot of pH versus mL of acid added for the titration of 50 mL of 1.0 M NaOH with 1.0 M HCl. Indicate the pH at the following points, and tell how many mL of acid are required to reach the equivalence point.

(a) at the start of the titration
(b) at the equivalence point
(c) after addition of a large excess of acid

16.95 A 0.10 L sample of a solution that is 0.30 M in HF and 0.30 M in NaF is diluted with 0.40 L of water. How does the added water affect the pH? Explain.

16.96 How many mL of 3.0 M NH_4Cl must be added to 250 mL of 0.20 M NH_3 to obtain a buffer solution having pH 9.40?

16.97 A 300 mL sample of hydrochloric acid having pH 2.00 is titrated with 0.100 M NaOH. How many mL of NaOH is required to reach the equivalence point?

16.98 What is the pH of a saturated solution of $Ca(OH)_2$ in water?

16.99 Use the Henderson-Hasselbalch equation to calculate the pH of a solution prepared by mixing equal volumes of 0.40 M NH_3 and 0.60 M NH_4Cl.

16.100 A saturated solution of $Mg(OH)_2$ in water has pH 10.35. Calculate K_{sp} for $Mg(OH)_2$.

16.101 A 10 mL volume of 0.44 M KOH is added to 30 mL of 0.20 M $HClO_4$. What is the pH of the solution?

16.102 The mercurous ion, Hg_2^{2+}, reacts with Cl^- to give a white precipitate of Hg_2Cl_2. How much Hg_2^{2+} remains in solution after addition of 1 drop (about 0.05 mL) of 6 M HCl to 1.0 mL of 0.010 M $Hg_2(NO_3)_2$? Express your answer in (a) mol/L and (b) g/L.

16.103 Calculate the molar solubility of $Fe(OH)_3$ in a buffer solution that is 0.10 M in NH_4Cl and 0.10 M in NH_3.

16.104 Calculate the concentrations of HF and F^- and the pH in a solution prepared by mixing 0.100 L of 0.600 M HCl and 0.500 L of 0.200 M NaF.

16.105 Dissolution of 5.0×10^{-3} mol of $Cr(OH)_3$ in 1.0 L of 1.0 M NaOH gives a solution of the complex ion $Cr(OH)_4^-$ ($K_f = 8 \times 10^{29}$). What fraction of the chromium in such a solution is present as uncomplexed Cr^{3+}?

16.106 In qualitative analysis, Ag^+, Hg_2^{2+}, and Pb^{2+} are separated from other cations by addition of HCl. Calculate the concentration of Cl^- required to just begin precipitation of (a) AgCl, (b) Hg_2Cl_2, and (c) $PbCl_2$ in a solution having metal ion concentrations of 0.030 M. What fraction of the Pb^{2+} remains in solution when the Ag^+ just begins to precipitate?

16.107 Calculate the concentrations of NH_4^+ and NH_3 and the pH in a solution prepared by mixing 20 g of NaOH and 500 mL of 1.5 M NH_4Cl. Assume that the volume remains constant.

16.108 Calculate the molar solubility of MnS in a 0.30 M NH_4Cl–0.50 M NH_3 buffer solution that is saturated with H_2S ([H_2S] ≈ 0.10 M). What is the solubility of MnS in g/L? K_{spa} for MnS is 3×10^{10}.

663

chapter 17

THERMODYNAMICS: ENTROPY, FREE ENERGY, AND EQUILIBRIUM

W hat factors determine the direction and extent of a chemical reaction? Some reactions, such as the combustion of hydrocarbon fuels, go almost to completion. Other reactions, such as the combination of gold and oxygen, hardly occur at all. Still other reactions—for example, the industrial synthesis of ammonia from N_2 and H_2—result in an equilibrium mixture that contains appreciable amounts of both reactants and products.

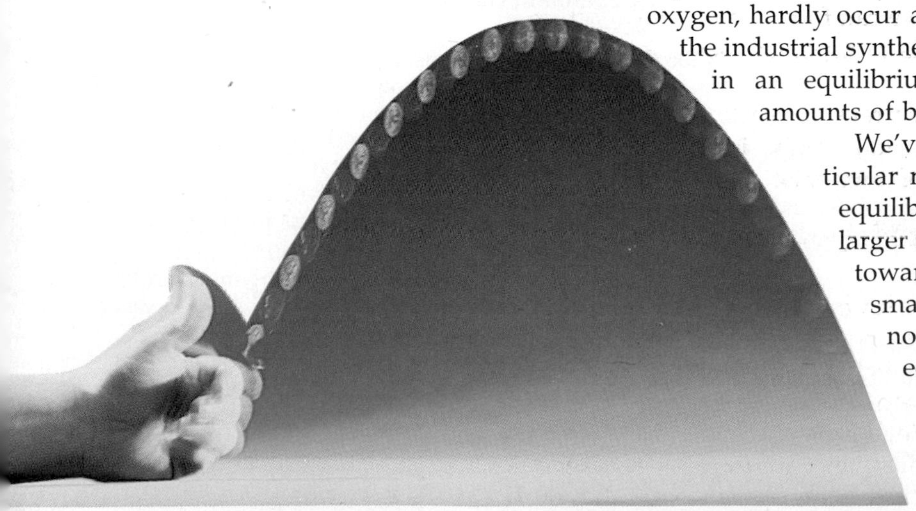

We've already seen that the extent of any particular reaction is described by the value of its equilibrium constant K: A value of K much larger than 1 indicates that the reaction goes far toward completion, and a value of K much smaller than 1 means that the reaction does not proceed very far before reaching an equilibrium state. But what determines the value of the equilibrium constant, and can we predict its value without measuring it? Put another way, what fundamental properties of nature determine the direction and extent of a particular chemical reaction? For answers to these questions, we turn to **thermodynamics**, the area of science that deals with the interconversion of heat and other forms of energy.

The flip of a coin. Heads and tails result with equal probability.

17.1 ►SPONTANEOUS PROCESSES

We have defined a **spontaneous process** as one that proceeds on its own without any external influence (Section 8.12). The reverse of a spontaneous process is always nonspontaneous and takes place only in the presence of some continuous external influence. Consider, for example, the expansion of an ideal gas into a vacuum. When the stopcock in the apparatus shown in Figure 17.1 is opened, the gas in bulb A expands spontaneously into the evacuated bulb B until the gas pressure in the two bulbs is the same. The reverse process, migration of all the gas molecules into one bulb, does not occur spontaneously. To compress a gas from a larger to a smaller volume, we have to push on the gas with a piston.

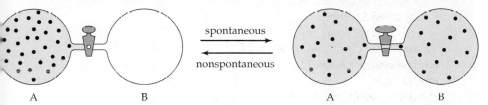

FIGURE 17.1 When the stopcock is opened, an ideal gas in bulb A expands spontaneously into evacuated bulb B to fill all the available volume. The reverse process, compression of the gas, is nonspontaneous.

As a second example, consider the combination of hydrogen and oxygen in the presence of a platinum catalyst:

$$2\ H_2(g) + O_2(g) \xrightarrow{\text{catalyst}} 2\ H_2O(l)$$

The forward reaction occurs spontaneously, but the reverse reaction, decomposition of water into its elements, does not occur no matter how long we wait. We'll see in the next chapter that we can force the reverse reaction to occur by electrolysis, but the process is nonspontaneous and requires a continuous input of electrical energy.

In general, whether the forward or reverse reaction is spontaneous depends on the temperature, pressure, and composition of the reaction mixture. Consider the Haber synthesis of ammonia:

$$N_2(g) + 3\ H_2(g) \xrightarrow{\text{catalyst}} 2\ NH_3(g)$$

A mixture of gaseous N_2, H_2, and NH_3, each at a partial pressure of 1 atm, reacts spontaneously at 300 K to convert some of the N_2 and H_2 to NH_3. We can predict the direction of spontaneous reaction from the relative values of the equilibrium constant K and the reaction quotient Q (Section 13.5). Since $K_p = 4.4 \times 10^5$ at 300 K and $Q_p = 1$ for partial pressures of 1 atm, the reaction will proceed in the forward direction because Q is less than K. Under these conditions the reverse reaction is nonspontaneous. At 700 K, however, $K_p = 8.8 \times 10^{-5}$, and the reverse reaction is spontaneous because Q is greater than K.

A spontaneous reaction always moves a system toward equilibrium. By contrast, a nonspontaneous reaction moves the composition of a mixture away from the equilibrium composition. Whereas a spontaneous reaction occurs naturally, a nonspontaneous reaction requires a continuous external influence. Remember that the word "spontaneous" doesn't mean the same thing as "fast." A spontaneous reaction can be slow—for example, the gradual rusting of iron metal. Thermodynamics tells us where a reaction is headed, but it says nothing about how long it takes to get there. As discussed in Section 12.9, the rate at which equilibrium is achieved depends on chemical kinetics, especially on the height of the activation energy barrier between the reactants and products (Figure 17.2).

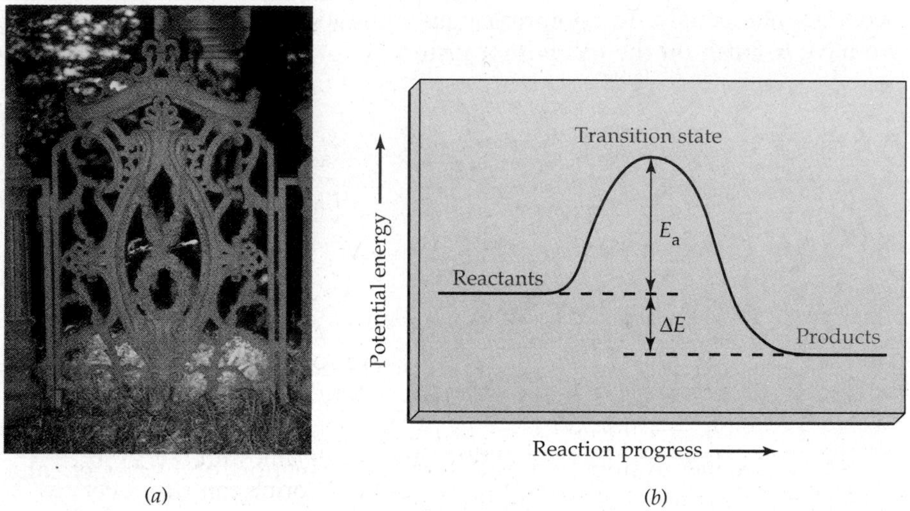

FIGURE 17.2 **(a)** The rusting of this iron gate is a spontaneous reaction, but it occurs slowly. **(b)** A spontaneous reaction occurs slowly if it has a high activation energy (Section 12.9).

PROBLEM 17.1 Which of the following processes are spontaneous, and which are nonspontaneous?

(a) diffusion of perfume molecules from one side of a room to the other
(b) heat flow from a cold object to a hot object
(c) decomposition of rust ($Fe_2O_3 \cdot H_2O$) to iron metal, oxygen, and water
(d) decomposition of solid $CaCO_3$ to solid CaO and gaseous CO_2 at 25°C and 1 atm pressure ($K_p = 1.4 \times 10^{-23}$)

17.2 ENTHALPY, ENTROPY, AND SPONTANEOUS PROCESSES: A BRIEF REVIEW

Let's look more closely at spontaneous processes and at the thermodynamic driving forces that cause them to occur. We saw in Chapter 8 that most spontaneous chemical reactions are accompanied by conversion of potential energy to heat. For example, when methane (natural gas) burns in air, the potential energy stored in the chemical bonds of CH_4 and O_2 is partly converted to heat, which flows from the system (reactants plus products) to the surroundings.

$$CH_4(g) + 2\,O_2(g) \rightarrow CO_2(g) + 2\,H_2O(l) \qquad \Delta H° = -890.3\text{ kJ}$$

Because heat is lost by the system, the reaction is exothermic and the standard enthalpy of reaction is negative ($\Delta H° = -890.3$ kJ). Of course, the total energy is conserved, and all the 890.3 kJ of energy lost by the system shows up as heat gained by the surroundings.

Because spontaneous reactions so often give off heat, the nineteenth-century French scientist Marcellin Berthelot proposed that spontaneous chemical or physical changes are *always* exothermic. But Berthelot's proposal can't be correct. Common sense tells you, for instance, that ice spontaneously absorbs heat and melts at temperatures above 0°C. Similarly, liquid water absorbs heat and spontaneously vaporizes at temperatures above 100°C. As further examples, gaseous N_2O_4 absorbs heat when it decomposes to NO_2 at 400 K, and table salt absorbs heat when it dissolves in water at room temperature:

The combustion of CH_4 in air is a spontaneous, exothermic reaction.

$$H_2O(s) \rightarrow H_2O(l) \qquad \Delta H_{fusion} = +6.01 \text{ kJ}$$

$$H_2O(l) \rightarrow H_2O(g) \qquad \Delta H_{vap} = +40.7 \text{ kJ}$$

$$N_2O_4(g) \rightarrow 2\ NO_2(g) \qquad \Delta H° = +57.1 \text{ kJ}$$

$$NaCl(s) \rightarrow Na^+(aq) + Cl^-(aq) \qquad \Delta H° = +3.88 \text{ kJ}$$

All these processes are endothermic, yet all are spontaneous. In all cases, the system moves spontaneously to a state of *higher* potential energy by absorbing heat from the surroundings.

Since some spontaneous reactions are exothermic and others are endothermic, it's clear that enthalpy alone can't account for the direction of spontaneous change; a second factor must be involved. This second thermodynamic driving force is nature's tendency to move to a condition of maximum randomness or disorder (Section 8.12).

The tendency of things to get "messed up" is common in everyday life. You may rake the leaves on a lawn into an orderly pile, but after a few windy days the leaves are scattered randomly. The reverse process is non-spontaneous; the wind never blows the randomly disordered leaves into a neatly arranged pile. Molecular systems behave similarly: *Molecular systems tend to move spontaneously to a state of maximum randomness or disorder.*

Why aren't these leaves ever blown into a neat pile?

Molecular randomness or disorder is called **entropy** and is denoted by the symbol S. Entropy is a state function (Section 8.3), and the entropy change ΔS for a process thus depends only on the initial and final states of the system:

$$\Delta S = S_{\text{final}} - S_{\text{initial}}$$

When the randomness or disorder of a system increases, ΔS has a positive value; when randomness decreases, ΔS is negative.

If you analyze the four spontaneous endothermic processes mentioned above, you'll see that each one involves an increase in the randomness of the system. When ice melts, for example, randomness increases because the highly ordered crystalline arrangement of rigidly held water molecules collapses, and the molecules become free to move about in the liquid. When liquid water vaporizes, randomness further increases because the molecules can now move independently in the much larger volume of the gas. In general, processes that convert a solid to a liquid or a liquid to a gas involve an increase in randomness and thus an increase in entropy (Figure 17.3).

FIGURE 17.3 Molecular randomness, and thus entropy, increases when a solid melts and when a liquid vaporizes. Conversely, randomness and entropy decrease when a vapor condenses and when a liquid freezes. Note the sign of ΔS for each process.

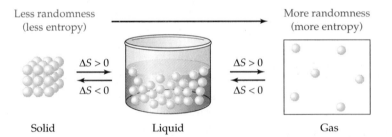

The decomposition of N_2O_4 ($O_2N–NO_2$) is accompanied by an increase in randomness because breaking the N–N bond allows the two gaseous NO_2 fragments to move independently. Whenever a molecule breaks into two or more pieces, the amount of molecular randomness increases. More specifically, randomness, and thus entropy, increases whenever a reaction results in an increase in the number of gaseous molecules (Figure 17.4).

FIGURE 17.4 Molecular randomness, and thus entropy, increases when a reaction results in an increase in the number of gaseous particles. For example, ΔS is positive for decomposition of N_2O_4 to NO_2 and is negative for formation of N_2O_4 from NO_2.

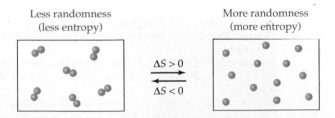

The increase in entropy on dissolving sodium chloride in water involves disruption of the crystal structure of solid NaCl and hydration of the Na^+ and Cl^- ions. Disruption of the crystal increases randomness, since the Na^+ and Cl^- ions are rigidly held in the solid but are free to move about in the liquid. The hydration process, however, *decreases* randomness because it puts the hydrating water molecules into an orderly arrangement about the Na^+ and Cl^- ions. It turns out that the overall dissolution process results in a net increase in randomness, and ΔS is thus positive (Figure 17.5). This is usually the case for dissolution of molecular solids, such as $HgCl_2$, and salts that contain +1 cations and −1 anions. For salts such as $CaSO_4$, which contain more highly charged ions, the hydrating water molecules are more strongly ordered about the ions, and the dissolution process often results in a net decrease in entropy. The following dissolution reactions illustrate the point:

$$HgCl_2(s) \rightarrow HgCl_2(aq) \qquad \Delta S = +9 \text{ J/(K} \cdot \text{mol)}$$

$$NaCl(s) \rightarrow Na^+(aq) + Cl^-(aq) \qquad \Delta S = +43 \text{ J/(K} \cdot \text{mol)}$$

$$CaSO_4(s) \rightarrow Ca^{2+}(aq) + SO_4{}^{2-}(aq) \qquad \Delta S = -140 \text{ J/(K} \cdot \text{mol)}$$

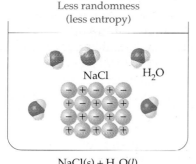

Less randomness
(less entropy)

NaCl   H₂O

$\Delta S > 0$
$\Delta S < 0$

NaCl(s) + H₂O(l)

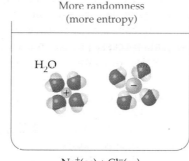

More randomness
(more entropy)

H₂O

Na⁺(aq) + Cl⁻(aq)

FIGURE 17.5 When NaCl dissolves in water, the crystal breaks up, and the Na^+ and Cl^- ions are surrounded by hydrating water molecules. The dipolar H_2O molecules are oriented such that the partially negative O atoms are near the cations and the partially positive H atoms are near the anions. Disruption of the crystal increases the entropy, but the hydration process decreases the entropy. For dissolution of NaCl, the net effect is an entropy increase.

EXAMPLE 17.1

Predict the sign of ΔS in the system (reactants plus products) for each of the following processes.
(a) $CO_2(s) \longrightarrow CO_2(g)$ (sublimation of dry ice)
(b) $CaSO_4(s) \longrightarrow CaO(s) + SO_3(g)$
(c) $N_2(g) + 3 H_2(g) \longrightarrow 2 NH_3(g)$
(d) $I_2(s) \longrightarrow I_2(aq)$ (dissolution of iodine in water)

SOLUTION   **(a)** A gas is much more disordered than a solid. Therefore, ΔS is positive.
(b) There is 1 mol of gaseous molecules on the product side of the equation and none on the reactant side. Since the reaction increases the number of gaseous molecules, the entropy change is positive.
(c) The entropy change is negative because the reaction decreases the number of gaseous molecules from 4 mol to 2 mol. Fewer particles can move independently after reaction than before.
(d) Iodine molecules are electrically neutral and form a molecular solid. The dissolution process destroys the order of the crystal and enables the iodine molecules to move about randomly in the liquid. Therefore, ΔS is positive.

┌ **PROBLEM 17.2** Predict the sign of ΔS in the system (reactants plus products) for each of the following processes.
(a) $H_2O(g) \longrightarrow H_2O(l)$ (formation of rain droplets)
(b) $I_2(g) \longrightarrow 2 I(g)$
(c) $CaCO_3(s) \longrightarrow CaO(s) + CO_2(g)$
(d) $Ag^+(aq) + Br^-(aq) \longrightarrow AgBr(s)$

Shaking a box that contains 20 quarters gives a random arrangement of heads and tails.

17.3 ►ENTROPY AND PROBABILITY

Why do systems tend to move spontaneously to a state of maximum randomness? The answer is that a random state is more probable than an ordered state because the random state can be achieved in more ways. A familiar example illustrates the point: Suppose that you shake a box containing 20 identical coins and then count the numbers of heads (H) and tails (T). Not surprisingly, it's very unlikely that all 20 coins will come up heads; that is, a perfectly ordered arrangement is much less probable than one in which heads and tails come up randomly.

The probabilities of the ordered and random states are proportional to the number of ways that the states can be achieved. The perfectly ordered state can be achieved in only one way because it consists of a single configuration (20 H). In how many ways, though, can a random state be achieved? If there were just two coins in the box, each of them could come up in two ways (H or T), and the two together could come up in $2 \times 2 = 2^2 = 4$ ways (HH, HT, TH, or TT). Three coins could come up in $2 \times 2 \times 2 = 2^3 = 8$ ways (HHH, HTH, THH, TTH, HHT, HTT, THT, or TTT), and so on. For the case of 20 coins, the number of possible combinations is $2^{20} = 1,048,576$.

Because the perfectly ordered state of 20 heads can be achieved in only one way, and the random state can be achieved in 2^{20} ways, the random state (the state with higher entropy) is 2^{20} times more probable than the perfectly ordered state. If you begin with an ordered state of 20 heads and shake the box, the system will move spontaneously to the random state of higher entropy.

The Austrian physicist Ludwig Boltzmann proposed in 1896 that the entropy of a particular state is related to the number of ways that the state can be achieved, according to the formula

$$S = k \ln W$$

where S is the entropy of the state, $\ln W$ is then the natural logarithm of the number of ways that the state can be achieved, and k, now known as Boltzmann's constant, is a universal constant equal to the gas constant R divided by Avogadro's number ($k = R/N_A = 1.38 \times 10^{-23}$ J/K). Because a logarithm is dimensionless, it follows from the Boltzmann equation that entropy has the same units as the constant k, namely, joules per kelvin.

Before considering some chemical examples, let's apply Boltzmann's formula to our 20 coins in a box. Because a perfectly ordered state consisting of 20 heads can be achieved in only one way ($W = 1$ in the Boltzmann equation) and because $\ln 1 = 0$, the entropy of the perfectly ordered state is zero:

$$S = k \ln W = k \ln 1$$
$$= 0$$

There is no disorder in this system. The more probable random state can be achieved in 2^{20} ways and thus has a higher entropy:

$$S = k \ln W = k \ln 2^{20}$$

$$= (1.38 \times 10^{-23} \text{ J/K}) (20) (\ln 2)$$

$$= 1.91 \times 10^{-22} \text{ J/K}$$

where we have made use of the relation $\ln x^a = a \ln x$ (Appendix A.2).

If our box contained 1 mol of coins, the entropy of the perfectly ordered state (6.02×10^{23} heads) would still be zero, but the entropy of the random state would be much higher because Avogadro's number of coins can be arranged randomly in a huge number of ways ($W = 2^{N_A} = 2^{6.02 \times 10^{23}}$). According to Boltzmann's formula, the entropy of the random state is

$$S = k \ln W = k \ln 2^{N_A} = kN_A \ln 2$$

Because $k = R/N_A$,

$$S = R \ln 2 = (8.314 \text{ J/K}) (0.693)$$

$$= 5.76 \text{ J/K}$$

An analogous chemical example is a crystal containing diatomic molecules such as carbon monoxide in which the two distinct ends of the molecule correspond to the heads and tails of a coin. Let's suppose that the temperature is 0 K—the temperature at which molecular motion ceases—and that the long dimensions of the molecules are oriented vertically (Figure 17.6). If the molecules pack together in a perfectly ordered "heads-up" arrangement, the entropy of the crystal is zero because the perfectly ordered structure can be achieved in only one way. If the molecules are arranged randomly with respect to the vertical direction, then the crystal has higher entropy: $k \ln 2^{20} = 1.91 \times 10^{-22}$ J/K for 20 molecules and $k \ln 2^{N_A} = R \ln 2 = 5.76$ J/K for 1 mol of molecules. Based on experimental measurements, the entropy of 1 mol of solid carbon monoxide near 0 K is about 5 J/K, indicating that the CO molecules adopt a nearly random arrangement.

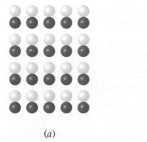 

(a) (b)

FIGURE 17.6 A hypothetical crystal containing 20 CO molecules. In **(a)**, the molecules are arranged in a perfectly ordered "heads-up" structure having entropy $S = 0$. In **(b)**, the molecules are arranged randomly in one of the 2^{20} ways in which the random structure can be obtained. The entropy of the random structure is $S = k \ln 2^{20} = 1.91 \times 10^{-22}$ J/K.

Boltzmann's formula also explains why a gas expands into a vacuum. If the two bulbs in Figure 17.1 have equal volumes, each molecule has one chance in two of being in bulb A (heads, in our coin example) and one chance in two of being in bulb B (tails) when the stopcock is opened. It's exceedingly unlikely that all the molecules in 1 mol of gas will be in bulb A, since that state can be achieved in only one way. The state in which Avogadro's number of molecules are randomly distributed between bulbs A and B can be achieved in $2^{6.02 \times 10^{23}}$ ways, and the entropy of the random state is therefore higher than the entropy of the ordered state by the now familiar amount, $R \ln 2 = 5.76$ J/K. Thus, a gas expands spontaneously because the state of greater volume is more probable.

In general, when the volume of 1 mol of an ideal gas changes from $V_{initial}$ to V_{final} at constant temperature, the entropy of the gas changes by an amount

$$\Delta S = R \ln \frac{V_{final}}{V_{initial}}$$

Because the pressure and volume of an ideal gas are related inversely ($P = nRT/V$), we can also write

$$\Delta S = R \ln \frac{P_{initial}}{P_{final}}$$

Thus, the entropy of a gas *increases* when its pressure *decreases* at constant temperature, and the entropy *decreases* when pressure *increases*. Common sense tells us that the more we squeeze the gas, the less space the gas molecules have and so the more ordered they will be.

┌ **PROBLEM 17.3** Which state has the higher entropy? Justify your answers in terms of probability.

(a) a completely ordered deck of cards or a shuffled deck in which the cards are arranged randomly
(b) a perfectly ordered crystal of solid nitrous oxide (N–N–O) or a disordered crystal in which the molecules are oriented randomly
(c) 1 mol of N_2 gas at STP or 1 mol of N_2 gas at 273 K in a volume of 11.2 L
(d) 1 mol of N_2 gas at STP or 1 mol of N_2 gas at 273 K and 0.25 atm

17.4 ➤ENTROPY AND TEMPERATURE

We've seen thus far that entropy is associated with the orientation and distribution of molecules in space. Disordered crystals have higher entropy than ordered crystals, and expanded gases have higher entropy than compressed gases. Entropy is also associated with molecular motion.

As the temperature of a substance increases, random molecular motion increases, and there is a corresponding increase in the average kinetic energy of the molecules. But not all the molecules have the same energy. As we saw in Section 9.6, there is a distribution of molecular speeds in a gas, a distribution that broadens and shifts to higher speeds with increasing temperature (Figure 9.13). In solids, liquids, and gases, the total energy of a substance can be distributed among the individual molecules in a number of ways, a number that increases as the total energy increases. According to Boltzmann's formula, the more ways (W) that the energy can be distributed, the greater the randomness of the state and the higher its entropy. Therefore, the entropy of a substance increases with increasing temperature (Figure 17.7).

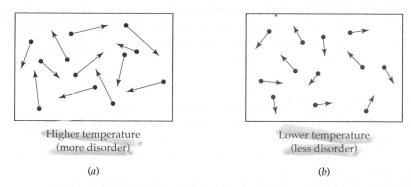

Higher temperature
(more disorder)

(a)

Lower temperature
(less disorder)

(b)

FIGURE 17.7 **(a)** A substance at a higher temperature has greater average molecular motion, more disorder, and greater entropy than **(b)** the same substance at a lower temperature.

A typical plot of entropy versus temperature is shown in Figure 17.8. At absolute zero, every substance is a solid whose particles are rigidly fixed in a crystalline structure. If there is no residual orientational disorder, like that in carbon monoxide [Figure 17.6(b)], the entropy of the substance at 0 K will be zero, a general result summarized in the **third law of thermodynamics**[1]:

THIRD LAW OF THERMODYNAMICS	The entropy of a perfectly ordered crystalline substance at 0 K is zero.

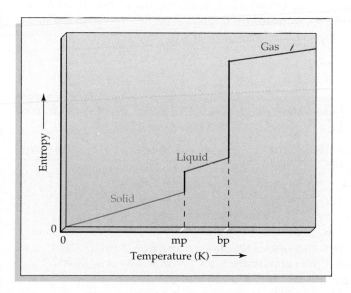

FIGURE 17.8 The entropy of a pure substance, equal to zero at 0 K, shows a steady increase with rising temperature, punctuated by discontinuous jumps in entropy at the temperatures of the phase transitions (mp = melting point, bp = boiling point).

[1] The first law of thermodynamics was discussed in Section 8.3. We'll review the first law and discuss the second law in Section 17.6.

As the temperature is raised, the molecules begin to vibrate. The number of ways in which the kinetic energy can be distributed increases with rising temperature, and the entropy of the solid thus increases steadily as the temperature goes up.

At the melting point of a solid, there is a discontinuous jump in entropy because there are many more ways of arranging the molecules in the liquid than in the solid. An even greater jump in entropy is observed at the boiling point because molecules in the gas are free to occupy a much larger volume. Between the melting point and the boiling point, the entropy of a liquid increases steadily as molecular motion becomes increasingly chaotic and the number of ways of distributing the total energy among the individual molecules increases. For the same reason, the entropy of a gas rises steadily above the boiling point.

17.5 ➤ STANDARD MOLAR ENTROPIES AND STANDARD ENTROPIES OF REACTION

We won't describe how the entropy of a substance is determined, except to note that two approaches are available: (1) calculations based on Boltzmann's formula and (2) experimental measurements of heat capacities (Section 8.7) down to very low temperatures. Suffice it to say that *standard molar entropies* are known for many substances.

The **standard molar entropy** of a substance, denoted by $S°$, is the entropy of 1 mol of the pure substance at 1 atm pressure and a specified temperature, usually 25°C. Values of $S°$ for some common substances at 25°C are listed in Table 17.1, and additional values are given in Appendix B. Note that the units of $S°$ are joules (not kilojoules) per kelvin-mole [J/(K · mol)]. Standard molar entropies are often called *absolute entropies* because they are measured with respect to an absolute reference point—the entropy of the substance at 0 K [$S° = 0$ J/(K · mol) at $T = 0$ K].

TABLE 17.1   Standard Molar Entropies for Some Common Substances at 25°C

Substance	Formula	$S°$ [J/(K · mol)]	Substance	Formula	$S°$ [J/(K · mol)]
Gases			*Liquids*		
Acetylene	C_2H_2	200.8	Acetic acid	CH_3COOH	160
Ammonia	NH_3	192.3	Ethanol	CH_3CH_2OH	161
Carbon dioxide	CO_2	213.6	Methanol	CH_3OH	127
Carbon monoxide	CO	197.6	Water	H_2O	69.9
Ethylene	C_2H_4	219.5	*Solids*		
Hydrogen	H_2	130.6	Calcium carbonate	$CaCO_3$	92.9
Methane	CH_4	186.2	Calcium oxide	CaO	39.7
Nitrogen	N_2	191.5	Diamond	C	2.4
Nitrogen dioxide	NO_2	240.0	Graphite	C	5.7
Dinitrogen tetroxide	N_2O_4	304.2	Iron	Fe	27.3
Oxygen	O_2	205.0	Iron(III) oxide	Fe_2O_3	87.4

Standard molar entropies make it possible to compare the entropies of different substances under the same conditions of temperature and pressure. It's apparent from Table 17.1, for example, that the entropies of gaseous substances tend to be larger than those of liquids, which, in turn, tend to be larger than those of solids. Table 17.1 also shows that $S°$ values increase with increasing molecular complexity. Compare, for example, CH_3OH, which has $S° = 127$ J/(K · mol), to CH_3CH_2OH, which has $S° = 161$ J/(K · mol).

Once we have values for standard molar entropies, it's easy to calculate the entropy change for a chemical reaction. The **standard entropy of reaction**, $\Delta S°$, can be obtained simply by subtracting the standard molar entropies of the reactants from the standard molar entropies of the products:

$$\Delta S° = S°(\text{products}) - S°(\text{reactants})$$

Because $S°$ values are quoted on a per-mole basis, we must multiply the $S°$ value for each substance by the stoichiometric coefficient of that substance in the balanced chemical equation. Thus, for the general reaction

$$a \, A + b \, B \rightarrow c \, C + d \, D$$

the standard entropy of reaction is

$$\Delta S° = [c \, S°(C) + d \, S°(D)] - [a \, S°(A) + b \, S°(B)]$$

where the units of the coefficients are mol, the units of $S°$ are J/(K · mol), and the units of $\Delta S°$ are J/K.

As an example, let's calculate the standard entropy change for the reaction

$$N_2O_4(g) \rightarrow 2 \, NO_2(g)$$

Using the appropriate $S°$ values obtained from Table 17.1, we find $\Delta S° = 175.8$ J/K:

$$\Delta S° = 2 \, S°(NO_2) - S°(N_2O_4)$$

$$= (2 \text{ mol})\left(240.0 \frac{J}{K \cdot mol}\right) - (1 \text{ mol})\left(304.2 \frac{J}{K \cdot mol}\right)$$

$$= 175.8 \text{ J/K}$$

Although the standard molar entropy of N_2O_4 is larger than that of NO_2, as expected for a more complex molecule, $\Delta S°$ for the reaction is positive because 1 mol of N_2O_4 is converted to 2 mol of NO_2. As noted earlier, we expect an increase in entropy whenever a molecule breaks into two or more pieces.

EXAMPLE 17.2

Calculate the standard entropy of reaction at 25°C for the industrial synthesis of ammonia:

$$N_2(g) + 3\,H_2(g) \rightarrow 2\,NH_3(g)$$

SOLUTION   The standard entropy change for the reaction is

$$\Delta S° = 2\,S°(NH_3) - [S°(N_2) + 3\,S°(H_2)]$$

Substituting into this equation the appropriate $S°$ values from Table 17.1, we obtain

$$\Delta S = (2\;mol)\left(192.3\,\frac{J}{K \cdot mol}\right) - \left[(1\;mol)\left(191.5\,\frac{J}{K \cdot mol}\right) + (3\;mol)\left(130.6\,\frac{J}{K \cdot mol}\right)\right]$$

$$= -198.7\;J/K$$

Note that the entropy change is negative, as predicted in Example 17.1, because the number of moles of gas decreases.

PROBLEM 17.4   Calculate the standard entropy of reaction at 25°C for the decomposition of calcium carbonate:

$$CaCO_3(s) \rightarrow CaO(s) + CO_2(g)$$

17.6 ➤ ENTROPY AND THE SECOND LAW OF THERMODYNAMICS

We've seen thus far that molecular systems tend to move spontaneously toward a state of minimum enthalpy and maximum entropy. In any particular reaction, however, the enthalpy of the system (reactants plus products) can either increase or decrease. Similarly, the entropy of the system can either increase or decrease. How, then, can we decide whether a reaction will occur spontaneously? In Section 8.13, we said that it is the value of the *free-energy* change, ΔG, that is the criterion for spontaneity, where $\Delta G = \Delta H - T\Delta S$. If $\Delta G < 0$, the reaction is spontaneous; if $\Delta G > 0$, the reaction is nonspontaneous; and if $\Delta G = 0$, the reaction is at equilibrium. In this section and the next, we'll see how that conclusion was reached. Let's begin by looking at the first two laws of thermodynamics:

FIRST LAW OF THERMODYNAMICS	In any process, spontaneous or nonspontaneous, the total energy of a system and its surroundings is constant.
SECOND LAW OF THERMODYNAMICS	In any *spontaneous* process, the total entropy of a system and its surroundings always increases.

The first law (Section 8.3) is simply a statement of the conservation of energy. It says that energy (or enthalpy) can flow between a system and its surroundings but that the total energy of the system plus the surroundings always remains constant. In an exothermic reaction, the system loses enthalpy to the surroundings; in an endothermic reaction, the system gains enthalpy from the surroundings. Since energy is conserved in *all* processes, spontaneous and nonspontaneous, the first law helps us keep track of energy flow between the system and the surroundings, but it doesn't tell us whether a particular reaction will be spontaneous or nonspontaneous.

The second law, however, provides a clear-cut criterion of spontaneity. It says that the direction of spontaneous change is always determined by the sign of the total entropy change:

$$\Delta S_{total} = \Delta S_{system} + \Delta S_{surroundings}$$

Specifically,

If $\Delta S_{total} > 0$, the reaction is spontaneous.

If $\Delta S_{total} < 0$, the reaction is nonspontaneous.

If $\Delta S_{total} = 0$, the reaction mixture is at equilibrium.

All reactions proceed spontaneously in the direction that increases the entropy of the system plus surroundings. A reaction that is nonspontaneous in the forward direction is spontaneous in the reverse direction because ΔS_{total} for the reverse reaction equals $-\Delta S_{total}$ for the forward reaction. If ΔS_{total} is zero, the reaction doesn't go spontaneously in either direction, and the reaction mixture is therefore at equilibrium.

To determine the value of ΔS_{total}, we need values for the entropy changes in the system and the surroundings. The entropy change in the system, ΔS_{sys}, is just the entropy of reaction, which can be calculated from standard molar entropies (Table 17.1), as described in the previous section. For a reaction that occurs at constant pressure, the entropy change in the surroundings is directly proportional to the enthalpy change for the reaction (ΔH) and inversely proportional to the kelvin temperature (T) of the surroundings, according to the equation

$$\Delta S_{surr} = \frac{-\Delta H}{T}$$

Although we won't derive this equation to calculate ΔS_{surr}, we can nevertheless justify its form. To see why ΔS_{surr} is proportional to $-\Delta H$, recall that for an exothermic reaction ($\Delta H < 0$), the system loses heat to the surroundings (Figure 17.9). As a result, the random, chaotic motion of the molecules in the surroundings increases, and the entropy of the surroundings also increases ($\Delta S_{surr} > 0$). Conversely, for an endothermic reaction ($\Delta H > 0$), the system gains heat from the surroundings, and the entropy of the surroundings therefore decreases ($\Delta S_{surr} < 0$). Because ΔS_{surr} is positive when ΔH is negative, and vice versa, ΔS_{surr} is proportional to $-\Delta H$.

$$\Delta S_{surr} \propto -\Delta H$$

The reason why ΔS_{surr} is inversely proportional to the absolute temperature T is more subtle. We can think of the surroundings as an infinitely

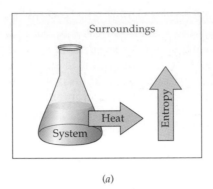

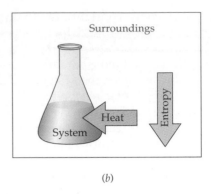

(a)   (b)

FIGURE 17.9 **(a)** When an exothermic reaction occurs in the system ($\Delta H < 0$), the surroundings gain heat and their entropy increases ($\Delta S_{surr} > 0$). **(b)** When an endothermic reaction occurs in the system ($\Delta H > 0$), the surroundings lose heat and their entropy decreases ($\Delta S_{surr} < 0$).

large constant-temperature bath to which heat can be added without changing its temperature. If the surroundings have a low temperature, they have only a small amount of disorder, in which case addition of a given quantity of heat results in a substantial increase in the amount of disorder (a relatively large value of ΔS_{surr}). If the surroundings have a high temperature, they already have a large amount of disorder, and addition of the same quantity of heat produces only a marginal increase in the amount of disorder (a relatively small value of ΔS_{surr}). Thus, we might expect ΔS_{surr} to vary inversely with temperature.

$$\Delta S_{surr} \propto \frac{1}{T}$$

Adding heat to the surroundings is somewhat analogous to tossing a rock into a lake. If the lake has little disorder (calm, smooth surface), the rock's impact produces considerable disorder, evident in a circular pattern of waves. If the lake is already appreciably disordered (rough, choppy surface), the additional disorder produced when the rock hits the water is hardly noticeable.

Adding heat to cold surroundings is analogous to tossing a rock into calm waters. Both processes produce considerable disorder and thus a relatively large increase in entropy.

Adding heat to hot surroundings is analogous to tossing a rock into rough waters. Both processes produce relatively little disorder and thus a relatively small increase in entropy.

EXAMPLE 17.3

Consider the oxidation of iron metal:

$$4 \ Fe(s) + 3 \ O_2(g) \rightarrow 2 \ Fe_2O_3(s)$$

By determining the sign of ΔS_{total}, show that the reaction is spontaneous at 25°C.

SOLUTION   To determine the sign of $\Delta S_{total} = \Delta S_{sys} + \Delta S_{surr}$, we must calculate the values of ΔS_{sys} and ΔS_{surr}. The entropy change in the system is equal to the standard entropy of reaction and can be calculated using the standard molar entropies in Table 17.1:

$$\Delta S_{sys} = \Delta S° = 2 \ S°(Fe_2O_3) - [4 \ S°(Fe) + 3 \ S°(O_2)]$$

$$= (2 \ mol)\left(87.4 \frac{J}{K \cdot mol}\right) - \left[(4 \ mol)\left(27.3 \frac{J}{K \cdot mol}\right) + (3 \ mol)\left(205.0 \frac{J}{K \cdot mol}\right)\right]$$

$$= -549.5 \ J/K$$

ΔS_{sys} is negative, as expected for a reaction that consumes 3 mol of gas.
To obtain $\Delta S_{surr} = -\Delta H°/T$, first calculate $\Delta H°$ for the reaction from standard enthalpies of formation (Section 8.9):

$$\Delta H° = 2 \ \Delta H_f°(Fe_2O_3) - [4 \ \Delta H_f°(Fe) + 3 \ \Delta H_f°(O_2)]$$

Because $\Delta H_f° = 0$ for elements and $\Delta H_f° = -824.2 \ kJ/mol$ for Fe_2O_3 (Appendix B), $\Delta H°$ for the reaction is

$$\Delta H° = 2 \ \Delta H_f°(Fe_2O_3) = (2 \ mol)(-824.2 \ kJ/mol) = -1648.4 \ kJ$$

The reaction is highly exothermic, and accordingly there is a large positive entropy change in the surroundings:

$$\Delta S_{surr} = \frac{-\Delta H°}{T} = \frac{-(-1,648,400 \ J)}{298 \ K} = 5532 \ J/K$$

The total entropy change is positive, and the reaction is therefore spontaneous under standard-state conditions at 25°C:

$$\Delta S_{total} = \Delta S_{sys} + \Delta S_{surr} = -549.5 \ J/K + 5532 \ J/K = 4982 \ J/K$$

PROBLEM 17.5 By determining the sign of ΔS_{total}, show whether the decomposition of calcium carbonate is spontaneous under standard-state conditions at 25°C.

$$CaCO_3(s) \rightarrow CaO(s) + CO_2(g)$$

17.7 ►FREE ENERGY

Chemists are generally more interested in the system (the reaction mixture) than the surroundings, and it's therefore convenient to restate the second law in terms of the thermodynamic properties of the system, without regard to the surroundings. For this purpose, we use the thermodynamic property called **free energy**, denoted by G in honor of J. Willard Gibbs (1839–1903), the American mathematical physicist who laid the foundations of chemical thermodynamics. As discussed in Section 8.13, the free energy G of a system is defined as

$$\textit{Free energy} \qquad G = H - TS$$

where H is the enthalpy, T is the temperature in kelvins, and S is the entropy. As you would expect from its name, the free energy has units of energy (J or kJ).

Why is G called the *free* energy? If you think of TS as the part of the system's energy that is already disordered, then $H - TS \ (= G)$ is the part of the system's energy that is still ordered and therefore free (available) to cause spontaneous change by becoming disordered.

Free energy, like enthalpy and entropy, is a state function, and the change in free energy (ΔG) for a process is therefore independent of path. For a reaction at constant temperature, ΔG is equal to the change in enthalpy minus the product of temperature times the change in entropy:

$$\Delta G = \Delta H - T\Delta S$$

To see what this equation for free-energy change has to do with spontaneity, let's return to the relationship

$$\Delta S_{\text{total}} = \Delta S_{\text{sys}} + \Delta S_{\text{surr}} = \Delta S + \Delta S_{\text{surr}}$$

where we have now dropped the subscript "sys". (It's generally understood that symbols without a subscript refer to the system, not the surroundings.) Since $\Delta S_{\text{surr}} = -\Delta H/T$, where ΔH is the heat gained by the system at constant pressure, we can also write

$$\Delta S_{\text{total}} = \Delta S - \frac{\Delta H}{T}$$

Multiplying both sides by $-T$ gives

$$-T\Delta S_{\text{total}} = \Delta H - T\Delta S$$

But the right-hand side of this equation is just ΔG, the change in the free energy of the system at constant temperature and pressure. Therefore,

$$-T\Delta S_{\text{total}} = \Delta G$$

Note that ΔG and ΔS_{total} have opposite signs because the absolute temperature T is always positive.

According to the second law of thermodynamics, a reaction is spontaneous if ΔS_{total} is positive, nonspontaneous if ΔS_{total} is negative, and at equilibrium if ΔS_{total} is zero. Since $-T\Delta S_{\text{total}} = \Delta G$ and since ΔG and ΔS_{total}

have opposite signs, we can restate the thermodynamic criterion for the spontaneity of a reaction carried out at constant temperature and pressure in the following way:

> If $\Delta G < 0$, the reaction is spontaneous.
>
> If $\Delta G > 0$, the reaction is nonspontaneous.
>
> If $\Delta G = 0$, the reaction mixture is at equilibrium.

In other words, *in any spontaneous process at constant temperature and pressure, the free energy of the system always decreases.*

As discussed in Section 8.13, the temperature T acts as a weighting factor that determines the relative importance of the enthalpy and entropy contributions to ΔG in the free-energy equation $\Delta G = \Delta H - T\Delta S$. If ΔH and ΔS are either both negative or both positive, the sign of ΔG (and therefore the spontaneity of the reaction) depends on the temperature (Table 17.2). If ΔH and ΔS are both negative, the reaction will be spontaneous only if the absolute value of ΔH is larger than the absolute value of $T\Delta S$. This is most likely at low temperatures, where the weighting factor T in $T\Delta S$ is small. If ΔH and ΔS are both positive, the reaction will be spontaneous only if $T\Delta S$ is larger than ΔH, which is most likely at high temperatures. We've already seen how these considerations apply to changes of state (Section 10.4).

TABLE 17.2   Signs of Enthalpy, Entropy, and Free-Energy Changes and Reaction Spontaneity for a Reaction at Constant Temperature and Pressure

ΔH	ΔS	$\Delta G = \Delta H - T\Delta S$	Reaction Spontaneity
−	+	−	Spontaneous at all temperatures
−	−	− or +	Spontaneous at low temperatures where $\|\Delta H\| > \|T\Delta S\|$ Nonspontaneous at high temperatures where $\|\Delta H\| < \|T\Delta S\|$
+	−	+	Nonspontaneous at all temperatures
+	+	− or +	Spontaneous at high temperatures where $T\Delta S > \Delta H$ Nonspontaneous at low temperatures where $T\Delta S < \Delta H$

EXAMPLE 17.4

Iron metal can be produced by reducing iron(III) oxide with hydrogen:

$$Fe_2O_3(s) + 3\ H_2(g) \rightarrow 2\ Fe(s) + 3\ H_2O(g) \qquad \Delta H° = 98.8\ kJ;\ \Delta S° = 141.5\ J/K$$

(a) Show whether this reaction is spontaneous at 25°C.
(b) Estimate the temperature at which the reaction becomes spontaneous.

SOLUTION   **(a)** At 25°C (298 K), ΔG for the reaction is

$$\Delta G = \Delta H - T\Delta S = (98.8 \text{ kJ}) - (298 \text{ K})(0.1415 \text{ kJ/K})$$

$$= (98.8 \text{ kJ}) - (42.2 \text{ kJ})$$

$$= 56.6 \text{ kJ}$$

Because the positive ΔH term is larger than the positive $T\Delta S$ term, ΔG is positive and the reaction is nonspontaneous at 298 K.

(b) At sufficiently high temperatures, $T\Delta S$ becomes larger than ΔH, ΔG becomes negative, and the reaction becomes spontaneous. We can estimate the temperature at which ΔG changes from positive to negative by setting $\Delta G = \Delta H - T\Delta S = 0$. Solving for T, we find that the reaction becomes spontaneous at 698 K.

$$T = \frac{\Delta H}{\Delta S} = \frac{98.8 \text{ kJ}}{0.1415 \text{ kJ/K}} = 698 \text{ K}$$

Note the assumption that the values of ΔH and ΔS are unchanged on going from 298 K to 698 K. In general, the temperature dependence of ΔH and ΔS is small and can usually be neglected.

PROBLEM 17.6 Consider the decomposition of gaseous N_2O_4:

$$N_2O_4(g) \rightarrow 2 \text{ NO}_2(g) \qquad \Delta H° = 57.1 \text{ kJ}; \Delta S° = 175.8 \text{ J/K}$$

(a) Show whether this reaction is spontaneous at 25°C.
(b) Estimate the temperature at which the reaction becomes spontaneous.

PROBLEM 17.7 The following data apply to the vaporization of mercury: $\Delta H_{vap} = 58.5 \text{ kJ/mol}; \Delta S_{vap} = 92.9 \text{ J/(K} \cdot \text{mol)}$.
(a) Does mercury boil at 325°C and 1 atm pressure?
(b) What is the normal boiling point of mercury?

17.8 ➤ STANDARD FREE-ENERGY CHANGES FOR REACTIONS

The free energy of a substance, like its enthalpy and entropy, depends on temperature, pressure, the physical state of the substance (solid, liquid, or gas), and its concentration, in the case of solutions. As a result, free-energy changes for chemical reactions must be compared under a well-defined set of standard-state conditions:

$$\textit{Standard-state conditions} \begin{cases} \text{Solids, liquids, and gases in pure form at 1 atm pressure} \\ \text{Solutes at 1 M concentration} \\ \text{A specified temperature, usually 25°C} \end{cases}$$

The **standard free-energy change**, $\Delta G°$, for a reaction is the change in free energy that occurs when reactants in their standard states are converted to products in their standard states. As with $\Delta H°$ (Section 8.9), the value of

$\Delta G°$ is an extensive property and refers to the number of moles indicated in the chemical equation. For example, $\Delta G°$ at 25°C for the reaction

$$Na(s) + H_2O(l) \rightarrow 1/2 \ H_2(g) + Na^+(aq) + OH^-(aq)$$

is the change in free energy that occurs when 1 mol of solid sodium reacts completely with 1 mol of liquid water to give 0.5 mol of hydrogen gas at 1 atm pressure and to give an aqueous solution that contains 1 mol of Na^+ ions and 1 mol of OH^- ions at concentrations of 1 M, with all reactants and products at a temperature of 25°C. For this reaction, $\Delta G° = -182$ kJ, where the superscript ° indicates that reactants and products are in their standard states.

Because the free-energy change for any process at constant temperature and pressure is $\Delta G = \Delta H - T\Delta S$, we can calculate the standard free-energy change $\Delta G°$ for a reaction from the standard enthalpy change $\Delta H°$ and the standard entropy change $\Delta S°$. Consider again the Haber synthesis of ammonia:

$$N_2(g) + 3 \ H_2(g) \rightarrow 2 \ NH_3(g) \qquad \Delta H° = -92.2 \text{ kJ}; \ \Delta S° = -198.7 \text{ J/K}$$

The standard free-energy change at 25°C (298 K) is

$$\Delta G° = \Delta H° - T\Delta S° = (-92.2 \times 10^3 \text{ J}) - (298 \text{ K})(-198.7 \text{ J/K})$$

$$= (-92.2 \times 10^3 \text{ J}) - (-59.2 \times 10^3 \text{ J})$$

$$= -33.0 \text{ kJ}$$

Because the negative enthalpy term is larger than the negative entropy term at 25°C, $\Delta G°$ is negative and the reaction is spontaneous under standard-state conditions.

Note that the standard free-energy change applies to a hypothetical process in which separate reactants in their standard states are *completely* converted to separate products in their standard states. For the Haber synthesis, the hypothetical process is

| 1 mol $N_2(g)$ (1 atm, 25°C) | + | 3 mol $H_2(g)$ (1 atm, 25°C) | $\xrightarrow{\Delta G° = -33.0 \text{ kJ}}$ | 2 mol $NH_3(g)$ (1 atm, 25°C) |

————— State 1 —————   State 2

$\Delta G° = -33.0$ kJ is the change in the free energy of the system on going from state 1 to state 2.

When we carry out an *actual* synthesis of ammonia, the reactants, N_2 and H_2, are not separate but are mixed together. Moreover, the reaction doesn't go to completion; it reaches an equilibrium state in which both reactants and products are present together. How, then, should we think about the meaning of $\Delta G°$ in the context of an actual reaction? One way would be to suppose that we have a mixture of N_2, H_2, and NH_3, with each substance present at a partial pressure of 1 atm. Suppose further that the mixture behaves as an ideal gas so that the free energy of each component in the mixture is the same as the free energy of the pure substance. Finally, suppose that the number of moles of each component—say x, y, and z—is very large so that the partial pressures don't change appreciably when 1 mol of N_2 and 3 mol of H_2 are converted to 2 mol of NH_3. In other words, we are imagining the following real process:

x mol $N_2(g)$ (1 atm, 25°C) y mol $H_2(g)$ (1 atm, 25°C) z mol $NH_3(g)$ (1 atm, 25°C)	$\xrightarrow{\Delta G° = -33.0 \text{ kJ}}$	$(x-1)$ mol $N_2(g)$ (1 atm, 25°C) $(y-3)$ mol $H_2(g)$ (1 atm, 25°C) $(z+2)$ mol $NH_3(g)$ (1 atm, 25°C)
State 1		State 2

The free-energy change for this process is the standard free-energy change $\Delta G°$, since each reactant and product is present at 1 atm pressure. If $\Delta G°$ is negative, the reaction will proceed spontaneously to give more products. If $\Delta G°$ is positive, the reaction will proceed in the reverse direction to give more reactants. Of course, the value of $\Delta G°$ provides no information concerning the rate of the reaction.

EXAMPLE 17.5

Methanol (CH_3OH), an important alcohol used in the manufacture of adhesives, fibers, and plastics, is synthesized industrially by the reaction

$$CO(g) + 2 H_2(g) \rightarrow CH_3OH(g)$$

(a) Calculate the standard free-energy change for this reaction at 25°C.
(b) Is the reaction spontaneous at 25°C?
(c) If it is spontaneous at 25°C, estimate the temperature at which the reverse reaction becomes spontaneous.

SOLUTION   We can calculate the standard free-energy change from the relation $\Delta G° = \Delta H° - T\Delta S°$, but first we must find $\Delta H°$ and $\Delta S°$ from standard enthalpies of formation ($\Delta H_f°$) and standard molar entropies ($S°$). The following values from Appendix B,

	$CO(g)$	$H_2(g)$	$CH_3OH(g)$
$\Delta H_f°$ (kJ/mol)	−110.5	0	−201.2
$S°$ (J/(K · mol))	197.6	130.6	238

give

$$\Delta H° = \Delta H_f°(CH_3OH) - [\Delta H_f°(CO) + 2 \Delta H_f°(H_2)]$$

$$= (1 \text{ mol})(-201.2 \text{ kJ/mol}) - [(1 \text{ mol})(-110.5 \text{ kJ/mol}) + (2 \text{ mol})(0 \text{ kJ/mol})]$$

$$= -90.7 \text{ kJ}$$

and

$$\Delta S° = (1 \text{ mol})\left(238 \frac{J}{K \cdot mol}\right) - \left[(1 \text{ mol})\left(197.6 \frac{J}{K \cdot mol}\right) + (2 \text{ mol})\left(130.6 \frac{J}{K \cdot mol}\right)\right]$$

$$= -221 \text{ J/K}$$

Therefore, $\Delta G° = \Delta H° - T\Delta S° = (-90.7 \times 10^3 \text{ J}) - (298 \text{ K})(-221 \text{ J/K})$

$$= (-90.7 \times 10^3 \text{ J}) + (65.9 \times 10^3 \text{ J})$$

$$= -24.8 \text{ kJ}$$

(b) Because $\Delta G°$ is negative, the reaction is spontaneous at 25°C. This means that a mixture of $CO(g)$, $H_2(g)$, and $CH_3OH(g)$, each at a partial pressure of 1 atm, will react at 25°C to produce more methanol.

(c) The reverse reaction becomes spontaneous at the temperature at which the sign of $\Delta G°$ for the forward reaction changes from negative to positive. We can estimate that temperature by setting

$$\Delta G° = \Delta H° - T\Delta S° = 0$$

Solving for T and assuming that the temperature dependence of $\Delta H°$ and $\Delta S°$ can be neglected gives

$$T = \frac{\Delta H°}{\Delta S°} = \frac{-90.7 \times 10^3 \text{ J}}{-221 \text{ J/K}} = 410 \text{ K}$$

⌐ PROBLEM 17.8 Consider the thermal decomposition of calcium carbonate:

$$CaCO_3(s) \rightarrow CaO(s) + CO_2(g)$$

(a) Using the data in Appendix B, calculate the standard free-energy change for this reaction at 25°C.

(b) Will a mixture of solid $CaCO_3$, solid CaO, and gaseous CO_2 at 1 atm pressure react spontaneously at 25°C to produce more CaO and CO_2?

(c) Assuming that $\Delta H°$ and $\Delta S°$ are temperature independent, estimate the temperature at which the reaction becomes spontaneous.

17.9 ►STANDARD FREE ENERGIES OF FORMATION

The **standard free energy of formation**, $\Delta G_f°$, of a substance is the free-energy change for formation of 1 mol of the substance in its standard state from the most stable form of the constituent elements in their standard states. For example, we found in the previous section that the standard free-energy change $\Delta G°$ for synthesis of 2 mol of NH_3 from its constituent elements is -33.0 kJ:

$$N_2(g) + 3 H_2(g) \rightarrow 2 NH_3(g) \qquad \Delta G° = -33.0 \text{ kJ}$$

Therefore, $\Delta G_f°$ for ammonia is -33.0 kJ/2 mol, or -16.5 kJ/mol.

Values of $\Delta G_f°$ at 25°C for some common substances are listed in Table 17.3, and additional values are given in Appendix B. Note that $\Delta G_f°$ for an element in its most stable form at 25°C is defined to be zero. Thus, solid graphite has $\Delta G_f° = 0$, but diamond, a less stable form of solid carbon at 25°C, has $\Delta G_f° = 2.9$ kJ/mol. As with standard enthalpies of formation, $\Delta H_f°$, a zero value of $\Delta G_f°$ for elements in their most stable form establishes a thermochemical "sea level," or reference point, with respect to which the standard

TABLE 17.3 Standard Free Energies of Formation for Some Common Substances at 25°C

Substance	Formula	ΔG_f° (kJ/mol)	Substance	Formula	ΔG_f° (kJ/mol)
Gases			*Liquids*		
Acetylene	C_2H_2	209.2	Acetic acid	CH_3COOH	−390
Ammonia	NH_3	−16.5	Ethanol	C_2H_5OH	−174.9
Carbon dioxide	CO_2	−394.4	Methanol	CH_3OH	−166.4
Carbon monoxide	CO	−137.2	Water	H_2O	−237.2
Ethylene	C_2H_4	68.1	*Solids*		
Hydrogen	H_2	0	Calcium carbonate	$CaCO_3$	−1128.8
Methane	CH_4	−50.8	Calcium oxide	CaO	−604.0
Nitrogen	N_2	0	Diamond	C	2.9
Nitrogen dioxide	NO_2	51.3	Graphite	C	0
Dinitrogen tetroxide	N_2O_4	97.8	Iron(III) oxide	Fe_2O_3	−742.2

free energies of other substances are measured. We can't measure the absolute value of a substance's free energy (as we can the entropy), but that's not a problem because we are interested only in free-energy *differences* between reactants and products.

The standard free energy of formation of a substance measures its thermodynamic stability with respect to its constituent elements. Substances that have a negative value of ΔG_f°, such as carbon dioxide and water, are stable in that they do not decompose to their constituent elements under standard-state conditions. Substances that have a positive value of ΔG_f°, such as ethylene and nitrogen dioxide, are thermodynamically unstable with respect to their constituent elements. Nevertheless, such substances, once prepared, can exist for long periods of time if the rate of their decomposition is slow.

Clearly, there's no point in trying to synthesize a substance from its elements under standard-state conditions if the substance has a positive value of ΔG_f°. Such a substance would have to be prepared at other temperatures and/or pressures, or it would have to be made from alternative starting materials using a reaction that has a negative free-energy change. Thus, thermodynamics can save us considerable time in chemical synthesis.

In the previous section, we calculated standard free-energy changes for reactions from the equation $\Delta G^\circ = \Delta H^\circ - T\Delta S^\circ$, using tabulated values of ΔH_f° and S° to find ΔH° and ΔS°. Alternatively, we can calculate ΔG° more directly by subtracting the standard free energies of formation of all the reactants from the standard free energies of formation of all the products:

$$\Delta G^\circ = \Delta G_f^\circ(\text{products}) - \Delta G_f^\circ(\text{reactants})$$

For the general reaction

$$a\,A + b\,B \rightarrow c\,C + d\,D$$

the standard free-energy change is

$$\Delta G^\circ = [c\,\Delta G_f^\circ(C) + d\,\Delta G_f^\circ(D)] - [a\,\Delta G_f^\circ(A) + b\,\Delta G_f^\circ(B)]$$

To illustrate, let's calculate the standard free-energy change for reduction of iron(III) oxide by carbon monoxide, a reaction that occurs in the commercial production of iron metal from iron ore:

$$Fe_2O_3(s) + 3\ CO(g) \rightarrow 2\ Fe(s) + 3\ CO_2(g)$$

Using the ΔG_f° values in Table 17.3, we obtain

$$\Delta G^\circ = [2\ \Delta G_f^\circ(Fe) + 3\ \Delta G_f^\circ(CO_2)] - [\Delta G_f^\circ(Fe_2O_3) + 3\ \Delta G_f^\circ(CO)]$$

$$= [(2\ mol)(0\ kJ/mol) + (3\ mol)(-394.4\ kJ/mol)]$$
$$- [(1\ mol)(-742.2\ kJ/mol) + (3\ mol)(-137.2\ kJ/mol)]$$

$$= -29.4\ kJ$$

EXAMPLE 17.6

(a) Calculate the standard free energy change for the oxidation of ammonia to give nitric oxide (NO) and water. Is it worth trying to find a catalyst for this reaction?

$$4\ NH_3(g) + 5\ O_2(g) \rightarrow 4\ NO(g) + 6\ H_2O(l)$$

(b) Is it worth trying to find a catalyst for the synthesis of NO from gaseous N_2 and O_2 at 25°C?

SOLUTION   **(a)** We can calculate ΔG° most easily from tabulated standard free energies of formation (Appendix B):

$$\Delta G^\circ = [4\ \Delta G_f^\circ(NO) + 6\ \Delta G_f^\circ(H_2O)] - [4\ \Delta G_f^\circ(NH_3) + 5\ \Delta G_f^\circ(O_2)]$$

$$= [(4\ mol)(86.6\ kJ/mol) + (6\ mol)(-237.2\ kJ/mol)]$$
$$- [(4\ mol)(-16.5\ kJ/mol) + (5\ mol)(0\ kJ/mol)]$$

$$= -1010.8\ kJ$$

It is worth looking for a catalyst because the negative value of ΔG° indicates that the reaction is spontaneous. (This reaction is the first step in the Ostwald process for production of nitric acid.) In industry, the reaction is carried out using a platinum–rhodium catalyst.
(b) It's not worth looking for a catalyst for the reaction $N_2(g) + O_2(g) \longrightarrow 2\ NO(g)$ because the standard free energy of formation of NO is positive ($\Delta G_f^\circ = 86.6\ kJ/mol$). This means that NO is unstable with respect to decomposition to N_2 and O_2 at 25°C. A catalyst could only increase the rate of decomposition.

⌐ **PROBLEM 17.9** **(a)** Using values of ΔG_f° in Appendix B, calculate the standard free-energy change for the reaction of calcium carbide (CaC_2) with water. Might this reaction be used for synthesis of acetylene (C_2H_2)?

$$CaC_2(s) + 2\ H_2O(l) \rightarrow C_2H_2(g) + Ca(OH)_2(s)$$

(b) Is it possible to synthesize acetylene from solid graphite and gaseous H_2 at 25°C and 1 atm pressure?

17.10 ➤FREE-ENERGY CHANGES AND COMPOSITION OF THE REACTION MIXTURE

The sign of the standard free-energy change $\Delta G°$ tells the direction of spontaneous reaction when both reactants and products are present at standard-state conditions. In actual reactions, however, the composition of the reaction mixture seldom corresponds to standard-state pressures and concentrations. Moreover, the partial pressures and concentrations change as a reaction proceeds. So how do we calculate the free-energy change ΔG for a reaction when the reactants and products are present at non-standard-state pressures and concentrations?

The answer is given by the relation

$$\Delta G = \Delta G° + RT \ln Q$$

where ΔG is the free-energy change under non-standard-state conditions, $\Delta G°$ is the free-energy change under standard-state conditions, R is the gas constant [$8.314 \text{ J/(K} \cdot \text{mol)}$], T is the absolute temperature in kelvins, and Q is the reaction quotient (Q_c or Q_p). Recall that the reaction quotient Q_c is an expression having the same form as the equilibrium constant expression K_c except that the concentrations do not necessarily have equilibrium values (Section 13.5). Similarly, Q_p has the same form as K_p except that the partial pressures have arbitrary values. For example, for the Haber synthesis of ammonia:

$$N_2(g) + 3 H_2(g) \rightarrow 2 NH_3(g) \qquad Q_p = \frac{(P_{NH_3})^2}{(P_{N_2})(P_{H_2})^3}$$

We won't derive the equation for ΔG under non-standard-state conditions, but we show how to use it in Example 17.7.

EXAMPLE 17.7

Calculate the free-energy change for ammonia synthesis at 25°C (298 K) given the following sets of partial pressures: **(a)** 1.0 atm N_2, 3.0 atm H_2, 0.020 atm NH_3; **(b)** 0.010 atm N_2, 0.030 atm H_2, 2.0 atm NH_3.

$$N_2(g) + 3 H_2(g) \rightarrow 2 NH_3(g) \qquad \Delta G° = -33.0 \text{ kJ}$$

SOLUTION   **(a)** First calculate the value of the reaction quotient Q_p:

$$Q_p = \frac{(P_{NH_3})^2}{(P_{N_2})(P_{H_2})^3} = \frac{(0.020)^2}{(1.0)(3.0)^3} = 1.5 \times 10^{-5}$$

Then substitute this value into the equation for ΔG:

$$\Delta G = \Delta G° + RT \ln Q = \Delta G° + 2.303RT \log Q_p$$

$$= (-33.0 \times 10^3 \text{ J}) + (2.303)(8.314 \text{ J/K})(298 \text{ K})(\log 1.5 \times 10^{-5})$$

$$= (-33.0 \times 10^3 \text{ J}) + (-27.5 \times 10^3 \text{ J})$$

$$= -60.5 \text{ kJ}$$

Note that ΔG is more negative than $\Delta G°$ because Q_p is less than 1. This means that the reaction has a greater thermodynamic driving force than it does under standard-state conditions, not surprising since the reaction mixture is rich in reactants and poor in the product. Note also that when each reactant and product is present at a partial pressure of 1 atm, $Q_p = 1$, $\log Q_p = 0$, and $\Delta G = \Delta G°$.

(b) In this case, the reaction mixture is rich in the product and poor in the reactants. Therefore, Q_p will be greater than 1, and ΔG should be more positive than $\Delta G°$. The value of Q_p is

$$Q_p = \frac{(P_{NH_3})^2}{(P_{N_2})(P_{H_2})^3} = \frac{(2.0)^2}{(0.010)(0.030)^3} = 1.5 \times 10^7$$

The corresponding value of ΔG is

$$\Delta G = \Delta G° + 2.303RT \log Q_p$$

$$= (-33.0 \times 10^3 \text{ J}) + (2.303)(8.314 \text{ J/K})(298 \text{ K})(\log 1.5 \times 10^7)$$

$$= (-33.0 \times 10^3 \text{ J}) + (40.9 \times 10^3 \text{ J})$$

$$= 7.9 \text{ kJ}$$

Because Q_p is large enough to give a positive value for ΔG, the reaction is nonspontaneous in the forward direction but spontaneous in the reverse direction. Thus, the direction of spontaneous reaction depends on the composition of the reaction mixture.

┌ **PROBLEM 17.10** Calculate ΔG for the formation of ethylene (C_2H_4) from carbon and hydrogen at 25°C when the partial pressures are 100 atm H_2 and 0.10 atm C_2H_4.

$$C(s) + 2 H_2(g) \rightarrow C_2H_4(g) \qquad \Delta G° = 68.1 \text{ kJ}$$

Is the reaction spontaneous in the forward or the reverse direction? ┘

17.11 ➤FREE ENERGY AND CHEMICAL EQUILIBRIUM

Now that we've seen how ΔG for a reaction depends on composition, we can understand how the total free energy of a reaction mixture changes as the reaction progresses toward equilibrium. Look again at the expression for calculating ΔG:

$$\Delta G = \Delta G° + RT \ln Q$$

If the reaction mixture contains mainly reactants and almost no products, Q will be much less than 1 and $RT \ln Q$ will be a very large negative number (minus infinity when $Q = 0$). Consequently, no matter what the value of $\Delta G°$ (positive or negative), the negative $RT \ln Q$ term will dominate the $\Delta G°$ term, and ΔG will be negative. This means that the forward reaction is always spontaneous when the concentration of products is very small. Conversely, if the reaction mixture contains mainly products and almost no reactants, Q will be much greater than 1, and $RT \ln Q$ will be a very large positive number (plus infinity when no reactants are present). Consequently, the positive $RT \ln Q$ term will dominate the $\Delta G°$ term, and ΔG will be positive. Thus, the reverse reaction is always spontaneous when the concentration of reactants is very small. These conditions are summarized as follows:

1. When the reaction mixture is mostly reactants:

$$Q \ll 1 \qquad RT \ln Q \ll 0 \qquad \Delta G < 0$$

The total free energy decreases as the reaction proceeds spontaneously in the forward direction.

2. When the reaction mixture is mostly products:

$$Q \gg 1 \qquad RT \ln Q \gg 0 \qquad \Delta G > 0$$

The total free energy decreases as the reaction proceeds spontaneously in the reverse direction.

Figure 17.10 shows how the total free energy of a reaction mixture changes as the reaction progresses. Because the free energy decreases as pure reactants form products and also decreases as pure products form reactants, the free-energy curve must go through a minimum somewhere

FIGURE 17.10 The total free energy of a reaction mixture as a function of the progress of the reaction. Beginning with either pure reactants or pure products, the free energy decreases (ΔG is negative) as the system moves toward equilibrium. The graph is drawn assuming that the pure reactants and pure products are in their standard states and that $\Delta G°$ for the reaction is negative. When $\Delta G°$ is negative, the equilibrium composition is rich in products.

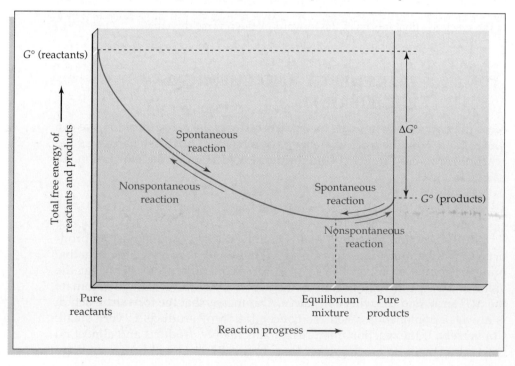

between pure reactants and pure products. At that minimum-free-energy composition, the system is at equilibrium because conversion either of reactants to products or of products to reactants would involve an increase in free energy. The equilibrium composition persists indefinitely unless the system is disturbed by an external influence.

The sign of ΔG for the reaction is the same as the sign of the slope of the free-energy curve (Figure 17.10). To the left of the equilibrium composition, ΔG and the slope of the curve are negative, and the free energy decreases as reactants are converted to products. To the right of the equilibrium composition, ΔG and the slope of the curve are positive. At the equilibrium composition, ΔG and the slope of the curve are zero, and no reaction occurs.

We can now derive a relationship between free energy and the equilibrium constant. At equilibrium, ΔG for a reaction is zero and the reaction quotient Q is equal to the equilibrium constant K. Substituting $\Delta G = 0$ and $Q = K$ into the equation

$$\Delta G = \Delta G^\circ + RT \ln Q$$

gives
$$0 = \Delta G^\circ + RT \ln K$$

or
$$\Delta G^\circ = -RT \ln K$$

This equation is one of the most important relationships in chemical thermodynamics because it allows us to calculate the equilibrium constant for a reaction from the standard free-energy change, or vice versa.

The relationship between ΔG° and $\ln K$ is especially useful when K is difficult to measure. Consider a reaction so slow that it takes more than an experimenter's lifetime to reach equilibrium or a reaction that goes essentially to completion, so that the equilibrium concentrations of the reactants are extremely small and hard to measure. We can't measure K directly in such cases, but we can calculate its value from ΔG°.

The relationship between ΔG° and the equilibrium constant K is summarized in Table 17.4. A reaction with a negative value of ΔG° has an equilibrium constant greater than 1, which corresponds to a minimum in the free-energy curve of Figure 17.10 at a composition rich in products. Conversely, a reaction that has a positive value of ΔG° has an equilibrium constant less than 1, and a minimum in the free-energy curve at a composition rich in reactants. Try redrawing Figure 17.10 for the case where $\Delta G^\circ > 0$.

TABLE 17.4 Relationship between the Standard Free-Energy Change and the Equilibrium Constant for a Reaction: $\Delta G^\circ = -RT \ln K$

ΔG°	$\ln K$	K	Comment
$\Delta G^\circ < 0$	$\ln K > 0$	$K > 1$	The equilibrium mixture is mainly products
$\Delta G^\circ > 0$	$\ln K < 0$	$K < 1$	The equilibrium mixture is mainly reactants
$\Delta G^\circ = 0$	$\ln K = 0$	$K = 1$	The equilibrium mixture contains comparable amounts of reactants and products ($K = 1$ for 1 M concentrations and 1 atm partial pressures)

EXAMPLE 17.8

Use the thermodynamic data in Appendix B to calculate the equilibrium constant at 25°C for the following reaction:

$$N_2(g) + 3\,H_2(g) \rightleftarrows 2\,NH_3(g)$$

SOLUTION   First calculate $\Delta G°$ for the reaction from the tabulated value of $\Delta G_f°$ for ammonia (-16.5 kJ/mol). Because values of $\Delta G_f°$ for elements in their standard states are zero, $\Delta G°$ for the reaction is just twice $\Delta G_f°(NH_3)$:

$$\Delta G° = (2\text{ mol})(-16.5\text{ kJ/mol}) = -33.0\text{ kJ}$$

Solving the equation $\Delta G° = -RT \ln K = -2.303RT \log K$

for $\log K$ gives

$$\log K = \frac{-\Delta G°}{2.303RT} = \frac{-(-33.0 \times 10^3\text{ J})}{(2.303)(8.314\text{ J/K})(298\text{ K})} = 5.78$$

Therefore,

$$K = K_p = 10^{5.78} = 6.0 \times 10^5$$

The equilibrium constant obtained by this procedure is K_p because reactants and products are gases, and their standard states are defined in terms of pressure. If we want the value of K_c, we must calculate it from the relation $K_p = K_c(RT)^{\Delta n}$ (Section 13.3), where R must be expressed in the proper units [$R = 0.0821$ L · atm/(K · mol)].

EXAMPLE 17.9

The value of $\Delta G_f°$ at 25°C for gaseous mercury is 31.85 kJ/mol. What is the vapor pressure of mercury at 25°C?

SOLUTION   The vapor pressure in atm is equal to K_p for the reaction

$$Hg(l) \rightleftarrows Hg(g)   K_p = P_{Hg}$$

Note that $Hg(l)$ is omitted from the equilibrium constant expression because it is a pure liquid. Since the standard state for elemental mercury is the pure liquid, $\Delta G_f° = 0$ for $Hg(l)$, and $\Delta G°$ for the vaporization reaction is simply equal to $\Delta G_f°$ for 1 mol of $Hg(g)$ (31.85 kJ). We can calculate K_p from $\Delta G°$ as in Example 17.8:

$$\log K_p = \frac{-\Delta G°}{2.303RT} = \frac{-(31.85 \times 10^3\text{ J})}{(2.303)(8.314\text{ J/K})(298\text{ K})} = -5.58$$

$$K_p = 10^{-5.58} = 2.6 \times 10^{-6}$$

Since K_p is defined in units of atm pressure, the vapor pressure of mercury at 25°C is 2.6×10^{-6} atm (0.0020 mm Hg). Because the vapor pressure is appreciable and mercury is toxic in the lungs, mercury should not be handled without adequate ventilation.

Mercury has an appreciable vapor pressure at room temperature, and its handling requires adequate ventilation.

EXAMPLE 17.10

At 25°C, K_{sp} for PbCrO$_4$ is 2.8×10^{-13}. Calculate the standard free-energy change at 25°C for the reaction PbCrO$_4(s) \rightleftarrows$ Pb$^{2+}(aq)$ + CrO$_4{}^{2-}(aq)$.

SOLUTION   We can calculate $\Delta G°$ directly from the equilibrium constant K_{sp}:

$$\Delta G° = -2.303RT \log K_{sp} = (-2.303)(8.314 \text{ J/K})(298 \text{ K})(\log 2.8 \times 10^{-13})$$

$$= 71.6 \times 10^3 \text{ J} = 71.6 \text{ kJ}$$

Note that $\Delta G°$ is a large positive number, in accord with a K_{sp} value much less than 1.

⌐ **PROBLEM 17.11**   Given the data in Appendix B, calculate K_p at 25°C for the reaction CaCO$_3(s) \rightleftarrows$ CaO(s) + CO$_2(g)$.

⌐ **PROBLEM 17.12**   Use the data in Appendix B to calculate the vapor pressure of water at 25°C.

⌐ **PROBLEM 17.13**   At 25°C, K_w for the dissociation of water is 1.0×10^{-14}. Calculate $\Delta G°$ for the reaction 2 H$_2$O$(l) \rightleftarrows$ H$_3$O$^+(aq)$ + OH$^-(aq)$.

interlude—SOME RANDOM THOUGHTS ABOUT ENTROPY

Cosmologists believe that entropy and the directionality of time arise from the expansion of the universe.

The increase in entropy that occurs during nuclear reactions on the sun makes possible the evolution of life on earth.

The idea of entropy has intrigued thinkers for more than a century. Poets, philosophers, physicists, and biologists have all struggled to understand the consequences of entropy.

Poets and philosophers have spoken of entropy as "time's arrow," a metaphor that arises out of the second law of thermodynamics. According to the second law, all spontaneously occurring processes are accompanied by an increase in the disorder of the universe. Every time a person breathes, every time a mountain range crumbles, and every time a star goes through its life cycle the amount of disorder in the universe increases. At some far distant time, when there is no order or available energy left and all is disorder, the universe, as we know it, must end.

Physicists speak of entropy as giving a directionality to time. There is a symmetry to basic physical laws that makes them equally valid when the signs of the quantities are reversed. The attraction between a positively charged proton and a negatively charged electron, for example, is exactly the same as the attraction between a negatively charged proton (a so-called antiproton) and a positively charged electron (a positron). Time, however, cannot be reversed because of entropy. Any process that takes place spontaneously over time must increase the entropy of the universe. The reverse process, which would have to go backward over time, can't occur because it would decrease the entropy of the universe. Thus there is a one-way nature to time that is not shared by other physical quantities. Cosmologists believe, in fact, that time's directionality is due ultimately to the expansion of the universe resulting from the "big bang" at the moment of creation.

Biologists, too, are intrigued by entropy and its consequences. Their problem is that, if the disorder of the universe is always increasing, how is it possible for enormously complex and increasingly sophisticated life forms to evolve? After all, the more complex the organism, the greater the amount of order required and the *lower* the entropy. The answer to the biologists' question again arises from the second law of thermodynamics: All spontaneous processes increase the disorder of the *universe*—that is, the total disorder of both system and surroundings. It's perfectly possible, however, for the disorder of any system to *decrease* spontaneously as long as the disorder of the surroundings *increases* by an even greater amount.

In nature, all life on earth can be thought of as the system, and our entire solar system as the surroundings. The energy used to drive evolution and to power all living organisms on earth comes from sunlight, caused ultimately by nuclear reactions in the sun. Evolution and the organization of small molecules into living organisms proceed with a decrease in entropy, but the nuclear reactions in the sun proceed with an even larger increase in entropy. Thus, the sun has been called "that great source of negative entropy in the sky."

Thermodynamics deals with the interconversion of heat and other forms of energy. It is important in chemistry because it allows us to predict the direction and extent of chemical reactions and other spontaneous processes. A **spontaneous process** proceeds on its own without any external influence. All spontaneous reactions move toward equilibrium, although thermodynamics tells us nothing about the rate at which equilibrium is achieved.

Entropy, denoted by S, is a state function that measures molecular disorder or randomness. The entropy of a system (reactants plus products) increases (ΔS is positive) for the following processes: phase transitions that convert a solid to a liquid or a liquid to a gas; reactions that increase the number of gaseous molecules; dissolution of molecular solids and certain salts in water; raising the temperature of a substance; expansion of a gas at constant temperature.

A disordered state of a system (state of high entropy) can be achieved in more ways (W) than an ordered state and is therefore more probable. The entropy of a state can be calculated from Boltzmann's formula, $S = k \ln W$. Entropy is associated with randomness in the orientation and distribution of molecules in space and with randomness in the distribution of the total energy of a substance. According to the **third law of thermodynamics**, the entropy of a pure, perfectly ordered crystalline substance at 0 K is zero.

The **standard molar entropy**, $S°$, of a substance is the absolute entropy of 1 mol of the pure substance at 1 atm pressure and a specified temperature, usually 25°C. The entropy change for a reaction, the **standard entropy of reaction**, $\Delta S°$, can be calculated from the relation $\Delta S° = S°(\text{products}) - S°(\text{reactants})$.

The **first law of thermodynamics** states that in any process (spontaneous or nonspontaneous) the total energy of a system and its surroundings remains constant. The **second law of thermodynamics** says that in any *spontaneous* process the total entropy of a system and its surroundings ($\Delta S_{\text{total}} = \Delta S_{\text{sys}} + \Delta S_{\text{surr}}$) always increases. A chemical reaction is spontaneous if $\Delta S_{\text{total}} > 0$, nonspontaneous if $\Delta S_{\text{total}} < 0$, and at equilibrium if $\Delta S_{\text{total}} = 0$. Reactions that are nonspontaneous in the forward direction are spontaneous in the reverse direction. For a reaction at constant pressure, $\Delta S_{\text{surr}} = -\Delta H/T$, and ΔS_{sys} is $\Delta S°$ for the reaction.

Free energy, $G = H - TS$, is a state function that indicates whether a reaction is spontaneous or nonspontaneous without regard to the surroundings. A reaction at constant temperature and pressure is spontaneous if $\Delta G < 0$, nonspontaneous if $\Delta G > 0$, and at equilibrium if $\Delta G = 0$. In the equation $\Delta G = \Delta H - T\Delta S$, temperature is a weighting factor that determines the relative importance of the enthalpy and entropy contributions to ΔG. When ΔH and ΔS have the same sign, the direction of spontaneous reaction depends on the temperature.

The **standard free-energy change**, $\Delta G°$, for a reaction is the change in free energy that occurs when reactants in their standard states are converted to products in their standard states. The standard state of a substance is the pure substance at 1 atm pressure (1 M concentration for solutes in solution) at a specified temperature, usually 25°C. The **standard free energy of formation**, $\Delta G_f°$, of a substance is the free-energy change for formation of 1 mol of the substance in its standard state from the most stable form of the constituent elements in their standard states. Substances with a negative value of ΔG_f° are thermodynamically stable with respect to the constituent elements. We can calculate ΔG° for a reaction in either of two ways: (1) $\Delta G^\circ = \Delta G_f^\circ(\text{products}) - \Delta G_f^\circ(\text{reactants})$ or (2) $\Delta G^\circ = \Delta H^\circ - T\Delta S^\circ$.

The free-energy change, ΔG, for a reaction under non-standard-state conditions is given by $\Delta G = \Delta G^\circ + RT \ln Q$, where Q is the reaction quotient. A plot of the total free energy of a reaction mixture versus reaction progress goes through a minimum somewhere between pure reactants and pure products. At that minimum, the system is in equilibrium, with $\Delta G = 0$ and $Q = K$. As a result, $\Delta G^\circ = -RT \ln K$, an important relation that allows us to calculate the equilibrium constant from ΔG° and vice versa.

UNDERSTANDING KEY CONCEPTS

1. Ideal gases A (red spheres) and B (blue spheres) occupy two separate bulbs. The contents of both bulbs constitute the initial state of an isolated system. Consider the process that occurs when the stopcock is opened.

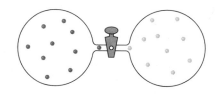

   (a) Draw a picture that represents the final (equilibrium) state of the system.
   (b) What are the signs (+, −, or 0) of ΔH, ΔS, and ΔG for this process? Explain.
   (c) How does this process illustrate the second law of thermodynamics?
   (d) Is the reverse process spontaneous or nonspontaneous? Explain.

2. What are the signs (+, −, or 0) of ΔH, ΔS, and ΔG for the spontaneous sublimation of a crystalline solid? Explain.

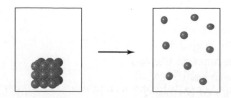

3. Consider the dissociation reaction $A_2(g) \rightleftarrows 2\,A(g)$. The following pictures represent two possible initial states and the equilibrium state of the system.

Initial state 1

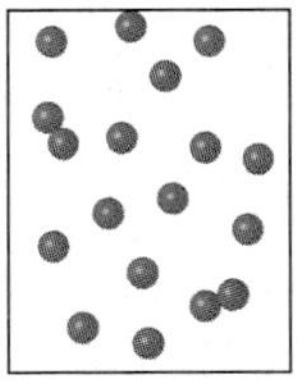

Initial state 2

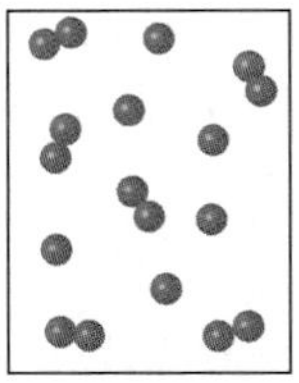

Equilibrium state

   (a) Is the reaction quotient Q_p for initial state 1 greater than, less than, or equal to the equilibrium constant K_p? Is Q_p for initial state 2 greater than, less than, or equal to K_p?
   (b) What are the signs (+, −, or 0) of ΔH, ΔS, and ΔG when the system goes from initial state 1 to the equilibrium state? Explain. Is this a spontaneous process?
   (c) What are the signs (+, −, or 0) of ΔH, ΔS, and ΔG when the system goes from initial state 2 to the equilibrium state? Explain. Is this a spontaneous process?
   (d) Relate each of the preceding pictures to the graph in Figure 17.10.

4. Consider again the dissociation reaction $A_2(g) \rightleftarrows 2\,A(g)$.

(a) What are the signs (+, −, or 0) of the standard enthalpy change, $\Delta H°$, and the standard entropy change, $\Delta S°$, for the forward reaction?
(b) Distinguish between the meaning of $\Delta S°$ for the dissociation reaction and ΔS for the process (Problem 3) in which the system goes from initial state 1 to the equilibrium state.
(c) Can you say anything about the sign of $\Delta G°$ for the dissociation reaction? How does $\Delta G°$ depend

on temperature. Will $\Delta G°$ increase, decrease, or remain the same if the temperature increases?
(d) Will the equilibrium constant K_p increase, decrease, or remain the same if the temperature increases? How will the picture for the equilibrium state (Problem 3) change if the temperature increases?
(e) What is the value of ΔG for the dissociation reaction when the system is at equilibrium?

ADDITIONAL PROBLEMS

Problems 17.1–17.13 appear within the chapter.

SPONTANEOUS PROCESSES

17.14 Distinguish between a spontaneous process and a nonspontaneous process, and give an example of each.

17.15 Which of the following processes are spontaneous, and which are nonspontaneous?
(a) freezing of water at 2°C
(b) corrosion of iron metal
(c) expansion of a gas to fill the available volume
(d) separation of an unsaturated aqueous solution of potassium chloride into solid KCl and liquid water

17.16 Tell whether the following processes are spontaneous or nonspontaneous:
(a) dissolution of sugar in hot coffee
(b) decomposition of NaCl to solid sodium and gaseous chlorine at 25°C and 1 atm pressure
(c) uniform mixing of bromine vapor and nitrogen gas
(d) boiling of gasoline at 25°C and 1 atm pressure

17.17 Assuming that gaseous reactants and products are present at 1 atm partial pressure, which of the fol-

lowing reactions are spontaneous in the forward direction?
(a) $N_2(g) + 2 H_2(g) \longrightarrow N_2H_4(l)$
$$K_p = 7 \times 10^{-27}$$
(b) $2 Mg(s) + O_2(g) \longrightarrow 2 MgO(s)$
$$K_p = 2 \times 10^{198}$$
(c) $MgCO_3(s) \longrightarrow MgO(s) + CO_2(g)$
$$K_p = 9 \times 10^{-10}$$
(d) $2 CO(g) + O_2(g) \longrightarrow 2 CO_2(g)$
$$K_p = 1 \times 10^{90}$$

17.18 Assuming that dissolved reactants and products are present at 1 M concentrations, which of the following reactions are nonspontaneous in the forward direction?
(a) $HCN(aq) + H_2O(l) \longrightarrow H_3O^+(aq) + CN^-(aq)$
$$K = 4.9 \times 10^{-10}$$
(b) $H_3O^+(aq) + OH^-(aq) \longrightarrow 2 H_2O(l)$
$$K = 1.0 \times 10^{14}$$
(c) $Ba^{2+}(aq) + CO_3^{2-}(aq) \longrightarrow BaCO_3(s)$
$$K = 3.8 \times 10^8$$
(d) $AgCl(s) \longrightarrow Ag^+(aq) + Cl^-(aq)$
$$K = 1.8 \times 10^{-10}$$

ENTROPY

17.19 Define entropy, and give an example of a process in which the entropy of a system increases.

17.20 Comment on the following statement: Exothermic reactions are spontaneous, but endothermic reactions are nonspontaneous.

17.21 Predict the sign of the entropy change in the system for each of the following processes:

(a) a solid sublimes
(b) a liquid freezes
(c) AgI precipitates from a solution containing Ag^+ and I^- ions
(d) gaseous CO_2 bubbles out of a carbonated beverage

17.22 Predict the sign of ΔS in the system for each of the following reactions.
   (a) $PCl_5(g) \longrightarrow PCl_3(g) + Cl_2(g)$
   (b) $CH_4(g) + 2\ O_2(g) \longrightarrow CO_2(g) + 2\ H_2O(l)$
   (c) $2\ H_3O^+(aq) + CO_3^{2-}(aq) \longrightarrow CO_2(g) + 3\ H_2O(l)$
   (d) $Mg(s) + Cl_2(g) \longrightarrow MgCl_2(s)$

17.23 Predict the sign of ΔS for each process in Problem 17.15.

17.24 Predict the sign of ΔS for each process in Problem 17.16.

17.25 Use Boltzmann's formula to calculate the entropy of the following arrangements of six quarters:
   (a) all heads
   (b) all tails
   (c) a random arrangement of heads and tails

17.26 Use Boltzmann's formula to calculate the entropy of the following arrangements of 100 pennies:
   (a) all heads
   (b) all tails
   (c) random arrangement of heads and tails

17.27 Consider a disordered crystal of monodeuteriomethane in which each tetrahedral CH_3D molecule is oriented randomly in one of four possible ways. Use Boltzmann's formula to calculate the entropy of the disordered state of the crystal if the crystal contains
   (a) 12 molecules
   (b) 120 molecules
   (c) 1 mol of molecules

What is the entropy of the crystal if the C–D bond of each of the CH_3D molecules points in the same direction?

17.28 Consider the distribution of ideal-gas molecules among three bulbs (A, B, and C) of equal volume. For each of the following states, determine the number of ways (W) that the state can be achieved, and use Boltzmann's formula to calculate the entropy of each of the following states.
   (a) 2 molecules in bulb A
   (b) 2 molecules randomly distributed among bulbs A, B, and C
   (c) 3 molecules in bulb A
   (d) 3 molecules randomly distributed among bulbs A, B, and C
   (e) 1 mol of molecules in bulb A
   (f) 1 mol of molecules randomly distributed among bulbs A, B, and C

What is ΔS on going from **(e)** to **(f)**? Compare that result with ΔS calculated from the equation $\Delta S = R \ln (V_{final}/V_{initial})$.

17.29 Which state in each of the following pairs has the higher entropy per mole of substance?
   (a) H_2 at 25°C in a volume of 10 L or H_2 at 25°C in a volume of 50 L
   (b) O_2 at 25°C and 1 atm or O_2 at 25°C and 10 atm
   (c) H_2 at 25°C and 1 atm or H_2 at 100°C and 1 atm
   (d) CO_2 at STP or CO_2 at 100°C and 0.1 atm

17.30 Which state in each of the following pairs has the higher entropy per mole of substance?
   (a) ice at −40°C or ice at 0°C
   (b) N_2 at STP or N_2 at 0°C and 10 atm
   (c) N_2 at STP or N_2 at 0°C in a volume of 50 L
   (d) water vapor at 150°C and 1 atm or water vapor at 100°C and 2 atm

STANDARD MOLAR ENTROPIES AND STANDARD ENTROPIES OF REACTION

17.31 What is meant by the standard molar entropy of a substance? Explain how standard molar entropies are used to calculate standard entropies of reaction.

17.32 Which substance in each of the following pairs would you expect to have the higher standard molar entropy? Justify your answer in each case.
   (a) $C_2H_2(g)$ or $C_2H_6(g)$
   (b) $CO_2(g)$ or $CO(g)$
   (c) $I_2(s)$ or $I_2(g)$
   (d) $CH_3OH(g)$ or $CH_3OH(l)$

17.33 Which substance in each of the following pairs would you expect to have the higher standard molar entropy? Explain your reasons in each case.
   (a) $P_4(s)$ or $P_4O_{10}(s)$ **(b)** $SO_2(g)$ or $SO_3(g)$
   (c) $Hg(l)$ or $Hg(s)$ **(d)** $CO_2(g)$ or $CO_2(s)$

17.34 Use the standard molar entropies in Appendix B to calculate $\Delta S°$ at 25°C for each of the following reactions. Account for the sign of the entropy change in each case.
   (a) $2\ H_2O_2(l) \longrightarrow 2\ H_2O(l) + O_2(g)$
   (b) $2\ Na(s) + Cl_2(g) \longrightarrow 2\ NaCl(s)$
   (c) $2\ O_3(g) \longrightarrow 3\ O_2(g)$
   (d) $4\ Al(s) + 3\ O_2(g) \longrightarrow 2\ Al_2O_3(s)$

17.35 Use the $S°$ values in Appendix B to calculate $\Delta S°$ at 25°C for each of the following reactions. Suggest a reason for the sign of $\Delta S°$ in each case.
   (a) $2\ S(s) + 3\ O_2(g) \longrightarrow 2\ SO_3(g)$
   (b) $SO_3(g) + H_2O(l) \longrightarrow H_2SO_4(aq)$
   (c) $AgCl(s) \longrightarrow Ag^+(aq) + Cl^-(aq)$
   (d) $NH_4NO_3(s) \longrightarrow N_2O(g) + 2\ H_2O(g)$

ENTROPY AND THE SECOND LAW OF THERMODYNAMICS

17.36 State the second law of thermodynamics.

17.37 An isolated system is one that exchanges neither matter nor energy with the surroundings. What is the entropy criterion for spontaneous change in an isolated system? Give an example of a spontaneous process in an isolated system.

17.38 Give an equation that relates the entropy change in the surroundings to the enthalpy change in the system. What is the sign of ΔS_{surr} for **(a)** an exothermic reaction and **(b)** an endothermic reaction?

17.39 When heat is added to the surroundings, the entropy of the surroundings increases. How does ΔS_{surr} depend on the temperature of the surroundings? Explain.

17.40 Use the data in Appendix B to calculate ΔS_{sys}, ΔS_{surr}, and ΔS_{total} at 25°C for the reaction

$$N_2(g) + 2\ O_2(g) \rightarrow N_2O_4(g)$$

Is this reaction spontaneous under standard-state conditions at 25°C?

17.41 Copper metal is obtained by smelting copper (I) sulfide ores:

$$Cu_2S(s) + O_2(g) \rightarrow 2\ Cu(s) + SO_2(g)$$

Use the data in Appendix B to calculate ΔS_{sys}, ΔS_{surr}, and ΔS_{total} at 25°C for this reaction. Is the reaction spontaneous under standard-state conditions at 25°C?

17.42 For vaporization of benzene, $\Delta H_{vap} = 30.7$ kJ/mol and $\Delta S_{vap} = 87.0$ J/(K · mol). Calculate ΔS_{surr} and ΔS_{total} at **(a)** 70°C **(b)** 80°C **(c)** 90°C. Does benzene boil at 70°C and 1 atm pressure? Calculate the normal boiling point of benzene.

17.43 For melting of sodium chloride, $\Delta H_{fusion} = 30.2$ kJ/mol and $\Delta S_{fusion} = 28.1$ J/(K · mol). Calculate ΔS_{surr} and ΔS_{total} at **(a)** 1050 K **(b)** 1075 K **(c)** 1100 K. Does NaCl melt at 1100 K? Calculate the melting point of NaCl.

FREE ENERGY

17.44 Define free energy, G, and tell how ΔG is related to ΔH and ΔS.

17.45 Explain how ΔG can be used to decide whether a reaction at constant temperature and pressure is spontaneous, nonspontaneous, or at equilibrium.

17.46 Describe how the signs of ΔH and ΔS determine whether a reaction is spontaneous or nonspontaneous at constant temperature and pressure.

17.47 What determines the direction of spontaneous reaction when ΔH and ΔS are both positive or both negative? Explain.

17.48 Given the data in Problem 17.42, calculate ΔG for vaporization of benzene at **(a)** 70°C **(b)** 80°C **(c)** 90°C. Use the values of ΔG to predict whether benzene boils at each of these temperatures and 1 atm pressure.

17.49 Given the data in Problem 17.43, calculate ΔG for melting of sodium chloride at **(a)** 1050 K **(b)** 1075 K **(c)** 1100 K. Use the values of ΔG to predict whether NaCl melts at each of these temperatures and 1 atm pressure.

17.50 Calculate the melting point of benzoic acid (C_6H_5COOH) given the following data: $\Delta H_{fusion} = 17.3$ kJ/mol and $\Delta S_{fusion} = 43.8$ J/K · mol.

17.51 Calculate the enthalpy of fusion of naphthalene ($C_{10}H_8$) given that its melting point is 128°C and its entropy of fusion is 47.7 J/K · mol.

STANDARD FREE-ENERGY CHANGES AND STANDARD FREE ENERGIES OF FORMATION

17.52 Define **(a)** the standard free-energy change, $\Delta G°$, for a reaction and **(b)** the standard free energy of formation, $\Delta G_f°$, of a substance.

17.53 What is meant by the standard state of a substance?

17.54 Use the data in Appendix B to calculate $\Delta H°$ and $\Delta S°$ for each of the following reactions. From the values of $\Delta H°$ and $\Delta S°$, calculate $\Delta G°$ at 25°C, and predict whether each reaction is spontaneous under standard-state conditions.

(a) $N_2(g) + 2\ O_2(g) \longrightarrow 2\ NO_2(g)$
(b) $2\ KClO_3(s) \longrightarrow 2\ KCl(s) + 3\ O_2(g)$
(c) $CH_3CH_2OH(l) + O_2(g) \longrightarrow CH_3COOH(l) + H_2O(l)$

17.55 Use the data in Appendix B to calculate $\Delta H°$ and $\Delta S°$ for each of the following reactions. From the values of $\Delta H°$ and $\Delta S°$, calculate $\Delta G°$ at 25°C, and predict whether each reaction is spontaneous under standard-state conditions.

(a) $2 SO_2(g) + O_2(g) \longrightarrow 2 SO_3(g)$
(b) $N_2(g) + 2 H_2(g) \longrightarrow N_2H_4(l)$
(c) $CH_3OH(l) + O_2(g) \longrightarrow HCOOH(l) + H_2O(l)$

17.56 Use the standard free energies of formation in Appendix B to calculate $\Delta G°$ at 25°C for each reaction in Problem 17.54.

17.57 Use the standard free energies of formation in Appendix B to calculate $\Delta G°$ at 25°C for each reaction in Problem 17.55.

17.58 Use the data in Appendix B to tell which of the following compounds are thermodynamically stable with respect to the constituent elements at 25°C:

(a) $BaCO_3(s)$ **(b)** $HBr(g)$ **(c)** $N_2O(g)$ **(d)** $C_2H_4(g)$

17.59 Use the data in Appendix B to decide whether synthesis of the following compounds from the constituent elements is thermodynamically feasible at 25°C:

(a) $N_2O_5(g)$ **(b)** $H_2S(g)$ **(c)** $H_2Se(g)$ **(d)** $CCl_4(l)$

17.60 Ethanol is manufactured in industry by hydration of ethylene.

$$CH_2{=}CH_2(g) + H_2O(l) \rightarrow CH_3CH_2OH(l)$$

Using the data in Appendix B to calculate $\Delta G°$, show that this reaction is spontaneous at 25°C. Why does this reaction become nonspontaneous at higher temperatures? Estimate the temperature at which the reaction becomes nonspontaneous.

17.61 Sulfur dioxide in the effluent gases from coal-burning electric power plants is one of the principal causes of acid rain. One method for reducing SO_2 emissions involves partial reduction of SO_2 to H_2S, followed by catalytic conversion of the H_2S and the remaining SO_2 to elemental sulfur:

$$2 H_2S(g) + SO_2(g) \rightarrow 3 S(s) + 2 H_2O(g)$$

Using the data in Appendix B to calculate $\Delta G°$, show that this reaction is spontaneous at 25°C. Why does this reaction become nonspontaneous at high temperatures? Estimate the temperature at which the reaction becomes nonspontaneous.

17.62 Consider the conversion of acetylene to benzene:

$$3 C_2H_2(g) \rightarrow C_6H_6(l)$$

Is it worth looking for a catalyst for this reaction? Is it possible to synthesize benzene from graphite and gaseous H_2 at 25°C and 1 atm pressure?

17.63 Consider the conversion of dichloroethane to vinyl chloride, the starting material for manufacturing polyvinyl chloride (PVC) plastics:

$$\underset{\text{Dichloroethane}}{CH_2ClCH_2Cl(l)} \rightarrow \underset{\text{Vinyl chloride}}{CH_2{=}CHCl(g)} + HCl(g)$$

Is this reaction spontaneous under standard-state conditions? Would it help to carry out the reaction in the presence of base to remove HCl? Explain. Is it possible to synthesize vinyl chloride from graphite, gaseous H_2, and gaseous Cl_2 at 25°C and 1 atm pressure?

FREE ENERGY, COMPOSITION, AND CHEMICAL EQUILIBRIUM

17.64 What is the relationship between the free-energy change under non-standard-state conditions, ΔG, the free-energy change under standard-state conditions, $\Delta G°$, and the reaction quotient, Q?

17.65 Compare the values of ΔG and $\Delta G°$ when **(a)** $Q < 1$, **(b)** $Q = 1$, and **(c)** $Q > 1$. Does the thermodynamic driving force increase or decrease as Q increases?

17.66 Sulfuric acid is produced in larger amounts by weight than any other chemical. It is used in manufacturing fertilizers, in oil refining, and in hundreds of other processes. An intermediate step in the industrial process for synthesis of H_2SO_4 involves catalytic oxidation of sulfur dioxide:

$$2 SO_2(g) + O_2(g) \rightarrow 2 SO_3(g) \qquad \Delta G° = -141.8 \text{ kJ}$$

Calculate ΔG at 25°C given the following sets of partial pressures:

(a) 100 atm SO_2, 100 atm O_2, 1.0 atm SO_3
(b) 2.0 atm SO_2, 1.0 atm O_2, 10 atm SO_3
(c) each reactant and product at a partial pressure of 1.0 atm

17.67 Urea (NH_2CONH_2), an important nitrogen fertilizer, is produced industrially by the reaction

$$2 NH_3(g) + CO_2(g) \rightarrow NH_2CONH_2(aq) + H_2O(l)$$

Given that $\Delta G° = -13.6$ kJ, calculate ΔG at 25°C for the following sets of conditions.

(a) 10 atm NH_3, 10 atm CO_2, 1 M NH_2CONH_2
(b) 0.10 atm NH_3, 0.10 atm CO_2, 1 M NH_2CONH_2.
Is the reaction spontaneous for the conditions in **(a)** and/or **(b)**?

17.68 What is the relationship between the standard free-energy change, $\Delta G°$, for a reaction and the equilibrium constant, K? What is the sign of $\Delta G°$ when
(a) $K > 1$ (b) $K = 1$ (c) $K < 1$

17.69 Do you expect a large or a small value of the equilibrium constant for a reaction when
(a) $\Delta G°$ is positive (b) $\Delta G°$ is negative

17.70 Calculate the equilibrium constant at 25°C for the reaction in Problem 17.66.

17.71 Calculate the equilibrium constant at 25°C for the reaction in Problem 17.67.

17.72 Given values of $\Delta G_f°$ at 25°C for liquid ethanol (-174.9 kJ/mol) and gaseous ethanol (-168.6 kJ/mol), calculate the vapor pressure of ethanol at 25°C.

17.73 At 25°C, K_a for acid dissociation of aspirin ($C_9H_8O_4$) is 3.0×10^{-4}. Calculate $\Delta G°$ for the reaction $C_9H_8O_4(aq) + H_2O(l) \rightleftarrows H_3O^+(aq) + C_9H_7O_4^-(aq)$

17.74 Used for making antifreeze (ethylene glycol, $HOCH_2CH_2OH$), ethylene oxide is produced industrially by catalyzed air oxidation of ethylene:

$$2 CH_2{=}CH_2 (g) + O_2 (g) \rightarrow 2 CH_2{-}CH_2 (g)$$
$$\underset{O}{\diagdown\diagup}$$

Ethylene oxide

Use the data in Appendix B to calculate $\Delta G°$ and K_p for this reaction at 25°C.

17.75 Use the data in Appendix B to calculate K_p at 25°C for the reaction

$$CO(g) + 2 H_2(g) \rightleftarrows CH_3OH(g)$$

What is ΔG for this reaction at 25°C when each reactant and product is present at a partial pressure of 20 atm?

GENERAL PROBLEMS

17.76 The oxidation of iron metal, $4 Fe(s) + 3 O_2(g) \longrightarrow 2 Fe_2O_3(s)$, is a spontaneous process, even though $\Delta S°$ for the reaction is negative. Is this consistent with the second law of thermodynamics? Explain.

17.77 The Haber synthesis of ammonia, $N_2(g) + 3 H_2(g) \longrightarrow 2 NH_3(g)$, requires a catalyst. Which of the following quantities are affected by the catalyst?
(a) rate of the forward reaction ✓
(b) rate of the reverse reaction ✓
(c) spontaneity of the reaction
(d) $\Delta H°$
(e) $\Delta S°$
(f) $\Delta G°$
(g) the equilibrium constant
(h) time required to reach equilibrium ✓

17.78 The standard free-energy change at 25°C for dissociation of water is 79.9 kJ:

$$2 H_2O(l) \rightleftarrows H_3O^+(aq) + OH^-(aq) \qquad \Delta G° = 79.9 \text{ kJ}$$

For each of the following sets of concentrations, calculate ΔG at 25°C, and indicate whether the reaction is spontaneous in the forward or reverse direction.
(a) $[H_3O^+] = [OH^-] = 1.0$ M
(b) $[H_3O^+] = [OH^-] = 1.0 \times 10^{-7}$ M
(c) $[H_3O^+] = 1.0 \times 10^{-7}$ M, $[OH^-] = 1.0 \times 10^{-10}$ M

Are your results consistent with Le Châtelier's principle? Use the thermodynamic data to calculate the equilibrium constant for the reaction.

17.79 Calculate the normal boiling point of ethanol (CH_3CH_2OH) given that its enthalpy of vaporization is 38.6 kJ and its entropy of vaporization is 110 J/K.

17.80 Indicate whether the following processes are spontaneous or nonspontaneous:
(a) heat transfer from a block of ice to a room maintained at 25°C
(b) evaporation of water from an open beaker
(c) conversion of iron(III) oxide to iron metal and oxygen
(d) uphill motion of an automobile

17.81 Do you agree with the following statements? If not, explain.
(a) Spontaneous reactions are always fast. no, just able
(b) In any spontaneous process, the entropy of the system always increases. no, sys + surr
(c) An endothermic reaction is always nonspontaneous. no ice melting
(d) A reaction that is nonspontaneous in the forward direction is always spontaneous in the reverse direction. yes

17.82 When rolling a pair of dice, there are two ways of getting a point total of 3 (1 + 2; 2 + 1) but only one way of getting a point total of 2 (1 + 1). How many ways are there of getting point totals of 4 through 12? What is the most probable point total? Calculate the entropy of each point total using Boltzmann's formula.

17.83 Chloroform ($CHCl_3$) has a normal boiling point of 61°C and an enthalpy of vaporization of 29.24 kJ. What are its values of ΔG_{vap} and ΔS_{vap} at 61°C?

17.84 Make a rough, qualitative plot of standard molar entropy versus temperature for methane from 0 K to 298 K. Incorporate the following data into your plot: mp = -182°C; bp = -164°C; $S° = 186.2$ J/K at 25°C.

17.85 Use the data in Appendix B to calculate $\Delta H°$, $\Delta S°$, and $\Delta G°$ at 25°C for each of the following reactions.
(a) $2 Mg(s) + O_2(g) \longrightarrow 2 MgO(s)$
(b) $MgCO_3(s) \longrightarrow MgO(s) + CO_2(g)$
(c) $Fe_2O_3(s) + 2 Al(s) \longrightarrow Al_2O_3(s) + 2 Fe(s)$
(d) $2 NaHCO_3(s) \longrightarrow Na_2CO_3(s) + CO_2(g) + H_2O(g)$

Are these reactions spontaneous or nonspontaneous at 25°C and 1 atm pressure? How does $\Delta G°$ change when the temperature is raised?

17.86 Ammonium nitrate is dangerous because it decomposes (sometimes explosively) when heated:

$$NH_4NO_3(s) \rightarrow N_2O(g) + 2 H_2O(g)$$

(a) Using the data in Appendix B, show that this reaction is spontaneous at 25°C.
(b) How does $\Delta G°$ for the reaction change when the temperature is raised?
(c) Calculate the equilibrium constant at 25°C.
(d) Calculate ΔG for the reaction when the partial pressure of each gas is 30 atm.

17.87 Make a qualitative plot of free energy versus reaction progress for a reaction that has a positive value of $\Delta G°$. Account for the shape of the curve, and identify the point at which $\Delta G = 0$. What is the significance of that point?

17.88 The melting point of benzene is 5.5°C. Predict the signs of ΔH, ΔS, and ΔG for melting of benzene at
(a) 0 °C   (b) 15°C

17.89 Consider a twofold expansion of an ideal gas at 25°C in the isolated system shown in Figure 17.1.
(a) What are the values of ΔH, ΔS, and ΔG for the process?
(b) How does this process illustrate the second law of thermodynamics?

17.90 The entropy change for a certain nonspontaneous reaction at 50°C is 104 J/K.
(a) Is the reaction endothermic or exothermic?
(b) What is the minimum value of ΔH (in kJ) for the reaction?

17.91 Trouton's rule says that the ratio of the molar heat of vaporization of a liquid to its normal boiling point in kelvins is approximately the same for all liquids: $\Delta H_{vap}/T_{bp} \approx 88$ J/K.
(a) Check the reliability of Trouton's rule for the liquids listed in the following table.
(b) Explain why liquids tend to have the same value of $\Delta H_{vap}/T_{bp}$.

(c) Which of the liquids in the table deviates from Trouton's rule? Explain.

Liquid	bp (°C)	ΔH_{vap} (kJ/mol)
Ammonia	-77.7	23.4
Benzene	80.1	30.8
Carbon tetrachloride	76.8	29.8
Chloroform	61.1	29.2
Mercury	356.6	56.9

17.92 Use the data in Appendix B to calculate the equilibrium pressure of CO_2 in a closed vessel that contains each of the following samples.
(a) 15 g of $MgCO_3$ and 1.0 g of MgO at 25°C
(b) 15 g of $MgCO_3$ and 1.0 g of MgO at 280°C
(c) 30 g of $MgCO_3$ and 1.0 g of MgO at 280°C
You may assume that $\Delta H°$ and $\Delta S°$ are temperature independent.

17.93 K_b for dissociation of aqueous ammonia is 1.710×10^{-5} at 20°C and 1.892×10^{-5} at 50°C. What are the values of $\Delta H°$ and $\Delta S°$ for the reaction?

$$NH_3(aq) + H_2O(l) \rightleftharpoons NH_4^+(aq) + OH^-(aq)$$

17.94 The temperature dependence of the equilibrium constant is given by the linear equation

$$\ln K = \frac{-\Delta H°}{R}\left(\frac{1}{T}\right) + \frac{\Delta S°}{R}$$

where $\Delta H°$ and $\Delta S°$ are assumed to be independent of temperature.
(a) Derive this equation from equations given in this chapter.
(b) Explain how this equation can be used to determine experimental values of $\Delta H°$ and $\Delta S°$ from values of K at several different temperatures.
(c) Use this equation to predict the sign of $\Delta H°$ for a reaction whose equilibrium constant increases with increasing temperature. Is the reaction endothermic or exothermic? Is your prediction in accord with Le Châtelier's principle?

17.95 The normal boiling point of bromine is 58.8°C, and the standard entropies of the liquid and vapor are $S°[Br_2(l)] = 152.2$ J/K · mol; $S°[Br_2(g)] = 245.4$ J/K · mol. At what temperature does bromine have a vapor pressure of 227 mm Hg?

17.96 A mixture of NO_2 and N_2O_4, each at an initial partial pressure of 1 atm, is heated to 100°C. Use the data in Appendix B to calculate the partial pressure of each gas at equilibrium. You may assume that $\Delta H°$ and $\Delta S°$ are temperature independent.

$$N_2O_4(g) \rightleftharpoons 2 NO_2(g)$$

17.97 One step in the commercial synthesis of sulfuric acid involves catalytic oxidation of sulfur dioxide:

$$2\ SO_2(g) + O_2(g) \rightleftarrows 2\ SO_3(g)$$

(a) A mixture of 192 g of SO_2, 48.0 g of O_2, and a V_2O_5 catalyst is heated to 800 K in a 15.0 L vessel. Use the data in Appendix B to calculate the partial pressures of SO_3, SO_2, and O_2 at equilibrium. You may assume that $\Delta H°$ and $\Delta S°$ are temperature independent.

(b) Does the percent yield of SO_3 increase or decrease on raising the temperature from 800 K to 1000 K? Explain.

(c) Does the total pressure increase or decrease on raising the temperature from 800 K to 1000 K? Calculate the total pressure in atm at 1000 K.

chapter 18 ELECTROCHEMISTRY

Batteries are everywhere in modern technological societies. They provide the electric current to start our automobiles and to power a host of products such as pocket calculators, digital watches, heart pacemakers, radios, and tape recorders. A battery is an example of an **electrochemical cell**, a device for interconverting chemical and electrical energy. A battery takes the energy released by a spontaneous chemical reaction and uses it to produce electricity.

Electrochemistry, the area of chemistry concerned with the interconversion of chemical and electrical energy, is enormously important in modern science and technology, not only because of batteries, but also because it makes possible the manufacture of essential industrial chemicals and materials. Sodium hydroxide, for example, which is used in the manufacture of paper, textiles, soaps, and detergents, is produced by passing an electric current through an aqueous solution of sodium chloride. Chlorine, essential to the manufacture of plastics such as polyvinyl chloride (PVC), is obtained in the same process. Aluminum is also produced in an electrochemical process, as is pure copper for use in electrical wiring.

In this chapter, we'll look at the principles involved in the design and operation of electrochemical cells. In addition, we'll explore some important connections between electrochemistry and thermodynamics.

Corrosion of this sunken ship's propeller occurs by an electrochemical mechanism.

18.1 ►GALVANIC CELLS

Electrochemical cells are of two basic types: **galvanic cells** (also called **voltaic cells**) and **electrolytic cells**. In a galvanic cell, a spontaneous chemical reaction generates an electric current. In an electrolytic cell, an electric current drives a nonspontaneous reaction. The two types are therefore the reverse of each other. The names "galvanic" and "voltaic" honor the Italian scientists Luigi Galvani (1737–1798) and Alessandro Volta (1745–1827), who conducted pioneering work in the field of electrochemistry. We'll take up galvanic cells in this section and will come to electrolytic cells later. First, though, let's review some of the language of oxidation–reduction, or redox, reactions (Section 4.6).

If you immerse a strip of zinc metal in an aqueous solution of copper sulfate, you find that a dark-colored solid deposits on the surface of the zinc and that the blue color characteristic of the Cu^{2+} ion slowly disappears from the solution (Figure 18.1). Chemical analysis shows that the dark-colored deposit is finely divided copper metal and that the solution now contains zinc ions. Therefore, the reaction is

$$Zn(s) + Cu^{2+}(aq) \rightarrow Zn^{2+}(aq) + Cu(s)$$

This is a redox reaction in which Zn is oxidized to Zn^{2+} and Cu^{2+} is reduced to Cu. Recall that an *oxidation* involves a loss of electrons (an increase in oxidation number) and a *reduction* involves a gain of electrons (a decrease in oxidation number).

(a) (b)

FIGURE 18.1 **(a)** A strip of zinc is immersed in an aqueous copper sulfate solution. **(b)** As time passes, a dark-colored deposit of copper appears on the zinc, and the blue color due to Cu^{2+} fades from the solution.

We can represent the oxidation and reduction aspects of the reaction by separating the overall process into **half-reactions**:

Oxidation half-reaction: $Zn(s) \rightarrow Zn^{2+}(aq) + 2\ e^-$

Reduction half-reaction: $Cu^{2+}(aq) + 2\ e^- \rightarrow Cu(s)$

We say that Cu^{2+} is the *oxidizing agent* because, in gaining electrons from Zn, it causes the oxidation of Zn to Zn^{2+}. Similarly, we say that Zn is the *reducing agent* because, in losing electrons to Cu^{2+}, it causes the reduction of Cu^{2+} to Cu.

When the reaction is carried out as shown in Figure 18.1, electrons are transferred directly from Zn to Cu^{2+}, and the enthalpy of reaction is lost to the surroundings as heat. If, however, the reaction is carried out using the electrochemical cell depicted in Figure 18.2, then some of the chemical energy released by the reaction is converted to electrical energy, which can be used to light a light bulb or run an electric motor.

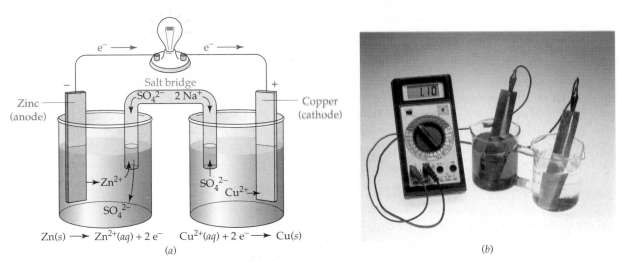

FIGURE 18.2 (a) A galvanic cell that uses oxidation of zinc metal to Zn^{2+} ions and the reduction of Cu^{2+} ions to copper metal. Note that the negative particles (electrons in the wire and anions in solution) travel around the circuit in the same direction. (b) An operating Daniell cell. The salt bridge in (a) is replaced by a porous glass frit that allows ion flow between anode and cathode compartments but prevents bulk mixing, which would bring Cu^{2+} ions into direct contact with zinc and short-circuit the cell. A digital voltmeter is a part of the external circuit (Section 18.3).

The apparatus shown in Figure 18.2 is a galvanic cell called a *Daniell cell*, after John Frederick Daniell, the English chemist who first constructed it in 1836. It consists of two *half-cells*, a beaker containing a strip of zinc that dips into an aqueous solution of zinc sulfate, and a second beaker containing a strip of copper that dips into aqueous copper sulfate. The strips of zinc and copper are called **electrodes** and are connected by a wire. In addition, the two solutions are connected by a **salt bridge**, a U-shaped tube that contains a gel permeated with a solution of an inert electrolyte, such as Na_2SO_4. The ions of the inert electrolyte do not react with the other ions in the solutions, and they are not oxidized or reduced at the electrodes.

The reaction that occurs in the Daniell cell is the same one that occurs when Zn reacts directly with Cu^{2+}, but now, because the Zn metal and Cu^{2+} ions are separated from each other, the electrons can be transferred from Zn to Cu^{2+} only through the wire. Consequently, the oxidation and reduction half-reactions occur at separate electrodes and an electric current flows through the wire. The electrode at which oxidation takes place is called the **anode** (the zinc strip in this example), and the electrode at which reduction takes place is called the **cathode** (the copper strip). Of course, the anode and cathode half-reactions must add to give the overall cell reaction:

Anode (oxidation) half-reaction:	$Zn(s) \rightarrow Zn^{2+}(aq) + 2\,e^-$
Cathode (reduction) half-reaction:	$Cu^{2+}(aq) + 2\,e^- \rightarrow Cu(s)$
Overall cell reaction:	$Zn(s) + Cu^{2+}(aq) \rightarrow Zn^{2+}(aq) + Cu(s)$

The salt bridge is necessary to complete the electrical circuit. Without it, the solution in the anode compartment would become positively charged as Zn^{2+} ions appeared in it, and the solution in the cathode compartment would become negatively charged as Cu^{2+} ions were removed from it. Because of the charge imbalance, the electrode reactions would quickly come to a halt, and electron flow through the wire would cease.

With the salt bridge in place, electrical neutrality is maintained in both compartments by a flow of ions. Anions (in this case SO_4^{2-}) flow through the salt bridge from the cathode to the anode, and cations migrate through the salt bridge from the anode to the cathode. For the cell shown in Figure 18.2, Na^+ ions move out of the salt bridge into the cathode compartment and Zn^{2+} ions move into the salt bridge from the anode compartment. (It's interesting to note that the anode and cathode get their names from the direction of ion flow between the two compartments: *An*ions move toward the *an*ode, and *cat*ions move toward the *cat*hode.)

The electrodes of commercial galvanic cells (batteries) are generally labeled with plus (+) and minus (−) signs, although the sign of the charge associated with each electrode depends on the point of view. From the perspective of the wire, the anode looks negative because a stream of negatively charged electrons comes from it. From the perspective of the solution, the anode looks positive because positively charged Zn^{2+} ions move from it. Because galvanic cells are used to supply electric current to an external circuit, it makes sense to adopt the perspective of the wire. Consequently, we regard the anode as the negative (−) electrode and the cathode as the positive (+) electrode. Thus, electrons move through the external circuit from the negative electrode, where they are produced by the anode half-reaction, to the positive electrode, where they are consumed by the cathode half-reaction.

Cathode:
- is where reduction occurs
- is where electrons are consumed
- is where cations migrate to
- has positive sign

Anode:
- is where oxidation occurs
- is where electrons are produced
- is where anions migrate to
- has negative sign

EXAMPLE 18.1

Design a galvanic cell that uses the redox reaction

$$Fe(s) + 2\ Fe^{3+}(aq) \rightarrow 3\ Fe^{2+}(aq)$$

Identify the anode and cathode half-reactions, and sketch the experimental setup. Label the anode and cathode, indicate the direction of electron and ion flow, and identify the sign of each electrode.

SOLUTION   In the overall cell reaction, iron metal is oxidized to iron(II) ions, and iron(III) ions are reduced to iron(II) ions. Therefore, the cell half-reactions are

Anode (oxidation): $Fe(s) \rightarrow Fe^{2+}(aq) + 2\ e^-$

Cathode (reduction): $2 \times [Fe^{3+}(aq) + e^- \rightarrow Fe^{2+}(aq)]$

Overall cell reaction: $Fe(s) + 2\ Fe^{3+}(aq) \rightarrow 3\ Fe^{2+}(aq)$

The cathode half-reaction has been multiplied by a factor of 2 so that the two half-reactions will add to give the overall cell reaction. When half-reactions are added, the electrons must cancel, since no electrons appear in the overall reaction.

A possible experimental setup is shown. The anode compartment consists of an iron metal electrode dipping into an aqueous solution of $Fe(NO_3)_2$. (Note, though, that *any inert electrolyte* can be used to carry the current in the anode compartment; Fe^{2+} need not be present initially because it's not a reactant in the anode half-reaction.) Since Fe^{3+} is a reactant in the cathode half-reaction, $Fe(NO_3)_3$ would be a good electrolyte for the cathode compartment. The cathode can be any electrical conductor that doesn't react with the ions in the solution. (Iron metal can't be used because it would react directly with Fe^{3+}, thus short-circuiting the cell.) A platinum wire is a common inert electrode. The salt bridge contains $NaNO_3$, but almost any inert electrolyte would do. Electrons flow through the wire from the iron anode ($-$) to the platinum cathode ($+$). Anions move from the cathode compartment toward the anode while cations migrate from the anode compartment toward the cathode.

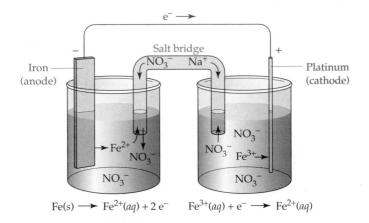

$$Fe(s) \longrightarrow Fe^{2+}(aq) + 2\,e^- \qquad Fe^{3+}(aq) + e^- \longrightarrow Fe^{2+}(aq)$$

PROBLEM 18.1 Describe a galvanic cell that uses the reaction

$$2\,Ag^+(aq) + Ni(s) \rightarrow 2\,Ag(s) + Ni^{2+}(aq)$$

Identify the anode and cathode half-reactions, and sketch the experimental setup. Label the anode and cathode, indicate the direction of electron and ion flow, and identify the sign of each electrode.

18.2 ▸SHORTHAND NOTATION FOR GALVANIC CELLS

It's convenient to have a shorthand notation for representing a galvanic cell. For the Daniell cell in Figure 18.2, which uses the reaction

$$Zn(s) + Cu^{2+}(aq) \rightarrow Zn^{2+}(aq) + Cu(s)$$

we can write the following expression:

$$Zn(s)\,|\,Zn^{2+}(aq)\,\|\,Cu^{2+}(aq)\,|\,Cu(s)$$

In this notation, each single vertical line (|) represents a phase boundary, such as that between a solid electrode and an aqueous solution, and the double vertical line (‖) denotes a salt bridge. The shorthand for the anode half-cell is always written on the left of the salt-bridge symbol, followed on the right of the symbol by the shorthand for the cathode half-cell. With that convention, electrons move through the external circuit from left to right (from anode to cathode). Reading the shorthand thus suggests the overall cell reaction: Zn is oxidized to Zn^{2+}, and Cu^{2+} is reduced to Cu.

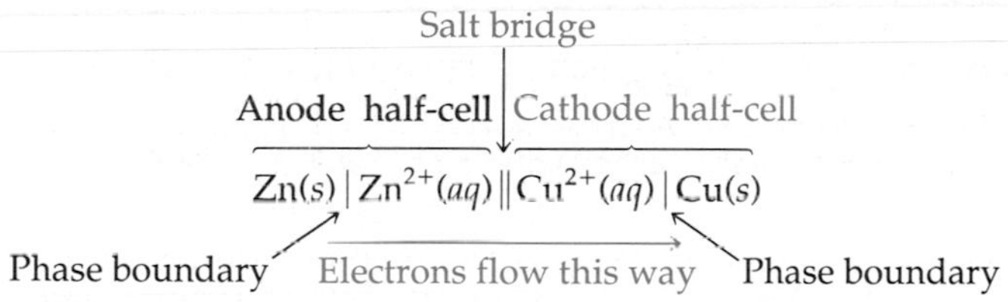

Salt bridge

Anode half-cell | Cathode half-cell

$$Zn(s) \,|\, Zn^{2+}(aq) \,\|\, Cu^{2+}(aq) \,|\, Cu(s)$$

Phase boundary   Electrons flow this way   Phase boundary

For the galvanic cell in Example 18.1 based on the reaction

$$Fe(s) + 2\,Fe^{3+}(aq) \rightarrow 3\,Fe^{2+}(aq)$$

the shorthand notation is

$$Fe(s) \,|\, Fe^{2+}(aq) \,\|\, Fe^{3+}(aq),\ Fe^{2+}(aq) \,|\, Pt(s)$$

The shorthand for the cathode half-cell includes both reactant (Fe^{3+}) and product (Fe^{2+}) as well as the electrode (Pt).

The notation for a cell involving a gas usually lists the gas immediately adjacent to the appropriate electrode. Thus, the notation

$$Cu(s) \,|\, Cu^{2+}(aq) \,\|\, Cl^-(aq) \,|\, Cl_2(g) \,|\, C(s)$$

specifies a cell in which copper is oxidized to Cu^{2+} at a copper anode and Cl_2 gas is reduced to Cl^- at a graphite (carbon) cathode. The cell reaction is

$$Cu(s) + Cl_2(g) \rightarrow Cu^{2+}(aq) + 2\,Cl^-(aq)$$

A more detailed notation would include ion concentrations and gas pressures, for example:

$$Cu(s) \,|\, Cu^{2+}(1.0\ M) \,\|\, Cl^-(1.0\ M) \,|\, Cl_2(1\ atm) \,|\, C(s)$$

EXAMPLE 18.2

Given the following cell notation

$$Pt(s) \,|\, Sn^{2+}(aq),\ Sn^{4+}(aq) \,\|\, Ag^+(aq) \,|\, Ag(s)$$

write a balanced equation for the cell reaction, and give a brief description of the cell.

SOLUTION Since the anode appears at the left in the shorthand notation, the anode (oxidation) half-reaction is

$$Sn^{2+}(aq) \rightarrow Sn^{4+}(aq) + 2\ e^-$$

The cathode (reduction) half-reaction is

$$2\ Ag^+(aq) + 2\ e^- \rightarrow 2\ Ag(s)$$

Note that we multiply the cathode half-reaction by a factor of 2 so that the electrons will cancel when we sum the two half-reactions to give the cell reaction:

$$Sn^{2+}(aq) + 2\ Ag^+(aq) \rightarrow Sn^{4+}(aq) + 2\ Ag(s)$$

The cell consists of a platinum wire anode dipping into an Sn^{2+} solution—say, $Sn(NO_3)_2(aq)$—and a silver cathode dipping into an Ag^+ solution—say, $AgNO_3(aq)$. As usual, the anode and cathode half-cells must be connected by a wire and a salt bridge.

⌐ PROBLEM 18.2 Write the shorthand notation for a galvanic cell that uses the reaction

$$Fe(s) + Sn^{2+}(aq) \rightarrow Fe^{2+}(aq) + Sn(s)$$

⌐ PROBLEM 18.3 Write a balanced equation for the overall cell reaction, and give a brief description of a galvanic cell represented by the following shorthand notation:

$$Pb(s) \mid Pb^{2+}(aq) \parallel Br_2(l) \mid Br^-(aq) \mid Pt(s)$$

18.3 ►CELL POTENTIALS AND FREE-ENERGY CHANGES FOR CELL REACTIONS

Let's return to the Daniell cell shown in Figure 18.2 to find an electrical measure of the driving force of a cell reaction. Electrons move through the external circuit from the zinc anode to the copper cathode because they have lower energy when on copper than on zinc. The driving force that pushes the negatively charged electrons away from the anode (− electrode) and pulls them toward the cathode (+ electrode) is an electrical potential called the **electromotive force (emf)**, the **cell potential (E)**, or the **cell voltage**. The SI unit of electrical potential is the volt (V). By definition, the potential of a galvanic cell is a positive quantity.

The relationship between the volt and the SI units of energy (joule, J) and electrical charge (coulomb, C) is given by the equation

$$1\ J = 1\ C \times 1\ V$$

where 1 C is the amount of charge transferred when a current of 1 ampere (A) flows for 1 s. (The current passing through a 100 W household light bulb is about 1 A, which means that the electrical charge of the electrons passing through the bulb in 1 s is 1 C.) When 1 C of charge moves between two electrodes that differ in electrical potential by 1 V, 1 J of energy is released by the cell and can be used to do electrical work.

A cell potential is measured with an electronic instrument called a *voltmeter*, which is designed to give a positive reading when the + and − terminals of the voltmeter are connected to the + (cathode) and − (anode) electrodes of the cell, respectively. Thus, the voltmeter–cell connections required to get a positive reading on the voltmeter tell us which electrode is the anode and which is the cathode.

This voltmeter shows a positive reading when its positive terminal is connected to the battery cathode (+) and its negative terminal is connected to the battery anode (−).

We've now seen two quantitative measures of the driving force of a chemical reaction: the cell potential E (an electrochemical quantity) and the free-energy change ΔG (a thermochemical quantity, Section 17.7). The values of ΔG and E are directly proportional and are related by the equation

$$\Delta G = -nFE$$

where n is the number of moles of electrons transferred in the reaction and F is the **faraday** (or *Faraday constant*), the electrical charge on 1 mol of electrons (96,485 C/mol e$^-$). In our calculations, we'll round the value of F to three significant figures:

$$1\ F = 96{,}500\ \text{C/mol e}^-$$

The faraday is named in honor of Michael Faraday (1791–1867), the nineteenth-century English scientist who laid the foundations for our current understanding of electricity.

Two features of the equation, $\Delta G = -nFE$, are worth noting: the units and the minus sign. When we multiply the charge transferred (nF) in coulombs by the cell potential (E) in volts, we obtain an energy (ΔG) in joules, in accord with the relationship 1 J = 1 C × 1 V. The minus sign is required because E and ΔG have opposite signs: The spontaneous reaction in a galvanic cell has a positive cell potential but a negative free-energy change (Section 17.7).

Later in this chapter, we'll see that cell potentials, like free-energy changes, depend on the composition of the reaction mixture. The **standard cell potential** $E°$ is the cell potential when both reactants and products are in their standard states—solutes at 1 M concentrations, gases at a partial pressure of 1 atm, solids and liquids in pure form, all at a specified temperature, usually 25°C. For example, $E°$ for the reaction

$$Zn(s) + Cu^{2+}(aq) \rightarrow Zn^{2+}(aq) + Cu(s)$$

is the cell potential measured at 25°C for a cell that has pure metal electrodes and 1 M concentrations of Zn^{2+} and Cu^{2+}. The standard free-energy change and the standard cell potential are related by the equation

$$\Delta G° = -nFE°$$

Because $\Delta G°$ and $E°$ are directly proportional, a voltmeter can be regarded as a "free-energy meter." When a voltmeter measures $E°$, it also measures $\Delta G°$.

EXAMPLE 18.3

Calculate the standard free-energy change at 25°C for the following reaction. The standard cell potential is 1.10 V at 25°C.

$$Zn(s) + Cu^{2+}(aq) \rightarrow Zn^{2+}(aq) + Cu(s)$$

SOLUTION   Two moles of electrons are transferred from Zn to Cu^{2+} in this reaction, and the standard free-energy change is therefore

$$\Delta G° = -nFE° = -(2 \text{ mol e}^-)\left(\frac{96{,}500 \text{ C}}{\text{mol e}^-}\right)(1.10 \text{ V})\left(\frac{1 \text{ J}}{1 \text{ C} \cdot \text{V}}\right)$$

$$= -212{,}000 \text{ J} = -212 \text{ kJ}$$

PROBLEM 18.4   The standard cell potential at 25°C is 0.92 V for the reaction

$$Al(s) + Cr^{3+}(aq) \rightarrow Al^{3+}(aq) + Cr(s)$$

What is the standard free-energy change for this reaction at 25°C?

18.4 ▶STANDARD REDUCTION POTENTIALS

The standard potential of any galvanic cell is the sum of the standard half-cell potentials for oxidation at the anode and for reduction at the cathode:

$$E°_{cell} = E°_{anode} + E°_{cathode}$$

Consider, for example, a cell in which zinc metal is oxidized to Zn^{2+} ions at the anode and H^+ ions are reduced to H_2 gas at the cathode (Figure 18.3):

Anode (oxidation):	$Zn(s) \rightarrow Zn^{2+}(aq) + 2 \text{ e}^-$
Cathode (reduction):	$2 H^+(aq) + 2 \text{ e}^- \rightarrow H_2(g)$
Overall cell reaction:	$Zn(s) + 2 H^+(aq) \rightarrow Zn^{2+}(aq) + H_2(g)$

FIGURE 18.3 A galvanic cell consisting of a Zn/Zn^{2+} (1 M) half-cell and a standard hydrogen electrode (S.H.E.). The S.H.E. is a piece of platinum foil that is in contact with bubbles of $H_2(g)$ at 1 atm pressure and with $H^+(aq)$ at 1 M concentration. Electrons flow from the zinc anode to the S.H.E. cathode. The measured standard cell potential of 0.76 V at 25°C is displayed on the voltmeter.

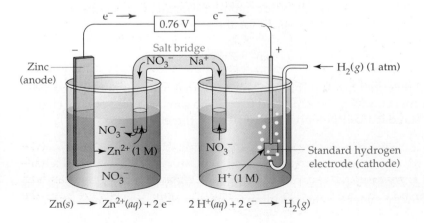

(We represent the hydrated proton as $H^+(aq)$ rather than $H_3O^+(aq)$ because we're interested here in electron transfer, not proton transfer as in Chapter 15). The standard potential for this cell—0.76 V at 25°C—is a measure of the combined driving forces of the oxidation and reduction half-reactions:

$$E^\circ_{cell} = E^\circ_{anode} + E^\circ_{cathode} = E^\circ_{Zn \to Zn^{2+}} + E^\circ_{H^+ \to H_2} = 0.76\ V$$

If we could determine E° values for individual half-reactions, we could combine those values to predict E° values for a host of cell reactions. Unfortunately, it's not possible to measure the potential of a single electrode; we can measure only a potential *difference* by placing a voltmeter between *two* electrodes. Nevertheless, we can develop a set of standard half-cell potentials by choosing an arbitrary standard half-cell as a reference point, assigning it an arbitrary potential, and then expressing the potential of all other half-cells relative to the reference half-cell. (Recall that this same approach was used in Section 8.9 for determining standard enthalpies of formation ΔH°_f.)

To define an electrochemical "sea level," chemists have chosen a reference half-cell called the **standard hydrogen electrode** (S.H.E., Figure 18.3). It consists of a platinum electrode in contact with H_2 gas and aqueous H^+ ions at standard-state conditions (1 atm $H_2(g)$, 1 M $H^+(aq)$, 25°C). The corresponding half-reaction is assigned an arbitrary potential of exactly 0 V:

$$2\ H^+(aq,\ 1\ M) + 2\ e^- \to H_2(g,\ 1\ atm)\qquad E^\circ = 0\ V$$

With this choice of standard reference electrode, the entire potential of the cell

$$Zn(s)\ |\ Zn^{2+}(1\ M)\ \|\ H^+(1\ M)\ |\ H_2(1\ atm)\ |\ Pt(s)$$

can be attributed to the Zn/Zn^{2+} half-cell:

$$E^\circ_{cell} = E^\circ_{Zn \to Zn^{2+}} + E^\circ_{H^+ \to H_2}$$
$$\uparrow\qquad\qquad \uparrow\qquad\qquad\qquad \uparrow$$
$$0.76\ V\quad 0.76\ V\qquad\qquad 0\ V$$

Since the Zn/Zn^{2+} half-reaction is an oxidation, the corresponding half-cell potential, $E^\circ = 0.76\ V$, is called a **standard oxidation potential**:

$$Zn(s) \to Zn^{2+}(aq) + 2\ e^-\qquad E^\circ = 0.76\ V\qquad \textit{Standard oxidation potential}$$

In a cell in which this half-reaction occurs in the opposite direction, the corresponding half-cell potential has the same magnitude but opposite sign:

$$Zn^{2+}(aq) + 2\ e^- \to Zn(s)\qquad E^\circ = -0.76\ V\qquad \textit{Standard reduction potential}$$

Because the Zn^{2+}/Zn half-reaction is a reduction, the potential, $E^\circ = -0.76$ V, is called a **standard reduction potential**. Whenever the direction of a half-reaction is reversed, the sign of E° must be reversed. That is, the standard oxidation potential and the standard reduction potential always have the same magnitude, but they have opposite signs.

We can determine standard potentials for other half-cells simply by constructing a galvanic cell in which the half-cell of interest is paired up with the standard hydrogen electrode. For example, to find the potential of a half-cell consisting of a copper electrode dipping into a 1 M Cu^{2+} solution, we would build the cell shown in Figure 18.4. The voltmeter–electrode connections required to get a positive reading on the voltmeter (0.34 V) would tell us that the standard hydrogen electrode is the anode and the copper electrode is the cathode. Therefore, the half-cell reactions involve oxidation of H_2 and reduction of Cu^{2+}. (Alternatively, we could identify the direction of the half-reactions by noting that the blue color due to Cu^{2+} fades as the reaction progresses.)

Anode (oxidation):	$H_2(g) \rightarrow 2\,H^+(aq) + 2\,e^-$	$E° = 0\ V$
Cathode (reduction):	$Cu^{2+}(aq) + 2\,e^- \rightarrow Cu(s)$	$E° = ?$
Overall cell reaction:	$H_2(g) + Cu^{2+}(aq) \rightarrow 2\,H^+(aq) + Cu(s)$	$E° = 0.34\ V$

Since the anode and cathode half-cell potentials must sum to give the overall cell potential, the standard *reduction* potential for the Cu^{2+}/Cu half-cell must be 0.34 V.

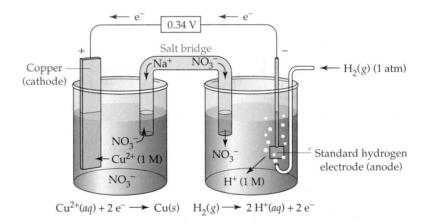

FIGURE 18.4 A galvanic cell consisting of a Cu^{2+}(1 M)/Cu half-cell and a standard hydrogen electrode. Electrons flow from the S.H.E. anode to the copper cathode. The measured standard cell potential at 25°C is 0.34 V.

We could also have calculated the standard reduction potential for the Cu^{2+}/Cu half-cell from the observed standard potential of the Daniell cell (1.10 V) and the standard oxidation potential for the Zn/Zn^{2+} half-cell (0.76 V):

Anode (oxidation):	$Zn(s) \rightarrow Zn^{+2}(aq) + 2\,e^-$	$E° = 0.76\ V$
Cathode (reduction):	$Cu^{2+}(aq) + 2\,e^- \rightarrow Cu(s)$	$E° = ?$
Overall cell reaction:	$Zn(s) + Cu^{2+}(aq) \rightarrow Zn^{+2}(aq) + Cu(s)$	$E° = 1.10\ V$

Since 1.10 V = 0.76 V + 0.34 V, we would again conclude that the standard reduction potential for the Cu^{2+}/Cu half-cell must be 0.34 V.

From experiments of the sort just described, hundreds of half-cell potentials have been determined. A short list is presented in Table 18.1, and a more complete tabulation is given in Appendix D. The following conventions are observed in constructing a table of half-cell potentials:

1. *The half-reactions are written as reductions* rather than as oxidations. This means that oxidizing agents and electrons are on the left side of each half-reaction and reducing agents are on the right side.

2. *The listed half-cell potentials are standard reduction potentials,* rather than standard oxidation potentials. Standard reduction potentials are also known as **standard electrode potentials**.

3. *The half-reactions are listed in order of decreasing standard reduction potential* (decreasing tendency to occur in the forward direction; increasing tendency to occur in the reverse direction). Consequently, the strongest oxidizing agents are located in the upper left of the table (F_2, H_2O_2, MnO_4^-, and so forth), and the strongest reducing agents are found in the lower right of the table (Li, Na, Mg, and so forth).

By choosing $E° = 0$ V for the standard hydrogen electrode, we obtain standard reduction potentials that range from about $+3$ V to -3 V.

TABLE 18.1   Standard Reduction Potentials at 25° C

	Reduction Half-Reaction		$E°$ (V)	
Stronger oxidizing agent	$F_2(g) + 2\,e^-$	$\rightarrow 2\,F^-(aq)$	2.87	Weaker reducing agent
	$H_2O_2(aq) + 2\,H^+(aq) + 2\,e^-$	$\rightarrow 2\,H_2O(l)$	1.78	
	$MnO_4^-(aq) + 8\,H^+(aq) + 5\,e^-$	$\rightarrow Mn^{2+}(aq) + 4\,H_2O(l)$	1.51	
	$Cl_2(g) + 2\,e^-$	$\rightarrow 2\,Cl^-(aq)$	1.36	
	$Cr_2O_7^{2-}(aq) + 14\,H^+(aq) + 6\,e^-$	$\rightarrow 2\,Cr^{3+}(aq) + 7\,H_2O(l)$	1.33	
	$O_2(g) + 4\,H^+(aq) + 4\,e^-$	$\rightarrow 2\,H_2O(l)$	1.23	
	$Br_2(l) + 2\,e^-$	$\rightarrow 2\,Br^-(aq)$	1.09	
	$Ag^+(aq) + e^-$	$\rightarrow Ag(s)$	0.80	
	$Fe^{3+}(aq) + e^-$	$\rightarrow Fe^{2+}(aq)$	0.77	
	$O_2(g) + 2\,H^+(aq) + 2\,e^-$	$\rightarrow H_2O_2(aq)$	0.70	
	$I_2(s) + 2\,e^-$	$\rightarrow 2\,I^-(aq)$	0.54	
	$O_2(g) + 2\,H_2O(l) + 4\,e^-$	$\rightarrow 4\,OH^-(aq)$	0.40	
	$Cu^{2+}(aq) + 2\,e^-$	$\rightarrow Cu(s)$	0.34	
	$Sn^{4+}(aq) + 2\,e^-$	$\rightarrow Sn^{2+}(aq)$	0.15	
	$2\,H^+(aq) + 2\,e^-$	$\rightarrow H_2(g)$	0	
	$Pb^{2+}(aq) + 2\,e^-$	$\rightarrow Pb(s)$	-0.13	
	$Ni^{2+}(aq) + 2\,e^-$	$\rightarrow Ni(s)$	-0.26	
	$Cd^{2+}(aq) + 2\,e^-$	$\rightarrow Cd(s)$	-0.40	
	$Fe^{2+}(aq) + 2\,e^-$	$\rightarrow Fe(s)$	-0.45	
	$Zn^{2+}(aq) + 2\,e^-$	$\rightarrow Zn(s)$	-0.76	
	$2\,H_2O(l) + 2\,e^-$	$\rightarrow H_2(g) + 2\,OH^-(aq)$	-0.83	
	$Al^{3+}(aq) + 3\,e^-$	$\rightarrow Al(s)$	-1.66	
Weaker oxidizing agent	$Mg^{2+}(aq) + 2\,e^-$	$\rightarrow Mg(s)$	-2.37	Stronger reducing agent
	$Na^+(aq) + e^-$	$\rightarrow Na(s)$	-2.71	
	$Li^+(aq) + e^-$	$\rightarrow Li(s)$	-3.04	

⌐ **PROBLEM 18.5** The standard potential for the following galvanic cell is 0.92 V:

$$Al(s) \mid Al^{3+}(aq) \parallel Cr^{3+}(aq) \mid Cr(s)$$

Look up the standard reduction potential for the Al^{3+}/Al half-cell in Table 18.1, and calculate the standard reduction potential for the Cr^{3+}/Cr half-cell. ⌐

18.5 ➤ USING STANDARD REDUCTION POTENTIALS

A table of standard reduction potentials summarizes an enormous amount of chemical information in a very small space. It enables us to arrange any two or more oxidizing or reducing agents in order of increasing strength, and it permits us to predict the spontaneity or nonspontaneity of thousands of redox reactions. Suppose, for example, that the table contains just 100 half-reactions. We can pair each reduction half-reaction with any one of the remaining 99 oxidation half-reactions to give a total of $100 \times 99 = 9900$ cell reactions. By calculating the $E°$ values for these cell reactions we would find that half of them are spontaneous and the other half are nonspontaneous.

To illustrate how to use the tabulated $E°$ values, let's calculate $E°$ for the oxidation of $Zn(s)$ by $Ag^+(aq)$:

$$2\ Ag^+(aq) + Zn(s) \rightarrow 2\ Ag(s) + Zn^{2+}(aq)$$

First, we find the relevant half-reactions in Table 18.1 and write them in the appropriate direction for reduction of Ag^+ and oxidation of Zn. Next, we add the half-reactions to get the overall reaction. Before adding, though, we must multiply the Ag^+/Ag half-reaction by a factor of 2 so that the electrons will cancel:

Reduction:	$2 \times [Ag^+(aq) + e^- \rightarrow Ag(s)]$	$E° = 0.80$ V
Oxidation:	$Zn(s) \rightarrow Zn^{2+}(aq) + 2\ e^-$	$E° = -(-0.76$ V$)$
Overall reaction:	$2\ Ag^+(aq) + Zn(s) \rightarrow 2\ Ag(s) + Zn^{2+}(aq)$	$E° = 1.56$ V

Then we tabulate the $E°$ values for the half-reactions, remembering that $E°$ for *oxidation* of zinc is the negative of the standard *reduction* potential $(-0.76$ V$)$. We do *not* multiply the $E°$ value for reduction of Ag^+ (0.80 V) by a factor of 2 because an electrical potential does not depend on how much reaction occurs.

The reason why $E°$ values are independent of the amount of reaction can be understood by looking at the equation $\Delta G° = -nFE°$. Free energy is an *extensive* property (Section 1.13) because it depends on the amount of substance. If we double the amount of Ag^+ reduced, the free-energy change, $\Delta G°$, doubles. But, of course, the number of electrons transferred, n, also doubles, and the ratio $E° = -\Delta G°/nF$ is therefore constant. Electrical potential is therefore an *intensive* property, one that does not depend on the amount of substance.

The $E°$ value for the overall reaction of $Zn(s)$ with $Ag^+(aq)$ is the sum of the $E°$ values for the two half-reactions: 0.80 V $+ 0.76$ V $= 1.56$ V. Because $E°$ is positive ($\Delta G°$ is negative), oxidation of zinc by Ag^+ is a spontaneous reaction under standard-state conditions. Just as Ag^+ can oxidize Zn, it's evident from Table 18.1 that Ag^+ can oxidize *any* reducing agent that lies below it in the table (Fe^{2+}, H_2O_2, I^-, and so forth). The sum of $E°$ for the Ag^+/Ag reduction (0.80 V) and $-E°$ for any half-reaction that lies below the Ag^+/Ag half-reaction always gives a positive $E°$ for the overall reaction.

In general, an oxidizing agent can oxidize any reducing agent that lies below it in the table but can't oxidize a reducing agent above it in the table. Thus, Ag^+ can't oxidize Br^-, H_2O, Cr^{3+}, and so forth because $E°$ for the overall reaction is negative. Simply by glancing at the location of the oxidizing and reducing agent in the table, we can predict whether a reaction is spontaneous or nonspontaneous.

EXAMPLE 18.4

(a) Arrange the following oxidizing agents in order of increasing strength under standard-state conditions: $Br_2(l)$, $Fe^{3+}(aq)$, $Cr_2O_7^{2-}(aq)$.
(b) Arrange the following reducing agents in order of increasing strength under standard-state conditions: $Al(s)$, $Na(s)$, $Zn(s)$.

SOLUTION   **(a)** Pick out the half-reactions in Table 18.1 that involve Br_2, Fe^{3+}, and $Cr_2O_7^{2-}$ and list them in the order in which they occur in the table:

$$Cr_2O_7^{2-}(aq) + 14\ H^+(aq) + 6\ e^- \rightarrow 2\ Cr^{3+}(aq) + 7\ H_2O(l) \qquad E° = 1.33\ V$$

$$Br_2(l) + 2\ e^- \rightarrow 2\ Br^-(aq) \qquad E° = 1.09\ V$$

$$Fe^{3+}(aq) + e^- \rightarrow Fe^{2+}(aq) \qquad E° = 0.77\ V$$

$Cr_2O_7^{2-}$ has the greatest tendency to be reduced (largest $E°$), and Fe^{3+} has the least tendency to be reduced (smallest $E°$). The species that has the greatest tendency to be reduced is the strongest oxidizing agent. Oxidizing strength therefore increases in the order $Fe^{3+} < Br_2 < Cr_2O_7^{2-}$. As a shortcut, simply note that the strength of the oxidizing agents (listed on the left side of Table 18.1) increases on moving up in the table ($Fe^{3+} < Br_2 < Cr_2O_7^{2-}$).
(b) List the half-reactions that involve $Al(s)$, $Na(s)$, $Zn(s)$ in the order in which they occur in Table 18.1:

$$Zn^{2+}(aq) + 2\ e^- \rightarrow Zn(s) \qquad E° = -0.76\ V$$

$$Al^{3+}(aq) + 3\ e^- \rightarrow Al(s) \qquad E° = -1.66\ V$$

$$Na^+(aq) + e^- \rightarrow Na(s) \qquad E° = -2.71\ V$$

The last half-reaction has the least tendency to occur in the forward direction (most negative $E°$) and the greatest tendency to occur in the reverse direction. Therefore, Na is the strongest reducing agent, and reducing strength increases in the order $Zn < Al < Na$. As a shortcut, note that the strength of the reducing agents (listed on the right side of Table 18.1) increases on moving down in the table ($Zn < Al < Na$).

EXAMPLE 18.5

From Table 18.1 predict whether $Pb^{2+}(aq)$ can oxidize Fe(s) or Cu(s) under standard-state conditions. Calculate $E°$ for each reaction at 25°C.

SOLUTION $Pb^{2+}(aq)$ is above Fe(s) in the table but below Cu(s). Therefore, $Pb^{2+}(aq)$ can oxidize Fe(s) but can't oxidize Cu(s). To confirm these predictions, we can calculate $E°$ values for the overall reactions.

(a) For oxidation of Fe by Pb^{2+}, $E°$ is positive (0.32 V), and the reaction is therefore spontaneous:

$$Pb^{2+}(aq) + 2\,e^- \rightarrow Pb(s) \qquad E° = -0.13 \text{ V}$$
$$\underline{Fe(s) \rightarrow Fe^{2+}(aq) + 2\,e^- \qquad E° = 0.45 \text{ V}}$$
$$Pb^{2+}(aq) + Fe(s) \rightarrow Pb(s) + Fe^{2+}(aq) \qquad E° = 0.32 \text{ V}$$

(b) For oxidation of Cu by Pb^{2+}, $E°$ is negative (-0.47 V), and the reaction is therefore nonspontaneous:

$$Pb^{2+}(aq) + 2\,e^- \rightarrow Pb(s) \qquad E° = -0.13 \text{ V}$$
$$\underline{Cu(s) \rightarrow Cu^{2+}(aq) + 2\,e^- \qquad E° = -0.34 \text{ V}}$$
$$Pb^{2+}(aq) + Cu(s) \rightarrow Pb(s) + Cu^{2+}(aq) \qquad E° = -0.47 \text{ V}$$

⌐ **PROBLEM 18.6** Which is the stronger oxidizing agent, $Cl_2(g)$ or $Ag^+(aq)$? Which is the stronger reducing agent, Fe(s) or Mg(s)?

⌐ **PROBLEM 18.7** From Table 18.1 predict whether each of the following reactions can occur under standard-state conditions.
(a) $2\,Fe^{3+}(aq) + 2\,I^-(aq) \rightarrow 2\,Fe^{2+}(aq) + I_2(s)$
(b) $3\,Ni(s) + 2\,Al^{3+}(aq) \rightarrow 3\,Ni^{2+}(aq) + 2\,Al(s)$

Confirm your predictions by calculating the value of $E°$ for each reaction. Which reaction(s) can occur in the reverse direction under standard-state conditions? ⌐

18.6 ►CELL POTENTIALS AND COMPOSITION
OF THE REACTION MIXTURE:
THE NERNST EQUATION

It's known from experiment that cell potentials depend on temperature and on the composition of the reaction mixture—that is, on the concentrations of solutes and the partial pressures of gases. This dependence can be derived from the equation

$$\Delta G = \Delta G° + RT \ln Q$$

Recall from Section 17.10 that ΔG is the free-energy change for a reaction under non-standard-state conditions, $\Delta G°$ is the free-energy change under standard-state conditions, and Q is the reaction quotient. Since $\Delta G = -nFE$, and $\Delta G° = -nFE°$, we can rewrite the equation for ΔG in the form

$$-nFE = -nFE° + RT \ln Q$$

Dividing by $-nF$, we obtain the **Nernst equation**, named after Walther Nernst (1864–1941), the German chemist who first derived it:

Nernst equation $$E = E° - \frac{RT}{nF} \ln Q \quad \text{or} \quad E = E° - \frac{2.303RT}{nF} \log Q$$

At 25°C, $2.303RT/F$ has a value of 0.0592 V, and we can therefore write the Nernst equation in the simplified form

$$E = E° - \frac{0.0592}{n} \log Q \quad \text{(in volts, at 25°C)}$$

In actual galvanic cells, the concentrations and partial pressures of reactants and products seldom have standard-state values, and the values change as the cell reaction proceeds. The Nernst equation is useful because it enables us to calculate cell potentials under non-standard-state conditions, as shown in Example 18.6.

EXAMPLE 18.6

Consider a galvanic cell using the reaction

$$\text{Zn}(s) + 2\,\text{H}^+(aq) \rightarrow \text{Zn}^{2+}(aq) + \text{H}_2(g)$$

Calculate the cell potential when $[\text{H}^+] = 1.0$ M, $[\text{Zn}^{2+}] = 0.0010$ M, and $P_{\text{H}_2} = 0.10$ atm.

SOLUTION   Qualitatively, we expect that the reaction will have a greater tendency to occur under the cited conditions than under standard-state conditions because the product concentrations are lower than standard-state values. We therefore anticipate that the cell potential E will be greater than the standard cell potential $E°$.

Quantitatively, the cell potential at 25°C is

$$E = E° - \frac{0.0592}{n} \log Q$$

$$= E° - \left(\frac{0.0592}{n}\right) \log \frac{[\text{Zn}^{2+}](P_{\text{H}_2})}{[\text{H}^+]^2}$$

where, as usual, zinc has been omitted from the reaction quotient because it's a pure solid. We can calculate the value of $E°$ from the standard potentials in Table 18.1:

$$E° = E°_{\text{Zn} \rightarrow \text{Zn}^{2+}} + E°_{\text{H}^+ \rightarrow \text{H}_2} = -(-0.76 \text{ V}) + 0 \text{ V} = 0.76 \text{ V}$$

For this reaction, 2 mol of electrons are transferred, and so $n = 2$. Substituting into the Nernst equation the appropriate values of $E°$, n, $[\text{H}^+]$, $[\text{Zn}^{2+}]$, and P_{H_2} gives

$$E = 0.76 \text{ V} - \left(\frac{0.0592 \text{ V}}{2}\right) \log \frac{(0.0010)(0.10)}{(1.0)^2} = 0.76 \text{ V} - \left(\frac{0.0592 \text{ V}}{2}\right)(-4.0)$$

$$= 0.76 \text{ V} + 0.12 \text{ V}$$

$$= 0.88 \text{ V} \quad \text{(at 25°C)}$$

⌐ **PROBLEM 18.8** Consider a galvanic cell that uses the reaction

$$Cu(s) + 2\ Fe^{3+}(aq) \rightarrow Cu^{2+}(aq) + 2\ Fe^{2+}(aq)$$

What is the potential of a cell at 25°C that has the following ion concentrations?

$$[Fe^{3+}] = 1.0 \times 10^{-4}\ M \qquad [Cu^{2+}] = 0.25\ M \qquad [Fe^{2+}] = 0.20\ M \qquad ◢$$

18.7 ►ELECTROCHEMICAL DETERMINATION OF pH

The electrochemical determination of pH using a pH meter is a particularly important application of the Nernst equation. Consider, for example, a cell with a hydrogen electrode as the anode and a second reference electrode as the cathode:

$$Pt\,|\,H_2(1\ atm)\,|\,H^+(?\ M)\,\|\,\text{reference cathode}$$

The hydrogen electrode consists of a platinum wire that is in contact with H_2 at 1 atm and dips into a solution of unknown pH. The potential of this cell is

$$E_{cell} = E_{H_2 \rightarrow H^+} + E_{ref}$$

We can calculate the potential for the hydrogen electrode half-reaction

$$H_2(g) \rightarrow 2\ H^+(aq) + 2\ e^-$$

by applying the Nernst equation:

$$E_{H_2 \rightarrow H^+} = E°_{H_2 \rightarrow H^+} - \left(\frac{0.0592}{n}\right) \log \frac{[H^+]^2}{P_{H_2}}$$

Since $E° = 0$ V for the standard hydrogen electrode, $n = 2$, and $P_{H_2} = 1$ atm, we can rewrite this equation as

$$E_{H_2 \rightarrow H^+} = -\left(\frac{0.0592}{2}\right) \log [H^+]^2$$

Further, because $\log [H^+]^2 = 2 \log [H^+]$ and $-\log [H^+] = pH$, the half-cell potential for the hydrogen electrode is directly proportional to the pH:

$$E_{H_2 \rightarrow H^+} = -\left(\frac{0.0592}{2}\right)(2)(\log [H^+]) = 0.0592\,(-\log [H^+])$$

$$E_{H_2 \rightarrow H^+} = 0.0592\ pH$$

The overall cell potential is

$$E_{cell} = 0.0592\ pH + E_{ref}$$

and the pH is therefore a linear function of the cell potential:

$$pH = \frac{E_{cell} - E_{ref}}{0.0592}$$

A higher cell potential indicates a higher pH. In other words, we can measure the pH of a solution simply by measuring E_{cell}.

In actual pH measurements, a *glass* electrode replaces the cumbersome hydrogen electrode and a *calomel* electrode is used as the reference. A glass electrode consists of a silver wire coated with silver chloride that dips into a reference solution of dilute hydrochloric acid (Figure 18.5). The hydrochloric acid is separated from the test solution of unknown pH by a thin glass membrane. A calomel electrode consists of mercury(I) chloride (Hg_2Cl_2, calomel) in contact with liquid mercury. The cathode half-reaction is

$$Hg_2Cl_2(s) + 2\ e^- \rightarrow 2\ Hg(l) + 2\ Cl^-(aq) \qquad E^\circ_{ref} = 0.28\ V$$

The cell potential is measured with a pH meter, a voltage-measuring device that electronically converts E_{cell} to pH and displays the result in pH units.

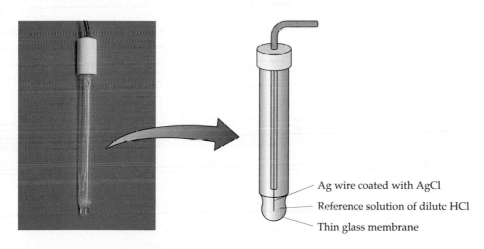

Ag wire coated with AgCl
Reference solution of dilute HCl
Thin glass membrane

FIGURE 18.5 A glass electrode consists of a silver wire coated with silver chloride that dips into a reference solution of dilute hydrochloric acid. The hydrochloric acid is separated from the test solution of unknown pH by a thin glass membrane. When a glass electrode is immersed in the test solution, its electrical potential depends linearly on the difference in the pH of the solutions on the two sides of the membrane.

EXAMPLE 18.7

The following cell has a potential of 0.61 V at 25°C:

$$Pt(s) \mid H_2(1\ atm) \mid H^+(?\ M) \parallel Cu^{2+}(1\ M) \mid Cu(s)$$

What is the pH of the solution in the anode compartment?

SOLUTION The cell reaction is

$$H_2(g) + Cu^{2+}(aq) \rightarrow 2\ H^+(aq) + Cu(s)$$

and the standard cell potential can be calculated from the data in Table 18.1:

$$E^\circ = E^\circ_{H_2 \rightarrow H^+} + E^\circ_{Cu^{2+} \rightarrow Cu} = 0\ V + 0.34\ V = 0.34\ V$$

The cell potential and the standard cell potential are related by the Nernst equation:

$$E = E^\circ - \left(\frac{0.0592}{n}\right) \log \frac{[H^+]^2}{(P_{H_2})[Cu^{2+}]}$$

Substituting in the values of E, E°, n, $[Cu^{2+}]$, and P_{H_2} gives

$$0.61\ V = 0.34\ V - \left(\frac{0.0592\ V}{2}\right) \log \frac{[H^+]^2}{(1)(1)} = 0.34\ V + (0.0592\ V)(pH)$$

Therefore, the pH is

$$pH = \frac{0.61\ V - 0.34\ V}{0.0592\ V} = 4.5$$

PROBLEM 18.9 What is the pH of the solution in the anode compartment of the following cell if the measured cell potential at 25°C is 0.28 V?

$$Pt(s)\ |\ H_2(1\ atm)\ |\ H^+(?\ M)\ \|\ Pb^{2+}(1\ M)\ |\ Pb(s)$$

18.8 ➤ STANDARD CELL POTENTIALS AND EQUILIBRIUM CONSTANTS

We saw in Section 18.3 that the standard free-energy change for a reaction is related to the standard cell potential by the equation

$$\Delta G^\circ = -nFE^\circ$$

In the previous chapter (Section 17.11), we showed that the standard free-energy change is also related to the equilibrium constant for the reaction:

$$\Delta G^\circ = -RT \ln K$$

Combining these two equations, we obtain

$$-nFE^\circ = -RT \ln K$$

or

$$E^\circ = \frac{RT}{nF} \ln K = \frac{2.303\ RT}{nF} \log K$$

Since 2.303 RT/F has a value of 0.0592 V at 25°C, we can rewrite this equation in the simplified form

$$E° = \frac{0.0592}{n} \log K \quad \text{(in volts, at 25°C)}$$

Because very small concentrations are difficult to measure, the determination of an equilibrium constant from concentration measurements is not feasible when K is either very large or very small. Standard cell potentials, however, are relatively easy to measure. Consequently, the most common use of the preceding equation is in calculating equilibrium constants from standard cell potentials.

As an example, let's calculate the value of K for the reaction in the Daniell cell:

$$Zn(s) + Cu^{2+}(aq) \rightarrow Zn^{2+}(aq) + Cu(s)$$

Solving the equation for $\log K$ and substituting in the appropriate values of $E°$ and n, we obtain:

$$E° = \left(\frac{0.0592}{n}\right) \log K$$

$$\log K = \frac{nE°}{0.0592} = \frac{(2)(1.10 \text{ V})}{0.0592 \text{ V}} = 37.2$$

$$K = 10^{37.2} = 2 \times 10^{37} \quad \text{(at 25°C)}$$

Because the equilibrium constant is a very large number, the reaction goes essentially to completion. When $[Zn^{2+}] = 1$ M, for example, $[Cu^{2+}]$ is less than 10^{-37} M.

Redox reactions typically go either essentially to completion (K is very large) or almost not at all (K is very small). This conclusion follows from the standard reduction potentials in Table 18.1, which span a range of about 6 V. Thus, $E°$ for a redox reaction can range from +6 V to −6 V, although values outside the range +3 V to −3 V are uncommon. For the case of $n = 2$, the correspondence between the value of $E°$ and the equilibrium constant K is indicated in Figure 18.6. Equilibrium constants for redox reactions tend to be

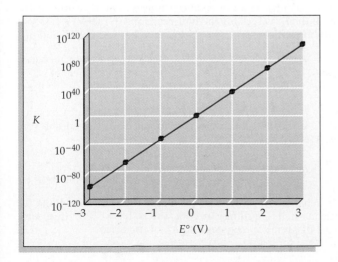

FIGURE 18.6 The relationship between the equilibrium constant K for a redox reaction with $n = 2$ and the standard cell potential $E°$. Note that K is plotted on a logarithmic scale.

either very large or very small in comparison with equilibrium constants for acid–base reactions, which are in the range of 10^{14} to 10^{-14}. Note that a positive value of $E°$ corresponds to $K > 1$, and a negative value of $E°$ corresponds to $K < 1$.

In the past several chapters, we've discussed three different ways to determine the value of an equilibrium constant K: from concentration data (Section 13.2), from thermochemical data (Section 17.11), and from electrochemical data. The following are the key relationships needed for each approach:

1. K from concentration data: $K = \dfrac{[C]^c[D]^d}{[A]^a[B]^b}$

2. K from thermochemical data: $\Delta G° = -RT \ln K$; $\ln K = \dfrac{-\Delta G°}{RT}$

3. K from electrochemical data: $E° = \dfrac{RT}{nF} \ln K$; $\ln K = \dfrac{nFE°}{RT}$

EXAMPLE 18.8

Using the standard reduction potentials in Table 18.1, calculate the equilibrium constant at 25°C for the reaction

$$6\,Br^-(aq) + Cr_2O_7^{2-}(aq) + 14\,H^+(aq) \rightarrow 3\,Br_2(l) + 2\,Cr^{3+}(aq) + 7\,H_2O(l)$$

SOLUTION   Find the relevant half-reactions in Table 18.1, and write them in the proper direction for oxidation of Br^- and reduction of $Cr_2O_7^{2-}$. Before adding the half-reactions to get the overall reaction, multiply the Br^-/Br_2 half-reaction by a factor of 3 so that the electrons will cancel:

$3 \times [2\,Br^-(aq) \rightarrow Br_2(l) + 2\,e^-]$	$E° = -1.09\text{ V}$
$Cr_2O_7^{2-}(aq) + 14\,H^+(aq) + 6\,e^- \rightarrow 2\,Cr^{3+}(aq) + 7\,H_2O(l)$	$E° = 1.33\text{ V}$
$6\,Br^-(aq) + Cr_2O_7^{2-}(aq) + 14\,H^+(aq) \rightarrow 3\,Br_2(l) + 2\,Cr^{3+}(aq) + 7\,H_2O(l)$	$E° = 0.24\text{ V}$

Note that $E°$ for the Br^-/Br_2 *oxidation* is the negative of the tabulated standard *reduction* potential (1.09 V), and remember that we don't multiply this $E°$ value by a factor of 3 because electrical potential is an intensive property. The $E°$ value for the overall reaction is the sum of the $E°$ values for the half-reactions: $-1.09\text{ V} + 1.33\text{ V} = 0.24\text{ V}$. To calculate the equilibrium constant, use the relation between $\log K$ and $nE°$, with $n = 6$:

$$\log K = \frac{nE°}{0.0592} = \frac{(6)(0.24\text{ V})}{0.0592\text{ V}} = 24 \qquad K = 10^{24} \text{ at } 25°C$$

⌐ **PROBLEM 18.10** Use the data in Table 18.1 to calculate the equilibrium constant at 25°C for the reaction

$$4\,Fe^{2+}(aq) + O_2(g) + 4\,H^+(aq) \rightarrow 4\,Fe^{3+}(aq) + 2\,H_2O(l)$$

⌐ **PROBLEM 18.11** What is the value of $E°$ for a redox reaction involving the transfer of 2 mol of electrons if its equilibrium constant is 1.8×10^{-5} (the same numerical value as the acid-dissociation constant K_a for acetic acid)?

18.9 ➤BATTERIES

By far the most important practical application of galvanic cells is their use as *batteries*. In multicell batteries, such as those used in automobiles, the individual galvanic cells are linked in series, with the anode of each cell connected to the cathode of the adjacent cell. The voltage provided by the battery is the sum of the individual cell voltages. The features required in a battery depend on the application. In general, however, a commercially successful battery should be compact, lightweight, physically rugged, and inexpensive, and it must provide a stable source of power for relatively long periods of time. Battery design is an active area of research that requires considerable ingenuity as well as a solid understanding of electrochemistry.

Let's look at several of the most common types of commercial batteries.

Lead Storage Battery

The *lead storage battery* is perhaps the most familiar of all galvanic cells because it has been used as a reliable source of power for starting automobiles for more than three quarters of a century. A typical 12 V battery consists of six cells connected in series, each cell providing a potential of about 2 V. The cell design is illustrated in Figure 18.7. The anode, a series of lead grids packed with spongy lead, and the cathode, a second series of grids packed with lead dioxide, dip into the electrolyte, an aqueous solution of sulfuric acid (38% by weight). When the cell is discharging (providing current), the electrode half-reactions and the overall cell reaction are:

Anode: $Pb(s) + HSO_4^-(aq) \rightarrow PbSO_4(s) + H^+(aq) + 2 e^-$ $E^\circ = 0.296$ V

Cathode: $PbO_2(s) + 3 H^+(aq) + HSO_4^-(aq) + 2 e^- \rightarrow PbSO_4(s) + 2 H_2O(l)$ $E^\circ = 1.628$ V

Overall: $Pb(s) + PbO_2(s) + 2 H^+(aq) + 2 HSO_4^-(aq) \rightarrow 2 PbSO_4(s) + 2 H_2O(l)$ $E^\circ = 1.924$ V

(These equations contain HSO_4^- ions because SO_4^{2-} is protonated in strongly acidic solutions.)

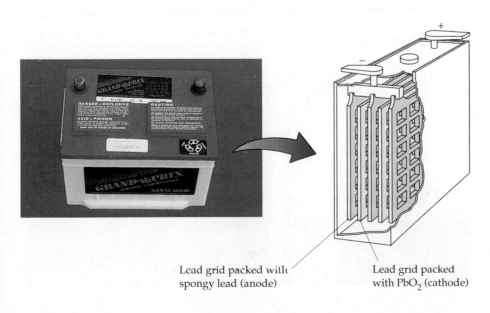

Lead grid packed with spongy lead (anode)

Lead grid packed with PbO_2 (cathode)

FIGURE 18.7 A lead storage battery and a cutaway view of one cell. Each electrode consists of several grids with a large surface area so that the battery can deliver the high currents required to start an automobile engine. The electrolyte (not shown) is aqueous sulfuric acid.

Lead is oxidized to lead sulfate at the anode, and lead dioxide is reduced to lead sulfate at the cathode. There is no need for the cell to have separate anode and cathode compartments because the oxidizing and reducing agents are both solids (PbO_2 and Pb) that don't come in contact. Insulating spacers between the grids keep them from touching.

Because the reaction product (solid $PbSO_4$) adheres to the surface of the electrodes, a "run-down" lead storage battery can be recharged by using an external source of direct current to drive the cell reaction in the reverse, nonspontaneous direction. In an automobile, the battery is continuously recharged by the alternator, which is driven by the engine. In older batteries, the charging process was accompanied by appreciable evolution of gaseous hydrogen and oxygen, owing to electrolysis of water. Not only did that electrolysis necessitate periodic water replacement, the accompanying gas evolution tended to loosen the Pb, PbO_2, and $PbSO_4$ from the electrode surfaces, thus shortening the life of the battery. Newer "maintenance-free" batteries use electrodes made from a calcium-containing lead alloy that inhibits the electrolytic decomposition of water. These batteries are sealed because no opening is needed for escape of gases or addition of water.

A lead storage battery should provide good service for several years, but eventually mechanical shock from driving over rough roads dislodges the $PbSO_4$ from the electrodes. Then it's no longer possible to recharge the battery, and it must be replaced.

Dry-Cell Batteries

The common household batteries used as power sources for flashlights, portable radios, and tape recorders use the *dry cell* or *Leclanché cell*, patented in 1866 by the Frenchman George Leclanché. A dry-cell battery has a zinc metal can, which serves as the anode, and an inert graphite rod surrounded by a paste of solid manganese dioxide (MnO_2) and carbon black, which functions as the cathode (Figure 18.8). Surrounding the MnO_2-containing paste is the electrolyte, a moist paste of ammonium chloride (NH_4Cl) and zinc chloride ($ZnCl_2$) in starch. The dry cell is not completely dry but gets its name from the fact that the electrolyte is a viscous, aqueous paste rather than a liquid solu-

FIGURE 18.8 Leclanché dry cell and a cutaway view.

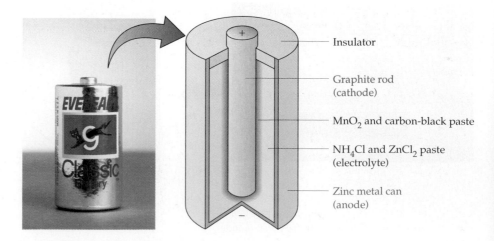

tion. The electrode reactions, which are rather complicated, can be represented in simplified form by the following equations:

Anode: $Zn(s) \rightarrow Zn^{2+}(aq) + 2\ e^-$

Cathode: $2\ MnO_2(s) + 2\ NH_4^+(aq) + 2\ e^- \rightarrow Mn_2O_3(s) + 2\ NH_3(aq) + H_2O(l)$

This cell provides a potential of about 1.5 V, but the potential deteriorates to about 0.8 V as the cell is used.

The *alkaline dry cell* is a modified version of the Leclanché cell in which the acidic electrolyte (NH_4Cl) of the Leclanché cell is replaced by an alkaline (basic) electrolyte, either NaOH or KOH. As in the Leclanché cell, the electrode reactions involve oxidation of zinc and reduction of manganese dioxide, but the oxidation product is zinc oxide, as is appropriate to the basic conditions:

Anode: $Zn(s) + 2\ OH^-(aq) \rightarrow ZnO(s) + H_2O(l) + 2\ e^-$

Cathode. $2\ MnO_2(s) + H_2O(l) + 2\ e^- \rightarrow Mn_2O_3(s) + 2\ OH^-(aq)$

Corrosion of the zinc anode is a significant side reaction under acidic conditions because zinc reacts with $H^+(aq)$ to give $Zn^{2+}(aq)$ and $H_2(g)$. Under basic conditions, however, the cell has a longer life because zinc corrodes more slowly. The alkaline cell also produces higher power and more stable current and voltage because of more efficient ion transport in the alkaline electrolyte.

Closely related to the alkaline dry cell is the *mercury battery*, often used In watches, heart pacemakers, and other devices where a battery of small size is required (Figure 18.9). The anode of the mercury battery is zinc, as in the alkaline dry cell, but the cathode is steel in contact with mercury(II) oxide (HgO) in an alkaline medium of KOH and $Zn(OH)_2$. Zinc is oxidized at the anode, and HgO is reduced at the cathode:

Anode: $Zn(s) + 2\ OH^-(aq) \rightarrow ZnO(s) + H_2O(l) + 2\ e^-$

Cathode: $HgO(s) + H_2O(l) + 2\ e^- \rightarrow Hg(l) + 2\ OH^-(aq)$

Even though mercury batteries can be made very small, they still produce a stable potential (about 1.3 V) for long periods of time. When possible, used mercury batteries should be recycled to recover the mercury because mercury is toxic and poses a threat to the environment.

FIGURE 18.9 A small mercury battery and a cutaway view.

Steel (cathode)

Insulator

HgO in KOH and $Zn(OH)_2$

Zinc container (anode)

Nickel–Cadmium Batteries

Nickel–cadmium or *"ni–cad"* batteries are popular for use in calculators and portable power tools because, unlike most other dry-cell batteries, they are rechargeable (Figure 18.10). The anode of a ni–cad battery is cadmium metal, and the cathode is a hydrated nickel oxide, NiO(OH), supported on nickel metal. The electrode reactions are:

$$\text{Anode:} \quad Cd(s) + 2\,OH^-(aq) \rightarrow Cd(OH)_2(s) + 2\,e^-$$

$$\text{Cathode:} \quad NiO(OH)(s) + H_2O(l) + e^- \rightarrow Ni(OH)_2(s) + OH^-(aq)$$

Ni–cad batteries can be recharged many times because the solid products of the electrode reactions adhere to the surface of the electrodes.

FIGURE 18.10 Rechargeable nickel–cadmium, or "ni–cad," storage batteries.

Fuel Cells

A **fuel cell** is a galvanic cell in which one of the reactants is a traditional fuel such as methane or hydrogen. A fuel cell differs from an ordinary battery in that the reactants are not self-contained within the cell but instead are continuously supplied from an external reservoir. Perhaps the best-known example is the hydrogen–oxygen fuel cell, which is used as a source of electric power in space vehicles (Figure 18.11). This cell contains porous carbon electrodes impregnated with metallic catalysts and an electrolyte consisting of hot, aqueous KOH. The fuel (gaseous H_2) and the oxidizing

FIGURE 18.11 A hydrogen–oxygen fuel cell. Gaseous H_2 is oxidized to water (steam) at the anode, and gaseous O_2 is reduced to water at the cathode.

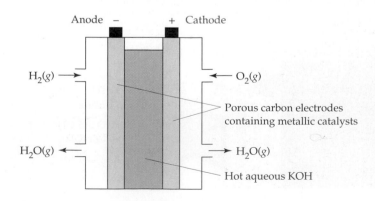

agent (gaseous O_2) don't react directly but instead flow into separate cell compartments where H_2 is oxidized at the anode and O_2 is reduced at the cathode:

Anode: $2\,H_2(g) + 4\,OH^-(aq) \rightarrow 4\,H_2O(l) + 4\,e^-$

Cathode: $O_2(g) + 2\,H_2O(l) + 4\,e^- \rightarrow 4\,OH^-(aq)$

Overall: $2\,H_2(g) + O_2(g) \rightarrow 2\,H_2O(l)$

The overall cell reaction is simply the conversion of hydrogen and oxygen to water. To date, large-scale applications of fuel cells have been limited because of cost, although the Tokyo Electric Power Company in Japan is now operating an 11 MW fuel-cell power plant capable of supplying power to about 4000 households.

Tokyo Electric Power Company 11 MW fuel-cell power plant at Goi, Japan.

┌ **PROBLEM 18.12**   Write a balanced equation for the overall cell reaction in each of the following batteries.

(a) Leclanché dry cell   **(b)** alkaline dry cell   **(c)** mercury battery
(d) nickel–cadmium battery

18.10 ►CORROSION

Corrosion is the oxidative deterioration of a metal, such as the conversion of iron to rust (a hydrated iron(III) oxide of approximate composition $Fe_2O_3 \cdot H_2O$). The rusting of iron has enormous economic consequences: It has been estimated that as much as one-fourth of the steel produced in the United States goes to replace steel structures and products that have been destroyed by corrosion.

To prevent corrosion, we first have to understand how it occurs. One important fact is that the rusting of iron requires both oxygen and water; it doesn't occur in oxygen-free water or in dry air. Another clue is the observation that rusting involves pitting of the metal surface, but the rust is deposited at a location physically separated from the pits. This suggests that rust does not form by direct reaction of iron and oxygen, but rather by an electrochemical process in which iron is oxidized in one region of the surface and oxygen is reduced in another region.

This iron bridge has corroded because the iron has been oxidized by moist air, yielding $Fe_2O_3 \cdot H_2O$ (rust).

A possible mechanism for rusting, consistent with the known facts, is illustrated in Figure 18.12. The surface of the iron and a droplet of surface water constitute a tiny galvanic cell in which different regions of the surface act as anode and cathode while the aqueous phase serves as the electrolyte. Iron is oxidized more readily in some regions (anode regions) than in others (cathode regions) because the composition of the metal is somewhat inhomogeneous and the surface is irregular. Factors such as impurities, phase boundaries, and mechanical stress may influence the ease of oxidation in a particular region of the surface.

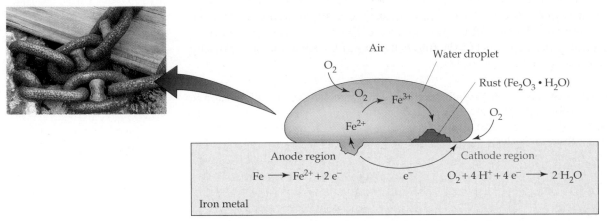

FIGURE 18.12 An electrochemical mechanism for corrosion of iron. The metal and a surface water droplet constitute a tiny galvanic cell in which iron is oxidized to Fe^{2+} in a region of the surface (anode region) remote from atmospheric O_2, and O_2 is reduced near the edge of the droplet at another region of the surface (cathode region). Electrons flow from anode to cathode through the metal, while ions flow through the water droplet. Fe^{2+} is further oxidized to Fe^{3+} by dissolved O_2 before being deposited as rust ($Fe_2O_3 \cdot H_2O$).

At an anode region, iron is oxidized to Fe^{2+} ions

$$Fe(s) \rightarrow Fe^{2+}(aq) + 2\ e^{-} \qquad E^{\circ} = 0.45\ \text{V}$$

while at a cathode region, oxygen is reduced to water:

$$O_2(g) + 4\ H^{+}(aq) + 4\ e^{-} \rightarrow 2\ H_2O(l) \qquad E^{\circ} = 1.23\ \text{V}$$

The actual potential for the reduction half-reaction is less than the standard potential (1.23 V) because the water droplet is not 1 M in H^{+} ions. (In fact, the water is only slightly acidic because the main source of H^{+} ions is the reaction of water with dissolved atmospheric carbon dioxide.) Even at pH 7, however, the potential for the reduction half-reaction is 0.81 V, which means that the cell potential is highly positive, indicative of a spontaneous reaction.

The electrons required for reduction of O_2 at the cathode region are supplied by a current that flows through the metal from the more easily oxidized anode region (Figure 18.12). The electrical circuit is completed by migration of ions in the water droplet. When Fe^{2+} ions migrate away from the pitted, anode region, they come in contact with O_2 (dissolved in the surface portion of the water droplet) and are further oxidized to Fe^{3+} ions:

$$4\ Fe^{2+}(aq) + O_2(g) + 4\ H^{+}(aq) \rightarrow 4\ Fe^{3+}(aq) + 2\ H_2O(l)$$

Iron(III) forms a very insoluble hydrated oxide even in moderately acidic solutions, and the iron(III) is deposited as the familiar red-brown material that we call rust:

$$2\ Fe^{3+}(aq) + 4\ H_2O(l) \rightarrow Fe_2O_3 \cdot H_2O(s) + 6\ H^+(aq)$$
$$\text{Rust}$$

An electrochemical mechanism for corrosion also explains nicely why automobiles rust more rapidly in parts of the country where road salt (NaCl or CaCl$_2$) is used to melt snow and ice. Dissolved salts in the water droplet greatly increase the conductivity of the electrolyte, thus accelerating the pace of corrosion.

A glance at a table of standard reduction potentials indicates that the O$_2$/H$_2$O half-reaction lies above the M^{n+}/M half-reaction for nearly all metals, meaning that O$_2$ should be able to oxidize all metals except a few, such as gold and platinum. Aluminum, for example, has $E° = -1.66$ V for the Al^{3+}/Al half-reaction and should be oxidized more readily than iron. In other words, the corrosion of aluminum products such as aircraft and automobile parts, window frames, cooking utensils, and soda cans should be a serious problem. Fortunately it isn't, because oxidation of aluminum gives a very hard, almost impenetrable film of Al$_2$O$_3$ that adheres to the surface of the metal and protects it from further attack. Other metals such as magnesium, chromium, titanium, and zinc form similar protective oxide coatings. In the case of iron, however, rust is too porous to shield the underlying metal from further oxidation.

This titanium bicycle doesn't corrode because of a hard, impenetrable layer of TiO$_2$ that adheres to the surface and protects the metal from further oxidation.

Prevention of Corrosion

Corrosion of iron can be prevented, or at least minimized, by shielding the metal surface from oxygen and moisture. A coat of paint is effective for a while, but rust begins to form as soon as the paint is scratched or chipped. Metals such as chromium, tin, or zinc, afford a more durable surface coating for iron. The steel used in making automobiles, for example, is coated by dipping into a bath of molten zinc, a process known as **galvanizing**. As the potentials indicate, zinc is oxidized more easily than iron and thus protects iron from oxidation even if the zinc layer becomes scratched (Figure 18.13).

$$Fe^{2+}(aq) + 2\ e^- \rightarrow Fe(s) \qquad E° = -0.45\ V$$

$$Zn^{2+}(aq) + 2\ e^- \rightarrow Zn(s) \qquad E° = -0.76\ V$$

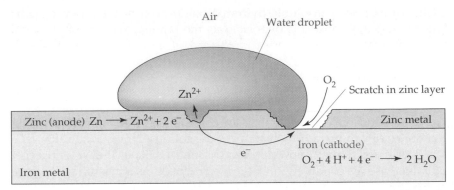

FIGURE 18.13 A layer of zinc protects iron from oxidation, even when the zinc layer becomes scratched. The zinc (anode), iron (cathode), and water droplet (electrolyte) constitute a tiny galvanic cell. Oxygen is reduced at the cathode, and zinc is oxidized at the anode, thus protecting the iron from oxidation.

The technique of protecting a metal from corrosion by connecting it to a second metal that is more easily oxidized is called **cathodic protection**. It's not necessary to cover the entire surface of the metal with a second metal, as in galvanizing iron. All that's required is electrical contact with the second metal. Underground steel pipeline, for example, is protected by connecting it through an insulated wire to a stake of magnesium, which acts as a **sacrificial anode** and corrodes instead of iron. In effect, the arrangement is a galvanic cell in which the easily oxidized magnesium acts as the anode, the pipeline behaves as the cathode, and moist soil is the electrolyte. The cell half-reactions are:

Magnesium is attached to the steel hull of this ship to protect the steel from rusting.

Anode: $Mg(s) \rightarrow Mg^{2+}(aq) + 2\ e^-$ $E° = 2.37$ V

Cathode: $O_2(g) + 4\ H^+(aq) + 4\ e^- \rightarrow 2\ H_2O(l)$ $E° = 1.23$ V

For large steel structures such as pipelines, storage tanks, bridges, and ships, cathodic protection is the best defense against premature rusting.

18.11 ▶ELECTROLYSIS AND ELECTROLYTIC CELLS

Thus far, we've been concerned only with *galvanic cells*—electrochemical cells in which a spontaneous redox reaction produces an electric current. A second important kind of electrochemical cell is the **electrolytic cell**, in which an electric current is used to drive a nonspontaneous reaction. Thus, the processes occurring in galvanic and electrolytic cells are the reverse of each other: A galvanic cell converts chemical energy to electrical energy when a reaction with a positive value of E (and a negative value of ΔG) proceeds toward equilibrium; an electrolytic cell converts electrical energy to chemical energy when an electric current drives a reaction with a negative value of E (and a positive value of ΔG) in a direction away from equilibrium. The process of using an electric current to bring about chemical change is called **electrolysis**. The opposite signs of E and $\Delta G = -nFE$ for the two kinds of cells are summarized in Table 18.2, along with the situation for a reaction that has reached equilibrium (a dead battery!).

TABLE 18.2	Relationship between Cell Potentials E and Free-Energy Changes ΔG		
Reaction Type	E	ΔG	**Cell Type**
Spontaneous	+	−	Galvanic
Nonspontaneous	−	+	Electrolytic
Equilibrium	0	0	Dead battery

Electrolysis of Molten Sodium Chloride

An electrolytic cell has two electrodes that dip into an electrolyte and are connected to a battery or some other source of direct electric current. A cell for electrolysis of molten sodium chloride, for example, is illustrated in Figure 18.14. The battery serves as an electron pump, pushing electrons into one electrode and pulling them out of the other electrode. The negative electrode attracts Na^+ cations, which combine with the electrons supplied by the battery and are thereby reduced to liquid sodium metal. Similarly, the positive electrode attracts Cl^- anions, which replenish the electrons removed by the battery and are thereby oxidized to chlorine gas. The electrode reactions and overall cell reaction are:

Anode (oxidation): $2\ Cl^-(l) \rightarrow Cl_2(g) + 2\ e^-$

Cathode (reduction): $\underline{2\ Na^+(l) + 2\ e^- \rightarrow 2\ Na(l)}$

Overall cell reaction: $2\ Na^+(l) + 2\ Cl^-(l) \rightarrow 2\ Na(l) + Cl_2(g)$

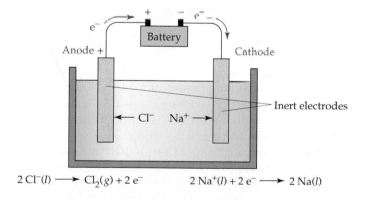

$2\ Cl^-(l) \longrightarrow Cl_2(g) + 2\ e^-$ $2\ Na^+(l) + 2\ e^- \longrightarrow 2\ Na(l)$

FIGURE 18.14 Electrolysis of molten sodium chloride. Chloride ions are oxidized to Cl_2 gas at the anode, and Na^+ ions are reduced to sodium metal at the cathode.

As for a galvanic cell, the anode is the electrode where oxidation takes place, and the cathode is the electrode where reduction takes place. The signs of the electrodes, however, are opposite for the two kinds of cells. In a galvanic cell, the anode is negative because it supplies electrons to the external circuit, but in an electrolytic cell, the anode is considered to be positive because electrons are pulled out of it by the battery.

Electrolysis of Aqueous Sodium Chloride

When an aqueous salt solution is electrolyzed, the electrode reactions can differ from those for electrolysis of a molten salt because water can be involved. In the electrolysis of aqueous sodium chloride, for example, the cathode half-reaction might involve either the reduction of Na^+ to sodium metal, as in the case of molten sodium chloride, or the reduction of water to hydrogen gas:

$$Na^+(aq) + e^- \rightarrow Na(s) \qquad\qquad E° = -2.71 \text{ V}$$

$$2\ H_2O(l) + 2\ e^- \rightarrow H_2(g) + 2\ OH^-(aq) \qquad E° = -0.83 \text{ V}$$

Because the standard potential is much less negative for reduction of water than for reduction of Na^+, we find that water is reduced and that bubbles of hydrogen gas are produced at the cathode.

The anode half-reaction might involve either the oxidation of Cl^- to Cl_2 gas, as in the case of molten sodium chloride, or oxidation of water to oxygen gas:

$$2\ Cl^-(aq) \rightarrow Cl_2(g) + 2\ e^- \qquad\qquad E° = -1.36 \text{ V}$$

$$2\ H_2O(l) \rightarrow O_2(g) + 4\ H^+(aq) + 4\ e^- \qquad E° = -1.23 \text{ V}$$

Based on the $E°$ values, we might expect a slight preference for oxidation of water in a solution having 1 M ion concentrations. For a neutral solution ($[H^+] = 10^{-7}$ M), the preference for water oxidation will be even greater because its oxidation potential at pH 7 is -0.81 V. The observed product at the anode, however, is Cl_2, not O_2, because of a phenomenon called *overvoltage*.

Experiments indicate that the applied voltage required for an electrolysis is always greater than the voltage calculated from standard oxidation and reduction potentials. The additional voltage required is called **overvoltage**. The overvoltage is needed to surmount the activation energy barrier for slow electron transfer at the electrode–solution interface. For electrode half-reactions involving metals, the overvoltage is quite small, but for half-reactions involving gases, such as H_2 and O_2, the overvoltage can be as large as 0.6 V. As a result, it's sometimes difficult to predict which half-reaction will occur when $E°$ values for the competing half-reactions are similar. In such cases, only experiment can tell what actually happens.

The observed electrode reactions and overall cell reaction for electrolysis of aqueous sodium chloride are:

Anode (oxidation):	$2\ Cl^-(aq) \rightarrow Cl_2(g) + 2\ e^-$	$E° = -1.36 \text{ V}$
Cathode (reduction):	$2\ H_2O(l) + 2\ e^- \rightarrow H_2(g) + 2\ OH^-(aq)$	$E° = -0.83 \text{ V}$
Cell reaction:	$2\ Cl^-(aq) + 2\ H_2O(l) \rightarrow Cl_2(g) + H_2(g) + 2\ OH^-(aq)$	$E° = -2.19 \text{ V}$

Sodium ion acts as a spectator ion and is not involved in the electrode reactions. Thus, the sodium chloride solution is converted to a sodium hydroxide solution as the electrolysis proceeds. The minimum potential required to force this nonspontaneous reaction to occur under standard-state conditions is 2.19 V plus the overvoltage.

Electrolysis of Water

The electrolysis of any aqueous solution requires the presence of an electrolyte to carry the current in solution. But if the ions of the electrolyte are less easily oxidized and reduced than water is, then water will react at both electrodes. Consider, for example, the electrolysis of an aqueous solution of the inert electrolyte Na_2SO_4. Water is oxidized at the anode in preference to SO_4^{2-} ions and is reduced at the cathode in preference to Na^+ ions. The electrode and overall cell reactions are:

Anode (oxidation): $2 H_2O(l) \rightarrow O_2(g) + 4 H^+(aq) + 4 e^-$

Cathode (reduction): $4 H_2O(l) + 4 e^- \rightarrow 2 H_2(g) + 4 OH^-(aq)$

Overall cell reaction: $6 H_2O(l) \rightarrow 2 H_2(g) + O_2(g) + 4 H^+(aq) + 4 OH^-(aq)$

If the anode and cathode solutions are mixed, the H^+ and OH^- ions react to form water:

$$4 H^+(aq) + 4 OH^-(aq) \rightarrow 4 H_2O(l)$$

The net electrolysis reaction is therefore the decomposition of water:

$$2 H_2O(l) \rightarrow 2 H_2(g) + O_2(g)$$

⌐ **PROBLEM 18.13** Predict the half-cell reactions that occur when aqueous solutions of the following salts are electrolyzed in a cell with inert electrodes: **(a)** LiCl **(b)** $CuSO_4$. What is the overall cell reaction in each case?   ⌐

18.12 ►COMMERCIAL APPLICATIONS OF ELECTROLYSIS

Electrolysis is used in the manufacture of many important chemicals and in numerous processes for purification and electroplating of metals. Let's look at some examples.

Manufacture of Sodium

Sodium metal is produced commercially in a *Downs cell* by electrolysis of a molten mixture of sodium chloride and calcium chloride (Figure 18.15). The presence of $CaCl_2$ allows the cell to be operated at a lower temperature, since the melting point of the $NaCl-CaCl_2$ mixture (about 580°C) is depressed well below that of pure NaCl (801°C). The liquid sodium produced at the cylindrical steel cathode is less dense than the molten salt and thus floats to the top part of the cell, where it is drawn off into a suitable container. Chlorine gas forms at the graphite anode, which is separated from the cathode by an iron screen. The cell design keeps the highly reactive sodium and chlorine away from each other and out of contact with air. Because the Downs process requires high currents, typically 25,000 to 40,000 A, plants for production of sodium are located near sources of inexpensive hydroelectric power, such as Niagara Falls, New York.

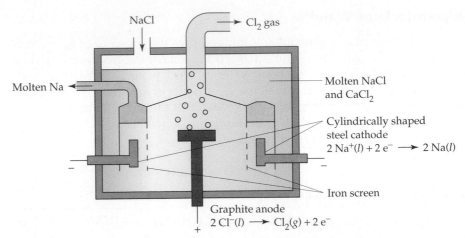

FIGURE 18.15 Cross-sectional view of a Downs cell for commercial production of sodium metal by electrolysis of molten sodium chloride. The cell design keeps the sodium and chlorine apart so that they can't react with each other.

Manufacture of Chlorine and Sodium Hydroxide

Production of chlorine and sodium hydroxide by electrolysis of aqueous sodium chloride is the basis of the **chlor-alkali industry**, a business that generates annual sales of approximately $4 billion in the United States alone. Both chlorine and sodium hydroxide rank among the top 10 chemicals in terms of production: Annual output in the United States is about 12 million tons of chlorine and 13 million tons of sodium hydroxide. Chlorine is used in water and sewage treatment and in the manufacture of plastics such as polyvinyl chloride (PVC). Sodium hydroxide is employed in making paper, textiles, soaps, and detergents.

Figure 18.16 shows the essential features of a membrane cell for commercial production of chlorine and sodium hydroxide. A saturated aqueous

FIGURE 18.16 A membrane cell for electrolytic production of Cl_2 and NaOH. Chloride ion is oxidized to Cl_2 gas at the anode, and water is converted to H_2 gas and OH^- ions at the cathode. Sodium ions move from the anode compartment to the cathode compartment through a cation-permeable membrane. Reactants (brine and water) enter the cell and products (Cl_2 gas, H_2 gas, aqueous NaOH, and depleted brine) leave through appropriately placed pipes.

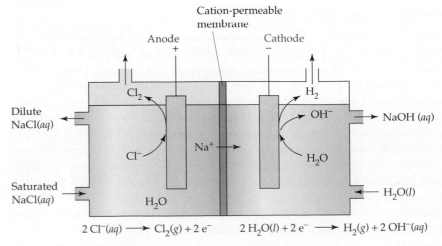

solution of sodium chloride (brine) flows into the anode compartment, where Cl^- is oxidized to Cl_2 gas, and water enters the cathode compartment, where it is converted to H_2 gas and OH^- ions. Between the anode and cathode compartments is a special plastic membrane that is permeable to cations but not to anions or water. The membrane keeps the Cl_2 and OH^- ions apart but allows a current of Na^+ ions to flow into the cathode compartment, thus carrying the current in solution and maintaining electrical neutrality in both compartments. The Na^+ and OH^- ions flow out of the cathode compartment as an aqueous solution of NaOH.

Manufacture of Aluminum

Although aluminum is the third most abundant element in the earth's crust (8.3% by weight), it remained a rare and expensive metal until 1886, when a 22 year old American, Charles Martin Hall, and a 23 year old Frenchman, Paul Heroult, independently devised a practical process for electrolytic production of aluminum. Still used today, the **Hall-Heroult process** involves electrolysis of a molten mixture of aluminum oxide (Al_2O_3) and cryolite (Na_3AlF_6) at about 1000°C in a cell with graphite electrodes (Figure 18.17). Electrolysis of pure Al_2O_3 is impractical since it melts at a very high temperature (2045°C), and electrolysis of aqueous Al^{3+} solutions is not feasible because water is reduced in preference to Al^{3+} ions. Thus, the use of cryolite as a solvent for Al_2O_3 is the key to the success of the Hall-Heroult process.

The electrode reactions are still not understood fully, but they probably involve complex anions of the type $AlF_xO_y^{+3-x-2y}$, formed by reaction of Al_2O_3 and Na_3AlF_6. The complex anions are reduced at the cathode to molten aluminum metal and are oxidized at the anode to O_2 gas, which reacts, in turn, with the graphite anodes to give CO_2 gas. As a result, the anodes are chewed up rapidly and must be replaced frequently. The cell operates at a low voltage (5 to 6 V) but with very high currents (up to 250,000 A) because 1 mol of electrons produces only 9 g of aluminum. Electrolytic production of aluminum is the largest single consumer of electricity in the

FIGURE 18.17   An electrolytic cell for production of aluminum by the Hall-Heroult process. Molten aluminum metal forms at the graphite cathode that lines the cell. Since molten aluminum is more dense than the Al_2O_3–Na_3AlF_6 mixture, it collects at the bottom of the cell and is drawn off periodically.

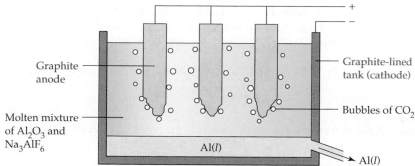

United States today, making recycling of aluminum products highly desirable.

Electrorefining and Electroplating

The purification of a metal by means of electrolysis is called **electrorefining**. For example, impure copper obtained from ores is converted to pure copper in an electrolytic cell that has impure copper as the anode and pure copper as the cathode (Figure 18.18). The electrolyte is an aqueous solution of copper sulfate.

FIGURE 18.18 Electrorefining of copper metal. **(a)** Alternating slabs of impure copper and pure copper serve as the electrodes in electrolytic cells for the refining of copper. **(b)** Copper is transferred through the $CuSO_4$ solution from the impure Cu anode to the pure Cu cathode. More easily oxidized impurities (Zn, Fe) remain in solution as cations, but noble metal impurities (Ag, Au, Pt) are not oxidized and collect as anode mud.

(a)

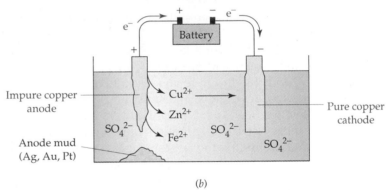

(b)

At the impure Cu anode, copper is oxidized along with more easily oxidized metallic impurities such as zinc and iron. Less easily oxidized impurities such as silver, gold, and platinum fall to the bottom of the cell as *anode mud*, which is reprocessed to recover the precious metals. At the pure Cu cathode, Cu^{2+} ions are reduced to pure copper metal, but the less easily reduced metal ions (Zn^{2+}, Fe^{2+}, and so forth) remain in the solution.

Anode (oxidation): $M(s) \rightarrow M^{2+}(aq) + 2\ e^-$ (M = Cu, Zn, Fe)

Cathode (reduction): $Cu^{2+}(aq) + 2\ e^- \rightarrow Cu(s)$

Thus, the net cell reaction simply involves transfer of copper metal from the impure anode to the pure cathode. The process typically takes about 4 weeks, and the copper obtained is 99.95% pure.

Closely related to electrorefining is **electroplating**, the coating of one metal on the surface of another using electrolysis. For example, steel automobile bumpers are plated with chromium to protect them from corrosion, and silver-plating is commonly used in making items of fine table service. The object to be plated is carefully cleaned and then made the cathode of an electrolytic cell that contains a solution of ions of the metal to be deposited.

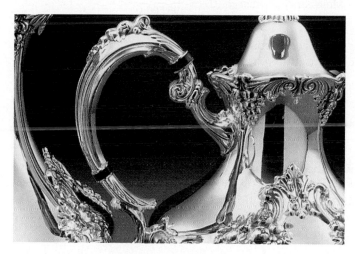

Silver-plated objects such as this teapot are made by electroplating pure silver metal onto steel.

18.13 ➤QUANTITATIVE ASPECTS OF ELECTROLYSIS

Michael Faraday showed in the 1830s that the amount of substance produced at an electrode by electrolysis depends on the quantity of charge passed through the cell. For example, passage of 1 mol of electrons through a Downs cell yields 1 mol (23.0 g) of sodium at the cathode:

$$Na^+(l) + e^- \rightarrow Na(l)$$
$$\text{1 mol   1 mol   1 mol (23.0 g)}$$

Similarly, passage of 1 mol of electrons in the Hall-Heroult process produces 1/3 mol (9.0 g) of aluminum, because 3 mol of electrons are required to reduce 1 mol of Al^{3+} to aluminum metal:

$$Al^{3+}(l) + 3\ e^- \rightarrow Al(l)$$
$$\text{1/3 mol   1 mol   1/3 mol (9.0 g)}$$

In general, the amount of product formed in an electrode reaction follows directly from the stoichiometry of the reaction and the atomic weight of the product.

To find out how many moles of electrons pass through a cell in a particular experiment, we need to measure the electric current and the time that the current flows. The number of coulombs of charge passed through the cell is equal to the product of the current in amperes (coulombs per second) and the time in seconds:

$$\text{Charge (C)} = \text{current (A)} \times \text{time (s)}$$

Since the charge on 1 mol of electrons is 96,500 C (Section 18.3), the number of moles of electrons passed through the cell is

$$\text{Moles of } e^- = \text{charge (C)} \times \frac{1 \text{ mol } e^-}{96{,}500 \text{ C}}$$

The following sequence of conversions is used to calculate the mass or volume of product produced by passing a known current for a fixed period of time:

$$\begin{matrix} \text{current} \\ \text{and} \\ \text{time} \end{matrix} \rightarrow \text{charge} \rightarrow \text{moles of } e^- \rightarrow \begin{matrix} \text{moles} \\ \text{of} \\ \text{product} \end{matrix} \rightarrow \begin{matrix} \text{grams or} \\ \text{liters of} \\ \text{product} \end{matrix}$$

The key is to think of the electrons as a "reactant" in a balanced chemical equation and then to proceed as with any other stoichiometry problem. Example 18.9 illustrates the calculations. Alternatively, we can calculate the current (or time) required to produce a given amount of product by working through this sequence in the reverse direction, as shown in Example 18.10.

EXAMPLE 18.9

A constant current of 30.0 A is passed through an aqueous solution of NaCl for a time of 1.00 h. How many grams of NaOH and how many liters of Cl_2 gas at STP are produced?

SOLUTION   Since electrons can be thought of as a reactant in the electrolysis process, the first step is to calculate the charge and the number of moles of electrons passed through the cell:

$$\text{Charge} = \left(30.0\frac{C}{s}\right)(1.00 \text{ h})\left(\frac{60 \text{ min}}{h}\right)\left(\frac{60 \text{ s}}{\text{min}}\right) = 1.08 \times 10^5 \text{ C}$$

$$\text{Moles of } e = (1.08 \times 10^5 \text{ C})\left(\frac{1 \text{ mol } e^-}{96{,}500 \text{ C}}\right) = 1.12 \text{ mol } e^-$$

The cathode reaction yields 2 mol of OH^- per 2 mol of electrons, so 1.12 mol of NaOH will be obtained:

$$2 \text{ H}_2\text{O}(l) + 2 \text{ e}^- \rightarrow \text{H}_2(g) + 2 \text{ OH}^-(aq)$$

$$\text{Moles of NaOH} = (1.12 \text{ mol } e^-)\left(\frac{2 \text{ mol NaOH}}{2 \text{ mol } e^-}\right) = 1.12 \text{ mol NaOH}$$

Converting the number of moles of NaOH to grams of NaOH gives

$$\text{Grams of NaOH} = (1.12 \text{ mol NaOH})\left(\frac{40.0 \text{ g NaOH}}{\text{mol NaOH}}\right) = 44.8 \text{ g NaOH}$$

The anode reaction gives 1 mol of Cl_2 per 2 mol of electrons, so 0.560 mol of Cl_2 will be obtained:

$$2\ Cl^-(aq) \rightarrow Cl_2(g) + 2\ e^-$$

$$\text{Moles of } Cl_2 = (1.12\ \text{mol } e^-)\left(\frac{1\ \text{mol } Cl_2}{2\ \text{mol } e^-}\right) = 0.560\ \text{mol } Cl_2$$

Since 1 mol of an ideal gas occupies 22.4 L at STP, the volume of Cl_2 obtained is

$$\text{Liters of } Cl_2 = (0.560\ \text{mol } Cl_2)\left(\frac{22.4\ \text{L } Cl_2}{\text{mol } Cl_2}\right) = 12.5\ \text{L } Cl_2$$

As a shortcut, the entire sequence of conversions can be carried out in one step. For example, the volume of Cl_2 produced at the anode is

$$\left(30.0\ \frac{C}{s}\right)(1.00\ h)\left(\frac{3600\ s}{h}\right)\left(\frac{1\ \text{mol } e^-}{96{,}500\ C}\right)\left(\frac{1\ \text{mol } Cl_2}{2\ \text{mol } e^-}\right)\left(\frac{22.4\ \text{L } Cl_2}{\text{mol } Cl_2}\right) = 12.5\ \text{L } Cl_2$$

EXAMPLE 18.10

What electric current must be passed through a Downs cell to produce sodium metal at a rate of 30.0 kg/h?

SOLUTION   Proceed through a sequence of conversions similar to that in Example 18.9, but in reverse order. Since the atomic weight of sodium is 23.0 g/mol, the number of moles of sodium produced per hour is

$$\text{Moles of Na} = (30.0\ \text{kg Na})\left(\frac{1000\ g}{1\ kg}\right)\left(\frac{1\ \text{mol Na}}{23.0\ \text{g Na}}\right) = 1.30 \times 10^3\ \text{mol Na}$$

To produce each mole of sodium, 1 mol of electrons must be passed through the cell.

$$Na^+(l) + e^- \rightarrow Na(l)$$

Therefore, the charge passed per hour is

$$\text{Charge} = (1.30 \times 10^3\ \text{mol Na})\left(\frac{1\ \text{mol } e^-}{1\ \text{mol Na}}\right)\left(\frac{96{,}500\ C}{\text{mol } e^-}\right) = 1.25 \times 10^8\ C$$

Since there are 3600 s in 1 h, the current required is

$$\text{Current} = \left(\frac{1.25 \times 10^8\ C}{3600\ s}\right) = 3.47 \times 10^4\ C/s = 34{,}700\ A$$

⌐ **PROBLEM 18.14**   How many kilograms of aluminum can be produced in 8 h by passing a constant current of 100,000 A through a molten mixture of aluminum oxide and cryolite?

⌐ **PROBLEM 18.15**   A layer of silver is electroplated on a coffee server using a constant current of 0.100 A. How much time is required to deposit 3.00 g of silver? ⌐

interlude—ELECTROCHEMICAL ART

These colored pieces are all made by anodizing aluminum, a silvery metal. The background is a bed of aluminum shavings.

The iridescent colors on this piece of metal are produced by anodizing different parts of the surface at different voltages.

If aluminum is a silvery metal, then why are some aluminum objects brightly colored? Aluminum bicycle parts, for instance, come in a spectrum of colors including red, blue, purple, yellow, and black.

Aluminum, chromium, titanium, and several other metals can be colored by an electrochemical process called **anodizing**. Unlike electroplating, in which a metal ion in the electrolyte is *reduced* and the metal is coated onto the surface of the cathode, anodizing involves *oxidation* of a metal anode to yield a metal oxide coat. In the oxidation of aluminum, for instance, the electrode reactions are as follows:

Cathode (reduction): $\qquad 6\,H^+(aq) + 6\,e^- \rightarrow 3\,H_2(g)$

Anode (oxidation): $\qquad 2\,Al(s) + 3\,H_2O(l) \rightarrow Al_2O_3(s) + 6\,H^+(aq) + 6\,e^-$

Overall cell reaction: $\quad 2\,Al(s) + 3\,H_2O(l) \rightarrow Al_2O_3(s) + 3\,H_2(g)$

The thickness of the aluminum oxide coating that forms on the anode can be controlled by varying the current flow during electrolysis. Typically, the coating is about 0.01 mm thick, which corresponds to about 4×10^4 atomic layers of Al_2O_3 on top of the underlying aluminum metal. The porosity of the Al_2O_3 coating can also be controlled by varying the electrolysis conditions, and it's this porosity that makes coloring possible. When an organic dye is added to the electrolyte, dye molecules soak into the spongy surface coating as it forms and become trapped as the surface hardens.

Titanium anodizing proceeds much like that of aluminum, but the resultant coat of TiO_2 is much thinner (10^{-4} mm) than the corresponding coat of Al_2O_3 (10^{-2} mm). Furthermore, the iridescent colors of anodized titanium result not from the absorption of organic dyes but from interference of light as it is reflected by the anodized surface. When a beam of white light strikes the anodized surface, part of the light is reflected from the outer TiO_2, while part penetrates through the semitransparent TiO_2 and is reflected from the inner metal. If the two reflections of a particular wavelength are out of phase, they interfere destructively, and that wavelength is canceled from the reflected light. As a result, the light that remains is colored. (Similar effects are responsible for the colors in oil slicks and peacock feathers.)

The exact color of anodized titanium depends on the thickness of the TiO_2 layer, which in turn varies with the voltage used to produce the layer. A voltage of 5 V, for example, gives a TiO_2 coating 3×10^{-5} mm thick with a yellow appearance, and a voltage of 30 V gives a coating 6×10^{-5} mm thick with a light blue appearance. Artists can use these effects to produce striking metal sculptures whose vibrant colors appear to change depending on the viewing angle.

Electrochemistry is the area of chemistry concerned with the interconversion of chemical and electrical energy. Chemical energy is converted to electrical energy in a **galvanic cell**, a device in which a spontaneous redox reaction is used to produce an electric current. Electrical energy is converted to chemical energy in an **electrolytic cell**, a cell in which an electric current drives a nonspontaneous reaction. It's convenient to separate the overall cell reaction into **half-reactions** because oxidation and reduction occur at separate **electrodes**. For both galvanic and electrolytic cells, the electrode at which oxidation occurs is called the **anode**, and the electrode at which reduction occurs is called the **cathode**.

The **cell potential** E (also called the **cell voltage** or **electromotive force**) is an electrical measure of the driving force of the cell reaction. Cell potentials depend on temperature, ion concentrations, and gas pressures. The **standard cell potential** $E°$ is the cell potential when reactants and products are in their standard states. Cell potentials are related to free-energy changes by the equations $\Delta G = -nFE$ and $\Delta G° = -nFE°$, where n is the number of moles of electrons transferred and F is the **faraday**, the electrical charge on 1 mol of electrons ($1\ F = 96{,}500\ \text{C/mol e}^-$).

The standard cell potential is the sum of the **standard oxidation potential** for the anode half-reaction and the **standard reduction potential** for the cathode half-reaction. Standard half-cell potentials are defined relative to an arbitrary value of 0 V for the **standard hydrogen electrode (S.H.E.):**

$$2\ \text{H}^+(aq, 1\ \text{M}) + 2\ \text{e}^- \rightarrow \text{H}_2(g, 1\ \text{atm}) \qquad E° = 0\ \text{V}$$

Tables of standard reduction potentials—also called **standard electrode potentials**—are used to arrange oxidizing and reducing agents in order of increasing strength, to calculate $E°$ values for cell reactions, and to decide whether a particular redox reaction is spontaneous.

Cell potentials under non-standard-state conditions can be calculated using the **Nernst equation**,

$$E = E° - \frac{0.0592}{n} \log Q \qquad \text{(in volts, at 25°C)}$$

where Q is the reaction quotient. At equilibrium, $E = 0$ V and $Q = K$. Therefore, the equilibrium constant K and the standard cell potential $E°$ are related by the equation

$$E° = \frac{0.0592}{n} \log K \qquad \text{(in volts, at 25°C)}$$

A battery is a convenient, portable source of electrical energy consisting of one or more galvanic cells. A **fuel cell**, such as the hydrogen–oxygen fuel cell, differs from an ordinary battery in that the reactants are continuously supplied to the cell.

Corrosion of iron (rusting) is an electrochemical process in which iron is oxidized in an anode region of the metal surface and oxygen is reduced in a cathode region. Corrosion can be prevented by covering iron with paint or another metal, such as zinc in the process called **galvanizing**, or by putting the iron in electrical contact with a second metal that is more easily oxidized, called **cathodic protection**.

Electrolysis, the process of using an electric current to bring about chemical change, is employed to produce several important industrial chemicals, including sodium, chlorine, sodium hydroxide (**chlor-alkali industry**), and aluminum (**Hall-Heroult process**). Additional electrolytic processes of importance are **electrorefining** of metals such as copper and **electroplating**. The product obtained at an electrode depends on reduction potentials and **overvoltage**. The amount of product obtained is related to the number of moles of electrons passed through the cell, which depends on the current and the time that the current flows.

UNDERSTANDING KEY CONCEPTS

1. The following galvanic cell has lead and zinc electrodes.

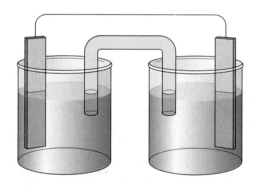

(a) Label the electrodes and identify the ions present in the solutions.
(b) Label the anode and cathode.
(c) Indicate the direction of electron flow in the wire and ion flow in the solutions.
(d) Identify the electrolyte and the salt bridge, and indicate the direction of ion flow.
(e) Write balanced equations for the electrode and overall cell reactions.

2. Sketch a cell with inert electrodes suitable for electrolysis of aqueous $CuBr_2$.
(a) Label the anode and cathode.
(b) Indicate the direction of electron and ion flow.
(c) Write balanced equations for the anode, cathode, and overall cell reactions.

3. Consider the following substances: $Fe(s)$, $PbO_2(s)$, $H^+(aq)$, $Al(s)$, $Ag(s)$, and $Cr_2O_7^{2-}(aq)$.
(a) Look at the $E°$ values in Appendix D, and classify each substance as an oxidizing agent or as a reducing agent.
(b) Which is the strongest oxidizing agent? Which is the weakest oxidizing agent?
(c) Which is the strongest reducing agent? Which is the weakest reducing agent?
(d) Which substance can be oxidized by $Cu^{2+}(aq)$? Which can be reduced by $H_2O_2(aq)$?

4. The following cell reactions occur spontaneously:

$$B + A^+ \rightarrow B^+ + A$$

$$C + A^+ \rightarrow C^+ + A$$

$$B + C^+ \rightarrow B^+ + C$$

(a) Arrange the three reduction half-reactions— $A^+ + e^- \longrightarrow A$, $B^+ + e^- \longrightarrow B$, and $C^+ + e^- \longrightarrow C$—in order of decreasing tendency to occur.
(b) Which of the preceding substances is the strongest oxidizing agent? Which is the strongest reducing agent?
(c) Which of the three cell reactions delivers the highest voltage?

5. "Tin cans" are made of steel that is protected with a thin layer of tin. Scratched tin cans rust rapidly, but scratched galvanized iron does not. Explain.

ADDITIONAL PROBLEMS

Problems 18.1–18.15 appear within the chapter.

GALVANIC CELLS

18.16 Define anode and cathode with reference to a specific galvanic cell.

18.17 Does the oxidizing agent react at the anode or at the cathode in a galvanic cell? Explain.

18.18 Why is the cathode of a galvanic cell considered to be the positive electrode?

18.19 What is the function of a salt bridge in a galvanic cell?

18.20 Describe galvanic cells that use each of the following reactions.

   (a) $Cd(s) + Sn^{2+}(aq) \longrightarrow Cd^{2+}(aq) + Sn(s)$
   (b) $2\ Al(s) + 3\ Cd^{2+}(aq) \longrightarrow 2\ Al^{3+}(aq) + 3\ Cd(s)$
   (c) $Pb(s) + 2\ H^{+}(aq) \longrightarrow Pb^{2+}(aq) + H_2(g)$
   (d) $6\ Fe^{2+}(aq) + Cr_2O_7{}^{2-}(aq) + 14\ H^{+}(aq) \longrightarrow$
      $6\ Fe^{3+}(aq) + 2\ Cr^{3+}(aq) + 7\ H_2O(l)$

In each case, write the anode and cathode half-reactions, and sketch the experimental setup. Label the anode and cathode, identify the sign of each electrode, and indicate the direction of electron and ion flow.

18.21 Write the standard shorthand notation for each of the cells in Problem 18.20.

18.22 An H_2/H^{+} half-cell (anode) and an Ag^{+}/Ag half-cell (cathode) are connected by a wire and a salt bridge.

   (a) Sketch the cell, indicating the direction of electron and ion flow.
   (b) Write balanced equations for the electrode and overall cell reactions.
   (c) Give the shorthand notation for the cell.

18.23 A galvanic cell is constructed from a Zn/Zn^{2+} half-cell (anode) and a Cl_2/Cl^{-} half-cell (cathode). Give the **(a)** sketch, **(b)** equations, and **(c)** shorthand requested in **(a)**, **(b)**, and **(c)** of Problem 18.22.

18.24 Write balanced equations for the electrode and overall cell reactions for each of the following galvanic cells. Make a sketch of each cell, labeling the anode and cathode and showing the direction of electron and ion flow.

   (a) $Co(s) \,|\, Co^{2+}(aq) \,\|\, Cu^{2+}(aq) \,|\, Cu(s)$
   (b) $Fe(s) \,|\, Fe^{2+}(aq) \,\|\, H_2O(l),\ H^{+}(aq) \,|\, O_2(g) \,|\, Pt(s)$

CELL POTENTIALS AND FREE-ENERGY CHANGES; STANDARD REDUCTION POTENTIALS

18.25 What are the SI units of electrical potential, electrical charge, and energy? How are they related?

18.26 Define all terms in the equation $\Delta G = -nFE$.

18.27 What conditions must be met for a cell potential E to qualify as a standard cell potential $E°$?

18.28 How are standard reduction potentials defined? How is a standard oxidation potential related to the corresponding standard reduction potential?

18.29 The silver oxide–zinc battery used in hearing aids delivers a voltage of 1.60 V. Calculate the free-energy change in kJ for the cell reaction:

$$Zn(s) + Ag_2O(s) \rightarrow ZnO(s) + 2\ Ag(s)$$

18.30 The standard cell potential for a lead storage battery is 1.924 V. Calculate $\Delta G°$ in kJ for the cell reaction:

$$Pb(s) + PbO_2(s) + 2\ H^{+}(aq) + 2\ HSO_4{}^{-}(aq) \rightarrow$$
$$2\ PbSO_4(s) + 2\ H_2O(l)$$

18.31 Using the standard free energies of formation in Appendix B, calculate the standard cell potential for the reaction in the hydrogen–oxygen fuel cell:

$$2\ H_2(g) + O_2(g) \rightarrow 2\ H_2O(l)$$

18.32 Consider a fuel cell that uses the reaction

$$CH_4(g) + 2\ O_2(g) \rightarrow CO_2(g) + 2\ H_2O(l)$$

Given the standard free energies of formation in Appendix B, what is the value of $E°$ for the cell reaction?

18.33 The standard potential for the following galvanic cell is 0.40 V:

$$Zn(s) \,|\, Zn^{2+}(aq) \,\|\, Eu^{3+}(aq),\ Eu^{2+}(aq) \,|\, Pt(s)$$

(Eu is the symbol for europium, one of the lanthanide elements.) Use the data in Table 18.1 to calculate the standard reduction potential for the Eu^{3+}/Eu^{2+} half-cell.

18.34 The following reaction has an $E°$ value of 0.27 V:

$$Cu^{2+}(aq) + 2\ Ag(s) + 2\ Br^{-}(aq) \rightarrow Cu(s) + 2\ AgBr(s)$$

Use the data in Table 18.1 to calculate the standard reduction potential for the half-reaction

$$AgBr(s) + e^{-} \rightarrow Ag(s) + Br^{-}(aq)$$

18.35 Arrange the following oxidizing agents in order of increasing strength under standard-state conditions: $O_2(g)$, $Cl_2(g)$, $Cu^{2+}(aq)$.

18.36 List the following reducing agents in order of increasing strength under standard-state conditions: $Mg(s)$, $Ni(s)$, $Zn(s)$.

18.37 Consider the following substances: $Br_2(l)$, $Fe^{2+}(aq)$, $MnO_4^-(aq)$. Which is the strongest oxidizing agent? Which is the weakest oxidizing agent?

18.38 Consider the following substances: $Fe^{2+}(aq)$, $Sn^{2+}(aq)$, $I^-(aq)$. Identify the strongest reducing agent and the weakest reducing agent.

18.39 Calculate the standard cell potential and the standard free-energy change in kJ for each of the reactions in Problem 18.20.

18.40 Calculate $E°$ and $\Delta G°$ in kJ for the cell reactions in Problem 18.24. (See Appendix D for standard reduction potentials.)

18.41 Calculate $E°$ for each of the following reactions. Which of these reactions are spontaneous under standard-state conditions?

(a) $2\ Fe^{2+}(aq) + Pb^{2+}(aq) \longrightarrow 2\ Fe^{3+}(aq) + Pb(s)$

(b) $Mg(s) + Ni^{2+}(aq) \longrightarrow Mg^{2+}(aq) + Ni(s)$

(c) $5\ Ag^+(aq) + Mn^{2+}(aq) + 4\ H_2O(l) \longrightarrow 5\ Ag(s) + MnO_4^-(aq) + 8\ H^+(aq)$

(d) $2\ H_2O_2(aq) \longrightarrow O_2(g) + 2\ H_2O(l)$

18.42 Use the data in Table 18.1 to predict whether the following reactions can occur under standard-state conditions.

(a) oxidation of $Sn^{2+}(aq)$ by $Br_2(l)$

(b) reduction of $Ni^{2+}(aq)$ by $Sn^{2+}(aq)$

(c) oxidation of $Ag(s)$ by $Pb^{2+}(aq)$

(d) reduction of $I_2(s)$ by $Cu(s)$

18.43 What reaction can occur, if any, when the following experiments are carried out under standard-state conditions?

(a) A strip of zinc is dipped into an aqueous solution of lead nitrate.

(b) An acidic solution of iron(II) sulfate is exposed to oxygen.

(c) A silver wire is immersed in an aqueous solution of nickel chloride.

(d) Hydrogen gas is bubbled through aqueous cadmium nitrate.

THE NERNST EQUATION

18.44 Consider a galvanic cell that uses the reaction

$$2\ Ag^+(aq) + Sn(s) \rightarrow 2\ Ag(s) + Sn^{2+}(aq)$$

Calculate the potential at 25°C for a cell that has the following ion concentrations: $[Ag^+] = 0.010$ M, $[Sn^{2+}] = 0.020$ M.

18.45 Consider a galvanic cell based on the reaction

$$2\ Fe^{2+}(aq) + Cl_2(g) \rightarrow 2\ Fe^{3+}(aq) + 2\ Cl^-(aq)$$

Calculate the cell potential at 25°C when $[Fe^{2+}] = 1.0$ M, $[Fe^{3+}] = 1.0 \times 10^{-3}$ M, $[Cl^-] = 3.0 \times 10^{-3}$ M, and $P_{Cl_2} = 0.50$ atm.

18.46 What is the cell potential at 25°C for the following galvanic cell?

$$Pb(s)\,|\,Pb^{2+}(1.0\ M)\,\|\,Cu^{2+}(1.0 \times 10^{-4}\ M)\,|\,Cu(s)$$

If the Pb^{2+} concentration is maintained at 1.0 M, what is the Cu^{2+} concentration when the cell potential drops to zero?

18.47 The Nernst equation applies to both cell reactions and half-reactions. For the conditions specified, calculate the potential for the following half-reactions at 25°C.

(a) $I_2(s) + 2\ e^- \longrightarrow 2\ I^-(aq)$; ($[I^-] = 0.020$ M)

(b) $Fe^{3+}(aq) + e^- \longrightarrow Fe^{2+}(aq)$; ($[Fe^{3+}] = [Fe^{2+}] = 0.10$ M)

(c) $Sn^{2+}(aq) \longrightarrow Sn^{4+}(aq) + 2\ e^-$; ($[Sn^{2+}] = 1.0 \times 10^{-3}$ M, $[Sn^{4+}] = 0.40$ M)

(d) $2\ Cr^{3+}(aq) + 7\ H_2O(l) \longrightarrow Cr_2O_7^{2-}(aq) + 14\ H^+(aq) + 6\ e^-$; ($[Cr^{3+}] = [Cr_2O_7^{2-}] = 1.0$ M, $[H^+] = 0.010$ M)

18.48 What is the reduction potential at 25°C for the hydrogen electrode in each of the following solutions? The half-reaction is

$$2\ H^+(aq) + 2\ e^- \rightarrow H_2(g,\ 1\ atm)$$

(a) 1.0 M HCl   **(b)** a solution having pH 4.00
(c) pure water   **(d)** 1.0 M NaOH

18.49 The following cell has a potential of 0.27 V at 25°C:

$$Pt(s)\,|\,H_2(1\ atm)\,|\,H^+(?\ M)\,\|\,Ni^{2+}(1\ M)\,|\,Ni(s)$$

What is the pH of the solution in the anode compartment?

18.50 What is the pH of the solution in the cathode compartment of the following cell if the measured cell potential at 25°C is 0.58 V?

$$Zn(s)\,|\,Zn^{2+}(1\ M)\,\|\,H^+(?\ M)\,|\,H_2(1\ atm)\,|\,Pt(s)$$

18.51 A galvanic cell has an iron electrode in contact with 0.10 M $FeSO_4$ and a copper electrode in contact with a $CuSO_4$ solution. If the measured cell potential at 25°C is 0.67 V, what is the concentration of Cu^{2+} in the $CuSO_4$ solution?

STANDARD CELL POTENTIALS AND EQUILIBRIUM CONSTANTS

18.52 Use the data in Table 18.1 to calculate the equilibrium constant at 25°C for the reaction

$$Ni(s) + 2 Ag^+(aq) \rightarrow Ni^{2+}(aq) + 2 Ag(s)$$

18.53 From standard reduction potentials, calculate the equilibrium constant at 25°C for the reaction

$$2 MnO_4^-(aq) + 10 Cl^-(aq) + 16 H^+(aq) \rightarrow$$
$$2 Mn^{2+}(aq) + 5 Cl_2(g) + 8 H_2O(l)$$

18.54 Beginning with the equations that relate $E°$, $\Delta G°$, and K, show that $\Delta G°$ is negative and $K > 1$ for a reaction that has a positive value of $E°$.

18.55 If a reaction has an equilibrium constant $K < 1$, is $E°$ positive or negative? What is the value of K when $E° = 0$ V?

18.56 Calculate the equilibrium constant at 25°C for each of the reactions in Problem 18.20.

18.57 Calculate the equilibrium constant at 25°C for disproportionation of Hg_2^{2+}:

$$Hg_2^{2+}(aq) \rightarrow Hg(l) + Hg^{2+}(aq)$$

See Appendix D for standard reduction potentials.

BATTERIES; CORROSION

18.58 For a lead storage battery **(a)** draw a diagram of one cell that shows the anode, cathode, electrolyte, direction of electron and ion flow, and sign of the electrodes. **(b)** Write the anode, cathode, and overall cell reactions. **(c)** Calculate the equilibrium constant for the cell reaction ($E° = 1.924$ V). **(d)** What is the cell voltage when the cell reaction reaches equilibrium?

18.59 Calculate the values of $E°$, $\Delta G°$ (in kJ), and K at 25°C for the cell reaction in a hydrogen–oxygen fuel cell: $2 H_2(g) + O_2(g) \longrightarrow 2 H_2O(l)$.

18.60 How many grams of HgO react at the cathode of a mercury battery when 2.00 g of zinc is consumed at the anode?

18.61 Write a balanced equation for the reaction that

occurs when a nickel–cadmium battery is recharged. If 10.0 g of $Ni(OH)_2$ is oxidized in the charging process, how many grams of cadmium are formed?

18.62 What is rust? What causes it to form? What can be done to prevent its formation?

18.63 What is meant by cathodic protection? Which of the following metals can offer cathodic protection to iron? Zn   Ni   Al   Sn

18.64 What is a sacrificial anode? Give an example.

18.65 The standard oxidation potential is 0.74 V for the reaction $Cr(s) \rightarrow Cr^{3+}(aq) + 3 e^-$. Despite the large, positive oxidation potential, chromium is used as a protective coating on steel automobile bumpers. Why doesn't the chromium corrode?

ELECTROLYSIS

18.66 Magnesium metal is produced by electrolysis of molten magnesium chloride using inert electrodes. **(a)** Make a sketch of the cell, label the anode and cathode, indicate the sign of the electrodes, and show the direction of electron and ion flow. **(b)** Write balanced equations for the anode, cathode, and overall cell reactions.

18.67 **(a)** Sketch a cell with inert electrodes suitable for electrolysis of an aqueous solution of sulfuric acid. Label the anode and cathode, and indicate the direction of electron and ion flow. Identify the positive and negative electrodes. **(b)** Write balanced equations for the anode, cathode, and overall cell reactions.

18.68 List the anode and cathode half-reactions that might occur when an aqueous solution of $MgCl_2$ is electro-

lyzed in a cell having inert electrodes. Predict which half-reactions will occur, and justify your answers.

18.69 What products should be formed when **(a)** molten KCl and **(b)** aqueous KCl are electrolyzed in a cell having inert electrodes. Account for any differences.

18.70 Predict the anode, cathode, and overall cell reactions when an aqueous solution of each of the following salts is electrolyzed in a cell having inert electrodes. **(a)** NaBr   **(b)** $CuCl_2$ **(c)** LiOH   **(d)** Ag_2SO_4

18.71 How many grams of silver will be obtained when an aqueous silver nitrate solution is electrolyzed for 20.0 min with a constant current of 2.40 A?

18.72 A constant current of 100.0 A is passed through an electrolytic cell having an impure copper anode, a pure copper cathode, and an aqueous $CuSO_4$ electrolyte. How many kilograms of copper are refined by transfer from the anode to the cathode in a 24.0 h period?

18.73 How many hours are required to produce 1000 kg of sodium by electrolysis of molten NaCl with a constant current of 30,000 A? How many liters of Cl_2 at STP will be obtained as a byproduct?

18.74 What constant current in amperes is required to produce aluminum by the Hall-Heroult process at a rate of 40.0 kg/h?

18.75 How many grams of $PbSO_4$ are reduced at the cathode if you charge a lead storage battery for 1.50 h with a constant current of 10.0 A?

18.76 A layer of chromium is electroplated on an automobile bumper by passing a constant current of 200.0 A through a cell that contains $Cr^{3+}(aq)$. How many minutes are required to deposit 125 g of chromium?

GENERAL PROBLEMS

18.77 Distinguish between each of the following terms.
(a) oxidizing agent and reducing agent
(b) galvanic cell and electrolytic cell
(c) battery and fuel cell
(d) anode and cathode

18.78 Briefly define each of the following terms.
(a) Hall-Heroult process   (b) Downs cell
(c) electrorefining   (d) galvanizing

18.79 Consider a galvanic cell that uses the following half-reactions:

$$MnO_4^-(aq) + 8\,H^+(aq) + 5\,e^- \rightarrow Mn^{2+}(aq) + 4\,H_2O(l)$$

$$Sn^{4+}(aq) + 2\,e^- \rightarrow Sn^{2+}(aq)$$

(a) Write a balanced equation for the overall cell reaction.
(b) What is the oxidizing agent and what is the reducing agent?
(c) Calculate the standard cell potential.

18.80 Given the following half-reactions and $E°$ values,

$$Mn^{3+}(aq) + e^- \rightarrow Mn^{2+}(aq) \qquad E° = 1.54\ V$$

$$MnO_2(s) + 4\,H^+(aq) + e^- \rightarrow Mn^{3+}(aq) + 2\,H_2O(l) \qquad E° = 0.95\ V$$

write a balanced equation for disproportionation of $Mn^{3+}(aq)$, and calculate the value of $E°$ for this reaction. Is the reaction spontaneous under standard-state conditions?

18.81 Consider the following half-reactions and $E°$ values:

$$Ag^+(aq) + e^- \rightarrow Ag(s) \qquad E° = 0.80\ V$$

$$Cu^{2+}(aq) + 2\,e^- \rightarrow Cu(s) \qquad E° = 0.34\ V$$

$$Pb^{2+}(aq) + 2\,e^- \rightarrow Pb(s) \qquad E° = -0.13\ V$$

(a) Which of these metals or ions is the strongest oxidizing agent? Which is the strongest reducing agent?
(b) The half-reactions can be used to construct three different galvanic cells. Sketch the cell that delivers the highest voltage. Label the anode and cathode, and indicate the direction of electron and ion flow.

(c) Write the cell reaction in (b), and calculate the values of $E°$, $\Delta G°$ (in kJ), and K for this reaction.
(d) Calculate the voltage for the cell in (b) if both ion concentrations are 0.010 M.

18.82 Consider the following substances: $Fe^{2+}(aq)$, $Fe^{3+}(aq)$, $H_2O_2(aq)$, $Cl^-(aq)$. Which is the strongest oxidizing agent? Which is the strongest reducing agent?

18.83 Consider the following substances: Al(s), Pb(s), $Br_2(l)$, $I^-(aq)$, $Cu^{2+}(aq)$, $MnO_4^-(aq)$. Which can be oxidized by $Fe^{2+}(aq)$? Which can be reduced by $Fe^{2+}(aq)$?

18.84 *Standard* reduction potentials for the Pb^{2+}/Pb and Cd^{2+}/Cd half-reactions are -0.13 and -0.40 V, respectively. At what relative concentrations of Pb^{2+} and Cd^{2+} will these half-reactions have the same reduction potential?

18.85 Consider a galvanic cell that uses the following half-reactions:

$$2\,H^+(aq) + 2\,e^- \rightarrow H_2(g)$$

$$Al^{3+}(aq) + 3\,e^- \rightarrow Al(s)$$

(a) Sketch the cell. Tell what materials are used for the electrodes. Label the anode and cathode, and indicate the direction of electron and ion flow.
(b) Write a balanced equation for the cell reaction, and calculate the standard cell potential.
(c) Calculate the cell potential at 25°C if the ion concentrations are 0.10 M and the partial pressure of H_2 is 10.0 atm.
(d) Calculate $\Delta G°$ (in kJ) and K for the cell reaction at 25°C.
(e) Calculate the mass change in grams of the aluminum electrode after the cell has supplied a constant current of 10.0 A for 25.0 min.

18.86 A Daniell cell delivers a constant current of 0.100 A for 200.0 h. How many grams of zinc are oxidized at the anode?

18.87 Approximately 12 million tons of Cl_2 are produced annually in the United States by electrolysis of brine. How many kilowatt-hours (kWh) of energy are required for the electrolysis if the cells operate at a potential of 4.5 V? (1 ton = 2000 lb; 1 kWh = 3.6 × 10^6 J.)

18.88 The reaction of MnO_4^- with oxalic acid ($H_2C_2O_4$) is widely used to determine the concentration of permanganate solutions:

$$2\,MnO_4^-(aq) + 5\,H_2C_2O_4(aq) + 16\,H^+(aq) \rightarrow$$
$$2\,Mn^{2+}(aq) + 10\,CO_2(g) + 8\,H_2O(l)$$

(a) Use the data in Appendix D to calculate $E°$ for the reaction.
(b) Show that the reaction goes to completion by calculating the values of $\Delta G°$ and K at 25°C.
(c) A 1.200 g sample of sodium oxalate ($Na_2C_2O_4$) is dissolved in dilute H_2SO_4 and then titrated with a $KMnO_4$ solution. If 32.50 mL of the $KMnO_4$ solution is required to reach the equivalence point, what is the molarity of the $KMnO_4$ solution?

18.89 When suspected drunk drivers are tested with a breath analyzer, the alcohol (ethanol) in the exhaled breath is oxidized to acetic acid with an acidic solution of potassium dichromate:

$$3\,CH_3CH_2OH(aq) + 2\,Cr_2O_7^{2-}(aq) + 16\,H^+(aq) \rightarrow$$
Ethanol
$$3\,CH_3COOH(aq) + 4\,Cr^{3+}(aq) + 11\,H_2O(l)$$
Acetic Acid

The color of the solution changes because some of the orange $Cr_2O_7^{2-}$ is converted to the green Cr^{3+}. The breath analyzer measures the color change and produces a meter reading calibrated in terms of blood alcohol content.

(a) What is $E°$ for the reaction if the standard half-cell potential for reduction of acetic acid to ethanol is 0.058 V?
(b) What is the value of E for the reaction when the concentrations of ethanol, $Cr_2O_7^{2-}$, and Cr^{3+} are 1 M and the pH is 4.00?

18.90 The standard half-cell potential for reduction of quinone to hydroquinone is 0.699 V:

Quinone, $C_6H_4O_2$

$E° = 0.699$ V

Hydroquinone, $C_6H_4(OH)_2$

When a photographic film is developed, silver bromide is reduced by hydroquinone (the developer) in a basic aqueous solution to give quinone and tiny black particles of silver metal:

$$2\,AgBr(s) + C_6H_4(OH)_2(aq) + 2\,OH^-(aq) \rightarrow$$
$$2\,Ag(s) + 2\,Br^-(aq) + C_6H_4O_2(aq) + 2\,H_2O(l)$$

(a) What is the value of $E°$ for this reaction under acidic conditions ($[H^+] = 1$ M)? Is the reaction spontaneous in 1 M acid?
(b) What is the value of $E°$ for the reaction under basic conditions ($[OH^-] = 1$ M)?

18.91 Adiponitrile, a key intermediate in the manufacture of nylon, is made industrially by an electrolytic process involving reduction of acrylonitrile:

Anode (oxidation): $2\,H_2O \rightarrow O_2 + 4\,H^+ + 4\,e^-$

Cathode (reduction):
$$2\,CH_2{=}CHCN + 2\,H^+ + 2\,e^- \rightarrow NC(CH_2)_4CN$$
Acrylonitrile   Adiponitrile

(a) Write a balanced equation for the overall cell reaction.
(b) How many kilograms of adiponitrile are produced in 10.0 h in a cell that has a constant current of 3.00×10^3 A?
(c) How many liters of O_2 at 740 mm Hg and 25°C are produced as a byproduct?

18.92 The following galvanic cell has a potential of 0.578 V at 25°C:

$$Ag(s) \mid AgCl(s) \mid Cl^-(1.0\,M) \parallel Ag^+(1.0\,M) \mid Ag(s)$$

Use this information to calculate K_{sp} for AgCl at 25°C.

18.93 A galvanic cell has a silver electrode in contact with 0.050 M $AgNO_3$ and a copper electrode in contact with 1.0 M $Cu(NO_3)_2$.

(a) Write a balanced equation for the cell reaction, and calculate the cell potential at 25°C.
(b) Excess $NaBr(aq)$ is added to the $AgNO_3$ solution to precipitate AgBr. What is the cell potential at 25°C after precipitation of AgBr if the concentration of excess Br^- is 1.0 M? Write a balanced equation for the cell reaction under these conditions. (K_{sp} for AgBr at 25°C is 5.4×10^{-13}.)
(c) Use the result in (b) to calculate the standard reduction potential $E°$ for the half-reaction $AgBr(s) + e^- \rightarrow Ag(s) + Br^-(aq)$.

THE MAIN-GROUP ELEMENTS

T he main-group elements are the 44 elements that occupy groups 1A–8A of the periodic table. They are subdivided into the *s*-block elements of groups 1A and 2A, with valence electron configuration ns^1 or ns^2, and the *p*-block elements of groups 3A–8A, with valence configurations ns^2np^{1-6}. Main-group elements are important because of their high natural abundance and their presence in commercially valuable chemicals. Eight of the 10 most abundant elements in the earth's crust and all 10 of the most abundant elements in the human body are main-group elements (Figure 19.1). In addition, the 10 most widely used industrial chemicals contain only main-group elements (Table 19.1).

The dark blue crystals are benitoite ($BaTiSi_3O_9$). Silicate minerals and silica account for about 95% of the mass of the earth's crust.

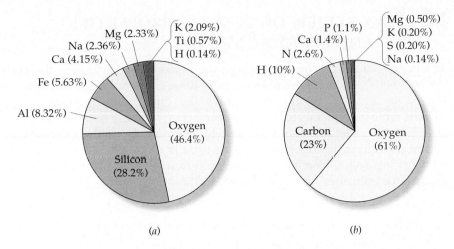

FIGURE 19.1 The 10 most abundant elements by mass **(a)** in the earth's crust and **(b)** in the human body.

TABLE 19.1	The Top 10 Chemicals (1993 U.S. Production)	
Chemical	**Millions of tons**	**Principal Uses**
Sulfuric acid (H_2SO_4)	40.2	Fertilizers, chemicals, oil refining
Nitrogen (N_2)	32.6	Inert atmospheres, low temperatures
Oxygen (O_2)	23.3	Steelmaking, welding, medical uses
Ethylene ($CH_2\!=\!CH_2$)	20.6	Plastics, antifreeze
Lime (CaO)	18.4	Steelmaking, chemicals, water treatment
Ammonia (NH_3)	17.2	Fertilizers, nitric acid
Sodium hydroxide (NaOH)	12.9	Chemicals, textiles, soaps
Chlorine (Cl_2)	12.0	Chemicals, plastics, water treatment
Methyl *tert*-butyl ether	12.0	Automobile fuel additive
Phosphoric acid (H_3PO_4)	11.5	Fertilizers, detergents

In Sections 6.5–6.9 we surveyed the alkali and alkaline-earth metals (groups 1A and 2A), aluminum (group 3A), the halogens (group 7A), and the noble gases (group 8A). In Chapter 14 we looked at the chemistry of hydrogen and oxygen. We'll now examine the remaining main-group elements, paying special attention to boron (group 3A), carbon and silicon (group 4A), nitrogen and phosphorus (group 5A), and sulfur (group 6A).

Sodium hydroxide is used in making soaps and textiles.

19.1 ▶A REVIEW OF GENERAL PROPERTIES AND PERIODIC TRENDS

Let's begin our survey of the main-group elements by reviewing the periodic trends that make it possible to classify these elements as metals, nonmetals, or semimetals. Figure 19.2 shows the main-group regions of the periodic table with metals to the left of the heavy staircase line, nonmetals to the right of the line, and semimetals—elements with intermediate properties—along the line. The elements usually classified as semimetals are boron (group 3A), silicon and germanium (group 4A), arsenic and antimony (group 5A), tellurium (group 6A), and astatine (group 7A).

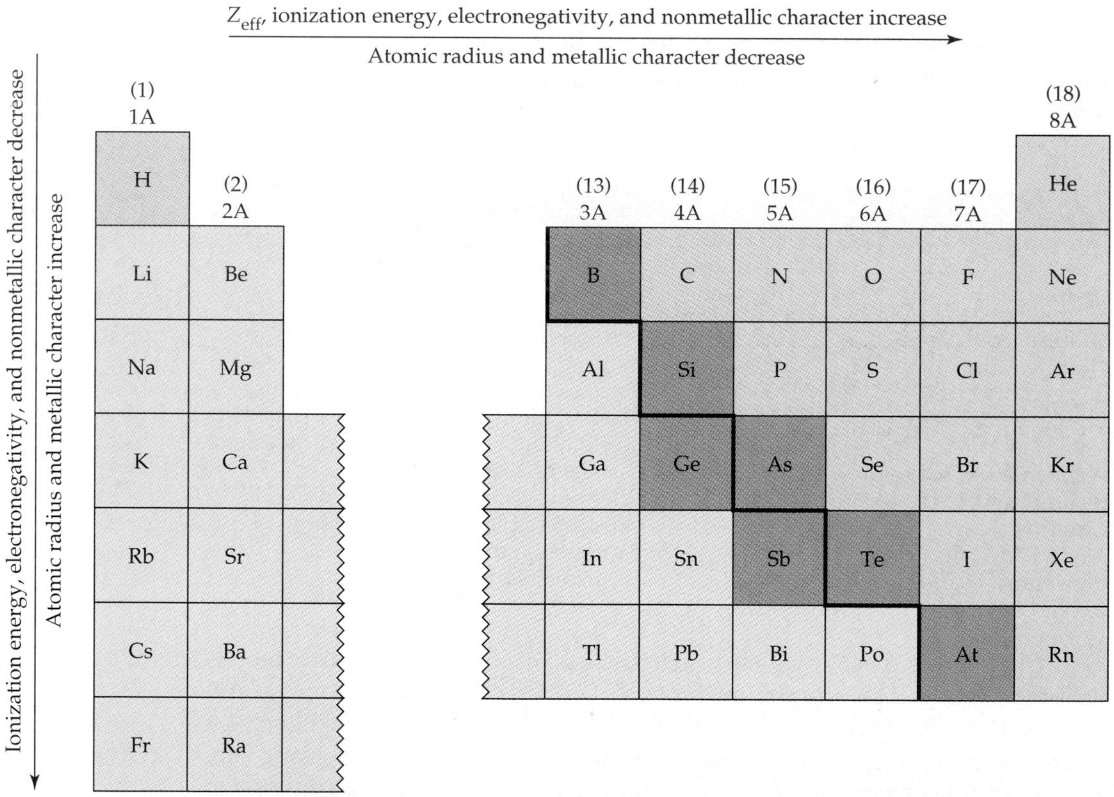

FIGURE 19.2 Periodic trends in the properties of the main-group elements. The metallic elements (blue) and the nonmetallic elements (red) are separated by the heavy staircase line. The semimetals, shown in violet, lie along the line.

From left to right across the periodic table, the effective nuclear charge Z_{eff} increases, because each additional valence electron does not completely shield the additional nuclear charge (Section 5.10). As a result, the atom's electrons are more strongly attracted to the nucleus, ionization energy increases, atomic radius decreases, and electronegativity increases. The elements on the left side of the table tend to form cations by losing electrons, and those on the right tend to form anions by gaining electrons. Thus, metallic character decreases and nonmetallic character increases across the table from left to right. In the third row, for example, sodium, magnesium, and aluminum are metals, silicon is a semimetal, and phosphorus, sulfur, and chlorine are nonmetals.

From the top to the bottom of a group in the periodic table, additional shells of electrons are occupied, and atomic radius therefore increases. Because the valence electrons are farther from the nucleus, though, ionization energy and electronegativity generally decrease. As a result, metallic character increases and nonmetallic character decreases down a group from top to bottom. In group 4A, for example, carbon is a nonmetal, silicon and germanium are semimetals, and tin and lead are metals. The horizontal and vertical periodic trends combine to locate the element with the most metallic character (cesium) in the lower left of the periodic table, the element with the most nonmetallic character (fluorine) in the upper right, and the semimetals along the diagonal staircase that stretches across the middle.

In earlier chapters, we saw examples of how the metallic or non-metallic character of an element affects its chemistry. Metals tend to form ionic compounds with nonmetals, whereas nonmetals tend to form covalent, molecular compounds with each other. Thus, binary metallic hydrides, such as NaH and CaH_2, are ionic solids with high melting points, and binary nonmetallic hydrides, such as CH_4, NH_3, H_2O, and HF, are covalent, molecular compounds that exist at room temperature as gases or volatile liquids (Section 14.5). Oxides exhibit similar trends. In the third row, for example, Na_2O and MgO are typical high-melting, ionic solids, and P_4O_{10}, SO_3, and Cl_2O_7 are volatile, covalent, molecular compounds (Section 14.9). Of course, the metallic or nonmetallic character of an oxide also affects its acid–base properties: Na_2O and MgO are basic, but P_4O_{10}, SO_3, and Cl_2O_7 are acidic. Table 19.2 summarizes some of the properties that distinguish metallic and nonmetallic elements.

TABLE 19.2   **Properties of Metallic and Nonmetallic Elements**

Metals	Nonmetals
All are solids at 25°C except Hg, which is a liquid	Eleven are gases at 25°C, one is a liquid (Br), and five are solids (C, P, S, Se, and I)
Most have a silvery shine	Most lack a metallic luster
Malleable and ductile	Nonmalleable and brittle
Good conductors of heat and electricity	Poor conductors of heat and electricity, except graphite
Relatively low ionization energies	Relatively high ionization energies
Relatively low electronegativities	Relatively high electronegativities
Lose electrons to form cations	Gain electrons to form anions; share electrons to form oxoanions
Hydrides are ionic (or interstitial)	Hydrides are covalent and molecular
Oxides are ionic and basic	Oxides are covalent, molecular, and acidic

EXAMPLE 19.1

Use the periodic table to predict which element in each of the following pairs has the more metallic character. **(a)** Ga or As **(b)** P or Bi **(c)** Sb or S

SOLUTION **(a)** Ga and As are in the same row of the periodic table, but Ga (group 3A) lies to the left of As (group 5A). Therefore, Ga is more metallic. **(b)** Bi lies below P in group 5A and is therefore more metallic. **(c)** Sb (group 5A) has more metallic character because it lies below and to the left of S (group 6A).

┌ **PROBLEM 19.1** Predict which element in each of the following pairs has the more nonmetallic character.

(a) B or Al   **(b)** Ge or Br   **(c)** In or Se

19.2 ➤DISTINCTIVE PROPERTIES OF THE SECOND-ROW ELEMENTS

The properties of the elements in the second row of the periodic table differ markedly from those of the heavier elements in the same periodic group. The second-row atoms have especially small sizes and especially high electronegativities. In group 5A, for example, the electronegativity of N is 3.0, whereas the electronegativities of P, As, Sb, and Bi are all in the range 2.1–1.9. Figure 7.4 graphically depicts the discontinuity in electronegativity that distinguishes the second-row elements from other elements of the same periodic group.

The small sizes and high electronegativities of the second-row elements accentuate their nonmetallic behavior. Thus, BeO is amphoteric (both acidic and basic), but the oxides of the other group 2A elements are basic. Boron differs from the metallic elements of group 3A in forming mainly covalent, molecular compounds. For example, BF_3 (bp $-100°C$) is a gaseous, molecular halide, but AlF_3 (mp 1290°C) is a typical high-melting, ionic solid. Furthermore, hydrogen bonding interactions are generally restricted to compounds of the highly electronegative second-row elements N, O, and F (Section 10.2). Recall also that HF contrasts with HCl, HBr, and HI in being the only weak hydrohalic acid (Section 15.2).

Another factor that distinguishes the second-row elements from the heavier elements is their lack of valence d orbitals. Because the second-row elements have only four valence orbitals ($2s$, $2p_x$, $2p_y$, and $2p_z$), they form a maximum of four covalent bonds. By contrast, the third-row elements can use d orbitals to form more than four bonds. Thus, nitrogen forms only NCl_3, but phosphorus forms both PCl_3 and PCl_5:

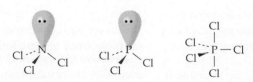

A further consequence of the small size of the second-row atoms C, N, and O is their ability to form multiple bonds involving π overlap of the $2p$

orbitals. By contrast, the $3p$ orbitals of the corresponding third-row atoms Si, P, and S are more diffuse, and the longer bond distances for these larger atoms result in poor π overlap.

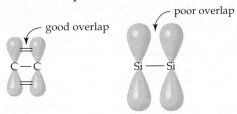

As a result of this poor overlap, π bonds involving p orbitals are rare for elements of the third and higher rows. Although compounds with C=C double bonds are common, molecules with Si=Si double bonds are uncommon and have been synthesized only recently. In group 5A, elemental nitrogen contains triply bonded N_2 molecules, whereas white phosphorus contains tetrahedral P_4 molecules, in which each P atom forms three single bonds rather than one triple bond. Similarly, O_2 contains an O=O double bond, whereas elemental sulfur contains crown-shaped S_8 rings, in which each S atom forms two single bonds rather than one double bond.

$$N\equiv N \qquad \text{(P}_4\text{ tetrahedron)} \qquad O=O \qquad \text{(S}_8\text{ ring)}$$

EXAMPLE 19.2

Account for the following observations.

(a) CO_2 is a gaseous molecular substance, but SiO_2 is a covalent network solid in which SiO_4 tetrahedra are linked to four neighboring SiO_4 tetrahedra by shared oxygen atoms.

$$O=C=O \qquad \text{(SiO}_4\text{ tetrahedron)}$$

(b) Glass made of SiO_2 is attacked by hydrofluoric acid with formation of SiF_6^{2-} anions. The analogous CF_6^{2-} anion does not exist.

SOLUTION **(a)** Because of its small size and good π overlap with other small atoms, carbon forms double bonds with two oxygens to give discrete CO_2 molecules. Because the larger Si atom does not form strong π bonds, it uses its four valence electrons to form four single bonds rather than two double bonds.
(b) Silicon has $3d$ orbitals and can use octahedral sp^3d^2 hybrid orbitals to bond to six F^- ions. With just $2s$ and $2p$ valence orbitals, carbon forms a maximum of four bonds.

⌈ **PROBLEM 19.2** **(a)** Draw Lewis electron-dot structures for HNO_3 and H_3PO_4, and suggest a reason for the difference in the formulas of these acids.
(b) Sulfur forms SF_6, but oxygen bonds to a maximum of two F atoms, yielding OF_2. Explain.

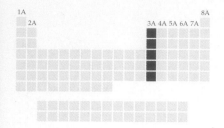

19.3 ➤ THE GROUP 3A ELEMENTS

The group 3A elements are boron, aluminum, gallium, indium, and thallium. We discussed aluminum in Section 6.7 and will take up boron in the next section. Gallium is remarkable for its unusually low melting point of 30°C, which gives it the largest liquid range of any metal—30 to 2403°C. Its most important use is in making gallium arsenide (GaAs), a semiconductor material employed in the manufacture of diode lasers for laser printers, compact-disc players, and fiber-optic communication devices. (We'll say more about semiconductors in Section 21.5.) Indium is also used in making semiconductor devices, such as transistors and thermistors. Thallium is extremely toxic and has no commercial uses.

A sample of gallium metal.

A sample of indium metal.

The semiconductor material gallium arsenide is prepared by heating gallium and arsenic in a furnace. The tube contains bars of gallium metal (left) and pieces of arsenic (right).

The valence electron configuration of the group 3A elements is $ns^2\,np^1$, and their primary oxidation state is +3. In addition, the heavier elements exhibit a +1 state, which is uncommon for gallium and indium but is the most stable oxidation state for thallium.

Despite some irregularities, the properties of the group 3A elements are generally consistent with increasing metallic character down the group (Table 19.3). All the group 3A elements are metals, except boron. Boron has a much smaller atomic radius and a higher electronegativity than the other elements of the group; it therefore shares its valence electrons in covalent bonds rather than transferring them to another element. Accordingly, boron has nonmetallic properties.

TABLE 19.3   Properties of the Group 3A Elements

Property	Boron	Aluminum	Gallium	Indium	Thallium
Valence electron configuration	$2s^2\,2p^1$	$3s^2\,3p^1$	$4s^2\,4p^1$	$5s^2\,5p^1$	$6s^2\,6p^1$
Common oxidation states	+3	+3	+3	+3	+3, +1
Atomic radius (pm)	83	143	135	167	170
M^{3+} ionic radius (pm)		51	62	81	95
First ionization energy (kJ/mol)	801	578	579	558	589
Electronegativity	2.0	1.5	1.6	1.7	1.8
Redox potential, $E°$ (V) for $M^{3+}(aq) + 3\,e^- \rightarrow M(s)$	-0.87^a	-1.66	-0.56	-0.34	-0.34^b

[a] $E°$ for the reaction $B(OH)_3(aq) + 3\,H^+(aq) + 3\,e^- \longrightarrow B(s) + 3\,H_2O(l)$.
[b] $E°$ for the reaction $Tl^+(aq) + e^- \longrightarrow Tl(s)$

19.4 ➤BORON

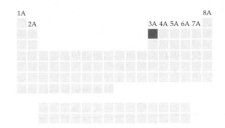

Boron is a relatively rare element, accounting for only about 0.001% of the earth's crust by mass. Nevertheless, boron is readily available because it occurs in concentrated deposits of borate minerals such as borax, $Na_2B_4O_7 \cdot 10\ H_2O$.

Elemental boron can be prepared by high-temperature reduction of B_2O_3 with magnesium, but the product is impure and amorphous:

$$B_2O_3(l) + 3\ Mg(s) \rightarrow 2\ B(s) + 3\ MgO(s)$$

High-purity, crystalline boron is best obtained by reaction of boron tribromide and hydrogen on a heated tantalum filament at temperatures above 1000°C:

$$2\ BBr_3(g) + 3\ H_2(g) \xrightarrow[1200°C]{Ta\ wire} 2\ B(s) + 6\ HBr(g)$$

Crystalline boron is a strong, hard, high-melting substance (mp 2300°C) that is chemically inert at room temperature, except for reaction with fluorine. These properties make boron fibers a desirable component in high-strength composite materials used in making sports equipment and military aircraft (Section 21.8). Unlike Al, Ga, In, and Tl, which are metallic conductors, boron is a semiconductor.

The borate minerals in this open pit borax mine near Boron, California, are believed to have been formed by evaporation of water from hot springs that were once present there.

Boron Compounds

BORON HALIDES The boron halides are highly reactive, volatile, covalent compounds that consist of trigonal planar BX_3 molecules. At room temperature, BF_3 and BCl_3 are gases, BBr_3 is a liquid, and BI_3 is a low-melting solid (mp 50°C). In their most important reactions, the boron halides behave as Lewis acids. For example, BF_3 reacts with ammonia to give the Lewis acid–base adduct F_3B-NH_3 (Section 15.16); it reacts with metal fluorides, yielding salts that contain the tetrahedral BF_4^- anion; and it acts as a catalyst in many industrially important organic reactions. In all these reactions, the boron atom uses its vacant $2p$ orbital in accepting a share in a pair of electrons from a Lewis base.

BORON HYDRIDES The boron hydrides, or **boranes**, are volatile, molecular compounds with formulas B_nH_m. The simplest is diborane (B_2H_6), the dimer of the unstable BH_3. Diborane can be prepared by reaction of sodium borohydride, $NaBH_4$, and iodine in an appropriate organic solvent:

$$2\ NaBH_4 + I_2 \rightarrow B_2H_6 + H_2 + 2\ NaI$$

Because boranes have high heats of combustion, they were once considered as potential lightweight, high-energy rocket fuels, but they proved to be too expensive to prepare. Nevertheless, boranes continue to be of interest to chemists because of their unusual structures and bonding.

The diborane molecule has a structure in which two BH_2 groups are connected by two bridging H atoms. The geometry about the B atoms is roughly tetrahedral, and the bridging B–H bonds are significantly longer than the terminal B–H bonds, 133 pm versus 119 pm. The structure differs from that of ethane (C_2H_6) and is unusual because hydrogen normally forms only one bond.

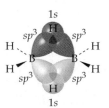

Diborane   Ethane

If each line in the structural formula of diborane were an ordinary two-center, two-electron bond (2c-2e bond), there would be 8 pairs, or 16 valence electrons. But diborane has a total of only 12 valence electrons—3 electrons from each boron and 1 from each hydrogen. Thus, diborane is said to be *electron-deficient*: It doesn't have enough electrons to form a 2c-2e bond between each pair of bonded atoms.

Because the geometry about the B atoms is roughly tetrahedral, we can assume that each boron uses sp^3 hybrid orbitals to bond to the four neighboring H atoms. The four terminal B–H bonds are assumed to be ordinary 2c-2e bonds, formed by overlap of a boron sp^3 hybrid orbital and a hydrogen 1s orbital, thereby using four of the six pairs of electrons. Each of the two remaining pairs of electrons forms a **three-center, two-electron bond** (3c-2e bond), which joins each bridging H atom to *both* B atoms. Each electron pair occupies a three-center molecular orbital formed by overlap of three atomic orbitals—one sp^3 hybrid orbital from each B atom and the 1s orbital on one bridging H atom (Figure 19.3). Because the two electrons in the B–H–B bridge are spread out over three atoms, the electron density between adjacent atoms is less than in an ordinary 2c-2e bond. Thus, this model nicely explains why the bridging B–H bonds are longer than the terminal B–H bonds.

FIGURE 19.3 Three-center bonding molecular orbitals in diborane. Each MO, shown in a different color, is formed by overlap of an sp^3 hybrid orbital on each B atom and the 1s orbital on one bridging H atom. Each MO contains one pair of electrons and accounts for the bonding in one B–H–B bridge.

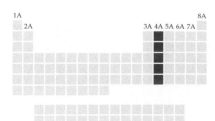

19.5 ►THE GROUP 4A ELEMENTS

The group 4A elements—carbon, silicon, germanium, tin, and lead—are especially important, both in industry and in living organisms. Carbon is present in all plants and animals, accounts for 23% of the mass of the human body, and is an essential constituent of the molecules on which life is based. Silicon is equally important in the mineral world: It is present in numerous silicate minerals and is the second most abundant element in the earth's crust. Both silicon and germanium are used in making modern solid-state electronic devices. Tin and lead have been known and used since ancient times.

The group 4A elements further illustrate the increase in metallic character down a group in the periodic table: Carbon is a nonmetal; silicon and germanium are semimetals; and tin and lead are metals. The usual periodic trends in atomic size, ionization energy, and electronegativity are evident in the data of Table 19.4.

TABLE 19.4   Properties of the Group 4A Elements

Property	Carbon	Silicon	Germanium	Tin	Lead
Valence electron configuration	$2s^2\,2p^2$	$3s^2\,3p^2$	$4s^2\,4p^2$	$5s^2\,5p^2$	$6s^2\,6p^2$
Melting point (°C)	>3550[a]	1410	937	232[b]	328
Boiling point (°C)		2355	2830	2260	1740
Density (g/cm^3)	3.51[a]	2.33	5.35	7.28[b]	11.3
Abundance in earth's crust (mass percent)	0.020	28.2	0.0005	0.0002	0.0013
Common oxidation states	+2, +4	+4	+4	+2, +4	+2, +4
Atomic radius (pm)	77	117	122	140	175
First ionization energy (kJ/mol)	1086	786	762	709	716
Electronegativity	2.5	1.8	1.8	1.8	1.9
Redox potential, $E°$ (V) for $M^{2+}(aq) + 2\,e^- \rightarrow M(s)$				−0.14	−0.13

[a] Diamond
[b] White Sn

Because the group 4A elements have the valence electron configuration $ns^2\,np^2$, their most common oxidation state is +4, as in CCl_4, $SiCl_4$, $GeCl_4$, $SnCl_4$, and $PbCl_4$. These compounds are volatile, molecular liquids in which the group 4A atom uses tetrahedral sp^3 hybrid orbitals to form covalent bonds to the Cl atoms. The +2 oxidation state occurs for tin and lead and is the most stable oxidation state for lead. Both $Sn^{2+}(aq)$ and $Pb^{2+}(aq)$ are common solution species, but there are no simple $M^{4+}(aq)$ ions for any of the group 4A elements. Instead, M(IV) species exist in solution as covalently bonded complex ions—for example, $SiF_6{}^{2-}$, $GeCl_6{}^{2-}$, $Sn(OH)_6{}^{2-}$, and $Pb(OH)_6{}^{2-}$. In general, the +4 oxidation-state compounds are covalent, and the compounds with tin and lead in the +2 oxidation state are largely ionic.

19.6 ➤CARBON

Carbon, although the second most abundant element in living organisms, accounts for only 0.02% of the mass of the earth's crust. It is present in carbonate minerals, such as limestone ($CaCO_3$), and in fossil fuels such as coal, petroleum, and natural gas. In uncombined form, carbon is found as diamond and graphite.

Recall from Section 10.11 that diamond has a covalent network structure in which each C atom uses sp^3 hybrid orbitals to form a tetrahedral

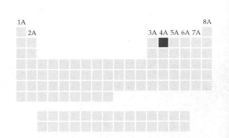

array of σ bonds, with bond lengths of 154 pm. The interlocking, three-dimensional network of strong bonds makes diamond the hardest known substance and gives it the highest known melting point for an element (>3550°C). Because the valence electrons are localized in the σ bonds, diamond is an electrical insulator.

Graphite has a two-dimensional sheetlike structure (Figure 10.28) in which each C atom uses sp^2 hybrid orbitals to form trigonal planar σ bonds to three neighboring C atoms. In addition, each C atom uses its remaining p orbital, perpendicular to the plane of the sheet, to form a π bond. Because each C atom must share its π bond with its three neighbors, the π electrons are delocalized and are free to move in the plane of the sheet. As a result, the electrical conductivity of graphite in a direction parallel to the sheets is about 10^{20} times greater than the conductivity of diamond, which makes graphite useful as an electrode material.

The carbon sheets in graphite are separated by a distance of 335 pm and are held together by weak London dispersion forces. Consequently, the sheets can easily slide over one another, thus accounting for the slippery feel of graphite and for its use as a lubricant. Because the sheets are so far apart, it's relatively difficult for an electron to hop from one sheet to the next, and the electrical conductivity in the direction perpendicular to the sheets is therefore about 10^4 times smaller than the conductivity parallel to the sheets.

A third allotrope of carbon, called *fullerene*, occurs in the soot formed by vaporizing graphite. It contains spherical C_{60} molecules with the shape of a soccer ball [Figure 10.28(b)].

Carbon also exists in more than 40 amorphous (noncrystalline) forms that resemble graphite: *coke*, *charcoal*, and *carbon black* among others. Coke is made by heating coal in the absence of air and is used as a reducing agent in the manufacture of steel (Section 21.3). Charcoal is formed when wood is heated in the absence of air. Because charcoal has a porous structure with an immense surface area (~25 m^2/g), it has strong adsorbent properties and is widely used in filters for removing malodorous molecules from air and water. Carbon black, used in the manufacture of printing inks and automobile tires, is made by heating hydrocarbons such as natural gas (CH_4) in a limited supply of oxygen:

$$CH_4(g) + O_2(g) \rightarrow C(s) + 2\,H_2O(g)$$

Carbon Compounds

Carbon forms millions of compounds, most of which are classified as *organic*; only CO_2, $CaCO_3$, HCN, and a handful of others are considered to be *inorganic*. The distinction is a historical one rather than a scientific one, though, and we'll deal with it in more detail in Chapter 23. For the present, we'll look only at some simple inorganic compounds of carbon.

OXIDES OF CARBON   The most important oxides of carbon are carbon monoxide (CO) and carbon dioxide (CO_2). Carbon monoxide is a colorless, odorless, toxic gas that forms when carbon or hydrocarbon fuels are burned in a limited supply of oxygen. In an excess of oxygen, CO burns to give CO_2:

$$2 \, C(s) + O_2(g) \rightarrow 2 \, CO(g) \qquad \Delta H° = -221 \text{ kJ}$$

$$2 \, CO(g) + O_2(g) \rightarrow 2 \, CO_2(g) \qquad \Delta H° = -566 \text{ kJ}$$

Carbon monoxide is synthesized industrially, along with H_2, by heating steam with hydrocarbons or coke (Section 14.3). It is used, among other purposes, for the industrial synthesis of methanol, CH_3OH:

$$CO(g) + 2 \, H_2(g) \xrightarrow[\text{ZnO/Cr}_2\text{O}_3 \text{ catalyst}]{400°C} CH_3OH(g)$$

The high toxicity of CO results from its ability to bond strongly to the iron(II) atom of hemoglobin, the oxygen-carrying protein in red blood cells. Because hemoglobin has a greater affinity for CO than for O_2, even small concentrations of CO in the blood can convert a substantial fraction of the O_2-bonded hemoglobin, called *oxyhemoglobin*, to the CO-bonded form, called *carboxyhemoglobin*, thus impairing the ability of hemoglobin to carry O_2 to the tissues:

$$\underset{\text{Oxyhemoglobin}}{Hb\text{—}O_2} + CO \rightleftarrows \underset{\text{Carboxyhemoglobin}}{Hb\text{—}CO} + O_2$$

A CO concentration in air of only 200 ppm can produce symptoms such as headache, dizziness, and nausea, and a concentration of 1000 ppm can cause death within 4 h.

Carbon dioxide is a colorless, odorless, nonpoisonous gas. It is produced when fuels are burned in an excess of oxygen and is an end product of food metabolism in humans and animals. Commercially, carbon dioxide is obtained as a byproduct of the yeast-catalyzed fermentation of sugar in the manufacture of alcoholic beverages:

$$\underset{\text{Glucose}}{C_6H_{12}O_6(aq)} \xrightarrow{\text{yeast}} \underset{\text{Ethanol}}{2 \, CH_3CH_2OH(aq)} + 2 \, CO_2(g)$$

Sodium carbonate reacts with acids, yielding CO_2 gas.

CO_2 can also be obtained by heating metal carbonates, and it is produced in the laboratory when metal carbonates are treated with acid:

$$CaCO_3(s) \xrightarrow{\text{heat}} CaO(s) + CO_2(g)$$

$$Na_2CO_3(s) + 2 \, H^+(aq) \rightarrow 2 \, Na^+(aq) + CO_2(g) + H_2O(l)$$

Carbon dioxide is used in beverages and in fire extinguishers. The "bite" of carbonated beverages is due to the mild acidity of CO_2 solutions (pH $\approx$ 4), which results from approximately 0.3% conversion of the dissolved CO_2 to carbonic acid (H_2CO_3), a weak diprotic acid (Section 15.11):

$$CO_2(aq) + H_2O(l) \rightleftarrows H_2CO_3(aq) \rightleftarrows H^+(aq) + HCO_3{}^-(aq)$$

CO_2 is useful in fighting fires because it is nonflammable and is about 1.5 times more dense than air. It therefore settles over a small fire like a blanket, separating the fire from its source of oxygen. Solid CO_2 (dry ice), which sublimes at $-78°C$, is used primarily as a refrigerant.

CARBONATES   Carbonic acid, H_2CO_3, has never been isolated as a pure substance, but it forms two series of salts: carbonates, which contain the trigonal planar CO_3^{2-} ion, and hydrogen carbonates or bicarbonates, which contain the HCO_3^- ion. Na_2CO_3, called *soda ash*, is used in multimillion-ton quantities in making glass, and $Na_2CO_3 \cdot 10\ H_2O$, known as *washing soda*, is used in laundering textiles. Its cleansing properties derive from the basicity of the carbonate ion:

$$CO_3^{2-}(aq) + H_2O(l) \rightleftarrows HCO_3^-(aq) + OH^-(aq)$$

$NaHCO_3$ is called *baking soda* because it reacts with acidic substances in food to yield bubbles of CO_2 gas that cause dough to rise:

$$NaHCO_3(s) + H^+(aq) \rightarrow Na^+(aq) + CO_2(g) + H_2O(l)$$

HYDROGEN CYANIDE AND CYANIDES   Hydrogen cyanide is a highly toxic, volatile substance (bp 26°C) produced when metal cyanide solutions are acidified:

$$CN^-(aq) + H^+(aq) \rightarrow HCN(aq)$$

Aqueous solutions of HCN, known as hydrocyanic acid, are very weakly acidic ($K_a = 4.9 \times 10^{-10}$).

The cyanide ion is called a *pseudohalide* ion because it behaves like Cl^- in forming an insoluble, white silver salt, AgCN. In complex ions such as $Fe(CN)_6^{3-}$, CN^- acts as a Lewis base (Section 15.16), bonding to transition metals through the lone pair of electrons on carbon. In fact, the toxicity of HCN and cyanides is due to the strong bonding of CN^- to iron(III) in cytochrome oxidase, an important enzyme in food metabolism. With CN^- attached to the iron, the enzyme is unable to function.

The bonding of CN^- to gold and silver is exploited in the extraction of these metals from their ores. The crushed rock containing small amounts of the precious metals is treated with an aerated cyanide solution, and the metals are then recovered from their $M(CN)_2^-$ complex ions by reduction with zinc. For gold, the reactions are

$$4\ Au(s) + 8\ CN^-(aq) + O_2(g) + 2\ H_2O(l) \rightarrow 4\ Au(CN)_2^-(aq) + 4\ OH^-(aq)$$

$$2\ Au(CN)_2^-(aq) + Zn(s) \rightarrow 2\ Au(s) + Zn(CN)_4^{2-}(aq)$$

CARBIDES   Carbon forms a number of binary compounds called *carbides*, in which the carbon atom has a negative oxidation state. Examples include ionic carbides of active metals such as CaC_2 and Al_4C_3, interstitial carbides of transition metals such as Fe_3C, and covalent network carbides such as SiC. Calcium carbide is a high-melting, colorless solid that has an

NaCl–type structure with Ca^{2+} ions in place of Na^+, and C_2^{2-} ions in place of Cl^-. CaC_2 is prepared by heating lime (CaO) and coke (C) at high temperatures, and it is used to prepare acetylene (C_2H_2) for oxyacetylene welding:

$$CaO(s) + 3\ C(s) \xrightarrow{2200°} CaC_2(s) + CO(g)$$

$$CaC_2(s) + 2\ H_2O(l) \rightarrow C_2H_2(g) + Ca(OH)_2(s)$$

Carborundum (SiC) is used as an industrial abrasive.

Iron carbide (Fe_3C) is an important constituent of steel, and silicon carbide (SiC) is the industrial abrasive carborundum. Almost as hard as diamond, SiC has a diamondlike structure with alternating Si and C atoms.

PROBLEM 19.3 HCN is a linear triatomic molecule. Draw its Lewis electron-dot structure, and indicate which hybrid orbitals are used by the carbon atom.

PROBLEM 19.4 The equilibrium between oxyhemoglobin and carboxy-hemoglobin suggests an approach to treating mild cases of carbon monoxide poisoning. Explain.

19.7 ►SILICON

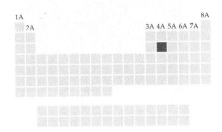

Silicon is a hard, gray, semiconducting solid that melts at 1410°C. It crystallizes in the diamond structure but does not form a graphitelike allotrope because of the relatively poor overlap of silicon π orbitals. In nature, silicon is generally found combined with oxygen in SiO_2 and in various silicate minerals. It is obtained in elemental form by reduction of silica sand (SiO_2) with coke (C) in an electric furnace:

$$SiO_2(l) + 2\ C(s) \xrightarrow{heat} Si(l) + 2\ CO(g)$$

The silicon used for making solid-state semiconductor devices such as transistors, computer chips, and solar cells must be ultrapure, with impurities at a level of less than $10^{-7}\%$ (1 ppb). For electronic applications, silicon is purified by converting it to $SiCl_4$, a volatile liquid (bp 58°C) that can be separated from impurities by fractional distillation and then converted back to elemental silicon by reduction with hydrogen:

$$Si(s) + 2\ Cl_2(g) \rightarrow SiCl_4(l)$$

$$SiCl_4(g) + 2\ H_2(g) \xrightarrow{heat} Si(s) + 4\ HCl(g)$$

The silicon is purified further by **zone refining** (Figure 19.4a). In this process, a heater melts a narrow zone at the top of a silicon rod. Because the impurities are more soluble in the liquid phase than in the solid, they concentrate in the molten zone. As the heater sweeps slowly down the rod, ultrapure silicon crystallizes at the trailing edge of the molten zone, and the impurities are dragged to the rod's lower end. Figure 19.4b shows some samples of ultrapure silicon.

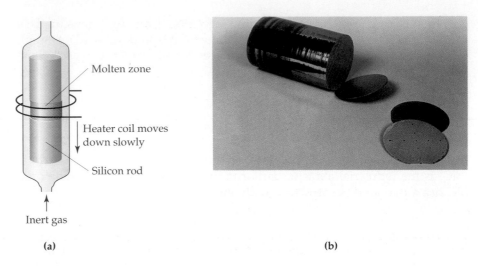

FIGURE 19.4 **(a)** Purification of silicon by zone refining. The heater coil sweeps the molten zone and the impurities to the lower end of the rod. After the rod has cooled, the impurities are removed by cutting off the rod's lower end. **(b)** A rod of ultrapure silicon and silicon wafers cut from such a rod. Silicon wafers are used in producing the integrated circuit chips found in solid-state electronic devices.

Molten zone

Heater coil moves down slowly

Silicon rod

Inert gas

(a)

(b)

The mineral zircon ($ZrSiO_4$) is a relatively inexpensive gemstone.

Silicates

Approximately 90% of the earth's crust consists of **silicates**, ionic compounds that contain silicon oxoanions along with cations, such as Na^+, K^+, Mg^{2+}, or Ca^{2+}, to balance the negative charge of the anions. As shown in Figure 19.5, the basic structural building block in silicates is the SiO_4 tetrahedron, a unit that occurs as the simple orthosilicate ion (SiO_4^{4-}) in the mineral zircon, $ZrSiO_4$. If two SiO_4 tetrahedra share a common O atom, the disilicate anion, $Si_2O_7^{6-}$, found in $Sc_2Si_2O_7$ results.

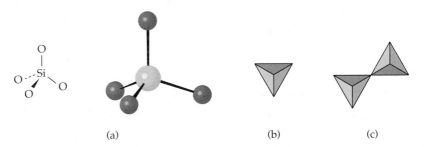

(a)   (b)   (c)

FIGURE 19.5 **(a)** A view of the SiO_4^{4-} anion showing the tetrahedral SiO_4 structural unit. **(b)** A tetrahedron is used as a shorthand representation of the SiO_4^{4-} anion. It is understood that an O atom is located at each corner of the tetrahedron and a Si atom is at the center. **(c)** A shorthand representation of the $Si_2O_7^{6-}$ anion. The corner shared by the two tetrahedra represents a shared O atom.

Simple anions such as SiO_4^{4-} and $Si_2O_7^{6-}$ are relatively rare in silicate minerals. More common are larger anions in which two or more O atoms bridge between Si atoms to give rings, chains, layers, and extended three-dimensional structures. Sharing of two O atoms per SiO_4 tetrahedron gives either cyclic anions, such as $Si_6O_{18}^{12-}$, or infinitely extended chain anions with repeating unit $Si_2O_6^{4-}$ (Figure 19.6). $Si_6O_{18}^{12-}$ is present in the mineral beryl, $Be_3Al_2Si_6O_{18}$, and in the gemstone emerald, a beryl in which about 2% of the Al^{3+} is replaced by the green-colored Cr^{3+}. Chain anions are found in pyroxene minerals such as diopside, $CaMgSi_2O_6$.

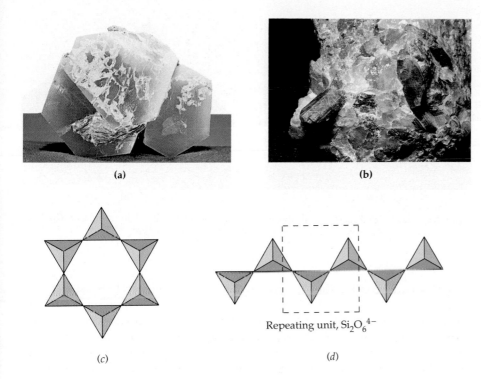

(a)

(b)

Repeating unit, $Si_2O_6^{4-}$

(c)

(d)

FIGURE 19.6 Samples of the silicate minerals **(a)** beryl ($Be_3Al_2Si_6O_{18}$) and **(b)** diopside ($CaMgSi_2O_6$). A shorthand representation of the structure of **(c)** the cyclic anion $Si_6O_{18}^{12-}$ in beryl and **(d)** the infinitely extended chain anion $(Si_2O_6^{4-})_n$ in diopside. Note that the number of negative charges on the $Si_2O_6^{4-}$ repeating unit equals the number of terminal (unshared) O atoms in that unit (4).

As shown in Figure 19.7, additional sharing of O atoms gives the double-stranded chain anions, $(Si_4O_{11}^{6-})_n$, found in asbestos minerals such as tremolite, $Ca_2Mg_5(Si_4O_{11})_2(OH)_2$, and the infinitely extended two-dimensional layer anions, $(Si_4O_{10}^{4-})_n$, found in clay minerals, micas, and talc. Talc has the formula $Mg_3(OH)_2(Si_4O_{10})$, where $Si_4O_{10}^{4-}$ is the formula

FIGURE 19.7 (a) A shorthand representation of the double-stranded chain anion $(Si_4O_{11}^{6-})_n$. Two of the single-stranded chains of Figure 19.6 **(d)** are laid side by side, and half of the SiO_4 tetrahedra share an additional O atom. **(b)** The layer anion $(Si_4O_{10}^{4-})_n$ is formed by the sharing of three O atoms per SiO_4 tetrahedron. Note again that the number of negative charges on each repeating unit equals the number of terminal O atoms in that unit.

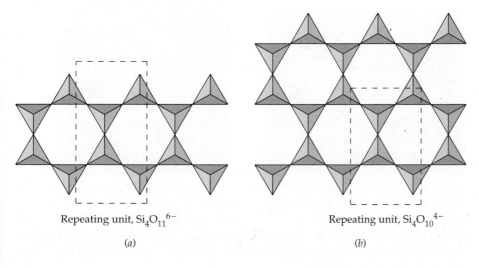

Repeating unit, $Si_4O_{11}^{6-}$

(a)

Repeating unit, $Si_4O_{10}^{4-}$

(b)

(a)   (b)

FIGURE 19.8 **(a)** Asbestos is a fibrous material because of its chain structure. **(b)** Mica cleaves into thin sheets because of its two-dimensional layer structure.

of the repeating unit in the layer anion. As shown in Figure 19.8, asbestos is a fibrous material because the bonds between chains are relatively weak and easily broken. Mica is a sheetlike material because the bonds between two-dimensional layers are relatively weak and easily broken.

If the layer anions of Figure 19.7(b) are stacked on top of each other and the terminal O atoms are shared, an infinitely extended three-dimensional structure is obtained in which all four O atoms of each SiO_4 tetrahedron are shared between two Si atoms, resulting in *silica* (SiO_2). The mineral quartz is one of many crystalline forms of SiO_2.

Partial substitution of the Si^{4+} of SiO_2 with Al^{3+} gives **aluminosilicates** called feldspars, the most abundant of all minerals. An example is orthoclase, $KAlSi_3O_8$, which has a three-dimensional structure like that of SiO_2. The $(AlSi_3O_8^-)_n$ framework consists of SiO_4 and AlO_4 tetrahedra that share all four of their corners with neighboring tetrahedra. K^+ balances the negative charge.

Crystalline quartz, one form of SiO_2.   A sample of orthoclase, $KAlSi_3O_8$.

PROBLEM 19.5 The mineral benitoite, $BaTiSi_3O_9$, contains cyclic $Si_3O_9^{6-}$ anions. Draw a two-dimensional representation of the structure of the $Si_3O_9^{6-}$ anion using Figure 19.6 as your model.

19.8 ➤GERMANIUM, TIN, AND LEAD

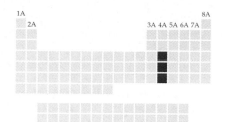

Germanium, tin, and lead have relatively low abundances in the earth's crust (Table 19.4), but tin and lead are concentrated in workable deposits and are readily extracted from their ores. Tin is obtained from the mineral cassiterite (SnO_2) by reduction of the purified oxide with carbon:

$$SnO_2(s) + 2\ C(s) \xrightarrow{\text{heat}} Sn(l) + 2\ CO(g)$$

Tin is used as a protective coating over steel in making tin cans and is an important component of alloys such as bronze (10% Sn, 90% Cu), pewter (90% Sn, with Cu and Sb), and solder (33% Sn, 67% Pb).

Lead is obtained from its ore, galena (PbS), by roasting the sulfide in air and reducing the resulting PbO with carbon monoxide in a blast furnace:

$$2\ PbS(s) + 3\ O_2(g) \rightarrow 2\ PbO(s) + 2\ SO_2(g)$$

$$PbO(s) + CO(g) \rightarrow Pb(l) + CO_2(g)$$

Lead is used in making pipes, cables, pigments, and, most importantly, electrodes for storage batteries (Section 18.9).

(a)   (b)

Samples of **(a)** cassiterite (SnO_2) and **(b)** galena (PbS).

Although tin and lead have been known for over 5000 y, germanium was not discovered until 1886, by C. A. Winkler in Germany. In fact, germanium was one of the "holes" in Mendeleev's periodic table (Figure 5.2). Germanium is used in making transistors and special glasses for infrared devices.

The physical properties of the heavier group 4A elements nicely illustrate the gradual transition from semimetal to metallic character. Germanium is a relatively high-melting, brittle semiconductor that has the same crystal structure as diamond and silicon. Tin exists in two allotropic forms, the usual silvery-white metallic form called white tin and a brittle, semiconducting form with the diamond structure called gray tin. White tin is the stable form at room temperature, but when kept for long periods of time

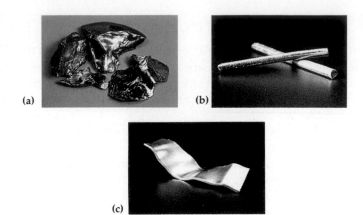

Samples of **(a)** germanium, **(b)** tin, and **(c)** lead.

below the transition temperature of 13°C, it slowly crumbles to gray tin, a phenomenon known as "tin disease."

$$\text{white tin} \underset{}{\overset{13°C}{\rightleftharpoons}} \text{gray tin}$$

Only the metallic form occurs for lead. Both white tin and lead are soft, malleable, low-melting metals.

19.9 ➤THE GROUP 5A ELEMENTS

The group 5A elements are nitrogen, phosphorus, arsenic, antimony, and bismuth. As shown in Table 19.5, these elements exhibit the expected trends of increasing atomic size, decreasing ionization energy, and decreasing electronegativity down the periodic group from N to Bi. Accordingly, metallic character increases in the same order. N and P are typical nonmetals, As and Sb are semimetals, and Bi is a metal. Thus, nitrogen is a gaseous substance made up of N_2 molecules, but bismuth is a silvery solid having an extended three-dimensional structure. The increasing metallic character of the heavier elements is also evident in the acid–base properties of their oxides: Most nitrogen and phosphorus oxides are acidic, arsenic and antimony oxides are amphoteric, and Bi_2O_3 is basic.

TABLE 19.5	Properties of the Group 5A Elements				
Property	**Nitrogen**	**Phosphorus**	**Arsenic**	**Antimony**	**Bismuth**
Valence electron configuration	$2s^2\,2p^3$	$3s^2\,3p^3$	$4s^2\,4p^3$	$5s^2\,5p^3$	$6s^2\,6p^3$
Melting point (°C)	−210	44[a]	613[b]	630	271
Boiling point (°C)	−196	280		1750	1560
Atomic radius (pm)	70	110	120	140	150
First ionization energy (kJ/mol)	1402	1012	947	834	703
Electronegativity	3.0	2.1	2.0	1.9	1.9

[a] White phosphorus.
[b] Sublimes.

(a)

(b)

(c)

Samples of **(a)** arsenic, **(b)** antimony, and **(c)** bismuth.

The valence electron configuration of the group 5A elements is ns^2np^3. They exhibit a maximum oxidation state of $+5$ in compounds such as HNO_3 and PF_5, in which they share all five valence electrons with a more electronegative element. They show a minimum oxidation state of -3 in compounds such as NH_3 and PH_3, where they share three valence electrons with a less electronegative element. The -3 state also occurs in ionic compounds such as Li_3N and Mg_3N_2, which contain the N^{3-} anion.

Nitrogen and phosphorus are unusual in that they exhibit all oxidation states between -3 and $+5$. For arsenic and antimony, the most important oxidation states are $+3$, as in $AsCl_3$, As_2O_3, and H_3AsO_3, and $+5$, as in AsF_5, As_2O_5, and H_3AsO_4. The $+5$ state becomes increasingly less stable from As to Sb to Bi. Another indication of increasing metallic character down the group is the existence of Sb^{3+} and Bi^{3+} cations in salts such as $Sb_2(SO_4)_3$ and $Bi(NO_3)_3$. By contrast, no simple cations are found in compounds of N or P.

The natural abundances of As, Sb, and Bi in the earth's crust are relatively low—about $2 \times 10^{-4}\%$ for As and about $2 \times 10^{-5}\%$ for Sb and Bi. All three elements are found in sulfide ores and are used in making various metal alloys. Arsenic is also used in making pesticides and semiconductors, such as GaAs. Bismuth compounds are present in some pharmaceuticals, such as Pepto-Bismol.

19.10 ➤NITROGEN

Elemental nitrogen is a colorless, odorless, tasteless gas that makes up 78% of the earth's atmosphere by volume. Because nitrogen (bp $-196°C$) is the most volatile component of liquid air, it is readily separated from the less volatile oxygen (bp $-183°C$) and argon (bp $-186°C$) by fractional distillation. As shown in Table 19.1, the annual U.S. production of nitrogen is 32.6 million tons, greater than that of any other industrial chemical except sulfuric acid. Nitrogen gas is used as a protective inert atmosphere in manufacturing processes, and the liquid is used as a refrigerant. By far the most important use of nitrogen, however, is in the Haber process for the manufacture of ammonia used in nitrogen fertilizers.

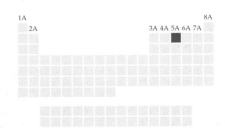

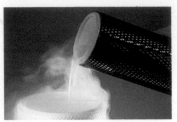

Liquid nitrogen is used as a coolant in scientific laboratories.

Under most conditions, the N_2 molecule is unreactive because a large amount of energy is required to break its strong nitrogen–nitrogen triple bond:

$$:N\equiv N: \rightarrow 2\ :\overset{\cdot}{N}\cdot \quad \Delta H° = 945\ kJ$$

As a result, reactions involving N_2 often have a high activation energy and/or an unfavorable equilibrium constant. For example, N_2 and O_2 do not combine to form nitric oxide at 25°C because the equilibrium constant for the reaction is 4.5×10^{-31}:

$$N_2(g) + O_2(g) \rightleftarrows 2\ NO(g) \quad \Delta H = 180.5\ kJ;\ K_c = 4.5 \times 10^{-31}\ at\ 25°C$$

At higher temperatures, however, the reaction does occur because it is endothermic, and the equilibrium shifts to the right with increasing temperature (Section 13.9). Indeed, high-temperature formation of NO from air in automobile engines is a major source of air pollution. Atmospheric N_2 and O_2 also react to form NO during electrical storms, where lightning discharges provide the energy required for the highly endothermic reaction.

N_2 reacts with H_2 in the Haber process for synthesis of NH_3, but the reaction requires high temperatures, high pressures, and a catalyst (Sections 13.8–13.10).

$$N_2(g) + 3\ H_2(g) \xrightarrow[\text{Fe/K}_2\text{O/Al}_2\text{O}_3\ \text{catalyst}]{400–500°C,\ 300–1000\ atm} 2\ NH_3(g)$$

It's convenient to classify nitrogen compounds by oxidation state, as shown in Table 19.6.

TABLE 19.6   Oxidation States of Nitrogen and Representative Compounds

Oxidation State	Compound	Formula	Lewis Electron-Dot Structure
−3	Ammonia	NH_3	H—N̈—H, H
−2	Hydrazine	N_2H_4	H—N̈—N̈—H, H H
−1	Hydroxylamine	NH_2OH	H—N̈—Ö—H, H
+1	Nitrous oxide	N_2O	:N≡N—Ö:
+2	Nitric oxide	NO	:N̈=Ö:
+3	Nitrous acid	HNO_2	H—Ö—N̈=Ö:
+4	Nitrogen dioxide	NO_2	:Ö—N̈=Ö:
+5	Nitric acid	HNO_3	H—Ö—N̈=Ö:, :Ö:

Nitrogen Compounds

AMMONIA   Ammonia and its synthesis by the Haber process serve as the gateway to nitrogen chemistry because ammonia is the starting material for industrial synthesis of other important nitrogen compounds, such as nitric acid. Used in agriculture as a fertilizer, ammonia is the most commercially important compound of nitrogen.

Ammonia is a colorless, pungent-smelling gas, consisting of polar, trigonal pyramidal NH_3 molecules that have a lone pair of electrons on the N atom. Because of hydrogen bonding (Section 10.2), gaseous NH_3 is extremely soluble in water and is easily condensed to liquid NH_3, which boils at $-33°C$. Like water, liquid ammonia is an excellent solvent for ionic compounds. It also dissolves alkali metals, as mentioned in Section 6.5.

Because ammonia is a Brønsted-Lowry base (Section 15.1), its aqueous solutions are weakly alkaline:

$$NH_3(aq) + H_2O(l) \rightleftharpoons NH_4^+(aq) + OH^-(aq) \qquad K_b = 1.8 \times 10^{-5}$$

Neutralization of aqueous ammonia with acids yields ammonium salts, which resemble alkali metal salts in their solubility.

HYDRAZINE   Hydrazine (H_2NNH_2) can be regarded as a derivative of NH_3 in which one H atom is replaced by an amino (NH_2) group. It can be prepared by reaction of ammonia with basic solutions of sodium hypochlorite (NaOCl):

$$2\ NH_3(aq) + OCl^-(aq) \rightarrow N_2H_4(aq) + H_2O(l) + Cl^-(aq)$$

No doubt you've heard that household cleaners should never be mixed because of possible exothermic reactions or formation of dangerous products. Formation of hydrazine on mixing household ammonia and hypochlorite-containing chlorine bleaches is a case in point.

Pure hydrazine is a poisonous, colorless liquid that smells like ammonia, freezes at $2°C$, and boils at $114°C$. It is violently explosive in the presence of air or other oxidizing agents and is used as a rocket fuel. For example, the *Apollo* lunar-landing module used a fuel composed of hydrazine and a derivative of hydrazine, along with dinitrogen tetroxide (N_2O_4) as the oxidizer. The highly exothermic reaction is

$$2\ N_2H_4(l) + N_2O_4(l) \rightarrow 3\ N_2(g) + 4\ H_2O(g) \qquad \Delta H° = -1049\ kJ$$

Hydrazine can be handled safely in aqueous solutions, where it behaves as a weak base ($K_b = 8.9 \times 10^{-7}$) and a versatile reducing agent. It reduces Fe^{3+} to Fe^{2+}, I_2 to I^-, and Ag^+ to metallic silver, for example.

OXIDES OF NITROGEN   Nitrogen forms a large number of oxides, but we'll discuss only three: nitrous oxide (N_2O), nitric oxide (NO), and nitrogen dioxide (NO_2).

Nitrous oxide (N_2O) is a colorless, sweet-smelling gas obtained when molten ammonium nitrate is heated gently at about $270°C$:

$$NH_4NO_3(l) \rightarrow N_2O(g) + 2\ H_2O(g)$$

(Strong heating can cause an explosion.) Known as "laughing gas" because small doses are mildly intoxicating, nitrous oxide is used as a dental anesthetic and as a propellant for dispensing whipped cream.

Nitric oxide (NO) is a colorless gas, produced in the laboratory when copper metal is treated with dilute nitric acid:

$$3 \ Cu(s) + 2 \ NO_3^-(aq) + 8 \ H^+(aq) \rightarrow 3 \ Cu^{2+}(aq) + 2 \ NO(g) + 4 \ H_2O(l)$$

As we'll discuss later, NO is prepared in large quantities by catalytic oxidation of ammonia, the first step in the industrial synthesis of nitric acid. Recently, it has been found that nitric oxide is important in biological processes: It plays a role in transmitting messages between nerve cells and in killing harmful microorganisms. It is also able to protect the heart from insufficient oxygen levels by widening blood vessels.

Nitrogen dioxide (NO_2) is the highly toxic, reddish brown gas that forms rapidly when nitric oxide is exposed to air:

$$2 \ NO(g) + O_2(g) \rightarrow 2 \ NO_2(g)$$

NO_2 is also produced when copper reacts with concentrated nitric acid (Figure 19.9):

$$Cu(s) + 2 \ NO_3^-(aq) + 4 \ H^+(aq) \rightarrow Cu^{2+}(aq) + 2 \ NO_2(g) + 2 \ H_2O(l)$$

FIGURE 19.9 Copper reacts with concentrated HNO_3 yielding noxious, red-brown fumes of NO_2. The blue color of the solution is due to Cu^{2+} ions.

NO_2 has an odd number of electrons (23) and is therefore paramagnetic. It tends to dimerize, forming colorless, diamagnetic N_2O_4, in which the unpaired electrons of two NO_2 molecules pair up to give an N–N bond:

$$O_2N\cdot + \cdot NO_2 \rightleftarrows O_2N\!-\!NO_2 \qquad \Delta H° = -57.2 \ kJ$$
$$\underset{\text{brown}}{} \qquad\qquad \underset{\text{colorless}}{}$$

In the gas phase, NO_2 and N_2O_4 are present in equilibrium. Because dimer formation is exothermic, N_2O_4 predominates at lower temperatures, and NO_2 predominates at higher temperatures. Thus, the color of the mixture fades on cooling and darkens on warming, as was illustrated in Figure 13.11. It's interesting to note that gaseous NO, which also has an odd number of electrons, has little tendency to dimerize.

NITROUS ACID   Nitrogen dioxide reacts with water to yield a mixture of nitrous acid (HNO_2) and nitric acid (HNO_3):

$$2 NO_2(g) + H_2O(l) \rightarrow HNO_2(aq) + H^+(aq) + NO_3^-(aq)$$

This is a disproportionation reaction in which nitrogen goes from the +4 oxidation state in NO_2 to the +3 state in HNO_2 and the +5 state in HNO_3. Nitrous acid is a weak acid ($K_a = 4.5 \times 10^{-4}$) that tends to disproportionate to nitric oxide and nitric acid. It has not been isolated as a pure compound:

$$3 HNO_2(aq) \rightleftarrows 2 NO(g) + H_3O^+(aq) + NO_3^-(aq)$$

NITRIC ACID   Nitric acid, one of the most important inorganic acids, is used mainly in making ammonium nitrate (NH_4NO_3) for fertilizers, but it is also used in manufacturing explosives, plastics, and dyes. Annual U.S. production of HNO_3 is approximately 9 million tons.

Nitric acid is produced industrially by the multistep **Ostwald process**, which involves (1) air oxidation of ammonia to nitric oxide at about 850°C over a platinum–rhodium catalyst, (2) rapid oxidation of the nitric oxide to nitrogen dioxide, and (3) disproportionation of NO_2 in water:

(1) $4 NH_3(g) + 5 O_2(g) \xrightarrow[\text{Pt/Rh catalyst}]{850°C} 4 NO(g) + 6 H_2O(g)$

(2) $2 NO(g) + O_2(g) \rightarrow 2 NO_2(g)$

(3) $3 NO_2(g) + H_2O(l) \rightarrow 2 HNO_3(aq) + NO(g)$

The catalyst used in oxidizing NH_3 to NO in a nitric acid plant is a large platinum–rhodium gauze.

Distillation of the resulting aqueous HNO_3 removes some of the water and gives concentrated (15 M) nitric acid, an HNO_3–H_2O mixture that is 68.5% HNO_3 by mass.

Further removal of water is required to obtain pure nitric acid, a colorless liquid that boils at 83°C. In the laboratory, you've probably noticed that concentrated nitric acid often has a yellow-brown color. The color is due to NO_2, produced by a slight amount of decomposition:

$$4 HNO_3(aq) \rightarrow 4 NO_2(aq) + O_2(g) + 2 H_2O(l)$$

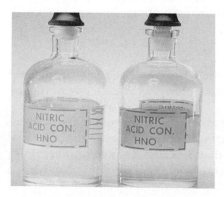

Concentrated nitric acid is colorless (left), but it turns yellow-brown on standing (right).

Nitric acid is a strong acid, essentially 100% dissociated in water. It's also a strong oxidizing agent, as indicated by large, positive $E°$ values for reduction to lower oxidation states:

$$NO_3^-(aq) + 2\,H^+(aq) + e^- \rightarrow NO_2(g) + H_2O(l) \qquad E° = 0.79\ V$$

$$NO_3^-(aq) + 4\,H^+(aq) + 3\,e^- \rightarrow NO(g) + 2\,H_2O(l) \qquad E° = 0.96\ V$$

Thus, nitric acid is a stronger oxidizing agent than $H^+(aq)$ and can oxidize relatively inactive metals like copper and silver that are not oxidized by aqueous HCl. The reduction product of HNO_3 in a particular reaction depends on the nature of the reducing agent and on the reaction conditions. We've already seen, for example, that copper reduces dilute HNO_3 to NO, but it reduces concentrated HNO_3 to NO_2.

An even more potent oxidizing agent than HNO_3 is *aqua regia*, a mixture of concentrated HCl and concentrated HNO_3 in a 3:1 ratio by volume. Aqua regia can dissolve even inactive metals like gold, which are insoluble in HCl and in HNO_3 separately:

$$Au(s) + 3\,NO_3^-(aq) + 6\,H^+(aq) + 4\,Cl^-(aq) \rightarrow AuCl_4^-(aq) + 3\,NO_2(g) + 3\,H_2O(l)$$

In this reaction, NO_3^- serves as the oxidizing agent, and Cl^- provides an additional driving force by converting the Au(III) oxidation product to the $AuCl_4^-$ complex ion.

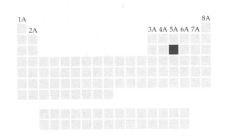

19.11 ▶PHOSPHORUS

Phosphorus is the most abundant element of group 5A, accounting for 0.10% of the mass of the earth's crust. It is found in phosphate rock, which is mostly calcium phosphate, $Ca_3(PO_4)_2$, and in fluorapatite, $Ca_5(PO_4)_3F$. The apatites are phosphate minerals with the formula $3\,Ca_3(PO_4)_2 \cdot CaX_2$, where X^- is usually F^- or OH^-. Phosphorus is also important in living systems and is the sixth most abundant element in the human body (Figure 19.1). Our bones are mostly $Ca_3(PO_4)_2$, and tooth enamel is almost pure $Ca_5(PO_4)_3OH$. Phosphate groups are present along with sugars and amine bases in the nucleic acids DNA and RNA (Section 24.12), which are both important in passing along genetic information.

Elemental phosphorus is produced industrially by heating phosphate rock, coke, and silica sand at about 1500°C in an electric furnace. The reaction can be represented by the simplified equation

$$2\,Ca_3(PO_4)_2(s) + 10\,C(s) + 6\,SiO_2(s) \rightarrow P_4(g) + 10\,CO(g) + 6\,CaSiO_3(l)$$

To condense the phosphorus, the gaseous reaction products are passed through water. Phosphorus is used to make phosphoric acid, one of the top 10 industrial chemicals (Table 19.1).

Phosphorus exists in two common allotropic forms: white phosphorus and red phosphorus. White phosphorus, the form produced in the industrial synthesis, is a toxic, waxy, white solid that contains discrete tetrahedral P_4 molecules. Red phosphorus, by contrast, is essentially nontoxic and has a polymeric structure:

A sample of the mineral apatite, $Ca_5(PO_4)_3OH$.

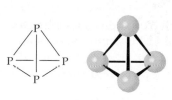
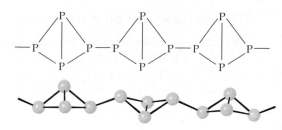

White phosphorus   Red phosphorus

As expected for a molecular solid that contains small, nonpolar molecules, white phosphorus has a low melting point (44°C) and is soluble in nonpolar solvents, such as carbon disulfide, CS_2. It is highly reactive, bursting into flames when exposed to air, and is thus stored under water. When white phosphorus is heated in the absence of air at about 300°C, it is converted to the more stable red form. Consistent with its polymeric structure, red phosphorus is higher melting (mp ~600°C), less soluble, and less reactive than white phosphorus, and it does not ignite on contact with air (Figure 19.10).

Finely divided white phosphorus, deposited on a piece of filter paper by evaporation of a carbon disulfide solution of P_4, bursts into flames in air.

FIGURE 19.10   Red phosphorus (left). White phosphorus, stored under water (right).

The high reactivity of white phosphorus is due to an unusual bonding that produces considerable strain in the P_4 molecules. If each P atom uses three $3p$ orbitals to form three P–P bonds, all the bond angles are expected to be 90°. The geometry of P_4, however, requires that all the bonds have 60° angles, which means that the p orbitals can't overlap in a head-on fashion. As a result, the P–P bonds are "bent," relatively weak, and highly reactive (Figure 19.11).

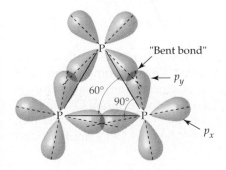

FIGURE 19.11   One equilateral triangular face of a tetrahedral P_4 molecule showing the 60° bond angles and the 90° angles between the p orbitals. The relatively poor orbital overlap in the bent bonds accounts for the high reactivity of white phosphorus.

Phosphorus Compounds

Like nitrogen, phosphorus forms compounds in all oxidation states between -3 and $+5$, but the $+3$ state, as in PCl_3, P_4O_6, and H_3PO_3, and the $+5$ state, as in PCl_5, P_4O_{10}, and H_3PO_4, are the most important. In comparison with nitrogen, phosphorus is more likely to be found in a positive oxidation state because of its lower electronegativity (Table 19.5).

PHOSPHINE   Phosphine (PH_3), a colorless, extremely poisonous gas, is the most important hydride of phosphorus. Like NH_3, phosphine has a trigonal pyramidal structure and has the group 5A atom in the -3 oxidation state. Unlike NH_3, however, its aqueous solutions are neutral, indicating that PH_3 is a poor proton acceptor. In accord with the low electronegativity of phosphorus, phosphine is easily oxidized, burning in air to form phosphoric acid:

$$PH_3(g) + 2\ O_2(g) \rightarrow H_3PO_4(l)$$

PHOSPHORUS HALIDES   Phosphorus reacts with all the halogens, forming phosphorus(III) halides PX_3 or phosphorus(V) halides PX_5 ($X = F$, Cl, Br, or I), depending on the relative amounts of the reactants:

Limited amount of X_2: $P_4 + 6\ X_2 \rightarrow 4\ PX_3$

Excess amount of X_2: $P_4 + 10\ X_2 \rightarrow 4\ PX_5$

All these halides are gases, volatile liquids, or low-melting solids. For example, phosphorus trichloride is a colorless liquid that boils at 76°C, and phosphorus pentachloride is an off-white solid that melts at 167°C. Both fume on contact with moist air due to hydrolysis, which breaks the P–Cl bonds, converting PCl_3 to phosphorous acid (H_3PO_3), and PCl_5 to phosphoric acid (H_3PO_4):

$$PCl_3(l) + 3\ H_2O(l) \rightarrow H_3PO_3(aq) + 3\ HCl(aq)$$
$$PCl_5(s) + 4\ H_2O(l) \rightarrow H_3PO_4(aq) + 5\ HCl(aq)$$

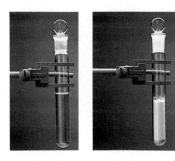

Samples of PCl_3 (left) and PCl_5 (right).

OXIDES AND OXOACIDS OF PHOSPHORUS   When phosphorus burns in air or oxygen, it yields phosphorus(III) oxide (P_4O_6, mp 24°C) or phosphorus(V) oxide (P_4O_{10}, mp 420°C), depending on the amount of oxygen present:

Limited amount of O_2: $P_4(s) + 3\ O_2(g) \rightarrow P_4O_6(s)$

Excess amount of O_2: $P_4(s) + 5\ O_2(g) \rightarrow P_4O_{10}(s)$

Both oxides are molecular compounds and have structures with a tetrahedral array of P atoms, as in white phosphorus. One O atom bridges each of the six edges of the P_4 tetrahedron, and an additional, terminal O atom is bonded to each P atom in P_4O_{10}. The P–O bonds to the terminal O atoms are shorter than the bonds to the bridging O atoms, 143 pm versus 160 pm, indicating that the terminal bonds have considerable P=O double bond character.

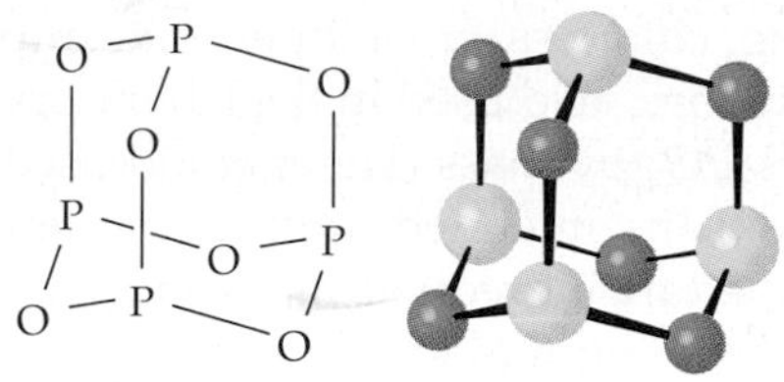

P_4O_6 and P_4O_{10} are acidic oxides and react with water to form aqueous solutions of phosphorous acid and phosphoric acid, respectively:

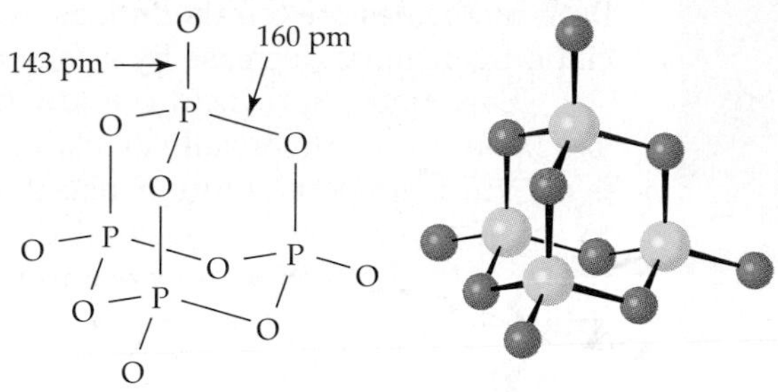

$$P_4O_6(s) + 6\,H_2O(l) \rightarrow 4\,H_3PO_3(aq)$$

$$P_4O_{10}(s) + 6\,H_2O(l) \rightarrow 4\,H_3PO_4(aq)$$

Because P_4O_{10} has a great affinity for water, it is widely used as a drying agent for gases and organic solvents.

The reaction of P_4O_{10} with water is vigorous and produces enough heat to convert some of the water to steam.

It's interesting to note that phosphorous acid (H_3PO_3) is a weak *diprotic* acid because only two of its three H atoms are bonded to oxygen. The H atom bonded directly to phosphorus is not acidic. In phosphoric acid, all three hydrogens are attached to oxygen, and thus phosphoric acid is a weak triprotic acid.

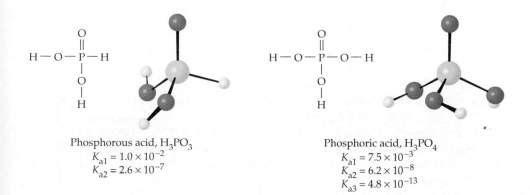

Phosphorous acid, H_3PO_3
$K_{a1} = 1.0 \times 10^{-2}$
$K_{a2} = 2.6 \times 10^{-7}$

Phosphoric acid, H_3PO_4
$K_{a1} = 7.5 \times 10^{-3}$
$K_{a2} = 6.2 \times 10^{-8}$
$K_{a3} = 4.8 \times 10^{-13}$

Soft drinks contain phosphoric acid.

Both molecules are tetrahedral, as expected. Note that the successive dissociation constants decrease by a factor of about 10^5 (Section 15.11).

Pure phosphoric acid is a low-melting, colorless, crystalline solid (mp 42°C), but the commercially available phosphoric acid used in the laboratory is a syrupy aqueous solution, about 82% H_3PO_4 by mass. The method used to manufacture phosphoric acid depends on its intended application. For use as a food additive—for example, as the tart ingredient in various soft drinks—pure phosphoric acid is made by burning molten phosphorus in a mixture of air and steam:

$$P_4(l) \xrightarrow{O_2} P_4O_{10}(s) \xrightarrow{H_2O} H_3PO_4(aq)$$

For use in making fertilizers, an impure form of phosphoric acid is produced by treating phosphate rock with sulfuric acid:

$$Ca_3(PO_4)_2(s) + 3\ H_2SO_4(aq) \rightarrow 2\ H_3PO_4(aq) + 3\ CaSO_4(aq)$$

Reaction of the phosphoric acid with phosphate rock gives $Ca(H_2PO_4)_2$, a water-soluble fertilizer known as triple superphosphate:

$$Ca_3(PO_4)_2(s) + 4\ H_3PO_4(aq) \rightarrow 3\ Ca^{2+}(aq) + 6\ H_2PO_4^-(aq)$$

Phosphoric acid is sometimes called *orthophosphoric acid* to distinguish it from other phosphoric acids that are obtained when H_3PO_4 is heated. For example, *diphosphoric acid*, $H_4P_2O_7$, also called *pyrophosphoric acid*, is obtained when two molecules of H_3PO_4 combine, with elimination of one water molecule:

Orthophosphoric acid, H_3PO_4

Diphosphoric acid, $H_4P_2O_7$

The next in the series is triphosphoric acid, $H_5P_3O_{10}$, and the one with an infinitely long chain of phosphate groups is called polymetaphosphoric acid, $(HPO_3)_n$:

Triphosphoric acid, $H_5P_3O_{10}$

Polymetaphosphoric acid, $(HPO_3)_n$

Repeating unit, HPO_3

All these acids have phosphorus in the +5 oxidation state, and all have structures in which PO_4 tetrahedra share bridging O atoms.

The sodium salt of triphosphoric acid, $Na_5P_3O_{10}$, is a component of some synthetic detergents. It is made by heating a stoichiometric mixture of the powdered orthophosphate salts Na_2HPO_4 and NaH_2PO_4:

$$2\ Na_2HPO_4 + NaH_2PO_4 \rightarrow Na_5P_3O_{10} + 2\ H_2O$$

The triphosphate ion $P_3O_{10}^{5-}$ acts as a water softener, bonding to ions such as Ca^{2+} and Mg^{2+} that are responsible for formation of soap scum in hard water. Unfortunately, $P_3O_{10}^{5-}$ can also contribute to excessive fertilization and rampant growth of algae when phosphate-rich waste water is discharged into lakes. Subsequent decomposition of the algae can deplete a lake of oxygen and cause fish to die off.

Sodium triphosphate, $Na_5P_3O_{10}$, is a common component of synthetic detergents and household cleaners.

EXAMPLE 19.3

Write balanced equations for conversion of orthophosphoric acid to **(a)** triphosphoric acid and **(b)** polymetaphosphoric acid.

SOLUTION Both reactions involve combination of H_3PO_4 molecules, with elimination of water. For phosphate chains of finite length, the number of water molecules eliminated is one less than the number of P atoms in the chain. For the infinitely long chain $(HPO_3)_n$, the number of water molecules eliminated equals the number of P atoms in the chain.

(a) $3\ H_3PO_4(aq) \longrightarrow H_5P_3O_{10}(aq) + 2\ H_2O(l)$
(b) $n\ H_3PO_4(aq) \longrightarrow (HPO_3)_n(aq) + n\ H_2O(l)$

⌐ PROBLEM 19.6 Draw structural formulas for H_3PO_4 and tetraphosphoric acid ($H_6P_4O_{13}$), and show how elimination of water from H_3PO_4 leads to formation of $H_6P_4O_{13}$. Write a balanced equation for the reaction.

⌐ PROBLEM 19.7 The structures of the various phosphate anions and silicate anions (Section 19.7) have many similarities. List some.

19.12 ▶THE GROUP 6A ELEMENTS

The group 6A elements are oxygen, sulfur, selenium, tellurium, and polonium. As shown in Table 19.7, their properties exhibit the usual periodic trends. Both oxygen and sulfur are typical nonmetals. Selenium and tellurium are primarily nonmetallic in character, though the most stable allotrope of selenium, gray selenium, is a lustrous semiconducting solid. Tellurium is also a semiconductor and is usually classified as a semimetal. Polonium, a radioactive element that occurs in trace amounts in uranium ores, is a silvery white metal.

With valence electron configuration $ns^2\ np^4$, the group 6A elements are just two electrons short of an octet configuration, and the −2 oxidation state is therefore a common one. The stability of the −2 state decreases, however, with increasing metallic character, as indicated by the $E°$ values in Table 19.7. Thus, oxygen is a powerful oxidizing agent, but $E°$ values for reduction

TABLE 19.7	Properties of the Group 6A Elements				
Property	Oxygen	Sulfur	Selenium	Tellurium	Polonium
Valence electron configuration	$2s^2\ 2p^4$	$3s^2\ 3p^4$	$4s^2\ 4p^4$	$5s^2\ 5p^4$	$6s^2\ 6p^4$
Melting point (°C)	−218	113[a]	217[b]	452	254
Boiling point (°C)	−183	445	685	1390	962
Atomic radius (pm)	66	104	116	143	167
X^{2-} ionic radius (pm)	132	184	191	211	
First ionization energy (kJ/mol)	1314	1000	941	869	812
Electron affinity (kJ/mol)	−141	−200	−195	−190	−183
Electronegativity	3.5	2.5	2.4	2.1	2.0
Redox potential, $E°$ (V) for $X + 2\ H^+ + 2\ e^- \rightarrow H_2X$	1.23	0.14	−0.40	−0.79	

[a] Rhombic S.
[b] Gray Se.

of Se and Te are negative, which means that H_2Se and H_2Te are reducing agents. Because S, Se, and Te are much less electronegative than oxygen, they are commonly found in positive oxidation states, especially +4, as in SF_4, SO_2, and H_2SO_3, and +6, as in SF_6, SO_3, and H_2SO_4.

Commercial uses of Se, Te, and Po are limited, though selenium is used in making red-colored glass and in photocopiers (see Interlude). Tellurium is used in alloys to improve their machinability, and polonium (^{210}Po) has been used as a heat source in space equipment and as a source of alpha particles in scientific research.

Elemental selenium

Elemental tellurium

The color of the red glass in these traffic signals is due to cadmium selenide, CdSe.

19.13 ▸SULFUR

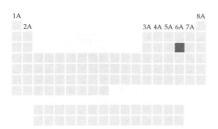

Sulfur is the sixteenth most abundant element in the earth's crust, accounting for 0.026% of its mass. It occurs in elemental form in large underground deposits and is present in numerous minerals, such as pyrite (FeS_2, which contains the S_2^{2-} ion), galena (PbS), cinnabar (HgS), and gypsum ($CaSO_4 \cdot 2\ H_2O$). Sulfur is also present in natural gas as H_2S and in crude oil as organic sulfur compounds. In plants and animals, sulfur occurs in various proteins, and it is one of the 10 most abundant elements in the human body (Figure 19.1).

Pyrite (FeS$_2$), often called fool's gold because of its golden yellow color. FeS$_2$ contains the disulfide ion (S$_2^{2-}$), the sulfur analog of the peroxide ion (O$_2^{2-}$).

Elemental sulfur is obtained from underground deposits. Because of concern about air pollution, an increasing amount of sulfur is now being recovered from natural gas and crude oil. The sulfur compounds in these sources are first converted to H$_2$S, one-third of which is then burned to give SO$_2$. Subsequent reaction of the SO$_2$ with the remaining H$_2$S yields elemental sulfur:

$$2\ H_2S(g) + 3\ O_2(g) \rightarrow 2\ SO_2(g) + 2\ H_2O(g)$$

$$SO_2(g) + 2\ H_2S(g) \xrightarrow[\text{Fe}_2\text{O}_3\ \text{catalyst}]{300°C} 3\ S(g) + 2\ H_2O(g)$$

In the United States, 88% of the sulfur produced is used to manufacture sulfuric acid.

Sulfur exists in many allotropic forms, but the most stable at 25°C is rhombic sulfur, a yellow crystalline solid (mp 113°C) that contains crown-shaped S$_8$ rings.

A sample of rhombic sulfur, the most stable allotrope of sulfur.

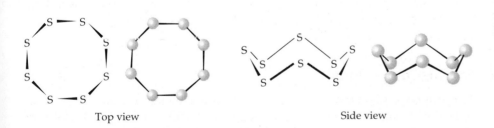

Top view   Side view

Above 95°C, rhombic sulfur is less stable than monoclinic sulfur (mp 119°C), an allotrope in which the cyclic S$_8$ molecules pack differently in the crystal. The phase transition from rhombic to monoclinic sulfur is very slow, however, and rhombic sulfur simply melts at 113°C when heated at an ordinary rate.

As shown in Figure 19.12, molten sulfur exhibits some striking changes when its temperature is increased. Just above its melting point, sulfur is a fluid, straw-colored liquid, but between 160 and 195°C its color becomes dark reddish brown, and its viscosity increases by a factor of more than 10,000. At still higher temperatures, the liquid becomes more fluid again and

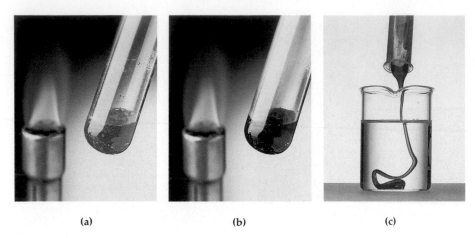

FIGURE 19.12 Effect of temperature on the properties of sulfur. **(a)** Fluid, straw-colored liquid sulfur at about 120°C. **(b)** Viscous, reddish brown liquid sulfur at about 180°C. **(c)** Plastic sulfur obtained by pouring liquid sulfur into water. Plastic sulfur is unstable and reverts to rhombic sulfur on standing at room temperature.

(a)   (b)   (c)

then boils at 445°C to give a vapor that contains mostly S_8 molecules along with smaller amounts of other S_n molecules ($2 \leq n \leq 10$). If the liquid is cooled rapidly by pouring it into water, the sulfur forms an amorphous, rubbery material called plastic sulfur.

The dramatic increase in the viscosity of molten sulfur at 160–195°C is due to the opening of the S_8 rings, yielding S_8 chains that form long polymers with more than 200,000 S atoms in the chain:

$$S_8 \text{ rings} \xrightarrow{\text{heat}} \cdot S - S_6 - S \cdot \text{ chains}$$

$$\cdot S - S_6 - S \cdot \text{ chains} + S_8 \text{ rings} \rightarrow \cdot S - S_{14} - S \cdot \text{ chains}$$

$$\xrightarrow{\text{etc.}} S_n \text{ chains} \quad (n > 200{,}000)$$

Whereas the small S_8 rings easily slide over one another in the liquid, the long polymer chains become entangled, thus accounting for the increase in viscosity. Above 200°C, the polymer chains begin to fragment into smaller pieces, and the viscosity therefore decreases. On rapid cooling, the chains are temporarily frozen in a disordered, tangled arrangement, which accounts for the elastic properties of plastic sulfur.

Sulfur Compounds

HYDROGEN SULFIDE   Hydrogen sulfide is a colorless gas (bp −61°C) with the strong, foul odor of rotten eggs, in which it occurs as a result of bacterial decomposition of sulfur-containing proteins. Hydrogen sulfide is extremely toxic, causing headaches and nausea at concentrations of 10 ppm, and sudden paralysis and death at 100 ppm. On initial exposure, the odor of H_2S can be detected at about 0.02 ppm, but unfortunately the gas tends to dull the sense of smell. It is thus an extremely insidious poison, even more dangerous than HCN.

In the laboratory, H_2S can be prepared by treating iron(II) sulfide with dilute sulfuric acid:

$$FeS(s) + 2\,H^+(aq) \rightarrow H_2S(g) + Fe^{2+}(aq)$$

For use in qualitative analysis (Section 16.15), H_2S is usually generated in solution by hydrolysis of thioacetamide:

$$\underset{\text{Thioacetamide}}{CH_3 \overset{\overset{\displaystyle S}{\|}}{-}C-NH_2(aq)} + H_2O(l) \rightarrow \underset{\text{Acetamide}}{CH_3 \overset{\overset{\displaystyle O}{\|}}{-}C-NH_2(aq)} + H_2S(aq)$$

H_2S is a very weak diprotic acid ($K_{a1} = 1.0 \times 10^{-7}$; $K_{a2} \approx 10^{-19}$) and a mild reducing agent. In reactions with mild oxidizing agents, it is oxidized to a milky white suspension of elemental sulfur:

$$H_2S(aq) + 2\ Fe^{3+}(aq) \rightarrow S(s) + 2\ Fe^{2+}(aq) + 2\ H^+(aq)$$

OXIDES AND OXOACIDS OF SULFUR   Sulfur dioxide (SO_2) and sulfur trioxide (SO_3) are the most important of the various oxides of sulfur. Sulfur dioxide, a colorless, toxic gas (bp $-10°C$) with a pungent, choking odor is formed when sulfur burns in air:

$$S(s) + O_2(g) \rightarrow SO_2(g)$$

Sulfur burns in air, yielding SO_2 gas.

SO_2 is slowly oxidized in the atmosphere to SO_3, which dissolves in rainwater to give sulfuric acid. The burning of sulfur-containing fuels is thus a major cause of acid rain (Section 9.9). In the laboratory, SO_2 is conveniently prepared by treating sodium sulfite with dilute acid:

$$Na_2SO_3(s) + 2\ H^+(aq) \rightarrow SO_2(g) + H_2O(l) + 2\ Na^+(aq)$$

Because SO_2 is toxic to microorganisms, it is used for sterilizing wine and dried fruit.

Both SO_2 and SO_3 are acidic, though aqueous solutions of SO_2 contain mainly dissolved SO_2 and little, if any, sulfurous acid (H_2SO_3):

$$SO_2(aq) + H_2O(l) \rightleftharpoons H_2SO_3(aq)$$

Nevertheless, it's convenient to write the acidic species as H_2SO_3, a weak diprotic acid:

$$H_2SO_3(aq) + H_2O(l) \rightleftharpoons H_3O^+(aq) + HSO_3^-(aq) \qquad K_{a1} = 1.5 \times 10^{-2}$$

$$HSO_3^-(aq) + H_2O(l) \rightleftharpoons H_3O^+(aq) + SO_3^{2-}(aq) \qquad K_{a2} = 6.3 \times 10^{-8}$$

Like carbonic acid, sulfurous acid has never been isolated in pure form.

Sulfuric acid (H_2SO_4), the world's most important industrial chemical, is manufactured by the **contact process**, a three-step reaction sequence in which (1) sulfur burns in air to give SO_2, (2) SO_2 is oxidized to SO_3 in the presence of a vanadium(V) oxide catalyst, and (3) SO_3 reacts with water to give H_2SO_4:

(1) $S(s) + O_2(g) \rightarrow SO_2(g)$

(2) $2 SO_2(g) + O_2(g) \xrightarrow[\text{V}_2\text{O}_5 \text{ catalyst}]{\text{heat}} 2 SO_3(g)$

(3) $SO_3(g) + H_2O$ (in conc H_2SO_4) $\rightarrow H_2SO_4(l)$

In the third step, the SO_3 is absorbed in concentrated sulfuric acid rather than in water because dissolution of SO_3 in water is slow. Water is then added to achieve the desired concentration. Commercial concentrated sulfuric acid is 98% H_2SO_4 by mass (18 M H_2SO_4). Anhydrous (100%) H_2SO_4 is a viscous, colorless liquid that freezes at 10.4°C and boils above 300°C.

The H_2SO_4 molecule is tetrahedral, with S–O bond lengths (142 pm) that are appreciably shorter than the S–OH bond lengths (157 pm). The geometry is therefore consistent with a Lewis structure that has considerable S=O double bonding:

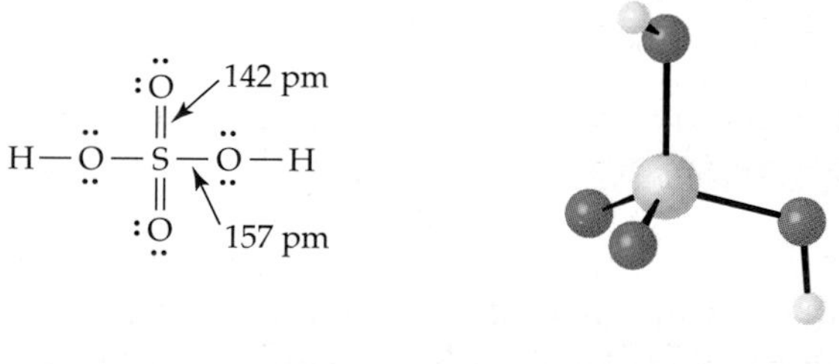

Sulfuric acid is a strong acid in the dissociation of its first proton and has $K_{a2} = 1.2 \times 10^{-2}$ for the dissociation of its second proton. As a diprotic acid, it forms two series of salts: hydrogen sulfates (bisulfates), such as $NaHSO_4$, and sulfates, such as Na_2SO_4.

The oxidizing properties of sulfuric acid depend on its concentration and on the temperature. In dilute solutions at room temperature, H_2SO_4 behaves like HCl, oxidizing metals that stand above hydrogen in the activity series (Table 4.3):

$$Fe(s) + 2 H^+(aq) \rightarrow Fe^{2+}(aq) + H_2(g)$$

Hot, concentrated H_2SO_4 is a better oxidizing agent than the dilute, cold acid is and can oxidize metals like copper, which are not oxidized by $H^+(aq)$. In the process, H_2SO_4 is reduced to SO_2:

$$Cu(s) + 2\ H_2SO_4(l) \rightarrow Cu^{2+}(aq) + SO_4{}^{2-}(aq) + SO_2(g) + 2\ H_2O(l)$$

U.S. production of sulfuric acid in 1993 was 40.2 million tons, far exceeding that of any other chemical (Table 19.1). It is used mostly in manufacturing soluble phosphate and ammonium sulfate fertilizers, but it is essential to many other industries (Figure 19.13). So widespread is the use of sulfuric acid in industrial countries that the amount produced is sometimes regarded as an indicator of economic activity.

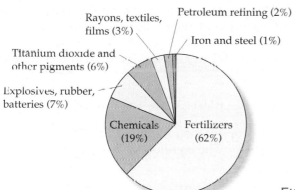

FIGURE 19.13 Uses of sulfuric acid in the United States.

19.14 ➤ THE HALOGENS: OXOACIDS AND OXOACID SALTS

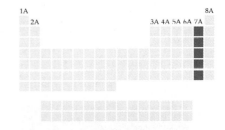

The halogens (group 7A) have valence electron configuration $ns^2\ np^5$ and are the most electronegative group of elements in the periodic table. We saw in Section 6.8 that the halogens tend to achieve an octet configuration by gaining one electron in reactions with metals, thus forming ionic compounds such as NaCl. They also share one electron with nonmetals to give molecular compounds such as HCl, BCl_3, PF_5, and SF_6. In all these compounds, the halogen is in the -1 oxidation state.

Among the most important compounds of halogens in positive oxidation states are the oxoacids of Cl, Br, and I (Table 19.8) and the corresponding oxoacid salts. In these compounds, the halogen shares its valence

TABLE 19.8	Oxoacids of the Halogens			

Oxidation State	Generic Name (Formula)	Chlorine	Bromine	Iodine
+1	Hypohalous acid (HXO)	HClO	HBrO	HIO
+3	Halous acid (HXO$_2$)	HClO$_2$	——	——
+5	Halic acid (HXO$_3$)	HClO$_3$	HBrO$_3$	HIO$_3$
+7	Perhalic acid (HXO$_4$)	HClO$_4$	HBrO$_4$	HIO$_4$, H$_5$IO$_6$

electrons with oxygen, a more electronegative element. (Electronegativities are O, 3.5; Cl, 3.0; Br, 2.8; I, 2.5.) The general formula for a halogen oxoacid is HXO_n, and the oxidation state of the halogen is +1, +3, +5, or +7, depending on the value of n.

Only four of the acids listed in Table 19.8 have been isolated in pure form: perchloric acid ($HClO_4$), iodic acid (HIO_3), and the two periodic acids, metaperiodic acid (HIO_4) and paraperiodic acid (H_5IO_6). The others are stable only in aqueous solution or in the form of their salts. Note that chlorous acid ($HClO_2$) is the only halous acid known.

The acid strength of the halogen oxoacids increases with increasing oxidation state of the halogen (Section 15.15). For example, acid strength increases from HClO, a weak acid ($K_a = 3.5 \times 10^{-8}$), to $HClO_4$, a very strong acid ($K_a \gg 1$). Note that the acidic proton is bonded to oxygen, not to the halogen, even though we usually write the molecular formula of these acids as HXO_n. All the halogen oxoacids and their salts are strong oxidizing agents.

Hypohalous acids are formed when Cl_2, Br_2, or I_2 dissolve in cold water:

$$X_2(g,\ l,\ \text{or}\ s) + H_2O(l) \rightleftharpoons HOX(aq) + H^+(aq) + X^-(aq)$$

In this reaction, the halogen disproportionates, going to the +1 oxidation state in HOX and the −1 state in X^-. The equilibrium lies to the left but is shifted to the right in basic solution:

$$X_2(g,\ l,\ \text{or}\ s) + 2\ OH^-(aq) \rightleftharpoons OX^-(aq) + X^-(aq) + H_2O(l)$$

Large amounts of aqueous sodium hypochlorite (NaOCl) are produced in the chlor-alkali industry (Section 18.12) when the Cl_2 gas and aqueous NaOH from electrolysis of aqueous NaCl are allowed to mix. Aqueous NaOCl is a strong oxidizing agent and is sold in 5% solution as chlorine bleach.

Further disproportionation of OCl^- to ClO_3^- and Cl^- is slow at room temperature but becomes fast at higher temperatures. Thus, when Cl_2 gas reacts with hot aqueous NaOH, it gives a solution that contains sodium chlorate ($NaClO_3$) rather than NaOCl:

$$3\ Cl_2(g) + 6\ OH^-(aq) \rightarrow ClO_3^-(aq) + 5\ Cl^-(aq) + 3\ H_2O(l)$$

Chlorate salts are used as weed killers and as strong oxidizing agents. Potassium chlorate, for example, is an oxidizer in matches, fireworks, and explosives. It reacts vigorously with organic matter, as shown in Figure 19.14.

Sodium perchlorate ($NaClO_4$) is produced commercially by electrolytic oxidation of aqueous sodium chlorate and is converted to perchloric acid by reaction with concentrated HCl:

$$ClO_3^-(aq) + H_2O(l) \xrightarrow{\text{electrolysis}} ClO_4^-(aq) + H_2(g)$$

$$NaClO_4(s) + HCl(aq) \rightarrow HClO_4(aq) + NaCl(s)$$

The $HClO_4$ is then concentrated by distillation at reduced pressure.

$KClO_3$ is an oxidizing agent in matches.

Pure, anhydrous perchloric acid is a colorless, shock-sensitive liquid that decomposes explosively on heating. It is a powerful and dangerous oxidizing agent, violently oxidizing organic matter and rapidly oxidizing even silver and gold. Perchlorate salts are also strong oxidants, and they too must be handled with caution. Ammonium perchlorate (NH_4ClO_4) is the oxidizer in the solid booster rockets used to propel the space shuttle.

Iodine differs from the other halogens in forming more than one perhalic acid. Paraperiodic acid (H_5IO_6) is obtained as white crystals (mp 128°C) when periodic acid solutions are evaporated. When heated to 100°C at reduced pressure, these crystals lose water and are converted to meta-periodic acid (HIO_4):

$$H_5IO_6(s) \xrightarrow[\text{12 mm Hg}]{\text{100°C}} HIO_4(s) + 2\ H_2O(g)$$

Metaperiodic acid is a strong monoprotic acid, whereas paraperiodic acid is a weak polyprotic acid ($K_{a1} = 5.1 \times 10^{-4}$; $K_{a2} = 4.9 \times 10^{-9}$). It has an octahedral structure in which a central iodine atom is bonded to one O atom and five OH groups:

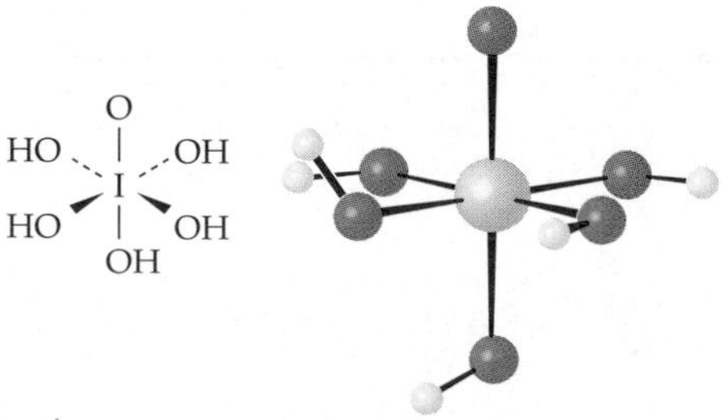

The oxidizer in the space shuttle's solid booster rockets is ammonium perchlorate, NH_4ClO_4.

$$
\begin{array}{c}
\text{HO} \diagdown \underset{|}{\overset{O}{}} \diagup \text{OH} \\
\text{HO} \diagup \overset{|}{I} \diagdown \text{OH} \\
\text{OH}
\end{array}
$$

Paraperiodic acid, H_5IO_6

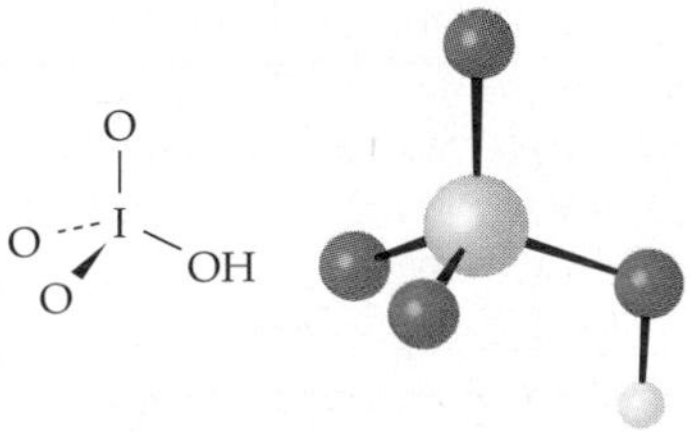

$$
\begin{array}{c}
\overset{O}{\overset{\|}{}} \\
O \cdots I \\
\diagup \diagdown \\
O \text{OH}
\end{array}
$$

Metaperiodic acid, HIO_4

Chlorine and bromine do not form perhalic acids of the type H_5XO_6 because their smaller sizes favor a tetrahedral structure over an octahedral one.

interlude—PHOTOCOPIERS

Not too many years ago, copies of documents were made either by using various wet photographic methods or by typing the original using "carbon paper." The introduction of the plain-paper photocopy machine in the mid 1950s revolutionized the way offices handle paper. Now, the most widely used method of document copying is based on *xerography*, a term derived from the Greek words for "dry writing."

Dry photocopiers make use of an unusual property of selenium, the group 6A element below sulfur and oxygen in the periodic table. Selenium is a *photoconductor*, a substance that is a poor electrical conductor when dark but whose conductivity increases (by a factor of 1000) when exposed to light. When the light is removed, the conductivity again drops.

As illustrated in Figure 19.15, the xerographic process begins when a selenium-coated drum is given a uniform positive charge and is then exposed through a lens to a brightly illuminated document. Those areas on the drum that correspond to a light part of the document become conducting when exposed and lose their charge, whereas those areas on the drum that correspond to a dark part of the document remain nonconducting and retain their charge. Thus, an image of the document is formed on the drum as an array of positive electrical charges.

Following its formation, the image is developed by exposing the drum to negatively charged dry ink particles (*toner*), which are attracted to the positively charged areas of the drum. The developed image is then transferred to paper by passing a sheet of paper between the drum and a positively charged development electrode, which induces the negatively charged toner particles to jump from the drum to the paper.

The toner particles, in addition to serving as pigment, are made of a resinous plastic material that fuses to the paper when heated, thereby fixing the image. The final copy then rolls out of the machine, and the drum is restored to its original condition by flooding it with light to remove all remaining charges and gently scraping off any bits of excess toner.

FIGURE 19.15 The photocopying process. A selenium-coated rotating drum is given a uniform positive charge (step 1) and is then exposed to an image (step 2). Negatively charged toner particles are attracted to the charged areas of the drum (step 3), and the image is transferred from the drum to a sheet of paper (step 4). Heating then fixes the image, and the drum is flooded with light and cleaned to ready the machine for another cycle (step 5).

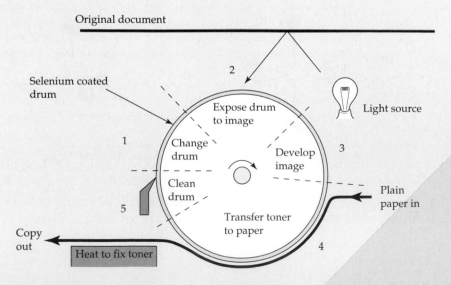

SUMMARY

The main-group elements are the s-block elements of groups 1A and 2A and the p-block elements of groups 3A–8A. From left to right across the periodic table, ionization energy, electronegativity, and nonmetallic character increase, whereas atomic radius and metallic character decrease. From top to bottom of a group in the periodic table, ionization energy, electronegativity, and nonmetallic character decrease, whereas atomic radius and metallic character increase. The second-row elements form strong multiple bonds but cannot form more than four bonds because they lack valence d orbitals.

The group 3A elements—B, Al, Ga, In, and Tl—are metals except for boron, which is a semimetal. Boron is a semiconductor and forms molecular compounds. **Boranes**, such as diborane (B_2H_6), are electron-deficient molecules that contain **three-center, two-electron bonds** (B–H–B).

The group 4A elements—C, Si, Ge, Sn, and Pb—exhibit the usual increase in metallic character down the group. Their most common oxidation state is +4, but the +2 state becomes increasingly more stable from Ge to Sn to Pb. In elemental form, carbon exists as diamond, graphite, fullerene, coke, charcoal, and carbon black. Its most important inorganic compounds include carbon monoxide (CO), carbon dioxide (CO_2), carbonates, such as Na_2CO_3 (soda ash, used in making glass), and $NaHCO_3$ (baking soda).

Silicon, the second most abundant element in the earth's crust, is obtained by reducing silica sand (SiO_2) with coke. It is purified for use in the semiconductor industry by **zone refining**. In the **silicates**, SiO_4 tetrahedra share common O atoms to give silicon oxoanions with ring, chain, layer, and extended three-dimensional structures. In **aluminosilicates**, such as $KAlSi_3O_8$, Al^{3+} replaces some of the Si^{4+}.

Molecular nitrogen (N_2) is unreactive because of its strong N-N triple bond. Nitrogen exhibits all oxidation states between -3 and $+5$. Its most important compounds are ammonia (NH_3), a nitrogen fertilizer, and nitric acid (HNO_3), used to make fertilizers, explosives, plastics, and dyes. Nitric acid is manufactured by the **Ostwald process**.

Phosphorus, the most abundant group 5A element, exists in two common allotropic forms—white phosphorus, which contains highly reactive tetrahedral P_4 molecules, and red phosphorus, which is polymeric. The most common oxidation states of P are -3 as in phosphine (PH_3), $+3$ as in PCl_3, P_4O_6, and H_3PO_3, and $+5$ as in PCl_5, P_4O_{10}, and H_3PO_4. On heating, H_3PO_4 molecules combine, with elimination of water molecules, yielding polyphosphoric acids (for example, $H_5P_3O_{10}$) or polymetaphosphoric acids $(HPO_3)_n$. Phosphates are used in fertilizers and detergents.

Sulfur is obtained from underground deposits and is recovered from natural gas and crude oil. The properties of sulfur change dramatically on heating as the S_8 rings of rhombic sulfur open and polymerize to give long chains, which then fragment at higher temperatures. The most common oxidation states of S are -2 as in H_2S, $+4$ as in SO_2 and H_2SO_3, and $+6$ as in SO_3 and H_2SO_4. Sulfuric acid, the world's most important industrial chemical, is manufactured by the **contact process**.

Chlorine, bromine, and iodine form a series of oxoacids: hypohalous acid (HXO), halous acid (HXO_2), halic acid (HXO_3), and perhalic acid (HXO_4). Acid strength increases as the oxidation state of the halogen increases from $+1$ to $+7$. Iodine forms two perhalic acids, HIO_4 and H_5IO_6. Halogen oxoacids and their salts are strong oxidizing agents.

1. Locate each of the following groups of elements on the periodic table.

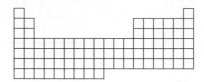

(a) main-group elements   (b) s-block elements
(c) p-block elements   (d) main-group metals
(e) nonmetals   (f) semimetals

2. Locate each of the following elements on the periodic table.

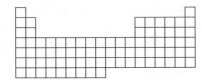

(a) element with the lowest ionization energy
(b) most electronegative element
(c) group 4A element with the largest atomic radius
(d) group 6A element with the smallest atomic radius
(e) group 3A element that is a semiconductor
(f) group 5A element that forms the strongest π bonds

3. Consider the six second- and third-row elements in groups 5A–7A of the periodic table.

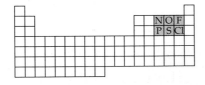

The molecular structures of common allotropes of these elements are shown below.

(a) What is the molecular structure of each of the six elements?
(b) Using Lewis electron-dot structures, explain the particular molecular structure of each element.
(c) Explain why nitrogen and phosphorus have different molecular structures, why oxygen and sulfur have different molecular structures, but why fluorine and chlorine have the same molecular structure.

4. The following pictures represent various silicate anions. Give the formula and charge of each anion.

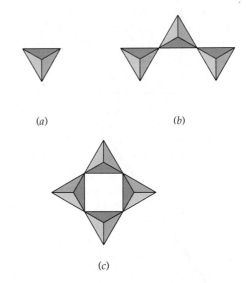

(a)   *(b)*

(c)

ADDITIONAL PROBLEMS

Problems 19.1–19.7 appear within the chapter.

GENERAL PROPERTIES AND PERIODIC TRENDS

19.8 Which of the following are main-group elements?
(a) Se   (b) Ca   (c) Cr   (d) In   (e) Au

19.9 Which of the following are p-block elements?
(a) K   (b) P   (c) Fe   (d) Be   (e) Br

19.10 Classify each of the following elements as a metal, nonmetal, or semimetal.
(a) Cl   (b) Ga   (c) As   (d) Pb   (e) S

19.11 Which element in each of the following pairs has the higher ionization energy?
(a) S or Cl   (b) Si or Ge   (c) In or O

19.12 Which element in each of the following pairs has the larger atomic radius?
(a) B or Al   (b) P or S   (c) Pb or Br

19.13 Which element in each of the following pairs has the higher electronegativity?

(a) Te or I   (b) N or P   (c) In or F

19.14 Which element in each of the following pairs has more metallic character?

(a) Si or Sn   (b) Ge or Se   (c) Bi or I

19.15 Arrange the following elements in order of increasing ionization energy.

(a) S   (b) Na   (c) Si   (d) O

19.16 Which compound in each of the following pairs is more ionic?

(a) CaH_2 or NH_3 (b) P_4O_6 or Ga_2O_3
(c) $SiCl_4$ or KCl

19.17 Which compound in each of the following pairs is more covalent?

(a) PBr_3 or AlF_3 (b) MgO or CO
(c) NaH or PH_3

19.18 Which of the following compounds are ionic and which are molecular?

(a) B_2H_6 (b) $BaCl_2$ (c) SO_3 (d) $GeCl_4$ (e) In_2O_3

19.19 Which compound in each of the following pairs is the more acidic oxide?

(a) Al_2O_3 or P_4O_{10} (b) B_2O_3 or Ga_2O_3
(c) SO_2 or SnO_2

19.20 Which compound in each of the following pairs has the higher melting point?

(a) LiCl or PCl_3 (b) CO_2 or SiO_2
(c) P_4O_{10} or NO_2

19.21 Which element in each of the following pairs is the better electrical conductor?

(a) B or Ga   (b) In or S   (c) Pb or P

19.22 Consider the following list of elements: C, Se, B, Sn, Cl. Identify the element on this list that

(a) has the largest atomic radius
(b) is most electronegative
(c) is the best electrical conductor
(d) has a maximum oxidation state of +6
(e) forms a hydride with the empirical formula XH_3

19.23 Consider the following list of elements: N, Si, Al, S, F. Identify the element on this list that

(a) has the highest ionization energy
(b) has the most metallic character
(c) forms the strongest π bonds
(d) is a semiconductor
(e) forms 2− anions

19.24 BF_3 reacts with F^- to give BF_4^-, whereas AlF_3 reacts with F^- to give AlF_6^-. Explain.

19.25 At ordinary temperatures sulfur exists as S_8, whereas oxygen exists as O_2. Explain.

19.26 The organic solvent acetone has the molecular formula $(CH_3)_2CO$. The silicon analogue, a thermally stable lubricant, is a polymer, $[(CH_3)_2SiO]_n$. Account for the difference in structure:

Acetone   Polydimethylsiloxane (silicone oil)

THE GROUP 3A ELEMENTS

19.27 What is the most common oxidation state for each of the group 3A elements?

19.28 List three ways in which the properties of boron differ from those of the other group 3A elements.

19.29 Explain why the properties of boron differ so markedly from the properties of the other group 3A elements.

19.30 What is borax, and why is it important?

19.31 How is crystalline boron prepared? Write a balanced equation for the reaction.

19.32 Tell what is meant by (a) an electron-deficient molecule and (b) a three-center two-electron bond. Illustrate each definition with an example.

19.33 Describe the structure of diborane (B_2H_6), and explain why the bridging B–H bonds are longer than the terminal B–H bonds.

19.34 Identify the group 3A element that best fits each of the following descriptions.

(a) is the most abundant element of the group
(b) is stable in the +1 oxidation state
(c) has an unusually low melting point
(d) is a semiconductor
(e) is extremely toxic
(f) forms an acidic oxide

THE GROUP 4A ELEMENTS

19.35 Identify the group 4A element that best fits each of the following descriptions.

(a) prefers the +2 oxidation state

(b) forms the strongest π bonds
(c) is the second most abundant element in the earth's crust
(d) forms the most acidic oxide

19.36 Select the group 4A element that best fits each of the following descriptions.

(a) forms the most basic oxide
(b) is the least dense semimetal
(c) is the second most abundant element in the human body
(d) is the most electronegative

19.37 Describe the geometry of each of the following molecules or ions, and tell which hybrid orbitals are used by the central atom.

(a) $GeBr_4$ (b) CO_2 (c) CO_3^{2-}
(d) $Sn(OH)_6^{2-}$ (e) $SnCl_2$

19.38 List three properties of diamond, and account for them in terms of structure and bonding.

19.39 Describe the structure and bonding in graphite, and explain why graphite is a good lubricant.

19.40 Why is graphite a better electrical conductor than diamond, and why does the conductivity of graphite depend on direction?

19.41 Using Le Châtelier's principle, explain why high pressures favor the formation of diamond (density $3.51\ g/cm^3$) from graphite (density $2.27\ g/cm^3$).

19.42 Give the name and formula of a compound in which carbon exhibits an oxidation state of

(a) +4   (b) +2   (c) −4

19.43 List three commercial uses for carbon dioxide, and relate each use to one of carbon dioxide's properties.

19.44 Draw Lewis electron-dot structures for CO and CN^-, and explain why these species are said to be isoelectronic.

19.45 Why are CO and CN^- so toxic to humans?

19.46 Write balanced net ionic equations for two reactions of the cyanide ion.

19.47 Give an example of an ionic carbide. What is the oxidation state of carbon in this substance?

19.48 Give the chemical formula for each of the following substances.

(a) baking soda   (b) cassiterite
(c) washing soda   (d) galena

19.49 Describe the preparation of silicon from silica sand, and tell how silicon is purified for use in semiconductor devices. Write balanced equations for all reactions.

19.50 What is meant by each of the following terms?

(a) zone refining
(b) silicate chain anion
(c) aluminosilicate

19.51 Using the notation of Figure 19.5, draw the structure of the silicate anion in (a) K_4SiO_4 and (b) $Ag_{10}Si_4O_{13}$. What is the relationship between the charge on the anion and the number of terminal O atoms?

19.52 Using the notation of Figure 19.6, draw the structure of the cyclic silicate anion in which four SiO_4 tetrahedra share O atoms to form an eight-membered ring of alternating Si and O atoms. Give the formula and charge of the anion.

19.53 The silicate anion in the mineral kinoite is a chain of three SiO_4 tetrahedra that share corners with adjacent tetrahedra. The mineral also contains Ca^{2+} ions, Cu^{2+} ions, and water molecules in a $1:1:1$ ratio.

(a) Give the formula and charge of the silicate anion.
(b) Give the complete formula for the mineral.

19.54 Suggest a plausible structure for the silicate anion in each of the following minerals.

(a) spodumene, $LiAlSi_2O_6$
(b) wollastonite, $Ca_3Si_3O_9$
(c) thortveitite, $Sc_2Si_2O_7$
(d) albite, $NaAlSi_3O_8$

19.55 What minerals are the starting materials for preparation of (a) silicon (b) tin (c) lead? Give the name and chemical formula for each mineral, and write a balanced equation for the preparation of each element.

19.56 List the main uses of silicon, germanium, tin, and lead.

THE GROUP 5A ELEMENTS

19.57 Identify the group 5A element(s) that best fits each of the following descriptions.

(a) makes up part of bones and teeth
(b) forms stable salts containing M^{3+} ions
(c) is the most abundant element in the atmosphere
(d) is the most abundant group 5A element in the earth's crust
(e) forms a basic oxide

19.58 Why is the N_2 molecule so unreactive?

19.59 Describe the process used for industrial production of the following chemicals.

(a) nitrogen   (b) ammonia
(c) nitric acid   (d) phosphoric acid

Write balanced equations for all chemical reactions.

19.60 Give the chemical formula for each of the following compounds, and indicate the oxidation state of the group 5A element.
(a) nitrous oxide
(b) hydrazine
(c) calcium phosphide
(d) phosphorous acid
(e) arsenic acid

19.61 Give the chemical formula for each of the following compounds, and indicate the oxidation state of the group 5A element.
(a) nitric oxide
(b) nitrous acid
(c) phosphine
(d) phosphorus(V) oxide
(e) triphosphoric acid

19.62 Draw Lewis electron-dot structures for (a) nitrous oxide (b) nitric oxide (c) nitrogen dioxide. Predict the molecular geometry of these molecules, and indicate which are expected to be paramagnetic.

19.63 Predict the geometry of each of the following molecules or ions, and tell which hybrid orbitals are used by the central atom.
(a) NO_2^-
(b) PH_3
(c) PF_5
(d) N_2O
(e) PCl_4^+
(f) PCl_6^-

19.64 Compare and contrast the properties of ammonia and phosphine.

19.65 Describe the structures of the white and red allotropes of phosphorus, and explain why white phosphorus is so reactive.

19.66 Account for each of the following observations.
(a) Phosphorous acid is a diprotic acid.
(b) Nitric acid is a strong oxidizing agent, but phosphoric acid is not.
(c) Nitrogen doesn't exist as a four-atom molecule like P_4.
(d) Phosphorus, arsenic, and antimony form trichlorides and pentachlorides, but nitrogen forms only NCl_3.

19.67 Write a balanced equation to account for each of the following observations.
(a) Nitric oxide turns brown when exposed to air.
(b) Nitric acid turns yellow-brown on standing.
(c) Silver dissolves in dilute HNO_3, yielding a colorless gas.
(d) Hydrazine reduces iodine to I^- and in the process is oxidized to N_2 gas.

19.68 Draw the structure of each of the following compounds.
(a) phosphorus(III) oxide
(b) phosphorus(V) oxide
(c) phosphorous acid
(d) orthophosphoric acid
(e) polymetaphosphoric acid

19.69 Write a balanced equation for the conversion of orthophosphoric acid to diphosphoric acid.

THE GROUP 6A ELEMENTS

19.70 Identify the group 6A element that best fits each of the following descriptions.
(a) is the most electronegative
(b) is a semimetal
(c) is radioactive
(d) has the most negative electron affinity
(e) is the most abundant element in the earth's crust
(f) is the most abundant element in the human body

19.71 How is sulfur recovered from natural sources, and what is its principal use?

19.72 Describe the structure of the sulfur molecules in
(a) rhombic sulfur
(b) monoclinic sulfur
(c) plastic sulfur
(d) liquid sulfur above 160°C

19.73 The viscosity of liquid sulfur increases sharply at about 160°C and then decreases again above 200°C. Explain.

19.74 Give the name and formula of two compounds in which sulfur exhibits an oxidation state of
(a) −2   (b) +4   (c) +6

19.75 What is the oxidation state of sulfur in each of the following compounds?
(a) HgS   (b) $Ca(HSO_4)_2$ (c) H_2SO_3
(d) FeS_2 (e) SF_4

19.76 Describe the contact process for manufacture of sulfuric acid, and write balanced equations for all reactions.

19.77 Describe a convenient laboratory method for preparation of each of the following compounds, and write balanced equations for all reactions.
(a) sulfur dioxide
(b) hydrogen sulfide
(c) sodium hydrogen sulfate

19.78 Write a balanced net ionic equation for each of the following reactions.

(a) $Zn(s)$ + dilute $H_2SO_4(aq)$ $\longrightarrow$

(b) $BaSO_3(s)$ + $HCl(aq)$ $\longrightarrow$

(c) $Cu(s)$ + hot, conc $H_2SO_4(l)$ $\longrightarrow$

(d) $H_2S(aq)$ + $I_2(aq)$ $\longrightarrow$

19.79 Account for each of the following observations.

(a) H_2SO_4 is a stronger acid than H_2SO_3.

(b) SF_4 exists, but OF_4 does not.

(c) The S_8 ring is nonplanar.

HALOGEN OXOACIDS AND OXOACID SALTS

19.80 Write the chemical formula for each of the following compounds, and indicate the oxidation state of the halogen.

(a) bromic acid

(b) hypoiodous acid

(c) sodium chlorite

(d) potassium metaperiodate

19.81 Write the chemical formula for each of the following compounds, and indicate the oxidation state of the halogen.

(a) chlorous acid

(b) potassium iodate

(c) paraperiodic acid

(d) sodium hypobromite

19.82 Name each of the following compounds.

(a) HIO_3

(b) $HClO_2$

(c) $NaOBr$

(d) $LiClO_4$

19.83 Name each of the following compounds.

(a) $KClO_2$

(b) HIO_4

(c) $HOBr$

(d) $NaBrO_3$

19.84 Write a Lewis electron-dot structure for each of the following molecules or ions, and predict the molecular geometry.

(a) HIO_3

(b) ClO_2^-

(c) $HOCl$

(d) IO_6^{5-}

19.85 Write a Lewis electron-dot structure for each of the following molecules or ions, and predict the molecular geometry.

(a) BrO_4^-

(b) ClO_3^-

(c) HIO_4

(d) $HOBr$

19.86 Explain why acid strength increases in the order $HClO < HClO_2 < HClO_3 < HClO_4$.

19.87 Write a balanced net ionic equation for each of the following reactions.

(a) $Br_2(l)$ + cold $NaOH(aq)$ $\longrightarrow$

(b) $Cl_2(g)$ + cold $H_2O(l)$ $\longrightarrow$

(c) $Cl_2(g)$ + hot $NaOH(aq)$ $\longrightarrow$

19.88 Write a balanced chemical equation for the reaction of potassium chlorate and sucrose (Figure 19.14). The products are $KCl(s)$, $CO_2(g)$, and $H_2O(g)$.

GENERAL PROBLEMS

19.89 What is the chemical formula for the most important mineral sources of each of the following elements?

(a) B   (b) P   (c) S

19.90 Carbon is an essential element in the molecules on which life is based. Would silicon be equally satisfactory? Explain.

19.91 Which of the nonmetallic elements are gases at 25°C and which is a liquid?

19.92 What sulfur-containing compound is present in acid rain, and how is it formed as a result of the burning of sulfur-containing fossil fuels?

19.93 How many of the four most abundant elements in the earth's crust and in the human body can you identify without consulting Figure 19.1?

19.94 Identify as many of the 10 most important industrial chemicals as you can without consulting Table 19.1.

19.95 Which of the group 4A elements have allotropes with the diamond structure? Which have metallic allotropes? How does the variation in the structure of the group 4A elements illustrate how metallic character varies down a periodic group?

19.96 Write the chemical formula for the following substances.

(a) pyrite   (b) soda ash
(c) gypsum   (d) cinnabar

19.97 Write a balanced chemical equation for a laboratory preparation of each of the following compounds.

(a) NH_3 (b) CO_2
(c) B_2H_6 (diborane)   (d) C_2H_2 (acetylene)
(e) N_2O (f) NO_2

19.98 Write balanced equations for the reactions of (a) H_3PO_4 and (b) $B(OH)_3$ with water. Classify each acid as a Brønsted-Lowry acid or as a Lewis acid.

19.99 What are the principal uses of the following industrial chemicals?

(a) H_2SO_4 (b) N_2 (c) NH_3 (d) HNO_3 (e) H_3PO_4

19.100 Give the name and formula of three compounds for each of the common oxidation states of phosphorus.

19.101 Could the strain in the P_4 molecule be reduced by using sp^3 hybrid orbitals in bonding instead of pure p orbitals? Explain.

19.102 Draw two plausible structures for H_3PO_3. For each one, predict the shape of the pH titration curve for titration of the H_3PO_3 ($K_{a1} \approx 10^{-2}$) with aqueous NaOH.

19.103 Account for each of the following observations.

(a) Diamond is extremely hard and high melting, whereas graphite is very soft and high melting.
(b) The terminal P–O bonds in P_4O_{10} are shorter than the bridging P–O bonds.
(c) Chlorine does not form a perhalic acid, H_5ClO_6.

19.104 Give one example from main-group chemistry that illustrates each of the following descriptions.

(a) covalent network solid
(b) disproportionation reaction
(c) paramagnetic oxide
(d) polar molecule that violates the octet rule
(e) Lewis acid
(f) amphoteric oxide
(g) semiconductor
(h) strong oxidizing agent
(i) allotropes

chapter 20

TRANSITION ELEMENTS AND COORDINATION CHEMISTRY

T he **transition elements** occupy the center part of the periodic table, bridging the gap between the active *s*-block metals of groups 1A and 2A on the left and the *p*-block metals, semimetals, and nonmetals of groups 3A–8A on the right (Figure 20.1). Because the *d* subshells are being filled in this region of the periodic table, the transition elements are called the ***d*-block elements**.

Because each *d* subshell consists of five orbitals and can accommodate 10 electrons, each transition series consists of 10 elements. The first series extends from scandium through zinc and includes many familiar metals, such as chromium, iron, and copper. The second series runs from yttrium through cadmium, and the third series runs from lanthanum through mercury. In addition, there is an incomplete fourth transition series made up of actinium through element 109.

Tucked into the periodic table between lanthanum (atomic number 57) and hafnium (atomic number 72) are the **lanthanides**. In this inner transition series of 14 metallic elements, the seven 4*f* orbitals are progressively filled. Following actinium (atomic number 89) is a second inner transition series of 14 elements, the **actinides**. In this series, the 5*f* subshell is pro-

Native copper

Main groups

(1) 1A	(2) 2A	Transition-metal groups										(13) 3A	(14) 4A	(15) 5A	(16) 6A	(17) 7A	(18) 8A
1 H																	2 He
3 Li	4 Be	(3) 3B	(4) 4B	(5) 5B	(6) 6B	(7) 7B	(8)	(9) 8B	(10)	(11) 1B	(12) 2B	5 B	6 C	7 N	8 O	9 F	10 Ne
11 Na	12 Mg											13 Al	14 Si	15 P	16 S	17 Cl	18 Ar
19 K	20 Ca	21 Sc	22 Ti	23 V	24 Cr	25 Mn	26 Fe	27 Co	28 Ni	29 Cu	30 Zn	31 Ga	32 Ge	33 As	34 Se	35 Br	36 Kr
37 Rb	38 Sr	39 Y	40 Zr	41 Nb	42 Mo	43 Tc	44 Ru	45 Rh	46 Pd	47 Ag	48 Cd	49 In	50 Sn	51 Sb	52 Te	53 I	54 Xe
55 Cs	56 Ba	57 La	72 Hf	73 Ta	74 W	75 Re	76 Os	77 Ir	78 Pt	79 Au	80 Hg	81 Tl	82 Pb	83 Bi	84 Po	85 At	86 Rn
87 Fr	88 Ra	89 Ac	104 Db	105 Jl	106 Rf	107 Bh	108 Hn	109 Mt									

Lanthanides	58 Ce	59 Pr	60 Nd	61 Pm	62 Sm	63 Eu	64 Gd	65 Tb	66 Dy	67 Ho	68 Er	69 Tm	70 Yb	71 Lu
Actinides	90 Th	91 Pa	92 U	93 Np	94 Pu	95 Am	96 Cm	97 Bk	98 Cf	99 Es	100 Fm	101 Md	102 No	103 Lr

FIGURE 20.1 The transition elements (*d*-block elements) are located in the central region of the periodic table between the *s*-block and *p*-block main-group elements. The two series of inner transition elements (*f*-block elements) follow lanthanum and actinium

gressively filled. The lanthanides and actinides together comprise the **f-block elements**, or **inner transition elements**.

The transition metals iron and copper have been known since antiquity and have played an important role in the development of civilization. Iron, the main constituent of steel, is still important as a structural material. Worldwide production of steel amounts to some 700 million tons per year. In newer technologies, other transition elements are useful. For example, the strong, lightweight metal titanium is a major component in modern jet aircraft. Transition metals are also used as heterogeneous catalysts in automobile catalytic converters and in the industrial synthesis of essential chemicals such as sulfuric acid, nitric acid, and ammonia.

The beams used in constructing this building are made of steel, an iron alloy.

The role of the transition elements in living systems is equally important. Iron is present in biomolecules such as hemoglobin, which transports oxygen from our lungs to other parts of the body. Cobalt is an essential component of vitamin B_{12}. Nickel, copper, and zinc are vital constituents of many enzymes, the large protein molecules that act as catalysts for biological reactions.

In this chapter, we'll look at the properties and chemical behavior of transition-metal compounds, paying special attention to *coordination compounds,* in which a central metal ion—usually a transition metal—is attached to a group of surrounding molecules or ions by coordinate covalent bonds (Section 7.4).

20.1 ►ELECTRON CONFIGURATIONS

Look at the electron configurations of potassium and calcium, the *s*-block elements immediately preceding the first transition series. These atoms have valence 4*s* electrons, but no *d* electrons:

$$K \ (Z = 19): [Ar] \ 3d^0 \ 4s^1 \qquad Ca \ (Z = 20): [Ar] \ 3d^0 \ 4s^2$$

The filling of the 3*d* subshell begins at atomic number 21 (scandium) and continues until the subshell is completely filled at atomic number 30 (zinc):

$$Sc \ (Z = 21): [Ar] \ 3d^1 \ 4s^2 \ \dashrightarrow Zn \ (Z = 30): [Ar] \ 3d^{10} \ 4s^2$$

The filling of the 3*d* subshell generally proceeds according to Hund's rule (Section 5.12)—one electron is added to each of the five 3*d* orbitals before a second electron is added to any one of them. There are just two exceptions to this expected filling pattern, chromium and copper:

	Expected configuration	Observed configuration

Cr (Z = 24):   ↑ ↑ ↑ ↑ __   ↑↓   ↑ ↑ ↑ ↑ ↑   ↑
                    $3d^4$ $4s^2$ $3d^5$ $4s^1$

Cu (Z = 29):   ↑↓ ↑↓ ↑↓ ↑↓ ↑   ↑↓   ↑↓ ↑↓ ↑↓ ↑↓ ↑↓   ↑
                    $3d^9$ $4s^2$ $3d^{10}$ $4s^1$

Electron configurations depend on both orbital energies and electron–electron repulsions. Consequently, it's not always possible to predict configurations when two valence subshells have similar energies. It's often found, however, that exceptions from the expected orbital filling pattern result in either half-filled or completely filled subshells. In the case of chromium, for example, the 3*d* and 4*s* subshells have similar energies. It's evidently advantageous to shift one electron from the 4*s* to the 3*d* subshell, which decreases electron–electron repulsions and gives two half-filled subshells. Because each valence electron is in a separate orbital, the electron–electron repulsion that would otherwise occur between the two 4*s* electrons in the expected configuration is eliminated. A similar shift of one electron from 4*s* to 3*d* in copper gives a completely filled 3*d* subshell and a half-filled 4*s* subshell.

In contrast to the situation for the neutral atoms, the electron configurations of transition-metal cations are easy to predict: All the valence electrons occupy the d orbitals. Iron, for example, which forms $+2$ and $+3$ cations, has the following valence electron configurations:

$$Fe: 3d^6 4s^2 \qquad Fe^{2+}: 3d^6 \qquad Fe^{3+}: 3d^5$$

When a neutral atom loses one or more electrons, the remaining electrons are less shielded, and the effective nuclear charge (Z_{eff}) increases. Consequently, the remaining electrons are more strongly attracted to the nucleus, and their orbital energies decrease. It turns out that the $3d$ orbitals experience a steeper drop in energy with increasing Z_{eff} than does the $4s$ orbital, making the $3d$ orbitals in cations lower in energy than the $4s$ orbital. As a result, all the valence electrons occupy the $3d$ orbitals, and the $4s$ orbital is vacant.

Even in neutral molecules and complex anions, all the metal's valence electrons are in d orbitals because the metal usually has a positive oxidation state. The metal atom in both VCl_4 and $MnO_4{}^{2-}$, for example, has valence configuration $3d^1$. Electron configurations and other properties for atoms and common ions of first-series transition elements are summarized in Table 20.1.

TABLE 20.1   Selected Properties of First-Series Transition Elements

Group:	3B	4B	5B	6B	7B		8B		1B	2B
Element:	Sc	Ti	V	Cr	Mn	Fe	Co	Ni	Cu	Zn
Valence electron configuration										
M Atom	$3d^1 4s^2$	$3d^2 4s^2$	$3d^3 4s^2$	$3d^5 4s^1$	$3d^5 4s^2$	$3d^6 4s^2$	$3d^7 4s^2$	$3d^8 4s^2$	$3d^{10} 4s^1$	$3d^{10} 4s^2$
M^{2+} Ion		$3d^2$	$3d^3$	$3d^4$	$3d^5$	$3d^6$	$3d^7$	$3d^8$	$3d^9$	$3d^{10}$
M^{3+} Ion	$3d^0$	$3d^1$	$3d^2$	$3d^3$	$3d^4$	$3d^5$	$3d^6$			
Elec. conductivity*	3	4	8	11	1	17	24	24	96	27
Melting point (°C)	1541	1660	1890	1857	1244	1535	1495	1455	1083	420
Boiling point (°C)	2836	3287	3380	2672	1962	2750	2870	2730	2567	907
Density (g/cm³)	2.99	4.5	5.96	7.20	7.20	7.86	8.9	8.90	8.92	7.14
Atomic radius (pm)	162	147	134	128	127	126	125	124	128	134
E_i†(kJ/mol)										
First	631	658	650	653	717	759	758	737	745	906
Second	1235	1310	1414	1592	1509	1561	1646	1753	1958	1733
Third	2389	2652	2828	2987	3248	2957	3232	3393	3554	3833

*Electrical conductivity relative to an arbitrary value of 100 for silver.
†Ionization energy.

EXAMPLE 20.1

Give the electron configuration of the metal in each of the following atoms or ions.
(a) Ni   **(b)** Cr^{3+} **(c)** $FeO_4{}^{2-}$ (ferrate ion)

SOLUTION   **(a)** Nickel ($Z = 28$) has a total of 28 electrons, including the argon core of 18. In neutral atoms of the first transition series, the 4s orbital is usually filled with 2 electrons, and the remaining electrons occupy the 3d orbitals. The electron configuration is therefore [Ar] $3d^8 4s^2$. (It's convenient to list the orbitals in order of principal quantum number because the 4s electrons are lost first in forming ions.)

(b) A neutral Cr atom ($Z = 24$) has a total of 24 electrons; a Cr^{3+} ion has 3 fewer, or a total of 21. In transition-metal ions, all the valence electrons occupy the d orbitals. Thus, the electron configuration is [Ar] $3d^3$.

(c) First, determine the oxidation state of iron in $FeO_4{}^{2-}$. Because each of the four oxygens has an oxidation number of -2 and the overall charge on the oxoanion is -2, the oxidation number of the iron must be $+6$. An iron(VI) atom has a total of 20 electrons, 6 less than a neutral iron atom ($Z = 26$). Because the valence electrons occupy the 3d orbitals, the electron configuration of Fe(VI) is [Ar] $3d^2$.

⌐ **PROBLEM 20.1**   Give the electron configuration of the metal in each of the following atoms or ions.

(a) V   **(b)** Co^{2+} **(c)** MnO_2

20.2 ➤PROPERTIES OF TRANSITION ELEMENTS

Let's look at some of the trends in the properties of the transition elements shown in Table 20.1 and try to understand them in terms of electron configurations.

Metallic Properties

All the transition elements are metals. Like the metals of groups 1A and 2A, the transition metals are malleable, ductile, lustrous, and good conductors of heat and electricity. Silver has the highest electrical conductivity of any element at room temperature, with copper a close second. The transition metals are harder, have higher melting and boiling points, and are more dense than the group 1A and 2A metals, largely because the sharing of d, as well as s, electrons results in stronger metallic bonding.

Crystals of chromium metal

Elemental mercury

From left to right across the first transition series, melting points increase from 1541°C for Sc to a maximum of 1890°C for V and then decrease to 1083°C for Cu (Table 20.1 and Figure 20.2). The second- and third-series elements exhibit a similar maximum in melting point at group 6B (2610°C for Mo; 3410°C for W, the metal with the highest melting point). The melting points increase as the number of unpaired d electrons available for metallic bonding increases and then decrease as the d electrons pair up and become less available for bonding. Note that zinc ($3d^{10}4s^2$), in which all the d and s electrons are paired, has a relatively low melting point (420°C) and that mercury ($4f^{14}5d^{10}6s^2$) is a liquid at room temperature (mp −39°C).

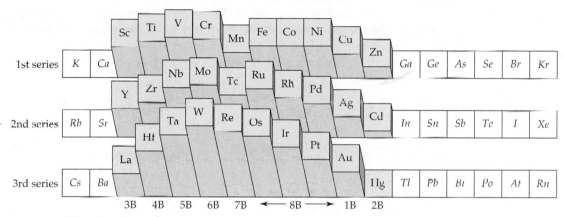

FIGURE 20.2 Relative melting points of the transition elements. Melting points reach a maximum value in the middle of each series.

Atomic Radii and Densities

Atomic radii are plotted in Figure 20.3. From left to right across a transition series, the atomic radii decrease, at first markedly and then more gradually after group 6B. Toward the end of each series, the radii increase again. The decrease in radii with increasing atomic number is expected, since the added d electrons only partially shield the added nuclear charge. As a result, the

FIGURE 20.3 Atomic radii (in pm) of the transition elements. The radii decrease with increasing atomic number and then increase again toward the end of each transition series. Note that the second- and third-series transition elements have nearly identical radii.

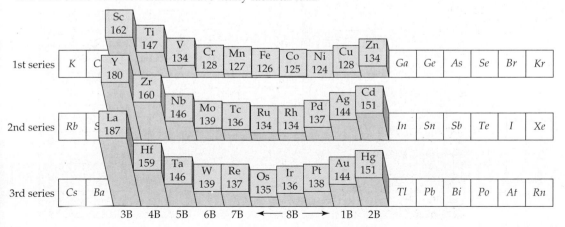

effective nuclear charge Z_{eff} increases. With increasing Z_{eff}, the electrons are more strongly attracted to the nucleus, and atomic size decreases. The upturn in radii toward the end of each series is probably due to more effective shielding and increasing electron–electron repulsion as double occupation of the *d* orbitals is completed. In contrast to the large variation in radii for main-group elements, transition-metal atoms have quite similar radii, which accounts for their ability to blend together in forming alloys such as steels.

It's interesting to note that the atomic radii of the second- and third-series transition elements from group 4B on are nearly identical, though we would expect an increase in size on adding an entire principal quantum shell of electrons. The smaller-than-expected sizes of the third-series atoms are associated with what is called the **lanthanide contraction**, the general decrease in atomic radii of the *f*-block lanthanide elements between the second and third transition series (Figure 20.4).

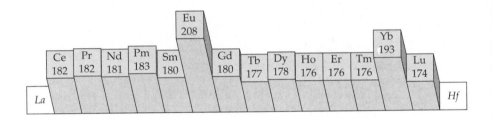

FIGURE 20.4 Atomic radii (in pm) of the lanthanide elements. The radii generally decrease with increasing atomic number.

The lanthanide contraction is due to the increase in effective nuclear charge as the 4*f* subshell is filled. By the end of the lanthanides, the size *decrease* due to a larger Z_{eff} almost exactly compensates for the expected size *increase* due to an added quantum shell of electrons. Consequently, atoms of the third transition series have very similar radii to those of the second transition series.

As you might expect, the densities of the transition metals are inversely related to their atomic radii (Figure 20.5). The densities initially increase from left to right across each transition series and then decrease toward the end of each series. Because the second- and third-series elements have

FIGURE 20.5 Relative densities of the transition metals. There is an initial increase in density across each series and then a decrease.

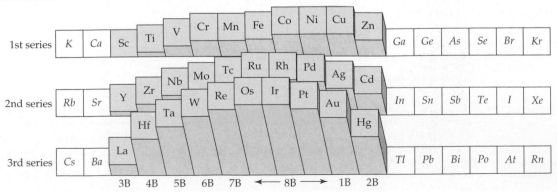

nearly the same atomic volume, the much heavier third-series elements have unusually high densities (22.5 g/cm^3 for osmium and 22.4 g/cm^3 for iridium, the most dense elements).

Ionization Energies and Standard Oxidation Potentials

Ionization energies generally increase from left to right across a transition series, though there are some irregularities, as indicated in Table 20.1 for the atoms of the first transition series. The general trend correlates with an increase in effective nuclear charge and a decrease in atomic radius.

Table 20.2 lists standard *oxidation* potentials $E°$ for the first-series transition metals. Note that the sign of these potentials is the opposite of the sign of standard *reduction* potentials (Table 18.1). Except for copper, all the $E°$ values are positive, which means that the solid metal is oxidized to the aqueous cation more readily than H$_2$ gas is oxidized to H$^+$(aq).

$$M(s) \rightarrow M^{2+}(aq) + 2 e^- \qquad E° > 0 \text{ V} \quad (\text{Product is } M^{3+} \text{ for } M = Sc)$$

$$H_2(g) \rightarrow 2 H^+(aq) + 2 e^- \qquad E° = 0 \text{ V}$$

In other words, the first-series metals, except for copper, are better reducing agents than H$_2$ gas and can therefore be oxidized by "nonoxidizing" acids like HCl that lack an oxidizing anion:

$$M(s) + 2 H^+(aq) \rightarrow M^{2+}(aq) + H_2(g) \qquad E° > 0 \text{ V} \quad (\text{except for } M = Cu)$$

Oxidation of copper requires a stronger oxidizing agent, such as HNO$_3$.

TABLE 20.2　　Standard Oxidation Potentials for First-Series Transition Elements

Oxidation Half-Reaction	$E°$ (V)
Sc(s) → Sc^{3+}(aq) + 3 e$^-$	2.08
Ti(s) → Ti^{2+}(aq) + 2 e$^-$	1.63
V(s) → V^{2+}(aq) + 2 e$^-$	1.18
Cr(s) → Cr^{2+}(aq) + 2 e$^-$	0.91
Mn(s) → Mn^{2+}(aq) + 2 e$^-$	1.18
Fe(s) → Fe^{2+}(aq) + 2 e$^-$	0.45
Co(s) → Co^{2+}(aq) + 2 e$^-$	0.28
Ni(s) → Ni^{2+}(aq) + 2 e$^-$	0.26
Cu(s) → Cu^{2+}(aq) + 2 e$^-$	−0.34
Zn(s) → Zn^{2+}(aq) + 2 e$^-$	0.76

A standard oxidation potential is a composite property that depends on $\Delta G°$ for sublimation of the metal, the ionization energies of the metal atom, and $\Delta G°$ for hydration of the metal ion:

$$M(s) \xrightarrow{\Delta G°_{subl}} M(g) \xrightarrow{E_i(-2 e^-)} M^{2+}(g) \xrightarrow{\Delta G°_{hydr}} M^{2+}(aq)$$

Nevertheless, the general trend in the $E°$ values shown in Table 20.2 correlates with the general trend in the ionization energies in Table 20.1. The ease of oxidation of the metal decreases as the ionization energies increase across the transition series from Sc to Zn. (Only Mn and Zn deviate from the trend of decreasing $E°$ values.) Thus, the early transition metals are oxidized most easily and are therefore the best reducing agents.

20.3 ▶OXIDATION STATES OF TRANSITION ELEMENTS

The transition elements differ from the main-group metals in that they exhibit a variety of oxidation states. Sodium, magnesium, and aluminum, for example, have a single oxidation state equal to their periodic group number (Na^+, Mg^{2+}, and Al^{3+}), but the transition elements frequently have oxidation states less than their group number. For example, manganese in group 7B shows oxidation states of $+2$ in $Mn^{2+}(aq)$, $+3$ in $Mn(OH)_3(s)$, $+4$ in $MnO_2(s)$, $+6$ in $MnO_4^{2-}(aq)$ (manganate ion), and $+7$ in $MnO_4^-(aq)$ (permanganate ion). Figure 20.6 summarizes the common oxidation states for elements of the first transition series, with the most frequently encountered ones indicated in red.

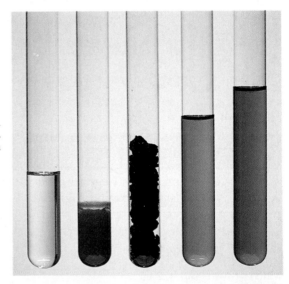

Manganese has different colors in $Mn^{2+}(aq)$, $Mn(OH)_3(s)$, $MnO_2(s)$, $MnO_4^{2-}(aq)$, and $MnO_4^-(aq)$ (from left to right).

FIGURE 20.6 Common oxidation states for first-series transition elements. The states encountered most frequently are in red. The highest oxidation state for the group 3B–7B metals is their periodic group number, but the later transition metals have a maximum oxidation state less than their group number. Most transition elements have more than one common oxidation state.

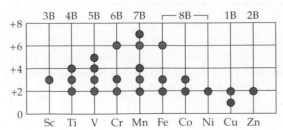

All the first-series transition elements except scandium form a +2 cation, corresponding to loss of the two $4s$ valence electrons. Because the $3d$ and $4s$ orbitals have similar energies, loss of a $3d$ electron is also possible, yielding +3 cations such as $V^{3+}(aq)$, $Cr^{3+}(aq)$, and $Fe^{3+}(aq)$. The energy required to remove the third electron is provided by the larger (more negative) $\Delta G°$ of hydration of the more highly charged +3 cation. Still higher oxidation states result from loss or sharing of additional d electrons. In their highest oxidation states, the transition elements are combined with the most electronegative elements (F, O, and sometimes Cl): for example, $VF_5(l)$ and $V_2O_5(s)$ for vanadium in group 5B; $CrO_2Cl_2(l)$, $CrO_4^{2-}(aq)$ (chromate ion), and $Cr_2O_7^{2-}(aq)$ (dichromate ion) for chromium in group 6B; $MnO_4^-(aq)$ for manganese in group 7B.

Note in Figure 20.6 that the highest oxidation state for the group 3B–7B metals is the group number, corresponding to loss of all the valence s and d electrons. For the later transition metals, though, loss of all the valence electrons is energetically prohibitive because of the increasing value of Z_{eff}. Consequently, only lower oxidation states are accessible for later transition metals—for example, +6 in $FeO_4^{2-}(aq)$ and +3 in $Co^{3+}(aq)$. Even these species have a great tendency to be reduced to still lower oxidation states. For example, the aqueous Co^{3+} ion oxidizes water to O_2 gas and is reduced to Co^{2+}:

$$4\ Co^{3+}(aq) + 2\ H_2O(l) \rightarrow 4\ Co^{2+}(aq) + O_2(g) + 4\ H^+(aq)$$

In general, ions that have the transition metal in a high oxidation state tend to be good oxidizing agents—for example, $Cr_2O_7^{2-}$, MnO_4^-, and FeO_4^{2-}. Conversely, early transition metal ions with the metal in a low oxidation state are good reducing agents—for example, V^{2+} and Cr^{2+}. Divalent ions of the later metals, such as Co^{2+}, Ni^{2+}, Cu^{2+}, and Zn^{2+}, are poor reducing agents because of the larger value of Z_{eff}. In fact, zinc has only one oxidation state (+2).

The elements of the second and third transition series also exhibit a variety of oxidation states. In general, the stability of the higher oxidation states increases down a periodic group. In group 8B, for example, the highest-oxidation-state oxide of iron is iron(III) oxide, Fe_2O_3. Ruthenium and osmium, though, form volatile tetroxides, RuO_4 and OsO_4, in which the metals have an oxidation state of +8.

20.4 ▶CHEMISTRY OF SELECTED TRANSITION ELEMENTS

Experimental work in transition-metal chemistry is particularly enjoyable because most transition-metal compounds have brilliant colors. In this section, we'll look at the chemistry of some representative elements commonly encountered in the laboratory.

Chromium

Chromium, which gets its name from the Greek word for color (*chroma*), is obtained from the ore chromite, a mixed metal oxide with the formula

This sculpture, *Herakles in Ithaka* by Jason Seley, was made from automobile bumpers plated with chromium.

$FeO \cdot Cr_2O_3$, or $FeCr_2O_4$. Reduction of chromite with carbon gives the alloy ferrochrome, used in making stainless steel, a hard, corrosion-resistant steel that contains up to 30% chromium.

$$FeCr_2O_4(s) + 4\ C(s) \rightarrow \underbrace{Fe(s) + 2\ Cr(s)}_{\text{Ferrochrome}} + 4\ CO(g)$$

Pure chromium is obtained by reducing chromium(III) oxide with aluminum:

$$Cr_2O_3(s) + 2\ Al(s) \rightarrow 2\ Cr(s) + Al_2O_3(s)$$

In addition to its use in making steels, chromium is widely used to electroplate metallic objects with an attractive, protective coating (Section 18.12). Chromium is hard and lustrous, takes a high polish, and is resistant to corrosion because an invisible, microscopic film of chromium(III) oxide shields the surface from further oxidation.

The aqueous solution chemistry of chromium can be systematized according to its oxidation states and the species that exist under acidic and basic conditions (Table 20.3). The common oxidation states are +2, +3, and +6, with the +3 state the most stable.

TABLE 20.3	Chromium Species in Common Oxidation States		
	Oxidation State		
	+2	**+3**	**+6**
Acidic solution:	$Cr^{2+}(aq)$ chromium(II) ion (chromous ion) blue	$\xrightarrow{+0.41\text{ V}}$ $Cr^{3+}(aq)$ chromium(III) ion (chromic ion) violet	$\xleftarrow{+1.33\text{ V}}$ $Cr_2O_7^{2-}(aq)$ dichromate ion orange
Basic solution:	$Cr(OH)_2(s)$ light blue	$Cr(OH)_3(s)$ pale green	$\xleftarrow{-0.13\text{ V}}$ $CrO_4^{2-}(aq)$ chromate ion yellow
		$Cr(OH)_4^{-}(aq)$ chromite ion deep green	

Chromium metal reacts with aqueous acids in the absence of oxygen to give $Cr^{2+}(aq)$, the beautiful blue chromium(II) (chromous) ion, in which Cr^{2+} is bound to six water molecules, $Cr(H_2O)_6^{2+}$ [Figure 20.7(a)].

$$Cr(s) + 2\ H^+(aq) \rightarrow Cr^{2+}(aq) + H_2(g)$$

In air, the reaction of chromium metal with acids yields chromium(III) because chromium(II) is rapidly oxidized by atmospheric oxygen:

$$4\ Cr^{2+}(aq) + O_2(g) + 4\ H^+(aq) \rightarrow 4\ Cr^{3+}(aq) + 2\ H_2O(l)$$

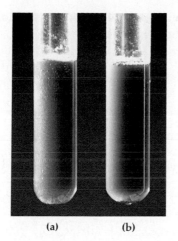

(a) **(b)**

FIGURE 20.7 **(a)** Reaction of chromium metal with 6 M H_2SO_4 in the absence of air gives bubbles of H_2 gas and a solution containing the blue chromium(II) ion. **(b)** Reaction of chromium metal with 6 M HCl in the presence of air yields a green solution containing chromium(III) species.

Although the $Cr(H_2O)_6^{3+}$ ion is violet, chromium(III) solutions are often green because anions replace some of the bound water molecules to give green complex ions such as $Cr(H_2O)_5Cl^{2+}$ and $Cr(H_2O)_4Cl_2^{+}$ [Figure 20.7(b)].

In basic solution, chromium(III) precipitates as chromium(III) hydroxide, a pale green solid that dissolves both in acid and in excess base (Figure 20.8):

In acid: $Cr(OH)_3(s) + 3\ H_3O^+(aq) \rightleftarrows Cr^{3+}(aq) + 6\ H_2O(l)$

In excess base: $Cr(OH)_3(s) + OH^-(aq) \rightleftarrows Cr(OH)_4^-(aq)$

Recall that this behavior is typical of amphoteric metal hydroxides (Section 16.12).

(a) **(b)** **(c)**

FIGURE 20.8 **(a)** Slow addition of aqueous NaOH to a solution of the violet $Cr(H_2O)_6^{3+}$ ion gives a pale green precipitate of chromium(III) hydroxide, $Cr(OH)_3$. **(b)** The $Cr(OH)_3$ dissolves on addition of H_2SO_4, reforming the $Cr(H_2O)_6^{3+}$ ion. **(c)** $Cr(OH)_3$ also dissolves on addition of excess NaOH, yielding the deep green chromite ion, $Cr(OH)_4^-$.

In contrast to the amphoteric $Cr(OH)_3$, chromium(II) hydroxide is a typical basic hydroxide. It dissolves in acid, but not in excess base. Conversely, the chromium(VI) compound, $CrO_2(OH)_2$, is a strong acid (chromic acid, H_2CrO_4). Recall that acid strength increases with increasing polarity of the O–H bonds (Section 15.15), which increases, in turn, with increasing oxidation state of the chromium atom.

$$Cr^{II}(OH)_2 \qquad\qquad Cr^{III}(OH)_3 \qquad\qquad Cr^{VI}O_2(OH)_2$$

basic   amphoteric   acidic

——————— *Increasing strength as a proton donor* ———→

In the +6 oxidation state, the most important solution species are the yellow chromate ion (CrO_4^{2-}) and the orange dichromate ion ($Cr_2O_7^{2-}$). These ions are interconverted by the rapid equilibrium reaction

$$2\,CrO_4^{2-}(aq) + 2\,H^+(aq) \rightleftarrows Cr_2O_7^{2-}(aq) + H_2O(l) \qquad K = 10^{14}$$

Because the equilibrium constant is 10^{14}, CrO_4^{2-} ions predominate in basic solutions, and $Cr_2O_7^{2-}$ ions predominate in acidic solutions (Figure 20.9).

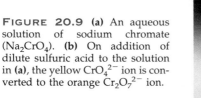

FIGURE 20.9 **(a)** An aqueous solution of sodium chromate (Na_2CrO_4). **(b)** On addition of dilute sulfuric acid to the solution in **(a)**, the yellow CrO_4^{2-} ion is converted to the orange $Cr_2O_7^{2-}$ ion.

(a)   (b)

The $Cr_2O_7^{2-}$ ion is a powerful oxidizing agent in acidic solution and is widely used as an oxidant in analytical chemistry.

$$Cr_2O_7^{2-}(aq) + 14\,H^+(aq) + 6\,e^- \rightarrow 2\,Cr^{3+}(aq) + 7\,H_2O(l) \qquad E° = +1.33\ V$$

In basic solution, where CrO_4^{2-} is the predominant species, chromium(VI) is a much weaker oxidizing agent:

$$CrO_4^{2-}(aq) + 4\,H_2O(l) + 3\,e^- \rightarrow Cr(OH)_3(s) + 5\,OH^-(aq) \qquad E° = -0.13\ V$$

Iron

Iron, the fourth most abundant element in the earth's crust (5.6% by mass), is immensely important both in human civilization and in living systems. Because iron is relatively soft and easily corroded, it is combined with carbon and other metals, such as vanadium, chromium, and manganese, to make alloys (steels) that are harder and less reactive than pure iron. Iron is obtained from its most important ores, hematite (Fe_2O_3) and magnetite (Fe_3O_4), by reduction with coke in a blast furnace.

The most important oxidation states of iron are +2 (ferrous) and +3 (ferric). When iron metal reacts in the absence of air with an acid such as HCl, which lacks an oxidizing anion, the product is the light green iron(II) ion, $Fe(H_2O)_6^{2+}$.

$$Fe(s) + 2\,H^+(aq) \rightarrow Fe^{2+}(aq) + H_2(g)$$

The oxidation stops at the iron(II) stage because the standard potential for oxidation of iron(II) ion is negative:

$$Fe^{2+}(aq) \rightarrow Fe^{3+}(aq) + e^- \qquad E° = -0.77\ \text{V}$$

In air, the iron(II) ion is slowly oxidized by atmospheric oxygen to the iron(III) ion, $Fe(H_2O)_6^{3+}$:

$$4\,Fe^{2+}(aq) + O_2(g) + 4\,H^+(aq) \rightarrow 4\,Fe^{3+}(aq) + 2\,H_2O(l) \qquad E° = +0.46\ \text{V}$$

When iron is treated with an acid that has an oxidizing anion—for example, dilute nitric acid—the metal is oxidized directly to iron(III):

$$Fe(s) + NO_3^-(aq) + 4\,H^+(aq) \rightarrow Fe^{3+}(aq) + NO(g) + 2\,H_2O(l)$$

Addition of base to iron(III) solutions precipitates the gelatinous, red-brown hydrous oxide, $Fe_2O_3 \cdot x\,H_2O$, usually written as $Fe(OH)_3$ (Figure 20.10):

$$Fe^{3+}(aq) + 3\,OH^-(aq) \rightarrow Fe(OH)_3(s)$$

Because $Fe(OH)_3$ is very insoluble ($K_{sp} = 2.6 \times 10^{-39}$), it forms as soon as the pH rises above pH 2. The red-brown rust stains that you've seen in sinks and bathtubs are due to air oxidation of $Fe^{2+}(aq)$ followed by deposition of hydrous iron(III) oxide. Unlike $Cr(OH)_3$, $Fe(OH)_3$ is not appreciably amphoteric. It dissolves in acid, but not in excess base.

FIGURE 20.10 When dilute NaOH is added to a solution of iron(III) sulfate, $Fe_2(SO_4)_3$, a gelatinous, red-brown precipitate of $Fe(OH)_3$ forms.

Copper

Copper, a reddish colored metal, is a relatively rare element, accounting for only $6.8 \times 10^{-3}\%$ of the earth's crust by mass. Like the other group 1B elements silver and gold, copper is found in the elemental state. Its most important ores are sulfides, such as chalcopyrite, $CuFeS_2$. In a multistep process, copper sulfides are concentrated, separated from iron, and converted to molten copper(I) sulfide, which is then reduced to elemental copper by blowing air through the hot liquid:

$$Cu_2S(l) + O_2(g) \rightarrow 2\ Cu(l) + SO_2(g)$$

The product, about 99% copper, is purified by electrolysis (Section 18.12).

Because of its high electrical conductivity and negative oxidation potential, copper is widely used to make electrical wiring and corrosion-resistant water pipes. Copper is also used in coins and is combined with other metals to make alloys such as brass (mostly copper and zinc) and bronze (mostly copper and tin). Though less reactive than other first-series transition metals, copper is oxidized on prolonged exposure to moist air and carbon dioxide, forming basic copper(II) carbonate, $Cu_2(OH)_2CO_3$. Subsequent reaction with dilute H_2SO_4 in acid rain then forms $Cu_2(OH)_2SO_4$, the green patina seen on bronze monuments.

$$2\ Cu(s) + O_2(g) + CO_2(g) + H_2O(g) \rightarrow Cu_2(OH)_2CO_3(s)$$

Copper is used to make electrical wiring, water pipes, and coins.

The green surface coating on this bronze sculpture is a basic copper(II) sulfate.

In its compounds, copper exists in two common oxidation states, +1 (cuprous) and +2 (cupric). Because the oxidation potential for the Cu^+/Cu^{2+} half-reaction is less negative than that for the Cu/Cu^+ half-reaction, any oxidizing agent strong enough to oxidize copper to the copper(I) ion is also able to oxidize the copper(I) ion to the copper(II) ion.

$$Cu(s) \rightarrow Cu^+(aq) + e^- \qquad E^\circ = -0.52\ V$$

$$Cu^+(aq) \rightarrow Cu^{2+}(aq) + e^- \qquad E^\circ = -0.15\ V$$

Dilute nitric acid, for example, oxidizes copper to the +2 oxidation level:

$$3\ Cu(s) + 2\ NO_3^-(aq) + 8\ H^+(aq) \rightarrow 3\ Cu^{2+}(aq) + 2\ NO(g) + 4\ H_2O(l)$$

It follows from the $E°$ values that $Cu^+(aq)$ can undergo a dispropor-tionation reaction, oxidizing and reducing itself:

$$Cu^+(aq) + e^- \rightarrow Cu(s) \qquad E° = +0.52 \text{ V}$$

$$\underline{Cu^+(aq) \rightarrow Cu^{2+}(aq) + e^- \qquad E° = -0.15 \text{ V}}$$

$$2\,Cu^+(aq) \rightarrow Cu(s) + Cu^{2+}(aq) \qquad E° = +0.37 \text{ V}$$

The positive $E°$ value for the disproportionation corresponds to a large equilibrium constant, indicating that the reaction proceeds far toward completion:

$$2\,Cu^+(aq) \rightleftarrows Cu(s) + Cu^{2+}(aq) \qquad K = 1.8 \times 10^6$$

Thus, the copper(I) ion is not an important species in aqueous solution, though copper(I) does exist in solid compounds, such as $CuCl$. In the presence of Cl^- ions, the disproportionation equilibrium is reversed because precipitation of the insoluble, white copper(I) chloride drives the reaction to completion:

$$Cu(s) + Cu^{2+}(aq) + 2\,Cl^-(aq) \rightarrow 2\,CuCl(s)$$

The more common +2 oxidation state is found in the blue aqueous ion, $Cu(H_2O)_6^{2+}$, and in numerous solid compounds and complex ions. Addition of base (aqueous ammonia) to a solution of a copper(II) salt gives a blue precipitate of copper(II) hydroxide, which dissolves in excess aqueous ammonia, yielding the dark blue complex ion $Cu(NH_3)_4^{2+}$ (Figure 20.11):

$$Cu^{2+}(aq) + 2\,OH^-(aq) \rightarrow Cu(OH)_2(s)$$

$$Cu(OH)_2(s) + 4\,NH_3(aq) \rightarrow Cu(NH_3)_4^{2+}(aq) + 2\,OH^-(aq)$$

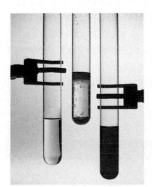

FIGURE 20.11 When an aqueous solution of $CuSO_4$ (left) is treated with aqueous ammonia (center), a blue precipitate of $Cu(OH)_2$ forms. On addition of excess ammonia (right), the precipitate dissolves, yielding the deep blue $Cu(NH_3)_4$ ion.

Perhaps the most common of all copper compounds is the blue-colored copper(II) sulfate pentahydrate, $CuSO_4 \cdot 5\,H_2O$. It has a structure in which four water molecules are bound to the copper(II) ion, and the fifth is hydrogen bonded to the sulfate ion. When heated, $CuSO_4 \cdot 5\,H_2O$ loses its water and its color (Figure 14.12), suggesting that the blue color of the pentahydrate is due to bonding of Cu^{2+} to the water molecules.

20.5 ►COORDINATION COMPOUNDS

A **coordination compound** (also called a **metal complex**) is a compound in which a central metal ion is attached to a group of surrounding molecules or ions by coordinate covalent bonds. A good example is the anticancer drug cisplatin, $Pt(NH_3)_2Cl_2$, in which two NH_3 molecules and two Cl^- ions use lone pairs of electrons to bond to the platinum(II) ion:

$$\ddot{:}\overset{\displaystyle :\ddot{Cl}:}{\underset{\displaystyle NH_3}{:\ddot{Cl}:Pt:NH_3}} \qquad \text{Cisplatin}$$

The molecules or ions that surround the central metal ion in a complex are called **ligands**, and the atoms that are attached directly to the metal ion are called **ligand donor atoms**. In cisplatin, for example, the ligands are NH_3 and Cl^-, and the ligand donor atoms are N and Cl. Note that complex formation is a Lewis acid–base interaction (Section 15.16) in which the ligands act as Lewis bases (electron-pair donors) and the central metal ion behaves as a Lewis acid (an electron-pair acceptor).

Not all coordination compounds are neutral molecules like $Pt(NH_3)_2Cl_2$. Many are salts, such as $[Ni(NH_3)_6]Cl_2$ and $K_3[Fe(CN)_6]$, which contain a complex cation or anion along with enough ions of opposite charge to give a compound that is electrically neutral overall. To indicate that the complex ion is a discrete structural unit, it is always enclosed in brackets in the formula of the salt. Thus, $[Ni(NH_3)_6]Cl_2$ contains $[Ni(NH_3)_6]^{2+}$ cations and Cl^- anions. The term *metal complex* (or simply complex) refers both to neutral molecules, such as $Pt(NH_3)_2Cl_2$, and to complex ions, such as $[Ni(NH_3)_6]^{2+}$ and $[Fe(CN)_6]^{3-}$.

The number of ligand donor atoms that surround a central metal ion in a complex is called the **coordination number** of the metal. Thus, platinum(II) has a coordination number of 4 in $Pt(NH_3)_2Cl_2$, and iron(III) has a coordination number of 6 in $[Fe(CN)_6]^{3-}$. The most common coordination numbers are 4 and 6, but others are well known (Table 20.4). The coordination number of a metal ion in a particular complex depends on the metal ion's size, charge, and electron configuration, and on the size and shape of the ligands.

Metal complexes have characteristic shapes, depending on the metal ion's coordination number. Two-coordinate complexes, such as

TABLE 20.4	Examples of Complexes with Various Coordination Numbers
Coordination Number	**Complex**
2	$[Ag(NH_3)_2]^+$, $[CuCl_2]^-$
3	$[HgI_3]^-$
4	$[Zn(NH_3)_4]^{2+}$, $[Ni(CN)_4]^{2-}$
5	$[Ni(CN)_5]^{3-}$, $Fe(CO)_5$
6	$[Cr(H_2O)_6]^{3+}$, $[Fe(CN)_6]^{3-}$
7	$[ZrF_7]^{3-}$
8	$[Mo(CN)_8]^{4-}$

$[Ag(NH_3)_2]^+$, are linear. Four-coordinate complexes are either tetrahedral or square planar; for example, $[Zn(NH_3)_4]^{2+}$ is tetrahedral, and $[Ni(CN)_4]^{2-}$ is square planar. Nearly all six-coordinate complexes are octahedral. The more common coordination geometries are illustrated in Figure 20.12. Coordination geometries were first deduced by the Swiss chemist Alfred Werner, who was awarded the 1913 Nobel Prize in chemistry for his pioneering studies of coordination compounds.

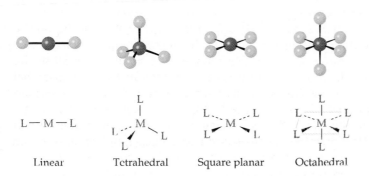

$$L-M-L$$

Linear Tetrahedral Square planar Octahedral

FIGURE 20.12 The arrangement of ligand donor atoms (L) in ML_n complexes with coordination number 2, 4, and 6. In the octahedral arrangement, four ligands are at the corners of a square, with one more above and one below the plane of the square.

The charge on a metal complex is equal to the charge on the metal ion (its oxidation state) plus the sum of the charges on the ligands. Thus, if we know the charge on each ligand and the charge on the complex, we can easily find the oxidation state of the metal.

EXAMPLE 20.2

A cobalt(III) ion forms a complex with four ammonia molecules and two chloride ions. What is the formula of the complex?

SOLUTION   The charge on the complex is the sum of the charges on the Co^{3+} ion, the four NH_3 ligands, and the two Cl^- ligands:

$$Co^{3+} \quad 4\ NH_3 \quad 2\ Cl^-$$
$$+3 \ + (4 \times 0) + [2 \times (-1)] = +1$$

Therefore, the formula of the complex is $[Co(NH_3)_4Cl_2]^+$.

EXAMPLE 20.3

What is the oxidation state of platinum in the coordination compound $K[Pt(NH_3)Cl_5]$?

SOLUTION   Because the compound is electrically neutral overall and contains one K^+ cation per complex anion, the anion must be $[Pt(NH_3)Cl_5]^-$. Since ammonia is neutral, and chloride has a charge of -1, the oxidation state of platinum must be $+4$:

$$K^+ \quad Pt^{n+} \quad NH_3 \quad 5\ Cl^-$$
$$+1\ +\ n\ +\ 0\ +\ [5 \times (-1)] = 0;\ \ n = +4$$

⌐ **PROBLEM 20.2** What is the formula of the chromium(III) complex that contains two ammonia and four thiocyanate (SCN^-) ligands?

⌐ **PROBLEM 20.3** What is the oxidation state of iron in $Na_4[Fe(CN)_6]$? ◢

20.6 ►LIGANDS

Structural formulas for some typical ligands are shown in Figure 20.13. Because all ligands are Lewis bases, they have at least one unshared pair of electrons, which can be used to form a coordinate covalent bond to a metal ion. They can be classified as *monodentate* or *polydentate*, depending on the number of ligand donor atoms that bond to the metal. Ligands such as H_2O, NH_3, or Cl^-, which bond using the electron pair of a single donor atom, are called *monodentate* ligands (literally, "one-toothed" ligands). Those that bond through electron pairs on more than one donor atom are termed **polydentate** ligands ("many-toothed" ligands). For example, ethylenediamine ($NH_2CH_2CH_2NH_2$, abbreviated en) is a **bidentate** ligand because it bonds to a metal using an electron pair on each of its two nitrogen atoms. The *hexadentate* ligand ethylenediaminetetraacetate ion ($EDTA^{4-}$) bonds to a metal ion through electron pairs on six donor atoms (two N atoms and four O atoms).

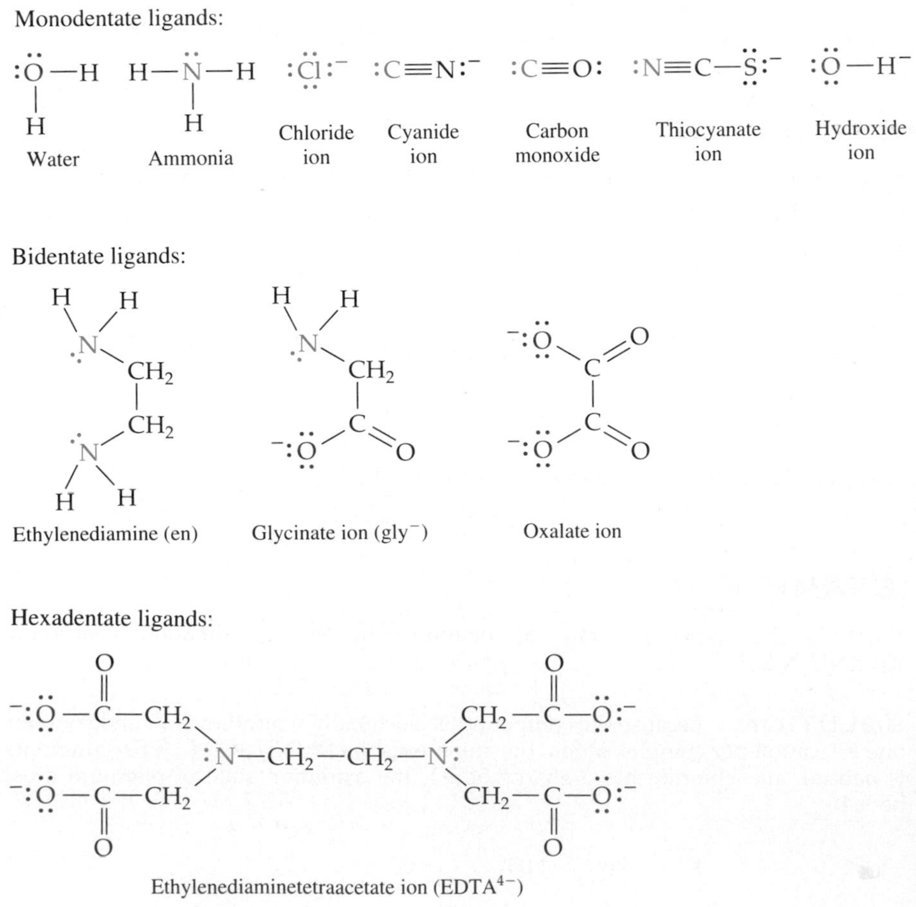

FIGURE 20.13 Structural formulas for some common ligands. Ligand donor atoms are in color.

Monodentate ligands:

Water   Ammonia   Chloride ion   Cyanide ion   Carbon monoxide   Thiocyanate ion   Hydroxide ion

Bidentate ligands:

Ethylenediamine (en)   Glycinate ion (gly⁻)   Oxalate ion

Hexadentate ligands:

Ethylenediaminetetraacetate ion ($EDTA^{4-}$)

Polydentate ligands are known as **chelating agents** (pronounced **key**-late-ing, from the Greek word *chele,* meaning "claw") because their multipoint attachment to a metal ion resembles the grasping of an object by the claws of a crab. For example, ethylenediamine holds a cobalt(III) ion with two claws, its two nitrogen donor atoms (Figure 20.14). The resulting five-membered ring consisting of the Co ion, two N atoms, and two C atoms of the ligand is called a **chelate ring**. A complex such as $[Co(en)_3]^{3+}$ or $[Co(EDTA)]^-$ that contains one or more chelate rings is known as a metal **chelate**. Because $EDTA^{4-}$ bonds to a metal ion through six donor atoms, it forms especially stable complexes and is often used to hold metal ions in solution. For example, in the treatment of lead poisoning, $EDTA^{4-}$ bonds to Pb^{2+}, which is then excreted by the kidneys as the soluble chelate $[Pb(EDTA)]^{2-}$.

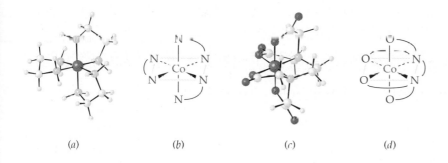

(a) (b) (c) (d)

FIGURE 20.14 (a) Computer-drawn model and **(b)** shorthand representation of the $[Co(en)_3]^{3+}$ ion. The complex contains three cobalt–ethylenediamine chelate rings. In **(b)**, N⌢N represents a bidentate $NH_2CH_2CH_2NH_2$ ligand, which spans adjacent corners of the octahedron. **(c)** Computer-drawn model and **(d)** shorthand representation of the $[Co(EDTA)]^-$ ion. The hexadentate EDTA ligand uses its two N atoms and four O atoms to bond to the metal, thus forming five chelate rings.

20.7 ➤ NAMING COORDINATION COMPOUNDS

In the early days of coordination chemistry, coordination compounds were named after their discoverer or according to their color. Nowadays, we use systematic names that specify the number of ligands of each particular type, the metal, and its oxidation state. Before listing the rules used to name coordination compounds, let's consider a few examples that will illustrate how to apply the rules:

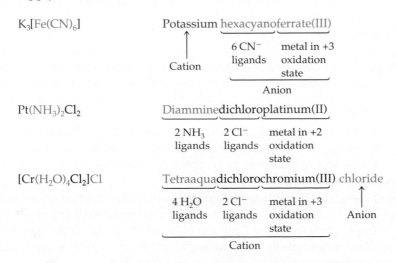

The following list summarizes the nomenclature rules recommended by the International Union of Pure and Applied Chemistry:

1. If the compound is a salt, name the cation first and then the anion, just as in naming simple salts (Section 2.10). $K_3[Fe(CN)_6]$, for example, is potassium hexacyanoferrate(III).

2. In naming a complex ion or a neutral complex, name the ligands first and then the metal. The names of anionic ligands end in *-o*. As shown in Table 20.5, they are usually obtained by changing the anion endings *-ide* to *-o* and *-ate* to *-ato*. Neutral ligands are specified by their usual names, except for H_2O, NH_3, and CO, which are called aqua, ammine, and carbonyl, respectively. The name of a complex is one word, with no space between the various ligand names and no space between the names of the last ligand and the metal.

TABLE 20.5   Names of Some Common Ligands

Anionic Ligand	Ligand Name	Neutral Ligand	Ligand Name
Bromide, Br^-	Bromo	Ammonia, NH_3	Ammine
Carbonate, $CO_3{}^{2-}$	Carbonato	Water, H_2O	Aqua
Chloride, Cl^-	Chloro	Carbon monoxide, CO	Carbonyl
Cyanide, CN^-	Cyano	Ethylenediamine, en	Ethylenediamine
Fluoride, F^-	Fluoro		
Glycinate, gly^-	Glycinato		
Hydroxide, OH^-	Hydroxo		
Oxalate, $C_2O_4{}^{2-}$	Oxalato		
Thiocyanate, SCN^-	Thiocyanato		

3. If the complex contains more than one ligand of a particular type, indicate the number with the appropriate Greek prefix: *di-, tri-, tetra-, penta-, hexa-,* and so forth. The ligands are listed in alphabetical order, and the prefixes are ignored in determining the order. Thus, tetra*a*qua precedes di*c*hloro in the name for $[Cr(H_2O)_4Cl_2]Cl$: tetraaquadichlorochromium(III) chloride.

4. If the name of a ligand itself contains a Greek prefix—for example, ethylenediamine—put the ligand name in parentheses and use one of the following alternate prefixes to specify the number of ligands: *bis-* (2), *tris-* (3), *tetrakis-* (4), and so forth. Thus, the name of $[Co(en)_3]Cl_3$ is tris(ethylenediamine)cobalt(III) chloride.

5. Use a Roman numeral in parentheses, immediately following the name of the metal, to indicate the metal's oxidation state. As shown by the preceding examples, there is no space between the name of the metal and the parenthesis.

6. In naming the metal, use the ending *-ate* if the metal is in an anionic complex. Thus, $[Fe(CN)_6]^{3-}$ is the hexacyanoferrate(III) anion. There are no simple rules for going from the name of the metal to the name of the metallate anion, partly because some of the anions have Latin names. Some common examples are given in Table 20.6.

TABLE 20.6	Names of Some Common Metallate Anions		
Metal	**Anion Name**	**Metal**	**Anion Name**
Aluminum	Aluminate	Iron	Ferrate
Chromium	Chromate	Manganese	Manganate
Cobalt	Cobaltate	Nickel	Nickelate
Copper	Cuprate	Platinum	Platinate
Gold	Aurate	Zinc	Zincate

The rules for naming coordination compounds make it possible to go from a formula to the systematic name or from a systematic name to the appropriate formula. The following problems give some practice.

EXAMPLE 20.4

Name each of the following.
(a) $[Co(NH_3)_6]Cl_3$, generally considered to be the first coordination compound, prepared by B. M. Tassaert in 1798
(b) $[Rh(NH_3)_5I]I_2$, a yellow compound obtained by heating $[Rh(NH_3)_5(H_2O)]I_3$ at 100°C
(c) $Fe(CO)_5$, a highly toxic, volatile liquid
(d) $[Fe(C_2O_4)_3]^{3-}$, the ion formed when Fe_2O_3 rust stains are dissolved in oxalic acid $(H_2C_2O_4)$

SOLUTION   (a) Because the chloride ion has a charge of -1 and ammonia is neutral, the oxidation state of cobalt is $+3$. Use the prefix *hexa-* to indicate that the cation contains six NH_3 ligands, and use a Roman numeral III in parentheses to indicate the oxidation state of cobalt. The name of $[Co(NH_3)_6]Cl_3$ is hexaamminecobalt(III) chloride.
(b) Because the iodide ion has a charge of -1, the complex cation is $[Rh(NH_3)_5I]^{2+}$ and the rhodium has an oxidation state of $+3$. List the ammine ligands before the iodo ligand, and use the prefix *penta* to indicate the presence of five NH_3 ligands. The name of $[Rh(NH_3)_5I]I_2$ is pentaammineiodorhodium(III) iodide.
(c) Because the carbonyl ligand is neutral, the oxidation state of iron is zero. The systematic name of $Fe(CO)_5$ is pentacarbonyliron(0), but the common name iron pentacarbonyl is often used.
(d) Because each oxalate ligand $(C_2O_4{}^{2-})$ has a charge of -2, and because $[Fe(C_2O_4)_3]^{3-}$ has an overall charge of -3, iron must have an oxidation state of $+3$. Use the name ferrate(III) for the metal because the complex is an anion. The name of $[Fe(C_2O_4)_3]^{3-}$ is the trioxalatoferrate(III) ion.

EXAMPLE 20.5

Write the formula for each of the following.
(a) potassium tetracyanonickelate(II)
(b) aquachlorobis(ethylenediamine)cobalt(III) chloride
(c) sodium hexafluoroaluminate
(d) diamminesilver(I) ion

SOLUTION   (a) Yellow potassium tetracyanonickelate(II) is obtained when an excess of KCN is added to an aqueous solution of a nickel(II) salt. Bonding of four CN^- to Ni^{2+} gives an $[Ni(CN)_4]^{2-}$ anion, which must be balanced by two K^+ cations. The formula for the compound is therefore $K_2[Ni(CN)_4]$.
(b) Because the complex cation contains one H_2O, one Cl^-, and two neutral en ligands, and because the metal is Co^{3+}, the cation is $[Co(en)_2(H_2O)Cl]^{2+}$. The $+2$ charge of the cation must be balanced by two Cl^- anions, so the formula of the

compound is [Co(en)$_2$(H$_2$O)Cl]Cl$_2$. The cation is the first product formed when [Co(en)$_2$Cl$_2$]$^+$ reacts with water.

(c) Sodium hexafluoroaluminate, also called cryolite, is used in the electrolytic production of aluminum metal (Section 18.12) and is an example of a coordination compound of a main-group element. The oxidation state of aluminum has been omitted from the name because aluminum has only one oxidation state (+3). Because F$^-$ has a charge of -1, the anion is [AlF$_6$]$^{3-}$. The charge of the anion must be balanced by three Na$^+$ cations; therefore, the formula for the compound is Na$_3$[AlF$_6$].

(d) Diamminesilver(I) ion, formed when silver chloride dissolves in an excess of aqueous ammonia, has the formula [Ag(NH$_3$)$_2$]$^+$.

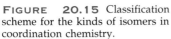 **PROBLEM 20.4** Name each of the following.

(a) [Cu(NH$_3$)$_4$]SO$_4$, a deep blue compound obtained when CuSO$_4$ is treated with an excess of ammonia

(b) Na[Cr(OH)$_4$], the compound formed when Cr(OH)$_3$ is dissolved in an excess of aqueous NaOH

(c) Co(gly)$_3$, a complex that contains the anion of the amino acid glycine

(d) [Fe(H$_2$O)$_5$(NCS)]$^{2+}$, the red complex formed in a qualitative analysis test for iron

PROBLEM 20.5 Write the formula for each of the following.

(a) tetraamminezinc(II) nitrate, the compound formed when zinc nitrate is treated with an excess of ammonia

(b) tetracarbonylnickel(0), the first metal carbonyl, prepared in 1888, and an important compound in the industrial process for refining nickel metal

(c) potassium amminetrichloroplatinate(II), a compound that contains a square planar anion

(d) the dicyanoaurate(I) ion, an ion important in the extraction of gold from its ores

20.8 ►ISOMERS

One of the more interesting aspects of coordination chemistry is the existence of **isomers,** compounds that have the same formula but a different arrangement of their constituent atoms. Because their atoms are arranged differently, isomers are different compounds, with different physical and chemical properties such as color, solubility, melting point, and chemical reactivity. Figure 20.15 shows a scheme for classifying some of the kinds of isomers in coordination chemistry. As we'll see in Chapters 23 and 24, isomers are also important in organic chemistry and biochemistry.

FIGURE 20.15 Classification scheme for the kinds of isomers in coordination chemistry.

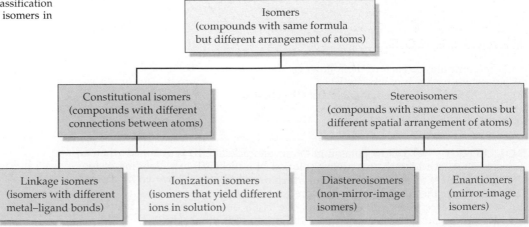

Constitutional Isomers

Isomers that have different connections among their constituent atoms are called **constitutional isomers**. Of the various types of constitutional isomers, we'll mention just two: *linkage isomers* and *ionization isomers*.

Linkage isomers arise when a ligand can bond to a metal through either of two different donor atoms. For example, the nitrite (NO_2^-) ion forms two different pentaamminecobalt(III) complexes, a yellow nitro complex that contains a Co–N bond, $[Co(NH_3)_5(NO_2)]^{2+}$, and a red nitrito complex that contains a Co–O bond, $[Co(NH_3)_5(ONO)]^{2+}$. The ligand in the nitrito complex is written as ONO to emphasize that it's linked to the cobalt through the oxygen atom. The thiocyanate (SCN^-) ion is another ligand that gives linkage isomers because it can bond to metals through either the sulfur atom or the nitrogen atom.

Ionization isomers are isomers that differ in the anion that is bonded to the metal ion. An example is the pair $[Co(NH_3)_5Br]SO_4$, a violet compound that has a Co–Br bond and a free sulfate anion, and $[Co(NH_3)_5SO_4]Br$, a red compound that has a Co-sulfate bond and a free bromide ion. Ionization isomers get their name from the fact that they yield different ions in solution.

Stereoisomers

Isomers that have the same connections among atoms but have a different arrangement of the atoms in space are called **stereoisomers.** In coordination chemistry, there are two kinds of stereoisomers: *diastereoisomers* and *enantiomers.*

Diastereoisomers, also called *geometric isomers*, have different relative orientations of their metal–ligand bonds. For example, in the square planar complex $Pt(NH_3)_2Cl_2$, the two Pt–Cl bonds can be oriented either adjacent (a 90° angle) or opposite (a 180° angle), as shown in Figure 20.16. The isomer in which identical ligands occupy adjacent corners of the square is called the **cis isomer**; that in which identical ligands are across from one another is called the **trans isomer.** (The word *trans* means across; *cis* means next to.) Because cis and trans isomers are different compounds, they have different properties. *cis*-$Pt(NH_3)_2Cl_2$ is a polar molecule and relatively more soluble in water than *trans*-$Pt(NH_3)_2Cl_2$, which is nonpolar because the two Pt–Cl and the two Pt–NH_3 bond dipoles point in opposite directions and therefore cancel. It's also interesting that *cis*-$Pt(NH_3)_2Cl_2$ (cisplatin) is an effective anticancer drug, whereas the trans isomer is physiologically inactive.

FIGURE 20.16 Diastereoisomers of the square planar complex $Pt(NH_3)_2Cl_2$. The two compounds have the same connections among atoms but have a different arrangement of atoms in space.

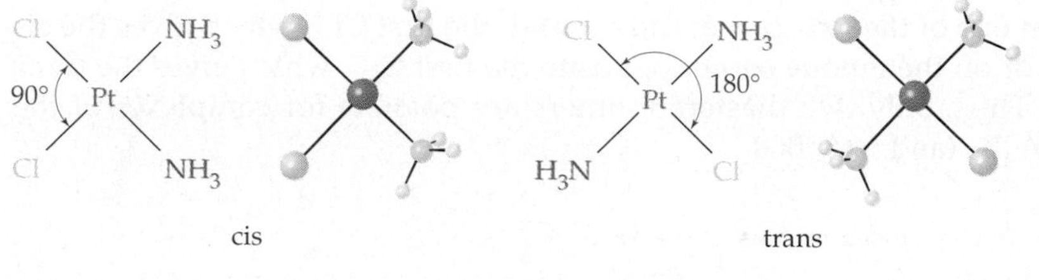

cis   trans

In general, square planar complexes of the type MA_2B_2 and MA_2BC—where M is a metal ion and A, B, and C are ligands—can exist as cis-trans isomers. No cis-trans isomers are possible, however, for four-coordinate *tetrahedral* complexes because all four corners of a tetrahedron are adjacent to one another.

Octahedral complexes of the type MA_4B_2 can also exist as diastereoisomers because the two B ligands can be either on adjacent or on opposite corners of the octahedron. An example is the violet compound *cis*-[Co(NH$_3$)$_4$Cl$_2$]Cl and the green compound *trans*-[Co(NH$_3$)$_4$Cl$_2$]Cl. As Figure 20.17 shows, there are several ways of drawing the cis and trans isomers because each complex can be rotated in space, changing the perspective but not the identity of the isomer.

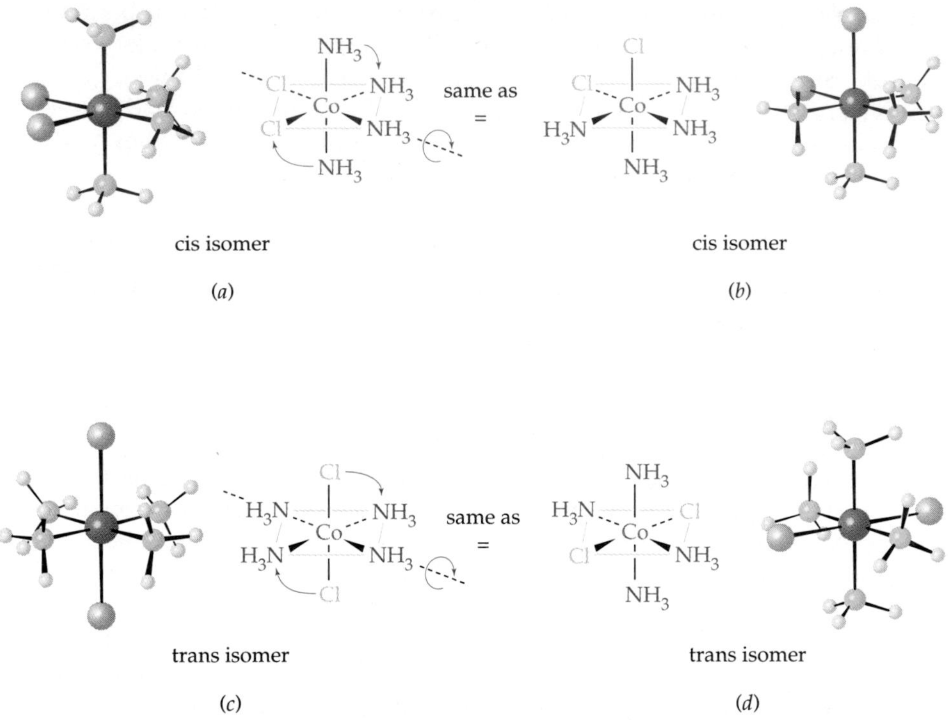

FIGURE 20.17 Diastereoisomers of the [Co(NH$_3$)$_4$Cl$_2$]$^+$ ion. Although they may appear different, cis isomers **(a)** and **(b)** are identical, as can be seen by rotating the entire complex by 90° about a Cl–Co–NH$_3$ axis. Similarly, trans isomers **(c)** and **(d)** are identical.

How can we be sure that there are only two diastereoisomers of the [Co(NH$_3$)$_4$Cl$_2$]$^+$ ion? The first Cl$^-$ ligand can be located at any one of the six corners of the octahedron. Once one Cl$^-$ is present, however, the five corners remaining are no longer equivalent. The second Cl$^-$ can be located either on one of the four corners adjacent to the first Cl$^-$, which gives the cis isomer, or on the unique corner opposite the first Cl$^-$, which gives the trans isomer. Thus, only two diastereoisomers are possible for complexes of the type MA_4B_2 (and MA_4BC).

EXAMPLE 20.6

Platinum(II) forms square planar complexes, and platinum(IV) gives octahedral complexes. How many diastereoisomers are possible for each of the following complexes? Describe their structures.
(a) $[Pt(NH_3)_3Cl]^+$ **(b)** $[Pt((NH_3)Cl_5]^-$ **(c)** $Pt(NH_3)_2Cl(NO_2)$ **(d)** $[Pt(NH_3)_4ClBr]^{2+}$

SOLUTION   **(a)** No isomers are possible for a square planar complex of the type MA_3B.
(b) No isomers are possible for an octahedral complex of the type MAB_5.
(c) Cis and trans isomers are possible for a square planar complex of the type MA_2BC. The Cl^- and NO_2^- ligands can be either on adjacent or on opposite corners of the square.
(d) Cis and trans isomers are possible for an octahedral complex of the type MA_4BC. The Cl^- and Br^- ligands can be either on adjacent or on opposite corners of the octahedron.

EXAMPLE 20.7

Draw the structures of all possible diastereoisomers of $Co(NH_3)_3Cl_3$.

SOLUTION   There are two diastereoisomers possible for an octahedral complex of the type MA_3B_3. One isomer **(a)** has the three Cl^- ligands in adjacent positions on one triangular face of the octahedron; the other isomer **(b)** has all three Cl^- ligands in a plane that contains the metal ion.

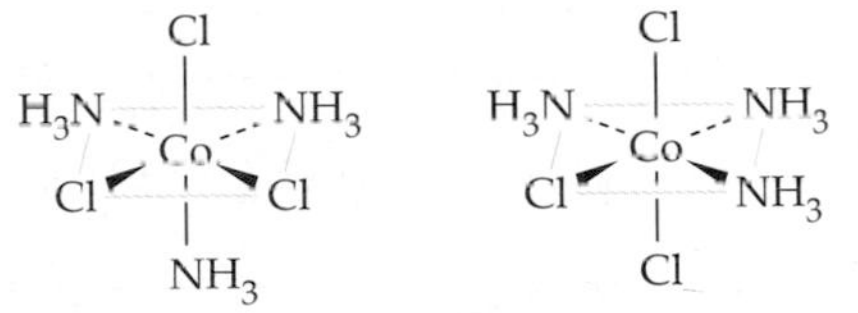

To convince yourself that there are only two diastereoisomers of $Co(NH_3)_3Cl_3$, look at Figure 20.17 and consider the products obtained from *cis*- and *trans*-$[Co(NH_3)_4Cl_2]^+$ when one of the four NH_3 ligands is replaced by a Cl^-. There are two kinds of NH_3 ligands in *cis*-$[Co(NH_3)_4Cl_2]^+$—two NH_3 ligands that are trans to each other and two NH_3 ligands that are trans to a Cl^-. Replacing an NH_3 trans to another NH_3 with Cl^- gives $Co(NH_3)_3Cl_3$ isomer **(a)**, while replacing an NH_3 trans to a Cl^- with another Cl^- gives isomer **(b)**. All four NH_3 ligands in *trans*-$[Co(NH_3)_4Cl_2]^+$ are equivalent, and replacing any one of them with a Cl^- gives isomer **(b)**. Thus, there are only two diastereoisomers for a complex of the type MA_3B_3.

⌐ PROBLEM 20.6 How many diastereoisomers are possible for each of the following complexes? Draw the structure of each diastereoisomer.
(a) $Pt(NH_3)_2(SCN)_2$ **(b)** $[CoCl_2Br_2]^{2-}$ (tetrahedral)   **(c)** $Co(NH_3)_3(NO_2)_3$
(d) $Pt(en)Cl_2$ **(e)** $[Cr(en)_2Br_2]^+$ **(f)** $[Rh(en)_3]^{3+}$

20.9 ►ENANTIOMERS AND MOLECULAR HANDEDNESS

Diastereoisomers are easy to distinguish because the various bonds in the cis and trans isomers point in different directions. *Enantiomers*, however, are stereoisomers that differ from one another in a more subtle way: **Enantiomers** are *molecules or ions that are nonidentical mirror images of one another;* they differ from one another because of their handedness.

Handedness affects almost everything we do. Anyone who has played much softball knows that the last available glove always fits the wrong hand. Any left-handed person sitting next to a right-handed person in a lecture knows that taking notes always means bumping elbows. The reason for these difficulties is that our hands aren't identical. Rather, they're mirror images. When you hold your left hand up to a mirror, the image you see looks like your right hand. (Try it.)

Not all objects are handed, of course. There's no such thing as a "right-handed" tennis ball or a "left-handed" coffee mug. When a tennis ball or a coffee mug is held up to a mirror, the image reflected is identical to the ball or mug itself. Objects that have a handedness to them are said to be **chiral** (pronounced **ky**-ral, from the Greek *cheir*, meaning hand), and objects that lack handedness are said to be nonchiral, or **achiral**.

Why is it that some objects are chiral but others aren't? In general, an object is not chiral if, like the coffee mug, it has a **symmetry plane** cutting through its middle so that one half of the object is a mirror image of the other half. If you were to cut the mug in half, one half of the mug would be the mirror image of the other half. A hand, however, has no symmetry plane and is therefore chiral. If you were to cut a hand in two, one "half" of the hand would not be a mirror image of the other half (Figure 20.18).

Symmetry
plane

 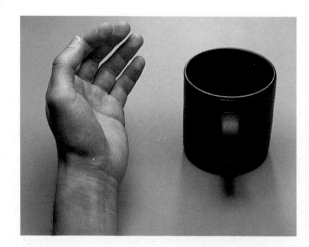

FIGURE 20.18 The meaning of a symmetry plane: An achiral object like the coffee mug has a symmetry plane passing through it, making the two halves mirror images. A chiral object like the hand has no symmetry plane because the two "halves" of the hand are not mirror images.

Just as certain objects like a hand are chiral, certain molecules and ions are also chiral. Consider the tris(ethylenediamine)cobalt(III) ion, $[Co(en)_3]^{3+}$, for example (Figure 20.19). $[Co(en)_3]^{3+}$ has no symmetry plane because its two "halves" aren't mirror-images. Like a hand, $[Co(en)_3]^{3+}$ is chiral and can exist in two nonidentical mirror image forms—a "right-handed" enantiomer in which the three ethylenediamine ligands spiral to the right (clockwise) and a "left-handed" enantiomer in which the ethylenediamine ligands spiral to the left (counterclockwise). By contrast, the analogous ammonia complex $[Co(NH_3)_6]^{3+}$ is achiral because, like a coffee mug, it has a symmetry plane. (Actually, $[Co(NH_3)_6]^{3+}$ has several symmetry planes, though only one is shown in Figure 20.19.) Thus, $[Co(NH_3)_6]^{3+}$ exists in a single form and does not have enantiomers.

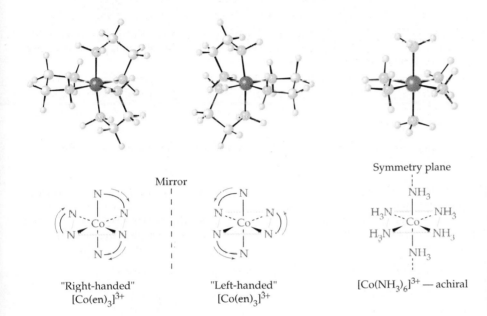

Mirror

"Right-handed"
$[Co(en)_3]^{3+}$

"Left-handed"
$[Co(en)_3]^{3+}$

Symmetry plane

$[Co(NH_3)_6]^{3+}$ — achiral

FIGURE 20.19 The structures of the $[Co(en)_3]^{3+}$ enantiomers and the achiral $[Co(NH_3)_6]^{3+}$ ion. $[Co(en)_3]^{3+}$ has a helical structure in which the three en ligands lie along the threads of a screw. As the arrows show, one enantiomer is "right-handed" in the sense that the screw would advance into the page as you rotated it to the right. The other enantiomer is "left-handed" because it would advance into the page as you rotated it to the left. In contrast, $[Co(NH_3)_6]^{3+}$ has several symmetry planes and is achiral. (Only one of the symmetry planes is identified.)

Enantiomers have identical properties except for their reactions with other chiral substances and their effect on *plane-polarized light*, light in which the electric vibrations of the light wave are restricted to a single plane, as shown in Figure 20.20. (In ordinary light, the electric vibrations occur in all planes parallel to the direction in which the light wave is traveling.) Plane-polarized light is obtained by passing ordinary light through a polarizing filter, like that found in certain kinds of sunglasses. When the plane-polarized light is passed through a solution of one of the enantiomers, the plane of polarization is rotated, either to the right or to the left. When the light is passed through a solution of the other enantiomer, its plane of polarization is rotated through an equal angle, but in the opposite direction. Enantiomers are sometimes called *optical isomers* because of their effect on plane-polarized light.

Enantiomers are labeled (+) or (−), depending on the direction of rotation of the plane of polarization. For example, the isomer of $[Co(en)_3]^{3+}$ that rotates the plane of polarization to the right is labeled (+)-$[Co(en)_3]^{3+}$. The isomer that rotates the plane of polarization to the left is designated (−)-$[Co(en)_3]^{3+}$. A 50:50 mixture of the (+) and (−) isomers, called a **racemic mixture**, produces no net optical rotation because the rotations produced by the individual enantiomers exactly cancel.

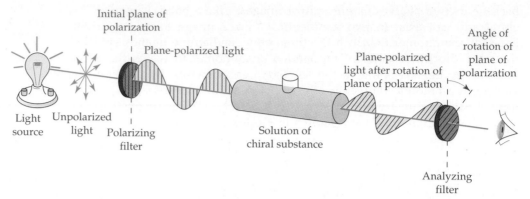

FIGURE 20.20 The essential features of a polarimeter, a device for measuring the angle through which the plane of plane-polarized light is rotated when the light is passed through a solution of a chiral substance.

EXAMPLE 20.8

Draw the structures of all possible diastereoisomers and enantiomers of $[Co(en)_2Cl_2]^+$.

SOLUTION   Ethylenediamine always spans adjacent corners of an octahedron, but the Cl^- ligands can be either on adjacent or on opposite corners. Therefore, there are two diastereoisomers, cis and trans. Because the trans isomer has several symmetry planes—one cuts through the Co and the en ligands—it is achiral and has no enantiomers. The cis isomer, however, is chiral and exists as a pair of enantiomers that are nonidentical mirror images.

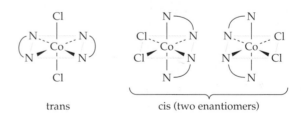

trans   cis (two enantiomers)

▢ **PROBLEM 20.7**   Which of the following objects are chiral?

(a) a glove   **(b)** a baseball   **(c)** a chair   **(d)** a foot   **(e)** a pencil

▢ **PROBLEM 20.8**   Which of the following complexes can exist as enantiomers? Draw the structure of each enantiomer.

(a) $[Fe(C_2O_4)_3]^{3-}$ **(b)** $[Co(NH_3)_4(en)]^{3+}$ **(c)** $[Co(NH_3)_2(en)_2]^{3+}$
(d) $[Cr(H_2O)_4Cl_2]^+$

20.10 ▶COLOR AND MAGNETIC PROPERTIES

Most transition-metal complexes have beautiful colors that depend on the identity of the metal and the ligands. The color of an aqua complex, for example, depends on the metal: $[Co(H_2O)_6]^{2+}$ is pink, $[Ni(H_2O)_6]^{2+}$ is green, $[Cu(H_2O)_6]^{2+}$ is blue, but $[Zn(H_2O)_6]^{2+}$ is colorless (Figure 20.21). If

FIGURE 20.21 Aqueous solutions of the nitrate salts of cobalt(II), copper(II), nickel(II), and zinc(II) (from left to right).

we keep the metal constant but vary the ligand, the color also changes. For example, $[Ni(H_2O)_6]^{2+}$ is green, $[Ni(NH_3)_6]^{2+}$ is blue, and $[Ni(en)_3]^{2+}$ is violet (Figure 20.22).

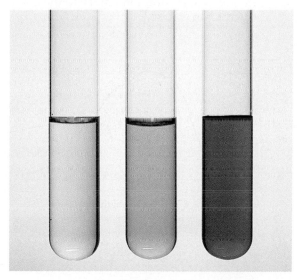

FIGURE 20.22 Aqueous solutions that contain $[Ni(H_2O)_6]^{2+}$, $[Ni(NH_3)_6]^{2+}$, and $[Ni(en)_3]^{2+}$ (from left to right). The two solutions on the right were prepared by adding ammonia and ethylenediamine, respectively, to aqueous nickel(II) nitrate.

How can we account for the color of transition-metal complexes? Let's begin by recalling that white light consists of a continuous spectrum of wavelengths corresponding to different colors (Section 5.2). When white light strikes a colored substance, some wavelengths are transmitted while others are absorbed. Just as an atom can absorb light by undergoing an electronic transition between atomic energy levels, thereby giving rise to atomic spectra (Section 5.3), so a metal complex can absorb light by undergoing an electronic transition from its lowest (ground) energy state to a higher (excited) energy state.

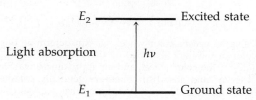

The wavelength λ of the light absorbed by a metal complex depends on the energy separation $\Delta E = E_2 - E_1$ between the two states, as given by the Planck equation $\Delta E = h\nu = hc/\lambda$, where h is Planck's constant, ν is the frequency of the light, and c is the speed of light:

$$\Delta E = E_2 - E_1 = h\nu = \frac{hc}{\lambda}$$

or

$$\lambda = \frac{hc}{\Delta E}$$

The measure of the amount of light absorbed by a substance is called the *absorbance*, and a plot of absorbance versus wavelength is called an **absorption spectrum**. For example, the absorption spectrum of the red-violet $[Ti(H_2O)_6]^{3+}$ ion has a broad absorption band at about 500 nm, a wavelength in the green part of the visible spectrum (Figure 20.23). Because the absorbance is smaller in the red and violet regions of the spectrum, these colors are largely transmitted, and we perceive the color of $[Ti(H_2O)_6]^{3+}$ to be red-violet. In general, the color that we see is complementary to the color absorbed (Figure 20.24).

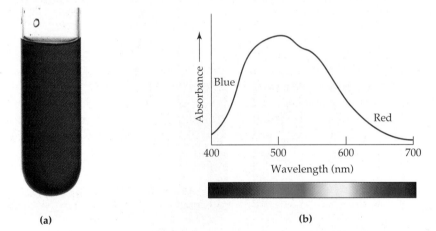

FIGURE 20.23 **(a)** A solution that contains the $[Ti(H_2O)_6]^{3+}$ ion. **(b)** Visible absorption spectrum of the $[Ti(H_2O)_6]^{3+}$ ion.

(a)

(b)

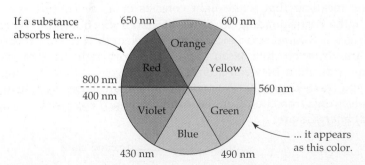

FIGURE 20.24 Using an artist's color wheel, we can determine the observed color of a substance from the color of the light absorbed by the substance. Complementary colors are shown on opposite sides of the color wheel, and an approximate wavelength range for each color is indicated. Observed and absorbed colors are generally complementary. Thus, if a complex absorbs only red light of 720 nm wavelength, it has a green color.

A second characteristic of complexes is their magnetic behavior. Recall that paramagnetic substances contain unpaired electrons and are attracted by magnetic fields, but diamagnetic substances contain only paired electrons and are weakly repelled by magnetic fields (Section 7.14). The number of unpaired electrons in a transition-metal complex can be determined by quantitative measurement of the force exerted on the complex by a magnetic field.

20.11 ➤ BONDING IN COMPLEXES: VALENCE BOND THEORY

According to the valence bond theory (Section 7.10), the bonding in metal complexes arises when a filled ligand orbital containing a pair of electrons overlaps a vacant hybrid orbital on the metal ion to give a coordinate covalent bond:

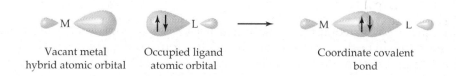

| Vacant metal hybrid atomic orbital | Occupied ligand atomic orbital | | Coordinate covalent bond |

The hybrid orbitals (Sections 7.11 and 7.12) that a metal ion uses to accept a share in the ligand electrons are those that point in the directions of the ligands. Remember that geometry and hybridization go together. Once you know the geometry of a complex, you automatically know which hybrid orbitals the metal ion uses. The relationship between the geometry of a complex and the hybrid orbitals used by the metal ion is summarized in Table 20.7.

TABLE 20.7   Hybrid Orbitals for Common Coordination Geometries

Coordination Number	Geometry	Hybrid Orbitals	Example
2	Linear	sp	$[Ag(NH_3)_2]^+$
4	Tetrahedral	sp^3	$[FeCl_4]^-$
4	Square planar	dsp^2	$[Ni(CN)_4]^{2-}$
6	Octahedral	d^2sp^3 or sp^3d^2	$[Cr(H_2O)_6]^{3+}$, $[Co(H_2O)_6]^{2+}$

To illustrate the relationship between geometry and hybridization, let's consider the tetrahedral complex $[CoCl_4]^{2-}$. A free Co^{2+} ion has electron configuration [Ar] $3d^7$, and its orbital diagram is

$$Co^{2+}: [Ar] \quad \underset{3d}{\uparrow\downarrow \ \uparrow\downarrow \ \uparrow \ \uparrow \ \uparrow} \qquad \underset{4s}{\underline{}} \qquad \underset{4p}{\underline{} \ \underline{} \ \underline{}}$$

Because the geometry of $[CoCl_4]^{2-}$ is tetrahedral, the hybrid orbitals that Co^{2+} uses to accept a share in the four pairs of ligand electrons must be sp^3

hybrids formed from the vacant $4s$ and $4p$ orbitals. It's convenient to represent the bonding in the complex by an orbital diagram that shows the hybridization of the metal orbitals and the four pairs of ligand electrons, now shared in the bonds between the metal and the ligands:

$[CoCl_4]^{2-}$: [Ar] $\underline{\uparrow\downarrow}$ $\underline{\uparrow\downarrow}$ $\underline{\uparrow}$ $\underline{\uparrow}$ $\underline{\uparrow}$ $\boxed{\underline{\uparrow\downarrow} \quad \underline{\uparrow\downarrow}\ \underline{\uparrow\downarrow}\ \underline{\uparrow\downarrow}}$

                          $3d$ $4s$ $4p$

Four sp^3 bonds to the ligands

In accord with this description, $[CoCl_4]^{2-}$ is paramagnetic and has three unpaired electrons.

As an example of a square planar complex, consider $[Ni(CN)_4]^{2-}$. A free Ni^{2+} ion has eight $3d$ electrons, two of which are unpaired in accord with Hund's rule:

Ni^{2+}: [Ar] $\underline{\uparrow\downarrow}$ $\underline{\uparrow\downarrow}$ $\underline{\uparrow\downarrow}$ $\underline{\uparrow}$ $\underline{\uparrow}$ $\underline{}$ $\underline{}\ \underline{}\ \underline{}$

                  $3d$ $4s$ $4p$

In square planar complexes, the metal uses a set of four hybrid orbitals called dsp^2 hybrids, which point toward the four corners of a square. By pairing up the two unpaired d electrons in one d orbital we obtain a vacant $3d$ orbital that can be hybridized with the $4s$ orbital and two of the $4p$ orbitals to give the square planar dsp^2 hybrids. These hybrids form bonds to the ligands by accepting a share in the four pairs of ligand electrons:

$[Ni(CN)_4]^{2-}$: [Ar] $\underline{\uparrow\downarrow}$ $\underline{\uparrow\downarrow}$ $\underline{\uparrow\downarrow}$ $\underline{\uparrow\downarrow}$ $\boxed{\underline{\uparrow\downarrow} \quad\quad \underline{\uparrow\downarrow} \quad\quad \underline{\uparrow\downarrow}\ \underline{\uparrow\downarrow}}$ $\underline{}$

                           $3d$ $4s$ $4p$

Four dsp^2 bonds to the ligands

In agreement with this description, $[Ni(CN)_4]^{2-}$ is diamagnetic.

In octahedral complexes, the metal ion uses either sp^3d^2 or d^2sp^3 hybrid orbitals. To see the difference between these two kinds of hybrids, let's consider the cobalt(III) complexes $[CoF_6]^{3-}$ and $[Co(CN)_6]^{3-}$. A free Co^{3+} ion has six $3d$ electrons, four of which are unpaired:

Co^{3+}: [Ar] $\underline{\uparrow\downarrow}$ $\underline{\uparrow}$ $\underline{\uparrow}$ $\underline{\uparrow}$ $\underline{\uparrow}$ $\underline{}$ $\underline{}\ \underline{}\ \underline{}$ $\underline{}\ \underline{}\ \underline{}\ \underline{}\ \underline{}$

            $3d$ $4s$ $4p$ $4d$

Magnetic measurements indicate that $[CoF_6]^{3-}$ is paramagnetic and contains four unpaired electrons. Evidently, none of the $3d$ orbitals is available to accept a share in the ligand electrons because each is already at least partially occupied. Consequently, the octahedral hybrids that Co^{3+} uses are formed from the vacant $4s$, $4p$, and $4d$ orbitals. These orbitals, called sp^3d^2 hybrids, share in the six pairs of ligand electrons, as shown in the following orbital diagram:

$[CoF_6]^{3-}$: [Ar] $\underline{\uparrow\downarrow}$ $\underline{\uparrow}$ $\underline{\uparrow}$ $\underline{\uparrow}$ $\underline{\uparrow}$ $\boxed{\underline{\uparrow\downarrow} \quad \underline{\uparrow\downarrow}\ \underline{\uparrow\downarrow}\ \underline{\uparrow\downarrow} \quad \underline{\uparrow\downarrow}\ \underline{\uparrow\downarrow}}$ $\underline{}\ \underline{}\ \underline{}$

                $3d$ $4s$ $4p$ $4d$

Six sp^3d^2 bonds to the ligands

In contrast to $[CoF_6]^{3-}$, magnetic measurements indicate that $[Co(CN)_6]^{3-}$ is diamagnetic. All six $3d$ electrons are therefore paired and occupy just three of the five $3d$ orbitals. That leaves two vacant $3d$ orbitals, which combine with the vacant $4s$ and $4p$ orbitals to give a set of six octahedral hybrid orbitals called d^2sp^3 hybrids. The d^2sp^3 hybrids form bonds to the ligands by accepting a share in the six pairs of ligand electrons:

$[Co(CN)_6]^{3-}$: [Ar] ⇅ ⇅ ⇅ | ⇅ ⇅　⇅　⇅ ⇅ ⇅ | — — — — — —

               $3d$ $4s$ $4p$ $4d$

Six d^2sp^3 bonds to the ligands

Note that the difference between d^2sp^3 and sp^3d^2 hybrids lies in the principal quantum number of the d orbitals. In d^2sp^3 hybrids the principal quantum number of the d orbitals is one less than the principal quantum number of the s and p orbitals. In sp^3d^2 hybrids the s, p, and d orbitals have the same principal quantum number. To determine which set of hybrids is used in any given complex, we must know the magnetic properties of the complex.

Complexes of metals, such as Co^{3+}, that exhibit more than one spin state are classified as *high-spin* or *low-spin*. A **high-spin complex**, such as $[CoF_6]^{3-}$, is one in which the d electrons are arranged according to Hund's rule to give the maximum number of unpaired electrons. A **low-spin complex**, such as $[Co(CN)_6]^{3-}$, is one in which the d electrons are paired up to give a maximum number of doubly occupied d orbitals and a minimum number of unpaired electrons.

EXAMPLE 20.9

Give a valence bond description of the bonding in $[V(NH_3)_6]^{3+}$. Include orbital diagrams for the free metal ion and the metal ion in the complex. Tell which hybrid orbitals the metal ion uses and the number of unpaired electrons present.

SOLUTION   The free V^{3+} ion has electron configuration $[Ar]3d^2$ and orbital diagram

V^{3+}: [Ar] ↑ ↑ — — —   —   — — —

              $3d$ $4s$ $4p$

Because $[V(NH_3)_6]^{3+}$ is octahedral, the V^{3+} ion must use either d^2sp^3 or sp^3d^2 hybrid orbitals in accepting a share in six pairs of electrons from the six NH_3 ligands. The preferred hybrids are d^2sp^3 because several $3d$ orbitals are vacant and d^2sp^3 hybrids have lower energy than sp^3d^2 hybrids. Thus, $[V(NH_3)_6]^{3+}$ has the following orbital diagram:

$[V(NH_3)_6]^{3+}$: [Ar] ↑ ↑ — | ⇅ ⇅　⇅　⇅ ⇅ ⇅ |

              $3d$ $4s$ $4p$

Six d^2sp^3 bonds to the ligands

The complex is paramagnetic because of its two unpaired electrons.

> **PROBLEM 20.9** Give a valence bond description of the bonding in each of the following complexes. Include orbital diagrams for the free metal ion and the metal ion in the complex. Tell which hybrid orbitals the metal ion uses and the number of unpaired electrons in each complex.
>
> **(a)** $[Fe(CN)_6]^{3-}$ (low-spin)   **(b)** $[Co(H_2O)_6]^{2+}$ (high-spin)
>
> **(c)** $[VCl_4]^-$ (tetrahedral)   **(d)** $[PtCl_4]^{2-}$ (square planar)

20.12 ➤CRYSTAL FIELD THEORY

The valence bond theory is useful in helping to visualize the bonding in complexes, but it doesn't account for their color or magnetic properties. To explain these properties, we turn to the **crystal field theory**, a model that views the bonding in complexes as arising from electrostatic interactions and considers the effect of the ligand charges on the energies of the metal ion *d* orbitals. This model was first applied to transition metal ions in ionic crystals—hence the name crystal field theory—but it is also applicable to metal complexes where the "crystal field" is the electric field due to the charges or dipoles of the ligands.

Octahedral Complexes

Let's first consider an octahedral complex such as $[TiF_6]^{3-}$ (Figure 20.25). According to the crystal field theory, the bonding is ionic and involves electrostatic attraction between the positively charged Ti^{3+} ion and the negatively charged F^- ligands. Of course, the F^- ligands repel one another, which is why they adopt the geometry (octahedral) that locates them as far apart from one another as possible. Because the metal–ligand attractions are greater than the ligand–ligand repulsions, the complex is more stable than the separated ions, thus accounting for the bonding.

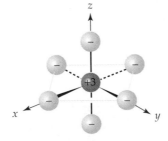

FIGURE 20.25 Crystal field model of the octahedral $[TiF_6]^{3-}$ complex. The metal ion and ligands are regarded as point charges held together by electrostatic attraction. The ligands lie along the $\pm x, \pm y,$ and $\pm z$ directions.

Note the differences between the crystal field theory and the valence bond theory. In crystal field theory, there are no covalent bonds, no shared electrons, and no hybrid orbitals—just electrostatic interactions within an array of ions. In complexes that contain neutral dipolar ligands, such as H_2O or NH_3, the electrostatic interactions are of the ion–dipole type (Section 10.2). For example, in $[Ti(H_2O)_6]^{3+}$, the Ti^{3+} ion attracts the negative end of the water dipoles.

To explain why complexes are colored, we need to look at the effect of the ligand charges on the energies of the *d* orbitals. Recall that four of the *d* orbitals are shaped like a cloverleaf, and the fifth (d_{z^2}) is shaped like a dumbbell inside a donut (Section 5.9). Figure 20.26 shows the spatial orientation of the *d* orbitals with respect to an octahedral array of ligand point charges located along the *x*, *y*, and *z* coordinate axes.

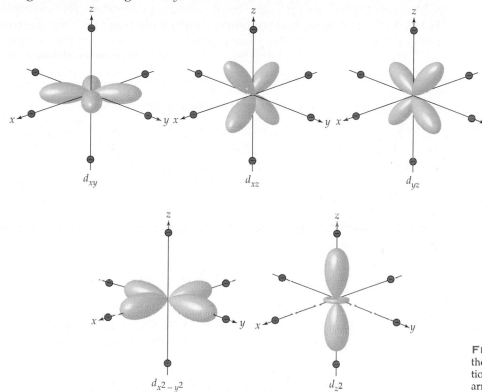

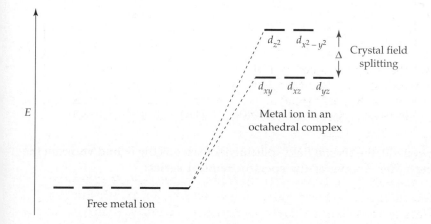

FIGURE 20.26 The shapes of the five *d* orbitals and their orientation with respect to an octahedral array of ligand point charges.

Because the *d* electrons are negatively charged, they are repelled by the negatively charged ligands. Thus, their orbital energies are higher in the complex than in the free metal ion. But not all the *d* orbitals are raised in energy by the same amount. Figure 20.27 shows that the d_{z^2} and $d_{x^2-y^2}$ orbitals, which point directly at the ligands, are raised in energy to a greater extent than the d_{xy}, d_{xz}, and d_{yz} orbitals, which point between the ligands. This energy splitting between the two sets of *d* orbitals is called the **crystal field splitting** and is represented by the Greek letter Δ.

FIGURE 20.27 A *d*-orbital energy level diagram for a free metal ion and a metal ion in an octahedral complex. In the absence of ligands, the five *d* orbitals have the same energy. When the metal ion is surrounded by an octahedral array of ligands, the *d* orbitals increase in energy and split into two sets that are separated in energy by the crystal field splitting Δ. The *d* orbitals whose lobes point directly toward the ligands (d_{z^2} and $d_{x^2-y^2}$) are higher in energy than the *d* orbitals whose lobes point between the ligands (d_{xy}, d_{xz}, and d_{yz}).

In general, the crystal field splitting energy Δ corresponds to wavelengths of light in the visible region of the spectrum, and the color of complexes can therefore be attributed to electronic transitions between the lower and higher energy sets of d orbitals. Consider, for example, $[Ti(H_2O)_6]^{3+}$, a complex that contains a single d electron (Ti^{3+} has electron configuration [Ar] $3d^1$). In the ground energy state, the d electron occupies one of the lower energy orbitals—xy, xz, or yz (from now on we'll denote the d orbitals by their subscripts). When $[Ti(H_2O)_6]^{3+}$ absorbs green light with a wavelength of about 500 nm, the absorbed energy is used to promote the d electron to one of the higher-energy orbitals, z^2 or $x^2 - y^2$.

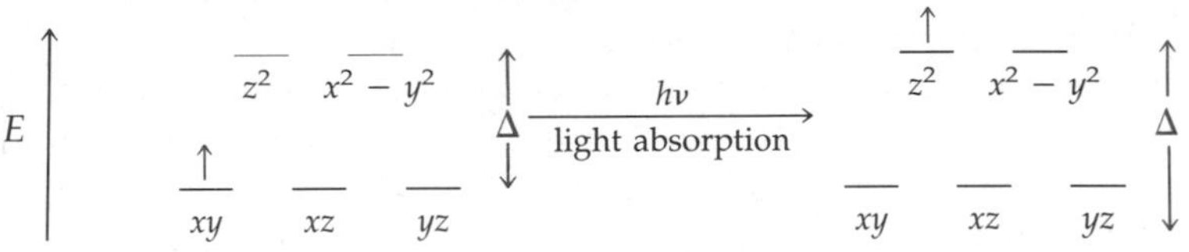

We can calculate the value of Δ from the wavelength of the absorbed light, about 500 nm for $[Ti(H_2O)_6]^{3+}$:

$$\Delta = h\nu = \frac{hc}{\lambda} = \frac{(6.626 \times 10^{-34} \text{ J} \cdot \text{s})(3.00 \times 10^8 \text{ m/s})}{500 \times 10^{-9} \text{ m}} = 3.98 \times 10^{-19} \text{ J}$$

This is the energy needed to excite a single $[Ti(H_2O)_6]^{3+}$ ion. To express Δ on a per-mole basis, we must multiply by Avogadro's number:

$$\Delta = \left(3.98 \times 10^{-19} \frac{\text{J}}{\text{ion}}\right)\left(6.02 \times 10^{23} \frac{\text{ions}}{\text{mol}}\right)$$

$$= 2.40 \times 10^5 \text{ J/mol} = 240 \text{ kJ/mol}$$

The absorption spectra of different complexes indicate that the size of the crystal field splitting depends on the nature of the ligands. For example, Δ for Ni^{2+} ([Ar] $3d^8$) complexes increases as the ligand varies from H_2O to NH_3 to ethylenediamine (en). Accordingly, the electronic transitions shift to higher energy (shorter wavelength) as the ligand varies from H_2O to NH_3 to en, thus accounting for the observed variation in color (Figure 20.22):

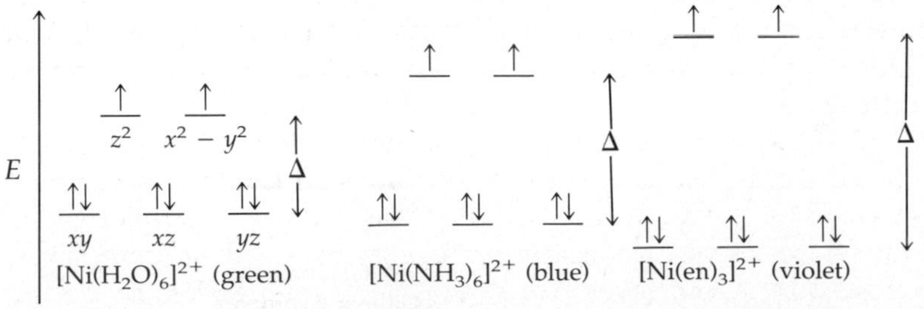

In general, the crystal field splitting increases as the ligand varies in the following order, known as the **spectrochemical series:**

Weak-field $I^- < Br^- < Cl^- < F^- < H_2O < NH_3 < en < CN^-$ Strong-field
ligands   ligands

Increasing Δ

Ligands such as halides and H_2O, which give a relatively small value of Δ, are called **weak-field ligands**. Ligands such as NH_3, en, and CN^-, which produce a relatively large value of Δ, are known as **strong-field ligands**. Different metal ions have different values of Δ, which explains why their complexes with the same ligand have different colors (Figure 20.21). Because d^0 ions like Ti^{4+}, d^{10} ions like Zn^{2+}, and main-group ions don't have partially filled d shells, they can't undergo d–d electronic transitions, and most of their compounds are therefore colorless.

   The crystal field theory accounts for the magnetic properties of complexes as well as for their color. It explains, for example, why complexes with weak-field ligands, such as $[CoF_6]^{3-}$, are high-spin, whereas related complexes with strong-field ligands, such as $[Co(CN)_6]^{3-}$, are low-spin. In $[CoF_6]^{3-}$, the six d electrons of Co^{3+} ([Ar] $3d^6$) occupy both the higher and lower energy d orbitals, whereas in $[Co(CN)_6]^{3-}$, all six d electrons are spin-paired in the lower energy orbitals.

$[CoF_6]^{3-}$ (high-spin) $[Co(CN)_6]^{3-}$ (low-spin)

   What determines which of the two spin states has the lower energy? In general, when an electron moves from a z^2 or $x^2 - y^2$ orbital to one of the lower-energy orbitals, the orbital energy decreases by Δ. But because of electron–electron repulsion, it costs energy to put the electron into an orbital that already contains another electron. The energy required is called the spin-pairing energy P. If Δ is greater than P, as it is for $[Co(CN)_6]^{3-}$, then the low-spin arrangement has lower energy. If Δ is less than P, as it is for $[CoF_6]^{3-}$, the high-spin arrangement has lower energy. Thus, the observed spin state depends on the relative values of Δ and P. In general, strong-field ligands give low-spin complexes, and weak-field ligands give high-spin complexes.

   Note that a choice between high-spin and low-spin electron configurations arises only for complexes of metal ions with four to seven d electrons, so-called d^4–d^7 complexes. For d^1–d^3 and d^8–d^{10} complexes, only one ground-state electron configuration is possible. In d^1–d^3 complexes, all the electrons occupy the lower-energy d orbitals, independent of the value of Δ. In d^8–d^{10} complexes, the lower-energy set of d orbitals is filled with three pairs of electrons, while the higher-energy set contains two, three, or four electrons, again independent of the value of Δ.

EXAMPLE 20.10

Draw a crystal field orbital energy level diagram, and predict the number of unpaired electrons for each of the following complexes.

(a) $[Cr(en)_3]^{3+}$ **(b)** $[Mn(CN)_6]^{3-}$ **(c)** $[Co(H_2O)_6]^{2+}$

SOLUTION **(a)** Cr^{3+} ([Ar] $3d^3$) has three unpaired electrons. In the complex, they occupy the lower energy set of d orbitals.

(b) Mn^{3+} ([Ar] $3d^4$) can have a high-spin or a low-spin configuration. Because CN^- is a strong-field ligand, all four d electrons go into the lower energy d orbitals. The complex is low-spin, with two unpaired electrons.

(c) Co^{2+} ([Ar] $3d^7$) has a high-spin configuration with three unpaired electrons because H_2O is a weak-field ligand. The orbital energy level diagrams are as follows:

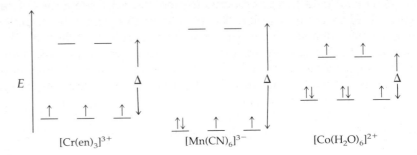

┌ **PROBLEM 20.10** Draw a crystal field orbital energy level diagram and predict the number of unpaired electrons for each of the following complexes.

(a) $[Fe(H_2O)_6]^{2+}$ **(b)** $[Fe(CN)_6]^{4-}$ **(c)** $[VF_6]^{3-}$

Tetrahedral and Square Planar Complexes

As you might expect, different geometric arrangements of the ligands give different energy splittings for the d orbitals. Figure 20.28 shows d-orbital energy level diagrams for tetrahedral and square planar complexes.

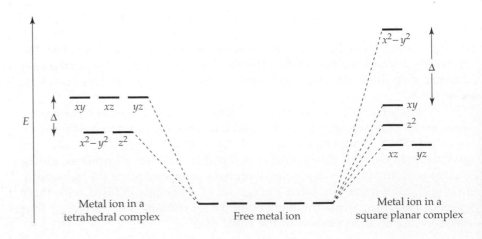

FIGURE 20.28 Energies of the d orbitals in tetrahedral and square planar complexes relative to their energy in the free metal ion. The crystal field splitting energy Δ is small in tetrahedral complexes but much larger in square planar complexes.

The splitting pattern in tetrahedral complexes is just the opposite of that in octahedral complexes; that is, the xy, xz, and yz orbitals have higher energy than the $x^2 - y^2$ and z^2 orbitals. (As with octahedral complexes, the energy ordering follows from the relative orientation of the orbital lobes and the ligands, but we won't try to derive the result.) Because none of the orbitals point directly at the ligands in tetrahedral geometry, and because there are only four ligands instead of six, the crystal field splitting in tetrahedral complexes is only about half of that in octahedral complexes. Consequently, Δ is almost always smaller than the spin-pairing energy P, and nearly all tetrahedral complexes are high-spin.

Square planar complexes look like octahedral ones (Figure 20.26) except that two of the trans ligands—say, those along the z axis—are missing. In square planar complexes, the $x^2 - y^2$ orbital is high in energy (Figure 20.28) because it points directly at the ligands, which lie along the x and y axes. The splitting pattern is more complicated here than for the octahedral and tetrahedral cases, but the main point to remember is that there is a large energy gap between the $x^2 - y^2$ orbital and the four lower-energy orbitals. Square planar geometry is most common for metal ions with electron configuration d^8 because this configuration favors low-spin complexes in which all four lower-energy orbitals are filled and the higher-energy $x^2 - y^2$ orbital is vacant. Common examples are $[Ni(CN)_4]^{2-}$, $[PdCl_4]^{2-}$, and $Pt(NH_3)_2Cl_2$.

EXAMPLE 20.11

Draw crystal field energy level diagrams and predict the number of unpaired electrons for the following complexes.

(a) $[FeCl_4]^-$ (tetrahedral) **(b)** $[PtCl_4]^{2-}$ (square planar)

SOLUTION **(a)** Take the tetrahedral energy level diagram from Figure 20.28. Because nearly all tetrahedral complexes are high-spin (small Δ), the five d electrons of Fe^{3+} ([Ar] $3d^5$) are distributed between the higher- and lower-energy orbitals as shown below. The complex is high-spin with five unpaired electrons.
(b) Pt^{2+} ([Xe] $4f^{14}5d^8$) has eight d electrons. Because Δ is large in square planar complexes, all the electrons occupy the four lower-energy d orbitals. There are no unpaired electrons, and the complex is diamagnetic.

PROBLEM 20.11 Draw a crystal field energy level diagram and predict the number of unpaired electrons for the following complexes.

(a) $[NiCl_4]^{2-}$ (tetrahedral) **(b)** $[Ni(CN)_4]^{2-}$ (square planar)

interlude—TITANIUM: A HIGH-TECH METAL

Named after the Titans, Greek mythological figures symbolic of power and strength, titanium is the ninth most abundant element in the earth's crust (0.57% by mass). It occurs in rutile, TiO_2, in the black beach sands of eastern Australia, and in the mineral ilmenite, $FeTiO_3$, in the United States, Canada, Malaysia, and elsewhere. Despite its abundance, titanium is difficult and expensive to produce, and it remained a curiosity until about 1950, when its potential applications in aerospace technology were recognized.

Titanium is a superb structural material because of its hardness, strength, heat resistance (mp 1660°C), and relatively low density (4.5 g/cm^3). Titanium is just as strong as steel, but 45% lighter; titanium is twice as strong as aluminum but only 60% heavier. When alloyed with a few percent aluminum and vanadium, titanium has a higher strength-to-weight ratio than any other engineering metal. These properties make titanium an ideal choice in aerospace applications, such as airframes and jet engines, as well as for recreational use. (Those who can afford one rave about the ride and feel of a titanium bike frame.)

Although titanium has a large positive oxidation potential, and Ti dust will burn in air, the bulk metal is remarkably immune to corrosion because its surface is covered with a thin, protective oxide film. Titanium objects are inert to sea water, nitric acid, hot aqueous NaOH, and even to aqueous chlorine gas. Titanium is therefore used in chemical plants, in desalination equipment, and in numerous other industrial processes that demand inert, noncorrosive materials. Because it is nontoxic and inert to body fluids, titanium is even used for manufacturing artificial joints.

Pure titanium is obtained commercially from rutile (TiO_2) by an indirect route that involves initial reaction of TiO_2 with Cl_2 gas and coke to yield liquid $TiCl_4$ (bp 136°C), which is purified by fractional distillation. Subsequent reduction to Ti metal is then carried out by reaction with molten magnesium at 900°C, and further purification is effected by melting the titanium in an electric arc under an atmosphere of argon.

$$TiO_2(s) + 2\ Cl_2(g) + 2\ C(s) \rightarrow TiCl_4(l) + 2\ CO(g)$$

$$TiCl_4(g) + 2\ Mg(l) \rightarrow Ti(s) + 2\ MgCl_2(l)$$

Although the process is extremely expensive and energy-intensive, the cost of producing titanium is justified because of its unique properties. Worldwide production of titanium now exceeds 100,000 tons per year.

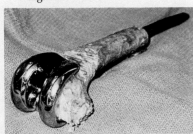
This artificial knee joint is made of titanium, a nontoxic, noncorrosive, high-tech metal.

SUMMARY

Transition elements (*d*-block elements) are the metallic elements in the central part of the periodic table. Most of the neutral atoms have valence electron configuration $(n-1)d^{1-10} ns^2$, and the cations have configuration $(n-1)d^{1-10}$, where n is the principal quantum number. Transition metals exhibit a variety of oxidation states. Ions with the metal in a high oxidation state tend to be good oxidizing agents ($Cr_2O_7^{2-}$, MnO_4^-). Early transition-metal ions with the metal in a low oxidation state are good reducing agents (V^{2+}, Cr^{2+}). The **lanthanides** and **actinides** comprise the *f*-block, or inner transition, elements.

Coordination compounds (metal complexes) are compounds in which a central metal ion is attached to a group of surrounding **ligands** by coordinate covalent bonds. The ligands act as electron-pair donors (Lewis bases), and the metal ion acts as an electron-pair acceptor (Lewis acid). Ligands can be **monodentate** or **polydentate** depending on the number of donor atoms attached to the metal. The **bidentate** ligand ethylenediamine ($NH_2CH_2CH_2NH_2$, en) uses two N donor atoms. Polydentate ligands are also called **chelating agents**. They form complexes (metal **chelates**) that contain rings of atoms known as **chelate rings**.

The number of ligand donor atoms bonded to a metal is called the **coordination number** of the metal. Common coordination numbers and associated geometries are 2 (linear), 4 (tetrahedral or square planar), and 6 (octahedral). Systematic names for complexes specify the number of ligands of each particular type, the metal, and its oxidation state.

Many complexes exist as **isomers**, compounds that have the same formula but a different arrangement of the constituent atoms. **Constitutional isomers**, such as **linkage isomers** and **ionization isomers**, have different bonds between their constituent atoms. **Stereoisomers**, such as **diastereoisomers** and **enantiomers**, have the same bonds but a different arrangement of the atoms in space. The most common diastereoisomers are **cis** and **trans isomers** of square planar MA_2B_2 complexes and octahedral MA_4B_2 complexes. Enantiomers, such as "right-handed" and "left-handed" $[Co(en)_3]^{3+}$, are nonidentical mirror images. One isomer rotates the plane of plane-polarized light to the right, and the other rotates this plane through an equal angle but in the opposite direction. A 50:50 mixture of the two isomers is called a **racemic mixture**. Molecules that have handedness are said to be **chiral**.

Valence bond theory describes the bonding in complexes in terms of two-electron, coordinate covalent bonds resulting from overlap of filled ligand orbitals with vacant metal hybrid orbitals that point in the directions of the ligands: sp (linear), sp^3 (tetrahedral), dsp^2 (square planar), and d^2sp^3 or sp^3d^2 (octahedral).

Crystal field theory assumes that the metal–ligand bonding is entirely ionic. Because of electrostatic repulsions between the d electrons and the ligands, the d orbitals are raised in energy and are differentiated by an energy separation called the **crystal field splitting**, Δ. In octahedral complexes, the z^2 and $x^2 - y^2$ orbitals have higher energy than the xy, xz, and yz orbitals. Tetrahedral and square planar complexes exhibit different splitting patterns. The color of complexes is due to electronic transitions from one set of d orbitals to another, and the transition energies depend on the position of the ligand in the **spectrochemical series**. **Weak-field ligands** give small Δ

KEY WORDS

absorption spectrum, *826*
achiral, *822*
actinides, *796*
bidentate, *814*
chelate, *815*
chelate ring, *815*
chelating agent, *815*
chiral, *822*
cis isomer, *819*
constitutional isomer, *819*
coordination compound, *812*
coordination number, *812*
crystal field splitting, *831*
crystal field theory, *830*
d-block element, *796*
diastereoisomer, *819*
enantiomer, *822*
f-block element, *797*
high-spin complex, *829*
inner transition element, *797*
ionization isomer, *819*
isomer, *818*
lanthanide, *796*
lanthanide contraction, *802*
ligand, *812*
ligand donor atom, *812*
linkage isomers, *819*
low-spin complex, *829*
metal complex, *812*
monodentate, *814*
polydentate, *814*
racemic mixture, *824*
spectrochemical series, *833*
stereoisomer, *819*
strong-field ligand, *833*
symmetry plane, *822*
trans isomer, *819*
transition element, *796*
weak-field ligand, *833*

values, and **strong-field ligands** give large Δ values. Crystal field theory accounts for the magnetic properties of complexes in terms of the relative values of Δ and the spin-pairing energy P. Small Δ values favor **high-spin complexes** (those with a maximum number of unpaired d electrons), and large Δ values favor **low-spin complexes**.

UNDERSTANDING KEY CONCEPTS

1. Locate on the periodic table the transition elements that have atoms with each of the following electron configurations. Identify each element.
   (a) [Ar] $3d^7 4s^2$ **(b)** [Ar] $3d^5 4s^1$
   (c) [Kr] $4d^2 5s^2$ **(d)** [Xe] $4f^3 6s^2$

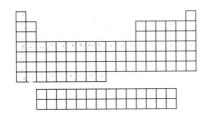

2. What is the general trend in each of the following properties from left to right across the first transition series (Sc to Zn)? Explain each trend in terms of effective nuclear charge and other relevant factors.
   (a) atomic radius   **(b)** density
   (c) ionization energy
   (d) standard oxidation potential

3. Classify the following ligands as monodentate, bidentate, or tridentate. Which can form chelate rings?
   (a) NH_2—CH_2—CH_2—NH_2
   (b) CH_3—CH_2—CH_2—NH_2
   (c) NH_2—CH_2—CH_2—NH—CH_2—CO_2^-
   (d) NH_2—CH_2—CH_2—NH_3^+

4. Give the oxidation state, coordination number, and coordination geometry of the transition metal in each of the following compounds.
   (a) $Na[Au(CN)_2]$ **(b)** $[Co(NH_3)_5Br]SO_4$
   (c) $Pt(en)Cl_2$ **(d)** $(NH_4)_2[PtCl_2(C_2O_4)_2]$

5. Consider the following isomers of $[Cr(NH_3)_2Cl_4]^-$.

   ● = Cr   ○ = NH_3 ◐ = Cl

(1)

(2)

(3)

(4)

(a) Label the isomers as cis or trans.
(b) Which isomers are identical, and which are different?
(c) Do any of these isomers exist as enantiomers? Explain.

6. Consider the following ethylenediamine complexes.

   ● = Cr   ◠○ = $NH_2CH_2CH_2NH_2$ ○ = Cl

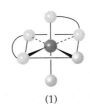

(1)

(2)

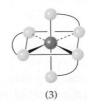

(3)

(4)

(a) Which complexes are chiral, and which are achiral?
(b) Draw the enantiomer of each chiral complex.
(c) Which, if any, of the chiral complexes are enantiomers of one another?

7. Predict the crystal field energy level diagram for a square pyramidal ML_5 complex that has two ligands along the $\pm x$ and $\pm y$ axes but only one ligand along the z axis. Your diagram should be intermediate between that for an octahedral ML_6 complex and that for a square planar ML_4 complex.

ADDITIONAL PROBLEMS

Problems 20.1–20.11 appear within the chapter.

ELECTRON CONFIGURATIONS AND PROPERTIES OF TRANSITION ELEMENTS

20.12 Define each of the following terms, and illustrate each with an example.

(a) transition element (b) lanthanide element
(c) actinide element (d) d-block element

20.13 Classify each of the following as an s-block, a p-block, or a d-block metal.

(a) Al (b) Mo (c) Ba (d) Hf (e) Ru

20.14 Use the periodic table to write the electron configuration for each of the following atoms.

(a) Ni (b) Cr (c) Zr (d) Zn

20.15 What is the electron configuration for each of the following atoms?

(a) V (b) Y (c) Cu (d) Hg

20.16 Use the periodic table to give the electron configuration for each of the following.

(a) Co^{2+} (b) Fe^{3+} (c) Mo^{3+}
(d) Cr(VI) in CrO_4^{2-}

20.17 Specify the electron configuration for each of the following.

(a) V^{3+} (b) Cu^+ (c) Rh^{3+} (d) Fe(VI) in FeO_4^{2-}

20.18 Predict the number of unpaired electrons for each of the following.

(a) Cu^{2+} (b) Ti^{2+} (c) Zn^{2+}

20.19 Predict the number of unpaired electrons for each of the following.

(a) Co^{3+} (b) Mn^{2+} (c) Sc^{3+}

20.20 Briefly account for each of the following observations.

(a) Titanium, used in making jet aircraft engines, is much harder than potassium or calcium.
(b) Molybdenum (mp 2610°C) has a higher melting point than yttrium (mp 1522°C) or cadmium (331°C).
(c) Atomic radii decrease in the order Sc > Ti > V.
(d) Densities increase in the order Ti < V < Cr.

20.21 Arrange the following atoms in order of decreasing atomic radii, and account for the trend.

(a) Cr (b) Ti (c) Mn (d) V

20.22 What is the lanthanide contraction, and why does it occur?

20.23 The atomic radii of zirconium and hafnium are nearly identical. Explain in terms of the lanthanide contraction.

20.24 What is the general trend in standard oxidation potentials across the first transition series from Sc to Zn? What is the reason for the trend?

20.25 Write a balanced net ionic equation for reaction of each of the following metals with hydrochloric acid in the absence of air. If no reaction occurs, indicate N.R.

(a) Cr (b) Zn (c) Cu (d) Fe

OXIDATION STATES

20.26 Which of the following metals have more than one oxidation state?

(a) Zn (b) Mn (c) Sr (d) Cu

20.27 Which of the following metals have only one oxidation state?

(a) Co (b) Fe (c) Sc (d) Al

20.28 What is the highest oxidation state for each of the elements from Sc to Zn?

20.29 The highest oxidation state for the early transition metals Sc, Ti, V, Cr, and Mn is the periodic group number. The highest oxidation state for the later transition elements Fe, Co, and Ni is less than the periodic group number. Explain.

20.30 Which is the stronger oxidizing agent, Cr^{2+} or Cu^{2+}? Explain.

20.31 Which is the stronger reducing agent, V^{2+} or Co^{2+}? Explain.

20.32 Which is more easily oxidized, Cr^{2+} or Ni^{2+}? Explain.

20.33 Which is more easily reduced, Ti^{3+} or Co^{3+}? Explain.

20.34 Arrange the following substances in order of increasing strength as an oxidizing agent, and account for the trend.

(a) Mn^{2+} (b) MnO_2 (c) MnO_4^-

20.35 Arrange the following ions in order of increasing strength as a reducing agent, and account for the trend.

(a) Cr^{2+} (b) Cr^{3+} (c) $Cr_2O_7^{2-}$

CHEMISTRY OF SELECTED TRANSITION ELEMENTS

20.36 Write a balanced equation for the industrial production of

(a) chromium metal from chromium(III) oxide
(b) copper metal from copper(I) sulfide.

20.37 Write a balanced net ionic equation for reaction of nitric acid with

(a) iron metal   (b) copper metal
(c) chromium metal.

20.38 What product is formed when dilute H_2SO_4, a non-oxidizing acid, is added to each of the following ions or compounds?

(a) CrO_4^{2-} (b) $Cr(OH)_3$ (c) $Cr(OH)_2$
(d) $Fe(OH)_2$ (e) $Cu(OH)_2$

20.39 What product is formed when excess aqueous NaOH is added to a solution of each of the following ions?

(a) Cr^{2+} (b) Cr^{3+} (c) $Cr_2O_7^{2-}$
(d) Fe^{2+} (e) Fe^{3+}

20.40 Arrange the following hydroxy compounds in order of increasing acid strength, and account for the trend.

(a) $CrO_2(OH)_2$ (b) $Cr(OH)_2$ (c) $Cr(OH)_3$

20.41 Explain how $Cr(OH)_3$ can act both as an acid and as a base.

20.42 Which of the following compounds is amphoteric?

(a) $Cr(OH)_2$ (b) $Fe(OH)_2$
(c) $Cr(OH)_3$ (d) $Fe(OH)_3$

20.43 Which of the following ions disproportionates in aqueous solution? Write a balanced net ionic equation for the reaction.

(a) Cr^{3+} (b) Fe^{3+} (c) Cu^+ (d) Cu^{2+}

20.44 What is a method for separating the following pairs of ions by addition of a single reagent? Include formulas for the major products of the reactions.

(a) Fe^{3+} and Na^+ (b) Cr^{3+} and Fe^{3+}
(c) Fe^{3+} and Cu^{2+}

20.45 Complete and balance the net ionic equation for each of the following reactions in acidic solution.

(a) $Cr_2O_7^{2-}(aq) + Fe^{2+}(aq) \longrightarrow$
(b) $Fe^{2+}(aq) + O_2(g) \longrightarrow$
(c) $Cu_2O(s) + H^+(aq) \longrightarrow$
(d) $Fe(s) + H^+(aq) \longrightarrow$

20.46 Write a balanced net ionic equation for each of the following reactions.

(a) A CrO_4^{2-} solution turns from yellow to orange upon addition of acid.
(b) $Fe^{3+}(aq)$ reacts with aqueous KSCN to give a deep red solution.
(c) Copper metal reacts with nitric acid to give NO gas and a blue solution.
(d) A deep green solution of $Cr(OH)_3$ in excess base turns yellow on addition of hydrogen peroxide.

20.47 Write a balanced net ionic equation for each of the following reactions.

(a) A $CuSO_4$ solution becomes dark blue when excess ammonia is added.
(b) A solution of $Na_2Cr_2O_7$ turns from orange to yellow on addition of base.
(c) When base is added to a solution of $Fe(NO_3)_3$, a red-brown precipitate forms.
(d) Dissolution of CuS in hot HNO_3 gives NO gas, a blue solution, and a yellow solid.

20.48 List the common oxidation states for chromium, iron, and copper, and give an example of an ion that has the metal in each of these oxidation states.

20.49 Draw the structure of each of the following ions.

(a) $Cr(H_2O)_5Cl^{2+}$ (b) $Cr_2O_7^{2-}$ (c) FeO_4^{2-}

COORDINATION COMPOUNDS; LIGANDS

20.50 Give a definition and an example for each of the following terms.

(a) coordination compound
(b) ligand
(c) ligand donor atom
(d) coordination number

20.51 Tell what is meant by each of the following terms, and illustrate each with an example.

(a) monodentate ligand
(b) bidentate ligand
(c) chelating agent
(d) chelate ring

20.52 Forming $[Ni(en)_3]^{2+}$ from Ni^{2+} and ethylenediamine is a Lewis acid-base reaction. Explain.

20.53 Identify the Lewis acid and the Lewis base in the reaction of oxalate ions ($C_2O_4^{2-}$) with Fe^{3+} to give $[Fe(C_2O_4)_3]^{3-}$.

20.54 Give an example of a coordination compound in which the metal exhibits a coordination number of
(a) 2   (b) 4   (c) 6.

20.55 What is the coordination number of the metal in each of the following complexes?
(a) $[HgCl_4]^{2-}$ (b) $[Cr(H_2O)_4Cl_2]^+$
(c) $[Au(CN)_2]^-$ (d) $[ZrF_8]^{4-}$
(e) $[Mo(CO)_5Br]^-$

20.56 What is the oxidation state of the metal in each of the complexes in Problem 20.55?

20.57 What is the formula of a complex that has each of the following geometries?
(a) tetrahedral   (b) linear
(c) octahedral   (d) square planar

20.58 Draw the structure of the iron oxalate complex $[Fe(C_2O_4)_3]^{3-}$. Describe the coordination geometry, and identify any chelate rings. What is the coordination number and the oxidation number of the iron?

20.59 What is the formula for each of the following complexes?
(a) an iridium(III) complex with three ammonia and three chloride ligands
(b) a chromium(III) complex with two water and two oxalate ligands
(c) a platinum(IV) complex with two ethylenediamine and two thiocyanate ligands

20.60 Identify the oxidation state of the metal in each of the following complexes
(a) $[Ni(CN)_5]^{3-}$ (b) $Ni(CO)_4$
(c) $[Co(en)_2(H_2O)Br]^{2+}$ (d) $[Cu(H_2O)_2(C_2O_4)_2]^{2-}$

20.61 What is the oxidation state of the metal in each of the following compounds?
(a) $(NH_4)_3[RhCl_6]$ (b) $[Cr(NH_3)_4(SCN)_2]Br$
(c) $[Cu(en)_2]SO_4$ (d) $Na_2[Mn(EDTA)]$

NAMING COORDINATION COMPOUNDS

20.62 What is the systematic name for each of the following ions?
(a) $[MnCl_4]^{2-}$ (b) $[Ni(NH_3)_6]^{2+}$
(c) $[Co(CO_3)_3]^{3-}$ (d) $[Pt(en)_2(SCN)_2]^{2+}$

20.63 Assign a systematic name to each of the following ions.
(a) $[AuCl_4]^-$ (b) $[Fe(CN)_6]^{4-}$
(c) $[Fe(H_2O)_5NCS]^{2+}$ (d) $[Cr(NH_3)_2(C_2O_4)_2]^-$

20.64 What is the systematic name for each of the following coordination compounds?
(a) $Cs[FeCl_4]$ (b) $[V(H_2O)_6](NO_3)_3$
(c) $[Co(NH_3)_4Br_2]Br$ (d) $Cu(gly)_2$

20.65 What is the systematic name for each of the following compounds?

(a) $[Cu(NH_3)_4]SO_4$ (b) $Cr(CO)_6$
(c) $K_3[Fe(C_2O_4)_3]$ (d) $[Co(en)_2(NH_3)CN]Cl_2$

20.66 Write the formula for each of the following compounds.
(a) tetraammineplatinum(II) chloride
(b) sodium hexacyanoferrate(III)
(c) tris(ethylenediamine)platinum(IV) sulfate
(d) triamminetrithiocyanatorhodium(III)

20.67 Write the formula for each of the following compounds.
(a) diamminesilver(I) nitrate
(b) potassium diaquadioxalatocobaltate(III)
(c) hexacarbonylmolybdenum(0)
(d) diamminebis(ethylenediamine)chromium(III) chloride

ISOMERS

20.68 Tell what is meant by each of the following terms, and illustrate each with an example.
(a) linkage isomers   (b) ionization isomers
(c) diastereoisomers   (d) enantiomers

20.69 Distinguish between the terms in each of the following pairs.
(a) constitutional isomers and stereoisomers
(b) cis and trans isomers
(c) chiral and achiral molecules

20.70 Draw all possible constitutional isomers of $[Ru(NH_3)_5(NO_2)]Cl$. Label the isomers as linkage isomers or ionization isomers.

20.71 There are four possible isomers for a square planar palladium(II) complex that contains two Cl^- and two SCN^- ligands. Sketch the structures of all four, and label them according to the isomer classification in Figure 20.15.

20.72 Which of the following complexes can exist as diastereoisomers?
(a) $[Cr(NH_3)_2Cl_4]^-$
(b) $[Co(NH_3)_5Br]^{2+}$
(c) $[FeCl_2(NCS)_2]^{2-}$ (tetrahedral)
(d) $[PtCl_2Br_2]^{2-}$ (square planar)

20.73 Tell how many diastereoisomers are possible for each of the following complexes, and draw their structures.
(a) $Pt(NH_3)_2(CN)_2$ (b) $[Co(en)(SCN)_4]^-$
(c) $[Cr(H_2O)_4Cl_2]^+$ (d) $Ru(NH_3)_3I_3$

20.74 Which of the following complexes are chiral?
  (a) [Pt(en)Cl$_2$]   **(b)** *cis*-[Co(NH$_3$)$_4$Br$_2$]$^+$
  (c) *cis*-[Cr(en)$_2$(H$_2$O)$_2$]$^{3+}$ **(d)** [Cr(C$_2$O$_4$)$_3$]$^{3-}$

20.75 Which of the following complexes can exist as enantiomers? Draw their structures.
  (a) [Cr(en)$_3$]$^{3+}$
  (b) *cis*-[Co(en)$_2$(NH$_3$)Cl]$^{2+}$
  (c) *trans*-[Co(en)$_2$(NH$_3$)Cl]$^{2+}$
  (d) [Pt(NH$_3$)$_3$Cl$_3$]$^+$

20.76 Draw all possible diastereoisomers and enantiomers of each of the following complexes.

  (a) Ru(NH$_3$)$_4$Cl$_2$ **(b)** [Pt(en)$_3$]$^{4+}$
  (c) [Pt(en)$_2$ClBr]$^{2+}$

20.77 Draw all possible diastereoisomers and enantiomers of each of the following complexes.
  (a) [Rh(C$_2$O$_4$)$_2$I$_2$]$^{3-}$ **(b)** [Cr(NH$_3$)$_2$Cl$_4$]$^-$
  (c) [Co(EDTA)]$^-$

20.78 How does plane-polarized light differ from ordinary light? Draw the structure of a chromium complex that rotates the plane of plane-polarized light.

20.79 What is a racemic mixture? Does it affect plane-polarized light? Explain.

COLOR AND MAGNETIC PROPERTIES; VALENCE BOND AND CRYSTAL FIELD THEORIES

20.80 What is an absorption spectrum? If the absorption spectrum of a complex has just one band at 455 nm, what is the color of the complex?

20.81 A red-colored complex has just one absorption band in the visible region of the spectrum. Predict the approximate wavelength of this band.

20.82 Give a valence bond description of the bonding in each of the following complexes. Include orbital diagrams for the free metal ion and the metal ion in the complex. Indicate which hybrid orbitals the metal ion uses for bonding, and specify the number of unpaired electrons.
  (a) [Ti(H$_2$O)$_6$]$^{3+}$
  (b) [NiBr$_4$]$^{2-}$ (tetrahedral)
  (c) [Fe(CN)$_6$]$^{3-}$ (low-spin)
  (d) [MnCl$_6$]$^{3-}$ (high-spin)

20.83 For each of the following complexes, describe the bonding in terms of the valence bond theory. Include orbital diagrams for the free metal ion and the metal ion in the complex. Indicate which hybrid orbitals the metal ion uses for bonding, and specify the number of unpaired electrons.
  (a) [AuCl$_4$]$^-$ (square planar)
  (b) [Ag(NH$_3$)$_2$]$^+$
  (c) [Fe(H$_2$O)$_6$]$^{2+}$ (high-spin)
  (d) [Fe(CN)$_6$]$^{4-}$ (low-spin)

20.84 Draw a crystal field energy level diagram for the 3d orbitals of titanium in [Ti(H$_2$O)$_6$]$^{3+}$. Indicate what is meant by the crystal field splitting, and tell why [Ti(H$_2$O)$_6$]$^{3+}$ is colored.

20.85 Explain using a sketch why the d_{xy} and $d_{x^2-y^2}$ orbitals have different energies in an octahedral complex. Which of the two orbitals has higher energy?

20.86 The [Ti(NCS)$_6$]$^{3-}$ ion exhibits a single absorption band at 544 nm. Calculate the crystal field splitting energy Δ in kJ/mol. Is NCS$^-$ a stronger or a weaker field ligand than water? Predict the color of [Ti(NCS)$_6$]$^{3-}$.

20.87 [Cr(H$_2$O)$_6$]$^{3+}$ is violet in color; [Cr(CN)$_6$]$^{3-}$ is yellow. Explain this difference in terms of crystal field theory. Use the colors to order H$_2$O and CN$^-$ in the spectrochemical series.

20.88 For each of the following complexes, draw a crystal field energy level diagram, assign the electrons to orbitals, and predict the number of unpaired electrons.
  (a) [CrF$_6$]$^{3-}$ **(b)** [V(H$_2$O)$_6$]$^{3+}$ **(c)** [Fe(CN)$_6$]$^{3-}$

20.89 Draw a crystal field energy level diagram, assign the electrons to orbitals, and predict the number of unpaired electrons for each of the following.
  (a) [Cu(en)$_3$]$^{2+}$ **(b)** [FeF$_6$]$^{3-}$
  (c) [Co(en)$_3$]$^{3+}$ (low-spin)

20.90 Ni^{2+}(*aq*) is green in color, but Zn^{2+}(*aq*) is colorless. Explain.

20.91 Cr^{3+}(*aq*) is violet in color, but Y^{3+}(*aq*) is colorless. Explain.

20.92 Weak-field ligands tend to give high-spin complexes, but strong-field ligands tend to give low-spin complexes. Explain.

20.93 Explain why nearly all tetrahedral complexes are high-spin.

20.94 Explain why square planar geometry is especially common for d^8 complexes.

20.95 For each of the following complexes, draw a crystal field energy level diagram, assign the electrons to orbitals, and predict the number of unpaired electrons.
  (a) [Pt(NH$_3$)$_4$]$^{2+}$ (square planar)
  (b) [MnCl$_4$]$^{2-}$ (tetrahedral)
  (c) [Co(NCS)$_4$]$^{2-}$ (tetrahedral)
  (d) [Cu(en)$_2$]$^{2+}$ (square planar)

GENERAL PROBLEMS

20.96 Write the electron configuration for each of the following atoms or ions.
(a) Mn   (b) Fe^{2+}
(c) V(IV) in VO^{2+} (d) Mn(VII) in MnO_4^-

20.97 Which of the following ions are paramagnetic?
(a) Sc^{3+} (b) Co^{2+} (c) V^{3+}
(d) CrO_4^{2-} (e) MnO_4^{2-}

20.98 Which of the following complexes are paramagnetic?
(a) $[Mn(CN)_6]^{3-}$ (b) $[Zn(NH_3)_4]^{2+}$
(c) $[Fe(CN)_6]^{4-}$ (d) $[FeF_6]^{4-}$

20.99 Which of the following complexes are diamagnetic?
(a) $[Ni(H_2O)_6]^{2+}$ (b) $[Co(CN)_6]^{3-}$
(c) $[HgI_4]^{2-}$ (d) $[Cu(NH_3)_4]^{2+}$

20.100 For each of the following reactions in acidic solution, predict the products and write a balanced net ionic equation.
(a) $Co^{3+}(aq) + H_2O(l) \longrightarrow$
(b) $Cr^{2+}(aq) + O_2(g) \longrightarrow$
(c) $Cu(s) + Cr_2O_7^{2-}(aq) \longrightarrow$
(d) $CrO_4^{2-}(aq) + H^+(aq) \longrightarrow$

20.101 Complete and balance the net ionic equation for each of the following reactions in acidic, basic, or neutral solution.
(a) $MnO_4^-(aq) + C_2O_4^{2-}(aq) \xrightarrow{\text{acidic}} Mn^{2+}(aq) + CO_2(g)$
(b) $Cr_2O_7^{2-}(aq) + Ti^{3+}(aq) \xrightarrow{\text{acidic}} Cr^{3+}(aq) + TiO^{2+}(aq)$
(c) $MnO_4^-(aq) + SO_3^{2-}(aq) \xrightarrow{\text{basic}} MnO_4^{2-}(aq) + SO_4^{2-}(aq)$
(d) $Fe(OH)_2(s) + O_2(g) \xrightarrow{\text{neutral}} Fe(OH)_3(s)$

20.102 Calculate the concentrations of Fe^{3+}, Cr^{3+}, and $Cr_2O_7^{2-}$ in a solution prepared by mixing 100 mL of 0.100 M $K_2Cr_2O_7$ and 100 mL of 0.400 M $FeSO_4$. The initial solutions are strongly acidic.

20.103 In basic solution, $Cr(OH)_4^-$ is oxidized to CrO_4^{2-} by hydrogen peroxide. Calculate the concentration of CrO_4^{2-} in a solution prepared by mixing 40 mL of 0.030 M $Cr_2(SO_4)_3$, 10 mL of 0.20 M H_2O_2, and 50 mL of 1.0 M NaOH.

20.104 Name each of the following compounds.
(a) $Na[Pt(H_2O)Br(C_2O_4)_2]$
(b) $[Cr(NH_3)_6][Co(C_2O_4)_3]$
(c) $[Co(NH_3)_6][Cr(C_2O_4)_3]$
(d) $[Rh(en)_2(NH_3)_2]_2(SO_4)_3$

20.105 $[Rh(en)_2(NO_2)(SCN)]^+$ can exist in 12 isomeric forms, including constitutional isomers and stereoisomers. Sketch the structures of all 12 isomers.

20.106 The glycinate anion, $gly^- = NH_2CH_2CO_2^-$, bonds to metal ions through the N atom and one of the O atoms. Using N⌒O to represent gly^-, sketch the structures of the four stereoisomers of $Co(gly)_3$.

20.107 Sketch the five possible diastereoisomers of $Pt(gly)_2Cl_2$, where gly^- is the bidentate ligand described in Problem 20.106. Which of the isomers has a dipole moment? Which can exist as a pair of enantiomers?

20.108 What is the difference between the valence bond theory and the crystal field theory? What are the advantages and disadvantages of each model?

20.109 Describe the bonding in $[Mn(CN)_6]^{3-}$ in terms of both the crystal field theory and the valence bond theory. Include the appropriate crystal field d orbital energy level diagram and the valence bond orbital diagram. Which model allows you to predict the number of unpaired electrons? How many do you expect?

20.110 $[FeCl_6]^{3-}$ is more paramagnetic than $[Fe(CN)_6]^{3-}$. Explain.

20.111 Although Cl^- is a weak-field ligand and CN^- is a strong-field ligand, $[CrCl_6]^{3-}$ and $[Cr(CN)_6]^{3-}$ exhibit approximately the same amount of paramagnetism. Explain.

20.112 In octahedral complexes, the choice between high-spin and low-spin electron configurations arises only for d^4–d^7 complexes. Explain.

20.113 Draw a crystal field energy level diagram and predict the number of unpaired electrons for each of the following.
(a) $[Mn(H_2O)_6]^{2+}$ (b) $Pt(NH_3)_2Cl_2$
(c) $[FeO_4]^{2-}$ (d) $[Ru(NH_3)_6]^{2+}$ (low-spin)

20.114 Aqueous solutions of $Fe(NO_3)_3$ are acidic. Explain.

20.115 Explain why $[CoCl_4]^{2-}$ (blue) and $[Co(H_2O)_6]^{2+}$ (pink) have different colors. Which complex has its absorption bands at longer wavelengths?

chapter *21* METALS AND SOLID-STATE
MATERIALS

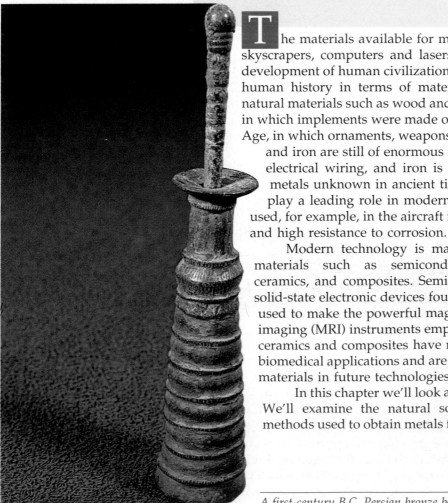

T he materials available for making tools and weapons, houses and skyscrapers, computers and lasers have had a profound effect on the development of human civilization. Indeed, archaeologists organize early human history in terms of materials—the Stone Age, in which only natural materials such as wood and stone were available; the Bronze Age, in which implements were made of copper alloyed with tin; and the Iron Age, in which ornaments, weapons, and tools were made of iron. Copper and iron are still of enormous importance. Copper is used in making electrical wiring, and iron is the main constituent of steel. Today, metals unknown in ancient times, such as aluminum and titanium, play a leading role in modern technology. These metals are widely used, for example, in the aircraft industry because of their low densities and high resistance to corrosion.

Modern technology is made possible by a host of solid-state materials such as semiconductors, superconductors, advanced ceramics, and composites. Semiconductors are used in the miniature solid-state electronic devices found in computers. Superconductors are used to make the powerful magnets found in the magnetic resonance imaging (MRI) instruments employed in medical diagnosis. Advanced ceramics and composites have numerous engineering, electronic, and biomedical applications and are likely to be among the more important materials in future technologies.

In this chapter we'll look at both metals and solid-state materials. We'll examine the natural sources of the metallic elements, the methods used to obtain metals from their ores, and the models used to

A first-century B.C. Persian bronze bottle for eye make-up.

844

Aluminum and titanium are used in making this jet airliner.

to describe the bonding in metals. We'll also look at the structure, bonding, properties, and applications of semiconductors, superconductors, ceramics, and composites.

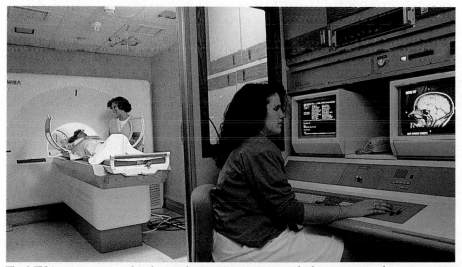
The MRI instruments used in hospitals contain magnets made from superconductors.

21.1 ▶SOURCES OF THE METALLIC ELEMENTS

Most metals occur in nature as **minerals**, the crystalline, inorganic constituents of the rocks that make up the earth's crust. Silicates and aluminosilicates (Section 19.7) are the most abundant minerals, but they are difficult to concentrate and reduce and are therefore generally unimportant as commercial sources of metals. More important are oxides and sulfides, such as hematite (Fe_2O_3), rutile (TiO_2), and cinnabar (HgS) (Figure 21.1), which yield iron, titanium, and mercury, respectively. Mineral deposits from which metals can be produced economically are called **ores** (Table 21.1).

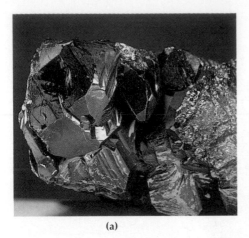

(a)

(b)

(c)

FIGURE 21.1 Samples of **(a)** hematite (Fe_2O_3), **(b)** star-shaped needles of rutile (TiO_2) on a quartz matrix, and **(c)** cinnabar (HgS).

TABLE 21.1   Principal Ores of Some Important Metals

Metal	Ore	Formula	Location of Important Deposits
Aluminum	Bauxite	$Al_2O_3 \cdot x\ H_2O$	Jamaica, Surinam
Chromium	Chromite	$FeCr_2O_4$	Russia, South Africa
Copper	Chalcopyrite	$CuFeS_2$	U.S., Chile, Canada
Iron	Hematite	Fe_2O_3	Australia, Ukraine, U.S.
Lead	Galena	PbS	U.S., Australia, Canada
Manganese	Pyrolusite	MnO_2	Brazil, Gabon, South Africa
Mercury	Cinnabar	HgS	Spain, Algeria, Mexico
Tin	Cassiterite	SnO_2	Malaysia, Bolivia
Titanium	Rutile	TiO_2	Australia
	Ilmenite	$FeTiO_3$	Canada, U.S., Australia
Zinc	Sphalerite	ZnS	U.S., Canada, Australia

Gold is sufficiently unreactive to occur in nature in elemental form.

It's interesting to note how the chemical composition of the most common ores correlates with the location of the metal in the periodic table (Figure 21.2). The early transition metals on the left side of the *d* block generally occur as oxides, and the more electronegative, late transition metals on the right side of the *d* block occur as sulfides. This pattern makes sense because the less electronegative metals tend to form ionic compounds by losing electrons to highly electronegative nonmetals such as oxygen. By contrast, the more electronegative metals tend to form compounds with more covalent character by bonding to the less electronegative nonmetals such as sulfur. Accordingly, we might expect to find oxide ores for the *s*-block metals and sulfide ores for the more electronegative *p*-block metals. In fact, sulfide ores are common for the *p*-block metals, except for Al and Sn, but oxides of the *s*-block metals are strongly basic and far too reactive to exist in an environment that contains acidic oxides such as CO_2 and SiO_2. Consequently, *s*-block metals are found in nature as carbonates, as silicates, and, in the case of Na and K, as chlorides (Sections 6.5 and 6.6). Only gold and the platinum-group metals (Ru, Os, Rh, Ir, Pd, and Pt) are sufficiently unreactive to occur commonly in uncombined form as the free metals.

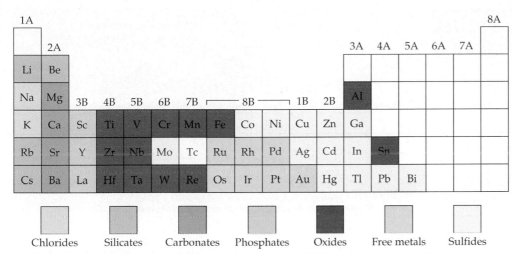

FIGURE 21.2 Primary mineral sources of metals. The *s*-block metals occur as chlorides, silicates, and carbonates. The *d*- and *p*-block metals are found as oxides and sulfides, except for the group 3B metals, which occur as phosphates, and the platinum-group metals and gold, which occur in uncombined form. There is no mineral source of technetium (Tc in group 7B), a radioactive element that is made in nuclear reactors.

21.2 ➤METALLURGY

An ore is a complex mixture of a metal-containing mineral and economically worthless material called **gangue** (pronounced "gang"), consisting of sand, clay, and other impurities. The extraction of a metal from its ore requires several steps: (1) concentration of the ore and, if necessary, chemical treatment prior to reduction; (2) reduction of the mineral to the free metal; (3) refining or purification of the metal. These processes are a part of **metallurgy**, the science and technology of extracting metals from their ores. Another aspect of metallurgy is the making of alloys, metallic materials composed of two or more elements. Steels and bronze, for example, are alloys.

In panning for gold, prospectors use density differences to separate gold from the gangue.

Concentration and Chemical Treatment of Ores

Ores are concentrated by separating the mineral from the gangue. The mineral and the gangue have different properties, which are exploited in various separation methods. Density differences, for example, are important in panning for gold, a procedure in which prospectors flush water over gold-bearing earth in a pan. The less dense gangue is washed away, and the more dense gold particles remain at the bottom of the pan. Differences in magnetic properties are used in concentrating the iron ore magnetite (Fe_3O_4). The Fe_3O_4 is strongly attracted by magnets, but the gangue is unaffected.

Metal sulfide ores are concentrated by **flotation**, a process that exploits differences in the ability of water and oil to wet the surfaces of the mineral and the gangue. A powdered ore such as chalcopyrite ($CuFeS_2$) is mixed with water, oil, and a detergent, and the mixture is vigorously agitated in a tank by blowing air through the liquid (Figure 21.3). The gangue, which

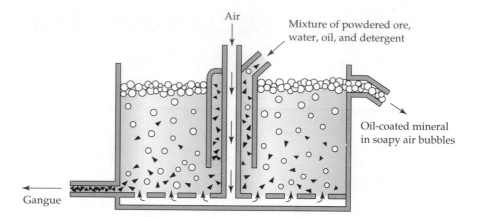

FIGURE 21.3 The flotation process for concentrating metal sulfide ores. Mineral particles float to the top of the tank along with the soapy air bubbles, while the gangue sinks to the bottom.

contains ionic silicates, is moistened by the polar water molecules and sinks to the bottom of the tank. The mineral particles, which contain the less polar metal sulfide, are coated by the oil and become attached to the soapy air bubbles created by the detergent. The metal sulfide particles are thus carried to the surface in the soapy froth, which is skimmed off at the top of the tank.

Sometimes, ores are concentrated by chemical treatment. In the *Bayer process*, for instance, the Al_2O_3 in bauxite ($Al_2O_3 \cdot x\ H_2O$) is separated from Fe_2O_3 impurities by treating the ore with hot aqueous NaOH. The amphoteric Al_2O_3 dissolves as the aluminate ion, $Al(OH)_4^-$, but the basic Fe_2O_3 does not:

$$Al_2O_3(s) + 2\ OH^-(aq) + 3\ H_2O(l) \rightarrow 2\ Al(OH)_4^-(aq)$$

The Bayer process is described in Section 16.12.

Chemical treatment is also used to convert minerals to compounds that are more easily reduced to the metal. For example, sulfide minerals, such as sphalerite (ZnS), are converted to oxides by **roasting**, a process that involves heating the mineral in air:

$$2\ ZnS(s) + 3\ O_2(g) \xrightarrow{\text{heat}} 2\ ZnO(s) + 2\ SO_2(g)$$

Formerly, the sulfur dioxide byproduct was a source of acid rain because it is oxidized in the atmosphere to SO_3, which reacts with water vapor to yield sulfuric acid. In modern roasting facilities, however, the SO_2 is converted on site to sulfuric acid rather than being vented to the atmosphere.

Reduction

Once an ore has been concentrated, it is reduced to the free metal, either by chemical reduction or by electrolysis. The method used (Table 21.2) depends on the activity of the metal as measured by its standard reduction potential (Table 18.1). The most active metals have the most negative standard reduction potentials and are the most difficult to reduce; the least active metals have the most positive standard reduction potentials and are the easiest to reduce.

TABLE 21.2	Reduction Methods for Producing Some Common Metals	
	Metal	**Reduction Method**
Least active	Au, Pt	None; found in nature as the free metal
	Cu, Ag, Hg	Roasting of the metal sulfide
	V, Cr, Mn, Fe, Ni, Zn, W, Pb	Reduction of the metal oxide with carbon, hydrogen, or a more active metal
	Al	Electrolysis of molten Al_2O_3 in cryolite
Most active	Li, Na, Mg	Electrolysis of the molten metal chloride

Gold and platinum are so inactive that they are usually found in nature in uncombined form, but copper and silver, which are slightly more active, are found in both combined and uncombined form. Copper, silver, and mercury commonly occur in sulfide ores that are easily reduced by roasting. Cinnabar (HgS), for example, yields elemental mercury when the ore is heated at 600°C in a stream of air:

$$HgS(s) + O_2(g) \xrightarrow{600°C} Hg(g) + SO_2(g)$$

Although it may seem strange that heating a substance in oxygen reduces it to the metal, both oxidation and reduction of HgS occur in this reaction: Sulfide ions are oxidized and mercury(II) is reduced.

More active metals, such as chromium, zinc, and tungsten, are obtained by reducing their oxides with a chemical reducing agent such as carbon, hydrogen, or a more active metal (Na, Mg, or Al). Pure chromium, for example, is produced by reducing Cr_2O_3 with aluminum (Section 20.4). Zinc, used in the automobile industry for galvanizing steel, is obtained by reducing ZnO with coke, a form of carbon produced by heating coal in the absence of air:

$$ZnO(s) + C(s) \xrightarrow{heat} Zn(g) + CO(g)$$

Although carbon is the cheapest available reducing agent, it is unsatisfactory for reducing oxides of metals like tungsten that form very stable carbides. (Tungsten carbide, WC, is an extremely hard material used to make high-speed cutting tools.) The preferred method for producing tungsten involves reducing tungsten(VI) oxide with hydrogen:

$$WO_3(s) + 3 H_2(g) \xrightarrow{850°C} W(s) + 3 H_2O(g)$$

Because of its high strength, high melting point (3410°C), low volatility, and high efficiency for converting electrical energy into light, tungsten is used to make filaments for electric light bulbs (Figure 21.4).

There are no chemical reducing agents strong enough to reduce compounds of the most active metals, so these metals are produced by electrolytic reduction (Section 18.12). Lithium, sodium, and magnesium, for

FIGURE 21.4 The familiar incandescent light bulb emits white light when an electric current passes through the tungsten wire filament, heating it to a high temperature. The glass bulb contains gases such as argon and nitrogen, which carry away heat from the filament. Although tungsten has the highest boiling point of any element (5660°C), the tungsten slowly evaporates from the hot filament and condenses as the black spot of tungsten metal usually visible on the inside surface of a burned-out bulb.

example, are obtained by electrolysis of their molten chlorides. Aluminum is manufactured by electrolysis of purified Al_2O_3 in molten cryolite (Na_3AlF_6).

Refining

For most applications, the metals obtained from processing and reducing ores need to be purified. The methods used include distillation, chemical purification, and electrorefining. Zinc (bp 907°C), for example, is volatile enough to be separated from cadmium, lead, and other impurities by distillation. In this way, the purity of the zinc obtained from reduction of ZnO is increased from 99% to 99.99%. Nickel, used as a catalyst and as a battery material, is purified by the **Mond process**, a chemical method involving formation and subsequent decomposition of the volatile compound nickel tetracarbonyl, $Ni(CO)_4$ (bp 43°C). Carbon monoxide is passed over impure nickel at about 150°C and 20 atm pressure, forming $Ni(CO)_4$ and leaving metal impurities behind. The $Ni(CO)_4$ is then decomposed at higher temperatures (about 230°C) on pellets of pure nickel. The process works because the equilibrium shifts to the left with increasing temperature:

$$\text{Ni}(s) + 4\,\text{CO}(g) \underset{\text{higher temp.}}{\overset{\text{lower temp.}}{\rightleftharpoons}} \text{Ni(CO)}_4(g) \quad \Delta H° = -160.8 \text{ kJ};\ \Delta S° = -410 \text{ J/K}$$

A similar strategy is employed to purify zirconium, which is used as cladding for fuel rods in nuclear reactors. The crude metal is heated at about 200°C with a small amount of iodine in an evacuated container to form the volatile ZrI_4. The ZrI_4 is then decomposed to pure zirconium by letting the vapor come in contact with an electrically heated tungsten or zirconium filament at about 1300°C:

$$\text{Zr}(s) + 2\,\text{I}_2(g) \underset{1300°C}{\overset{200°C}{\rightleftharpoons}} \text{ZrI}_4(g)$$

Zirconium is used as cladding for fuel rods in nuclear reactors.

Copper from the reduction of ores must be purified for use in making electrical wiring because impurities increase its electrical resistance. The method used is electrorefining, an electrolytic process in which copper is oxidized to Cu^{2+} at an impure copper anode, and Cu^{2+} from an aqueous copper sulfate solution is reduced to copper at a pure copper cathode. The process is described in Section 18.12.

21.3 ➤ IRON AND STEEL

The metallurgy of iron is of special technological importance because iron is the major constituent of steel, the most widely used of all metallic materials. Worldwide production of steel amounts to some 700 million tons per year. Iron is produced by carbon monoxide reduction of iron ore, usually hematite (Fe_2O_3), in a huge reactor called a blast furnace (Figure 21.5). A charge of iron ore, coke, and limestone ($CaCO_3$) is introduced at the top of the furnace, and a blast of hot air is sent in at the bottom, where the coke burns, yielding carbon monoxide and producing temperatures of about 2000°C:

$$2\ C(s) + O_2(g) \rightarrow 2\ CO(g) \qquad \Delta H° = -221\ \text{kJ}$$

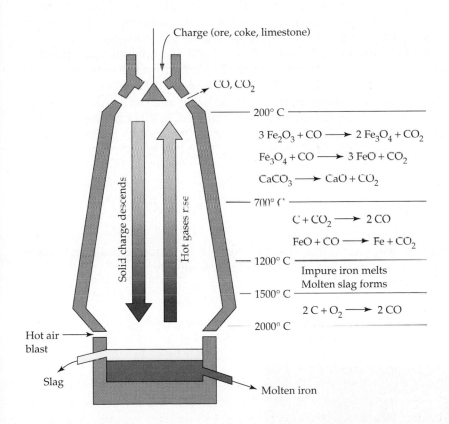

FIGURE 21.5 A diagram of a blast furnace for reduction of iron ore. Modern blast furnaces are as large as 60 m high and 14 m in diameter. They are designed for continuous operation and produce up to 10,000 tons of iron per day. Note the approximate temperatures and the chemical reactions that occur in the various regions of the furnace.

As the charge descends and the hot carbon monoxide rises, a complex series of high-temperature reactions occur in the various regions of the furnace, as shown in Figure 21.5. The key overall reaction is the reduction of Fe_2O_3 to iron metal (mp 1535°C), which is obtained as an impure liquid at the bottom of the furnace:

$$Fe_2O_3(s) + 3\ CO(g) \rightarrow 2\ Fe(l) + 3\ CO_2(g)$$

The purpose of the limestone is to remove the gangue from the iron ore. At the high temperatures of the furnace, the limestone decomposes to lime (CaO), a basic oxide that reacts with SiO_2 and other acidic oxides present in the gangue. The product, called **slag**, is a molten material consisting mainly of calcium silicate:

$$CaCO_3(s) \rightarrow CaO(s) + CO_2(g)$$

$$\underset{\text{Lime}}{CaO(s)} + \underset{\text{Sand}}{SiO_2(s)} \rightarrow \underset{\text{Slag}}{CaSiO_3(l)}$$

The slag, which is less dense than the molten iron, floats on the surface of the iron, thus allowing the iron and the slag to be removed from the bottom of the furnace through separate taps.

The iron obtained from a blast furnace, a brittle material called *cast iron*, or *pig iron*, contains about 4% elemental carbon and smaller amounts of other impurities such as elemental silicon, phosphorus, sulfur, and manganese. The latter elements are formed from their compounds in the reducing atmosphere of the furnace. The most important of several methods for purifying the iron and converting it to steel is the **basic oxygen process**, shown in Figure 21.6. Molten iron from the blast furnace is exposed to a jet of pure oxygen gas for about 20 min in a furnace that is lined with basic oxides such as CaO. The impurities are oxidized, and the acidic oxides that form react with the basic CaO to yield a molten slag that can be poured off. Phosphorus, for example, is oxidized to P_4O_{10}, which then reacts with CaO to give molten calcium phosphate:

$$P_4(l) + 5\ O_2(g) \rightarrow P_4O_{10}(l)$$

$$\underset{\text{Basic oxide}}{6\ CaO(s)} + \underset{\text{Acidic oxide}}{P_4O_{10}(l)} \rightarrow \underset{\text{Slag}}{2\ Ca_3(PO_4)_2(l)}$$

Manganese also passes into the slag because its oxide is basic and reacts with added SiO_2, yielding molten manganese silicate:

$$2\ Mn(l) + O_2(g) \rightarrow 2\ MnO(s)$$

$$\underset{\text{Basic oxide}}{MnO(s)} + \underset{\text{Acidic oxide}}{SiO_2(s)} \rightarrow \underset{\text{Slag}}{MnSiO_3(l)}$$

FIGURE 21.6 Using the basic oxygen process to make steel.

The basic oxygen process produces steels that contain about 1% carbon but only very small amounts of phosphorus and sulfur. Usually, the composition of the liquid steel is monitored by chemical analysis, and the amounts of oxygen and impure iron used are adjusted to achieve the desired concentrations of carbon and other impurities. The hardness, strength, and malleability of the steel depend on its chemical composition, on the rate at which the liquid steel is cooled, and on subsequent heat treatment of the solid. The mechanical and chemical properties of a steel can also be altered by adding other metals. Stainless steel, for example, is a corrosion-resistant iron alloy that contains up to 30% chromium along with smaller amounts of nickel.

The bodies of these DeLorean cars are made of stainless steel.

21.4 ➤BONDING IN METALS

Thus far, we've discussed the sources, production, and properties of some important metals. Some properties, such as hardness and melting point, vary considerably among metals, but other properties are characteristic of metals in general. All metals are malleable, ductile, lustrous, and good conductors of heat and electricity. By *malleable*, we mean that a metal can be hammered into sheets; by *ductile*, we mean that it can be drawn into wires. Both properties reflect a metal's ability to be deformed without breaking into pieces like glass or an ionic crystal. The *luster* of a metal is its shiny appearance, most evident when light is reflected from a well-polished metallic surface. The high thermal and electrical conductivity of metals are familiar properties. When you touch a metal, it feels cold because the metal efficiently conducts heat away from your hand, and when you connect a metal wire to the terminals of a battery, it conducts an electric current.

To understand these properties, we need to look at the bonding in metals. We'll consider two theoretical models that are commonly used: the electron-sea model and the molecular orbital theory.

Electron-Sea Model of Metals

If you try to draw a Lewis electron-dot structure for a metal, you'll quickly realize that there aren't enough valence electrons available to form an electron-pair bond between every pair of adjacent atoms. Sodium, for example, which has just one valence electron per Na atom ($3s^1$), crystallizes in a body-centered cubic structure in which each Na atom is surrounded by eight nearest neighbors (Sections 10.8 and 10.9). Consequently, the valence electrons can't be localized in a bond between any particular pair of atoms. Instead, they are delocalized and belong to the crystal as a whole. In the **electron-sea model**, we visualize the crystal as a three-dimensional array of metal cations immersed in a sea of delocalized electrons that are free to move throughout the crystal (Figure 21.7). The continuum of delocalized, mobile valence electrons acts as an electrostatic glue that holds the metal cations together.

The electron-sea model affords a simple qualitative explanation for the electrical and thermal conductivity of metals. Because the electrons are mobile, they are free to move away from a negative electrode and toward a positive electrode when a metal is subjected to an electrical potential. The mobile electrons can also conduct heat by carrying kinetic energy from one

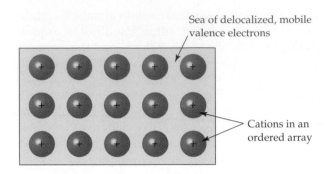

Sea of delocalized, mobile valence electrons

Cations in an ordered array

FIGURE 21.7 Two-dimensional representation of the electron-sea model of a metal. An ordered array of cations is immersed in a continuous distribution of delocalized, mobile valence electrons. The valence electrons do not belong to any particular metal ion but to the crystal as a whole.

part of the crystal to another. The malleability and ductility of metals follows from the fact that the delocalized bonding extends in all directions; that is, it is not confined to oriented bond directions as in covalent network solids like SiO_2. When a metallic crystal is deformed, no localized bonds are broken. Instead, the electron sea simply adjusts to the new distribution of cations, and the energy of the deformed structure is similar to that of the original. Thus, the energy required to deform a metal like sodium is relatively small. Of course, the energy required to deform a transition metal like iron is greater because iron has more valence electrons ($4s^2\ 3d^6$), and the electrostatic glue is therefore more dense.

Molecular Orbital Theory for Metals

A more detailed understanding of the bonding in metals is provided by the molecular orbital theory, a model that is a logical extension of the molecular orbital description of small molecules discussed in Sections 7.13–7.15. Recall that in the H_2 molecule the $1s$ orbitals of the two H atoms overlap to give a σ bonding MO and a higher-energy, σ^* antibonding MO. The bonding in the gaseous Na_2 molecule is similar: The $3s$ orbitals of the two Na atoms combine to give a σ and a σ^* MO. Because each Na atom has just one $3s$ valence electron, the lower-energy, bonding orbital is filled and the higher-energy, antibonding orbital is empty.

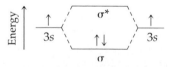

Now consider how the energy level diagram is altered if we bring together an increasingly large number of Na atoms and so build up a crystal of sodium metal. The key idea to remember from Section 7.13 is that the number of molecular orbitals formed is the same as the number of atomic orbitals combined. Thus, there will be three MOs for a triatomic Na_3 molecule, four MOs for Na_4, and so on. A cubic crystal of sodium metal, 1.5 mm on an edge, contains about 10^{20} Na atoms and therefore has about 10^{20} MOs, each of which is delocalized over all the atoms in the crystal. As shown in Figure 21.8, the difference in energy between successive MOs in an Na_n molecule decreases as the number of Na atoms increases, so that the MOs merge into an almost continuous band of energy levels. Consequently, MO

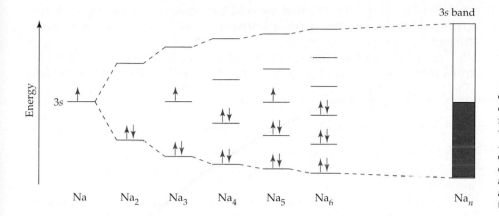

FIGURE 21.8 Molecular orbital energy levels for Na$_n$ molecules. A crystal of sodium metal can be regarded as a giant Na$_n$ molecule, where n has a value of about 10^{20}. As the value of n increases, the energy levels merge into an almost continuous band. Because each Na atom has one 3s valence electron and each MO can hold two electrons, the 3s band is half-filled.

theory for metals is often called **band theory**. The bottom half of the band consists of bonding MOs and is filled, whereas the top half of the band consists of antibonding MOs and is empty.

How does band theory account for the electrical conductivity of metals? According to quantum mechanics, each electron in a metal has a discrete kinetic energy and a discrete velocity. These values depend on the particular MO energy level and increase from the bottom of a band to the top of a band. For a one-dimensional metal wire, electrons traveling in opposite directions at the same speed have the same value of the kinetic energy. Thus, the energy levels within a band occur in degenerate pairs; one set of energy levels applies to electrons moving to the right, and the other set applies to electrons moving to the left. In the absence of an electrical potential, the two sets of levels are equally populated. That is, for each electron moving to the right, there is another electron moving to the left with exactly the same speed (Figure 21.9). As a result, there is no net electric current in either direction. In the presence of an electrical potential, however, those electrons moving to the right (toward the positive terminal of a battery) are accelerated, those moving to the left (toward the negative terminal) are slowed down, and those moving to the left with very slow speeds undergo a change

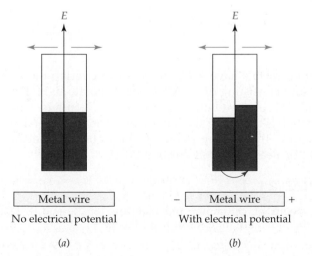

Metal wire
No electrical potential
(a)

− Metal wire +
With electrical potential
(b)

FIGURE 21.9 Half-filled 3s band of MO energy levels for sodium metal. The direction of electron motion for the two degenerate sets of energy levels is indicated by the horizontal arrows. **(a)** In the absence of an electrical potential, the two sets of levels are equally populated, and there is no electric current through the wire. **(b)** In the presence of an electrical potential (positive electrode on the right), some of the electrons shift from one set of energy levels to the other, and there is a net current of electrons from left to right.

of direction. Thus, the number of electrons moving to the right is now greater than the number moving to the left, and so there is a net electric current.

It's evident from Figure 21.9 that an electrical potential can shift electrons from one set of energy levels to the other only if the band is partially filled. If the band is completely filled, there are no available vacant energy levels to which electrons can be excited, and therefore the two sets of levels must remain equally populated, even in the presence of an electrical potential. This means that an electrical potential can't accelerate the electrons in a completely filled band. *Materials that have only completely filled bands are therefore electrical insulators.* By contrast, *materials that have partially filled bands are metals.*

Based on the preceding analysis, we would predict that magnesium should be an insulator because it has electron configuration [Ar] $3s^2$ and should therefore have a completely filled $3s$ band. This prediction is wrong, however, because we have not yet considered the $3p$ valence orbitals. Just as the $3s$ orbitals combine to form a $3s$ band, so the $3p$ orbitals can combine to form a $3p$ band. If the $3s$ and $3p$ bands were widely separated in energy, the $3s$ band would be filled, the $3p$ band would be empty, and magnesium would be an insulator. In fact, though, the $3s$ and $3p$ bands overlap in energy, and the resulting composite band is only partially filled (Figure 21.10). Thus, magnesium is a conductor.

FIGURE 21.10 In magnesium metal, the $3s$ and $3p$ bands have similar energies and overlap to give a composite band consisting of four MOs per Mg atom. The composite band can accommodate eight electrons per Mg atom but is only partially filled, since each Mg atom has just two valence electrons. (In this and subsequent figures, we don't show the separate sets of energy levels for the right- and left-moving electrons.)

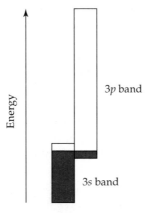

Transition metals have a d band that can overlap the s band to give a composite band consisting of six MOs per metal atom. Half of the MOs are bonding and half are antibonding, and thus we might expect maximum bonding for metals that have six valence electrons per metal atom. In accord with this picture, the melting points of the transition metals go through a maximum at or near group 6B (Section 20.2).

EXAMPLE 21.1

The melting points of chromium and zinc are 1857°C and 420°C, respectively. Account for the difference in terms of band theory.

SOLUTION   The electron configurations are [Ar] $3d^5 4s^1$ for Cr and [Ar] $3d^{10} 4s^2$ for Zn. Assume that the $3d$ and $4s$ bands overlap. The composite band, which can accommodate 12 valence electrons per metal atom, will be half-filled for Cr and completely filled for Zn. Strong bonding and a high melting point are expected for Cr because all the bonding MOs are occupied and all the antibonding MOs are empty. Weak bonding and a low melting point are expected for Zn because both the bonding and the antibonding MOs are occupied. (The fact that Zn is a metal suggests that the $4p$ orbitals also contribute to the composite band.)

PROBLEM 21.1   Mercury metal is a liquid at room temperature. Using band theory, suggest a reason for its low melting point ($-39°C$).

21.5 ➤ SEMICONDUCTORS

A **semiconductor**, such as silicon or germanium, is a material that has an electrical conductivity intermediate between that of a metal and that of an insulator. To understand the electrical properties of semiconductors, let's look first at the bonding in insulators. Consider diamond, for example, a covalent network solid in which each C atom is bonded tetrahedrally to four other C atoms (Figure 10.17). In a localized description of the bonding, C–C electron-pair bonds result from overlap of sp^3 hybrid orbitals. In a delocalized description, the $2s$ and $2p$ valence orbitals of all the C atoms combine to give bands of bonding and antibonding MOs. As is generally the case for insulators, the bonding MOs, called the **valence band**, and the higher-energy, antibonding MOs, called the **conduction band**, are separated in energy by a large **band gap**. The band gap in diamond is about 520 kJ/mol.

Each of the two bands in diamond can accommodate four electrons per C atom. Because carbon has just four valence electrons ($2s^2 2p^2$), the valence band is completely filled and the conduction band is completely empty. Diamond is therefore an insulator because there are no vacant MOs in the valence band to which electrons can be excited by an electrical potential and because excitation to the vacant MOs of the conduction band is prevented by the large band gap. By contrast, metallic conductors have no energy gap between the highest occupied and lowest unoccupied MOs. This fundamental difference between the energy levels of metals and insulators is illustrated in Figure 21.11.

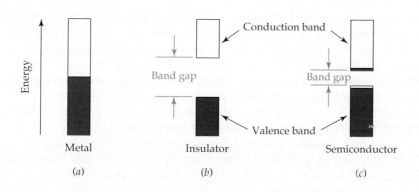

FIGURE 21.11 Bands of MO energy levels for **(a)** a metallic conductor, **(b)** an electrical insulator, and **(c)** a semiconductor. A metallic conductor has a partially filled band. An electrical insulator has a completely filled valence band and a completely empty conduction band, which are separated in energy by a large band gap. In a semiconductor, the band gap is smaller. As a result, the conduction band is partially occupied with a few electrons, and the valence band is partially empty. Electrical conductivity in metals and semiconductors results from the presence of partially filled bands.

The MOs of a semiconductor are similar to those of an insulator, but the band gap in a semiconductor is smaller (Figure 21.11). As a result, a few electrons have enough energy to jump the gap and occupy the higher-energy, conduction band. The conduction band is thus partially filled, and the valence band is also partially filled because it now contains a few unoccupied MOs. When an electrical potential is applied to a semiconductor, it conducts a small amount of current because the potential can accelerate the electrons in the partially filled bands. Table 21.3 shows how the electrical properties of the group 4A elements vary with the size of the band gap.

TABLE 21.3	Band Gaps for the Group 4A Elements	
Element[a]	**Band Gap (kJ/mol)**	**Type of Material**
C (diamond)	520	Insulator
Si	107	Semiconductor
Ge	65	Semiconductor
Sn (gray tin)	8	Semiconductor
Sn (white tin)	0	Metal
Pb	0	Metal

[a] Si, Ge, and gray Sn have the same structure as diamond.

The electrical conductivity of a semiconductor *increases* with increasing temperature because the number of electrons with sufficient energy to occupy the conduction band increases as the temperature rises. At higher temperatures, there are more charge carriers (electrons) in the conduction band and more vacancies in the valence band. By contrast, the electrical conductivity of a metal *decreases* with increasing temperature. At higher temperatures, the metal cations undergo increased vibrational motion about their lattice sites, and vibration of the cations disrupts the flow of electrons through the crystal.

The conductivity of a semiconductor can be greatly increased by adding certain impurities in small (ppm) amounts, a process called **doping**. Consider, for example, the addition of a group 5A element such as phosphorus to a group 4A semiconductor such as silicon. Like diamond, silicon has a structure in which each Si atom is surrounded tetrahedrally by four others. The added P atoms occupy normal Si positions in the structure, but each P atom has five valence electrons and therefore introduces an extra electron not needed for bonding. In the MO picture, the extra electrons occupy the conduction band, as shown in Figure 21.12(a). The number of electrons in the conduction band of the doped silicon is much greater than in pure silicon, and the conductivity of the doped semiconductor is therefore correspondingly higher. Because the charge carriers are *negative* electrons, the group 5A–doped silicon is called an ***n*-type semiconductor**.

Now let's consider a semiconductor in which silicon is doped with a group 3A element such as boron. Each B atom has just three valence electrons and therefore does not have enough electrons to form bonds to its four Si neighbors. In the MO picture, the bonding MOs of the valence band

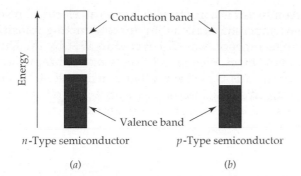

(a)   (b)

n-Type semiconductor   *p*-Type semiconductor

FIGURE 21.12   MO energy levels for doped semiconductors. **(a)** An *n*-type semiconductor, such as silicon doped with phosphorus, has more electrons than needed for bonding and thus has *negative* electrons in the partially filled conduction band. **(b)** A *p*-type semiconductor, such as silicon doped with boron, has fewer electrons than needed for bonding and thus has vacancies—*positive* holes—in the valence band.

are only partially filled, as shown in Figure 21.12(b). The vacancies in the valence band can be regarded as *positive* holes in a filled band, and the group 3A–doped semiconductor is therefore called a **p-type semiconductor**.

Doped semiconductors are essential components in the modern solid-state electronic devices found in radios, television sets, pocket calculators, and computers. Devices such as transistors, which control electrical signals in these products, are made from *n*-type and *p*-type semiconductors. In modern integrated circuits, an amazing number of extremely small devices can be packed into a small space, thus decreasing the size and increasing the speed of electrical equipment. For example, computer microprocessors now contain up to 100,000 transistors on a silicon chip with a surface area less than 1 cm² and are able to execute as many as 40 million instructions per second.

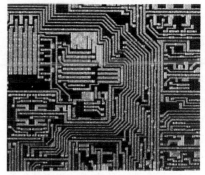

A computer microprocessor chip.

EXAMPLE 21.2

Consider a crystal of germanium that has been doped with a small amount of aluminum. Is the doped crystal an *n*-type or a *p*-type semiconductor? Compare the conductivity of the doped crystal with that of pure germanium.

SOLUTION   Germanium, like silicon, is a group 4A semiconductor, and aluminum, like boron, is a group 3A element. The doped germanium is therefore a *p*-type semiconductor because each Al atom has one less valence electron than needed for bonding to the four neighboring Ge atoms. (Like silicon, germanium has the diamond structure.) The valence band is thus partially filled, which accounts for the electrical conductivity. The conductivity will be greater than that of pure germanium because the doped germanium has many more positive holes in the valence band. That is, it has more vacant MOs available to which electrons can be excited by an electrical potential.

⌐ **PROBLEM 21.2**   Is germanium doped with arsenic an *n*-type or a *p*-type semiconductor? Why is its conductivity greater than that of pure germanium?   ⌐

21.6 ➤SUPERCONDUCTORS

The discovery of high-temperature superconductors is surely one of the most exciting scientific developments in the last 15 years. It has stimulated an enormous amount of research in chemistry, physics, and materials science that could some day lead to a world of superfast computers, magnetically levitated trains, and power lines that carry electric current without loss of energy.

A **superconductor** is a material that loses all electrical resistance below a characteristic temperature called the **superconducting transition temperature**, T_c. This phenomenon was discovered in 1911 by the Dutch physicist Heike Kamerlingh Onnes, who found that mercury abruptly loses its electrical resistance when it is cooled with liquid helium to 4.2 K (Figure 21.13). Below its T_c, a superconductor becomes a perfect conductor, and an electric current, once started, flows indefinitely without loss of energy.

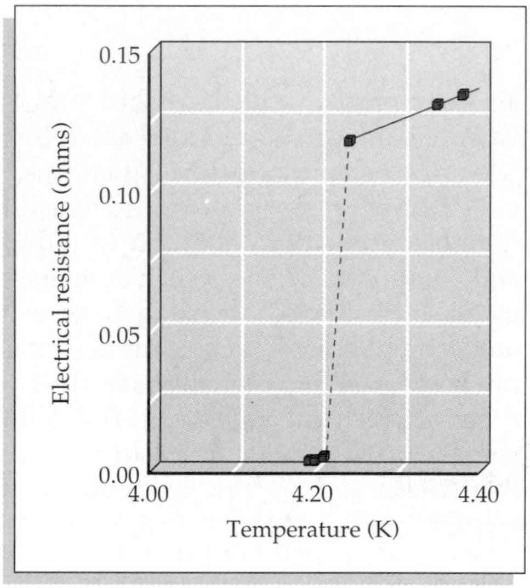

FIGURE 21.13 The electrical resistance of mercury falls to zero at its superconducting transition temperature, T_c = 4.2 K. Above T_c, mercury is a metallic conductor: Its resistance increases (conductivity decreases) with increasing temperature. Below T_c, mercury is a superconductor.

FIGURE 21.14 A computer graphics representation of the crystal structure of $YBa_2Cu_3O_7$. Layers of Y atoms (yellow) and Ba atoms (blue) are sandwiched between layers of CuO_5 square pyramids (red) and chains of square planar CuO_4 groups (green).

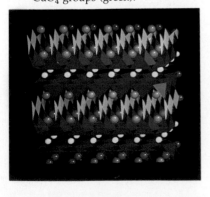

Since 1911, scientists have been searching for materials that superconduct at higher temperatures, and more than 6000 superconductors are now known. Until 1986, however, the record value of T_c was only 23.2 K, for the compound Nb_3Ge. The situation changed dramatically in 1986 when K. Alex Müller and J. Georg Bednorz of the IBM Zürich Research Laboratory reported a T_c of 35 K for the nonstoichiometric barium lanthanum copper oxide $Ba_xLa_{2-x}CuO_4$, where x has a value of about 0.1. Soon thereafter, scientists found even higher values of T_c for other copper-containing oxides: 90 K for $YBa_2Cu_3O_7$, 125 K for $Tl_2Ca_2Ba_2Cu_3O_{10}$, and 133 K for the 1993 record holder, $HgCa_2Ba_2Cu_3O_{8+x}$. High values of T_c for these compounds were completely unexpected because most metal oxides—nonmetallic inorganic solids called *ceramics*—are electrical insulators. Within just one year of discovering the first ceramic superconductor, Müller and Bednorz were awarded the 1987 Nobel Prize in physics.

The crystal structure of $YBa_2Cu_3O_7$, the so-called 1-2-3 compound (one yttrium, two bariums, and three coppers), is illustrated in Figure 21.14, and one unit cell of the structure is shown in Figure 21.15. The crystal contains parallel planes of Y, Ba, and Cu atoms. Two-thirds of the Cu atoms are surrounded by a square pyramid of five O atoms, some of which are shared with neighboring CuO_5 groups to give two-dimensional layers of square pyramids. The remaining Cu atoms are surrounded by a square of four O atoms, two of which are shared with neighboring CuO_4 squares to give chains of CuO_4 groups. It's interesting to note that the Cu atoms have a

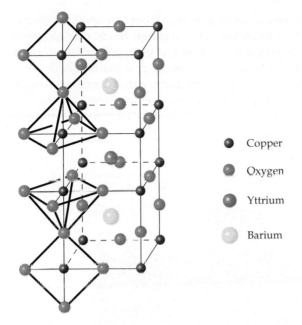

Copper
Oxygen
Yttrium
Barium

FIGURE 21.15 One unit cell of the crystal structure of $YBa_2Cu_3O_7$ contains one Y atom, two Ba atoms, three Cu atoms, and seven O atoms. In counting the Cu and O atoms, recall that the unit cell contains one-eighth of each corner atom, one-fourth of each edge atom, and one-half of each face atom (Section 10.10). The figure includes eight O atoms from neighboring unit cells to show the square pyramidal CuO_5 groups and the square planar CuO_4 groups.

fractional oxidation number of +2.33, based on the usual oxidation numbers of +3 for Y, +2 for Ba, and −2 for O. Both the infinitely extended layers of Cu and O atoms and the fractional oxidation number of Cu appear to play a role in the current flow, but a generally accepted theory of superconductivity in ceramic superconductors is not yet available. This is a field where experiment is far ahead of theory.

One of the most dramatic properties of a superconductor is its ability to levitate a magnet (Figure 21.16). When a superconductor is cooled below its T_c and a magnet is lowered toward it, the superconductor and the magnet repel each other, and the magnet hovers above the superconductor as though suspended in midair. The repulsive force arises in the following way: When the magnet moves toward the superconductor, it induces a supercurrent in the surface of the superconductor that continues to flow even after the magnet stops moving. The supercurrent, in turn, induces a magnetic field in the superconductor that exactly cancels the applied field from the magnet. Thus, the net magnetic field within the bulk of the super-

FIGURE 21.16 Levitation of a magnet above a pellet of $YBa_2Cu_3O_7$ cooled to 77 K with liquid nitrogen. $YBa_2Cu_3O_7$ becomes a superconductor at approximately 90 K.

conductor is zero, a phenomenon called the *Meissner effect*. Outside of the superconductor, however, the magnetic fields due to the magnet and the supercurrent repel each other, just as the north poles of two bar magnets do. The magnet therefore experiences an upward magnetic force as well as the usual downward gravitational force, and it remains suspended above the superconductor at the point where the two forces are equal. Potential applications of the Meissner effect include high-speed, magnetically levitated trains (Figure 21.17).

FIGURE 21.17 An experimental magnetically levitated train in Japan. The train is suspended above superconducting magnets that are cooled with liquid helium.

Some applications of superconductors already exist. For example, powerful superconducting magnets are essential components in the magnetic resonance imaging (MRI) instruments widely used in medical diagnosis (Section 22.10). All present applications, however, use conventional superconductors ($T_c \leq 20$ K) that are cooled to 4.2 K with liquid helium, an expensive substance that requires sophisticated cryogenic (cooling) equipment. Much of the excitement surrounding the new, high-temperature superconductors arises because their T_c values are above the boiling point of liquid nitrogen (bp 77 K), an abundant refrigerant that is cheaper than milk. Of course, the search goes on for materials with still higher values of T_c. For applications such as long-distance electric-power transmission, the goal is a material that superconducts at room temperature.

Several serious problems must be surmounted before applications of high-temperature superconductors can become a reality. Presently known ceramic superconductors are brittle powders with high melting points and are not easily fabricated into the wires and coils needed for electrical equipment. Also, the currents that these materials are able to carry at 77 K are still too low for practical applications. Thus, applications are likely in the future but are not right around the corner.

In 1991, scientists at AT&T Bell Laboratories discovered a new class of high-temperature superconductors based on fullerene, the allotrope of carbon that contains C_{60} molecules (Section 10.11). Called "buckyballs," after the architect R. Buckminster Fuller, these soccer-ball-shaped C_{60} molecules react with potassium to give K_3C_{60}, a stable crystalline solid that contains a face-centered-cubic array of buckyballs, with K^+ ions in the cavities between the buckyballs (Figure 21.18). K_3C_{60} is a metallic conductor at room temperature, but it becomes a superconductor at 18 K. The rubidium fulleride,

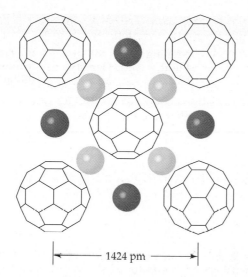

|← 1424 pm →|

FIGURE 21.18 A portion of one unit cell of the face-centered-cubic structure of K_3C_{60} viewed perpendicular to a cube face. The C_{60} "buckyballs" are located at the cube corners and face centers, and the K^+ ions (red and blue spheres) are in two kinds of holes between the C_{60}^{3-} ions. The K^+ ions shown in red lie in the plane of the cube face (the plane of the paper) and are surrounded octahedrally by six C_{60}^{3-} ions. Those shown in blue lie in a plane one-fourth of a cell edge length below the plane of the paper and are surrounded tetrahedrally by four C_{60}^{3-} ions.

Rb_3C_{60}, and a rubidium–thallium–C_{60} compound of unknown stoichiometry have higher T_c values of 30 K and 45–48 K, respectively. Because the fullerides are three-dimensional superconductors, whereas the copper oxide ceramics are two-dimensional superconductors, the fullerides may prove to be better materials for making superconducting wires. The ceramic superconductors conduct a current parallel to the layers of Cu and O atoms but not between layers, whereas the fullerides can conduct in any direction.

PROBLEM 21.3 Show that one unit cell of $YBa_2Cu_3O_7$ (Figure 21.15) contains three Cu atoms and seven O atoms.

21.7 ►CERAMICS

Ceramics are inorganic, nonmetallic, nonmolecular solids, including both crystalline and amorphous materials, such as glasses. Known since ancient times, traditional silicate ceramics, such as pottery and porcelain, are made by heating aluminosilicate clays to high temperatures. Modern, so-called **advanced ceramics**—materials that have high-tech engineering, electronic, and biomedical applications—include *oxide ceramics*, such as alumina (Al_2O_3), and *nonoxide ceramics*, such as silicon carbide (SiC) and silicon nitride (Si_3N_4). Additional examples are listed in Table 21.4, which compares properties of ceramics with those of metallic materials such as aluminum and steel. (Note that oxide ceramics are named by replacing the *um* ending of the element name with an *a*; thus, BeO is beryllia, ZrO_2 is zirconia, and so forth.)

In many respects, the properties of ceramics are superior to those of metals: Ceramics have higher melting points, and they are stiffer, harder, and more resistant to wear and corrosion. Moreover, they maintain much of their strength at high temperatures, where metals either melt or corrode because of oxidation. Silicon nitride and silicon carbide, for example, are stable to oxidation in air up to 1400–1500°C, and oxide ceramics don't react with oxygen because they are already fully oxidized. Because ceramics are

Prototype silicon nitride rotor for use in gas-turbine engines.

TABLE 21.4	Properties of Some Ceramic and Metallic Materials			
Material	Melting Point (°C)	Density (g/cm³)	Elastic Modulus (GPa)[a]	Hardness (mohs)[b]
Oxide ceramics				
Alumina, Al_2O_3	2072	4.0	380	9
Beryllia, BeO	2530	3.0	370	8
Zirconia, ZrO_2	2700	5.9	210	8
Nonoxide ceramics				
Boron carbide, B_4C	2350	2.5	280	9
Silicon carbide, SiC	2700	3.2	400	9
Silicon nitride, Si_3N_4	1900	3.4	310	9
Metals				
Aluminum	660	2.7	70	3
Plain carbon steel	1515	7.9	210	5

[a] The elastic modulus, measured in units of pressure (1 gigapascal = 1 GPa = 10^9 Pa) indicates the stiffness of a material when it is subjected to a load. The larger the value, the stiffer the material.

[b] Numbers on the Mohs hardness scale range from 1 for talc, a very soft material, to 10 for diamond, the hardest known substance.

less dense than steel, they are attractive lightweight, high-temperature materials for replacing metal components in aircraft, space vehicles, and automotive engines.

Unfortunately, ceramics are brittle, as anyone who has dropped a teacup well knows. The brittleness, hardness, stiffness, and high melting points of ceramics are due to strong chemical bonding. Take silicon carbide, for example, a covalent network solid that crystallizes in the diamond structure (Figure 21.19). Each Si atom is bonded tetrahedrally to four C atoms, and each C atom is bonded tetrahedrally to four Si atoms. The strong, highly directional covalent bonds ($D_{Si-C} = 435$ kJ/mol) prevent the planes of atoms from sliding over one another when the solid is subjected to the stress of a load or an impact. As a result, the solid can't deform to relieve the stress.

FIGURE 21.19 One unit cell of the cubic form of silicon carbide, SiC. The Si atoms are located at the corners and face centers of a face-centered-cubic unit cell, while the C atoms occupy cavities (tetrahedral holes) between four Si atoms. Each C atom is bonded tetrahedrally to four Si atoms, and each Si atom is bonded tetrahedrally to four C atoms. To visualize the tetrahedral bonding of the Si atoms, consider the location of the C atoms in adjacent unit cells (not shown). The crystal can't deform under stress because the bonds are strong and highly directional.

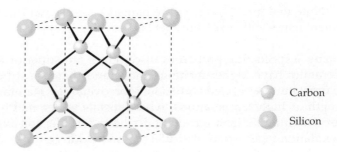

Carbon

Silicon

It maintains its shape up to a point, but then the bonds give way suddenly, and the material fails catastrophically when the stress exceeds a certain threshold value. Oxide ceramics, in which the bonding is largely ionic, behave similarly. By contrast, metals are able to deform under stress because their planes of metal cations can slide easily in the electron sea (Section 21.4). As a result, metals dent but ceramics shatter.

Ceramic processing, the series of steps that leads from raw material to the finished ceramic object, is important in determining the strength and the resistance to fracture of the product. Processing often begins with a fine powder, which is combined with an organic binder, shaped, compacted, and finally *sintered* at temperatures of 1300 to 2000°C. **Sintering**, which occurs below the melting point, is a process in which the particles of the powder are "welded" together without completely melting. During sintering, the crystal grains grow larger and the density of the material increases as the void spaces between particles disappear. Unfortunately, impurities and remaining voids can lead to microscopic cracks that cause the material to fail under stress. It is therefore important to minimize impurities and voids by beginning with high-purity, fine powders that can be tightly compacted prior to sintering.

One approach to making such powders is the **sol–gel method**, which involves synthesis of a metal oxide powder from a metal alkoxide, a compound derived from a metal and an alcohol. In the synthesis of titania (TiO_2) from titanium ethoxide, $Ti(OCH_2CH_3)_4$, for example, the $Ti(OCH_2CH_3)_4$ starting material is made by reacting titanium(IV) chloride with ethanol and ammonia in a benzene solution:

$$TiCl_4 + 4\ HOCH_2CH_3 + 4\ NH_3 \xrightarrow{\ C_6H_6\ } Ti(OCH_2CH_3)_4 + 4\ NH_4Cl$$

Ethanol   Titanium ethoxide

Pure $Ti(OCH_2CH_3)_4$ is then dissolved in an appropriate organic solvent, and water is added to bring about a *hydrolysis* reaction, represented by the following simplified equation:

$$Ti(OCH_2CH_3)_4 + 4\ H_2O \xrightarrow{\ \text{solution}\ } Ti(OH)_4(s) + 4\ HOCH_2CH_3$$

In this reaction, $Ti-OCH_2CH_3$ bonds are broken, $Ti-OH$ bonds are formed, and ethanol is regenerated. The $Ti(OH)_4$ forms in the liquid as a colloidal dispersion called a *sol* consisting of extremely fine particles having a diameter of only 0.001 to 0.1 μm (1 $\mu m = 10^{-3}$ mm). Subsequent reactions eliminate water molecules and form oxygen bridges between Ti atoms:

$$(HO)_3Ti-O\!-\!\boxed{H + H-O}\!-\!Ti(OH)_3 \rightarrow (HO)_3Ti-O-Ti(OH)_3 + H_2O$$

Because all the OH groups can undergo the preceding reaction, the particles of the sol link together through a three-dimensional network of oxygen bridges, and the sol is converted to a more rigid, gelatin-like material called a *gel*. The remaining water and solvent are then removed by heating the gel, and TiO_2 is obtained as a fine powder consisting of high-purity particles with a diameter less than 1 μm (Figure 21.20).

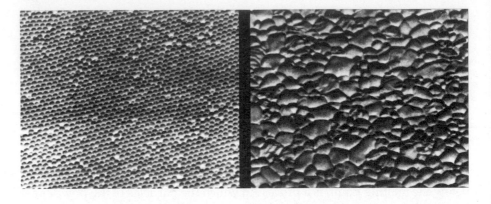

FIGURE 21.20 Electron micrographs of a titania powder consisting of tightly packed particles of TiO_2 with a diameter less than 1 μm (left) and the dense ceramic produced by sintering the powder (right). The powder was made by the sol–gel method.

The insulating material in this spark plug is made of alumina.

Oxide ceramics have many important uses. Alumina, for example, is the material of choice for making spark-plug insulators because of its high electrical resistance, high strength, high thermal stability, and chemical inertness. It is nontoxic and essentially inert in biological systems, so it is used in constructing dental crowns and the heads of artificial hips. Alumina is also used as a substrate material for electronic circuit boards.

EXAMPLE 21.3

Alumina (Al_2O_3) powders can be prepared from aluminum ethoxide by the sol–gel method. Write a balanced equation for the hydrolysis of aluminum ethoxide.

SOLUTION   First, determine the chemical formula of aluminum ethoxide. Because aluminum is a group 3A element, it has an oxidation number of +3. The ethoxide ligand is the anion of ethanol, $HOCH_2CH_3$, and has a charge of −1. Since aluminum ethoxide is a neutral compound, its formula must be $Al(OCH_2CH_3)_3$. The hydrolysis reaction, which breaks the $Al–OCH_2CH_3$ bonds and forms $Al–OH$ bonds, requires one H_2O molecule for each ethoxide ligand:

$$Al(OCH_2CH_3)_3 + 3\ H_2O \rightarrow Al(OH)_3 + 3\ HOCH_2CH_3$$

Subsequent heating converts the aluminum hydroxide to alumina.

EXAMPLE 21.4

The 1-2-3 ceramic superconductor $YBa_2Cu_3O_7$ has been synthesized by the sol–gel method from a stoichiometric mixture of yttrium ethoxide, barium ethoxide, and copper(II) ethoxide in an appropriate organic solvent. The oxide product, before heating in oxygen, has the formula $YBa_2Cu_3O_{6.5}$. Write a balanced equation for hydrolysis of the stoichiometric mixture of metal ethoxides.

SOLUTION   First, determine the chemical formulas of the metal ethoxides. Because yttrium is a group 3B element and has an oxidation number of +3, the formula of yttrium ethoxide must be $Y(OCH_2CH_3)_3$. Similarly, because both barium in group 2A and copper(II) have an oxidation number of +2, the formulas of barium ethoxide and copper ethoxide must be $Ba(OCH_2CH_3)_2$ and $Cu(OCH_2CH_3)_2$, respectively. If the three metal ethoxides were present separately, hydrolysis would give $Y(OH)_3$, $Ba(OH)_2$, and $Cu(OH)_2$. Because they are present together in a 1:2:3 ratio, write the product as $Y(OH)_3 \cdot 2\,Ba(OH)_2 \cdot 3\,Cu(OH)_2$, or $YBa_2Cu_3(OH)_{13}$. Thus, the hydrolysis reaction requires 13 H_2O molecules for 13 OCH_2CH_3 ligands, and the balanced equation is

$$Y(OCH_2CH_3)_3 + 2\,Ba(OCH_2CH_3)_2 + 3\,Cu(OCH_2CH_3)_2 + 13\,H_2O$$
$$\rightarrow YBa_2Cu_3(OH)_{13} + 13\,HOCH_2CH_3$$

Subsequent heating removes water, converting the mixed-metal hydroxide to the oxide $YBa_2Cu_3O_{6.5}$, which is then oxidized to $YBa_2Cu_3O_7$ by heating in O_2 gas.

PROBLEM 21.4 Silica glasses used in making lenses, laser mirrors, and other optical components can be made by the sol–gel method. One step in the process involves hydrolysis of $Si(OCH_3)_4$. Write a balanced equation for the reaction.

PROBLEM 21.5 Crystals of the oxide ceramic barium titanate, $BaTiO_3$, have an unsymmetrical arrangement of ions, and as a result, the crystals have an electric dipole moment. Such materials are called ferroelectrics and are used in making various electronic devices. $BaTiO_3$ can be made by the sol–gel method, which involves hydrolysis of a mixture of metal alkoxides. Write a balanced equation for the hydrolysis of a 1:1 mixture of barium isopropoxide and titanium isopropoxide, and explain how the resulting sol is converted to $BaTiO_3$. (The isopropoxide ligand is the anion of isopropanol, $HOCH(CH_3)_2$, also known as rubbing alcohol.)

21.8 ►COMPOSITES

We saw in the previous section that ceramics are brittle and prone to fracture. They can be strengthened and toughened, though, by mixing the ceramic powder prior to sintering with fibers of a second ceramic material, such as carbon, boron, or silicon carbide. The resulting hybrid material, called a **ceramic composite**, combines the advantageous properties of both components. An example is the composite consisting of fine grains of alumina reinforced with whiskers of silicon carbide. Whiskers are tiny, fiber-shaped particles, about 0.5 μm in diameter and 50 μm long, that are very strong because they are single crystals. Silicon carbide-reinforced alumina possesses high strength and high shock resistance, even at high temperatures, and has therefore been used to make high-speed cutting tools for machining very hard steels.

How do fibers and whiskers increase the strength and fracture toughness of a composite material? First, fibers have great strength along the fiber axis because most of the chemical bonds are aligned in that direction. Second, there are several ways in which fibers can prevent microscopic cracks from propagating to the point that they lead to fracture of the material. The fibers can deflect cracks, thus preventing them from moving cleanly in one direction, and they can bridge cracks, thus holding the two sides of a crack together.

Silicon carbide-reinforced alumina is an example of a composite in which both the fibers and the surrounding matrix are ceramics. There are other composites, however, in which the two phases are different types of materials. Examples are **ceramic–metal composites**, or *cermets*, such as aluminum metal reinforced with boron fiber, and **ceramic–polymer composites**, such as boron/epoxy and carbon/epoxy. (Epoxy is a resin consisting of long-chain organic molecules.) These materials are popular for aerospace and military applications because of their high strength-to-weight ratios. Boron-reinforced aluminum, for example, is used as a lightweight structural material in the space shuttle, and boron/epoxy and carbon/epoxy skins are used on military aircraft (Figure 21.21). Increased use of composite materials in commercial aircraft could result in weight savings of 20 to 30% and corresponding savings in fuel, which accounts for more than 50% of the cost of operating an airline.

FIGURE 21.21 The skin of the B-2 advanced technology aircraft is a strong, lightweight composite material that contains carbon fibers.

Ceramic fibers used in composites are usually made by high-temperature methods. Carbon (graphite) fiber, for example, can be made by oxidizing and carbonizing fibers of polyacrylonitrile, a long-chain organic molecule also used to make the textile Orlon:

$$-CH_2-CH \underset{CN}{-} \boxed{CH_2-CH \underset{CN}{-}} CH_2-CH \underset{CN}{-} \qquad \text{Polyacrylonitrile}$$

Repeating unit

In the final step of the multistep process, the carbon in the fiber is converted to graphite by heating at 1400 to 2500°C. Similarly, silicon carbide fiber can be made by heating fibers that contain long-chain molecules with alternating silicon and carbon atoms:

$$-SiH_2-CH_2 \boxed{-SiH_2-CH_2-} SiH_2-CH_2- \xrightarrow[-H_2]{\text{heat}} SiC \text{ fiber}$$

Repeating unit

PROBLEM 21.6 Classify each of these composites as ceramic–ceramic, ceramic–metal, or ceramic–polymer:

(a) cobalt/tungsten carbide   (b) silicon carbide/zirconia
(c) boron nitride/epoxy   (d) boron carbide/titanium

DIAMONDS AND DIAMOND FILMS — *interlude*

Their sparkling beauty has made diamonds an object of fascination and desire for millennia. Far more important—at least industrially—is their hardness. Diamond powder is unsurpassed for polishing, diamond-tipped bits are unequalled for drilling, and diamond-tipped sawblades are unparalled for cutting. In fact, the industrial demand for diamonds dwarfs that of the jewelry trade and far outstrips the supply of natural stones. As a result of this demand, an intensive effort to produce synthetic diamonds began in the 1940s and culminated in the mid-1950s with independent announcements by the General Electric company in the United States and the Allemanna Svenska Elektriska company in Sweden of a method for preparing diamond from graphite.

The original General Electric method for producing diamond from graphite used a combination of high temperature (2500°C) and pressure (100,000 atm) and yielded a black gritty material suitable for use as an industrial abrasive. Subsequent technical improvements have made possible the synthesis of gem-quality stones of up to 2 carats in mass (about 0.4 g) at a price competitive with natural diamond. So inexpensive (relatively speaking) have synthetic diamonds become that the electronics industry uses them as heat sinks to help cool heat-producing semiconductor components. This application arises from another important property of diamond: its ability to conduct heat about four times better than either copper or silver.

These diamonds, fine enough to be of gem quality, were made from graphite at high temperature and high pressure.

The most exciting recent advance in diamond-producing technology was the development by several research groups in the 1980s of methods for depositing diamond *films* on a variety of surfaces at low pressure. When a dilute mixture (about 1% by volume) of CH_4 in H_2 gas at a pressure of about 40 mm Hg is heated in a microwave discharge to a temperature near 1000°C, free carbon and hydrogen atoms are produced. The hydrogen atoms evidently impede the formation of graphite, and the carbon atoms therefore deposit on a nearby surface in the form of diamond. The thin, filmlike coating that results is made up of tiny diamond crystals of about 10 to 20 nm in diameter.

Imagine the possibilities, some of which are just now becoming reality: Tools, knifeblades, and scalpels coated with diamond remain forever sharp. Eyeglass lenses and wristwatches coated with diamond remain scratch-free. High-fidelity loudspeakers coated with diamond give a nearly perfect, undistorted sound at high frequencies. Hip joints and other biological implants coated with diamond are not rejected by the body's immune system.

Further improvements in deposition technology are still needed—only substrates stable at a temperature of 1000°C can be coated at present—but the time may not be too far off before diamond coatings on a variety of consumer products become commonplace.

The diaphragm in this high-frequency speaker is coated with a thin diamond film, allowing distortion-free sound reproduction.

Most metals occur in nature as **minerals**, the silicates, carbonates, oxides, sulfides, phosphates, and chlorides that make up the rocks of the earth's crust. The type of mineral found for a particular metal can be correlated with the metal's position in the periodic table. Minerals from which metals can be produced economically are called **ores**.

Metallurgy, the science and technology of extracting metals from their ores, involves three steps: (1) concentration and chemical treatment to separate the mineral from impurities called **gangue**; (2) reduction of the mineral to the free metal; and (3) refining or purification of the metal. The first step uses processes such as **flotation** and **roasting**. The reduction method depends on the activity of the metal: The least active metals can be obtained by roasting the metal sulfide. Metals of intermediate activity are obtained by reducing the metal oxide with C, H_2, or a more active metal (Na, Mg, or Al). The most active metals require electrolysis. The refining step involves distillation, chemical purification (for example, the **Mond process** for nickel), or electrorefining.

Iron is produced in a blast furnace by reducing iron oxide ores with CO, formed from coke (carbon) and hot air. Added limestone ($CaCO_3$) decomposes to CaO, which reacts with SiO_2 and other acidic oxide impurities to yield a molten **slag** (mainly $CaSiO_3$) that is separated from the molten iron at the bottom of the furnace. The iron is converted to steel by the **basic oxygen process**, which removes impurities such as elemental Si, P, S, and most of the C. The impurities are oxidized by a jet of O_2 gas, and the resulting oxides react with CaO and are removed as slag.

Two bonding models are used for metals. The **electron-sea model** pictures a metal as an array of metal cations immersed in a sea of delocalized, mobile valence electrons that act as an electrostatic glue holding the cations together. In the **molecular orbital theory** for metals, also called **band theory**, the delocalized valence electrons occupy a vast number of MO energy levels that are so closely spaced in energy that they merge into an almost continuous band. Both theories account for properties such as malleability, ductility, and high thermal and electrical conductivity, but band theory better explains how the number of valence electrons affects properties such as melting point and hardness.

Band theory also accounts for the electrical properties of **metals, insulators**, and **semiconductors**. Because electrical conductivity involves excitation of valence electrons to readily accessible higher-energy MOs, materials with partially filled bands are metallic conductors, and materials with only completely filled bands are electrical insulators. In insulators, the bonding MOs, called the **valence band**, and the antibonding MOs, called the **conduction band**, are separated in energy by a large **band gap**. In semiconductors, the band gap is smaller, and a few electrons have enough energy to occupy the conduction band. The resulting partially filled valence and conduction bands give rise to a small conductivity. The conductivity can be increased by **doping**—adding a group a 5A impurity to a group 4A element, which gives an *n*-**type semiconductor**, or adding a group 3A impurity to a group 4A element, which gives a *p*-**type semiconductor**.

A **superconductor** is a material that loses all electrical resistance below a characteristic temperature called the **superconducting transition temperature**, T_c. Examples include mixed-metal oxides such as $YBa_2Cu_3O_7$ ($T_c =$

90 K) and fullerene-based compounds such as Rb_3C_{60} ($T_c = 30$ K). Below T_c, a superconductor is able to levitate a magnet, a phenomenon called the Meissner effect.

   Ceramics are inorganic, nonmetallic, nonmolecular solids. Modern **advanced ceramics** include oxide ceramics such as Al_2O_3 and $YBa_2Cu_3O_7$ and nonoxide ceramics such as SiC and Si_3N_4. Ceramics are generally lighter, stiffer, harder, and more resistant to wear and corrosion than metals. Ceramics also have higher melting points. One approach to ceramic processing is the **sol–gel** method, in which high-purity, fine metal oxide powders are prepared by hydrolysis of a metal alkoxide. Unfortunately, ceramics are brittle and prone to fracture. Their strength and fracture toughness can be increased by making **ceramic composites**, hybrid materials such as Al_2O_3/SiC in which fine grains of Al_2O_3 are reinforced by whiskers of SiC. Other types of composites include **ceramic–metal composites** and **ceramic–polymer composites**.

UNDERSTANDING KEY CONCEPTS

1. Look at the location of elements A, B, C, and D in the following periodic table.

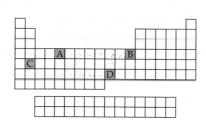

   Without looking at Figure 21.2, predict whether these elements are likely to be found in nature as carbonates, oxides, sulfides, or in uncombined form. Explain.

2. Among the methods of extracting metals from their ores are (i) roasting a metal sulfide, (ii) chemical reduction of a metal oxide, and (iii) electrolysis. The preferred method depends on the $E°$ value for the reduction half-reaction $M^{n+}(aq) + ne^- \longrightarrow M(s)$.
   (a) Which method is appropriate for the metals with the most negative $E°$ values?
   (b) Which method is appropriate for the metals with the most positive $E°$ values?

3. The following pictures show the electron populations of the bands of MO energy levels for four different materials.

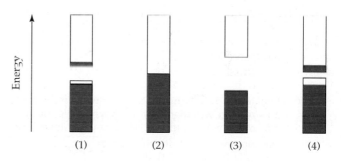

   (a) Classify each of the materials as an insulator, a semiconductor, or a metal.
   (b) Arrange the four materials in order of increasing electrical conductivity. Explain.
   (c) Tell whether the conductivity of each material increases or decreases when the temperature increases.

4. The following pictures show the electron population of the composite *s-d* band for three different transition metals.

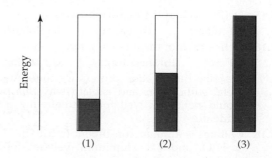

(a) Which metal has the highest melting point? Explain.

(b) Which metal has the lowest melting point? Explain.

(c) Arrange the metals in order of increasing hardness. Explain.

5. The following picture represents the electron population of the bands of MO energy levels for elemental silicon.

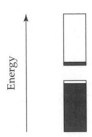

(a) Identify the valence band, the conduction band, and the band gap.

(b) Draw a picture that shows how the electron population changes when the silicon is doped with gallium.

(c) Draw a picture that shows how the electron population changes when the silicon is doped with arsenic.

(d) Compare the electrical conductivity of the doped silicon samples with that of pure silicon. Account for the differences.

ADDITIONAL PROBLEMS

Problems 21.1–21.6 appear within the chapter.

SOURCES OF THE METALLIC ELEMENTS

21.7 List three examples of metals that are found in nature as oxide ores.

21.8 List three examples of metals that are found in nature in uncombined form.

21.9 Locate the following metals in the periodic table, and predict whether they are likely to be found in nature as oxides, sulfides, or in uncombined form.

(a) copper   (b) zirconium
(c) palladium   (d) bismuth

21.10 Locate the following metals in the periodic table, and predict whether they are likely to be found in nature as oxides, sulfides, or in uncombined form.

(a) iridium   (b) chromium
(c) mercury   (d) titanium

21.11 Explain why early transition metals occur in nature as oxides but late transition metals occur as sulfides.

21.12 Explain why iron occurs in nature as an oxide but calcium occurs as a carbonate.

21.13 Name the following minerals.

(a) Fe_2O_3 (b) PbS   (c) TiO_2 (d) $CuFeS_2$

21.14 Write chemical formulas for the following minerals.

(a) cinnabar   (b) bauxite
(c) sphalerite   (d) chromite

METALLURGY

21.15 Define each of the following terms.

(a) mineral   (b) gangue   (c) flotation

21.16 Define each of the following terms.

(a) ore   (b) metallurgy   (c) roasting

21.17 Describe the flotation process for concentrating a metal sulfide ore, and explain why this process would not work well for concentrating a metal oxide ore.

21.18 Gallium is a minor constituent in bauxite, $Al_2O_3 \cdot x\ H_2O$, and like aluminum, gallium forms an amphoteric oxide, Ga_2O_3. Explain how the Bayer process makes it possible to separate Ga_2O_3 from Fe_2O_3, which is also present in bauxite. Write a balanced net ionic equation for the reaction.

21.19 The metals of the zinc subgroup are found in nature as sulfide ores. Values of $E°$ for the reduction half-reaction $M^{2+}(aq) + 2\ e^- \rightarrow M(s)$ are -0.76 V for M = Zn, -0.40 V for M = Cd, and $+0.85$ V for M = Hg. Explain why roasting sphalerite (ZnS) gives ZnO but roasting cinnabar (HgS) gives elemental mercury. What product would you expect from roasting CdS?

21.20 Zinc metal is produced by reducing ZnO with coke, and magnesium metal is produced by electrolysis of molten $MgCl_2$. Look at the standard reduction potentials in Appendix D, and explain why magnesium can't be prepared by the method used for zinc.

21.21 Complete and balance each of the following equations.

(a) $V_2O_5(s) + Ca(s) \longrightarrow$
(b) $PbS(s) + O_2(g) \longrightarrow$
(c) $MoO_3(s) + H_2(g) \longrightarrow$
(d) $MnO_2(s) + Al(s) \longrightarrow$
(e) $MgCl_2(l) \xrightarrow{\text{electrolysis}}$

21.22 Complete and balance each of the following equations.

(a) $Fe_2O_3(s) + H_2(g) \longrightarrow$
(b) $Cr_2O_3(s) + Al(s) \longrightarrow$
(c) $Ag_2S(s) + O_2(g) \longrightarrow$
(d) $TiCl_4(g) + Mg(l) \longrightarrow$
(e) $LiCl(l) \xrightarrow{\text{electrolysis}}$

21.23 Use the data in Appendix B to calculate $\Delta H°$ and $\Delta G°$ for the roasting of sphalerite, ZnS:

$$2\,ZnS(s) + 3\,O_2(g) \rightarrow 2\,ZnO(s) + 2\,SO_2(g)$$

Explain why $\Delta H°$ and $\Delta G°$ have different values, and account for the sign of $(\Delta H° - \Delta G°)$.

21.24 Use the data in Appendix B to calculate $\Delta H°$ and $\Delta G°$ for the roasting of cinnabar, HgS:

$$HgS(s) + O_2(g) \rightarrow Hg(l) + SO_2(g)$$

Explain why $\Delta H°$ and $\Delta G°$ have different values, and account for the sign of $(\Delta H° - \Delta G°)$.

21.25 Ferrochrome, an iron–chromium alloy used in making stainless steel, is produced by reducing chromite ($FeCr_2O_4$) with coke:

$$FeCr_2O_4(s) + 4\,C(s) \rightarrow \underbrace{Fe(s) + 2\,Cr(s)}_{\text{Ferrochrome}} + 4\,CO(g)$$

(a) How many kilograms of chromium can be obtained by reacting 236 kg of chromite with an excess of coke?
(b) How many liters of carbon monoxide at 25°C and 740 mm Hg are obtained as a byproduct?

21.26 One step in the industrial process for producing copper from chalcopyrite ($CuFeS_2$) involves reducing molten copper(I) sulfide with a blast of hot air:

$$Cu_2S(l) + O_2(g) \rightarrow 2\,Cu(l) + SO_2(g)$$

(a) How many kilograms of Cu_2S must be reduced to account for the world's annual copper production of about 8×10^9 kg?
(b) How many liters of SO_2 at STP are produced as a byproduct?
(c) If all the SO_2 escaped into the atmosphere and was converted to sulfuric acid in acid rain, how many kg of H_2SO_4 would there be in the rain?

21.27 Nickel, used in making stainless steel, can be purified by electrorefining. The electrolysis cell has an impure nickel anode, a pure nickel cathode, and an aqueous solution of nickel sulfate as the electrolyte. How many kg of nickel can be refined in an 8.00 h day if the current passed through the cell is held constant at 52.5 A?

21.28 Pure copper for use in electrical wiring is obtained by electrorefining. How many hours are required to transfer 7.50 kg of copper from an impure copper anode to a pure copper cathode if the current passed through the electrolysis cell is held constant at 40.0 A.

IRON AND STEEL

21.29 Write a balanced equation for the overall reaction that occurs in a blast furnace when iron ore is reduced to iron metal. Identify the oxidizing agent and the reducing agent.

21.30 What is the role of coke in the commercial process for producing iron? Write balanced equations for the relevant chemical reactions.

21.31 What is slag, and what is its role in the commercial process for producing iron?

21.32 Why is limestone added to the blast furnace in the commercial process for producing iron? Write balanced equations for the chemical reactions.

21.33 Briefly describe the basic oxygen process. Write balanced equations for the relevant chemical reactions.

21.34 Why does slag form in the basic oxygen process?

Write balanced equations for the relevant reactions.

21.35 When iron ore is reduced in a blast furnace, some of the SiO_2 impurity is also reduced as it reacts with carbon to give elemental silicon and carbon monoxide. The silicon is subsequently reoxidized in the basic oxygen process, and the resulting SiO_2 reacts with CaO, yielding slag, which is then separated from the molten steel. Write balanced equations for the three reactions involving SiO_2.

21.36 In a blast furnace, some of the $CaSO_4$ impurity in iron ore is reduced by carbon, yielding elemental sulfur and carbon monoxide. The sulfur is subsequently oxidized in the basic oxygen process, and the product reacts with CaO to give a molten slag. Write balanced equations for the reactions.

BONDING IN METALS

21.37 List three properties that are common to all metals.

21.38 List two properties of metals that vary considerably depending on the particular metal.

21.39 Potassium metal crystallizes in a body-centered-cubic structure. Draw one unit cell, and try to draw a Lewis structure for bonding of the central K atom to its nearest neighbor K atoms. What is the problem?

21.40 Describe the electron-sea model of the bonding in cesium metal. Cesium has a body-centered-cubic structure.

21.41 How does the electron-sea model account for the malleability and ductility of metals?

21.42 How does the electron-sea model account for the electrical and thermal conductivity of metals?

21.43 Cesium metal is very soft, and tungsten metal is very hard. Explain the difference in terms of the electron-sea model.

21.44 Sodium melts at 98°C, and magnesium melts at 649°C. Account for the higher melting point of magnesium in terms of the electron-sea model.

21.45 Why is the molecular orbital theory for metals called band theory?

21.46 Draw an MO energy level diagram that shows the population of the 4s band for potassium metal.

21.47 How does band theory account for the electrical conductivity of metals?

21.48 Materials with partially filled bands are metallic conductors, and materials with only completely filled bands are electrical insulators. Explain why the population of the bands affects the conductivity.

21.49 Draw an MO energy level diagram for beryllium metal, and show the population of the MOs for the following two cases: **(a)** the 2s and 2p bands are well separated in energy; **(b)** the 2s and 2p bands overlap in energy. Which diagram agrees with the fact that beryllium has a high electrical conductivity? Explain.

21.50 Draw an MO energy level diagram for calcium metal, and show the population of the MOs for the following two cases: **(a)** the 4s and 3d bands are well separated in energy; **(b)** the 4s and 3d bands overlap in energy. Which diagram agrees with the fact that calcium has a high electrical conductivity? Explain.

21.51 The melting points for the second-series transition elements increase from 819°C for yttrium to 2610°C for molybdenum and then decrease again to 321°C for cadmium. Account for the trend in terms of band theory.

21.52 Copper has a Mohs hardness value of 3, and iron has a Mohs hardness value of 5. Use band theory to explain why copper is softer than iron.

SEMICONDUCTORS

21.53 Define a semiconductor, and give three examples.

21.54 Explain each of the following terms.
  (a) valence band   **(b)** conduction band
  (c) band gap   **(d)** doping

21.55 Make a drawing of the bands of MO energy levels and the electron population for **(a)** a semiconductor and **(b)** an electrical insulator. Explain why a semiconductor has the higher electrical conductivity.

21.56 Make a drawing of the bands of MO energy levels and the electron population for **(a)** a semiconductor and **(b)** a metallic conductor. Explain why a semiconductor has the lower electrical conductivity.

21.57 How does the electrical conductivity of a semiconductor change as the size of the band gap increases? Explain.

21.58 How does the electrical conductivity of a semiconductor change as the temperature increases? Explain.

21.59 Explain what an *n*-type semiconductor is, and give an example. Draw an MO energy level diagram, and show the population of the valence band and the conduction band for an *n*-type semiconductor.

21.60 Explain what a *p*-type semiconductor is, and give an example. Draw an MO energy level diagram, and show the population of the valence band and the conduction band for a *p*-type semiconductor.

21.61 Explain why germanium doped with phosphorus has a higher electrical conductivity than pure germanium.

21.62 Explain why silicon doped with gallium has a higher electrical conductivity than pure silicon.

21.63 Classify the following semiconductors as *n*-type or *p*-type.
  (a) Si doped with In   **(b)** Ge doped with Sb
  (c) gray Sn doped with As

21.64 Classify the following semiconductors as *n*-type or *p*-type.
  (a) Ge doped with As   **(b)** Ge doped with B
  (c) Si doped with Sb

21.65 Arrange the following materials in order of increasing electrical conductivity.

(a) Cu
(b) Al_2O_3
(c) Fe
(d) pure Ge
(e) Ge doped with In

21.66 Arrange the following materials in order of increasing electrical conductivity.

(a) pure gray Sn
(b) gray Sn doped with Sb
(c) NaCl
(d) Ag
(e) pure Si

SUPERCONDUCTORS

21.67 Define a superconductor, and give an example.

21.68 Define the superconducting transition temperature, T_c.

21.69 What are the two most characteristic properties of a superconductor?

21.70 $YBa_2Cu_3O_7$ is a superconductor below its T_c of 90 K and a metallic conductor above 90 K. Make a rough plot of electrical resistance versus temperature for $YBa_2Cu_3O_7$.

21.71 Compare the structure and properties of ceramic superconductors, such as $YBa_2Cu_3O_7$, and fullerene-based superconductors, such as Rb_3C_{60}.

21.72 Look at Figure 21.15, and identify the coordination numbers of the Cu, Y, and Ba atoms.

CERAMICS AND COMPOSITES

21.73 What is a ceramic, and what properties distinguish a ceramic from a metal?

21.74 Contrast the bonding in ceramics with the bonding in metals.

21.75 Why are ceramics more wear resistant than metals?

21.76 Why are oxide ceramics more corrosion-resistant than metals?

21.77 Silicon nitride (Si_3N_4), a high-temperature ceramic useful for making engine components, is a covalent network solid in which each Si atom is bonded to four N atoms and each N atom is bonded to three Si atoms. Explain why silicon nitride is more brittle than a metal like copper.

21.78 Magnesia (MgO), used as an insulator for electrical heating devices, has a face-centered-cubic structure like that of NaCl. Draw one unit cell of the structure of MgO, and explain why MgO is more brittle than magnesium metal. (*Hint:* What happens if you try to slide planes of ions in a direction parallel to an edge of the unit cell?)

21.79 What is ceramic processing?

21.80 Describe what happens when a ceramic powder is sintered.

21.81 Describe the differences between a sol and a gel.

21.82 What is the difference between a hydrolysis reaction and the reaction that occurs when a sol is converted to a gel?

21.83 Zirconia (ZrO_2), an unusually tough oxide ceramic, has been used to make very sharp table knives.

Write a balanced equation for the hydrolysis of zirconium isopropoxide in the sol–gel method for making zirconia powders. The isopropoxide ligand is the anion of isopropanol, $HOCH(CH_3)_2$.

21.84 Zinc oxide (ZnO) is a semiconducting ceramic used to make varistors (variable resistors). Write a balanced equation for the hydrolysis of zinc ethoxide in the sol–gel method for making ZnO powders.

21.85 Describe the reactions that occur when an $Si(OH)_4$ sol becomes a gel. What is the formula of the ceramic obtained when the gel is dried and sintered?

21.86 Describe the reactions that occur when a $Y(OH)_3$ sol becomes a gel. What is the chemical formula of the ceramic obtained when the gel is dried and sintered?

21.87 Silicon nitride powder can be made by reacting silicon tetrachloride vapor with gaseous ammonia. The byproduct is gaseous hydrogen chloride. Write a balanced equation for the reaction.

21.88 Boron, which is used in making composites, is deposited on a tungsten wire when the wire is heated electrically in the presence of boron trichloride vapor and gaseous hydrogen. Write a balanced equation for the reaction.

21.89 Explain why graphite/epoxy composites are good materials for making tennis rackets and golf clubs.

21.90 Explain why silicon carbide-reinforced alumina is stronger and tougher than pure alumina.

GENERAL PROBLEMS

21.91 List three examples of metals that are found in nature as sulfide ores.

21.92 List three examples of metals that are commonly prepared by electrolysis.

21.93 What metals are the main constituents of stainless steel?

21.94 Imagine a planet with an atmosphere that contains O_2 and SO_2 but no CO_2. Give the chemical composition of the minerals you would expect to find for the alkaline-earth metals on such a planet.

21.95 Define each of the following terms.

   (a) sol   **(b)** gel
   (c) sintering   **(d)** ceramic processing

21.96 Describe two applications of ceramics. How do these applications depend on the properties of the materials used?

21.97 Superconductors with values of T_c above 77 K are of special interest. What's so special about 77 K?

21.98 What properties of metals are better explained by band theory than by the electron-sea model?

21.99 Tungsten is hard and has a very high melting point (3410°C), and gold is soft and has a relatively low melting point (1064°C). Are these facts in better agreement with the electron-sea model or with the MO model (band theory)? Explain.

21.100 Explain why the enthalpy of vaporization of vanadium (460 kJ/mol) is much larger than that of zinc (114 kJ/mol).

21.101 Classify each of the following materials as a metallic conductor, an *n*-type semiconductor, a *p*-type semiconductor, or an electrical insulator.

   (a) MgO   **(b)** Si doped with Sb
   (c) white tin   **(d)** Ge doped with Ga
   (e) stainless steel

21.102 Gallium arsenide, a material used in the manufacture of laser printers and compact disc players, has a band gap of 130 kJ/mol. Is GaAs a metallic conductor, a semiconductor, or an electrical insulator? With what group 4A element is GaAs isoelectronic? Draw an MO energy level diagram for GaAs, and show the population of the bands.

21.103 A 3.4×10^3 kg batch of cast iron contains 0.45% by mass of phosphorus as an impurity. In the conversion of cast iron to steel by the basic oxygen process, the phosphorus is oxidized to P_4O_{10}, which then reacts with CaO and is removed as slag.

   (a) Write balanced equations for the oxidation of P_4 and for the formation of slag.

   (b) How many kg of CaO are required to react with all the P_4O_{10}?

21.104 The $YBa_2Cu_3O_7$ superconductor can be synthesized by the sol–gel method from a stoichiometric mixture of metal ethoxides. How many grams of $Y(OCH_2CH_3)_3$ and how many grams of $Ba(OCH_2CH_3)_2$ are required for reaction with 75.4 g of $Cu(OCH_2CH_3)_2$ and an excess of water? Assuming a 100% yield, how many grams of $YBa_2Cu_3O_7$ are obtained?

21.105 Calculate the percent composition for $YBa_2Cu_3O_7$.

21.106 Mullite, $3 Al_2O_3 \cdot 2 SiO_2$, a high-temperature ceramic being considered for use in engines, can be made by the sol–gel method.

   (a) Write a balanced equation for formation of the sol by hydrolysis of a stoichiometric mixture of aluminum ethoxide and tetraethoxysilane, $Si(OCH_2CH_3)_4$.

   (b) Describe the reactions that convert the sol to a gel.

   (c) What additional steps are required to convert the gel to the ceramic product?

21.107 The glass in photochromic sunglasses is called a nanocomposite because it contains tiny silver halide particles with a diameter of less than 10 nm (1 nm $= 10^{-9}$ m $= 10^{-6}$ mm). This glass darkens and becomes less transparent outdoors in sunlight but returns to its original transparency indoors, where there is less light. Speculate on how these reversible changes occur.

21.108 The Mond process for purifying nickel involves the reaction of impure nickel with carbon monoxide to give nickel tetracarbonyl at about 150°C. The nickel tetracarbonyl then decomposes to pure nickel at about 230°C:

$$Ni(s) + 4 CO(g) \rightleftharpoons Ni(CO)_4(g)$$
$$\Delta H° = -160.8 \text{ kJ}; \Delta S° = -410 \text{ J/K}$$

The values of $\Delta H°$ and $\Delta S°$ apply at 25°C, but they are relatively independent of temperature and can be used at 150°C and 230°C.

   (a) Calculate $\Delta G°$ and the equilibrium constant K_p at 150°C.

   (b) Calculate $\Delta G°$ and the equilibrium constant K_p at 230°C.

   (c) Why does the reaction have a large negative value for $\Delta S°$? Show that the change in $\Delta G°$ with increasing temperature is consistent with a negative value of $\Delta S°$.

   (d) Show that the change in K_p with increasing temperature is consistent with a negative value of $\Delta H°$.

21.109 Figure 21.5 indicates that carbon reacts with carbon dioxide to form carbon monoxide at temperatures above approximately 700°C:

$$C(s) + CO_2(g) \rightarrow 2 CO(g)$$

(a) Use the thermodynamic data in Appendix B to calculate values of $\Delta H°$, $\Delta S°$, and $\Delta G°$ for this reaction at 25°C.

(b) Is the reaction endothermic or exothermic?

(c) Account for the sign of the entropy change.

(d) Assuming that $\Delta H°$ and $\Delta S°$ do not depend on temperature, calculate the values of $\Delta G°$ at 500°C and at 1000°C. Note the sign of $\Delta G°$ at the two temperatures.

(e) Calculate values of the equilibrium constant K_p at 500°C and at 1000°C.

21.110 At high temperatures, coke reduces silica impurities in iron ore to elemental silicon:

$$SiO_2(s) + 2\ C(s) \rightarrow Si(s) + 2\ CO(g)$$

(a) Use the data in Appendix B to calculate values of $\Delta H°$, $\Delta S°$, and $\Delta G°$ for this reaction at 25°C.

(b) Is the reaction endothermic or exothermic?

(c) Account for the sign of the entropy change.

(d) Is the reaction spontaneous at 25°C and 1 atm pressure of CO?

(e) Assuming that $\Delta H°$ and $\Delta S°$ do not depend on temperature, estimate the temperature at which the reaction becomes spontaneous at 1 atm pressure of CO.

chapter 22 NUCLEAR CHEMISTRY

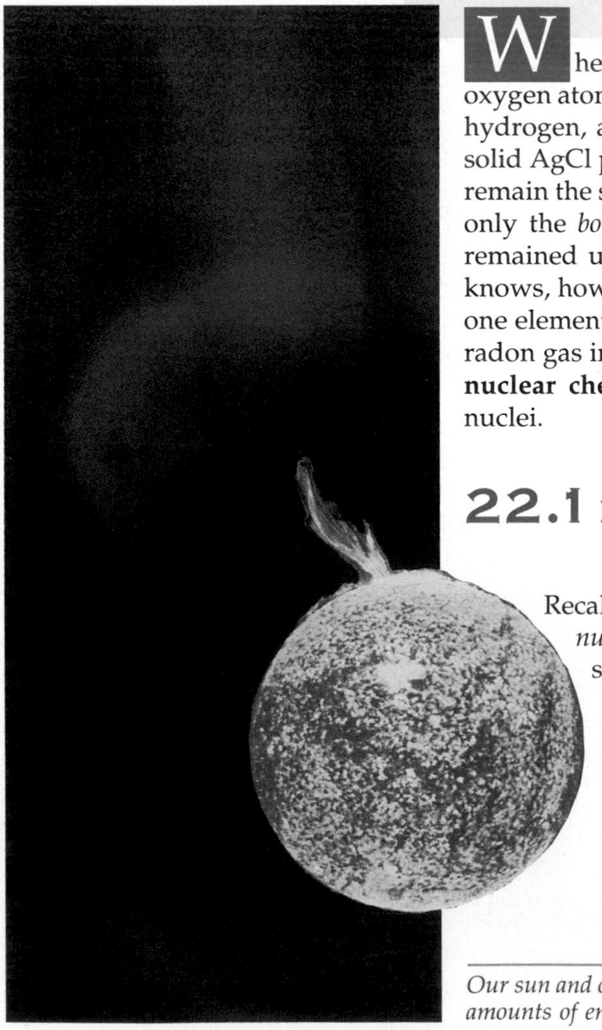

W hen methane (CH_4) burns in oxygen, the carbon, hydrogen, and oxygen atoms recombine to yield CO_2 and H_2O, but they still remain carbon, hydrogen, and oxygen. When aqueous NaCl reacts with aqueous $AgNO_3$, solid AgCl precipitates from solution, but the silver and chlorine themselves remain the same. In fact, in all the reactions we've discussed up to this point, only the *bonds* between atoms have changed; the atoms themselves have remained unchanged. Anyone who reads the paper or watches television knows, however, that atoms *can* change, often resulting in the conversion of one element into another. Atomic weapons, nuclear energy, and radioactive radon gas in our homes are all topics of societal importance, and all involve **nuclear chemistry**—the study of the properties and reactions of atomic nuclei.

22.1 ➤ NUCLEAR REACTIONS AND THEIR CHARACTERISTICS

Recall from Section 2.5 that an atom is characterized by its *atomic number*, Z, and its *mass number*, A. The atomic number, written as a subscript to the left of the element symbol, gives the number of protons in the nucleus. The mass number, written as a super-script to the left of the element symbol, gives the total number of **nucleons**, a general term for both protons (*p*) and neutrons (*n*).

Our sun and other stars are powered by nuclear reactions, which release almost unimaginable amounts of energy.

A nuclear power plant for generating electricity.

The most common isotope of carbon, for example, has 12 nucleons: 6 protons and 6 neutrons.

Mass number

$^{12}_{6}C$

6 protons
6 neutrons
————————
12 nucleons

Atomic number

Atoms with identical atomic numbers (the same number of protons) but different mass numbers (different numbers of neutrons) are called *isotopes*, or **nuclides**. There are 13 known nuclides of carbon, 2 of which occur naturally (^{12}C and ^{13}C) and 11 of which have been produced artificially.[1] Only the 2 naturally occurring isotopes are indefinitely stable; the remaining 11 undergo spontaneous **nuclear reactions** that change the nucleus. Carbon-14, for example, slowly decomposes to give nitrogen-14 plus an electron, a process we can write as

$$^{14}_{6}C \rightarrow {}^{14}_{7}N + {}^{0}_{-1}e$$

The electron is often written as $^{0}_{-1}e$, where the superscript 0 indicates that the mass of an electron is essentially zero when compared with that of a proton or neutron, and the subscript -1 indicates that the charge is -1. (The subscript in this instance is not a true atomic number.)

Nuclear reactions, such as the spontaneous decomposition of ^{14}C, are distinguished from chemical reactions in several ways:

[1] In a strict sense, carbon-14 can also be said to occur naturally, since it is produced in small amounts in the upper atmosphere by the action of neutrons (produced by cosmic rays) on nitrogen-14.

$$^{14}_{7}N + {}^{1}_{0}n \rightarrow {}^{14}_{6}C + {}^{1}_{1}H$$

1. *Nuclear* reactions involve a change in an atom's nucleus, usually producing a different element. *Chemical* reactions, by contrast, involve only a change in distribution of the outer-shell electrons around an atom and never change the nucleus itself or produce a different element.

2. Different nuclides of an element have essentially the same behavior in chemical reactions but often have completely different behavior in nuclear reactions.

3. The rates of nuclear reactions are unaffected by a change in temperature or pressure or by the addition of a catalyst.

4. The nuclear reaction of an atom is essentially the same, regardless of whether the atom is in a chemical compound or is in uncombined elemental form.

5. The energy changes accompanying nuclear reactions are far greater than those accompanying chemical reactions. The nuclear transformation of 1.0 g of uranium-235 releases 8.2×10^7 kJ of energy, for example, whereas the chemical combustion of 1.0 g of methane (natural gas) releases only 56 kJ.

22.2 ➤ NUCLEAR REACTIONS AND RADIOACTIVITY

Scientists have known since 1896 that many naturally occurring nuclei are **radioactive**—that is, they spontaneously emit radiation. Early studies of radioactive nuclei by the British physicist Ernest Rutherford in 1897 showed that there are three common types of radiation with markedly different properties: alpha (α), beta (β), and gamma (γ) radiation, named after the first three letters of the Greek alphabet.

Alpha (α) Radiation

Using the simple experiment shown in Figure 22.1, Rutherford found that α **radiation** consists of a stream of particles that are repelled by a positively charged electrode, are attracted by a negatively charged electrode, and have a mass-to-charge ratio identifying them as helium nuclei, $^4_2\text{He}^{2+}$. Alpha particles thus consist of two protons and two neutrons.

Because the emission of an alpha particle from a nucleus results in a loss of two protons and two neutrons, it reduces the mass number of the nucleus by four and reduces the atomic number by two. Alpha emission is particularly common for heavy radioactive isotopes, also called **radio-**

FIGURE 22.1 The effect of an electric field on α, β, and γ radiation. The radioactive source in the shielded box emits radiation, which passes between two electrodes. Alpha radiation is deflected toward the negative electrode, β radiation is strongly deflected toward the positive electrode, and γ radiation is undeflected.

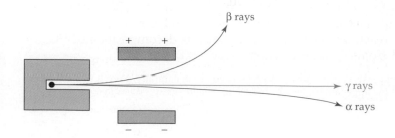

isotopes, or **radionuclides**: Uranium-238, for example, spontaneously emits an alpha particle and forms thorium-234.

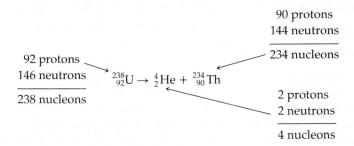

$$^{238}_{92}\text{U} \rightarrow \,^{4}_{2}\text{He} + \,^{234}_{90}\text{Th}$$

92 protons
146 neutrons
——————
238 nucleons

90 protons
144 neutrons
——————
234 nucleons

2 protons
2 neutrons
——————
4 nucleons

Note the way that the **nuclear equation** for the radioactive decay of uranium-238 into lighter elements is written. The equation is not balanced in the usual chemical sense because the kinds of nuclei are not the same on both sides of the arrow. Instead, we say that a nuclear equation is balanced when the sums of the nucleons are the same on both sides of the equation and when the sums of the charges on the nuclei and any elementary particles (protons, neutrons, and electrons) are the same on both sides. In the decay of $^{238}_{92}\text{U}$ to give $^{4}_{2}\text{He}$ and $^{234}_{90}\text{Th}$, for example, there are 238 nucleons and 92 nuclear charges on both sides of the nuclear equation.

Note also that we are concerned only with charges on elementary particles and on nuclei when we write nuclear equations, not with ionic charges on atoms. The alpha particle is thus written as $^{4}_{2}\text{He}$ rather than as $^{4}_{2}\text{He}^{2+}$, and the thorium resulting from radioactive decay of $^{238}_{92}\text{U}$ is written as $^{234}_{90}\text{Th}$ rather than as $^{234}_{90}\text{Th}^{2-}$. We ignore the ionic charges both because they are irrelevant to nuclear disintegration and because they soon disappear. The $^{4}_{2}\text{He}^{2+}$ ion immediately picks up two electrons from whatever it strikes, yielding a neutral helium atom, and the $^{234}_{90}\text{Th}^{2-}$ ion immediately gives two electrons to whatever it is in electrical contact with, yielding a neutral thorium atom.

Beta (β) Radiation

Further work by Rutherford in the late 1800s showed that **β radiation** consists of a stream of particles that are attracted to a positive electrode (Figure 22.1), are repelled by a negative electrode, and have a mass-to-charge ratio identifying them as electrons, $^{0}_{-1}\text{e}$ or β^{-}. Beta emission occurs when a neutron in the nucleus spontaneously decays into a proton plus an electron, which is then ejected. The product nucleus has the same mass number as the starting nucleus because a neutron has turned into a proton, but it has a higher atomic number because of the newly created proton. The reaction of ^{131}I to give ^{131}Xe is an example:

$$^{131}_{53}\text{I} \rightarrow \,^{131}_{54}\text{Xe} + \,^{0}_{-1}\text{e}$$

53 protons
78 neutrons
——————
131 nucleons

0 nucleons
but −1 charge

54 protons
77 neutrons
——————
131 nucleons

Writing the emitted β particle as $_{-1}^{0}e$ in the nuclear equation makes clear the charge balance of the nuclear reaction: The subscript in the $_{53}^{131}I$ nucleus on the left (53) is balanced by the sum of the two subscripts on the right $(54 - 1 = 53)$.

Gamma (γ) Radiation

Gamma radiation is unaffected by electric fields (Figure 22.1), has no mass, and is simply electromagnetic radiation of very high energy and short wavelength $(\lambda = 10^{-11}\text{--}10^{-14}$ m). In modern terms, we would say that **γ rays** are a stream of high-energy photons. Gamma radiation almost always accompanies α and β emission as a mechanism for the release of energy, but because it changes neither the mass number nor the atomic number of the product nucleus, γ emission is often not shown when writing nuclear equations.

Positron Emission and Electron Capture

In addition to α, β, and γ radiation, two other common types of radioactive decay processes occur. **Positron emission** involves the conversion of a proton in the nucleus into a neutron plus an ejected *positron*, $_{1}^{0}e$ or β^{+}, a particle that can be thought of as a "positive electron."[2] A positron has the same mass as an electron but an opposite charge. The result of positron emission is a decrease in the atomic number of the product nucleus but no change in the mass number. Potassium-40, for example, undergoes positron emission to yield argon-40, a nuclear reaction important in geology for dating rocks.

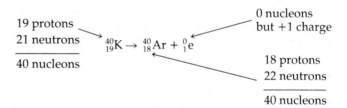

Note once again that the sum of the two subscripts on the right of the nuclear equation $(18 + 1 = 19)$ is equal to the subscript in $_{19}^{40}K$ the nucleus on the left.

Electron capture is a process in which a proton in the nucleus captures an inner-shell electron and is thereby converted into a neutron. The mass number of the product nucleus is unchanged, but the atomic number decreases by one, just as in positron emission. The conversion of mercury-197 into gold-197 is an example:

---

[2] A positron is an example of a so-called *antiparticle*, the exact opposite of a particle. When an electron and a positron collide, the particle and antiparticle annihilate each other, producing only gamma rays.

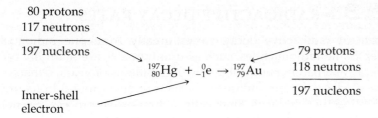

Characteristics of the different kinds of radioactive decay processes are summarized in Table 22.1.

TABLE 22.1	A Summary of Radioactive Decay Processes			
Process	Symbol	Change in Atomic Number	Change in Mass Number	Change in Neutron Number
α Emission	$^4_2$He or α	-2	-4	-2
β Emission	$^0_{-1}$e or β^-	$+1$	0	-1
γ Emission	$^0_0\gamma$	0	0	0
Positron emission	$^0_1$e or β^+	-1	0	$+1$
Electron capture	E. C.	-1	0	$+1$

EXAMPLE 22.1

Write balanced nuclear equations for each of the following.
(a) alpha emission from curium-242: $^{242}_{96}\text{Cm} \rightarrow {}^4_2\text{He} + ?$
(b) beta emission from magnesium-28: $^{28}_{12}\text{Mg} \rightarrow {}^0_{-1}\text{e} + ?$
(c) positron emission from xenon-118: $^{118}_{54}\text{Xe} \rightarrow {}^0_1\text{e} + ?$

SOLUTION   The key to writing nuclear equations is to make sure that the number of nucleons is the same on both sides of the equation and that the number of elementary and nuclear charges is the same.

(a) In alpha emission, the mass number decreases by four, and the atomic number decreases by two, giving plutonium-238: $^{242}_{96}\text{Cm} \rightarrow {}^4_2\text{He} + {}^{238}_{94}\text{Pu}$
(b) In beta emission, the mass number is unchanged, and the atomic number increases by one, giving aluminum-28: $^{28}_{12}\text{Mg} \rightarrow {}^0_{-1}\text{e} + {}^{28}_{13}\text{Al}$
(c) In positron emission, the mass number is unchanged, and the atomic number decreases by one, giving iodine-118: $^{118}_{54}\text{Xe} \rightarrow {}^0_1\text{e} + {}^{118}_{53}\text{I}$

⌐ **PROBLEM 22.1**   Write balanced nuclear equations for each of the following.
(a) beta emission from ruthenium-106: $^{106}_{44}\text{Ru} \rightarrow {}^0_{-1}\text{e} + ?$
(b) alpha emission from bismuth-189: $^{189}_{83}\text{Bi} \rightarrow {}^4_2\text{He} + ?$
(c) electron capture by polonium-204: $^{204}_{84}\text{Po} + {}^0_{-1}\text{e} \rightarrow ?$

⌐ **PROBLEM 22.2**   What particle is produced by decay of thorium-214 to radium-210? $^{214}_{90}\text{Th} \rightarrow {}^{210}_{88}\text{Ra} + ?$

22.3 ▶RADIOACTIVE DECAY RATES

The rate of radioactive decay varies greatly from one radionuclide to another. Some radionuclides, such as uranium-238, are naturally occurring and decay at a barely perceptible rate over billions of years. Others, such as carbon-17, decay within milliseconds and are never found in nature.

Radioactive decay is kinetically a first-order process (Section 12.3) whose rate is proportional to the number of radioactive nuclei N in a sample:

$$\text{Decay rate} = k \times N$$

where k is a first-order rate constant called the **decay constant**. As we saw in Section 12.4, a first-order rate law can be converted into an integrated rate law of the form

$$\ln\left(\frac{N}{N_0}\right) = -kt$$

where N_0 is the number of radioactive nuclei originally present in a sample, and N is the number remaining at time t. Note that the decay constant k has units of time^{-1}, so that the quantity kt is unitless.

Like all first-order processes, radioactive decay is characterized by a *half-life, $t_{1/2}$,* the time required for the number of radioactive nuclei in a sample to drop to one-half of the initial value (Section 12.5). For example, the half-life of iodine-131, a radionuclide used in thyroid testing, is 8.04 d. If today you have a certain amount of $^{131}_{53}\text{I}$—say 1.000 g—then 8.04 d from now you will have only 0.500 g of $^{131}_{53}\text{I}$ remaining because one-half of the sample will have decayed by beta emission, yielding 0.500 g of $^{131}_{54}\text{Xe}$. After 8.04 more days (16.08 total) only 0.250 g of $^{131}_{53}\text{I}$ will remain; after a further 8.04 d (24.12 total) only 0.125 g will remain; and so on. Each passage of a half-life causes the decay of one-half of whatever sample remains, as shown graphically by the curve in Figure 22.2. The half-life is the same no matter what the size of the sample, the temperature, or any other external condition.

$$^{131}_{53}\text{I} \rightarrow {}^{131}_{54}\text{Xe} + {}^{0}_{-1}\text{e} \qquad t_{1/2} = 8.04 \text{ d}$$

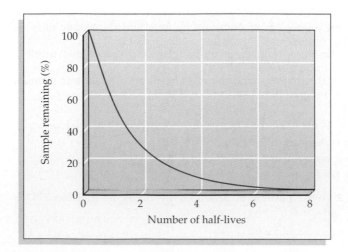

FIGURE 22.2 The decay of a radioactive nuclide over time. No matter what the value of the half-life, 50% of the sample remains after one half-life, 25% remains after two half-lives, 12.5% remains after three half-lives, and so on.

Mathematically, the value of a half-life can be calculated from the integrated rate law by setting $N = 1/2\,N_0$ at time $t_{1/2}$:

$$\ln\left(\frac{1/2N_0}{N_0}\right) = -kt_{1/2} \quad \text{or} \quad \ln\frac{1}{2} = -\ln 2 = -kt_{1/2}$$

so $\quad t_{1/2} = \dfrac{\ln 2}{k} \quad$ and $\quad k = \dfrac{\ln 2}{t_{1/2}}$

If we now substitute the value $\ln 2 = 0.693$, we can arrive at simple expressions for $t_{1/2}$ and k:

$$t_{1/2} = \frac{0.693}{k} \quad \text{and} \quad k = \frac{0.693}{t_{1/2}}$$

These equations say that if we know the value of either the decay constant k or the half-life, $t_{1/2}$, then we can calculate the value of the other. Furthermore, if we know the value of $t_{1/2}$, we can calculate the ratio of remaining and initial amounts of radioactive sample N/N_0 at any time t by substituting the expression for k into the integrated rate law:

Since $\quad \ln\left(\dfrac{N}{N_0}\right) = -kt \quad$ and $\quad k = \dfrac{0.693}{t_{1/2}}$

then $\quad \ln\left(\dfrac{N}{N_0}\right) = -0.693\left(\dfrac{t}{t_{1/2}}\right)$

Example 22.2 shows how to calculate a half-life from a decay constant, and Example 22.4 shows how to determine the percentage of radioactive sample remaining at time t.

The half-lives of some useful radionuclides are given in Table 22.2. As you might expect, radionuclides used internally in medical applications have fairly short half-lives so that they decay rapidly and cause no long-term health hazards.

Technetium-99*m*, a short-lived radionuclide used for brain scans, is obtained by neutron bombardment of molybdenum-99 and then stored in this "molybdenum cow" in the form of MoO_4^{2-}. Small amounts are removed by passing a saline solution through the cylinder.

TABLE 22.2	Half-lives of Some Useful Radionuclides			
Radionuclide	**Symbol**	**Radiation**	**Half-life**	**Use**
Tritium	^3_1H	β^-	12.32 y	Biochemical tracer
Carbon-14	$^{14}_6\text{C}$	β^-	5715 y	Archaeological dating
Phosphorus-32	$^{32}_{15}\text{P}$	β^-	14.28 d	Leukemia therapy
Potassium-40	$^{40}_{19}\text{K}$	β^-	1.26×10^9 y	Geological dating
Cobalt-60	$^{60}_{27}\text{Co}$	β^-, γ	5.27 y	Cancer therapy
Technetium-99*m*[a]	$^{99\text{m}}_{43}\text{Tc}$	γ	6.01 h	Brain scans
Iodine-123	$^{123}_{53}\text{I}$	γ	13.1 h	Thyroid therapy
Uranium-235	$^{235}_{92}\text{U}$	α, γ	7.04×10^8 y	Nuclear reactors

[a] The *m* in technetium-99*m* stands for *metastable*, meaning that it undergoes gamma emission but does not change its mass number or atomic number.

EXAMPLE 22.2

The decay constant for sodium-24, a radionuclide used medically in blood studies, is 4.63×10^{-2} h^{-1}. What is the half-life of ^{24}Na?

SOLUTION   Half-life can be calculated from the decay constant with the equation

$$t_{1/2} = \frac{0.693}{k}$$

Substituting the value $k = 4.63 \times 10^{-2}$ h^{-1} into the equation gives

$$t_{1/2} = \frac{0.693}{4.63 \times 10^{-2}\ \text{h}^{-1}} = 15.0\ \text{h}$$

The half-life of sodium-24 is 15.0 h.

EXAMPLE 22.3

The half-life of radon-222, a radioactive gas of concern as a health hazard in some homes, is 3.823 d. What is the decay constant of ^{222}Rn?

SOLUTION   A decay constant can be calculated from the half-life with the equation

$$k = \frac{0.693}{t_{1/2}}$$

Substituting the value $t_{1/2} = 3.823$ d into the equation gives

$$k = \frac{0.693}{3.823\ \text{d}} = 0.181\ \text{d}^{-1}$$

EXAMPLE 22.4

Phosphorus-32, a radionuclide used in leukemia therapy, has a half-life of 14.28 days. What percent of a sample remains after 35.0 days?

SOLUTION   The ratio of remaining (N) and initial (N_0) amounts of a radioactive sample at time t is given by the equation

$$\ln\left(\frac{N}{N_0}\right) = -0.693\left(\frac{t}{t_{1/2}}\right)$$

Substituting values for t and for $t_{1/2}$ into the equation gives

$$\ln\left(\frac{N}{N_0}\right) = -0.693\left(\frac{35.0\ \text{d}}{14.28\ \text{d}}\right) = -1.70$$

Taking the natural antilog of -1.70 by using the INV ln x or the ANTILN key on a calculator then gives the ratio N/N_0.

$$\frac{N}{N_0} = \text{antiln } (-1.70) = 0.183$$

Since the initial amount of ^{32}P was 100%, we can set $N_0 = 100\%$ and solve for N

$$\frac{N}{100\%} = 0.183$$

so $N = 0.183 \times 100\% = 18.3\%$

After 35.0 d, 18.3% of a ^{32}P sample remains, and $100\% - 18.3\% = 81.7\%$ has decayed.

EXAMPLE 22.5

A sample of ^{41}Ar, a radionuclide used to measure the flow of gases from smoke-stacks, decays initially at a rate of 34,500 disintegrations/min, but the decay rate falls to 21,500 disintegrations/min after 75.0 min. What is the half-life of ^{41}Ar?

SOLUTION   For a first-order process like radioactive decay, in which rate = kN, the ratio of the decay rate at any time t to the decay rate at time $t = 0$ is the same as the ratio of N and N_0.

$$\frac{\text{Decay rate at time } t}{\text{Decay rate at time } t = 0} = \frac{kN}{kN_0} = \frac{N}{N_0}$$

Thus, we need to find $t_{1/2}$ in the equation

$$\ln \left(\frac{N}{N_0}\right) = -0.693 \left(\frac{t}{t_{1/2}}\right)$$

Substituting the proper values into the equation gives

$$\ln \left(\frac{21,500}{34,500}\right) = -0.693 \left(\frac{75.0 \text{ min}}{t_{1/2}}\right) \quad \text{or} \quad -0.473 = \frac{-52.0 \text{ min}}{t_{1/2}}$$

so $t_{1/2} = \dfrac{-52.0 \text{ m}}{-0.473} = 110 \text{ min}$

The half-life of ^{41}Ar is 110 min.

┌ **PROBLEM 22.3** The decay constant for mercury-197, a radionuclide used medically in kidney scans, is $1.08 \times 10^{-2} \text{ h}^{-1}$. What is the half-life of mercury-197?

┌ **PROBLEM 22.4** The half-life of carbon-14 is 5715 y. What is its decay constant?

┌ **PROBLEM 22.5** What percentage of $^{14}_{6}C$ ($t_{1/2} = 5715$ y) remains in a sample estimated to be 16,230 y old?

┌ **PROBLEM 22.6** What is the half-life of iron-59, a radionuclide used medically in the diagnosis of anemia, if a sample with an initial decay rate of 16,800 disintegrations/min decays at a rate of 10,860 disintegrations/min after 28.0 d? ┘

22.4 ➤NUCLEAR STABILITY

Why do some nuclei undergo radioactive decay while others do not? Why, for example, does a carbon-*14* nucleus, with six protons and *eight* neutrons in its nucleus, spontaneously emit a β particle, whereas a carbon-*13* nucleus, with six protons and *seven* neutrons in its nucleus, is stable indefinitely? Before answering these questions, it's important to define what we mean by *stable*. In the context of nuclear chemistry, we'll use the word "stable" to refer to isotopes that can be prepared and whose half-lives can be measured, even if that half-life is only a fraction of a second. Those isotopes that can't be prepared or that decay too rapidly for their half-lives to be measured we'll call *unstable*. Those isotopes that do not undergo radioactive decay we'll call *nonradioactive*, or *stable indefinitely*.

The answer to the question about why some isotopes are radioactive while others are not has to do with the neutron/proton ratio in the nucleus and with the forces holding the nucleus together. To see the effect of the neutron/proton ratio on nuclear stability, look at the grid pictured in Figure 22.3. Along the side of the grid are divisions representing the number of neutrons in nuclei, a number arbitrarily cut off at 200. Along the bottom of the grid are divisions representing the number of protons in nuclei—the first 92 divisions represent the naturally occurring elements,[3] the next 17 represent the artificially produced **transuranium elements**, and the divisions beyond 109 represent elements yet unknown.

When the more than 3000 known, stable nuclides are plotted on the neutron/proton grid in Figure 22.3, they fall in a curved band sometimes

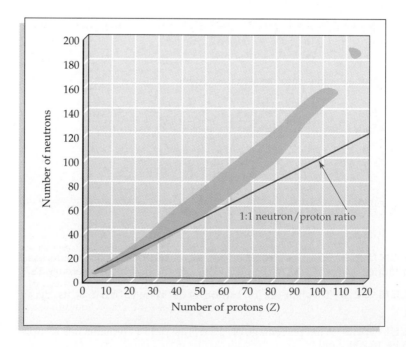

FIGURE 22.3 The peninsula of nuclear stability, indicating various neutron–proton combinations that give rise to observable nuclei with measurable half-lives. Combinations outside the peninsula are too unstable to exist. The "island of stability" near 114 protons and 184 neutrons corresponds to a group of superheavy nuclei that have not yet been observed but that are predicted to be stable.

---

[3] Actually, only 90 of the first 92 elements occur naturally. Neither Tc nor Pm occurs naturally because all their isotopes are radioactive and have very short half-lives. Fr and At occur on earth only in very tiny amounts.

called the "peninsula of nuclear stability." Even within the peninsula, all but 264 of the nuclides disintegrate spontaneously, though their rates of decay vary enormously. On either side of the peninsula is a "sea of instability" representing the large number of unstable neutron/proton combinations that have never been seen. Particularly interesting is the "island of stability" predicted to exist for a few superheavy nuclides near 114 protons and 184 neutrons. None of these nuclides has yet been made, though the search is being actively pursued.

Several generalizations can be made from the data in Figure 22.3:

1. Every element in the periodic table has at least one radioactive isotope.
2. Hydrogen is the only element whose most abundant stable isotope (^1_1H) contains more protons (one) than neutrons (zero). The most abundant stable isotopes of other elements lighter than calcium ($Z = 20$) usually have either the same number of protons and neutrons (^4_2He, $^{12}_6C$, $^{16}_8O$, and $^{28}_{14}Si$, for example), or have only one more neutron than protons ($^{11}_5B$, $^{19}_9F$, and $^{23}_{11}Na$, for example).
3. The ratio of neutrons to protons gradually increases for elements heavier than calcium, giving a curved appearance to the peninsula of stability. The most abundant stable isotope of bismuth, for example, has 126 neutrons and 83 protons ($^{209}_{83}Bi$).
4. All isotopes beyond bismuth-209 are radioactive, even though they may occur naturally.
5. Nonradioactive isotopes generally have an even number of neutrons. Of the 264 nonradioactive isotopes, 156 have even numbers of both protons and neutrons, and 51 have an even number of neutrons but an odd number of protons. Only 4 nonradioactive isotopes have an odd number of both protons and neutrons (Figure 22.4).

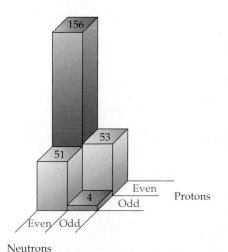

FIGURE 22.4 Numbers of nonradioactive isotopes with various even/odd combinations of neutrons and protons. The majority of nonradioactive isotopes have both an even number of protons and an even number of neutrons. Only four nonradioactive isotopes have both an odd number of protons and an odd number of neutrons.

Observations such as those shown in Figure 22.3 lead to the suggestion that neutrons function as a kind of nuclear "glue" that holds nuclei together by overcoming proton–proton repulsions. The more protons there are in the nucleus, the more glue is needed. Furthermore, there appear to be certain "magic numbers" of protons or neutrons—2, 8, 20, 28, 50, 82, 126—that give rise to particularly stable nuclei. A nucleus with a magic number of either protons or neutrons is likely to be unusually stable. There are ten naturally occurring isotopes of tin, which has a magic number of protons ($Z = 50$), but there are only two naturally occurring isotopes of its neighbor indium ($Z = 49$) and two naturally occurring isotopes of its neighbor antimony ($Z = 51$). Lead-208 is especially stable because it has a *double* magic number of nucleons: 126 neutrons and 82 protons.

The correlation of nuclear stability with special numbers of nucleons is reminiscent of the correlation of chemical stability with special numbers of electrons—the octet rule discussed in Section 6.10. As a result, a shell model of nuclear structure has been proposed, analogous to the shell model of electronic structure. The magic numbers of nucleons evidently correspond to filled nuclear-shell configurations.

A close-up look at a segment of the peninsula of nuclear stability shown in Figure 22.5 reveals some more interesting trends. One trend is that elements with an even atomic number have a larger number of nonradioactive nuclides than do elements with an odd atomic number. Tungsten ($Z = 74$) has five nonradioactive nuclides, for example, while its neighbor rhe-

FIGURE 22.5 A close-up look at the peninsula of nuclear stability in the region from $Z = 66$ (dysprosium) through $Z = 79$ (gold) shows the types of radioactive processes undergone by various nuclei. Nuclei with a lower neutron/proton ratio tend to undergo positron emission, electron capture, or alpha emission, whereas nuclei with a higher neutron/proton ratio tend to undergo beta emission.

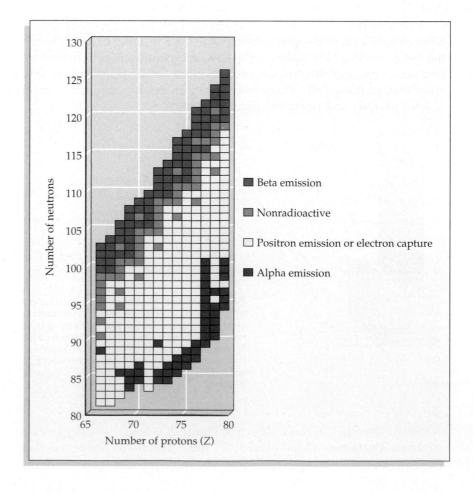

nium (Z = 75) has only one. Another trend is that radioactive nuclei with lower neutron/proton ratios (right side of the peninsula) tend to undergo nuclear disintegration by positron emission, by electron capture, or by alpha emission, while nuclei with higher neutron/proton ratios (left side of the peninsula) tend to emit beta particles. This makes sense if you think about it: The nuclei on the right side of the peninsula are neutron poor and therefore undergo processes that *increase* the neutron/proton ratio; the nuclei on the left side of the peninsula, by contrast, are neutron rich and therefore undergo a process that *decreases* the neutron/proton ratio.

These processes increase the neutron/proton ratio:

Positron emission: proton $\rightarrow$ neutron + β^+

Electron capture: proton + electron $\rightarrow$ neutron

Alpha emission: $^A_Z X \rightarrow {}^{A-4}_{Z-2} X + {}^4_2 He$

This process decreases the neutron/proton ratio:

Beta emission: neutron $\rightarrow$ proton + β^-

One further point about nuclear decay is that some nuclides, particularly those of the heavy elements above bismuth, can't reach a nonradioactive nucleus in a single emission. The product nucleus resulting from the first decay is itself radioactive and therefore undergoes a further disintegration. In fact, such nuclei must often undergo a whole **decay series** of nuclear disintegrations before they ultimately reach a nonradioactive product. Uranium-238, for example, undergoes a series of 14 sequential nuclear reactions, ultimately ending up at lead-206 (Figure 22.6).

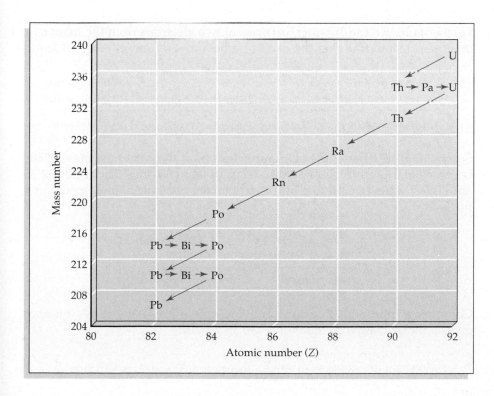

FIGURE 22.6 The decay series from to $^{238}_{92}U$ to $^{206}_{82}Pb$. Each nuclide except for the last is radioactive and undergoes nuclear decay. The diagonal arrows represent alpha emissions, and the horizontal arrows represent beta emissions.

22.5 ►ENERGY CHANGES DURING NUCLEAR REACTIONS

We said in the previous section that neutrons appear to act as a kind of nuclear glue by overcoming the proton–proton repulsions that would otherwise cause the nucleus to fly apart. In principle, it should be possible to measure the strength of the forces holding a nucleus together by measuring the amount of heat released on forming the nucleus from isolated protons and neutrons. For example, the energy change associated with the combining of two neutrons and two protons to yield an $_2^4\text{He}$ nucleus should be a direct measure of nuclear stability.

$$2\,_1^1\text{H} + 2\,_0^1\text{n} \rightarrow\,_2^4\text{He} \qquad \Delta E = \text{?}$$

Unfortunately, there are problems with actually carrying out the measurement because temperatures rivaling those in the interior of the sun (10^7 K) are necessary to get the reaction to occur. Nevertheless, the energy change for the process can be *calculated* by using the Einstein equation $\Delta E = \Delta mc^2$, which relates the energy change (ΔE) of a nuclear process to a corresponding mass change Δm (Section 5.5).

Take a helium-4 nucleus for example. We know from Table 2.1 that the mass of two neutrons and two protons is 4.031 88 amu:

Mass of two neutrons	=	(2) (1.008 66 amu) =	2.017 32 amu
Mass of two protons	=	(2) (1.007 28 amu) =	2.014 56 amu
Total mass of 2 n + 2 p		=	4.031 88 amu

Furthermore, we can subtract the mass of two electrons from the mass of a helium-4 *atom* to find that the mass of a helium-4 *nucleus* is 4.001 50 amu:

Mass of helium-4 atom		=	4.002 60 amu
− Mass of two electrons	= −(2) (5.486 × 10^{-4} amu) =	−0.001 10 amu
Mass of helium-4 nucleus		=	4.001 50 amu

Subtracting the mass of the helium nucleus from the combined mass of the neutrons and protons shows a difference of 0.030 38 amu. That is, 0.030 38 amu is *lost* when two protons and two neutrons combine to form a helium-4 nucleus:

Mass of 2 n + 2 p	=	4.031 88 amu
− Mass of ^{4}He nucleus	=	−4.001 50 amu
Mass difference	=	0.030 38 amu

Converting from amu to grams and multiplying by Avogadro's number gives a molar mass difference of 0.030 38 g/mol:

$$(0.030\,38\text{ amu})\left(1.660\,54 \times 10^{-24}\frac{\text{g}}{\text{amu}}\right)(6.022 \times 10^{23}\text{ mol}^{-1}) = 0.030\,38\frac{\text{g}}{\text{mol}}$$

The loss in mass that occurs when protons and neutrons combine to form a nucleus is called the **mass defect** of the nucleus. The lost mass does not simply disappear, though. It is converted into energy—a **binding energy** that holds the nucleons together and that we can calculate using the Einstein equation:

$$\Delta E = \Delta m c^2 \qquad \text{where } c = 3.00 \times 10^8 \text{ m/s}$$

$$= (3.038 \times 10^{-5} \text{ kg/mol}) (3.00 \times 10^8 \text{ m/s})^2$$

$$= 2.73 \times 10^{12} \text{ kg} \cdot \text{m}^2/(\text{mol} \cdot \text{s}^2)$$

$$= 2.73 \times 10^{12} \text{ J/mol} = 2.73 \times 10^9 \text{ kJ/mol}$$

For a helium-4 nucleus, the binding energy is 2.73×10^9 kJ/mol. In other words, 2.73×10^9 kJ/mol of energy is released when helium-4 nuclei are formed, and 2.73×10^9 kJ/mol of energy must be added to decompose helium-4 nuclei into isolated protons and neutrons. This enormous amount of energy is more than *10 million times* the energy change associated with a typical chemical process!

To make comparisons among different isotopes easier, binding energies are usually expressed on a per-nucleon basis using *electron volts* as the energy unit. One electron volt (eV) = 1.60×10^{-19} J, and 1 million electron volts (1 MeV) = 1.60×10^{-13} J.[4] Thus, the helium-4 binding energy is 7.08 MeV/nucleon:

$$\begin{aligned}\text{Helium-4 binding} \\ \text{energy per nucleon}\end{aligned} = \frac{2.73 \times 10^{12} \text{ J/mol}}{6.022 \times 10^{23} \text{ nuclei/mol}} \times \frac{1 \text{ MeV}}{1.60 \times 10^{-13} \text{ J}} \times \frac{1 \text{ nucleus}}{4 \text{ nucleons}}$$

$$= 7.08 \text{ MeV/nucleon}$$

A plot of binding energy per nucleon for the most stable isotope of each element is shown in Figure 22.7. Higher stability in this context corre-

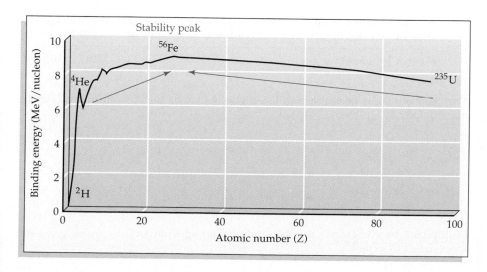

FIGURE 22.7 The binding energy per nucleon for the most stable isotope of each naturally occurring element. Note that binding energy reaches a maximum of 8.79 MeV/nucleon at ^{56}Fe. As a result, there is an increase in stability when much lighter elements fuse together to yield heavier elements up to ^{56}Fe and when much heavier elements split apart to yield lighter elements down to ^{56}Fe, as indicated by the arrows.

[4] When 1 eV is multiplied by Avogadro's number and is thus expressed on a per-mole basis, 1 eV/mol = 96.5 kJ/mol.

sponds to a higher binding energy per nucleon, so that the most stable nuclei are at the top of the curve. Iron-56, with a binding energy of 8.79 MeV/nucleon, is the most stable isotope known.

The idea that mass and energy are interconvertible is a potentially disturbing one because it seems to overthrow two of the fundamental principles on which chemistry is based—the law of mass conservation and the law of energy conservation. In fact, what the mass/energy interconversion means is that the two individual laws must be *combined*. Neither mass nor energy is conserved separately; only the combination of the two is conserved. Every time we do a chemical reaction, mass and energy are interconverted, but the combination of the two is conserved. For the energy changes involved in ordinary chemical reactions, however, the effects are so small that the mass change can't be seen by even the best analytical balances. Example 22.7 illustrates a mass–energy calculation for a chemical reaction.

EXAMPLE 22.6

Helium-6 is a radioactive isotope with $t_{1/2} = 0.81$ s. Calculate the mass defect in g/mol for the formation of a ^{6}He nucleus, and calculate the binding energy in MeV/nucleon. Is a ^{6}He nucleus more stable or less stable than a ^{4}He nucleus? (The mass of a ^{6}He atom is 6.018 89 amu.)

SOLUTION   First, calculate the total mass of the nucleons ($4\,n + 2\,p$)

$$
\begin{aligned}
\text{Mass of four neutrons} &= (4)\,(1.008\,66 \text{ amu}) = 4.034\,64 \text{ amu} \\
\text{Mass of two protons} &= (2)\,(1.007\,28 \text{ amu}) = 2.014\,56 \text{ amu} \\
\hline
\text{Mass of } 4\,n + 2\,p &\phantom{= (2)\,(1.007\,28 \text{ amu})} = 6.049\,20 \text{ amu}
\end{aligned}
$$

Next, calculate the mass of a ^{6}He nucleus by subtracting the mass of two electrons from the mass of a ^{6}He atom.

$$
\begin{aligned}
\text{Mass of helium-6 atom} &\phantom{= -(2)\,(5.486 \times 10^{-4} \text{ amu})} = 6.018\,89 \text{ amu} \\
-\text{ Mass of two electrons} &= -(2)\,(5.486 \times 10^{-4} \text{ amu}) = -0.001\,10 \text{ amu} \\
\hline
\text{Mass of helium-6 nucleus} &\phantom{= -(2)\,(5.486 \times 10^{-4} \text{ amu})} = 6.017\,79 \text{ amu}
\end{aligned}
$$

Then subtract the mass of the nucleons from the mass of the ^{6}He nucleus to find the mass defect:

$$\text{Mass defect} = \text{mass of nucleons} - \text{mass of nucleus}$$

$$= (6.049\,20 \text{ amu}) - (6.017\,79 \text{ amu})$$

$$= 0.031\,41 \text{ amu}$$

Multiplying by Avogadro's number and converting from amu to grams gives the mass defect in g/mol:

$$(0.031\,41 \text{ amu})\left(1.660\,54 \times 10^{-24}\,\frac{\text{g}}{\text{amu}}\right)(6.022 \times 10^{23} \text{ mol}^{-1}) = 0.031\,41\,\frac{\text{g}}{\text{mol}}$$

Now, use the Einstein equation to convert the mass defect into the binding energy.

$$\Delta E = \Delta mc^2 = \left(0.031\,41\,\frac{\text{g}}{\text{mol}}\right)\left(10^{-3}\,\frac{\text{kg}}{\text{g}}\right)\left(3.00\times10^8\,\frac{\text{m}}{\text{s}}\right)^2$$

$$= 2.83\times10^{12}\,\text{J/mol} = 2.83\times10^9\,\text{kJ/mol}$$

This value can be expressed in the desired units of MeV/nucleon by dividing by Avogadro's number, converting to MeV, and dividing by the number of nucleons:

$$\frac{2.83\times10^{12}\,\dfrac{\text{J}}{\text{mol}}}{6.022\times10^{23}\,\dfrac{\text{nuclei}}{\text{mol}}} \times \frac{1\,\text{MeV}}{1.60\times10^{-13}\,\text{J}} \times \frac{1\,\text{nucleus}}{6\,\text{nucleons}} = 4.89\,\text{MeV/nucleon}$$

The binding energy of a radioactive ^6He nucleus is 4.89 MeV/nucleon, making it less stable than a ^4He nucleus, whose binding energy is 7.08 MeV/nucleon.

EXAMPLE 22.7

What is the change in mass in grams when hydrogen atoms combine to form a mole of hydrogen molecules?

$$2\,\text{H} \rightarrow \text{H}_2 \qquad \Delta E = -436\,\text{kJ}$$

SOLUTION   The problem asks us to calculate a mass defect Δm when the energy change ΔE is known. To do this, we have to rearrange the Einstein equation to solve for mass:

$$\Delta m = \frac{\Delta E}{c^2} = \frac{(-436\,\text{kJ})\left(10^3\,\dfrac{\text{J}}{\text{kJ}}\right)\left(\dfrac{1\,\text{kg}\cdot\text{m}^2/\text{s}^2}{1\,\text{J}}\right)}{\left(3.00\times10^8\,\dfrac{\text{m}}{\text{s}}\right)^2}$$

$$= -4.84\times10^{-12}\,\text{kg} = -4.84\times10^{-15}\,\text{g}$$

The loss in mass accompanying formation of 1 mol of H_2 molecules from its constituent atoms is 4.84×10^{-15} g, far too small an amount to be detectable by any balance.

PROBLEM 22.7 Calculate the mass defect in g/mol for the formation of an oxygen-16 nucleus, and calculate the binding energy in MeV/nucleon. The atomic mass of an ^{16}O atom is 15.994 92 amu.

PROBLEM 22.8 What is the mass change in g/mol for the "thermite" reaction of aluminum with iron(III) oxide?

$$2\,\text{Al}(s) + \text{Fe}_2\text{O}_3(s) \rightarrow \text{Al}_2\text{O}_3(s) + 2\,\text{Fe}(s) \qquad \Delta E = -852\,\text{kJ/mol}$$

22.6 ►NUCLEAR FISSION AND FUSION

A careful look at the plot of atomic number versus binding energy per nucleon in Figure 22.7 leads to some interesting and enormously important conclusions. The fact that binding energy per nucleon begins at a relatively low value for $^{2}_{1}H$, reaches a maximum at $^{56}_{26}Fe$, and then gradually tails off implies that both lighter and heavier elements are less stable than midweight elements near iron-56. Very heavy nuclei can therefore gain stability and release energy if they fragment to yield midweight elements, while very light nuclei can gain stability and release energy if they fuse together. The two resultant processes—**fission** for the fragmenting of heavy nuclei and **fusion** for the joining together of light nuclei—have changed the world since their discovery in the late 1930s and early 1940s.

Nuclear Fission

Certain nuclei—uranium-233, uranium-235, and plutonium-239, for example—do more than undergo simple radioactive decay; they break into fragments when struck by neutrons. As illustrated in Figure 22.8, an incoming neutron causes the nucleus to split into two smaller pieces of roughly similar size.

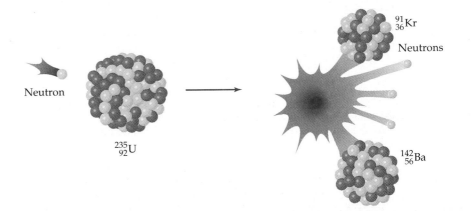

FIGURE 22.8 A representation of nuclear fission. A uranium-235 nucleus fragments when struck by a neutron, yielding two smaller nuclei and releasing an enormous amount of energy.

The fission of a nucleus does not occur in exactly the same way each time: More than 100 different fission pathways have been identified for uranium-235, yielding more than 200 different fission products. One of the more frequently occurring pathways generates barium-142 and krypton-91, along with two additional neutrons plus the one neutron that initiated the fission.

$$^{1}_{0}n + ^{235}_{92}U \rightarrow ^{142}_{56}Ba + ^{91}_{36}Kr + 3\ ^{1}_{0}n$$

The three neutrons released by fission of a ^{235}U nucleus can induce three more fissions yielding nine neutrons, which can induce nine more fissions yielding 27 neutrons, and so on indefinitely. The result is a **chain**

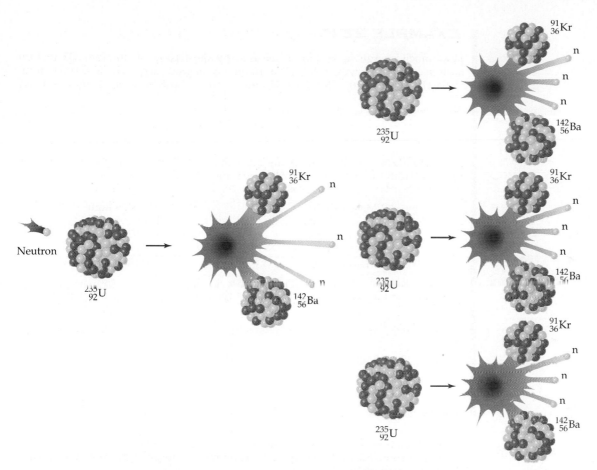

FIGURE 22.9 A chain reaction, in which each fission event produces additional neutrons that induce more fissions. The rate of the process increases at each stage.

reaction that continues to occur even if the supply of neutrons from outside is cut off (Figure 22.9). If the sample size is small, many of the neutrons escape before initiating additional fission events, and the chain reaction stops. If a sufficient amount of ^{235}U is in the sample, however, an amount called the **critical mass**, then the chain reaction becomes self-sustaining and can occur so rapidly that a nuclear explosion results. For ^{235}U the critical mass is about 56 kg, although the amount can be reduced to 15 kg by placing a "tamper" of ^{238}U around the ^{235}U to reflect back some of the escaping neutrons.

The amount of energy released during nuclear fission can be calculated as in Example 22.8 by finding the accompanying mass change and then using the Einstein mass–energy relationship discussed in the previous section. When calculating the mass change, it's simplest to use the masses of the *atoms* corresponding to the relevant nuclei, rather than the masses of the nuclei themselves, since the number of electrons is the same in both reactants and products.

Enormous amounts of energy are released in the explosion that accompanies an uncontrolled nuclear chain reaction.

EXAMPLE 22.8

How much energy in kJ/mol is released by the fission of uranium-235 to form barium-142 and krypton-91? [The fragment masses are ^{238}U (235.0439 amu), ^{142}Ba (141.9164 amu), ^{91}Kr (90.9234 amu), and n (1.008 66 amu); 1 amu = 1.6605 $\times$ 10^{-27} kg.]

$$\ce{^1_0n + ^{235}_{92}U -> ^{142}_{56}Ba + ^{91}_{36}Kr + 3\ ^1_0n}$$

SOLUTION   First calculate the mass change by subtracting the masses of the products from the masses of the ^{235}U reactant:

Mass of ^{235}U:	235.0439 amu
− Mass of ^{142}Ba:	−141.9164 amu
− Mass of ^{91}Kr	−90.9234 amu
− Mass of 2 n: (−2 × 1.00866 amu) =	−2.0173 amu
Mass change:	0.1868 amu

Next, use the Einstein equation to convert the mass loss into an energy change:

$$\Delta E = \Delta mc^2$$

$$\Delta E = (0.1868 \text{ amu})\left(1.6605 \times 10^{-27} \frac{\text{kg}}{\text{amu}}\right)(6.022 \times 10^{23} \text{ mol}^{-1})\left(3.00 \times 10^8 \frac{\text{m}}{\text{s}}\right)^2$$

$$\Delta E = 1.68 \times 10^{13} \text{ kg} \cdot \text{m}^2/(\text{s}^2 \cdot \text{mol}) = 1.68 \times 10^{10} \text{ kJ/mol}$$

Nuclear fission of 1 mol of ^{235}U releases 1.68×10^{10} kJ of energy.

PROBLEM 22.9 An alternate pathway for the nuclear fission of ^{235}U produces tellurium-137 and zirconium-97. How much energy in kJ/mol is released in this fission pathway?

$$\ce{^1_0n + ^{235}_{92}U -> ^{137}_{52}Te + ^{97}_{40}Zr + 2\ ^1_0n}$$

The atomic masses are ^{235}U (235.0439 amu), ^{137}Te (136.9254 amu), ^{97}Zr (96.9110 amu), and n (1.00866 amu); 1 amu = 1.6605 $\times$ 10^{-27} kg.

Nuclear Reactors

The same fission process that leads to a nuclear explosion when uncontrolled can be used for the generation of electric power when carried out in a controlled manner in a **nuclear reactor** (Figure 22.10). The principle behind a nuclear reactor is simple: A subcritical amount of uranium fuel is placed in a containment vessel surrounded by circulating coolant, and *control rods* are added. Made of substances such as boron and cadmium that absorb and thus regulate the flow of neutrons, the control rods are raised and lowered as necessary to maintain the fission at a controlled rate so that overheating is prevented. Energy from the controlled fission heats the circulating coolant, which in turn produces steam to drive a turbine and produce electricity.

Naturally occurring uranium is a mixture of two isotopes. The nonfissionable ^{238}U isotope has a natural abundance of 99.3%, while the fissionable ^{235}U isotope is present only to the extent of 0.7%. The fuel used in nuclear reactors is typically made of compressed pellets of UO_2 that have been

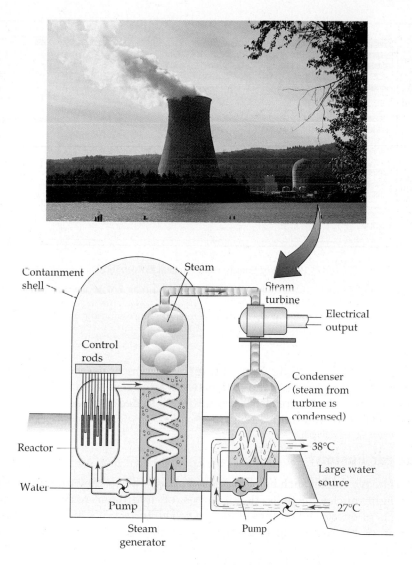

Energy from the fission of uranium atoms is used to produce steam and generate electricity in this nuclear power plant. The characteristically shaped stacks are cooling towers for the steam.

FIGURE 22.10 A diagram of a nuclear power plant. Heat produced in the reactor core is transferred by coolant circulating in a closed loop to a steam generator, and the steam then drives a turbine to generate electricity.

isotopically enriched to a 3% concentration of ^{235}U and then encased in zirconium rods. The rods are placed in a pressure vessel filled with water, which acts as a moderator to slow the neutrons so they can be captured more readily. No nuclear explosion can occur in a reactor because the amount of fuel is well below the critical mass. In a worst-case accident, however, uncontrolled fission could lead to enormous overheating that could melt the reactor and surrounding containment vessel, thereby releasing large amounts of radioactivity to the environment.

Twenty-four countries around the world now obtain some of their electricity from nuclear energy (Figure 22.11). France leads the way with 73%, followed by a number of other European countries that have also made a substantial commitment to the technology. The United States has been more cautious, with only 22% of its power coming from nuclear plants. The

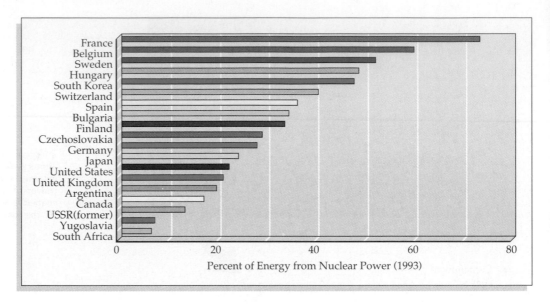

FIGURE 22.11 Percentage of electricity generated by nuclear power in 1993.

primary problem holding back future development is the yet unsolved matter of how to dispose of the radioactive wastes generated by the plants. It will take at least 600 y for waste strontium-90 to decay to safe levels, and at least 20,000 y for plutonium-239.

Nuclear Fusion

Just as heavy nuclei such as ^{235}U release energy when they undergo *fission*, very light nuclei such as the isotopes of hydrogen release enormous amounts of energy when they undergo *fusion*. In fact, it's just such a fusion reaction of hydrogen nuclei to produce helium that powers our sun and other stars. Among the processes thought to occur in the sun are those in the following sequence leading to helium-4:

$$^1_1\text{H} + ^1_1\text{H} \;\longrightarrow\; ^2_1\text{H} + ^0_1\text{e}$$

$$^1_1\text{H} + ^2_1\text{H} \;\longrightarrow\; ^3_2\text{He}$$

$$^3_2\text{He} + ^3_2\text{He} \longrightarrow ^4_2\text{He} + 2\,^1_1\text{H}$$

$$^3_2\text{He} + ^1_1\text{H} \;\longrightarrow\; ^4_2\text{He} + ^0_1\text{e}$$

The main appeal of nuclear fusion as a potential power source is that the hydrogen isotopes used as fuel are cheap and plentiful and that the fusion products are nonradioactive and nonpolluting. The technical problems that must be solved before achieving a practical and controllable fusion method are staggering, however. Not the least of the problems is that a temperature of approximately 40 *million* kelvins is needed to initiate the fusion process.

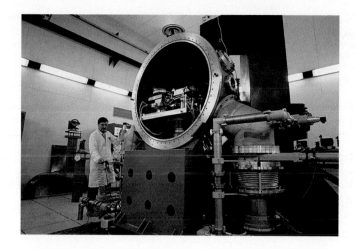

The search for controllable nuclear fusion is going on with the help of this Nova reactor, in which glass-encased pellets of a deuterium/tritium mixture are heated by a laser pulse to a temperature of 10^8 K.

⌐ PROBLEM 22.10   Calculate the amount of energy released in kJ/mol for the fusion reaction of ^{1}H and ^{2}H atoms to yield a ^{3}He atom:

$$^1_1H + {}^2_1H \rightarrow {}^3_2He$$

The atomic masses are ^{1}H (1.007 83 amu), ^{2}H (2.014 10 amu), and ^{3}He (3.016 03 amu); 1 amu = 1.6605×10^{-27} kg.

22.7 ➤ NUCLEAR TRANSMUTATION

Only about 300 of the more than 3000 known isotopes occur naturally. The remainder have been made by **nuclear transmutation**, the change of one element into another brought about by bombardment of an atom with a high energy particle such as a proton, a neutron, or an α particle. In the ensuing collision between particle and atom, an unstable nucleus is momentarily created, a nuclear change occurs, and a different element is produced. The first nuclear transmutation was accomplished in 1917 by Rutherford, who bombarded ^{14}N nuclei with α particles and found that ^{17}O was produced:

$$^{14}_7N + {}^4_2He \rightarrow {}^{17}_8O + {}^1_1H$$

Other nuclear transmutations can lead to the synthesis of entirely new elements never before seen on earth. In fact, all the transuranium elements—those elements with atomic numbers greater than 92—have been produced by bombardment reactions. Plutonium (Pu), for example, can be made by bombardment of uranium-238 with α particles:

$$^{238}_{92}U + {}^4_2He \rightarrow {}^{241}_{94}Pu + {}^1_0n$$

The plutonium-241 that results from uranium-238 bombardment is itself radioactive with a half-life of 14.4 y, decaying by β emission to yield americium-241. (If the name sounds familiar, it's because americium is used commercially in making smoke detectors). Americium-241 is also radioactive, decaying by α emission with a half-life of 432 y.

$$^{241}_{94}\text{Pu} \rightarrow {}^{241}_{95}\text{Am} + {}^{0}_{-1}\text{e}$$

$$^{241}_{95}\text{Am} \rightarrow {}^{237}_{93}\text{Np} + {}^{4}_{2}\text{He}$$

Still other nuclear transmutations are carried out using neutrons, protons, or other particles for bombardment. The cobalt-60 used in radiation therapy for cancer patients can be prepared by neutron bombardment of iron-58 according to a three-step sequence of events: Iron-58 first adds a neutron to yield iron-59, the iron-59 undergoes β decay to yield cobalt-59, and the cobalt-59 adds a second neutron to yield cobalt-60.

$$^{58}_{26}\text{Fe} + {}^{1}_{0}\text{n} \rightarrow {}^{59}_{26}\text{Fe}$$

$$^{59}_{26}\text{Fe} \rightarrow {}^{59}_{27}\text{Co} + {}^{0}_{-1}\text{e}$$

$$^{59}_{27}\text{Co} + {}^{1}_{0}\text{n} \rightarrow {}^{60}_{27}\text{Co}$$

The overall change can be written as

$$^{58}_{26}\text{Fe} + 2\,{}^{1}_{0}\text{n} \rightarrow {}^{60}_{27}\text{Co} + {}^{0}_{-1}\text{e}$$

EXAMPLE 22.9

The element berkelium, first prepared at the University of California at Berkeley in 1949, is made by α bombardment of $^{241}_{95}\text{Am}$. Two neutrons are also produced during the reaction. What isotope of berkelium results from this transmutation? Write a balanced nuclear equation.

SOLUTION   According to the periodic table, berkelium has $Z = 97$. Since the sum of the reactant mass numbers is $241 + 4 = 245$, and since there are two neutrons produced, the berkelium isotope must have a mass number of 243.

$$^{241}_{95}\text{Am} + {}^{4}_{2}\text{He} \rightarrow {}^{243}_{97}\text{Bk} + 2\,{}^{1}_{0}\text{n}$$

┌ **PROBLEM 22.11** Write a balanced nuclear equation for the reaction of argon-40 with a proton: $^{40}_{18}\text{Ar} + {}^{1}_{1}\text{p} \rightarrow ? + {}^{1}_{0}\text{n}$

┌ **PROBLEM 22.12** Write a balanced nuclear equation for the reaction of of uranium-238 with a deuteron ($^{2}_{1}\text{H}$): $^{238}_{92}\text{U} + {}^{2}_{1}\text{H} \rightarrow ? + 2\,{}^{1}_{0}\text{n}$

22.8 ►DETECTING AND MEASURING RADIOACTIVITY

Radioactive emissions are invisible. We can't see, hear, smell, touch, or taste them, no matter how high the dose. We can, however, detect radiation by measuring its *ionizing* properties. High-energy radiation of all kinds is usually grouped under the name **ionizing radiation** because interaction of the radiation with a molecule knocks an electron from the molecule, thereby ionizing it.

$$\text{Molecule} \xrightarrow[\text{radiation}]{\text{ionizing}} \text{ion} + e^-$$

Ionizing radiation includes not only α particles, β particles, and γ rays, but also X rays and **cosmic rays**. X rays, like γ rays, are high-energy photons ($\lambda = 10^{-8}$–10^{-11} m) rather than particles; cosmic rays are energetic particles coming from interstellar space. They consist primarily of protons, along with some alpha and beta particles.

The simplest device for detecting radiation is the photographic film badge worn by people who routinely work with radioactive materials. Any radiation striking the badge causes it to fog. Perhaps the best known method for measuring radiation is the **Geiger counter**, an argon-filled tube containing two electrodes (Figure 22.12). The inner walls of the tube are coated with an electrically conducting material and given a negative charge, and a wire in the center of the tube is given a positive charge. As radiation enters the tube through a thin window, it strikes and ionizes argon atoms, releasing electrons that briefly conduct a tiny electric current from the negatively charged walls to the positively charged center electrode. The passage of the current is detected, amplified, and used to produce a clicking sound or to register on a meter. The more radiation that enters the tube, the more frequent the clicks.

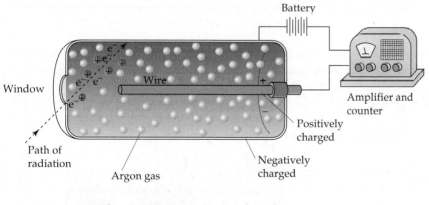

FIGURE 22.12 A Geiger counter for measuring radiation. As radiation enters the tube through a thin window, it ionizes argon atoms and produces electrons that conduct a tiny electric current from the negatively charged walls to the positively charged center electrode. The current flow then registers on the meter.

Radiation is conveniently detected and measured using this scintillation counter, which electronically counts the flashes produced when radiation strikes a phosphor.

The most versatile method for measuring radiation in the laboratory is the *scintillation counter*, in which a substance called a *phosphor* (often a sodium iodide crystal containing a small amount of thallium iodide) emits a flash of light when it is struck by radiation. The number of flashes are counted electronically and converted into an electrical signal.

Radiation intensity is expressed in different ways, depending on what is being measured. Some units measure the number of nuclear decay occurrences; others measure the amount of exposure to radiation or the biological consequences of radiation (Table 22.3).

TABLE 22.3 Units for Measuring Radiation

Unit	Quantity Measured	Description
Becquerel (Bq)	Decay events	Amount of sample that undergoes 1 disintegration/s
Curie (Ci)	Decay events	Amount of sample that undergoes 3.7×10^{10} disintegrations/s
Gray (Gy)	Energy absorbed per kg of tissue	1 Gy = 1 J/kg tissue
Rad	Energy absorbed per kg of tissue	1 rad = 0.01 Gy
Sievert (Sv)	Tissue damage	1 Sv = 1 J/kg
Rem	Tissue damage	1 rem = 0.01 Sv

- The *becquerel* (Bq) is the SI unit for measuring the number of radioactive disintegrations occurring each second in a sample: 1 Bq = 1 disintegration/s. The *curie* (Ci) and *millicurie* (mCi) also measure disintegrations per unit time, but they are far larger units than the becquerel and are more often used, particularly in medicine and biochemistry. One curie is the decay rate of 1 g of radium, equal to 3.7×10^{10} Bq.

$$1 \text{ Bq} = 1 \text{ disintegration/s}$$

$$1 \text{ Ci} = 3.7 \times 10^{10} \text{ Bq} = 3.7 \times 10^{10} \text{ disintegrations/s}$$

For example, a 1.5 mCi sample of tritium is equal to 5.6×10^7 Bq, meaning that it undergoes 5.6×10^7 disintegrations/s.

$$(1.5 \text{ mCi})\left(10^{-3} \ \frac{\text{Ci}}{\text{mCi}}\right)\left(3.7 \times 10^{10} \ \frac{\text{Bq}}{\text{Ci}}\right) = 5.6 \times 10^7 \text{ Bq}$$

- The *gray* (Gy) is the SI unit for measuring the amount of energy absorbed per kilogram of tissue exposed to a radiation source: 1 Gy = 1 J/kg. The *rad* (*radiation absorbed dose*) also measures tissue exposure and is more often used in medicine.

$$1 \text{ Gy} = 1 \text{ J/kg}$$

$$1 \text{ rad} = 0.01 \text{ Gy}$$

• The *sievert* (Sv) is the SI unit that measures the amount of tissue damage caused by radiation. It takes into account not just the energy absorbed per kilogram of tissue but also the different biological effects of different kinds of radiation. For example, 1 Gy of α radiation causes 20 times more tissue damage than 1 Gy of γ rays, but 1 Sv of α radiation and 1 Sv of γ rays cause the same amount of damage. The *rem* (*roentgen equivalent for man*) is an analogous non-SI unit that is more frequently used in medicine.

$$1 \text{ rem} = 0.01 \text{ Sv}$$

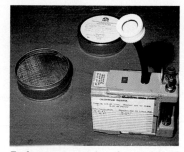

Radon gas can be detected and measured by this home detection kit.

22.9 ►BIOLOGICAL EFFECTS OF RADIATION

The effects of ionizing radiation on the human body vary with the kind of radiation and its energy, the length of exposure, and whether the source is outside or inside the body. When coming from outside the body, γ and X radiation are potentially more harmful than α and β particles because they pass through clothing and skin and into the body. When coming from inside the body, α and β particles are particularly dangerous because all their radiation energy is given up to the immediately surrounding tissue. Alpha emitters are especially hazardous internally and are almost never used in medical applications.

Because of their relatively large mass, α particles move slowly (up to only one-tenth the speed of light) and can be stopped by a few sheets of paper or by the top layer of skin. Beta particles, because they are much lighter, move at up to nine-tenths the speed of light and have about 100 times the penetrating power of alpha particles. A block of wood or heavy protective clothing is necessary to stop beta radiation, which would otherwise penetrate and burn the skin. Gamma rays and X rays move at the speed of light and have about 1000 times the penetrating power of alpha particles. A lead block several inches thick is needed to stop gamma and X radiation, which would otherwise penetrate and damage the body's internal organs. Some properties of different kinds of ionizing radiation are summarized in Table 22.4.

TABLE 22.4	**Some Properties of Ionizing Radiation**	
Type of Radiation	**Energy Range**[a]	**Penetrating Distance in Water**
α	3–9 MeV	0.02–0.04 mm
β	0–3 MeV	0–1 mm
X[b]	100 ev–10 keV	0.01 cm–1 cm
γ[b]	10 keV–10 MeV	1 cm–20 cm

[a] Energies are measured in electron volts (eV): $1 \text{ eV} = 1.602 \times 10^{-19} \text{ J}$.

[b] Distances at which one half of the radiation has been stopped.

The biological effects of different radiation doses are given in Table 22.5. Although the effects sound fearful, the average radiation dose received annually by most people is only about 120 mrem. About 70% of this radiation comes from natural sources (rocks and cosmic rays); the remaining

TABLE 22.5	Biological Effects of Short-term Radiation on Humans
Dose (rem)	**Biological Effects**
0–25	No detectable effects
25–100	Temporary decrease in white blood cell count
100–200	Nausea, vomiting, longer term decrease in white blood cells
200–300	Vomiting, diarrhea, loss of appetite, listlessness
300–600	Vomiting, diarrhea, hemorrhaging, eventual death in some cases
Above 600	Eventual death in nearly all cases

30% comes from medical procedures like taking X rays. The amount due to emissions from nuclear power plants and to fallout from atmospheric testing of nuclear weapons in the 1950s is barely detectable.

⌐ **PROBLEM 22.13** Initial estimates of radiation loss from the 1986 Chernobyl nuclear power plant disaster in Russia predicted a worldwide increase in background radiation of about 5 mrem. (People near the plant received a far greater dose.) By what percentage will this amount increase the annual dose of most people?

22.10 ➤APPLICATIONS OF NUCLEAR CHEMISTRY

Dating with Radionuclides

Biblical scrolls are found in a cave near the Dead Sea—are they authentic? A mummy is discovered in an Egyptian tomb—how old is it? The burned bones of a man are dug up near Lubbock, Texas—how long have humans been on the North American continent? These and many other questions can be answered by archaeologists using a technique called **radiocarbon dating**. (The Dead Sea Scrolls are 1900 y old and authentic, the mummy is 3100 y old, and the human remains found in Texas are 9900 y old.)

Radiocarbon dating places the age of this Egyptian mummy at 3100 y.

Radiocarbon dating of archaeological artifacts depends on the slow and constant production of radioactive carbon-14 in the upper atmosphere by neutron bombardment of nitrogen atoms. (The neutrons come from the bombardment of other atoms by cosmic rays.)

$$\ce{^{14}_7N + ^1_0n -> ^{14}_6C + ^1_1H}$$

Carbon-14 atoms produced in the upper atmosphere combine with oxygen to yield $^{14}CO_2$, which slowly diffuses into the lower atmosphere, where it mixes with ordinary $^{12}CO_2$ and is taken up by plants during photosynthesis. When these plants are eaten, carbon-14 enters the food chain and is ultimately distributed evenly throughout all living organisms.

As long as a plant or animal is living, a dynamic equilibrium exists in which an organism excretes or exhales the same amount of ^{14}C that it takes in. As a result, the ratio of ^{14}C to ^{12}C in the living organism is the same as that in the atmosphere—about 1 part in 10^{12}. When the plant or animal dies, however, it no longer takes in more ^{14}C, and the $^{14}C/^{12}C$ ratio in the organism slowly decreases as ^{14}C undergoes radioactive decay by beta emission ($t_{1/2} = 5715$ y).

$$\ce{^{14}_6C -> ^{14}_7N + ^0_{-1}e}$$

At 5715 y (one ^{14}C half-life) after the death of the organism, the $^{14}C/^{12}C$ ratio has decreased by a factor of 2; at 11,430 y after death, the $^{14}C/^{12}C$ ratio has decreased by a factor of 4; and so on. By measuring the amount of ^{14}C remaining in the traces of any once-living organism, archaeologists can determine how long ago the organism died. Human or animal hair from well-preserved remains, charcoal or wood fragments from once-living trees, and cotton or linen from once-living plants are all useful sources for radiocarbon dating. The technique becomes less accurate as samples get older and the amount of ^{14}C they contain diminishes, but artifacts with an age of 1000 to 20,000 y can be dated with reasonable accuracy.

Just as radiocarbon measurements allow dating of once-living organisms, similar measurements on other radionuclides make possible the dating of rocks. Uranium-238, for example, has a half-life of 4.46×10^9 y and decays through the series of events shown previously in Figure 22.6 to yield lead-206. The age of a uranium-containing rock can therefore be determined by measuring the $^{238}U/^{206}Pb$ ratio. Similarly, potassium-40 has a half-life of 1.26×10^9 y and decays through electron capture and through positron emission to yield argon-40. (Both processes yield the same product.)

$$\ce{^{40}_{19}K + ^0_{-1}e -> ^{40}_{18}Ar}$$

$$\ce{^{40}_{19}K -> ^{40}_{18}Ar + ^0_1e}$$

The age of a rock can be found by crushing a sample, measuring the amount of ^{40}Ar gas that escapes, and comparing the amount of argon-40 with the amount of ^{40}K remaining in the sample. It is through techniques such as these that the age of the earth has been estimated at approximately 4.5 billion y.

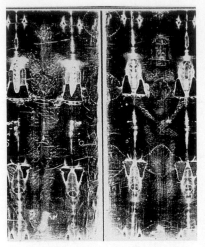

The Shroud of Turin, thought by some to be the burial shroud of Christ, is actually a medieval artifact dating from approximately A.D. 1380.

EXAMPLE 22.10

Radiocarbon measurements made in 1988 on the Shroud of Turin, a religious artifact thought by some to be the burial shroud of Christ, showed a ^{14}C decay rate of 14.2 disintegrations per minute per gram of carbon. What is the age of the shroud if currently living organisms decay at the rate of 15.3 disintegrations per minute per gram of carbon? The half-life of ^{14}C is 5715 y.

SOLUTION   As we saw in Example 22.5, the ratio of the decay rate at any time t to the decay rate at time $t = 0$ is the same as the ratio of N and N_0.

$$\frac{\text{Decay rate at time } t}{\text{Decay rate at time } t = 0} = \frac{kN}{kN_0} = \frac{N}{N_0}$$

To date the shroud, we need to calculate the time t that corresponds to the observed decay rate. This can be done with the equation

$$\ln\left(\frac{N}{N_0}\right) = -0.693\left(\frac{t}{t_{1/2}}\right)$$

Substituting the proper values into the equation gives.

$$\ln\left(\frac{14.2}{15.3}\right) = -0.693\left(\frac{t}{5715 \text{ y}}\right) \quad \text{or} \quad -0.0746 = -0.693\left(\frac{t}{5715 \text{ y}}\right)$$

so

$$t = \frac{(-0.0746)(5715 \text{ y})}{-0.693} = 615 \text{ y}$$

The Shroud of Turin is only 615 y old, showing it to be from medieval times.

⌐ **PROBLEM 22.14** Charcoal found in the Lascaux cave in France, site of many prehistoric cave paintings, was observed to decay at a rate of 2.4 disintegrations per minute per gram of carbon. What is the age of the charcoal if currently living organisms decay at the rate of 15.3 disintegrations per minute per gram of carbon? The half-life of ^{14}C is 5715 y.

What is the age of these paintings?

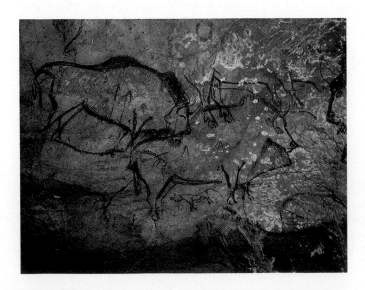

Medical Uses of Radioactivity

The origins of nuclear medicine date to 1901, when the French physician Henri Danlos first used radium in the treatment of a tuberculous skin lesion. Since that time, the use of radioactivity has become a crucial part of modern medical care, both diagnostic and therapeutic. Currently used nuclear techniques can be grouped into four classes: (1) in vivo procedures, (2) in vitro procedures, (3) radiation therapy, and (4) imaging procedures.

IN VIVO PROCEDURES   In vivo studies—those that take place *inside* the body—are carried out to assess the functioning of a particular organ or body system. A radiopharmaceutical agent is administered, and its path in the body—whether it is absorbed, excreted, diluted, or concentrated—is determined by analysis of blood or urine samples.

An example of the many in vivo procedures using radioactive agents is the determination of whole-blood volume by injecting a known quantity of red blood cells labeled with radioactive chromium-51. After a suitable interval to allow the labeled cells to be distributed evenly throughout the body, a blood sample is taken and blood volume is calculated by comparing the concentration of labeled cells in the blood with the quantity of labeled cells injected.

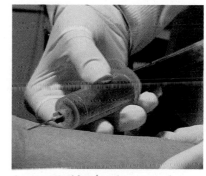

A person's blood volume can be found by injecting a small amount of radioactive chromium-51 and measuring the dilution factor.

$$\text{Blood volume} = \frac{C_0\, V_0}{C_{\text{blood}}}$$

where C_0 = concentration of labeled cells injected (μCi/mL)
   V_0 = volume of labeled cells injected (mL)
   C_{blood} = concentration of labeled cells in blood (μCi/mL)

IN VITRO PROCEDURES: RADIOIMMUNOASSAY   In vitro studies —those that take place *outside* the body—often involve radioimmunoassay (RIA) techniques to measure small concentrations of substances such as drugs or hormones in body fluids. As shown in Figure 22.13, RIA takes

FIGURE 22.13 Radioimmunoassay (RIA) involves measuring the binding of antigen to an antibody. **(a)** Antibodies of known binding ability are prepared and **(b)** added to a sample of biological fluid containing the antigen to be measured. A known amount of radioactively labeled antigen is also added. **(c)** The unknown amount of antigen in the fluid can be determined by counting the amount of labeled antigen bound to the antibodies and comparing the result with a set of known standards. The relative numbers of labeled and unlabeled antigens that are bound are the same as the relative numbers combined (1 to 1 in this example).

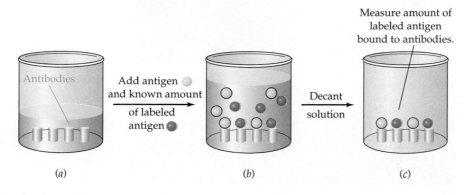

Measure amount of labeled antigen bound to antibodies.

Antibodies

Add antigen and known amount of labeled antigen

Decant solution

(a)   *(b)*   *(c)*

advantage of the ability of large protein molecules called *antibodies* to bind specifically to smaller molecules called *antigens* (the drug or hormone of interest).

To carry out an RIA analysis, a blood or urine sample containing an unknown concentration of antigen is combined in a test tube with small known amounts of both the antibody and a radioactively labeled version of the antigen. The labeled and unlabeled antigen molecules then compete for binding to the limited amount of antibody until an equilibrium is established. At that point, the relative numbers of labeled and unlabeled antigen molecules bound to the antibody reflect their original concentrations. The distribution between bound and unbound labeled molecules is measured with a radiation detector, and the antigen concentration is then determined by comparing the observed distribution to a set of standard distributions measured using known amounts of antigen. The RIA technique can measure drug or hormone concentrations as low as 10^{-14} M and can be carried out on a single drop of blood.

THERAPEUTIC PROCEDURES   Therapeutic procedures—those in which radiation is used as a weapon to kill diseased tissue—can involve either external or internal sources of radiation. External radiation therapy for the treatment of cancer is often carried out with gamma rays from a cobalt-60 source. The highly radioactive source is shielded by a thick lead container and has a small opening directed toward the site of the tumor. By focusing the radiation beam on the tumor and rotating the patient's body, the tumor receives the full exposure while the exposure of surrounding parts of the body is minimized. Nevertheless, sufficient exposure occurs so that most patients treated in this manner suffer some effects of radiation sickness.

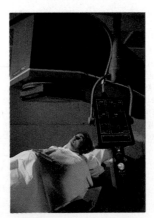

Cancerous tumors can be treated by irradiation with gamma rays from this cobalt-60 source.

Internal radiation therapy is a much more selective technique than external therapy. In the treatment of thyroid disease, for example, iodine-131, a powerful beta emitter known to localize in the target tissue, is administered internally. Since β particles penetrate no farther than several millimeters, the localized ^{131}I produces a high radiation dose that destroys only the surrounding diseased tissue.

IMAGING PROCEDURES   Imaging procedures give diagnostic information about the health of body organs by analyzing the distribution pattern of radionuclides introduced into the body. A radiopharmaceutical agent that is known to concentrate in a specific tissue or organ is injected into the body, and its distribution pattern is monitored by external radiation detectors. Depending on the disease and organ, a diseased organ might concentrate more of the radiopharmaceutical than a normal organ and thus show up as a radioactive hot spot against a cold background. Alternatively, the diseased organ might concentrate less of the radiopharmaceutical than a normal organ and thus show up as a cold spot on a hot background.

A different kind of imaging procedure makes use of a technique called *magnetic resonance imaging*, or MRI.[5] MRI uses no radionuclides and introduces no known hazards. Instead, MRI uses radio waves to stimulate certain nuclei in the presence of an extremely powerful magnetic field. Those stimulated nuclei (normally the hydrogen nuclei in H_2O molecules) then give off a signal that can be measured, interpreted, and correlated with their environment in the body. Figure 22.14 shows a brain scan carried out by MRI.

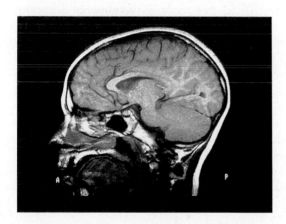

FIGURE 22.14 An MRI brain scan.

[5] The original name of the technique is nuclear magnetic resonance (NMR) imaging, but the word *nuclear* is usually dropped because of the unpleasant connotations it has for many people.

interlude—THE ORIGIN OF CHEMICAL ELEMENTS

The stars in the Milky Way galaxy condensed from gas clouds under gravitational attraction.

Astronomers tell us that our universe began some 15 billion years ago in an extraordinary moment they call the big bang. One second later, the temperature had dropped to about 10^{10} K, and elementary particles began to form: protons, neutrons, and electrons, as well as more exotic particles called *positrons* (positively charged electrons) and *neutrinos* (neutral particles with a mass much less than that of an electron). Three minutes later, the temperature had dropped still further, to 10^9 K, and protons and neutrons fused to form helium nuclei, $^4_2\text{He}^{2+}$. After another few minutes, helium synthesis ceased, and the remaining neutrons decayed to yield protons and electrons.

Matter is thought to have remained in this form for many millions of years until the expanding universe had cooled to about 10,000 K. Electrons were then able to bind to protons and to helium nuclei, forming stable hydrogen and helium atoms. The attractive forces of gravity acting on regions of higher-than-average density slowly produced massive local concentrations of matter, ultimately forming billions of galaxies, each with many billions of what were soon to become stars. As the gas clouds of hydrogen and helium condensed under gravitational attraction and stars formed, the temperatures in their cores reached 10^7 K, and their densities reached 100 g/cm^3. Under these extreme conditions, protons in the cores began to fuse to yield helium nuclei, generating vast amounts of heat and light in the process—about 6×10^8 kJ per mole of protons undergoing fusion.

Most of these early stars burned for perhaps billions of years and then died. A few, however, were so massive that, as helium nuclei accumulated in their cores and new core conditions developed, gravitational attraction caused a rapid contraction leading to still higher core temperatures and higher density—up to 5×10^8 K and 5×10^5 g/cm^3. Even larger nuclei were now formed: carbon, oxygen, silicon, magnesium, and especially iron, the most stable of all elements. But the stars were becoming unstable. As their hydrogen fuel ran out, they underwent an instantaneous gravitational collapse in a time as short as one second. The resultant conditions in the core led to the conversion of iron into still heavier elements and ultimately produced a cataclysmic explosion of the star, an event visible throughout the universe as a *supernova*.

Matter released by exploding supernovae was blown throughout the galaxy, ultimately acting as raw material for the formation of a new generation of stars and planets. Our own sun and solar system, for instance, are thought to have been formed only about 4.5 billion years ago from the matter of former supernovae. Except for hydrogen and helium, all the atoms in our bodies, our planet, and our solar system were created more than 5 billion years ago in exploding stars. We are truly made of stardust.

Instantaneous gravitational collapse of a massive star resulted in this supernova explosion observed in 1987.

Nuclear chemistry is the study of the properties and reactions of atomic nuclei. **Nuclear reactions** differ from chemical reactions in that they involve a change in an atom's nucleus, often producing a different element. The rates of nuclear reactions are unaffected by the addition of a catalyst or by a change in temperature or pressure, and the energy changes accompanying nuclear reactions are far greater than those accompanying chemical reactions. Of the more than 3000 known isotopes, most have been made by **nuclear transmutation,** the change of one element into another brought about by bombardment of an atom with a high energy particle such as a proton, a neutron, or an α particle.

Radioactivity is the spontaneous emission of radiation from an unstable nucleus. **Alpha (α) radiation** consists of helium nuclei, small particles containing two protons and two neutrons (^4_2He). **Beta (β) radiation** consists of electrons ($^0_{-1}\text{e}$), and **gamma (γ) radiation** consists of high-energy photons that have no mass. **Positron emission** is the conversion of a proton in the nucleus into a neutron plus an ejected *positron*, ^0_1e or β^+, a particle that has the same mass as an electron but an opposite charge. **Electron capture** is the capture of an inner-shell electron by a proton in the nucleus. The process is accompanied by emission of gamma rays and results in the conversion of a proton in the nucleus into a neutron. Every element in the periodic table has at least one radioactive isotope **(radionuclide)**. Radioactive decay is characterized kinetically by a first-order **decay constant** and by a half-life, $t_{1/2}$, the time required for the number of radioactive nuclei in a sample to drop to one-half of their initial value. Different radionuclides have half-lives of anywhere from milliseconds to billions of years.

The stability of a given nucleus is related to its neutron/proton ratio. Neutrons appear to act as a kind of nuclear "glue" that holds nuclei together by overcoming proton-proton repulsions. The strength of the forces involved can be measured by calculating an atom's **mass defect**—the difference in mass between a given nucleus and the total mass of its individual nucleons (protons and neutrons). Applying the Einstein equation, $\Delta E = \Delta mc^2$, then allows calculation of the nuclear **binding energy**.

Certain heavy nuclei such as uranium-235 undergo **nuclear fission** when struck by neutrons, breaking apart into fragment nuclei and releasing enormous amounts of energy. Light nuclei such as the isotopes of hydrogen undergo **nuclear fusion** when heated to sufficiently high temperatures, forming heavier nuclei and releasing energy.

High-energy radiation of all types—α particles, β particles, γ rays, X rays, and cosmic rays—is called **ionizing radiation**. When any of these kinds of radiation strikes a molecule, it dislodges an electron and gives an ion. Radiation intensity is expressed in different ways according to what property is being measured. The becquerel (Bq) and the curie (Ci) measure the number of radioactive disintegrations per second in a sample. The gray (Gy) and the rad measure the amount of radiation absorbed per kilogram of tissue. The sievert (Sv) and the rem measure the amount of tissue damage caused by radiation. Radiation effects become noticeable with a human exposure of 25 rem and become lethal at an exposure above 600 rem.

1. $^{40}_{19}$K decays by positron emission to give $^{40}_{18}$Ar. If yellow spheres represent $^{40}_{19}$K atoms, and blue spheres represent $^{40}_{18}$Ar atoms, how many half-lives have passed in the following sample?

2. Make a drawing similar to that in Problem 1 to represent the decay of a sample of $^{40}_{19}$K after approximately 0.5 half-life has passed.

3. Write the symbol of the isotope represented by the following drawing. Blue spheres represent neutrons, and red spheres represent protons.

4. The nucleus shown in Problem 3 is radioactive. Would you expect it to decay by positron emission or β emission? Explain.

ADDITIONAL PROBLEMS

Problems 22.1–22.14 appear within the chapter.

NUCLEAR REACTIONS AND RADIOACTIVITY

22.15 Positron emission and electron capture both give a product nuclide whose atomic number is one less than the starting nuclide. Explain.

22.16 What is the difference between an alpha particle and a helium atom?

22.17 Why is an alpha emitter more hazardous to an organism internally than externally, whereas a gamma emitter is equally hazardous internally and externally?

22.18 Why do nuclides that are "neutron rich" emit beta particles, but nuclides that are "neutron poor" emit alpha particles or positrons or undergo electron capture?

22.19 Harmful chemical spills can often be cleaned up by treatment with another chemical. A spill of H_2SO_4, for example, can be neutralized by addition of $NaHCO_3$. Why can't harmful radioactive wastes from nuclear power plants be cleaned up just as easily?

22.20 Complete and balance the following nuclear equations.
(a) $^{126}_{50}$Sn $\longrightarrow$ $^{0}_{-1}$e + ?
(b) $^{210}_{88}$Ra $\longrightarrow$ $^{4}_{2}$He + ?
(c) $^{77}_{37}$Rb $\longrightarrow$ $^{0}_{1}$e + ?
(d) $^{76}_{36}$Kr + $^{0}_{-1}$e $\longrightarrow$?

22.21 Complete and balance the following nuclear equations.
(a) $^{90}_{38}$Sr $\longrightarrow$ $^{0}_{-1}$e + ?
(b) $^{247}_{100}$Fm $\longrightarrow$ $^{4}_{2}$He + ?
(c) $^{49}_{25}$Mn $\longrightarrow$ $^{0}_{1}$e + ?
(d) $^{37}_{18}$Ar + $^{0}_{-1}$e $\longrightarrow$?

22.22 What particle is produced in each of the following decay reactions?
(a) $^{188}_{80}$Hg $\longrightarrow$ $^{188}_{79}$Au + ?
(b) $^{218}_{85}$At $\longrightarrow$ $^{214}_{83}$Bi + ?
(c) $^{234}_{90}$Th $\longrightarrow$ $^{234}_{91}$Pa + ?

22.23 What particle is produced in each of the following decay reactions?
(a) $^{24}_{11}$Na $\longrightarrow$ $^{24}_{12}$Mg + ?
(b) $^{135}_{60}$Nd $\longrightarrow$ $^{135}_{59}$Pr + ?
(c) $^{170}_{78}$Pt $\longrightarrow$ $^{166}_{76}$Os + ?

22.24 Write balanced nuclear equations for the following processes.
(a) alpha emission of ^{162}Re
(b) electron capture of ^{138}Sm
(c) beta emission of ^{188}W
(d) positron emission of ^{165}Ta

22.25 Write balanced nuclear equations for the following processes.
  (a) beta emission of ^{157}Eu
  (b) electron capture of ^{126}Ba
  (c) alpha emission of ^{146}Sm
  (d) positron emission of ^{125}Ba

22.26 Of the two isotopes of iodine, ^{136}I and ^{122}I, one decays by beta emission and one decays by positron emission. Which is which? Explain.

22.27 Of the two isotopes of tungsten, ^{160}W and ^{185}W, one decays by beta emission and one decays by alpha emission. Which is which? Explain.

22.28 Americium-241, a radionuclide used in smoke detectors, decays by a series of 12 reactions involving sequential loss of $\alpha, \alpha, \beta, \alpha, \alpha, \beta, \alpha, \alpha, \alpha, \beta, \alpha$, and β particles. Identify each intermediate nuclide and the final stable product nucleus.

22.29 Radon-222 decays by a series of three alpha emissions and two beta emissions. What is the final stable nuclide?

22.30 Thorium-232 decays by a 10-step series, ultimately yielding lead-208. How many alpha particles and how many beta particles are emitted?

RADIOACTIVE DECAY RATES

22.31 What does it mean when we say that the half-life of iron-59 is 44.5 d?

22.32 What is the difference between a half-life and a decay constant, and what is the relationship between them?

22.33 The half-life of indium-111, a radionuclide used in studying the distribution of white blood cells, is $t_{1/2} = 2.806$ d. What is the decay constant of ^{111}In?

22.34 What is the decay constant of gallium-67, a radionuclide used for imaging soft-tissue tumors? The half-life of ^{67}Ga is $t_{1/2} = 78.25$ h.

22.35 The decay constant of thallium-201, a radionuclide used for parathyroid imaging, is 0.227 d^{-1}. What is the half-life of ^{201}Tl?

22.36 The decay constant of plutonium-239, a waste product from nuclear reactors, is 2.88×10^{-5} y^{-1}. What is the half-life of ^{239}Pu?

22.37 The half-life of ^{241}Am is 432.2 y. What percentage of an ^{241}Am sample remains after 65 d? after 65 y? after 650 y?

22.38 Fluorine-18 has $t_{1/2} = 109.8$ min. What percentage of an ^{18}F sample remains after 24 min? After 24 h? After 24 d?

22.39 How old is a sample of wood whose ^{14}C content is found to be 43% that of a living tree? The half-life of ^{14}C is 5715 y.

22.40 What is the age of a rock whose $^{40}Ar/^{40}K$ ratio is 1.15? The half-life of ^{40}K is 1.26×10^9 y.

22.41 The decay constant of ^{35}S is 7.95×10^{-3} d^{-1}. What percentage of an ^{35}S sample remains after 185 d?

22.42 Plutonium-239 has a decay constant of 2.88×10^{-5} y^{-1}. What percentage of a ^{239}Pu sample remains after 1000 y? After 25,000 y? After 100,000 y?

22.43 Polonium-209, an α emitter, has a half-life of 105 y. How many alpha particles are emitted in 1.0 s from a 1.0 ng sample of ^{209}Po?

22.44 Chlorine-36 is a β emitter, with a half-life of 3×10^5 y. How many beta particles are emitted in 1.0 min from a 5.0 mg sample of ^{36}Cl? How many curies of radiation does the 5.0 mg sample represent?

22.45 What is the age of the linen shroud around a mummy if the carbon in the shroud has an average of 9.2 disintegrations per minute per gram? The carbon in living organisms undergoes an average of 15.3 disintegrations per minute per gram, and the half-life of ^{14}C is 5715 y.

22.46 A 1.0 mg sample of ^{79}Se decays initially at a rate of 2.8×10^6 disintegrations/s. What is the half-life of ^{79}Se in years?

22.47 What is the half-life in years of ^{44}Ti if a 1.0 ng sample decays initially at a rate of 6.4×10^3 disintegrations/s?

22.48 A sample of a ^{37}Ar undergoes 8540 disintegrations/min initially but undergoes 6990 disintegrations/min after 10.0 d. What is the half-life of ^{37}Ar?

22.49 A sample of ^{28}Mg decays initially at a rate of 53,500 disintegrations/min, but the decay rate falls to 10,980 disintegrations/min after 48.0 h. What is the half-life of ^{28}Mg?

ENERGY CHANGES DURING NUCLEAR REACTIONS

22.50 Why does a given nucleus have less mass than the sum of its constituent protons and neutrons?

22.51 What is the wavelength of gamma rays whose energy is 1.50 MeV?

22.52 What is the frequency of X rays whose energy is 6.82 keV?

22.53 Calculate the mass defect in g/mol for the following nuclei:

(a) ^{52}Fe (atomic mass = 51.948 11 amu)

(b) ^{92}Mo (atomic mass = 91.906 81 amu)

22.54 Calculate the mass defect in g/mol for the following nuclei.

(a) ^{32}S (atomic mass = 31.972 07 amu)

(b) ^{40}Ca (atomic mass = 39.962 59 amu)

22.55 Calculate the binding energy in MeV/nucleon for the following nuclei.

(a) ^{58}Ni (atomic mass = 57.935 35 amu)

(b) ^{84}Kr (atomic mass = 83.911 51 amu)

22.56 Calculate the binding energy in MeV/nucleon for the following nuclei.

(a) ^{63}Cu (atomic mass = 62.939 60 amu)

(b) ^{84}Sr (atomic mass = 83.913 43 amu)

22.57 How much energy in kJ/mol is released when an alpha particle is emitted from ^{174}Ir? The atomic mass of ^{174}Ir is 173.966 66 amu, the atomic mass of ^{170}Re is 169.958 04 amu, and the atomic mass of a ^{4}He atom is 4.002 60 amu.

$$^{174}_{77}\text{Ir} \rightarrow {}^{170}_{75}\text{Re} + {}^4_2\text{He} \qquad \Delta E = ?$$

22.58 Magnesium-28 is a beta emitter that decays to aluminum-28. How much energy is released in kJ/mol? The atomic mass of ^{28}Mg is 27.983 88 amu, and the atomic mass of ^{28}Al is 27.981 91 amu.

22.59 What is the mass change accompanying the formation of ammonia from hydrogen and nitrogen?

$$N_2(g) + 3\,H_2(g) \rightarrow 2\,NH_3(g) \qquad \Delta H° = -92.2 \text{ kJ}$$

22.60 A positron has the same mass as an electron (9.109×10^{-31} kg) but an opposite charge. When the two particles encounter each other, annihilation occurs and only gamma rays are produced. How much energy in kJ/mol is produced?

22.61 How much energy is released in the fusion reaction of ^{2}H to yield ^{3}He? The atomic mass of ^{2}H is 2.0141 amu, and the atomic mass of ^{3}He is 3.0160 amu.

$$2\,{}^2_1\text{H} \rightarrow {}^3_2\text{He} + {}^1_0\text{n}$$

NUCLEAR TRANSMUTATION

22.62 Give the products of the following nuclear reactions.

(a) $^{109}_{47}\text{Ag} + {}^4_2\text{He} \longrightarrow$?

(b) $^{10}_5\text{B} + {}^4_2\text{He} \longrightarrow$? $+ {}^1_0\text{n}$

22.63 Balance the following equations for the nuclear fission of ^{235}U.

(a) $^{235}_{92}\text{U} \longrightarrow {}^{160}_{62}\text{Sm} + {}^{72}_{30}\text{Zn} + ?\,{}^1_0\text{n}$

(b) $^{235}_{92}\text{U} \longrightarrow {}^{87}_{35}\text{Br} + ?\, + 2\,{}^1_0\text{n}$

22.64 Element 109 ($^{266}_{109}\text{Une}$), the heaviest known element, was prepared in 1982 by bombardment of ^{209}Bi atoms with ^{58}Fe atoms. Identify the other product that must have formed, and write a balanced nuclear equation.

22.65 Molybdenum-99 is formed by neutron bombardment of a naturally occurring nuclide. If one neutron is absorbed and no byproducts are formed, what is the starting nuclide?

22.66 Californium-246 is formed by bombardment of uranium-238 atoms. If four neutrons are formed as byproducts, what particle is used for the bombardment?

22.67 Balance the following transmutation reactions.

(a) $^{246}_{96}\text{Cm} + {}^{12}_6\text{C} \longrightarrow$? $+ 4\,{}^1_0\text{n}$

(b) $^{253}_{99}\text{Es} + ?\ \longrightarrow {}^{256}_{101}\text{Md} + {}^1_0\text{n}$

(c) $^{250}_{98}\text{Cf} + {}^{11}_5\text{B} \longrightarrow$? $+ 4\,{}^1_0\text{n}$

GENERAL PROBLEMS

22.68 Potassium ion, K^+, is present in most foods and is an essential nutrient in the human body. Potassium-40, however, which has a natural abundance of 0.0117%, is radioactive with $t_{1/2} = 1.26 \times 10^9$ y. What is the decay constant of ^{40}K? How many ^{40}K$^+$ ions are present in 1.00 g of KCl? How many disintegrations/s does 1.00 g of KCl undergo?

22.69 Chlorine-34 has a half-life of only 1.53 s. How long does it take for 99.99% of a ^{34}Cl sample to decay?

22.70 The decay constant for ^{20}F is 0.063 s^{-1}. How long does it take for 99.99% of a ^{20}F sample to decay?

22.71 Calculate the mass defect and the binding energy in MeV/nucleon for the following nuclides. Which of the two is more stable?

(a) ^{50}Cr (atomic mass = 49.946 05 amu

(b) ^{64}Zn (atomic mass = 63.929 15 amu)

22.72 What is the age of a bone fragment that shows an average of 2.9 disintegrations per minute per gram of carbon? The carbon in living organisms undergoes an average of 15.3 disintegrations per minute per gram, and the half-life of ^{14}C is 5715 y.

22.73 How much energy is released in the fusion reaction of ^{2}H with ^{3}He?

$$^2_1H + ^3_2He \rightarrow ^4_2He + ^1_1H$$

The relevant masses are ^{2}H (2.0141 amu); ^{3}He (3.0160 amu); ^{4}He (4.0026 amu); ^{1}H (1.0078 amu); 1 amu = 1.6605×10^{-27} kg.

22.74 Boron is used in control rods for nuclear reactors because it can absorb neutrons and emit alpha particles: $^{10}_5B + ^1_0n \rightarrow ? + ^4_2He$. Balance the equation.

22.75 The longest half-life yet measured for radioactive decay is the double beta emission of selenium-82, for which $t_{1/2} = 1.1 \times 10^{20}$ y.

$$^{82}_{34}Se \rightarrow ^{82}_{36}Kr + 2\ ^0_{-1}e$$

How many disintegrations per day occur initially in a 1.0 mol sample of ^{82}Se?

22.76 The most abundant isotope of uranium, ^{238}U, does not undergo fission. In a *breeder reactor*, however, a ^{238}U atom captures a neutron and emits two beta particles to make a fissionable isotope of plutonium, which can then be used as fuel in a nuclear reactor. Write a balanced nuclear equation.

22.77 A sample of ^{25}Na decays initially at a rate of 44,500 disintegrations/min, but the decay rate falls to 1350 disintegrations/min after 5.00 min. What is the half-life of ^{25}Na?

22.78 It has been estimated that 3.9×10^{23} kJ/s of energy is radiated into space by the sun. What is the rate of the sun's mass loss in kg/s?

chapter 23 ORGANIC CHEMISTRY

As the systematic study of chemistry slowly began in the 1700s, differences were noted between compounds obtained from animals and those from minerals. Chemicals from animal sources were often more difficult to isolate, purify, and work with than were those from mineral sources. To express this difference, the term *organic chemistry* was introduced to mean the study of compounds from living organisms, while *inorganic chemistry* was used to refer to the study of compounds from minerals. Today we know that there are no fundamental differences between organic and inorganic compounds—the same principles apply to both. The only common characteristic of compounds from living sources is that all contain the element carbon. Thus, **organic chemistry** is now defined as the study of carbon compounds.

Why is carbon special, and why do chemists still treat organic chemistry as a separate branch? The answers to these questions involve the ability of carbon atoms to bond together, forming long chains and rings. Of all the elements, only carbon is able to form such an immense array of compounds, from methane, with one carbon atom, to deoxyribonucleic acid (DNA), with tens of billions of carbon atoms. More than 11 million organic compounds have been made, and thousands of new ones are made each week in chemical laboratories throughout the world.

The chemistry of this plant differs from that of the nearby rock, because it is based on the element carbon.

23.1 ➤THE NATURE OF ORGANIC MOLECULES

Let's review what we've seen in earlier chapters about organic molecules:

1. Carbon is tetravalent (Section 7.5); it has four outer-shell electrons ($1s^2\ 2s^2\ 2p^2$) and forms four bonds. In methane, for example, carbon is connected to four hydrogen atoms:

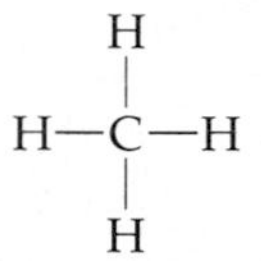

$$
\begin{array}{c}
\text{H} \\
| \\
\text{H---C---H} \qquad \text{Methane, } CH_4 \\
| \\
\text{H}
\end{array}
$$

2. Organic molecules have *covalent bonds* (Section 7.1). In ethane, for example, the bonds result from the sharing of two electrons, either between C and C or between C and H:

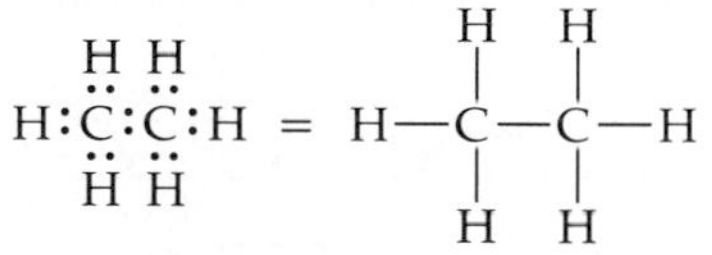

$$
\text{H:C:C:H} \;=\; \text{H---C---C---H} \qquad \text{Ethane, } C_2H_6 \text{ or } CH_3CH_3
$$

3. Organic molecules have *polar covalent bonds* when carbon bonds to an element on the right or left side of the periodic table (Section 7.7). In chloromethane, the electronegative chlorine atom attracts electrons more strongly than carbon, resulting in polarization of the carbon-chlorine bond so that carbon has a partial positive charge, δ^+. In methyllithium, the lithium attracts electrons less strongly than carbon, resulting in polarization of the carbon-lithium bond so that carbon has a partial negative charge, δ^-:

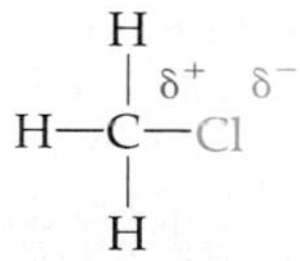

$$
\text{H---C---Cl} \qquad\qquad \text{H---C---Li}
$$

Chloromethane, CH_3Cl (electron-poor carbon)   Methyllithium, CH_3Li (electron-rich carbon)

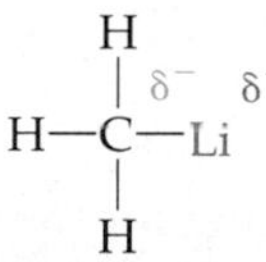

4. Carbon can form *multiple covalent bonds* by sharing more than two electrons with a neighboring atom (Section 7.4). In ethylene, the two carbon atoms share four electrons in a double bond. In acetylene, the two carbons share six electrons in a triple bond:

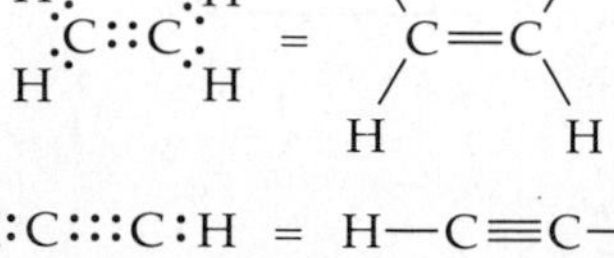

$$
\text{C::C} \;=\; \text{C}=\text{C} \qquad \text{Ethylene, } C_2H_4
$$

$$
\text{H:C:::C:H} \;=\; \text{H---C}\equiv\text{C---H} \quad \text{Acetylene, } C_2H_2
$$

5. Organic molecules have specific three-dimensional shapes, as predicted by the VSEPR model (Section 7.9). When carbon is bonded to four atoms, as in methane, the bonds are oriented toward the four corners of

an imagined tetrahedron with carbon in the center and with H–C–H angles of 109.5°.

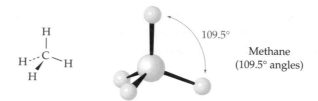

Methane
(109.5° angles)

When carbon bonds to three atoms, as in ethylene, the bonds are at angles of 120° to one another. When carbon bonds to two atoms, as in acetylene, the bonds are at angles of 180°. (Note that in the computer-generated pictures of these molecules, the multiple bonds aren't specified. Only the connections between atoms are shown.)

Ethylene (120° angles)

H—C≡C—H
Acetylene (180° angles)

6. Carbon uses *hybrid atomic orbitals* for bonding to other atoms (Sections 7.11 and 7.12). Carbons that bond to four atoms use sp^3 orbitals formed by the combination of an atomic *s* orbital with three atomic *p* orbitals. These sp^3 orbitals point toward the corners of an imagined tetrahedron, accounting for the observed geometry of carbon. Each of the four equivalent C–H bonds in methane is formed by overlap of a carbon sp^3 orbital with a hydrogen 1*s* orbital (Figure 23.1).

FIGURE 23.1 The carbon atom in methane is sp^3 hybridized and forms four equivalent bonds pointing toward the corners of a tetrahedron. Each bond is formed by overlap of a carbon sp^3 orbital with a hydrogen 1*s* orbital.

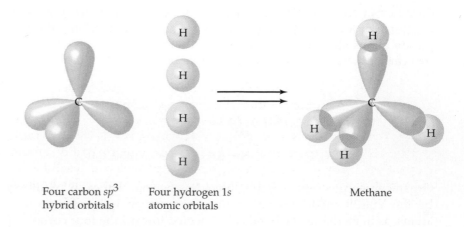

Four carbon sp^3
hybrid orbitals

Four hydrogen 1*s*
atomic orbitals

Methane

Doubly bonded carbons are sp^2 hybridized. Carbon has three sp^2 hybrid orbitals, which lie in a plane and point toward the corners of an equilateral triangle, and one unhybridized p orbital, which is oriented at a 90° angle to the plane of the sp^2 hybrids. When two sp^2 hybridized carbon atoms approach each other with sp^2 orbitals aligned head-on for sigma bonding, the unhybridized p orbitals on each carbon overlap to form a pi bond, resulting in a net carbon–carbon double bond. The two remaining sp^2 hybrids on each carbon bond to hydrogen (Figure 23.2).

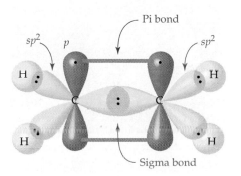

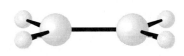

Ethylene - C_2H_4
(The carbon atoms are sp^2 hybridized.)

FIGURE 23.2 The carbon atoms in ethylene are sp^2 hybridized and form three bonds at angles of 120°. The carbon–carbon double bond is formed by head-on (σ) overlap of carbon sp^2 orbitals and sideways (π) overlap of p orbitals.

Triply bonded carbons are sp hybridized. Carbon has two sp hybrid orbitals, which are at 180° to each other, and two unhybridized p orbitals, which are oriented 90° from the sp hybrids and 90° from each other. When two sp hybridized carbon atoms approach each other with sp orbitals aligned head-on for σ bonding, the p orbitals on each carbon overlap to form two π bonds, resulting in a net carbon–carbon triple bond. The remaining sp hybrid on each carbon bonds to a hydrogen (Figure 23.3).

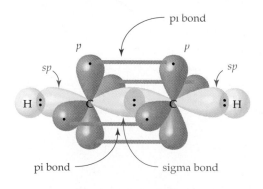

Acetylene C_2H_2
(The carbons are sp hybridized.)

FIGURE 23.3 The carbon atoms in acetylene are sp hybridized and form two bonds at angles of 180°. The carbon–carbon triple bond is formed by σ overlap of carbon sp orbitals and two sets of π overlap of carbon p orbitals.

Covalent bonding gives organic compounds properties that are quite different from those of ionic, inorganic salts. Intermolecular forces between individual organic molecules are weak, and organic compounds therefore have lower melting and boiling points than do ionic compounds. In fact, many simple organic compounds are liquid at room temperature. In addition, most organic compounds are insoluble in water and don't conduct electricity. Only a few small polar organic molecules such as glucose and ethyl alcohol dissolve in water.

23.2 ►ALKANES AND THEIR ISOMERS

How can there be so many organic compounds? The answer is that a relatively small number of atoms can bond together in a great many ways. Take molecules that contain only carbon and hydrogen (**hydrocarbons**) and that have only single bonds. Such compounds belong to the family of organic molecules called **saturated** hydrocarbons, or **alkanes**.

The paraffin wax used in home canning is a mixture of alkanes.

If you imagine ways that one carbon and four hydrogens can combine, only methane, CH_4 is possible. If you imagine ways that two carbons and six hydrogens can combine, only ethane is possible; and if you imagine the combination of three carbons with eight hydrogens, only propane is possible.

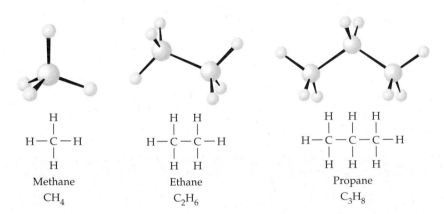

Methane	Ethane	Propane
CH_4	C_2H_6	C_3H_8

If larger numbers of carbons and hydrogens combine, though, *more than one kind of molecule can result*. There are two ways in which molecules with the formula C_4H_{10} can result: Either the four carbons can be in a row, or they can have a branched arrangement. Similarly, there are three ways in which molecules with the formula C_5H_{12} can result, and even more ways for larger alkanes. Compounds with all their carbons connected in a row are called **straight-chain alkanes** (or *n*-alkanes; *n* for *normal*), and those with a branching connection of carbons are called **branched-chain alkanes**.

C_4H_{10}

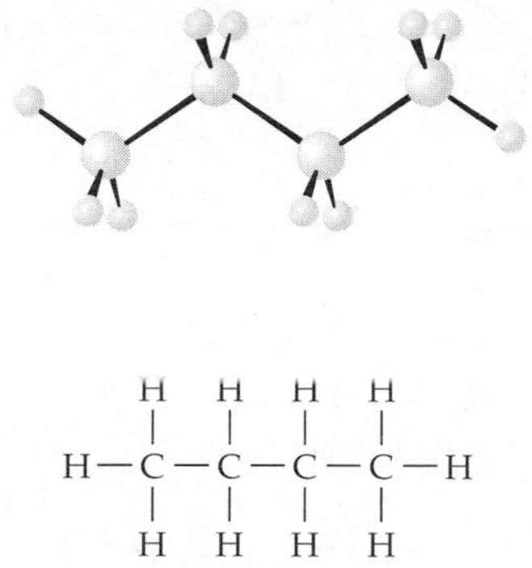

$$H-\underset{\underset{H}{|}}{\overset{\overset{H}{|}}{C}}-\underset{\underset{H}{|}}{\overset{\overset{H}{|}}{C}}-\underset{\underset{H}{|}}{\overset{\overset{H}{|}}{C}}-\underset{\underset{H}{|}}{\overset{\overset{H}{|}}{C}}-H$$

Butane (straight chain)

H—C—H   ⟩ Branch point

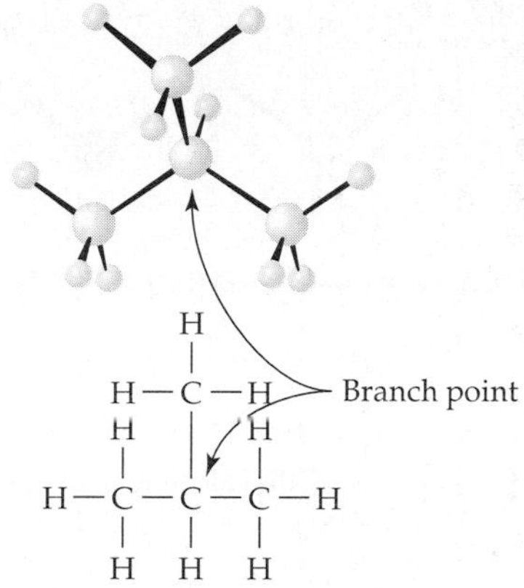

2-Methylpropane (branched chain)

C_5H_{12}

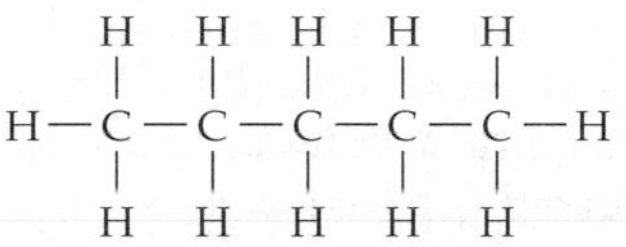

Pentane
(straight chain)

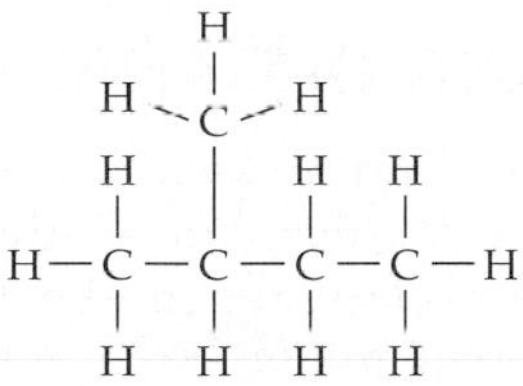

2-Methylbutane
(branched chain)

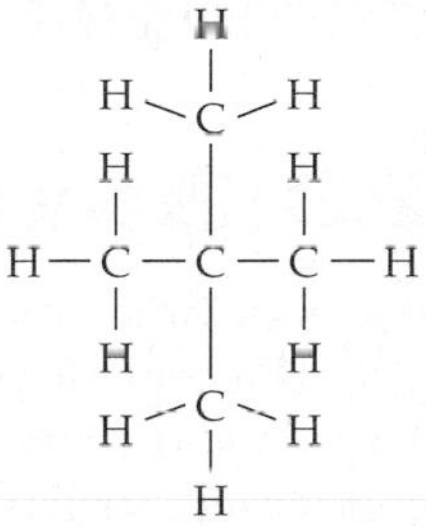

2,2-Dimethylpropane
(branched chain)

Compounds like the two different C_4H_{10} molecules and the three different C_5H_{12} molecules that have the same molecular formula but different structures are called *isomers*, as we saw in Section 20.8. As Table 23.1 shows, the number of possible alkane isomers grows rapidly as the number of carbon atoms increases.

TABLE 23.1   Numbers of Possible Alkane Isomers

Formula	No. of Isomers	Formula	No. of Isomers
C_6H_{14}	5	$C_{10}H_{22}$	75
C_7H_{16}	9	$C_{20}H_{42}$	366,319
C_8H_{18}	18	$C_{30}H_{62}$	4,111,846,763
C_9H_{20}	35	$C_{40}H_{82}$	62,491,178,805,831

It's important to realize that different isomers are different chemical compounds. They have different structures and different physical properties, such as melting point and boiling point. For example, ethyl alcohol (ethanol, or grain alcohol) and dimethyl ether both have the formula C_2H_6O, yet ethyl alcohol is a liquid with a boiling point of 78.5°C, whereas dimethyl ether is a gas with a boiling point of −23°C.

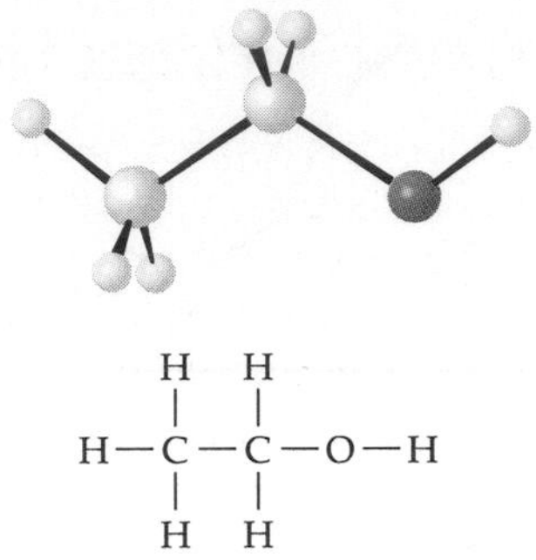

H H
| |
H—C—C—O—H
| |
H H

Ethyl alcohol

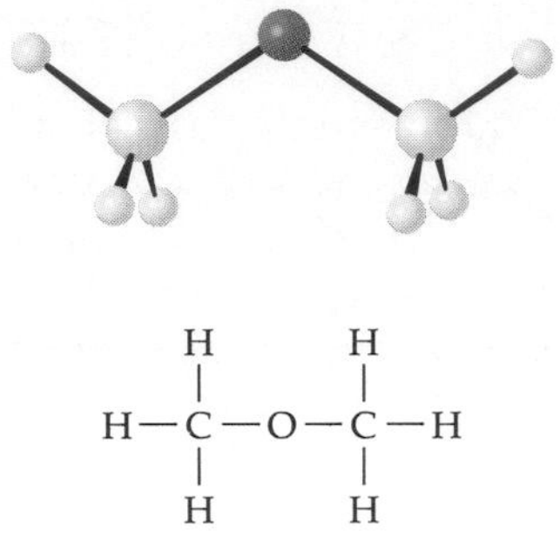

H H
| |
H—C—O—C—H
| |
H H

Dimethyl ether

PROBLEM 23.1 Draw the straight-chain isomer with the formula C_7H_{16}.

PROBLEM 23.2 Draw the five alkane isomers with the formula C_6H_{14}.

23.3 ▸WRITING ORGANIC STRUCTURES

It's both time consuming and awkward to draw all the bonds and all the atoms in organic compounds, even for relatively small molecules like C_4H_{10}. Thus, a shorthand way of drawing **condensed structures** is often used. In condensed structures, carbon–hydrogen and carbon–carbon single bonds aren't shown; rather, they're "understood." If a carbon atom has three hydrogens bonded to it, we write CH_3; if the carbon has two hydrogens bonded to it, we write CH_2; and so on. For example, the four-carbon, straight-chain alkane (called *butane*) and its branched-chain isomer (called *2-methylpropane*, or *isobutane*) can be written in the following way:

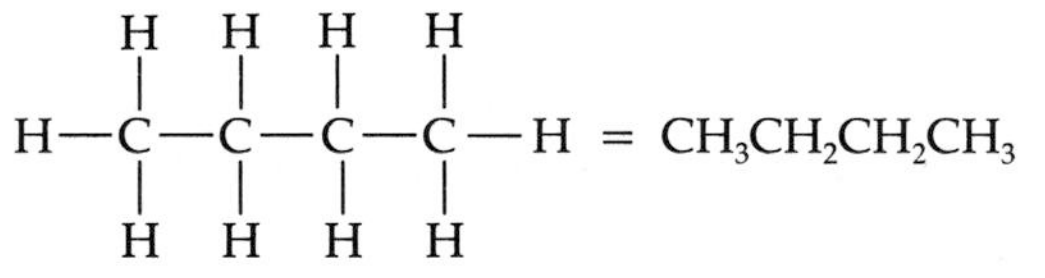

H H H H
| | | |
H—C—C—C—C—H = $CH_3CH_2CH_2CH_3$ Butane
| | | |
H H H H

H
|
H—C—H
H | H CH_3
| | | |
H—C—C—C—H = CH_3CHCH_3 2-Methylpropane
| | | (Isobutane)
H H H

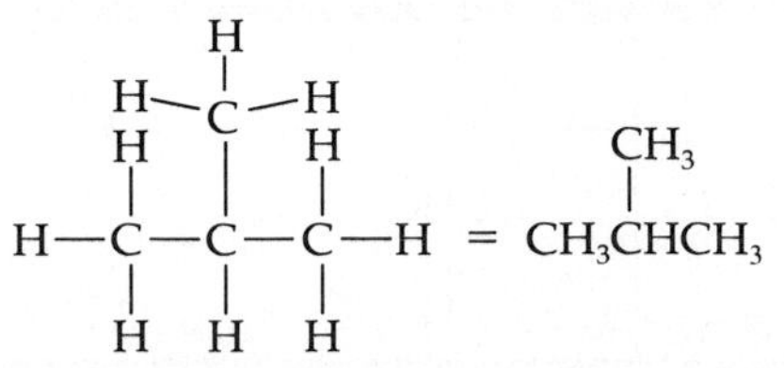

Note that the horizontal bonds between carbons aren't shown—the CH_3 and CH_2 units are simply placed next to each other—but that the vertical bond in 2-methylpropane is shown for clarity.

PROBLEM 23.3 Draw the three isomers of C_5H_{12} as condensed structures.

23.4 ►THE SHAPES OF ORGANIC MOLECULES

Chemists don't usually worry about exact three-dimensional shapes when writing the condensed structure of an organic molecule, and a molecule can be arbitrarily shown in a great many ways. Butane, for example, might be represented by any of the following structures. These structures don't imply any particular three-dimensional shape for butane; they indicate only the connections among atoms.

$$CH_3-CH_2-CH_2-CH_3 \qquad CH_3-CH_2-\overset{\overset{\displaystyle CH_3}{|}}{CH_2} \qquad CH_3-\overset{\overset{\displaystyle CH_2-CH_3}{|}}{CH_2}$$

$$\overset{\displaystyle CH_2-CH_2-CH_3}{\underset{\displaystyle CH_3}{|}} \qquad CH_3(CH_2)_2CH_3 \qquad CH_3CH_2CH_2CH_3$$

Some representations of butane, C_4H_{10}

In fact, butane has no one single shape because *rotation* is possible around carbon–carbon single bonds. The two parts of a molecule joined by a carbon–carbon single bond are free to spin around the bond, giving rise to an infinite number of possible three-dimensional structures, or **conformations**. A given butane molecule might be fully extended at one instant but be twisted an instant later. In fact, a large sample of butane contains a great many molecules, which are constantly changing their shape. At any given instant, though, most of the molecules have an extended, zigzag conformation, which is slightly more stable than other possibilities. The same is true for other alkanes.

Butane

EXAMPLE 23.1

The following structures have the same formula, C_8H_{18}. Which of them represent the same molecule?

(a) $$\overset{\overset{\displaystyle CH_3}{|}}{CH_3CH}CH_2\overset{\overset{\displaystyle CH_3}{|}}{CH}CH_2CH_3$$ (b) $$CH_3CH_2\overset{\overset{\displaystyle CH_3}{|}}{CH}CH_2\overset{\overset{\displaystyle CH_3}{|}}{CH}CH_3$$ (c) $$CH_3\overset{\overset{\displaystyle CH_3}{|}}{CH}CH_2CH_2\overset{\overset{\displaystyle CH_3}{|}}{CH}CH_3$$

SOLUTION   Pay attention to the order of connection between atoms. Don't get confused by the apparent differences caused by writing a structure right-to-left versus left-to-right. In this example, structure **(a)** has a straight chain of six carbons with –CH_3 branches on the second and fourth carbons from the end. Structure **(b)** also has a straight chain of six carbons with –CH_3 branches on the second and fourth carbons from the end and is therefore identical to **(a)**. The only difference between **(a)** and **(b)** is that one is written "forwards" and one is written "backwards." Structure **(c)** has a straight chain of six carbons with –CH_3 branches on the second and fifth carbons from the end, and is an isomer of **(a)** and **(b)**.

PROBLEM 23.4 Which of the following structures are identical?

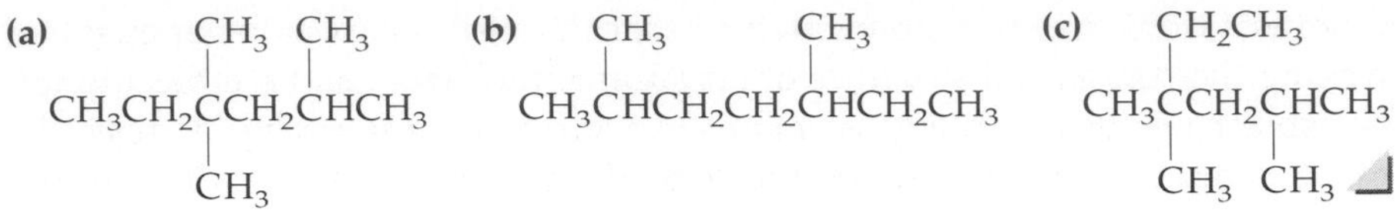

23.5 ➤NAMING ALKANES

In earlier times, when relatively few pure organic chemicals were known, new compounds were named at the whim of their discoverer. Thus, urea is a crystalline substance first isolated from urine, and the barbiturates are a group of tranquilizing agents named by their discoverer in honor of his friend Barbara. As more and more compounds became known, however, the need for a systematic method of naming organic compounds became apparent.

The system of naming now used was devised by the International Union of Pure and Applied Chemistry, IUPAC. In the IUPAC system, a chemical name has three parts: parent, suffix, and prefix. The parent name specifies the overall size of the molecule by telling how many carbon atoms are present in the longest continuous chain; the suffix identifies what family the molecule belongs to; and the prefix specifies the location of various substituent groups attached to the parent chain:

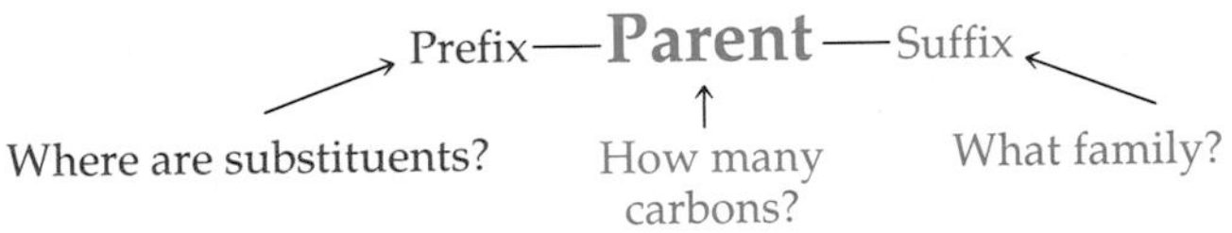

Straight-chain alkanes are named by counting the number of carbon atoms in the chain and adding the family suffix *-ane*. With the exception of the first four compounds—methane, ethane, propane, and butane—whose names have historical origins, the alkanes are named from Greek numbers according to the number of carbons present. Thus, *pent*ane is the five-carbon alkane, *hex*ane is the six-carbon alkane, and so on, as shown in Table 23.2.

TABLE 23.2 Names of Straight-Chain Alkanes

Number of Carbons	Structure	Name	Number of Carbons	Structure	Name
1	CH_4	*Meth*ane	6	$CH_3CH_2CH_2CH_2CH_2CH_3$	*Hex*ane
2	CH_3CH_3	*Eth*ane	7	$CH_3CH_2CH_2CH_2CH_2CH_2CH_3$	*Hept*ane
3	$CH_3CH_2CH_3$	*Prop*ane	8	$CH_3CH_2CH_2CH_2CH_2CH_2CH_2CH_3$	*Oct*ane
4	$CH_3CH_2CH_2CH_3$	*But*ane	9	$CH_3CH_2CH_2CH_2CH_2CH_2CH_2CH_2CH_3$	*Non*ane
5	$CH_3CH_2CH_2CH_2CH_3$	*Pent*ane	10	$CH_3CH_2CH_2CH_2CH_2CH_2CH_2CH_2CH_2CH_3$	*Dec*ane

Branched-chain alkanes are named by the following four steps.

Step 1. Name the main chain.   Find the *longest continuous chain of carbons* present in the molecule, and use the name of that chain as the parent name. The longest chain may not always be obvious from the manner of writing; you may have to "turn corners" to find it.

$$CH_3-CH_2$$
$$|$$
$$CH_3-CH-CH_2-CH_2-CH_3$$

Named as a hexane, not as a pentane, because the longest chain has six carbons.

If you prefer, you can redraw the structure so that the longest chain is on one line:

$$CH_3-CH_2 \qquad\qquad CH_3$$
$$| \qquad\qquad\qquad\qquad\qquad\qquad |$$
$$CH_3 \quad CH \quad CH_2 \quad CH_2 \quad CH_3 \qquad\qquad CH_3 \quad CH_2 \quad CH \quad CH_2 \quad CH_2 \quad CH_3$$

same as

Step 2. Number the carbon atoms in the main chain.   Beginning at the end nearer the first branch point, number each carbon atom in the parent chain:

$$CH_3$$
$$|$$
$$CH_3-CH_2-CH-CH_2-CH_2-CH_3$$
$$\quad 1 \qquad 2 \qquad 3 \qquad 4 \qquad 5 \qquad 6$$
$$[\ 6 \qquad 5 \qquad 4 \qquad 3 \qquad 2 \qquad 1\ ]$$

wrong numbering

The first (and only) branch occurs at C3 if we start numbering from the left, but would occur at C4 if we started from the right by mistake

Step 3. Identify and number the branching substituent.   Assign a number to each branching substituent group on the parent chain according to its point of attachment.

$$CH_3$$
$$|$$
$$CH_3-CH_2-CH-CH_2-CH_2-CH_3$$
$$\quad 1 \qquad 2 \qquad 3 \qquad 4 \qquad 5 \qquad 6$$

The main chain is a hexane. There is a $-CH_3$ substituent group connected to C3 of the chain.

If there are two substituent groups on the same carbon, assign the same number to both. There must always be as many numbers in the name as there are substituents.

$$CH_2-CH_3$$
$$|$$
$$CH_3-CH_2-C-CH_2-CH_2-CH_3$$
$$\quad 1 \qquad 2 \qquad 3\ | \qquad 4 \qquad 5 \qquad 6$$
$$\qquad\qquad\qquad CH_3$$

The main chain is a hexane. There are two substituents, a $-CH_3$ and a $-CH_2CH_3$, both connected to C3 of the chain.

The $-CH_3$ and $-CH_2CH_3$ substituents that branch off the main chain in the previous two compounds are called **alkyl groups**. You can think of an alkyl group as the part of an alkane that remains when a hydrogen is removed. For example, removal of a hydrogen from methane, CH_4, gives the *methyl group*, $–CH_3$, and removal of a hydrogen from ethane, CH_3CH_3, gives the *ethyl group*, $–CH_2CH_3$. Alkyl groups are named by replacing the *-ane* ending of the parent alkane with an *-yl* ending.

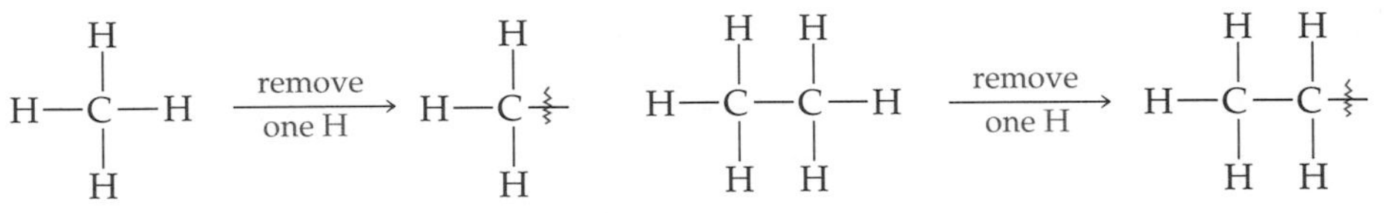

Methane   A methyl group   Ethane   An ethyl group

Step 4. Write the name as a single word. Use hyphens to separate the different prefixes and use commas to separate numbers if necessary. If two or more different substituent groups are present, cite them in alphabetical order. If two or more identical substituents are present, use one of the prefixes *di-*, *tri-*, *tetra-*, and so forth, but don't use these prefixes for alphabetizing purposes. Look at the following examples to see how names are written:

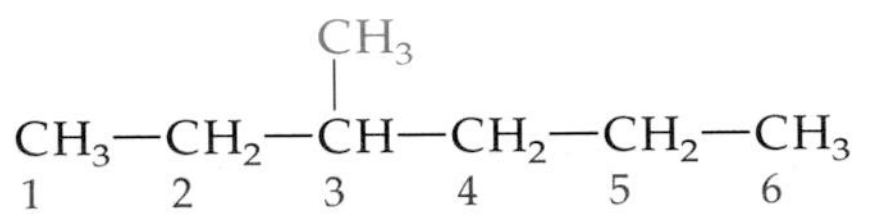

3-Methylhexane—a six-carbon main chain with a 3-methyl substituent

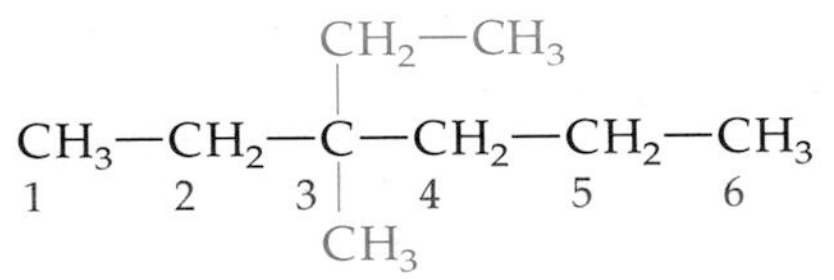

3-Ethyl-3-methylhexane—a six-carbon main chain with 3-ethyl and 3-methyl substituents

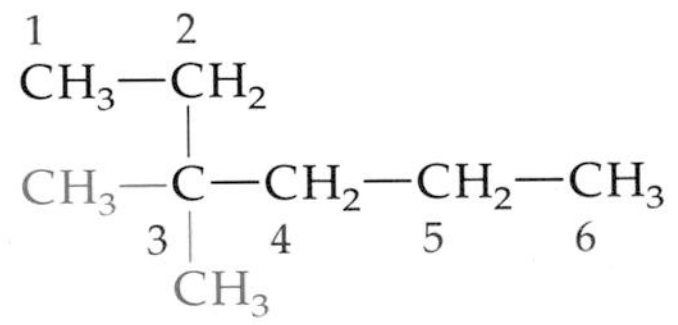

3,3-Dimethylhexane—a six-carbon main chain with two 3-methyl substituents

More About Alkyl Groups

It doesn't matter which hydrogen is removed from CH_4 to form a methyl group or which hydrogen is removed from CH_3CH_3 to form an ethyl group because all the hydrogen atoms in each molecule are equivalent to one another. The eight hydrogens in $CH_3CH_2CH_3$, however, are not all equivalent. Propane has two "kinds" of hydrogens—six on the end carbons and two on the middle carbon. Depending on which kind of hydrogen is removed, two different propyl groups can result. Removal of one of the six hydrogens attached to an end carbon yields a straight-chain propyl group called *n-propyl*. Removal of one of the two hydrogens attached to the middle carbon yields a branched-chain propyl group called *isopropyl*.

Similarly, there are four different kinds of butyl groups. Two (*n*-butyl and *sec*-butyl) are derived from straight-chain butane, and two (isobutyl and *tert*-butyl) are derived from branched-chain isobutane. [The prefixes *sec-* (for *secondary*) and *tert-* (for *tertiary*) refer to the number of other carbon atoms attached to the branching carbon. There are two other carbons attached to the branch point in a *sec*-butyl group and three other carbons attached to the branch point in a *tert*-butyl group.]

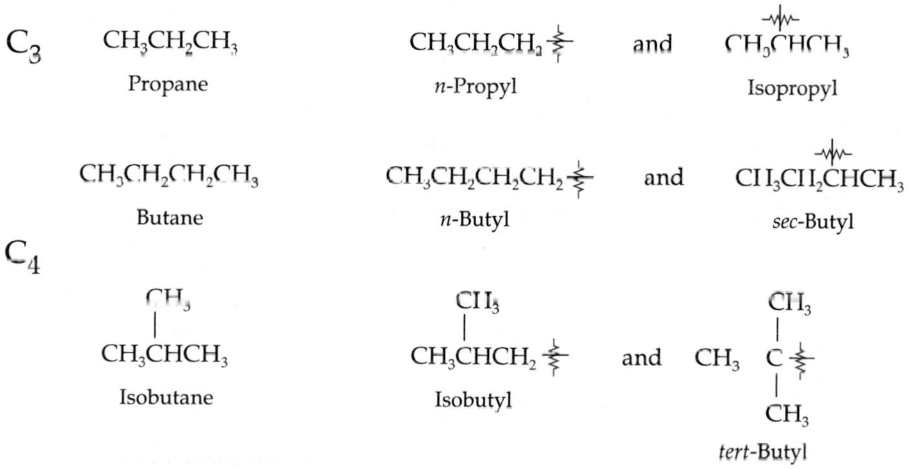

C₃   CH₃CH₂CH₃   CH₃CH₂CH₂⧸   and   CH₃CHCH₃

        Propane   *n*-Propyl   Isopropyl

        CH₃CH₂CH₂CH₃   CH₃CH₂CH₂CH₂⧸   and   CH₃CH₂CHCH₃

        Butane   *n*-Butyl   *sec*-Butyl

C₄

             CH₃   CH₃   CH₃
              | | |
        CH₃CHCH₃   CH₃CHCH₂⧸   and   CH₃ C⧸
                                                           |
        Isobutane   Isobutyl   CH₃

                                                     tert-Butyl

It's important to realize that alkyl groups themselves are not compounds and that the "removal" of a hydrogen from an alkane is just a useful way of looking at things, not a chemical reaction. Alkyl groups are simply parts of molecules that help us to name compounds.

EXAMPLE 23.2

What is the IUPAC name of the following alkane?

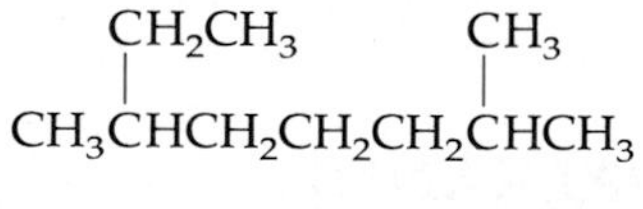

          CH₂CH₃   CH₃
           | |
    CH₃CHCH₂CH₂CH₂CHCH₃

SOLUTION   The molecule has a chain of eight carbons (octane) with two methyl substituents. Numbering from the end nearer the first methyl substituent indicates that the methyls are at C2 and C6, giving the name 2,6-dimethyloctane. The numbers are separated by a comma and are set off from the rest of the name by a hyphen.

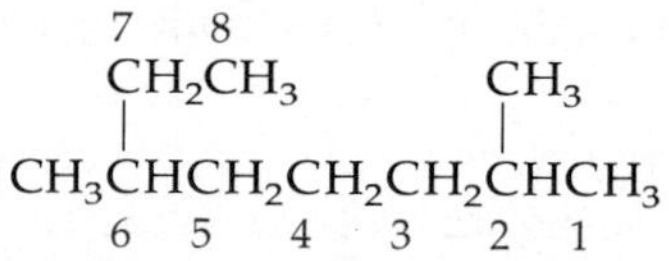

          7   8
          CH₂CH₃   CH₃
           | |
    CH₃CHCH₂CH₂CH₂CHCH₃   2,6-Dimethyloctane
    6   5   4   3   2   1

EXAMPLE 23.3

Draw the structure of 3-isopropyl-2-methylhexane.

SOLUTION   First, look at the parent name (hexane) and draw its carbon structure:

$$C-C-C-C-C-C \qquad \text{Hexane}$$

Next, find the substituents (3-isopropyl and 2-methyl), and place them on the proper carbons:

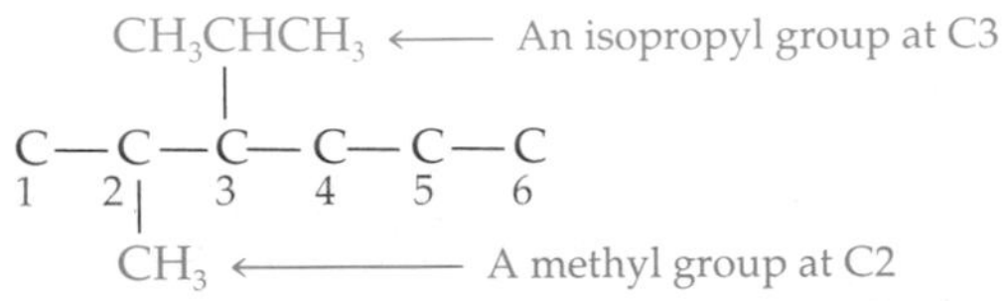

$$CH_3CHCH_3 \quad \longleftarrow \text{ An isopropyl group at C3}$$

$$\underset{1 \quad 2 \quad 3 \quad 4 \quad 5 \quad 6}{C-C-C-C-C-C}$$

$$CH_3 \quad \longleftarrow \text{ A methyl group at C2}$$

Finally, add hydrogens to complete the structure:

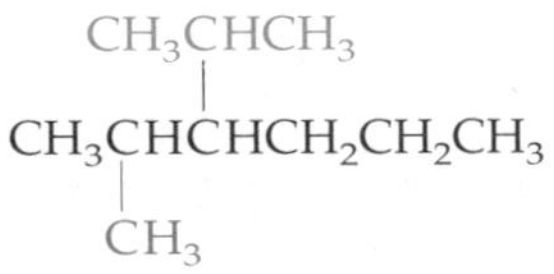

$$CH_3CHCH_3$$
$$CH_3CHCHCH_2CH_2CH_3 \qquad \text{3-Isopropyl-2-methylhexane}$$
$$CH_3$$

⌐ PROBLEM 23.5   What are the IUPAC names of the following alkanes?
(a) the three isomers of C_5H_{12}

(b)
$$CH_3$$
$$CH_3CH_2CHCHCH_3$$
$$CH_2CH_3$$

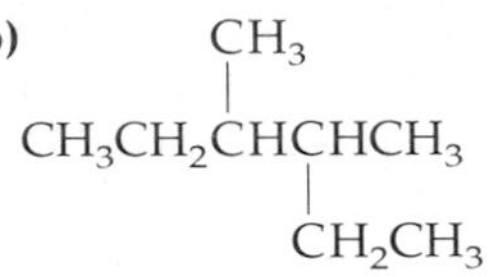

(c)
$$CH_3 \quad CH_3$$
$$CH_3CHCH_2CHCH_3$$

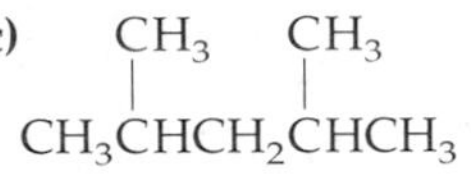

(d)
$$CH_3 \quad CH_2CH_3$$
$$CH_3CCH_2CH_2CHCH_3$$
$$CH_3$$

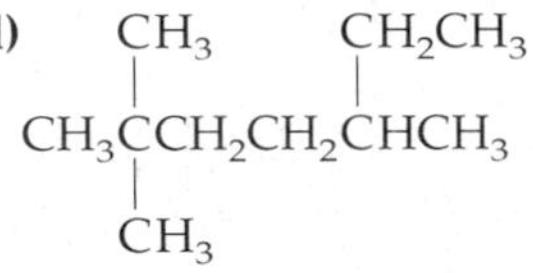

⌐ PROBLEM 23.6   Draw structures corresponding to the following IUPAC names:
(a) 3,4-dimethylnonane
(b) 3-ethyl-4,4-dimethylheptane
(c) 2,2-dimethyl-4-propyloctane
(d) 2,2,4-trimethylpentane

23.6 ➤CYCLOALKANES

The compounds we've been dealing with thus far have all been open-chain, or **acyclic**, alkanes. **Cycloalkanes**, which contain rings of carbon atoms, are also well known and are widespread throughout nature. Compounds of all ring sizes from 3 through 30 carbons and beyond have been prepared. The four simplest cycloalkanes having 3 carbons (cyclopropane), 4 carbons (cyclobutane), 5 carbons (cyclopentane), and 6 carbons (cyclohexane) are shown.

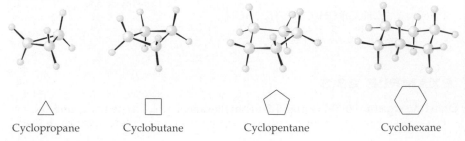

Cyclopropane   Cyclobutane   Cyclopentane   Cyclohexane

Even condensed structures are awkward for cyclic molecules, and a streamlined way of drawing structures is often used in which cycloalkanes are represented by polygons. A triangle represents cyclopropane, a square represents cyclobutane, and so on. Notice that carbon and hydrogen atoms aren't shown in these **line structures**. A carbon atom is simply understood to be at every junction of lines, and the proper number of hydrogen atoms needed to fill out carbon's valency is supplied mentally. Methylcyclohexane looks like this in a line structure:

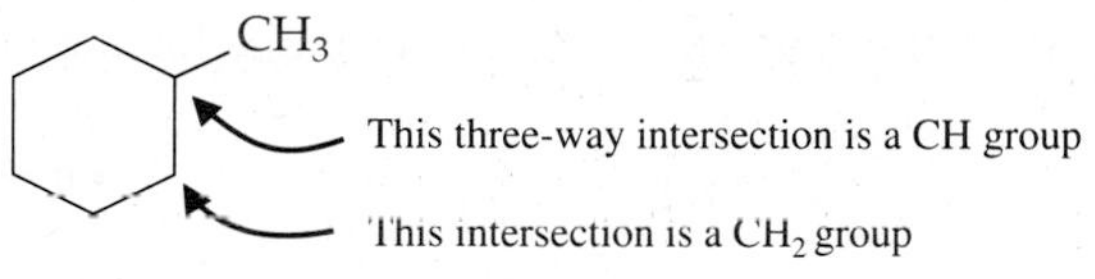

CH$_3$

This three-way intersection is a CH group

This intersection is a CH$_2$ group

Methylcyclohexane

As you might expect, the C–C bonds in cyclopropane and cyclobutane are considerably distorted from the ideal 109.5° value. Cyclopropane, for example, has the shape of an equilateral triangle, with C–C–C angles of 60°. As a result, the bonds in three- and four-membered rings are weaker than normal, and the molecules are more reactive than other alkanes. Cyclopentane, cyclohexane, and larger cycloalkanes, however, pucker into shapes that allow bond angles to be near their normal tetrahedral value, as shown in the computer-generated models at the bottom of the preceding page.

Substituted cycloalkanes are named using the cycloalkane as the parent name and identifying the positions on the ring where substituents are attached. We start numbering at the group that has alphabetical priority, and proceed in the direction that gives the second substituent the lower possible number. For example,

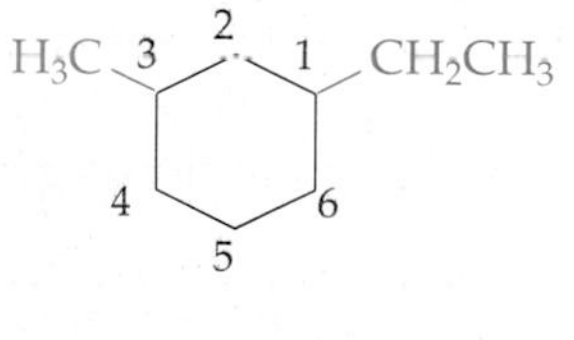

H$_3$C 3 2 1 CH$_2$CH$_3$
        4   6
           5

1-Ethyl-3-methylcyclohexane

not 1-Methyl-3-ethylcyclohexane
or 1-Ethyl-5-methylcyclohexane
or 1-Methyl-5-ethylcyclohexane

EXAMPLE 23.4

What is the IUPAC name of the following cycloalkane?

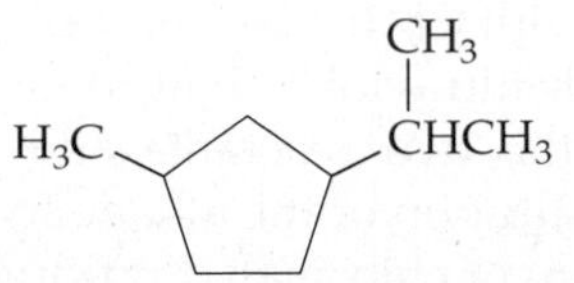

CH$_3$
|
H$_3$C   CHCH$_3$

SOLUTION   First, identify the parent cycloalkane (cyclopentane) and the two substituents (a methyl group and an isopropyl group). Then, number the ring beginning at the group having alphabetical priority (isopropyl rather than methyl), and proceed in a direction that gives the second group the lower possible number.

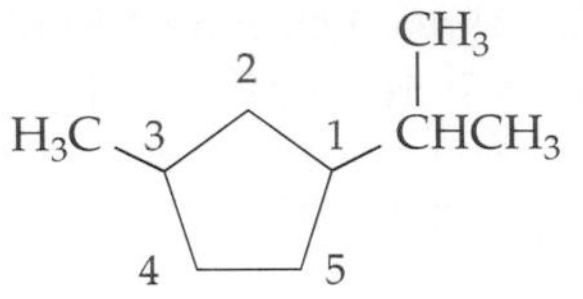

1-Isopropyl-3-methylcylopentane

PROBLEM 23.7 Give IUPAC names for the following cycloalkanes.

(a)

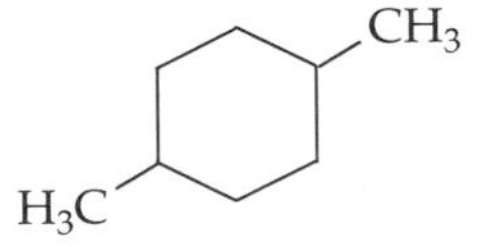

(b)

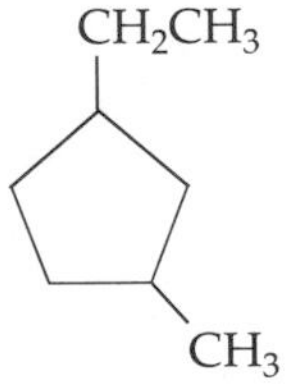

(c)

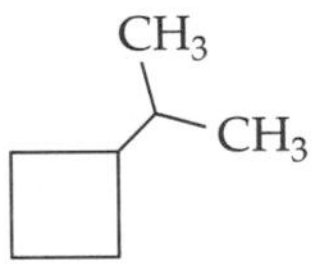

PROBLEM 23.8 Draw structures corresponding to the following IUPAC names. Use simplified line structures rather than condensed structures.

(a) 1,1-dimethylcyclobutane   (b) 1-*tert*-butyl-2-methylcyclopentane
(c) 1,3,5-trimethylcycloheptane

23.7 ►REACTIONS OF ALKANES

Alkanes have relatively low chemical reactivity and are inert to acids, bases, and most other common laboratory reagents. They do, however, react with oxygen and with halogens under appropriate conditions. The chemical reaction of alkanes with oxygen occurs during combustion in an engine or furnace when the alkane is burned as fuel. Carbon dioxide and water are formed as products, and a large amount of heat is released. For example, methane, the main component of natural gas, reacts with oxygen to release 890 kJ per mole of methane burned.

$$CH_4(g) + 2\,O_2(g) \rightarrow CO_2(g) + 2\,H_2O(l) \qquad \Delta H° = -890 \text{ kJ}$$

Propane (the LP gas used in campers and rural homes), gasoline (a mixture of C_5–C_{11} alkanes), kerosene (a mixture of C_{11}–C_{14} alkanes), and other alkanes burn similarly.

Reaction of alkanes with Cl_2 (or Br_2) occurs when a mixture of the two reagents is irradiated with ultraviolet light, denoted by $h\nu$. Depending on the relative amounts of the two reactants and on the time allowed for reaction, a sequential substitution of the alkane hydrogen atoms by chlorine occurs, leading to a mixture of chlorinated products. Methane, for example, reacts with chlorine to yield a mixture of chloromethane (CH_3Cl), dichloromethane (CH_2Cl_2), trichloromethane ($CHCl_3$), and tetrachloromethane (CCl_4):

Methane gas is burned off on the tops of these oil wells.

$$CH_4 + Cl_2 \xrightarrow{\;h\nu\;} CH_3Cl + HCl$$

$$\xrightarrow{\;Cl_2\;} CH_2Cl_2 + HCl$$

$$\xrightarrow{\;Cl_2\;} CHCl_3 + HCl$$

$$\xrightarrow{\;Cl_2\;} CCl_4 + HCl$$

PROBLEM 23.9 Draw all the monochlorinated substitution products you would expect to obtain from chlorination of 2-methylbutane.

23.8 ➤ FAMILIES OF ORGANIC MOLECULES: FUNCTIONAL GROUPS

Chemists have learned through experience that organic compounds can be classified into families according to their structural features and that the chemical behavior of the members of a family is often predictable. The structural features that make it possible to class compounds together are called *functional groups*. A **functional group** is a part of a larger molecule and is composed of an atom or group of atoms that has characteristic chemical behavior. A given functional group undergoes the same kinds of reactions in every molecule it's a part of. Look at the carbon–carbon double-bond functional group, for example. Ethylene (C_2H_4), the simplest compound with a double bond, undergoes many chemical reactions similar to those of α-pinene ($C_{10}H_{16}$), a larger and more complex compound (and major constituent of turpentine). Both, for example, react with hydrogen in the same way (Figure 23.4).

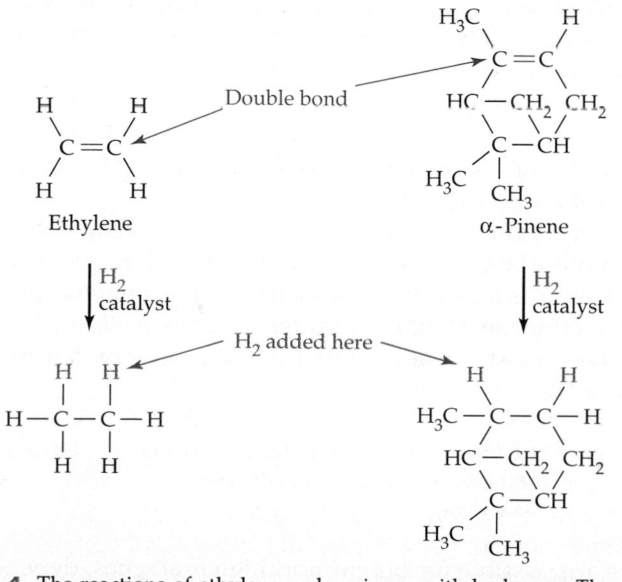

FIGURE 23.4 The reactions of ethylene and α-pinene with hydrogen. The carbon–carbon double-bond functional group adds two hydrogen atoms in each case, regardless of the complexity of the rest of the molecule.

The example shown in Figure 23.4 is typical: *The chemistry of an organic molecule, regardless of its size and complexity, is determined by the functional groups it contains.* Table 23.3 lists some of the most common functional groups and gives examples of where they occur. Look carefully at this table to see the many types of functional groups found in organic compounds. Some functional groups, such as alkenes, alkynes, and aromatic rings, have only carbon–carbon double or triple bonds; others contain oxygen, nitrogen, or halogen atoms.

⌐ PROBLEM 23.10 Locate and identify the functional groups in the following molecules.

(a) lactic acid, from sour milk

(b) styrene, used to make polystyrene

$$CH_3-\overset{\overset{\displaystyle H}{|}}{\underset{\underset{\displaystyle OH}{|}}{C}}-\overset{\overset{\displaystyle O}{\|}}{C}-OH$$

⌐ PROBLEM 23.11 Propose structures for molecules that fit the following descriptions.

(a) C_2H_4O containing an aldehyde functional group
(b) $C_3H_6O_2$ containing a carboxylic acid functional group

23.9 ►ALKENES AND ALKYNES

In contrast with alkanes, which have only single bonds, alkenes and alkynes have *multiple* bonds: **Alkenes** are hydrocarbons that contain a carbon–carbon double bond, and **alkynes** are hydrocarbons that contain a carbon–carbon triple bond. Both groups of compounds are **unsaturated**, meaning that they have fewer hydrogens per carbon than the related alkanes. Ethylene ($H_2C=CH_2$), for example, has the formula C_2H_4, whereas ethane (CH_3CH_3) has the formula C_2H_6.

Alkenes are named by counting the longest chain of carbons that contains the double bond and adding the family suffix *-ene*. Thus, ethylene, the simplest alkene, is followed by propene, butene, pentene, hexene, and so on. Note that ethylene should properly be called *ethene*, but the name ethylene has been used for so long that it is accepted by IUPAC. Similarly, the name *propylene* is often used for propene.

$H_2C=CH_2$	$CH_3CH=CH_2$	$CH_3CH_2CH=CH_2$	$CH_3CH=CHCH_3$
Ethene	Propene	1-Butene	2-Butene
(Ethylene)	(Propylene)		

Isomers are possible for butene and higher alkenes, depending on the position of the double bond in the chain, which must be specified by a numerical prefix. Numbering starts from the chain end nearer the double bond, and only the first of the double-bond carbons is cited. If a substituent is present on the chain, its identity is noted and its position of attachment is

TABLE 23.3   Some Important Families of Organic Molecules

Family Name	Functional Group Structure[a]	Simple Example	Name	Name Ending
Alkane	(contains only C—H and C—C single bonds)	CH_3CH_3	Ethane	*-ane*
Alkene	$\diagdown C{=}C \diagup$	$H_2C{=}CH_2$	Ethene (Ethylene)	*-ene*
Alkyne	$-C{\equiv}C-$	$H-C{\equiv}C-H$	Ethyne (Acetylene)	*-yne*
Arene (aromatic)	(benzene ring structure)	(benzene ring structure)	Benzene	None
Alcohol	$-\overset{\vert}{\underset{\vert}{C}}-O-H$	CH_3OH	Methanol	*-ol*
Ether	$-\overset{\vert}{\underset{\vert}{C}}-O-\overset{\vert}{\underset{\vert}{C}}-$	CH_3OCH_3	Dimethyl ether	*-ether*
Amine	$-\overset{\vert}{\underset{\vert}{C}}-\overset{\vert}{N}-$	CH_3NH_2	Methylamine	*-amine*
Aldehyde	$-\overset{\vert}{\underset{\vert}{C}}-\overset{O}{\overset{\Vert}{C}}-H$	$CH_3\overset{O}{\overset{\Vert}{C}}H$	Ethanal (Acetaldehyde)	*-al*
Ketone	$-\overset{\vert}{\underset{\vert}{C}}-\overset{O}{\overset{\Vert}{C}}-\overset{\vert}{\underset{\vert}{C}}-$	$CH_3\overset{O}{\overset{\Vert}{C}}CH_3$	Propanone (Acetone)	*-one*
Carboxylic acid	$-\overset{\vert}{\underset{\vert}{C}}-\overset{O}{\overset{\Vert}{C}}-O-H$	$CH_3\overset{O}{\overset{\Vert}{C}}OH$	Ethanoic acid (Acetic acid)	*-oic acid*
Ester	$-\overset{\vert}{\underset{\vert}{C}}-\overset{O}{\overset{\Vert}{C}}-O-\overset{\vert}{\underset{\vert}{C}}-$	$CH_3\overset{O}{\overset{\Vert}{C}}OCH_3$	Methyl ethanoate (Methyl acetate)	*-oate*
Amide	$-\overset{\vert}{\underset{\vert}{C}}-\overset{O}{\overset{\Vert}{C}}-\overset{\vert}{N}-$	$CH_3\overset{O}{\overset{\Vert}{C}}NH_2$	Ethanamide (Acetamide)	*-amide*

[a] The bonds whose connections aren't specified are assumed to be attached to carbon or hydrogen atoms in the rest of the molecule.

given. If the double bond is equidistant from both ends of the chain, numbering starts at the end nearer the substituent.

$$CH_3CH\!=\!CHCH_2\overset{\overset{\displaystyle CH_3}{|}}{C}HCH_3$$
$$1 \quad 2 \qquad 3 \quad 4 \quad 5 \quad 6$$
5-Methyl-2-hexene

(numbered to give double bond the lower number)

$$CH_3CH_2CH\!=\!CH\overset{\overset{\displaystyle CH_3}{|}}{C}HCH_3$$
$$6 \quad 5 \quad 4 \qquad 3 \quad 2 \quad 1$$
2-Methyl-3-hexene

(numbered to give substituent the lower number when the double bond is equidistant from both ends)

In addition to the alkene isomers that exist because of double-bond *position*, alkene isomers can also exist because of double-bond *geometry*. For example, there are two **cis-trans isomers** of 2-butene that differ in their geometry about the double bond. The cis isomer has its two –CH_3 groups on the same side of the double bond, and the trans isomer has its two –CH_3 groups on opposite sides of the double bond. Like the other kinds of isomers we've discussed, the individual cis and trans isomers of an alkene are different substances with different physical properties and different (though often similar) chemical behavior. *cis*-2-Butene boils at 4°C, for example, but *trans*-2-butene boils at 0.9°C.

$$\overset{\displaystyle H_3C}{\diagdown}\underset{}{\overset{}{C}}\!=\!\overset{}{\underset{}{C}}\overset{\displaystyle CH_3}{\diagup}$$
H   H

cis-2-Butene
(methyl groups on
the same side)

(Top view)

(Side view)

$$\overset{\displaystyle H_3C}{\diagdown}\underset{}{\overset{}{C}}\!=\!\overset{}{\underset{}{C}}\overset{\displaystyle H}{\diagup}$$
H   CH₃

trans-2-Butene
(methyl groups on
opposite sides)

(Top view)

(Side view)

Cis-trans isomerism in alkenes arises because of the electronic structure of the carbon–carbon double bond. Because the π part of the double bond is formed by sideways overlap of two parallel p orbitals (Section 7.12), rotation would break the bond and is therefore energetically unfavorable (Figure 23.5). Experiments show, in fact, that it takes about 240 kJ/mol of energy to cause rotation around a C=C double bond. Since double-bond rotation can't occur, cis and trans isomers can't interconvert.

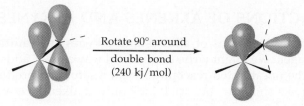

Rotate 90° around
double bond
(240 kj/mol)

pi bond—*p* orbital overlap   broken bond—no *p* orbital overlap

FIGURE 23.5 Rotation around a carbon–carbon double bond requires a large amount of energy because *p* orbital overlap is destroyed. Cis-trans alkene isomerism arises as a result.

Alkynes are similar in many respects to alkenes and are named using the family suffix -*yne*. The simplest alkyne, HC≡CH, is usually called by its alternative name *acetylene* rather than by its systematic name *ethyne*.

$$\underset{\substack{\text{Ethyne}\\ \text{(Acetylene)}}}{HC\text{=}CH} \qquad \underset{\text{Propyne}}{CH_3C\text{=}CH} \qquad \underset{\text{2-Butyne}}{\overset{1\ \ \ 2\ \ \ 3\ 4}{CH_3C\text{=}CCH_3}} \qquad \underset{\text{1-Butyne}}{\overset{4\ \ \ 3\ \ \ 2\ \ \ 1}{CH_3CH_2C\text{=}CH}}$$

As with alkenes, isomers are possible for butyne and higher alkynes, depending on the position of the triple bond in the chain. Unlike the alkenes, however, no cis-trans isomers are possible for alkynes because of their linear geometry.

EXAMPLE 23.5

Draw the structure of *cis*-3-heptene.

SOLUTION   The name 3-heptene indicates that the molecule has seven carbons (hept-) and has a double bond between carbons 3 and 4:

$$\overset{1\ \ \ 2\ \ \ \ 3\ \ \ \ \ 4\ \ \ 5\ \ \ 6\ \ \ 7}{CH_3CH_2CH\text{=}CHCH_2CH_2CH_3} \qquad \text{3-Heptene}$$

The prefix *cis*- indicates that the two alkyl groups attached to the double-bond carbons lie on the same side of the double bond:

$$\underset{H\qquad\ \ H}{\overset{CH_3CH_2\qquad CH_2CH_2CH_3}{C\text{=}C}} \qquad \textit{cis}\text{-3-Heptene}$$

PROBLEM 23.12 Give IUPAC names for the following alkenes and alkynes.

(a) $\underset{CH_3CHCH\text{=}CH_2}{\overset{CH_3}{|}}$

(b) $CH_3CH_2CH_2$
$\underset{CH_3C\text{=}CHCH_2CH_3}{|}$

(c) $\underset{HC\text{≡}CCHCH_2CH_2CH_3}{\overset{CH_2CH_3}{|}}$

PROBLEM 23.13 Draw structures corresponding to the following IUPAC names.

(a) 2,2-dimethyl-3-hexene   (b) 4-isopropyl-2-heptyne   (c) *trans*-3-heptene

23.10 ►REACTIONS OF ALKENES AND ALKYNES

The most important transformations of alkenes and alkynes are **addition reactions**. That is, a reagent we might write in a general way as X–Y adds to the multiple bond of the unsaturated reactant to yield a saturated product. Alkenes and alkynes react similarly, but we'll look only at alkenes because they're more common and more important.

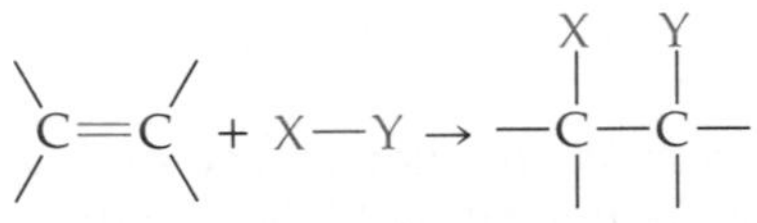

$$\underset{/}{\overset{\backslash}{C}}=\underset{\backslash}{\overset{/}{C}} + X—Y \rightarrow \overset{\overset{X}{|}}{-C}-\overset{\overset{Y}{|}}{C}- \qquad \textit{An addition reaction}$$

- **ADDITION OF HYDROGEN**   Alkenes react with hydrogen in the presence of a platinum or palladium catalyst to yield the corresponding alkane product. For example:

$$\underset{\text{1-Butene}}{CH_3CH_2CH{=}CH_2} + H_2 \xrightarrow[\text{catalyst}]{Pd} \underset{\text{Butane}}{CH_3CH_2CH_2CH_3}$$

The addition of hydrogen to an alkene, often called **hydrogenation**, is used commercially to convert unsaturated vegetable oils to the saturated fats used in margarine and cooking fats.

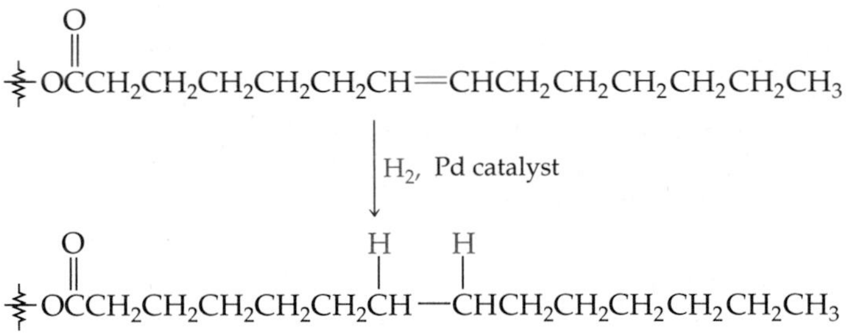

$$\overset{\overset{O}{\|}}{\xi OCCH_2CH_2CH_2CH_2CH_2CH}{=}CHCH_2CH_2CH_2CH_2CH_2CH_3 \qquad \text{Partial structure of a vegetable oil}$$

$$\downarrow H_2, \text{ Pd catalyst}$$

$$\overset{\overset{O}{\|}}{\xi OCCH_2CH_2CH_2CH_2CH_2CH}\overset{\overset{H}{|}}{-}\overset{\overset{H}{|}}{C}HCH_2CH_2CH_2CH_2CH_2CH_3 \qquad \text{Partial structure of a saturated cooking fat}$$

- **ADDITION OF Cl_2 AND Br_2**   Alkenes react with the halogens Cl_2 and Br_2 to give 1,2-dihaloalkane addition products, a process called **halogenation**. For example,

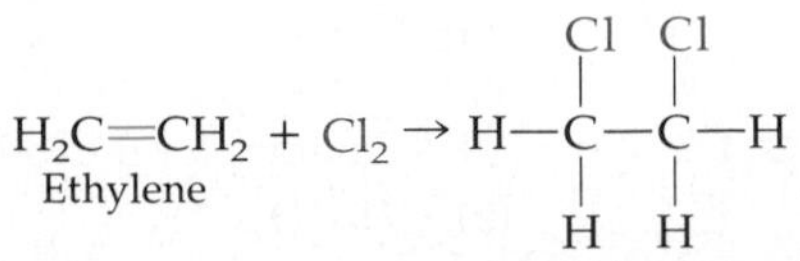

$$\underset{\text{Ethylene}}{H_2C{=}CH_2} + Cl_2 \rightarrow H-\overset{\overset{Cl}{|}}{\underset{\underset{H}{|}}{C}}-\overset{\overset{Cl}{|}}{\underset{\underset{H}{|}}{C}}-H$$
$$\text{1,2-Dichloroethane}$$

More than 8 million tons of 1,2-dichloroethane is manufactured each year in the United States by reaction between ethylene and chlorine as the first step in making PVC [poly(vinyl chloride)] plastics.

- **ADDITION OF WATER**   Alkenes don't react with pure water but, in the presence of a strong acid catalyst such as sulfuric acid, a **hydration** reaction takes place to yield an *alcohol*. For example,

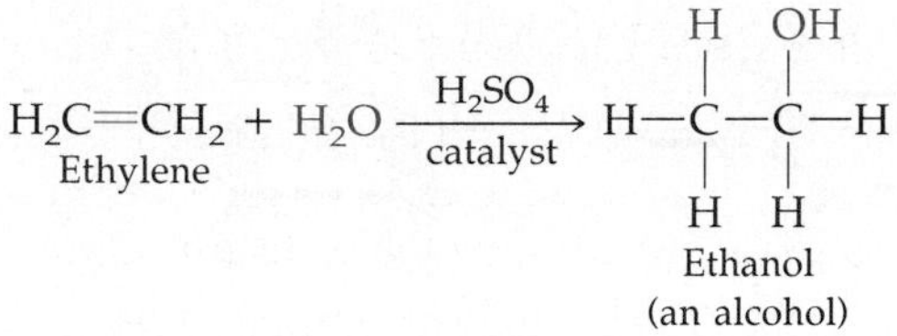

$$H_2C=CH_2 + H_2O \xrightarrow[\text{catalyst}]{H_2SO_4} \overset{\displaystyle H \quad OH}{\underset{\displaystyle H \quad H}{H-C-C-H}}$$

Ethylene

Ethanol
(an alcohol)

Nearly 300 million gallons of ethyl alcohol (ethanol) are produced each year in the United States by the acid-catalyzed addition of water to ethylene.

PROBLEM 23.14 Show the products of the reaction of 2-butene with the following reagents.

(a) H_2, Pd catalyst   (b) Br_2 (c) H_2O, H_2SO_4 catalyst

PROBLEM 23.15 Reaction of 2-pentene with H_2O in the presence of H_2SO_4 yields a mixture of two alcohol products. Draw their structures.

23.11 ►AROMATIC COMPOUNDS AND THEIR REACTIONS

In the early days of organic chemistry, the word *aromatic* was used to describe certain fragrant substances from fruits, trees, and other natural sources. Chemists soon realized, however, that substances grouped as aromatic behaved in a chemically different manner from most other organic compounds. Today, the term **aromatic** refers to the class of compounds containing a six-membered ring with three double bonds. Benzene is the simplest aromatic compound, but aspirin, diazepam (Valium), and many other important compounds also contain aromatic rings.

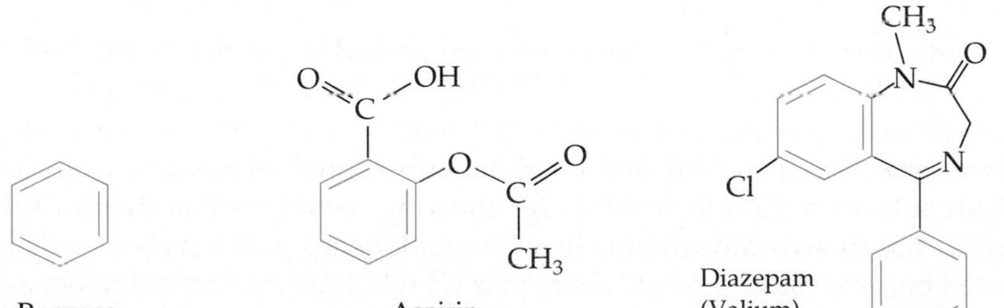

Benzene

Aspirin

Diazepam
(Valium)

Toluene (methylbenzene) is the aromatic compound responsible for the odor of balsam fir.

Benzene is a flat, symmetrical molecule that is often represented as a six-membered ring with three double bonds. The problem with this representation is that it gives the wrong impression about benzene's chemical reactivity. Since benzene appears to have three double bonds, we might expect it to react with H_2, Br_2, and H_2O to give the same kinds of addition products that alkenes do. But this expectation is wrong. Benzene and other aromatic compounds are much less reactive than alkenes and don't normally undergo addition reactions.

Benzene's stability is a consequence of its electronic structure. As shown by the orbital picture in Figure 23.6(b), each of the six carbons is sp^2

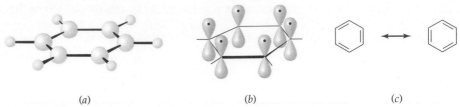

FIGURE 23.6 Some representations of benzene: **(a)** a computer-generated structure; **(b)** an orbital picture; and **(c)** two equivalent resonance structures.

hybridized and has a *p* orbital perpendicular to the ring. When these *p* orbitals overlap to form π bonds, there are two possibilities, shown in Figure 23.6(c).

Neither of the two equivalent structures in Figure 23.6(c) is correct by itself. Rather, each represents one resonance form of the true benzene structure, which is a resonance hybrid of the two (Section 7.6). The stability of benzene is thus explained by supposing that its six π electrons are spread around the *entire* ring. Because the six electrons aren't confined to specific double bonds in the normal way, benzene doesn't react to give addition products in the normal way. Such an idea is hard to represent using lines for covalent bonds, but we sometimes indicate the situation by representing the six electrons as a circle inside the six-membered ring. In this book, though, we'll indicate aromatic rings by showing just one of the individual resonance structures, because it's easier to keep track of electrons that way.

Substituted aromatic compounds are named using the suffix *-benzene*. Thus, C_6H_5Br is bromobenzene, $C_6H_5CH_2CH_3$ is ethylbenzene, and so on. Disubstituted aromatic compounds are named using one of the prefixes *ortho-*, *meta-*, or *para-*. An *ortho-* or *o*-disubstituted benzene has its two substituents in a 1,2 relationship on the ring; a *meta-* or *m*-disubstituted benzene has its two substituents in a 1,3 relationship; and a *para-* or *p*-disubstituted benzene has its substituents in a 1,4 relationship. When the benzene ring itself is a substituent, the name **phenyl** (pronounced **fen**-nil) is used.

ortho-Dimethylbenzene   *meta*-Dibromobenzene   *para*-Dinitrobenzene   A phenyl group

Unlike alkenes, which undergo addition reactions, aromatic compounds usually undergo **substitution reactions**. That is, a group Y *substi-*

tutes for one of the hydrogen atoms on the aromatic ring without changing the ring itself. Of course, it doesn't matter which of the six ring hydrogens is replaced since all six are equivalent.

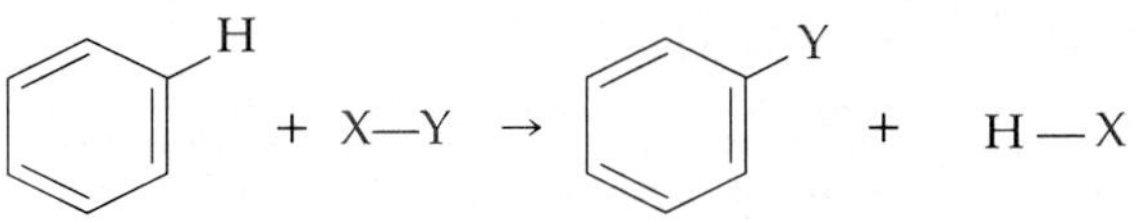

+ X—Y → + H — X   *A substitution reaction*

- **NITRATION**   Substitution of a nitro group ($-NO_2$) for a ring hydrogen occurs when benzene reacts with nitric acid in the presence of sulfuric acid as catalyst:

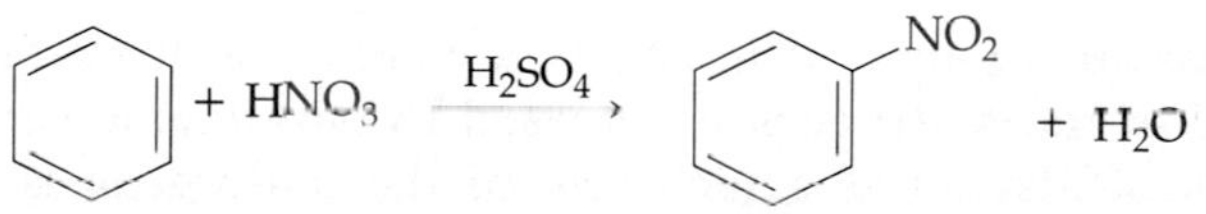

+ HNO_3 $\xrightarrow{H_2SO_4}$ + H_2O

Benzene   Nitrobenzene

Nitration of aromatic rings is a key step in the synthesis of explosives such as TNT (trinitrotoluene) and of many important pharmaceutical agents. Nitrobenzene itself is a starting material for preparing many of the brightly colored dyes used in clothing.

- **HALOGENATION**   Substitution of a bromine or chlorine for a ring hydrogen occurs when benzene reacts with Br_2 or Cl_2 in the presence of $FeBr_3$ or $FeCl_3$ as catalyst:

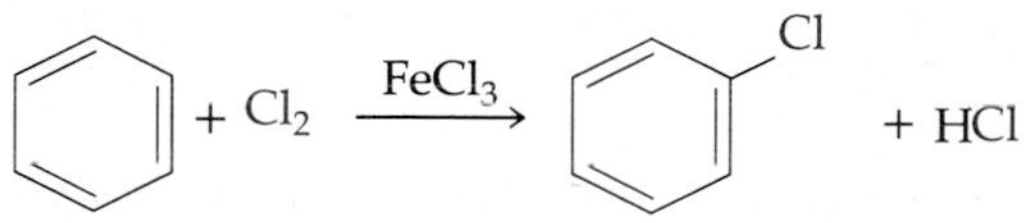

+ Cl_2 $\xrightarrow{FeCl_3}$ + HCl

Benzene   Chlorobenzene

The chlorination of an aromatic ring is a step used in the synthesis of numerous pharmaceutical agents such as the tranquilizer Valium.

- **SULFONATION**   Substitution of a sulfonic acid group ($-SO_3H$) for a ring hydrogen occurs when benzene reacts with concentrated sulfuric acid and sulfur trioxide:

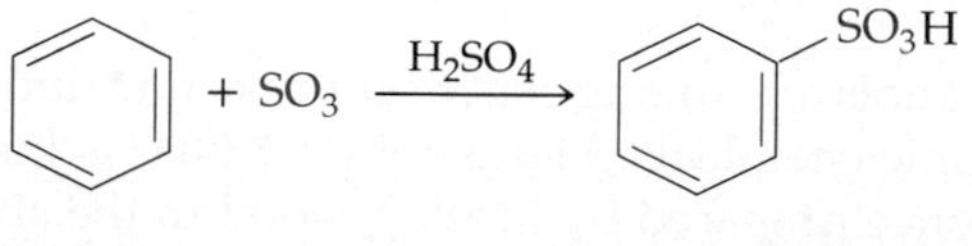

+ SO_3 $\xrightarrow{H_2SO_4}$ + Benzenesulfonic acid

Benzene   Benzenesulfonic acid

Aromatic-ring sulfonation is a key step in the synthesis of such compounds as aspirin and the sulfa-drug family of antibiotics.

⌐ **PROBLEM 23.16** Draw structures corresponding to the following names.
(a) *o*-dibromobenzene   **(b)** *p*-chloronitrobenzene   **(c)** *m*-diethylbenzene

⌐ **PROBLEM 23.17** Write the products from reaction of these reagents with *para*-dimethylbenzene.
(a) Br_2, $FeBr_3$ **(b)** HNO_3, H_2SO_4 **(c)** SO_3, H_2SO_4

⌐ **PROBLEM 23.18** Reaction of $Br_2/FeBr_3$ with toluene (methylbenzene) can lead to a mixture of *three* substitution products. Show the structure of each.   ⌐

23.12 ➤ALCOHOLS, ETHERS, AND AMINES

Alcohols

Alcohols can be thought of in two ways: either as derivatives of water in which one of the hydrogens is replaced by an organic substituent, or as derivatives of alkanes in which one of the hydrogens is replaced by a hydroxyl group (–OH). Because of their structural resemblance to water, many simple alcohols are water soluble.

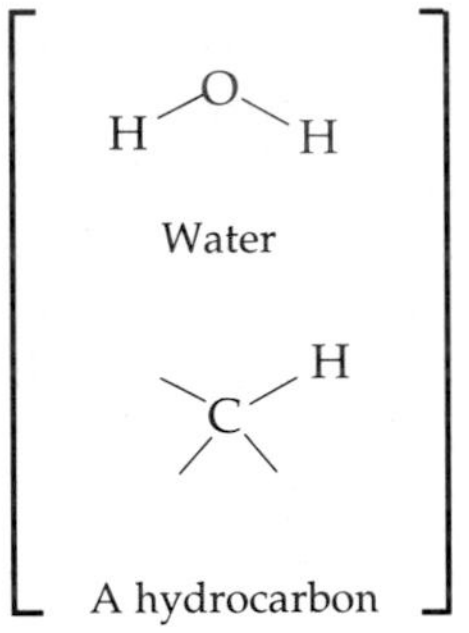

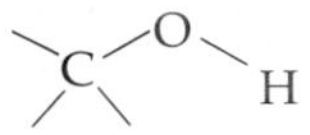

Water

A hydrocarbon

An alcohol

for example: CH_3CH_2OH

Ethanol

Alcohols are named by specifying the point of attachment of the –OH group to the hydrocarbon chain and using the suffix *-ol* to replace the terminal *-e* in the alkane name. Numbering of the chain begins at the end nearer the –OH group. For example,

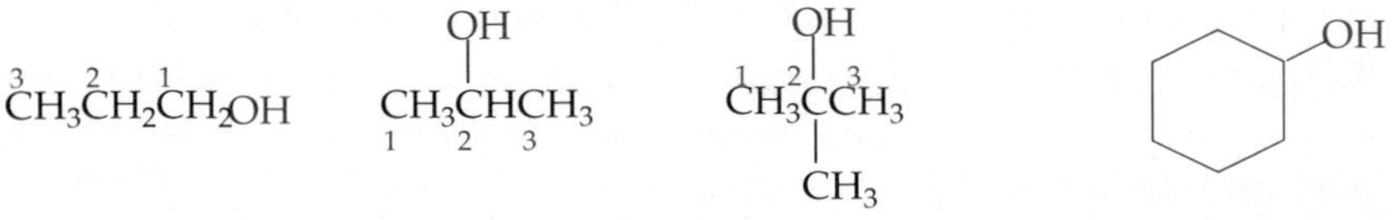

1-Propanol   2-Propanol   2-Methyl-2-propanol   Cyclohexanol

Simple alcohols are among the most important and commonly encountered of all organic chemicals. Methanol (CH_3OH) is known as *wood alcohol* because it was once prepared by heating wood in the absence of air. Today, approximately 1.7 billion gallons of methanol are manufactured each year in the United States by catalytic reduction of carbon monoxide with hydrogen gas.

$$CO + 2\,H_2 \xrightarrow[\text{ZnO/Cr}_2\text{O}_3 \text{ catalyst}]{400°C} CH_3OH$$

Though toxic to humans, causing blindness in low doses and death in larger amounts, methanol is an important industrial starting material for preparing formaldehyde (CH_2O), acetic acid (CH_3CO_2H), and other chemicals.

Ethanol (CH_3CH_2OH) is one of the oldest known pure organic chemicals. Its production by fermentation of grains and sugars, and its subsequent purification by distillation, go back at least as far as the twelfth century A.D. Sometimes called *grain alcohol*, ethanol is the "alcohol" present in all wines (10 to 13%), beers (3 to 5%), and distilled liquors (35 to 90%). Fermentation is carried out by adding yeast to an aqueous sugar solution and allowing enzymes in the yeast to break down carbohydrates into ethanol and CO_2.

$$\underset{\text{Glucose}}{C_6H_{12}O_6} \xrightarrow{\text{yeast}} 2\ CO_2 + 2\ \underset{\text{Ethanol}}{CH_3CH_2OH}$$

Only about 5% of the ethanol produced industrially comes from fermentation. Most is obtained by acid-catalyzed addition of water to ethylene (Section 23.10).

2-Propanol [$(CH_3)_2CHOH$], commonly called isopropyl alcohol or *rubbing alcohol*, is used primarily as a solvent. It is prepared industrially by addition of water to propene.

$$\underset{\text{Propene}}{CH_3CH{=}CH_2} + H_2O \xrightarrow[\text{catalyst}]{\text{acid}} \underset{\text{2-Propanol}}{CH_3\overset{\displaystyle OH}{\overset{|}{C}}HCH_3}$$

Still other important alcohols include 1,2-ethanediol (ethylene glycol), 1,2,3-propanetriol (glycerol), and the aromatic alcohol phenol. Ethylene glycol is the principal constituent of automobile antifreeze, glycerol is used as a moisturizing agent in many foods and cosmetics, and phenol is used for preparing nylon, epoxy adhesives, and heat-setting resins.

$$HOCH_2CH_2OH$$

1,2-Ethanediol
(Ethylene glycol)

$$HOCH_2\overset{\displaystyle OH}{\overset{|}{C}}HCH_2OH$$

1, 2, 3-Propanetriol
(Glycerol)

Phenol

Propylene glycol (1,2-propanediol) is sprayed on airplanes in winter to de-ice the exterior prior to takeoff.

Ethers

Ethers are compounds that have two organic groups bonded to the same oxygen atom. They are fairly inert chemically and are often used as reaction solvents for that reason. Diethyl ether, the most common member of the ether family, was used for many years as a surgical anesthetic agent but has now been replaced by safer nonflammable alternatives.

$$CH_3CH_2OCH_2CH_3$$

Diethyl ether

Tetrahydrofuran
(a cyclic ether)

Amines

Amines are organic derivatives of ammonia in the same way that alcohols and ethers are organic derivatives of water. That is, one or more of the ammonia hydrogens is replaced in amines by an organic substituent. As the following examples indicate, the suffix *-amine* is used in naming these compounds.

Ammonia   Methylamine   Dimethylamine   Trimethylamine   Benzeneamine
(Aniline)

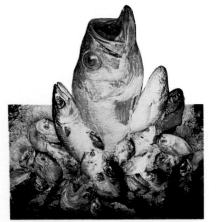

The characteristic aroma of ripe fish is due to methylamine, CH_3NH_2.

Like ammonia, amines are bases because they can use the lone pair of electrons on nitrogen to accept H^+ from an acid and give ammonium salts (Section 15.12). Because they're ionic, ammonium salts are much more soluble in water than are neutral amines. Thus, a water-insoluble amine such as triethylamine dissolves readily in water when converted into its ammonium salt by reaction with HCl.

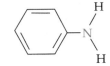

$$CH_3CH_2-\overset{..}{N}-CH_2CH_3 + HCl(aq) \rightarrow CH_3CH_2-\overset{H}{\underset{CH_2CH_3}{\overset{|}{N}}}{}^+-CH_2CH_3 \ Cl^- (aq)$$

Triethylamine
(water insoluble)

Triethylammonium chloride
(water soluble)

The increase in water solubility on conversion of an amine to its protonated salt has enormous practical consequences in drug delivery. Many important amine-containing drugs, such as morphine (a painkiller) and tetracycline (an antibiotic), are insoluble in aqueous body fluids and are thus difficult to deliver to the appropriate site within the body. Converting these drugs to their ammonium salts, however, increases their solubility to the point where delivery through the bloodstream becomes possible.

┌ **PROBLEM 23.19** Write the structures of the ammonium salts produced by
reaction of the following amines with HCl.

(a)

NHCH$_3$

(b)

CH$_3$CH$_2$CH$_2$NH$_2$

◢

23.13 ▸KETONES AND ALDEHYDES

Look back at the functional groups listed in Table 23.3 and you'll see that
many of them have a carbon–oxygen double bond (C=O), called a **carbonyl
group** (pronounced car-bo-**neel**). Carbonyl-containing compounds occur
everywhere. Carbohydrates, fats, proteins, and nucleic acids all contain
carbonyl groups; most pharmaceutical agents contain carbonyl groups; and
many of the synthetic polymers used for clothing and other applications
contain carbonyl groups.

It's useful to classify carbonyl compounds into two categories based on
their chemical properties. In one category are aldehydes and ketones.
Because the carbonyl group in these compounds is bonded to atoms (H and
C) that are not strongly electronegative, the bonds to the carbonyl-
group carbon of aldehydes and ketones are nonpolar, and the two groups of
compounds thus have similar properties. In the second category are carboxylic acids, esters, and amides. Since the carbonyl-group carbon in these
compounds is bonded to an atom (O or N) that is strongly electronegative,
the bonds are polar. Thus, carboxylic acids, esters, and amides behave
similarly.

| Aldehyde | Ketone | Carboxylic acid | Ester | Amide |

| Less polar | | More polar | | |

Simple aldehydes and ketones are used throughout chemistry and
biology. For example, an aqueous solution of formaldehyde (properly
named methanal) is used under the name *formalin* as a biological sterilant
and preservative. Formaldehyde is also used in the chemical industry as a
starting material for the manufacture of the plastics Bakelite and melamine,
as a component of the adhesives used to bind plywood, and as a part of the
foam insulation used in houses. Note that formaldehyde differs from other
aldehydes in having two hydrogens attached to the carbonyl group rather
than one. Acetone (properly named propanone) is perhaps the most widely
used of all organic solvents. You might have seen cans of acetone sold in
paint stores for general-purpose cleanup work. When naming these groups
of compounds, aldehydes take the suffix -*al* and ketones take the suffix -*one*.

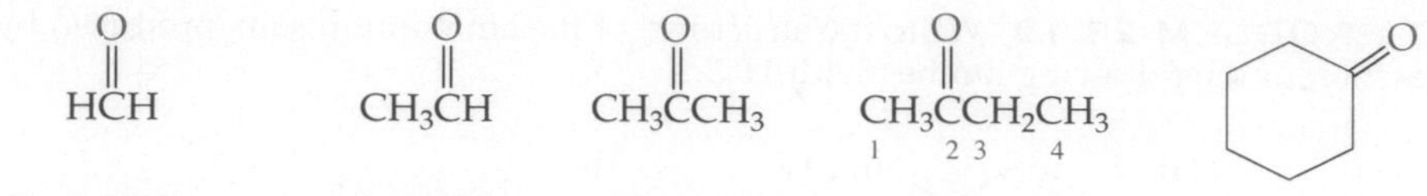

$$\underset{\text{Formaldehyde}\atop\text{(Methanal)}}{\overset{\overset{\displaystyle O}{\|}}{HCH}} \qquad \underset{\text{Acetaldehyde}\atop\text{(Ethanal)}}{\overset{\overset{\displaystyle O}{\|}}{CH_3CH}} \qquad \underset{\text{Acetone}\atop\text{(Propanone)}}{\overset{\overset{\displaystyle O}{\|}}{CH_3CCH_3}} \qquad \underset{\text{2-Butanone}}{\overset{\overset{\displaystyle O}{\|}}{\underset{1\ \ \ 2\ \ 3\ \ \ \ 4}{CH_3CCH_2CH_3}}}$$

Cyclohexanone

Aldehyde and ketone functional groups are also present in many biologically important compounds. Glucose and most other sugars contain aldehyde groups, for example. Testosterone and many other steroid hormones contain ketone groups.

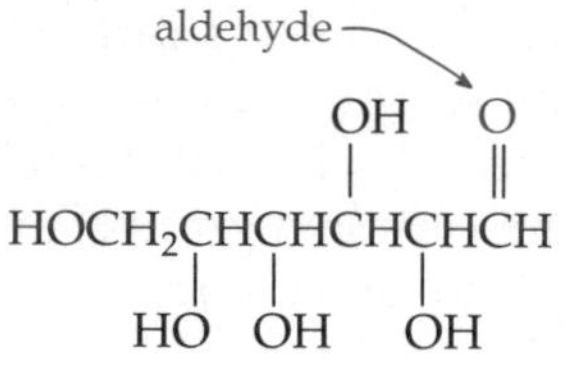

aldehyde

$$\underset{\underset{HO\ \ \ OH\ \ \ OH}{|\ \ \ \ \ |\ \ \ \ |}}{HOCH_2CHCHCHCHCH}$$

OH   O

Glucose—a pentahydroxyhexanal

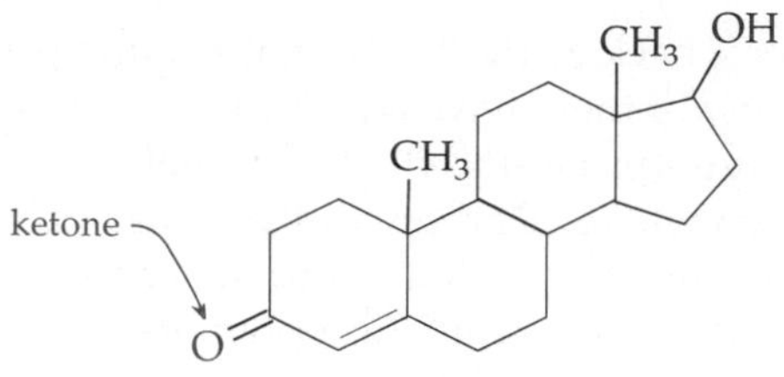

ketone

Testosterone—a steriod hormone

The industrial preparation of simple aldehydes and ketones usually involves an oxidation of the related alcohol. Thus, formaldehyde is prepared by oxidation of methanol, and acetone is prepared by oxidation of 2-propanol.

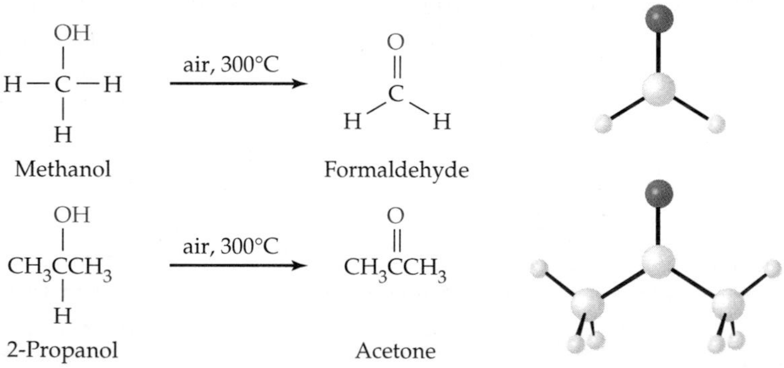

$$\underset{\text{Methanol}}{H-\overset{\overset{\displaystyle OH}{|}}{\underset{\underset{\displaystyle H}{|}}{C}}-H} \xrightarrow{\text{air, }300°C} \underset{\text{Formaldehyde}}{\overset{\overset{\displaystyle O}{\|}}{\underset{H\ \ \ \ \ \ H}{C}}}$$

$$\underset{\text{2-Propanol}}{\overset{\overset{\displaystyle OH}{|}}{\underset{\underset{\displaystyle H}{|}}{CH_3CCH_3}}} \xrightarrow{\text{air, }300°C} \underset{\text{Acetone}}{\overset{\overset{\displaystyle O}{\|}}{CH_3CCH_3}}$$

23.14 ➤CARBOXYLIC ACIDS, ESTERS, AND AMIDES

Carboxylic acids, esters, and amides have their carbonyl groups bonded to an atom (O or N) that strongly attracts electrons All three families undergo **carbonyl-group substitution reactions**, in which a group we can represent as –Y substitutes for the –OH, –OC, or –N group of the starting material.

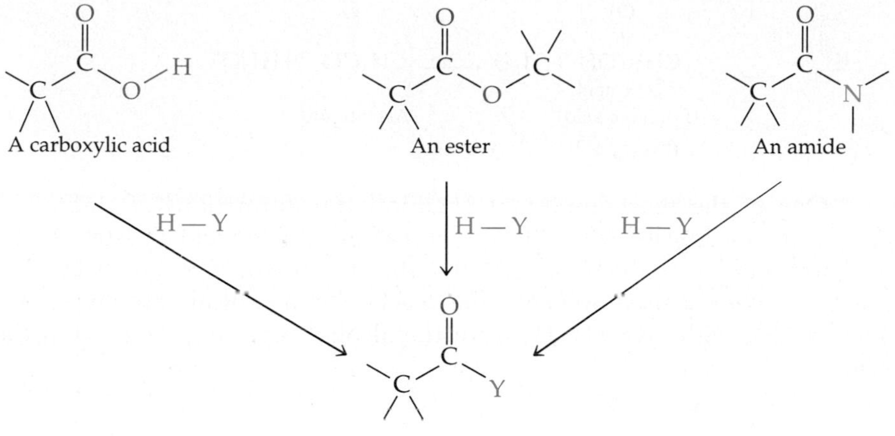

A carboxylic acid   An ester   An amide

A carbonyl-group substitution reaction

Carboxylic Acids

Carboxylic acids occur widely throughout the plant and animal kingdoms. Acetic acid (ethanoic acid), for example, is the principal organic constituent of vinegar, and butanoic acid is responsible for the odor of rancid butter. In addition, long-chain carboxylic acids such as stearic acid are components of all animal fats and vegetable oils. Although many carboxylic acids have common names—*acetic acid* instead of *ethanoic acid*, for instance—systematic names are derived by replacing the final *-e* of the corresponding alkane with *-oic acid*.

Acetic acid
(Ethanoic acid)

Butanoic acid

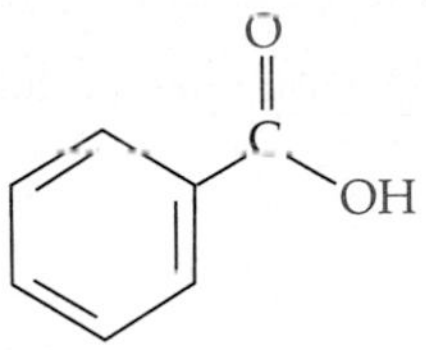

Benzoic acid

$$CH_3CH_2CH_2CH_2CH_2CH_2CH_2CH_2CH_2CH_2CH_2CH_2CH_2CH_2CH_2CH_2CH_2COH$$

Stearic acid
(Octadecanoic acid)

As their name implies, carboxylic acids are *acidic*; they dissociate slightly in aqueous solution to give H_3O^+ and a **carboxylate anion**. Carboxylic acids are much weaker than inorganic acids like HCl or H_2SO_4, however. The K_a of acetic acid, for example, is 1.78×10^{-5} ($pK_a = 4.75$), meaning that only about 1% of acetic acid molecules dissociate in a 1.0 M aqueous solution.

$$CH_3\overset{\displaystyle O}{\overset{\|}{C}}OH + H_2O \rightleftharpoons CH_3\overset{\displaystyle O}{\overset{\|}{C}}O^- + H_3O^+$$

Acetic acid
(Ethanoic acid)   Acetate ion
$pK_a = 4.75$

One of the most important chemical transformations of carboxylic acids is their acid-catalyzed reaction with an alcohol to yield an ester. Acetic acid, for example, reacts with ethanol in the presence of H_2SO_4 to yield ethyl acetate, a widely used solvent. The reaction is a typical carbonyl-group substitution, with $-OCH_2CH_3$ from the alcohol replacing $-OH$ from the acid.

$$CH_3-\overset{\displaystyle O}{\overset{\|}{C}}-\boxed{OH + H}-OCH_2CH_3 \xrightarrow[\text{catalyst}]{H^+} CH_3-\overset{\displaystyle O}{\overset{\|}{C}}-OCH_2CH_3 + H_2O$$

Acetic acid   Ethanol   Ethyl acetate

Esters

Esters have many uses in medicine, in industry, and in living systems. In medicine, a number of important pharmaceutical agents, including aspirin and the local anesthetic benzocaine, are esters. In industry, polyesters such as Dacron and Mylar are used to make synthetic fibers and films. In nature, many simple esters are responsible for the fragrant odors of fruits and flowers. For example, pentyl acetate is found in bananas, and octyl acetate is found in oranges.

The odors of these fruits are due to simple esters.

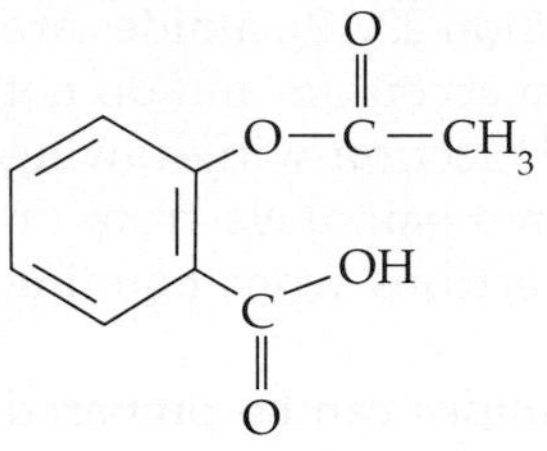

Aspirin

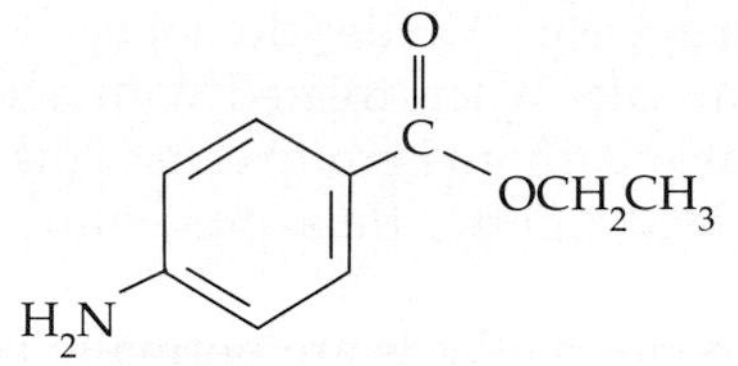

Benzocaine

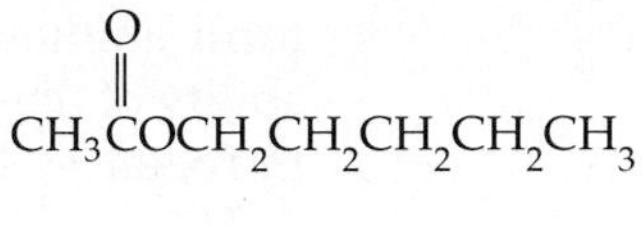

Pentyl acetate

The most important reaction of esters is their conversion by a carbonyl-group substitution reaction into carboxylic acids. Both in the laboratory and in the body, esters undergo a reaction with water—a **hydrolysis**—that splits the ester molecule into a carboxylic acid and an alcohol. The net effect is a substitution of –OC by –OH. Although the reaction is slow in pure water, it is catalyzed by both acid and base. Base-catalyzed ester hydrolysis is often called **saponification**, from the Latin word *sapo* meaning "soap." Soap, in fact, is a mixture of sodium salts of long-chain carboxylic acids and is produced by hydrolysis with aqueous NaOH of the naturally occurring esters in animal fat.

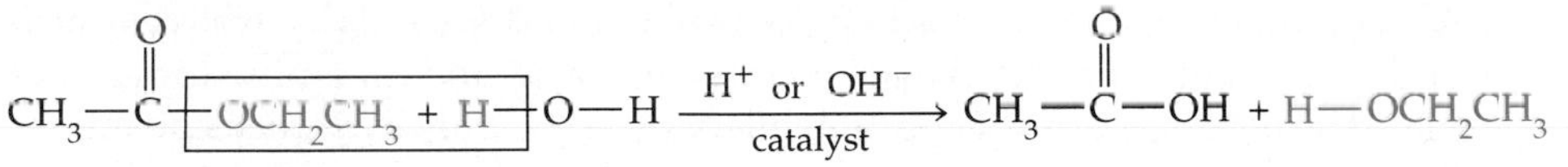

Since esters are derived from carboxylic acids and alcohols, they are named by first identifying the alcohol-related part and then the acid-related part using the *-ate* ending. Ethyl acetate, for example, is the ester derived from ethanol and acetic acid.

Amides

Without **amides**, there would be no life. As we'll see in the next chapter, the amide bond between nitrogen and a carbonyl-group carbon is the fundamental link used by organisms for forming proteins. In addition, some synthetic polymers such as nylon contain amide groups, and important pharmaceutical agents such as acetaminophen, the aspirin substitute found in Tylenol and Excedrin, are amides.

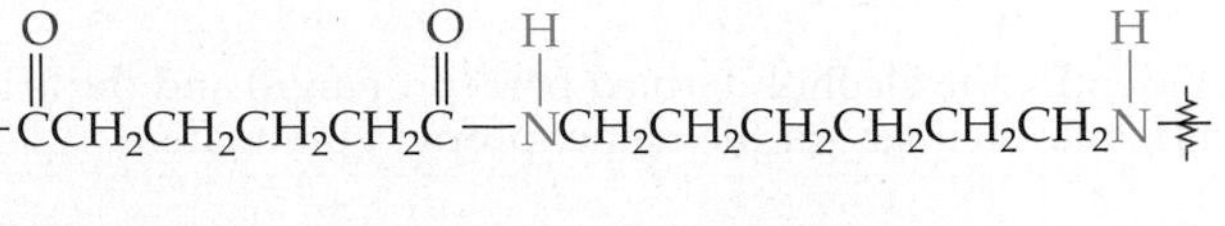

Repeating unit of nylon 66

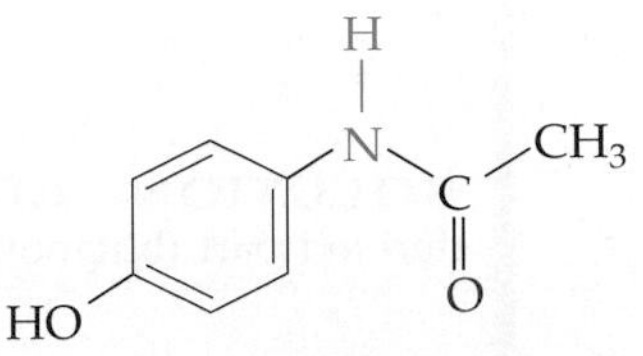

Acetaminophen

Unlike amines, which also contain nitrogen (Section 23.12), amides are neutral rather than basic. Amides do not act as proton acceptors and do not form ammonium salts when treated with acid. The electron-withdrawing ability of the nearby carbonyl group causes the unshared pair of electrons on nitrogen to be held tightly, thus preventing the electrons from bonding to H^+.

Although better methods are normally used, amides can be prepared by the reaction of a carboxylic acid with ammonia or an amine, just as esters are prepared by the reaction of a carboxylic acid with an alcohol. In both cases, water is a byproduct, and the –OH part of the carboxylic acid is replaced. Amides are named by first citing the *N*-alkyl group on the amine part (*N* because the group is attached to nitrogen) and then identifying the carboxylic acid part using the *-amide* ending.

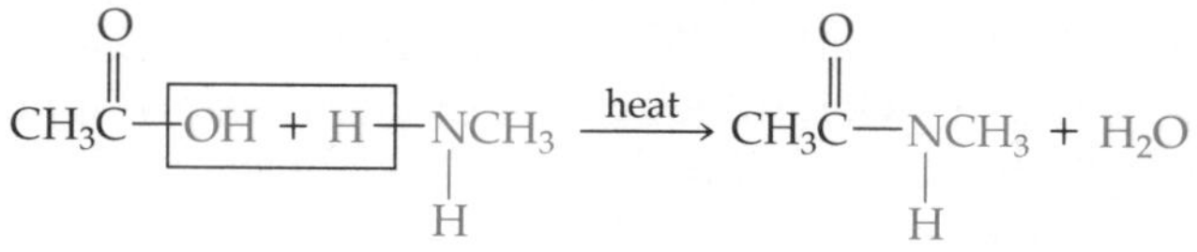

$$CH_3C\overset{O}{\overset{\|}{}}\boxed{OH + H}NCH_3 \xrightarrow{heat} CH_3C\overset{O}{\overset{\|}{}}-NCH_3 + H_2O$$

Acetic acid   Methylamine   *N*-Methylacetamide
(an amide)

Amides undergo an acid- or base-catalyzed hydrolysis reaction with water in the same way that esters do. Just as an ester yields a carboxylic acid and an alcohol, an amide yields a carboxylic acid and an amine (or ammonia). The net effect is a substitution of –N by –OH. This hydrolysis of amides is the key process that occurs in the stomach during digestion of proteins.

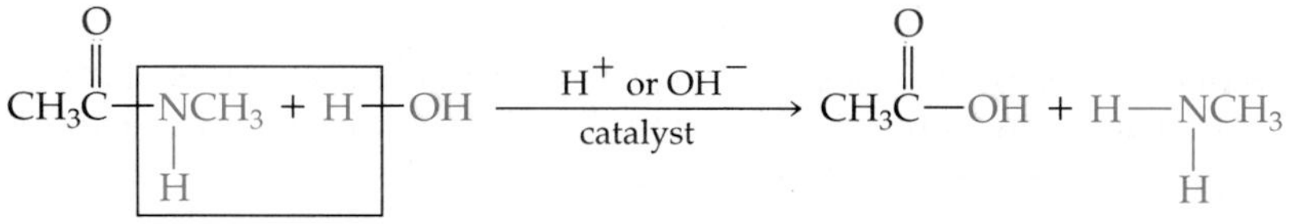

$$CH_3C\overset{O}{\overset{\|}{}}\boxed{NCH_3 + H}OH \xrightarrow[catalyst]{H^+\ or\ OH^-} CH_3C\overset{O}{\overset{\|}{}}-OH + H-NCH_3$$

N-Methylacetamide   Acetic acid   Methylamine

EXAMPLE 23.6

Give the systematic names of the following compounds.

(a)

$$CH_3CH_2CH_2\overset{O}{\overset{\|}{C}}OCH_2CH_2CH_3$$

(b)

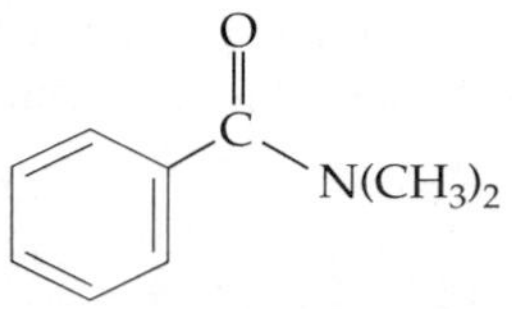

SOLUTION   **(a)** First identify the alcohol-derived part (propanol) and the acid-derived part (butanoic acid), and then assign the name: propyl butanoate.

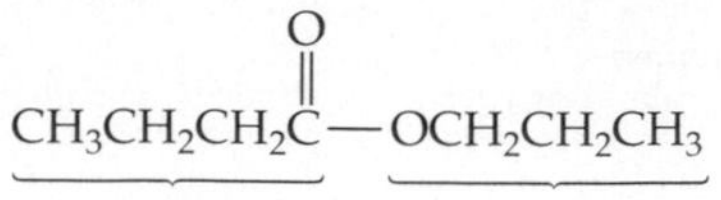

$$\underbrace{CH_3CH_2CH_2\overset{O}{\overset{\|}{C}}}_{Butanoic\ acid}-\underbrace{OCH_2CH_2CH_3}_{Propanol} \quad \text{Propyl butanoate}$$

(b) First identify the amine-derived part (dimethylamine) and the acid-derived part (benzoic acid), and then assign the name: *N,N*-dimethylbenzamide.

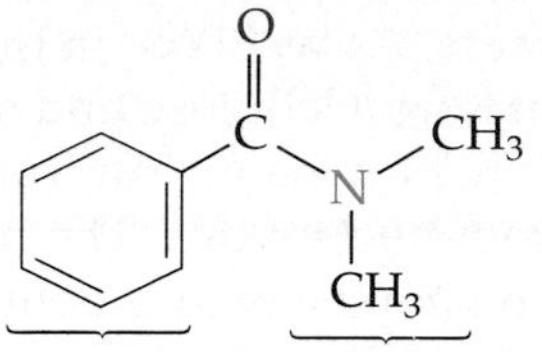

N, N-Dimethylbenzamide

Benzoic acid   Dimethylamine

EXAMPLE 23.7

Write the products of the following reactions.

(a)

$$CH_3CH(CH_3)CH_2C(=O)OH + HOCH(CH_3)CH_3 \longrightarrow ?$$

(b)

$$CH_3CH_2CH(Br)CH_2C(=O)NH_2 + H_2O \longrightarrow ?$$

SOLUTION **(a)** The reaction of a carboxylic acid with an alcohol yields an ester plus water. Write the reagents to show how H_2O is removed, and then connect the remaining fragments to complete the substitution reaction:

$$CH_3CH(CH_3)CH_2C(=O)\boxed{OH + H}OCH(CH_3)CH_3 \rightarrow CH_3CH(CH_3)CH_2C(=O){-}OCH(CH_3)CH_3 + H_2O$$

(b) The reaction of an amide with water yields a carboxylic acid and an amine (or ammonia). Write the reagents to show how NH_3 is removed, and then connect the remaining fragments to complete the substitution reaction:

$$CH_3CH_2CH(Br)CH_2C(=O)\boxed{NH_2 + H}OH \rightarrow CH_3CH_2CH(Br)CH_2C(=O)OH + NH_3$$

PROBLEM 23.20 Draw structures corresponding to the following names.

(a) 4-methylpentanoic acid **(b)** isopropyl benzoate **(c)** *N*-ethylpropanamide

PROBLEM 23.21 Write the products of the following reactions.

(a)

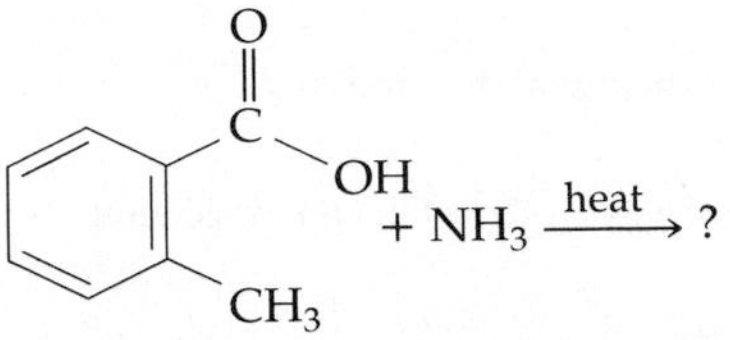

(b)

$$CH_3CH(Cl)CH_2C(=O)OH + CH_3CH_2CH(CH_3)OH \xrightarrow[\text{catalyst}]{H^+} ?$$

23.15 ►SYNTHETIC POLYMERS

Polymers are large molecules formed by the repetitive bonding together of many smaller molecules, called **monomers**. As we'll see in the next chapter, biological polymers occur throughout nature. Cellulose and starch are polymers built from small sugars, proteins are polymers built from amino acids, and nucleic acids are polymers built from nucleotides. The basic idea is the same, but synthetic polymers are much simpler than biopolymers because the starting monomer units are usually smaller and simpler.

Many simple alkenes, called **vinyl monomers**, undergo polymer-forming (*polymerization*) reactions: Ethylene yields polyethylene, propylene (propene) yields polypropylene, styrene yields polystyrene, and so forth. The polymer molecules that result may have anywhere from a few hundred to a few thousand monomer units incorporated into a long chain. Some commercially important polymers are shown in Table 23.4.

TABLE 23.4   Some Alkene Polymers and Their Uses

Monomer Name	Structure	Polymer Name	Uses
Ethylene	$H_2C{=}CH_2$	Polyethylene	Packaging , bottles
Propylene	$H_2C{=}CH{-}CH_3$	Polypropylene	Bottles, rope, pails, medical tubing
Vinyl chloride	$H_2C{=}CH{-}Cl$	Poly(vinyl chloride)	Insulation, plastic pipe
Styrene	$H_2C{=}CH{-}$⬡ 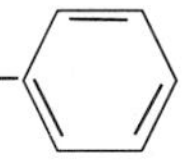	Polystyrene	Foams and molded plastics
Acrylonitrile	$H_2C{=}CH{-}C{\equiv}N$	Orlon, Acrilan	Fibers, outdoor carpeting

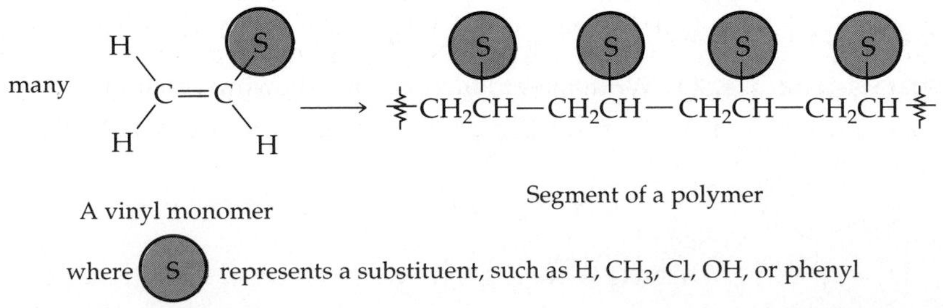

A vinyl monomer

Segment of a polymer

where **S** represents a substituent, such as H, CH_3, Cl, OH, or phenyl

The fundamental process in alkene polymerization is an addition reaction to the double bonds. A species called an *initiator*, In, first adds to the double bond of an alkene, yielding a reactive intermediate that in turn adds to a second alkene molecule to produce another reactive intermediate, and so on.

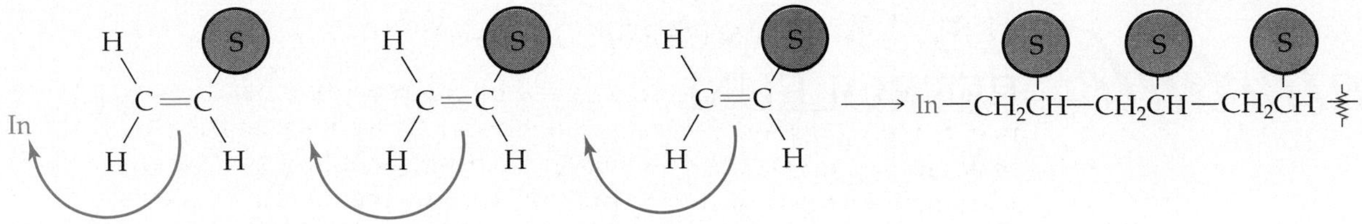

$$\text{In}-\text{CH}_2\text{CH}-\text{CH}_2\text{CH}-\text{CH}_2\text{CH} \not\succeq$$

A second kind of polymerization process occurs when molecules with *two* functional groups react. We've seen, for example, that a carboxylic acid reacts with an amine to yield an amide (Section 23.14). If a molecule with two carboxylic acid groups reacts with a molecule having two amino groups, an initial reaction joins the two molecules together, and further reactions then link more and more molecules together until a giant poly-amide chain results. Nylon 66, one of the most important such polymers, is prepared by heating adipic acid (hexanedioic acid) with 1,6-hexanediamine at 280°C.

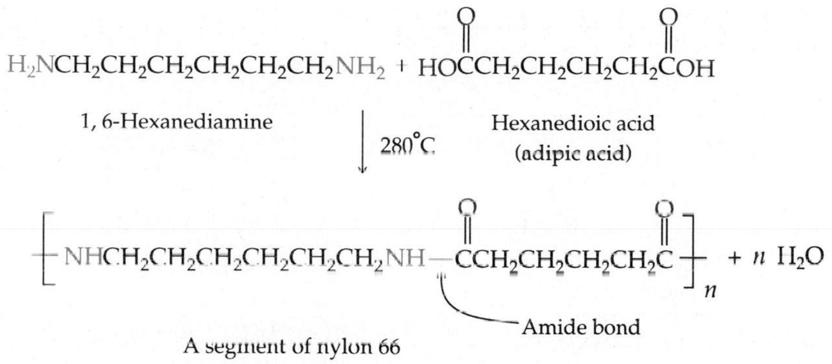

$$H_2NCH_2CH_2CH_2CH_2CH_2CH_2NH_2 + HOCCH_2CH_2CH_2CH_2COH$$

1, 6-Hexanediamine $\qquad\qquad$ Hexanedioic acid
$\qquad\qquad\qquad$ 280°C $\qquad$ (adipic acid)

$$\left[NHCH_2CH_2CH_2CH_2CH_2CH_2NH-CCH_2CH_2CH_2CH_2C \right]_n + n\ H_2O$$

Amide bond

A segment of nylon 66

Nylons have many uses, both in engineering applications and in fibers. High impact strength and resistance to abrasion make nylon an excellent material for bearings and gears. High tensile strength makes it suitable as fibers for a range of applications from clothing to mountaineering ropes to carpets.

Just as diacids and diamines react to give *polyamides*, diacids and dialcohols react to give *polyesters*. The most industrially important polyester, made from reaction of terephthalic acid (1,4-benzenedicarboxylic acid) with ethylene glycol (1,2-ethanediol), is used under the trade name Dacron to make clothing fiber and under the name Mylar to make plastic film and recording tape.

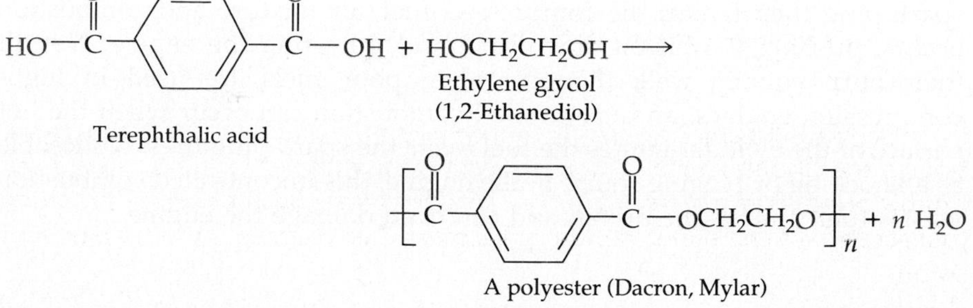

$$HO-C-\bigcirc-C-OH + HOCH_2CH_2OH \longrightarrow$$

$\qquad\qquad\qquad\qquad$ Ethylene glycol
$\qquad\qquad\qquad\qquad$ (1,2-Ethanediol)
Terephthalic acid

$$\left[C-\bigcirc-C-OCH_2CH_2O \right]_n + n\ H_2O$$

A polyester (Dacron, Mylar)

interlude—PETROLEUM

The large columns in this oil refinery are where crude oil is distilled into fractions to yield gasoline.

Laid down eons ago by the decomposition of once living matter, the world's natural gas and petroleum deposits are by far the most abundant source of organic chemicals. Natural gas consists largely of methane but also contains ethane, propane, and butane. Petroleum is an enormously complex mixture of substances, primarily hydrocarbons, that must be refined into different fractions before it can be used. Refining begins by fractional distillation (Section 11.10) of crude petroleum into three principal cuts: straight-run gasoline (boiling point 30–200°C), kerosene (bp 175–300°C), and gas oil (bp 275–400°C). Further distillation under reduced pressure yields lubricating oils and waxes, and leaves an undistillable tarry residue of asphalt.

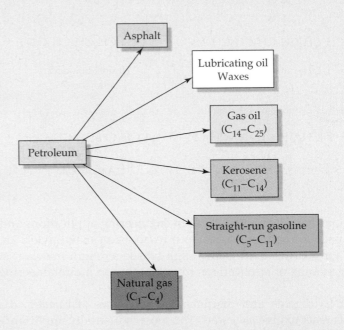

The straight-run gasoline that comes directly from petroleum is a particularly poor fuel because it causes *engine knock*. In the typical automobile engine, a piston draws a mixture of fuel and air into a cylinder on its downward stroke and compresses the mixture on its upward stroke. A spark plug then ignites the compressed fuel/air mixture and combustion occurs, pushing the piston downward and powering the engine. Not all fuels burn equally well, though. When poor fuels are used in high-compression engines, an uncontrolled combustion can occur when the hot surface of the cylinder ignites the fuel *before* the spark plug fires. Noticeable as a knocking or pinging sound in the engine, this uncontrolled combustion wastes fuel, annoys the driver, and can even damage the engine.

High-performance engines need high-octane gasoline to prevent engine knock.

A fuel's knock-inducing behavior is rated by its *octane number*—the higher the rating, the better the fuel. Heptane, a straight-chain hydrocarbon and a particularly bad fuel, is assigned a base value octane number of 0. 2,2,4-trimethylpentane, a branched-chain hydrocarbon commonly known as isooctane, has a rating of 100.

$$CH_3CH_2CH_2CH_2CH_2CH_2CH_3$$

$$CH_3\overset{\overset{\displaystyle CH_3}{|}}{\underset{\underset{\displaystyle CH_3}{|}}{C}}CH_2\overset{\overset{\displaystyle CH_3}{|}}{C}HCH_3$$

Heptane
octane number = 0

2,2,4-Trimethylpentane (isooctane)
octane number = 100

Straight-run gasoline, with its high percentage of unbranched alkanes, is a poor fuel, and petroleum chemists have therefore devised ways of improving it. In a process called *catalytic cracking*, the kerosene cut (C_{11}–C_{14}) is cracked into smaller molecules at high temperature on a silica–alumina catalyst. The major gasoline-range products of cracking are C_7–C_{10} branched-chain molecules with high octane ratings. Another way of increasing a fuel's octane rating is to add an octane booster. Tetraethyllead, $(CH_3CH_2)_4Pb$ was for many years the most common booster, but fears of toxic effects from lead pollution have curtailed its use. Oxygenated organic compounds such as ethanol, CH_3CH_2OH, and methyl *tert*-butyl ether, $[CH_3OC(CH_3)_3]$, are now used instead of tetraethyllead to boost the octane rating of premium gasolines.

SUMMARY

Organic chemistry is the study of carbon compounds. More than 11 million organic compounds are known, organized into families according to the functional groups they contain. A **functional group** is an atom or group of atoms within a molecule that has characteristic chemical behavior and that undergoes the same kinds of reactions in every molecule it's a part of.

The simplest family of compounds are the **saturated hydrocarbons**, or **alkanes**, which contain only carbon and hydrogen and have only single bonds. **Straight-chain alkanes** have all their carbons connected in a row, **branched-chain alkanes** have a branching connection of atoms in their chain, and **cycloalkanes** have a ring of carbon atoms. Isomerism is possible in alkanes having four or more carbons. Isomers are compounds that have the same molecular formula but different structures. Straight-chain alkanes are named in the IUPAC system by adding the family ending -*ane* to the Greek number that tells how many carbon atoms are present. Branched-chain alkanes are named by identifying the longest continuous chain of carbon atoms and then telling what **alkyl groups** are present as branches off the main chain. Alkanes are chemically rather inert, although they undergo **combustion** with oxygen and undergo a **substitution reaction** with chlorine.

Alkenes are hydrocarbons that contain a carbon–carbon double bond, and **alkynes** are hydrocarbons that contain a carbon–carbon triple bond. **Cis-trans isomers** are possible for substituted alkenes because of the lack of rotation about the carbon–carbon double bond. The cis isomer has two substituents on the same side of the double bond, and the trans isomer has two substituents on opposite sides. The most important transformations of alkenes and alkynes are **addition reactions**, in which a reagent adds to the multiple bond to yield a saturated product.

Aromatic compounds contain a six-membered ring with three double bonds. These compounds usually undergo substitution reactions, in which a group substitutes for one of the hydrogen atoms on the aromatic ring. **Alcohols** can be thought of as derivatives of water in which one of the hydrogens is replaced by an organic substituent. Similarly, **amines** are derivatives of ammonia in which one or more of the ammonia hydrogens are replaced by an organic substituent. Amines are bases and can be protonated by acids to yield ammonium salts.

Compounds that contain a **carbonyl group**, C=O, can be classified into two categories based on their chemical properties. In aldehydes and ketones, the carbonyl-group carbon is bonded to atoms (H and C) that don't attract electrons strongly. In **carboxylic acids**, **esters**, and **amides**, the carbonyl-group carbon is bonded to an atom (O or N) that *does* attract electrons strongly. As a result, these latter three families of compounds undergo **carbonyl-group substitution reactions**, in which a group –Y substitutes for the –OH, –OC, or –N group of the starting material.

Polymers are large molecules formed by the repetitive bonding together of many smaller **monomers**. Alkene polymers such as polyethylene result from the polymerization of simple substituted alkenes. Nylons and polyesters result from the sequential reaction of two difunctional molecules.

ADDITIONAL PROBLEMS

Problems 23.1–23.21 appear within the chapter.

FUNCTIONAL GROUPS AND ISOMERS

23.22 What are functional groups, and why are they important?

23.23 Give examples of compounds that are members of the following families.
(a) alcohol   (b) amine
(c) carboxylic acid   (d) aromatic

23.24 Propose structures for molecules that meet the following descriptions.
(a) a ketone with the formula $C_5H_{10}O$
(b) an ester with the formula $C_6H_{12}O_2$
(c) a compound with formula $C_2H_5NO_2$ that is both an amine and a carboxylic acid

23.25 There are three isomers with the formula C_3H_8O. Draw their structures.

23.26 Write structures for each of the following molecular formulas. You may have to use rings and/or multiple bonds in some instances.
(a) C_2H_7N (b) C_4H_8 (c) C_2H_4O (d) CH_2O_2

23.27 How many isomers can you write that fit the following descriptions?
(a) alcohols with formula $C_4H_{10}O$
(b) amines with formula C_3H_9N
(c) ketones with formula $C_5H_{10}O$
(d) aldehydes with formula $C_5H_{10}O$

23.28 Identify the functional groups in the following molecules.

(a)

Retinal (Vitamin A)

(b)

Estrone, a female sex hormone

ALKANES

23.29 What is the difference between a straight-chain alkane and a branched-chain alkane?

23.30 What kind of hybrid orbitals does carbon use in forming alkanes?

23.31 If someone reported the preparation of a compound with the formula C_3H_9, most chemists would be skeptical. Why?

23.32 What is wrong with each of the following structures?
(a) CH_3=$CHCH_2CH_2OH$
(b)

$$CH_3CH_2CH=\overset{\overset{\displaystyle O}{\|}}{C}CH_3$$

(c) $CH_3CH_2C{\equiv}CH_2CH_3$

23.33 What are the IUPAC names of the following alkanes?

(a)

$$CH_3CH_2CH_2CH_2\overset{\overset{\displaystyle CH_2CH_3}{|}}{C}H\underset{\underset{\displaystyle CH_3}{|}}{C}HCH_2CH_3$$

(b)

$$CH_3CH_2CH_2\overset{\overset{\displaystyle CH_3CHCH_3}{|}}{C}HCH_2\underset{\underset{\displaystyle CH_3}{|}}{C}HCH_3$$

(c)

$$CH_3\overset{\overset{\displaystyle CH_3}{|}}{C}CH_2CH_2CH_2\underset{\underset{\displaystyle CH_3}{|}}{C}HCH_3$$

(d)

$$CH_3CH_2CH_2\overset{\overset{\displaystyle CH_2CH_2CH_2CH_3}{|}}{C}CH_3$$
$$\underset{\underset{\displaystyle CH_2CH_3}{|}}{}$$

23.34 Write condensed structures for each of the following compounds:
(a) 3-ethylhexane
(b) 2,2,3-trimethylpentane
(c) 3-ethyl-3,4-dimethylheptane
(d) 5-isopropyl-2-methyloctane

23.35 The following compound, known commonly as isooctane, is important as a reference substance for determining the octane rating of gasoline. What is the IUPAC name of isooctane?

$$CH_3\overset{\overset{\displaystyle CH_3}{|}}{C}CH_2\overset{\overset{\displaystyle CH_3}{|}}{C}HCH_3 \quad \text{Isooctane}$$
$$\underset{\underset{\displaystyle CH_3}{|}}{}$$

23.36 Give IUPAC names for each of the five isomers with the formula C_6H_{14}.

23.37 Draw structures corresponding to the following IUPAC names.
(a) cyclooctane
(b) 1,1-dimethylcyclopentane
(c) 1,2,3,4-tetramethylcyclobutane
(d) 4-ethyl-1,1-dimethylcyclohexane

23.38 Give IUPAC names for each of the following cycloalkanes.

(a)

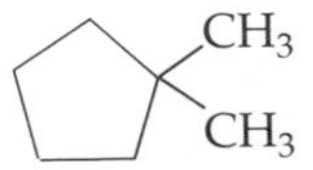

(b)

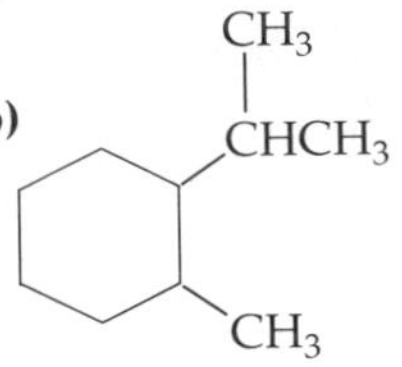

(c)

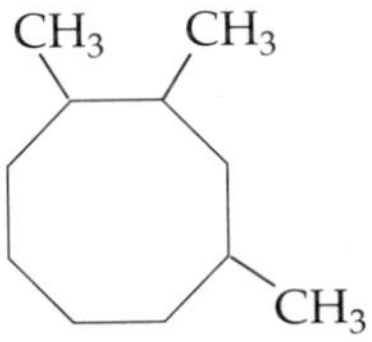

23.39 The following names are incorrect. What is wrong with each, and what are the correct names?

(a)

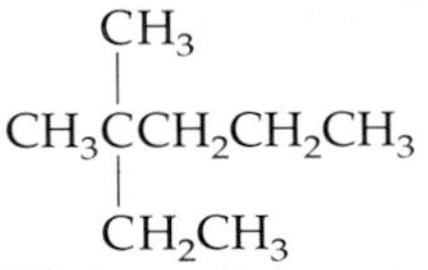

$$CH_3$$
$$CH_3CCH_2CH_2CH_3$$
$$CH_2CH_3$$

4-Ethyl-4-methylpentane

(b)

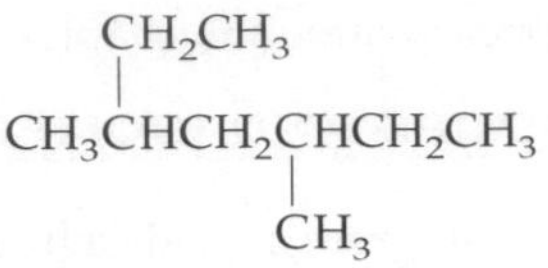

$$CH_2CH_3$$
$$CH_3CHCH_2CHCH_2CH_3$$
$$CH_3$$

5-Ethyl-3-methylhexane

(c)

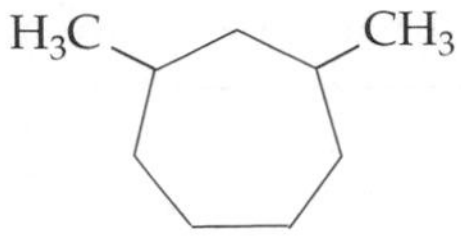

1,4-Dimethylcyclooctane

23.40 Draw structures and give IUPAC names for the nine isomers of C_7H_{16}.

23.41 Write a balanced equation for the combustion of butane.

23.42 Write the formulas of all monochlorinated substitution products that might result from a substitution reaction of the following substances with Cl_2.
(a) hexane
(b) 3-methylpentane
(c) methylcyclohexane

23.43 Which of the following reactions is likely to have a higher yield? Explain.

(a)

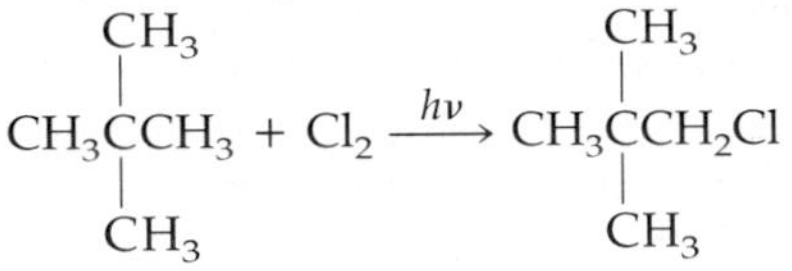

$$CH_3 \qquad\qquad CH_3$$
$$CH_3CCH_3 + Cl_2 \xrightarrow{h\nu} CH_3CCH_2Cl$$
$$CH_3 \qquad\qquad CH_3$$

(b)

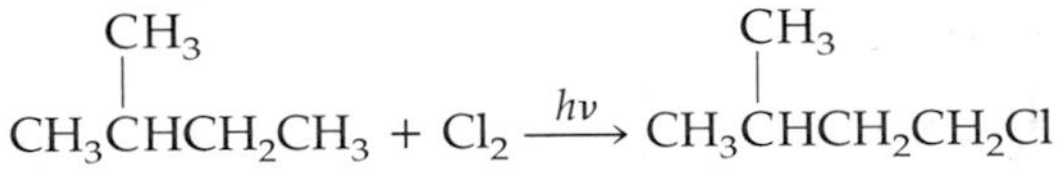

$$CH_3 \qquad\qquad\qquad CH_3$$
$$CH_3CHCH_2CH_3 + Cl_2 \xrightarrow{h\nu} CH_3CHCH_2CH_2Cl$$

ALKENES, ALKYNES, AND AROMATIC COMPOUNDS

23.44 What kind of hybrid orbitals does carbon use in forming (a) double bonds, (b) triple bonds, and (c) aromatic rings?

23.45 Why are alkenes, alkynes, and aromatic compounds said to be unsaturated?

23.46 Not all compounds that smell nice are called "aromatic," and not all compounds called "aromatic" smell nice. Explain.

23.47 What is meant by the term *addition reaction*?

23.48 Write structural formulas for compounds that meet the following descriptions:
(a) an alkene with five carbons
(b) an alkyne with four carbons
(c) a substituted aromatic compound with eight carbons

23.49 How many dienes (compounds with two double bonds) are there with the formula C_5H_8? Draw structures of as many as you can.

23.50 Give IUPAC names for the following compounds:

(a)

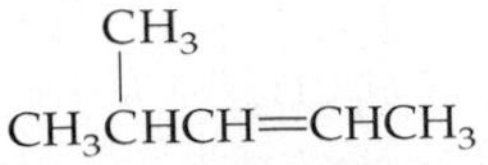

$$CH_3$$
$$CH_3CHCH=CHCH_3$$

(b)

$$CH=CH_2$$
$$CH_3CH_2CHCH_3$$

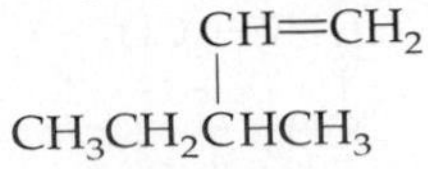

(c)

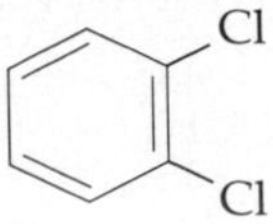

(d)

$$CH_3CH=CCH_3$$
with CH_3 above the second carbon

(e)

$$CH_3CH_2C\equiv CCH_2CH_2CHCH_3$$
with CH_3 above the terminal carbon

23.51 Draw structures corresponding to the following IUPAC names:

(a) cis-2-hexene   (b) 2-methyl-3-hexene

(c) 2-methyl-1,3-butadiene

23.52 Excluding cis-trans isomers, there are five alkenes with the formula C_5H_{10}. Draw structures for as many as you can, and give their IUPAC names.

23.53 Which of the alkenes in Problem 23.52 can exist as cis-trans isomers?

23.54 There are three alkynes with the formula C_5H_8. Draw and name them.

23.55 The following names are incorrect by IUPAC rules. Draw the structures represented and give the correct names.

(a) 2-methyl-4-hexene
(b) 5,5-dimethyl-3-hexyne
(c) 2-butyl-1-propene
(d) 1,5-diethylbenzene

23.56 Why is cis-trans isomerism possible for alkenes but not for alkanes or alkynes?

23.57 Which of the following compounds are capable of cis-trans isomerism?

(a) 1-hexene   (b) 2-hexene   (c) 3-hexene

23.58 Which of the following compounds are capable of cis-trans isomerism?

(a)

$$CH_3CHCH=CHCH_3$$
with CH_3 above the second carbon

(b)

$$CH_3CH_2CHCH_3$$
with $CH=CH_2$ above the third carbon

(c)

$$CH_3CH=CHCHCH_2CH_3$$
with Cl above the fourth carbon

23.59 Draw structures of the following compounds, indicating the cis or trans geometry of the double bond if necessary.

(a) *cis*-3-heptene
(b) *cis*-4-methyl-2-pentene
(c) *trans*-2,5-dimethyl-3-hexene

23.60 Why do you suppose small-ring cycloalkenes such as cyclohexene don't exist as cis-trans isomers?

23.61 Write equations for the reaction of 2,3-dimethyl-2-butene with each of the following reagents.

(a) H_2 and Pd catalyst
(b) Br_2
(c) H_2O and H_2SO_4 catalyst

23.62 Write equations for the reaction of 2-methyl-2-butene with the reagents given in Problem 23.61.

23.63 Write equations for the reaction of *p*-dichlorobenzene with the following reagents.

(a) Br_2 and $FeBr_3$ catalyst
(b) HNO_3 and H_2SO_4 catalyst
(c) H_2SO_4 and SO_3
(d) Cl_2 and $FeCl_3$ catalyst

23.64 Benzene and other aromatic compounds don't normally react with hydrogen in the presence of a palladium catalyst. If very high pressures (200 atm) and high temperatures are used, however, benzene will add three molecules of H_2 to give an addition product. What is a likely structure for the product?

23.65 The explosive trinitrotoluene (TNT) is made by three successive nitration reactions on toluene (methylbenzene). If these nitrations take place in the ortho and para positions relative to the methyl group, what is the structure of TNT?

ALCOHOLS, AMINES, AND CARBONYL COMPOUNDS

23.66 Draw structures corresponding to the following names:

(a) 2,4-dimethyl-2-pentanol
(b) 2,2-dimethylcyclohexanol
(c) 5,5-diethyl-1-heptanol
(d) 3-ethyl-3-hexanol

23.67 Draw structures corresponding to the following names.

(a) propylamine
(b) diethylamine
(c) *N*-methylpropylamine

23.68 Show the products from reaction of the amines in Problem 23.67 with aqueous HCl.

23.69 Assume that you had samples of quinine, an amine, and menthol, an alcohol. What simple chemical test could you do to distinguish between them?

23.70 Assume that you're given samples of pentanoic acid and methyl butanoate, both of which have the formula $C_5H_{10}O_2$. Describe how you can tell them apart.

23.71 What is the structural difference between an aldehyde and a ketone?

23.72 How do aldehydes and ketones differ from carboxylic acids, esters, and amides?

23.73 What general kind of reaction do carboxylic acids, esters, and amides undergo?

23.74 Identify the kinds of carbonyl groups in the following molecules (aldehyde, amide, ester, or ketone).

(a)

(b) $CH_3CH_2CH_2CHO$

(c)
$$CH_3CHCH_2COCH_3$$
with CH_3 above the second carbon

(d)
a cyclopentane ring with $CONH_2$ substituent

(e)
$$CH_3CHCH_2COOCH_3$$
with CH_3 above the second carbon

23.75 Write the equation for the dissociation of benzoic acid in water. If the K_a of benzoic acid is 6.5×10^{-5}, what is the extent of its dissociation in a 1.0 M solution?

23.76 How many grams of NaOH would it take to neutralize 5.0 g of benzoic acid?

23.77 Assume that you have a sample of acetic acid ($pK_a = 4.75$) dissolved in water:
   (a) Draw the structure of the major species present in the water solution.
   (b) Now assume that aqueous HCl is added to the acetic acid solution until pH 2 is reached. Draw the structure of the major species present.
   (c) Finally, assume that aqueous NaOH is added to the acetic acid solution until pH 12 is reached. Draw the structure of the major species present.

23.78 Draw and name compounds that meet the following descriptions:
   (a) three different amides with the formula $C_5H_{11}NO$
   (b) three different esters with the formula $C_6H_{12}O_2$

23.79 Give the IUPAC names of the following compounds.
   (a)
$$CH_3CHCH_2CH_2COCH_3$$
with CH_3 above the second carbon and O above the carbonyl

   (b)
$$CH_3CCH_2CH_2COH$$
with CH_3 above and below the second carbon, O above the carbonyl

(c)
$$CH_3CH_2CH_2CHCNH_2$$
with O above the carbonyl and CH_3 below

23.80 Draw structures corresponding to the following IUPAC names.
   (a) methyl pentanoate
   (b) isopropyl 2-methylbutanoate
   (c) cyclohexyl acetate

23.81 Write equations showing how you could prepare each of the esters in Problem 23.80 from the appropriate alcohols and carboxylic acids.

23.82 Draw structures corresponding to the following IUPAC names.
   (a) 3-methylpentanamide
   (b) *N*-phenylacetamide
   (c) *N*-ethyl-*N*-methylbenzamide

23.83 Write equations showing how you could prepare each of the amides in Problem 23.82 from the appropriate amines and carboxylic acids.

23.84 Novocaine, a local anesthetic, has the following structure. Identify the functional groups present in novocaine, and show the structures of the alcohol and carboxylic acids you would use to prepare it.

$$H_2N-\!\!\!\bigcirc\!\!\!-C-OCH_2CH_2NCH_2CH_3$$
with O above the carbonyl and CH_2CH_3 above the N

Novocaine

23.85 Ordinary soap is a mixture of the sodium or potassium salts of long-chain carboxylic acids that arise from saponification of animal fat. Draw the structures of soap molecules produced in the following reaction:

$$
\begin{aligned}
&CH_2OC(CH_2)_{14}CH_3\\
&CHOC(CH_2)_7CH\!=\!CH(CH_2)_7CH_3 + 3\ KOH \rightarrow\\
&CH_2OC(CH_2)_{16}CH_3
\end{aligned}
$$
A fat

POLYMERS

23.86 What is the difference between a monomer and a polymer?

23.87 Show the structure of poly(vinyl chloride) by drawing several repeating units. Vinyl chloride is $H_2C\!=\!CHCl$.

23.88 Show the monomer units you would use to prepare the following alkene polymers:

(a)

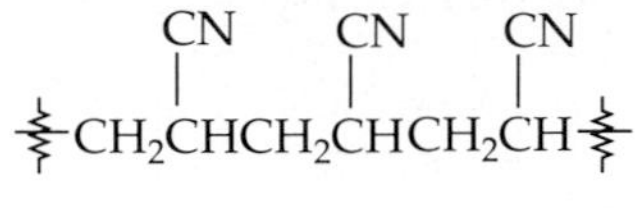

$$\text{+CH}_2\text{CHCH}_2\text{CHCH}_2\text{CH+}$$
with CN, CN, CN substituents

(b)

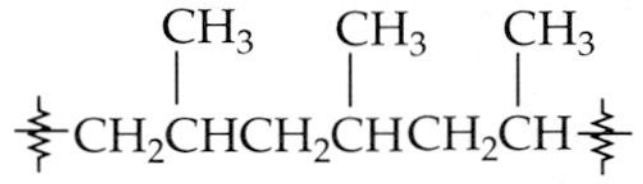

$$\text{+CH}_2\text{CHCH}_2\text{CHCH}_2\text{CH+}$$
with CH_3, CH_3, CH_3 substituents

(c)

$$\text{+CH}_2\text{CCH}_2\text{CCH}_2\text{C+}$$
with Cl, Cl, Cl above and Cl, Cl, Cl below

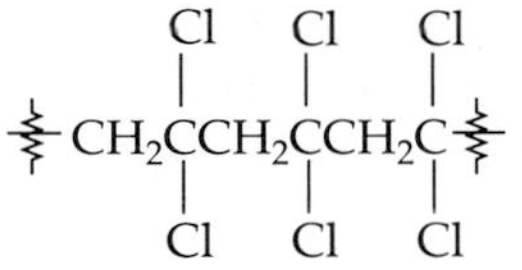

23.89 Show the structures of the polymers you would obtain from the following monomers:
 (a) $F_2C=CF_2$ gives Teflon
 (b) $H_2C=CHCOOCH_3$ gives Lucite

23.90 Kevlar, a nylon polymer prepared by reaction of 1,4-benzenedicarboxylic acid (terephthalic acid) with *p*-diaminobenzene, is so strong that it's used to make bulletproof vests. Draw the structure of a segment of Kevlar.

23.91 Draw the structure of the polyester that results from reaction of ethylene glycol ($HOCH_2CH_2OH$) with butanedioic acid ($HOOCCH_2CH_2COOH$).

GENERAL PROBLEMS

23.92 Draw structural formulas for the following:
 (a) 2-methylheptane
 (b) 4-ethyl-2-methylhexane
 (c) 4-ethyl-3,4-dimethyloctane
 (d) 2,4,4-trimethylheptane
 (e) 1,1-dimethylcyclopentane
 (f) 4-isopropyl-3-methylheptane

23.93 Give IUPAC names for the following alkanes:
 (a)
$$CH_3CH_2CH_2CHCHCH_3$$
with CH_3 and CH_3

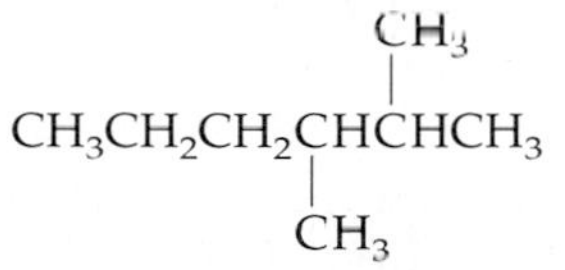

 (b)
$$CH_3CH_2CH_2CHCHCH_3$$
with CH_3 and $CH_2CH_2CH_2CH_3$

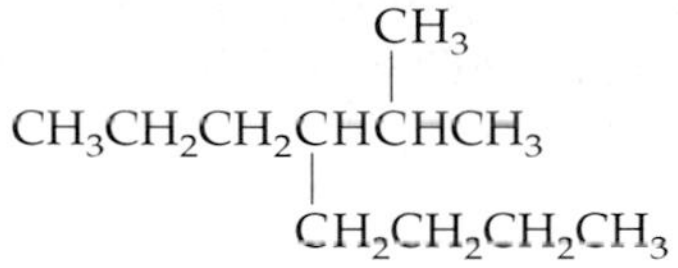

 (c)
$$CH_3CHCH_2-C-CH_3$$
with CH_3, CH_2CH_3, CH_2CH_3

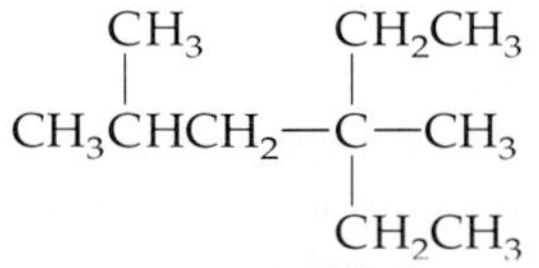

 (d)
$$CH_3CH_2-C-CH_2CH_3$$
with CH_2CH_3 and CH_2CH_3

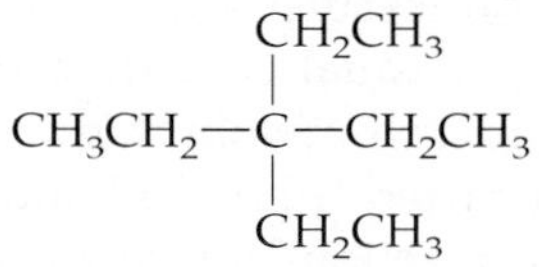

23.94 Assume that you have two unlabeled bottles, one with cyclohexane and one with cyclohexene. How could you tell them apart by doing chemical reactions?

23.95 Assume you have two unlabeled bottles, one with cyclohexene and one with benzene. How could you tell them apart by doing chemical reactions?

23.96 Write the products of the following reactions.
 (a)
$$CH_3CH_2CH=CHCHCH_3 \xrightarrow[\text{Pd catalyst}]{H_2}$$
with CH_3

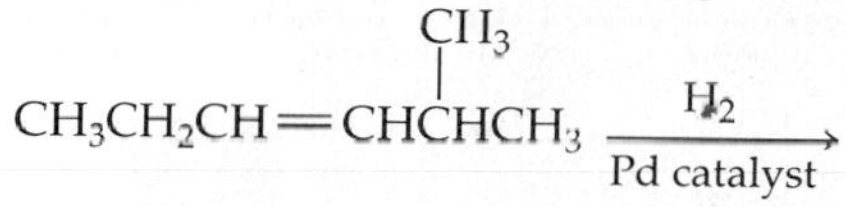

 (b)
$$Br-\text{⬡}-Br \xrightarrow[H_2SO_4]{HNO_3}$$

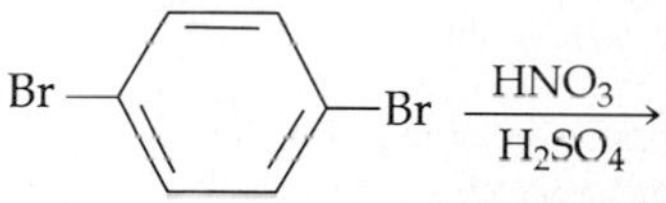

 (c)
ring with CH_3, CH_3 $\xrightarrow[H_2SO_4]{H_2O}$

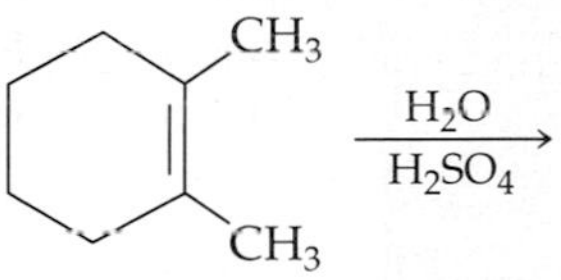

23.97 Show the structure of the nylon polymer that results from heating 6-aminohexanoic acid.

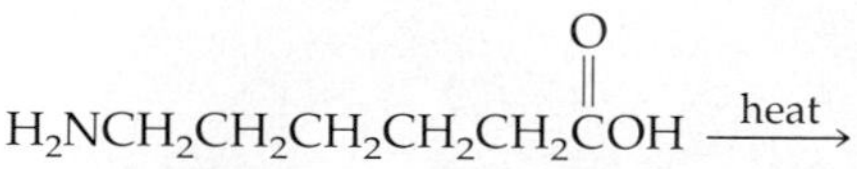

$$H_2NCH_2CH_2CH_2CH_2CH_2\overset{\displaystyle O}{\overset{\|}{C}}OH \xrightarrow{\text{heat}}$$

6-Aminohexanoic acid

chapter 24 BIOCHEMISTRY

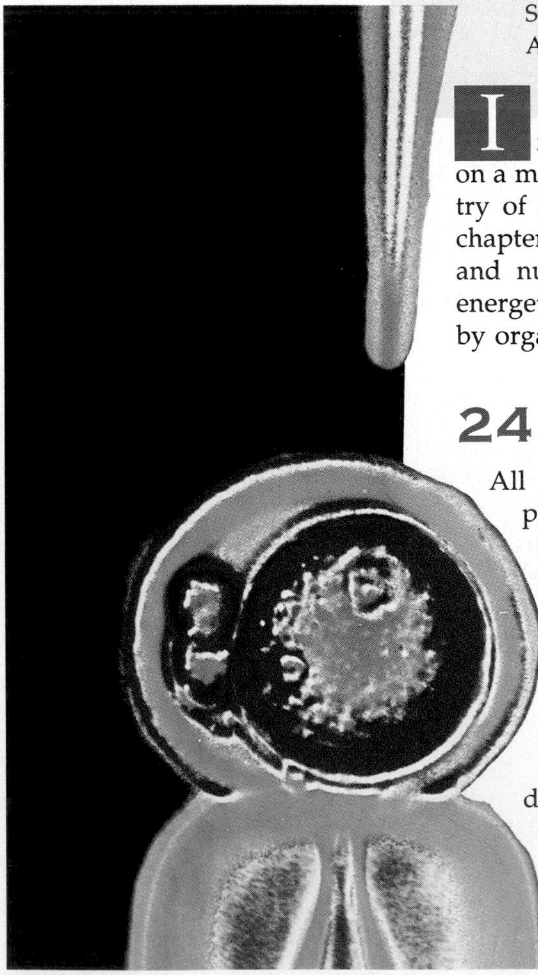

I f the ultimate goal of chemistry is to understand the world around us on a molecular level, then surely a knowledge of **biochemistry**, the chemistry of living organisms, is a crucial part of that goal. We'll look in this chapter at the main classes of biomolecules: proteins, carbohydrates, lipids, and nucleic acids. First, though, we'll take a brief look at biochemical energetics to see where biochemical energy comes from and how it is used by organisms.

24.1 ► BIOCHEMICAL ENERGY

All living things do mechanical work. Microorganisms engulf food, plants bend toward the sun, animals move about. Organisms also do the chemical work of synthesizing biomolecules needed for energy storage, growth, and repair. Even individual cells do work when they move molecules and ions across cell membranes.

In animals, it is the energy extracted from food and released in the exquisitely interconnected reactions of *metabolism* that allows work to be done. All animals are powered by the oxidation in cells of biomolecules containing mainly carbon, hydrogen, and oxygen. The end products of this biological oxidation are carbon dioxide, water, and energy, just as they are when an organic fuel such as methane is burned with oxygen in a furnace.

The microinjection of DNA into a single mouse embryo is a powerful technique in biochemical investigations.

$$C, H, O \text{ (food molecules)} + O_2 \rightarrow CO_2 + H_2O + \text{energy}$$

Ultimately, the energy used by living organisms comes from the sun. Photosynthesis in plants converts sunlight to potential energy stored mainly in the chemical bonds of carbohydrates. Plant-eating animals use some of this energy for living and store the rest of it, mainly in the chemical bonds of fats. Other animals, including ourselves, then eat plants and other animals and make use of the chemical energy these organisms have stored.

The energy used by living organisms ultimately comes from the sun. Plants use solar energy for the photosynthesis of glucose from CO_2, and animals then eat the plants.

The sum of the many organic reactions that go on in cells is called **metabolism**. These reactions usually occur in long sequences, which may be either linear or cyclic. In a linear sequence, the product of one reaction serves as the starting material for the next. In a cyclic sequence, a series of reactions regenerates the first reactant.

A linear
sequence

$$A \rightarrow B \rightarrow C \rightarrow D \rightarrow \cdots$$

A cyclic
sequence

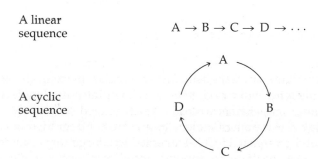

Those reaction sequences that break molecules apart are known collectively as **catabolism**, while those that put building blocks back together to assemble larger molecules are known as **anabolism**. Catabolic reactions generally release energy that is used to power living organisms, whereas anabolic reactions generally absorb energy. The overall picture of catabolism and energy production can be roughly divided into the four stages shown in Figure 24.1.

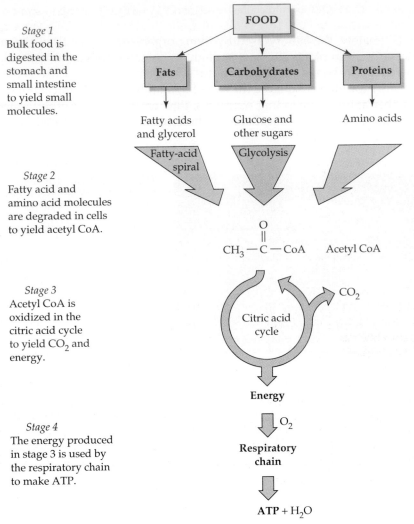

Stage 1
Bulk food is digested in the stomach and small intestine to yield small molecules.

Stage 2
Fatty acid and amino acid molecules are degraded in cells to yield acetyl CoA.

Stage 3
Acetyl CoA is oxidized in the citric acid cycle to yield CO_2 and energy.

Stage 4
The energy produced in stage 3 is used by the respiratory chain to make ATP.

FIGURE 24.1 An overview of catabolic pathways in the four stages of food degradation and the production of biochemical energy.

The first stage of catabolism, *digestion*, takes place in the stomach and small intestine when bulk food is broken down into small molecules such as simple sugars, long-chain carboxylic acids (called *fatty acids*), and amino acids. In stage 2, these small molecules are further degraded in cells to yield 2-carbon acetyl groups (CH_3CO-) attached to a large carrier molecule called *coenzyme A*. The resultant compound, *acetyl coenzyme A (acetyl CoA)*, is an intermediate in the breakdown of all main classes of food molecules.

Acetyl groups are oxidized in the third stage of catabolism, the *citric acid cycle*, to yield carbon dioxide and water. This stage releases a great deal of energy that is used in stage 4, the *respiratory chain*, to make molecules of **adenosine triphosphate (ATP)**. ATP, the final product of food catabolism, plays a pivotal role in the production of biological energy. As the molecule that stores the energy needed to power all life processes, ATP has been called the "energy currency of the living cell." Catabolic reactions "pay off" in ATP by synthesizing it from adenosine diphosphate (ADP) plus hydrogen phosphate ion, $HPO_4{}^{2-}$. Anabolic reactions "spend" ATP by transferring a phosphate group to other molecules, thereby regenerating ADP. The entire process of energy production thus revolves around the $ATP \rightleftarrows ADP$ interconversion.

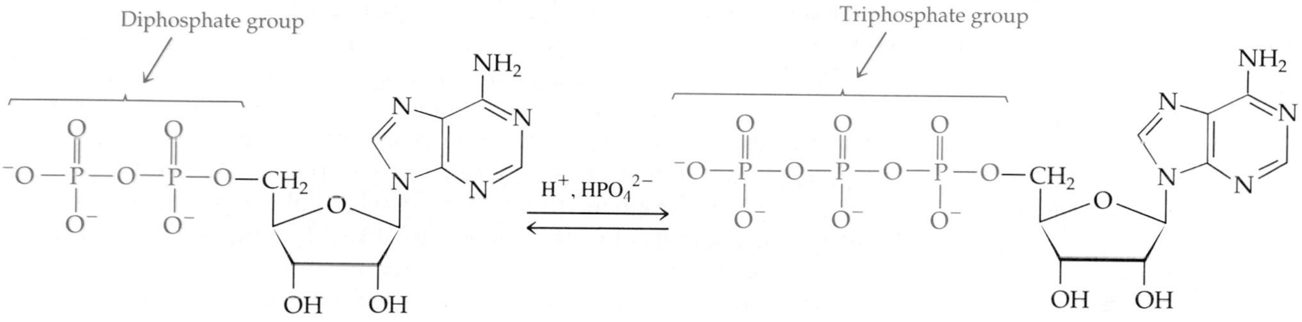

Adenosine diphosphate (ADP)   Adenosine triphosphate (ATP)

Because the primary metabolic function of ATP is to store energy, biochemists often refer to it as a "high-energy molecule" or an "energy storehouse." We don't mean by this that ATP is somehow different from other compounds; we mean only that ATP releases a large amount of energy when its P–O–P (*phosphoric anhydride*) bonds are broken and a phosphate group is transferred.

What does the body do with ATP? Recall from Section 8.13 that a chemical reaction is favorable, or spontaneous, if free energy, G, is released during the process. Conversely, a reaction is unfavorable if free energy is absorbed. The change in free energy (ΔG) depends on two factors: the release or absorption of heat (ΔH) and the increase or decrease in entropy (ΔS). The larger the amount of heat released and the greater the increase in entropy, the more favorable the reaction.

$$\Delta G = \Delta H - T\Delta S$$

Reactions in living organisms are no different from laboratory reactions in test tubes. Both follow the same laws, and both have the same kinds of energy requirements. For any biochemical reaction to occur spontaneously, free energy must be released, meaning that ΔG must be negative. For example, oxidation of glucose, the principal source of energy for animals, releases 2870 kJ per mole of glucose.

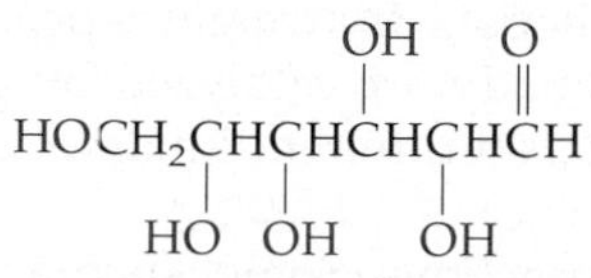

$$HOCH_2CHCHCHCHCH + 6\ O_2 \rightarrow 6\ CO_2 + 6\ H_2O \qquad \Delta G° = -2870\ kJ$$

Glucose ($C_6H_{12}O_6$)

Reactions in which free energy is absorbed can also occur, but such reactions can't be spontaneous and must have positive values of ΔG. An example is the conversion of glucose to glucose 6-phosphate, an important step in the breakdown of glucose derived from carbohydrates in the diet:

$$\underset{\substack{| \quad \quad | \quad \quad |\\ \text{HO} \quad \text{OH} \quad \text{OH}}}{\overset{\substack{\text{OH} \quad \text{O}\\ | \quad \quad ||}}{\text{HOCH}_2\text{CHCHCHCHCH}}} \xrightarrow[-\text{H}_2\text{O}]{\text{HPO}_4{}^{2-}} \underset{\substack{| \quad \quad \quad \quad | \quad \quad | \quad \quad |\\ \text{O}^- \quad \quad \text{HO} \quad \text{OH} \quad \text{OH}}}{\overset{\substack{\text{O} \quad \quad \quad \quad \text{OH} \quad \text{O}\\ || \quad \quad \quad \quad | \quad \quad ||}}{{}^-\text{OPOCH}_2\text{CHCHCHCHCH}}} \qquad \Delta G° = +13.8 \text{ kJ}$$

Glucose   Glucose 6-phosphate

What usually happens for an energetically unfavorable reaction to occur is that it is "coupled" to an energetically favorable reaction so that the *overall* free-energy change for the two reactions together is favorable. For example, the reaction of glucose with hydrogen phosphate ion ($HPO_4{}^{2-}$) to yield glucose 6-phosphate plus water doesn't take place spontaneously because it is energetically unfavorable by 13.8 kJ. When ATP is involved, however, the reaction of ATP with water to yield ADP plus hydrogen phosphate ion is energetically *favorable* by about 30.5 kJ. Thus, if the two reactions are coupled, the overall process for the synthesis of glucose 6-phosphate is favorable by about 16.7 kJ. That is, the reaction of ATP with water releases more than enough free energy to "drive" the unfavorable reaction of glucose with hydrogen phosphate:

glucose + $HPO_4{}^{2-} \rightarrow$ glucose 6-phosphate + H_2O	$\Delta G° = +13.8$ kJ
ATP + $H_2O \rightarrow$ ADP + $HPO_4{}^{2-}$ + H^+	$\Delta G° = -30.5$ kJ
glucose + ATP $\rightarrow$ glucose 6-phosphate + ADP + H^+	$\Delta G° = -16.7$ kJ

It's this ability to drive otherwise unfavorable reactions that makes ATP so useful. In fact, most of the thousands of reactions going on in your body every minute are powered by energy from ATP. It's no exaggeration to say that the transfer of a phosphate group from ATP is the one chemical reaction that makes life possible.

PROBLEM 24.1 One of the steps in fat metabolism is the reaction of glycerol ($HOCH_2$–$CHOH$–CH_2OH) with ATP to yield glycerol 1-phosphate. Formulate the reaction, and draw the structure of glycerol 1-phosphate.

24.2 ➤AMINO ACIDS AND PEPTIDES

Taken from the Greek *proteios*, meaning "primary," the name *protein* aptly describes a group of biological molecules that are of primary importance to all living organisms. Approximately 50% of the body's dry weight is protein, and almost all the reactions that occurs in the body are catalyzed by proteins. In fact, a human body contains over *100,000* different kinds of proteins.

Proteins have many different biological functions. Some, such as the keratin in skin, hair, and fingernails, serve a structural purpose. Others, such as the insulin that controls glucose metabolism, act as hormones to regulate specific body processes. And still other proteins, such as DNA polymerase, are enzymes, the biological catalysts that carry out body chemistry. Regardless of their function, all **proteins** are biological polymers made up of many amino acids linked together to form a long chain. As their name implies,

Bird feathers are made largely of the protein *keratin*.

amino acids are molecules that contain two functional groups, a basic amino group ($-NH_2$) and an acidic $-COOH$ group. Alanine is one of the simplest examples.

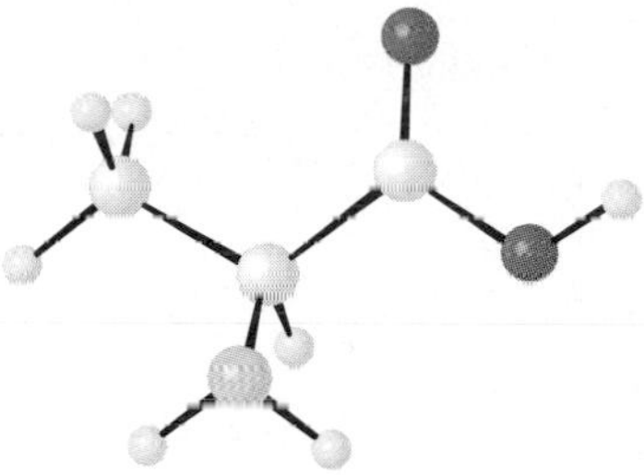

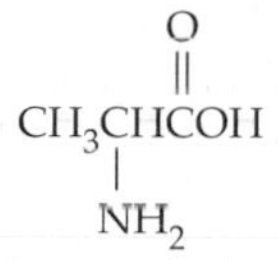

$$\underset{\substack{| \\ NH_2}}{CH_3CHCOH}\overset{\displaystyle O}{\overset{\displaystyle \|}{}}$$

Alanine   an amino acid

Two or more amino acids can link together by forming amide bonds (Section 23.14), usually called **peptide bonds**. A *dipeptide* results from the linking together of two amino acids by formation of a peptide bond between the $-NH_2$ group of one and the $-COOH$ group of the second. In the same way, a *tripeptide* results from linking three amino acids by two peptide bonds, and so on. Shorter chains of up to 100 amino acids are usually called **polypeptides**, while the term *protein* is reserved for longer chains.

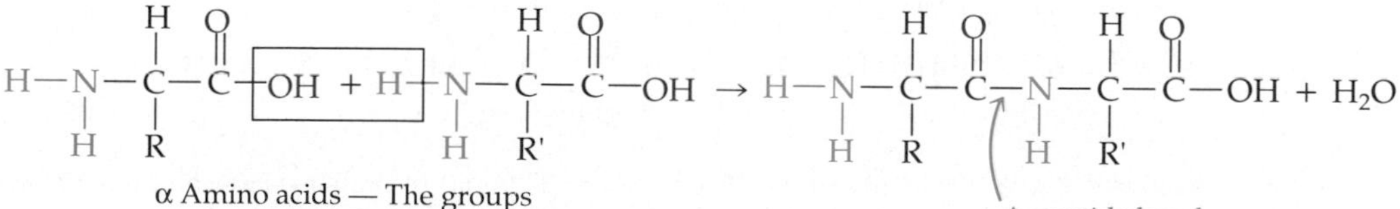

α Amino acids — The groups symbolized by R and R' represent different amino acid side chains.

A peptide bond

There are 20 different amino acids commonly found in proteins, as shown in Figure 24.2. Each amino acid is referred to by a three-letter shorthand code, such as Ala (alanine), Gly (glycine), Pro (proline), and so on. All 20 are called **alpha (α) amino acids** because the amino group in each is connected to the carbon atom *alpha to* (next to) the carboxylic acid group. Nineteen of the 20 have an $-NH_2$ amino group, but one (proline) has an $-NH-$ amino group as part of a ring.

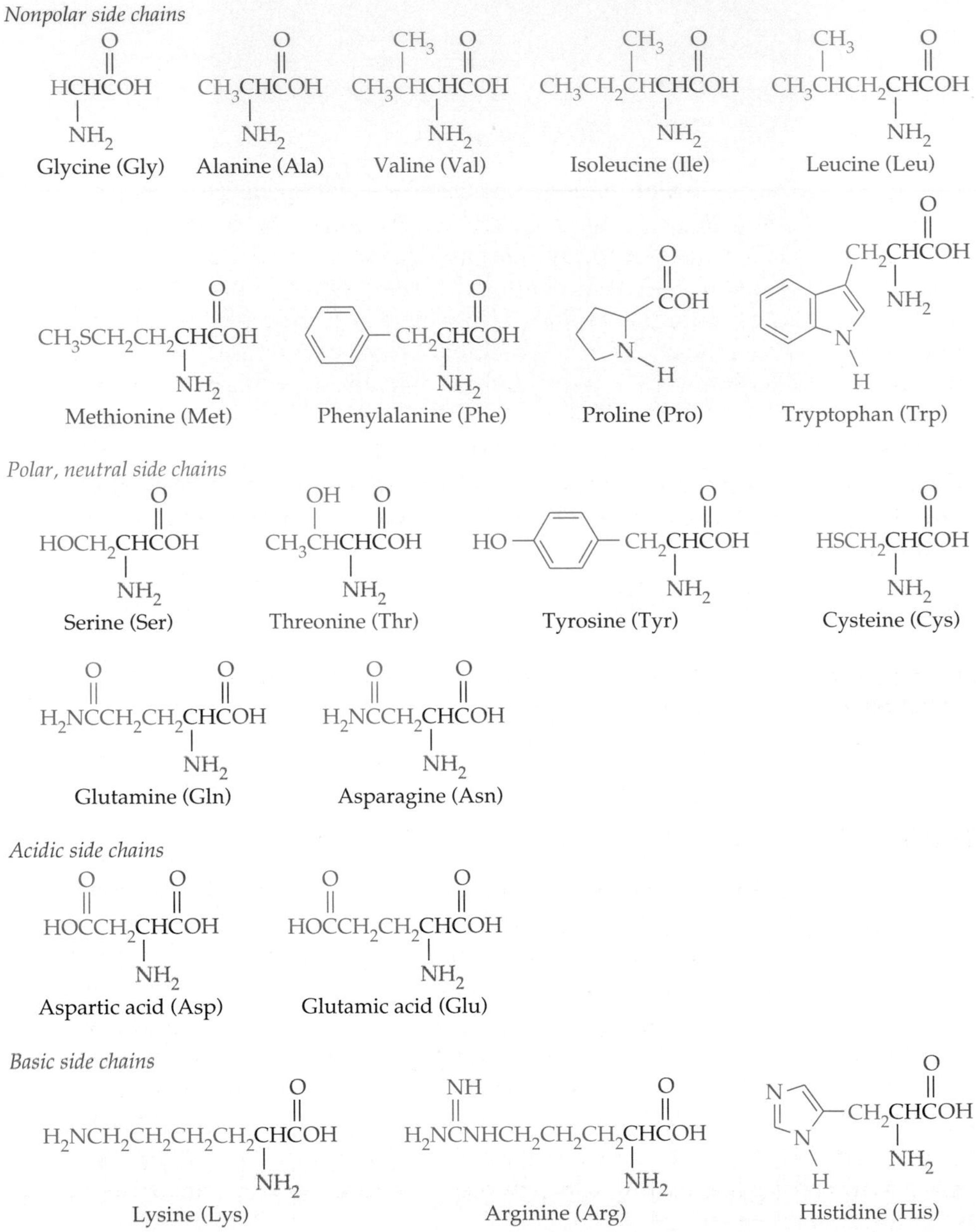

Nonpolar side chains

Glycine (Gly)   Alanine (Ala)   Valine (Val)   Isoleucine (Ile)   Leucine (Leu)

Methionine (Met)   Phenylalanine (Phe)   Proline (Pro)   Tryptophan (Trp)

Polar, neutral side chains

Serine (Ser)   Threonine (Thr)   Tyrosine (Tyr)   Cysteine (Cys)

Glutamine (Gln)   Asparagine (Asn)

Acidic side chains

Aspartic acid (Asp)   Glutamic acid (Glu)

Basic side chains

Lysine (Lys)   Arginine (Arg)   Histidine (His)

FIGURE 24.2 Structures of the 20 α-amino acids found in proteins. Fifteen of the 20 have neutral side chains, two have acidic side chains, and three have basic side chains.

The 20 amino acids differ only in the nature of the group attached to the α carbon. Usually called the *side chain*, this group can be symbolized in a general way by the letter R. (In a broader context, the symbol R is used throughout organic chemistry to refer to an organic fragment of unspecified structure.) Our bodies can synthesize only 10 of the 20 amino acids. The remaining 10, named in red in Figure 24.2, are called essential amino acids because they must be obtained from the diet.

α Carbon

R—CH—COH

Side chain

NH₂

O
‖

Generalized structure of
an α-amino acid

The 20 common amino acids can be classified as *neutral*, *basic*, or *acidic*, depending on the structure of their side chains. Fifteen of the 20 have neutral side chains. Two (aspartic acid and glutamic acid) have an additional carboxylic acid group in their side chains and are classified as acidic amino acids. Three (lysine, arginine, and histidine) have an additional amine function in their side chains and are classified as basic amino acids. The 15 neutral amino acids can be further divided into those with nonpolar side chains and those with polar functional groups such as amide or hydroxyl groups. Nonpolar side chains are often described as *hydrophobic* (water fearing) because they are repelled by water, while polar side chains are described as *hydrophilic* (water loving) because they are attracted to water.

PROBLEM 24.2 Which common amino acids contain an aromatic (benzenelike) ring? Which contain sulfur? Which are alcohols? Which have alkyl-group side chains?

PROBLEM 24.3 Draw alanine using wedges and dashes (Section 7.9) to show the tetrahedral geometry of its α carbon.

24.3 ➤AMINO ACIDS AND MOLECULAR HANDEDNESS

We saw in Section 20.9 that certain molecules, like a hand, lack a plane of symmetry and are therefore *chiral*. When held up to a mirror, a chiral molecule is not identical to its reflected image. Instead, the molecule and its mirror image have a right-hand/left-hand relationship. Compare alanine and propane, for example (Figure 24.3). An alanine molecule is chiral; it has no symmetry plane and can exist in two forms—a "right-handed" form called D-alanine (from *dextro*, Latin for "right") and a "left-handed" form called L-alanine (from *levo*, Latin for "left"). Propane, however, is achiral. It has a symmetry plane cutting through the three carbons and thus exists in only a single form.

Why are some organic molecules chiral but others aren't? The answer has to do with the three-dimensional nature of organic compounds. We've seen that carbon forms four bonds that are oriented toward the four corners of a regular tetrahedron. Whenever a carbon atom is bonded to four *different* atoms or groups of atoms, chirality results. If a carbon is bonded to two or more of the same groups, however, no chirality is possible. In alanine, for example, carbon 2 is bonded to four different groups: a –COOH group, an –H atom, an –NH₂ group, and a –CH₃ group. Thus, alanine is chiral. In propane, however, each of the three carbons is bonded to at least two groups—the –H atoms—that are identical. Thus, propane is achiral.

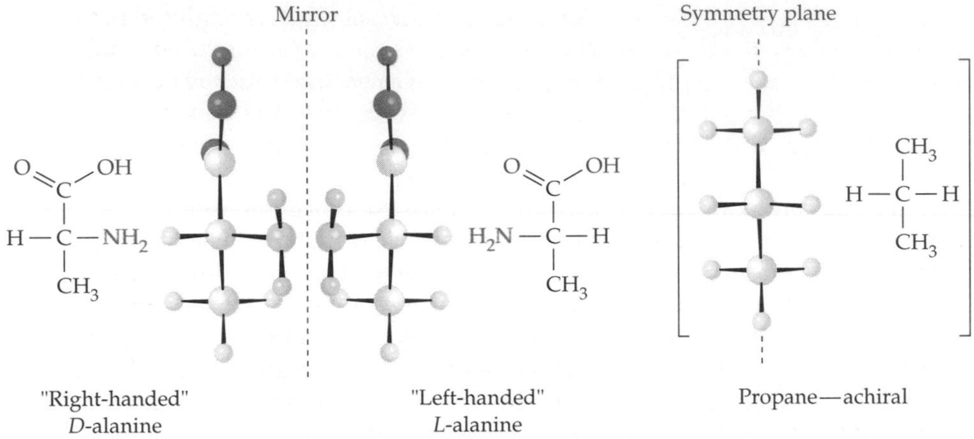

Mirror

Symmetry plane

$$O = C \diagup OH$$
$$H - C - NH_2$$
$$CH_3$$

$$O = C \diagup OH$$
$$H_2N - C - H$$
$$CH_3$$

$$CH_3$$
$$H - C - H$$
$$CH_3$$

"Right-handed"
D-alanine

"Left-handed"
L-alanine

Propane—achiral

FIGURE 24.3 Alanine (2-aminopropanoic acid) has no symmetry plane; the two "halves" of the molecule are not mirror images. Thus, alanine can exist in two forms—a "right-handed" form and a "left-handed" form. Propane, however, has a symmetry plane and is achiral.

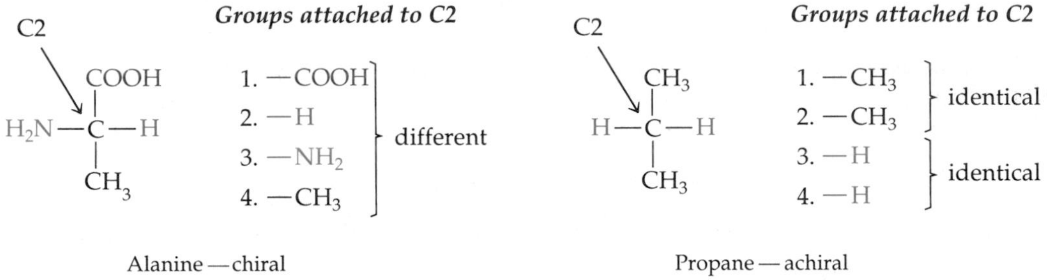

C2

$$\diagdown COOH$$
$$H_2N - C - H$$
$$CH_3$$

Groups attached to C2

1. —COOH
2. —H
3. —NH$_2$
4. —CH$_3$
} different

C2

$$\diagdown CH_3$$
$$H - C - H$$
$$CH_3$$

Groups attached to C2

1. —CH$_3$
2. —CH$_3$
} identical
3. —H
4. —H
} identical

Alanine—chiral

Propane—achiral

As mentioned in Section 20.9, the two mirror-image forms of a chiral molecule are called *enantiomers*, or *optical isomers*. The mirror-image relationship of the enantiomers of a compound with four different groups on a chiral carbon atom is shown in Figure 24.4.

Of the 20 common amino acids, 19 are chiral because they have four different groups bonded to their α carbons, –H, –NH$_2$, –COOH, and –R (the

FIGURE 24.4 A carbon atom that is bonded to four different groups is chiral and is not identical to its mirror image.

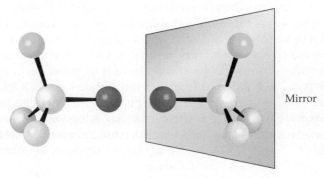

Mirror

side chain). Only glycine, H_2NCH_2COOH, is achiral. Even though the naturally occurring chiral α-amino acids can exist as pairs of enantiomers, only L-amino acids are found in proteins. When drawn with the –COOH group at the top and the side-chain R group at the bottom, an L-amino acid has its –NH_2 group coming out of the plane of the paper on the left side of the structure:

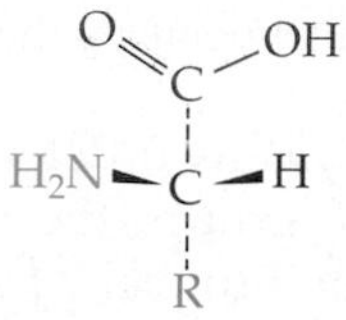

$$
\begin{array}{c}
O{=}C{-}OH \\
| \\
H_2N{-}C{-}H \\
| \\
R
\end{array}
$$
An L-amino acid

EXAMPLE 24.1

Lactic acid can be isolated from sour milk. Is lactic acid chiral?

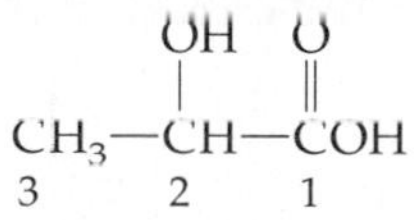

$$
\underset{3 \quad\quad 2 \quad\quad 1}{CH_3{-}\underset{\underset{OH}{|}}{CH}{-}\underset{\underset{O}{\|}}{C}OH}
\quad \text{Lactic acid}
$$

SOLUTION   To find out if lactic acid is chiral, list the groups attached to each carbon:

Groups on carbon 1	*Groups on carbon 2*	*Groups on carbon 3*
1. –OH	1. –COOH	1. –CH(OH)COOH
2. =O	2. –OH	2. –H
3. –CH(OH)CH₃	3. –H	3. –H
	4. –CH₃	4. –H

Next, look at the lists to see if any carbon is attached to four *different* groups. Of the three carbons, carbon 2 has four different groups, and lactic acid is therefore chiral.

▶ **PROBLEM 24.4** Which of the following objects are chiral?
(a) a glove   **(b)** a baseball   **(c)** a screw   **(d)** a nail

▶ **PROBLEM 24.5** 2-Aminopropane is an achiral molecule, but 2-aminobutane is chiral. Explain.

▶ **PROBLEM 24.6** Which of these molecules are chiral?
(a) 3-chloropentane   **(b)** 2-chloropentane   **(c)**

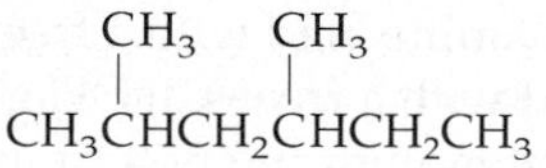

$$
\underset{}{CH_3\underset{\underset{CH_3}{|}}{CH}CH_2\underset{\underset{CH_3}{|}}{CH}CH_2CH_3}
$$

▶ **PROBLEM 24.7** Two of the 20 common amino acids have two chiral carbon atoms in their structures. Identify them.

24.4 ▶PROTEINS

We saw in Section 24.2 that proteins are amino acid polymers in which the individual amino acids, often called *residues*, are linked together by peptide (amide) bonds. The repeating chain of amide linkages to which the side chains are attached is called the *backbone*.

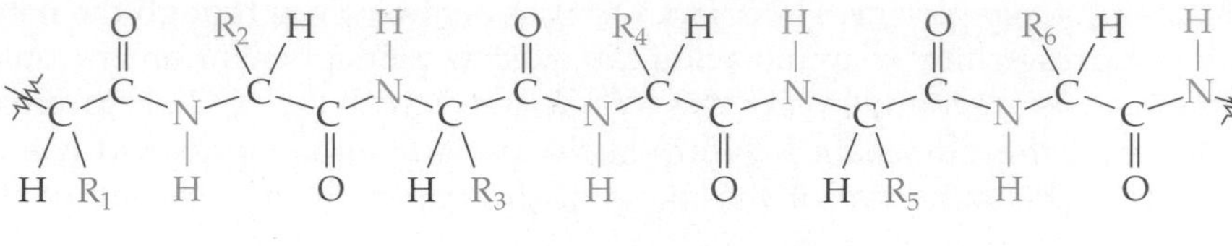

A segment of a protein backbone. The side-chain R groups
of the individual amino acids are substituents on the backbone.

Since amino acids can be assembled in any order, depending on which
–COOH group joins with which –NH$_2$ group, the number of possible
isomeric peptides increases rapidly as the number of amino acid residues
increases. There are six ways in which three amino acids can be joined, more
than 40,000 ways in which the eight amino acids present in the blood
pressure-regulating hormone angiotensin II can be joined (Figure 24.5), and
an inconceivably vast number of ways in which the *1800* amino acids in
myosin, the major component of muscle filaments, can be arranged.

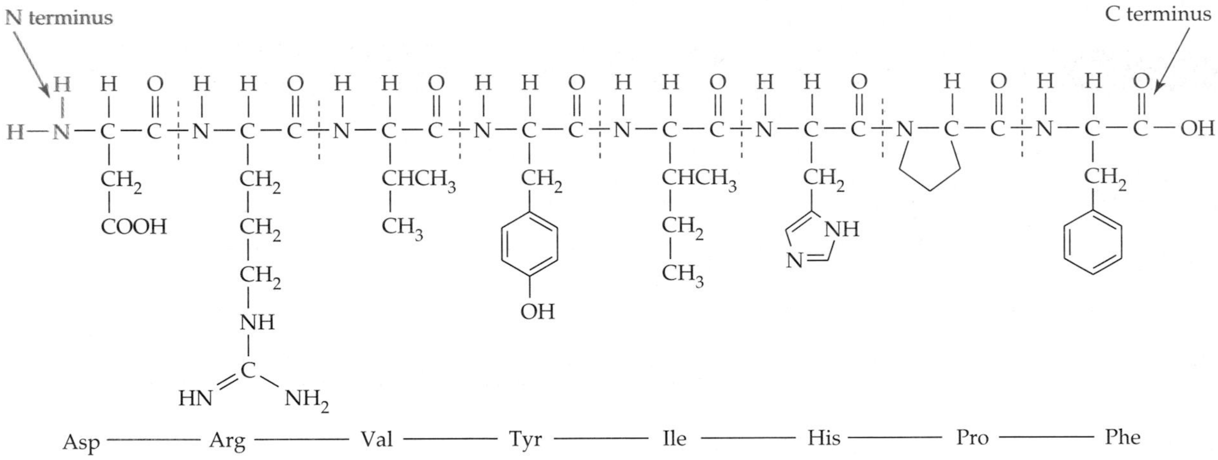

FIGURE 24.5 The structure of angiotensin II, an octapeptide in blood plasma that regulates
blood pressure.

No matter how long the chain, all noncyclic proteins have an **N-terminal amino acid** with a free –NH$_2$ group on one end and a **C-terminal amino
acid** with a free –COOH group on the other end. By convention, a protein is
written with the N-terminal residue on the left and the C-terminal residue
on the right, and its name is indicated with the three-letter abbreviations
listed in Figure 24.2. Thus, angiotensin II (Figure 24.5) is abbreviated Asp-
Arg-Val-Tyr-Ile-His-Pro-Phe.

Proteins are classified according to their three-dimensional shape as
either *fibrous* or *globular*. Fibrous proteins, such as collagen and the keratins,
consist of polypeptide chains arranged side by side in long filaments.
Because these proteins are tough and insoluble in water, they are used in
nature for structural materials like skin, tendons, hair, ligaments, and muscle. Globular proteins, by contrast, are usually coiled into compact, nearly
spherical shapes, as seen in the computer-generated picture of the peni-

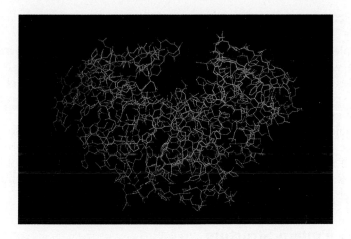

FIGURE   24.6 A computer-generated view of the penicillopepsin enzyme, a globular protein containing 323 amino acids.

cillopepsin enzyme in Figure 24.6. Most of the 2000 or so known enzymes present inside cells are globular proteins.

Another way to classify proteins is according to biological function. As indicated in Table 24.1, proteins have a remarkable diversity of roles.

TABLE 24.1   Some Biological Functions of Proteins

Type	Function	Example
Enzymes	Catalyze biological processes	Pepsin
Hormones	Regulate body processes	Insulin
Storage proteins	Store nutrients	Ferritin
Transport proteins	Transport oxygen and other substances through the body	Hemoglobin
Structural proteins	Form an organism's structure	Collagen
Protective proteins	Help fight infection	Antibodies
Contractile proteins	Form muscles	Actin and myosin
Toxic proteins	Serve as a defense for the plant or animal	Snake venoms

EXAMPLE 24.2

Draw the structure of the dipeptide Ala-Ser.

SOLUTION   First, look up the names and structures of the two amino acids, Ala (alanine) and Ser (serine). Since alanine is *N*-terminal and serine is *C*-terminal, Ala-Ser must have an amide bond between the alanine –COOH and the serine –NH$_2$.

N-terminal $\longrightarrow$ $\underset{\underset{CH_3}{|}}{H_2NCHC}\overset{\overset{O}{\|}}{}-\underset{\underset{CH_2OH}{|}}{NHCHC}\overset{\overset{O}{\|}}{OH}$ $\longleftarrow$ C-terminal

Alanine   Serine   Ala-Ser

PROBLEM 24.8 Use the three-letter shorthand notations to name the two isomeric dipeptides that can be made from valine and cysteine. Draw both structures.

PROBLEM 24.9 Name the six tripeptides that contain valine, tyrosine, and glycine.

24.5 ►LEVELS OF PROTEIN STRUCTURE

With molecular weights of up to one-half *million*, many proteins are so large that the word *structure* takes on a broader meaning with these immense molecules than it does with simple organic molecules. In fact, chemists usually speak about four levels of structure when describing proteins. The **primary structure** of a protein specifies the sequence in which the various amino acids are linked together. **Secondary structure** refers to how segments of the protein chain are oriented into a regular pattern. **Tertiary structure** refers to how the entire protein chain is coiled and folded into a specific three-dimensional shape, and **quaternary structure** refers to how several protein chains can aggregate to form a larger unit.

Primary Protein Structure

Primary structure is the most important of the four structural levels because it is the protein's amino acid sequence that determines its overall shape and function. So crucial is primary structure to function that the change of only one amino acid out of several hundred can drastically alter biological properties. The disease sickle-cell anemia, for example, is caused by a genetic defect in blood hemoglobin whereby valine is substituted for glutamic acid at only one position in a chain of 146 amino acids.

Secondary Protein Structure

When looking at the primary structure of a protein like angiotensin II in Figure 24.5, you might get the idea that the molecule is long and threadlike. In fact, though, most proteins are not threadlike. Most proteins fold in such a way that segments of the protein chain orient into regular patterns, called *secondary structures*. There are two common kinds of patterns: the α-helix and the β-pleated sheet.

α-Keratin, a fibrous protein found in wool, hair, fingernails, and feathers, wraps into a helical coil, much like the cord on a telephone (Figure 24.7). Called an α-**helix**, this secondary structure is stabilized by the formation of hydrogen bonds (Section 10.2) between the N–H group of one amino acid and the C=O group of another amino acid four residues away. Each coil of the helix contains 3.6 amino acid residues, with a distance between coils of 0.54 nm (5.4 Å).

Lamb's wool is made largely of keratin, a fibrous protein with an α-helical secondary structure.

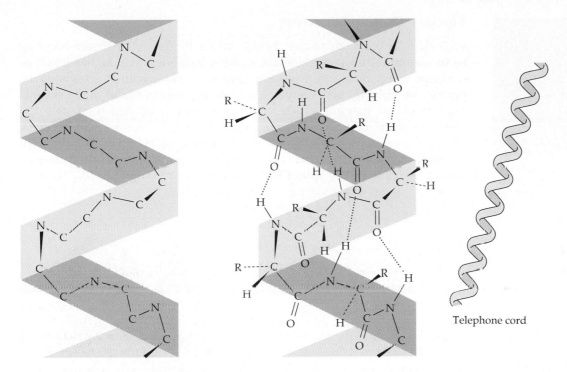

Telephone cord

FIGURE 24.7 The helical secondary structure of α-keratin. The amino acid backbone winds in a spiral, much like that of a telephone cord.

Fibroin, the fibrous protein found in silk, has a secondary structure called a β-**pleated sheet**, in which polypeptide chains line up in a parallel arrangement held together by hydrogen bonds (Figure 24.8). Although not as common as the α-helix, small β-pleated sheet regions are often found in proteins where sections of peptide chains double back on themselves.

Silk is made of fibroin, a fibrous protein with a pleated-sheet secondary structure.

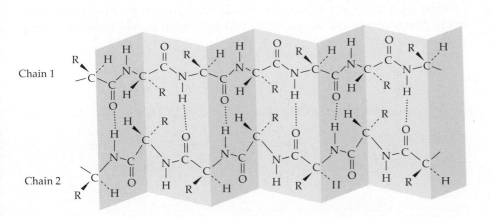

Chain 1

Chain 2

FIGURE 24.8 The β-pleated sheet secondary structure of silk fibroin.

The muscles of sea mammals contain myoglobin, a globular protein that stores oxygen.

Tertiary Protein Structure

Secondary protein structures result primarily from hydrogen-bonding between amide linkages along the protein backbone, but higher levels of structure result primarily from interactions of side-chain R groups in the protein. Myoglobin, for example, is a globular protein found in the skeletal muscles of sea mammals, where it stores oxygen needed to sustain the animals during long dives. With a single chain of 153 amino acid residues, myoglobin consists of eight straight segments, each of which adopts an α-helical secondary structure. These helical sections then fold up further to form a compact, nearly spherical, tertiary structure (Figure 24.9).

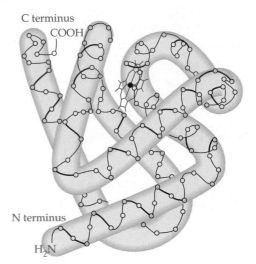

FIGURE 24.9 Secondary and tertiary structure of myoglobin, a globular enzyme found in the muscles of sea mammals.

The most important force stabilizing a protein's tertiary structure results from the hydrophobic interactions of hydrocarbon side chains on amino acids. Those amino acids with neutral, nonpolar side chains have a strong tendency to congregate on the hydrocarbon-like interior of a protein molecule, away from the aqueous medium. Those amino acids with polar side chains are usually found on the exterior of the protein, where they can be solvated by water. Also important for stabilizing a protein's tertiary structure are *disulfide bridges* (covalent S–S bonds formed between nearby cysteine residues), *salt bridges* (ionic attractions between positively and negatively charged sites on the protein), and hydrogen bonds between nearby amino acids.

$$O=C-CHCH_2SH \; + \; HSCH_2CH-C=O \quad \xrightarrow{[O]} \quad O=C-CHCH_2S-SCH_2CH-C=O$$

Two cysteine residues   A disulfide bridge

24.6 ➤ENZYMES

Enzymes are large proteins that act as catalysts for biological reactions. A *catalyst*, as we saw in Section 12.11, is an agent that speeds up the rate of a chemical reaction without itself undergoing change. For example, sulfuric acid catalyzes the reaction of a carboxylic acid with an alcohol to yield an ester (Section 23.14). Enzymes are similar to sulfuric acid in that they catalyze reactions that might otherwise occur very slowly, but they differ in two important respects. First, enzymes are far larger, more complicated molecules than simple inorganic catalysts. Second, enzymes are far more specific in their action. Sulfuric acid catalyzes the reaction of nearly *every* carboxylic acid with nearly *every* alcohol, but enzymes often catalyze only a *single* reaction of a single compound, called the enzyme's **substrate**. For example, the enzyme *amylase* found in human digestive systems is able to catalyze the breakdown of starch to yield glucose but has no effect on cellulose, even though starch and cellulose are structurally similar. Thus, humans can digest potatoes (starch) but not grass (cellulose).

$$\text{Starch} + H_2O \xrightarrow{\text{amylase}} \text{many glucose molecules}$$

$$\text{Cellulose} + H_2O \xrightarrow{\text{amylase}} \text{no reaction}$$

The catalytic activity of an enzyme is measured by its **turnover number**, which is defined as the number of substrate molecules acted on by one molecule of enzyme per unit time. As indicated in Table 24.2, enzymes vary greatly in their turnover number. Most enzymes have values in the 1 to 20,000 range, but carbonic anhydrase, which catalyzes the reaction of CO_2 with water to yield bicarbonate ion, acts on *600,000* substrate molecules per second.

TABLE 24.2   Turnover Numbers of Some Enzymes

Enzyme	Turnover Number (s^{-1})
Carbonic anhydrase	600,000
Acetylcholinesterase	25,000
β-Amylase	18,000
Penicillinase	2,000
DNA polymerase I	15

In addition to their protein part, many enzymes contain small, nonprotein parts called **cofactors**. In such enzymes, the protein part is called an **apoenzyme**, and the entire assembly of apoenzyme plus cofactor is called a **holoenzyme**. Only holoenzymes are active as catalysts; neither apoenzyme nor cofactor alone can catalyze a reaction.

$$\text{Holoenzyme} = \text{apoenzyme} + \text{cofactor}$$

An enzyme cofactor can be either an inorganic ion (usually a metal ion) or a small organic molecule called a **coenzyme**. (The requirement of many enzymes for metal-ion cofactors is the main reason behind our dietary need for trace minerals. Iron, zinc, copper, manganese, molybdenum, cobalt, nickel, and selenium are all essential trace elements that function as enzyme cofactors.) A large number of different organic molecules serve as coenzymes. Often, though not always, the coenzyme is a *vitamin*. Thiamine (vitamin B_1), for example, is a coenzyme required in the metabolism of carbohydrates.

Thiamine (vitamin B_1); an enzyme cofactor

How Enzymes Work: The Lock-And-Key Model

Our current picture of how enzymes work uses the **lock-and-key model** to explain both their specificity and their catalytic activity. In this model, an enzyme is pictured as a large, irregularly shaped molecule with a cleft or crevice in its middle. Inside the crevice is an **active site**, a small three-dimensional region with the specific shape necessary to bind the substrate and catalyze the appropriate reaction. In other words, the shape of the active site acts like a lock into which only a specific key (substrate) can fit (Figure 24.10). An enzyme's active site is lined by various acidic, basic, and neutral amino acid side chains, all properly positioned for maximum interaction with the substrate.

FIGURE 24.10 According to the lock-and-key model, an enzyme is a large, three-dimensional molecule containing a crevice with an active site. Only a substrate whose shape and structure are complementary to those of the active site can fit into the enzyme. The active site of the enzyme penicillopepsin is clearly visible as the cleft at the top in this computer-generated structure of penicillopepsin, as is the fit of the substrate (blue) in the active site.

Enzyme-catalyzed reactions begin when the substrate migrates into the active site to form an enzyme–substrate complex. No covalent bonds are formed; the enzyme and substrate are held together by hydrogen bonds and by weak electrical attractions between functional groups. With enzyme and substrate thus held together in a precisely defined arrangement, the appropriately positioned functional groups in the active site carry out a chemical reaction on the substrate molecule, and enzyme plus product then separate:

$$E + S \quad \rightarrow \quad E\text{–}S \quad \rightarrow \quad E + P$$

Enzyme Substrate   Complex   Enzyme Product

As an example of enzyme action, look in Figure 24.11 at the enzyme hexose kinase, which catalyzes the reaction of adenosine triphosphate (ATP) with glucose to yield glucose 6-phosphate and adenosine diphosphate (ADP). The enzyme first binds a molecule of ATP cofactor at a position near the active site, and glucose then bonds to the active site with its C6 hydroxyl group held rigidly in position next to the ATP molecule. Reaction ensues, and the two products are released from the enzyme.

Glucose   Glucose 6-phosphate

FIGURE 24.11   The hexose kinase-catalyzed reaction of glucose with ATP. Glucose enters the cleft of the enzyme and binds to the active site, where it reacts with a molecule of ATP cofactor held nearby.

Compare the enzyme-catalyzed reaction shown in Figure 24.11 with the same reaction in the absence of enzyme. Without enzyme present, the two reagents would spend most of their time surrounded by solvent molecules, far away from each other and only occasionally bumping together. With an enzyme present, however, the two reagents are forced into close contact. Enzymes act as catalysts because of their ability to bring reagents together, to hold them at the exact distance and with the exact orientation necessary for reaction, and to provide proton donating or accepting groups as required.

24.7 ►CARBOHYDRATES

Carbohydrates occur in every living organism. The starch in food and the cellulose in grass are nearly pure carbohydrate. Modified carbohydrates form part of the coating around all living cells, and other carbohydrates are found in the DNA that carries genetic information from one generation to the next. The word *carbohydrate* was used originally to describe glucose, which has the formula $C_6H_{12}O_6$ and was once thought to be a "hydrate of carbon"—$C_6(H_2O)_6$. This view was soon abandoned, but the word persisted and is now used to refer to a large class of polyhydroxylated aldehydes and ketones. Glucose, for example, is a six-carbon aldehyde with five hydroxyl groups.

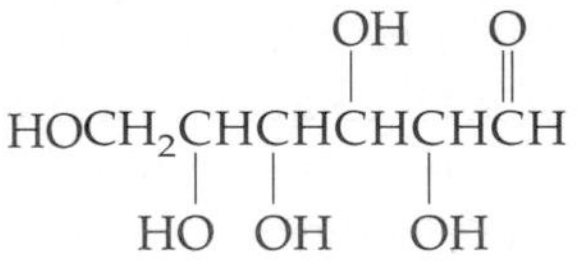

$$\underset{\substack{|\\HO}}{\underset{|}{HOCH_2CH}}\underset{\substack{|\\OH}}{CH}\underset{\substack{|\\OH}}{CH}\overset{OH}{\underset{|}{CH}}\overset{O}{\underset{\|}{CH}}\qquad \text{Glucose—a pentahydroxy aldehyde}$$

Carbohydrates are classified as either simple or complex. Simple sugars, or **monosaccharides**, are carbohydrates such as glucose and fructose that can't be broken down into smaller molecules by hydrolysis with aqueous acid. Complex carbohydrates, or **polysaccharides**, are compounds such as cellulose and starch that are made of many simple sugars linked together. On hydrolysis, polysaccharides are cleaved to yield many molecules of simple sugars.

Monosaccharides are further classified as either aldoses or ketoses. An *aldose* contains an aldehyde carbonyl group; a *ketose* contains a ketone carbonyl group. The *-ose* suffix indicates a sugar, and the number of carbon atoms in the sugar is specified by using one of the prefixes *tri-*, *tetr-*, *pent-*, or *hex-*. Thus, glucose is an aldohexose (a six-carbon aldehyde sugar), fructose is a ketohexose (a six-carbon ketone sugar), and ribose is an aldopentose (a five-carbon aldehyde sugar). Most commonly occurring sugars are either aldopentoses or aldohexoses.

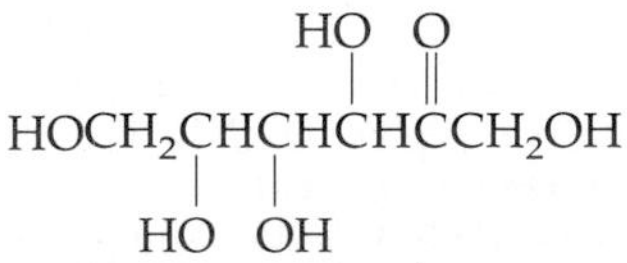

$$HOCH_2CHCHCHCCH_2OH \qquad HOCH_2CHCH-CHCH$$

Fructose—a ketohexose 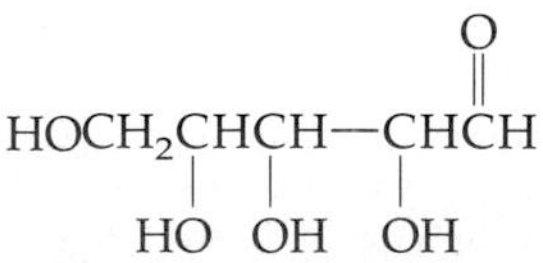   Ribose—an aldopentose

□ **PROBLEM 24.10** Classify each of the following monosaccharides.

(a) $HOCH_2CHCHCHCH$ **(b)** $HOCH_2CCH_2OH$ **(c)** $HOCH_2CHCHCH$

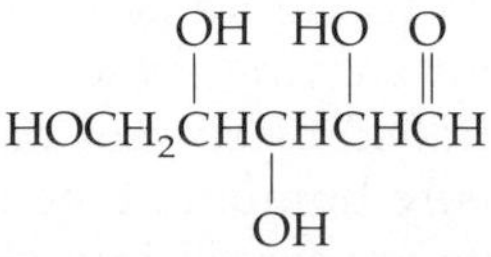

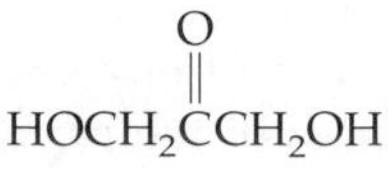

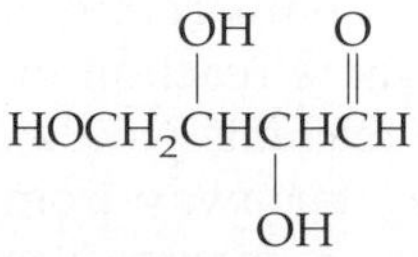

24.8 ►HANDEDNESS OF CARBOHYDRATES

We saw in Section 24.3 that compounds are chiral if they have a carbon atom bonded to four different atoms or groups of atoms. Such compounds lack a plane of symmetry and can exist as a pair of enantiomers in either a "right-handed" D form or a "left-handed" L form. For instance, the simple triose glyceraldehyde is chiral because it has four different groups bonded to C2: –CHO, –H, –OH, and –CH₂OH. Only D-glyceraldehyde occurs naturally.

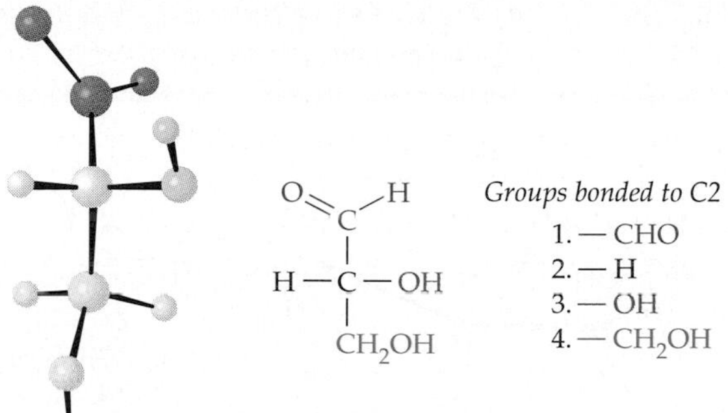

$$O{=}C{-}H$$
$$H{-}C{-}OH$$
$$CH_2OH$$

D-Glyceraldehyde

Groups bonded to C2
1. — CHO
2. — H
3. — OH
4. — CH₂OH

Glyceraldehyde has only one chiral carbon atom and can exist as two enantiomers, but other sugars have two, three, four, or even more chiral carbons. In general, a compound with n chiral carbon atoms has a maximum of 2^n possible forms. Glucose, for example, has four chiral carbon atoms and a total of $2^4 = 16$ forms that differ in the spatial arrangements of the substituents around the chiral carbon atoms.

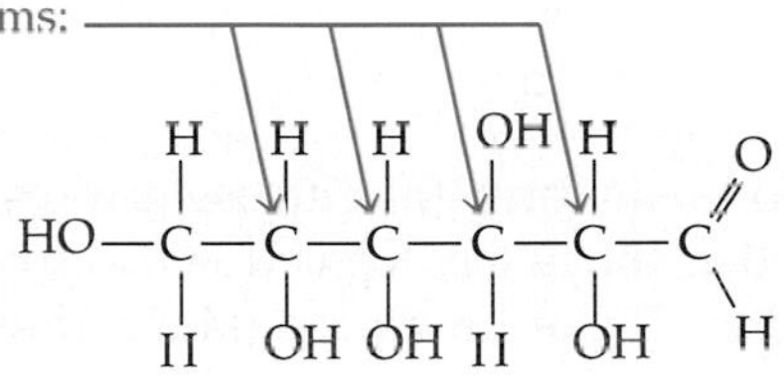

Four different groups are
attached to these atoms:

$$HO{-}C{-}C{-}C{-}C{-}C{-}C{\Big<}^O_H$$

┌ **PROBLEM 24.11** Draw tetrahedral representations of the two glyceraldehyde enantiomers using wedged, dashed, and normal lines to show three-dimensionality.

┌ **PROBLEM 24.12** Ribose has three chiral carbon atoms. What is the maximum number of ribose forms?

24.9 ►CYCLIC STRUCTURES
OF MONOSACCHARIDES

Glucose and other monosaccharides are often shown for convenience as having open-chain structures. They actually exist, however, primarily as cyclic molecules in which an –OH group near the bottom of the chain adds to the carbonyl group near the top of the chain to form a ring. In glucose itself, ring formation occurs between the –OH group on C5 and the C=O group at C1 (Figure 24.12).

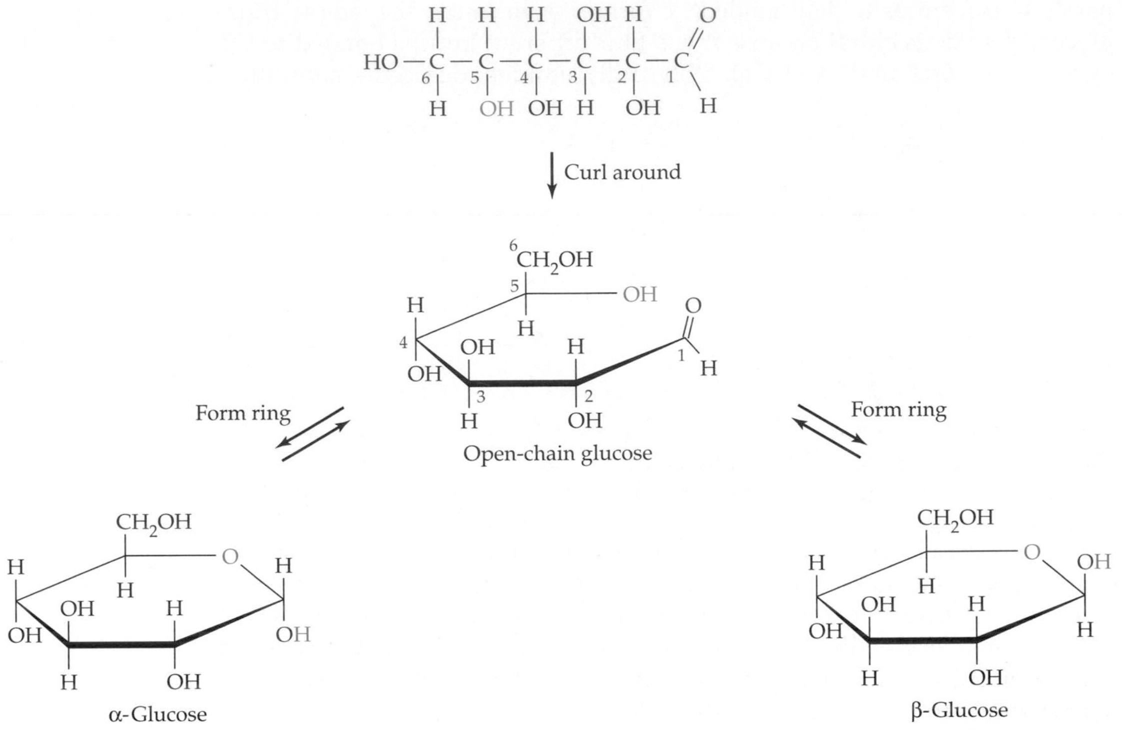

FIGURE 24.12 The cyclic α and β forms of glucose that result from ring formation between the –OH group at C5 and the C=O group at C1. The α form has the –OH group at C1 on the bottom side of the ring; the β form has the –OH group on the top.

Note that two cyclic forms of glucose can result from ring formation, depending on whether the newly formed –OH group at C1 is on the bottom or top side of the ring. The ordinary crystalline glucose you might take from a bottle is entirely the cyclic α form, in which the C1 –OH group is on the bottom side of the ring. In water solution, however, all three forms are present in an equilibrium mixture in the proportion 0.02% open-chain form, 36% α form, and 64% β form.

24.10 ►SOME COMMON DISACCHARIDES AND POLYSACCHARIDES

Lactose

Lactose, or *milk sugar*, is the major carbohydrate present in mammalian milk. Human milk, for example, is about 7% lactose. Structurally, lactose is a disaccharide whose hydrolysis with aqueous acids yields one molecule of glucose and one molecule of a different monosaccharide called galactose. The two sugars are bonded together by what is called a *1,4 link*, a bridging oxygen atom between C1 of β-galactose and C4 of β-glucose.

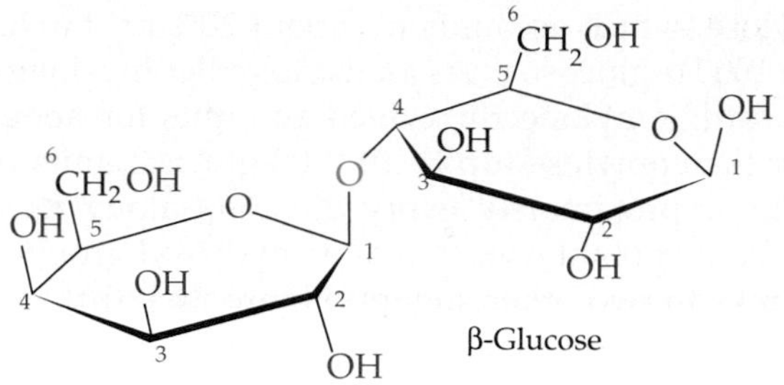

Lactose

β-Glucose

β-Galactose

Sucrose

Sucrose, or plain table sugar, is probably the most common pure organic chemical in the world. Although sucrose is found in many plants, sugar beets (20% by weight) and sugar cane (15% by weight) are the most common sources. Hydrolysis of sucrose yields one molecule of glucose and one molecule of fructose. The 50:50 mixture of glucose and fructose that results, often called *invert sugar*, is commonly used as a food additive.

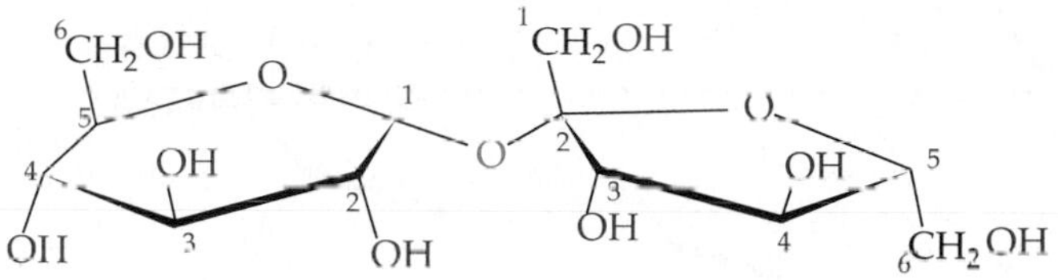

Sucrose

α-Glucose   β-Fructose

Cellulose and Starch

Cellulose, the fibrous substance used by plants as a structural material in grasses, leaves, and stems, consists of several thousand β-glucose molecules joined together by 1,4 links to form an immense polysaccharide.

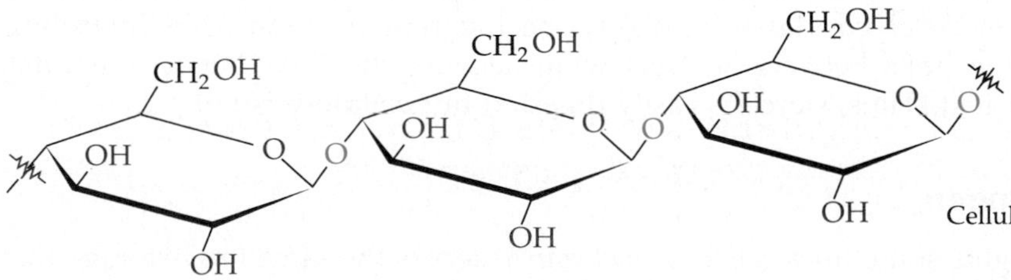

Cellulose

β-Glucose units

Starch, like cellulose, is also a polymer of glucose. Unlike cellulose, though, starch is edible. Indeed, the starch in such vegetables as beans, rice, and potatoes is an essential part of the human diet. Structurally, starch differs from cellulose in that it contains α- rather than β-glucose units. Starch is also more structurally complex than cellulose and is of two types: *amylose* and *amylopectin*. Amylose, which accounts for about 20% of starch, consists of several hundred to 1000 α-glucose units joined together in a long chain by 1,4 links [Figure 24.13(a)]. Amylopectin, which accounts for about 80% of starch, is much larger than amylose (up to 100,000 glucose units per molecule) and has *branches* approximately every 25 units along its chain. A glucose molecule at a branch point uses two of its hydroxyl groups (those at C4 and C6) to form links to two other sugars [Figure 24.13(b)].

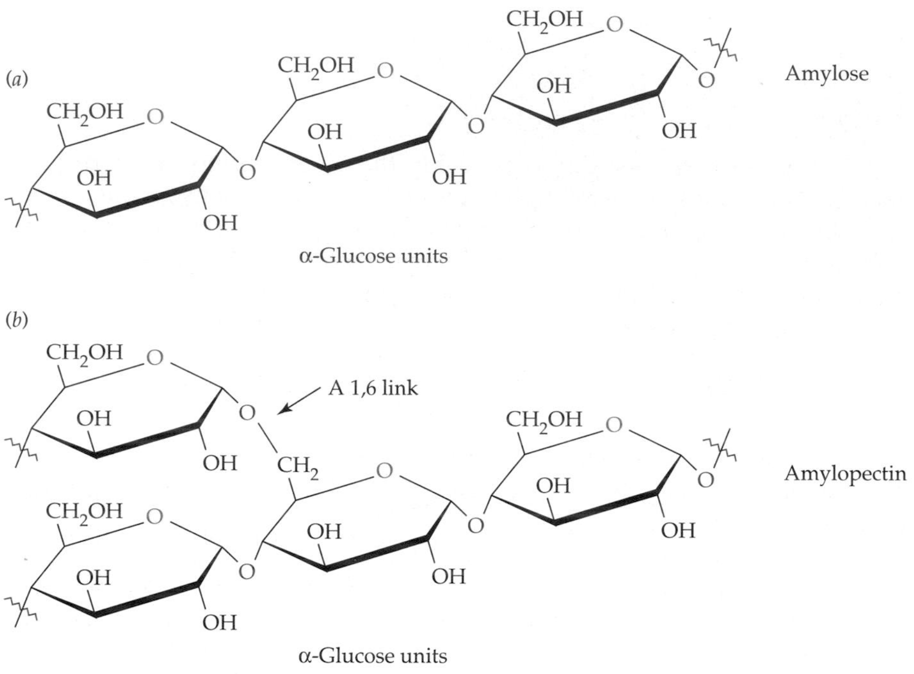

FIGURE 24.13 Glucose polymers in starch. **(a)** Amylose consists only of linear chains of α-glucose units linked by 1,4 bonds. **(b)** Amylopectin has branch points about every 25 sugars in the chain. A glucose unit at a branch point uses two of its hydroxyls (at C4 and C6) to form 1,4 and 1,6 links to two other sugars.

Starch molecules are digested in the stomach by enzymes called α-*glycosidases*, which break down the polysaccharide chain and release individual glucose molecules. As is usually the case with enzyme-catalyzed reactions, α-glycosidases are highly specific in their action. They hydrolyze only the links between α units while leaving the links between β units untouched. Thus, starch is easily digested but cellulose is not.

Glycogen

Glycogen, sometimes called *animal starch*, serves the same food-storage role in animals that starch serves in plants. After we eat starch and the body breaks it down into simple glucose units, some of the glucose is used immediately as fuel and some is stored in the body as glycogen for later use. Structurally, glycogen is similar to amylopectin in being a long polymer of α-glucose units with branch points in its chain. Glycogen has many more branches than amylopectin, however, and is much larger—up to 1 million glucose units per molecule.

24.11 ►LIPIDS

Lipids are less well known to most people than proteins or carbohydrates, yet they are just as essential to life. Lipids have many important biological roles, serving as sources of fuel storage, as protective coatings around many plants and insects, and as major components of the membranes that surround every living cell.

Apples and other fruits have a waxy coating of lipids.

Chemically, **lipids** are the naturally occurring organic molecules that dissolve in nonpolar organic solvents when a sample of plant or animal tissue is crushed or ground. Because they're defined by solubility, a physical property, rather than by chemical structure, it's not surprising that there are a great many different kinds of lipids (Figure 24.14). Note that all the lipids in Figure 24.14 contain large hydrocarbon portions, which accounts for their solubility behavior.

$$CH_2OCCH_2CH_2CH_2CH_2CH_2CH_2CH_2CH_2CH_2CH_2CH_2CH_2CH_2CH_3$$

$$CHOCCH_2CH_2CH_2CH_2CH_2CH_2CH_2CH=CHCH_2CH_2CH_2CH_2CH_2CH_2CH_2CH_3$$

$$CH_2OCCH_2CH_2CH_2CH_2CH_2CH_2CH_2CH_2CH_2CH_2CH_2CH_2CH_3$$

An animal fat or vegetable oil

Cholesterol—a steroid

$PGF_{2\alpha}$—a prostaglandin

FIGURE 24.14 Structures of some representative lipids isolated from plant and animal tissue by extraction with nonpolar organic solvents. All have large hydrocarbon portions.

Fats and Oils

Animal fats and vegetable oils are the most plentiful lipids in nature. Although they appear different—animal fats like butter and lard are usually solid, whereas vegetable oils like corn and peanut oil are liquid—their structures are similar. All fats and oils are **triacylglycerols**, or *triglycerides*, triesters of glycerol (1,2,3-propanetriol) with three long-chain carboxylic acids called **fatty acids**. The fatty acids are usually unbranched and have an even number of carbon atoms in the range 12–22.

As shown by the triacylglycerol structure in Figure 24.14, the three fatty acids of a given molecule need not be the same. Furthermore, the fat or oil from a given source is a complex mixture of many different triacylglycerols. Table 24.3 shows the structures of some commonly occurring fatty acids, and Table 24.4 lists the composition of several fats and oils. Note that vegetable oils are largely unsaturated, but animal fats contain a much higher percentage of saturated fatty acids.

TABLE 24.3   Structures of Some Common Fatty Acids

Name	No. of Carbons	No. of Double Bonds	Structure	Melting Point (°C)
Saturated				
Myristic	14	0	$CH_3(CH_2)_{12}COOH$	53
Palmitic	16	0	$CH_3(CH_2)_{14}COOH$	63
Stearic	18	0	$CH_3(CH_2)_{16}COOH$	70
Unsaturated				
Oleic	18	1	$CH_3(CH_2)_7CH{=}CH(CH_2)_7COOH$	16
Linoleic	18	2	$CH_3(CH_2)_4CH{=}CHCH_2CH{=}CH(CH_2)_7COOH$	−5
Linolenic	18	3	$CH_3CH_2CH{=}CHCH_2CH{=}CHCH_2CH{=}CH(CH_2)_7COOH$	−11

TABLE 24.4   Approximate Composition of Some Common Fats and Oils

Source	Saturated Fatty Acids (%)			Unsaturated Fatty Acids (%)	
	C_{14} Myristic	C_{16} Palmitic	C_{18} Stearic	C_{18} Oleic	C_{18} Linoleic
Animal Fat					
Butter	10	25	10	25	5
Human fat	3	25	8	46	10
Whale blubber	8	12	3	35	10
Vegetable Oil					
Corn	1	8	4	46	41
Olive	1	5	5	82	7
Peanut	–	7	5	60	20

About 40 different fatty acids occur naturally. Palmitic acid (C_{16}) and stearic acid (C_{18}) are the most abundant saturated acids; oleic and linoleic acids (both C_{18}) are the most abundant unsaturated ones. Oleic acid is *monounsaturated* because it has only the one double bond, but linoleic and linolenic acids are *polyunsaturated fatty acids* (called *PUFAs*) because they have more than one carbon–carbon double bond. Although the reasons are not yet clear, it appears that a diet rich in saturated fats leads to a higher level of blood cholesterol and consequent higher risk of heart attack than a diet rich in unsaturated fat.

The data in Table 24.3 show that unsaturated fatty acids generally have lower melting points than their saturated counterparts, a trend that's also true for triacylglycerols. Since vegetable oils have a higher proportion of unsaturated fatty acids than do animal fats, they have lower melting points and appear as liquids rather than solids. This behavior arises because the carbon–carbon double bonds in unsaturated vegetable oils introduce bends and kinks into the hydrocarbon chains, making it difficult for the chains to nestle closely together to become solid.

The carbon–carbon double bonds in vegetable oils can be hydrogenated to yield saturated fats in the same way that any alkene can react with hydrogen to yield an alkane (Section 23.10). By carefully controlling the extent of hydrogenation, the final product can have any desired consistency. Margarine, for example, is prepared so that only about two-thirds of the double bonds present in the starting vegetable oil are hydrogenated.

$$\overset{\text{O}}{\underset{\text{||}}{\text{OCCH}_2\text{CH}_2\text{CH}_2\text{CH}_2\text{CH}_2\text{CH}_2\text{CH}_2\text{CH}=\text{CHCH}_2\text{CH}=\text{CHCH}_2\text{CH}=\text{CHCH}_2\text{CH}_3}}$$

Partial structure of a vegetable oil

H_2,   Pd catalyst

$$\overset{\text{O}}{\underset{\text{||}}{\text{OCCH}_2\text{CH}_2\text{CH}_2\text{CH}_2\text{CH}_2\text{CH}_2\text{CH}_2\text{CH}_2\text{CH}_2\text{CH}_2\text{CH}_2\text{CH}_2\text{CH}_2\text{CH}_2\text{CH}_2\text{CH}_2\text{CH}_3}}$$

Saturated product

The purple foxglove, a common backyard plant, contains the steroidal heart stimulant digitoxigenin.

PROBLEM 24.13 Show the structure of glyceryl trioleate, a fat molecule whose components are glycerol and three oleic acid units.

Steroids

A **steroid** is a lipid whose structure is based on the tetracyclic (four-ring) system shown in the following examples. Three of the rings are six-membered, while the fourth is five-membered. Steroids have many diverse roles throughout both the plant and animal kingdoms. Some steroids, such as digitoxigenin, isolated from the purple foxglove (*Digitalis purpurea*), are used in medicine as heart stimulants. Others, such as hydrocortisone, are hormones, and still others have a wide variety of different physiological functions.

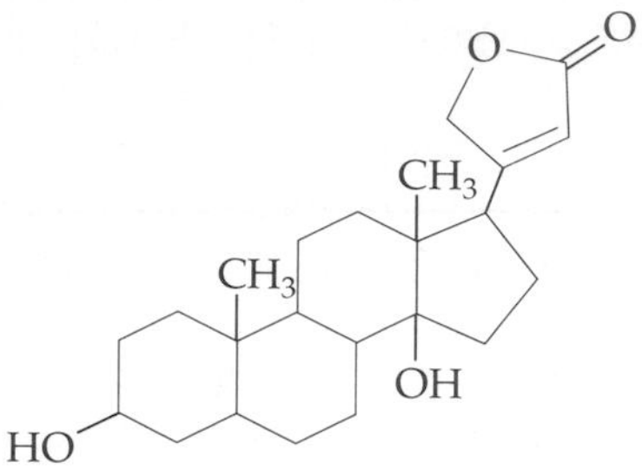

Digitoxigenin

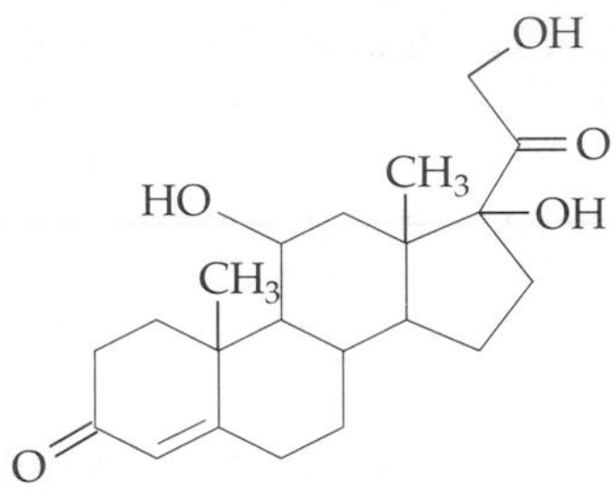

Hydrocortisone

Cholesterol, an unsaturated alcohol whose structure is shown in Figure 24.14, is the most abundant animal steroid. It's been estimated that a 60 kg (130 lb) person has a total of about 175 g of cholesterol distributed throughout the body. Much of this cholesterol is bonded through ester links to fatty acids, but some is found as the free alcohol. Gallstones, for example, are nearly pure cholesterol. Cholesterol serves two important functions in the body. First, it is a minor component of cell membranes, where it helps to keep the membranes fluid. Second, it serves as the body's starting material for the synthesis of all other steroids, including the sex hormones. Although news reports sometimes make cholesterol sound dangerous, there would be no life without it. The human body obtains its cholesterol both by synthesis in the liver and by ingestion of food. Even on a strict no-cholesterol diet, an adult is able to synthesize approximately 800 mg per day.

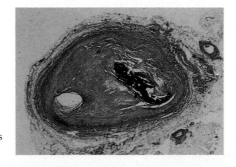

The plaque clogging this artery is nearly pure cholesterol.

24.12 ►NUCLEIC ACIDS

How does a seed "know" what kind of plant to become? How does a fertilized ovum know how to grow into a human being? How does a cell know what part of the body it's in so that it can carry out the right reactions? The answers to such questions involve the biological molecules called *nucleic acids*. Deoxyribonucleic acid (DNA) and ribonucleic acid (RNA) are the chemical carriers of an organism's genetic information. Coded in an organism's DNA is all the information that determines the nature of the organism and all the directions that are needed for producing the thousands of different proteins required by the organism.

Just as proteins are polymers made of amino acid units, nucleic acids are polymers made up of **nucleotide** units linked together to form a long chain. Each nucleotide is composed of a **nucleoside** plus phosphoric acid, H_3PO_4, and each nucleoside is composed of an aldopentose sugar plus an amine base.

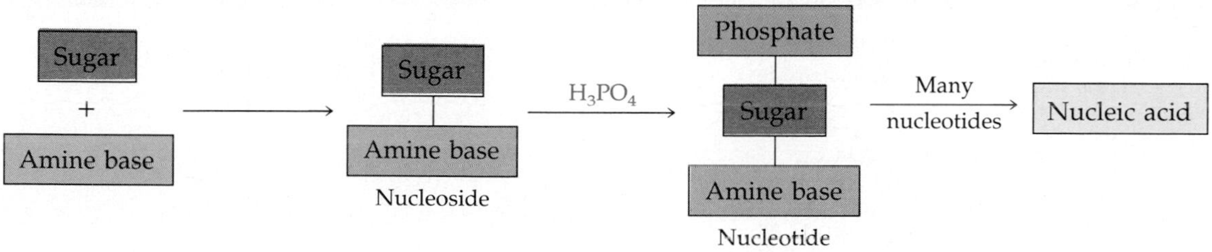

The sugar component in RNA is ribose, and the sugar in DNA is 2-deoxyribose (2-deoxy means that oxygen is missing from C2 of ribose).

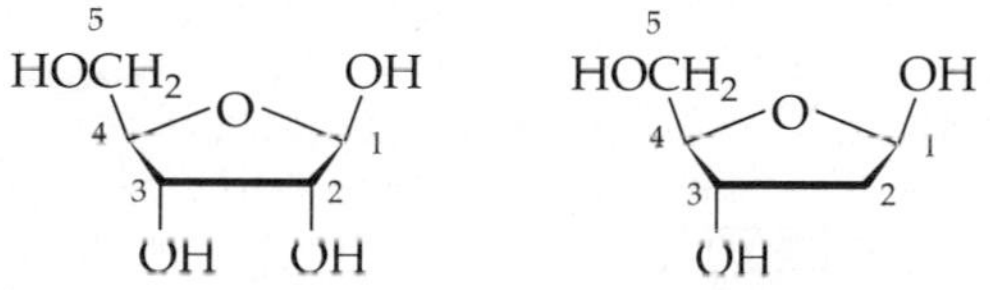

There are four different cyclic amine bases in DNA: adenine, guanine, cytosine, and thymine. Adenine, guanine, and cytosine also occur in RNA, but thymine is replaced in RNA by a related base called uracil.

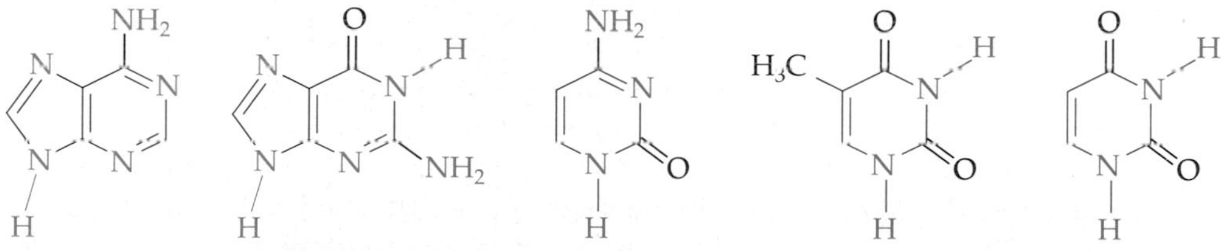

Adenine (A)	Guanine (G)	Cytosine (C)	Thymine (T)	Uracil (U)
DNA	DNA	DNA	DNA	RNA
RNA	RNA	RNA		

In both DNA and RNA, the cyclic amine base is bonded to C1′ of the sugar, and the phosphoric acid is bonded to the C5′ sugar position. Thus, nucleosides and nucleotides have the general structure shown in Figure 24.15. (In discussions of RNA and DNA, numbers with a prime superscript refer to positions on the sugar component of a nucleotide; numbers without a prime refer to positions on the cyclic amine base.)

Nucleotides join together in nucleic acids by forming a phosphate ester bond between the phosphate group at the 5′ end of one nucleotide and the hydroxyl group on the sugar component at the 3′ end of another nucleotide (Figure 24.16).

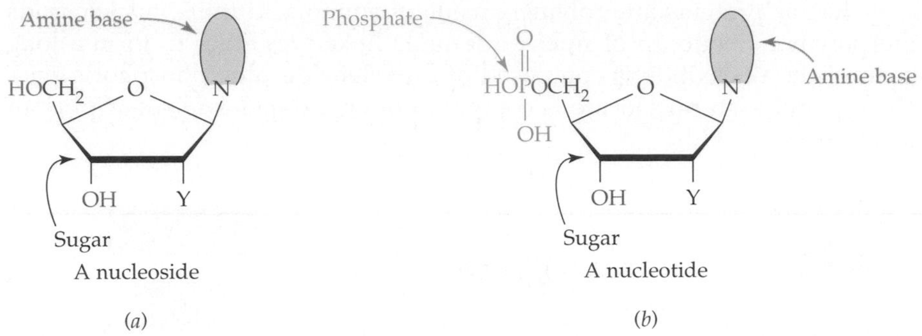

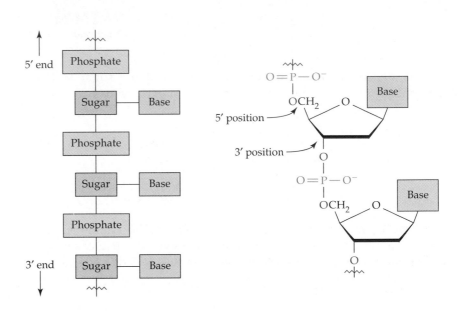

FIGURE 24.15 General structures of **(a)** a nucleoside and **(b)** a nucleotide. When Y = H, the sugar is deoxyribose; when Y = OH, the sugar is ribose.

FIGURE 24.16 Generalized structure of a nucleic acid.

Just as the structure of a protein depends on its sequence of individual amino acids, the structure of a nucleic acid depends on its sequence of individual nucleotides. To carry the analogy further, just as a protein has a polyamide backbone with different side chains attached to it, a nucleic acid has an alternating sugar-phosphate backbone with different amine base side chains attached.

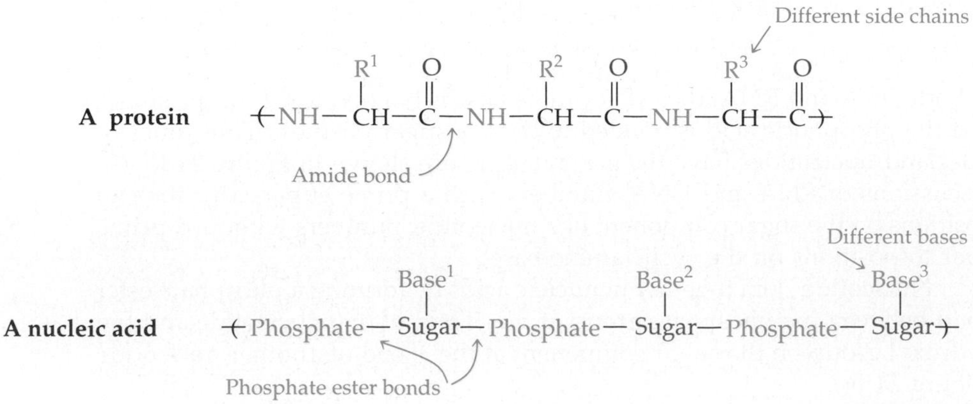

The sequence of nucleotides is described by starting at the 5′ phosphate end and identifying the bases in order. Abbreviations are used for each nucleotide: A for adenosine, T for thymidine, G for guanosine, C for cytidine, and U for uracil. Thus, a typical DNA sequence might be written as –T–A–G–G–C–T–.

EXAMPLE 24.3

Draw the full structure of the DNA dinucleotide C–T.

SOLUTION

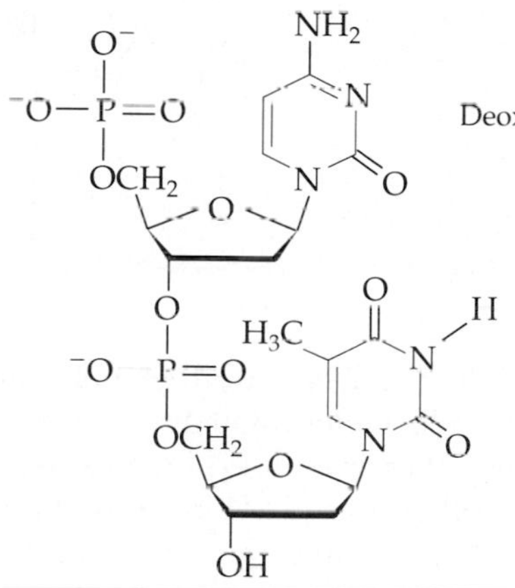

Deoxycytidine (C)

Deoxythymidine (T)

PROBLEM 24.14 Draw the full structure of the DNA dinucleotide A–G.

PROBLEM 24.15 Draw the full structure of the RNA dinucleotide U–A.

24.13 ▶BASE PAIRING IN DNA: THE WATSON-CRICK MODEL

Molecules of DNA isolated from different tissues of the same species have the same proportions of nucleotides, but molecules from different species can have quite different proportions. For example, human DNA contains about 30% each of A and T and about 20% each of G and C, but the bacterium *Clostridium perfringens* contains about 37% each of A and T and only 13% each of G and C. Note that in both cases, the bases occur in pairs. Adenine and thymine are usually present in equal amounts, as are guanine and cytosine. Why should this be?

According to the **Watson-Crick model**, DNA consists of two polynucleotide strands coiled around each other in a *double helix*. The sugar–phosphate backbone is on the outside of the helix, and the heterocyclic bases are on the inside, so that a base on one strand points directly in toward a base on the second strand. The two strands run in opposite directions and are held together by hydrogen bonds between pairs of bases. A and T form two strong hydrogen bonds to each other but not to G or C; G and C form three strong hydrogen bonds to each other but not to A or T.

Adenine   Thymine   Guanine   Cytosine

The two strands of the DNA double helix aren't identical; rather, they're complementary. Whenever a G base occurs in one strand, a C base occurs opposite it in the other strand. When an A base occurs in one strand, a T base occurs in the other strand. This complementary pairing of bases explains why the pair A and T and the pair G and C are always found in equal amounts. Figure 24.17 illustrates this base pairing, showing how the two complementary strands coil into the double helix. X-ray measurements show that the DNA double helix is 2.0 nm (20 Å) wide, that there are exactly 10 base pairs in each full turn, and that each turn is 3.4 nm (34 Å) high.

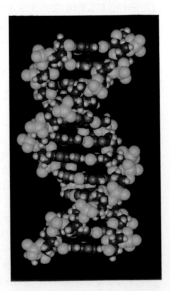

FIGURE 24.17 The coil of the sugar-phosphate backbone is visible on the outside of this computer-generated DNA model, while the pairs of hydrogen-bonded amine bases lie flat on the inside of the helix. (yellow = P, light blue = O, red = C, dark blue = N)

EXAMPLE 24.4

What sequence of bases on one strand of DNA is complementary to the sequence G–C–A–T–T–A–T on another strand?

SOLUTION   Since A and G form complementary pairs with T and C, respectively, go through the given sequence replacing A with T, G with C, T with A, and C with G.

Original: G—C—A—T—T—A—T

Complement: C—G—T—A—A—T—A

PROBLEM 24.16   What sequence of bases on one strand of DNA is complementary to the following sequence on another strand?

G—G—C—C—C—G—T—A—A—T

24.14 ➤NUCLEIC ACIDS AND HEREDITY

Most DNA of higher organisms, both plant and animal, is found in the nucleus of cells in the form of threadlike strands that are coated with proteins and wound into complex assemblies called **chromosomes**. Each chromosome is made up of several thousand **genes**, where a gene is a segment of a DNA chain that contains the instructions necessary to make a specific protein. By decoding the right genes at the right time, an organism uses genetic information to synthesize the thousands of proteins needed for living. Thus, the function of DNA is to act as a storage medium for an organism's genetic information. The function of RNA is to read, decode, and use the information received from DNA to make proteins.

Three processes take place in the transfer and use of genetic information: **Replication** is the means by which identical copies of DNA are made, forming additional molecules and preserving genetic information for passing on to offspring. **Transcription** is the means by which information in the DNA is transferred to and decoded by RNA. **Translation** is the means by which RNA uses the information to build proteins.

$$\text{Replication} \left(\; \text{DNA} \xrightarrow{\text{transcription}} \text{RNA} \xrightarrow{\text{translation}} \text{Proteins}\right.$$

Replication

DNA replication is an enzyme-catalyzed process that begins with a partial unwinding of the double helix. As the DNA strands separate and bases are exposed, new nucleotides line up on each strand in a complementary manner, A to T and C to G, and two new strands begin to grow. Each new strand is complementary to its old template strand, and two new, identical DNA double helixes are produced (Figure 24.18).

The magnitude of the replication process is extraordinary. The nucleus of a human cell contains 46 chromosomes (23 pairs), each of which consists of one large DNA molecule. Each chromosome, in turn, is made up of several thousand genes, and the sum of all genes in a human cell (the *genome*) is approximately 3 billion base pairs. This immense base sequence is faithfully copied during replication, with an error occurring only about once each 10–100 billion bases.

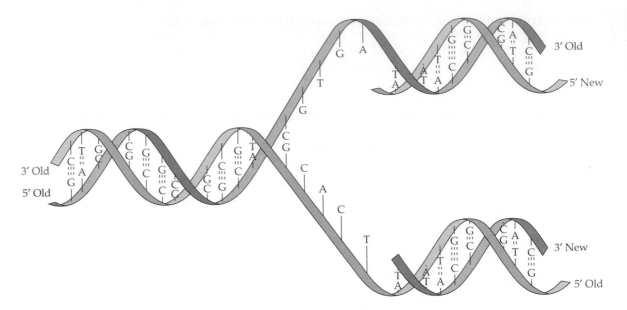

FIGURE 24.18 DNA replication. A portion of the DNA double helix unwinds, and complementary nucleotides line up for linking to yield two new DNA molecules. Each of the new DNA molecules contains one of the original strands and one new strand.

Transcription

The genetic instructions contained in DNA are transcribed into RNA when a small portion of the DNA double helix unwinds and one of the two DNA strands acts as a template for complementary *ribonucleotides* to line up, just as in DNA replication (Figure 24.19). The only difference is that uracil (U) rather than thymine lines up opposite adenine. Once completed, the RNA molecule separates from the DNA template, and the DNA rewinds to its stable double-helix conformation.

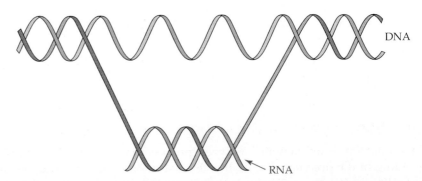

FIGURE 24.19 Transcription of DNA to synthesize RNA. A small portion of the DNA double helix unwinds, and one of the two DNA strands acts as a template on which ribonucleotides line up. The RNA produced is complementary to the DNA strand from which it is transcribed.

Translation

Protein biosynthesis is directed by a special kind of RNA called *messenger RNA,* or *mRNA.* It takes place on knobby protuberances within a cell called *ribosomes.* The specific ribonucleotide sequence in mRNA acts like a long coded sentence to specify the order in which different amino acid residues are to be joined. Each of the estimated 100,000 proteins in the human body is synthesized from a different mRNA that has been transcribed from a specific gene segment on DNA.

Each "word" along the mRNA chain consists of a series of three ribonucleotides that is specific for a given amino acid. For example, the series cytosine-uracil-guanine (C–U–G) on mRNA is a three-letter word directing that the amino acid leucine be incorporated into the growing protein. The words are read by another kind of RNA called *transfer RNA,* or *tRNA.* Each of the 60 or so different tRNAs contains a complementary base sequence that allows it to recognize a three-letter word on mRNA and act as a carrier to bring a specific amino acid into place for transfer to the growing peptide chain. When synthesis of the protein is completed, a "stop" word signals the end, and the protein is released from the ribosome. The process is illustrated in Figure 24.20.

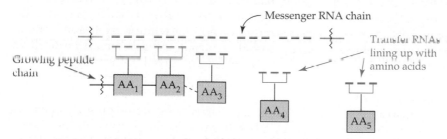

FIGURE 24.20 Protein biosynthesis. Messenger RNA is read by tRNA that contains complementary three-base sequences. Transfer RNA then assembles the proper amino acids (AA$_1$, AA$_2$, and so on) into position for incorporation into the peptide.

EXAMPLE 24.5

What RNA base sequence is complementary to the following DNA base sequence?

$$G—C—C—T—A—A—G—T—G$$

SOLUTION   Go through the sequence replacing A with U, G with C, T with A, and C with G.

Original DNA: G—C—C—T—A—A—G—T—G

Complementary RNA: C—G—G—A—U—U—C—A—C

PROBLEM 24.17 Show how uracil can form strong hydrogen bonds to adenine.

PROBLEM 24.18 What RNA base sequence is complementary to the following DNA base sequence?

$$C—G—T—G—A—T—T—A—C—A$$

PROBLEM 24.19 From what DNA base sequence was the following RNA sequence transcribed?

$$U—G—C—A—U—C—G—A—G—U$$

interlude—DIETS, BABIES, AND HIBERNATING BEARS

Brown fat tissue helps bears and other mammals hibernate over the winter.

Most warmblooded animals generate enough heat to maintain body temperature through normal metabolic reactions. In certain cases, however, normal metabolism is not sufficient to satisfy the body's need for warmth, and a supplemental method of heat generation is needed. This occurs, for example, in newborn infants and in hibernating animals that must survive for long periods without food. Babies, bears, and many other creatures have therefore been provided with *brown fat tissue*. Brown fat tissue, so called because its color contrasts with that of normal white fatty tissue, contains high concentrations of blood vessels and *mitochondria*, small intracellular structures that act as the main site of energy generation in organisms. In addition, brown fat tissue contains a special protein that acts as an *uncoupler* of ATP synthesis.

We saw in Section 24.1 that the reactions of the respiratory chain are coupled to the synthesis of ATP. In other words, the purpose of the respiratory chain is to harness the energy released during catabolism to synthesize ATP. In the presence of an uncoupling agent like the protein in brown fat, however, the reactions in the respiratory chain continue to take place, but no ATP is synthesized. The energy that would otherwise be used for ATP synthesis is released simply as heat, and the temperature of the organism rises. In essence, brown fat acts as a furnace to produce heat energy rather than stored energy.

The uncoupling of ATP synthesis from the respiratory chain can be caused by chemical agents as well as by brown-fat protein. It has been found, for example, that 2,4-dinitrophenol can cause uncoupling even in the normal white fat tissue of adults.

2, 4-Dinitrophenol

Imagine the possibilities for a simple chemical agent that causes food to be converted directly into heat rather than into ATP. The body, in critical need of ATP to drive its many reactions, would continue to burn fat at a frantic rate. Pound after pound of fat would literally be burned away, fulfilling the dream of every previously unsuccessful dieter. In fact, the idea works, and weight-control pills containing 2,4-dinitrophenol were marketed for a brief period in the 1930s. Unfortunately, the dose needed to *slow* ATP synthesis is very close to the lethal dose that *stops* ATP synthesis, a fact that was discovered only after several fatalities had occurred. Thus, there's still no magic pill for dieters. Exercise and sensible eating remain the only safe way to control weight.

Fats, carbohydrates, and proteins are metabolized in the body to yield acetyl CoA, which is further degraded in the citric acid cycle to yield two molecules of carbon dioxide plus a large amount of energy. The energy output of the various steps in the citric acid cycle is coupled to the respiratory chain, a series of enzyme-catalyzed reactions whose ultimate purpose is to synthesize **adenosine triphosphate (ATP)**.

Proteins are large biomolecules consisting of α-amino acid residues linked together by amide bonds. Twenty amino acids are commonly found in proteins, and all except glycine have a handedness. In general, any carbon atom that is bonded to four different groups has a handedness and is said to be chiral. The right-handed and left-handed forms of a chiral substance are called enantiomers, or optical isomers. Proteins can be classified either by shape or biological function. Fibrous proteins such as α keratin are tough, threadlike, and water insoluble; globular proteins such as myoglobin are compact, water soluble, and mobile within cells. Some proteins are enzymes, some are hormones, and some serve to store nutrients. Others act as structural, protective, or transport agents.

A protein's **primary structure** is its amino acid sequence. Its **secondary structure** is the orientation of segments of the protein chain into a regular pattern, such as an α-helix or a β-pleated sheet. Its **tertiary structure** is the three-dimensional shape into which the entire protein molecule is coiled.

Enzymes are large proteins that function as biological catalysts. Enzyme specificity is due to a lock-and-key fit between enzyme and substrate. Enzymes contain a crevice, inside which is an **active site,** a small three-dimensional region of the enzyme with the specific shape necessary to bind the proper substrate. The catalytic action of enzymes is due to their ability to bring reagents together and hold them in the position necessary for reaction.

Carbohydrates are polyhydroxy aldehydes and ketones. Simple carbohydrates such as glucose can't be hydrolyzed to smaller molecules; complex carbohydrates such as starch and cellulose contain many simple sugars linked together. **Monosaccharides** such as glucose and fructose exist as a mixture of an open-chain form and two cyclic forms called the α form and the β form. Disaccharides such as lactose and sucrose contain two simple sugars joined by a linking oxygen atom. Important **polysaccharides** include cellulose, starch, and glycogen. All sugars are chiral.

Lipids are the naturally occurring organic molecules that dissolve when a plant or animal sample is crushed or ground with a nonpolar solvent. Because they are defined by a physical property rather than by structure, there are a great many different kinds of lipids. Animal fats and vegetable oils are **triacylglycerols**—triesters of glycerol with three long-chain **fatty acids**. The fatty acids are unbranched, have an even number of carbon atoms, and may be either saturated or unsaturated. Vegetable oils contain a higher proportion of unsaturated fatty acids than do animals fats and consequently have lower melting points.

Deoxyribonucleic acid (DNA) and ribonucleic acid (RNA) are the chemical carriers of an organism's genetic information. Nucleic acids are made up of many individual building blocks, called **nucleotides**, linked together to form a long chain. Each nucleotide consists of a cyclic amine base linked to C1 of a sugar, with the sugar in turn linked through its C5 –OH group to phosphoric acid. The sugar component in RNA is ribose; the sugar in DNA is 2-deoxyribose. The bases in DNA are adenine, guanine, cytosine and thymine; the bases in RNA are adenine, guanine, cytosine, and uracil. Molecules of DNA consist of two complementary polynucleotide strands, held together by hydrogen bonds between bases on the two strands and coiled into a double helix. Adenine (A) and thymine (T) form hydrogen bonds only to each other, as do cytosine (C) and guanine (G).

Three processes take place in the transfer of genetic information: **Replication** is the process by which identical copies of DNA are made and genetic information is preserved. **Transcription** is the process by which messenger RNA is produced. **Translation** is the process by which mRNA directs protein synthesis. Each mRNA has three-base segments that are recognized by small amino acid–carrying molecules of transfer RNA (tRNA), which deliver the appropriate amino acids in the correct order for protein synthesis.

ADDITIONAL PROBLEMS

Problems 24.1–24.19 appear within the chapter.

AMINO ACIDS, PEPTIDES, AND PROTEINS

24.20 What does the prefix "α" mean when referring to α-amino acids?

24.21 Why are the naturally occurring amino acids called L-amino acids?

24.22 What amino acids do the following abbreviations stand for?

(a) Ser (b) Thr (c) Pro (d) Phe (e) Cys

24.23 Name and draw the structures of amino acids that fit the following descriptions.

(a) contains an isopropyl group
(b) contains an alcohol group
(c) contains a thiol (–SH) group
(d) contains an aromatic ring

24.24 Much of the chemistry of amino acids is the familiar chemistry of carboxylic acid and amine functional groups. What products would you expect to obtain from the following reactions of glycine?

(a)

$$H_2NCH_2\overset{\displaystyle O}{\overset{\|}{C}}OH + CH_3OH \xrightarrow[\text{catalyst}]{H_2SO_4}$$

(b)

$$H_2NCH_2\overset{\displaystyle O}{\overset{\|}{C}}OH + HCl \rightarrow$$

24.25 Identify the amino acids present in the following hexapeptide.

24.26 Look at the structure of angiotensin II in Figure 24.5, and identify both the N-terminal and C-terminal amino acids.

24.27 What is meant by the following terms as they apply to proteins?

(a) primary structure
(b) secondary structure
(c) tertiary structure

24.28 What is the difference between fibrous and globular proteins?

24.29 Name three different biological functions that proteins have in the body.

24.30 What kinds of intramolecular interactions are important in stabilizing a protein's tertiary structure?

24.31 What kind of bonding stabilizes helical and β-pleated-sheet secondary protein structures?

24.32 Why is cysteine such an important amino acid for defining the tertiary structure of proteins?

$$H_2N-CH-\overset{\displaystyle O}{\overset{\|}{C}}-NH-CH-\overset{\displaystyle O}{\overset{\|}{C}}-NH-CH-\overset{\displaystyle O}{\overset{\|}{C}}-NH-CH-\overset{\displaystyle O}{\overset{\|}{C}}-NH-CH-\overset{\displaystyle O}{\overset{\|}{C}}-NH-CH-\overset{\displaystyle O}{\overset{\|}{C}}-OH$$

CH(CH₃)₂   CH₂OH   CH₂   CH₂SCH₃   CHOH   CH₃

CH₃

24.33 Which of the following amino acids are most likely to be found on the outside of a globular protein and which on the inside? Explain.

(a) valine   (b) leucine
(c) aspartic acid   (d) asparagine

24.34 Use the three-letter abbreviations to name all tripeptides containing methionine, isoleucine, and lysine.

24.35 Write structural formulas for the two dipeptides containing phenylalanine and glutamic acid.

24.36 *Aspartame*, marketed under the trade name Nutra-Sweet for use as a nonnutritive sweetener, is the methyl ester of a simple dipeptide. Identify the two amino acids present in aspartame, and show all the products of digestion, assuming that both amide and ester bonds are hydrolyzed in the stomach.

Aspartame

24.37 Cytochrome *c* is an important enzyme found in the cells of all aerobic organisms. Elemental analysis of cytochrome *c* shows that it contains 0.43% iron. What is the minimum molecular weight of this enzyme?

MOLECULAR HANDEDNESS

24.38 Which of the following objects are chiral?

(a) a shoe
(b) a bed
(c) a light bulb
(d) a flower pot

24.39 Give two examples of chiral objects and two examples of achiral objects.

24.40 Which of the following compounds are chiral?

(a) 2,4-dimethylheptane
(b) 5-ethyl-3,3-dimethylheptane

24.41 Draw chiral molecules that meet the following descriptions.

(a) a chloroalkane, $C_5H_{11}Cl$
(b) an alcohol, $C_6H_{14}O$
(c) an alkene, C_6H_{12}
(d) an alkane, C_8H_{18}

24.42 There are eight alcohols with the formula $C_5H_{12}O$. Draw them and tell which are chiral.

24.43 Propose structures for compounds that meet the following descriptions.

(a) a chiral alcohol with four carbons
(b) a chiral aldehyde
(c) a compound with two chiral centers

CARBOHYDRATES

24.44 What is the structural difference between an aldose and a ketose?

24.45 Classify each of the following carbohydrates by indicating the nature of its carbonyl group and the number of carbon atoms present. For example, glucose is an aldohexose.

(a)

$$HOCH_2\overset{OH}{\underset{}{CH}}\overset{}{CH}\overset{OH}{\underset{}{CH}}\overset{OH}{\underset{}{CH}}\overset{O}{\underset{}{CH}}$$

(b)

$$HOCH_2\overset{OH}{\underset{}{CH}}CHCH\overset{}{CH}CHCH$$ OH HO OH

(c)

$$HOCH_2\overset{OH}{\underset{}{CH}}CH\overset{}{CH}CHCCH_2OH$$ HO OH

24.46 Starch and cellulose are both polymers of glucose. What is the main structural difference between

them, and what different roles do they serve in nature?

24.47 How are amylose and amylopectin similar, and how are they different?

24.48 Starch and glycogen are both α-linked polymers of glucose. What is the structural difference between them, and what different roles do they serve in nature?

24.49 Write the open-chain structure of a ketotetrose.

24.50 Write the open-chain structure of a four-carbon deoxy sugar.

24.51 D-Mannose, an aldohexose found in orange peels, has the following structure in open-chain form. Coil mannose around, and draw it in cyclic α- and β-forms.

$$HOCH_2\overset{HO}{\underset{}{CH}}\overset{OH}{\underset{}{CH}}CH\overset{O}{\underset{}{CH}}\!-\!CH$$ HO OH   Mannose

24.52 Show two D-mannose molecules (Problem 24.51) attached by an α-1,4 link.

LIPIDS

24.53 What is a fatty acid?

24.54 What does it mean to say that fats and oils are triacylglycerols?

24.55 Draw the structure of glycerol myristate, made from glycerol and three myristic acid molecules (Table 24.3).

24.56 Spermaceti, a fragrant substance isolated from sperm whales, was a common ingredient in cosmetics until its use was banned in 1976 to protect the whales from extinction. Chemically, spermaceti is cetyl palmitate, the ester of palmitic acid with cetyl alcohol (the straight-chain C_{16} alcohol). Show the structure of spermaceti.

24.57 There are two isomeric fat molecules whose components are glycerol, one palmitic acid, and two stearic acids. Draw the structures of both, and explain how they differ.

24.58 One of the two molecules you drew in Problem 24.57 is chiral. Which molecule is chiral, and why?

24.59 Draw the structures of all products you would obtain by reaction of the following lipid with aqueous KOH. What are the names of the products?

$$CH_2-O-\overset{\displaystyle O}{\overset{\|}{C}}(CH_2)_{16}CH_3$$
$$CH-O-\overset{\displaystyle O}{\overset{\|}{C}}(CH_2)_7CH=CH(CH_2)_7CH_3$$
$$CH_2-O-\overset{\displaystyle O}{\overset{\|}{C}}(CH_2)_7CH=CHCH_2CH=CHCH_2CH=CHCH_2CH_3$$

24.60 Draw the structure of the product you would obtain on hydrogenation of the lipid in Problem 24.59. What is its name?

24.61 Would the product in Problem 24.60 have a higher or lower melting point than the original lipid? Why?

24.62 What products would you obtain by treating oleic acid with the following reagents?

(a) Br_2 (b) H_2, Pd catalyst
(c) CH_3OH, HCl catalyst

24.63 One of the constituents of the carnauba wax used in floor and furniture polish is an ester of a C_{32} straight-chain alcohol with a C_{20} straight-chain carboxylic acid. Draw the structure of this ester.

NUCLEIC ACIDS

24.64 What is a nucleotide, and what three kinds of components does it contain?

24.65 What are the names of the sugars in DNA and RNA, and how do they differ?

24.66 Where in the cell is most DNA found?

24.67 What is meant by the following terms as they apply to nucleic acids?
(a) base pairing
(b) replication
(c) translation
(d) transcription

24.68 What is the difference between a gene and a chromosome?

24.69 What genetic information does a single gene contain?

24.70 Show by drawing structures how the phosphate and sugar components of a nucleic acid are joined.

24.71 Show by drawing structures how the sugar and heterocyclic base components of a nucleic acid are joined.

24.72 Draw the complete structure of deoxycytidine 5'-phosphate, one of the four deoxyribonucleotides.

24.73 The DNA from sea urchins contains about 32% A and about 18% G. What percentages of T and C would you expect in sea urchin DNA? Explain.

24.74 If the sequence T–A–C–C–G–A appeared on one strand of DNA, what sequence would appear opposite it on the other strand?

24.75 What sequence would appear on the mRNA molecule transcribed from the DNA in Problem 24.74?

24.76 Human insulin is composed of two polypeptide chains. One chain contains 21 amino acids, and the other contains 30 amino acids. How many nucleotides are present in the DNA to code for each chain?

GENERAL PROBLEMS

24.77 The catabolism of glucose to yield carbon dioxide and water has $\Delta G° = -2870$ kJ/mol. What is the value of $\Delta G°$ for the photosynthesis of glucose from carbon dioxide and water in green plants?

24.78 The *endorphins* are a group of naturally occurring neuroproteins that act in a manner similar to morphine to control pain. Research has shown that the biologically active part of the endorphin molecule is a simple pentapeptide called an *enkephalin*, with the structure Tyr–Gly–Gly–Phe–Met. Draw the complete structure of this enkephalin.

24.79 Write full structures for the following peptides, and indicate the positions of the amide bonds.

(a) Val–Phe–Cys   **(b)** Glu–Pro–Ile–Leu

24.80 Why is proline never encountered in a protein α-helix? The α-helical segments of myoglobin and other proteins stop when a proline residue is encountered in the chain.

24.81 Why do you suppose diabetics must receive insulin by injection rather than by mouth?

24.82 Jojoba wax, used in candles and cosmetics, is partially composed of the ester of stearic acid and a straight-chain 22-carbon alcohol. Draw the structure of this ester.

24.83 Write representative structures for the following.

(a) a fat   **(b)** a vegetable oil   **(c)** a steroid

24.84 What DNA sequence is complementary to the following sequence?

A—G—T—T—C—A—T—C—G

MATHEMATICAL OPERATIONS

appendix

A.1 ➤ SCIENTIFIC NOTATION

The numbers that you encounter in chemistry are often either very large or very small. For example, there are about 33,000,000,000,000,000,000,000 H_2O molecules in 1.0 mL of water, and the distance between the H and O atoms in an H_2O molecule is 0.000 000 000 095 7 m. These quantities are more conveniently written in scientific notation as 3.3×10^{22} molecules and 9.57×10^{-11} m, respectively. In scientific notation, numbers are written in the exponential format $A \times 10^n$, where A is a number between 1 and 10, and the exponent n is a positive or negative integer.

How do you convert a number from ordinary notation to scientific notation? If the number is greater than or equal to 10, shift the decimal point to the *left* by n places until you obtain a number between 1 and 10. Then, multiply the result by 10^n. For example, the number 8137.6 is written in scientific notation as 8.1376×10^3:

$$8137.6 = 8.1376 \times 10^3$$

Number of places decimal point was shifted to the left

Shift decimal point to the left by 3 places to get a number between 1 and 10

When you shift the decimal point to the left by three places, you are in effect dividing the number by $10 \times 10 \times 10 = 1000 = 10^3$. Therefore, you must multiply the result by 10^3 so that the value of the number is unchanged.

To convert a number less than 1 to scientific notation, shift the decimal point to the *right* by n places until you obtain a number between 1 and 10. Then, multiply the result by 10^{-n}. For example, the number 0.012 is written in scientific notation as 1.2×10^{-2}:

$$0.012 = 1.2 \times 10^{-2}$$

Number of places decimal point was shifted to the right

Shift decimal point to the right by 2 places to get a number between 1 and 10

When you shift the decimal point to the right by two places, you are in effect multiplying the number by $10 \times 10 = 100 = 10^2$. Therefore, you must multiply the result by 10^{-2} so that the value of the number is unchanged. ($10^2 \times 10^{-2} = 10^0 = 1$.)

The following table gives some additional examples. To convert from scientific notation to ordinary notation, simply reverse the preceding process. Thus, to write the number 5.84×10^4 in ordinary notation, drop the factor of 10^4 and move the decimal point by 4 places to the *right* ($5.84 \times 10^4 = 58{,}400$). To write the number 3.5×10^{-1} in ordinary notation, drop the factor of 10^{-1} and move the decimal point by 1 place to the *left* ($3.5 \times 10^{-1} = 0.35$). Note that you don't need scientific notation for numbers between 1 and 10 because $10^0 = 1$.

Number	Scientific Notation
58,400	5.84×10^4
0.35	3.5×10^{-1}
7.296	$7.296 \times 10^0 = 7.296 \times 1$

Addition and Subtraction

To add or subtract two numbers expressed in scientific notation, both numbers must have the same exponent. Thus, to add 7.16×10^3 and 1.32×10^2, first write the latter number as 0.132×10^3 and then add:

$$\begin{array}{r} 7.16\ \times 10^3 \\ +\ 0.132 \times 10^3 \\ \hline 7.29\ \times 10^3 \end{array}$$

The answer has three significant figures. (Significant figures are discussed in Section 1.10.) Alternatively, you can write the first number as 71.6×10^2 and then add:

$$\begin{array}{r} 71.6\ \times 10^2 \\ +\ \ 1.32 \times 10^2 \\ \hline 72.9\ \times 10^2 = 7.29 \times 10^3 \end{array}$$

Multiplication and Division

To multiply two numbers expressed in scientific notation, multiply the factors in front of the powers of 10 and then add the exponents:

$$(A \times 10^n)(B \times 10^m) = AB \times 10^{n+m}$$

For example,

$$(2.5 \times 10^4)(4.7 \times 10^7) = (2.5)(4.7) \times 10^{4+7} = 12 \times 10^{11} = 1.2 \times 10^{12}$$

$$(3.46 \times 10^5)(2.2 \times 10^{-2}) = (3.46)(2.2) \times 10^{5+(-2)} = 7.6 \times 10^3$$

Both answers have two significant figures.

To divide two numbers expressed in scientific notation, divide the factors in front of the powers of 10 and then subtract the exponent in the denominator from the exponent in the numerator:

$$\frac{A \times 10^n}{B \times 10^m} = \frac{A}{B} \times 10^{n-m}$$

For example,

$$\frac{3 \times 10^6}{7.2 \times 10^2} = \frac{3}{7.2} \times 10^{6-2} = 0.4 \times 10^4 = 4 \times 10^3 \quad \text{(1 significant figure)}$$

$$\frac{7.50 \times 10^{-5}}{2.5 \times 10^{-7}} = \frac{7.50}{2.5} \times 10^{-5-(7)} = 3.0 \times 10^2 \quad \text{(2 significant figures)}$$

Powers and Roots

To raise a number $A \times 10^n$ to a power m, raise the factor A to the power m and then multiply the exponent n by the power m:

$$(A \times 10^n)^m = A^m \times 10^{n \times m}$$

For example, 3.6×10^2 raised to the 3rd power is 4.7×10^7:

$$(3.6 \times 10^2)^3 = (3.6)^3 \times 10^{2 \times 3} = 47 \times 10^6 = 4.7 \times 10^7 \quad \text{(2 significant figures)}$$

To take the mth root of a number $A \times 10^n$, raise the number to the power $1/m$. That is, raise factor A to the power $1/m$ and then divide the exponent n by the root m:

$$\sqrt[m]{A \times 10^n} = (A \times 10^n)^{1/m} = A^{1/m} \times 10^{n/m}$$

For example, the square root of 9.0×10^8 is 3.0×10^4:

$$\sqrt[2]{9.0 \times 10^8} = (9.0 \times 10^8)^{1/2} = (9.0)^{1/2} \times 10^{8/2} = 3.0 \times 10^4 \quad \text{(2 significant figures)}$$

Because the exponent in the answer (n/m) is an integer we must sometimes rewrite the original number by shifting the decimal point so that the exponent n is an integral multiple of the root m. For example, to take the cube root of 6.4×10^{10}, we first rewrite this number as 64×10^9 so that the exponent (9) is an integral multiple of the root 3:

$$\sqrt[3]{6.4 \times 10^{10}} = \sqrt[3]{64 \times 10^9} = (64)^{1/3} \times 10^{9/3} = 4.0 \times 10^3$$

Scientific Notation and Electronic Calculators

With a scientific calculator you can carry out calculations in scientific notation. You should consult the instruction manual for your particular calculator to learn how to enter and manipulate numbers expressed in an exponential format. On most calculators, you enter the number $A \times 10^n$ by (i) entering the number A, (ii) pressing a key labeled EXP or EE, and (iii) entering the exponent n. If the exponent is negative, you press a key labeled $+/-$ before entering the value of n. (Note that you do not enter the number

10.) The calculator displays the number $A \times 10^n$ with the number A on the left followed by some space and then the exponent n. For example,

$$4.625 \times 10^2 \quad \text{is displayed as} \quad 4.625 \quad 02$$

To add, subtract, multiply, or divide exponential numbers, use the same sequence of key strokes as you would in working with ordinary numbers. When you add or subtract on a calculator, the numbers need not have the same exponent; the calculator automatically takes account of the different exponents. Remember, though, that the calculator often gives more digits in the answer than the allowed number of significant figures. It's sometimes helpful to outline the calculation on paper, as in the preceding examples, to keep track of the number of significant figures.

Most calculators have x^2 and $\sqrt{x}$ keys for squaring a number and finding its square root. Just enter the number, and press the appropriate key. You probably have a y^x (or a^x) key for raising a number to a power. To raise 4.625×10^2 to the 3rd power, for example, use the following keystrokes: (i) Enter the number 4.625×10^3 in the usual way, (ii) press the y^x key, (iii) enter the power 3, and (iv) press the = key. The result is displayed as 9.8931641 07, but it must be rounded to 4 significant figures. Therefore, $(4.625 \times 10^2)^3 = 9.893 \times 10^7$.

To take the mth root of a number, raise the number to the power $1/m$. For example, to take the 5th root of 4.52×10^{11}, use the following keystrokes: (i) Enter the number 4.52×10^{11}, (ii) press the y^x key, (iii) enter the number 5 (for the 5th root), (iv) press the $1/x$ key (to convert the 5th root to the power $1/5$), and (v) press the = key. The result is

$$\sqrt[5]{4.52 \times 10^{11}} = (4.52 \times 10^{11})^{1/5} = 2.14 \times 10^2$$

The calculator is able to handle the nonintegral exponent $11/5$, so there is no need to enter the number as 45.2×10^{10} to make the exponent an integral multiple of the root 5.

⌐ **PROBLEM A.1** Perform the following calculations, expressing the results in scientific notation with the correct number of significant figures. (You don't need a calculator for these.)

(a) $(1.50 \times 10^4) + (5.04 \times 10^3)$ **(b)** $(2.5 \times 10^{-2}) - (5.0 \times 10^{-3})$ **(c)** $(4.0 \times 10^4)^2$
(d) $\sqrt[3]{8 \times 10^{12}}$ **(e)** $\sqrt{2.5 \times 10^5}$

ANSWERS: **(a)** 2.00×10^4 **(b)** 2.0×10^{-2} **(c)** 1.6×10^9
(d) 2×10^4 **(e)** 5.0×10^2

⌐ **PROBLEM A.2** Perform the following calculations, expressing the results in scientific notation with the correct number of significant figures. (Use a calculator for these.)

(a) $(9.72 \times 10^{-1}) + (3.4823 \times 10^2)$ **(b)** $(3.772 \times 10^3) - (2.891 \times 10^4)$
(c) $(7.62 \times 10^{-3})^4$ **(d)** $\sqrt[3]{8.2 \times 10^7}$ **(e)** $\sqrt[5]{3.47 \times 10^{-12}}$

ANSWERS: **(a)** 3.4920×10^2 **(b)** -2.514×10^4 **(c)** 3.37×10^{-9}
(d) 4.3×10^2 **(e)** 5.11×10^{-3}

A.2 ►LOGARITHMS

Common Logarithms

Any positive number x can be written as 10 raised to some power z—that is, $x = 10^z$. The exponent z is called the *common*, or *base 10*, *logarithm* of the number x and is denoted by $\log_{10} x$, or simply $\log x$:

$$x = 10^z \qquad \log x = z$$

For example, 100 can be written as 10^2, and $\log 100$ is therefore equal to 2:

$$100 = 10^? \qquad \log 100 = 2$$

Similarly,

$$10 = 10^1 \qquad \log 10 = 1$$
$$1 = 10^0 \qquad \log 1 = 0$$
$$0.1 = 10^{-1} \qquad \log 0.1 = -1$$

In general, the logarithm of a number x is the power z to which 10 must be raised to equal the number x.

As Figure A.1 shows, the logarithm of a number greater than 1 is positive, the logarithm of 1 is zero, and the logarithm of a positive number less than 1 is negative. The logarithm of a *negative* number is undefined because 10 raised to any power is always positive ($x = 10^z > 0$).

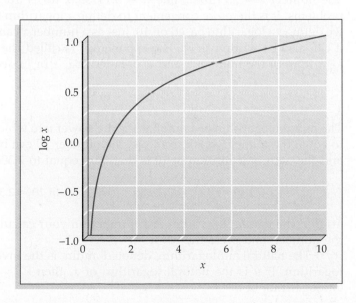

FIGURE A.1 Values of $\log x$ for values of x in the range 0.1 to 10.

You can use a calculator to find the logarithm of a number that is not an integral power of 10. For example, to find the logarithm of 60.2, simply enter 60.2, and press the LOG key. The logarithm should be between 1 and 2 because 60.2 is between 10^1 and 10^2. The calculator gives a value of 1.7795965, which must be rounded to 1.780 because 60.2 has three significant figures. The only significant figures in a logarithm are the digits beyond the decimal point, as explained in Section 15.5.

Antilogarithms

The antilogarithm, denoted antilog, is the inverse of the common logarithm. If z is the logarithm of x, then x is the antilogarithm of z. But since x can be written as 10^z , the antilogarithm of z is 10^z:

$$\text{If } z = \log x \quad \text{then} \quad x = \text{antilog } z = 10^z$$

In other words, the antilog of a number is 10 raised to a power equal to that number. For example, the antilog of 2 is $10^2 = 100$, and the antilog of 0.70 is $10^{0.70}$.

To find the value of antilog 0.70, use your calculator. If you have a 10^x key, enter 0.70 and press the 10^x key. If you have a y^x key, use the following keystrokes: (i) Enter 10, (ii) press the y^x key, (iii) enter the exponent 0.70, and (iv) press the = key. If you have an INV (inverse) key, enter 0.70, press the INV key, and then press the LOG key. The result is

$$\text{antilog } 0.70 = 5.0 \quad \text{(2 significant figures)}$$

Natural Logarithms

The number $e = 2.71828...$, like $\pi = 3.14159...$, turns up in many scientific problems. It is therefore convenient to define a logarithm based on e, just as we defined a logarithm based on 10. Just as a number x can be written as 10^z, it can also be written as e^u. The exponent u is called the *natural*, or *base e*, logarithm of the number x and is denoted $\log_e x$ or, more commonly, $\ln x$:

$$x = e^u \qquad \ln x = u$$

The natural logarithm of a number x is the power u to which e must be raised to equal the number x. For example, the number 10 can be written as $e^{2.303}$, and the natural logarithm of 10 is therefore equal to 2.303:

$$10 = e^{2.303} = (2.71828...)^{2.303} \qquad \ln 10 = 2.303$$

To find the natural logarithm of a number on your calculator, simply enter the number, and press the LN key.

The natural antilogarithm, denoted antiln, is the inverse of the natural logarithm. If u is the natural logarithm of x, then x ($= e^u$) is the natural antilogarithm of u:

$$\text{If } u = \ln x \quad \text{then} \quad x = \text{antiln } u = e^u$$

In other words, the natural antilogarithm of a number is e raised to a power equal to that number. For example, the natural antilogarithm of 0.70 is $e^{0.70}$, which equals 2.0:

$$\text{antiln } 0.70 = e^{0.70} = 2.0$$

Your calculator probably has an INV (inverse) key or an e^x key. To find the natural antilogarithm of a number—say, 0.70—enter 0.70, press the INV key, and then press the LN key. Alternatively, you can enter 0.70, and press the e^x key.

Some Mathematical Properties of Logarithms

Because logarithms are exponents, the algebraic properties of exponents can be used to derive the following useful relationships involving logarithms:

1. The logarithm (either common or natural) of a product xy equals the sum of the logarithm of x and the logarithm of y:

$$\log xy = \log x + \log y \qquad \ln xy = \ln x + \ln y$$

2. The logarithm of a quotient x/y equals the difference between the logarithm of x and the logarithm of y:

$$\log \frac{x}{y} = \log x - \log y \qquad \ln \frac{x}{y} = \ln x - \ln y$$

It follows from these relationships that

$$\log \frac{y}{x} = -\log \frac{x}{y} \qquad \ln \frac{y}{x} = -\ln \frac{x}{y}$$

Because $\log 1 = \ln 1 = 0$, it also follows that

$$\log \frac{1}{x} = -\log x \qquad \ln \frac{1}{x} = -\ln x$$

3. The logarithm of x raised to a power a equals a times the logarithm of x:

$$\log x^a = a \log x \qquad \ln x^a = a \ln x$$

Similarly,

$$\log x^{1/a} = \frac{1}{a} \log x \qquad \ln x^{1/a} = \frac{1}{a} \ln x$$

where $x^{1/a} = \sqrt[a]{x}$

What is the numerical relationship between the common logarithm and the natural logarithm? To derive it, we begin with the definitions of log x and ln x:

$$\log x = z \quad \text{where} \quad x = 10^z$$

$$\ln x = u \quad \text{where} \quad x = e^u$$

We then write ln x in terms of 10^z and make use of the property that $\ln x^a = a \ln x$:

$$\ln x = \ln 10^z = z \ln 10$$

Because $z = \log x$ and $\ln 10 = 2.303$, we find that the natural logarithm is 2.303 times the common logarithm:

$$\ln x = 2.303 \log x$$

┌ PROBLEM A.3 Use a calculator to evaluate the following expressions.
(a) log 705   **(b)** ln (3.4×10^{-6}) **(c)** antilog (-2.56) **(d)** antiln 8.1

ANSWERS: **(a)** 2.848   **(b)** −12.59   **(c)** 2.8×10^{-3} **(d)** 3×10^3 ┘

A.3 ➤ STRAIGHT-LINE GRAPHS AND LINEAR EQUATIONS

The results of a scientific experiment are often summarized in the form of a graph. Consider an experiment in which some property y is measured as a function of some variable x. (A real example would be measurement of the volume of a gas as a function of its temperature, but we'll use y and x to keep the discussion general.) Suppose that we obtain the following experimental data:

x	y
−1	−5
1	1
3	7
5	13

The graph in Figure A.2 shows values of x, called the independent variable, along the horizontal axis and values of y, the dependent variable, along the vertical axis. Each pair of experimental values of x and y is represented by a point on the graph. For this particular experiment, the four data points lie on a straight line.

The equation of a straight line can be written as

$$y = mx + b$$

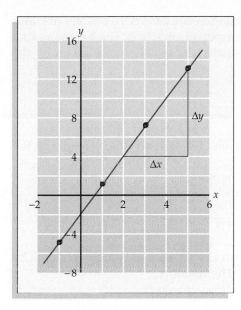

FIGURE A.2 A straight-line *y* versus *x* plot of the data in the table.

where *m* is the slope of the line and *b* is the intercept, the value of *y* at the point where the line crosses the *y* axis—that is, the value of *y* when *x* = 0. The slope of the line is the change in *y* (Δ*y*) for a given change in *x* (Δ*x*):

$$m = \text{slope } = \frac{\Delta y}{\Delta x}$$

The right triangle in Figure A.2 shows that *y* changes from 4 to 13 when *x* changes from 2 to 5. Therefore, the slope of the line is 3:

$$m = \text{slope } = \frac{\Delta y}{\Delta x} = \frac{13 - 4}{5 - 2} = \frac{9}{3} = 3$$

The graph shows a *y*-intercept of −2 (*b* = −2), and the equation of the line is therefore

$$y = 3x - 2$$

An equation of the form *y* = *mx* + *b* is called a *linear equation* because values of *x* and *y* that satisfy such an equation are the coordinates of points that lie on a straight line. We also say that *y* is a *linear function* of *x*, or that *y* is *directly proportional to x*. In our example, the rate of change of *y* is 3 times that of *x*.

A.4 ➤QUADRATIC EQUATIONS

A quadratic equation is an equation that can be written in the form

$$ax^2 + bx + c = 0$$

where a, b, and c are constants. The equation contains only powers of x and is called quadratic because the highest power of x is 2. The solutions to a quadratic equation (values of x that satisfy the equation) are given by the *quadratic formula*:

$$x = \frac{-b \pm \sqrt{b^2 - 4ac}}{2a}$$

The $\pm$ indicates that there are two solutions, one given by the $+$ sign, and the other given by the $-$ sign.

As an example, let's solve the equation

$$x^2 = \frac{2 - 6x}{3}$$

First, we put the equation into the form $ax^2 + bx + c = 0$ by multiplying it by 3 and moving $2 - 6x$ to the left side. The result is

$$3x^2 + 6x - 2 = 0$$

Then we apply the quadratic formula with $a = 3$, $b = 6$, and $c = -2$:

$$x = \frac{-6 \pm \sqrt{(6)^2 - 4(3)(-2)}}{2(3)}$$

$$= \frac{-6 \pm \sqrt{36 + 24}}{6} = \frac{-6 \pm \sqrt{60}}{6} = \frac{-6 \pm 7.746}{6}$$

The two solutions are

$$x = \frac{-6 + 7.746}{6} = \frac{1.746}{6} = 0.291 \quad \text{and} \quad x = \frac{-6 - 7.746}{6} = \frac{-13.746}{6} = -2.291$$

THERMODYNAMIC *appendix* B
PROPERTIES
AT 25°C

TABLE B.1 Inorganic Substances

Substance and state	ΔH_f° (kJ/mol)	ΔG_f° (kJ/mol)	S° [J/(K · mol)]
Aluminum			
$Al(s)$	0	0	28.3
$AlCl_3(s)$	−704.2	−628.9	110.7
$Al_2O_3(s)$	−1676	−1582	50.9
Barium			
$Ba(s)$	0	0	62.8
$BaCO_3(s)$	1216	−1138	112
$BaSO_4(s)$	−1473	−1362	132
Beryllium			
$Be(s)$	0	0	9.5
BeO	−609.6	−580.3	14.1
Bromine			
$Br(g)$	111.9	82.4	174.9
$Br^-(aq)$	−121.5	−104.0	82.4
$Br_2(l)$	0	0	152.2
$Br_2(g)$	30.9	3.14	245.4
$HBr(g)$	−36.4	−53.4	198.6
Calcium			
$Ca(s)$	0	0	41.4
$CaC_2(s)$	−59.8	−64.8	70.0
$CaCO_3(s)$	−1206.9	−1128.8	92.9
$Ca(OH)_2(s)$	−986.1	−898.6	83.4
$CaO(s)$	−635.1	−604.0	39.7
$CaSO_4(s)$	−1434.1	−1321.9	107
Carbon			
$C(s, graphite)$	0	0	5.7
$C(s, diamond)$	1.9	2.9	2.4
$CO(g)$	−110.5	−137.2	197.6
$CO_2(g)$	−393.5	−394.4	213.6
Chlorine			
$Cl(g)$	121.7	105.7	165.1
$Cl^-(aq)$	−167.2	−131.3	56.5
$Cl_2(g)$	0	0	223.0
$HCl(g)$	−92.3	−95.3	186.8
Chromium			
$Cr(s)$	0	0	23.8
$Cr_2O_3(s)$	−1140	−1058	81.2

TABLE B.1 continued

Substance and state	ΔH_f° (kJ/mol)	ΔG_f° (kJ/mol)	S° [J/(K · mol)]
Copper			
$Cu(s)$	0	0	33.1
$CuO(s)$	−157	−130	42.6
$Cu_2S(s)$	−79.5	−86.2	120.9
Fluorine			
$F(g)$	79.0	61.9	158.6
$F^-(aq)$	−332.6	−278.8	−13.8
$F_2(g)$	0	0	202.7
$HF(g)$	−271	−273	173.7
Hydrogen			
$H(g)$	218.0	203.3	114.6
$H^+(aq)$	0	0	0
$H_2(g)$	0	0	130.6
$OH^-(aq)$	−230.0	−157.3	−10.8
$H_2O(l)$	−285.8	−237.2	69.9
$H_2O(g)$	−241.8	−228.6	188.7
$H_2O_2(l)$	−187.8	−120.4	110
Iodine			
$I(g)$	106.8	70.3	180.7
$I^-(aq)$	−55.2	−51.6	111
$I_2(s)$	0	0	116.1
$I_2(g)$	62.4	19.4	260.6
$HI(g)$	26.5	1.7	206.5
Iron			
$Fe(s)$	0	0	27.3
$FeO(s)$	−272	−255	61
$Fe_2O_3(s)$	−824.2	−742.2	87.4
$Fe_3O_4(s)$	−1118	−1015	146
Lead			
$Pb(s)$	0	0	64.8
$PbO_2(s)$	−277	−217.4	68.6
$PbS(s)$	−100	−98.7	91.2
$PbSO_4(s)$	−919.9	−813.2	148.6
Lithium			
$Li(s)$	0	0	29.1
$Li(g)$	159.4	126.7	138.7
$Li^+(aq)$	−278.5	−293.3	13
Magnesium			
$Mg(s)$	0	0	32.7
$MgCO_3(s)$	−1096	−1012	65.7
$MgCl_2(s)$	−641.6	−591.8	89.6
$MgO(s)$	−601.7	−569.4	26.9
Manganese			
$Mn(s)$	0	0	32.0
$MnO(s)$	−385.2	−362.9	59.7
$MnO_2(s)$	−520.0	−465.1	53.1
Mercury			
$Hg(l)$	0	0	76.0
$Hg(g)$	61.32	31.85	174.8
$HgCl_2(s)$	−224	−179	146
$Hg_2Cl_2(s)$	−265.2	−210.8	192
$HgO(s)$	−90.8	−58.6	70.3
$HgS(s)$	−58.2	−50.6	82.4

TABLE B.1   continued

Substance and state	ΔH_f° (kJ/mol)	ΔG_f° (kJ/mol)	S° [J/(K · mol)]
Nickel			
Ni(s)	0	0	29.9
NiCl$_2$(s)	−305.3	−259.1	97.7
NiO(s)	−240	−212	38.0
NiS(s)	−82.0	−79.5	53.0
Nitrogen			
N(g)	472.7	455.6	153.2
N$_2$(g)	0	0	191.5
NH$_3$(g)	−46.1	−16.5	192.3
NH$_3$(aq)	−80.3	−26.6	111
NH$_4^+$(aq)	−132.5	−79.4	113
N$_2$H$_4$(l)	50.6	149.2	121.2
N$_2$H$_4$(g)	95.4	159.3	238.4
NO(g)	90.2	86.6	210.7
NO$_2$(g)	33.2	51.3	240.0
N$_2$O(g)	82.0	104.2	219.7
N$_2$O$_4$(g)	9.16	97.8	304.2
N$_2$O$_5$(g)	11	115	356
HNO$_3$(l)	−174.1	−80.8	155.6
HNO$_3$(aq)	−207.4	−111.3	146
NH$_4$NO$_3$(s)	−365.6	−184.0	151.1
Oxygen			
O(g)	249.2	231.7	160.9
O$_2$(g)	0	0	205.0
O$_3$(g)	143	163	238.8
Phosphorus			
P(s, white)	0	0	41.1
P(s, red)	−18	−12	22.8
P$_4$(g)	58.9	24.5	279.9
PCl$_3$(l)	−320	−272	217
PH$_3$(g)	5.4	13	210.1
P$_4$O$_{10}$(s)	−2984	−2698	228.9
H$_3$PO$_4$(s)	−1279	−1119	110.5
Potassium			
K(s)	0	0	64.2
K(g)	89.2	60.6	160.2
K$^+$(aq)	−252.4	−283.3	103
KCl(s)	−436.7	−409.2	82.6
KClO$_3$(s)	−397.7	−296.3	143
KOH(s)	−424.8	−379.1	78.9
KOH(aq)	−482.4	−440.5	91.6
Selenium			
Se(s, black)	0	0	42.44
H$_2$Se(g)	29.7	15.9	219.0
Silicon			
Si(s)	0	0	18.8
SiCl$_4$(l)	−687.0	−619.9	240
SiO$_2$(s, quartz)	−910.9	−856.7	41.8

TABLE B.1   concluded

Substance and state	ΔH_f° (kJ/mol)	ΔG_f° (kJ/mol)	S° [J/(K · mol)]
Silver			
Ag(s)	0	0	42.6
Ag$^+$(aq)	105.6	77.1	72.7
AgBr(s)	−100.4	−96.9	107
AgCl(s)	−127.1	−109.8	96.2
AgI(s)	−61.8	−66.2	115
Ag$_2$O(s)	−31.0	−11.2	121
Ag$_2$S(s)	−32.6	−40.7	144.0
Sodium			
Na(s)	0	0	51.2
Na(g)	107.3	76.8	153.6
Na$^+$(aq)	−240.1	−261.9	59.0
NaBr(s)	−361.1	−349.0	86.8
Na$_2$CO$_3$(s)	−1130.7	−1044.5	135.0
NaHCO$_3$(s)	−950.8	−851.0	102
NaCl(s)	−411.2	−384.2	72.1
NaOH(s)	−425.6	−379.5	64.5
NaOH(aq)	−470.1	−419.2	48.2
Na$_2$O(s)	−414.2	−375.5	75.1
Na$_2$O$_2$(s)	−510.9	−447.7	95.0
Sulfur			
S(s, rhombic)	0	0	31.8
H$_2$S(g)	−20.6	−33.6	205.7
SO$_2$(g)	−296.8	−300.2	248.1
SO$_3$(g)	−395.7	−371.1	256.6
H$_2$SO$_4$(l)	−814.0	−690.1	156.9
H$_2$SO$_4$(aq)	−909.3	−744.6	20
HSO$_4$$^-$(aq)	−887.3	−756.0	132
SO$_4$$^{2-}$(aq)	−909.3	−744.6	20
Tin			
Sn(s, white)	0	0	51.5
Sn(s, gray)	−2.1	0.1	44.1
SnO(s)	−286	−257	56.5
SnO$_2$(s)	−580.7	−519.7	52.3
Titanium			
Ti(s)	0	0	30.6
TiCl$_4$(l)	−804.2	−737.2	252.3
TiO$_2$(s)	−944.7	−889.5	50.3
Zinc			
Zn(s)	0	0	41.6
Zn(g)	130.7	95.2	160.9
ZnO(s)	−348.3	−318.3	43.6
ZnS(s)	−206.0	−201.3	57.7

TABLE B.2   Organic Substances

Substance and state	Formula	ΔH_f° (kJ/mol)	ΔG_f° (kJ/mol)	S° [J/(K · mol)]
Acetic acid(l)	CH_3COOH	−484.5	−390	160
Acetylene(g)	C_2H_2	226.7	209.2	200.8
Benzene(l)	C_6H_6	49.0	124.5	172.8
Butane(g)	C_4H_{10}	−126	−17	310
Carbon tetrachloride(l)	CCl_4	−135.4	−65.3	216.4
Dichloroethane(l)	CH_2ClCH_2Cl	−165.2	−79.6	208.5
Ethane(g)	C_2H_6	−84.7	−32.9	229.5
Ethanol(l)	C_2H_5OH	−277.7	−174.9	161
Ethanol(g)	C_2H_5OH	−235.1	−168.6	282.6
Ethylene(g)	C_2H_4	52.3	68.1	219.5
Ethylene oxide(g)	C_2H_4O	−52.6	−13.1	242.4
Formic acid(l)	$HCOOH$	−424.7	−361.4	129.0
Glucose(s)	$C_6H_{12}O_6$	−1260	−910	212.1
Methane(g)	CH_4	−74.8	−50.8	186.2
Methanol(l)	CH_3OH	−238.7	−166.4	127
Methanol(g)	CH_3OH	−201.2	−161.9	238
Propane(g)	C_3H_8	105	−25	270
Vinyl chloride	$CH_2{=}CHCl$	35	51.9	263.9

appendix C

EQUILIBRIUM CONSTANTS AT 25°C

TABLE C.1	Acid-Dissociation Constants at 25°C			
Acid	**Formula**	K_{a1}	K_{a2}	K_{a3}
Acetic	CH_3COOH	1.8×10^{-5}		
Acetylsalicylic	$C_9H_8O_4$	3.0×10^{-4}		
Arsenic	H_3AsO_4	5.6×10^{-3}	1.7×10^{-7}	4.0×10^{-12}
Arsenious	H_3AsO_3	6×10^{-10}		
Ascorbic	$C_6H_8O_6$	8.0×10^{-5}		
Benzoic	C_6H_5COOH	6.5×10^{-5}		
Boric	H_3BO_3	5.8×10^{-10}		
Carbonic	H_2CO_3	4.3×10^{-7}	5.6×10^{-11}	
Chloroacetic	$CH_2ClCOOH$	1.4×10^{-3}		
Citric	$C_6H_8O_7$	7.1×10^{-4}	1.7×10^{-5}	4.1×10^{-7}
Formic	$HCOOH$	1.8×10^{-4}		
Hydrazoic	HN_3	1.9×10^{-5}		
Hydrocyanic	HCN	4.9×10^{-10}		
Hydrofluoric	HF	3.5×10^{-4}		
Hydrogen peroxide	H_2O_2	2.4×10^{-12}		
Hydrosulfuric	H_2S	1.0×10^{-7}	$\sim 10^{-19}$	
Hypobromous	$HOBr$	2.0×10^{-9}		
Hypochlorous	$HOCl$	3.5×10^{-8}		
Hypoiodous	HOI	2.3×10^{-11}		
Iodic	HIO_3	1.7×10^{-1}		
Lactic	$HC_3H_5O_3$	1.4×10^{-4}		
Nitrous	HNO_2	4.5×10^{-4}		
Oxalic	$H_2C_2O_4$	5.9×10^{-2}	6.4×10^{-5}	
Phenol	C_6H_5OH	1.3×10^{-10}		
Phosphoric	H_3PO_4	7.5×10^{-3}	6.2×10^{-8}	4.8×10^{-13}
Phosphorous	H_3PO_3	1.0×10^{-2}	2.6×10^{-7}	
Saccharin	$C_7H_5NO_3S$	2.1×10^{-12}		
Selenic	H_2SeO_4	Very large	1.2×10^{-2}	
Selenious	H_2SeO_3	3.5×10^{-2}	5×10^{-8}	
Sulfuric	H_2SO_4	Very large	1.2×10^{-2}	
Sulfurous	H_2SO_3	1.5×10^{-2}	6.3×10^{-8}	
Tartaric	$C_4H_6O_6$	1.0×10^{-3}	4.6×10^{-5}	
Water	H_2O	1.8×10^{16}		

TABLE C.2   Acid-Dissociation Constants at 25°C for Hydrated Metal Cations

Cation	K_a	Cation	K_a
$Fe^{2+}(aq)$	3.2×10^{-10}	$Be^{2+}(aq)$	3×10^{-7}
$Co^{2+}(aq)$	1.3×10^{-9}	$Al^{3+}(aq)$	1.4×10^{-5}
$Ni^{2+}(aq)$	2.5×10^{-11}	$Cr^{3+}(aq)$	1.6×10^{-4}
$Zn^{2+}(aq)$	2.5×10^{-10}	$Fe^{3+}(aq)$	6.3×10^{-3}

Note: As an example, K_a for $Fe^{2+}(aq)$ is the equilibrium constant for the reaction

$$Fe(H_2O)_6{}^{2+}(aq) + H_2O(l) \rightleftarrows H_3O^+(aq) + Fe(H_2O)_5(OH)^+(aq)$$

TABLE C.3   Base-Dissociation Constants at 25°C

Base	Formula	K_b
Ammonia	NH_3	1.8×10^{-5}
Aniline	$C_6H_5NH_2$	4.3×10^{-10}
Codeine	$C_{18}H_{21}NO_3$	1.6×10^{-6}
Dimethylamine	$(CH_3)_2NH$	5.4×10^{-4}
Ethylamine	$C_2H_5NH_2$	6.4×10^{-4}
Hydrazine	N_2H_4	8.9×10^{-7}
Hydroxylamine	NH_2OH	9.1×10^{-9}
Methylamine	CH_3NH_2	3.7×10^{-4}
Morphine	$C_{17}H_{19}NO_3$	1.6×10^{-6}
Piperidine	$C_5H_{11}N$	1.3×10^{-3}
Propylamine	$C_3H_7NH_2$	5.1×10^{-4}
Pyridine	C_5H_5N	1.8×10^{-9}
Strychnine	$C_{21}H_{22}N_2O_2$	1.8×10^{-6}
Trimethylamine	$(CH_3)_3N$	6.5×10^{-5}

TABLE C.4 Solubility Product Constants at 25°C

Compound	Formula	K_{sp}
Aluminum hydroxide	$Al(OH)_3$	1.9×10^{-33}
Barium carbonate	$BaCO_3$	2.6×10^{-9}
Barium chromate	$BaCrO_4$	1.2×10^{-10}
Barium fluoride	BaF_2	1.8×10^{-7}
Barium hydroxide	$Ba(OH)_2$	5.0×10^{-3}
Barium sulfate	$BaSO_4$	1.1×10^{-10}
Cadmium carbonate	$CdCO_3$	6.2×10^{-12}
Cadmium hydroxide	$Cd(OH)_2$	5.3×10^{-15}
Calcium carbonate	$CaCO_3$	5.0×10^{-9}
Calcium fluoride	CaF_2	1.5×10^{-10}
Calcium hydroxide	$Ca(OH)_2$	4.7×10^{-6}
Calcium phosphate	$Ca_3(PO_4)_2$	2.1×10^{-33}
Calcium sulfate	$CaSO_4$	7.1×10^{-5}
Chromium(III) hydroxide	$Cr(OH)_3$	6.7×10^{-31}
Cobalt(II) hydroxide	$Co(OH)_2$	1.1×10^{-15}
Copper(I) bromide	$CuBr$	6.3×10^{-9}
Copper(I) chloride	$CuCl$	1.7×10^{-7}
Copper(II) carbonate	$CuCO_3$	2.5×10^{-10}
Copper(II) hydroxide	$Cu(OH)_2$	1.6×10^{-19}
Copper(II) phosphate	$Cu_3(PO_4)_2$	1.4×10^{-37}
Iron(II) hydroxide	$Fe(OH)_2$	4.9×10^{-17}
Iron(III) hydroxide	$Fe(OH)_3$	2.6×10^{-39}
Lead bromide	$PbBr_2$	6.6×10^{-6}
Lead chloride	$PbCl_2$	1.2×10^{-5}
Lead chromate	$PbCrO_4$	2.8×10^{-13}
Lead iodide	PbI_2	8.5×10^{-9}
Lead sulfate	$PbSO_4$	1.8×10^{-8}
Magnesium carbonate	$MgCO_3$	6.8×10^{-6}
Magnesium fluoride	MgF_2	7.4×10^{-11}
Magnesium hydroxide	$Mg(OH)_2$	5.6×10^{-12}
Manganese(II) carbonate	$MnCO_3$	2.2×10^{-11}
Manganese(II) hydroxide	$Mn(OH)_2$	2.1×10^{-13}
Mercury(I) bromide	Hg_2Br_2	6.4×10^{-23}
Mercury(I) chloride	Hg_2Cl_2	1.4×10^{-18}
Mercury(I) iodide	Hg_2I_2	5.3×10^{-29}
Mercury(II) hydroxide	$Hg(OH)_2$	3.1×10^{-26}
Nickel(II) hydroxide	$Ni(OH)_2$	5.5×10^{-16}
Silver bromide	$AgBr$	5.4×10^{-13}
Silver carbonate	Ag_2CO_3	8.4×10^{-12}
Silver chloride	$AgCl$	1.8×10^{-10}
Silver chromate	Ag_2CrO_4	1.1×10^{-12}
Silver cyanide	$AgCN$	6.0×10^{-17}
Silver iodide	AgI	8.5×10^{-17}
Silver sulfate	Ag_2SO_4	1.2×10^{-5}
Strontium carbonate	$SrCO_3$	5.6×10^{-10}
Tin(II) hydroxide	$Sn(OH)_2$	5.4×10^{-27}
Zinc carbonate	$ZnCO_3$	1.2×10^{-10}
Zinc hydroxide	$Zn(OH)_2$	4.1×10^{-17}

TABLE C.5	Solubility Products in Acid (K_{spa}) at 25°C	

Compound	Formula	K_{spa}
Cadmium sulfide	CdS	8×10^{-7}
Cobalt(II) sulfide	CoS	3
Copper(II) sulfide	CuS	6×10^{-16}
Iron(II) sulfide	FeS	6×10^2
Lead sulfide	PbS	3×10^{-7}
Manganese(II) sulfide	MnS	3×10^{10}
Mercury(II) sulfide	HgS	2×10^{-32}
Nickel(II) sulfide	NiS	8×10^{-1}
Silver sulfide	Ag_2S	6×10^{-30}
Tin(II) sulfide	SnS	1×10^{-5}
Zinc sulfide	ZnS	3×10^{-2}

Note: K_{spa} for MS is the equilibrium constant for the reaction

$$MS(s) + 2\,H_3O^+(aq) \rightleftarrows M^{2+}(aq) + H_2S(aq) + 2\,H_2O(l)$$

We use K_{spa} for metal sulfides rather than K_{sp} because K_{a2} for $H_2S \approx 10^{-19}$ is very small and its value is uncertain (see R. J. Myers, *J. Chem. Ed.*, **1986**, 63, 687–690).

TABLE C.6	Formation Constants for Complex Ions at 25°C		

Complex ion	K_f	Complex ion	K_f
$Ag(NH_3)_2^+$	1.7×10^7	$Cu(NH_3)_4^{2+}$	1.1×10^{13}
$Ag(CN)_2^-$	1×10^{21}	$Ni(NH_3)_6^{2+}$	5.6×10^8
$Ag(S_2O_3)_2^{3-}$	2.9×10^{13}	$Zn(NH_3)_4^{2+}$	2.9×10^9
$Al(OH)_4^-$	2.1×10^{34}	$Zn(OH)_4^{2-}$	2.8×10^{15}
$Cr(OH)_4^-$	8×10^{29}		

STANDARD REDUCTION POTENTIALS AT 25°C

Half-reaction	$E°$ (V)
$F_2(g) + 2\,e^- \rightarrow 2\,F^-(aq)$	2.87
$O_3(g) + 2\,H^+(aq) + 2\,e^- \rightarrow O_2(g) + 2\,H_2O(l)$	2.08
$S_2O_8^{2-}(aq) + 2\,e^- \rightarrow 2\,SO_4^{2-}(aq)$	2.01
$Co^{3+}(aq) + e^- \rightarrow Co^{2+}(aq)$	1.81
$H_2O_2(aq) + 2\,H^+(aq) + 2\,e^- \rightarrow 2\,H_2O(l)$	1.78
$MnO_4^-(aq) + 4\,H^+(aq) + 3\,e^- \rightarrow MnO_2(s) + 2\,H_2O(l)$	1.68
$PbO_2(s) + 3\,H^+(aq) + HSO_4^-(aq) + 2\,e^- \rightarrow PbSO_4(s) + 2\,H_2O(l)$	1.628
$MnO_4^-(aq) + 8\,H^+(aq) + 5\,e^- \rightarrow Mn^{2+}(aq) + 4\,H_2O(l)$	1.51
$Cl_2(g) + 2\,e^- \rightarrow 2\,Cl^-(aq)$	1.36
$Cr_2O_7^{2-}(aq) + 14\,H^+(aq) + 6\,e^- \rightarrow 2\,Cr^{3+}(aq) + 7\,H_2O(l)$	1.33
$O_2(g) + 4\,H^+(aq) + 4\,e^- \rightarrow 2\,H_2O(l)$	1.23
$MnO_2(s) + 4\,H^+(aq) + 2\,e^- \rightarrow Mn^{2+}(aq) + 2\,H_2O(l)$	1.22
$2\,IO_3^-(aq) + 12\,H^+(aq) + 10\,e^- \rightarrow I_2(s) + 6\,H_2O(l)$	1.20
$Br_2(l) + 2\,e^- \rightarrow 2\,Br^-(aq)$	1.09
$NO_3^-(aq) + 4\,H^+(aq) + 3\,e^- \rightarrow NO(g) + 2\,H_2O(l)$	0.96
$2\,Hg^{2+}(aq) + 2\,e^- \rightarrow Hg_2^{2+}(aq)$	0.92
$Hg^{2+}(aq) + 2\,e^- \rightarrow Hg(l)$	0.85
$Ag^+(aq) + e^- \rightarrow Ag(s)$	0.80
$Hg_2^{2+}(aq) + 2\,e^- \rightarrow 2\,Hg(l)$	0.80
$Fe^{3+}(aq) + e^- \rightarrow Fe^{2+}(aq)$	0.77
$O_2(g) + 2\,H^+(aq) + 2\,e^- \rightarrow H_2O_2(aq)$	0.70
$MnO_4^-(aq) + e^- \rightarrow MnO_4^{2-}(aq)$	0.56
$I_2(s) + 2\,e^- \rightarrow 2\,I^-(aq)$	0.54
$Cu^+(aq) + e^- \rightarrow Cu(s)$	0.52
$O_2(g) + 2\,H_2O(l) + 4\,e^- \rightarrow 4\,OH^-(aq)$	0.40
$Cu^{2+}(aq) + 2\,e^- \rightarrow Cu(s)$	0.34
$Hg_2Cl_2(s) + 2\,e^- \rightarrow 2\,Hg(l) + 2\,Cl^-(aq)$	0.28
$AgCl(s) + e^- \rightarrow Ag(s) + Cl^-(aq)$	0.22
$SO_4^{2-}(aq) + 4\,H^+(aq) + 2\,e^- \rightarrow H_2SO_3(aq) + H_2O(l)$	0.17
$Cu^{2+}(aq) + e^- \rightarrow Cu^+(aq)$	0.15
$Sn^{4+}(aq) + 2\,e^- \rightarrow Sn^{2+}(aq)$	0.15
$AgBr(s) + e^- \rightarrow Ag(s) + Br^-(aq)$	0.07
$2\,H^+(aq) + 2\,e^- \rightarrow H_2(g)$	0
$Fe^{3+}(aq) + 3\,e^- \rightarrow Fe(s)$	−0.04
$Pb^{2+}(aq) + 2\,e^- \rightarrow Pb(s)$	−0.13
$Sn^{2+}(aq) + 2\,e^- \rightarrow Sn(s)$	−0.14
$Ni^{2+}(aq) + 2\,e^- \rightarrow Ni(s)$	−0.26
$Co^{2+}(aq) + 2\,e^- \rightarrow Co(s)$	−0.28
$PbSO_4(s) + H^+(aq) + 2\,e^- \rightarrow Pb(s) + HSO_4^-(aq)$	−0.296

Half-reaction	$E°$ (V)
$Cd^{2+}(aq) + 2\ e^- \rightarrow Cd(s)$	-0.40
$Cr^{3+}(aq) + e^- \rightarrow Cr^{2+}(aq)$	-0.41
$Fe^{2+}(aq) + 2\ e^- \rightarrow Fe(s)$	-0.45
$2\ CO_2(g) + 2\ H^+(aq) + 2\ e^- \rightarrow H_2C_2O_4(aq)$	-0.49
$Cr^{3+}(aq) + 3\ e^- \rightarrow Cr(s)$	-0.74
$Zn^{2+}(aq) + 2\ e^- \rightarrow Zn(s)$	-0.76
$2\ H_2O(l) + 2\ e^- \rightarrow H_2(g) + 2\ OH^-(aq)$	-0.83
$Cr^{2+}(aq) + 2\ e^- \rightarrow Cr(s)$	-0.91
$Mn^{2+}(aq) + 2\ e^- \rightarrow Mn(s)$	-1.18
$Al^{3+}(aq) + 3\ e^- \rightarrow Al(s)$	-1.66
$Mg^{2+}(aq) + 2\ e^- \rightarrow Mg(s)$	-2.37
$Na^+(aq) + e^- \rightarrow Na(s)$	-2.71
$Ca^{2+}(aq) + 2\ e^- \rightarrow Ca(s)$	-2.87
$K^+(aq) + e^- \rightarrow K(s)$	-2.93
$Li^+(aq) + e^- \rightarrow Li(s)$	-3.04

$\mathscr{A}$NSWERS TO SELECTED PROBLEMS

Chapter 1

1.1 (a) Cu; (b) Pt; (c) Pu   **1.3** (a) Ti, metal; (b) Te, semimetal; (c) Se, nonmetal; (d) Sc, metal; (e) At, semimetal; (f) Ar, nonmetal   **1.5** (a) microgram; (b) decimeter; (c) picosecond; (d) kiloampere; (e) millimole   **1.7** (a) 195 K; (b) 316°F; (c) 215°F   **1.9** Results are both precise and accurate.   **1.11** (a) 3.774 L; (b) 255 K; (c) 55.26 kg   **1.13** (a) ~2000°F; (b) ~5 × 10^{-11} cm³   **1.15** 8.88 g, 0.313 oz   **1.17** 2.212 g/cm³   *Understanding Key Concepts*   **3.** red–gas; blue–42; green–sodium;   **4.** (a) poor accuracy, good precision; (b) good accuracy, good precision; (c) poor accuracy, poor precision   **5.** Weight = 50 lb; mass = 135 lb (61.2 kg); person would fly into space.   *Additional Problems*   **1.31** (a) tellurium; (b) rhenium; (c) beryllium; (d) argon; (e) plutonium   **1.33** (a) carbon is C; (b) sodium is Na; (c) nitrogen is N; (d) chlorine is Cl   **1.47** (a) deciliter (10^{-1} L); (b) decimeter (10^{-1} m); (c) micrometer (10^{-6} m); (d) nanoliter (10^{-9} L)   **1.49** 10^6 μL/L; 2 × 10^4 μL/20 mL   **1.51** (a) A liter is just slightly larger than a quart. (b) A mile is about twice as long as a kilometer. (c) An ounce is about 30 times larger than a gram. (d) An inch is about 2.5 times larger than a centimeter.   **1.53** 3.6665 × 10^6 m³   **1.55** (a) 0.003221 mm; (b) 894,000 m; (c) 0.000 000 000 001 350 82 m³   **1.57** (a) 35,670 m (4 sig. fig.), 35,670.1 m (6 sig. fig.); (b) 69 g (2 sig. fig.), 68.5 g (3 sig. fig.); (c) 5.00 × 10^3 cm; (d) 2.3098 × 10^{-4} kg   **1.59** (a) 0.125; (b) 3.1   **1.61** (a) 0.14 m; (b) 30.08 kg; (c) 2.090 × 10^{-3} m³; (d) 29 m/s; (e) 748.0 m³   **1.63** (a) 189 cm; (b) 0.2 m³   **1.65** 10 g   **1.67** 0.61 cm/shake   **1.69** 39.9°C = 103.8°F; 22.2°C = 72.0°F   **1.71** 3410°C = 3683 K   **1.73** (a) 2.26°A/°C; (b) 1.25°A/°F; (c) mp = 175°A; bp = 401°A; (d) 259°A   **1.75** 0.179 cm³; 162,000 cm³   **1.77** 11 g/cm³   **1.79** 2.33 g/cm³   **1.85** (a) selenium; (b) antimony; (c) barium; (d) cesium; (e) neptunium   **1.87** (a) centiliter; (b) square meter; (c) kelvin; (d) nanometer   **1.89** mp = 801°C = 1474°F; bp = 1413°C = 2575°F   **1.91** 75.85 mL   **1.93** 1.990 × 10^{10} L

Chapter 2

2.1 3/2   **2.3** ~40 times   **2.5** $^{35}_{17}$Cl (18 n), $^{37}_{17}$Cl (20 n)   **2.7** $^{109}_{47}$Ag, silver   **2.9** 2.04 × 10^{22} Cu atoms   **2.11** (a) ionic; (b) molecular; (c) molecular; (d) ionic   **2.13** (a) cesium fluoride; (b) potassium oxide; (c) copper(II) oxide; (d) barium sulfide; (e) beryllium bromide   **2.15** (a) nitrogen trichloride; (b) tetraphosphorus hexoxide; (c) disulfur difluoride   **2.17** (a) calcium hypochlorite; (b) silver thiosulfate or silver(I) thiosulfate; (c) sodium dihydrogen phosphate; (d) tin(II) nitrate; (e) lead(IV) acetate   **2.19** (a) periodic acid; (b) bromous acid; (c) chromic acid   *Understanding Key Concepts*   **1.** (a)   **3.** Na (b); Ca²⁺ (c); F⁻ (a)   **5.** (d)   **6.** (a) gallium, Ga; (b) chromium, Cr; (c) aluminum, Al   **9.**

```
    H         H
    |         |
H—C—O—C—H
    |         |
    H         H
```

10. $C_3H_7NO_2$; $C_2H_6O_2$; $C_2H_4O_2$ *Additional Problems*   **2.25** $\dfrac{\text{carbon suboxide}}{CO_2} = \dfrac{3}{1}$ **2.27** C_3O_2 **2.29** Zn/S = 1/1   **2.39** (a) $^{220}_{86}$Rn; (b) $^{210}_{84}$Po; (c) $^{197}_{79}$Au   **2.41** (a) 7 p, 7 e⁻; 8 n; (b) 27 p, 27 e⁻, 33 n; (c) 53 p, 53 e⁻, 78 n   **2.43** (a) $^{24}_{12}$Mg, magnesium; (b) $^{58}_{28}$Ni, nickel   **2.45** 10.8 amu   **2.47** 25.982 amu (^{26}Mg)   **2.57** (a) 4 p, 2 e⁻; (b) 37 p, 36 e⁻; (c) 34 p, 36 e⁻; (d) 79 p, 76 e⁻   **2.59** C_3H_8O **2.61**

```
              H              H
              |              |
         H—C—H     H—C—H
     H        |        H       |       H
     |        |        |        |       |
H—C—C—C—C—C—H
     |        |        |        |       |
     H        |        H       H      H
         H—C—H
              |
              H
```

2.63 (a) I⁻; (b) $H_2PO_4^-$; (c) HCO_3^- **2.65** (a) KCl; (b) $SnBr_2$; (c) CaO; (d) $BaCl_2$; (e) AlH_3 **2.67** (a) barium ion; (b) cesium ion; (c) vanadium(III) ion; (d) hydrogen carbonate ion; (e) ammonium ion; (f) nickel(II) ion; (g) nitrite ion; (h) chlorite ion; (i) manganese(II) ion; (j) perchlorate ion   **2.69** (a) SO_3^{2-}; (b) PO_4^{3-}; (c) Zr^{4+}; (d) CrO_4^{2-}; (e) $C_2H_3O_2^-$; (f) $S_2O_3^{2-}$ **2.71** (a) zinc(II) cyanide; (b) iron(III) nitrite; (c) titanium(IV) sulfate; (d) tin(II) phosphate; (e) mercury(I) sulfide; (f) manganese(IV) oxide; (g) potassium periodate; (h) copper(II) acetate   **2.73** (a) Na_2SO_4; (b) $Ba_3(PO_4)_2$; (c) $Ga_2(SO_4)_3$

2.75 (a) sodium sulfate; (b) barium phosphate; (c) gallium sulfate   **2.77** (a) sodium bromate; (b) phosphoric acid; (c) phosphorous acid; (d) vanadium(V) oxide   **2.79** For NH_3, 0.505 g H; for N_2H_4, 0.337 g H   **2.81** TeO_4^{2-}, tellurate; TeO_3^{2-}, tellurite; TeO_4^{2-} and TeO_3^{2-} are analogous to SO_4^{2-} and SO_3^{2-}.   **2.83** (a) I^-; (b) Au^{3+}; (c) Kr   **2.85** 39.9641 amu

Chapter 3

3.1 $2 KClO_3 \longrightarrow 2 KCl + 3 O_2$ **3.3** (a) 159.7 amu; (b) 98.1 amu; (c) 192.0 amu; (d) 334.1 amu   **3.5** 6.02×10^{23} HCl molecules/mole   **3.7** 3.3 g of acetic anhydride; 5.9 g of aspirin; 2.0 g of acetic acid   **3.9** 4230 g   **3.11** Li_2O is the limiting reactant. 41 kg H_2O **3.13** (a) 0.025 mol; (b) 1.62 mol   **3.15** 690 mL   **3.17** 0.656 M   **3.19** 10.0 mL   **3.21** 0.758 M   **3.23** $MgCO_3$ **3.25** (a) C_6H_6; (b) C_2H_2; (c) $C_2H_4O_2$ *Understanding Key Concepts*   **1.** (b)   **2.** (c) $2 A + B_2 \longrightarrow A_2B_2$ **3.** $C_2H_4 + 3 O_2 \longrightarrow 2 CO_2 + 2 H_2O$ **4.** reactants (d); products (c)   *Additional Problems*   **3.27** (a) not balanced; (b) balanced; (c) not balanced   **3.29** Adding a subscript "2" to the oxygen in H_2O would change water to a new compound, hydrogen peroxide. This is not allowed.   **3.31** (a) $2 NH_4NO_3 \longrightarrow 2 N_2 + O_2 + 4 H_2O$; (b) $C_2H_6O + O_2 \longrightarrow C_2H_4O_2 + H_2O$; (c) $C_2H_8N_2 + 2 N_2O_4 \longrightarrow 3 N_2 + 2 CO_2 + 4 H_2O$ **3.33** (a) 47.88 g; (b) 159.81 g; (c) 200.59 g; (d) 18.0 g   **3.35** 5.0 mol   **3.37** 1.27 mol   **3.39** 386.7 amu   **3.41** 5×10^{18} C atoms   **3.43** 6.44×10^{-4} mol; 3.88×10^{20} molecules   **3.45** (a) 0.14 mol; (b) 0.0051 mol; (c) 2.7×10^{-3} mol   **3.47** 167 kg   **3.49** (a) $2 Fe_2O_3 + 3 C \longrightarrow 4 Fe + 3 CO_2$; (b) 4.94 mol C; (c) 59.3 g C   **3.51** (a) $2 Mg + O_2 \longrightarrow 2 MgO$; (b) 16.5 g O_2, 41.5 g MgO; (c) 38.0 g Mg, 63.0 g MgO   **3.53** (a) $2 HgO \longrightarrow 2 Hg + O_2$; (b) 42.1 g Hg, 3.4 g O_2; (c) 451 g HgO   **3.55** Al_2O_3 **3.57** 15.8 g NH_3; 83.3 g N_2 left over   **3.59** 5.22 g $C_2H_4Cl_2$ **3.61** (a) 0.1 g NaCl; (b) 5.53 g H_2SO_4 **3.63** Percent yield = 86.8%   **3.65** 38 g   **3.67** (a) 6.90 g; (b) 2.06 g   **3.69** 0.0685 M   **3.71** Na^+, 0.147 M; Ca^{2+}, 0.00298 M; K^+, 0.00402 M; Cl^-, 0.157 M   **3.73** 15.5 g   **3.75** 166 mL   **3.77** 0.958 M   **3.79** (a) % Cu = 57.5%; % O = 36.2%; % C = 5.4%; % H = 0.92%   **3.79** (b) % C = 63.5%; % H = 6.0%; % N = 9.3%; % O = 21.2%; (c) % Fe = 45.5%; % C = 25.2%; % N = 29.3%   **3.81** (a) $C_{13}H_{18}O_2$; (b) PbC_8H_{20}; (c) $ZrSiO_4$ **3.83** $C_8H_{17}N$ **3.85** High-resolution mass spectrometry is capable of measuring the mass of a particular isotopic composition of the molecule.   **3.87** 220,000 g/mol   **3.89** Mo   **3.91** (a) $C_9H_8O_4$; (b) $FeTiO_3$; (c) $Na_2S_2O_3$ **3.93** 386.7 g/mol   **3.95** NaH is the limiting reactant. 13.5 g $NaBH_4$; 1.83 g B_2H_6 left over   **3.97** $FeC_{10}H_{10}$

Chapter 4

4.1 (a) precipitation; (b) redox; (c) acid–base neutralization   **4.3** (a) $2 Ag^+(aq) + CrO_4^{2-}(aq) \longrightarrow Ag_2CrO_4(s)$; (b) $2 H^+(aq) + MgCO_3(s) \longrightarrow H_2O(l) + CO_2(g) + Mg^{2+}(aq)$ **4.5** (a) $Ni^{2+}(aq) + S^{2-}(aq) \longrightarrow NiS(s)$; (b) $Pb^{2+}(aq) + CrO_4^{2-}(aq) \longrightarrow PbCrO_4(s)$; (c) $Ag^+(aq) +$

$Br^-(aq) \longrightarrow AgBr(s)$ **4.7** (a) $2 Cs^+(aq) + 2 OH^-(aq) + 2H^+(aq) + SO_4^{2-}(aq) \longrightarrow 2 Cs^+(aq) + SO_4^{2-}(aq) + 2 H_2O(l)$; $H^+(aq) + OH^-(aq) \longrightarrow H_2O(l)$; (b) $Ca^{2+}(aq) + 2 OH^-(aq) + 2 CH_3COOH(aq) \longrightarrow Co^{2+}(aq) + 2 CH_3COO^-(aq) + 2 H_2O(l)$; $CH_3COOH(aq) + OH^-(aq) \longrightarrow CH_3COO^-(aq) + H_2O(l)$ **4.9** $2 Cu^{2+}(aq) + 4 I^-(aq) \longrightarrow 2 CuI(s) + I_2(aq)$

Cu^{2+}	$+2$
I^-	-1
CuI	Cu $+1$, I -1
I_2	0

Oxidizing agent, Cu^{2+}; reducing agent, I^- **4.11** (a) N.R.; (b) N.R.   **4.13** $H_2O(l) + 2 MnO_4^-(aq) + Br^-(aq) \longrightarrow 2 MnO_2(s) + BrO_3^-(aq) + 2 OH^-(aq)$ **4.15** $2 NO_3^-(aq) + 8 H^+(aq) + 3 Cu(s) \longrightarrow 3 Cu^{2+}(aq) + 2 NO(g) + 4 H_2O(l)$ **4.17** 1.98 M   *Understanding Key Concepts*   **1.** (a) $Ba^{2+}(aq) + SO_4^{2-}(aq) \longrightarrow BaSO_4(s)$; (b) $2 H^+(aq) + CO_3^{2-}(aq) \longrightarrow CO_2(g) + H_2O(l)$ **3.** C > A > D > B   **4.** Reaction (a) will occur. Reaction (b) will not occur.   **5.** (a) 1; (b) 2; (c) 3   *Additional Problems*   **4.19** (a) redox; (b) precipitation; (c) acid–base neutralization   **4.21** (a) $S_8(s) + 8 O_2(g) \longrightarrow 8 SO_2(g)$; (b) $Ni^{2+}(aq) + S^{2-}(aq) \longrightarrow NiS(s)$; (c) $CH_3COOH(aq) + OH^-(aq) \longrightarrow H_2O(l) + CH_3COO^-(aq)$ **4.23** A strong electrolyte is a compound that completely dissociates into its ions when placed in water. When a weak electrolyte is placed in water, it dissociates only slightly into its ions.   **4.25** (a) 2.25 M; (b) 1.42 M   **4.27** H_2O is a polar molecule and allows the polar HCl to dissociate into ions in aqueous solution. $CHCl_3$ is not very polar and does not allow the polar HCl to dissociate into ions.   **4.29** (a) insoluble; (b) insoluble; (c) insoluble (soluble in hot water); (d) insoluble   **4.31** (a) Precipitation reaction; MnS(s) will form. (b) No precipitate will form. (c) Precipitation reaction; $Hg_3(PO_4)_2(s)$ will form.   **4.33** (a) $AlCl_3(aq) + 3 NaOH(aq) \longrightarrow Al(OH)_3(s) + 3 NaCl(aq)$; (b) $Fe(NO_3)_2(aq) + Na_2S(aq) \longrightarrow FeS(s) + 2 NaNO_3(aq)$; (c) $CoSO_4(aq) + K_2CO_3(aq) \longrightarrow CoCO_3(s) + K_2SO_4(aq)$ **4.35** Add $Na_2SO_4(aq)$; it will precipitate $BaSO_4(s)$.   **4.37** (b) NO_3^- **4.39** Add the solution to an active metal, such as magnesium. Bubbles of H_2 gas indicate the presence of an acid.   **4.41** (a) $2 H^+(aq) + 2 ClO_4^-(aq) + Ca^{2+}(aq) + 2 OH^-(aq) \longrightarrow Ca^{2+}(aq) + 2 ClO_4^-(aq) + 2 H_2O(l)$; (b) $CH_3COOH(aq) + Na^+(aq) + OH^-(aq) \longrightarrow CH_3COO^-(aq) + Na^+(aq) + H_2O(l)$ **4.43** (a) $H^+(aq) + OH^-(aq) \longrightarrow H_2O(l)$; (b) $H^+(aq) + OH^-(aq) \longrightarrow H_2O(l)$ **4.45** Best reducing agents, bottom left; best oxidizing agents, top right (excluding inert gases)   **4.47** (a) gains electrons; (b) loses electrons; (c) loses electrons; (d) gains electrons   **4.49** (a) NO_2 O -2, N $+4$; (b) SO_3 O -2, S $+6$; (c) $COCl_2$ O -2, Cl -1, C $+4$; (d) CH_2Cl_2 Cl -1, H $+1$, C 0; (e) $KClO_3$ O -2, K $+1$, Cl $+5$; (f) HNO_3 O -2, H $+1$, N $+5$ **4.51** (a) ClO_3^- O -2, Cl $+5$; (b) SO_3^{2-} O -2, S $+4$; (c) $C_2O_4^{2-}$ O -2, C $+3$; (d) NO_2^- O -2, N $+3$; (e) BrO^- O -2, Br $+1$ **4.53** (a) Ca(s) is oxidized; $Sn^{2+}(aq)$ is reduced. (b) not a redox reaction   **4.55** (a) N.R.; (b) N.R.; (c) N.R.; (d) $Au^{3+}(aq) + 3 Ag(s) \longrightarrow 3 Ag^+(aq) + Au(s)$ **4.57** (a) N.R.; (b) N.R.   **4.59** (a) reduction; (b) oxidation; (c) reduction; (d) oxidation   **4.61** (a) $4 e^- + 2 H_2O(l) + O_2(g) \longrightarrow 4 OH^-(aq)$;

(b) $H_2O_2(aq) + 2 OH^-(aq) \longrightarrow O_2(g) + 2 H_2O(l) + 2 e^-$; (c) $MnO_4^-(aq) + e^- \longrightarrow MnO_4^{2-}(aq)$; (d) $CH_3OH(aq) + 2 OH^-(aq) \longrightarrow CH_2O(aq) + 2 H_2O(l) + 2 e^-$ **4.63** (a) oxidation, $Mn(s) \longrightarrow Mn^{2+}(aq)$; reduction, $NO_3^-(aq) \longrightarrow NO_2(g)$; (b) oxidation, $Mn^{3+}(aq) \longrightarrow MnO_2(s)$; reduction, $Mn^{3+}(aq) \longrightarrow Mn^{2+}(aq)$ **4.65** (a) $2 H^+(aq) + VO^{2+}(aq) + e^- \longrightarrow V^{3+}(aq) + H_2O(l)$; (b) $2 Ni(OH)_2(s) + 2 OH^-(aq) \longrightarrow Ni_2O_3(s) + 3 H_2O(l) + 2 e^-$; (c) $2 H^+(aq) + NO_3^-(aq) + e^- \longrightarrow NO_2(aq) + H_2O(l)$; (d) $Br_2(aq) + 12 OH^-(aq) \longrightarrow 2 BrO_3^-(aq) + 6 H_2O(l) + 10 e^-$ **4.67** (a) $2 S_2O_3^{2-}(aq) + I_2(aq) \longrightarrow S_4O_6^{2-}(aq) + 2 I^-(aq)$; (b) $Mn^{2+}(aq) + H_2O_2(aq) + 2 OH^-(aq) \longrightarrow MnO_2(s) + 2 H_2O(l)$; (c) $6 H_2O(l) + 4 Zn(s) + NO_3^-(aq) + 7 OH^-(aq) \longrightarrow 4 Zn(OH)_4^{2-}(aq) + NH_3(aq)$; (d) $2 Bi(OH)_3(s) + 3 Sn(OH)_3^-(aq) + 3 OH^-(aq) \longrightarrow 2 Bi(s) + 3 Sn(OH)_6^{2-}(aq)$ **4.69** (a) $12 H^+(aq) + 4 MnO_4^-(aq) + 5 C_2H_5OH(aq) \longrightarrow 4 Mn^{2+}(aq) + 11 H_2O(l) + 5 CH_3COOH(aq)$; (b) $8 H^+(aq) + Cr_2O_7^{2-}(aq) + 3 H_2O_2(aq) \longrightarrow 2 Cr^{3+}(aq) + 7 H_2O(l) + 3 O_2(g)$; (c) $4 Sn^{2+}(aq) + 8 H^+(aq) + IO_4^-(aq) \longrightarrow 4 Sn^{4+}(aq) + I^-(aq) + 4 H_2O(l)$; (d) $2 PbO_2(s) + 4 H^+(aq) + 4 Cl^-(aq) \longrightarrow 2 PbCl_2(s) + 2 H_2O(l) + O_2(g)$ **4.71** 78.4 mL **4.73** 0.1829 M **4.75** 0.3902 M **4.77** 2.09 mg **4.79** $H_3C_6H_5O_7(aq) + 3 OH^-(aq) \longrightarrow C_6H_5O_7^{3-}(aq) + 3 H_2O(l)$ **4.81** 26.80% **4.83** (a) $4 H^+(aq) + 5 PbO_2(s) + 2 Mn^{2+}(aq) \longrightarrow 5 Pb^{2+}(aq) + 2 H_2O(l) + 2 MnO_4^-(aq)$; (b) $3 H_2O(l) + As_2O_3(s) + 2 H^+(aq) + 2 NO_3^-(aq) \longrightarrow 2 H_3AsO_4(aq) + 2 HNO_2(aq)$; (c) $2 H_2O(l) + Br_2(aq) + SO_2(g) \longrightarrow 2 Br^-(aq) + HSO_4^-(aq) + 3 H^+(aq)$; (d) $4 H^+(aq) + 2 NO_2^-(aq) + 2 I^-(aq) \longrightarrow 2 NO(g) + I_2(s) + 2 H_2O(l)$

Chapter 5

5.1 gamma ray, 8.43×10^{18} Hz; radar, 2.91×10^9 Hz **5.3** Violet has the highest frequency; red has the longest wavelength. **5.5** 1875 nm **5.7** 1310 kJ/mol **5.9** 2.34×10^{-38} m

5.11

n	l	m_l	Orbital designation	No. of orbitals
5	0	0	$5s$	1
	1	$-1, 0, +1$	$5p$	3
	2	$-2, -1, 0, +1, +2$	$5d$	5
	3	$-3, -2, -1, 0, +1, +2, +3$	$5f$	7
	4	$-4, -3, -2, -1, 0, +1, +2, +3, +4$	$5g$	9
				25

There are 25 possible orbitals in the fifth shell.
5.13 (a) $2p$; (b) $4f$; (c) $3d$ **5.15** 1.31×10^3 kJ/mol
5.17 (a) Ti $1s^2 2s^2 2p^6 3s^2 3p^6 4s^2 3d^2$; $\underset{1s}{\uparrow\downarrow}$ $\underset{2s}{\uparrow\downarrow}$ $\underset{2p}{\uparrow\downarrow \; \uparrow\downarrow \; \uparrow\downarrow}$

$\underset{3s}{\uparrow\downarrow}$ $\underset{3p}{\uparrow\downarrow \; \uparrow\downarrow \; \uparrow\downarrow}$ $\underset{4s}{\uparrow\downarrow}$ $\underset{3d}{\uparrow \; \uparrow \; \_ \; \_ \; \_}$; (b) Zn

$1s^2 2s^2 2p^6 3s^2 3p^6 4s^2 3d^{10}$; $\underset{1s}{\uparrow\downarrow}$ $\underset{2s}{\uparrow\downarrow}$ $\underset{2p}{\uparrow\downarrow \; \uparrow\downarrow \; \uparrow\downarrow}$ $\underset{3s}{\uparrow\downarrow}$

$\underset{3p}{\uparrow\downarrow \; \uparrow\downarrow \; \uparrow\downarrow}$ $\underset{4s}{\uparrow\downarrow}$ $\underset{3d}{\uparrow\downarrow \; \uparrow\downarrow \; \uparrow\downarrow \; \uparrow\downarrow \; \uparrow\downarrow}$;

(c) Sn $1s^2 2s^2 2p^6 3s^2 3p^6 4s^2 3d^{10} 4p^6 5s^2 4d^{10} 5p^2$ [Kr] $\underset{5s}{\uparrow\downarrow}$

$\underset{4d}{\uparrow\downarrow \; \uparrow\downarrow \; \uparrow\downarrow \; \uparrow\downarrow \; \uparrow\downarrow}$ $\underset{5p}{\uparrow \; \uparrow \; \_}$; (d) Pb [Xe]

$6s^2 4f^{14} 5d^{10} 6p^2$ **5.19** Cr, Cu, Nb, Mo, Ru, Rh, Pd, Ag, Gd, Pt, Au, Ac, Th, Pa, U, Np, Cm **5.21** (a) Ba; (b) W; (c) Sn; (d) Ce ***Understanding Key Concepts*** **2.** Ga **4.** K^+ $r = 133$ pm; Cl^- $r = 184$ pm; K $r = 227$ pm **5.** $Z = 118$ [Rn] $7s^2 5f^{14} 6d^{10} 7p^6$ **6.** $Z = 121$ **7.** (a) $3p_y$, $n = 3$; $l = 1$; (b) $4d_{z^2}$ $n = 4$; $l = 2$ ***Additional Problems*** **5.23** Violet has the higher frequency and the higher energy; red has the higher wavelength. **5.25** 5.5×10^{-8} m **5.27** 1.6×10^{-22} J/photon; 0.096 kJ/mol **5.29** 1.60×10^{-4} kJ/mol **5.31** (a) 1320 nm, near IR; (b) 0.149 m, radiowave; (c) 65.4 nm, UV **5.33** 5.28×10^{-15} m **5.35** 3.10×10^{-28} m **5.39** 364.6 nm **5.41** $n = 5$, $\lambda = 4051$ nm, $E = 29.55$ kJ/mol, IR; $n = 6$, $\lambda = 2625$ nm, $E = 45.60$ kJ/mol, IR **5.49** $l = 2$ for d subshells; $l = 3$ for f subshells **5.51** (a) $4s$ $n = 4$; $l = 0$; $m_l = 0$; $m_s = \pm\frac{1}{2}$; (b) $3p$ $n = 3$; $l = 1$; $m_l = -1, 0, +1$; $m_s = \pm\frac{1}{2}$; (c) $5f$ $n = 5$; $l = 3$; $m_l = -3, -2, -1, 0, +1, +2, +3$; $m_s = \pm\frac{1}{2}$; (d) $5d$ $n = 5$; $l = 2$; $m_l = -2, -1, 0, +1, +2$; $m_s = \pm\frac{1}{2}$ **5.53** (a) not allowed, for $l = 0$, $m_l = 0$; (b) allowed; (c) not allowed, for $n = 4$, $l = 0, 1, 2, 3$ **5.55** 20 e^- **5.61** (a) $2p < 3p < 5s < 4d$; (b) $2s < 4s < 3d < 4p$; (c) $3d < 4p < 5p < 6s$ **5.63** (a) $3s$; (b) $3d$; (c) $6s$; (d) $4f$ **5.65** (a) $Z = 55$, Cs, [Kr] $5s^2 4d^{10} 5p^6 6s^1$; (b) $Z = 41$, Nb, [Kr] $5s^2 4d^3$; (c) $Z = 80$, Hg, [Xe] $6s^2 4f^{14} 5d^{10}$; (d) $Z = 62$, Sm, [Xe] $6s^2 4f^6$ **5.67** (a) $Z = 25$, Mn, [Ar] $\underset{4s}{\uparrow\downarrow}$

$\underset{3d}{\uparrow \; \uparrow \; \uparrow \; \uparrow \; \uparrow}$; (b) $Z = 56$, Ba, [Xe] $\underset{6s}{\uparrow\downarrow}$; (c) $Z = 28$, Ni, [Ar] $\underset{4s}{\uparrow\downarrow}$ $\underset{3d}{\uparrow\downarrow \; \uparrow\downarrow \; \uparrow\downarrow \; \uparrow \; \uparrow}$; (d) $Z = 47$, Ag, [Kr]

$\underset{5s}{\uparrow}$ $\underset{4d}{\uparrow\downarrow \; \uparrow\downarrow \; \uparrow\downarrow \; \uparrow\downarrow \; \uparrow\downarrow}$ **5.69** $Z = 118$ [Rn] $7s^2 5f^{14} 6d^{10} 7p^6$ **5.71** $Z = 115$ [Rn] $7s^2 5f^{14} 6d^{10} 7p^3$ **5.73** (a) Ga; (b) Pd **5.75** Zn **5.79** F < O < S **5.81** Mg because of higher Z_{eff} and smaller size **5.83** 2A **5.85** $S^{2-} > Ca^{2+} > Sc^{3+} > Ti^{4+}$; Z_{eff} increases on going from S^{2-} to Ti^{4+}. **5.87** $n = 4$, $\lambda = 97.2$ nm; $n = 5$, $\lambda = 95.0$ nm; $n = 6$, $\lambda = 93.8$ nm; $n = 7$, $\lambda = 93.1$ nm **5.89** 2279 nm **5.91** (a) 0.151 kJ/mol; (b) 2.17×10^{-8} kJ/mol; (c) 2.91 kJ/mol **5.93** 4.17×10^{-33} m **5.95** 151 kJ/mol **5.97** K, $Z_{eff} = 2.26$; Kr, $Z_{eff} = 4.06$ **5.99** 8.3×10^{28} photons

Chapter 6

6.1 (a) Br; (b) S; (c) Se; (d) Ne **6.3** (b) Cl has the highest E_{i1} and the smallest E_{i4}. **6.5** +91 kJ/mol, unfavorable **6.7** (a) KCl; (b) CaF_2; (c) CaO **6.9** (a) $2 Cs(s) + 2 H_2O(l) \longrightarrow 2 Cs^+(aq) + 2 OH^-(aq) + H_2(g)$; (b) N.R.; (c) $Rb(s) + O_2(g) \longrightarrow RbO_2(s)$; (d) $2 K(s) + 2 NH_3(g) \longrightarrow 2 KNH_2(s) + H_2(g)$; (e) $2 Rb(s) + H_2(g) \longrightarrow 2 RbH(s)$ **6.11** $BeCl_2 + 2 K \longrightarrow Be + 2 KCl$ **6.13** oxidizing agent, H^+; reducing agent, Al **6.15** (a) $Br_2(l) + Cl_2(g) \longrightarrow 2 BrCl(g)$; (b) $2 Al(s) + 3 F_2(g) \longrightarrow 3 AlF_3(s)$; (c) $H_2(g) + I_2(s) \longrightarrow 2 HI(g)$ **6.17** (a) XeF_2 F -1, Xe $+2$; (b) XeF_4 F -1, Xe $+4$; (c) $XeOF_4$ F -1, O -2, Xe $+6$ **6.19** gain 2 e^- ***Understanding Key Concepts*** **1.** (a) solid; (b) dark, like I_2, maybe with a metallic sheen; (c) yes, NaAt **3.** (a)

ionic compound; (b) covalent compound   **6.** predicted
for Fr: mp 23°C; bp 650°C; density 2 g/cm³; atomic
radius 275 pm   ***Additional Problems***   **6.21** largest E_{i1},
8A because of largest values of Z_{eff}; smallest E_{i1}, 1A
because of smallest values of Z_{eff} **6.23** (a) 3p; (b) 4p;
(c) 4s   **6.25** (a) Sn; (b) Br   **6.27** (a) 5s; (b) 4p; (c) 6s
6.29 (a) lowest, K; highest, Li; (b) lowest, B; highest, Cl;
(c) lowest, Ca; highest, Cl   **6.31** (a) 496 kJ/mol; (b) 2189
kJ/mol; (c) 5140 kJ/mol; (d) 39,546 kJ/mol   **6.33** They
have the same magnitude but opposite sign.   **6.35** Br
6.39 AlBr₃ > CaO > MgBr₂ > LiBr   **6.41** +195 kJ/mol
6.43 (a) −606 kJ/mol; (b) −1215 kJ/mol   **6.45** 2390 kJ/
mol   **6.47** absorbs 29 kJ/mol   **6.49** CaCl, −176 kJ/mol;
CaCl₂, −799 kJ/mol   **6.61** (a) Ca(s) + H₂(g) ⟶
CaH₂(s); (b) Ca(s) + 2 H₂O(l) ⟶ Ca²⁺(aq) +
2 OH⁻(aq) + H₂(g); (c) N.R.; (d) Ca(s) + Br₂(l) ⟶
CaBr₂(s); (e) 2 Ca(s) + O₂(g) ⟶ 2 CaO(s)   **6.63** AlCl₃
+ 3 Na ⟶ Al + 3 NaCl; Na is oxidized. Al³⁺ is
reduced.   **6.65** 590 g   **6.67** 16.0 g
6.69   Oxidizing Agent   Reducing Agent
(a)   H⁺   Mg
(b)   F₂   Kr
(c)   Cl₂   I₂
6.71 403 kJ/mol   **6.73** MgF, −294 kJ/mol; MgF₂, −1114
kJ/mol   **6.77** (a) 2 Li(s) + H₂(g) ⟶ 2 LiH(s); (b)
2 Li(s) + 2 H₂O(l) ⟶ 2 Li⁺(aq) + 2 OH⁻(aq) + H₂(g);
(c) 2 Li(s) + 2 NH₃(l) ⟶ 2 LiNH₂(s) + H₂(g); (d)
2 Li(s) + Br₂(l) ⟶ 2 LiBr(s); (e) 6 Li(s) + N₂(g) ⟶ 2
Li₃N(s); (f) 4 Li(s) + O₂(g) ⟶ Li₂O(s)   **6.79** For metals,
values of Z_{eff}/r are similar for diagonal elements.

Chapter 7

7.1 (a), (b), **7.3** (a), (b), (c), **7.5**, **7.7** (a),(b),(c),(d) [structures]

7.9 :N̈=N=Ö: ⟷ :N≡N—Ö:
7.11 (a) polar covalent; (b) ionic; (c) polar covalent;
(d) polar covalent
7.13 [resonance structures]
7.15 (a) bent; (b) trigonal pyramidal; (c) linear;
(d) octahedral; (e) square pyramidal; (f) tetrahedral;
(g) tetrahedral; (h) tetrahedral; (i) square planar;
(j) trigonal planar
7.17 [structures]
7.21 sp^3d hybridization
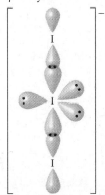
Understanding Key Concepts   **1.** (a) square pyramidal;
(b) trigonal pyramidal; (c) square planar; (d) trigonal
planar   **2.** (a) trigonal bipyramidal; (b) tetrahedral;
(c) square pyramidal   **4.** (a) sp^2; (b) sp^3d^2;
(c) sp^3 **6.** Every carbon is sp^2 hybridized. 18 σ bonds;
5 π bonds   ***Additional Problems***
7.27 (a),(b),(c),(d),(e),(f) [structures]
7.29 (a) [structure]

(b)

$$\left[:\ddot{O} - \ddot{C}l - \ddot{O}: \atop \underset{:\ddot{O}:}{|} \right]^-$$

(c)

$$\left[\underset{:\ddot{O}:}{\overset{:\ddot{O}:}{:\ddot{O} - \overset{|}{C}l - \ddot{O}:}} \right]^-$$

There are also structures with one Cl=O double bond.

7.31 (a) no; (b) no; (c) no; (d) yes

7.33 $\ddot{O}=C=\ddot{O}$ two double bonds

7.35

$$H-\underset{\underset{H}{|}}{\overset{\overset{H}{|}}{C}}-\ddot{N}=C=\ddot{O}: \longleftrightarrow H-\underset{\underset{H}{|}}{\overset{\overset{H}{|}}{C}}-N\equiv C-\ddot{O}:$$

7.37 (a) X = Se; (b) X = Kr

7.39 (a)

$$H-\overset{\overset{\displaystyle :O:}{\|}}{C}-\underset{\underset{\displaystyle H}{|}}{N}-H$$

(b)

$$H-\underset{\underset{\displaystyle H}{|}}{\overset{\overset{\displaystyle H}{|}}{C}}-C\equiv N-\ddot{O}:$$

7.43 (a) C—Cl; (b) Si—Cl; (c) N—Mg

7.45 $:\bar{C}\equiv\overset{+}{O}:$

7.47 (a)

$$H-\underset{\underset{\displaystyle H}{|}}{N}-\ddot{O}-H$$; no formal charges

(b)

$$\left[H-\underset{\cdot\cdot}{\overset{\overset{\displaystyle H}{|}}{N}}-\underset{\underset{\displaystyle H}{|}}{C}-H \right]^-$$

(c)

$$:\ddot{O}:^- \atop :\ddot{C}l-\overset{+}{P}-\ddot{C}l: \atop :\ddot{C}l:$$

7.49 (a)

$$\underset{H}{\overset{H}{>}}C=\overset{+}{N}=\overset{-}{\ddot{N}}:$$

(b)

$$\underset{H}{\overset{H}{>}}C-\overset{+}{\ddot{N}}=\overset{-}{\ddot{N}}:$$

structure (a) is more important   **7.51** (a) 4; (b) 6; (c) 3 or 4; (d) 2 or 5; (e) 6; (f) 4   **7.53** (a) 5; (b) 6; (c) 5; (d) 5; (e) 3; (f) 2 or 5   **7.55** (a) tetrahedral; (b) tetrahedral; (c) tetrahedral; (d) bent   **7.57** (a) trigonal planar; (b) linear; (c) bent   **7.59** (a) T-shaped; (b) trigonal pyramidal; (c) bent; (d) trigonal planar

7.61

$$\underset{\cdot\cdot}{\overset{H}{>}}N=\overset{\cdot}{\overset{\cdot}{N}} \qquad \overset{\cdot}{\overset{\cdot}{N}}=\overset{\cdot}{N}$$

They are geometric isomers, not resonance forms. In resonance forms the atoms have the same geometrical arrangement.

7.63

$$H-\underset{\underset{\displaystyle H}{|}}{\overset{\overset{\displaystyle H}{|}}{C}}-\underset{\cdot\cdot}{\overset{\overset{\displaystyle :O:}{\|}}{S}}-\underset{\underset{\displaystyle H}{|}}{\overset{\overset{\displaystyle H}{|}}{C}}-H$$ ~109° ~109°

7.65 All six C's are sp^2 hybridized, and the bond angle is ~120°. The geometry about each C is trigonal planar.

7.69 (a) sp^3; (b) sp^3d^2; (c) sp^2 or sp^3; (d) sp or sp^3d; (e) sp^3d^2 **7.71** (a) sp^2; (b) sp^3; (c) sp^3d^2; (d) sp^2 **7.73** C sp^2 N sp^3

$$H_2N-\underset{\underset{\displaystyle H}{|}}{\overset{\overset{\displaystyle O}{\|}}{C}}-\underset{\underset{\displaystyle H}{|}}{\ddot{N}}$$ ~120° ~109° ~109°

7.75

	O_2^+	O_2	O_2^-
σ^*_{2p}	—	—	—
π^*_{2p}	↑ —	↑ ↑	↑↓ ↑
σ_{2p}	↑↓	↑↓	↑↓
π_{2p}	↑↓ ↑↓	↑↓ ↑↓	↑↓ ↑↓
σ^*_{2s}	↑↓	↑↓	↑↓
σ_{2s}	↑↓	↑↓	↑↓

	Bond Order
O_2^+	2.5
O_2	2
O_2^-	1.5

All are stable. All have unpaired electrons.

7.77

σ^*_{2p} —	
π^*_{2p} — —	Bond order = 3
σ_{2p} ↑↓	
π_{2p} ↑↓ ↑↓	$[:C\equiv C:]^{2-}$
σ^*_{2s} ↑↓	
σ_{2s} ↑↓	

C_2^{2-}

7.81 B sp^2; N sp^2; All bond angles are 120°. The overall geometry of the molecule is planar.   **7.83** Triply bonded C's are sp hybridized. Theoretical bond angle for C—C≡C is 180°. Benzyne is so reactive because the C—C≡C bond angle is closer to 120° and is very strained.   **7.85** (a) H—C≡C—H; (b) $H-\ddot{N}=\ddot{N}-H$ (c)

$$:\ddot{O}: \atop Cl-\underset{\cdot\cdot}{S}-Cl$$

Chapter 8

8.1 16 J   **8.3** (a) and (b) are state functions; (c) is not
8.5 $W = +570$ J; $\Delta E = -483$ kJ   **8.7** (a) 779 kJ evolved;
(b) 1.24 kJ absorbed   **8.9** 0.13 J/(g · °C)   **8.11** -202 kJ
8.15 $+2816$ kJ   **8.17** -81 kJ   **8.19** $\Delta S° < 0$ because the
reaction decreases the number of moles of gaseous mol-
ecules   **8.21** $\Delta G° = -32.9$ kJ; reaction is spontaneous;
$T = 190$°C   ***Understanding Key Concepts***   **1.** (a) yes;
$w < 0$; (b) yes; $\Delta H < 0$; exothermic

3.

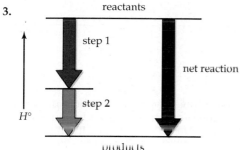

5. (a) positive; (b) negative   **6.** $\Delta G < 0$; $\Delta S > 0$; $\Delta H \approx 0$
for the mixing of gaseous molecules   ***Additional Prob-***
lems   **8.23** car, 7.1×10^5 J; truck, 6.7×10^5 J   **8.25** -70
J; energy change is negative   **8.31** for system, $\Delta H =$
-244 kJ; $\Delta E = -209$ kJ; for surroundings, $\Delta H = +244$
kJ; $\Delta E = +209$ kJ   **8.33** $w = 0.283$ kJ; $\Delta E = -314$ kJ
8.35 226 kJ   **8.37** 8.01 kJ evolved; exothermic
8.41 0.523 J/(g · °C); 25.0 J/mol · °C   **8.47** -395.7 kJ/
mol   **8.51** $+48.6$ kJ/mol   **8.53** cyclopropane; its more
positive $\Delta H_f°$ results in a more negative $\Delta H°$ for the combustion reaction.   **8.55**   17.1 kJ/mol   **8.57**   14 kJ
8.59 -3695 kJ   **8.65** (a) +; (b) −; (c) +; (d) +   **8.67** ΔS
< 0; the reaction decreases the number of gas molecules.
8.69 (a) spontaneous, exothermic; (b) nonspontaneous,
endothermic; (c) spontaneous, endothermic   **8.71** 569 K
8.73 (a) favored by enthalpy; $\Delta G° = -147.8$ kJ; (b) crossover temperature = 929.9 K   **8.75** (a) $+172.5$ kJ; (b)
-196.0 kJ; (c) -24.8 kJ   **8.77** 59°C   **8.79** (a) -238.6 kJ;
(b) & (c) $\Delta H < 0$ and $\Delta S > 0$; the reaction is spontaneous at all temperatures.   **8.81** (a) -128.2 kJ; (b) -29.3
kJ; (c) yes; (d) $\Delta H°$; (e) below 386 K; (f) $+163.9$ kJ; (g)
$+115.6$ kJ; (h) no; (i) $\Delta H°$; (j) above 1012 K; (k) $\Delta G° =$
$+86.3$ kJ, $\Delta H° = +35.7$ kJ, $\Delta S° = -170$ J/K; (l) no; (m)
Run the reactions separately, first reaction at low temperature, second reaction at high temperature.
8.83 1.73 °C

Chapter 9

9.1 1 atm = 14.7 psi; 1.0 mm Hg = 1.93×10^{-2} psi
9.3 0.650 atm   **9.5** 5.0 atm   **9.7** 28°C   **9.9** 190 L
9.11 $X(H_2) = 0.7281$; $X(N_2) = 0.2554$; $X(NH_3) = 0.0165$
9.13 0.0280 atm   **9.15** (a) O_2, 1.62; (b) C_2H_2, 1.04
9.17 3.8×10^{-5} m   ***Understanding Key Concepts***
2. (c)   **3.** (c)   **5.** The gas pressure in the bulb in mm is
equal to the difference in the height of the Hg in the
two arms of the manometer.   ***Additional Problems***
9.19 Temperature is a measure of the kinetic energy
content of a chemical system.   **9.23** (a) 4.87×10^5 Pa;

(b) 1.35 atm; (c) 230 Pa   **9.25** 930 mm Hg   **9.27** $1.36 \times$
10^6 Pa   **9.31** They all contain the same number of gas
molecules.   **9.33** At 1.02 atm, volume = 7210 L; at 35°C,
volume = 51.5 L.   **9.35** 2.1×10^4 mm Hg   **9.37** volume = 6×10^{15} L; density = 1×10^8 molecules/L
9.39 6920 L   **9.41** 167 balloons   **9.43** The containers are
identical. Both containers contain the same number of
gas molecules. Weigh the containers. Since the molecular
weight for O_2 is greater than the molecular weight for
H_2, the heavier container contains O_2.   **9.45** $n = 7.8 \times$
10^{-3} mol O_2; volume = 200 mL   **9.47** 34.0 amu
9.49 $X(Ar) = 0.3065$; $X(N_2) = 0.6935$ **9.51** 0.547 L
9.53 (a) 9.44 L; (b) 6.05 g Zn   **9.55** (a) 380 g; (b) 5.4
days   **9.57** $P(CH_4) = 1.4$ atm; $P(C_2H_6) = 0.059$ atm;
$P(C_3H_8) = 0.022$ atm; $P(C_4H_{10}) = 0.0074$ atm
9.59 $X(HCl) = 0.026$; $X(H_2) = 0.094$; $X(Ne) = 0.88$
9.61 $P(H_2) = 723$ mm Hg; mass of Mg = 3.36 g
9.67 -32°C   **9.69** 17.2 g/mol   **9.71** Since they both
have the same mass, they will have the same diffusion
rates.   **9.73** -268°C   **9.75** 504.3 g   **9.77** (a) A contains
$CO_2(g)$ and $N_2(g)$; B contains $CO_2(g)$, $N_2(g)$ and $H_2O(s)$;
(b) 0.0013 mol H_2O, (c) A contains $N_2(g)$, B contains
$N_2(g)$ and $H_2O(s)$; C contains $N_2(g)$ and $CO_2(s)$; (d)
0.01091 mol N_2; (e) 0.0181 mol CO_2 **9.79** From the
ideal gas law, $P = 72$ atm; from the van der Waals
equation, $P = 69$ atm.   **9.81** (a) 2 $C_8H_{18}(l) + 25\ O_2(g)$
$\longrightarrow$ 16 $CO_2(g) + 18\ H_2O(g)$; (b) 1.1×10^{11} kg; (c) $5.7 \times$
10^{13} L.   **9.83** fp of $H_2O = 492$°R; $R = 0.0456\ \dfrac{L \cdot atm}{°R \cdot mol}$
9.85 (a) 0.0290 mol; (b) 0.0100 mol A; A is H_2O; (c)
0.0120 mol B; B is CO_2; (d) 0.0010 mol C and 0.0060 mol
D; 28.0 g/mol D is N_2; C is O_2; (e) 4 $C_3H_5N_3O_9(s) \longrightarrow$
10 $H_2O(g) + 12\ CO_2(g) + O_2(g) + 6\ N_2(g)$ **9.87** 1.1 L

Chapter 10

10.1 41%; HF has more ionic character than HCl.
10.3 (a) no; (b) no;
(c)   (d)

10.5 H_2S, dipole–dipole, dispersion; CH_3OH, hydrogen
bonding, dipole–dipole, dispersion; CBr_4, dispersion;
Ne, dispersion; Ne < H_2S < CH_3OH < CBr_4 **10.7** 334
K   **10.9** 31.4 kJ/mol   **10.11** (a) 2 atoms; (b) 4 atoms
10.13 9.31 g/cm³   **10.15** For CuCl there are 4 pluses
and 4 minuses; for $BaCl_2$ there are 8 pluses and 8
minuses.   ***Understanding Key Concepts***   **1.** (a) dispersion; (b) dispersion; (c) hydrogen bonding   **3.** (a) cubic
closest-packed; (b) simple cubic; (c) hexagonal closest-
packed; (d) body-centered cubic   **5.** (a) cubic closest-
packed; (b) 4 S^{2-} and 4 Zn^{2+} **6.** (a) normal boiling
point ≈ 300 K, normal melting point ≈ 180 K; (b) (i)
solid; (ii) gas; (iii) supercritical fluid   ***Additional Prob-***
lems   **10.23** For CH_3OH and CH_4 dispersion forces are
small. CH_3OH can hydrogen bond; CH_4 cannot. This
accounts for the large difference in boiling points. For
1-decanol and decane dispersion forces are comparable
and relatively large along the C—H chain. 1-decanol can

hydrogen bond; decane cannot. This accounts for the 55°C higher boiling point for 1-decanol.

10.25 (a) (b)

(c) XeF_2 is linear with zero dipole moment. (d) PCl_5 is trigonal bipyramidal with zero dipole moment. **10.27** (a) F_2 and Kr have no dipole–dipole forces. (b) HF has the largest hydrogen bond forces. (c) Kr has the largest dispersion forces. **10.33** (a) $Hg(l) \longrightarrow$ $Hg(g)$; (b) no change of state; Hg remains a liquid; (c) $Hg(g) \longrightarrow Hg(l) \longrightarrow Hg(s)$ **10.37** 3.89 kJ of heat is released **10.41** 80.1°C **10.43** 7.12 J/(K · mol) **10.45** 294 mm Hg **10.47** 288 mm Hg **10.49** $\Delta H_{vap} =$ 30.1 kJ/mol; $C = 7.89$ **10.55** 128 pm **10.57** 175 pm; 11.3 g/cm³ **10.59** 549 pm **10.61** radius = 186 pm; edge = 429 pm **10.63** (a) 4 Ca atoms; (b) Since the unit cell contains 4 Ca atoms, the unit cell is the face-centered cube. **10.65** 244 pm **10.67** 357.1 pm **10.71** (a) gas; (b) liquid; (c) solid

10.73

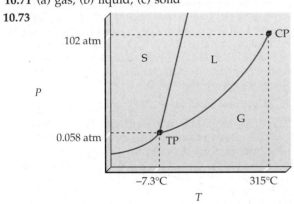

10.75

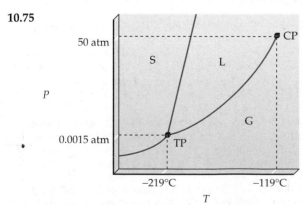

10.77

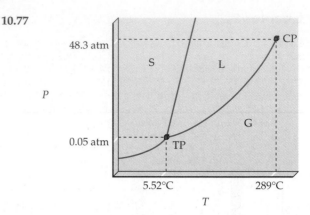

10.79 solid $\longrightarrow$ liquid $\longrightarrow$ supercritical fluid $\longrightarrow$ liquid $\longrightarrow$ solid $\longrightarrow$ gas **10.81** (a) C_8H_{18}, dispersion; (b) $C_2H_5NH_2$, dipole–dipole, hydrogen bonding, dispersion; (c) $C_6H_6(OH)_6$, dipole–dipole, hydrogen bonding, dispersion; (d) aqueous NaOH, ion–dipole, hydrogen bonding, dispersion **10.87** 650°C **10.89** −30.7°C **10.91** 23.3 kJ/mol

10.93

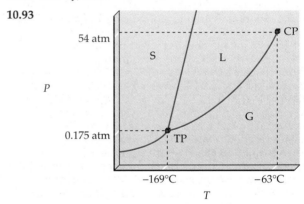

Kr cannot be liquefied at room temperature because room temperature is above T_c (−63°C). **10.95** 68% **10.97** 6.01×10^{23} atoms/mol **10.99** 556 pm

Chapter 11

11.1 toluene $<$ Br_2 $<$ KBr **11.3** 5.52% **11.5** 0.614 M **11.7** 312 g **11.9** 0.513 m **11.11** 3.2×10^{-2} mol/(L · atm) **11.13** 98.6 mm Hg **11.15** 27.1 mm Hg; 46.6 mm Hg **11.17** −2.15°C **11.19** 9.54 atm **11.21** 128 g/mol **11.23** (a) bp ≈ 107°C; (b) $X_{toluene}$ ≈ 0.64, $X_{benzene}$ ≈ 0.36 *Understanding Key Concepts* **1.** (a) $<$ (b) $<$ (c)

6. (b) ~95°C

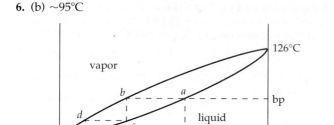

Additional Problems **11.33** The HBr solution will be warm to touch. The AgNO$_3$ solution will be cool to touch. **11.35** 17.8°C **11.41** Dissolve 3.27 g of C$_7$H$_6$O$_2$ in 1.000 kg of CHCl$_3$, and take 165 mL of the solution. **11.43** (a) 11.2%; (b) 0.00270%; (c) 3.65% **11.45** 5 mg **11.47** (a) 0.196 m; (b) X(C$_{10}$H$_{14}$N$_2$) = 0.0145, X(CH$_2$Cl$_2$) = 0.985 **11.49** 0.71 m, X(NaOCl) = 0.013 **11.51** 6.78 M **11.53** 3.7 g **11.55** X(C$_{12}$H$_{22}$O$_{11}$) = 0.00646; weight % = 11.0%; molality = 0.361 m **11.57** 0.195 mol/(L · atm) **11.59** 0.010 mol/(L · atm) **11.65** solid < liquid < solution < gas **11.69** (a) 100.57°C; (b) 101.6°C **11.71** 125 mm Hg **11.73** In the liquid, X(acetone) = 0.602 and X(ethyl acetate) = 0.398. In the vapor, X(acetone) = 0.785 and X(ethyl acetate) = 0.215. **11.75** In the liquid, X(CHCl$_3$) = 0.222 and X(CH$_2$Cl$_2$) = 0.778. In the vapor, X(CHCl$_3$) = 0.124 and X(CH$_2$Cl$_2$) = 0.876. **11.77** 3.6°C · kg/mol **11.79** 0.573 m **11.81** 5.41 mm Hg **11.83** 0.197 M **11.87** 342.5 g/mol **11.89** HCl is a strong electrolyte in H$_2$O. Each HCl molecule completely dissociate into two solute particles. HF is a weak electrolyte in H$_2$O. Only a few percent of the HF molecules dissociates into ions. **11.91** 146 g/mol; C$_6$H$_{14}$N$_2$O$_2$ **11.93** 47 g **11.95** The KCl solution has the higher boiling point because it is a strong electrolyte, yielding two ions per formula unit. **11.99** 0.002 atm **11.101** freezing point = −2.3°C; boiling point = 100.63°C **11.103** 11.7 g of rubbing alcohol contains 10.5 g of isopropyl alcohol. **11.105** 0.081 m

Chapter 12

12.1 Rate of disappearance of N$_2$O$_5$ = 2.2 × 10^{-5} M/s; rate of appearance of O$_2$ = 1.10 × 10^{-5} M/s. **12.3** rate = k[BrO$_3^-$][Br$^-$][H$^+$]^2: 1st order in BrO$_3^-$, 1st order in Br$^-$, 2nd order in H$^+$, 4th order overall; rate = k[H$_2$][I$_2$]: 1st order in H$_2$, 1st order in I$_2$, 2nd order overall; rate = k[CH$_3$CHO]$^{3/2}$: ½ order in CH$_3$CHO, ½ order overall **12.5** s^{-1}; s^{-1}; M^{-3} s^{-1}; M^{-1} s^{-1}; M$^{-1/2}$ s^{-1} **12.7** A plot of log [cyclopropane] vs. time is linear, indicating that the data fit the equation for a first-order reaction. k = 6.7 × 10^{-4}s (0.040/min) **12.9** (a) A plot of 1/[HI] vs. time is linear. The reaction is second order. (b) 0.0308 M^{-1} min^{-1}; (c) 260 min; (d) 81.2 min **12.11** (a) rate = k[O$_3$][O]; (b) rate = k[Br]^2[Ar]; (c) rate = k[Co(CN)$_5$(H$_2$O)$^{2-}$] **12.13** (a) 104 kJ/mol; (b) 1.4 ×

10^{-4}/s **Understanding Key Concepts** **1.** 2 : 1 : 4 : 2 **2.** The ks are all the same **3.** (b) 1 min **4.** (a) second order; (b) rate = k[A]^2 **5.** (a) bimolecular; (b) unimolecular; (c) termolecular **6.** (a) rate = k[B$_2$][C]; (b) B$_2$ + C ⟶ CB + B (slow); CB + A ⟶ AB + C (fast); (c) C is a catalyst. C doesn't appear in the chemical equation because it is used up in the first step and regenerated in the second step. **Additional Problems** **12.15** molecules/(cm^3 · time) **12.17** (a) 2.0 × 10^{-5} M/s; (b) 1.5 × 10^{-5} M/s **12.19** (a) 3 times faster; (b) 2 times faster **12.21** (a) 4.5 × 10^{-3} M/s; (b) 3.0 × 10^{-3} M/s **12.23** 1st order in CHCl$_3$; ½ order in Cl$_2$; ³⁄₂ order overall **12.25** rate = k[NO]^2[H$_2$]; units for k are 1/(M^2 · s) **12.27** (a) rate = k[CH$_3$NNCH$_3$]; (b) 2.5 × 10^{-4}/s; (c) 5.0 × 10^{-6} M/s **12.29** (a) rate = k[NO]^2[Cl$_2$]; (b) 3.0/(M^2 · s); (c) 5.2 × 10^{-3} M/s **12.31** t$_{1/2}$ = 17 min; t = 69 min **12.33** t$_{1/2}$ = 21 min; t = 42 min **12.35** The reaction is second order in NOBr; k = 0.80/(M · s) **12.37** 2.79 × 10^{-3}/s **12.39** The rate-determining step is the slowest step in a multistep reaction. The rate law shows the molecularity of the rate-determining step. The observed rate law is the rate law for the rate determining step. **12.41** (a) 2 NO(g) + Cl$_2$(g) ⟶ 2 NOCl(g); (b) NOCl$_2$ is a reaction intermediate. (c) Each elementary step is bimolecular. **12.43** (a) unimolecular; rate = k[I$_2$]; (b) termolecular; rate = k[NO]^2[Br$_2$]; (c) bimolecular; rate = k[CH$_3$Br][OH$^-$]; (d) unimolecular; rate = k[N$_2$O$_5$] **12.45** H$_2$(g) + ICl(g) ⟶ HI(g) + HCl(g) (slow) HI(g) + ICl(g) ⟶ I$_2$(g) + HCl(g) (fast) **12.47** 104 kJ/mol **12.49** (a) 134 kJ/mol; (b) 6.0/(M · s) **12.51** The reactant molecules may not have sufficient kinetic energy or correct orientation to react. **12.57** (a) O$_3$(g) + O(g) ⟶ 2 O$_2$(g); (b) Cl acts as a catalyst. (c) ClO is a reaction intermediate. (d) A catalyst is consumed in one step and regenerated in a subsequent step. A reaction intermediate is produced in one step and consumed in another. **12.59** Reactant concentration, temperature, and a catalyst. Reactant concentrations and a catalyst may show up in the rate law. If they do, changes in their concentrations will affect the rate. The rate constant changes with temperature; it increases with an increase in temperature and decreases with a decrease in temperature. **12.61** (a) 2 A + B ⟶ D; (b) 1st order in A; zero order in B; 1st order in C; (c) rate = k[A][C]; (d) C is a catalyst. C appears in the rate law, but it is not consumed in the reaction; (e) A + C ⟶ AC (slow) AC + B ⟶ AB + C (fast) A + AB ⟶ D (fast) (f) 3.4 × 10^{-4}/(M · s) **12.63** For E$_a$ = 50 kJ/mol, f = 2.0 × 10^{-9}; for E$_a$ = 100 kJ/mol, f = 3.9 × 10^{-18} **12.65** (a) 1/8; (b) 0.024/y; (c) 193 y **12.67** (a) 2 NO(g) + Br$_2$(g) ⟶ 2 NOBr(g); (b) NOBr$_2$ is a reaction intermediate. (c) rate = k[NO][Br$_2$]; (d) It can't be the first step. It must be the second step.

Chapter 13

13.1 0.0862 M **13.3** (a) 7.9 × 10^4; (b) 1.3 × 10^{-5} **13.5** At 500 K, K$_p$ = 1.7 × 10^4; at 1000 K, K$_c$ = 1.1

13.7 $K_c = 1.2 \times 10^{-42}$. Since K_c is very small, the equilibrium mixture contains mostly H_2 molecules. **13.9** (a) 3.5×10^{-22} M; (b) 210 H atoms and 6.0×10^{22} H_2 molecules **13.11** $[N_2O_4] = 0.0429$ M; $[NO_2] = 0.0141$ M **13.13** (a) remains the same; (b) increases; (c) decreases **13.15** (a) remains the same; (b) increases; (c) decreases; (d) remains the same; (e) decreases **Understanding Key Concepts** **1.** (a) (1) and (3); (b) $K_c = 1.5$; (c) Since the same number of molecules appear on both sides of the equation, the volume terms in K_c would all cancel. Therefore, we can calculate K_c without including the volume. **2.** (a) $A_2 + C_2 \rightleftharpoons 2$ AC; (b) $A_2 + B_2 \rightleftharpoons 2$ AB **3.** (a) (2); (b) (1), reverse; (3), forward **4.** (a) $A_2 + 2$ B $\rightleftharpoons 2$ AB; (b) The number of AB molecules will increase, because as the volume is decreased at constant temperature, the pressure will increase and the reaction will shift to the side of fewer molecules to reduce the pressure. **5.** When the stopcock is opened, the reaction will go in the reverse direction because there will initially be an excess of AB molecules. **6.** As the temperature is raised the reaction proceeds in the reverse direction. This is consistent with an exothermic reaction, where "heat" can be considered as a product. **Additional Problems** **13.17** (a) $K_c = \dfrac{[PCl_3][Cl_2]}{[PCl_5]}$; (b) $K_c = \dfrac{[NOCl]^2}{[NO]^2[Cl_2]}$; (c) $K_c = \dfrac{[CS_2][H_2]^4}{[CH_4][H_2S]^2}$ **13.19** $K_c = \dfrac{[C_2H_5OC_2H_5][H_2O]}{[C_2H_5OH]^2}$ **13.21** (a) $K_c = \dfrac{[\text{malic acid}]}{[\text{fumaric acid}]}$;

(b) $K_c = \dfrac{[\text{citric acid}]}{[\text{acetic acid}][\text{oxaloacetic acid}]}$ **13.23** 0.058 **13.25** 29.0 **13.27** (a) $K_c = \dfrac{[CH_3COOC_2H_5][H_2O]}{[CH_3COOH][C_2H_5OH]}$; (b) $K_c = 3.4$; Since the same number of molecules appear on both sides of the equation, the volume terms in K_c cancel. Therefore, we can calculate K_c without including the volume. **13.29** $K_p = K_c = 4.24$ **13.31** $K_p = 0.0313$; $K_c = 1.28 \times 10^{-3}$ **13.33** (a) $K_c = \dfrac{[CO_2]^3}{[CO]^3}$ $K_p = \dfrac{(P_{CO_2})^3}{(P_{CO})^3}$; (b) $K_c = \dfrac{1}{[O_2]^3}$ $K_p = \dfrac{1}{(P_{O_2})^3}$; (c) $K_c = [SO_3]$ $K_p = P_{SO_3}$; (d) $K_c = [Ba^{2+}][SO_4{}^{2-}]$ **13.35** (a) proceeds hardly at all toward completion; (b) goes almost all the way to completion **13.37** $K_c = 1.2 \times 10^{82}$ is very large. When equilibrium is reached, very little if any ethanol will remain. **13.39** Q_c has the same form as K_c but contains arbitrary (not necessarily equilibrium) concentrations, so it can have any value. **13.41** $Q_c = 0.69$; Since $Q_c < K_c$, the reaction will proceed from left to right to reach equilibrium. **13.43** 3.7×10^{-3} M **13.45** $[N_2] = 2.22$ M, $[O_2] = 0.54$ M, $[NO] = 0.045$ M **13.47** (a) 6.8 mol; (b) 0.03 mol CH_3COOH, 9.03 mol C_2H_5OH, 0.97 mol $CH_3COOC_2H_5$, 0.97 mol H_2O **13.49** (a) increases; (b) increases; (c) decreases; (d) decreases **13.51** (a) shift to the left; reaction product decreases; (b) no change; (c) shift to the right; reaction product increases **13.53** $[H_2]$ decreases when the temperature is increased. As the temperature is decreased, K_c increases. **13.55** [HI] decreases when the tempera-

ture is increased. **13.57** (a) amount of CH_3OH decreases; (b) amount of CH_3OH increases; (c) no change; (d) amount of CH_3OH increases; (e) no change **13.59** $k_f[A][B] = k_r[C]$ $\dfrac{k_f}{k_r} = \dfrac{[C]}{[A][B]} = K_c$ **13.61** 210 **13.63** (i) change concentrations; (ii) change volume; (iii) change temperature; A change in temperature will cause a change in the value of the equilibrium constant. **13.67** (a) 0.0400 M; (b) 0.122 **13.69** K_c is very large. The reaction goes to completion. $[CO_2] = 6.2 \times 10^{-25}$ M **13.71** 1.25 atm, 75.8% **13.73** $[H_2] = [I_2] = 0.067$ M; [HI] = 0.766 M **13.75** (a) The reaction is exothermic. (b) no change for all three **13.77** (a) no change; (b) $[CO_2]$ decreases; (c) $[CO_2]$ decreases; (d) $[CO_2]$ increases **13.79** (a) $K_c = \dfrac{[C_2H_6][C_2H_4]}{[C_4H_{10}]}$ $K_p = \dfrac{P_{C_2H_6}P_{C_2H_4}}{P_{C_4H_{10}}}$; (b) 0.19; (c) 38%, $P_{total} = 69$ atm; (d) A decrease in volume would decrease the percent conversion of C_4H_{10}. **13.81** (a) 0.0840; (b) $[(CH_3)_2C{=}CCH_2] = [HCl] = 0.0942$ M; $[(CH_3)_3CCl] = 0.106$ M; (c) $P_{t\text{-butyl chloride}} = 0.055$ atm; $P_{isobutylene} = 0.345$ atm; $P_{HCl} = 0.545$ atm

Chapter 14

14.1 $d(H_2) = 8.24 \times 10^{-5}$ g/cm^3; Air is 14 times more dense than H_2. **14.3** 2 Ga(s) + 6 H$^+$(aq) $\longrightarrow$ 3 H_2(g) + 2 Ga^{3+}(aq) **14.5** $PdH_{0.74}$; $d(H) = 0.0841$ g/cm^3; H molarity = 83.4 M **14.7** (a) Li_2O(s) + H_2O(l) $\longrightarrow$ 2 Li$^+$(aq) + 2 OH$^-$(aq); (b) SO_3(l) + H_2O(l) $\longrightarrow$ H$^+$(aq) + $HSO_4{}^-$(aq); (c) Cr_2O_3(s) + 6 H$^+$(aq) $\longrightarrow$ 2 Cr^{3+}(aq) + 3 H_2O(l); (d) Cr_2O_3(s) + 2 OH$^-$(aq) + 3 H_2O(l) $\longrightarrow$ 2 Cr(OH)$_4{}^-$(aq) **14.9** (a) Rb_2O_2(s) + H_2O(l) $\longrightarrow$ 2 Rb$^+$(aq) + HO$_2{}^-$(aq) + OH$^-$(aq); (b) CaO(s) + H_2O(l) $\longrightarrow$ Ca^{2+}(aq) + 2 OH$^-$(aq); (c) 2 CsO$_2$(s) + H_2O(l) $\longrightarrow$ O$_2$(g) + 2 Cs$^+$(aq) + HO$_2{}^-$(aq) + OH$^-$(aq); (d) SrO$_2$(s) + H_2O(l) $\longrightarrow$ Sr^{2+}(aq) + HO$_2{}^-$(aq) + OH$^-$(aq); (e) CO$_2$(g) + H_2O(l) $\longrightarrow$ H$^+$(aq) + HCO$_3{}^-$(aq) **14.11** PbS(s) + 4 H_2O_2(aq) $\longrightarrow$ PbSO$_4$(s) + 4 H_2O(l) **14.13** NiSO$_4 \cdot 7$ H_2O **Understanding Key Concepts** **1.** (a) (1) covalent (2) ionic (3) covalent (4) interstitial; (b) (1) H, +1; other element, -3 (2) H, -1; other element, +1 (3) H, +1; other element, -2 **2.** (a) A = NaH; B = PdH$_x$; C = H_2S; D = HI; (b) NaH (ionic); PdH$_x$ (interstitial); H_2S and HI (covalent); (c) H_2S and HI (molecular); NaH and PdH$_x$ (three-dimensional crystal); (d) NaH: Na, +1; H, -1, H_2S: S, -2; H, +1, HI: I, -1; H, +1 **3.** reducing agent, Ca or Al **4.** (a) 6; (b) 18 **5.** (a) A = CaO; B = Al$_2O_3$; C = SO$_3$; D = SeO$_3$; (b) CaO (basic); Al$_2O_3$ (amphoteric); SO$_3$ and SeO$_3$ (acidic); (c) CaO (most ionic); SO$_3$ (most covalent); (d) CaO and Al$_2O_3$ (three-dimensional crystal); SO$_3$ and SeO$_3$ (molecular); (e) CaO (highest melting point); SO$_3$ (lowest melting point) **6.** Water does not undergo a disproportionation reaction because H is already in its highest oxidation state and O is already in its lowest oxidation state. **7.** K is oxidized by water. F_2 is reduced by water. Cl_2 and Br_2 disproportionate when treated with water. **Additional Problems** **14.19** (a) 4.71×10^{16} kg D; (b) 5×10^2 kg T **14.21** There are 10. PH$_3$, PH$_2$D, PH$_2$T, PD$_3$, PD$_2$H, PD$_2$T, PT$_3$, PT$_2$H, PT$_2$D, PHDT **14.23** (a) Fe(s) + 2 H$^+$(aq) $\longrightarrow$

$Fe^{2+}(aq) + H_2(g)$; (b) $Ca(s) + 2 H_2O(l) \longrightarrow H_2(g) + Ca^{2+}(aq) + 2 OH^-(aq)$; (c) $2 Al(s) + 6 H^+(aq) \longrightarrow 2 Al^{3+}(aq) + 3 H_2(g)$; (d) $C_2H_6(g) + 2 H_2O(g) \longrightarrow 2 CO(g) + 5 H_2(g)$ **14.27** 9.67×10^4 kg **14.31** (a) KH, H^-; (b) PH_3, covalent; (c) $H_2S(g)$ covalent; (d) BaH_2, H^- **14.33** (a) H_2S, covalent; (b) NaH, ionic; (c) PdH_x, interstitial; (d) SrH_2, ionic; (e) CH_4, covalent **14.35** (a) SrH_2, $H -1$; (b) H_2Se, $H +1$; (c) HBr, $H +1$; (d) NaH, $H -1$ **14.37** (a) KH, ionic bonding; (b) PH_3, covalent bonding **14.39** (a) GeH_4, tetrahedral; (b) H_2S, bent; (c) NH_3, trigonal pyramidal **14.41** (a) $MgH_2(s) + 2 H^+(aq) \longrightarrow Mg^{2+}(aq) + 2 H_2(g)$; (b) $NaH(s) + H_2O(l) \longrightarrow H_2(g) + Na^+(aq) + OH^-(aq)$; (c) $2 Li(l) + H_2(g) \longrightarrow 2 LiH(s)$; (d) $BaH_2(s) + H^+(aq) + HSO_4^-(aq) \longrightarrow 2 H_2(g) + BaSO_4(s)$ **14.43** $d = 0.16$ g/cm³; volume $= 1.8 \times 10^3$ cm³ **14.47** 20.7 g of C_2H_2 and 44.6 L of O_2 **11.51** The highest occupied molecular orbital in O_2 is the doubly degenerate π^*2p orbital, which contains 2 unpaired electrons. The bond order is 2 because O_2 has four $\pi 2p$ and two $\sigma 2p$ bonding electrons and two π^*2p antibonding electrons. **14.53** (a) $2 Ca(s) + O_2(g) \longrightarrow 2 CaO(s)$; (b) $C(s) + O_2(g) \longrightarrow CO_2(g)$; (c) $4 As(s) + 5 O_2(g) \longrightarrow 2 As_2O_5(s)$; (d) $4 B(s) + 3 O_2(g) \longrightarrow 2 B_2O_3(s)$ **14.55** $Li_2O < BeO < B_2O_3 < CO_2 < N_2O_5$ **14.57** $Cl_2O_7 < Al_2O_3 < Na_2O < Cs_2O$ **14.59** (a) CrO_3; (b) N_2O_5; (c) SO_3 **14.61** (a) $Cl_2O_7(l) + H_2O(l) \longrightarrow 2 H^+(aq) + 2 ClO_4^-(aq)$; (b) $K_2O(s) + H_2O(l) \longrightarrow 2 K^+(aq) + 2 OH^-(aq)$; (c) $SO_3(l) + H_2O(l) \longrightarrow H^+(aq) + HSO_4^-(aq)$ **14.63** (a) $ZnO(s) + 2 H^+(aq) \longrightarrow Zn^{2+}(aq) + H_2O(l)$; (b) $ZnO(s) + 2 OH^-(aq) + H_2O(l) \longrightarrow Zn(OH)_4^{2-}(aq)$ **14.67** (a) BaO_2; (b) CaO; (c) CsO_2; (d) Li_2O; (e) Na_2O_2

14.69

	O_2		O_2^-		O_2^{2-}	
σ^*2p	—		—		—	
π^*2p	↑	↑	↑↓	↑	↑↓	↑↓
$\pi 2p$	↑↓	↑↓	↑↓	↑↓	↑↓	↑↓
$\sigma 2p$	↑↓		↑↓		↑↓	
	Bond order = 2		Bond order = 1.5		Bond order = 1	

(a) The O—O bond length increases because the bond order decreases. The bond order decreases because of the increased occupancy of antibonding orbitals. (b) O_2^- has one unpaired electron and is paramagnetic. O_2^{2-} has no unpaired electrons and is diamagnetic. **14.73** (a) $H_2O_2(aq) + 2 H^+(aq) + 2 I^-(aq) \longrightarrow I_2(aq) + 2 H_2O(l)$; (b) $3 H_2O_2(aq) + 8 H^+(aq) + Cr_2O_7^{2-}(aq) \longrightarrow 2 Cr^{3+}(aq) + 3 O_2(g) + 7 H_2O(l)$ **14.81** (a) $2 F_2(g) + 2 H_2O(l) \longrightarrow O_2(g) + 4 HF(aq)$; (b) $I_2(s) + H_2O(l) \longrightarrow HOI(aq) + H^+(aq) + I^-(aq)$; (c) $2 K(s) + 2 H_2O(l) \longrightarrow H_2(g) + 2 K^+(aq) + 2 OH^-(aq)$ **14.85** $CaSO_4 \cdot 2H_2O$ **14.89** 1.38 kg of Mg^{2+} **14.95** 3.0 million tons **14.97** (a) NH_3, $H -1$; (b) BaH_2, $H -1$; (c) $Ca(OH)_2$, $H +1$; (d) HBr, $H +1$; (e) H_3PO_4, $H +1$ **14.101** (a) $Ca(OH)_2$; (b) Cr_2O_3; (c) RbO_2; (d) Na_2O_2; (e) BaH_2; (f) H_2Se **14.103** (a) $2 H_2(g) + O_2(g) \longrightarrow 2 H_2O(l)$; (b) $5 H_2O_2(aq) + 2 MnO_4^-(aq) + 6 H^+(aq) \longrightarrow 5 O_2(g) + 2 Mn^{2+}(aq) + 8 H_2O(l)$; (c) $2 F_2(g) + 2 H_2O(l) \longrightarrow O_2(g) + 4 HF(aq)$

Chapter 15

15.1
(a) $H_2SO_4(aq) + H_2O(l) \longrightarrow H_3O^+(aq) + HSO_4^-(aq)$
conjugate base

(b) $HSO_4^-(aq) + H_2O(l) \longrightarrow H_3O^+(aq) + SO_4^{2-}(aq)$
conjugate base

(c) $H_3O^+(aq) + H_2O(l) \longrightarrow H_3O^+(aq) + H_2O(l)$
conjugate base

15.3 $[H_3O^+] = 2.0 \times 10^{-9}$ M; solution is basic **15.5** (a) 8.20; (b) 4.22 **15.7** (a) 1.30; (b) 0.00; (c) -0.78 **15.9** 11.81 **15.13** (a) 8.0%; (b) 2.6% **15.15** $[H_2SO_4] = 0$ M; $[HSO_4^-] = 0.49$ M; $[SO_4^{2-}] = 0.011$ M; $[H_3O^+] = 0.51$ M; $[OH^-] = 2.0 \times 10^{-14}$ M; pH = 0.29 **15.17** (a) 7.7×10^{-12}; (b) 2.9×10^{-7} **15.19** 8.32 **15.21** (a) neutral; (b) basic; (c) acidic; (d) acidic; (e) acidic **15.23** (a) Lewis acid, $AlCl_3$; Lewis base, Cl^-; (b) Lewis acid, Ag^+; Lewis base, NH_3; (c) Lewis acid, SO_2; Lewis base, OH^-; (d) Lewis acid, Cr^{3+}; Lewis base, H_2O *Understanding Key Concepts* **1.** (a) acids, HCO_3^- and H_3O^+; bases, H_2O and CO_3^{2-}; (b) acids, HF and H_2CO_3; bases, HCO_3^- and F^- **2.** (a) HX^-, Y^-, Z^-; (b) $Z^- < Y^- < HX^-$ **3.** (a) $HX < HZ < HY$; (b) HY; (c) HX; (d) 20% **4.** (c) represents a solution of a weak diprotic acid, H_2A. Because K_{a2} is always less than K_{a1}, (a) and (d) represent impossible situations. **5.** (a) $A^-(aq) + H_2O(l) \longrightarrow HA(aq) + OH^-(aq)$; basic; (b) $M(H_2O)_6^{3+}(aq) + H_2O(l) \longrightarrow H_3O^+(aq) + M(H_2O)_5(OH)^{2+}(aq)$; acidic; (c) $2 H_2O(l) \rightleftharpoons H_3O^+(aq) + OH^-(aq)$; neutral; (d) $M(H_2O)_6^{3+}(aq) + A^-(aq) \longrightarrow HA(aq) + M(H_2O)_5(OH)^{2+}(aq)$; acidic **6.** (a) H_2S, weakest; HBr, strongest; (b) H_2SeO_3, weakest; $HClO_3$, strongest **7.** (a) Brønsted-Lowry acids: NH_4^+, $H_2PO_4^-$; Brønsted-Lowry bases: SO_3^{2-}, OCl^-, $H_2PO_4^-$; (b) Lewis acids: Fe^{3+}, BCl_3; Lewis bases: SO_3^{2-}, OCl^-, $H_2PO_4^-$ *Additional Problems* **15.27** (a) OCl^-; (b) PO_4^{3-}; (c) OH^-; (d) CH_3NH_2; (e) HCO_3^-; (f) H^- **15.29**
(a) $CH_3COOH(aq) + NH_3(aq) \longrightarrow NH_4^+(aq) + CH_3COO^-(aq)$
acid   base————acid   base

(b) $CO_3^{2-}(aq) + H_3O^+(aq) \longrightarrow H_2O(l) + HCO_3^-(aq)$
base   acid————base   acid

(c) $HSO_3^-(aq) + H_2O(l) \longrightarrow H_3O^+(aq) + SO_3^{2-}(aq)$
acid   base————acid   base

(d) $HSO_3^-(aq) + H_2O(l) \longrightarrow H_2SO_3(aq) + OH^-(aq)$
base   acid   acid   base

15.33 (a) left; (b) left; (c) right; (d) right **15.35** $2 H_2O(l) \rightleftharpoons H_3O^+(aq) + OH^-(aq)$; $K_w = [H_3O^+][OH^-]$; $[H_2O]$ is omitted because it is a constant. **15.37** (a) 4.0×10^{-11} M; (b) 5.0×10^{-15} M; (c) 1.8×10^{-6} M; (d) 6.7×10^{-12} M; (e) 1.0×10^{-7} M **15.39** (a) 8×10^{-5} M; (b) 1.5×10^{-11} M; (c) 1.0 M; (d) 5.6×10^{-15} M; (e) 10 M **15.41** (a) 10; (b) 10^{10}; (c) 1.26 **15.43** (a) 0.70; (b) 11.80; (c) 11.90; (d) -0.18; (e) 14.18 **15.45** (a) 2.30; (b) 2.92; (c) 0.46 **15.47** (a) $HClO_2(aq) + H_2O(l) \rightleftharpoons H_3O^+(aq) + ClO_2^-(aq)$; $K_a = \dfrac{[H_3O^+][ClO_2^-]}{[HClO_2]}$; (b) $HOBr(aq) + H_2O(l) \rightleftharpoons H_3O^+(aq) + OBr^-(aq)$; $K_a = \dfrac{[H_3O^+][OBr^-]}{[HOBr]}$;

(c) $HCOOH(aq) + H_2O(l) \rightleftharpoons H_3O^+(aq) + HCOO^-(aq)$;

$K_a = \dfrac{[H_3O^+][HCOO^-]}{[HCOOH]}$ **15.49** 2.0×10^{-9} **15.51**

$[C_6H_5OH] = 0.10$ M; $[C_6H_5O^-] = [H_3O^+] = 3.6 \times 10^{-6}$ M; $[OH^-] = 2.8 \times 10^{-9}$ M; % dissociation = 0.0036%; pH = 5.44 **15.53** pH = 1.59; % dissociation = 1.7% **15.55** (a) $H_3PO_4(aq) + H_2O(l) \rightleftharpoons H_3O^+(aq) + H_2PO_4^-(aq)$; $K_{a1} = \dfrac{[H_3O^+][H_2PO_4^-]}{[H_3PO_4]}$; (b) $H_2PO_4^-(aq) + H_2O(l) \rightleftharpoons H_3O^+(aq) + HPO_4^{2-}(aq)$; $K_{a2} = \dfrac{[H_3O^+][HPO_4^{2-}]}{[H_2PO_4^-]}$; (c) $HPO_4^{2-}(aq) + H_2O(l) \rightleftharpoons H_3O^+(aq) + PO_4^{3-}(aq)$; $K_{a3} = \dfrac{[H_3O^+][PO_4^{3-}]}{[HPO_4^{2-}]}$ **15.57** pH = 1.08; $[C_2O_4^{2-}] = 6.4 \times 10^{-5}$ M **15.59** $[H_3O^+] = 0.4$ M; $[SO_4^{2-}] = 0.007$ M **15.61** 1×10^{-6} **15.63** $[C_6H_5NH_3^+] = [OH^-] = 8.0 \times 10^{-6}$ M; $[C_6H_5NH_2] = 0.15$ M; $[H_3O^+] = 1.2 \times 10^{-9}$ M; pH = 8.90 **15.65** (a) 2.0×10^{-11}; (b) 1.1×10^{-6}; (c) 2.3×10^{-5}; (d) 5.6×10^{-6} **15.67**

(a) $\underset{\text{acid}}{CH_3NH_3^+(aq)} + \underset{\text{base}}{H_2O(l)} \rightleftharpoons \underset{\text{acid}}{H_3O^+(aq)} + \underset{\text{base}}{CH_3NH_2(aq)}$

(b) $\underset{\text{acid}}{Cr(H_2O)_6^{3+}(aq)} + \underset{\text{base}}{H_2O(l)} \rightleftharpoons \underset{\text{acid}}{H_3O^+(aq)} + \underset{\text{base}}{Cr(H_2O)_5(OH)^{2+}(aq)}$

(c) $\underset{\text{base}}{CH_3COO^-(aq)} + \underset{\text{acid}}{H_2O(l)} \rightleftharpoons \underset{\text{acid}}{CH_3COOH(aq)} + \underset{\text{base}}{OH^-(aq)}$

(d) $\underset{\text{base}}{PO_4^{3-}(aq)} + \underset{\text{acid}}{H_2O(l)} \rightleftharpoons \underset{\text{acid}}{HPO_4^{2-}(aq)} + \underset{\text{base}}{OH^-(aq)}$

15.69 (a) acidic; (b) neutral; (c) basic; (d) acidic; (e) acidic **15.71** (a) $[NO_3^-] = [C_2H_5NH_3^+] = 0.10$ M; $[H_3O^+] = [C_2H_5NH_2] = 1.2 \times 10^{-6}$ M; $[OH^-] = 8.0 \times 10^{-9}$ M; pH = 5.90; (b) $[Na^+] = [CH_3COO^-] = 0.10$ M; $[CH_3COOH] = [OH^-] = 7.5 \times 10^{-6}$ M; $[H_3O^+] = 1.3 \times 10^{-9}$ M; pH = 8.89; (c) $[Na^+] = [NO_3^-] = 0.10$ M; $[H_3O^+] = [OH^-] = 1.0 \times 10^{-7}$ M; pH = 7.00 **15.73** (a) $PH_3 < H_2S < HCl$; (b) $NH_3 < PH_3 < AsH_3$; (c) $HBrO < HBrO_2 < HBrO_3$ **15.75** (a) H_2Te; (b) H_3PO_4; (c) $H_2PO_4^-$; (d) NH_4^+ **15.79** (a) Lewis acid, SiF_4; Lewis base, F^-; (b) Lewis acid, Zn^{2+}; Lewis base, NH_3; (c) Lewis acid, $Al(OH)_3$; Lewis base, OH^-; (d) Lewis acid, $AgCl$; Lewis base, Cl^- **15.81** (a) BF_3; (b) SO_3; (c) Sn^{4+}; (d) CH_3^+ **15.83** 9.45 **15.85** (a) $HCO_3^-(aq) + H_2O(l) \rightleftharpoons H_3O^+(aq) + CO_3^{2-}(aq)$; (b) $HCO_3^-(aq) + H_2O(l) \rightleftharpoons H_2CO_3(aq) + OH^-(aq)$ **15.87** $[Na^+] = [C_6H_5COO^-] = 0.050$ M; $[C_6H_5COOH] = [OH^-] = 2.8 \times 10^{-6}$ M; $[H_3O^+] = 3.6 \times 10^{-9}$ M; pH = 8.44

15.89

Additional H_2O molecules can hydrogen bond to the other two H atoms of H_3O^+. **15.91** fraction dissociated = 6.09×10^{-10}; % dissociation = 6.09×10^{-8}%; pH = 7.471 **15.93** $[H_3PO_4] = 0.076$ M; $[H_2PO_4^-] = [H_3O^+] =$ 0.024 M; $[HPO_4^{2-}] = 6.2 \times 10^{-8}$ M; $[PO_4^{3-}] = 1.2 \times 10^{-18}$ M; $[OH^-] = 4.2 \times 10^{-13}$ M; pH = 1.62 **15.95** $HCO_3^-(aq) + Al(H_2O)_6^{3+}(aq) \longrightarrow H_2O(l) + CO_2(g) + Al(H_2O)_5(OH)^{2+}(aq)$ **15.97** $HBr < NH_4Br < NaBr < (NH_4)_2CO_3 < Na_2CO_3$ **15.101** 2.54 **15.103** For $[HCl] = 1.0 \times 10^{-10}$ M, pH = 7.00, and the principal source of H_3O^+ is the dissociation of H_2O. For $[HCl] = 1.0 \times 10^{-7}$ M, pH = 6.79. **15.105** (a) $[NH_4^+] = [F^-] = 0.25$ M; $[NH_3] = [HF] = 3.1 \times 10^{-4}$ M; $[H_3O^+] = 4.4 \times 10^{-7}$ M; $[OH^-] = 2.3 \times 10^{-8}$ M; pH = 6.36; (b) $[NH_4^+] = [CH_3COO^-] = 0.25$ M; $[NH_3] = [CH_3COOH] = 1.4 \times 10^{-3}$ M; $[H_3O^+] = [OH^-] = 1.0 \times 10^{-7}$ M; pH = 7.00; (c) $[NH_4^+] = 0.47$ M; $[SO_3^{2-}] = 0.22$ M; $[NH_3] = [HSO_3^-] = 0.030$ M; $[H_3O^+] = 8.6 \times 10^{-9}$ M; $[OH^-] = 1.2 \times 10^{-6}$ M; pH = 8.06

Chapter 16

16.1 (a) $HNO_2(aq) + OH^-(aq) \rightleftharpoons NO_2^-(aq) + H_2O(l)$; pH > 7.00; (b) $H_3O^+(aq) + NH_3(aq) \rightleftharpoons NH_4^+(aq) + H_2O(l)$; pH < 7.00; (c) $OH^-(aq) + H_3O^+(aq) \rightleftharpoons 2 H_2O(l)$; pH = 7.00 **16.3** $[H_3O^+] = 1.2 \times 10^{-9}$ M, $[HCN] = 0.025$ M, $[Na^+] = [CN^-] = 0.010$ M, $[OH^-] = 8.2 \times 10^{-6}$ M, pH = 8.91 **16.5** Buffer pH = 3.76 (a) pH = 3.71; (b) pH = 3.87 **16.7** 9.95 **16.9** Suggested buffer system: $HOCl$ ($K_a = 3.5 \times 10^{-8}$) and $NaOCl$ **16.11** (a) 12.00; (b) 4.1; (c) 2.15 **16.13** Use thymolphthalein (pH 9.4–10.6). Bromthymol blue is unacceptable because it changes color halfway to the equivalence point. **16.15** (a) $K_{sp} = [Ag^+][Cl^-]$; (b) $K_{sp} = [Pb^{2+}][I^-]^2$; (c) $K_{sp} = [Ca^{2+}]^3[PO_4^{3-}]^2$ **16.17** 1.10×10^{-10} **16.19** 1.4×10^{-5} M **16.21** 5.4×10^{-14} M **16.23** (a) IP > K_{sp}, a precipitate of $BaCO_3$ will form. (b) IP < K_{sp}, no precipitate will form. *Understanding Key Concepts* **1.** (a) (1) and (3); (b) (3) **2.** (a) (2) has the highest pH; (3) has the lowest pH **3.** (4) **4.** (b) 50.0 mL; (c) pH > 7.00; (d) pH = 4 **5.** (a) (a) decreased; (b) unaffected; (c) decreased; (d) increased; (b) (a) unaffected; (b) increased; (c) decreased; (d) decreased *Additional Problems* **16.27** (a) 1.0×10^{14}; (b) 3.5×10^6; (c) 4.3×10^4; (d) 6.5×10^9 **16.29** $K_n = 2.3 \times 10^{-5}$; K_n is small. The neutralization reaction does not proceed very far to completion. **16.31** (a) $NaNO_2$; (b) HCl; (c) $Ba(NO_2)_2$ **16.33** (a) increases; (b) and (c) remains the same; (d) decreases **16.35** For 0.10 M HN_3, % dissociation = 1.4%; for 0.10 M HN_3 in 0.10 M HCl, % dissociation = 0.019% because of the common ion (H_3O^+) effect. **16.39** (a), (c), and (d) **16.41** When blood absorbs acid, the equilibrium shifts to the left, decreasing the pH, but not by much because the $[HCO_3^-]/[H_2CO_3]$ ratio remains nearly constant. When blood absorbs base, the equilibrium shifts to the right, increasing the pH, but not by much because the $[HCO_3^-]/[H_2CO_3]$ ratio remains nearly constant. **16.43** 9.09 **16.45** 9.25; (a) 9.30; (b) 9.08 **16.47** (a) 1.0×10^{-5}; (b) 2.0×10^{-9}; (b) is the weaker acid. **16.49** 0.093 **16.51** Solution should have 0.50 mol of CH_3COO^- per mole of CH_3COOH. For example, you might dissolve 41g of CH_3COONa in 1.00

L of 1.00 M CH_3COOH.   **16.53** (a) 9.00 mmol; (b) 20.0 mL; (c) 7.00   **16.55** (a) 0.92; (b) 1.77; (c) 3.5; (d) 7.00; (e) 10.5; (f) 12.60   **16.59** (a) 1.74; (b) 2.34; (c) 6.02; (d) 9.70   **16.63** (a) $K_{sp} = [Ca^{2+}][OH^-]^2$; (b) $K_{sp} = [Ag^+]^3[PO_4^{3-}]$; (c) $K_{sp} = [Ba^{2+}][CO_3^{2-}]$; (d) $K_{sp} = [Ca^{2+}]^5[PO_4^{3-}]^3[OH^-]$ **16.65** 8.39×10^{-12} **16.67** (a) 1.6×10^{-5} M; (b) 1.4×10^{-2} M; (c) 1.3×10^{-8} M   **16.69** $Ag_2CO_3(s) \rightleftharpoons 2\, Ag^+(aq) + CO_3^{2-}(aq)$ (a) $AgNO_3$, source of Ag^+; reaction shifts left; (b) HNO_3, source of H_3O^+; removes CO_3^{2-}; reaction shifts right; (c) Na_2CO_3, source of CO_3^{2-}; reaction shifts left; (d) NH_3, forms $Ag(NH_3)_2^+$; removes Ag^+; reaction shifts right   **16.71** (a) 1.0×10^{-4} M; (b) 2.1×10^{-7} M   **16.73** (b) $CaCO_3$, (c) $Ni(OH)_2$, and (d) $Ca_3(PO_4)_2$ **16.75** 1×10^{-22} **16.77** (a) 9.2×10^{-9} M; (b) 0.050 M   **16.79** IP < K_{sp}; no precipitate will form. A $[Cl^-]$ just greater than 0.049 M will result in precipitation.   **16.81** (a) IP < K_{sp}; no precipitate; (b) IP > K_{sp}; $CaCO_3$ will precipitate.   **16.83** Yes. For FeS, $Q_c < K_{spa}$, and no FeS will precipitate. For SnS, $Q_c > K_{spa}$, and SnS will precipitate.   **16.85** For $Mg(OH)_2$, IP < K_{sp}; no precipitate will form. For $Al(OH)_3$, IP > K_{sp}; $Al(OH)_3$ will precipitate.   **16.87** (a) Add Cl^- and precipitate Hg_2Cl_2; (b) Add $(NH_4)_2HPO_4$ and precipitate $MgNH_4PO_4$; (c) Add HCl and H_2S, and precipitate HgS; (d) Add Cl^- and precipitate $PbCl_2$.   **16.89** (a), (c), and (e)   **16.91** (a) $H_3O^+(aq) + NH_3(aq) \rightleftharpoons NH_4^+(aq) + H_2O(l)$; (b) $H_3O^+(aq) + CH_3NH_2(aq) \rightleftharpoons CH_3NH_3^+(aq) + H_2O(l)$; (c) $H_3O^+(aq) + OH^-(aq) \rightleftharpoons 2\, H_2O(l)$; Reaction (c) proceeds farthest to the right.   **16.93** (a), (b), and (d)   **16.95** A solution that is 0.30 M HF and 0.30 M NaF is a buffer solution. The pH of a buffer solution is not changed on dilution.   **16.97** 30 mL   **16.99** 9.07   **16.101** 1.40   **16.103** 4.5×10^{-25} M   **16.105** 1×10^{-30} **16.107** $[NH_4^+] = 0.50$ M; $[NH_3] = 1.0$ M; pH = 9.55

Chapter 17

17.1 (a) spontaneous; (b), (c), and (d) nonspontaneous   **17.3** (a) shuffled deck; (b) disordered crystal; (c) 1 mol N_2 at STP; (d) 1 mol N_2 at 273 K and 0.25 atm   **17.5** $\Delta S_{total} = -438$ J/K, not spontaneous   **17.7** (a) At 325°C, $\Delta G > 0$, and Hg does not boil; (b) 630 K = 357°C   **17.9** (a) $\Delta G° = -150.2$ kJ; This reaction can be used for the synthesis of C_2H_2, since $\Delta G < 0$. (b) No, since $\Delta G_f°(C_2H_2) > 0$.   **17.11** 1.4×10^{-23} **17.13** +80 kJ   *Understanding Key Concepts*   **1.** (b) $\Delta H = 0$; $\Delta S = +$; $\Delta G = -$; (c) In an isolated system $\Delta S_{surr} = 0$, and $\Delta S_{sys} = \Delta S_{total} > 0$ for the spontaneous process. (d) $\Delta G = +$, and process is nonspontaneous.   **2.** $\Delta H = +$; $\Delta S = +$; $\Delta G = -$ **3.** (a) For *initial state 1*, $Q_p < K_p$; for *initial state 2*, $Q_p > K_p$; (b) $\Delta H = +$; $\Delta S = +$; $\Delta G = -$; reaction is spontaneous; (c) $\Delta H = -$; $\Delta S = -$; $\Delta G = -$; reaction is spontaneous; (d) State 1 lies to the left of the minimum in Figure 17.10; state 2 lies to the right of the minimum.   **4.** (a) $\Delta H° = +$, and $\Delta S° = +$; (b) $\Delta S°$ is for the complete conversion of 1 mol of A_2 in its standard state to 2 mol of A in its standard state; (c) There is not enough information to say anything about the sign of

$\Delta G°$: $\Delta G°$ will decrease (becomes less positive or more negative) as the temperature is increased; (d) K_p will increase as the temperature increases. As the temperature increases there will be more A and less A_2; (e) $\Delta G = 0$ at equilibrium.   *Additional Problems*   **17.15** (a) and (d) nonspontaneous; (b) and (c) spontaneous   **17.17** (b) and (d) spontaneous   **17.21** (a) +; (b) −; (c) −; (d) +   **17.23** (a) −; (b) −; (c) + (d) −   **17.25** (a) 0; (b) 0; (c) 5.74×10^{-23} J/K   **17.27** (a) 2.30×10^{-22} J/K; (b) 2.30×10^{-21} J/K; (c) 11.5 J/K   **17.29** (a) H_2 at 25°C in 50 L; (b) O_2 at 25°C, 1 atm; (c) H_2 at 100°C, 1 atm; (d) CO_2 at 100°C, 0.1 atm   **17.33** (a) $P_4O_{10}(s)$; (b) $SO_3(g)$; (c) Hg(l); (d) $CO_2(g)$ **17.35** (a) −165.4 J/K; (b) −306 J/K; (c) +33.0 J/K; (d) +446.0 J/K   **17.37** For a spontaneous process $\Delta S_{total} = \Delta S_{sys} + \Delta S_{surr} > 0$. For an isolated system $\Delta S_{surr} = 0$ and $\Delta S_{sys} > 0$ is the criterion for spontaneous change. An example of a spontaneous process in an isolated system is the mixing of two gases.   **17.41** $\Delta S_{sys} = -11.6$ J/K; $\Delta S_{surr} = +728.8$ J/K; $\Delta S_{total} = +717.2$ J/K; reaction is spontaneous.   **17.43** (a) $\Delta S_{surr} = -28.8$ J/(K·mol); $\Delta S_{total} = -0.7$ J/(K·mol); (b) $\Delta S_{surr} = -28.1$ J/(K·mol); $\Delta S_{total} = 0$; (c) $\Delta S_{surr} = -27.5$ J/(K·mol); $\Delta S_{total} = +0.6$ J/(K·mol); NaCl melts at 1100 K because $\Delta S_{total} > 0$. The melting point of NaCl is 1075 K, where $\Delta S_{total} = 0$.   **17.49** (a) +0.7 kJ; does not melt; (b) 0; does melt; (c) −0.7 kJ; does melt   **17.51** 19.1 kJ/mol   **17.55** (a) $\Delta H° = -197.8$ kJ; $\Delta S° = -188.0$ J/K; $\Delta G° = -141.8$ kJ; spontaneous; (b) $\Delta H° = 50.6$ kJ; $\Delta S° = -331.5$ J/K; $\Delta G° = +149.4$ kJ; nonspontaneous; (c) $\Delta H° = -471.8$ kJ; $\Delta S° = -133.1$ J/K; $\Delta G° = -432.1$ kJ; spontaneous   **17.57** (a) −141.8 kJ; (b) +149.2 kJ; (c) −432.2 kJ   **17.59** (a) no; (b) yes; (c) no; (d) yes   **17.61** $\Delta G° = -90.0$ kJ; the reaction becomes nonspontaneous at high temperatures because $\Delta S°$ is negative. The reaction becomes nonspontaneous at 507°C.   **17.63** $\Delta G° = +36.2$ kJ; reaction is nonspontaneous under standard-state conditions. Using NaOH(aq), $\Delta G° = -79.7$ kJ; reaction is spontaneous. (More generally, base removes HCl, driving the reaction to the right.) $\Delta G_f° = +51.9$ kJ; the synthesis of vinyl chloride from its elements is not possible.   **17.65** (a) $\Delta G < \Delta G°$; (b) $\Delta G = \Delta G°$; (c) $\Delta G > \Delta G°$; As Q increases, the thermodynamic driving force decreases.   **17.67** (a) −30.7 kJ; spontaneous; (b) +3.5 kJ; nonspontaneous   **17.71** $K_p = 2.4 \times 10^2$ **17.73** +20.1 kJ   **17.75** $\Delta G° = -24.7$ kJ; $K_p = 2.1 \times 10^4$; $\Delta G = -39.5$ kJ   **17.77** (a), (b), and (h) are affected by catalyst. All others are not.   **17.79** 351 K = 78°C   **17.83** $\Delta S_{vap} = 87.5$ J/K; $\Delta G_{vap} = 0$ **17.85** (a) $\Delta H° = -1203.4$ kJ; $\Delta S° = -216.6$ J/K; $\Delta G° = -1138.8$ kJ; reaction is spontaneous at 25°C, and $\Delta G°$ will become less negative as the temperature is raised; (b) $\Delta H° = +101$ kJ; $\Delta S° = +174.8$ J/K; $\Delta G° = +48$ kJ; reaction is not spontaneous at 25°C, and $\Delta G°$ will become less positive as the temperature is raised; (c) $\Delta H° = -852$ kJ; $\Delta S° = -38.5$ J/K; $\Delta G° = -840$ kJ; reaction is spontaneous at 25°C, and $\Delta G°$ will become less negative as the temperature is raised; (d) $\Delta H° = +135.6$ kJ; $\Delta S° = +333$ J/K; $\Delta G° = +34.5$ kJ; reaction is not spontaneous at 25°C, and $\Delta G°$ will become less positive as the temperature is raised.   **17.89** (a) $\Delta H = 0$; $\Delta S = 5.76$ J/K; $\Delta G = -1.72$ kJ; (b) $\Delta S_{total} = \Delta S_{sys} > 0$

17.91 (a)

	$\Delta H_{vap}/T_{bp}$
ammonia	120 J/K
benzene	87 J/K
carbon tetrachloride	85 J/K
chloroform	87 J/K
mercury	90 J/K

(b) All processes are the conversion of a liquid to a gas at the boiling point. They should all have similar ΔS values. $\Delta H_{vap}/T_{bp}$ is equal to ΔS_{vap}; (c) NH_3 deviates from Trouton's rule because of hydrogen bonding. Because $NH_3(l)$ is more ordered than the other liquids, ΔS_{vap} is larger.   **17.93** $\Delta H^\circ = +2.66$ kJ and $\Delta S^\circ = -82.2$ J/K   **17.95** 26°C   **17.97** (a) $P_{SO_3} = 12.5$ atm; $P_{SO_2} = 0.63$ atm; $P_{O_2} = 0.32$ atm; (b) % yield of SO_3 decreases with increasing temperature because ΔS° is negative. ΔG° becomes less negative and K_p gets smaller as temperature increases; (c) P_{total} increases to 18.6 atm (because of gas expansion [Charles' Law] and because K_p decreases).

Chapter 18

18.3 $Pb(s) + Br_2(l) \longrightarrow Pb^{2+}(aq) + 2\ Br^-(aq)$ There is a Pb anode in an aqueous solution of Pb^{2+}. The cathode is a Pt wire that dips into a pool of liquid Br_2 and an aqueous solution that is saturated with Br_2. A salt bridge connects the anode and cathode compartments. The electrodes are connected through an external circuit.   **18.5** -0.74 V   **18.7** (a) Reaction can occur; $E^\circ = 0.23$ V; (b) Reaction cannot occur; $E^\circ = -1.40$ V; Reaction (b) can occur in the reverse direction.   **18.9** 6.9   **18.11** -0.140 V   **18.13** (a) cathode reaction: $2\ H_2O(l) + 2\ e^- \longrightarrow H_2(g) + 2\ OH^-(aq)$; anode reaction: $2\ Cl^-(aq) \longrightarrow Cl_2(g) + 2e^-$; overall reaction: $2\ Cl^-(aq) + 2\ H_2O(l) \longrightarrow Cl_2(g) + H_2(g) + 2\ OH^-(aq)$; (b) cathode reaction: $Cu^{2+}(aq) + 2\ e^- \longrightarrow Cu(s)$; anode reaction: $2\ H_2O(l) \longrightarrow O_2(g) + 4\ H^+(aq) + 4\ e^-$; overall reaction: $2\ Cu^{2+}(aq) + 2\ H_2O(l) \longrightarrow 2\ Cu(s) + O_2(g) + 4\ H^+(aq)$ **18.15** 7.45 h   *Understanding Key Concepts*   **1.** (e) anode reaction: $Zn(s) \longrightarrow Zn^{2+}(aq) + 2\ e^-$; cathode reaction: $Pb^{2+}(aq) + 2\ e^- \longrightarrow Pb(s)$; overall reaction: $Zn(s) + Pb^{2+}(aq) \longrightarrow Zn^{2+}(aq) + Pb(s)$ **2.** (c) anode reaction: $2\ Br^-(aq) \longrightarrow Br_2(aq) + 2\ e^-$; cathode reaction: $Cu^{2+}(aq) + 2\ e^- \longrightarrow Cu(s)$; overall reaction: $Cu^{2+}(aq) + 2\ Br^-(aq) \longrightarrow Cu(s) + Br_2(aq)$ **3.** (a) oxidizing agents: PbO_2, H^+, $Cr_2O_7^{2-}$; reducing agents: Al, Fe, Ag; (b) PbO_2 is the strongest oxidizing agent. H^+ is the weakest oxidizing agent; (c) Al is the strongest reducing agent. Ag is the weakest reducing agent; (d) oxidized by Cu^{2+}: Fe and Al; reduced by H_2O_2: PbO_2 and $Cr_2O_7^-$ **4.** (a) $A^+ + e^- \longrightarrow A$; $C^+ + e^- \longrightarrow C$; $B^+ + e^- \longrightarrow B$; (b) A^+ is the strongest oxidizing agent; B is the strongest reducing agent; (c) $A^+ + B \longrightarrow B^+ + A$ **5.** Sn is not as easily oxidized as Fe and it does not offer cathodic protection. Once the Fe is exposed (scratched), the tin can rusts. A galvanized can is Fe coated with Zn. The Zn offers cathodic protection even if the Fe is exposed. The Zn is sacrificially oxidized.   *Additional Problems*

18.17 The oxidizing agent gets reduced, and reduction takes place at the cathode.   **18.21** (a) $Cd(s) \mid Cd^{2+}(aq) \parallel Sn^{2+}(aq) \mid Sn(s)$; (b) $Al(s) \mid Al^{3+}(aq) \parallel Cd^{2+}(aq) \mid Cd(s)$; (c) $Pb(s) \mid Pb^{2+}(aq) \parallel H^+(aq) \mid H_2(g) \mid Pt(s)$; (d) $Pt(s) \mid Fe^{2+}(aq), Fe^{3+}(aq) \parallel Cr_2O_7^{2-}(aq), Cr^{3+}(aq) \mid Pt(s)$ **18.23** (b) anode reaction: $Zn(s) \longrightarrow Zn^{2+}(aq) + 2\ e^-$; cathode reaction: $Cl_2(g) + 2\ e^- \longrightarrow 2\ Cl^-(aq)$; overall reaction: $Zn(s) + Cl_2(g) \longrightarrow Zn^{2+}(aq) + 2\ Cl^-(aq)$; (c) $Zn(s) \mid Zn^{2+}(aq) \parallel Cl^-(aq) \mid Cl_2(g) \mid C(s)$ **18.29** -309 kJ   **18.31** $+1.23$ V   **18.33** -0.36 V   **18.35** $Cu^{2+} < O_2 < Cl_2$ **18.37** MnO_4^- is the strongest oxidizing agent. Fe^{2+} is the weakest oxidizing agent.   **18.39** (a) $E^\circ = +0.26$ V; $\Delta G^\circ = -50$ kJ; (b) $E^\circ = +1.26$ V; $\Delta G^\circ = -730$ kJ; (c) $E^\circ = +0.13$ V; $\Delta G^\circ = -25$ kJ; (d) $E^\circ = +0.56$ V; $\Delta G^\circ = -320$ kJ   **18.41** (a) $E^\circ = -0.90$ V; nonspontaneous; (b) $E^\circ = +2.11$ V; spontaneous; (c) $E^\circ = -0.71$ V; nonspontaneous; (d) $E^\circ = +1.08$ V; spontaneous   **18.43** (a) $Zn(s) + Pb^{2+}(aq) \longrightarrow Zn^{2+}(aq) + Pb(s)$; (b) $4\ Fe^{2+}(aq) + O_2(g) + 4\ H^+(aq) \longrightarrow 4\ Fe^{3+}(aq) + 2\ H_2O(l)$; (c) and (d), no reaction   **18.45** $+0.91$ V   **18.47** (a) $+0.64$ V; (b) $+0.77$ V; (c) -0.23 V; (d) -1.05 V   **18.49** 9.0   **18.51** 8.9×10^{-6} M   **18.53** 2×10^{25} **18.55** If $K < 1$, $E^\circ < 0$. When $E^\circ = 0$, $K = 1$.   **18.57** 9×10^{-3} **18.59** $E^\circ = +1.23$ V; $\Delta G^\circ = -475$ kJ; $K = 1 \times 10^{83}$ **18.61** 6.06 g

18.63 Cathodic protection is the attachment of a more easily oxidized metal to the metal you want to protect. This forces the metal you want to protect to be the cathode, hence the name, cathodic protection. Zn and Al can offer cathodic protection to Fe (Ni or Sn cannot).   **18.65** Cr forms a protective oxide coating similar to Al.   **18.69** (a) $K(l)$ and $Cl_2(g)$; (b) $H_2(g)$ and $Cl_2(g)$. Solvent H_2O is reduced in preference to K^+.   **18.71** 3.22 g   **18.73** 38.8 hours; volume $= 4.87 \times 10^5$ L of $Cl_2(g)$ **18.75** 84.9 g   **18.79** (a) $2\ MnO_4^-(aq) + 16\ H^+(aq) + 5\ Sn^{2+}(aq) \longrightarrow 2\ Mn^{2+}(aq) + 5\ Sn^{4+}(aq) + 8\ H_2O(l)$; (b) MnO_4^- is the oxidizing agent, Sn^{2+} is the reducing agent; (c) $+1.36$ V   **18.81** (a) Ag^+ is the strongest oxidizing agent. Pb is the strongest reducing agent. (c) $Pb(s) + 2\ Ag^+(aq) \longrightarrow Pb^{2+}(aq) + 2\ Ag(s)$; $E^\circ = +0.93$ V; $\Delta G^\circ = -180$ kJ; $K = 10^{31}$; (d) $+0.87$ V   **18.83** Fe^{2+} can oxidize Al. Fe^{2+} can reduce $Br_2(l)$ and MnO_4^-.   **18.85** (b) $2\ Al(s) + 6\ H^+(aq) \longrightarrow 2\ Al^{3+}(aq) + 3\ H_2(g)$; $E^\circ = +1.66$ V; (c) $+1.59$ V; (d) $\Delta G^\circ = -961$ kJ; $K = 10^{168}$; (e) 1.40 g   **18.87** 3.7×10^{10} kWh   **18.89** (a) $+1.27$ V; (b) $+0.95$ V   **18.91** (a) $4\ CH_2{=}CHCN + 2\ H_2O \longrightarrow 2\ NC(CH_2)_4CN + O_2$; (b) 60.4 kg; (c) 7030 L   **18.93** (a) $2\ Ag^+(aq) + Cu(s) \longrightarrow Cu^{2+}(aq) + 2\ Ag(s)$; $E = 0.38$ V; (b) $Cu^{2+}(aq) + 2\ Ag(s) + 2\ Br^-(aq) \longrightarrow 2\ AgBr(s) + Cu(s)$; $E = +0.27$ V; (c) $+0.07$ V

Chapter 19

19.1 (a) B; (b) Br; (c) Se   **19.3** H—C≡N: The carbon is sp hybridized.   **19.7** In silicate and phosphate anions both Si and P are surrounded by tetrahedra of O, which can link together to form chains.   *Understanding Key Concepts*   **3.** (a) N_2, O_2, F_2, P_4, S_8, Cl_2

(b) :N⋮⋮⋮N: :O::O: :F:F: :Cl:Cl:

(c) The smaller N and O can form effective π bonds, whereas P and S cannot. In both F_2 and Cl_2 the atoms are joined by a single bond. **4.** (a) SiO_4^{4-}; (b) $Si_3O_{10}^{8-}$; (c) $Si_4O_{12}^{8-}$ *Additional Problems* **19.9** (b) P and (e) Br **19.11** (a) Cl; (b) Si; (c) O **19.13** (a) I; (b) N; (c) F **19.15** Na < Si < S < O **19.17** (a) PBr_3; (b) CO; (c) PH_3 **19.19** (a) P_4O_{10}; (b) B_2O_3; (c) SO_2 **19.21** (a) Ga; (b) In; (c) Pb **19.23** (a) F; (b) Al; (c) N; (d) Si; (e) S **19.25** In O_2 a π bond is formed by $2p$ orbitals on each O. S does not form strong π bonds with its $3p$ orbitals, which leads to the S_8 ring structure. **19.35** (a) Pb; (b) C; (c) Si; (d) C **19.37** (a) tetrahedral, sp^3; (b) linear, sp; (c) trigonal planar, sp^2; (d) octahedral, sp^3d^2; (e) bent, sp^2 **19.41** The more dense form (diamond) is favored at high pressure.
19.51 (a)   (b)

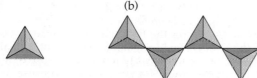

The charge on the anion is equal to the number of terminal O atoms. **19.53** (a) $Si_3O_{10}^{8-}$; (b) $Ca_2Cu_2Si_3O_{10} \cdot 2 H_2O$ **19.57** (a) P; (b) Sb and Bi; (c) N; (d) P; (e) Bi **19.61** (a) NO, +2; (b) HNO_2, +3; (c) PH_3, −3; (d) P_4O_{10}, +5; (e) $H_5P_3O_{10}$, +5 **19.63** (a) bent, sp^2; (b) trigonal pyramidal, sp^3; (c) trigonal bipyramidal, sp^3d; (d) linear, sp; (e) tetrahedral, sp^3; (f) octahedral, sp^3d^2 **19.67** (a) $2 NO(g) + O_2(g) \longrightarrow 2 NO_2(g)$; (b) $4 HNO_3(aq) \longrightarrow 4 NO_2(aq) + O_2(g) + 2 H_2O(l)$; (c) $3 Ag(s) + 4 H^+(aq) + NO_3^-(aq) \longrightarrow 3 Ag^+(aq) + NO(g) + 2 H_2O(l)$; (d) $N_2H_4(aq) + 2 I_2(aq) \longrightarrow N_2(g) + 4 H^+(aq) + 4 I^-(aq)$ **19.75** (a) −2; (b) +6; (c) +4; (d) −1; (e) +4 **19.77** (a) $Na_2SO_3(s) + 2 H^+(aq) \longrightarrow SO_2(g) + H_2O(l) + 2 Na^+(aq)$; (b) $CH_3C(S)NH_2(aq) + H_2O(l) \longrightarrow CH_3C(O)NH_2(aq) + H_2S(aq)$; (c) $NaOH(aq) + H_2SO_4(aq) \longrightarrow NaHSO_4(aq) + H_2O(l)$ **19.79** (a) Acid strength increases as the number of O's increases; (b) In comparison with S, O is much too electronegative to form compounds of O in the +4 oxidation state. Also, S uses sp^3d hybrid orbitals for bonding in SF_4, but O doesn't have valence d orbitals and so it can't form four bonds to F; (c) Each S is sp^3 hybridized with two lone pairs of electrons. The bond angles are therefore 109.5°. A planar ring would require bond angles of 135°. **19.81** (a) $HClO_2$, +3; (b) KIO_3, +5; (c) H_5IO_6, +7; (d) $NaBrO$, +1 **19.83** (a) potassium chlorite; (b) metaperiodic acid; (c) hypobromous acid; (d) sodium bromate **19.85** (a) BrO_4^-

$$\left[\begin{array}{c} :\ddot{O}: \\ :\ddot{O}:\overset{\displaystyle :\!\!\ddot{O}:}{\underset{\displaystyle :\!\!\ddot{O}:}{Br}}:\ddot{O}: \end{array} \right]^-$$ tetrahedral

(b) ClO_3^-

$$\left[\begin{array}{c} :\ddot{O}: \\ :\ddot{Cl}:\ddot{O}: \\ :\ddot{O}: \end{array} \right]^-$$ trigonal pyramidal

(c) HIO_4

$$H:\ddot{O}:\overset{\displaystyle :\!\!\ddot{O}:}{\underset{\displaystyle :\!\!\ddot{O}:}{I}}:\ddot{O}:$$ tetrahedral

(d) HOBr $H:\ddot{O}:\ddot{Br}:$ bent

19.87 (a) $Br_2(l) + 2 OH^-(aq) \longrightarrow OBr^-(aq) + Br^-(aq) + H_2O(l)$; (b) $Cl_2(g) + H_2O(l) \longrightarrow HOCl(aq) + H^+(aq) + Cl^-(aq)$; (c) $3 Cl_2(g) + 6 OH^-(aq) \longrightarrow ClO_3^-(aq) + 5 Cl^-(aq) + 3 H_2O(l)$ **19.89** (a) $Na_2B_4O_7 \cdot 10 H_2O$; (b) $Ca_3(PO_4)_2$; (c) elemental sulfur, FeS_2, PbS, HgS, $CaSO_4 \cdot 2 H_2O$ **19.91** Gases: H, He, Ne, Ar, Kr, Xe, Rn, F, Cl, O, N; Liquid: Br **19.95** C, Si, Ge, and Sn have allotropes with the diamond structure. Sn and Pb have metallic allotropes. C (nonmetal), Si (semimetal), Ge (semimetal), Sn (semimetal and metal), Pb (metal) **19.97** (a) $NH_4^+(aq) + OH^-(aq) \longrightarrow NH_3(g) + H_2O(l)$; (b) $CO_3^{2-}(aq) + 2 H^+(aq) \longrightarrow CO_2(g) + H_2O(l)$; (c) $2 NaBH_4 + I_2 \longrightarrow B_2H_6(g) + H_2(g) + 2 NaI$; (d) $CaC_2(s) + 2 H_2O(l) \longrightarrow C_2H_2(g) + Ca(OH)_2(aq)$; (e) $NH_4NO_3(l) \longrightarrow N_2O(g) + 2 H_2O(g)$; (f) $Cu(s) + 2 NO_3^-(aq) + 4 H^+(aq) \longrightarrow Cu^{2+}(aq) + 2 NO_2(g) + 2 H_2O(l)$ **19.101** The angle required by P_4 is 60°. The strain would not be reduced by using sp^3 hybrid orbitals, since their angle is ~109°. **19.103** (a) In diamond each C is covalently bonded to four additional C's in a rigid three-dimensional network solid. Graphite is a two-dimensional covalent network solid of carbon sheets that can slide over each other. Both are high melting because melting requires the breaking of C—C bonds; (b) There is some multiple bond character in terminal P—O bonds, utilizing empty d orbitals on the P. The bridging P—O bonds are single bonds; (c) Chlorine does not form perhalic acids of the type H_5XO_6 because its smaller size favors a tetrahedral structure over an octahedral one.

Chapter 20

20.1 (a) $[Ar]3d^34s^2$; (b) $[Ar]3d^7$; (c) $[Ar]3d^3$ **20.3** +2 **20.5** (a) $[Zn(NH_3)_4](NO_3)_2$; (b) $Ni(CO)_4$; (c) $K[Pt(NH_3)Cl_3]$; (d) $[Au(CN)_2]^-$ **20.7** (a) yes; (b) no; (c) no; (d) yes; (e) no

20.9 (a) Fe^{3+} [Ar] ↑ ↑ ↑ ↑ ↑ — — — —;
                               $3d$ $4s$ $4p$

$[Fe(CN)_6]^{3-}$ [Ar] ↑↓ ↑↓ ↑ ⟨↑↓ ↑↓ ↑↓ ↑↓ ↑↓⟩
                        $3d$ $4s$ $4p$

d^2sp^3; 1 unpaired e^-;

(b) Co^{2+} [Ar] $\underline{\uparrow\downarrow}$ $\underline{\uparrow\downarrow}$ $\underline{\uparrow}$ $\underline{\uparrow}$ $\underline{\uparrow}$ $\underline{\quad}$ $\underline{\quad}$ $\underline{\quad}$ $\underline{\quad}$
$\quad\quad\quad\quad\quad 3d \quad\quad\quad\quad 4s \quad\quad 4p$

$[Co(H_2O)_6]^{2+}$ [Ar] $\underline{\uparrow\downarrow}$ $\underline{\uparrow\downarrow}$ $\underline{\uparrow}$ $\underline{\uparrow}$ $\underline{\uparrow}$
$\quad\quad\quad\quad\quad\quad\quad 3d$

$\boxed{\underline{\uparrow\downarrow}\ \underline{\uparrow\downarrow}\ \underline{\uparrow\downarrow}\ \underline{\uparrow\downarrow}\ \underline{\uparrow\downarrow}\ \underline{\uparrow\downarrow}}$ $\underline{\quad}$ $\underline{\quad}$ $\underline{\quad}$; sp^3d^2;
$4s \quad\quad 4p \quad\quad\quad 4d$

3 unpaired e^-; (c) V^{3+} [Ar] $\underline{\uparrow}$ $\underline{\uparrow}$ $\underline{\quad}$ $\underline{\quad}$ $\underline{\quad}$ $\underline{\quad}$
$\quad\quad\quad\quad\quad\quad\quad 3d \quad\quad\quad\quad\quad 4s$

$\underline{\quad}$ $\underline{\quad}$ $\underline{\quad}$; $[VCl_4]^-$ [Ar] $\underline{\uparrow}$ $\underline{\uparrow}$ $\underline{\quad}$ $\underline{\quad}$ $\underline{\quad}$
$\quad\quad 4p \quad\quad\quad\quad\quad\quad\quad 3d$

$\boxed{\underline{\uparrow\downarrow}\ \underline{\uparrow\downarrow}\ \underline{\uparrow\downarrow}\ \underline{\uparrow\downarrow}}$ sp^3; 2 unpaired e^-;
$4s \quad\quad 4p$

(d) Pt^{2+} [Xe] $\underline{\uparrow\downarrow}$ $\underline{\uparrow\downarrow}$ $\underline{\uparrow\downarrow}$ $\underline{\uparrow}$ $\underline{\uparrow}$ $\underline{\quad}$ $\underline{\quad}$ $\underline{\quad}$ $\underline{\quad}$
$\quad\quad\quad\quad\quad 5d \quad\quad\quad\quad 6s \quad\quad 6p$

$[PtCl_4]^{2-}$ [Xe] $\underline{\uparrow\downarrow}$ $\underline{\uparrow\downarrow}$ $\underline{\uparrow\downarrow}$ $\underline{\uparrow\downarrow}$ $\boxed{\underline{\uparrow\downarrow}\ \underline{\uparrow\downarrow}\ \underline{\uparrow\downarrow}\ \underline{\uparrow\downarrow}}$ $\underline{\quad}$
$\quad\quad\quad\quad\quad\quad 5d \quad\quad\quad\quad 6s \quad\quad 6p$

dsp^2; no unpaired electrons

20.11

$\underline{\uparrow\downarrow}$ $\underline{\uparrow}$ $\underline{\uparrow}$ $\quad\quad\quad x^2 - y^2$ $\underline{\quad}$
$xy\ \ xz\ \ yz$

$\quad\quad\quad\quad\quad\quad\quad\quad z^2$ $\underline{\uparrow\downarrow}$
$\quad\quad\quad\quad\quad\quad\quad\quad xy$

$\underline{\uparrow\downarrow}$ $\underline{\uparrow\downarrow}$ $\quad\quad\quad\quad z^2$ $\underline{\uparrow\downarrow}$
$z^2\ \ x^2-y^2$

$\quad\quad\quad\quad\quad\quad\quad\quad\quad$ $\underline{\uparrow\downarrow}$ $\underline{\uparrow\downarrow}$
$\quad\quad\quad\quad\quad\quad\quad\quad xz\ \ yz$

2 unpaired e^- $\quad\quad$ no unpaired e^-
$\quad\quad$ (a) $\quad\quad\quad\quad\quad\quad$ (b)

Understanding Key Concepts 2.(a) The atomic radii decrease, at first markedly and then more gradually. Toward the end of the series, the radii increase again. The decrease in atomic radii is a result of an increase in Z_{eff}. The increase is due to electron-electron repulsions in doubly occupied d orbitals. (b) The densities of the transition metals are inversely related to their atomic radii. The densities initially increase from left to right and then decrease toward the end of the series. (c) Ionization energies generally increase from left to right across the series. The general trend correlates with an increase in Z_{eff} and a decrease in atomic radii. (d) The standard oxidation potentials generally decrease, which correlates with the general trend in ionization energies. 3. (a) bidentate; (b) monodentate; (c) tridentate; (d) monodentate; (a) and (c) can form chelates 4. (a) oxidation state = +1; coordination number = 2; linear; (b) oxidation state = +3; coordination number = 6; octahedral; (c) oxidation state = +2; coordination number = 4; square planar; (d) oxidation state = +4; coordination number = 6; octahedral 5. (a) (1) cis (2) trans (3) trans (4) cis; (b) (1) and (4) are the same. (2) and (3) are the same. (c) No, their mirror images are identical. 6. (a) (1) chiral (2) archiral (3) chiral (4) chiral; (c) (1) and (4) are enantiomers.

7.

E $\Big|$
$x^2 - y^2$
z^2
xy
$xz\ \ yz$

Additional Problems 20.13(a) p-block; (b) d-block; (c) s-block; (d) d-block; (e) d-block 20.15 (a) $[Ar]3d^34s^2$; (b) $[Kr]4d^15s^2$; (c) $[Ar]3d^{10}4s^1$; (d) $[Xe]4f^{14}5d^{10}6s$ 20.17 (a) $[Ar]3d^2$; (b) $[Ar]3d^{10}$; (c) $[Kr]4d^6$; (d) $[Ar]3d^2$ 20.19 (a) 4; (b) 5; (c) 0 20.21 Ti > V > Cr > Mn. Atomic radius decreases with increasing Z_{eff}. 20.23 Zr and Hf are about the same size. The lanthanide contraction for Hf is due to the increase in Z_{eff} as the $4f$ subshell is filled. 20.25 (a) $Cr(s) + 2\ H^+(aq) \longrightarrow Cr^{2+}(aq) + H_2(g)$; (b) $Zn(s) + 2\ H^+(aq) \longrightarrow Zn^{2+}(aq) + H_2(g)$; (c) N.R.; (d) $Fe(s) + 2\ H^+(aq) \longrightarrow Fe^{2+}(aq) + H_2(g)$ 20.27 (c) Sc; (d) Al 20.29 The highest oxidation state for the group 3B–7B metals is the group number, corresponding to the loss of all valence s and d electrons. For the later transition metals, loss of all valence electrons is energetically prohibitive because of the increasing Z_{eff}. Therefore, only lower oxidation states are accessible for the later transition metals. 20.31 V^{2+} is the stronger reducing agent because of a smaller Z_{eff}. 20.33 Co^{3+} is more easily reduced because of a higher Z_{eff}. 20.35 $Cr_2O_7^{2-} < Cr^{3+} < Cr^{2+}$ because of decreasing oxidation state of the Cr. 20.37 (a) $Fe(s) + NO_3^-(aq) + 4\ H^+(aq) \longrightarrow Fe^{3+}(aq) + NO(g) + 2\ H_2O(l)$; (b) $3\ Cu(s) + 2\ NO_3^-(aq) + 8\ H^+(aq) \longrightarrow 3\ Cu^{2+}(aq) + 2\ NO(g) + 4\ H_2O(l)$; (c) $Cr(s) + NO_3^-(aq) + 4\ H^+(aq) \longrightarrow Cr^{3+}(aq) + NO(g) + 2\ H_2O(l)$ 20.39 (a) $Cr(OH)_2$; (b) $Cr(OH)_4^-$; (c) CrO_4^{2-}; (d) $Fe(OH)_2$; (e) $Fe(OH)_3$ 20.41 acid: $Cr(OH)_3(s) + OH^-(aq) \longrightarrow Cr(OH)_4^-(aq)$; base: $Cr(OH)_3(s) + 3\ H_3O^+(aq) \longrightarrow Cr^{3+}(aq) + 6\ H_2O(l)$ 20.43 (c) Cu^+; $2\ Cu^+(aq) \longrightarrow Cu(s) + Cu^{2+}(aq)$ 20.45 (a) $Cr_2O_7^{2-}(aq) + 6\ Fe^{2+}(aq) + 14\ H^+(aq) \longrightarrow 2\ Cr^{3+}(aq) + 6\ Fe^{3+}(aq) + 7\ H_2O(l)$; (b) $4\ Fe^{2+}(aq) + O_2(g) + 4\ H^+(aq) \longrightarrow 4\ Fe^{3+}(aq) + 2\ H_2O(l)$; (c) $Cu_2O(s) + 2\ H^+(aq) \longrightarrow Cu(s) + Cu^{2+}(aq) + H_2O(l)$; (d) $Fe(s) + 2\ H^+(aq) \longrightarrow Fe^{2+}(aq) + H_2(g)$ 20.47 (a) $Cu^{2+}(aq) + 4\ NH_3(aq) \longrightarrow [Cu(NH_3)_4]^{2+}(aq)$; (b) $Cr_2O_7^{2-}(aq) + 2\ OH^-(aq) \longrightarrow 2\ CrO_4^{2-}(aq) + H_2O(l)$; (c) $Fe^{3+}(aq) + 3\ OH^-(aq) \longrightarrow Fe(OH)_3(s)$; (d) $3\ CuS(s) + 8\ H^+(aq) + 2\ NO_3^-(aq) \longrightarrow 3\ Cu^{2+}(aq) + 3\ S(s) + 2\ NO(g) + 4\ H_2O(l)$

20.49 (a)

$$\left[\begin{array}{c} Cl \\ H_2O \diagdown \ \big| \diagup OH_2 \\ Cr \\ H_2O \diagup \big| \diagdown OH_2 \\ OH_2 \end{array}\right]^{2-}$$

(b)

$$\left[\begin{array}{c} O\quad\quad O \\ \big|\quad\quad\big| \\ O\diagdown Cr \diagdown O \diagup Cr \diagup O \\ \big|\quad\quad\big| \\ O\quad\quad O \end{array}\right]^{2-}$$

(c)

$$\left[\begin{array}{c} O \\ \big| \\ O \diagdown Fe \diagup O \\ \big| \\ O \end{array}\right]^{2-}$$

20.53 Lewis acid = Fe^{3+}; Lewis base = $C_2O_4^{2-}$ **20.55** (a) 4; (b) 6; (c) 2; (d) 8; (e) 6   **20.59** (a) $Ir(NH_3)_3Cl_3$; (b) $[Cr(H_2O)_2(C_2O_4)_2]^-$; (c) $[Pt(en)_2(SCN)_2]^{2+}$ **20.61** (a) +3; (b) +3; (c) +2; (d) +2   **20.63** (a) tetrachloroaurate(III); (b) hexacyanoferrate(II); (c) pentaaquoisothiocyanatoiron(III); (d) diamminedi(oxalato)chromate(III)   **20.65** (a) tetraamminecopper(II) sulfate; (b) hexacarbonylchromium(0); (c) potassium tri(oxalato)ferrate(III); (d) amminecyanatobis(ethylenediamine)cobalt(III) chloride   **20.67** (a) $[Ag(NH_3)_2]NO_3$; (b) $K[Co(H_2O)_2(C_2O_4)_2]$; (c) $Mo(CO)_6$; (d) $[Cr(NH_3)_2(en)_2]Cl_3$ **20.73** (a) 2 (cis and trans); (b) none; (c) 2 (cis and trans); (d) 2   **20.75** (a) and (b) can exist as enantiomers   **20.81** ~525 nm

20.83
(a) Au^{3+} [Xe] (orbital diagram: 5d ↿⇂ ↿⇂ ↿⇂ ↿ ↿; 6s __; 6p __ __ __)

$[AuCl_4]^-$ [Xe] (orbital diagram: 5d ↿⇂ ↿⇂ ↿⇂ ↿⇂; boxed 6s ↿⇂ 6p ↿⇂ ↿⇂; __)

dsp^2; no unpaired e^-;
(b) Ag^+ [Kr] (4d ↿⇂ ↿⇂ ↿⇂ ↿⇂ ↿⇂; 5s __; 5p __)

$[Ag(NH_3)_2]^+$ [Kr] (4d ↿⇂ ↿⇂ ↿⇂ ↿⇂ ↿⇂; boxed 5s ↿⇂ 5p ↿⇂; __)

sp; no unpaired e^-;
(c) Fe^{2+} [Ar] (3d ↿⇂ ↿ ↿ ↿ ↿; 4s __; 4p __ __ __)

$[Fe(H_2O)_6]^{2+}$ [Ar]

(3d ↿⇂ ↿ ↿ ↿ ↿; boxed 4s ↿⇂ 4p ↿⇂ ↿⇂ ↿⇂ 4d ↿⇂ ↿⇂; __)

__ __ sp^3d^2; 4 unpaired e^-;
(d) Fe^{2+} [Ar] (3d ↿⇂ ↿ ↿ ↿ ↿; 4s __; 4p __ __ __)

$[Fe(CN)_6]^{4-}$ [Ar] (3d ↿⇂ ↿⇂ ↿⇂)

(boxed: ↿⇂ ↿⇂ | ↿⇂ | ↿⇂ ↿⇂ ↿⇂; 4s 4p) d^2sp^3; no unpaired e^-

20.85

$d_{x^2-y^2}$ d_{xy}

The $d_{x^2-y^2}$ orbital is higher in energy because its lobes are pointing directly at the ligands.   **20.87** $[Cr(H_2O)_6]^{3+}$ absorbs at ~580 nm, while $[Cr(CN)_6]^{3-}$ absorbs at ~ 415 nm. CN^- is a stronger-field ligand, while H_2O is a weaker-field ligand.   **20.89** Number of unpaired electrons: (a) 1; (b) 5; (c) 0.   **20.91** Cr^{3+} is $3d^3$ and should be colored. Y^{3+} is $4d^0$ and should be colorless.   **20.93** Ligands do not point directly at any d orbitals, and the crystal field splitting is small, leading to high-spin complexes.   **20.95** Number of unpaired electrons: (a) 0; (b) 5; (c) 3; (d) 1   **20.97** (b) Co^{2+}; (c) V^{3+}; (e) MnO_4^{2-} **20.99** (b) and (c)   **20.101** (a) $16\ H^+(aq) + 2\ MnO_4^-(aq) + 5\ C_2O_4^{2-}(aq) \longrightarrow 2\ Mn^{2+}(aq) + 8\ H_2O(l) + 10\ CO_2(g)$; (b) $6\ Ti^{3+}(aq) + Cr_2O_7^{2-}(aq) + 2\ H^+(aq) \longrightarrow 6\ TiO^{2+}(aq) + 2\ Cr^{3+}(aq) + H_2O(l)$; (c) $2\ MnO_4^-(aq) + SO_3^{2-}(aq) + 2\ OH^-(aq) \longrightarrow 2\ MnO_4^{2-}(aq) + SO_4^{2-}(aq) + H_2O(l)$; (d) $2\ H_2O(l) + 4\ Fe(OH)_2(s) + O_2(g) \longrightarrow 4\ Fe(OH)_3(s)$ **20.103** 0.013 M   **20.109** $[Mn(CN)_6]^{3-}$ (Mn^{3+} $3d^4$) valence bond theory:

[Ar] (3d ↿⇂ ↿ ↿; boxed 4s ↿⇂ ↿⇂ 4p ↿⇂ ↿⇂ ↿⇂) d^2sp^2

crystal field theory: __ __

(↿⇂ ↿ ↿; ↿ with Δ)

Crystal field model predicts 2 unpaired electrons.   **20.111** Cr^{3+} is a $3d^3$ ion. Regardless of the crystal field splitting energy, the three electrons singly occupy the three lower-energy d orbitals.   **20.115** $[CoCl_4]^{2-}$ is tetrahedral. $[Co(H_2O)_6]^{2+}$ is octahedral. Because $\Delta_{tet} < \Delta_{oct}$, these complexes have different colors. $[CoCl_4]^{2-}$ has absorption bands at longer wavelengths.

Chapter 21

21.1 The electron configuration for Hg is $[Xe]4f^{14}5d^{10}6s^2$. Assuming the 5d and 6s bands overlap, the composite band can accommodate 12 valence electrons per metal atom. Weak bonding and a low melting point are expected for Hg because both the bonding and antibonding MOs are occupied.

21.3 8 Cu at corners $8 \times \frac{1}{8} = 1$ Cu
8 Cu on edges $8 \times \frac{1}{4} = \underline{2\ Cu}$
Total   = 3 Cu

12 O on edges $12 \times \frac{1}{4} = 3$ O
8 O on faces $8 \times \frac{1}{2} = \underline{4\ O}$
Total   = 7 O

21.5 $Ba[OCH(CH_3)_2]_2 + Ti[OCH(CH_3)_2]_4 + 6\ H_2O \longrightarrow BaTi(OH)_6(s) + 6\ HOCH(CH_3)_2$; $BaTi(OH)_6(s) \xrightarrow{heat} BaTiO_3(s)$ ***Understanding Key Concepts*** **2.** (a) (iii) electrolysis; (b) (i) roasting a metal sulfide   **3.** (a) (1) and (4) are semiconductors; (2) is a metal; (3) is an insulator; (b) (3) < (1) < (4) < (2). The conductivity increases with decreasing band gap. (c) (1) and (4) increase; (2) decreases; (3) not much change.   **4.** (a) (2); bonding MOs are filled. (b) (3); bonding and antibonding MOs are filled. (c) (3) < (1) < (2). Hardness increases with increasing MO bond order.   ***Additional Problems*** **21.11** The less electronegative early transition metals tend to form ionic compounds by losing electrons to highly electronegative nonmetals such as oxygen. The more electronegative late transition metals tend to form compounds with more covalent character by bonding to the less electronegative nonmetals such as sulfur.   **21.17** The flotation process exploits the differences in the ability of water and oil to wet the surfaces of the mineral and the gangue. The gangue, which contains ionic silicates, is moistened by the polar water molecules and sinks to the bottom of the tank. The mineral particles, which contain the less polar metal sulfide, are coated by the oil and become attached to the soapy air bubbles created by the detergent. The metal sulfide particles are carried to the surface in the soapy froth, which is skimmed off at the top of the tank. This process would not work well for a metal oxide because it is too polar and would be wet by the water and sink with the gangue.   **21.19** Since $E° < 0$ for Zn^{2+}, the reduction of Zn^{2+} is not favored. Since $E° > 0$ for Hg^{2+}, the reduction of Hg^{2+} is favored. The roasting of CdS should yield CdO because, like the $E°$ for Zn^{2+}, $E° < 0$ for the reduction of Cd^{2+}.   **21.21** (a) $V_2O_5(s) + 5\ Ca(s)$ $\longrightarrow 2\ V(s) + 5\ CaO(s)$; (b) $2\ PbS(s) + 3\ O_2(g) \longrightarrow$ $2\ PbO(s) + 5\ SO_2(g)$; (c) $MoO_3(s) + 3\ H_2(g) \longrightarrow Mo(s)$ $+ 3\ H_2O(g)$; (d) $3\ MnO_2(s) + 4\ Al(s) \longrightarrow Mn(s) +$ $2\ Al_2O_3(s)$; (e) $MgCl_2(l) \xrightarrow{\text{electrolysis}} Mg(l) + Cl_2(g)$ **21.23** $\Delta H° = -878.2$ kJ; $\Delta G° = -834.4$ kJ; $\Delta H°$ and $\Delta G°$ are different because of the entropy change associated with the reaction. The minus sign for $(\Delta H° - \Delta G°)$ indicates that the entropy is negative, which is consistent with a decrease in the number of moles of gas from 3 mol to 2 mol.   **21.25** 110 kg Cr; 1.06×10^5 L CO **21.27** 0.460 kg   **21.29** $Fe_2O_3(s) + 3\ CO(g) \longrightarrow 2\ Fe(l) +$ $3\ CO_2(g)$; Fe_2O_3 is the oxidizing agent; CO is the reducing agent.   **21.33** Molten iron from a blast furnace is exposed to a jet of pure oxygen gas for about 20 minutes. The impurities are oxidized to yield a molten slag that can be poured off. $P_4(l) + 5\ O_2(g) \longrightarrow P_4O_{10}(l)$; $6\ CaO(s) + P_4O_{10}(l) \longrightarrow 2\ Ca_3(PO_4)_2(l)$ (slag); $2\ Mn(l) +$ $O_2(g) \longrightarrow 2\ MnO(s)$; $MnO(s) + SiO_2(s) \longrightarrow MnSiO_3(l)$ (slag)   **21.35** $SiO_2(s) + 2\ C(s) \longrightarrow Si(s) + 2\ CO(g)$; $Si(s)$ $+ O_2(g) \longrightarrow SiO_2(s)$; $CaO(s) + SiO_2(s) \longrightarrow CaSiO_3(l)$ **21.43** The energy required to deform a transition metal like W is greater than that for Cs because W has more valence electrons and hence more electrostatic "glue." **21.45** The difference in energy between successive MOs in a metal decreases as the number of metal atoms increases, so that the MOs merge into an almost continuous band of energy levels. Consequently, MO theory for metals is often called band theory.   **21.47** The energy levels within a band occur in degenerate pairs; one set of energy levels applies to electrons moving to the right, and the other set applies to electrons moving to the left. In the abence of an electrical potential, the two sets of levels are equally populated. As a result there is no net electric current. In the presence of an electrical potential those electrons moving to the right are accelerated, those moving to the left are slowed down, and some change direction. Thus, the two sets of energy levels are now unequally populated. The number of electrons moving to the right is now greater than the number moving to the left, so there is a net electric current.

21.49

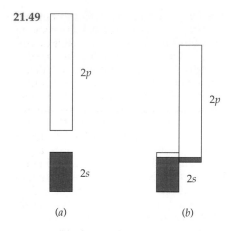

(a)   (b)

Diagram (b) shows the 2s and 2p bands overlapping in energy and the resulting composite band is only partially filled. Thus, Be is a good electrical conductor. **21.51** Transition metals have a *d* band that can overlap the *s* band to give a composite band consisting of six MOs per metal atom. Half of the MOs are bonding and half are antibonding; thus, maximum bonding is expected for metals that have six valence electrons per metal atom. Accordingly, the melting points of the transition metals go through a maximum at or near group 6B.   **21.57** As band gap increases, electrical conductivity decreases. As the band gap increases, the number of electrons able to jump the gap and occupy the higher energy conduction band decreases and thus conductivity decreases.   **21.59** An *n*-type semiconductor is a semiconductor doped with a substance with more valence electrons than the semiconductor itself. Si doped with P is an example.

21.61 In the MO picture, the extra electrons occupy the conduction band. The number of electrons in the conduction band of the doped Ge is much greater than for pure Ge, and the conductivity of the doped semiconductor is correspondingly higher.   **21.63** (a) *p*-type; (b) *n*-type; (c) *n*-type   **21.65** Al_2O_3 < Ge < Ge doped with In < Fe < Cu   **21.69** (1) A superconductor is able to levitate a magnet. (2) In a superconductor once an electric current is started, it flows indefinitely without loss of energy.   **21.71** The fullerides are three-dimensional superconductors, whereas the copper oxide ceramics are two-dimensional superconductors.   **21.75** Ceramics have higher melting points, and they are stiffer, harder, and more wear resistant than metals because they have stronger bonding. They maintain much of their strength at high temperatures, where metals either melt or corrode because of oxidation. **21.77** The brittleness of ceramics is due to strong chemical bonding. In silicon nitride each Si atom is bonded to four N atoms, and each N atom is bonded to three Si atoms. The strong, highly directional covalent bonds prevent the planes of atoms from sliding over one another when the solid is subjected to a stress. As a result, the solid can't deform to relieve the stress. It maintains its shape up to a point, but then the bonds give way suddenly and the material fails catastrophically when the stress exceeds a certain threshold value. By contrast, metals are able to deform under stress because their planes of metal cations can slide easily in the electron sea. **21.83** $Zr[OCH(CH_3)_2]_4$ + 4 $H_2O \longrightarrow Zr(OH)_4$ + 4 $HOCH(CH_3)_2$ **21.85** $(HO)_3Si—O—H + H—O—Si(OH)_3 \longrightarrow$ $(HO)_3Si—O—Si(OH)_3 + H_2O$. Further reactions of this sort give a three-dimensional network of Si—O—Si bridges. On heating, SiO_2 is obtained. **21.87** 3 $SiCl_4(g)$ + 4 $NH_3(g) \longrightarrow Si_3N_4(s)$ + 12 $HCl(g)$ **21.89** Graphite/epoxy composites are good materials for making tennis rackets and golf clubs because of their high strength-to-weight ratios. **21.97** 77 K is the boiling point of the readily available liquid N_2. **21.99** W $[Xe]4f^{14}5d^46s^2$; Au $[Xe]4f^{14}5d^{10}6s^1$; These facts are better explained with the MO band model. Transition metals have a d band that can overlap the s band to give a composite band consisting of six MOs per metal atom. Half of the MOs are bonding and half are antibonding; thus, maximum bonding is expected for metals that have six valence electrons per metal atom. Accordingly, the hardness and melting points of the transition metals go through a maximum at or near group 6B. **21.101** (a) insulator; (b) n-type semiconductor; (c) metallic conductor; (d) p-type semiconductor; (e) metallic conductor **21.103** (a) $P_4(l)$ + 5 $O_2(g) \longrightarrow P_4O_{10}(l)$; 6 $CaO(s) + P_4O_{10}(l) \longrightarrow$ 2 $Ca_3(PO_4)_2(l)$ (slag); (b) 42 kg **21.105** 13.35% Y; 41.23% Ba; 28.62% Cu; 16.81% O **21.107** $AgCl \underset{dark}{\overset{sunlight}{\rightleftharpoons}} Ag + Cl$ AgCl is transparent; Ag is opaque **21.109** (a) $\Delta H° =$ +172.5 kJ; $\Delta S° = +175.9$ J/K; $\Delta G° = +120.0$ kJ; (b) endothermic; (c) The number of moles of gas increases from 1 mol to 2 mol; therefore $\Delta S° > 0$; (d) at 500°C, $\Delta G° =$ +36.5 kJ; at 1000°C, $\Delta G° = -51.4$ kJ; (e) at 500°C, $K_p =$ 3.4×10^{-3}; at 1000°C, $K_p = 1.3 \times 10^2$

Chapter 22

22.1 (a) $^{106}_{45}Rh$; (b) $^{185}_{81}Tl$; (c) $^{204}_{83}Bi$ **22.3** 64.2 h **22.5** 14.0% **22.7** mass defect = 0.13699 g/mol; binding energy = 8.00 MeV/nucleon **22.9** 1.79×10^{10} kJ/mol **22.11** $^{40}_{19}K$ **22.13** 4% *Understanding Key Concepts* **1.** 2 half lives **3.** $^{14}_6C$ **4.** β emission because the n/p ratio is high *Additional Problems* **22.21** (a) $^{90}_{39}Y$; (b) $^{243}_{98}Cf$; (c) $^{49}_{24}Cr$; (d) $^{37}_{17}Cl$ **22.23** (a) $-^0_1e$; (b) ^0_1e; (c) ^4_2He **22.25** (a) $^{157}_{63}Eu \longrightarrow ^{157}_{64}Gd + -^0_1e$; (b) $^{126}_{56}Ba + -^0_1e \longrightarrow$ $^{126}_{55}Cs$; (c) $^{146}_{62}Sm \longrightarrow ^{142}_{60}Nd + ^4_2He$; (d) $^{125}_{56}Ba \longrightarrow ^{125}_{55}Cs$

+ $^0_{-1}e$ **22.27** ^{160}W is neutron poor and decays by α-emission. ^{185}W is neutron rich and decays by β-emission **22.29** $^{210}_{82}Pb$ **22.33** 0.247 d^{-1} **22.35** 3.05 d **22.37** after 65 d, 99.97%; after 65 y, 90.10%; after 650 y, 35.27% **22.39** 6960 y **22.41** 23.0% **22.43** 600 α-particles **22.45** 4200 y **22.47** 47 y **22.49** 21.0 h **22.51** 8.27 $\times 10^{-13}$ m **22.53** (a) 0.48059 g/mol; (b) 0.85499 g/mol **22.55** (a) 8.76 MeV/nucleon; (b) 8.74 MeV/nucleon **22.57** 5.42×10^8 kJ/mol **2.59** 1.02×10^{-9} g **22.61** 3.2 $\times 10^8$ kJ/mol **22.63** (a) 3; (b) $^{146}_{57}La$ **22.65** $^{98}_{42}Mo$ **22.67** (a) $^{254}_{102}No$; (b) ^4_2He; (c) $^{257}_{103}Lr$ **22.69** 20.3 s **22.71** (a) mass defect = 0.46700 amu; binding energy = 8.72 MeV/nucleon; (b) mass defect = 0.60015 amu; binding energy = 8.76 MeV/nucleon; ^{64}Zn is more stable. **22.73** 1.77×10^9 kJ/mol **22.75** 10 disintegrations/d **22.7** 0.991 min

Chapter 23

23.1

23.3 $CH_3CH_2CH_2CH_2CH_3$ $CH_3CH_2CHCH_3$ (with CH_3 substituent)

23.5 (a) pentane; 2-methylbutane; 2,2-dimethylpropane; (b) 3,4-dimethylhexane; (c) 2,4-dimethylpentane; (d) 2,2,5-trimethylheptane **23.7** (a) 1,4-dimethylcyclohexane; (b) 1-ethyl-3-methylcyclopentane; (c) isopropylcyclobutane

23.9 $ClCH_2CHCH_2CH_3$ $CH_3CCH_2CH_3$ (with CH_3 and Cl substituents)

$CH_3CHCHCH_3$ $CH_3CHCH_2CH_2Cl$ (with CH_3 and Cl substituents)

23.11 (a) $CH_3\overset{O}{\overset{\|}{C}}H$; (b) $CH_3CH_2\overset{O}{\overset{\|}{C}}OH$

23.13 (a) $CH_3\overset{CH_3}{\overset{|}{C}}CH=CHCH_2CH_3$ (with CH_3 and $CHCH_3$ groups)

(b) $CH_3C≡CCHCH_2CH_2CH_3$

(c)

$$CH_3CH_2 \quad H$$
$$C=C$$
$$H \quad CH_2CH_2CH_3$$

23.15

$$\begin{matrix} H & OH \\ CH_3C-CCH_2CH_3 \\ H & H \end{matrix} \qquad \begin{matrix} HO & H \\ CH_3C-CCH_2CH_3 \\ H & H \end{matrix}$$

23.17 (a)

(b)

(c)

23.19 (a)

$$\overset{+}{N}H_2CH_3 \qquad Cl^-$$

(b) $CH_3CH_2CH_2\overset{+}{N}H_3 \ Cl^-$

23.21

(a)

(b)

$$\begin{matrix} Cl & O & CH_3 \\ CH_3CHCH_2COCHCH_2CH_3 \end{matrix}$$

23.23 (a) CH_3CH_2OH; (b) CH_3NH_2;

(c) $CH_3CH_2\overset{O}{\overset{\|}{C}}OH$ (d)

23.25 $CH_3CH_2CH_2OH$; $CH_3\overset{OH}{\overset{|}{C}HCH_3}$; $CH_3CH_2OCH_3$

23.27 (a) $CH_3CH_2CH_2CH_2OH$ $CH_3CH_2\overset{OH}{\overset{|}{C}HCH_3}$

$$\overset{CH_3}{\overset{|}{CH_3CHCH_2OH}} \quad \overset{CH_3}{\underset{CH_3}{\overset{|}{CH_3COH}}}$$; (b) $CH_3CH_2CH_2NH_2$

$$\overset{NH_2}{\overset{|}{CH_3CHCH_3}} \quad CH_3CH_2NHCH_3 \quad \overset{CH_3}{\overset{|}{CH_3NCH_3}}$$

(c) $\overset{O}{\overset{\|}{CH_3CH_2CH_2CCH_3}}$ $\overset{O}{\overset{\|}{CH_3CH_2CCH_2CH_3}}$

$$\overset{O}{\overset{\|}{CH_3\overset{|}{\underset{CH_3}{CH}CCH_3}}};$$ (d) $\overset{O}{\overset{\|}{CH_3CH_2CH_2CH_2CH}}$ $\overset{CH_3 \ O}{\overset{| \quad \|}{CH_3CHCH_2CH}}$

$$\overset{O}{\overset{\|}{CH_3CH_2\overset{|}{\underset{CH_3}{CH}CH}}} \quad \overset{H_3C \ O}{\overset{| \quad \|}{CH_3\overset{|}{\underset{H_3C}{C}}-CH}}$$

23.31 C_3H_9 contains one less H than needed for an alkane and one more H than needed for an alkene.
23.33 (a) 4-ethyl-3-methyloctane; (b) 4-isopropyl-2-methylheptane; (c) 2,2,6-trimethylheptane; (d) 4-ethyl-4-methyloctane **23.35** 2,2,4-trimethylpentane **23.37**

(a)

(b)

(c)

(d)

23.39 (a) 3,3-dimethylhexane; (b) 3,5-dimethylheptane; (c) 1,3-dimethylcycloheptane **23.41** $2 \ C_4H_{10}(g) + 13 \ O_2(g) \longrightarrow 8 \ CO_2(g) + 10 \ H_2O(g)$ **23.49**
$CH_2{=}C{=}CHCH_2CH_3$ $CH_2{=}CHCH{=}CHCH_3$
$CH_2{=}CHCH_2CH{=}CH_2$ $CH_3CH{=}C{=}CHCH_3$

$$\overset{CH_3}{\overset{|}{CH_2{=}CCH{=}CH_2}} \quad \overset{CH_3}{\overset{|}{CH_2{=}C{=}CCH_3}}$$

23.51 (a)

$$H \quad H$$
$$C=C$$
$$H_3C \quad CH_2CH_2CH_3$$

(b) $\overset{CH_3}{\overset{|}{CH_3CHCH{=}CHCH_2CH_3}}$

(c) $\overset{CH_3}{\overset{|}{CH_2{=}CCH{=}CH_2}}$ **23.53** 2-pentene

23.55 (a) $CH_3\overset{CH_3}{\overset{|}{CH}CH_2CH{=}CHCH_3}$;
5-methyl-2-hexene

(b) 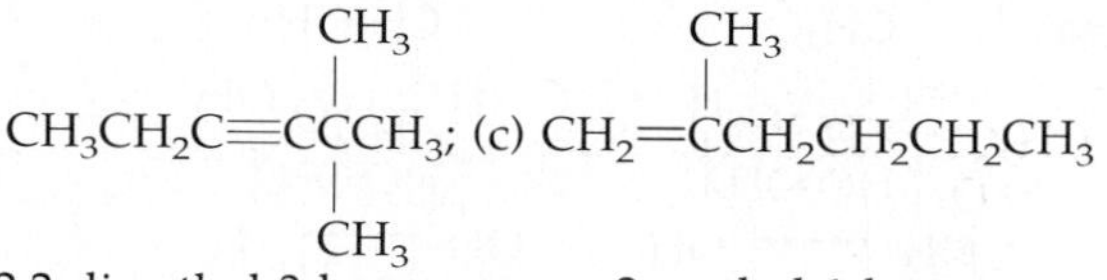 $CH_3CH_2C{\equiv}CCCH_3$ (with CH_3 groups); **(c)** $CH_2{=}CCH_2CH_2CH_2CH_3$ (with CH_3)

2,2-dimethyl-3-hexyne   2-methyl-1-hexene

(d) 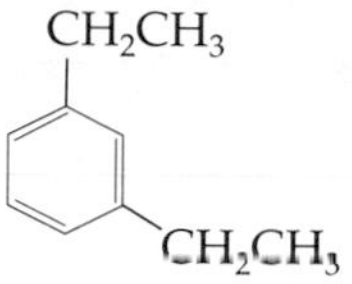   **23.57** (a) no; (b) yes; (c) yes

1,3-diethybenzene

23.59 (a)

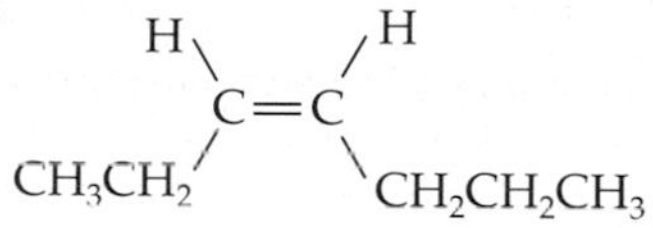

(b) 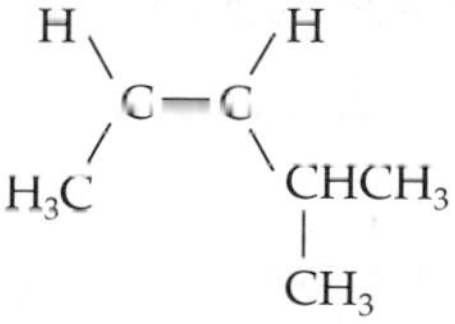   (c)

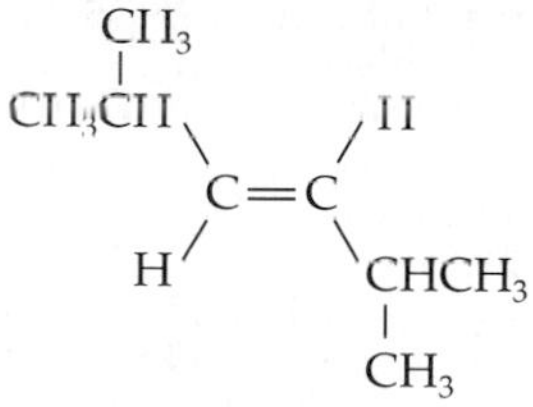

23.65

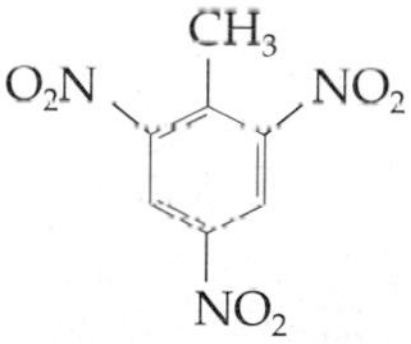

23.67 (a) $CH_3CH_2CH_2NH_2$; (b) $(CH_3CH_2)_2NH$, (c) $CH_3CH_2CH_2NHCH_3$ **23.75** $C_6H_5COOH\,(aq) + H_2O(l) \rightleftharpoons H_3O^+(aq) + C_6H_5COO^-(aq)$; % dissociation = 0.81%

23.77 (a) 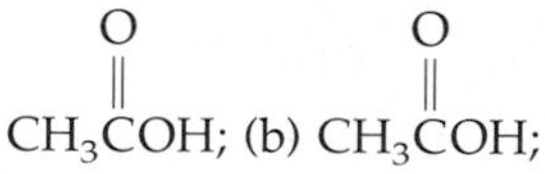 CH_3COH; (b) CH_3COH;

(c) 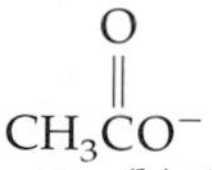 $CH_3CO^-\ Na^+$ **23.79** (a) methyl 4-methylpentanoate; (b) 4,4-dimethylpentanoic acid; (c) 2-methylpentanamide   **23.87** $-CH_2CHCH_2CHCH_2CHCH_2CH-$ (with Cl groups)

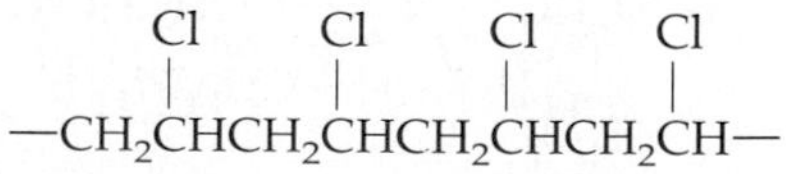

23.89 (a)

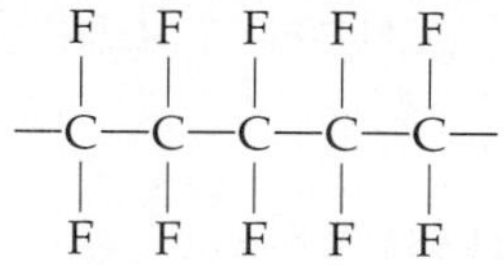

(b) 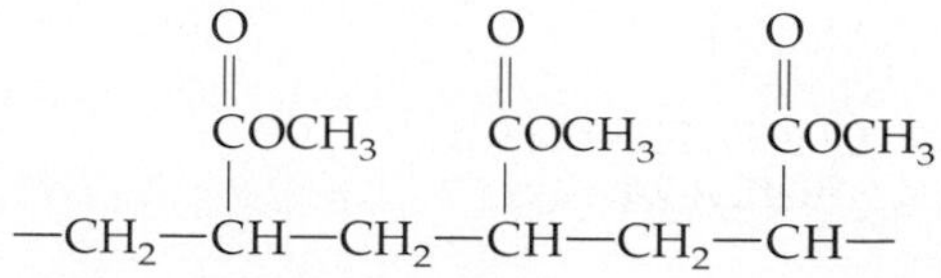 $-CH_2-CH-CH_2-CH-CH_2-CH-$ (with $COCH_3$ groups)

23.91

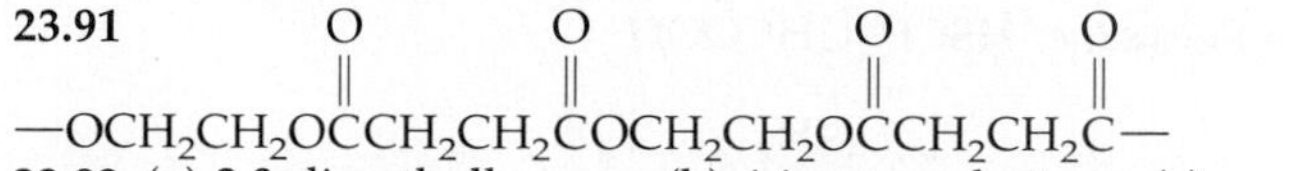

23.93 (a) 2,3-dimethylhexane; (b) 4-isopropyloctane; (c) 4-ethyl-2,4-dimethylhexane; (d) 3,3-diethylpentane

23.97

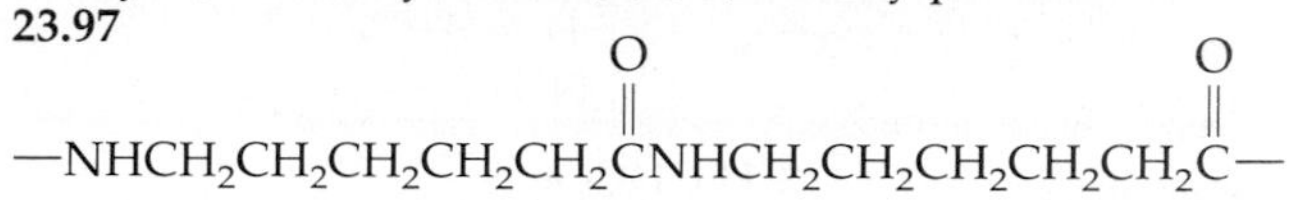

Chapter 24

24.1 $HOCH_2CH(OH)CH_2OH + ATP \longrightarrow HOCH_2CH(OH)CH_2OPO_3^{2-} + ADP$

24.3

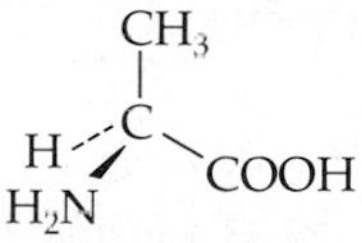

24.7 isoleucine and threonine   **24.9** Val–Tyr–Gly; Val–Gly–Tyr; Tyr–Gly–Val; Tyr–Val–Gly; Gly–Tyr–Val; Gly–Val–Tyr

24.11

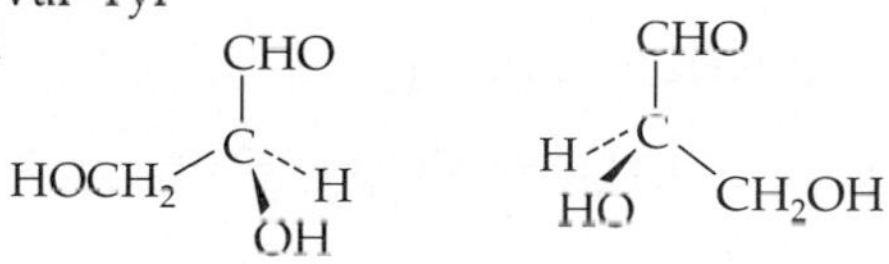

24.17

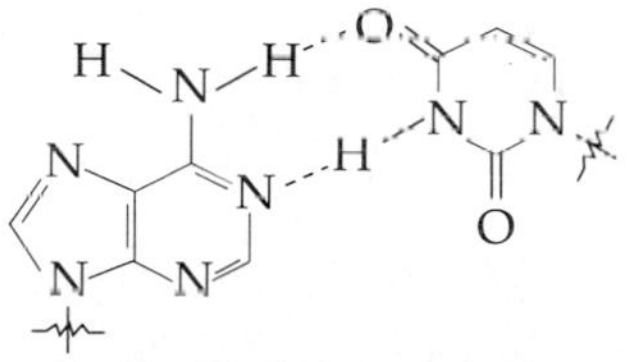

24.19 A–C–G–T–A–G–C–T–C–A   **24.23** (a)

valine $CH_3CHCHCOOH$ (with CH_3 and NH_2)

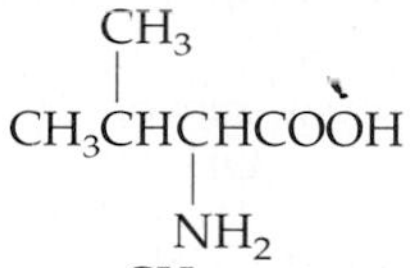

leucine $CH_3CHCH_2CHCOOH$ (with CH_3 and NH_2)

(b) serine $HOCH_2CHCOOH$ (with NH_2)

threonine $CH_3CHCHCOOH$ (with OH and NH_2)

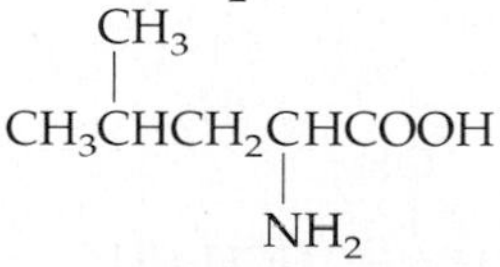

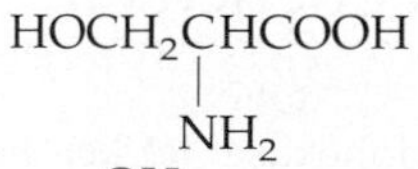

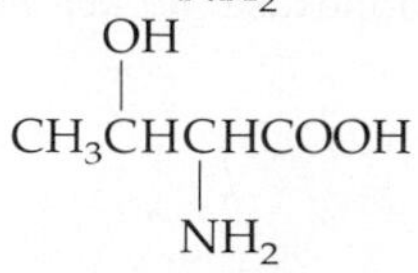

tyrosine

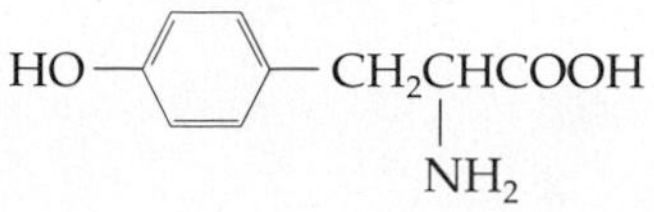

(c) cysteine $HSCH_2CHCOOH$
$\quad\quad\quad\quad\quad\quad\quad\quad |$
$\quad\quad\quad\quad\quad\quad\quad\quad NH_2$

(d) phenylalanine ⬡—$CH_2CHCOOH$
$\quad\quad\quad\quad\quad\quad\quad\quad\quad\quad |$
$\quad\quad\quad\quad\quad\quad\quad\quad\quad\quad NH_2$

tryptophan $CH_2CHCOOH$
$\quad\quad\quad\quad\quad\quad\quad\quad |$
$\quad\quad\quad\quad\quad\quad\quad\quad NH_2$

tyrosine HO—⬡—$CH_2CHCOOH$
$\quad\quad\quad\quad\quad\quad\quad\quad\quad\quad\quad |$
$\quad\quad\quad\quad\quad\quad\quad\quad\quad\quad\quad NH_2$

24.25 Val–Ser–Phe–Met–Thr–Ala **24.33** outside: (c) and (d), hydrophilic; inside: (a) and (b), hydrophobic
24.35

$\quad\quad\quad\quad\quad O \quad\quad\quad O$
$\quad\quad\quad\quad\quad \| \quad\quad\quad \|$
$H_2NCHCNHCHCOH$
$\quad\quad | \quad\quad\quad\quad |$
$\quad\quad CH_2 \quad\quad CH_2CH_2COOH$
$\quad\quad |$
$\quad\quad$⬡

$\quad\quad\quad\quad O \quad\quad\quad O$
$\quad\quad\quad\quad \| \quad\quad\quad \|$
$H_2NCHCNHCHCOH$
$\quad\quad | \quad\quad\quad\quad |$
$\quad\quad CH_2 \quad\quad CH_2$—⬡
$\quad\quad |$
$\quad\quad CH_2COOH$

24.37 13,000 amu **24.39** chiral: screw, glove; achiral: pencil, tire **24.43**

$\quad\quad\quad\quad\quad\quad\quad\quad\quad\quad OH$
$\quad\quad\quad\quad\quad\quad\quad\quad\quad\quad |$
(a) $CH_3CH_2CCH_3$
$\quad\quad\quad\quad\quad\quad\quad\quad\quad\quad |$
$\quad\quad\quad\quad\quad\quad\quad\quad\quad\quad H$

$\quad\quad\quad H \quad O \quad\quad\quad\quad OH$
$\quad\quad\quad | \quad \| \quad\quad\quad\quad |$
(b) CH_3CH_2C—CH (c) $CH_3CHCHCH_2CH_3$
$\quad\quad\quad\quad\quad |\quad\quad\quad\quad\quad\quad\quad\quad |$
$\quad\quad\quad\quad\quad CH_3 \quad\quad\quad\quad\quad\quad\quad OH$

24.45 (a) aldopentose; (b) aldohexose; (c) ketohexose
24.49

$\quad\quad\quad\quad\quad\quad\quad O$
$\quad\quad\quad\quad\quad\quad\quad \|$
$HOCH_2CHCCH_2OH$
$\quad\quad\quad\quad | $
$\quad\quad\quad\quad OH$

24.51

$\quad\quad\alpha \quad\quad\quad\quad\quad\quad\quad\quad \beta$

24.55

$\quad\quad\quad\quad O$
$\quad\quad\quad\quad \|$
$CH_2OC(CH_2)_{12}CH_3$
$\quad\quad\quad\quad O$
$\quad\quad\quad\quad \|$
$CHOC(CH_2)_{12}CH_3$
$\quad\quad\quad\quad O$
$\quad\quad\quad\quad \|$
$CH_2OC(CH_2)_{12}CH_3$

24.61 It would have a higher melting point because it is saturated.

$\quad\quad\quad\quad\quad\quad\quad\quad O$
$\quad\quad\quad\quad\quad\quad\quad\quad \|$
24.63 $CH_3(CH_2)_{18}CO(CH_2)_{31}CH_3$
24.71

24.73 32% A and T because they are complementary. 18% G and C because they are complementary.
24.75 A–U–G–G–C–U **24.77** $\Delta G° = +2870$ kJ/mol
24.77 $\Delta G° = +2870$ kJ/mol
24.79

$\quad\quad\quad\quad\quad\quad\quad\quad O \quad\quad\quad\quad O$
$\quad\quad\quad\quad\quad\quad\quad\quad \| \quad\quad\quad\quad \|$
$H_2NCHCNHCHCNHCHCOOH$
$\quad\quad | \quad\quad\quad | \quad\quad\quad |$
$\quad CHCH_3 \quad CH_2 \quad CH_2SH$
$\quad\quad | \quad\quad\quad\quad |$
$\quad CH_3 \quad\quad\quad$⬡

$\quad\quad\quad\quad O \quad\quad\quad\quad O \quad\quad\quad\quad O$
$\quad\quad\quad\quad \| \quad\quad\quad\quad \| \quad\quad\quad\quad \|$
H_2NCHC—N—$CNHCHCNHCHCOOH$
$\quad\quad | \quad\quad\quad\quad\quad\quad\quad\quad | \quad\quad\quad |$
$\quad CH_2 \quad\quad\quad\quad\quad\quad CH_3CH \quad CH_2CHCH_3$
$\quad\quad | \quad\quad\quad\quad\quad\quad\quad\quad | \quad\quad\quad |$
$\quad CH_2COOH \quad\quad\quad\quad CH_2CH_3 \quad CH_3$

 GLOSSARY

Absorption spectrum a plot of the amount of light absorbed versus wavelength (Section 20.10)

Accuracy a measure of how close to the true value a given measurement is (Section 1.10)

Achiral lacking handedness (Section 20.9)

Acid–base indicator a substance that changes color in a specific pH range (Section 15.6)

Acid–base neutralization reaction a process in which an acid reacts with a base to yield water plus an ionic compound called a salt (Sections 4.1, 4.5, 16.1)

Acid-dissociation constant (K_a) the equilibrium constant for the dissociation of an acid in water (Section 15.8)

Activation energy (E_a) the height of the energy barrier between reactants and products (Section 12.9)

Active site a small three-dimensional region of an enzyme with the specific shape necessary to bind the substrate and catalyze the appropriate reaction (Section 24.6)

Activity series a list of elements in order of their reducing ability in aqueous solution (Section 4.7)

Alcohol an organic molecule that contains an –OH group (Section 23.12)

Aldehyde an organic molecule that contains one alkyl group and one hydrogen bonded to a C=O carbon (Section 23.13)

Alkali metal an element in group 1A of the periodic table (Sections 1.4, 6.5)

Alkaline earth metal an element in group 2A of the periodic table (Sections 1.4, 6.6)

Alkane a compound that contains only carbon and hydrogen and has only single bonds (Section 23.2)

Alkene a hydrocarbon that has a carbon–carbon double bond (Section 23.9)

Alkyl group the part of an alkane that remains when a hydrogen is removed (Section 23.5)

Alkyne a hydrocarbon that has a carbon–carbon triple bond (Section 23.9)

Allotropes different structural forms of an element (Section 10.11)

Alpha (α) radiation a type of radioactive emission; a helium nucleus (Sections 2.4, 22.2)

Aluminosilicate a silicate mineral in which partial substitution of Si^{4+} by Al^{3+} has occurred (Section 19.7)

Amide an organic molecule that contains one alkyl group and one nitrogen bonded to a C=O carbon (Section 23.14)

Amine an organic derivative of ammonia (Section 23.12)

Amino acid the building block used to make proteins (Section 24.2)

Amorphous solid a solid whose constituent particles are randomly arranged and have no ordered long range structure (Section 10.6)

Amphoteric exhibiting both acidic and basic properties (Section 14.9)

Amplitude a wave's height measured from the midpoint between peak and trough (Section 5.2)

Anabolism metabolic reaction sequences that put building blocks together to assemble larger molecules (Section 24.1)

Anion a negatively charged particle (Section 2.8)

Anode the electrode at which oxidation takes place (Section 18.1)

Anodizing the oxidation of a metal anode to yield a protective metal oxide coat (Chapter 18 Interlude)

Apoenzyme the protein part of an enzyme (Section 24.6)

Aromatic referring to the class of compounds related to benzene (Section 23.11)

Arrhenius acid a substance that provides H^+ ions when dissolved in water (Sections 2.9, 4.5)

Arrhenius base a substance that provides OH^- ions when dissolved in water (Sections 2.9, 4.5)

Arrhenius equation an equation relating reaction rate, temperature, and activation energy; $k = Ae^{-E_a/RT}$ (Section 12.9)

Atom the smallest particle that retains the chemical properties of an element (Chapter 2)

Atomic mass unit (amu) a convenient unit of mass; 1/12th the mass of a $^{12}_{6}C$ atom (Section 2.6)

Atomic number (Z) the number of protons in an atom's nucleus (Section 2.5)

Atomic weight the weighted average mass of an element's atoms (Section 2.6)

Aufbau principle a set of rules that guides the filling order of orbitals in atoms (Section 5.12)

Avogadro's law the volume of a gas at a fixed pressure and temperature is proportional to its molar amount (Section 9.2)

Avogadro's number   the number of units in a mole; 6.02 × 10²³ (Section 3.3)

Balanced equation   an equation in which the numbers and kinds of atoms on both sides of the arrow are the same (Section 3.1)

Band gap   the energy difference between the valence band and the conduction band in a metal (Section 21.5)

Band theory   the molecular orbital theory for metals (Section 21.4)

Base-dissociation constant (K_b)   the equilibrium constant for the reaction of a base with water (Section 15.12)

Basic oxygen process   a method for purifying iron and converting it to steel (Section 21.3)

Battery   *see* **Galvanic cell** (Section 18.9)

Bayer process   purification of Al_2O_3 by treating bauxite with hot aqueous NaOH (Section 16.12)

Beta (β) radiation   a type of radioactive emission consisting of electrons (Section 22.2)

Bidentate ligand   a ligand that bonds to a metal using electron pairs on two donor atoms (Section 20.6)

Bimolecular reaction   an elementary reaction that results from collisions between two reactant molecules (Section 12.7)

Binary hydride   a compound that contains hydrogen and one other element (Section 14.5)

Binding energy   the energy that holds nucleons together in the nucleus (Section 22.5)

Biochemistry   the chemistry of living organisms (Chapter 24)

Body-centered cubic packing   a packing arrangement of spheres into a body-centered cubic unit cell (Section 10.9)

Body-centered cubic unit cell   a cubic unit cell with an atom at each of its eight corners and an additional atom in the center of the cube (Section 10.8)

Boiling point   the temperature at which liquid and vapor coexist in equilibrium (Section 10.4)

Bond angle   the angle at which two adjacent bonds intersect (Section 7.9)

Bond dissociation energy (D)   the amount of energy needed to break a chemical bond in an isolated molecule in the gaseous state (Section 7.2)

Bond length   the optimum distance between nuclei in a covalent bond (Section 7.1)

Bond order   the number of electron pairs shared between two bonded atoms (Section 7.4)

Bonding electron pair   a pair of valence electrons in a covalent bond (Section 7.4)

Born-Haber cycle   a pictorial way of viewing the steps in the formation of an ionic solid from its elements (Section 6.4)

Boyle's law   the volume of a fixed amount of gas at a constant temperature varies inversely with its pressure (Section 9.2)

Branched-chain alkane   an alkane with a branching connection of carbons (Section 23.2)

Brønsted-Lowry acid   a substance that can transfer H^+ (Section 15.1)

Brønsted-Lowry base   a substance that can accept H^+ (Section 15.1)

Buffer capacity   a measure of the amount of acid or base that a buffer can absorb without a significant change in pH (Section 16.3)

Buffer solution   a solution of a weak acid and its conjugate base that resists drastic changes in pH (Section 16.3)

Carbohydrate   a large class of organic molecules related to glucose (Section 24.7)

Carbonyl group   the C=O group (Section 23.13)

Carboxylic acid   an organic molecule that contains the –COOH group (Section 23.14)

Catabolism   metabolic reaction sequences that break molecules apart (Section 24.1)

Catalyst   a substance that increases the rate of a reaction without itself being consumed (Section 12.11)

Cathode   the electrode at which reduction takes place (Section 18.1)

Cathode ray   the visible glow emitted when an electric potential is applied between two electrodes in an evacuated chamber (Section 2.3)

Cathodic protection   a technique for protecting a metal from corrosion by connecting it to a second metal that is more easily oxidized (Section 18.10)

Cation   a positively charged particle (Section 2.8)

Cell potential (E)   *see* **Electromotive force** (Section 18.3)

Cell voltage   *see* **Electromotive force** (Section 18.3)

Celsius degree (°C)   a common unit of temperature; 0°C = 273.15 K (Section 1.9)

Centimeter (cm)   a common unit of length; 1 cm = 0.01 m (Section 1.7)

Ceramic composite   a hybrid material made of two ceramics (Section 21.8)

Ceramic   an inorganic, nonmetallic, nonmolecular solid (Section 21.7)

Chain reaction   a reaction whose product initiates further reaction (Section 22.6)

Change of state   *see* **Phase change** (Section 10.4)

Charles' law   the volume of a fixed amount of gas at a constant pressure varies directly with its absolute temperature (Section 9.2)

Chelate   the complex formed by a metal atom and a polydentate ligand (Section 20.6)

Chelating agent   a polydentate ligand (Section 20.6)

Chemical bond   forces that hold atoms together in chemical compounds (Section 2.8)

Chemical energy   potential energy stored in the chemical bonds of molecules (Section 8.2)

Chemical equation   a format for writing a chemical reaction, listing reactants on the left, products on the right, and an arrow between them (Section 2.7)

Chemical equilibrium   the state reached when the concentrations of reactants and products remain constant in time (Chapter 13)

Chemical formula   a format for listing the number and kind of constituent elements in a compound (Section 2.7)

Chemical kinetics   the area of chemistry concerned with reaction rates and the sequence of steps by which reactions occur (Chapter 12)

Chemical property   a characteristic that results in a change in the chemical makeup of a sample (Section 1.13)

Chemical reaction   the transformation of one substance into another (Section 2.7)

Chemistry   the study of the composition, properties, and transformations of matter (Chapter 1)

Chiral   having handedness (Section 20.9)

Chlor-alkali industry   the commercial production method for Cl_2 and NaOH by electrolysis of aqueous sodium chloride (Section 18.12)

Chromosome   a threadlike strand of DNA in the nucleus of cells (Section 24.14)

Cis isomer   the isomer of a metal complex in which identical ligands are adjacent (Section 20.8)

Coefficient   a number placed before a formula in an equation to indicate how many formula units are required to balance the equation (Section 3.1)

Coenzyme   an inorganic ion or a small organic molecule that acts as an enzyme cofactor (Section 24.6)

Cofactor   the small, nonprotein part of an enzyme (Section 24.6)

Colligative property   a property that depends on the amount of dissolved solute but not on the chemical identity of the solute (Section 11.5)

Collision theory   a model in which bimolecular reactions occur when two properly oriented reactant molecules come together in a sufficiently energetic collision (Section 12.9)

Colloid   a homogeneous mixture containing particles with diameters in the range 2–1000 nm (Section 11.1)

Common-ion effect   the shift in the position of an equilibrium on addition of a substance that provides an ion in common with one of the ions already involved in the equilibrium (Section 16.2)

Compound   a chemical substance composed of more than one kind of element (Section 2.7)

Condensed structure   a shorthand method for drawing organic structures in which C–H and C–C single bonds are "understood" rather than shown (Section 23.3)

Conduction band   the antibonding molecular orbitals in a metal (Section 21.5)

Conformation   the three-dimensional structure of a molecule (Section 23.4)

Conjugate acid   the species HA formed by addition of H^+ to the base A^- (Section 15.1)

Conjugate acid–base pair   chemical species whose formulas differ by only one proton (Section 15.1)

Conjugate base   the species A^- formed by loss of H^+ from the acid HA (Section 15.1)

Constitutional isomers   isomers that have different connections among their constituent atoms (Section 20.8)

Contact process   the commercial process for making sulfuric acid from sulfur (Section 19.13)

Conversion factor   an expression to describe the relationship between units (Section 1.12)

Coordinate covalent bond   a bond formed when one atom donates two electrons to another atom that has a vacant valence orbital (Section 7.4)

Coordination compound   a compound in which a central metal ion is attached to a group of surrounding molecules or ions by coordinate covalent bonds (Section 20.5)

Coordination number   the number of ligand donor atoms that surround a central metal ion in a complex (Section 20.5)

Core electrons   inner-shell electrons (Section 6.1)

Corrosion   the oxidative deterioration of a metal, such as the conversion of iron to rust (Section 18.10)

Cosmic ray   a stream of energetic particles, primarily protons, coming from interstellar space (Section 22.8)

Covalent bond   a bond that occurs when two atoms share several (usually two) electrons (Section 2.8; Chapter 7)

Covalent network solid   a solid whose atoms are linked together by covalent bonds into a giant three-dimensional array (Section 10.6)

Critical mass   the amount of material necessary for a nuclear chain reaction to become self-sustaining (Section 22.6)

Critical point   a combination of temperature and pressure beyond which a gas cannot be liquefied (Section 10.12)

Crystal field splitting   the energy splitting between two sets of d orbitals in a metal complex (Section 20.12)

Crystal field theory   a model that views the bonding in metal complexes as arising from electrostatic interactions and considers the effect of the ligand charges on the energies of the metal-ion d orbitals (Section 20.12)

Crystalline solid   a solid whose atoms, ions, or molecules have an ordered arrangement extending over a long range (Section 10.6)

Cubic centimeter (cm^3)   a common unit of volume, equal in size to the milliliter; $1 \, cm^3 = 10^{-6} \, m^3$ (Section 1.8)

Cubic closest-packed   a packing arrangement of spheres into a face-centered cubic unit cell with three alternating layers (Section 10.9)

Cubic meter (m^3)   the SI unit of volume (Section 1.8)

Cycloalkane   an alkane that contains a ring of carbon atoms (Section 23.6)

d-Block element   a transition-metal element that results from d-orbital filling (Section 5.13; Chapter 20)

Dalton's law of partial pressures   the total pressure exerted by a mixture of gases in a container at constant V and T is equal to the sum of the pressures exerted by each individual gas in the container (Section 9.5)

Decay constant the first-order rate constant for radioactive decay (Section 22.3)

Decay series a sequence of nuclear disintegrations ultimately leading to a nonradioactive product (Section 22.4)

Degenerate having the same energy level (Section 5.12)

Density an intensive physical property that relates the mass of an object to its volume (Section 1.13)

Diamagnetic a substance that is weakly repelled by a magnetic field (Section 7.14)

Diastereoisomers non-mirror-image stereoisomers (Section 20.8)

Diffraction scattering of a beam by an object containing regularly spaced lines or points (Section 10.7)

Diffusion the mixing of different gases by random molecular motion and with frequent collisions (Section 9.7)

Dimensional-analysis method a method of problem solving whereby problems are set up so that unwanted units cancel (Section 1.12)

Dipole-dipole force an intermolecular force resulting from electrical interactions among dipoles on neighboring molecules (Section 10.2)

Dipole moment (μ) the measure of net molecular polarity; $\mu = Q \times r$ (Section 10.1)

Diprotic acid an acid that has two dissociable protons (Section 15.7)

Disproportionation reaction a reaction in which a substance is both oxidized and reduced (Section 14.10)

Dissociate to split apart to give ions when dissolved in water (Section 4.2)

Doping the addition of a small amount of impurities to increase the conductivity of a semiconductor (Section 21.5)

Double bond a covalent bond formed by sharing four electrons (Section 7.4)

Effective nuclear charge (Z_{eff}) the net nuclear charge actually felt by an electron (Section 5.10)

Effusion the escape of gas molecules through a tiny hole in a membrane, without collisions (Section 9.7)

Electrochemical cell a device for interconverting chemical and electrical energy (Section 18.1)

Electrochemistry the area of chemistry concerned with the interconversion of chemical and electrical energy (Chapter 18)

Electrolysis the process of using an electric current to bring about chemical change (Section 18.11)

Electrolyte a substance that dissolves in water to produce ions (Section 4.2)

Electrolytic cell an electrochemical cell in which an electric current drives a nonspontaneous reaction (Section 18.11)

Electromagnetic radiation radiant energy (Section 5.2)

Electromagnetic spectrum the range of different kinds of electromagnetic radiation (Section 5.2)

Electromotive force (emf) the electrical potential that pushes electrons away from the anode and pulls them toward the cathode (Section 18.3)

Electron a negatively charged, fundamental atomic particle (Section 2.3)

Electron affinity (E_{ea}) the energy change that occurs when an electron is added to an isolated atom in the gaseous state (Section 6.3)

Electron capture a process in which a proton in the nucleus captures an inner-shell electron and is thereby converted into a neutron (Section 22.2)

Electron configuration a description of the arrangement of electrons in the orbitals of an atom (Section 5.12)

Electron-sea model a model that visualizes metals as a three-dimensional array of metal cations immersed in a sea of delocalized electrons that are free to move about (Section 21.4)

Electronegativity (EN) the ability of an atom in a molecule to attract the shared electrons in a bond (Section 7.7)

Electroplating the coating of one metal on the surface of another using electrolysis (Section 18.12)

Electrorefining the purification of a metal by means of electrolysis (Section 18.12)

Element a fundamental substance that can't be chemically changed or broken down into anything simpler (Section 1.2)

Elementary reaction a single step in a reaction mechanism (Section 12.7)

Empirical formula a formula that gives the ratios of atoms in a compound (Section 3.11)

Enantiomers stereoisomers that are nonidentical mirror images of one another (Section 20.9)

Endothermic a chemical change in which heat is absorbed and the temperature of the surroundings falls (Section 8.6)

Energy the capacity to do work or supply heat (Section 8.1)

Enthalpy change (ΔH) the heat change at constant pressure; $\Delta H = \Delta E + P\Delta V$ (Section 8.5)

Entropy (S) the amount of molecular disorder or randomness in a system (Sections 8.12, 17.3)

Entropy of solution (ΔS_{soln}) the entropy change during formation of a solution (Section 11.2)

Enzyme a large protein that acts as a catalyst for a biological reaction (Section 24.6)

Equilibrium constant (K_c) the constant in the equilibrium equation (Section 13.2)

Equilibrium constant (K_p) the equilibrium constant for reaction of gases, defined using partial pressures (Section 13.3)

Equilibrium constant expression the expression on the right side of the equilibrium equation (Section 13.2)

Equilibrium mixture a mixture of reactants and products at equilibrium (Section 13.1)

Equivalence point the point in a titration at which stoichiometrically equivalent quantities of reactants have been mixed together (Section 16.5)

Ester an organic molecule that contains the –COOR group (Section 23.14)

Ether an organic molecule that contains two alkyl groups bonded to the same oxygen atom (Section 23.12)

Exothermic a chemical change in which heat is evolved and the temperature of the surroundings rises (Section 8.6)

Extensive property a property whose value depends on the sample size (Section 1.13)

***f*-Block element** a lanthanide or actinide element that results from *f*-orbital filling (Section 5.13)

Face-centered cubic unit cell a cubic unit cell with an atom at each of its eight corners and an additional atom on each of its six faces (Section 10.8)

Faraday the electrical charge on 1 mol of electrons (96,485 C/mol e⁻) (Section 18.3)

First law of thermodynamics the energy of the universe is constant (Section 8.2)

Flotation a metallurgical process that exploits differences in the ability of water and oil to wet the surfaces of mineral and gangue (Section 21.2)

Formal charge the result of a method of electron book-keeping that tells whether an atom in a molecule has gained or lost electrons compared to an isolated atom (Section 7.8)

Formation constant (K_f) the equilibrium constant for formation of a complex ion (Section 16.11)

Formula unit one unit (atom, ion, or molecule) corresponding to a given formula (Section 3.1)

Formula weight the sum of atomic weights of all atoms in one formula unit of a substance (Section 3.3)

Fractional distillation the separation of volatile liquids on the basis of boiling point (Section 11.10)

Free-energy change (ΔG) $\Delta G = \Delta H - T\Delta S$ (Sections 8.13, 17.7)

Frequency (ν) the number of wave maxima that pass by a fixed point per unit time (Section 5.2)

Fuel cell a galvanic cell in which one of the reactants is a traditional fuel such as methane or hydrogen (Section 18.9)

Functional group a part of a larger molecule; composed of an atom or group of atoms that has characteristic chemical behavior (Section 23.8)

Galvanic cell an electrochemical cell in which a spontaneous chemical reaction generates an electric current (Section 18.1)

Galvanizing a process for protecting steel from corrosion by coating it with zinc (Section 18.10)

Gamma (γ) radiation a type of radioactive emission consisting of a stream of high-energy photons (Section 22.2)

Gangue the economically worthless material consisting of sand, clay, and other impurities that accompanies an ore (Section 21.2)

Gas laws relationships among the variables P, V, n, and T for a gas sample (Section 9.2)

Gas shift reaction the reaction of CO with H_2O to yield hydrogen (Section 14.3)

Gene a segment of a DNA chain that contains the instructions necessary to make a specific protein (Section 24.14)

Graham's law the rate of effusion of a gas is inversely proportional to the square root of its molar mass (Section 9.7)

Gram (g) a common unit of mass; 1 g = 0.001 kg (Section 1.6)

Ground-state configuration the lowest-energy electron configuration of an atom (Section 5.12)

Group a column of elements in the periodic table (Section 1.3)

Half-life ($t_{1/2}$) the time required for a reactant concentration to drop to one-half of its initial value (Section 12.5)

Half-reaction method a method for balancing redox equations (Section 4.9)

Hall-Heroult process the commercial production method for aluminum by electrolysis of a molten mixture of aluminum oxide and cryolite (Section 18.12)

Halogen an element in group 7A of the periodic table (Sections 1.4, 6.8)

Halogenation the addition of halogen (Cl_2 or Br_2) to an alkene (Section 23.20)

Heat the energy transferred from one object to another as the result of a temperature difference between them (Section 8.2)

Heat capacity (C) the amount of heat required to raise the temperature of an object or substance a given amount (Section 8.7)

Heat of combustion the amount of energy released on burning a substance (Section 8.11)

Heat of fusion (ΔH_{fusion}) the amount of heat required for melting (Section 8.6)

Heat of reaction the enthalpy change for a reaction (Section 8.6)

Heat of solution (ΔH_{soln}) the enthalpy change during formation of a solution (Section 11.2)

Heat of sublimation (ΔH_{subl}) the amount of heat required for sublimation (Section 8.6)

Heat of vaporization (ΔH_{vap}) the amount of heat required for evaporation (Section 8.6)

Heisenberg uncertainty principle we can never know both the position and the velocity of an electron beyond a certain level of precision (Section 5.6)

Henderson-Hasselbalch equation an equation relating the pH of a solution to the pK_a of the weak acid; pH = pK_a + log ([base]/[acid]) (Section 16.4)

Henry's law the solubility of a gas in a liquid at a given temperature is directly proportional to the partial pressure of the gas over the solution (Section 11.4)

Hertz (Hz) a unit of frequency; 1 Hz = 1 s⁻¹ (Section 5.2)

Hess's law the overall enthalpy change for a reaction is equal to the sum of the enthalpy changes for the individual steps in the reaction (Section 8.8)

Heterogeneous catalyst one that exists in a different phase than the reactants (Section 12.12)

Heterogeneous equilibria equilibria in which reactants and products are present in more than one phase (Section 13.4)

Heterogeneous mixture a mixture having regions with differing compositions (Section 2.7)

Hexagonal closest-packed a packing arrangement of spheres into a noncubic unit cell with two alternating layers (Section 10.9)

High-spin complex a metal complex in which the *d* electrons are arranged to give the maximum number of unpaired electrons (Section 20.11)

Holoenzyme the assembly of apoenzyme plus cofactor (Section 24.6)

Homogeneous catalyst one that exists in the same phase as the reactants (Section 12.12)

Homogeneous equilibria equilibria in which all reactants and products are in a single phase, usually either gaseous or solution (Section 13.4)

Homogeneous mixture a mixture having a constant composition throughout (Section 2.7)

Hund's rule if two or more degenerate orbitals are available, one electron goes in each until all are half full (Section 5.12)

Hybrid atomic orbital a set of wave functions derived by combination of atomic wave functions (Section 7.11)

Hydrate a solid compound that contains water molecules (Section 14.14)

Hydration the addition of water to an alkene (Section 23.10)

Hydrocarbon a compound that contains only carbon and hydrogen (Section 23.2)

Hydrogen bond an attractive interaction between a hydrogen atom bonded to an electronegative O, N, or F atom and an unshared electron pair on another nearby electronegative atom (Section 10.2)

Hydrogenation the addition of H_2 to an alkene to yield an alkane (Section 23.10)

Hydronium ion H_3O^+ (Section 4.5)

Hygroscopic absorbing water from the air (Section 14.14)

Ideal-gas law a description of how the volume of a gas is affected by changes in pressure, temperature, and amount; $PV = nRT$ (Section 9.3)

Initial rate the instantaneous rate of a reaction at the beginning (Section 12.1)

Inner transition-metal element an element in the fourteen groups shown separately at the bottom of the periodic table (Section 1.4)

Instantaneous rate the rate of a reaction at a particular time (Section 12.1)

Intensive property a property whose value does not depend on the sample size (Section 1.13)

Intermolecular force an attraction between molecules that holds them together (Section 10.2)

Internal energy (*E*) the sum of kinetic and potential energies for each particle in the system (Section 8.3)

International System of Units (SI) the seven base units, along with others derived from them, used for all scientific measurements (Section 1.5)

Interstitial hydride a metallic hydride that consists of a crystal lattice of metal atoms with the smaller hydrogen atoms occupying holes between the larger metal atoms (Section 14.5)

Ion product (IP) a number defined in the same way as K_{sp}, except that the concentrations in the expression for IP are nonequilibrium values (Section 16.13)

Ion a charged particle (Section 2.8)

Ion-dipole force an intermolecular force resulting from electrical interactions between an ion and the partial charges on a polar molecule (Section 10.2)

Ion-product constant for water (K_w) $[H_3O^+][OH^-] = 1.0 \times 10^{-14}$ (Section 15.4)

Ionic bond a bond that results from a complete transfer of one or more electrons between atoms (Sections 2.8, 6.4)

Ionic equation a reaction written so that the ions are explicitly shown (Section 4.3)

Ionic solid a solid whose constituent particles are ions ordered into a regular three-dimensional arrangement held together by ionic bonds (Sections 2.8, 10.6)

Ionization energy (E_i) the amount of energy necessary to remove the outermost electron from an isolated neutral atom in the gaseous state (Section 6.1)

Ionization isomers isomers that differ in the anion bonded to the metal ion (Section 20.8)

Ionizing radiation radiation that knocks an electron from a molecule, thereby ionizing it (Section 22.8)

Isomers compounds that have the same formula but a different arrangement of their constituent atoms (Section 20.8)

Isotope effect differences in properties that arise from the differences in isotopic mass (Section 14.2)

Isotopes atoms with identical atomic numbers but different mass numbers (Section 2.5)

Kelvin (K) the SI unit of temperature; 0 K = absolute zero (Section 1.9)

Ketone an organic molecule that contains two alkyl groups bonded to a C=O carbon (Section 23.13)

kilogram (kg) the SI unit of mass; 1 kg = 2.205 U.S. lb (Section 1.6)

Kinetic energy (E_K) the energy of motion; $E_K = (1/2)mv^2$ (Section 8.1)

Kinetic-molecular theory a theory describing the behavior of gases (Section 9.6)

Lanthanide contraction the decrease in atomic radii of the *f*-block lanthanide elements (Section 20.2)

Lattice energy (*U*) the sum of the electrostatic interactions between ions in a solid that must be overcome to break a crystal into individual ions (Section 6.4)

Law of conservation of energy energy can be neither created nor destroyed (Section 8.2)

Law of definite proportions different samples of a pure chemical substance always contain the same proportion of elements by mass (Section 2.1)

Law of mass conservation mass is neither created nor destroyed in chemical reactions (Section 2.1)

Law of multiple proportions the mass ratios are small, whole-number multiples of one another when two elements combine in different ways to form different substances (Section 2.2)

Le Châtelier's Principle if a stress is applied to a reaction mixture at equilibrium, reaction occurs in the direction that relieves the stress (Section 13.6)

Lewis acid an electron-pair acceptor (Section 15.16)

Lewis base an electron-pair donor (Section 15.16)

Lewis structure a representation of an atom that shows valence electrons as dots (Section 7.4)

Ligand a molecule or ion that surrounds the central metal ion in a complex (Section 20.5)

Ligand donor atom an atom attached directly to the metal ion in a metal complex (Section 20.5)

Limiting reactant the reactant present in limiting amount that controls the extent to which reaction occurs (Section 3.6)

Line spectrum the wavelengths of light emitted by an energetically excited atom (Section 5.3)

Line structure a shorthand structure in which only bonds (lines) are shown; carbon atoms are understood to be at every junction of lines (Section 23.6)

Linkage isomers isomers that arise when a ligand bonds to a metal through either of two different donor atoms (Section 20.8)

Lipid a naturally occurring organic molecule that dissolves in nonpolar organic solvents when a sample of plant or animal tissue is crushed or ground (Section 24.11)

Liquid crystal a state of matter in which molecules move around as if in a viscous liquid, but have a restricted range of motion as if in a solid (Chapter 10 Interlude)

Liter (L) a common unit of volume; $1 \text{ L} = 10^{-3} \text{ m}^3$ (Section 1.8)

Lock-and-key model a model that pictures an enzyme as a large, irregularly shaped molecule with a cleft into which substrate can fit (Section 24.6)

London dispersion force an intermolecular force resulting from the motion of electrons around atoms (Section 10.2)

Lone pair electrons a pair of valence electrons not used for bonding (Section 7.4)

Low-spin complex a metal complex in which the *d* electrons are paired to give a maximum number of doubly occupied *d* orbitals and a minimum number of unpaired electrons (Section 20.11)

Main-group element an element in the two larger groups on the left and the six larger groups on the right of the periodic table (Section 1.4, Chapters 6, 19)

Mass the amount of matter in an object (Section 1.6)

Mass defect the loss in mass that occurs when protons and neutrons combine to form a nucleus (Section 22.5)

Mass number the total number of protons and neutrons in an atom (Section 2.5)

Matter a catchall term used to describe anything you can touch, taste, or smell (Section 1.6)

Melting point the temperature at which solid and liquid coexist in equilibrium (Section 10.4)

Metabolism the sum of the many organic reactions that go on in cells (Section 24.1)

Metal complex *see* **Coordination compound** (Section 20.5)

Metal an element on the left side of the periodic table, bounded on the right by a zigzag line running from boron to astatine (Sections 1.4, 21.4)

Metallic solid a solid consisting of metal atoms, whose crystals have metallic properties such as electrical conductivity (Section 10.6)

Metalloid *see* **Semimetal** (Section 1.4)

Metallurgy the science and technology of extracting metals from their ores (Section 21.2)

Meter (m) the SI unit of length (Section 1.7)

Milligram (mg) a common unit of mass; $1 \text{ mg} = 0.001 \text{ g} = 10^{-6} \text{ kg}$ (Section 1.6)

Milliliter (mL) a common unit of volume; $1 \text{ mL} = 1 \text{ cm}^3$ (Section 1.8)

Millimeter (mm) a common unit of length; $1 \text{ mm} = 0.001 \text{ m}$ (Section 1.7)

Mineral a crystalline, inorganic constituent of the rocks that make up the earth's crust (Section 21.1)

Miscible mutually soluble in all proportions (Section 11.4)

Molality (*m*) a unit of concentration; the number of moles of solute per kilogram of solvent (mol/kg) (Section 11.3)

Molar heat capacity (C_m) the amount of heat necessary to raise the temperature of 1 mol of a substance 1°C (Section 8.7)

Molar mass the mass of 1 mol of substance; equal to the molecular or formula weight of the substance in grams (Section 3.3)

Molarity (M) a common unit of concentration; the number of moles of solute per liter of solution (Sections 3.7, 11.3)

Mole the SI unit for amount of substance; the quantity of a substance that contains as many molecules or formula units as there are atoms in exactly 12 g of carbon-12 (Section 3.3)

Mole fraction (*X*) a unit of concentration; the number of moles of a component divided by the total number of moles in the mixture (Section 9.5)

Molecular equation a reaction written using the full formulas of reactants and products (Section 4.3)

Molecular formula a formula that tells the actual numbers of atoms in a compound (Section 3.11)

Molecular orbital theory a quantum mechanical description of bonding in which electrons occupy molecular orbitals that belong to the entire molecule rather than to an individual atom (Section 7.13)

Molecular solid a solid whose constituent particles are molecules held together by intermolecular forces (Section 10.6)

Molecular weight (MW)   the sum of atomic weights of the atoms in a molecule (Section 3.3)

Molecularity   the number of molecules (or atoms) on the reactant side of the chemical equation for an elementary reaction (Section 12.7)

Molecule   the unit of matter that results when two or more atoms are joined by covalent bonds (Section 2.8)

Mond process   a chemical method for purification of nickel from its ore (Section 21.2)

Monodentate ligand   a ligand that bonds to a metal using the electron pair of a single donor atom (Section 20.6)

Monomer   a small molecule that when bonded to itself many times, forms a polymer (Section 23.15)

Monoprotic acid   an acid that has a single dissociable proton (Section 15.7)

Monosaccharide   a carbohydrate such as glucose that can't be broken down into smaller molecules by hydrolysis (Section 24.7)

n-Type semiconductor   a semiconductor doped with an impurity that has more electrons than necessary for bonding (Section 21.5)

Nanometer (nm)   a common unit of length; 1 nm = 10^{-9} m (Section 1.7)

Natural gas   largely methane, but also contains ethane, propane, and butane (Chapter 23 Interlude)

Nernst equation   an equation for calculating cell potentials under non-standard-state conditions; $E = E° - (RT \ln Q)/nF$ (Section 18.6)

Net ionic equation   a reaction written so that the spectator ions are removed (Section 4.3)

Neutralization reaction   *see* **Acid–base neutralization reaction** (Section 4.1)

Neutron   a neutral, fundamental atomic particle in the nucleus of atoms (Section 2.5)

Noble gas   an element in group 8A of the periodic table (Sections 1.4, 6.9)

Node   a region where a wave has zero amplitude (Section 5.9)

Nonelectrolyte   a substance that does not produce ions when dissolved in water (Section 4.2)

Nonmetal   an element on the right side of the periodic table, bounded on the left by a zigzag line running from boron to astatine (Section 1.4)

Nonstoichiometric compound   a compound whose atomic composition can't be expressed as a ratio of small whole numbers (Section 14.5)

Normal boiling point   the temperature at which boiling occurs when there is exactly 1 atm of external pressure (Section 10.5)

Normal melting point   the temperature at which melting occurs when there is exactly 1 atm of external pressure (Section 10.12)

Nuclear chemistry   the study of the properties and reactions of atomic nuclei (Chapter 22)

Nuclear equation   an equation for a nuclear reaction in which the sums of the nucleons are the same on both sides and the sums of the charges on the nuclei and any elementary particles are the same on both sides (Section 22.2)

Nuclear fission   the fragmenting of heavy nuclei (Section 22.6)

Nuclear fusion   the joining together of light nuclei (Section 22.6)

Nuclear reaction   a reaction that changes an atomic nucleus (Section 22.1)

Nuclear transmutation   the change of one element into another (Section 22.7)

Nucleic acid   a polymer made up of nucleotide units linked together to form a long chain (Section 24.12)

Nucleon   a general term for both protons and neutrons (Section 22.1)

Nucleus   the central core of an atom consisting of protons and neutrons (Section 2.4)

Nuclide   an isotope (Section 22.1)

Octet rule   main-group elements tend to undergo reactions that leave them with eight valence electrons (Section 6.10)

Orbital   a solution to the Schrödinger wave equation; describes a region of space where an electron is likely to be found (Section 5.6)

Ore   a mineral deposit from which a metal can be produced economically (Section 21.1)

Organic chemistry   the study of carbon compounds (Chapter 23)

Osmosis   the passage of solvent through a membrane from the less concentrated side to the more concentrated side (Section 11.8)

Osmotic pressure   the amount of pressure necessary to cause osmosis to stop (Section 11.8)

Ostwald process   the commercial process for making nitric acid from ammonia (Section 19.10)

Overvoltage   the additional voltage required above that calculated for an electrolysis reaction (Section 18.11)

Oxidation   the loss of one or more electrons by a substance (Section 4.6)

Oxidation number   a value that measures whether an atom in a compound is neutral, electron-rich, or electron-poor compared with an isolated atom (Section 4.6)

Oxidation-number method   a method for balancing redox equations (Section 4.8)

Oxidation-reduction (redox) reaction   a process in which one or more electrons are transferred between reaction partners (Sections 4.1, 4.6)

Oxide   a binary compound with oxygen in the −2 oxidation state (Section 14.9)

Oxidizing agent   a substance that causes an oxidation by accepting an electron (Section 4.6)

Oxoacid   an acid that contains oxygen in addition to hydrogen and another element (Section 2.10)

Oxoanion   an anion in which an atom is combined with oxygen (Section 2.10) *p*-**Block element** an element in groups 3A–8A that results from *p*-orbital filling (Section 5.13)

p-**Type semiconductor** a semiconductor doped with an impurity that has fewer electrons than necessary for bonding (Section 21.5)

Paramagnetic a substance that is attracted by a magnetic field (Section 7.14)

Pauli exclusion principle no two electrons in an atom can have the same four quantum numbers (Section 5.11)

Peptide *see* **Polypeptide** (Section 24.2)

Percent composition a list of elements present in a compound and the mass percent of each (Section 3.11)

Percent dissociation the concentration of the acid that dissociates divided by the initial concentration of the acid times 100% (Section 15.10)

Percent yield the amount of product actually formed in a reaction divided by the amount theoretically possible and multiplied by 100% (Section 3.5)

Period a row of elements in the periodic table (Section 1.3)

Periodic table a chart of the elements arranged by increasing atomic number so that elements in a given group have similar chemical properties (Section 1.3)

Peroxide a binary compound with oxygen in the −1 oxidation state (Sections 14.9, 14.10)

Petroleum a complex mixture of substances, primarily hydrocarbons (Chapter 23 Interlude)

pH the negative base-10 logarithm of the molar hydronium ion concentration (Section 15.5)

pH titration curve a plot of the pH of a solution as a function of the volume of added titrant (Section 16.5)

Phase a state of matter (Section 10.4)

Phase change a process in which the physical form but not the chemical identity of a substance changes (Section 10.4)

Phase diagram a plot showing the effects of pressure and temperature on the physical state of a substance (Section 10.12)

Photoelectric effect the ejection of electrons from a metal on exposure to radiant energy (Section 5.4)

Photon the smallest possible amount of radiant energy (Section 5.4)

Physical property a characteristic that can be determined without changing the chemical makeup of a sample (Section 1.13)

Pi (π) bond a bond in which shared electrons occupy a region above and below a line connecting the two nuclei (Section 7.12)

Picometer (pm) a common unit of length; 1 pm = 10^{-12} m (Section 1.7)

Polar covalent bond a bond in which the bonding electrons are attracted somewhat more strongly by one atom than by the other (Section 7.7)

Polarizability the ease with which a molecule's electron cloud can be distorted by a nearby electric field (Section 10.2)

Polyatomic ion a charged, covalently bonded group of atoms (Section 2.8)

Polydentate ligand a ligand that bonds to a metal through electron pairs on more than one donor atom (Section 20.6)

Polymer a large molecule formed by the repetitive bonding together of many smaller molecules (Section 23.15)

Polypeptide a chain of up to 100 amino acids (Section 24.2)

Polyprotic acid an acid that contains more than one dissociable proton (Section 15.11)

Polysaccharide a compound such as cellulose that is made of many simple sugars linked together (Section 24.7)

Positron emission the conversion of a proton in the nucleus into a neutron plus an ejected positron (Section 22.2)

Potential energy (E_P) energy that is stored, either in an object because of its position or in a molecule because of its chemical composition (Section 8.1)

Precipitation reaction a process in which an insoluble solid precipitate forms and drops out of solution (Section 4.1)

Precision a measure of how well a number of independent measurements agree with one another (Section 1.10)

Primary protein structure the sequence in which amino acids are linked together (Section 24.5)

Primitive-cubic unit cell a cubic unit cell with an atom at each of its eight corners (Section 10.8)

Property any characteristic that can be used to describe or identify matter (Section 1.13)

Protein a biological polymer made up of many amino acids linked together to form a long chain (Section 24.2)

Proton a positively charged, fundamental atomic particle in the nucleus of atoms (Section 2.5)

Qualitative analysis a procedure for identifying the ions present in an unknown solution (Section 16.15)

Quantum the smallest possible amount of radiant energy (Section 5.4)

Quantum mechanical model a model of atomic structure that concentrates on an electron's wavelike properties (Section 5.6)

Quantum number a variable in the Schrödinger wave equation; describes the energy level and position in space where an electron is most likely to be found (Section 5.7)

Quaternary protein structure aggregation of several protein chains to form a larger unit (Section 24.5)

Racemic mixture a 50 : 50 mixture of enantiomers (Section 20.9)

Radioactivity the spontaneous emission of radiation (Section 22.2)

Radiocarbon dating a technique for dating archaeological artifacts by measuring the amount of ^{14}C in the sample (Section 22.10)

Radioisotope a radioactive isotope (Section 22.2)

Raoult's law the vapor pressure of a solution containing a nonvolatile solute is equal to the vapor pressure of pure solvent times the mole fraction of the solvent (Section 11.6)

Rate constant the proportionality constant in a rate law (Section 12.2)

Rate-determining step the slowest step in a reaction mechanism (Section 12.8)

Rate law an equation that tells how reaction rate depends on the concentration of each reactant (Section 12.2)

Reaction intermediate a species formed in one step of a reaction mechanism and consumed in a subsequent step (Section 12.7)

Reaction mechanism the sequence of molecular events that defines the pathway from reactants to products (Section 12.7)

Reaction order the value of the exponents of concentration terms in the rate law (Section 12.2)

Reaction quotient (Q_c) similar to the equilibrium constant K_c except that the concentrations in the equilibrium constant expression are not necessarily equilibrium values (Section 13.5)

Reaction rate the increase in the concentration of a product per unit time or the decrease in the concentration of a reactant per unit time (Section 12.1)

Redox reaction *see* **Oxidation–reduction reaction** (Section 4.6)

Reducing agent a substance that causes a reduction by donating an electron (Section 4.6)

Reduction the gain of one or more electrons by a substance (Section 4.6)

Replication the process by which identical copies of DNA are made (Section 24.14)

Resonance hybrid an average of several valid Lewis structures for a molecule (Section 7.6)

Reverse osmosis the passage of solvent through a membrane from the more concentrated side to the less concentrated side (Section 11.9)

Roasting a metallurgical process that involves heating a mineral in air (Section 21.2)

Rounding off deleting digits to keep only the correct number of significant figures (Section 1.11)

s-Block element an element in groups 1A or 2A that results from s-orbital filling (Section 5.13)

Sacrificial anode an easily oxidized metal that corrodes instead of a less reactive metal to which it is connected (Section 18.10)

Salt an ionic compound formed in an acid–base neutralization reaction (Section 4.1)

Salt bridge a tube that contains a gel permeated with a solution of an inert electrolyte connecting the two sides of an electrochemical cell (Section 18.1)

Saponification the base-catalyzed hydrolysis of an ester to yield a carboxylic acid and an alcohol (Section 23.14)

Saturated containing only single bonds (Section 23.2)

Saturated solution a solution containing the maximum possible amount of dissolved solute at equilibrium (Section 11.4)

Scientific notation a system in which a large or small number is written as a number between 1 and 10 times a power of 10 (Section 1.5)

Second law of thermodynamics in any spontaneous process, the total entropy of a system and its surroundings always increases (Section 17.6)

Secondary protein structure the orientation of segments of a protein chain into a regular pattern (Section 24.5)

Semiconductor a material that has an electrical conductivity intermediate between that of a metal and that of an insulator (Section 21.5)

Semimetal an element adjacent to the zigzag boundary between metals and nonmetals (Section 1.4)

Semipermeable membrane a membrane that allows passage of water or other small molecules but not the passage of large solute molecules or ions (Section 11.8)

Shell a grouping of orbitals according to principal quantum number (Section 5.7)

Sigma (σ) bond a bond in which the shared electrons are centered about the axis between the two nuclei (Section 7.12)

Significant figures the total number of digits in a measurement (Section 1.10)

Silicate an ionic compound that contains silicon oxoanions along with cations, such as Na^+, K^+, Mg^{2+}, or Ca^{2+} (Section 19.7)

Simple cubic packing a packing arrangement of spheres into a primitive-cubic unit cell (Section 10.9)

Single bond a covalent bond formed by sharing two electrons (Section 7.4)

Sintering a process in which the particles of a powder are "welded" together without completely melting (Section 21.7)

Slag a byproduct of iron production, consisting mainly of calcium silicate (Section 21.3)

Solubility the amount of a substance that dissolves in a given volume of solvent (Section 4.4)

Solubility product constant (K_{sp}) the equilibrium constant for a dissolution reaction (Section 16.10)

Solute the dissolved substance in a solution (Section 11.1)

Solution a homogeneous mixture containing particles the size of a typical ion or covalent molecule (Section 11.1)

Solvent the major component in a solution (Section 11.1)

Sol–gel method a method of preparing ceramics, involving synthesis of a metal oxide powder from a metal alkoxide (Section 21.7)

sp Hybrid orbital a hybrid orbital formed by combination of one atomic s orbital with one p orbital (Section 7.12)

sp^2 Hybrid orbital a hybrid orbital formed by combination of one s and two p atomic orbitals (Section 7.12)

sp^3 Hybrid orbital a hybrid orbital formed by combination of one s and three p atomic orbitals (Section 7.11)

sp^3d Hybrid orbital a hybrid orbital formed by combination of one s, three p, and one d atomic orbitals (Section 7.12)

sp^3d^2 Hybrid orbital   a hybrid orbital formed by combination of one s, three p, and two d atomic orbitals (Section 7.12)

Specific heat   the amount of heat necessary to raise the temperature of 1 g of a substance 1°C (Section 8.7)

Spectator ion   an ion that appears on both sides of the reaction arrow (Section 4.3)

Spectrochemical series   an ordered list of ligands in which crystal field splitting increases (Section 20.12)

Spontaneous process   one that proceeds on its own without any continuous external influence (Section 8.12)

Standard cell potential ($E°$)   the cell potential when both reactants and products are in their standard states (Section 18.3)

Standard electrode potential   *see* **Standard reduction potential** (Section 18.4)

Standard enthalpy of reaction ($\Delta H°$)   enthalpy change under standard-state conditions (Section 8.5)

Standard entropy of reaction ($\Delta S°$)   the entropy change for a chemical reaction under standard-state conditions (Section 17.5)

Standard free-energy change ($\Delta G°$)   the free-energy change that occurs when reactants in their standard states are converted to products in their standard states (Section 17.8)

Standard free energy of formation ($\Delta G_f°$)   the free-energy change for formation of 1 mol of the substance in its standard state from the most stable form of the constituent elements in their standard states (Section 17.9)

Standard heat of formation ($\Delta H_f°$)   the enthalpy change $\Delta H_f°$ for the hypothetical formation of 1 mol of a substance in its standard state from the most stable forms of its constituent elements in their standard states (Section 8.9)

Standard hydrogen electrode (S.H.E.)   a reference half-cell consisting of a platinum electrode in contact with H_2 gas and aqueous H^+ ions at standard-state conditions (Section 18.4)

Standard molar entropy ($S°$)   the entropy of 1 mol of the pure substance at 1 atm pressure and a specified temperature, usually 25°C (Section 17.5)

Standard molar volume   the volume of 1 mol of a gas at 0°C and 1 atm pressure; 22.4 L (Section 9.2)

Standard oxidation potential   the standard potential for an oxidation half-cell (Section 18.4)

Standard reduction potential   the standard potential for a reduction half-cell (Section 18.4)

Standard temperature and pressure (STP) $T = 273.15$ K; $P = 1$ atm (Section 9.3)

State function   a function or property whose value depends only on the present condition of the system, not on the path used to arrive at that condition (Section 8.3)

Steam–hydrocarbon reforming process   an important industrial method for producing hydrogen from methane (Section 14.3)

Stereoisomers   isomers that have the same connections among atoms but have a different arrangement of the atoms in space (Section 20.8)

Steric factor   the fraction of collisions having proper orientation for conversion of reactants to products (Section 12.9)

Steroid   a lipid whose structure is similar to that of cholesterol (Section 24.11)

Stoichiometry   mole/mass relationships between reactants and products (Section 3.4)

Straight-chain alkane   an alkane with all its carbons connected in a row (Section 23.2)

Strong acid   an acid that dissociates completely in water and is a strong electrolyte (Sections 4.5, 15.2)

Strong electrolyte   a compound that dissociates completely into ions when dissolved in water (Section 4.2)

Strong-field ligand   ligand that has a large crystal field splitting (Section 20.12)

Structural formula   a representation that shows the specific connections between atoms in a molecule (Section 2.8)

Sublimation   the direct conversion of a solid to a vapor without going through a liquid state (Section 8.6)

Subshell   a grouping of orbitals according to angular-momentum quantum number (Section 5.7)

Superconducting transition temperature (T_c)   the temperature below which a superconductor loses all electrical resistance (Section 21.6)

Superconductor   a material that loses all electrical resistance below a certain temperature (Section 21.6)

Supercritical fluid   a state of matter beyond the critical point that is neither liquid nor gas (Section 10.12)

Superoxide   a binary compound with oxygen in the $-1/2$ oxidation state (Sections 14.9, 14.10)

Supersaturated solution   a solution containing a greater-than-equilibrium amount of solute (Section 11.4)

Surface tension   the resistance of a liquid to spreading out and increasing its surface area (Section 10.3)

Suspension   a homogeneous mixture containing particles greater than about 1000 nm in diameter that are visible with a low-power microscope (Section 11.1)

Symmetry plane   a plane that cuts through an object so that half of the object is a mirror image of the other half (Section 20.9)

Temperature   a measure of the kinetic energy of molecular motion (Section 8.2)

Tertiary protein structure   folding of a protein chain into a specific three-dimensional shape (Section 24.5)

Theory   a consistent explanation of known observations (Section 1.1)

Thermal energy   heat (Section 8.2)

Thermochemistry   a study of the heat changes that take place during reactions (Chapter 8)

Thermodynamic standard state   conditions under which thermodynamic measurements are reported; 298.15 K (25°C), 1 atm pressure for each gas, 1 M concentration for solutions (Section 8.5)

Thermodynamics   the study of the interconversion of heat and other forms of energy (Chapter 17)

Third law of thermodynamics   the entropy of a perfectly ordered crystalline substance at 0 K is zero (Section 17.4)

Three-center, two-electron bond   a covalent bond in which three atoms share two electrons (Section 19.4)

Titration   a procedure for determining the concentration of a solution (Section 3.10)

Trans isomer   the isomer of a metal complex in which identical ligands are opposite one another (Section 20.8)

Transcription   the process by which information in DNA is transferred to and decoded by RNA (Section 24.14)

Transition-metal element   an element in the 10 smaller groups in the middle of the periodic table (Section 1.4; Chapter 20)

Transition state   the configuration of atoms at the maximum in the potential energy profile for a reaction (Section 12.9)

Translation   the process by which RNA builds proteins (Section 24.14)

Transuranium elements   the 17 artificially produced elements beyond uranium in the periodic table (Section 22.4)

Triacylglycerol   a triester of glycerol (1,2,3-propanetriol) with three long-chain carboxylic acids (Section 24.11)

Triple bond   a covalent bond formed by sharing six electrons (Section 7.4)

Triple point   a unique combination of pressure and temperature at which all three phases coexist in equilibrium (Section 10.12)

Turnover number   the number of substrate molecules acted on by one molecule of enzyme per unit time (Section 24.6)

Unimolecular reaction   an elementary reaction that involves a single reactant molecule (Section 12.7)

Unit cell   the small repeating units that make up a crystal (Section 10.8)

Unsaturated   an organic molecule that has a double or triple bond (Section 23.9)

Valence band   the bonding molecular orbitals in a metal (Section 21.5)

Valence bond theory   a quantum mechanical description of bonding that pictures covalent bond formation as the overlap of two singly occupied atomic orbitals (Section 7.10)

Valence shell   the outermost electron shell (Section 5.12)

Valence-shell electron-pair repulsion (VSEPR) model   a model for predicting the approximate geometry of a molecule (Section 7.9)

Van der Waals forces   an alternative name for intermolecular forces (Section 10.2)

Vapor pressure   the partial pressure of a gas in equilibrium with liquid (Section 10.5)

Viscosity   the measure of a liquid's resistance to flow (Section 10.3)

Voltaic cell   *see* Galvanic cell (Section 18.1)

Volume   the amount of space occupied by an object (Section 1.8)

Watson-Crick model   a model of DNA, consisting of two polynucleotide strands coiled around each other in a double helix (Section 24.13)

Wave function   a solution to the Schrödinger wave equation (Section 5.6)

Wavelength (λ)   the length of a wave from one maximum to the next (Section 5.2)

Weak acid   an acid that dissociates incompletely in water and is a weak electrolyte (Sections 4.5, 15.2)

Weak electrolyte   a molecular compound that dissociates incompletely when dissolved in water (Section 4.2)

Weak-field ligand   ligand that has a small crystal field splitting (Section 20.12)

Weight percent (wt %)   a unit of concentration; the mass of one component divided by the total mass of the solution times 100% (Section 11.3)

Work (w)   the distance (d) moved times the force (F) that opposes the motion (Section 8.4)

Zone refining   a purification technique in which a heater melts a narrow zone at the top of a rod of some material and then sweeps slowly down the rod, bringing impurities with it (Section 19.7)

INDEX

Molar heat capacity *Cont.*
  table of, 297
Molar mass, **76**
  determination of, 436
Molarity, **86**
  advantages of, 413
Mole, **75**
Mole fraction, **339, 413**
  advantages of, 413
Mole-to-gram conversion, 79–81
Molecular compound, names of, 56–57
Molecular equation, **110**
Molecular formula, **92**
Molecular geometry, summary of, 254–55
Molecular orbital, **265**
  antibonding, **265**
  bond order in, 267
  bonding, **265**
  properties of, 265
Molecular orbital theory, **264–70**
  diatomic molecules and, 266–69
  insulators and, 857–59
  metals and, 854–56
  oxygen and, 267–69
  semiconductors and, 857–59
  summary of, 267
  valence bond theory and, 270–71
Molecular solid, **382**
Molecular weight, **75**
  determination of, 99
Molecularity, **470**
  rate law and, 472–73
Molecule, **49**
Mond process, **850**
Monodentate ligand, **814**
Monomer, **952**
Monoprotic acid, **582**
Monosaccharide, **980**
Morton, William, 354
MRI, *see* Magnetic resonance imaging,
  910–11
Mt. Everest, pressure on, 329
Müller, K. Alex, 860
Multielectron atom, atomic orbitals in,
  169–70
  electron configurations of, 174–75
Multiple bond, 234
Multiple proportions, law of, **37**
Mylar, structure of, 953
Myoglobin, structure of, 976
Myosin, 656

N-terminal amino acid, **972**
n-type semiconductor, **859**
Names, acids, 60–61
  alkanes, 926–30
  alkenes, 934–36
  alkyl groups, 928–29
  anions, 54–55
  cations, 54–55
  coordination compounds, 815–18
  ionic compounds, 54–56
  ligands, 816
  molecular compounds, 56–57
  oxoacids, 60–61
  oxoanions, 59
  polyatomic ions, 58–59

Nano-, as prefix, 10
Naphthalene, analysis of, 95–97
Neon, orbital-filling diagram for, 173
  properties of, 218
Nernst, Walther, 719
Nernst equation, **719**
  cell potential and, 718–19
  equilibrium constant and, 722–24
  pH determination and, 720–21
  reaction quotient and, 718–19
Net ionic equation, **111**
Network solid, **382**
Neutral solution, 578
Neutralization reaction, **108, 115,** 617–20
  calculations for, 617–20
  characteristics of, 114–16
  common-ion effect and, 620–22
  driving force in, 108
  net ionic equation for, 116
Neutron, **43**
  mass of, 43
Newton, definition of, 326
Nickel, manufacture of, 850
Nickel-cadmium battery, electrode reactions of, 728
  operation of, 728
Nickel tetracarbonyl, Mond process and,
  850
Nicotine, K_b of, 615
Nitrate, resonance structures of, 242
Nitration, aromatic compounds and, 941
Nitric acid, manufacture of, 773
  oxidizing properties of, 774
Nitric oxide, air pollution and, 349
  oxidation of, 456–57
Nitrogen, bond strength of, 235
  chemistry of, 769–74
  combustion of, 770
  inorganic compounds of, 771–74
  manufacture of, 769
  molar volume of, 334
  molecular orbital theory of, 267–69
  orbital-filling diagram for, 173
  oxides of, 771–72
  triple bond in, 234
  uses of, 769
Nitrogen dioxide, dimerization of, 772
Nitrogen-14, transmutation of, 901
Nitrogenase enzyme, 484
Nitroglycerin, explosive use of, 489
Nitrous acid, acidity of, 773
  biological properties of, 772
  formal charges in, 246
  K_a of, 585
  manufacture of, 771–73
  resonance in, 246
  uses of, 772
Nobel, Alfred, 489
Noble gas(es), **8**
  electron affinity of, 197
  occurrence of, 218–19
  properties of, 8, 218
  reactions of, 219
  uses of, 218–19
Node, **166**
Nonbonding electrons, 233
Nonelectrolyte, **109**

Nonmetal, **8**
  characteristics of, 753
  properties of, 8
Nonspontaneous process, 311, 666
  entropy and, 312
Nonstoichiometric compound, **545**
Normal boiling point, **379, 395**
Normal melting point, **395**
Northern lights, 184
Novocaine, structure of, 960
Nuclear chemistry, **878–911**
  applications of, 906–11
Nuclear equation, **881**
Nuclear fuel, purification of, 346
Nuclear reaction, **879**
  balancing, 881
  characteristics of, 878–80
  energy changes during, 892–95
  rates of, 884–87
Nuclear reactor, **898–900**
  electricity from, 900
Nuclear transmutation, **901**
Nucleic acid, **988–91**
  see also DNA, RNA
  structure of, 990
Nucleon, **878**
Nucleoside, **989**
  structure of, 990
Nucleosynthesis, 912
Nucleotide, **989**
  structure of, 990
Nucleus, **41**
  discovery of, 41–42
  shell model of, 890
  size of, 42
Nuclide(s), **871**
  number of, 901
  stability of, 888
Nylon, structure of, 953
  uses of, 953

Ocean, composition of, 561
  salt concentration of, 203
  volume of, 560
Octahedral complexes, isomers of, 819–20
  crystal field theory of, 830–33
  valence bond theory of, 828–29
Octahedral molecules, 252–53
  hybrid orbitals in, 262
Octane rating, 955
Octet rule, **220**
  effective nuclear charge and, 220
  exceptions to, 220–21
  explanation of, 220
  main-group elements and, 233–34
Oleic acid, structure of, 986
Olive oil, composition of, 986
Onnes, H. K. 860
Optical isomer, 823
  see also Enantiomer
Orbital, **158**
  energy levels of, 163, 169–70
  hybridization of, 257–64
d-Orbital, crystal field theory and, 831–35
  nodes in, 168
  shape of, 168

PHOTO CREDITS

All photographs not listed: © Richard Megna/Fundamental Photographs

CHAPTER 1 CO1 Runk/Schoenberger/Grant Heilman Photography **p. 1 (top)** SmithKline Beecham **p. 1 (bot. left)** NASA/Phototake **p. 1, (bot. rt.)** Al Hamdan/The Image Bank **p. 2 (top)** Karl Hentz/The Image Bank **p. 2 (ctr.)** Link/Visuals Unlimited **p. 2 (bot.)** McCracken Photographers **Fig. 1.1(a)** Superstock **(b)** Terje Rakke/The Image Bank **Fig. 1.2** Granger Collection **p. 4** John Cancalosi/Peter Arnold; AC/GM/Peter Arnold **p. 7 (bot.)** McCracken Photographers **p. 8 (top left)** McCracken Photographers **p. 8 (top rt.)** Michael Dalton/Fundamental Photographs **p. 8 (bot.)** McCracken Photographers **p. 9 (top)** McCracken Photographers **p. 9 (bot.)** Courtesy Yonex Corp. **p. 11 (top)** McCracken Photographers **Fig. 1.5** McCracken Photographers **Fig. 1.6** McCracken Photographers **p. 12 (bot.)** Dr. Tony Brain/David Parker/Science Photo Library/Photo Researchers **Fig. 1.8** McCracken Photographers **p. 14** Photo Researchers **p. 16** McCracken Photographers **p. 17** McCracken Photographers **p. 19** AAA Photo/Phototake **p. 23 (top)** Art Wolfe/Allstock **p. 23 (bot.)** McCracken Photographers **p. 24 (ctr.)** Frans Lanting/Allstock **p. 24 (bot.)** McCracken Photographers **Fig. 1.10** Barry L. Runk/Grant Heilman Photography **p. 26 (top)** Bob Daemmrich/Stock Boston **p. 26 (bot.)** Chris Noble/Allstock **Fig. 1.11** Michael Melford/The Image Bank **p. 27** McCracken Photographers

CHAPTER 2 CO2 Bill Ross/WESTLIGHT **p. 35** McCracken Photographers **p. 38 (top)** McCracken Photographers **p. 38 (bot.)** IBM Research/Peter Arnold **p. 42 (top)** Bob Daemmrich/Stock Boston **p. 42 (bot.)** McCracken Photographers **p. 45** Will McIntyre/Allstock **p. 48 (rt.)** Paul Silverman/Fundamental Photographs **p. 49 (top)** Larry Cameron/Photo Researchers **p. 49 (ctr.)** Dr. E. R. Degginger **p. 52** McCracken Photographers **p. 58** Paul Silverman/Fundamental Photographs **Fig. 2.11(a)** Courtesy IBM Corp. **Fig. 2.11(b)** Dr. E. R. Degginger **Fig. 2.11(c)** IBM **Fig. 2.13** Tom Pantages

CHAPTER 3 CO3 Lee Lockwood/Black Star **p. 72 (top)** McCracken Photographers **p. 72 (bot.)** Yoav Levy/Phototake **p. 73** McCracken Photographers **p. 78** Bethlehem Steel **p. 80** T. J. Florian/Rainbow **p. 81 (top)** Diane Schiumo/Fundamental Photographs **p. 81 (bot.)** Phototake **p. 83 (top)** Gianalberto Cigolini/The Image Bank **p. 83 (bot.)** Dan McCoy/Rainbow **p. 84** Larry Mulvehill/Photo Researchers **p. 85** NASA **p. 87 (top)** Paul Silverman/Fundamental Photographs **p. 87 (bot.)** Dan McCoy/Rainbow **Fig. 3.8** McCracken Photographers **p. 93** Phil Degginger **p. 95** NASA **p. 97** McCracken Photographers **p. 100** Phil Degginger

CHAPTER 4 CO4 Runk/Schoenberger/Grant Heilman Photography **p. 106** McCracken Photographers **p. 107** McCracken Photographers **p. 113** McCracken Photographers **p. 114** McCracken Photographers **p. 115** McCracken Photographers **p. 117** McCracken Photographers **p. 122** McCracken Photographers **p. 123** McCracken Photographers **p. 124 (top)** McCracken Photographers **p. 124 (ctr. rt.)** McCracken Photographers **p. 129** McCracken Photographers **Fig. 4.2 p. 135** McCracken Photographers

CHAPTER 5 CO5 Runk/Schoenberger/Grant Heilman Photography **p. 147** Grant Heilman/Grant Heilman Photography **p. 148** Phil Degginger **Fig. 5.5(a)** Dr. E. R. Degginger/Photo Researchers **Fig. 5.5(b)** Malcolm Boulton/Photo Researchers **p. 152** Jack Fields/Photo Researchers **p. 153 (top)** Larry Mulvehill/Science Source/Photo Researchers **p. 153 (bot.)** R. L. Kaylin/Allstock **p. 154** Michael Smelter/Photo Researchers **p. 155 (top)** L. Steinmark/Custom Medical Stock Photo; **(bot)** Gerry Davis/Phototake **p. 157** Donald Deitz/Stock Boston **p. 158** Science Photo Library/Photo Researchers **p. 159** Dr. Harold E. Edgerton/Palm Press, Inc. **p. 160** Dr. E. R. Degginger **p. 161** Yoav Levy/Phototake **p. 184 (top)** Robert Slobins/Phototake; **(bot.)** McCracken Photographers

CHAPTER 6 CO6 Barry L. Runk/Grant Heilman Photography **p. 197** McCracken Photographers **p. 200** Dr. E. R. Degginger **p. 202 (rt.)** Robert Goldstein/Photo Researchers **p. 203 (ctr. left)** John Coletti/Stock Boston **p. 203 (bot.)** Grant Heilman/Grant Heilman Photography **p. 204** McCracken Photographers **p. 206 (rt.)** McCracken Photographers **p. 208** Paul Silverman/Fundamental Photographs **p. 209 (top)** McCracken Photographers **p. 209 (ctr. left)** Matthew McVay/Allstock **p. 209 (ctr. rt.)** McCracken Photographers **p. 209 (bot. rt.)** Dean Hulse/Rainbow **p. 209 (bot. left)** Tim Thompson/Allstock **p. 210 (top)** Dr. E. R. Degginger **p. 210 (bot.)** McCracken Photographers **p. 212**

(rt.) David Barnes/Allstock **p. 212 (bot.)** Stephen Frisch/Stock Boston **p. 213** McCracken Photographers **p. 214 (top)** Matthew McVay/Stock Boston **p. 214 (ctr.)** McCracken Photographers **p. 214 (ctr. left)** Stephen Frisch/Stock Boston **p. 214 (bot.)** McCracken Photographers **p. 215 (bot.)** Guy Marche/Allstock **p. 216** The Image Bank **p. 219 (top)** Merlin Metalworks, Inc. **p. 222** Martin Miller/Visuals Unlimited **p. 223** Co Rentmeester/The Image Bank

CHAPTER 7 CO7 Ralph Clevenger/WESTLIGHT **p. 229** George Disario/Stock Market **p. 232** McCracken Photographers **p. 241** McCracken Photographers **Fig. 7.16** McCracken Photographers **p. 272** Phil Degginger **p. 273** Phil Degginger

CHAPTER 8 CO8 Grant Heilman Photography **Fig. 8.3** Eric Kamp/Phototake **p. 281** George H. Harrison/Grant Heilman Photography **p. 282 (top)** Jim Caccavo/Stock Boston **p. 282 (bot.)** Larry Lefever/Grant Heilman Photography **p. 284** D.O.E./Science Source/Photo Researchers **p. 287** George H. Harrison/Grant Heilman Photography **p. 290** McCracken Photographers **p. 293** Paul Silverman/Fundamental Photographs **p. 298 (top)** McCracken Photographers **p. 299** McCracken Photographers **p. 304** Peter M. Fisher/Stock Market **p. 305** Dravo Lime **p. 306** Liane Enkelis/Stock Boston **p. 308** Francois Gohier/Photo Researchers **Fig. 8.14** Spencer Grant/Stock Boston **p. 309** NASA **p. 311 (top)** David W. Hamilton/The Image Bank **p. 311 (bot.)** Bruce Forster/Allstock **p. 314 (left)** Kristen Brochmann/Fundamental Photographs **p. 317** Springer/Bettmann

CHAPTER 9 CO9 Runk/Schoenberger/Grant Heilman Photography **Tb. 9-1** Kaz Mori/The Image Bank **Fig. 9.2(a)** NASA/Stock Boston **Fig. 9.2(b)** NASA Robbi Newman/The Image Bank **p. 326** Phil Degginger **p. 329** Keren Su/Stock Boston **p. 335** Link/Visuals Unlimited **p. 336** BMW **p. 340** Fred Bavendam/Allstock **p. 346** U.S. Department of Energy/Mark Marten/Photo Researchers **p. 349** NASA/Phototake **p. 350** A.J. Copley/Visuals Unlimited Superstock **p. 351 (top)** John D. Cunningham/Visuals Unlimited; **(bot.)** Dallas & John Heaton/Stock Boston **Fig. 9.19** NASA **p. 354** Stacy Pick/Stock Boston

CHAPTER 10 CO10 Runk/Schoenberger/Grant Heilman Photography **Fig. 10.8** Phil Degginger **p. 371** Michael Dalton/Fundamental Photographs **Fig. 10.21** Cornell University **p. 373** Nik Kleinberg/Stock Boston **Fig. 10.12** Dr. E.R. Degginger **p. 374** Gary E. Holscher/Allstock **p. 379 (top)** Larry Lefever/Grant Heilman; **(bot.)** C. Seghers/Photo Researchers **Fig. 10.16** Runk/Schoenberger/Grant Heilman Photography **p. 385** Michael Dalton/Fundamental Photographs **p. 392** Paul Silverman/Fundamental Photographs **p. 393** Stacy Pick/Stock Boston **p. 394** Michael Dalton/Fundamental Photographs **Fig. 10.32** Paul Silverman/Fundamental Photographs **p. 399** Bob Daemmrich/Stock Boston

CHAPTER 11 CO11 Greg Gawlowski/Impact Images **p. 407 (left)** Yoav Levy/Phototake **p. 407 (ctr.)** Michael Dalton/Fundamental Photographs **p. 407 (rt.)** Stockphotos **p. 408 (left)** Jim Pickerell/Stock Boston **p. 408 (rt.)** Erci Kamp/Phototake **p. 409 (top)** Runk/Schoenberger/Grant Heilman Photography **p. 409 (bot.)** Paul Silverman/Fundamental Photographs **p. 418** Eurgen Gordon **p. 420** Douglas Faulkner/Photo Researchers **p. 421** Yoav Levy/Phototake **p. 432 (left)** McCracken Photographers **p. 432 (rt.)** McCracken Photographers **p. 434** George Riley/Stock Boston **p. 435** Grant Heilman/Grant Heilman Photography **Fig. 11-16** Joseph P. Sinnot/Fundamental Photographs **p. 441** Diane Schiumo/Fundamental Photographs

CHAPTER 12 CO12 Max Dunham/Photo Researchers **Fig. 12.2** McCracken Photographers **Fig. 12.5** McCracken Photographers **Fig. 12.11** McCracken Photographers **Fig. 12.15** McCracken Photographers **Fig. 12.17** Harshaw/Filtrol Partnership **p. 449 (top)** Steve McCutcheon/Visuals Unlimited **p. 449 (bot. rt.)** Grant Heilman Photography **p. 453** Phil Degginger **p. 463 (bot.)** Aaron Haupt/Stock Boston **p. 471** Robert Slobins/Phototake **p. 474** Jeffry W. Myers/West Stock, Inc. **p. 486** McCracken Photographers **p. 488** Hank Morgan/Rainbow **Fig. 12.18** General Motors/AC Rochester **p. 489** John English

CHAPTER 13 CO13 Martin Rogers/Stock Boston **p. 498** Nada Penik/Visuals Unlimited **p. 504** Dan McCoy/Rainbow **p. 506** McCracken Photographers **p. 514** Grant Heilman/Grant Heilman Photography **Fig. 13.11** McCracken Photographers **Fig. 13.13** Courtesy of United Catalysts, Inc., Louisville, Kentucky

CHAPTER 14 CO14 Runk/Schoenberger/Grank Heilman Photography **Fig. 14.5(b)** McCracken Photographers **Fig. 14.5(c)** Tom Bochsler,

Photography Ltd. **Fig. 14.14** NASA **p.** 538 Bill Iburg/Visuals Unlimited **p.** 539 Ken Kay/Fundamental Photographs **p.** 541 DiAnn Lein **p.** 542 AP/Wide World Photos **p.** 546 Carl Purcell/Photo Researchers **p.** 547 McCracken Photographers **p.** 548 **(top)** G. Bruce Forster/ Allstock **p. 548, bot.** Liane Enkelis/Stock Boston **p.** 558 **(top)** Paul Silverman/Fundamental Photographs **p.** 558 **(bot.)** NASA/Mark Marten/ Science Source/Photo Researchers **p.** 560 **(top)** Paul Silverman/Fundamental Photographs **p.** 560 **(bot.)** Grant Heilman/Grant Heilman Photography

CHAPTER 15 CO15 Paul Silverman/Fundamental Photographs **Fig. 15.3** Courtesy Hach Company **p.** 571 Runk/Schoenberger/Grant Heilman Photography **p.** 608 **(top)** David Nunuk/Science Photo Library/ Photo Researchers **Fig. 15.5** Runk/Schoenberger/Grant Heilman Photography **p.** 595 Michael Tamborrino/FPG International **p.** 608 **(bot.)** Ray Pfortner/Peter Arnold

CHAPTER 16 CO16 Richard Megna/Fundamental Photographs **Fig. 16.4** Donald Clegg and Roxy Wilson **p.** 641 **(top)** Alexander Tsiaras/ Science Source/Photo Researchers; **(ctr.)** Manfred Kage/Peter Arnold **p.** 644 Grant Heilman Photography **p.** 647 Robert Mathena/Fundamental Photographs **Fig. 16.16** Tom Bochsler, Photography Ltd. **p.** 656 Runk/ Schoenberger/Grant Heilman Photography

CHAPTER 17 CO17 Barry L. Runk/Grant Heilman Photography **Fig. 17.2** John D. Cunningham/Visuals Unlimited **Fig. 17.8** Dr. Rudolph Schild/Science Photo Library/Photo Researchers **p.** 667 **(top)** John Kaprielian/Photo Researchers; **(bot.)** Larry Lefever/Grant Heilman Photography **p.** 670 Paul Silverman/Fundamental Photographs **p.** 678 **(left)** James H. Karales/Peter Arnold; **(right)** W. A. Banaszewski/Visuals Unlimited **p.** 694 **(top)** Photo Researchers; **(bot.)** Dale E. Boyer/Photo Researchers

CHAPTER 18 CO18 David & Doris Krumholz/Fundamental Photographs **Fig. 18.1(a)** Richard Megna/Fundamental Photographs **Fig. 18.1(b)** Richard Megna/Fundamental Photographs **Fig. 18.8** Dr. E. R. Degginger **Fig. 18.5** Tom Pantages **Fig. 18.7** Tom Pantages **Fig. 18.8** Michael Dalton/Fundamental Photographs **Fig. 18.9** Tom Pantages **Fig. 18.10** Michael Dalton/Fundamental Photographs **Fig. 18.12** Zandria Muench/Allstock **Fig. 18.18** Visuals Unlimited **p.** 707 Tom Pantages **p.** 711 Dr. E. R. Degginger **p.** 729 **(top)** The Tokyo Electric Power Company, Inc.; **(bot.)** Phil Degginger **p.** 731 Merlin Metalworks, Inc. **p.** 732 Science VU-NSRDC/Visuals Unlimited **p.** 739 Bill Gallery/ Stock Boston **p.** 742 **(top)** Paul Silverman/Fundamental Photographs; **(bot.)** Bill Seeley

CHAPTER 19 CO19 Karl Hartmann/Sachs/Phototake **p.** 751 Comstock **p.** 756 **(ctr.)** Stephen Frisch/Stock Boston **Fig. 19.4(b)** Texas Instruments **Fig. 19.6(a)** Barry L. Runk/Grant Heilman Photography; **(b)** William E. Ferguson **Fig. 19.10(a),(b)** Tom Bochsler, Photography Ltd. **p.** 766 **(top left)** Runk/Schoenberger/Grant Heilman Photography; **(top rt.)** Barry L. Runk/Grant Heilman Photography **p.** 768 Dr. E.R. Degginger **p.** 757 **(top)** U.S. Borax **p.** 756 **(left)** Paul Silverman/Fundamental Photographs **p.** 756 **(rt.)** Hank Morgan/Science Source/Photo Researchers **p.** 761 Richard Megna/Fundamental Photographs **p.** 763 Paul Silverman/Fundamental Photographs **p.** 764 William McCoy/Rainbow **p.** 766 **(bot.)** Paul Silverman/Paul Silverman/Fundamental Photographs **p.** 767 **(left)** Runk Schoenberger/Grant Heilman Photography **p.** 767 **(rt.)** Barry L. Runk/Grant Heilman Photography **p.** 768 **(ctr.)** Stephen Frisch **p.** 768 **(rt.)** Stephen Frisch **p.** 769 **(left)** Dr. E.R. Degginger **p.** 769 **(ctr.)** Russ Lappa/Science/Photo Researchers **p.** 769 **(rt.)** Paul Silverman/Fundamental Photographs **p.** 770 Tom Bochsler,

Photography Ltd. **p.** 773 **(top)** Johnson-Matthey Metals, Ltd. **p.** 773 **(bot.)** Tom Bochsler, Photography Ltd. **p.** 774 Runk/Schoenberger/Grant Heilman Photography **p.** 778 James Knowles/Stock Boston **p.** 779 Diane Schiumo/Fundamental Photographs **p.** 780 **(top)** Stephen Frisch; **(bot.)** Stephen Frisch; **(rt.)** Kent Knudson/FPG International **p.** 781 **(top)** Runk/Schoenberger/Grant Heilman Photography **p.** 781 **(bot.)** Manfred Kage/Peter Arnold **p.** 786 Joseph P. Sinnot/Fundamental Photographs **p.** 787 NASA

CHAPTER 20 CO20 Paul Silverman/Fundamental Photographs **p.** 797 Foto Werner H. Muller/Peter Arnold **p.** 800 **(left)** Paul Silverman/ Fundamental Photographs **p.** 806 Chris Hildreth/Cornell University **Fig. 20.7** Donald Clegg and Roxy Wilson **p.** 810 **(left)** Paul Silverman/Fundamental Photographs **p.** 810 **(rt.)** Michael Dalton/Fundamental Photographs **p.** 811 Paul Silverman/Fundamental Photographs **p.** 836 Visuals Unlimited

CHAPTER 21 CO21 E.R. Degginger/Bruce Coleman **Fig. 21.1(a)** Barry L. Runk/Grant Heilman Photography **Fig. 21.1(b)** Karl Hartmann/Sachs/ Phototake **Fig. 21.1(c)** Runk/Schoenberger/Grant Heilman Photography **Fig. 21.4** Rich Treptow/Photo Researchers **Fig. 21.6** Science VU/ Visuals Unlimited **Fig. 21.14** Chemical Design Ltd/Science Photo Library/Photo Researchers **Fig. 21.16** Runk/Schoenberger/Grant Heilman Photography **Fig. 21.17** Marcello Bertinetti/Photo Researchers **Fig. 21.20** Corning Corporation/Scientific American **Fig. 21.21** Northrup Corporation **p.** 845 **(top)** Courtesy of Boeing **(bot.)** Dan McCoy/Rainbow **p.** 846 Arnold Fisher/Science Photo Library/Photo Researchers **p.** 847 Richard Hutchings/Photo Researchers **p.** 850 Science VU/Visuals Unlimited **p.** 859 Astrid & Hanns-Frieder Michler/Science Photo Library/ Photo Researchers **p.** 863 Courtesy of GTE Laboratories, Inc. **p.** 866 Dr. E.R. Degginger **p.** 869 **(top)** J. & L. Weber/Peter Arnold **p.** 869 **(bot.)** Courtesy JVC Corp.

CHAPTER 22 CO22 NASA/Phototake **p.** 879 Jean-Claude Lejeune/ Stock Boston **p.** 885 David Parker/Science Photo Library/Photo Researchers **p.** 897 Phototake **p.** 899 Earl Scott/Photo Researchers **p.** 901 Phototake **Fig. 22.12** Rennie Van Munchow/Phototake **p.** 904 Kevin Schafer/Peter Arnold **p.** 905 Runk/Schoenberger/Grant Heilman Photography **p.** 906 John D. Cunningham/Visuals Unlimited **p.** 908 **(top)** NASA/Science Photo Library/Photo Researchers **p.** 908 **(bot.)** Yan/ Photo Researchers **p.** 909 Tom McCarthy Photos **p.** 910 Bill Pierce/ Rainbow **Fig. 22.14** VU/SIU/Visuals Unlimited **Fig. 22.15** Leonard Lessin/Peter Arnold **p.** 912 **(top)** Dennis Di Cicco/Peter Arnold **p.** 912 **(bot.)** Royal Observatory, Edinburgh/Science Photo Library/Photo Researchers

CHAPTER 23 CO23 Rod Planck/Photo Researchers **p.** 922 Runk/ Schoenberger/Grant Heilman Photography **p.** 932 Ken Graham/Allstock **p.** 939 Grant Heilman Photography **p.** 943 Chris Johns/Allstock **p.** 944 Kristen Brochmann/Fundamental Photographs **p.** 948 Larry Lefever/Grant Heilman Photography **p.** 955 Joe Sohm/Chromosohm/ Allstock

CHAPTER 24 CO24 Jon Gordon/Phototake **Fig. 24.6** Courtesy Dr. Christine Humblet, Parke & Davis **Fig. 24.17** Philippe Plailly/Science Photo Library/Photo Researchers **p.** 963 Dr. E.R. Degginger **p.** 967 Richard Pasley/Stock Boston **p.** 974 Gregory G. Dimijian/Photo Researchers **p.** 975 Christopher Arnesen/Allstock **p.** 976 Richard Ellis/ Photo Researchers **p.** 985 Larry Lefever/Grant Heilman Photography **p.** 987 Darrell Gulin/Allstock **p.** 988 F. Sloop/W. Ober/Visuals Unlimited **p.** 996 Tom J. Ulrich/Visuals Unlimited

Physical Constants

Avogadro's number	$N_A = 6.022\ 137 \times 10^{23}/\text{mol}$
Boltzmann constant	$k = 1.380\ 66 \times 10^{-23}\ \text{J/K}$
Electron charge (magnitude)	$e = 1.602\ 177\ 3 \times 10^{-19}\ \text{C}$
Electron mass	$m_e = 9.109\ 390 \times 10^{-31}\ \text{kg}$
Faraday constant	$F = 96{,}485.31\ \text{C/mol}$
Gas constant	$R = 0.082\ 0578\ (\text{L} \cdot \text{atm})/(\text{K} \cdot \text{mol})$
	$\quad = 8.314\ 51\ \text{J}/(\text{K} \cdot \text{mol})$
	$\quad = 1.987\ 22\ \text{cal}/(\text{K} \cdot \text{mol})$
Neutron mass	$m_n = 1.674\ 929 \times 10^{-27}\ \text{kg}$
Planck's constant	$h = 6.626\ 076 \times 10^{-34}\ \text{J} \cdot \text{s}$
Proton mass	$m_p = 1.672\ 623 \times 10^{-27}\ \text{kg}$
Speed of light (in vacuum)	$c = 2.997\ 924\ 58 \times 10^8\ \text{m/s}$

Conversion Factors

Mass [SI unit: kilogram (kg)]
   $1\ \text{kg} = 10^3\ \text{g} = 2.2046\ \text{lb}$
   $1\ \text{lb} = 16\ \text{oz} = 453.59\ \text{g}$
   $1\ \text{ton} = 2000\ \text{lb} = 907.185\ \text{kg}$
   $1\ \text{amu} = 1.660\ 540 \times 10^{-27}\ \text{kg}$

Length [SI unit: meter (m)]
   $1\ \text{m} = 10^2\ \text{cm} = 1.0936\ \text{yd}$
   $1\ \text{km} = 0.621\ 37\ \text{mi}$
   $1\ \text{mi} = 5280\ \text{ft} = 1760\ \text{yd} = 1.6093\ \text{km}$
   $1\ \text{in} = 2.54\ \text{cm}$
   $1\ \text{pm} = 10^{-12}\ \text{m} = 10^{-10}\ \text{cm} = 10^{-2}\ \text{Å}$

Volume [SI unit: cubic meter (m^3)]
   $1\ \text{L} = 10^{-3}\ \text{m}^3 = 1\ \text{dm}^3 = 1.0567\ \text{qt}$
   $1\ \text{mL} = 1\ \text{cm}^3$
   $1\ \text{gal} = 4\ \text{qt} = 3.7854\ \text{L}$

Temperature [SI unit: kelvin (K)]
   $0\ \text{K} = -273.15°\text{C}$
   $\text{K} = °\text{C} + 273.15$

$$°\text{C} = \frac{5}{9}(°\text{F} - 32)$$

$$°\text{F} = \left(\frac{9}{5} \times °\text{C}\right) + 32$$

Pressure [SI unit: pascal (Pa)]
   $1\ \text{Pa} = 1\ \text{N/m}^2 = 1\ \text{kg}/(\text{m} \cdot \text{s}^2)$
   $1\ \text{atm} = 760\ \text{mm Hg (torr)}$
   $\quad\quad = 101{,}325\ \text{Pa}$
   $\quad\quad = 14.70\ \text{lb/in}^2$

Energy [SI unit: joule (J)]
   $1\ \text{J} = 1\ (\text{kg} \cdot \text{m}^2)/\text{s}^2$
   $1\ \text{cal} = 4.184\ \text{J}$
   $1\ \text{eV/molecule} = 96.485\ 31\ \text{kJ/mol}$
   $1\ \text{MeV} = 1.602\ 18 \times 10^{-13}\ \text{J}$